FIELD REFERENCE MANUAL

Specifications for Structural Concrete
ACI 301-05

with selected ACI references

Hartness Library
Vermont Technical College
One Main St.
Randolph Center, VT 05061

American Concrete Institute®
Advancing concrete knowledge

PUBLICATION SP-15 (05)
American Concrete Institute, Farmington Hills

Copyright © 2005
American Concrete Institute
P.O. Box 9094
Farmington Hills, MI 48333-9094

All rights reserved including rights of reproduction and use in any form or by any means, including the making of copies by any photo process, or by any electronic or mechanical device, printed, written, or oral, or recording for sound or visual reproduction or for use in any knowledge or retrieval system or device, unless permission in writing is obtained from the copyright proprietors.

Printed in the United States of America

November 2005

ISBN: 0-87031-194-8
Library of Congress Control Number: 2005935052

Contents

1	Specifications for Structural Concrete	ACI 301-05
51	Cement and Concrete Terminology	ACI 116R-00
125	Standard Specifications for Tolerances for Concrete Construction and Materials and Commentary	ACI 117/117R-90 (Reapproved 2002)
149	Chemical Admixtures for Concrete	ACI 212.3R-04
179	Guide for Structural Lightweight-Aggregate Concrete	ACI 213R-03
217	Evaluation of Strength Test Results of Concrete	ACI 214R-02
237	Guide for Obtaining Cores and Interpreting Compressive Strength Results	ACI 214.4R-03
253	Causes, Evaluation and Repair of Cracks in Concrete Structures	ACI 224.1R-93 (Reapproved 1998)
275	Controlled Low-Strength Materials	ACI 229R-99 (Reapproved 2005)
291	Guide for Concrete Floor and Slab Construction	ACI 302.1R-04
369	Guide to Cast-In-Place Architectural Concrete Practice	ACI 303R-04
401	Guide for Measuring, Mixing, Transporting, and Placing Concrete	ACI 304R-00
443	Hot Weather Concreting	ACI 305R-99
463	Cold Weather Concreting	ACI 306R-88 (Reapproved 2002)
487	Guide to Curing Concrete	ACI 308R-01
519	Guide for Consolidation of Concrete	ACI 309R-96
559	Building Code Requirements for Structural Concrete (Chapters 3-7)	ACI 318-05
619	Guide to Formwork for Concrete	ACI 347-04
651	Design Responsibility for Architectural Precast-Concrete Projects	ACI 533.1R-02

ACI 301-05

Specifications for Structural Concrete
An ACI Standard

Reported by ACI Committee 301

W. Calvin McCall	Colin L. Lobo
Chair	Secretary

Jon B. Ardahl	Marwan A. Daye	Clifford Gordon[†]	David K. Maxwell
Domingo J. Carreira[*]	Mario R. Diaz	David P. Gustafson[*]	Timothy L. Moore
Oleh B. Ciuk	James A. Farny[*]	Jerry A. Holland	Jerry Parnes
Steven R. Close[*]	W. Bryant Frye	Roy H. Keck[*]	Aimee Pergalsky
D. Gene Daniel	Richard D. Gaynor	James A. Lee	James M. Shilstone, Sr.

Voting Subcommittee Members

James E. Anderson	Gene Hightower	G. Michael Robinson	Daniel J. Stanley
Ramon L. Carrasquillo	Narendra V. Jadhav	Edward D. Russell	Bruce A. Suprenant
Paul A. Decker	Michael L. Leming	Mehmet A. Samee	Robert L. Teerman
Dan Ellery[†]	William M. Klorman	W. Thomas Scott	Michael A. Whisonant
Alphonse E. Engelman	Mark A. Payne	William C. Sherman	Michelle L. Wilson
Thomas M. Greene	Kenneth B. Rear	Douglas J. Sordyl	Richard M. Wing

Consulting Members

Jeffrey W. Coleman	Gilbert J. Haddad	Ross S. Martin	Joseph A. McElroy
Steven H. Gebler	Atilano Lamana	Bryant Mather[†]	Carlos Videla

[*]Subcommittee chair.
[†]Deceased.

This specification is a Reference Specification that the Engineer or Architect can make applicable to any construction project by citing it in the Project Specifications. The Architect/Engineer supplements the provisions of this Reference Specification as needed by designating or specifying individual project requirements.

The document covers materials and proportioning of concrete; reinforcing and prestressing steels; production, placing, finishing, and curing of concrete; and formwork design and construction. Methods of treatment of joints and embedded items, repair of surface defects, and finishing of formed and unformed surfaces are specified. Separate sections are devoted to architectural concrete, lightweight concrete, mass concrete, prestressed concrete, and shrinkage-compensating concrete. Provisions governing testing, evaluation, and acceptance of concrete as well as acceptance of the structures are included.

Keywords: admixture; aggregate; air entrainment; architectural concrete; cement; cementitious materials; cold weather; compressive strength; concrete; concrete construction; concrete durability; concrete slab; consolidation; conveyor; curing; density; exposed-aggregate finish; finish; floors; formwork; grout; grouting; hot-weather; inspection; joint (construction, contraction, and isolation); lightweight concrete; mix; mixture proportion; placing; prestressed concrete; prestressing steel; reinforced concrete; reinforcement; repair; reshoring; shoring; shrinkage-compensating concrete; specification; subgrade; temperature; test; tolerance; water-cementitious material ratio; welded wire reinforcement.

NOTES TO SPECIFIER
This specification is incorporated by reference in the project specifications using the wording in P3 of the preface and including the information from the mandatory, optional, and submittal checklists following the specification.

PREFACE
P1. ACI Specification 301 is intended to be used by reference or incorporation in its entirety in the Project Specification. Do not copy individual Parts, Sections, Articles, or Paragraphs into the Project Specification, because taking them out of context may change their meaning.

P2. If Sections or Parts of ACI Specification 301 are copied into the Project Specification or any other document,

---

ACI 301-05 supersedes ACI 301-99 and became effective April 20, 2005.
Copyright © 2005, American Concrete Institute.
All rights reserved including rights of reproduction and use in any form or by any means, including the making of copies by any photo process, or by electronic or mechanical device, printed, written, or oral, or recording for sound or visual reproduction or for use in any knowledge or retrieval system or device, unless permission in writing is obtained from the copyright proprietors.

do not refer to them as an ACI Specification, because the specification has been altered.

P3. A statement such as the following will serve to make ACI Specification 301 a part of the Project Specification:

"Work on (Project Title) shall conform to all requirements of ACI 301-05 published by the American Concrete Institute, Farmington Hills, Michigan, except as modified by these Contract Documents."

P4. Each technical Section of ACI Specification 301 is written in the three-part Section format of the Construction Specifications Institute, as adapted for ACI requirements. The language is imperative and terse.

P5. The Specification is written to the Contractor. When a provision of this specification requires action on the Contractor's part, the verb "shall" is used. If the Contractor is allowed to exercise an option, the verb "may" or, when limited alternatives are available, the conjunctive phrase "shall either... or..." is used. Statements provided in the specification as information to the contractor use the verbs "may" or "will." Informational statements typically identify activities or options that "will" be taken or "may" be taken by the Owner or the Architect/Engineer.

CONTENTS
Preface, p. 1

SPECIFICATION:
Section 1—General requirements, p. 301-3
- 1.1—Scope
  - 1.1.1— Work specified
  - 1.1.2— Work not specified
- 1.2—Definitions
- 1.3—Reference standards and cited publications
  - 1.3.1—Reference standards
  - 1.3.2—Cited publications
  - 1.3.3—Field references
- 1.4—Standards-producing organizations
- 1.5—Submittals
  - 1.5.1—General
  - 1.5.2—Testing agency reports
- 1.6—Quality assurance
  - 1.6.1—General
  - 1.6.2—Testing agencies
  - 1.6.3—Testing responsibilities of Contractor
  - 1.6.4—Testing responsibilities of Owner's testing agency
  - 1.6.5—Tests on hardened concrete in-place
  - 1.6.6—Evaluation of concrete strength tests
  - 1.6.7—Acceptance of concrete strength
  - 1.6.8—Field acceptance of concrete
- 1.7—Acceptance of structure
  - 1.7.1—General
  - 1.7.2—Dimensional tolerances
  - 1.7.3—Appearance
  - 1.7.4—Strength of structure
  - 1.7.5—Durability
- 1.8—Protection of in-place concrete
  - 1.8.1—Loading and support of concrete
  - 1.8.2—Protection from mechanical injury

Section 2—Formwork and formwork accessories, p. 301-10
- 2.1—General
  - 2.1.1—Description
  - 2.1.2—Submittals
- 2.2—Products
  - 2.2.1—Materials
  - 2.2.2—Performance and design requirements
  - 2.2.3—Fabrication and manufacture
- 2.3—Execution
  - 2.3.1—Construction and erection of formwork
  - 2.3.2—Removal of formwork
  - 2.3.3—Reshoring and backshoring
  - 2.3.4—Strength of concrete required for removal of formwork

Section 3—Reinforcement and reinforcement supports, p. 301-13
- 3.1—General
  - 3.1.1—Submittals, data, and drawings
  - 3.1.2—Materials delivery, storage, and handling
- 3.2—Products
  - 3.2.1—Materials
  - 3.2.2—Fabrication
- 3.3—Execution
  - 3.3.1—Preparation
  - 3.3.2—Placement

Section 4—Concrete mixtures, p. 301-16
- 4.1—General
  - 4.1.1—Description
  - 4.1.2—Submittals
  - 4.1.3—Quality control
  - 4.1.4—Materials storage and handling
- 4.2—Products
  - 4.2.1—Materials
  - 4.2.2—Performance and design requirements
  - 4.2.3—Proportioning
- 4.3—Execution
  - 4.3.1—Measuring, batching, and mixing
  - 4.3.2—Delivery

Section 5—Handling, placing, and constructing, p. 301-20
- 5.1—General
  - 5.1.1—Description
  - 5.1.2—Submittals
  - 5.1.3—Delivery, storage, and handling
- 5.2—Products
  - 5.2.1—Materials
  - 5.2.2—Performance and design requirements
- 5.3—Execution
  - 5.3.1—Preparation
  - 5.3.2—Placement of concrete
  - 5.3.3—Finishing formed surfaces
  - 5.3.4—Finishing unformed surfaces
  - 5.3.5—Sawed contraction joints
  - 5.3.6—Curing and protection
  - 5.3.7—Repair of surface defects

Section 6—Architectural concrete, p. 301-26
6.1—General
  6.1.1—Description
  6.1.2—Submittals
  6.1.3—Quality assurance
  6.1.4—Product delivery, storage, and handling
  6.1.5—Project conditions
6.2—Products
  6.2.1—Materials
  6.2.2—Performance and design requirements
6.3—Execution
  6.3.1—Preparation
  6.3.2—Proportioning concrete mixtures
  6.3.3—Consolidation
  6.3.4—Formwork monitoring
  6.3.5—Formwork removal
  6.3.6—Repair of tie holes and surface defects
  6.3.7—Finishing

Section 7—Lightweight concrete, p. 301-28
7.1—General
  7.1.1—Description
  7.1.2—Submittals
  7.1.3—Product delivery, storage, and handling
7.2—Products
  7.2.1—Aggregates
  7.2.2—Performance and design requirements
  7.2.3—Mixtures
  7.2.4—Batching and mixing
7.3—Execution
  7.3.1—Consolidation
  7.3.2—Finishing
  7.3.3—Field quality control

Section 8—Mass concrete, p. 301-29
8.1—General
  8.1.1—Description
  8.1.2—Submittals
8.2—Products
  8.2.1—Materials
  8.2.2—Performance and design requirements
8.3—Execution
  8.3.1—Placement
  8.3.2—Curing and protection

Section 9—Prestressed concrete, p. 301-29
9.1—General
  9.1.1—Description
  9.1.2—Submittals
  9.1.3—Quality control
  9.1.4—Product delivery, handling, and storage
9.2—Products
  9.2.1—Materials
  9.2.2—Proportioning of concrete and grout mixtures
9.3—Execution
  9.3.1—Inspection
  9.3.2—Preparation
  9.3.3—Placement
  9.3.4—Tensioning

Section 10—Shrinkage-compensating concrete, p. 301-32
10.1—General
  10.1.1—Scope
  10.1.2—General requirements
  10.1.3—Submittals
10.2—Products
  10.2.1—Materials
  10.2.2—Performance and design requirements
  10.2.3—Proportioning
  10.2.4—Reinforcement
  10.2.5—Isolation-joint filler materials
10.3—Execution
  10.3.1—Reinforcement
  10.3.2—Placing
  10.3.3—Isolation joints
  10.3.4—Curing

NOTES TO SPECIFIER:
Foreword to checklists, p. 301-35

Mandatory requirements checklist, p. 301-37

Optional requirements checklist, p. 301-38

Submittals checklist, p. 301-45

SECTION 1—GENERAL REQUIREMENTS
1.1—Scope
1.1.1 *Work specified*—This Specification covers cast-in-place structural concrete.

Provisions of this Specification shall govern except where other provisions are specified in the Contract Documents.

1.1.2 *Work not specified*—The following subjects are not in the scope of this specification:
- Precast concrete products;
- Heavyweight shielding concrete;
- Slipformed paving concrete;
- Terrazzo;
- Insulating concrete;
- Refractory concrete;
- Shotcrete;
- Slipformed concrete walls; and
- Tilt-up concrete construction.

1.2—Definitions
acceptable or **accepted**—acceptable to or accepted by the Architect/Engineer.

ACI Concrete Field Testing Technician Grade 1—a person who has demonstrated knowledge and ability to perform and record the results of ASTM standard tests on freshly mixed concrete and to make and cure test specimens. Such knowledge and ability shall be demonstrated by passing prescribed written and performance examinations and having credentials that are current with the American Concrete Institute.

Architect/Engineer or **Engineer/Architect**—the Architect, Engineer, architectural firm, engineering firm, or architectural and engineering firm issuing project drawings and specifications or administering work under the Contract Documents.

architectural concrete—concrete that is exposed as an interior or exterior surface in the completed structure and is designated as architectural concrete in the Contract Documents; contributes to visual character of the completed structure and therefore requires special care in the selection of the concrete materials, forming, placing, and finishing to obtain the desired architectural appearance.

backshores—shores placed snugly under a concrete slab or structural member after the original formwork and shores have been removed from a small area at a time, without allowing the slab or member to deflect, or support its own weight or existing construction loads from above.

cement, expansive—a cement that, when mixed with water, produces a paste that, after setting, tends to increase in volume to a significantly greater degree than does portland cement paste; used to compensate for volume decrease due to shrinkage or to induce tensile stress in reinforcement.

cement, expansive Type K—a mixture of portland cement, anhydrous tetracalcium trialuminate sulfate ($C_4A_3S\bullet$), calcium sulfate ($CaSO_4$), and lime (CaO); the $C_4A_3\overline{S}$ is a constituent of a separately burned clinker that is interground with portland cement, or alternatively, is formed simultaneously with the portland-cement clinker compounds during the burning process.

Contract Documents—a set of documents supplied by Owner to Contractor as the basis for construction; these documents contain contract forms, contract conditions, specifications, drawings, addenda, and contract changes.

Contractor—the person, firm, or entity under contract for construction of the Work.

duct—a conduit (plain or corrugated) to accommodate prestressing steel for post-tensioned concrete.

exposed to public view—situated so that it can be seen from a public location after completion of the building.

high-early-strength concrete—concrete that is capable of attaining specified strength at an earlier age than 28 days through the use of high-early-strength cement or admixtures.

lightweight concrete—concrete of substantially lower density than normalweight concrete.

mass concrete—any volume of concrete with dimensions large enough to require that measures be taken to cope with generation of heat from hydration of the cement and attendant volume change to minimize cracking.

mass concrete, plain—Mass concrete containing no reinforcement or less reinforcement than necessary to be considered reinforced mass concrete.

mass concrete, reinforced—mass concrete containing adequate prestressed or nonprestressed reinforcement to act together with the concrete in resisting forces including those induced by temperature and shrinkage.

normalweight concrete—concrete having a density of approximately 150 lb/ft^3 made with gravel or crushed stone aggregates.

Owner—the corporation, association, partnership, individual, public body, or authority for whom the Work is constructed.

permitted—accepted or acceptable to the Architect/Engineer; usually pertains to a request by the Contractor, or to an item specified in the Contract Documents.

post-tensioning—a method of prestressing reinforced concrete in which tendons are tensioned after the concrete has hardened.

prestressed concrete—concrete in which internal stresses of sufficient magnitude and distribution are introduced to counteract to a desired degree the tensile stresses resulting from the service loads; in reinforced concrete, the prestress is commonly introduced by tensioning the tendons.

project drawings—graphic presentation of project requirements.

project specifications—the written document that details requirements for the Work in accordance with service parameters and other specific criteria.

reference specification—a standardized mandatory-language document prescribing materials, dimensions, and workmanship, incorporated by reference in Contract Documents, with information in the Mandatory Requirements Checklist required to be provided in the Project Specification.

reference standards—standardized mandatory-language documents of a technical society, organization, or association, including codes of local or federal authorities, which are incorporated by reference in Contract Documents.

required—required in this Specification or the Contract Documents.

reshores—shores placed snugly under a stripped concrete slab or other structural member after the original forms and shores have been removed from a large area, thus requiring the new slab or structural member to deflect and support its own weight and existing construction loads applied before the installation of the reshores.

sheathing, prestressing—a material encasing prestressing steel to prevent bonding of the prestressing steel with the surrounding concrete, to provide corrosion protection, and to contain the corrosion-inhibiting coating.

sheathing, wood formwork—the materials forming the contact face of forms; also called lagging or sheeting.

shop drawing—a drawing that provides details for a particular task that is developed by the Contractor and reviewed by the Engineer. The shop drawing is prepared to the requirements of the project drawings and project specifications.

shore—a temporary support designed to support formwork, fresh concrete, and construction loads from above for recently built structures that have not developed full design strength.

shrinkage-compensating concrete—a concrete made using an expansive cement that increases in volume after setting, designed to induce compressive stresses in elastically restrained concrete to approximately offset the tensile stresses resulting from drying shrinkage.

strength test—the average of the compressive strengths of two or more cylinders made from the same sample of concrete and tested at 28 days or at the specified test age.

structural lightweight concrete—Structural concrete made with lightweight aggregate; the equilibrium density, as calculated by ASTM C 567, usually is in the range of 90 to 115 lb/ft^3 with a minimum compressive strength of 2500 psi.

submitted—documents or materials provided to Architect/Engineer for review or acceptance.

Work—the entire construction or separately identifiable parts thereof required to be furnished under Contract Documents.

1.3—Reference standards and cited publications

1.3.1 *Reference standards*—Standards of ACI, ASTM, CRD, and AWS referred to in this Specification are listed with serial designation including year of adoption or revision and are part of this Specification.

1.3.1.1 *ACI standards*

ACI 117-90	Standard Specifications for Tolerances for Concrete Construction and Materials
ACI 423.6-01	Specification for Unbonded Single-Strand Tendons

1.3.1.2 *ASTM standards*

A 82-02	Standard Specification for Steel Wire, Plain, for Concrete Reinforcement
A 184/ A 184M-01	Standard Specification for Welded Deformed Steel Bar Mats for Concrete Reinforcement
A 185-02	Standard Specification for Steel Welded Wire Reinforcement, Plain, for Concrete
A 416/A 416M-02	Standard Specification for Steel Strand, Uncoated Seven-Wire, for Prestressed Concrete
A 421/A 421M-02	Standard Specification for Uncoated Stress-Relieved Steel Wire for Prestressed Concrete
A 496-02	Standard Specification for Steel Wire, Deformed, for Concrete Reinforcement
A 497/A 497M-02	Standard Specification for Steel Welded Wire Reinforcement, Deformed, for Concrete
A 615/A 615M-04b	Standard Specification for Deformed and Carbon Steel Bars for Concrete Reinforcement
A 706/A 706M-04b	Standard Specification for Low-Alloy Steel Deformed and Plain Bars for Concrete Reinforcement
A 722/A 722M-98 (2003)	Standard Specification for Uncoated High-Strength Steel Bars for Prestressing Concrete
A 767/ A 767M-00b	Standard Specification for Zinc-Coated (Galvanized) Steel Bars for Concrete Reinforcement
A 775/ A 775M-04a	Standard Specification for Epoxy-Coated Steel Reinforcing Bars
A 779/A 779M-00	Standard Specification for Steel Strand, Seven-Wire, Uncoated, Compacted, Stress-Relieved for Prestressed Concrete
A 780-01	Standard Practice for Repair of Damaged Hot-Dip Galvanized Coatings
A 882/A 882M-04a	Standard Specification for Filled Epoxy-Coated Seven-Wire Prestressing Steel Strand
A 884/A 884M-04	Standard Specification for Epoxy-Coated Steel Wire and Welded Wire Reinforcement
A 934/A 934M-04	Standard Specification for Epoxy-Coated Prefabricated Steel Reinforcing Bars
A 955/ A 955M-04a^{e1}	Standard Specification for Deformed and Plain Stainless Steel Bars for Concrete Reinforcement
A 970/ A 970M-04a^{e1}	Standard Specification for Welded or Forged Headed Bars for Concrete Reinforcement
A 996/A 996M-04	Standard Specification for Rail-Steel and Axle-Steel Deformed Bars for Concrete Reinforcement
C 31/C 31M-03a	Standard Practice for Making and Curing Concrete Test Specimens in the Field
C 33-03	Standard Specification for Concrete Aggregates
C 39/C 39M-03	Standard Test Method for Compressive Strength of Cylindrical Concrete Specimens
C 42/ C 42M-04	Standard Test Method for Obtaining and Testing Drilled Cores and Sawed Beams of Concrete
C 94/C 94M-04	Standard Specification for Ready-Mixed Concrete
C 138/C 138M-01a	Standard Test Method for Unit Weight, Yield, and Air Content (Gravimetric) of Concrete
C 143/ C143M-03	Standard Test Method for Slump of Hydraulic-Cement Concrete
C 150-04a	Standard Specification for Portland Cement
C 171-03	Standard Specification for Sheet Materials for Curing Concrete
C 172-04	Standard Practice for Sampling Freshly Mixed Concrete
C 173/ C 173M-01^{e1}	Standard Test Method for Air Content of Freshly Mixed Concrete by the Volumetric Method
C 192/C 192M-02	Standard Practice for Making and Curing Concrete Test Specimens in the Laboratory
C 231-04	Standard Test Method for Air Content of Freshly Mixed Concrete by the Pressure Method
C 260-01	Standard Specification for Air-Entraining Admixtures for Concrete
C 309-03	Standard Specification for Liquid Membrane-Forming Compounds for Curing Concrete
C 330-04	Standard Specification for Lightweight Aggregates for Structural Concrete

C 387-04	Standard Specification for Packaged, Dry, Combined Materials for Mortar and Concrete
C 404-03	Standard Specification for Aggregates for Masonry Grout
C 494/C 494-04	Standard Specification for Chemical Admixtures for Concrete
C 567-04	Standard Test Method for Determining Density of Structural Lightweight Concrete
C 595-03	Standard Specification for Blended Hydraulic Cements
C 597-02	Standard Test Method for Pulse Velocity Through Concrete
C 618-03	Standard Specification for Coal Fly Ash and Raw or Calcined Natural Pozzolan for Use in Concrete
C 684-99 (2003)	Standard Test Method for Making, Accelerated Curing, and Testing Concrete Compression Test Specimens
C 685/C 685M-01	Standard Specification for Concrete Made By Volumetric Batching and Continuous Mixing
C 803/C 803M-03	Standard Test Method for Penetration Resistance of Hardened Concrete
C 805-02	Standard Test Method for Rebound Number of Hardened Concrete
C 845-04	Standard Specification for Expansive Hydraulic Cement
C 873-04	Standard Test Method for Compressive Strength of Concrete Cylinders Cast in Place in Cylindrical Molds
C 878-03	Standard Test Method for Restrained Expansion of Shrinkage-Compensating Concrete
C 881/C 881M-02	Standard Specification for Epoxy-Resin-Base Bonding Systems for Concrete
C 900-01	Standard Test Method for Pullout Strength of Hardened Concrete
C 928-00	Standard Specification for Packaged, Dry, Rapid Hardening Cementitious Materials for Concrete Repairs
C 939-02	Standard Test Method for Flow of Grout for Preplaced-Aggregate Concrete (Flow Cone Method)
C 989-04	Standard Specification for Ground Granulated Blast-Furnace Slag for Use in Concrete and Mortars
C 1017/ C 1017M-03	Standard Specification for Chemical Admixtures for Use in Producing Flowing Concrete
C 1012-04	Standard Test Method for Length Change of Hydraulic-Cement Mortars Exposed to a Sulfate Solution
C 1059-99	Standard Specification for Latex Agents for Bonding Fresh to Hardened Concrete
C 1064/ C 1064M-04	Standard Test Methods for Temperature of Freshly Mixed Portland Cement Concrete
C 1074-04	Standard Practice for Estimating Concrete Strength by the Maturity Method
C 1077-02	Standard Practice for Laboratories Testing Concrete and Concrete Aggregates for Use in Construction and Criteria for Laboratory Evaluation
C 1107-02	Standard Specification for Packaged Dry, Hydraulic Cement Grout (Nonshrink)
C1157-03	Standard Performance Specification for Hydraulic Cement
C 1218/ C 1218M-99	Standard Test Method for Water-Soluble Chloride in Mortar and Concrete
C 1240-04	Standard Specification for Silica Fume Used in Cementitious Mixtures
C 1315-03	Standard Specification for Liquid Membrane-Forming Compounds Having Special Properties for Curing and Sealing Concrete
C 1602/ C 1602M-04	Standard Specification for Mixing Water Used in the Production of Hydraulic Cement Concrete
D 98-98	Standard Specification for Calcium Chloride
D 994-98 (2003)	Standard Specification for Preformed Expansion Joint Filler for Concrete (Bituminous Type)
D 1621-04	Standard Test Methods for Compressive Properties of Rigid Cellular Plastics
D 1751-99	Standard Specification for Preformed Expansion Joint Fillers for Concrete Paving and Structural Construction (Non-extruding and Resilient Bituminous Types)
D 1752-04	Standard Specification for Preformed Sponge Rubber and Cork Expansion Joint Fillers for Concrete Paving and Structural Construction
D 3575-00[e1]	Standard Test Methods for Flexible Cellular Materials Made from Olefin Polymers
E 329-03	Standard Specification for Agencies Engaged in the Testing and/or Inspection of Materials Used in Construction
E 1155-96(2001)	Standard Test Method for Determining FF Floor Flatness and FL Floor Levelness Numbers

1.3.1.3 *Other referenced standards*—Other standards referenced in this Specification:

ANSI/ AWS D1.4-98	Structural Welding Code—Reinforcing Steel
CRD-C 513-74	Specification for Rubber Waterstops
CRD-C 572-74	Specification for Polyvinyl chloride Waterstops

1.3.2 *Cited publications*—Publications cited in this Specification:

ACI 318-05	Building Code Requirements for Structural Concrete
ACI CP1-04	ACI Certification Concrete Field Testing Technician—Grade I
ACI CP10-95	ACI Certification Flatwork Technician and Flatwork Finisher
ACI SP-15	Field Reference Manual: Specifications for Structural Concrete (ACI 301-05) with Selected ACI and ASTM References
CRSI MSP-2-01	Manual of Standard Practice, 27th Edition, Voluntary Certification Program for Fusion-Bonded Epoxy Coating Applicator Plants

1.3.3 *Field references*—Keep in Contractor's field office a copy of the following reference:

ACI SP-15   Field Reference Manual: Specifications for Structural Concrete (ACI 301-05) with Selected ACI and ASTM References.

1.4—Standards-producing organizations

Abbreviations for and complete names and addresses of organizations issuing documents referred to in this Specification are listed:

American Concrete Institute (ACI)
P.O. Box 9094
Farmington Hills, MI 48333-9094

ASTM International
100 Barr Harbor Drive
West Conshohocken, PA 19428

American Welding Society (AWS)
550 Northwest 42nd Avenue
Miami, FL 33126

Concrete Reinforcing Steel Institute (CRSI)
933 N. Plum Grove Road
Schaumburg, IL 60173

U.S. Army Corps of Engineers [COE(CRD)]
Waterways Experiment Station
3909 Halls Ferry Road
Vicksburg, MS 39180

National Ready Mixed Concrete Association (NRMCA)
900 Spring Street
Silver Spring, MD 20910

1.5—Submittals

1.5.1 *General*—Unless otherwise specified, submittals required in this Specification shall be submitted for review and acceptance.

1.5.2 *Testing agency reports*—Testing agencies will report results of concrete and concrete materials tests and inspections performed during the course of the Work to the Owner, Architect/Engineer, Contractor, and the concrete supplier. Strength test reports will include location in the Work where the batch represented by test was deposited and the batch ticket number. Reports of strength tests will include detailed information of storage and curing of specimens before testing. Final reports will be provided within seven days of test completion.

1.6—Quality assurance

1.6.1 *General*—Concrete materials and operations may be tested and inspected by the Owner as work progresses. Failure to detect defective work or material will not prevent rejection if a defect is discovered later nor shall it obligate the Architect/Engineer for final acceptance.

1.6.2 *Testing agencies*—Agencies that test concrete materials shall meet the requirements of ASTM C 1077. Testing agencies that test reinforcing steel shall meet the requirements of ASTM E 329. Testing agencies shall be accepted by the Architect/Engineer before performing any work. Field tests of concrete required in 1.6.3 and 1.6.4 shall be made by an ACI Concrete Field Testing Technician Grade 1 certified in accordance with ACI CP1 or equivalent. Equivalent certification programs shall include requirements for written and performance examinations as stipulated in ACI publication CP1.

1.6.3 *Testing responsibilities of Contractor*

1.6.3.1 Submit data on qualifications of proposed testing agency for acceptance. Use of testing services will not relieve the Contractor of the responsibility to furnish materials and construction in compliance with the Contract Documents.

1.6.3.2 *Duties and responsibilities*—Unless otherwise specified in the Contract Documents, the Contractor shall assume the duties and responsibilities given in 1.6.3.2.a through 1.6.3.2.g:

1.6.3.2.a Qualify proposed materials and establish mixture proportions.

1.6.3.2.b Allow access to the project site or to the source of materials and assist Owner's testing agency in obtaining and handling samples at the project site or at the source of materials.

1.6.3.2.c Advise Owner's testing agency at least 24 h in advance of operations to allow for completion of quality tests and for assignment of personnel.

1.6.3.2.d Provide and maintain adequate facilities on the project site for safe storage and initial curing of concrete test specimens as required by ASTM C 31/C 31M for the sole use of the testing agency.

1.6.3.2.e Submit test data and documentation on concrete ingredient materials and mixture proportions.

1.6.3.2.f Submit quality-control program of the concrete supplier and provide copies of test reports pertaining to the Work.

1.6.3.2.g When specified or permitted to base concrete acceptance on accelerated strength testing, submit correlation data for the standard 28-day compressive strength based on at least 15 sets of test data in accordance with 1.6.4.2.e with concrete made with the same materials encompassing a range of at least the required average strength f'_{cr}, plus or minus 1000 psi.

1.6.3.3 *Tests required of Contractor's testing agency*—Unless otherwise specified in the Contract Documents, the

Contractor shall provide, at no cost to the Owner, the necessary testing services given in 1.6.3.3.a and 1.6.3.3.b:

1.6.3.3.a Qualification of proposed materials and establishment of concrete mixtures.

1.6.3.3.b Other testing services needed or required by Contractor.

1.6.4 *Testing responsibilities of Owner's testing agency*

1.6.4.1 Unless otherwise specified in the Contract Documents, the Owner's testing agency will provide the necessary services given in 1.6.4.1.a through 1.6.4.1.c:

1.6.4.1.a Representatives of the Owner's testing agency will inspect, sample, and test materials and production of concrete as required by the Architect/Engineer. When it appears that material furnished or work performed by the Contractor fails to conform to Contract Documents, the testing agency will immediately report such deficiency to the Architect/Engineer, Contractor, and concrete supplier.

1.6.4.1.b The testing agency and its representatives are not authorized to revoke, alter, relax, enlarge, or release any requirement of the Contract Documents, nor to accept or reject any portion of the Work.

1.6.4.1.c The testing agency will report test and inspection results that pertain to the Work to the Architect/Engineer, Contractor, and concrete supplier within seven days after tests and inspections are performed.

1.6.4.2 *Testing services*—When required by the Owner or the Architect/Engineer, the Owner's testing agency will perform the following testing services given in 1.6.4.2.a through 1.6.4.2.i at no cost to the Contractor:

1.6.4.2.a Review and check-test proposed materials for compliance with Contract Documents.

1.6.4.2.b Review and check-test proposed concrete mixture as required by the Architect/Engineer.

1.6.4.2.c Obtain production samples of materials at plants or stockpiles during the course of the Work and test for compliance with the Contract Documents.

1.6.4.2.d Obtain samples in accordance with ASTM C 172. Select the truckloads or batches of concrete to be tested on a random basis, using random numbers selected before commencement of concrete placement.

Obtain at least one composite sample for each 100 yd^3, or fraction thereof, of each concrete mixture placed in any one day. When the total quantity of a given concrete mixture is less than 50 yd^3, the strength tests may be waived by the Architect/Engineer.

1.6.4.2.e Conduct concrete strength tests during construction in accordance with the following procedures:

- Mold and cure a minimum of three cylinders from each sample in accordance with ASTM C 31/C 31M. Record any deviations from the ASTM requirements in the test report.
- Test cylinders in accordance with ASTM C 39/C39M. Test one specimen at seven days for information, and test a minimum of two specimens at 28 days for acceptance, unless otherwise specified. The compressive strength test results for acceptance shall be the average of the compressive strengths from the specimens tested at 28 days. If a specimen in a test shows evidence of improper sampling, molding, or testing, discard the specimen and consider the strength of the remaining cylinder or cylinders to be the test result. If all specimens in a test show defects, discard the entire test.
- When accelerated testing of concrete is specified or permitted as an alternative to standard testing, mold and cure two specimens from each composite sample in accordance with ASTM C 684, following the procedure specified by the Architect/Engineer. Make at least one accelerated strength test from each composite sample in 1.6.4.2.d and one standard cured 28-day compressive-strength test for at least every other accelerated strength test in accordance with ASTM C 31/C 31M. Use these test results to maintain and update the correlation between accelerated and standard 28-day compressive-strength tests.

1.6.4.2.f Determine slump of each composite sample taken in accordance with 1.6.4.2.d and whenever consistency of concrete appears to vary, using ASTM C 143/C 143M.

1.6.4.2.g Determine the temperature of each composite sample taken in accordance with 1.6.4.2.d using ASTM C 1064/C1064M.

1.6.4.2.h Determine the air content of normalweight concrete using ASTM C 231, C 173, or C 138 for each composite sample taken in accordance with 1.6.4.2.d or as directed by the Architect/Engineer. Additional tests may be performed as necessary.

1.6.4.2.i When the Contract Documents indicate concrete will be exposed to deicing salts, air content tests will be made on samples from the first three batches in the placement and until three consecutive batches have air contents within the range specified in 4.2.2.4, at which time every fifth batch will be tested. This test frequency will be maintained until a batch is not within the range specified in 4.2.2.4, at which time testing of each batch will be resumed until three consecutive batches have air contents within the range specified in 4.2.2.4. Additional tests may be performed as necessary for control. These air content tests may be taken on composite samples in 1.6.4.2.d or on samples from the batch at any time after discharge of 2 ft^3 of concrete.

1.6.4.3 *Additional testing services*—When required by the Architect/Engineer, the Owner's testing agency will perform the following testing services at no cost to the Contractor:

- Inspect the concrete batching, mixing, and delivery operations;
- Inspect forms, foundation preparation, reinforcing steel, embedded items, reinforcing steel placement, and concrete placing, finishing, and curing operations;
- Sample concrete at point of placement and other locations as directed by the Architect/Engineer and perform required tests;
- Review the manufacturer's report for each shipment of cement, reinforcing steel, and prestressing tendons, and conduct laboratory tests or spot checks of the materials received for compliance with specifications; and
- Other testing or inspection services as required by the Architect/Engineer.

1.6.4.4 *Other testing services as needed*—The Contractor shall pay for the following testing services performed, when necessary, by the Owner's testing agency:
- Additional testing and inspection required because of changes in materials or mixture proportions requested by the Contractor; and
- Additional testing of materials or concrete occasioned by failure to meet specification requirements.

1.6.5 *Tests on hardened concrete in-place*

1.6.5.1 *General*—When needed, tests on hardened concrete will be performed by the Owner's testing agency. Testing shall be at the Contractor's expense when this Specification requires such tests to verify the strength of the structure. The Owner will pay costs if tests are at the Owner's request and not required by this Specification.

1.6.5.2 *Nondestructive tests*—Use of the rebound hammer in accordance with ASTM C 805, pulse-velocity method in accordance with ASTM C 597, or other nondestructive tests may be permitted by the Architect/Engineer for evaluating the uniformity and relative concrete strength in place, or for selecting areas to be cored.

1.6.5.3 *Core tests*

1.6.5.3.a Where required by the Architect/Engineer, obtain cores in accordance with ASTM C 42/C42 M. Wipe cores surface-dry immediately after coring and allow to dry in air for a period not exceeding one hour after drilling. Seal cores in plastic bags or nonabsorbent containers until testing. End preparation of cores shall be completed within 48 h after drilling. Test cores not earlier than 48 h after drilling or last wetting and not later than seven days after the cores were drilled from the structure.

1.6.5.3.b At least three representative cores shall be taken from each area of in-place concrete that is considered potentially deficient. The location of cores shall be determined by the Architect/Engineer to impair the strength of the structure as little as possible. If, before testing, cores show evidence of having been damaged subsequent to or during removal from the structure, replacement cores shall be taken.

1.6.5.3.c Fill core holes with low-slump concrete or mortar of a strength equal to or greater than the original concrete.

1.6.6 *Evaluation of concrete strength tests*

1.6.6.1 *Standard molded and cured strength specimens*—Test results from standard molded and cured test cylinders shall be evaluated separately for each specified concrete mixture. Evaluation will be valid only if tests have been conducted in accordance with procedures specified. For evaluation, each specified mixture shall be represented by at least five tests.

1.6.6.2 *Nondestructive tests*—Test results will be evaluated by the Architect/Engineer and will be valid only if tests have been conducted using properly calibrated equipment in accordance with recognized standard procedures and an acceptable correlation between test results and concrete compressive strength has been established and is submitted.

1.6.6.3 *Core tests*—Core test results will be evaluated by the Architect/Engineer and will be valid only if tests have been conducted in accordance with specified procedures.

1.6.7 *Acceptance of concrete strength*

1.6.7.1 *Standard molded and cured strength specimens*—The strength level of concrete will be considered satisfactory when: the averages of all sets of three consecutive compressive strength test results molded and cured in accordance with the requirements of ASTM C 31/C 31M equal or exceed f'_c; and no individual strength test result falls below f'_c by more than 500 psi when f'_c is 5000 psi or less, or by more than $0.10f'_c$ when f'_c is more than 5000 psi. These criteria also apply to accelerated strength testing unless another basis for acceptance is specified in the Contract Documents.

1.6.7.2 *Nondestructive tests*—Nondestructive tests shall not be used as the sole basis for accepting or rejecting concrete, but may be used, when permitted, to evaluate concrete where standard molded and cured cylinders have yielded results not meeting the criteria in 1.6.7.1.

1.6.7.3 *Core tests*—Strength level of concrete in the area represented by core tests will be considered adequate when the average compressive strength of the cores is equal to at least 85% of f'_c, and if no single core is less than 75% of the specified compressive strength f'_c.

1.6.8 *Field acceptance of concrete*

1.6.8.1 *Air content*—Concrete not within the limits of air-entrainment indicated in 4.2.2.4 and tested in accordance with 1.6.4.2.h shall not be used in the Work.

1.6.8.2 *Slump*—Concrete not within the slump limits of 4.2.2.2 shall not be used in the Work.

1.6.8.3 *Temperature*—Concrete not within temperature limits of 4.2.2.8 shall not be used in the Work.

1.7—Acceptance of structure

1.7.1 *General*—Completed concrete work shall conform to applicable requirements of this Specification and the Contract Documents.

1.7.1.1 Concrete work that fails to meet one or more requirements of the Contract Documents but subsequently is repaired to bring the concrete into compliance will be accepted.

1.7.1.2 Concrete work that fails to meet one or more requirements of the Contract Documents and cannot be brought into compliance may be rejected.

1.7.1.3 Repair rejected concrete work by removing and replacing or by reinforcing with additional construction as required by the Architect/Engineer. To bring rejected work into compliance, use repair methods that will maintain specified strength and meet applicable requirements for function, durability, dimensional tolerances, and appearance as determined by the Architect/Engineer.

1.7.1.4 Submit for acceptance the proposed repair methods, materials, and modifications needed to repair the concrete work to meet the requirements of Contract Documents.

1.7.1.5 Contractor shall pay all costs to bring concrete work into compliance with requirements of Project Specification.

1.7.1.6 Concrete members cast in the wrong location may be rejected.

1.7.2 *Dimensional tolerances*

1.7.2.1 Formed surfaces resulting in concrete outlines smaller than permitted by the tolerances of ACI 117 may be considered deficient in strength and subject to the provisions of 1.7.4.

1.7.2.2 Formed surfaces resulting in concrete outlines larger than permitted by ACI 117 may be rejected. Remove excess materials when required by the Architect/Engineer.

1.7.2.3 Inaccurately formed concrete surfaces that exceed ACI 117 tolerances may be rejected.

1.7.2.4 Finished slabs exceeding the tolerances in 5.3.4.3 may be corrected provided they are brought into compliance with 1.7.3, 1.7.4, and 1.7.5.

1.7.2.5 Concrete with tolerances and defects exceeding the limitations of 2.2.2.4 may be rejected.

1.7.3 *Appearance*

1.7.3.1 Concrete not meeting the requirements of 5.3.3 or 5.3.4 shall be brought into compliance in accordance with 1.7.1.

1.7.4 *Strength of structure*

1.7.4.1 *Criteria for determining potential strength deficiency*—Strength will be considered deficient and concrete work will be rejected when the work fails to comply with requirements that control the strength of the structure including, but not limited to, the conditions given in 1.7.4.1.a through 1.7.4.1.f:

1.7.4.1.a Concrete strength failing to comply with requirements of 1.6.7.

1.7.4.1.b Reinforcing steel size, quantity, grade, position, or arrangement at variance with the requirements of Section 3 or other Contract Documents.

1.7.4.1.c Concrete elements that differ from the required dimensions or location.

1.7.4.1.d Curing not performed in accordance with Contract Documents.

1.7.4.1.e Inadequate protection of concrete from extreme temperature and other adverse environmental conditions during early stages of hardening and strength development.

1.7.4.1.f Mechanical injury, construction fires, or premature removal of formwork resulting in deficient strength.

1.7.4.2 *Action required when strength is potentially deficient*—When strength of the structure is considered potentially deficient, the actions given in 1.7.4.2.a through 1.7.4.2.e may be required by the Architect/Engineer:

1.7.4.2.a Structural analysis or additional testing, or both.

1.7.4.2.b Core tests.

1.7.4.2.c If testing is inconclusive or impractical or if structural analysis does not confirm the safety of the structure, load tests may be required and their results evaluated in accordance with ACI 318.

1.7.4.2.d Concrete work rejected by structural analysis or by results of a load test shall be strengthened with additional construction when required by the Architect/Engineer, or replaced.

1.7.4.2.e Document all repair work proposed to bring strength-deficient concrete work into compliance with Contract Documents, and submit the documentation to the Architect/Engineer for acceptance.

1.7.5 *Durability*

1.7.5.1 *Criteria for determining potential durability deficiency*—Durability of concrete work will be considered deficient and the concrete work will be rejected when it fails to comply with the requirements that control durability of the structure, including, but not limited to, the conditions given in 1.7.5.1a through 1.7.5.1.f:

1.7.5.1.a Strength failing to comply with 1.6.7.

1.7.5.1.b Materials for concrete not conforming to the requirements in 4.2.1.1, 4.2.1.2, 4.2.1.3, and 4.2.1.4.

1.7.5.1.c Concrete not conforming to the air-entrainment requirements in Contract Documents or the air content limits of Table 4.2.2.4.

1.7.5.1.d Curing not in accordance with Contract Documents.

1.7.5.1.e Inadequate protection of concrete from detrimental temperature and other detrimental environmental conditions during early stages of hardening and strength development.

1.7.5.1.f Concrete exceeding the maximum allowable chloride-ion content requirements in Table 4.2.2.6.

1.7.5.2 *Action required when durability is potentially deficient*—When durability of the structure is considered to be potentially deficient, the actions given in 1.7.5.2.a through 1.7.5.2.e may be required by the Architect/Engineer:

1.7.5.2.a Obtain and test samples of the ingredient materials used in the concrete.

1.7.5.2.b Obtain samples of concrete from the structure by coring, sawing, or other acceptable means.

1.7.5.2.c Laboratory evaluation of concrete and concrete materials to assess the ability of concrete to resist weathering action, chemical attack, abrasion, or other deterioration, and to protect reinforcement and embedments from corrosion.

1.7.5.2.d Repair or replace concrete rejected for durability deficiency as directed by the Architect/Engineer.

1.7.5.2.e Document repair work to bring concrete work into compliance with Contract Documents and submit the documentation to the Architect/Engineer for acceptance.

1.8—Protection of in-place concrete

1.8.1 *Loading and support of concrete*—Do not allow construction loads to exceed the superimposed load that the structural member, with necessary supplemental support, is capable of supporting safely and without damage or unacceptable deflections.

1.8.2 *Protection from mechanical injury*—During the curing period, protect concrete from damaging mechanical disturbances, including load-induced stresses, shock, and harmful vibration. Protect concrete surfaces from damage by construction traffic, equipment, materials, rain or running water, and other adverse weather conditions.

SECTION 2—FORMWORK AND FORMWORK ACCESSORIES

2.1—General

2.1.1 *Description*—This section covers design, construction, and treatment of formwork to confine and shape concrete to the required dimensions.

2.1.2 *Submittals*

2.1.2.1 Submit the data required in 2.1.2.1.a through 2.1.2.1.f unless otherwise specified:

2.1.2.1.a *Formwork facing materials*—Data on form-facing materials proposed for smooth-form finish if different from that specified in 2.2.1.1.

2.1.2.1.b *Construction and contraction joints*—Location of construction and contraction joints proposed if different from those indicated in the Contract Documents.

2.1.2.1.c *Testing for formwork removal*—Data on method for determining strength of concrete for removal of formwork in accordance with 2.3.4.2 when a method other than field-cured cylinders is proposed.

2.1.2.1.d *Formwork removal plans*—Detail plans for formwork removal operations when removal of forms at concrete strengths lower than that specified in 2.3.2.5 is proposed.

2.1.2.1.e *Reshoring and backshoring plans*—When reshoring or backshoring is required or permitted, submit procedures and plans of operations, before use, sealed by a professional Engineer licensed in the state where work will be performed. Indicate on shop drawings the magnitude of construction loads permitted during reshoring or backshoring.

2.1.2.1.f Data on formwork release agent or form liner proposed for use with each formed surface.

2.1.2.2 Submit data required in 2.1.2.2.a through 2.1.2.2.e when required by the Contract Documents:

2.1.2.2.a Shop drawings for formwork sealed by a professional Engineer licensed in the state where the work will be done.

2.1.2.2.b Calculations for formwork, reshoring and backshoring, sealed by a professional Engineer licensed in the state where the work will be done.

2.1.2.2.c Manufacturer's data and samples of form ties.

2.1.2.2.d Manufacturer's data and samples of expansion joint materials.

2.1.2.2.e Manufacturer's data and samples of waterstops.

2.2—Products

2.2.1 *Materials*

2.2.1.1 *Form-facing materials*—Materials for form faces in contact with concrete shall meet 5.3.3.5 and the following requirements unless otherwise specified in Contract Documents.
- For rough-form finish—No form-facing material is specified.
- For smooth-form finish—Use plywood, tempered concrete-form-grade hardboard, metal, plastic, paper, or other acceptable materials capable of producing the desired finish for form-facing materials. Form-facing materials shall produce a smooth, uniform texture on the concrete. Do not use form-facing materials with raised grain, torn surfaces, worn edges, dents, or other defects that will impair the texture of concrete surfaces.

2.2.1.2 *Formwork accessories*—Use commercially manufactured accessories for formwork accessories that are partially or wholly embedded in concrete, including ties and hangers. Do not use nonfabricated wire form ties. Where indicated in the Contract Documents, use form ties with integral water barrier plates in walls or other acceptable positive water barriers.

2.2.1.3 *Formwork release agents*—Use commercially manufactured formwork release agents that prevent formwork absorption of moisture, prevent bond with concrete, and do not stain the concrete surfaces.

2.2.1.4 *Expansion joint filler*—Premolded expansion joint filler shall conform to ASTM D 994, D 1751, or D 1752.

2.2.1.5 *Other embedded items*—Use waterstops, sleeves, inserts, anchors, and other embedded items of the material and design indicated in the Contract Documents. Waterstop materials shall meet requirements of CRD C 513 for rubber waterstop, or CRD C 572 for polyvinyl chloride waterstop. Splice the waterstops and use molded pieces as recommended by the manufacturer.

2.2.2 *Performance and design requirements*

2.2.2.1 Design and engineering of formwork shall be the responsibility of the Contractor. When required by the Contract Documents, design calculations for formwork and formwork drawings shall be sealed by a professional Engineer licensed in the state where the Work will be done.

2.2.2.2 Design formwork, shores, reshores, and backshores to support all loads transmitted to them and to comply with the requirements of the applicable building code. Design formwork to withstand the pressure resulting from placement and vibration of concrete and to maintain specified tolerances.

2.2.2.3 Do not use earth cuts as forms for vertical or sloping surfaces unless required or permitted by Contract Documents.

2.2.2.4 Maximum deflection of facing materials reflected on concrete surfaces exposed to public view shall be 1/240 of the span between structural members of the formwork. For architectural concrete, see 6.2.2.1.a.

2.2.2.5 *Formed construction and contraction joints*

2.2.2.5.a Locate and form construction joints that least impair strength of the structure and meet the requirements of 5.3.2.6.

2.2.2.5.b Unless otherwise specified or permitted, locate and detail formed construction joints to the following requirements:
- Locate construction joints within the middle third of the spans of slabs, beams, and girders. When a beam intersects a girder within this region, offset the joint in the girder a distance equal to or greater than twice the width of the beam;
- Locate joints in walls and columns at the underside of slabs, beams, or girders and at the tops of footings or slabs; and
- Make joints perpendicular to the main reinforcement.

2.2.2.5.c Provide keyways where indicated on Contract Documents. Unless otherwise specified, longitudinal keyways indicated on the Contract Documents, shall be a minimum of 1-1/2 in. deep in joints in walls and between walls and slabs or footings.

2.2.2.5.d Provide construction and contraction joints where indicated on the Contract Documents. Submit for acceptance the location of construction and contraction joints differing from those indicated on the Contract Documents.

2.2.2.6 For a smooth-form finish, set the facing materials in an orderly and symmetrical arrangement, and keep the number of seams to a practical minimum. Facing materials shall be supported with studs or other backing capable of maintaining deflections within the tolerances specified in 2.2.2.4.

2.2.3 *Fabrication and manufacture*

2.2.3.1 Formwork shall be tight to prevent loss of mortar from concrete.

2.2.3.2 Place 3/4 in. minimum chamfer strips in the corners of formwork to produce beveled edges on permanently exposed surfaces unless otherwise specified. Do not bevel reentrant corners or edges of formed joints of concrete unless specified in the Contract Documents.

2.2.3.3 Inspect formwork and remove deleterious material immediately before concrete is placed. Provide temporary openings where needed at the base of column and wall formwork to facilitate cleaning and inspection.

2.2.3.4 Fabricate form ties so ends or end fasteners can be removed with minimum spalling at the faces of concrete. After the ends or end fasteners of form ties have been removed, terminate the embedded portion of ties not less than two diameters, or twice the minimum cross-sectional dimension of the tie, from the formed concrete surface. This distance shall be a minimum of 3/4 in. Repair tie holes in accordance with 5.3.7.2.

2.2.3.5 Locate waterstops in joints where indicated on Contract Documents. Use pieces of premolded waterstop with a maximum practicable length to create the minimum number of end joints. Make joints in waterstops in accordance with the manufacturer's recommendations. Ensure that joints develop effective watertightness equal to the continuous waterstop material, permanently develop not less than 50% of the strength of the parent section and permanently retain flexibility.

2.3—Execution

2.3.1 *Construction and erection of formwork*

2.3.1.1 At construction joints, lap contact surface of the form sheathing for flush surfaces exposed to view over the hardened concrete in the previous placement. Ensure formwork is sealed against hardened concrete to prevent offsets or loss of mortar at construction joints and to maintain a true surface.

2.3.1.2 Unless otherwise specified in the Contract Documents, construct formwork so concrete surfaces conform to the tolerance limits of ACI 117. The class of surface for offset between adjacent pieces of formwork facing material shall be Class B for surfaces permanently exposed to public view and Class D for surfaces that will be permanently concealed, unless otherwise specified.

2.3.1.3 Provide positive means of adjustment (such as wedges or jacks) of shores and struts. Do not make adjustments in the formwork after concrete has reached its time of initial setting. Brace formwork securely against lateral deflection and lateral instability.

2.3.1.4 To maintain specified tolerances, camber formwork to compensate for anticipated deflections in formwork during concrete placement. Set formwork and intermediate screed strips for slabs accurately to produce designated elevations and contours of the finished surface before removal of formwork. Ensure that edge forms and screed strips are strong enough to support vibrating screeds or roller pipe screeds when the finish specified requires the use of such equipment.

2.3.1.5 When formwork is cambered, set screeds to the same camber to maintain specified concrete thickness.

2.3.1.6 Fasten form wedges in place after final adjustment of forms and before concrete placement.

2.3.1.7 Anchor formwork to shores, supporting surfaces, or members to prevent upward or lateral movement of the formwork system during concrete placement.

2.3.1.8 Construct formwork for wall openings to facilitate removal and to counteract swelling of wood formwork.

2.3.1.9 Provide runways for moving equipment and support runways directly on the formwork or structural member without resting on the reinforcing steel.

2.3.1.10 Place sleeves, inserts, anchors, and embedded items required for adjoining work or for support of adjoining work before concrete placement.

2.3.1.11 Position and support expansion joint materials, waterstops, and other embedded items to prevent displacement. Fill voids in sleeves, inserts, and anchor slots temporarily with readily removable material to prevent entry of concrete into voids.

2.3.1.12 Clean surfaces of formwork and embedded materials of mortar, grout, and foreign materials before concrete is placed.

2.3.1.13 Cover surfaces of formwork with an acceptable material that will prevent bond with the concrete. A field-applied formwork release agent or a factory-applied liner may be used. If a formwork release agent is used, apply to the surfaces of the formwork in accordance with the manufacturer's recommendations before placing reinforcing steel. Do not allow formwork release agent to puddle in the forms. Do not allow formwork release agent to contact reinforcing steel or hardened concrete against which fresh concrete is to be placed.

2.3.2 *Removal of formwork*

2.3.2.1 When formed surfaces require finishing, remove forms as soon as removal operations will not damage concrete.

2.3.2.2 Remove top forms on sloping surfaces of concrete as soon as removal will not allow concrete to sag. Perform needed repairs or treatment required at once and follow immediately with specified curing.

2.3.2.3 Loosen wood formwork for wall openings as soon as loosening operations will not damage concrete.

2.3.2.4 Do not damage concrete during removal of formwork for columns, walls, sides of beams, and other parts not supporting the weight of the concrete. Perform needed repair and treatment required on vertical surfaces at once and follow immediately with specified curing.

2.3.2.5 Unless otherwise specified, leave formwork and shoring in place to support the weight of concrete in beams, slabs, and in-place structural members until concrete has reached f_c', in accordance with 2.3.4. If a lower compressive strength is proposed for removal of formwork and shoring, submit detailed plans for review and acceptance. When shores and other vertical supports are arranged to allow the form-facing material to be removed without loosening or disturbing the shores and supports, the facing material may be removed at an earlier age unless otherwise specified.

2.3.2.6 Construct formwork to permit easy removal.

2.3.3 *Reshoring and backshoring*

2.3.3.1 Submittals for reshoring and backshoring operations shall comply with 2.1.2.1.e and 2.1.2.2.b.

2.3.3.2 During reshoring and backshoring, do not allow concrete in beam, slab, column, or any structural member to be loaded with combined dead and construction loads in excess of the loads permitted by the Architect/Engineer for the concrete compressive strength at the time of reshoring and backshoring.

2.3.3.3 Place reshores and backshores in sequence with stripping operations.

2.3.3.4 Tighten reshores and backshores to carry the required loads without overstressing the concrete members. Leave them in place until tests required by 2.3.4 indicate that the concrete compressive strength has attained the minimum value specified in 2.3.2.5.

2.3.3.5 For floors supporting shores under newly placed concrete, either leave the original supporting shores in place, or install reshores or backshores. The shoring system and the supporting slabs shall resist the anticipated loads. Locate reshores and backshores directly under a shore position or as indicated on formwork shop drawings.

2.3.3.6 In multistory buildings, place reshoring or backshoring over a sufficient number of stories to distribute the weight of newly placed concrete, forms, and construction live loads such that the design loads of the floors supporting the shores, reshores or backshores are not exceeded.

2.3.4 *Strength of concrete required for removal of formwork*

2.3.4.1 When removal of formwork or reshoring is based on concrete reaching a specified compressive strength, concrete will be presumed to have reached this strength when test cylinders, field cured the same as the concrete they represent, have reached the compressive strength specified for removal of formwork or reshoring. Mold cylinders in accordance with ASTM C 31/C 31M, and cure them under the same conditions for moisture and temperature as used for the concrete they represent. Test cylinders in accordance with ASTM C 39/C 39M.

2.3.4.2 Alternatively, when specified or permitted, use methods in 2.3.4.2 b through 2.3.4.2.d to evaluate concrete strength for formwork removal. Before using methods in 2.3.4.2.b through 2.3.4.2.d, submit data using project materials to demonstrate correlation of measurements on the structure with the compressive strength of laboratory-cured molded cylinders or drilled cores. Submit correlation data on the proposed alternative method for determining strength to the Architect/Engineer.

2.3.4.2.a Tests of cast-in-place cylinders in accordance with ASTM C 873. This is limited to slabs with concrete depths from 5 to 12 in.

2.3.4.2.b Penetration resistance in accordance with ASTM C 803/C 803M.

2.3.4.2.c Pullout strength in accordance with ASTM C 900.

2.3.4.2.d Maturity method in accordance with ASTM C 1074.

2.3.5 *Field quality control*

2.3.5.1 Establish and maintain survey controls and benchmarks in an undisturbed condition until final completion and acceptance of the project.

2.3.5.2 Variations from plumb and designated building lines shall not exceed the tolerances specified in ACI 117.

SECTION 3—REINFORCEMENT AND REINFORCEMENT SUPPORTS

3.1—General

This section covers materials, fabrication, placement, and tolerances of reinforcement and reinforcement accessories.

3.1.1 *Submittals, data, and drawings*—Unless otherwise required by Contract Documents, submit data and drawings specified in 3.1.1.1 through 3.1.1.3 for review and acceptance before fabrication and execution:

3.1.1.1 Submit the data specified in 3.1.1.1.a through 3.1.1.1g unless otherwise specified:

3.1.1.1.a *Reinforcement*—Submit manufacturer's certified test report.

3.1.1.1.b *Placing drawings*—Submit placing drawings showing fabrication dimensions and placement locations of reinforcement and reinforcement supports.

3.1.1.1.c *Splices*—Submit a list of splices and request to use splices not indicated in Contract Documents.

3.1.1.1.d *Mechanical splices*—Submit request to use mechanical splices not shown on the project drawings.

3.1.1.1.e *Column dowels*—Submit request to place column dowels without the use of templates.

3.1.1.1.f *Field bending*—Submit request and procedure to field bend or straighten reinforcement partially embedded in concrete.

3.1.1.1.g *Certification*—Submit copy of current CRSI Plant Certification Manual.

3.1.1.2 Submit the data specified in 3.1.1.2.a through 3.1.1.2.b when required:

3.1.1.2.a *Welding*—Submit description of reinforcement weld locations, welding procedures, and welder certification when welding is permitted in accordance with 3.2.2.2.

3.1.1.2.b *Supports*—If coated reinforcement is required, submit description of reinforcement supports and materials for fastening coated reinforcement not described in 3.3.2.4.

3.1.1.3 Submit the data specified in 3.1.1.3.a through 3.1.1.3.b when alternatives are proposed:

3.1.1.3.a *Reinforcement relocation*—Submit a request to relocate any reinforcement that exceeds specified placement tolerances.

3.1.1.3.b Inspection and quality-control program of plants applying epoxy coating if proposed plant is not certified in accordance with the CRSI Certification Program.

3.1.2 *Materials delivery, storage, and handling*

3.1.2.1 Prevent bending, coating with earth, oil, or other material, or otherwise damaging the reinforcement.

3.1.2.2 When handling coated reinforcement, use equipment having contact areas padded to avoid damaging the coating. Lift bundles of coated reinforcement at multiple pickup points to prevent bar-to-bar abrasion from sags in the bundles. Do not drop or drag coated reinforcement. Store coated reinforcement on cribbing that will not damage the coating.

3.2—Products

3.2.1 *Materials*

3.2.1.1 *Reinforcing bars*—Reinforcement shall be deformed bars, except spirals and welded wire reinforcement, which may be plain. Reinforcement shall be the grades, types, and sizes required by Contract Documents and shall conform to one of the following:
- ASTM A 615/A 615M;
- ASTM A 706/A 706M;
- ASTM A 970/A 970M; or
- ASTM A 996/A 996M, rail-steel bars shall be Type R.

3.2.1.2 *Coated reinforcing bars*—Use zinc or epoxy-coated reinforcing bars as specified in the Contract Documents.

3.2.1.2.a Zinc-coated (galvanized) reinforcing bars shall conform to ASTM A 767/A 767M. Repair coating damage due to shipping, handling, and placing in accordance with ASTM A 780. The maximum total damaged areas shall not exceed 2% of the surface area in each linear foot of each bar.

3.2.1.2.b Epoxy-coated reinforcing bars shall conform to ASTM A 775/A 775M or ASTM A 934/A 934M as specified in the Contract Documents.

Coatings shall be applied in plants that are certified in accordance with the Concrete Reinforcing Steel Institute (CRSI) Certification Program or an equivalent program acceptable to the Architect/Engineer.

Repair damaged areas with patching material conforming to ASTM A 775/A 775M or ASTM A 934/A 934M as applicable and in accordance with the material manufacturer's written recommendations. Repair coating damage due to shipping, handling, and placing. The maximum total damaged areas shall not exceed 2% of the surface area in each linear foot of each bar. Fading of the coating color will not be cause for rejection of epoxy-coated reinforcing bars.

3.2.1.3 *Stainless steel bars*—Stainless steel bars shall conform to ASTM A 955/A 955M.

3.2.1.4 *Bar mats*—Bar mats shall conform to ASTM A 184/A 184M:

3.2.1.5 *Wire*—Use plain or deformed wire as indicated on Contract Documents. Plain wire may be used for spirals.

3.2.1.5.a Plain wire shall conform to ASTM A 82.

3.2.1.5.b Deformed wire size D4 and larger shall conform to ASTM A 496.

3.2.1.5.c Epoxy-coated wire shall conform to ASTM A 884/A 884M. The maximum total damaged areas, including areas repaired at the manufacturing facility, shall not exceed 2% of the surface area in each linear foot of each wire. Repair all damaged areas.

3.2.1.5.d For wire with f_y exceeding 60,000 psi, f_y shall correspond to a strain of 0.35%.

3.2.1.6 *Welded wire reinforcement*—Use welded wire reinforcement specified in Contract Documents and conforming to one of the specifications given in 3.2.1.6.a through 3.2.1.6.c.

3.2.1.6.a *Plain welded wire reinforcement*—ASTM A 185, with welded intersections spaced not farther apart than 12 in. in the direction of principal reinforcement.

3.2.1.6.b *Deformed welded wire reinforcement*—ASTM A 497/A497M, with welded intersections spaced not farther apart than 16 in. in the direction of principal reinforcement.

3.2.1.6.c *Epoxy-coated welded wire reinforcement*—ASTM A 884/A 884M, the maximum total damaged areas, including areas repaired at the manufacturing facility, shall not exceed 2% of the surface area in each linear foot of each wire. Repair all damaged areas.

3.2.1.6.d For welded wire reinforcement with f_y exceeding 60,000 psi, f_y shall correspond to a strain of 0.35%.

3.2.1.7 *Wire-reinforcement supports*—Unless otherwise specified or permitted, use wire-reinforcement supports complying with Class 1, maximum protection, or Class 2, moderate protection, as indicated in Chapter 3 of the CRSI Manual of Standard Practice.

3.2.1.8 *Coated wire-reinforcement supports*

3.2.1.8.a For epoxy-coated reinforcement—Use wire-reinforcement supports coated with dielectric material, including epoxy or another polymer for a minimum distance of 2 in. from the point of contact with epoxy-coated reinforcement.

3.2.1.8.b For zinc-coated reinforcement—Use galvanized wire-reinforcement supports or wire-reinforcement supports coated with dielectric material.

3.2.1.9 *Precast concrete reinforcement supports*—Use concrete supports that have a surface area of not less than 4 in.^{2} and have a compressive strength equal to or greater than the specified compressive strength of the concrete being placed.

3.2.2 *Fabrication*

3.2.2.1 *Reinforcement*—Bend reinforcement cold unless heating is permitted. Fabricate reinforcement in accordance with fabricating tolerances of ACI 117.

3.2.2.2 *Welding*

3.2.2.2.a When welding of reinforcement is specified or permitted, comply with the requirements of ANSI/AWS D1.4. Do not weld crossing bars (tack welding) for assembly of reinforcement, supports, or embedded items.

3.2.2.2.b After completing welds on zinc-coated (galvanized) or epoxy-coated reinforcement, repair coating damage in accordance with requirements in 3.2.1.2.a or 3.2.1.2.b, respectively. Coat welds and steel splice devices used to splice reinforcement with the same material used for repair of coating damage.

3.3—Execution

3.3.1 *Preparation*

3.3.1.1 When concrete is placed, reinforcement shall be free of materials deleterious to bond. Reinforcement with rust, mill scale, or a combination of both will be considered satisfactory, provided the minimum nominal dimensions, nominal weight, and the minimum average height of deformations of a hand-wire-brushed test specimen are not less than the applicable ASTM specification requirements.

3.3.2 *Placement*

3.3.2.1 *Tolerances*—Place, support, and fasten reinforcement as shown on the project drawings. Do not exceed the placing tolerances specified in ACI 117 before concrete is placed. Placing tolerances shall not reduce cover requirements except as specified in ACI 117.

3.3.2.2 *Reinforcement relocation*—When it is necessary to move reinforcement beyond the specified placing tolerances to avoid interference with other reinforcement, conduits, or embedded items, submit the resulting reinforcement arrangement for acceptance.

3.3.2.3 *Concrete cover*—Unless otherwise specified, minimum concrete cover for reinforcement shall be as indicated in Table 3.3.2.3.

For bundled bars, minimum concrete cover shall be equal to the equivalent diameter of the bundle but need not be greater than 2 in.; except the minimum cover shall not be less than specified in Table 3.3.2.3. The equivalent diameter of the bundle shall be computed based on the total area of the bundle. Tolerances on minimum concrete cover shall meet the requirements of ACI 117.

3.3.2.4 *Reinforcement supports*—Unless otherwise permitted, use the reinforcement supports given in 3.3.2.4.a through 3.3.2.4.i:

3.3.2.4.a Use precast concrete reinforcement supports to support reinforcement from the ground or a mud mat.

3.3.2.4.b Use reinforcement supports made of concrete, metal, or plastic to support uncoated reinforcement.

3.3.2.4.c Use wire reinforcement supports that are galvanized, coated with dielectric material, or made of dielectric material to support zinc-coated (galvanized) reinforcement.

3.3.2.4.d Reinforcement and embedded steel items used with zinc-coated (galvanized) reinforcement shall be zinc-coated (galvanized) or coated with nonmetallic materials.

3.3.2.4.e Support epoxy-coated reinforcement on coated wire reinforcement supports or on reinforcement supports made of dielectric material. Use coatings or materials compatible with concrete.

3.3.2.4.f When precast concrete reinforcement supports with embedded tie wires or dowels are used with epoxy-coated reinforcement, use wires or dowels coated with dielectric material.

3.3.2.4.g Reinforcement used as supports with epoxy-coated reinforcement shall be epoxy coated.

3.3.2.4.h In walls reinforced with epoxy-coated reinforcement, use spreader bars that are epoxy coated. Proprietary combination bar clips and spreaders used in walls with epoxy-coated reinforcement shall be made of corrosion-resistant material or coated with dielectric material.

3.3.2.4.i Fasten epoxy-coated reinforcement with tie wires coated with epoxy or other polymer.

3.3.2.5 *Welded wire reinforcement*—For slabs on ground, extend welded wire reinforcement to within 2 in. of the concrete edge. Lap splice edges and ends of welded wire reinforcement sheets as shown on the project drawings. Unless otherwise specified or permitted, do not extend welded wire reinforcement through contraction joints. Support welded wire reinforcement during placing of concrete to maintain positioning in the slab. Do not place welded wire reinforcement on grade and subsequently raise into position in concrete.

3.3.2.6 *Column dowels*—Furnish and use templates for placement of column dowels unless otherwise permitted.

3.3.2.7 *Splices*—Make splices as indicated on the project drawings unless otherwise permitted. Mechanical splices for reinforcement not shown on the project drawings shall not be used unless accepted by the Architect/Engineer. Remove reinforcement coating in the area of the mechanical splice if required by the splice manufacturer. After installing mechanical splices on zinc-coated (galvanized) or epoxy-coated reinforcement, repair coating damage and areas of removed coating in accordance with 3.2.1.2.a or 3.2.1.2.b. Coat exposed parts of mechanical splices used on coated bars with the same material used to repair coating damage.

3.3.2.8 *Field bending or straightening*—When permitted, bend or straighten reinforcement partially embedded in

Table 3.3.2.3—Minimum concrete cover for reinforcement

Minimum concrete cover for reinforcement, except for extremely corrosive atmospheres, other severe exposures, or additional fire protection, shall be as follows:

	Minimum cover, in.
Slabs and joists	
Top and bottom bars for dry conditions	
No. 11 bars and smaller	3/4 in.
No. 14 and 18 bars	1-1/2 in.
Formed concrete surfaces exposed to earth, water, or weather, and over or in contact with sewage and for bottoms bearing on work mat, or slabs supporting earth cover	
No. 5 bars and smaller, W31 or D31 wire and smaller	1-1/2 in.
No. 6 through 18 bars, W45 or D45 wire	2 in.
Beams and columns, formed	
For dry conditions	
Stirrups, spirals, and ties	1-1/2 in.
Principal reinforcement	2 in.
Exposed to earth, water, sewage, or weather	
Stirrups and ties	2 in.
Principal reinforcement	2-1/2 in.
Walls	
For dry conditions	
No. 11 bars and smaller	3/4 in.
No. 14 and 18 bars	1-1/2 in.
Formed concrete surfaces exposed to earth, water, sewage, weather, or in contact with ground	2 in.
Footings and base slabs	
At formed surfaces and bottoms bearing on concrete work mat	2 in.
At unformed surfaces and bottoms in contact with earth	3 in.
Top of footings	Same as slabs
Over top of piles	2 in.

Table 3.3.2.8—Minimum diameter of bend

Bar size	Minimum inside bend diameter
No. 3 through 8	Six bar diameters
No. 9, 10, and 11	Eight bar diameters
No. 14 and 18	Ten bar diameters

concrete in accordance with procedures 3.3.2.8.a through 3.3.2.8.c. Reinforcing bar sizes No. 3 through 5 may be bent cold the first time, provided reinforcing bar temperature is above 32 °F. For other bar sizes, preheat reinforcing bars before bending.

3.3.2.8.a *Preheating*—Apply heat by any method that does not harm the reinforcing bar material or cause damage to the concrete. Preheat a length of reinforcing bar equal to at least five bar diameters in each direction from the center of the bend but do not extend preheating below the surface of the concrete. Do not allow the temperature of the reinforcing bar at the concrete interface to exceed 500 °F. The preheat temperature of the reinforcing bar shall be between 1100 and 1200 °F. Maintain the preheat temperature until bending or straightening is complete. Measure the preheat temperature by temperature measurement crayons, contact pyrometer, or other acceptable methods. Do not artificially cool heated reinforcing bars until the temperature of the bar is less than 600 °F.

3.3.2.8.b *Bend diameters*—Minimum inside bend diameters shall conform to the requirements of Table 3.3.2.8. In addition, beginning of the bend shall not be closer to the concrete surface than the minimum diameter of bend.

3.3.2.8.c *Repair of bar coatings*—After field bending or straightening zinc-coated (galvanized) or epoxy-coated reinforcing bars, repair coating damage in accordance with 3.2.1.2.a or 3.2.1.2.b.

3.3.2.9 *Field cutting of reinforcement*—Field cut reinforcement only when specifically permitted using cutting methods specified by or acceptable to the Architect/Engineer. Do not flame cut epoxy-coated reinforcement.

3.3.2.9.a When zinc-coated (galvanized) reinforcing bars are cut in the field, coat the ends of the bars with a zinc-rich formulation used in accordance with the manufacturer's recommendations, and repair any coating damage in accordance with 3.2.1.2.a.

3.3.2.9.b When epoxy-coated reinforcing bars are cut in the field, coat the ends of the bars with the same material used for repair of coating damage, and repair any coating damage in accordance with 3.2.1.2.b.

3.3.2.10 *Reinforcement through expansion joint*—Do not continue reinforcement or other embedded metal items bonded to concrete through expansion joints. Dowels bonded on only one side of a joint and waterstops shall extend through the joint.

SECTION 4—CONCRETE MIXTURES
4.1—General

4.1.1 *Description*—This section covers the requirements for materials, proportioning, production, and delivery of concrete.

4.1.2 *Submittals*

4.1.2.1 *Mixture proportions*—Submit concrete mixture proportions and characteristics.

4.1.2.2 *Mixture proportion data*—Submit field test records used to establish the required average strength in accordance with 4.2.3.3. Submit test data used to establish the average compressive strength of the mixture in accordance with 4.2.3.4.

4.1.2.3 *Concrete materials*—Submit the following information for concrete materials, along with evidence demonstrating compliance with 4.2.1:
- For cementitious materials: types, manufacturing locations, shipping locations, and certificates showing compliance with ASTM C 150, ASTM C 595, ASTM C 618, ASTM C 845, ASTM C 989, or ASTM C 1157.
- For aggregates: types, pit or quarry locations, producers' names, gradings, specific gravities, and evidence not more than 90 days old demonstrating compliance with 4.2.1;
- For admixtures: types, brand names, producers, manufacturer's technical data sheets, and certification data; and
- For water and ice: source of supply.

4.1.2.4 *Field test data basis*—When field test records are used as the basis for selecting proportions for a concrete mixture, submit data on materials and mixture proportions with supporting test results confirming conformance with specified requirements.

4.1.2.5 *Mixture proportion adjustments*—Submit any adjustments to mixture proportions or changes in materials, along with supporting documentation, made during the course of the Work.

4.1.2.6 *Concrete for floors*—Submit evaluations and test results verifying adequacy of concrete to be placed in floors when the cementitious materials content is less than that specified in Table 4.2.2.1.

4.1.2.7 *Calcium chloride*—When it is desired to use calcium chloride, submit a request including data demonstrating compliance with 4.2.2.5.

4.1.2.8 *Volumetric batching*—When it is desired to produce concrete by the volumetric batching method, submit request along with description of proposed method.

4.1.2.9 *Time of discharge*—When it is desired to exceed the maximum time for discharge of concrete permitted by ASTM C 94/C 94M, submit a request along with a description of the precautions to be taken.

4.1.3 *Quality control*

4.1.3.1 Maintain records verifying that materials used are of the specified and accepted types and sizes and are in conformance with the requirements of 4.2.1.

4.1.3.2 Ensure that production and delivery of concrete conform to the requirements of 4.3.1 and 4.3.2.

4.1.3.3 Ensure that the concrete produced has the specified characteristics in the freshly mixed state and that these characteristics are maintained during transport and delivery.

4.1.4 *Materials storage and handling*

4.1.4.1 *Cementitious materials*—Store cementitious materials in dry, weathertight buildings, bins, or silos that will exclude contaminants.

4.1.4.2 *Aggregates*—Store and handle aggregate in a manner that will avoid segregation and prevent contamination with other materials or other sizes of aggregates. Store aggregates in locations that will permit them to drain freely. Do not use aggregates that contain frozen lumps.

4.1.4.3 *Water and ice*—Protect mixing water and ice from contamination during storage and delivery.

4.1.4.4 *Admixtures*—Protect stored admixtures against contamination, evaporation, or damage. Provide agitating equipment for admixtures used in the form of suspensions or nonstable solutions to ensure uniform distribution of the ingredients. Protect liquid admixtures from freezing and from temperature changes that would adversely affect their characteristics.

4.2—Products
4.2.1 Materials

4.2.1.1 *Cementitious materials*—Use ASTM C 150 Type I or Type II cement unless one or a combination of the cementitious materials given in 4.2.1.1.a through 4.2.1.1.f are specified or permitted:

4.2.1.1.a Portland cement conforming to ASTM C 150.

4.2.1.1.b Blended hydraulic cement conforming to ASTM C 595 or C 1157. For the sections of the structure that are designated as subject to deicing chemicals, submit certification on the composition of the cement verifying that the concrete mixture meets the requirements of Table 4.2.2.9.

4.2.1.1.c Hydraulic cement conforming to ASTM C 1157. For sections of the structure that will be subjected to deicing chemicals, submit certification on the composition of the cement verifying that the concrete mixture meets the requirements of Table 4.2.2.9.

4.2.1.1.d Pozzolanic mineral admixture conforming to ASTM C 618. When fly ash is used, the minimum amount shall be 15% by weight of the total cementitious materials unless otherwise specified.

4.2.1.1.e Ground-granulated blast-furnace slag conforming to ASTM C 989.

4.2.1.1.f Silica fume conforming to ASTM C 1240.

4.2.1.1.g Use cementitious materials that are of the same brand and type and from the same plant of manufacture as the cementitious materials used in the concrete represented by the submitted field test records or used in the trial mixtures.

4.2.1.2 *Aggregates*—Aggregates shall conform to ASTM C 33 unless otherwise specified. When a single size or a combination of two or more sizes of coarse aggregates are used, the final grading shall conform to the grading requirements of ASTM C 33 unless otherwise specified or permitted. Aggregates used in concrete shall be obtained from the same sources and have the same size ranges as the aggregates used in the concrete represented by submitted historical data or used in trial mixtures.

4.2.1.3 *Water and ice*—Mixing water for concrete and water used to make ice shall meet the requirements of ASTM C 1602/C 1602M. Use potable water unless alternative sources of water complying with ASTM 1602/C 1602M are permitted.

4.2.1.4 *Admixtures*—When required or permitted, admixtures shall meet the requirements of the following:
- Air-entraining admixtures—ASTM C 260;
- Chemical admixtures—ASTM C 494;
- Chemical admixtures for use in producing flowing concrete—ASTM C 1017/C 1017M; and
- Calcium chloride—ASTM D 98.

Admixtures used in concrete shall be the same as those used in the concrete represented by submitted field test records or used in trial mixtures.

4.2.1.5 *Change of materials*—When changes in brand, type, size, or source of cementitious materials, aggregates, water, ice, or admixtures are proposed, submit new field data, data from new trial mixtures, or other evidence that the change will not adversely affect the relevant properties of the concrete. Data shall be submitted for acceptance before changes are made.

4.2.2 Performance and design requirements

4.2.2.1 *Cementitious-material content*—The cementitious-material content shall be adequate for concrete to satisfy the specified requirements for strength, water-cementitious material ratio, durability, and finishing ability. For concrete used in floors, cementitious-material content shall not be less than indicated in Table 4.2.2.1 unless otherwise accepted. Acceptance of a lower cementitious-material content will be contingent upon verification that concrete mixtures with the lower cementitious-material content will meet the specified strength requirements and will produce concrete with equal finish quality, appearance, durability, and surface hardness. When a history of finishing quality is not available, evaluate the proposed mixture by placing concrete in a slab at the project site using project materials, equipment, and personnel. The slab shall be at least 8 x 8 ft and have an acceptable thickness. Slump shall not exceed the specified slump. Submit evaluation results for acceptance.

4.2.2.2 *Slump*—Unless otherwise specified or permitted, concrete shall have, at the point of delivery, a slump of 4 in. Determine the slump by ASTM C 143/C 143M. Slump tolerances shall meet the requirements of ACI 117. When use of a Type I or II plasticizing admixture conforming to ASTM C 1017/C 1017M or when a Type F or G high-range water-reducing admixture conforming to ASTM C 494 is permitted to increase the slump of concrete, concrete shall have a slump of 2 to 4 in. before the admixture is added and a maximum slump of 8 in. at the point of delivery after the admixture is added unless otherwise specified.

4.2.2.3 *Size of coarse aggregate*—Unless otherwise specified or permitted, nominal maximum size of coarse aggregate shall not exceed three-fourths of the minimum clear spacing between reinforcing bars, one-fifth of the narrowest dimension between sides of forms, or one-third of the thickness of slabs or toppings.

4.2.2.4 *Air content*—Unless otherwise specified, concrete shall be air-entrained and the air content at the point of delivery shall conform to the requirements of Table 4.2.2.4 for severe exposure. For specified compressive strengths above 5000 psi, the air contents indicated in Table

Table 4.2.2.1—Minimum cementitious-materials content requirements for floors

Nominal maximum size of aggregate, in.	Minimum cementitious material content, lb/yd^3
1-1/2	470
1	520
3/4	540
3/8	610

Note: When fly ash is used, quantity shall not be less than 15% nor more than 25% by weight of total cementitious material.

Table 4.2.2.4—Air content* of concrete for various sizes of coarse aggregate

Nominal maximum size of aggregate, in.	Air content,† %		
	Severe exposure	Moderate exposure	Mild exposure
Less than 3/8	9	7	5
3/8	7.5	6	4.5
1/2	7	5.5	4
3/4	6	5	3.5
1	6	4.5	3
1-1/2	5.5	4.5	2.5
2	5	4	2
3	4.5	3.5	1.5
6	4	3	1.5

*Measured in accordance with ASTM C 138, C 173, or C 231.
†Air-content tolerance is ±1-1/2%.

4.2.2.4 may be reduced by 1%. Measure air content in accordance with either ASTM C 138, C 173, or C 231.

4.2.2.5 *Admixtures*—When admixtures are specified in Contract Documents for particular parts of the Work, use the types specified. Use of calcium chloride or other admixtures containing chloride ions shall be subject to the limitations in 4.2.2.6. When accepted, add calcium chloride into the concrete mixture in solution form only.

4.2.2.6 *Chloride-ion concentration*—Unless otherwise specified, maximum water-soluble chloride-ion concentrations in hardened concrete at ages from 28 to 42 days contributed from the ingredients including water, aggregates, cementitious materials, and admixtures shall not exceed the limits of Table 4.2.2.6. Measure water-soluble chloride-ion content in accordance with ASTM C 1218/C 1218M. The type of member described in Table 4.2.2.6 shall apply to the Work as indicated in the Contract Documents.

4.2.2.7 *Sulfate resistance*—For those portions of the structure designated as requiring sulfate resistance, provide concrete meeting the requirements specified in the Contract Documents. Submit documentation verifying compliance with specified requirements. Do not use calcium chloride admixture in sulfate-resistant concrete.

4.2.2.8 *Concrete temperature*—When the average of the highest and lowest temperature during the period from midnight to midnight is expected to drop below 40 °F for more than three successive days, deliver concrete to meet the following minimum temperatures immediately after placement:
- 55 °F for sections less than 12 in. in the least dimension;
- 50 °F for sections 12 to 36 in. in the least dimension;
- 45 °F for sections 36 to 72 in. in the least dimension; and

Table 4.2.2.6—Maximum allowable chloride-ion content

Type of member	Maximum water-soluble chloride ion (Cl$^-$) content in concrete, percent by weight of cement
Prestressed concrete	0.06
Reinforced concrete exposed to chloride in service	0.15
Reinforced concrete that will be dry or protected from moisture in service	1.00
Other reinforced concrete construction	0.30

- 40 °F for sections greater than 72 in. in the least dimension.

The temperature of concrete as placed shall not exceed these values by more than 20 °F. These minimum requirements may be terminated when temperatures above 50 °F occur during more than half of any 24 h duration.

Unless otherwise specified or permitted, the temperature of concrete as delivered shall not exceed 90 °F.

4.2.2.9 *Strength and water-cementitious material ratio*—The compressive strength and, when required, the water-cement or water-cementitious material ratio of the concrete for each portion of the Work shall be as specified in the Contract Documents.

4.2.2.9.a When required for concrete exposed to deicing chemicals, the maximum weight of fly ash, natural pozzolans, silica fume, or ground-granulated blast-furnace slag that is included in the concrete shall not exceed the percentages of the total weight of cementitious materials given in Table 4.2.2.9.

4.2.2.9.b Unless otherwise specified, strength requirements shall be based on a 28-day compressive strength determined on 6 x 12 in. cylindrical specimens made and tested in accordance with ASTM C 31/C 31M and C 39/C 39M, respectively.

4.2.3 *Proportioning*

4.2.3.1 Proportion concrete to comply with 4.2.2 to provide workability and consistency so concrete can be worked readily into forms and around reinforcement without segregation or bleeding, and to provide an average compressive strength adequate to meet acceptance requirements of 1.6.7.1. If the production facility has records of field tests performed within the past 12 months and spanning a period of not less than 60 calendar days for a class of concrete within 1000 psi of that specified for the Work, calculate a standard deviation and establish the required average compressive strength f'_{cr} in accordance with 4.2.3.2 and 4.2.3.3.a. If field test records are not available, select f'_{cr} from Table 4.2.3.3.b.

4.2.3.2 *Standard deviation*

4.2.3.2.a *Field test data*—Field test records used to calculate standard deviation shall represent materials, quality-control procedures, and climatic conditions similar to those expected in the Work. Changes in materials and proportions in concrete represented by the test records shall not have been more closely restricted than those in the proposed Work. Test records shall comply with one of the following:
- Data from a single group of at least 15 consecutive com-

Table 4.2.2.9—Maximum cementitious material requirements for concrete exposed to deicing chemicals

Cementitious material	Maximum percent of total cementitious material by weight*
Fly ash or other pozzolans conforming to ASTM C 618	25
Slag conforming to ASTM C 989	50
Silica fume conforming to ASTM C 1240	10
Total of fly ash or other pozzolans, slag, and silica fume	50†
Total of fly ash or other pozzolans and silica fume	35†

*Total cementitious material also includes ASTM C 150, C 595, and C 845 cement. The maximum percentages above shall include:
a) Fly ash or other pozzolans present in Type IP or I(PM) blended cement, ASTM C 595;
b) Slag used in manufacture of an IS or I(SM) blended cement, ASTM C 595; and
c) Silica fume, ASTM C 1240, present in blended cement.
†Fly ash or other pozzolans and silica fume shall constitute no more than 25 and 10%, respectively, of the total weight of cementitious material.

pressive-strength tests with the same mixture proportions.
- Data from two groups of consecutive compressive strength tests totaling at least 30. Neither of the two groups shall consist of less than 10 tests.

4.2.3.2.b *Standard deviation*—Calculate the standard deviation s of the strength test records as follows:
- For a single group of consecutive test results:

$$s = \left[\frac{\sum_{i=1}^{n} (X_i - \bar{X})^2}{(n-1)} \right]^{1/2} \quad (4\text{-}1)$$

where
s = standard deviation;
n = number of test results considered;
X = average of n test results considered; and
$\bar{X}_i$ = individual test result.

- For two groups of consecutive test results:

$$s = \left[\frac{(n_1 - 1)s_1^2 + (n_2 - 1)s_2^2}{(n_1 + n_2 - 2)} \right]^{1/2} \quad (4\text{-}2)$$

where
s = standard deviation for the two groups combined;
s_1, s_2 = standard deviations for Groups 1 and 2, respectively, calculated in accordance with Eq. (4-1); and
n_1, n_2 = number of test results in Groups 1 and 2, respectively.

4.2.3.3 *Required average compressive strength*—Calculate f'_{cr} for the specified class of concrete in accordance with 4.2.3.3.a or 4.2.3.3.b:

4.2.3.3.a Use the standard deviation calculated in accordance with 4.2.3.2 to establish f'_{cr} in accordance with Table 4.2.3.3.a. Use the larger of the two values of f'_{cr} calculated.

4.2.3.3.b When field test records are not available to establish a standard deviation, select the required average compressive strength f'_{cr} from Table 4.2.3.3.c.

Table 4.2.3.3.a—Required average compressive strength f'_{cr}, when data are available to establish a standard deviation, psi

f'_c, psi	f'_{cr}, psi	
	Use the larger of:	Equation
5000 or less	$f'_{cr} = f'_c + 1.34ks$	(4-3)
	$f'_{cr} = f'_c + 2.33ks - 500$	(4-4)
Over 5000	$f'_{cr} = f'_c + 1.34ks$	(4-3)
	$f'_{cr} = 0.90f'_c + 2.33ks$	(4-5)

Notes: f'_{cr} = required average compressive strength; f'_c = specified concrete strength; k = factor from Table 4.2.3.3.b to adjust standard deviation if total number of tests is less than 30; and s = standard deviation calculated in accordance with 4.2.3.2

Table 4.2.3.3.b—k-factor for increasing standard deviation for number of tests considered

Total no. of tests considered	k-factor for increasing standard deviation
15	1.16
20	1.08
25	1.03
30 or more	1.00

Note: Linear interpolation for intermediate number of tests is acceptable.

Table 4.2.3.3.c—Required average compressive strength f'_{cr}*

f'_c, psi	f'_{cr}, psi
Less than 3000	$f'_c + 1000$
3000 to 5000	$f'_c + 1200$
Over 5000	$1.1f'_c + 700$

*When data are not available to establish standard deviation.

4.2.3.4 *Documentation of required average compressive strength*—Documentation indicating the proposed concrete proportions will produce an average compressive strength equal to or greater than the required average compressive strength, and shall consist of field strength records or trial mixture.

4.2.3.4.a *Field test data*—If field test data are available and represent a single group of at least 10 consecutive strength tests for one mixture, using the same materials, under the same conditions, and encompassing a period of not less than 60 days, verify that the average of the field test results equals or exceeds f'_{cr}. Submit for acceptance the mixture proportions along with the field test data.

If the field test data represent two groups of compressive strength tests for two mixtures, plot the average strength $\bar{X}_1$ and $\bar{X}_2$ of each group versus the water-cementitious material ratio of the corresponding mixture proportions and interpolate between them to establish the required water-cementitious material ratio. Establish mixture proportions for f'_{cr} based on the required water-cementitious material ratio.

4.2.3.4.b *Trial mixtures*—Establish mixture proportions based on trial mixtures in accordance with the following requirements:
- Use materials and material combinations listed in 4.2.1.1 through 4.2.1.4 proposed for the Work.
- Determine f'_{cr} according to 4.2.3.3.a if suitable field test data are available, or use Table 4.2.3.3.b.

- Make at least three trial mixtures complying with 4.2.2. Each trial mixture shall have a different cementitious material content. Select water-cementitious material ratios that will produce a range of compressive strengths encompassing f'_{cr}.
- Proportion trial mixtures to produce a slump within 3/4 in. of the maximum specified, and for air-entrained concrete, an air content within 0.5% of the required air content indicated in Table 4.2.2.4. The temperature of the freshly mixed concrete shall be recorded and shall be within 10 °F of the intended maximum temperature of the concrete as mixed and delivered.
- For each trial mixture, make and cure three compressive strength cylinders for each test age in accordance with ASTM C 192/C 192M. Test for compressive strength in accordance with ASTM C 39/C 39M at 28 days or at the test age specified in the Contract Documents.
- From results of these tests, plot a curve showing the relationship between water-cementitious material ratio and compressive strength.
- From the curve of water-cementitious material ratio versus compressive strength, select the water-cementitious material ratio corresponding to f'_{cr}. This is the maximum water-cementitious material ratio that may be used to establish mixture proportions, unless a lower water-cementitious material ratio is specified in 4.2.2.8.
- Establish mixture proportions so that the maximum water-cementitious material ratio is not exceeded when slump is at the maximum specified.

4.2.3.5 *Field verification of adequacy of selected mixture proportions*—Using materials and mixture proportions accepted for use in the Work, verify that the concrete can be adequately placed using the intended placing method. Place the concrete mixture using project equipment and personnel. Verify that the slump and air content obtained at the form are acceptable. Make suitable corrections to the placing methods or to the mixture proportions, if needed. Submit any adjustments to the mixture proportions to the Architect/Engineer for review and acceptance.

4.2.3.6 *Revisions to concrete mixtures*—When 15 consecutive compressive strength test results become available from the field, calculate the actual average compressive strength and standard deviation. Calculate a revised value for f'_{cr} in accordance with 4.2.3.3.a. Verify that both of the requirements of 1.6.7.1 are met.

4.2.3.6.a When the actual average compressive strength X exceeds the revised value of f'_{cr} and requirements of 1.6.7.1 are met, f'_{cr} may be decreased. The revised mixture shall meet the requirements of 4.2.2.

4.2.3.6.b If the actual average compressive strength $\overline{X}$ is less than the revised value of f'_{cr}, or if either of the two requirements in 1.6.7.1 are not met, take immediate steps to increase average compressive strength of the concrete.

4.2.3.6.c Submit revised mixture proportions for acceptance before placing in the Work.

4.3—Execution

4.3.1 *Measuring, batching, and mixing*—Production facilities shall produce concrete of the specified quality and conforming to this Specification.

4.3.1.1 *Ready-mixed and site-produced concrete*—Unless otherwise specified, measure, batch, and mix concrete materials and concrete in conformance with ASTM C 94/C 94M.

4.3.1.2 *Concrete produced by volumetric batching and continuous mixing*—When concrete made by volumetric batching and continuous mixing is acceptable, it shall conform to the requirements of ASTM C 685/C 685M and shall satisfy the requirements of this Specification.

4.3.1.3 *Prepackaged dry materials used in concrete*—If packaged dry-combined materials are used, they shall conform to the requirements of ASTM C 387 and shall satisfy the requirements of this Specification.

4.3.2 *Delivery*—Concrete shall possess the specified characteristics in the freshly mixed state at the point of placing. Transport and deliver concrete in equipment conforming to the requirements of ASTM C 94/C 94M.

4.3.2.1 *Slump adjustment*—When concrete arrives at the point of delivery with a slump below that which will result in the specified slump at the point of placement and is unsuitable for placing at that slump, the slump may be adjusted to the required value by adding water up to the amount allowed in the accepted mixture proportions unless otherwise specified by the Architect/Engineer. Addition of water shall be in accordance with ASTM C 94/C 94M. Do not exceed the specified water-cementitious material ratio or slump. Do not add water to concrete delivered in equipment not acceptable for mixing. After plasticizing or high-range water-reducing admixtures are added to the concrete at the site to achieve flowable concrete, do not add water to the concrete. Measure slump and air content of air-entrained concrete after slump adjustment to verify compliance with specified requirements.

4.3.2.2 *Time of discharge*—Time for completion of discharge shall comply with ASTM C 94/C 94M unless otherwise permitted. When discharge is permitted after more than 90 min have elapsed since batching or after the drum has revolved 300 revolutions, verify that air content of air-entrained concrete, slump, and temperature of concrete are as specified.

SECTION 5—HANDLING, PLACING, AND CONSTRUCTING

5.1—General

5.1.1 *Description*—This section covers the production of cast-in-place structural concrete. Included are methods and procedures for obtaining quality concrete through proper handling, placing, finishing, curing, and repair of surface defects.

5.1.2 *Submittals*

5.1.2.1 Submit the following data unless otherwise specified:

5.1.2.1.a *Field control test reports*—Maintain and submit accurate records of test and inspection reports.

5.1.2.1.b *Conveying equipment*—Submit description of conveying equipment.

5.1.2.1.c *Temperature measurement*—Submit proposed method of measuring concrete surface temperature changes.

5.1.2.1.d *Repair methods*—When stains, rust, efflorescence, and surface deposits must be removed as described in 5.3.7.7, submit the proposed method of removal.

5.1.2.1.e *Qualifications of finishers*—Submit qualifications of the finishing contractor and of flatwork finishers who will perform the Work as stipulated in 5.3.4.1.

5.1.2.2 Submit the data specified in 5.1.2.2.a through 5.1.2.2.g when required:

5.1.2.2.a *Drawings and data*—Submit shop drawings and data for review as required by the Contract Documents.

5.1.2.2.b *Placement notification*—When Contract Documents require advance notification of concrete placement, submit notification at least 24 h in advance.

5.1.2.2.c *Preplacement requirements*—Submit, when required, request for acceptance of preplacement activities.

5.1.2.2.d *Wet-weather placement*—When placement is scheduled during wet weather, submit, when required, request for acceptance of protection.

5.1.2.2.e *Hot-weather placement*—When placement of concrete exceeding 90 °F is desired as described in 5.3.2.1.c, submit, when required, request for placement along with proposed precautions.

5.1.2.2.f *Matching sample finish*—When required by Contract Documents, submit sample finish as described in 5.3.3.

5.1.2.2.g *Exposed-aggregate surface*—When an exposed-aggregate surface is specified and a surface retarder is proposed to be used, submit specification and manufacturer's data for the retarder and the proposed method of using retarder.

5.1.2.3 When alternatives are proposed, submit the data specified in 5.1.2.3.a through 5.1.2.3.g:

5.1.2.3.a *Construction joints*—Submit information for acceptance of proposed location and treatment of construction joints not indicated on the project drawings.

5.1.2.3.b *Two-course slabs*—When a bonding agent other than cement grout is proposed, submit specification and manufacturer's data for bonding agent.

5.1.2.3.c *Underwater placement*—When underwater placement is planned, submit request for acceptance of proposed method.

5.1.2.3.d *Contraction joints*—When contraction joints other than those indicated on the Contract Documents are proposed, submit request of location.

5.1.2.3.e *Moisture-preserving method*—When a moisture-preserving method other than specified in 5.3.6.4. is proposed, submit request of the proposed method.

5.1.2.3.f *Coated ties*—When coated form ties described in 5.3.7.2 are proposed to preclude the requirement to patch tie holes, submit proposed coated tie description.

5.1.2.3.g *Repair materials*—When a repair material described in 5.2.1.3 is proposed, submit the repair material specification, manufacturer's data on the proposed patching material, and the proposed preparation and application procedure.

5.1.3 *Delivery, storage, and handling*

5.1.3.1 *Delivery*—Place concrete within the time limits required in 4.3.2.2.

5.1.3.2 *Storage and handling*—Store and handle products to retain original quality. Do not use products stored beyond the manufacturer's recommended shelf life.

5.2—Products

5.2.1 *Materials*

5.2.1.1 *Curing compounds*—Use curing compounds that conform to ASTM C 309 or ASTM C 1315.

5.2.1.2 *Waterproof sheet materials*—Use waterproof sheet materials that conform to ASTM C 171.

5.2.1.3 *Proprietary patching materials*—Use acceptable proprietary patching materials complying with 5.3.7.6.

5.2.1.4 *Bonding grout*—Use bonding grout in accordance with 5.3.7.4.

5.2.1.5 *Site-mixed portland-cement repair mortar*—Use repair mortar in accordance with 5.3.7.5.

5.2.2 *Performance and design requirements*

5.2.2.1 *Construction and contraction joints*—Make and locate construction and contraction joints that are proposed, but not indicated on the project drawings, in accordance with 2.2.2.5. Do not impair strength of the structure with joints.

5.3—Execution

5.3.1 *Preparation*

5.3.1.1 Do not place concrete until data on materials and mixture proportions are accepted.

5.3.1.2 Remove hardened concrete and foreign materials from the inner surfaces of conveying equipment.

5.3.1.3 Before placing concrete in forms, complete the following:

- Comply with formwork requirements specified in Section 2;
- Remove snow, ice, frost, water, and other foreign materials from surfaces, including reinforcement and embedded items, against which concrete will be placed;
- Comply with reinforcing steel placement requirements specified in Section 3;
- Position and secure in place expansion joint materials, anchors, and other embedded items; and
- Obtain acceptance of finished preparation.

5.3.1.4 Before placing a concrete slab on ground, remove foreign materials from the subgrade and complete the following:

- Subgrade shall be well drained and of uniform load-bearing nature;
- In-place density of subgrade soils shall be uniform throughout the area and at least the minimum required by Contract Documents;
- Subgrade shall be free from frost and ice; and
- Subgrade shall be moist with no free water and no muddy or soft spots.

5.3.1.5 When high evaporative conditions necessitate protection of concrete immediately after placing or finishing, make provisions in advance of concrete placement for windbreaks, shading, fogging, sprinkling, ponding, or wet covering.

5.3.1.6 During ambient temperature conditions described in 4.2.2.8, make provisions in advance of concrete placement to maintain the temperature of the concrete as specified in 5.3.2.1.b. Use heating, covering, or other means adequate to maintain required temperature without overheating or drying of concrete due to concentration of heat. Do not use combustion heaters unless precautions are taken to prevent exposure of the concrete to exhaust gases containing carbon dioxide.

5.3.2 *Placement of concrete*

5.3.2.1 *Weather considerations*

5.3.2.1.a *Wet weather*—Do not begin to place concrete while rain, sleet, or snow is falling unless adequate protection is provided and, when required, acceptance of protection is obtained. Do not allow rain water to increase mixing water or to damage the surface of the concrete.

5.3.2.1.b *Cold weather*—Concrete temperatures and ambient temperatures shall meet minimum temperature requirements of 4.2.2.8.

5.3.2.1.c *Hot weather*—The temperature of concrete as placed shall not exceed 90 °F unless otherwise specified or permitted. Loss of slump, flash set, or cold joints due to temperature of concrete as placed will not be acceptable. When temperature of concrete exceeds 90 °F, obtain acceptance, when required, of proposed precautionary measures. When temperature of steel reinforcement, embedments, or forms is greater than 120 °F, fog steel reinforcement, embedments, and forms with water immediately before placing concrete. Remove standing water before placing concrete.

5.3.2.2 *Conveying*—Rapidly convey concrete from mixer to the place of final deposit by methods that prevent segregation or loss of ingredients and ensure the required quality of concrete. Do not use aluminum pipes or chutes.

5.3.2.3 *Conveying equipment*—Use acceptable conveying equipment of a size and design that will prevent cold joints from occurring. Clean conveying equipment before each placement.

5.3.2.3.a Use belt conveyors that are horizontal or at a slope that will not cause excessive segregation or loss of ingredients. Protect concrete to minimize drying and effects of temperature rise. Use an acceptable discharge baffle or hopper at the discharge end to prevent segregation. Do not allow mortar to adhere to the return length of the belt.

5.3.2.3.b Use metal or metal-lined chutes having rounded bottoms, and sloped between one vertical to two horizontal and one vertical to three horizontal. Chutes longer than 20 ft long and chutes not meeting slope requirements may be used provided the discharge is into a hopper before distributing into the forms.

5.3.2.3.c Use pumping equipment that permits placement rates that avoid cold joints and prevents segregation in discharge of pumped concrete.

5.3.2.4 *Depositing*—Deposit concrete continuously in one layer or in layers to have fresh concrete deposited on in-place concrete that is still plastic. Do not deposit fresh concrete on concrete that has hardened sufficiently to cause formation of seams or planes of weakness within the section, unless construction joint requirements of 5.3.2.6 are met. Do not use concrete that has surface-dried, partially hardened, or contains foreign material. When temporary spreaders are used in the forms, remove the spreaders as their service becomes unnecessary. Spreaders made of metal or concrete may be left in place if prior acceptance is obtained. Do not place concrete over columns and walls until concrete in columns and walls is no longer plastic and has been in place at least 1 h. Do not subject concrete to any procedure that will cause segregation. Deposit concrete as near as practicable to the final position to avoid segregation. Place concrete for beams, girders, brackets, column capitals, haunches, and drop panels at the same time as concrete for slabs. When underwater placement is required or permitted, place concrete by an acceptable method. Deposit fresh concrete so concrete enters the mass of previously placed concrete from within, displacing water with a minimum disturbance to the surface of concrete.

5.3.2.5 *Consolidating*—Consolidate concrete by vibration. Thoroughly work concrete around reinforcement and embedded items and into corners of forms, eliminating air and stone pockets that may cause honeycombing, pitting, or planes of weakness. Use internal vibrators of the largest size and power that can properly be used in the Work as described in Table 5.3.2.5. Use immersion-type vibrators with nonmetallic heads when consolidating concrete around epoxy-coated reinforcement. Workers shall be experienced in use of the vibrators. Do not use vibrators to move concrete within the forms.

5.3.2.6 *Construction joints and other bonded joints*—Locate construction joints as indicated on the project drawings or as accepted in accordance with 5.1.2.3.a. Formed construction joints shall meet requirements of 2.2.2.5. Remove laitance and thoroughly clean and dampen construction joints before placement of fresh concrete. When bond is required or permitted, use one of the following methods:

- Use an acceptable adhesive applied in accordance with the manufacturer's recommendations;
- Use an acceptable surface retarder in accordance with manufacturer's recommendations;
- Roughen the surface in an acceptable manner that exposes the aggregate uniformly and does not leave laitance, loosened particles of aggregate, or damaged concrete at the surface; or
- Use portland-cement grout of the same proportions as the mortar in the concrete in an acceptable manner.

5.3.3 *Finishing formed surfaces*

5.3.3.1 *General*—After removal of forms, give each formed surface one or more of the finishes described in 5.3.3.2, 5.3.3.3, or 5.3.3.4. When Contract Documents do not specify a finish, finish surfaces as required by 5.3.3.5.

5.3.3.2 *Matching sample finish*—When the finish is required by the Contract Documents to match a sample panel furnished to the Contractor, reproduce the sample finish on an area at least 100 ft^2 in a location designated by the Architect/Engineer. Obtain acceptance before proceeding with that finish in the specified locations.

Table 5.3.2.5—Range of characteristics, performance, and applications of internal vibrators

Column 1	Column 2	Column 3	Column 4	Column 5	Column 6	Column 7	Column 8	Column 9
Group	Diameter of head, in.	Frequency, vibrations per min.	Eccentric moment, in.-lb	Average amplitude, in.	Centrifugal force, lb	Radius of action, in.	Rate of concrete placement, yd^3/h per vibrator	Application
1	3/4 to 1-1/2	9000 to 15,000	0.03 to 0.10	0.015 to 0.03	100 to 400	3 to 6	1 to 5	Plastic and flowing concrete in very thin members and confined places
2	1-1/4 to 2-1/2	8500 to 12,500	0.08 to 0.25	0.02 to 0.04	300 to 900	5 to 10	3 to 10	Plastic concrete in thin walls, columns, beams, precast piles, thin slabs, and along construction joints
3	2 to 3-1/2	8000 to 12,000	0.20 to 0.70	0.025 to 0.05	700 to 2000	7 to 14	6 to 20	Stiff plastic concrete (<3 in. slump) in general construction such as walls, columns, beams, prestressed piles, and heavy slabs
4	3 to 6	7000 to 10,500	0.70 to 2.5	0.03 to 0.06	1500 to 4000	12 to 20	15 to 40	Mass and structural concrete of 0 to 2 in. slump deposited in quantities up to 4 yd^3 in relatively open forms of heavy construction
5	5 to 7	5500 to 8500	2.25 to 3.50	0.04 to 0.08	2500 to 6000	16 to 24	25 to 50	Mass concrete in gravity dams, large piers, massive walls, etc.

Column 3—While vibrator is operating in concrete.
Column 4—Computed eccentric moment ef, in.-lb, where e = distance from center of gravity of eccentric to its center of rotation, in., and f = force of gravity of eccentric, lb.
Column 5—Measured or computed peak amplitude while operating in air (deviating from point of rest), $a = ew/(W + w)$, in., where W = mass of shell and other nonmoving parts, lb, and w = mass of eccentric, lb.
Column 6—Computed centrifugal force of vibrator, $F = 4\pi^2 n^2 ew/g$, lb, where n = frequency of vibrator while operating in concrete, cycles/s, and g = acceleration due to gravity, 386.1 in./s^2.
Column 7—Radius over which concrete is fully consolidated.
Column 8—Assumes insertion spacing is 1-1/2 times radius of action, and that vibrator operates 2/3 of time concrete is being placed.
Column 7 and 8—These ranges reflect capacity of vibrator, mixture workability, degree of consolidation desired, and other construction conditions.

5.3.3.3 *As-cast finishes*—Use form-facing materials meeting the requirements of 2.2.1.1. Unless otherwise specified, produce as-cast form finishes in accordance with the requirements given in 5.3.3.3.a through 5.3.3.3.c:

5.3.3.3.a *Rough-form finish*—Patch tie holes and defects. Chip or rub off fins exceeding 1/2 in. in height. Leave surfaces with the texture imparted by the forms.

5.3.3.3.b *Smooth-form finish*—Patch tie holes and defects. Remove fins exceeding 1/8 in. in height.

5.3.3.3.c *Architectural finishes*—Produce architectural finishes including special textured finishes, exposed-aggregate finish, and aggregate transfer finish in accordance with Section 6.

5.3.3.4 *Rubbed finishes*—Remove forms as early as permitted by 2.3.2. Produce one of the finishes given in 5.3.3.4.a through 5.3.3.4.c on concrete specified to have a smooth form finish:

5.3.3.4.a *Smooth-rubbed finish*—Remove forms as early as permitted by 2.3.2, and perform necessary patching. Produce finish on hardened concrete no later than the day following formwork removal. Wet the surface and rub it with carborundum brick or other abrasive until uniform color and texture are produced. Use no cement grout other than cement paste drawn from the concrete itself by the rubbing process.

5.3.3.4.b *Grout-cleaned finish*—Begin cleaning operations after contiguous surfaces to be cleaned are completed and accessible. Do not clean surfaces as work progresses. Wet the surface and apply grout consisting of one part portland cement and one and one-half parts fine sand with enough water to produce the consistency of thick paint. Match color of surrounding concrete. Scrub grout into voids, and remove excess grout. When grout whitens, rub the surface and keep the surface damp for 36 h afterward.

5.3.3.4.c *Cork-floated finish*—Perform necessary repairs. Remove ties, burrs, and fins. Wet the surface and apply stiff grout of one part portland cement and one part fine sand, filling voids. Match color of surrounding concrete. Use enough water to produce a stiff consistency. Compress grout into voids by grinding the surface with a slow-speed grinder. Produce the final finish with cork float, using a swirling motion.

5.3.3.5 *Unspecified finishes*—When a specific finish is not specified in Contract Documents for a concrete surface, apply the following finishes:

- Rough-form finish on concrete surfaces not exposed to public view; and
- Smooth-form finish on concrete surfaces exposed to public view.

5.3.4 *Finishing unformed surfaces*

5.3.4.1 *Placement*—Place concrete at a rate that allows spreading, straightedging, and darbying or bullfloating before bleed water appears. Strike smooth the top of walls, buttresses, horizontal offsets, and other similar unformed surfaces and float them to a texture consistent with finish of adjacent formed surface. Finish slab surfaces in accordance with one of the finishes in 5.3.4.2, as designated in the Contract Documents. Use qualified flatwork finishers acceptable to the Architect/Engineer. Unless otherwise permitted, a minimum of one finisher or finishing supervisor shall be a certified ACI Flatwork Concrete Finisher/Technician or a certified ACI Flatwork Technician as defined in ACI CP 10 or equivalent.

5.3.4.2 *Finishes and tolerances*

5.3.4.2.a *Scratch finish*—Place, consolidate, strikeoff, and level concrete, eliminating high spots and low spots. Roughen the surface with stiff brushes or rakes before the final set. Produce a finish that will meet conventional bull-floated tolerance requirements of ACI 117.

5.3.4.2.b *Float finish*—Place, consolidate, strikeoff, and level concrete, eliminating high spots and low spots. Do not work concrete further until it is ready for floating. Begin floating with a hand float, a bladed power float equipped with float shoes, or a powered disk float when the bleedwater sheen has disappeared, and the surface has stiffened sufficiently to permit the operation. Produce a finish that will meet conventional straightedged tolerance requirements of ACI 117, then refloat the slab immediately to a uniform texture.

5.3.4.2.c *Trowel finish*—Float concrete surface, then power-trowel the surface. Hand-trowel the surface smooth and free of trowel marks. Continue hand-troweling until a ringing sound is produced as the floor is troweled. Tolerance for concrete floors shall be conventional straightedged tolerance in accordance with ACI 117 unless otherwise specified.

5.3.4.2.d *Broom or belt finish*—Immediately after concrete has received a floated finish, give the concrete surface a coarse transverse scored texture by drawing a broom or burlap belt across the surface.

5.3.4.2.e *Dry-shake finish*—Blend metallic or mineral aggregate specified in Contract Documents with portland cement in the proportions recommended by the aggregate manufacturer, or use bagged, premixed material specified in Contract Documents as recommended by the manufacturer. Float-finish the concrete surface. Apply approximately two-thirds of the blended material required for coverage to the surface by a method that ensures even coverage without segregation. Float-finish the surface after application of the first dry-shake. Apply the remaining dry-shake material at right angles to the first application and in locations necessary to provide the specified minimum thickness. Begin final floating and finishing immediately after application of the dry-shake. After selected material is embedded by the two floatings, complete operation with a broomed, floated, or troweled finish, as specified in the Contract Documents.

5.3.4.2.f *Heavy-duty topping for two-course slabs*—For heavy-duty topping mixture, use the materials and methods specified in Contract Documents. Place and consolidate concrete for the base slab, and screed concrete to the specified depth below the top of the finished surface. Topping placed the same day as the base slab shall be placed as soon as bleedwater in the base slab has disappeared and the surface will support a person without appreciable indentation. When topping placement is deferred, brush the surface with a coarse wire broom to remove laitance and scratch the surface when concrete is plastic. Wet-cure the base slab at least three days. Before placing the topping, clean the base slab surface thoroughly of contaminants and loose mortar or aggregate. Dampen the surface, leaving it free of standing water. Immediately before placing topping, scrub into the slab surface a coat of bonding grout consisting of equal parts of cement and fine sand with enough water to make a creamy mixture. Do not allow grout to set or dry before topping is placed. Bonding agents other than cement grout may be used with prior acceptance. Spread, compact, and float the topping mixture. Check for flatness of surface and complete operation with a floated, troweled, or broom finish as specified in the Contract Documents.

5.3.4.2.g *Topping for two-course slab not intended for heavy-duty service*—Preparation of base slab, selection of topping material, mixing, placing, consolidating, and finishing operations shall be as specified in Section 5.3.4.2.f, except that the aggregate need not be selected for special wear resistance.

5.3.4.2.h *Nonslip finish*—Where a nonslip finish is required, give the surface a broom or belt finish or a dry-shake application of crushed aluminum oxide or other abrasive particles, as specified in the Contract Documents. Rate of application shall be not less than 25 lb/100 ft^2.

5.3.4.2.i *Exposed-aggregate finish*—Immediately after surface of the concrete has been leveled to meet the conventional straightedged tolerance requirements of ACI 117 and the bleedwater sheen has disappeared, spread aggregate of the color and size specified in Contract Documents uniformly over the surface to provide complete coverage to a depth of one stone. Tamp the aggregate lightly to embed aggregate in the surface. Float the surface until the embedded stone is fully coated with mortar and the surface has been finished to meet the conventional straightedged tolerance requirements of ACI 117. After the matrix has hardened sufficiently to prevent dislodgment of the aggregate, apply water carefully and brush the surface with a fine-bristled brush to expose the aggregate without dislodging it. When specified or permitted, a surface retarder sprayed on freshly floated concrete surface may be used to extend the working time for the exposure of aggregate.

5.3.4.2.j *Nonspecified finish*—When the type of finish is not specified in Contract Documents, use one of the following appropriate finishes and accompanying tolerances.
- Scratch finish—For surfaces intended to receive bonded cementitious mixtures;
- Float finish—For walks, drives, steps, ramps, and for surfaces intended to receive waterproofing, roofing, insulation, or sand-bed terrazzo; or
- Trowel finish—For floors intended as walking surfaces, floors in manufacturing, storage, and warehousing areas, or for reception of floor coverings.

5.3.4.3 *Measuring tolerances for slabs*

5.3.4.3.a Measure slabs for suspended floors and slabs-on-ground to verify compliance with the tolerance requirements of ACI 117 as specified in 5.3.4.2.a through 5.3.4.2.c. Measure floor finish tolerances within 72 h after slab finishing and before removal of supporting formwork or shoring.

5.3.4.3.b Unless otherwise specified in the Contract Documents, for residential floors and nonresidential floor installations 10,000 ft^2 or less in total project area, measure floor finish tolerances in accordance with the "10-ft straight edge method" in ACI 117.

5.3.4.3.c Unless otherwise specified in the Contract Documents, for nonresidential floor installations exceeding 10,000 ft^2 in total project area, measure floor finish tolerances in accordance with ASTM E 1155 and the F-number system in ACI 117.

5.3.5 *Sawed contraction joints*—Where saw-cut joints are required or permitted, start cutting as soon as concrete has hardened sufficiently to prevent dislodgment of aggregates. Saw a continuous slot to a depth of one-fourth the thickness of the slab but not less than 1 in. Complete sawing within 12 h after placement. If an alternative method, timing, or depth is proposed for saw cutting, submit detailed procedure plans for review and acceptance.

5.3.6 *Curing and protection*

5.3.6.1 *Curing*—Cure concrete in accordance with 5.3.6.2 or 5.3.6.3 for a minimum of seven days after placement. Cure high-early-strength concrete for a minimum of three days after placement. Alternatively, moisture retention measures may be terminated when:

a. Tests made on at least two cylinders kept adjacent to the structure and cured by the same methods as the structure indicate that 70% of f_c', as determined in accordance with ASTM C 39/C 39M, has been attained;

b. The compressive strength of laboratory-cured cylinders, representative of the in-place concrete, exceeds 85% f_c', provided the temperature of the in-place concrete has been maintained at 50 °F or higher during curing; or

c. Strength of concrete reaches f_c' as determined by accepted nondestructive test methods meeting the requirements of 2.3.4.2.

When one of the curing procedures in 5.3.6.4 is used initially, the curing procedure may be replaced by one of the other procedures when concrete is one day old, provided the concrete is not permitted to become surface-dry at any time. Use a curing procedure of 5.3.6.4 that supplies additional water during the entire curing period for concrete containing silica fume and when specified in the Contract Documents.

5.3.6.2 *Unformed concrete surfaces*—Apply one of the procedures in 5.3.6.4 after completion of placement and finishing of concrete surfaces that are not in contact with forms.

5.3.6.3 *Formed concrete surfaces*—Keep absorbent wood forms wet until they are removed. After formwork removal, cure concrete by one of the methods in 5.3.6.4.

5.3.6.4 *Preservation of moisture*—After placing and finishing, use one or more of the following methods to preserve moisture in concrete:

a. Ponding, continuous fogging, or continuous sprinkling;

b. Application of mats or fabric kept continuously wet;

c. Continuous application of steam (under 150 °F);

d. Application of sheet materials conforming to ASTM C 171;

e. Application of a curing compound conforming to ASTM C 309 or C 1315. Apply the compound in accordance with manufacturer's recommendation as soon as water sheen has disappeared from the concrete surface and after finishing operations. The application rate shall not be less than 1 gal./200 ft^2. For rough surfaces such as those specified in 5.3.4.2.a and 5.3.4.2.d, apply curing compound in two applications at right angles to each other. The material applied in each coat shall not be less than 1 gal./200 ft^2 of area. Do not use curing compound on any surface where concrete or other material will be bonded, unless the curing compound will not prevent bond or unless measures are to be taken to completely remove the curing compound from areas to receive bonded applications; or

f. Application of other accepted moisture-retaining method.

5.3.6.5 *Protection*—Immediately after placement, protect concrete from premature drying, excessively hot or cold temperatures, and mechanical injury. Protect concrete during the curing period such that the concrete temperature does not fall below the requirements of 4.2.2.8.

Maintain the concrete protection to prevent freezing of the concrete and to ensure the necessary strength development for structural safety. Remove protection in such a manner that the maximum decrease in temperature measured at the surface of the concrete in a 24 h period shall not exceed the following:

- 50 °F for sections less than 12 in. in the least dimension;
- 40 °F for sections from 12 to 36 in. in the least dimension;
- 30 °F for sections 36 to 72 in. in the least dimension; or
- 20 °F for sections greater than 72 in. in the least dimension.

Measure concrete temperature using a method acceptable to the Architect/Engineer, and record the concrete temperature. When the surface temperature of the concrete is within 20 °F of the ambient or surrounding temperature, protection measures may be removed.

5.3.7 *Repair of surface defects*

5.3.7.1 *General*—Repair tie holes and surface defects immediately after formwork removal. Where the concrete surface will be textured by sandblasting or bush-hammering, repair surface defects before texturing.

5.3.7.2 *Repair of tie holes*—Plug tie holes except where stainless steel ties, noncorroding ties, or acceptably coated ties are used. When portland-cement patching mortar conforming to 5.3.7.5 is used for plugging, clean and dampen tie holes before applying the mortar. When other materials are used, apply them in accordance with manufacturer's recommendations.

5.3.7.3 *Repair of surface defects other than tie holes*—Outline honeycombed or otherwise defective concrete with a 1/2 to 3/4 in. deep saw cut and remove such concrete down to sound concrete. When chipping is necessary, leave chipped edges perpendicular to the surface or slightly undercut. Do not feather edges. Dampen the area to be patched plus another 6 in. around the patch area perimeter. Prepare bonding grout according to 5.3.7.4. Thoroughly brush grout into the surface. When the bond coat begins to lose water sheen, apply patching mortar prepared in accordance with 5.3.7.5, and thoroughly consolidate mortar into place. Strikeoff mortar, leaving the patch slightly higher than the surrounding surface to compensate for shrinkage. Leave the patch undisturbed for 1 h before finishing. Keep the patch damp for seven days.

5.3.7.4 *Preparation of bonding grout*—For bonding grout, mix approximately one part cement and one part fine sand with water to the consistency of thick cream.

5.3.7.5 *Site-mixed portland-cement repair mortar*—Mix repair mortar using the same materials as concrete to be patched with no coarse aggregate. Do not use more than one part cement to two and one-half parts sand by damp loose volume. For repairs in exposed concrete, make a trial batch and check color compatibility of repair material with surrounding concrete. When the repair is too dark, substitute white portland cement for a part of the gray cement to produce a color closely matching surrounding concrete. Use a repair mortar at a stiff consistency with no more mixing water than is necessary for handling and placing. Mix the repair mortar and manipulate the mortar frequently with a trowel without adding water.

5.3.7.6 *Repair materials other than site-mixed portland-cement mortar*—Acceptable repair materials other than site-mixed portland-cement mortar may be used for repair. Use repair materials in accordance with manufacturer's recommendations. Materials include, but are not limited to, 5.3.7.6.a and 5.3.7.6.b:

   5.3.7.6.a Shotcrete;

   5.3.7.6.b Commercial patching products, including:

- Portland-cement mortar modified with a latex bonding agent conforming to ASTM C 1059 Type II;
- Epoxy mortars and epoxy compounds that are moisture-insensitive during application and after curing, that embody an epoxy binder conforming to ASTM C 881/C 881M, Type III. The type, grade, and class shall be appropriate for the application as specified in ASTM C 881/C 881M;
- Shrinkage-compensating or nonshrink portland-cement grout conforming to ASTM C 1107; and
- Packaged, dry concrete repair materials conforming to ASTM C 928.

5.3.7.7 *Removal of stains, rust, efflorescence, and surface deposits*—Remove stains, rust, efflorescence, and surface deposits considered objectionable by the Architect/Engineer by acceptable methods.

SECTION 6—ARCHITECTURAL CONCRETE
6.1—General

6.1.1 *Description*

6.1.1.1 *Scope*—This section covers construction of architectural concrete as designated in Contract Documents.

6.1.1.2 *Coordination*—Provide coordination between this Work and work of other trades, and other concrete work on the structure. Integrate this Work into the structure. Prevent damage or defects that will lessen the quality of the surface.

6.1.1.3 *General requirements*—Architectural concrete shall comply with the requirements of Sections 1 through 5 unless otherwise indicated in Contract Documents and in this section.

6.1.2 *Submittals*

6.1.2.1 Submit the data specified in 6.1.2.1.a unless otherwise specified:

   6.1.2.1.a *Drawings and data*—Submit shop drawings of forms for architectural concrete. Show jointing of facing panels; locations and details of form ties and recesses; and details of joints, anchorages, and other accessories.

6.1.2.2 Submit the data specified in 6.1.2.2.a through 6.1.2.2.c when required:

   6.1.2.2.a *Mock-ups*—When Contract Documents require full-scale mock-ups of structural items, submit a request for acceptance of the proposed location at the project site.

   6.1.2.2.b *Special finishes*—Submit, when required, mock-ups or sample panels of aggregate transfer and other special finishes.

   6.1.2.2.c *Exposed-aggregate finishes*—Submit, when required, the proposed method of producing exposed-aggregate finishes.

6.1.2.3 *Review of submittals*—Do not construct forms until submittals have been accepted. Do not place concrete until submitted plans for batching, mixing, placing, and curing have been accepted.

6.1.3 *Quality assurance*

6.1.3.1 *Concrete construction technical specialists*—For architectural concrete operations listed in Project Specifications, provide a technical specialist trained or approved by the specialty item manufacturer. The specialist shall be on the project site during the first three days of construction operations using the specialty item and at other times required by the Project Specifications to provide technical assistance.

6.1.3.2 *Preconstruction conference*—A preconstruction conference shall be held for this phase of the Work. Organization and procedures shall be established and agreed to by all individuals involved with this phase of the Work.

6.1.3.3 *Samples and mock-up*—Make full-scale mock-ups of structural items when specified in Contract Documents. Use the same equipment, materials, and procedures that will be used in the final work. Make mock-ups at acceptable locations on the project site. Use mock-ups as samples of required quality of finished construction.

6.1.4 *Product delivery, storage, and handling*

6.1.4.1 *Aggregates*—Deliver each size of aggregate to the mixer at uniform moisture content throughout each day's concrete production.

6.1.5 *Project conditions*

6.1.5.1 *Environmental conditions*—Protect architectural concrete from damage, disfigurement, and discoloration from construction to acceptance.

6.2—Products

6.2.1 *Materials*

6.2.1.1 *Curing water and coverings*—Use curing water and coverings that will not stain the concrete.

6.2.1.2 *Reinforcement supports and spacers*—Use stainless steel, plastic, or plastic-coated reinforcement supports and spacers near exposed surfaces, except that plastic-coated products shall not be used near surfaces that are to be sandblasted.

6.2.1.3 *Formwork*—Use formwork that is watertight.

6.2.2 *Performance and design requirements*

6.2.2.1 *Formwork*

   6.2.2.1.a Design forms to produce the required finish. Limit deflection of facing materials between studs and deflection of studs and walers to 0.0025 times the clear span ($L/400$).

   6.2.2.1.b Where natural plywood form finish, grout-cleaned finish, smooth-rubbed finish, or other finish is required, form faces shall be smooth and forms shall be true to line and grade. Surfaces produced shall require only minor dressing to arrive at true surfaces. Where an as-cast finish is required, construct and install the forms so that no dressing will be required in the finishing operation to match the accepted sample.

6.2.2.1.c Where as-cast surfaces, including natural plywood form finish, are specified, ensure that the panels are orderly in arrangement, with joints planned in an acceptable relation to openings, building corners, and other architectural features.

6.2.2.1.d Where panels for as-cast surfaces are separated by recessed or emphasized joints, provide in the structural design of the forms the locations of ties within the joints so patches of tie holes will be in the recessed or emphasized joints, unless otherwise specified.

6.2.2.1.e Do not reuse forms with surface wear, tears, or defects that lessen the quality of the surface. Thoroughly clean and properly coat forms before reuse.

6.3—Execution

6.3.1 *Preparation*—Thoroughly clean and inspect formwork and batching, mixing, conveying, and placing equipment before use. Do not use equipment for other concrete construction during architectural concrete operations.

6.3.2 *Proportioning concrete mixtures*—Maintain designated colors and uniformity of color, except when not required by Contract Documents. For a concrete mixture of a specified color, use the same materials and proportions throughout the project. Avoid changes in quantity of cementitious materials per unit volume of concrete. Use only one type and one brand of cement from one mill, only one source and one nominal maximum size of coarse aggregate, only one source of fine aggregate, and only one placing consistency. For architectural concrete with exterior exposure, use air-entrained concrete with a water-cementitious material ratio not exceeding 0.45 by weight. Air content shall comply with Table 4.2.2.4.

6.3.3 *Consolidation*—Do not allow vibrators to contact formwork for exposed concrete surfaces. Where a smooth-rubbed or similar finish is specified, work the coarse aggregate back from the forms by spading or form vibration, leaving a full surface of mortar but avoiding surface voids.

6.3.4 *Formwork monitoring*—During concrete placement, continuously observe formwork. If deviations from desired elevation, alignment, plumbness, or camber are observed, or if weakness develops and the falsework shows undue settlement or distortion, stop work, remove the affected construction if it is unacceptably damaged, and strengthen the falsework.

6.3.5 *Formwork removal*—Prevent damage to concrete from formwork removal. Do not pry against face of concrete. Use only wooden wedges to separate forms from concrete.

6.3.6 *Repair of tie holes and surface defects*

6.3.6.1 *Repair area*—Where as-cast finishes are specified, the total area requiring repair shall not exceed 2 ft^2 in each 1000 ft^2 of as-cast surface. This is in addition to tie-hole patches, if Contract Documents permit ties to fall within as-cast areas.

6.3.6.2 *Color match*—Repairs in as-cast architectural concrete shall match color and texture of surrounding surfaces. Determine by trial the mixture of repair mortar to obtain a color match with the concrete when both repair and concrete are cured and dry. After initial set, dress surfaces of repairs manually to obtain texture matching the surrounding surfaces.

6.3.6.3 *Exposed aggregate*—Any finishing process intended to expose aggregate on the surface shall show aggregate faces in patched areas. The outer 1 in. of patch shall contain the same aggregates as the surrounding concrete. In aggregate transfer finish, the patching mixture shall contain the same selected colored aggregates. After patches have been cured thoroughly, expose the aggregates together with the aggregates of adjoining surfaces by the same process of mortar removal.

6.3.6.4 *Curing of patches*—Cure patches in architectural concrete surfaces for seven days. Protect patches from premature drying.

6.3.7 *Finishing*—Finishes shall comply with one of the finishes specified in 6.3.7.1 through 6.3.7.4 or other finishes as indicated in the Contract Documents:

6.3.7.1 *Textured finishes*—Use textured forms or textured form liners of plastic, wood, or sheet metal. Secure liner panels in forms by cementing or stapling. Do not permit impressions of nail heads, screw heads, or washers to be imparted to the surface of the concrete. Seal edges of textured panels to each other or to divider strips to prevent bleeding of cement paste. Use a sealant that will not stain the concrete surface.

6.3.7.2 *Aggregate transfer finishes*—Produce aggregate transfer and special finishes that duplicate the mockups or sample panels that were prepared in advance and accepted.

6.3.7.3 *Exposed-aggregate finishes*—Expose aggregate by an acceptable method including blasting, bush-hammering, or a surface retarder. Provide a concrete surface that will duplicate the mockups or sample panels that were prepared in advance and accepted.

6.3.7.3.a *Scrubbed finish*—Provide a scrubbed finish on partially hardened concrete. Wet the concrete surface thoroughly and scrub with fiber or wire brushes, using water freely, until surface mortar is removed and aggregate is uniformly exposed. Then rinse with clear water. If portions of the surface have become too hard to permit uniform aggregate exposure, use dilute hydrochloric acid (1 part commercial muriatic acid diluted with four to 10 parts water) to remove the excess surface mortar after the concrete has been in place at least 14 days. Remove the acid from the finished surface with clean water within 15 min after application. To facilitate aggregate exposure, cast concrete against form faces coated with a surface retarder applied in accordance with the manufacturer's recommendations.

6.3.7.3.b *Blast finish*—Sandblast or waterblast the concrete surface to a degree sufficient to expose aggregates. Blast surfaces with the same specified blast finish at the same age of the concrete. Use stainless steel or plastic reinforcement supports and spacers near concrete surfaces to be blasted. Protect adjacent materials and inserts during abrasive blasting operations. Unless otherwise specified in the Contract Documents, degree of blasting shall be light and shall expose fine aggregate with occasional exposure of coarse aggregate, to produce a uniform color, and not exceed a reveal of 1/16 in.

6.3.7.3.c *Tooled finish*—Dress the thoroughly cured concrete surface with electric, air, or hand tools to a uniform texture, as specified by Contract Documents.

6.3.7.3.d When blasted or tooled finishes are specified, remove surface mortar to the degree specified in Contract Documents.

6.3.7.4 *Applied finishes*—When finishes of stucco, cementitious coatings, or similar troweled materials are specified, prepare the surface of concrete to ensure permanent adhesion of the finish. When concrete is less than 24 h old, roughen it with a heavy wire brush or scoring tool. When concrete is older than 24 h, roughen the surface mechanically or by etching with acid. After roughening, wash the surface free of dust, acid, surface retarder, and other foreign material before final finish is applied.

SECTION 7—LIGHTWEIGHT CONCRETE
7.1—General

7.1.1 *Description*—This section covers requirements for structural lightweight concrete. Portions of structures to be lightweight concrete under the provisions of this section shall be designated in Contract Documents. Lightweight concrete shall comply with the requirements of Sections 1 through 5 unless otherwise specified in this section.

7.1.2 *Submittals*

7.1.2.1 *Review of submittals*—Obtain the Architect/Engineer's acceptance of required submittals before placing concrete.

7.1.3 *Product delivery, storage, and handling*

7.1.3.1 *Aggregate storage*—Unless otherwise specified or permitted, prewet dry lightweight aggregate and leave aggregates in the stockpile after prewetting for at least 12 h before using. Follow lightweight aggregate supplier's recommendations for storage, handling, prewetting, and draining, if applicable.

7.1.3.2 *Aggregate handling*—Do not allow machinery to run over lightweight aggregates.

7.2—Products

7.2.1 *Aggregates*—Fine and coarse lightweight aggregates for lightweight concrete shall conform to ASTM C 330. Normalweight aggregate used in lightweight concrete shall conform to 4.2.1.2.

7.2.2 *Performance and design requirements*

7.2.2.1 *Concrete exposed to weather*—When specified in the Contract Documents, entrain air in lightweight concrete subject to potentially destructive exposure (other than wear or loading), including exposure to freezing-and-thawing, severe weather, or deicer chemicals. Use 6 ± 1.5% air content when the nominal maximum size of aggregate is greater than 3/8 in. Use 7.5 ± 1.5% air content when the nominal maximum size is 3/8 in. or less. Determine the air content by the volumetric methods of ASTM C 173. Select concrete mixture proportions for air-entrained concrete to provide the specified compressive strength f_c', specified in the Contract Documents.

7.2.2.2 *Floors*—For troweled floors, the slump of structural lightweight concrete placed by pump shall not exceed 5 in. at the point of placement. For other floors, slump of lightweight concrete shall not exceed 4 in. at the point of placement.

7.2.3 *Mixtures*

7.2.3.1 *Density*—Proportion lightweight concrete mixtures to meet the specified equilibrium density determined by the calculated equilibrium method in ASTM C 567 unless otherwise specified or permitted. Correlate equilibrium density with the fresh bulk density of concrete. Use the fresh bulk density as the basis for acceptance during construction. Submit test results and correlation for review.

7.2.3.2 *Proportioning*—Determine the quantity of cementitious materials needed to attain the specified strength for lightweight concrete in accordance with 4.2.3. Relate strength to cementitious materials content of the concrete.

7.2.4 *Batching and mixing*

7.2.4.1 *Procedure*—When batching and mixing procedures are at variance with this Specification, submit recommendations to the Architect/Engineer for acceptance.

7.2.4.2 *Low-absorption aggregate*—Batch and mix aggregate that has been shown to absorb less than 2% water by weight during the first hour after inundation, as required by 4.3.1. Test aggregate for water absorption with the minimum moisture content likely to occur on the project. Predampening may be used to achieve this condition.

7.2.4.3 *High-absorption aggregate*—Batch and mix concrete made with lightweight aggregates absorbing 2% or more water by weight as specified in 7.2.4.3.a and 7.2.4.3.b:

7.2.4.3.a First add aggregate to approximately 80% of the mixing water and mix for a minimum of 1-1/2 min in a stationary mixer or 15 revolutions at mixing speed in a transit mixer.

7.2.4.3.b Then add admixtures, cement, and the withheld portion of mixing water and complete the mixing in accordance with 4.3.1.

7.2.4.4 *Slump adjustment*—When permitted, add water or air-entraining admixture to the mixture to bring the mixture to the specified slump after transport. For pumped concrete, increase slump of concrete entering the pump to maintain the specified slump at point of placement, as long as the requirements of 4.3.2.1 are met.

Prewet lightweight aggregate in accordance with 7.1.3.1 unless otherwise specified. For pumped concrete, prewetting shall be sufficient to ensure that slump loss through the pump line does not exceed 4 in.

7.3—Execution

7.3.1 *Consolidation*—Do not vibrate lightweight concrete to the extent that large particles of aggregate float to the surface.

7.3.2 *Finishing*—Do not work lightweight concrete to the extent that mortar is driven down and lightweight aggregate appears at the surface.

7.3.3 *Field quality control*

7.3.3.1 *Additional testing*

7.3.3.1.a *Density*—Acceptance of lightweight concrete in the field will be based on fresh bulk density measured in accordance with ASTM C 138. The fresh bulk density required in the field shall be that corresponding to the specified equilibrium density. When the fresh bulk density varies by more than plus or minus 3 lb/ft^3 from the required fresh bulk density, adjust the mixture as promptly as conditions permit to bring the density to the desired level. Do not use concrete for which fresh bulk density varies by more than plus or minus 4 lb/ft^3 from the required fresh bulk density.

7.3.3.1.b *Air content*—Determine the air content of the lightweight concrete sample for each strength test in accordance with ASTM C 173.

SECTION 8—MASS CONCRETE
8.1—General
8.1.1 *Description*

8.1.1.1 *Scope*—This section covers requirements for mass concrete as designated in Contract Documents.

8.1.1.2 *General requirements*—Mass concrete, either plain or reinforced, shall comply with the requirements of Sections 1 through 5 unless otherwise specified in this section.

8.1.2 *Submittals*—Comply with 4.1.2 and the following requirements:

8.1.2.1 *Cementitious materials*—Submit brand names, manufacturer's certifications, and test data on heat of hydration.

8.2—Products
8.2.1 *Materials*

8.2.1.1 *Cementitious materials*—Comply with 4.2.1.1, 8.2.1.1.a, and 8.2.1.1.b:

8.2.1.1.a Do not use ASTM C 150 Type III cement.

8.2.1.1.b Unless otherwise specified or permitted, use moderate heat of hydration portland cement, blended hydraulic cement with moderate or low heat of hydration properties, or portland cement with fly ash, pozzolan, or ground-granulated blast-furnace slag.

8.2.1.2 *Admixtures*—Comply with 4.2.1.4 and the following requirements:

8.2.1.2.a Do not use calcium chloride or other accelerating admixtures unless specifically permitted.

8.2.1.2.b Use an acceptable retarding admixture, pretested with project materials under project conditions, whenever prevailing temperature conditions make it necessary to prevent cold joints due to the quantity of concrete placed, to offset the effects of high concrete temperature, to permit revibration of the concrete, or to reduce the maximum temperature and rate of temperature rise.

8.2.2 *Performance and design requirements*

8.2.2.1 *Cementitious materials content*—Use the minimum cementitious material content required to attain f_c', desired durability, and properties as specified in 4.2.2.

8.2.2.2 *Slump*—Unless otherwise permitted or specified, the slump of mass concrete shall conform to the following:
- For plain mass concrete, a maximum slump of 3 in.; and
- For reinforced mass concrete, the requirements of 4.2.2.2.

8.3—Execution
8.3.1 *Placement*

8.3.1.1 *Placing temperatures*—Unless otherwise permitted or specified, the temperature of concrete at the point of placement shall not exceed 70 °F or be less than 35 °F. Concrete placed in cold weather shall meet the requirements of 4.2.2.8.

8.3.1.2 *Slump*—Slump of the concrete when placed shall meet the tolerances of ACI 117.

8.3.1.3 *Consolidation*—Place concrete in layers not more than 18 in. thick. Extend vibrator heads into the previously placed layer of plastic concrete.

8.3.2 *Curing and protection*

8.3.2.1 *Preservation of moisture*

8.3.2.1.a Cure mass concrete for the minimum curing period specified in 5.3.6 unless Contract Documents require longer curing.

8.3.2.1.b When a specific curing method is not specified in the Contract Documents, preserve the moisture either by maintaining the forms in place or, for surfaces not in contact with forms, by applying one of the procedures specified in 5.3.6.4.

8.3.2.2 *Cold weather concrete placement*—Protect the concrete from freezing and moisture loss for the required curing period in accordance with 5.3.6.1. Do not use steam or other curing methods that will add heat to the concrete.

8.3.2.3 *Hot-weather concrete placement*—Keep forms and exposed concrete continuously wet during the curing period whenever the surrounding air temperature is above 90 °F.

8.3.2.4 *Control of concrete surface temperature*—Unless otherwise specified, cool the concrete gradually so that the drop in concrete surface temperature during and at the conclusion of the specified curing period does not exceed 20 °F in any 24 h period.

SECTION 9—PRESTRESSED CONCRETE
9.1—General
9.1.1 *Description*—This section covers requirements for site-cast, post-tensioned, and prestressed structural concrete members. Post-tensioned concrete shall comply with the requirements in Sections 1 through 5 unless otherwise specified.

9.1.2 *Submittals*

9.1.2.1 Required submittals before execution of the Work are specified in 9.1.2.1.a through 9.1.2.1.c:

9.1.2.1.a *Drawings*—Installation drawings of post-tensioned concrete construction providing the following information in addition to that required by Section 2 and Section 3:
- Sizes and heights of tendon supports, including bars, and chairs;
- The location of tendons throughout their length;
- Size, details, location, materials, and stress grade (where applicable) for tendons and accessories, including anchorages and couplers;
- Jacking procedures, stressing sequence, and tensioning forces;
- Values of the wobble and curvature coefficients and anchorage set assumed, and calculated tendon elongations; and
- Details of reinforcing steel to prevent bursting and spalling.

The installation drawings shall be sealed by a professional engineer licensed in the state where the Work will be performed when specific tendon locations, stressing sequences, and elongations are not shown on the Contract Documents or when deviations from the tendon sizes or tendon locations shown on the Contract Documents are proposed.

9.1.2.1.b Gauge pressures and calibration curves for the specific jacks and gauges to be used. Calibrations shall be within 6 months of use.

9.1.2.1.c Grout mixture proportions and test data demonstrating compliance with 9.2.2.2.

9.1.2.2 *Optional submittals*—Submit the information specified in 9.1.2.2.a through 9.1.2.2.c when required by the Contract Documents:

9.1.2.2.a Test data substantiating the expected curvature and wobble coefficients and expected anchorage set.

9.1.2.2.b Results of tests required in 9.1.3.1, including demonstration of compliance with 9.2.1.5 through 9.2.1.7.

9.1.2.2.c Jack clearances.

9.1.2.3 Required submittals during the execution of the Work are specified in 9.1.2.3.a and 9.1.2.3.b.

9.1.2.3.a Certified mill test reports for a sample representing each production lot from which the prestressing steel was taken.

9.1.2.3.b Stressing records required in 9.3.4.3 for review before trimming extensions of tendons past anchorages.

9.1.3 *Quality control*

9.1.3.1 *Testing*—Test materials in accordance with the requirements of 9.1.3.1.a through 9.1.3.1.d. Include in the report a detailed description of test procedures and apparatus as well as test results.

9.1.3.1.a *Test assembly*—Test, in accordance with 9.1.3.1.b, two specimens of each tendon size at least 10 ft long and complete with standard production-quality anchorages. For unbonded tendons, test a third sample in accordance with 9.1.3.1.c.

9.1.3.1.b *Static test*—Test prestressed reinforcement specimens in accordance with the appropriate ASTM Specification of 9.2.1.1. Test tendon assembly with a method that will allow accurate determination of yield strength, breaking strength, and elongation of the specimen to ensure compliance with 9.2.1.5, or 9.2.1.6, and 9.2.1.7.

9.1.3.1.c *Cyclic test for unbonded tendons*—Perform a cyclic test on a representative tendon assembly that shall withstand, without failure, 500,000 cycles from 60 to 66% and back to 60% of its breaking strength. Test single element tendons using one strand, bar, or wire as a complete tendon assembly. Systems using multiple strands, wires, or bars may be tested using a prototype tendon provided the assembly has not less than 10% of the full-sized tendon.

9.1.3.1.d *Grout testing*—The Owner's testing agency will test the grout for strength and fluidity daily in accordance with ASTM C 1107.

9.1.3.2 *Tolerances*—Comply with the tolerances given in 9.1.3.2.a and 9.1.3.2.b.

9.1.3.2.a Bearing surface between anchorage and concrete shall be concentric with the tendon. The bearing plate or anchorage shall be perpendicular within plus or minus 1 degree to the direction of the tendon at the anchorage.

9.1.3.2.b Place tendons and anchorages within the tolerances of ACI 117 for reinforcement placement, distance between reinforcement, and concrete cover. These tolerances apply separately to both vertical and horizontal dimensions and may be different for each direction except that in slabs the horizontal tolerance shall not exceed 1 in. in 15 ft of tendon length.

9.1.4 *Product delivery, handling, and storage*—Deliver, handle, and store materials in a manner that prevents mechanical damage and corrosion. Store cement and premixed grout to prevent hydration during storage. Only use cement that has been properly stored for grouting.

9.2—Products

9.2.1 *Materials*—Use materials that comply with the requirements of 9.2.1.1 through 9.2.1.8:

9.2.1.1 *Prestressing tendons*—Prestressed reinforcement shall be of the type and strength required by the Contract Documents and shall conform to one of the following specifications:
- ASTM A 416/A 416M;
- ASTM A 421/A 421M;
- ASTM A 722/A 722M;
- ASTM A 779/A 779M; or
- ASTM A 882/A 882M.

Tendons shall be fabricated in Post Tensioning Institute (PTI) certified plants.

Prestressing steel shall be clean and free of excessive rust, scale, oil, dirt, and pitting. A light coating of rust is permissible.

9.2.1.2 *Protection of unbonded tendons*—Protect unbonded single-strand tendons against corrosion in accordance with ACI 423.6, "Specification for Unbonded Single-Strand Tendons." The protection shall be continuous over the entire length to be unbonded and shall prevent intrusion of cement paste or loss of coating materials during concrete placement.

9.2.1.3 *Ducts for bonded tendons*

9.2.1.3.a Duct-forming materials shall not react with alkalies in the cement, shall be strong enough to retain their shape and resist damage during construction, and shall prevent the intrusion of water from the cement paste. Duct-forming material left in place shall not directly or indirectly cause electrolytic action or deterioration. Ducts shall be corrugated or otherwise capable of transmitting forces from the grout to the surrounding concrete.

9.2.1.3.b The inside diameter of the duct shall be at least 1/4 in. larger than the wire, bar, or strand and shall have an inside cross-sectional area at least twice that of the net area of the prestressed reinforcement.

9.2.1.3.c Ducts shall have grout holes or vents at each end and at each intended high point. Provide drain holes at each intended low point if the tendon will be subjected to freezing after placing and before grouting.

9.2.1.4 *Sheathing for unbonded tendons*

9.2.1.4.a In accordance with ACI 423.6, sheathing for unbonded tendons shall have sufficient strength and water resistance to prevent damage or deterioration during transportation, storage at project site, installation, and concrete placement. The sheathing shall be continuous over the unbonded length of the tendons. The sheathing shall prevent the intrusion of water from the cement paste and the escape of coating material.

9.2.1.4.b When specified in the Contract Documents and for applications in corrosive environments, the sheathing shall be connected to stressing, intermediate, and fixed anchorages to provide full encapsulation of the prestressing steel in accordance with ACI 423.6.

9.2.1.5 *Anchorages for bonded tendons*—Anchorages for bonded tendons tested in an unbonded state shall develop 95% of the actual breaking strength of the prestressed reinforcement, without exceeding anticipated set. The actual breaking strength of the prestressed reinforcement shall not be less than specified in Section 9.2.1.1, and shall be determined by tests on representative samples of the material in accordance with its ASTM standard. Anchorages that develop less than 100% of the actual breaking strength shall be used only where the bond length provided is equal to or greater than the bond length required to develop 100% of the actual breaking strength of the tendon. Provide the required bond length between the anchorage and the zone where the full prestressing force is required under service and factored loads.

9.2.1.6 *Anchorages for unbonded tendons*—Anchorages for unbonded tendons shall develop at least 95% of the actual breaking strength of the prestressed reinforcement without exceeding anticipated set. Total elongation of the tendon under ultimate load shall be not less than 2% when measured over a minimum gauge length of 10 ft.

9.2.1.7 *Couplers*—Use couplers only where indicated on the Contract Documents or as acceptable. All couplers shall develop strength in excess of the actual breaking strength of the prestressing steel without exceeding anticipated set of either the couplers or the prestressed reinforcement, and shall not reduce the ductility of the tendon below the minimum 2% elongation specified in 9.2.1.6. Enclose couplers in housings that permit necessary movements during stressing. For bonded tendons, provide fittings to allow complete grouting of all the couplers components.

9.2.1.8 *Sleeves and gaskets*—Connect sheathing at joints with leak tight sleeves or gaskets.

9.2.2 *Proportioning of concrete and grout mixtures*—Comply with 9.2.2.1 through 9.2.2.2.f for concrete and grout mixtures.

9.2.2.1 *Concrete*—Proportion concrete mixtures in compliance with Section 4.

9.2.2.2 *Grout*

9.2.2.2.a Unless otherwise permitted or specified, grout shall consist of a mixture of cement and water unless the gross inside cross-sectional area of the duct exceeds four times the tendon cross-sectional area, in which case fine aggregate may be added to the mixture. Fly ash or other pozzolans conforming to ASTM C 618 or ground granulated blast-furnace slag (GGBFS) conforming to ASTM C 989 may be added in accordance with the strength requirements of 9.2.2.2.e. The maximum water-soluble chloride ion concentration in the hardened grout shall be in accordance the prestressed concrete requirements of 4.2.2.6.

9.2.2.2.b When required by the Contract Documents, add an acceptable shrinkage-compensating or expanding admixture to produce an unrestrained expansion of the mixture of between 0 and 5% by volume of the grout.

9.2.2.2.c Do not use admixtures containing more than trace (from impurities, not as an intended ingredient) amounts of chlorides, fluorides, aluminum, zinc, or nitrates. Other admixtures may be used, provided acceptable tests or performance records show conclusively that the admixtures will have no harmful effects on the tendons, accessories, or grout.

9.2.2.2.d Use fine aggregate conforming to ASTM C 404, Size No. 2, except that all material shall pass the No. 16 sieve.

9.2.2.2.e Proportion grout to achieve a minimum compressive strength of 2500 psi at seven days and 5000 psi at 28 days when tested in accordance with Section 12.5 of ASTM C 1107, and have a consistency that will facilitate placement. When required by the Contract Documents, the consistency of the grout shall be verified in accordance with ASTM C 939, and the efflux time of the grout sample immediately after mixing shall be less than 25 s. Water content shall be the minimum necessary for proper placement, and the water-cementitious material ratio shall not exceed 0.45 by weight.

9.2.2.2.f Mix the grout in a mechanical mixer capable of continuous mixing that will produce a grout free of lumps and undispersed cement. Pass the grout through the No. 16 sieve into pumping equipment that has provisions for recirculation. Begin pumping grout as soon after mixing as possible. Continue pumping as long as the grout retains the required consistency. Discard grout that has partially set.

9.3—Execution

9.3.1 *Inspection*—Conduct a visual inspection to ensure that the requirements of this Specification and the Contract Documents are met. Inspection shall include, but not be limited to, the following:
- Cleanliness of material and formwork;
- Location of materials and formwork;
- Proper tensioning of prestressing tendons; and
- Proper grouting of grouted tendons.

9.3.2 *Preparation*

9.3.2.1 *Tendons and concrete*

9.3.2.1.a Keep tendons dry and water out of the ducts. Maintain concrete around grouted tendons at a temperature of 40 °F or higher for at least three days before grouting.

9.3.2.1.b Keep ducts free of grease, oil, paint, and other foreign matter. A light coat of rust on the prestressed reinforcement is permissible, provided loose rust has been removed and the surface is not pitted.

9.3.2.1.c Keep tendons for use in unbonded construction clean and undamaged, and protect them with a permanent, continuous coating specified in 9.2.1.2.

9.3.2.1.d When parts of the tendon extend beyond the ends of the member, or when tendons are outside the concrete of the post-tensioned element, cover the exposed or specified parts of the tendon with an additional coating. The coating may be shop or field-applied, and can be plastic, epoxy, or other acceptable material.

9.3.2.1.e Keep end anchorages that will be permanently protected with concrete free of loose rust, grease, oil, and other foreign matter.

9.3.2.1.f Protect grout fittings and ducts for bonded tendons from collapse, obstructions, and other damage before and during concrete placement. Before placing concrete, examine the duct and grout fittings for holes, and repair any holes located. If the tendon remains ungrouted for more than 28 days from the time of tendon placement, provide acceptable temporary corrosion protection.

9.3.2.2 *Grouting*—Provide a dependable high-pressure water supply of sufficient volume before beginning the grouting operation.

9.3.3 *Placement*

9.3.3.1 *Tendons and accessories*—Place tendons and anchorages within the tolerances of 9.1.3.2. Firmly support tendons to prevent displacement during concrete placement.

9.3.3.2 *Grout*

9.3.3.2.a For bonded-tendon construction, grout shall fill all voids between prestressed reinforcement, ducts, and anchorage fittings. Continue injection until grout of the same consistency as the grout injected flows from vent and drain openings without the presence of air bubbles. Close vent and drain openings progressively in the direction of the flow. After vent and drain openings have been closed, raise the grouting pressure to at least 50 psi and plug the injection hole.

9.3.3.2.b In the event of a blockage or an interruption of grouting, remove grout from the duct by flushing with water.

9.3.4 *Tensioning*

9.3.4.1 *Sequence*—Stress tendons in the sequence at the concrete compressive strength and at the construction stage indicated in the Contract Documents.

9.3.4.2 *Tensioning multiple-strand tendons*—Tension tendons composed of multiple strands in a common duct simultaneously unless the tendon is designed for the strands to be stressed individually.

9.3.4.3 *Prestressing force*—Tension the prestressed reinforcement using hydraulic jacks equipped with a pressure gauge calibrated to the jack within an accuracy of plus or minus 2%. Calibrate the gauge or dynamometer within six months before use. The pressure gauge shall have graduations no larger than 100 psi. Apply the jacking force required to produce the prestressing force shown on the Contract Documents or Installation Drawings and measure the tendon elongation. Verify that the prestressing force is adequate by comparing the measured elongations to the calculated elongations. If the measured elongations differ from the calculated elongations by more than 7%, determine and correct the cause of the discrepancy. Elongation calculations shall be based on average values of load-elongation curves for the prestressed reinforcement used. For each tendon, keep and submit a record of the measured elongations and the gauge pressure readings. Do not remove stressing tails, grout ducts, or grout stressing pockets until the Architect/Engineer has reviewed the elongation records.

9.3.4.4 *Prestress loss*—The total loss of prestressing force in any post-tensioned element due to unreplaced broken tendons shall not exceed 2% of the total prestressing force.

9.3.4.5 *Formwork*

9.3.4.5.a Ensure that formwork does not restrain elastic shortening, deflection, or camber resulting from application of the prestressing force, and is sufficiently rigid to prevent displacement of the tendons beyond the tolerances of 9.1.3.2. Anchor tendon supports to the formwork to maintain the tendon profile during concrete placement.

9.3.4.5.b Do not remove formwork supports until sufficient prestressing force has been applied to support the dead load, formwork, and anticipated construction loads. When a structure will be post-tensioned in two directions, formwork shall support the load that is redistributed by the partially completed stressing operation.

9.3.4.6 *Prevention of damage to tendons*—Do not expose tendons to mechanical damage, welding sparks, flame, or electric ground currents. Do not conduct burning and welding operations in the vicinity of tendons, except as permitted by 9.3.4.7.

9.3.4.7 *Trimming of tendons*—Surplus lengths of tendons beyond anchorages may be removed by either rapid oxyacetylene burning, abrasive wheel, or shears unless the procedure is contrary to the recommendations of the prestressed reinforcement or anchorage manufacturer.

SECTION 10—SHRINKAGE-COMPENSATING CONCRETE

10.1—General

10.1.1 *Scope*—This section covers shrinkage-compensating concrete using expansive cement conforming to ASTM C 845, Type E-1.

10.1.2 *General requirements*—Portions of structures to be constructed using shrinkage-compensating concrete under the provisions of this section shall be designated in the Contract Documents. Shrinkage-compensating concrete shall comply with the requirements of Sections 1 through 5 unless otherwise specified in this section.

10.1.3 *Submittals*

10.1.3.1 *Review of submittals*—Obtain the Architect/Engineer's acceptance of required submittals before placing concrete.

10.1.3.2 Submit expansion test results measured in accordance with ASTM C 878 for the concrete mixture proportions.

10.1.3.3 Submit placing sequence.

10.2—Products

10.2.1 Materials

10.2.1.1 *Cementitious materials*

10.2.1.1.a Unless otherwise specified, the cement shall comply with ASTM C 845, Type E-1 (K).

10.2.1.1.b When permitted, silica fume shall comply with ASTM C 1240.

10.2.1.1.c Unless otherwise specified, do not use fly ash or ground-granulated blast-furnace slag.

10.2.1.2 *Admixtures*

10.2.1.2.a Do not use accelerating admixtures or admixtures containing calcium chloride unless otherwise specified or permitted.

10.2.1.2.b Do not change type, brand, or dosage rate of admixtures without evaluating the revised concrete mixture for expansion as measured in accordance with ASTM C 878 unless permitted.

10.2.2 *Performance and design requirements*—Comply with 4.2.2 and 10.2.2.1 through 10.2.2.3:

10.2.2.1 *Minimum cement content*—Cement content shall not be less than 564 lb/yd^3.

10.2.2.2 *Expansion*—Unless otherwise specified, the concrete expansion shall be a minimum of 0.03% and a maximum of 0.10%, measured in accordance with ASTM C 878.

10.2.2.3 *Slump*—Unless otherwise specified or permitted, the slump shall not exceed 6 in. at the point of placement.

10.2.3 *Proportioning*—Comply with 4.2.3 and 10.2.3.1 through 10.2.3.3.

10.2.3.1 When laboratory trial mixtures are used, stop the mixer after the initial mixing cycle and cover the laboratory concrete mixer for 20 min unless otherwise specified. After this time period, add water, as necessary, to produce the maximum specified slump within 3/4 in. The concrete shall then be mixed for an additional 2 min.

10.2.3.2 For the proposed concrete mixture, provide laboratory test results for three expansion bars cast and tested in accordance with ASTM C 878. Record the expansion test results and submit for acceptance.

10.2.3.3 *Revisions to concrete mixtures*—When concrete mixture proportions are revised in accordance with 4.2.3.6, evaluate the effect on expansion by performing laboratory tests on three expansion bars cast with the revised concrete mixture in accordance with ASTM C 878. Submit test results along with the revised mixture proportions.

10.2.4 *Reinforcement*—Use deformed bars or deformed welded-wire reinforcement meeting the requirements of 3.2 at the amounts specified in the Contract Documents.

10.2.5 *Isolation-joint filler materials*—Unless otherwise specified, use compressible isolation-joint filler material that does not develop a stress greater than 25 psi at 50% strain when tested in accordance with ASTM D 1621 or D 3575.

10.3—Execution

10.3.1 *Reinforcement*

10.3.1.1 Place reinforcement on supports that are rigid and spaced adequately to ensure proper positioning of the reinforcement during placement.

10.3.1.2 Unless otherwise specified, position reinforcement 2 in. from the top surface for reinforced slabs on ground.

10.3.2 *Placing*

10.3.2.1 *Placing sequence*—Sequence of concrete placements shall permit the previous placements to have two adjacent edges free to expand.

10.3.2.2 Unless otherwise specified or permitted, the minimum time between casting adjoining sections shall be 72 h.

10.3.3 *Isolation joints*—Provide isolation joints at junctions with columns, walls, drains, or any other rigid obstruction in the structure, in accordance with the Contract Documents.

10.3.4 *Curing*—Unless otherwise specified, wet-cure shrinkage-compensating concrete for a minimum of seven days in accordance with 5.3.6.4 a or b.

Flowchart for Selection of Concrete Mixtures

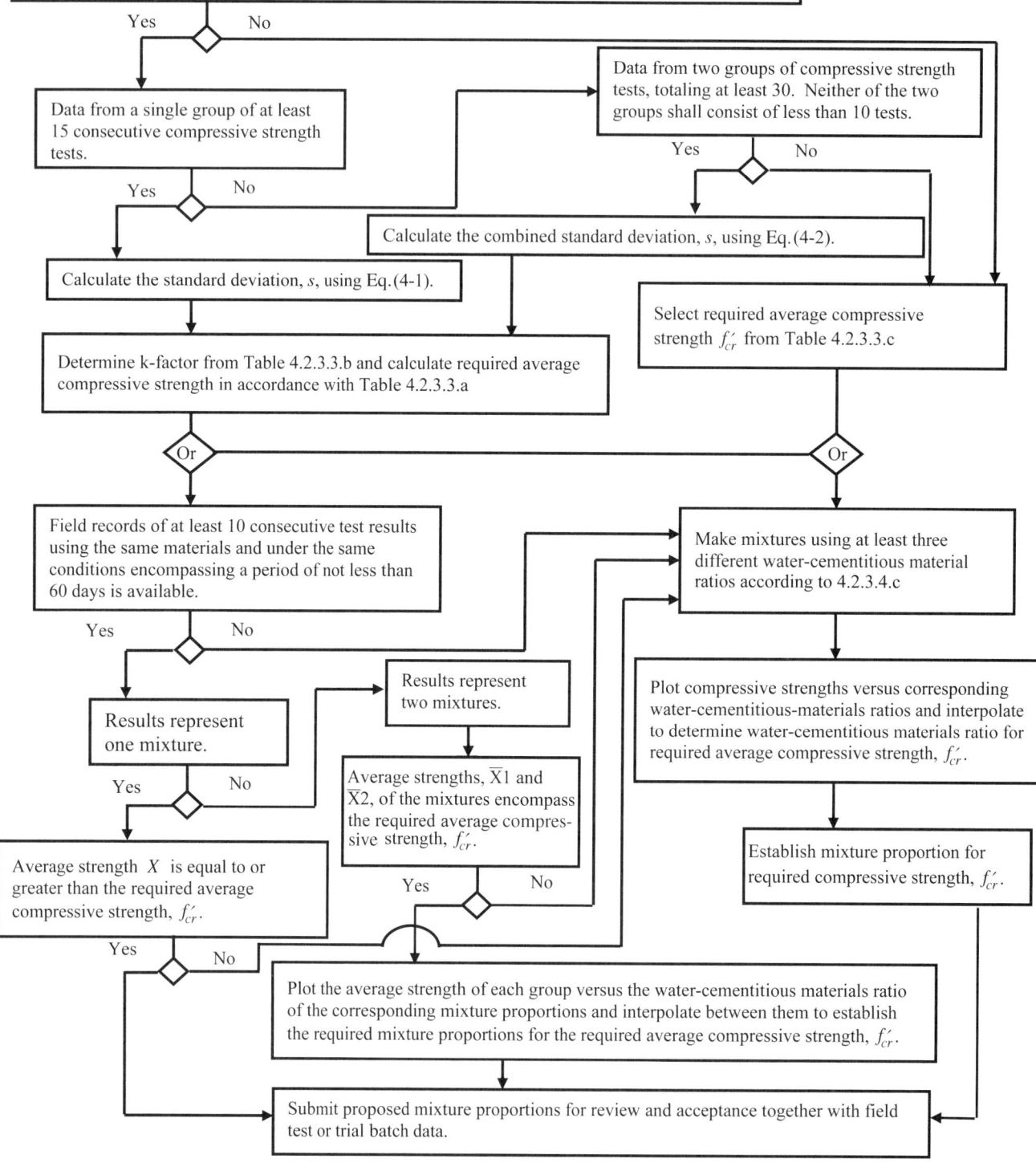

NOTES TO SPECIFIER:

FOREWORD TO CHECKLISTS

F1. This Foreword is included for explanatory purposes only; it does not form a part of Specification ACI 301.

F2. ACI Specification 301 may be referenced by the Specifier in the Project Specification for any building project, together with supplementary requirements for the specific project. Responsibilities for project participants must be defined in the Project Specification. The ACI Specification cannot and does not address responsibilities for any project participant other than the Contractor.

F3. Checklists do not form a part of ACI Specification 301. Checklists assist the Specifier in selecting and specifying project requirements in the Project Specification.

F4. Building codes set minimum requirements necessary to protect the public. ACI Specification 301 may stipulate requirements more restrictive than the minimum. The Specifier shall make adjustments to the needs of a particular project by reviewing each of the items in the checklists and including those the Specifier selects as mandatory requirements in the Project Specification.

F5. The Mandatory Requirements Checklist indicates work requirements regarding specific qualities, procedures, materials, and performance criteria that are not defined in ACI Specification 301.

F6. The Optional Requirements Checklist identifies Specifier choices and alternatives. The checklists identify the Sections, Parts, and Articles of the reference specification and the action required or available to the Specifier.

F7. *Recommended References*—Documents and publications that are referenced in the Checklists of ACI Specification 301 are listed. These references provide guidance to the Specifier and are not considered to be part of ACI Specification 301.

American Concrete Institute (ACI)

ACI 117R	Commentary on Standard Specifications for Tolerances for Concrete Construction and Materials
ACI 201.2R	Guide to Durable Concrete
ACI 207.2R	Effect of Restraint, Volume Change, and Reinforcement on Cracking of Mass Concrete
ACI 211.1	Standard Practice for Selecting Proportions for Normal, Heavyweight, and Mass Concrete
ACI 222R	Protection of Metals in Concrete Against Corrosion
ACI 223	Standard Practice for the Use of Shrinkage Compensating Concrete
ACI 225R	Guide to the Selection and Use of Hydraulic Cements
ACI 228.1R	In-Place Methods to Estimate Concrete Strength
ACI 302.1R	Guide for Concrete Floor and Slab Construction
ACI 303R	Guide to Cast-In-Place Architectural Concrete Practice
ACI 303.1	Standard Specification for Cast-In-Place Architectural Concrete
ACI 305R	Hot Weather Concreting
ACI 306.1	Standard Specification for Cold Weather Concreting
ACI 308.1	Standard Practice for Curing Concrete
ACI 311.1R	ACI Manual of Concrete Inspection—SP-2 (99)
ACI 311.4R	Guide for Concrete Inspection
ACI 311.5R	Guide for Plant Inspection and Field Testing of Ready-Mixed Concrete
ACI 318	Building Code Requirements for Structural Concrete (ACI 318-05) and Commentary (318R-05)
ACI 347	Guide to Formwork for Concrete
ACI CP 10	Craftsman Workbook for ACI Certification of Concrete Flat Work Technician/Finisher

ASTM International

ASTM C 441	Test Method for Effectiveness of Pozzolans or Ground Blast-Furnace Slag in Preventing Excessive Expansion of Concrete Due to the Alkali-Silica Reaction
ASTM D 698	Standard Test Methods for Laboratory Compaction Characteristics of Soil Using Standard Effort [12,400 ft-lbf/ft^3 (600 kN-m/m^3)]
ASTM D 1557	Standard Test Method for Laboratory Compaction Characteristics of Soil Using Modified Effort [56,000 ft-lbf/ft^3 (2,700 kN-m/m^3)]

National Ready Mixed Concrete Association (NRMCA)
Quality Control Manual, Section 3—Certification of Ready Mixed Concrete Production Facilities

Portland Cement Association (PCA)
PCA Design and Control of Concrete Mixtures, 14th Edition

Wire Reinforcement Institute (WRI)
WRI Manual of Standard Practice

Guidance for evaluating the degree of rusting on strand is given in "Evaluation of Degree of Rusting on Prestressed Concrete Strand," by A. S. Sason, *PCI Journal*, V. 37, No. 3, May-June 1992, pp. 25-30.

The above publications may be obtained from the following organizations (additional references can be found in Section 1.3 of the Specification):

American Concrete Institute (ACI)
P.O. Box 9094
Farmington Hills, MI 48333-9094

ASTM International
100 Barr Harbor Dr.
West Conshohocken, PA 19428

National Ready Mixed Concrete Association
90 Spring St.
Silver Spring, MD 20910

Prestressed Concrete Institute
209 W. Jackson Blvd Ste. 500
Chicago, IL 60606

Portland Cement Association
5420 Old Orchard Road
Skokie, IL 60076

Wire Reinforcement Institute, Inc.
942 Main St., Suite 300
Hartford, CT 06103

SPECIFICATIONS FOR STRUCTURAL CONCRETE (ACI 301-05)

MANDATORY REQUIREMENTS CHECKLIST

Section/Part/Article	Notes to Architect/Engineer
Reinforcement and reinforcement supports	
3.2.1.1	Specify required grades, types, and sizes of reinforcing steel.
3.3.2.7	Show splices on the project drawings.
Concrete mixtures	
4.2.2.6	Designate which portions of the structure are classified in accordance with Table 4.2.2.6 member types. Additional information on the effects of chlorides on corrosion of reinforcing steel is given in ACI 201.2R and ACI 222R. Test procedures must conform to those given in ASTM C 1218/C 1218M. An initial evaluation can be obtained by testing individual concrete ingredients for total chloride content. If total chloride-ion content, calculated on the basis of concrete proportions, exceeds that permitted in Table 4.2.2.6, it may be necessary to test samples of hardened concrete for water-soluble chloride-ion content as described in ASTM C 1218/C 1218M. Some of the total chloride ions present in the ingredients will either be insoluble or will react with the cement during hydration and become insoluble under the test procedure described. When concrete is tested for water-soluble chloride-ion content, the tests should be made at an age of 28 to 42 days. The limits in Table 4.2.2.6 are to be applied to chlorides contributed from the concrete ingredients, not those from the environment surrounding the concrete. The water-soluble chloride-ion limits in Table 4.2.2.6 differ from the acid-soluble chloride limits recommended in ACI 201.2R and ACI 222R. For reinforced concrete that will be dry in service, a limit of 1% has been included to control total soluble chlorides. Table 4.2.2.6 includes limits of 0.15 and 0.30% for reinforced concrete that will be exposed to chlorides or will be damp in service, respectively. These water-soluble chloride-ion limits compare with the recommended acid-soluble chloride-ion limits of 0.10 and 0.15 in ACI 201.2R while 222R recommends acid-soluble chloride-ion limits of 0.08 and 0.20% for prestressed and reinforced concrete, respectively.
4.2.2.7	Designate in the Contract Documents the portions of the structure requiring concrete resistant to sulfate attack. Specify requirements for concrete in these portions of the structure in accordance with Table 4.3.1 of ACI 318. Consult ACI 201.2R for guidance on establishing the degree of sulfate resistance needed. Criteria for evaluating effectiveness of materials against sulfate attack using test method ASTM C 1012 are provided: • in ASTM C 618 for fly ash and other pozzolans; • in ASTM C 1240 for silica fume; and • in ASTM C 989 for ground-granulated blast-furnace slag. Alternatively, consider accepting concrete mixtures that have shown a service record in excess of eight years in exposure at least as severe as that of the proposed structure.
4.2.2.9	Indicate the specified compressive strength of concrete f_c' for various portions of the Work. For most structural members, the requirements of the design will dictate the required strength. A higher compressive strength may be required for durability considerations. For floors, the specified compressive strength f_c' will generally depend upon the intended use and expected wear unless durability considerations dictate higher strengths. If the floor will be exposed to abrasive wear from early construction traffic, consider requiring a minimum compressive strength at three days of 1800 psi or higher. See ACI 302.1R for guidance on compressive strengths to specify for various classes of floors.
Handling, placing, and constructing	
5.3.1.4	Specify the required in-place density of subgrade soils for slabs-on-ground as a percentage of the maximum laboratory density. Specify the test methods to be used such as ASTM D 698 or ASTM D 1557.
Architectural concrete	
6.3.7	Specify which of the finishes from 6.3.7.1 through 6.3.7.3 (a through d) are required. Specify any special finishes that are required, but not covered by the above.
Prestressed concrete	
9.2.1.1	Specify type and minimum tensile strength of prestressed reinforcement.

OPTIONAL REQUIREMENTS CHECKLIST

Section/Part/Article	Notes to Architect/Engineer
General requirements	
1.6.3.2, 1.6.3.3, 1.6.4.1	Specify if other testing arrangements are required, such as Owner's testing agency establishing mixture proportions or any testing responsibilities of the Owner's testing agency that will be performed by the Contractor's testing agency.
1.6.3.2.g, 1.6.4.2.e	If accelerated testing of concrete is specified or permitted as an alternative to standard testing, specify the procedure from ASTM C 684 that is to be followed. Specify when compressive test specimens are to be tested if other than seven and 28 days.
1.6.4.3	Specify additional testing services desired for the Work, if applicable. Note that these additional testing services are to be performed by the Owner's testing laboratory; hence, the term "will" is used in place of "shall" in 1.6.4.3. Refer to ACI 311.1R (SP-2), ACI 311.4R, and 311.5R for specific inspection items that may be appropriate. When it is necessary or desirable to know properties of concrete at the point of placement or at locations other than the delivery point, specify that concrete is to be sampled at these other locations for testing. See the discussion under Optional Requirements in Section 4.2.2.2.
1.6.5.2	Specify if nondestructive tests will be permitted to evaluate uniformity or relative in-place strength of concrete. Refer to ACI 228.1R for guidance on nondestructive test methods.
1.6.7.1	If another basis for acceptance of concrete strength level is required for accelerated strength testing, specify the basis for acceptance.
Formwork and formwork accessories	
2.1.2.1	Review the list of submittal items and specify in Contract Documents the items that need not be submitted.
2.1.2.2	Review the list of submittal items and specify in Contract Documents the items to be submitted.
2.2.1.1	Specify other materials for form faces in contact with concrete.
2.2.1.2	Indicate where walls require form ties with a positive water-barrier.
2.2.2.1	Specify if calculations and drawings for formwork must be sealed by a licensed Engineer.
2.2.2.3	Specify if earth cuts will be permitted or required.
2.2.2.4	Specify more or less stringent limitations on deflection of facing materials when needed. Refer to ACI 347 for further guidance.
2.2.2.5.b	Specify or allow alternative locations for formed construction joints when necessary to facilitate formwork removal or accelerate construction, provided that the alternative joint locations do not adversely affect the strength of the structure.
2.2.2.5.c	Specify keyway depths other than 1-1/2 in. when required.
2.2.3.2	Specify if chamfer strips are not required on exterior corners of permanently exposed surfaces. Specify if bevels are required on re-entrant corners of concrete or on edges of formed concrete joints.
2.3.1.2	Specify tolerance limits required to be different than those of ACI 117. Specify when a more or less restrictive tolerance for abrupt offset is required. Refer to ACI 347 and the Commentary to ACI 117 for further guidance.
2.3.2.5	Specify the minimum compressive strength for removal of forms supporting the weight of concrete if different than f_c'. Specify if nonload-carrying form-facing material is not permitted to be removed at an earlier age than the load-carrying portion of the formwork.
2.3.4.2	Specify if the alternative methods for evaluating concrete strength for formwork removal are permitted.
Reinforcement and reinforcement supports	
3.1.1	Specify if the submittals listed in 3.1.1.1 through 3.1.1.3 are not required to be submitted. Otherwise, they will be required to be submitted.
3.2.1.1	For headed bars, specify type of steel for reinforcing bars: • Low-alloy steel (ASTM A 706/A 706M); • Carbon steel (ASTM A 615/A 615M). For carbon steel (ASTM A 615/A 615M) also specify grade; and • Rail steel or axle steel deformed bars (ASTM A 996/A 996M).
3.2.1.2	Specify if coated reinforcing bars are required and, if so, whether the coating is to be zinc or epoxy.
3.2.1.2.a	For zinc-coated reinforcing bars conforming to ASTM A 767/A 767M, specify the class of coating, whether galvanizing is to be performed before or after fabrication, and indicate which bars require special finished bend diameters (usually smaller sizes used for stirrup and ties). Avoid mixing galvanized and nongalvanized reinforcing steel or other embedded steel that could result in galvanic cells.
3.2.1.2.b	Specify the ASTM specification to which epoxy-coated reinforcing bars are to conform.
3.2.1.4	Specify which of the three combinations will apply.
3.2.1.5	Specify plain or deformed wire and, if required, epoxy-coated wire.

OPTIONAL REQUIREMENTS CHECKLIST (cont.)

Section/Part/Article	Notes to Architect/Engineer
3.2.1.6	Specify plain or deformed welded wire reinforcement and, if required, epoxy-coated wire reinforcement. Refer to "WRI Manual of Standard Practice" for additional guidance.
3.2.1.7	Specify if wire reinforcement supports are required or permitted.
3.2.2.2	Specify if bar welds are required or permitted. If required or permitted, specify any desired requirements for preparation for welding (such as removal of zinc or epoxy coating) more stringent than those in AWS D1.4. Specify desired requirements for chemical composition of reinforcing bars more stringent than those of the referenced ASTM specifications. Specify special heat treatment of welded assemblies, if required. Specify supplementary requirements for welding of wire to wire, and welding of wire or welded wire reinforcement to reinforcing bars or structural steels.
3.3.2.3	Specify special cover requirements for corrosive atmosphere, other severe exposures, or fire protection not covered in Table 3.3.2.3. Some concrete covers in Table 3.3.2.3 may exceed minimum concrete covers required by ACI 318. Concrete covers used for design must agree with the covers specified in Table 3.3.2.3.
3.3.2.4	Specify if the methods of support are to be other than those indicated in 3.3.2.4.a through 3.3.2.4.i.
3.3.2.5	Specify where reinforcement may extend through contraction joints, including saw-cut joints.
3.3.2.8	Specify if bending or straightening reinforcement partially embedded in concrete is permitted.
3.3.2.9	Specify if field cutting of reinforcement is permitted and specify cutting methods to be used.
Concrete mixtures	
4.2.1.1	Specify the other standards that the cementitious material may be required or permitted to conform to if cement other than ASTM C 150 Type I or Type II is required or permitted. Specify if ASTM C 150 cement with ASTM C 618 pozzolanic materials, ASTM C 989 ground-granulated blast-furnace slag, or ASTM C 1240 silica fume, is required. Specify the class of pozzolan or grade of slag that is required. Specify if ASTM C 595 blended hydraulic cement or C 1157 hydraulic cement are required or permitted. Use ACI 318 and 225R to determine what will be acceptable for the project conditions. ASTM C 1157 is a performance specification for hydraulic cement. Requirements for physical properties of the cement are specified in ASTM C 1157; however, there are no restrictions on the composition of the cement or its constituents. When using ASTM C 1157, the Architect/Engineer should consider whether any additional limits on physical properties or constituents should be specified. If it is suspected that concrete will be exposed to sulfates in service, evaluate the water-soluble sulfates in the soil and groundwater. Use the criteria of ACI 318, Section 4.3.1 and Table 4.3.1, to determine the cement type to use. Use any of the cements in ACI 318, Table 4.3.1 for concrete exposed to sea water. Verify the availability of the cement specified. Do not use ASTM C 595, Type S and SA. Specify if less than 15% fly ash is permitted. In some instances, using less than 15% fly ash can increase the concrete's susceptibility to excessive expansions caused by alkali-silica reactivity (ASR). If a smaller percentage of fly ash is proposed for use, the proposed project mixture of fly ash and portland cement from the same source should be tested and compared to a control mixture using only the portland cement in accordance with ASTM C 441. The project mixture should be considered acceptable, provided the average length increase of the project mixture does not exceed that of the control mixture. For projects where expansions due to ASR may be critical, consider requiring the test comparison at some frequency during the Work, such as every three months. If reactive aggregates are available and may be used, specify the use of natural pozzolan, fly ash, slag, or silica fume in an amount shown to be effective in mitigating harmful expansions due to alkali-silica reactivity. Alternatively, specify a low-alkali cement be used as described in the Optional Requirements Checklist for 4.2.1.2.
4.2.1.2	If aggregates are to conform to a specification other than ASTM C 33 for grading, deleterious substances, or soundness, specify the other requirements. Specify the test for determining conformance to requirements for cleanliness, and specify grading be performed on samples obtained from the aggregates at the point of batching. Specify any additional requirements for aggregate such as hardness, color, mineralogical composition, texture, or shape (crushed or gravel). If concrete will be exposed to wetting, extended exposure to humid atmosphere, or in contact with moist ground, specify the use of aggregates that do not contain materials deleteriously reactive with alkalies in the cement; however, such aggregates may be used with cement containing less than 0.60% alkalies such as (Na_2O + $0.658K_2O$) or with a material such as natural pozzolan, fly ash, slag, or silica fume in an amount shown to be effective in preventing harmful expansion due to alkali-aggregate reaction in accordance with ASTM C 441.
4.2.1.4	Specify the admixtures listed in 4.2.1.4 that are required. Indicate the parts of the Work in which each type of admixture should or may be used.
4.2.2.1	Specify if less than 15% or more than 25% fly ash is permitted in floors. If more than 25% is permitted, a history should be available demonstrating the finishing ability of the proposed concrete mixture.

OPTIONAL REQUIREMENTS CHECKLIST (cont.)

Section/Part/Article	Notes to Architect/Engineer
4.2.2.2	If slump is to be different than 4 in., specify the requirement. It might be necessary at times to specify that the slump of concrete be determined at the point of placement rather than at the point of delivery. For example, pumped concrete is often specified to have slump measured at the end of the pumpline to preclude problems encountered with varying slump loss during pumping. This would provide for a slump higher than 4 in. at the point of delivery to obtain 4 in. slump at the end of the pumpline. Once the slump loss during pumping can be determined, acceptance or rejection of concrete based on slump can then be determined at the delivery point. For example, if a 1-1/2 in. slump loss during pumping has been established and confirmed by comparative testing, a slump of 5-1/2 in. measured at the point of delivery will meet the 4 in. slump requirement at the end of the pumpline. Specify if a plasticizing or high-range water-reducing admixture is required or permitted to produce concrete with high slumps. If so, specify the required slump if different from those indicated in 4.2.2.2. For floors, refer to ACI 302.1R for guidance on slumps to specify for the various classes of floors. If a plasticizing or high-range water-reducing admixture is required or permitted to obtain high-strength concrete with a low water-cementitious material ratio, such as 0.25 to 0.30, modify the requirements accordingly for the slump before adding the admixture. Confer with concrete suppliers and admixture suppliers in the area where the project is located to determine their experience and input for such high-performance concrete.
4.2.2.3	If an aggregate size requirement differs from that specified by 4.2.2.3 (for example, smaller size in floor toppings), specify nominal maximum size of aggregate.
4.2.2.4	Specify if concrete is not required to be air-entrained. Intentionally entrained air should not be incorporated in normalweight concrete slabs that require a dense, polished, machine-troweled surface. Refer to ACI 302.1R for further guidance. For air-entrained concrete for other than severe exposure, specify the type of exposure as indicated in Table 4.2.2.4. Exposure is defined as follows: *Mild exposure*—Service in a climate where concrete will not be exposed to freezing, deicing agents, or other aggressive agents, but where air entrainment is desired for other beneficial effects, such as to improve workability or cohesion in concrete with a low cementitious material content. To improve strength, air contents lower than those needed for durability can be used. This exposure includes indoor or outdoor service. *Moderate exposure*—Service in a climate where freezing is expected, but where the concrete will not be continually exposed to moisture or free water for long periods before freezing and will not be exposed to deicing agents, other aggressive agents, or other aggressive chemicals. Examples include exterior beams, columns, walls, girders, and slabs that are not in contact with wet soil and are located so that they will not receive direct application of deicing salts. *Severe exposure*—Concrete that is exposed to deicing chemicals or other aggressive agents or that may become highly saturated by continual contact with moisture or free water before freezing. Examples include parking structures, pavements, bridge decks, curbs, gutters, sidewalks, canal linings, and exterior water tanks or sumps. Specify if a particular ASTM test method (ASTM C 231, C 138, or C 173) is required for measuring air content. For the same reasons as described in the Optional Requirements to 4.2.2.2, it may be necessary to specify that air content be measured at the point of placement to account for loss of air content during pumping. Once the loss of air content during pumping is established, acceptance limits at the point of placement can be determined.
4.2.2.5	Specify types of admixture required and indicate the parts of the Work in which each type should or may be used. Calcium chloride as an admixture shall not be used in concrete to be subjected to severe or very severe sulfate exposure as defined in Table 4.3.1 of ACI 318.
4.2.2.6	When epoxy- or zinc-coated bars are used, the limits in Table 4.2.2.6 may be more restrictive than necessary. Specify if higher limits are allowed. See the references given in the Mandatory Requirements Checklist for 4.2.2.6.
4.2.2.8	These requirements have been excerpted from ACI 306.1. For projects in cold climates, such as in northern winters, or in situations where it is prudent to require the Contractor to follow specific procedures to achieve the limits of 4.2.2.8, the temperature limits for cold weather may be deleted and ACI 306.1 can be referred to in its entirety. Options within ACI 306.1 must then be exercised. Also, see the Optional Requirements Checklist for 5.3.6.1. If concrete delivered in hot weather with a temperature higher than 90 °F has been used successfully in given climates or situations, the higher temperature may be specified in place of the 90 °F limit.
4.2.2.9	Concrete exposed to alternating cycles of freezing and thawing in a saturated condition; deicer salts; fresh, brackish, or seawater including the area in the splash or spray zone; sulfates; and concrete that is required to have low permeability to water should be specified to have a water-cementitious material ratio not exceeding the value in ACI 318 Tables 4.2.2 and 4.3.1, whichever is applicable.

OPTIONAL REQUIREMENTS CHECKLIST (cont.)

Section/Part/Article	Notes to Architect/Engineer
4.2.2.9.a	Specify those areas that will be exposed to deicing chemicals and must comply with the limitations in Table 4.2.2.9.
4.2.2.9.b	If the test specimen is to be other than a 6 x 12 in. cylindrical specimen, specify the size of the specimen. If age at test is to be other than 28 days, specify age at test. If a different test method is required, specify the test method.
4.2.3.4.b	Specify the test ages, if other than 28 days, for trial mixture proportioning.
4.3.1.1	If concrete materials are to be specified, measured, batched, or mixed other than in conformance with ASTM C 94/C 94M, specify how these procedures are to be accomplished. Specify if the ready-mixed concrete production facility must be certified by the NRMCA Program for Certification of Ready-Mixed Concrete Production Facilities or an equivalent program.
4.3.2.1	Specify when slump adjustment by addition of water at the project site is not permitted.
4.3.2.2	If shorter or longer time for completion of discharge is required or permitted, specify maximum time.
Handling, placing, and constructing	
5.1.2.1	Specify if submittals listed in 5.1.2.1.a through 5.1.2.1.e are not required to be submitted.
5.1.2.2.a	Specify if shop drawings must be submitted.
5.1.2.2.b	Specify if advance notification of concrete placement is required.
5.1.2.2.c	Specify if a request for acceptance of preplacement activities must be submitted. When necessary, specify a preconstruction conference be held before the start of construction activities.
5.1.2.2.d	Specify if a request for acceptance of wet-weather protection must be submitted.
5.1.2.2.e	Specify if a request for acceptance of hot-weather precautions must be submitted.
5.1.2.2.f	Specify if samples finished in accordance with 5.3.3.2 must be submitted.
5.1.2.2.g	Specify if an exposed-aggregate surface is required.
5.3.2.1.c	If concrete temperatures higher than 90 °F are acceptable, based on location, relative humidity, and past experience, specify a higher allowable concrete temperature in hot weather. Review the ACI 305R report for guidance to specify a higher temperature.
5.3.2.6	Specify if bond is required at construction joints.
5.3.3.2	Specify if the finish is required to match that of a sample panel to be furnished for comparison purposes. Specify the sample finish location and the in-place finish location.
5.3.3.3	Specify more restrictive tolerances for as-cast form finishes as needed based on importance of surface appearance. See Optional Requirements Checklist for 2.3.1.2 for additional guidance.
5.3.3.5	Specify if finishes other than those in 5.3.3.5 are required.
5.3.4.1	Specify which of the finishes or combination of finishes in 5.3.4.2 are required. If this is not done, the finishes will be as required in 5.3.4.2.j. Specify when more or less certified flatwork concrete finishers may be required or permitted. More stringent qualifications for the finishing contractor and finishers may be appropriate where floor serviceability is significant to the Owner and for large floor projects with specific requirements for flatness, heavy loading, frequent lift truck traffic, or automated warehouse truck traffic. For such projects, specify that the finishing contractors use qualified flatwork finishers who are skilled in the specific type of Work required.
5.3.4.2.c	Specify more restrictive tolerances if applicable. The conventional straightedged tolerance from ACI 117 applies to most general floor construction. For floors requiring tighter tolerances, such as in areas housing sensitive test or monitoring equipment, specify either "flat" or "very flat" floor tolerances from ACI 117. Refer to the commentary for ACI 117 and ACI 302.1R for more guidance. Specify tolerances that may be more or less restrictive when applicable.
5.3.4.2.e	For dry-shake finishes, specify the metallic or mineral aggregate, the final finishing method, and the location.
5.3.4.2.f	For heavy-duty topping for two-course slabs, specify the materials, the final finishing method, and the location.
5.3.4.2.h	For nonslip finishes, specify the location. Where abrasive particles other than crushed aluminum oxides are to be used, specify the other abrasive particles.
5.3.4.2.i	For exposed-aggregate finishes, specify the location, color, surface retarder, and size of aggregate. (Usually 3/8 to 5/8 in.)

OPTIONAL REQUIREMENTS CHECKLIST (cont.)

Section/Part/Article	Notes to Architect/Engineer
5.3.4.3.b, 5.3.4.3.c	Alternative floor finish tolerances, types of floors, and floor areas may be specified where applicable. The 10 ft straightedge method of measuring tolerances from ACI 117 applies to many small general floor construction applications. The F-number measuring system specified in ACI 117 applies to many large specialized and general floor construction applications. For floors requiring tighter tolerances, such as in areas with frequent lift-truck traffic, automated warehouse forklifts, or housing sensitive test and monitoring equipment, specify either "flat" or "very flat" floor tolerances from ACI 117 using the F-number measuring system. When specifying the F-number measuring system for unshored floors, specify only the F_F value, not the F_L value. Note that the commentary for ACI 117 contains cautions (per ASTM E 1155) to not use the F-number measuring system within 2 ft of an imbed or construction joint. Caution should also be used in specifying the F-number measuring system in floor areas that slope, unless a specific constant slope has been specified so that the F_F value is appropriate. Refer to the commentary for ACI 117 and ACI 302.1R for further guidance.
5.3.5	Specify if saw-cut contraction joints are required.
5.3.6.1	For concrete surfaces that require enhanced durability, such as high wear resistance, low permeability, or minimal cracking, a longer duration of curing may be needed than is required to meet compressive strength criteria alone. When such enhanced properties are required, minimum curing periods of 7 days for high-early-strength concrete, 14 days for concrete incorporating Type I or Type II cements, and 14 to 21 days for concrete incorporating pozzolan as one of the cementitious materials are recommended. See ACI 308.1 for additional guidance. Specify if a curing procedure of 5.3.6.4 that supplies additional water is required.
5.3.6.5	Requirements for rate of temperature change have been excerpted from ACI 306.1. For optional cold-weather concreting requirements, see the Optional Requirements Checklist for 4.2.2.8 and specify ACI 306.1 in its entirety, if appropriate.
5.3.7.7	Where stains, rust, efflorescence, and surface deposits are to be limited, describe the degree to which they are unacceptable.
Architectural concrete	
6.1.1.1	Designate areas to be treated as architectural concrete. Describe special color requirements. If necessary, specify a symmetrical array of formwork panels of a specified size.
6.1.1.3	Review Sections 1 through 5 and specify requirements to be omitted or added for architectural concrete. Designate any special cementitious materials, aggregates, or admixtures required for architectural concrete.
6.1.2.1, 6.1.2.2	Specify which submittals are required.
6.1.3.1	If the importance of the Work warrants it, list operations for which a technical specialist, trained or approved by the specialty item manufacturer, is to be on the project site to provide technical assistance during the first three days of construction operations using the specialty item. Specify any other times when a technical specialist is to be on the project site to provide technical assistance.
6.1.3.3	Specify for which structural items the Contractor is to make full-scale mock-ups as samples of finished construction.
6.2.2.1.d	Specify if it is permissible for ties to be located within exposed areas of architectural concrete.
6.3.2	Specify areas where designated colors and uniformity of color need not be maintained. Specify areas where stucco or cementitious coating is required. If so, specify applicable requirements or refer to the applicable part of the Contract Documents for such requirements. If applicable, specify required color.
6.3.3	Specify areas where a smooth-rubbed or similar finish is required.
6.3.6.1	Specify areas where as-cast finishes are permitted or required. Specify if ties are permitted within as-cast areas.
6.3.7.3.a	When acid washing or surface retarders are to be used to obtain a scrubbed finish, review ACI 303R, "Guide to Cast-in-Place Architectural Concrete Practice," and specify the recommendations and appropriate safety precautions that should be followed.

6.3.7.3.b	For a blast finish, if the degree of blasting is to be other than light, specify what degree of blasting is to be used based on the following:
	  Brush Sufficient to dull surface sheen but not to have any reveal
	  Light Maximum 1/16 in. aggregate exposure
	  Medium Maximum 1/4 in. aggregate exposure
	  Heavy Maximum 1/3 in. of the large aggregate diameter
	Refer to ACI 303.1 for additional guidance.

6.3.7.3.c	Where a tooled finish surface texture is specified, such as a hand-tooled, rough- or fine-pointed, or bush-hammered surface texture, provide description of specified surface texture.

OPTIONAL REQUIREMENTS CHECKLIST (cont.)

Section/Part/Article	Notes to Architect/Engineer
6.3.7.3.d	If blasted or tooled finishes are specified, specify the degree to which surface mortar is to be removed (removal of surface mortar only, removal of sufficient mortar to expose the surface of some coarse aggregate in relief to a specified depth, or removal with tools of sufficient material to abrade the coarse aggregate).
6.3.7.4	Specify areas where designated colors and uniformity of color need not be maintained. Specify areas where stucco or cementitious coating is required. If so, specify applicable requirements or refer to the applicable part of the Contract Documents for such requirements. If applicable, specify required color.
Lightweight concrete	
7.1.1	Designate portions of the structure to be constructed of lightweight concrete. Review Sections 1 through 5 and specify requirements to be omitted or added for lightweight concrete.
7.1.3.1	Specify if prewetting lightweight aggregate is not required.
7.2.2.1	Where lightweight concrete is subject to potentially destructive exposure other than wear or loading, specify that it be air-entrained. Destructive exposures include freezing and thawing, severe weather, or deicer chemicals. Specify the required compressive strength based on the requirements of ACI 318 Section 4.2.2 and Table 4.2.2.
7.2.3.1	For lightweight concrete, specify the equilibrium density. Specify method of determining equilibrium density of other than the calculated method in ASTM C 567.
7.2.4.4	Specify if presoaking lightweight aggregate by means other than vacuum saturation, ponding, or sprinkling is required.
Mass concrete	
8.1.1.1	Designate portions of the structure to be treated as either plain mass concrete or reinforced mass concrete. Whether or not concrete should be designated as mass concrete depends on many factors such as weather conditions, the volume-surface ratio, rate of hydration, degree of restraint to volume change, temperature and mass of surrounding materials, and functional and aesthetic effect of cracking. In general, heat generation should be considered when the minimum cross-sectional dimension approaches or exceeds 2-1/2 ft or when cement contents above 600 lb/yd^3 are used. The requirements for each project, however, should be evaluated on their own merits.
8.1.1.2	Review Sections 1 through 5 and specify additional requirements or any requirements to be omitted for mass concrete.
8.2.1.1.b	For mass concrete sections, cements such as ASTM C 150, Type II moderate heat; ASTM C 150, Type IV; ASTM C 595 (MH or LH) cements; or cement combinations with fly ash, pozzolans, or ground granulated blast-furnace slag should be used for the low heat-of-hydration benefits. Because low heat-of-hydration cementitious materials generally have lower early strengths, the compatibility of the concrete using such materials should be considered with the other work on the project. If the lower early concrete strength obtained using such cementitious materials is not acceptable, specify appropriate procedures to be used. The availability of cementitious materials should also be considered when specifying a particular cement or cementitious material combination.
8.2.2.1	When 28-day strength is not required for service conditions, a reduction in cement content can be achieved by requiring that concrete mixtures be proportioned for a strength at ages other than 28 days, such as at 56 or 90 days. Use of fly ash or other acceptable pozzolan may also reduce the required cement content. The Contract Documents should specify the use of pozzolans and later-age design strengths when acceptable.
8.2.2.2	Specify the maximum permissible slump if it is to be other than 3 in. for plain mass concrete or that required by 4.2.2.2 for reinforced mass concrete.
8.3.1.1	If the limits on temperature of concrete when deposited are to be other than as given in 8.3.1.1, specify maximum and minimum placing temperatures.
8.3.2.1.a	A curing period of 7 days is sufficient for mass concrete proportioned for a 28-day specified strength. When concrete strength is based on 56- or 90-day compressive strength, the curing period should be extended to a minimum of 14 days. Specify the duration of curing if longer than 7 days is required.
8.3.2.1.b	Mass concrete is best cured with water for the additional cooling benefit in warm weather. When water curing is impractical, such as when the surrounding air temperature is less than 32 °F, other methods such as the use of liquid membrane-forming compounds may be used. Specify if a particular curing method is desired.
8.3.2.4	Specify additional or optional temperature controls as appropriate to minimize thermal cracking. For example, limitations on temperature differentials between the center and surface of the concrete may be desirable for large structurally reinforced placements, such as large mat foundations, if the entire concrete section can be cast in one continuous placement and the external restraint from adjacent concrete elements can be avoided. Complying with limitations on temperature differentials will normally require keeping concrete warm with insulation. Additional reinforcing steel may also be needed to minimize crack widths from base restraint and the higher peak concrete temperatures. See PCA publication, *Design and Control of Concrete Mixtures*, 14th edition, and ACI 207.2R for additional guidance.

OPTIONAL REQUIREMENTS CHECKLIST (cont.)

Section/Part/Article	Notes to Architect/Engineer
Prestressed concrete	
9.1.1	Review Sections 1 through 5 and specify additional requirements or requirements to be omitted for post-tensioned concrete.
9.1.2.2	If required, specify that the Contractor submit test data substantiating expected coefficients and anchorage set.
9.2.1.1	Guidance for evaluating the degree of rusting on strand is given in "Evaluation of Degree of Rusting on Prestressed Concrete Strand," by A. S. Sason, *PCI Journal*, V. 37, No. 3, May-June 1992, pp. 25-30.
9.2.1.4.b	Indicate areas that are considered to be corrosive environments where encapsulation of the prestressing steel at stressing, intermediate, and fixed anchorages is required.
9.2.1.7	Indicate areas where couplers may be used.
9.2.2.2.a	Specify specific grout mixture for the project, if desired.
9.2.2.2.b	If a shrinkage-compensating or expanding admixture is required for the grout, so specify in the Contract Documents.
9.2.2.2.e	Specify when verification of grout consistency is desired.
9.3.2.1.d	Indicate areas that require an additional coating on the exposed or specified parts of the tendon.
9.3.4.1	Specify the sequence, the concrete strength, and the stages at which tendons should be stressed.
Shrinkage-compensating concrete	
10.1.2	Designate areas to be constructed using shrinkage-compensating concrete. Specify requirements of Sections 1 to 5 that do not apply.
10.2.1.1.a	If an expansive cement other than ASTM C 845, Type E-1 (K) is acceptable or required, specify the cement type.
10.2.1.1.c	Fly ash or ground granulated blast-furnace slag will affect the expansion and should not be used without adequate testing.
10.2.1.2.a	Accelerating admixtures, specifically ones that contain calcium chloride, may reduce the expansion of the concrete and should not be permitted for use in shrinkage-compensating concrete.
10.2.1.2.b	Admixtures may have an effect on the expansion of the specific concrete mixture. Do not permit changes in admixture dosage or type without additional testing. See ACI 223 for additional information.
10.2.2.2	If different minimum and maximum limits for expansion are desired, specify the requirements. Minimum expansion needed is based on the projected shrinkage for the particular concrete mixture and the amount of reinforcement used. Consult ACI 223 for guidance.
10.2.2.3	If slump is to be different than 6 in. maximum at the point of placement, specify the requirement. Refer to Optional Requirements Checklist 4.2.2.2 for guidance on slump loss between delivery and placements points.
10.2.3.1	Due to the initial slump loss of shrinkage-compensating concrete, it is necessary to proportion the concrete mixture to consider initial slump loss. If the concrete mixture used in the Work has a delivery time longer than 20 min, specify a longer hold time to be used in the trial mixture proportioning procedure. Consult ACI 223 for guidance.
10.2.4	Specify the grade of reinforcing bar and the amounts of reinforcement required. Shrinkage-compensating concrete must always be reinforced. The reinforcement should be determined in accordance with ACI 318. See ACI 223 for additional guidance.
10.2.5	Specify alternative compressible isolation-joint filler material if desired.
10.3.1.2	Specify position of bars in reinforced slabs on ground if different from 2 in. from top surface.
10.3.2.2	If a longer time between casting of adjoining sections is needed, specify the time required. See ACI 223 for guidance.
10.3.4	If shrinkage-compensating concrete is cured by a method other than wet curing, the expansion will be reduced significantly. The structure or slab should be designed to compensate for this reduced expansion. See ACI 223 for guidance. If curing is to be continued for a period longer than 7 days, or if a method other than water curing is acceptable, specify the requirements in the Contract Documents.

SUBMITTALS CHECKLIST

NOTE: The items listed will be submitted by the Contractor and reviewed by the Architect/Engineer.

Notify the Contractor of acceptance or rejection after review of submittals. All submittals and responses should be retained in files for future reference during the Work. Some submittal requirements shown will apply only when optional requirements are selected and written into the Project Specifications. Once optional requirements have been selected, review the Section/Part/Article indicated for the submittal item to see if it applies.

Section/Part/Article	Submittal items and notes to Architect/Engineer
General requirements	
1.6.3.1	Proposed testing agency.
1.6.3.2.e	Test data and documentation on materials and concrete mixtures.
1.6.3.2.f	Quality-control program of the concrete supplier.
1.6.3.2.g	Request to use accelerated testing and correlation data.
1.6.4.1.c	Test and inspection results.
1.7.1.4	Proposed repair methods, materials, and modifications to the Work.
1.7.4.2.e	Description of repair work performed to bring strength-deficient concrete into compliance with the Contract Documents.
1.7.5.2.e	Description of repair performed to bring potentially nondurable concrete into compliance with the Contract Documents.
Formwork and formwork accessories	
2.1.2.1.a	Data on formwork facing materials if different from that specified in 2.2.1.1.
2.1.2.1.b	Data on proposed departure from location or detail of construction and contraction joints shown on the project drawings.
2.1.2.1.c	Correlation data for alternative methods of determining strength of concrete for formwork removal. Refer to ACI 228.1R for recommendations on developing suitable correlation data.
2.1.2.1.d	Detailed plan for formwork removal at a lower compressive strength than specified.
2.1.2.1.e	Plan and procedures for installation and removal of reshoring and backshoring. See ACI 347 for guidance on items to consider.
2.1.2.1.f	Data on formwork release agent or formwork liners.
2.1.2.2.a	Shop drawings for formwork.
2.1.2.2.b	Calculations for formwork, reshoring, and backshoring.
2.1.2.2.c	Data and samples of form ties.
2.1.2.2.d	Data and samples of expansion joint materials.
2.1.2.2.e	Data and samples of waterstops.
2.2.2.3	Request to use earth cuts as form surfaces.
2.2.2.5.b	Alternative location and details of construction joints.
2.2.2.5.d	Alternative locations and details for formed construction and contraction joints.
2.3.2.5	Detailed plan for formwork removal at a lower compressive strength than f'_c.
2.3.4.2	Data correlating alternative concrete strength-measuring methods for formwork removal. Refer to ACI 228.1R for recommendations on developing suitable correlation data.
Reinforcement and reinforcement supports	
3.1.1.1.a	Certified test reports on materials.
3.1.1.1.b	Placing drawings showing fabrication dimensions and locations for placement of reinforcement and supports.
3.1.1.1.c	List of splices and request to use splices not shown on the project drawings.
3.1.1.1.d	Request to use mechanical splices not shown on the project drawings.
3.1.1.1.e	Request for placement of column dowels without the use of templates.
3.1.1.1.f	Request and procedure to field bend or straighten partially embedded reinforcement.
3.1.1.1.g	Copy CRSI Plant Certification.
3.1.1.2.a	Description of reinforcement weld locations, weld procedures, and welder qualifications.
3.1.1.2.b	Proposed supports for coated reinforcement and materials for fastening coated reinforcement not covered in 3.3.2.4.
3.1.1.3.a, 3.3.2.2	When the Contractor finds it necessary to move reinforcement beyond the specified placing tolerances to avoid interference with other reinforcement, conduits, or embedded items, review a submittal showing the resulting arrangement of reinforcement.

SUBMITTALS CHECKLIST (cont.)

Section/Part/Article	Submittal items and notes to Architect/Engineer
3.1.1.3.b	Inspection and quality-control program of plants that are not certified by the Concrete Reinforcing Steel Institute.
3.2.2.1	Request to heat and bend reinforcement.
3.3.2.6	Request to use alternate method for setting column dowels.
3.3.2.7	Request to use mechanical splices not shown on the project drawings.
3.3.2.8	Request and procedure to field bend or straighten partially embedded reinforcement.
3.3.2.9	Request to field cut reinforcement.
Concrete mixtures	
4.1.2.1	Mixture proportions and characteristics. Check that mixture proportions conform to the requirements of 4.2.2 for cementitious material content, water-cementitious material ratio, slump, nominal maximum size of coarse aggregate, air content, admixtures, and chloride-ion concentration, as well as compressive strength and yield.
4.1.2.2, 4.2.3.4.a	Method and test data used to establish mixture proportions. Several different methods can be used to select mixture proportions to produce the necessary placeability, density, strength, and durability of the concrete. Field experience of concrete mixtures previously used under similar conditions provides the best assurance that the proposed concrete mixture can be used satisfactorily and will have the specified properties. If there is no field experience, ACI 211.1 provides guidance for selection of the initial quantities of materials based on material properties and specified concrete properties. When a field test record is not available, ACI 211.1 recommends that mixture characteristics be checked by trial batches in the laboratory or in the field. Blending aggregates to meet criteria for a combined grading is another proportioning method that can be used. Listed below are some of the different procedures that have been used to determine proportions of blended aggregates: • Combined fineness modulus; • 8 to 18% retained on each of the standard sieves; • Coarseness factor chart; and • 0.45 power chart. When one of the above or other similar proportioning methods are used, the specific combined grading to which aggregate is to be blended, along with the tolerances for control, should be submitted. This proportioning method also requires concrete characteristics to be checked by trial batches.
4.1.2.3	Information on types, classes, producers' names, and plant locations for cementitious materials; types, pit or quarry locations, producers' names, gradings, and properties required by ASTM C 33 for aggregates; types, brand names, and producers' names for admixtures; and source of supply for water and ice. Except for admixtures and water, test results confirming conformance with applicable specifications shall not be more than 90 days old. Test results for aggregate soundness, abrasion, and reactivity may be older than 90 days, but not older than 1 year, provided test results for the other properties specified in ASTM C 33 indicate that the aggregate quality has not changed.
4.1.2.4	Materials, mixture proportions, and field strength-test data used for proportioning.
4.1.2.5, 4.2.3.5	Requests for adjustments to mixture proportions. Requests to adjust mixture proportions necessary for workability or consistency. If the Contractor desires to decrease the cementitious materials content of the concrete mixture after having satisfied the requirements of 4.2.3.6, review a request for acceptance of the proposed revised mixture with a lower cementitious materials content on a trial basis. If the Contractor finds it necessary to increase the cementitious materials content, review a request for acceptance of the proposed revised mixture with a higher cementitious materials content on a trial basis. Confirm adequacy of modified proportions has been verified from a set of new field test data.
4.1.2.6	Evaluation and test results required in 4.2.2.1 verifying the adequacy of concrete to be placed in floors if the cementitious materials content is less than the minimum specified in Table 4.2.2.1.
4.1.2.7	Request to use calcium chloride.
4.1.2.8, 4.3.1.1	Request to use the volumetric batching method.
4.1.2.9	Requests to exceed the ASTM C 94/C 94M required time of discharge.
4.2.1.1	Requests to use cementitious materials other than ASTM C 150 Type I or Type II. When ASTM C 595 or C 1157 cements are used in structures that will be subjected to deicing chemicals, verify compliance of the concrete with Table 4.2.2.9.

SUBMITTALS CHECKLIST (cont.)

Section/Part/Article	Submittal items and notes to Architect/Engineer
4.2.1.3	Request to use alternative sources of water.
4.2.1.4	Request to use admixtures.
4.2.1.5	Request to change materials and data verifying that properties of the concrete mixture conform to the requirements of 4.2.2.
4.2.2.1	Request to use a lower cementitious material content.
4.2.2.2	Request to use a slump other than that specified.
4.2.2.7	Documentation indicating compliance with the specified requirements for sulfate resistance. Documentation may include test results on the cementitious material to be used in the proposed concrete mixture in accordance with ASTM C 1012. Alternatively, evidence of adequate sulfate resistance of at least 8-year old existing structures in exposures at least as severe as the proposed structure may be accepted. The level of sulfate ions in the soil or water of the existing structure should be comparable to the exposure of the proposed structure as provided by the Architect/Engineer.
4.2.3.6.c	Revised mixture proportions based on revised value of f'_{cr}.
Handling, placing, and constructing	
5.1.2.1.a	Test and inspection records.
5.1.2.1.b	Description of conveying equipment.
5.1.2.1.c	Proposed method of measuring concrete surface temperature changes.
5.1.2.1.d	Proposed method for removal of stains, rust, efflorescence, and surface deposits.
5.1.2.1.e	Qualifications of finishing contractor and flatwork finishers.
5.1.2.2.a	Shop drawings of placing, handling, and constructing methods.
5.1.2.2.b	Advance notification of forthcoming placement. Arrange for tests and inspection to be properly coordinated.
5.1.2.2.c	Request for acceptance of preplacement activities to ensure the preplacement activities are properly inspected, if necessary.
5.1.2.2.d, 5.3.2.1.a	Proposed wet-weather protection activities.
5.1.2.2.e	Proposed precautions for placement of concrete hotter than 90 °F.
5.1.2.2.f	Sample finish.
5.1.2.2.g, 5.3.4.2.i	Specification and manufacturer's data for surface retarder used in producing exposed-aggregate finish along with method of use.
5.1.2.3.a	Proposed location and treatment of construction joints not shown on the project drawings. Review proposed methods for preparing the surface and the use of portland-cement grout.
5.1.2.3.b	Bonding agents other than cement grout for two-course slabs.
5.1.2.3.c, 5.3.2.4	Proposed method for underwater placement.
5.1.2.3.d	Proposed location of contraction joints not indicated on the project drawings.
5.1.2.3.e	Proposed methods of curing other than those of 5.3.6.4.
5.1.2.3.f	Description of proposed coated form ties.
5.1.2.3.g, 5.2.1.3, 5.3.7.6	Specification and data and methods of use for any proposed repair material other than site-mixed portland-cement mortar described in 5.3.7.5 (see 5.3.7.6). For patches in exposed concrete, exercise caution when using the materials described in 5.3.7.6, particularly with regard to both possible color changes from weathering and delamination due to differing coefficients of thermal expansion. Ensure that the material, including ASTM type or class, is appropriate for the moisture and thermal conditions of exposure.
5.3.2.1.c	Request to exceed 90 °F concrete temperature, along with proposed precautionary measures and supporting data.
5.3.2.6	Proposed materials and methods to prepare the concrete surface to achieve bond.
5.3.4.2.f	Request to use bonding agents other than cement grout.
5.3.5	Detailed plan for alternative saw-cutting method, such as shallow-cut and dry-cut method. See ACI 302.1R for further guidance.
5.3.6.4	Proposed methods of curing other than those listed in 5.3.6.4.a through e.
5.3.6.5	Method of measuring concrete surface temperature.

SUBMITTALS CHECKLIST (cont.)

Architectural concrete	
6.1.2.1.a	Shop drawings and fabricating drawings of formwork for architectural concrete. These drawings should show jointing of facing panels; locations and details of form ties and recesses; details of joints, anchorages, and other accessories; and any necessary alignment bracing. Review drawings for condition of finished surface, jointing, location of form tie holes and their treatment, types of form ties, location and details of rustication strips, leak-tightness, assembly, and removal.
6.1.2.2.a	Request for the proposed location of full-scale mock-ups at the project site.
6.1.2.2.b	Mock-ups or sample panels of aggregate transfer and other special finishes.
6.1.2.2.c	When an exposed-aggregate finish is required, review a description of the method (such as blasting, bush-hammering, or use of a surface retarder) the Contractor desires to use to expose aggregate.
Lightweight concrete	
7.1.3.1	Request for alternate prewetting methods or times for lightweight aggregate.
7.2.3.1	Test results or calculations correlating equilibrium density to the required fresh bulk density.
7.2.4.1	Batching and mixing procedure that varies from the specified requirements in Section 4.
7.2.4.4	If the Contractor needs additional water or air entrainment to bring the concrete to the specified slump, review the request and quantities to be added.
Mass concrete	
8.1.2	Documentation and test data on cementitious material, aggregates, admixtures, and water. If the Contractor deems it necessary to use a retarding, accelerating, or other admixture in mass concrete, review manufacturer's data on the admixture and the Contractor's test results on the admixture with the other project materials.
8.2.1.2.a	Request to use an accelerating admixture. As a general rule, accelerating admixtures should not be used in mass concrete because they contributed to early undesirable heat development. On rare occasions, such as when early formwork removal is critical, accelerating admixtures may be needed to accelerate strength development in reinforced mass concrete during winter conditions. Calcium chloride, if used, should not be permitted in excess of 1% by weight of cement. The Architect/Engineer must accept the use of any accelerating admixture.
8.2.2.2	Requests to use a slump greater than 3 in. for plain mass concrete or a slump differing from the requirements of 4.2.2.2 for reinforced mass concrete.
8.3.1.1	Requests to allow limits on concrete temperature at placement to exceed 70 °F or to be less than 35 °F.
Prestresssed concrete	
9.1.2.1.a	Installation drawings and data on: • Sizes and heights of tendon support bars and chairs; • Tendon locations; • Size, details, location, materials, and stress grade (where applicable) for tendons and accessories; • Jacking procedures, stressing sequence, and tensioning forces; • Wobble and curvature coefficients and anchorage set data; and • Details of reinforcement to prevent busting and spalling.
9.1.2.1.b	Gauge pressures and calibration curves for the rams and gauges.
9.1.2.1.c	Grout mixture proportions and test data demonstrating compliance with 9.2.2.2.
9.1.2.2.a	Test data substantiating the expected coefficient and anchorage set.
9.1.2.2.b	Results of tests required in 9.1.3.1.
9.1.2.2.c	Jack clearances.
9.1.2.3.a	Certified mill tests for a sample taken from the production lot of the prestressing tendon that will be used in the Work.
9.1.2.3.b	Stressing and elongation records.
9.2.1.7	Proposed locations of couplers at locations other than as indicated on the Contract Documents.
9.2.2.2.a	Request to use different grout mixtures.
9.3.2.1.d	Data on coating material for tendons extending outside the concrete, or otherwise specified to receive an additional coating.
9.3.2.1.f	Data on corrosion protection material for tendons ungrouted for more than 28 days after tendon placement.
9.3.4.3	Records of measured elongation and gauge pressure readings for the prestressing force.

SUBMITTALS CHECKLIST (cont.)

Shrinkage-compensating concrete	
10.1.3.2	Expansion test results for the proposed concrete mixtures.
10.1.3.3	Proposed sequences of concrete placements. It is critical that the concrete be placed in such manner that will permit the placement to expand. Consult ACI 223 for guidance.
10.2.1.1.c	Request to use silica fume.
10.2.3.2	Expansion test results for the proposed concrete mixtures.
10.2.3.3	Proportions and expansion test results for revised mixture proportions.

ACI 116R-00

Cement and Concrete Terminology

Reported by ACI Committee 116

Hamid Farzam
Chairman

Glen Bollin	Richard H. Howe	Joaquin Marin
Bernard J. Erlin	Henri L. Isabelle	Bryant Mather
Fred K. Gibbe	Lawrence J. Kaetzel	Alvaro G. Meseguer
Robert L. Henry	Tarek S. Khan	Richard C. Mielenz
Mark B. Hogan	James R. Libby	Austin H. Morgan, Jr.
Edward P. Holub	Mark D. Luther	Todd Rutenbeck

FOREWORD

This report is the authoritative glossary for cement and concrete technology. It is to be used generally and specifically in ACI technical communications, correspondence, and publications. One mission of Committee 116 is to produce and maintain a list of terms with their meaning in the field of cement and concrete technology.

Committee 116 has tried to produce a glossary that will be useful, comprehensive, and up-to-date. It recognizes, however, that the listing may not be complete and that some definitions may be at variance with some commonly accepted meanings.

Users of the glossary are invited to submit suggestions for changes and additions to ACI Headquarters for consideration by Committee 116 when preparing future editions. In the event that a user disagrees with any of the definitions, it is hoped that the reasons for such will be given to the committee.

The committee is aware that some of the definitions included may seem entirely self-evident to an expert in the concrete field. This occurs because no term has been discarded if there was reason to believe it would appear to be technical in nature to a casual reader of the ACI literature.

The committee voted to use the following procedural rules:

1. Each definition shall be stated in one sentence;
2. Each definition shall consist of the term printed in boldface, a dash, and the definition statement;
3. The definition statement shall not repeat the term and should state the class or group and identify the features unique to the term; as "**mathematics**—the science of numbers and spaces";
4. Verbs should be stated in the infinitive rather than the participle; for example the term to be defined should be "**abrade**" not "abrading";
5. Notes may be appended to definition statements;
6. Cross references may take the place of a definition as "**green concrete**—see **concrete, green**." They also may call attention to related items as "**flint**—a variety of chert. (See also **chert**)." Where the committee has found two or more terms with the same meaning, the definition is given where the preferred term appears, the synonyms are cross referenced to the preferred term, and in many cases, the fact is stated;
7. Generally, where there are a number of terms, the last word of which is the same, the definitions are given where the terms are listed in the inverted form, as "**cement, low-heat**" rather than "**low-heat cement**," but under the latter entry, there will be a cross reference "see **cement, low-heat**;" and
8. In selecting terms and definitions, there shall be coordination with the terminology subcommittees of ASTM Committees C-1 on Cement, and C-9 on Concrete and Concrete Aggregates.

The invaluable contributions of the past chairmen of Committee 116, B. J. Erlin, R. C. Mielenz, D. L. Bloem, W. H. Price, R. E. Davis, Jr., J. R. Dise, K. F. Gibbe, Robert L. Henry, M. D. Luther, B. Mather, and E. Senbetta, those of the present members of the committee, as well as the diligent efforts of William Lorman and Lewis H. Tuthill, are gratefully acknowledged.

For drafting this edition, all members, both associates and voting, participated.

ACI Committee Reports, Guides, Standard Practices, and Commentaries are intended for guidance in planning, designing, executing, and inspecting construction. This document is intended for the use of individuals who are competent to evaluate the significance and limitations of its content and recommendations and who will accept responsibility for the application of the material it contains. The American Concrete Institute disclaims any and all responsibility for the stated principles. The Institute shall not be liable for any loss or damage arising therefrom.

Reference to this document shall not be made in contract documents. If items found in this document are desired by the Architect/Engineer to be a part of the contract documents, they shall be restated in mandatory language for incorporation by the Architect/Engineer.

ACI 116R-00 supersedes ACI 116R-90 and became effective March 16, 2000.
Copyright © 2000, American Concrete Institute.
All rights reserved including rights of reproduction and use in any form or by any means, including the making of copies by any photo process, or by electronic or mechanical device, printed, written, or oral, or recording for sound or visual reproduction or for use in any knowledge or retrieval system or device, unless permission in writing is obtained from the copyright proprietors.

This document has been approved for use by agencies of the Department of Defense and for listing in the DoD Index of Specifications and Standards.

A

Abrams' law—see **law, Abrams'**.
abrasion damage—see **damage, abrasion**.
abrasion resistance—see **resistance, abrasion**.
absolute specific gravity—see **specific gravity, absolute**.
absolute volume—see **volume, absolute**.
absorbed moisture—see **moisture, absorbed**.
absorbed water—see **moisture, absorbed**.
absorption—the process by which a liquid is drawn into and tends to fill permeable voids in a porous solid body; also, the increase in mass of a porous solid body resulting from the penetration of a liquid into its permeable voids.
abutment—in bridges, the end structure (usually of concrete) that supports the beams, girders, and deck of the bridge, or combinations thereof, and sometimes retains the earthen bank or supports the end of the approach pavement slab; in prestressing, the structure against which the tendons are stressed in producing pretensioned precast members or post-tensioned pavement; and in dams, the side of the gorge or bank of the stream against which a dam abuts.
accelerating admixture—see **admixture, accelerating**.
acceleration—increase in velocity or in rate of change, especially the quickening of the natural progress of a process such as setting or strength development (hardening) of concrete. (See also **admixture, accelerating**.)
accelerator—see **admixture, accelerating**.
accidental air—see **air, entrapped**.
acrylic resin—see **resin, acrylic**.
addition—a material that is interground or blended in limited amounts into a hydraulic cement during manufacture either as a "processing addition" to aid in manufacturing and handling the cement or as a "functional addition" to modify the use properties of the finished product.
advancing-slope grouting—see **grouting, advancing-slope**.
additive—see **agent**.
adhesion—the state in which two surfaces are held together by interfacial effects that may consist of molecular forces, interlocking action, or both.
adhesives—the group of materials used to join or bond similar or dissimilar materials; for example, in concrete work, the epoxy resins.
adiabatic—a condition in which heat neither enters nor leaves a system.
adiabatic curing—see **curing, adiabatic**.
adjustment screw—see **screw, adjustment**.
admixture—a material other than water, aggregates, hydraulic cement, and fiber reinforcement, used as an ingredient of a cementitious mixture to modify its freshly mixed, setting, or hardened properties and that is added to the batch before or during its mixing.
  admixture, accelerating—an admixture that causes an increase in the rate of hydration of the hydraulic cement and thus shortens the time of setting, increases the rate of strength development, or both.
  admixture, air-entraining—an admixture that causes the development of a system of microscopic air bubbles in concrete, mortar, or cement paste during mixing, usually to increase its workability and resistance to freezing and thawing. (See also **air, entrained**.)
  admixture, retarding—an admixture that causes a decrease in the rate of hydration of the hydraulic cement and lengthens the time of setting.
  admixture, water-reducing—an admixture that either increases slump of freshly mixed mortar or concrete without increasing water content or maintains slump with a reduced amount of water, the effect being due to factors other than air entrainment.
  admixture, water-reducing (high-range)—a water-reducing admixture capable of producing large water reduction or great flowability without causing undue set retardation or entrainment of air in mortar or concrete.
adobe—unburnt brick dried in the sun.
adsorbed water—see **water, adsorbed**.
adsorption—development (at the surface of either a liquid or solid) of a higher concentration of a substance than exists in the bulk of the medium; especially formation of one or more layers of molecules of gases, of dissolved substances, or of liquids at the surface of a solid (such as cement, cement paste, or aggregates), or of air-entraining agents at the air-water interfaces; also, the process by which a substance is adsorbed. (See also **water, adsorbed**.)
advancing-slope grouting—see **grouting, advancing-slope**.
advancing-slope method—see **method, advancing-slope**.
aerated concrete—see **concrete, cellular** and **concrete, foamed**.
A/F ratio—see **ratio, A/F**.
afwillite—a mineral with composition $3CaO \cdot 2SiO_2 \cdot 3H_2O$ occurring naturally in South Africa, Northern Ireland, and California, and artificially in some hydrated portland cement mixtures.
agent—a general term for a material that may be used either as an addition to cement or an admixture in concrete; for example, an air-entraining agent.
  agent, air-entraining—see **admixture, air-entraining**.
  agent, bonding—a substance applied to a suitable substrate to create a bond between it and a succeeding layer.
  agent, parting—see **agent, release** (preferred term).
  agent, release—material used to prevent bonding of concrete to a surface. (See also **bond breaker** and **oil, form**.)
  agent, surface-active—a substance that markedly affects the interfacial or surface tension of solutions when present even in low concentrations.
  agent, wetting—a substance capable of lowering the surface tension of liquids, facilitating the wetting of solid surfaces, and permitting the penetration of liquid into the capillaries.
agglomeration—a gathering into a ball or mass.
aggregate—granular material, such as sand, gravel, crushed stone, crushed hydraulic-cement concrete, or iron blast-

furnace slag, used with a hydraulic cementing medium to produce either concrete or mortar. (See also **aggregate, heavyweight** and **aggregate, lightweight**.)

aggregate, angular—aggregate particles that possess well-defined edges formed at the intersection of roughly planar faces.

aggregate, coarse—aggregate predominantly retained on the 4.75 mm (No. 4) sieve, or that portion retained on the 4.75 mm (No. 4) sieve. (See also **aggregate**.)

aggregate, crusher-run—aggregate that has been mechanically broken and has not been subjected to subsequent screening.

aggregate, dense-graded—aggregates graded to produce low void content and maximum density when compacted. (See also **aggregate, well-graded**.)

aggregate, fine—aggregate passing the 9.5 mm (3/8 in.) sieve, almost entirely passing the 4.75 mm (No. 4) sieve, and predominantly retained on the 75 µm (No. 200) sieve; or that portion passing the 4.75 mm (No. 4) sieve and predominantly retained on the 75 µm (No. 200) sieve. (See also **aggregate** and **sand**.)

aggregate, gap-graded—aggregate graded so that certain intermediate sizes are substantially absent.

aggregate, heavyweight—aggregate of high density, such as barite, magnetite, hematite, limonite, ilmenite, iron, or steel, used in heavyweight concrete.

aggregate, lightweight—aggregate of low density, such as: a) expanded or sintered clay, shale, slate, diatomaceous shale, perlite, vermiculite, or slag; b) natural pumice, scoria, volcanic cinders, tuff, and diatomite; and c) sintered fly ash or industrial cinders, used in lightweight concrete.

aggregate, mineral—aggregate consisting essentially of inorganic nonmetallic rock materials, either natural or crushed and graded.

aggregate, normalweight—aggregate that is neither heavyweight nor lightweight.

aggregate, open-graded—aggregate in which the voids are relatively large when the aggregate is compacted.

aggregate, reactive—aggregate containing substances capable of reacting chemically with the products of solution or hydration of the portland cement in concrete or mortar under ordinary conditions of exposure, resulting in some cases in harmful expansion, cracking, or staining.

aggregate, refractory—aggregate having refractory properties that, when bound together into a conglomerate mass by a matrix, forms a refractory body.

aggregate, single-sized—aggregate in which a major portion of the particles is in a narrow size range.

aggregate, well-graded—aggregate having a particle-size distribution that produces maximum density, that is, minimum void space.

aggregate blending—the process of intermixing two or more aggregates to produce a different set of properties, generally, but not exclusively, to improve grading.

aggregate-cement ratio—see **ratio, aggregate-cement**.

aggregate gradation—see **grading** (preferred term).

aggregate interlock—the effect of portions of aggregate particles from one side of a joint or crack in concrete protruding into recesses in the other side of the joint or crack so as to transfer load in shear and maintain alignment.

aggregate transparency—discoloration of a concrete surface consisting of darkened areas over coarse aggregate particles immediately below the concrete surface.

agitating speed—see **speed, agitating**.

agitating truck—see **truck, agitating**.

agitation—
1. the process of providing motion in mixed concrete just sufficient to prevent segregation or loss of plasticity; and
2. the mixing and homogenization of slurries or finely ground powders by either mechanical means or injection of air. (See also **agitator**.)

agitator—a device for maintaining plasticity and preventing segregation of mixed concrete by agitation. (See also **agitation**.)

aids, grinding—materials used to expedite the process of grinding by eliminating ball coating, dispersing the finely ground product, or both.

air—

air, accidental—see **air, entrapped** (preferred term).

air, entrained—microscopic air bubbles intentionally incorporated in mortar or concrete during mixing, usually by use of a surface-active agent; typically between 10 and 1000 µm (1 mm) in diameter and spherical or nearly so. (See also **air entrainment**.)

air, entrapped—air voids in concrete that are not purposely entrained and that are larger, mainly irregular in shape, and less useful than those of entrained air; and 1 mm or larger in size.

air blow pipe—air jet used in shotcrete gunning to remove rebound or other loose material from the work area.

air-blown mortar—see **shotcrete** (preferred term).

air content—the volume of air voids in cement paste, mortar, or concrete, exclusive of pore space in aggregate particles; usually expressed as a percentage of total volume of the paste, mortar, or concrete.

air-cooled blast-furnace slag—see **blast-furnace slag**.

air entraining—the capability of a material or process to develop a system of microscopic bubbles of air in cement paste, mortar, or concrete during mixing. (See also **air entrainment**.)

air-entraining agent—see **admixture, air-entraining**.

air-entraining hydraulic cement—see **cement, air-entraining hydraulic**.

air entrainment—the incorporation of air in the form of microscopic bubbles (typically smaller than 1 mm) during the mixing of either concrete or mortar. (See also **air entraining** and **air, entrained**.)

air lift—equipment whereby slurry or dry powder is lifted through pipes by means of compressed air.

air meter—see **meter, air**.

air

air-permeability test—see **test, air-permeability** and **test, Blaine**.

air ring—see **ring, air**.

air separator—see **separator, air**.

air void—see **void, air**.

air-water jet—see **jet, air-water**.

akermanite—a mineral of the melilite group, $Ca_2MgSi_2O_7$. (See also **gehlenite**, **melilite**, and **merwinite**.)

alabaster—a compact, crystalline, weakly textured form of practically pure gypsum.

alignment wire—see **wire, ground** (preferred term).

alite—a name used to identify tricalcium silicate, including small amounts of MgO, Al_2O_3, Fe_2O_3, and other oxides; a principal constituent of portland-cement clinker. (See also **belite**, **celite**, and **felite**.)

alkali—salts of alkali metals, principally sodium and potassium; specifically, sodium and potassium occurring in constituents of concrete and mortar, usually expressed in chemical analyses as the oxides Na_2O and K_2O. (See also **cement, low-alkali**.)

alkali-aggregate reaction—see **reaction, alkali-aggregate**.

alkali-carbonate rock reaction—see **reaction, alkali-carbonate rock**.

alkali reactivity (of aggregate)—see **reactivity (of aggregate), alkali**.

alkali-silica reaction—see **reaction, alkali-silicate**.

alkyl aryl sulfonate—synthetic detergent used to entrain air in hydraulic cement mixtures.

allowable bearing capacity—the maximum pressure to which a soil or other material should be subjected to guard against shear failure or excessive settlement.

allowable load—see **load, service dead** and **load, service live**.

allowable stress—see **stress, allowable**.

alternate-lane construction—see **construction, alternate-lane**.

alumina—aluminum oxide (Al_2O_3).

aluminate cement—see **cement, calcium-aluminate**.

aluminate concrete—see **concrete, aluminate**.

aluminous cement—see **cement, calcium-aluminate**.

amount of mixing—the extent of mixer action employed in combining the ingredients for either concrete or mortar; in the case of stationary mixers, the mixing time; and in the case of truck mixers, the number of revolutions of the drum at mixing speed after the intermingling of the cement with water and aggregates. (See also **mixing time**.)

amplitude—the maximum displacement from the mean position in connection with vibration.

analysis, dynamic—analysis of stresses in framing as functions of displacement under transient loading.

analysis, mechanical—the process of determining particle-size distribution of an aggregate. (See also **analysis, sieve**.)

analysis, sieve—particle-size distribution; usually expressed as the mass percentage retained upon each of a series of standard sieves of decreasing size and the percentage passed by the sieve of finest size. (See also **grading**.)

anchor—in prestressed concrete, to lock the stressed tendon in position so that it will retain its stressed condition; in precast-concrete construction, to attach the precast units to the building frame; and in slabs on grade or walls, to fasten to rock or adjacent structures to prevent movement of the slab or wall with respect to the foundation, adjacent structure, or rock. (See also **anchor, form**.)

anchor, form—device used to secure formwork to previously placed concrete of adequate strength; the device is normally embedded in the concrete during placement.

anchor bolt—see **bolt, anchor**.

anchorage—in post-tensioning, a device used to anchor the tendon to the concrete member; in pretensioning, a device used to maintain the elongation of a tendon during the time interval between stressing and release; in precast-concrete construction, the devices for attaching precast units to the building frame; and in slab or wall construction, the device used to anchor the slab or wall to the foundation, rock, or adjacent structure.

   anchorage, dead-end—the anchorage at that end of a tendon that is opposite the jacking end.

   anchorage, end—

     1. length of reinforcement, mechanical anchor, hook, or combination thereof, beyond the point of nominal zero stress in the reinforcement of cast-in-place concrete; and

     2. mechanical device for transmitting prestressing force to the concrete in a post-tensioned member. (See also **anchorage**.)

   anchorage, mechanical—any mechanical device capable of developing the strength of the reinforcement without damage to the concrete.

   anchorage, threaded—an anchorage device that is provided with threads to facilitate attaching the jacking device and to effect the anchorage.

   anchorage, wedge—a device for anchoring a tendon by wedging.

anchorage bond stress—see **stress, anchorage bond**.

anchorage deformation—see **deformation, anchorage** or **slip**.

anchorage device—see **anchorage** (preferred term).

anchorage loss—see **deformation, anchorage**.

anchorage slip—see **deformation, anchorage** or **slip**.

anchorage zone—see **zone, anchorage**.

angle float—see **float, angle**.

angle of repose—the angle between the horizontal and the natural slope of loose material below which the material will not slide.

angular aggregate—see **aggregate, angular**.

anhydrite—a mineral, anhydrous calcium sulfate ($CaSO_4$); gypsum from which the water of crystallization has been removed, usually by heating above 325 F (160 C); natural anhydrite is less reactive than that obtained by calcination of gypsum.

anhydrous calcium chloride—see **calcium chloride, anhydrous**.

apparent specific gravity—see **specific gravity, absolute**.

architect-engineer or **engineer-architect**—the architect, engineer, architectural firm, engineering firm, or architectural and engineering firm issuing project drawings and specifications, or administering the work under contract specifications and drawings, or both.

architectural concrete—see **concrete, architectural**.

arc spectrography—spectrographic identification of elements in a sample of material heated to volatilization in an electric arc or spark.

area of steel—the cross-sectional area of the steel reinforcement. (See also **effective area of reinforcement**.)

arenaceous—composed primarily of sand; sandy.

argillaceous—composed primarily of clay or shale; clayey.

arris—the ridge formed by the meeting of two surfaces.

arrissing tool—see **tool, arrissing**.

artificial pozzolan—see **pozzolan, artificial**.

asbestos-cement products—products manufactured from rigid material composed essentially of asbestos fiber and portland cement.

ashlar—see **masonry, ashlar**.

ashlar, patterned—see **masonry, ashlar**.

ashlar masonry—see **masonry, ashlar**.

ashlar, random—see **masonry, ashlar**.

asphalt—a dark brown to black cementitious material in which the predominating constituents are bitumens that occur in nature or are obtained in petroleum processing.

asphalt cement—see **cement, asphalt**.

asphaltic concrete—see **concrete, asphaltic**.

atmospheric-pressure steam curing—see **curing, atmospheric-pressure steam**.

Atterberg limits—see **limits, Atterberg**.

Atterberg test—see **test, Atterberg**.

autoclave—a pressure vessel in which an environment of steam at high pressure may be produced; used in the curing of concrete products and in the testing of hydraulic cement.

autoclave curing—see **curing, autoclave**.

autoclave cycle—see **cycle, autoclave**.

autoclaved—see **curing, autoclave**.

autoclaving—see **curing, autoclave**.

autogenous healing—see **healing, autogenous**.

autogenous length change—see **length change, autogenous**.

autogenous volume change—see **volume change, autogenous**.

automatic batcher—see **batcher**.

auxiliary reinforcement—see **reinforcement, auxiliary**.

average bond stress—see **bond stress, average**.

average compressive strength—see **compressive strength, average**.

axis, neutral—a line in the plane of a structural member subject to bending where the longitudinal stress is zero.

axle load—see **load, axle**.

axle steel—see **steel, axle**.

axle-steel reinforcement—see **reinforcement, axle-steel**.

B

b/b_o—see **factor, coarse-aggregate** (preferred term).

bacillus, cement—see **ettringite** (preferred term).

backfill concrete—see **concrete, backfill**.

back form—see **form, top** (preferred term).

back plastering—plaster applied to one face of a lath system following application and subsequent hardening of plaster applied to the opposite face. (See also **parge**.)

back stay—see **brace** (preferred term).

backshores—shores placed snugly under a concrete slab or structural member after the original formwork and shores have been removed from a small area without allowing the entire slab or member to deflect or support its own mass or existing construction loads.

bacterial corrosion—see **corrosion, bacterial**.

bag (of cement; also **sack)**—a quantity of portland cement: 94 lb (43 kg) in the U.S.; for other kinds of cement, quantity indicated on the bag.

balanced load—see **load, balanced**.

balanced moment—moment capacity at simultaneous crushing of concrete and yielding of tension steel.

balanced reinforcement—an amount and distribution of reinforcement in a flexural member such that in working-stress design the allowable tensile stress in the steel and the allowable compressive stress in the concrete are attained simultaneously; or such that in strength design, the tensile reinforcement reaches its specified yield strength simultaneously with the concrete in compression reaching its assumed ultimate strain of 0.003.

ball mill—see **mill, ball**.

ball test—see **test, ball**.

band iron—thin metal strap used as a form tie, hanger, etc.

bar—an element, normally composed of steel, with a nominally uniform cross-sectional area used to reinforce concrete.

    bar, coated—a bar on which a coating has been applied, usually to increase resistance to corrosion.

    bar, deformed—a reinforcing bar with a manufactured pattern of surface ridges intended to reduce slip and increase pullout resistance of bars embedded in concrete.

    bar, epoxy-coated—a reinforcing bar coated by an epoxy-resin system, usually to increase resistance to corrosion.

    bar, high-bond—see **bar, deformed** (preferred term).

    bar, plain—a reinforcing bar without surface deformations, or one having deformations that do not conform to the applicable requirements.

    bar, reinforcement—see **reinforcement**.

    bar, standard hooked—a reinforcing bar with the end bent into a hook to provide anchorage.

    bar, tie—bar at right angles to, and tied to reinforcement to keep it in place.

bar bender—a tradesman who cuts and bends steel reinforcement; or a machine for bending steel reinforcement.

bar-end check—a check of the ends of reinforcing bars to determine whether they fit the devices intended for connecting the bars. (See also **mechanical connection**.)

bar mat—an assembly of steel reinforcement composed of two or more layers of bars placed at angles to each other and secured together either by welding or tying.

bar schedule—a list of the reinforcement, showing the shape, number, size, and dimensions of every different element required for a structure or a portion of a structure.

bar spacing—the distance between parallel reinforcing bars, measured center to center of the bars perpendicular to their longitudinal axes.

bar support—hardware used to support or hold reinforcing bars in proper position to prevent displacement before and during concreting. (See also **bat**; **bolster, slab**; **chair**.)

barite—a mineral, barium sulfate ($BaSO_4$), used in either pure or impure form as concrete aggregate primarily for the construction of high-density radiation shielding concrete; designated "barytes" in the UK.

barrage—a low dam erected to control the level of a stream.

barrel (of cement)—a quantity of portland cement: 376 lb (4 bags) in the U.S. (obsolete); also wood or metal container formerly used for shipping cement.

barrel-vault roof—see **roof, barrel-vault**.

barrier, moisture—a vapor barrier.

barrier, vapor—membranes located under concrete floor slabs that are placed on grade to retard transmission of water vapor from the subgrade.

bars, bundled—a group of not more than four parallel reinforcing bars in contact with each other, usually tied together.

bars, stem—bars used in the wall section of a cantilevered retaining wall or in the webs of a box; when a cantilevered retaining wall and its footing are considered as an integral unit, the wall is often referred to as the stem of the unit.

base—a subfloor slab or "working mat," either previously placed and hardened or freshly placed, on which floor topping is placed in a later operation; also the underlying stratum on which a concrete slab, such as a pavement, is placed. (See also **mud slab** and **subbase**.)

base bead—see **base screed** (preferred term).

base coat—any plaster coat or coats applied before application of the finish coat.

base course—a layer of specified select material of planned thickness constructed on the subgrade or subbase of a pavement to serve one or more functions, such as distributing loads, providing drainage, or minimizing frost action; also the lowest course of masonry in a wall or pier.

base plate—a plate of metal or other material formerly placed under pavement joints and the adjacent slab ends to prevent the infiltration of soil and moisture from the sides or bottom of the joint opening; also a steel plate used to distribute vertical loads, as for bridge beams, building columns, or machinery.

base screed—a preformed metal screed with perforated or expanded flanges to provide a guide for thickness and planeness of plaster and to provide a separation between plaster and other materials.

basic creep—see **creep, basic**.

basket—see **load-transfer assembly** (preferred term).

bassanite—calcium sulfate hemihydrate, $2CaSO_4 \cdot H_2O$. (See also **hemihydrate** and **plaster of paris**.)

bat—a broken brick sometimes used to support reinforcement. (See also **bar support**.)

batch—*n.* quantity of either concrete or mortar mixed at one time; *v.* to weigh or volumetrically measure and introduce into the mixer the ingredients for a quantity of either concrete or mortar.

batch, trial—a batch of concrete prepared to establish or check proportions of the constituents.

batch box—container of known volume used for measuring constituents of a batch of either concrete or mortar in proper proportions.

batch mixer—see **mixer, batch**.

batch plant—an installation for batching or for batching and mixing concrete materials.

batch weights—the quantities of the various ingredients (cement, water, the several sizes of aggregate, and admixtures if used) that compose a batch of concrete.

batched water—the mixing water added by a batcher to a cementitious mixture either before or during the initial stages of mixing (also called batch water).

batcher—a device for measuring ingredients for a batch of concrete.

batcher, automatic—a batcher equipped with gates or valves that, when actuated by a single starter switch, will open automatically at the start of the weighing operation of each material, and will close automatically when the designated quantity of each material has been reached, interlocked in such a manner that: a) the charging mechanism cannot be opened until the scale has returned to zero; b) the charging mechanism cannot be opened if the discharge mechanism is open; c) the discharge mechanism cannot be opened if the charging mechanism is open; d) the discharge mechanism cannot be opened until the designated quantity has been reached within the allowable tolerance; and e) if different kinds of aggregates or different kinds of cements are measured cumulatively in a single batcher, interlocked sequential controls are provided.

batcher, manual—a batcher equipped with gates or valves that are operated manually, with or without supplementary power (pneumatic, hydraulic, or electrical), the accuracy of the weighing operation being dependent on the operator's observation of the scale.

batcher, semiautomatic—a batcher equipped with gates or valves that are separately opened manually to allow the material to be weighed but that are closed automatically when the designated quantity of each material has been reached.

batching, cumulative—measuring more than one ingredient of a batch in the same container by bringing the batcher scale into balance at successive total weights as each ingredient is accumulated in the container.

batten (also **batten strip**)—a narrow strip of wood placed over the vertical joint of sheathing or paneling; also used to hold several boards together. (See also **cleat**.)

batter—inclination from the vertical or horizontal.

batter boards—pairs of horizontal boards nailed to wooden stakes adjoining an excavation; used as a guide to elevations and to outline the building.

batter pile—see **pile, batter**.

bauxite—a rock composed principally of hydrous aluminum oxides; the principal ore of aluminum and a raw material for manufacture of calcium-aluminate cement.

bay—the space, in plan, between the centerlines of adjacent piers, mullions, or columns; a small, well-defined area of concrete placed at one time in the course of placing large areas, such as floors, pavements, or runways.

beam—a structural member subjected to primarily flexure, but also to axial load; and, the graduated horizontal bar of a weighing scale on which the balancing poises ride. (See also **beam, spandrel**; **girder**; **girt**; **joist**; **ledger**; **purlin**; and **stringer**.)

beam, double-tee—a precast concrete member composed of two stems and a combined top flange, commonly used as a beam but also used vertically in exterior walls.

beam, drop-in—a precast element simply supported on adjacent cantilevered elements.

beam, edge—a stiffening beam at the edge of a slab.

beam, grade—a reinforced concrete beam, usually at ground level, that strengthens or stiffens the foundation or supports overlying construction.

beam, simple—a beam without rotational restraint or continuity at its supports; also known as a simply supported beam.

beam, slender—a beam that, if loaded to failure without lateral bracing of the compression flange, would fail by buckling rather than in flexure.

beam, spandrel—a beam in the perimeter of a building, spanning between columns and usually supporting a floor or roof.

beam-and-slab floor (roof)—a reinforced concrete system in which a slab is supported by and is often monolithic with reinforced-concrete beams.

beam bottom—soffit or bottom form for a beam.

beam-column—a structural member subjected to axial load and flexure forces but primarily axial load.

beam form—a retainer or mold so erected as to give the necessary shape, support, and finish to a concrete beam.

beam form-clamp—any of various types of tying or fastening units used to hold the sides of beam forms.

beam hanger—a wire, strap, or other hardware device that supports formwork from structural members.

beam pocket—opening left in a vertical member in which a beam is to rest; also an opening in the column or girder form where forms for an intersecting beam will be framed.

beam saddle—see **beam hanger** (preferred term).

beam side—vertical or sloping side of a beam.

beam test—a method of measuring the flexural strength (modulus of rupture) of concrete by testing a standard unreinforced beam.

bearing capacity—see **allowable bearing capacity**.

bearing stratum—the soil or rock stratum on which a concrete footing or mat bears or that carries the load transferred to it by a concrete pile, caisson, or similar deep foundation unit.

belite—a name used to identify one form of the constituent of portland-cement clinker now known when pure as dicalcium silicate ($2CaO \cdot SiO_2$). (See also **alite**; **celite**; and **felite**.)

bench—see **pretensioning bed**.

bending moment—see **moment, bending**.

bending moment diagram—a graphical representation of the variation of bending moment along the length of the member for a given stationary system of loads.

beneficiation—improvement of the chemical or physical properties of a raw material or intermediate product by the removal or modification of undesirable components or impurities.

bent, pile—two or more piles driven in a row transverse to the long dimension of the structure and fastened together by capping and (sometimes) bracing.

bent bar—a reinforcing bar bent to a prescribed shape. (See also **hook**; **bar, hooked**; **stirrup**; and **tie**.)

bentonite—a clay composed principally of minerals of the montmorillonoid group, characterized by high adsorption and very large volume change with wetting or drying.

Berliner—a type of terrazzo topping using small and large pieces of marble paving, usually with a standard terrazzo matrix between pieces, also called Palladiana.

billet steel—see **steel, billet**.

binder—a cementing material, either a hydrated cement or reaction products of cement or lime and reactive siliceous material, the kind of cement and curing conditions governing the characteristics of the product formed; also materials such as asphalt, resins, and other materials forming the matrix of concretes, mortars, and sanded grouts.

biological shielding—shielding provided to attenuate or absorb nuclear radiation, such as neutron, proton, alpha and beta particles, and gamma radiation; the shielding is provided mainly by the density of the concrete, except that in the case of neutrons the attenuation is achieved by compounds of some of the lighter elements (for example, hydrogen and boron). (See also **concrete, shielding**.)

bituminous cement—see **cement, bituminous**.

Blaine apparatus—air-permeability apparatus for measuring the surface area of a finely ground cement, raw material, or other product. (See ASTM C 204.)

Blaine fineness—the fineness of powdered materials such as cement and pozzolans, expressed as surface area per unit mass usually in square meters per kilogram, determined by the Blaine apparatus. (See also **surface, specific**.)

Blaine test—see **test, Blaine**.

blanket, curing—a covering of sacks, matting, burlap, straw, waterproof paper, or other suitable material placed over freshly finished concrete. (See also **burlap**.)

blast-furnace slag—the nonmetallic product consisting essentially of silicates and aluminosilicates of calcium and other bases that is developed in a molten condition simultaneously with iron in a blast furnace.
1. air-cooled blast-furnace slag is the material resulting from solidification of molten blast-furnace slag under atmospheric conditions; subsequent cooling may be accelerated by application of water to the solidified surface;
2. expanded blast-furnace slag is the low density, cellular material obtained by controlled processing of molten blast-furnace slag with water, or water and other agents, such as steam, compressed air, or both;
3. granulated blast-furnace slag is the glassy, granular material formed when molten blast-furnace slag is rapidly chilled, as by immersion in water; and
4. ground granulated blast-furnace slag is granulated blast-furnace slag that has been finely ground and is a hydraulic cement.

bleed—to undergo bleeding. (See **bleeding**.)

bleeding—the autogenous flow of mixing water within, or its emergence from, newly placed concrete or mortar; caused by the settlement of the solid materials within the mass; also called water gain.

bleeding capacity—the ratio of volume of water released by bleeding to the volume of paste or mortar.

bleeding rate—the rate at which water is released from a paste or mortar by bleeding.

blemish—any superficial defect that causes visible variation from a consistently smooth and uniformly colored surface of hardened concrete. (See also **bug holes**; **efflorescence**; **honeycomb**; **joint, lift**; **laitance**; **popout**; **rock pocket**; and **sand streak**.)

blended cement—see **cement, blended**.

blinding—the application of a layer of lean concrete or other suitable material to reduce surface voids or to provide a clean, dry working surface; also the filling or plugging of the openings in a screen or sieve by the material being separated. (See **concrete, lean**.)

blistering—the irregular raising of a thin layer at the surface of placed mortar or concrete during or soon after completion of the finishing operation, or in the case of pipe after spinning; also bulging of the finish plaster coat as it separates and draws away from the base coat.

bloated—swollen, as in certain lightweight aggregates as a result of processing.

block, concrete—a concrete masonry unit, usually containing hollow cores.

block, end—an enlarged end section of a member intended to reduce anchorage stresses to allowable values and provide space needed for post-tensioning anchorages.

block, wood—a solid piece of wood used in concrete formwork to fill space or prevent movement of the formwork.

block beam—a flexural member composed of individual blocks that are joined together by prestressing. (See also **member, segmental**.)

blockout—a space within a concrete structure under construction in which fresh concrete is not to be placed, called core in the UK.

blowdown period—time taken to reduce pressure in an autoclave from maximum to atmospheric.

blowholes—see **surface air voids** (preferred term).

blowup—the raising of two concrete slabs off the subgrade where they meet as a result of greater expansion than the joint between them will accommodate; typically occurs only in unusually hot weather where joints have become filled with incompressible material; often results in cracks on both sides of the joint and parallel to it.

board butt joint—construction joint in shotcrete formed by sloping the sprayed surface to a 1 in. (25 mm) board laid flat.

bolster, slab—continuous wire bar support used to support bars in the bottom of slabs; top wire is corrugated at 1 in. centers to hold bars in position. (See also **bar support**.)

bolt, anchor—a metal bolt or stud, headed or threaded, either cast in place, grouted in place, or drilled into finished concrete, used to hold various structural members or embedments in the concrete, and to resist shear, tension, and vibration loadings from various sources, such as wind and machine vibration; also known as a hold-down bolt or a foundation bolt.

bolt, foundation—see **bolt, anchor**.

bolt, hold-down—anchor bolt provided near the ends of shear walls for transferring boundary-member loads from the shear wall to the foundation. (See also **bolt, anchor**.)

bolt, she—a type of form tie and spreader bolt in which the end fastenings are threaded into the end of the bolt, thus eliminating cones and reducing the size of holes left in the concrete surface.

bolt sleeve—a tube surrounding a bolt in a concrete wall to prevent concrete from adhering to the bolt and acting as a spreader for the formwork.

bond—
1. adhesion of concrete or mortar to reinforcement or other surfaces against which it is placed, including friction due to shrinkage and longitudinal shear in the concrete engaged by the bar deformations;
2. adhesion of cement paste to aggregate;
3. adhesion or cohesion between plaster coats or between plaster and a substrate produced by adhesive or cohesive properties of plaster or supplemental materials; and
4. patterns formed by the exposed faces of masonry units, for example, running bond or flemish bond.

bond, ceramic—the development of fired strength as a result of thermochemical reactions between materials exposed to temperatures approaching the fusion point of the mixture such as that which may occur, under these conditions, between calcium-aluminate cement and a refractory aggregate.

bond, chemical—bond between materials that is the result of cohesion and adhesion developed by chemical reaction.

bond, flexural stress—in structural-concrete members, the stress between the concrete and the reinforcing element that results from the application of external load.

bond, mechanical—
1. in general concrete construction, the physical interlock between cement paste and aggregate, or between concrete and reinforcement (specifically, the sliding resistance, not the adhesive resistance, of an embedded bar); and
2. in plastering, the physical keying of a plaster coat to: a) another; b) to the plaster base by means of plaster keys to the lath; or c) through interlock with adjacent plaster casts created by means of scratching or cross raking.

bond, transfer—in pretensioning, the bond stress resulting from the transfer of stress from the tendon to the concrete.

bond area—the nominal area of interface between two elements across which adhesion develops or may develop, as between cement paste and aggregate.

bond breaker—a material used to prevent adhesion of newly placed concrete to the substrate. (See also **oil, form** and **agent, release**.)

bond length—see **length, development** (preferred term).

bond plaster—a specially formulated gypsum plaster designed as first-coat application over monolithic concrete.

bond prevention—measures taken to prevent adhesion of concrete or mortar to surfaces against which it is placed.

bond strength—see **strength, bond**.

bond stress—see **stress, bond**.

bond stress, average—the force in a bar divided by the product of the perimeter and the development length of the bar.

bond stress, development—see **stress, anchorage bond** (preferred term).

bonded hollow-wall masonry—see **masonry, bonded hollow-wall**.

bonded member—a prestressed-concrete member in which the tendons are bonded to the concrete either directly or through grouting.

bonded post-tensioning—see **post-tensioning, bonded**.

bonded tendon—see **tendon, bonded**.

bonder—a masonry unit that ties two or more wythes (leaves) of a wall together by overlapping. (See **also header** and **wythe [leaf]**.)

bonding agent—see **agent, bonding**.

bonding layer—see **layer, bonding**.

bored pile—see **pier, drilled**.

boring—the removal by drilling of rock; a sample of soil or concrete for tests.

boron frits—clear, colorless, synthetic glass produced by fusion and quenching, containing boron. (See also **concrete, boron-loaded**.)

boron-loaded concrete—see **concrete, boron-loaded**.

box out—to form an opening in concrete by a box-like form.

brace—a structural member used to provide lateral support for another member, generally for the purpose of ensuring stability or resisting lateral loads.

bracing—see **brace** (preferred term).

bracket—an overhanging member projecting from a wall or other body to support weight acting outside the wall, or a similar piece to strengthen an angle. (See also **corbel**.)

breccia—rock composed of angular fragments of older rock cemented together.

bredigite—a mineral, alpha prime dicalcium silicate ($2CaO \cdot SiO_2$), occurring naturally at Scawt Hill, Northern Ireland, and at the Isle of Muck, Scotland and is also present in slags and portland cement.

breeze—usually clinker; also fine, divided material from coke production.

brick, calcium-silicate—a concrete product made principally from sand and lime that is hardened by autoclave curing.

brick, concrete—solid concrete masonry units of relatively small prescribed dimensions.

brick, rubbing—a silicon-carbide brick used to smooth and remove irregularities from surfaces of hardened concrete.

brick, sand-lime—see **brick, calcium-silicate** (preferred term).

brick seat—ledge on wall or footing to support a course of masonry.

bridge deck—see **deck, bridge**.

briquette (also **briquet**)—a molded specimen of mortar with enlarged extremities and reduced center having a cross section of definite area, used for measurement of tensile strength.

broadcast—to toss granular material, such as sand, over a horizontal surface so that a thin, uniform layer is obtained.

broom finish—see **finish, broom**.

brown coat—see **coat, brown**.

brown out—to complete application of base coat plaster.

brown oxide—see **oxide, brown**.

brownmillerite—a ternary compound originally regarded as $4CaO \cdot Al_2O \cdot Fe_2O_3$ (C_4AF) occurring in portland and calcium-aluminate cement; now used to refer to a series of solid solutions between $2CaO \cdot Fe_2O_3$ (C_2F) and $2CaO \cdot Al_2O_3$ (C_2A).

brucite—a mineral having the composition magnesium hydroxide, $Mg(OH)_2$, and a specific crystal structure.

brushed surface—see **surface, brushed**.

buck—framing around an opening in a wall; a door buck encloses the opening in which a door is placed.

buckling—failure by lateral or torsional instability of a structural member, occurring with stresses below the yield or ultimate values.

bug holes—see **surface air voids** (preferred term).

buggy—a two-wheeled hand or motor-driven cart, usually rubber-tired, for transporting small quantities of concrete from hoppers or mixers to forms; sometimes called a concrete cart.

building

building official—the official charged with administration and enforcement of the applicable building code, the duly authorized representative of the official.

build-up—spraying of shotcrete in successive layers to form a thicker mass; also the accumulation of residual hardened concrete in a mixer.

bulk cement—see **cement, bulk**.

bulk density—see **density, bulk**.

bulk loading—see **loading, bulk**.

bulk modulus—see **modulus, bulk**.

bulk specific gravity—see **specific gravity, absolute** and **density, bulk**.

bulk specific gravity (saturated-surface dry)—see **specific gravity, absolute**.

bulkhead—a partition in formwork blocking fresh concrete from a section of the form, or a partition closing a section of the form, such as at a construction joint; a partition in a storage tank or bin, as for cement or aggregate.

bulking—increase in the volume occupied by a quantity of sand in a moist condition over the volume of the same quantity dry or completely inundated.

bulking curve—graph of change in volume of a quantity of sand due to change in moisture content.

bulking factor—see **factor, bulking**.

bull float—see **float, bull**.

bundled bars—see **bars, bundled**.

burlap—a coarse fabric of jute, hemp, or less commonly, flax, for use as a water-retaining covering in curing concrete surfaces; also called Hessian.

bush-hammer—a hammer having a serrated face, as rows of pyramidal points, used to roughen or dress a surface; to finish a concrete surface by application of a bush-hammer.

bush-hammer finish—see **finish, bush-hammer**.

butt joint—see **joint, butt**.

butter—to spread mortar on a masonry unit with a trowel; also the process by which the interior of a concrete mixer, transportation unit, or other item coming in contact with fresh concrete is provided with a mortar coating so that fresh concrete coming in contact with it will not be depleted of mortar.

buttress—a projecting structure to support either a wall or a building.

butyl stearate—a colorless, oily, and practically odorless material ($C_{17} H_{35} COOC_4 H_9$) used as an admixture for concrete to provide dampproofing.

C

cabinet, moist—an upright and compartmented case having doors and shelves of moderate dimensions for storing and curing small test specimens of cement paste, mortar, and concrete in an atmosphere of approximately 73 F (23 C) and at least 95% relative humidity. (See also **moist room**.)

cable—see **tendon** (preferred term).

cage—a rigid assembly of reinforcement ready for placing in position.

caisson—part of a foundation, a watertight chamber used in construction underwater, or a hollow floating box used as a floodgate for a dock or basin.

caisson pile—see **pile, caisson**.

calcareous—containing calcium carbonate or, less generally, containing the element calcium.

calcine—to alter composition or physical state by heating below the temperature of fusion.

calcite—a mineral having the composition calcium carbonate ($CaCO_3$) and a specific crystal structure; the principal constituent of limestone, chalk, and marble; a major constituent in the manufacture of portland cement.

calcium—a silver-white metallic element of the alkaline-earth group occurring naturally only in combination with other elements.

calcium-aluminate cement—see **cement, calcium-aluminate**.

calcium chloride—a crystalline solid, $CaCl_2$; in various technical grades, used as a drying agent, as an accelerator of concrete, as a deicing chemical, and for other purposes. (See also **admixture, accelerating**.)

calcium chloride, anyhdrous ($CaCl_2$)—a solid, usually 94% calcium chloride, typically in pellet form.

calcium chloride, hydrous ($CaCl_2 \cdot 2H_2O$)—a solid, usually 77% calcium chloride, in flake form.

calcium chloride solution—an aqueous solution of calcium chloride (usually at a specified concentration so that a given amount can be gauged to provide a specific concentration) usually expressed as a percent calcium chloride by mass of portland cement.

calcium hydroxide—see **lime, hydrated**.

calcium stearate—$Ca(C_{18}H_{35}O_2)_2$, commonly marketed in powder form, insoluble in water, used as a water repellent admixture in concrete.

calcium-silicate brick—see **brick, calcium-silicate**.

calcium-silicate hydrate—see **hydrate, calcium-silicate**.

caliche—gravel, sand, and desert debris cemented by calcium carbonate or other salts.

California bearing ratio (CBR)—the ratio of the force per unit area required to penetrate a soil mass with a 3 in.^{2} (1940 mm^2) circular piston at the rate of 0.05 in. (1.3 mm) per min to the force required for corresponding penetration of a standard material; the ratio is usually determined at 0.1 in. (2.5 mm) penetration.

calorimeter—an instrument for measuring heat exchange during a chemical reaction, such as the quantity of heat liberated by the combustion of a fuel or hydration of a cement.

camber—a deflection that is intentionally built into a structural element or form to improve appearance or to nullify the deflection of the element under the effects of loads, shrinkage, and creep.

cant strip—see **strip, chamfer** (preferred term).

cap—a smooth plane surface of suitable material bonded to the bearing surfaces of test specimens to distribute the load during strength testing.

cap cables—short cables (tendons) introduced to prestress the zone of negative moment only.

capacity—a measure of the rated volume of a particular concrete mixer or agitator, usually limited by specifications to a maximum percentage of total gross volume; also the output of concrete, aggregate, or other product per unit of time (as plant capacity or screen capacity); also load-carrying limit of a structure.

capacity-reduction factor—see **strength-reduction factor** (preferred term).

capillarity—the movement of a liquid in the interstices of concrete, soil, or other finely porous material due to surface tension. (See also **flow, capillary**.)

capillary flow—see **flow, capillary**.

capillary space—see **space, capillary**.

cap, pile—
1. a structural member that is placed on top of a group of piles and used to transmit loads from the structure through the pile group into the soil; the piles may be connected to the cap with reinforcement to resist uplift or with reinforcement to resist moment so as to form a bent; also known as a rider cap or girder; also a masonry, timber, or concrete footing resting on a group of piles; and
2. a metal cap or helmet temporarily fitted over the head of a precast pile to protect it during driving; some form of shock-absorbing material is often incorporated.

cap, rider—see **cap, pile** (preferred term).

carbon black—a finely divided form of carbon produced by the combustion or partial decomposition of hydrocarbon, used as an admixture to color concrete.

carbonation—reaction between carbon dioxide and a hydroxide or oxide to form a carbonate, especially in cement paste, mortar, or concrete; the reaction with calcium compounds to produce calcium carbonate.

carbonation shrinkage—see **shrinkage, carbonation**.

carriageway—in the UK, a term used in the same meaning as the word "road" in the U.S.

cast-in-place—referring to a cementitious mixture that is deposited in the place where it is required to harden as part of the structure, as opposed to precast concrete.

cast-in-place concrete—see **concrete, cast-in-place**.

cast-in-place pile—see **pile, cast-in-place**.

cast-in-situ—see **cast-in-place** (preferred term).

cast stone—see **stone, cast**.

castable refractory—see **refractory, castable**.

catalyst—a substance that accelerates a chemical reaction and enables it to proceed under conditions more mild than otherwise required and which is not, itself, permanently changed by the reaction. (See also **catalyst, negative**.)

catalyst, negative—a substance that slows a chemical reaction and which, itself, does not enter into the reaction; inhibitor.

catface—blemish or rough depression in the finish plaster coat caused by variations in the base coat thickness.

cathead—a notched wedge placed between two formwork members meeting at an oblique angle; a spindle on a hoist; the large, round retention nut used on she bolts.

cathodic protection—the form of corrosion protection wherein one metal is caused to corrode in preference to another, thereby protecting the latter from corrosion.

catwalk—a narrow elevated walkway.

caulk—to place a material in a crack or joint with the intent of retarding entry of dirt or water. (See also **joint filler** or **sealant, joint**.)

cavitation damage—see **damage, cavitation**.

celite—a name used to identify the calcium aluminoferrite constituent of portland cement. (See also **alite**; **belite**; **felite**; and **brownmillerite**.)

cellular concrete—see **concrete, cellular**.

cellular construction—see **construction, cellular**.

cement—see **cement, hydraulic**.

    cement, air-entraining hydraulic—hydraulic cement containing an air-entraining agent in sufficient amount to entrain air in mortar within specified limits.

    cement, aluminous—see **cement, calcium-aluminate** (preferred term).

    cement, asphalt—asphalt that is refined to meet specifications for use in the manufacture of bituminous pavements.

    cement, bituminous—a black solid, semisolid, or liquid substance at natural air temperatures and appreciably soluble only in carbon disulfide or some volatile liquid hydrocarbon, being composed of mixed indeterminate hydrocarbons mined from natural deposits, produced as a residue in the distillation of petroleum, or obtained by the destructive distillation of coal or wood.

    cement, blended—a hydraulic cement consisting essentially of an intimate and uniform blend of granulated blast-furnace slag and hydrated lime; or an intimate and uniform blend of portland cement and granulated blast-furnace slag, portland cement and pozzolan, or portland blast-furnace slag cement and pozzolan, produced by intergrinding portland cement clinker with the other materials or by blending portland cement with the other materials, or a combination of intergrinding and blending.

    cement, bulk—cement that is transported and delivered in bulk (usually in specially constructed vehicles) instead of in bags.

    cement, calcium-aluminate—the product obtained by pulverizing clinker consisting essentially of hydraulic calcium aluminates resulting from fusing or sintering a suitably proportioned mixture of aluminous and calcareous materials; called high-alumina cement in the UK.

    cement, chemically prestressing—a type of expansive cement containing a higher percentage of expansive component than a shrinkage-compensating cement, when used in concretes with adequate internal or external restraint, that will expand sufficiently due to chemical reactions within the matrix, to develop the stresses necessary for prestressing the concrete. (See also **cement, expansive**.)

    cement, expanding—see **cement, expansive** (preferred term).

cement

cement, expansive—a cement that, when mixed with water, produces a paste that, after setting, increases in volume to a significantly greater degree than does portland-cement paste; used to compensate for volume decrease due to shrinkage or to induce tensile stress in reinforcement (post-tensioning).

1. *cement, expansive, Type K*—a mixture of portland cement, anhydrous tetracalcium trialuminate sulfate ($C_4A_3\bar{S}$), calcium sulfate ($CaSO_4$), and lime (CaO); the $C_4A_3\bar{S}$ is a constituent of a separately burned clinker that is interground with portland cement or alternately, it may be formed simultaneously with the portland-cement clinker compounds during the burning process;
2. *cement, expansive, Type M*—interground or blended mixtures of portland cement, calcium-aluminate cement, and calcium sulfate suitably proportioned; and
3. *cement, expansive, Type S*—a portland cement containing a high computed tricalcium aluminate (C_3A) content and an amount of calcium sulfate above the usual amount found in portland cement

cement, high-alumina—see **cement, calcium-aluminate** (preferred term).

cement, high-early-strength—portland cement characterized by attaining a given level of strength in mortar or concrete earlier than does normal portland cement; referred to in the U.S. as Type III.

cement, high-fineness—a hydraulic cement of substantially higher specific surface and substantially smaller mean particle diameter than typical for products of similar composition, produced by additional grinding or by separation by particle size.

cement, hot—newly manufactured cement that has not had an opportunity to cool after burning and grinding of the component materials.

cement, hydraulic—a cement that sets and hardens by chemical interaction with water and is capable of doing so underwater, for example, portland cement and ground granulated blast-furnace slag are hydraulic cements.

cement, hydrophobic—unhydrated cement treated so as to have reduced tendency to take up moisture.

cement, Keene's—a cement composed of finely ground, anhydrous, calcined gypsum, the set of which is accelerated by the addition of other materials.

cement, low-alkali—a portland cement that contains a relatively small amount of sodium or potassium or both; in the U.S., a portland cement containing not more than 0.60% Na_2O equivalent, that is, percent Na_2O + 0.658 × percent K_2O.

cement, low-heat—a portland cement for use when a low heat of hydration is desired, referred to in U.S. as Type IV.

cement, masonry—a hydraulic cement for use in mortars for masonry construction; contains one or more of the following materials: portland cement, portland blast-furnace slag cement, portland-pozzolan cement, natural cement, slag cement or hydraulic lime; and in addition usually contains one or more materials, such as hydrated lime, limestone, chalk, calcareous shell, talc, slag, or clay in finely ground condition.

cement, moderate sulfate-resisting—a portland cement for use when either moderate sulfate resistance or moderate heat of hydration or both is desired, now referred to as Type II.

cement, modified—a portland cement for use when either moderate heat of hydration, moderate sulfate resistance, or both, is desired, now referred to as Type II (an obsolete term).

cement, natural—a hydraulic cement produced by calcining an argillaceous limestone at a temperature below the sintering point and then grinding to a fine powder.

cement, nonstaining—a masonry cement that contains not more than a stipulated amount of water-soluble alkali as measured by a stipulated test method.

cement, normal—general purpose portland cement, referred to in the U.S. as Type I.

cement, oil-well—hydraulic cement suitable for use under high pressure and temperature in sealing water and gas pockets, and setting casing during the drilling and repair of wells; often contains retarders to meet the requirements of use.

cement, ordinary portland—the term used in the UK and elsewhere to designate the equivalent of American normal portland cement or Type I cement; commonly abbreviated OPC.

cement, plastic—a special product manufactured for plaster and stucco application.

cement, portland—a hydraulic cement produced by pulverizing portland-cement clinker, usually in combination with calcium sulfate.

cement, portland blast-furnace slag—a hydraulic cement consisting of an intimately interground mixture of portland-cement clinker and granulated blast-furnace slag or an intimate and uniform blend of portland cement and fine granulated blast-furnace slag in which the amount of the slag constituent is within specified limits.

cement, portland-pozzolan—a hydraulic cement consisting of an intimate and uniform blend of portland cement or portland blast-furnace slag cement and fine pozzolan produced by intergrinding portland-cement clinker and pozzolan, by blending portland cement or portland blast-furnace slag cement and finely divided pozzolan, or a combination of intergrinding and blending, in which the pozzolan constituent is within specified limits.

cement, regulated-set—a hydraulic cement containing fluorine-substituted calcium aluminate, capable of very rapid setting.

cement, Roman—a misnomer for a hydraulic cement made by calcining a natural mixture of calcium carbonate and clay, such as argillaceous limestone, to a temperature below that required to sinter the material but high enough to decompose the calcium carbonate, followed by grinding; so named because its brownish color resembles ancient Roman cements produced by use of lime-pozzolan mixtures.

cement, self-stressing—see **cement, expansive**.

cement, shrinkage-compensating—see **cement, expansive**.

cement, slag—hydraulic cement consisting mostly of an intimate and uniform blend of granulated blast-furnace slag and portland cement, hydrated lime, or both, in which the slag constituent is at least 10% by mass of the finished product.

cement, sticky—finished cement that develops low or zero flowability during or after storage in silos, or after transportation in bulk containers, hopper-bottom cars, etc.; may be caused by: a) interlocking of particles; b) mechanical compaction; c) electrostatic attraction between particles. (See also **set, warehouse**.)

cement, sulfate-resistant—portland cement, low in tricalcium aluminate, that reduces susceptibility of concrete to attack by dissolved sulfates in water or soils, designated Type V in the U.S.

cement, sulfoaluminate—see **cement, expansive, Type K**.

cement, supersulfated—a hydraulic cement made by intimately intergrinding a mixture of granulated blast-furnace slag, calcium sulfate, and a small amount of lime, portland cement, or portland cement clinker; so named because the equivalent content of sulfate exceeds that for portland blast-furnace slag cement.

cement, white—portland cement that hydrates to a white paste; made from raw materials of low iron content, the clinker for which is fired by a reducing flame.

cement-aggregate ratio—see **ratio, aggregate-cement**.

cement bacillus—see **ettringite** (preferred term).

cement-bound macadam—see **macadam, cement-bound**.

cement content—quantity of cement contained in a concrete, mortar, or grout, preferably expressed as mass per unit volume of concrete, mortar, or grout.

cement factor—see **cement content** (preferred term).

cement gel—see **gel, cement**.

cement gun—see **gun, cement**.

cement kiln—see **kiln, cement**.

cement paint—see **paint, cement**.

cement paste—binder of concrete and mortar consisting essentially of cement, water, hydration products, and any admixtures together with very finely divided materials included in the aggregates. (See also **cement paste, neat**.)

cement paste, neat—a plastic mixture of hydraulic cement and water both before and after setting and hardening.

cement plaster—see **plaster** and **stucco**.

cement rock—natural impure limestone that contains the ingredients for production of portland cement in approximately the required proportions.

cementation process—the process of injecting cement grout under pressure into certain types of ground (for example, gravel, or fractured rock) to solidify it.

cementitious—having cementing properties.

cementitious materials—see **materials, cementitious**.

cementitious mixture—a mixture (mortar, concrete, or grout) containing hydraulic cement.

center matched—tongue-and-groove lumber with the tongue and groove at the center of the piece rather than offset as in standard matched. (See also **standard matched**.)

centering—falsework used in the construction of arches, shells, space structures, or any continuous structure where the entire falsework is lowered (struck or decentered) as a unit. (See also **falsework** and **formwork**.)

central-mixed concrete—see **concrete, central-mixed**.

central mixer—see **mixer, central**.

centrifugally cast concrete—see **concrete, centrifugally cast**.

centrifugal process—see **process, centrifugal**.

ceramic bond—see **bond, ceramic**.

chair—see **bar support** (preferred term), and **bat**.

chalk—a soft limestone composed chiefly of the calcareous remains of marine organisms.

chalking—formation of a loose powder resulting from the disintegration of the surface of concrete or of applied coating, such as cement paint.

chamfer—either a beveled edge or corner formed in concrete work by means of a chamfer strip.

chamfer strip—see **strip, chamfer**.

charge—to introduce, feed, or load materials into a concrete or mortar mixer, furnace, or other container or receptacle where they will be further treated or processed.

checking—development of shallow cracks at closely spaced but irregular intervals on the surface of plaster, cement paste, mortar, or concrete. (See also **cracks** and **crazing**.)

chemical bond—see **bond, chemical**.

chemically prestressing cement—see **cement, chemically prestressing**.

chemically prestressing concrete—see **concrete, chemically prestressing**.

chert—a very fine-grained siliceous rock characterized by a variety of colors, by hardness and conchoidal fracture in dense varieties, and the fracture becoming splintery and the hardness decreasing in porous varieties; it is composed of silica in the form of chalcedony, cryptocrystalline or microcrystalline quartz, opal, or combinations of any of these minerals.

chipping—treatment of a hardened concrete surface by chiseling.

chips—broken fragments of marble or other mineral aggregate screened to specified sizes.

chord modulus—see **modulus of elasticity**.

chute—a sloping trough or tube for conducting concrete, cement, aggregate, or other free flowing materials from a higher to a lower point.

clamp—see **coupler** (preferred term).

class (of concrete)—an arbitrary characterization of concrete of various qualities or usages, usually by compressive strength.

clay—natural mineral material having plastic properties and composed of very fine particles; the clay mineral fraction of a soil is usually considered to be the portion consisting of particles finer than 2 µm; clay minerals are essentially hydrous aluminum silicates or occasionally hydrous magnesium silicates.

clay, fire—an earthy or stony mineral aggregate that has as the essential constituent hydrous silicates of aluminum with or without free silica, and that is plastic when sufficiently pulverized and wetted, rigid when subsequently dried, and of suitable refractoriness for use in commercial refractory products.

clay content—mass fraction of clay of a heterogeneous material, such as a soil or a natural concrete aggregate or crushed stone.

cleanout—an opening in the forms for removal of refuse, to be closed before the concrete is placed; a port in tanks, bins, or other receptacles for inspection and cleaning.

cleanup—treatment of horizontal construction joints to remove surface material and contamination down to a condition of soundness corresponding to that of a freshly broken surface of hardened concrete.

cleat—small board used to connect formwork members or used as a brace. (See also **batten**.)

climbing form—see **form, climbing**.

clinker—a partially fused product of a kiln, which is ground to make cement; also other vitrified or burnt material. (See also **clinker, portand-cement**.)

clinker, portland-cement—a partially fused ceramic material consisting primarily of hydraulic calcium silicates and calcium aluminates. (See also **clinker**.)

clip—wire or sheet-metal device used to attach various types of lath to supports or to secure adjacent lath sheets.

closed-circuit grouting—see **grouting, closed-circuit**.

coarse aggregate—see **aggregate, coarse**.

coarse-aggregate factor—see **factor, coarse-aggregate**.

coarse-grained soil—see **soil, coarse-grained**.

coat—a film or layer as of paint or plaster applied in a single operation.

  coat, brown—the second coat in three-coat plaster application.

  coat, dash-bond—a thick slurry of portland cement, sand, and water flicked on surfaces with a paddle or brush to provide a base for subsequent portland cement plaster coats; sometimes used as a final finish on plaster.

  coat, finish—final thin coat of shotcrete preparatory to hand finishing; also exposed coat of plaster and stucco.

  coat, flash—a light coat of shotcrete used to cover minor blemishes on a concrete surface.

  coat, scratch—the first coat of plaster or stucco applied to a surface in three-coat work; usually cross-raked or scratched to form a mechanical key with the brown coat.

coated bar—see **bar, coated**.

coating—
1. *on concrete*—material applied to a surface by brushing, dipping, mopping, spraying, troweling, etc., to preserve, protect, decorate, seal, or smooth the substrate;
2. *on aggregate particles*—foreign or deleterious substances found adhering to the aggregate particles; or
3. *on architectural concrete*—material used to protect a concrete surface from atmospheric contaminants and those that penetrate slightly and leave a visible clear or pigmented film on the surface. (See also **sealer**.)

coating, polysulfide—a protective-coating system prepared by polymerizing a chlorinated alkyl polyether with an inorganic polysulfide.

coating, form—a liquid applied to formwork surfaces for a specific purpose, such as to promote easy release from the concrete, to preserve the form material, or to retard setting of the near-surface matrix for preparation of exposed-aggregate finishes.

cobble—in geology, a rock fragment between 2-1/2 and 10 in. (64 and 256 mm) in diameter; as applied to coarse aggregate for concrete, the material in the nominal size range (3 to 6 in. [75 to 150 mm]).

cobblestone—a rock fragment, usually rounded or semirounded, with an average dimension between 3 and 12 in. (75 and 300 mm).

coefficient of subgrade friction—the coefficient of friction between a slab and its subgrade, commonly used in design of slabs-on-grade to estimate the force induced in the slab due to volume changes and elastic shortening if prestressed.

coefficient of subgrade reaction—ratio of: a) load per unit area of horizontal surface of a mass of soil; to b) corresponding settlement of the surface; determined as the slope of the secant, drawn between the point corresponding to zero settlement and the point of 0.05 in. (1.3 mm) settlement, of a load-settlement curve obtained from a plate load test on a soil using a 30 in. (762 mm) or greater diameter loading plate; used in the design of concrete pavements by the Westergaard method; also called modulus of subgrade reaction or subgrade modulus.

coefficient of thermal expansion—change in linear dimension per unit length or change in volume per unit volume per degree of temperature change.

coefficient of variation (V)—the standard deviation expressed as a percentage of the average. (See also **standard deviation**.)

cold-drawn wire reinforcement—see **reinforcement, cold-drawn wire**.

cold face—the surface of a refractory section not exposed to the source of heat; surface of concrete or masonry exposed to low ambient temperatures.

cold joint—see **joint, cold**.

cold-joint lines—visible lines on the surfaces of formed concrete indicating the presence of discontinuities where one layer of concrete had hardened before subsequent concrete was placed. (See also **joint, cold**.)

cold strength—see **strength, cold**.

cold-water paint—see **paint, cold-water**.

cold weather—a period in which for more than three successive days the average daily outdoor temperature drops below 40 F (5 C). Note: The average daily temperature is the average of the highest and lowest temperature during the period from midnight to midnight. When temperatures above 50 F (10 C) occur during more than half of any 24-h duration, the period shall no longer be regarded as cold weather.

cold-worked steel reinforcement—see **reinforcement, cold-worked steel**.

colemanite—a mineral, hydrated calcium borate ($Ca_2B_6O_{11} \cdot 5H_2O$). (See also **concrete, boron-loaded**.)

colloid—a substance that is in a state of division preventing passage through a semipermeable membrane, consisting of particles ranging from 0.1 to 0.001 mm in diameter.

colloidal concrete—see **concrete, colloidal**.

colloidal mixer—see **mixer, colloidal**.

colloidal grout—see **grout, colloidal**.

colloidal particle—see **particle, colloidal**.

colorimetric value—an indication of the amount of organic impurities present in fine aggregate.

column—a member used primarily to support axial compression loads and with a height of at least three times its least lateral dimension.

column, composite—a concrete compression member reinforced longitudinally with structural steel shapes, pipe, or tubing with or without longitudinal reinforcing bars.

column, long—a column whose load capacity is limited by buckling rather than strength. (See also **column, slender**.)

column, pipe—a column made of steel pipe; often filled with concrete.

column, short—a column whose load capacity is limited by strength rather than buckling; a column that is customarily so stocky and sufficiently restrained that at least 95% of the cross-sectional strength can be developed.

column, slender—a column whose load capacity is reduced by the increased eccentricity caused by secondary deflection moments.

column, spirally reinforced—a column in which the vertical bars are enveloped by spiral reinforcement, that is, closely spaced continuous hooping.

column, tied—a column laterally reinforced with ties.

column capital—an enlargement of a column below a slab intended to increase the shearing resistance.

column clamp—any of various types of tying or fastening units to hold column form sides together.

column side—one of the vertical panel components of a column form.

column strip—the portion of a flat slab over the columns and consisting of the two adjacent quarter panels on each side of the column center line.

combined-aggregate grading—see **grading, combined-aggregate**.

combined footing—see **footing, combined**.

come-along —
  1. a hoe-like tool with a blade approximately 4 in. (100 mm) high and 20 in. (500 mm) wide and curved from top to bottom, used for spreading concrete; or
  2. a colloquial name for a device (load binder) used to tighten chains holding loads in place on a truck bed.

compacting factor—the ratio obtained by dividing the observed mass of concrete that fills a container of standard size and shape when allowed to fall into it under standard conditions of test, by the mass of fully compacted concrete which fills the same container.

compaction—see **consolidation** (preferred term).

component, expansive—the portion of an expansive cement that is responsible for the expansion, generally one of several anhydrous calcium aluminate or sulfoaluminate compounds and a source of sulfate, with or without free lime, (CaO); the expansive component may be produced separately and later ground or blended with a normal portland-cement clinker; in other instances, produced by firing in a kiln with the constituents of portland cement.

composite column—see **column, composite**.

composite concrete flexural members—concrete flexural members consisting of concrete elements constructed in separate placements but so interconnected that the elements respond to loads as a unit.

composite construction—see **construction, composite**.

composite pile—see **pile, composite**.

composite sample—see **sample, composite**.

compound, curing—a liquid that can be applied as a coating to the surface of newly placed concrete to retard the loss of water or, in the case of pigmented compounds, and also to reflect heat so as to provide an opportunity for the concrete to develop its properties in a favorable temperature and moisture environment. (See also **curing** and **curing, membrane**.)

compound, joint-sealing—an impervious material used to fill joints in pavements or structures.

compound, sealing—see **sealer**.

compound, waterproofing—material used to impart water repellency to a structure or a constructional unit.

compression flange—see **flange, compression**.

compression member—see **member, compression**.

compression reinforcement—see **reinforcement, compression**.

compression test—see **test, compression**.

compressive strength—see **strength, compressive**.

compressive-strength, average—the average compressive strength of a given class or strength level of concrete; in ACI 214, defined as average compressive strength required to statistically meet a designated specific strength.

compressive stress—see **stress**.

concentric tendons—see **tendons, concentric**.

concordant tendons—see **tendons, concordant**.

concrete—a composite material that consists essentially of a binding medium within which are embedded particles or fragments of aggregate, usually a combination of fine aggregate and coarse aggregate; in portland-cement concrete, the binder is a mixture of portland cement and water, with or without admixtures.

concrete, aerated—see **concrete, foamed** and **concrete, cellular**.

concrete, aluminate—concrete made with calcium-aluminate cement; used primarily where high-early-strength and refractory or acid-resistant concrete is required.

concrete, architectural—concrete that will be permanently exposed to view and therefore requires special care in selection of the concrete materials, forming, placing, and finishing to obtain the desired architectural appearance.

concrete, asphaltic—a mixture of asphalt cement and aggregate.

concrete, backfill—nonstructural concrete used to correct over-excavation, fill excavated pockets in rock, or prepare a surface to receive structural concrete.

concrete, boron-loaded—high density concrete including a boron-containing admixture or aggregate, such as the mineral colemanite, boron frits, or boron metal alloys, to act as a neutron attenuator. (See also **biological shielding** and **concrete, shielding**.)

concrete, cast-in-place—concrete that is deposited and allowed to harden in the place where it is required to be in the completed structure, as opposed to precast concrete.

concrete, cellular—a low-density product consisting of portland cement, cement-silica, cement-pozzolan, lime-pozzolan, lime-silica pastes, or pastes containing blends of these ingredients and having a homogeneous void or cell structure, attained with gas-forming chemicals or foaming agents (for cellular concretes containing binder ingredients other than, or in addition to, portland cement, autoclave curing is usually employed).

concrete, central-mixed—concrete that is completely mixed in a stationary mixer from which it is transported to the delivery point.

concrete, centrifugally cast—concrete compacted by centrifugal action, for example, in the manufacture of pipe and poles. (See also **centrifugal process**.)

concrete, chemically prestressing—concrete made with expansive cement and reinforcement under conditions such that the expansion of the cement induces tensile stress in the reinforcement so as to produce prestressed concrete.

concrete, colloidal—concrete in which the aggregate is bound by colloidal grout.

concrete, confined—concrete containing closely spaced special transverse reinforcement that is provided to restrain the concrete in directions perpendicular to the applied stress.

concrete, cyclopean—mass concrete in which large stones, each of 100 lb (50 kg) or more, are placed and embedded in the concrete as it is deposited. (See also **concrete, rubble**.)

concrete, dense—concrete containing a minimum of voids.

concrete, dry-mix—concrete of very low water content used in the dry-cast process. (See also **process, dry-cast**.)

concrete, dry-packed—concrete placed by dry packing.

concrete, epoxy—a mixture of epoxy resin and catalyst (binder), fine aggregate, and coarse aggregate. (See also **concrete, polymer, mortar, epoxy**; and **resins, epoxy**.)

concrete (mortar or grout), expansive-cement—concrete (mortar or grout) made with expansive cement.

concrete, exposed—concrete surfaces formed so as to yield an acceptable texture and finish for permanent exposure to view. (See also **concrete, architectural**.)

concrete, fair-face—a concrete surface that, on completion of the forming process, requires no further (concrete) treatment other than curing. (See also **concrete, architectural**.)

concrete, fat—concrete containing a relatively large amount of plastic and cohesive mortar.

concrete, fiber-reinforced—concrete containing dispersed, randomly oriented fibers.

concrete, fibrous—see **concrete, fiber-reinforced**.

concrete, field—concrete delivered or mixed, placed, and cured on the job site.

concrete, flowing—concrete that is characterized by a slump greater than 7-1/2 in. (190 mm) while remaining cohesive.

concrete, foamed—low-density concrete made by the addition of a prepared foam or by generation of gas within the unhardened mixture.

concrete, fresh—concrete that possesses enough of its original workability so that it can be placed and consolidated by the intended methods.

concrete, gap-graded—concrete containing a gap-graded aggregate.

concrete, gas—lightweight concrete produced by developing voids with gas generated within the fresh mixture (usually from the action of cement alkalies on aluminum powder used as an admixture). (See also **concrete, foamed**.)

concrete, granolithic—concrete suitable for use as a wearing surface finish to floors, made with specially selected aggregate of suitable hardness, surface texture, and particle shape.

concrete, green—concrete that has set but not hardened appreciably.

concrete, grouted-aggregate—see **concrete, preplaced-aggregate**.

concrete, gypsum—concrete in which the cementitious constituent is partially dehydrated calcium sulfate (plaster).

concrete, hardened—concrete that has developed sufficient strength to serve some purpose or resist breaking under stipulated loading.

concrete, heat-resistant—any concrete that will not disintegrate when exposed to constant or cyclic heating at any temperature below that at which a ceramic bond is formed.

concrete, heavy—see **concrete, high-density** (preferred term).

concrete, heavyweight—see **concrete, high-density** (preferred term).

concrete, high-density—concrete of substantially higher density than that made using normal-density aggregates, usually obtained by use of high-density aggregates and used especially for radiation shielding.

concrete, high-early-strength—concrete which, through the use of high-early-strength cement or admixtures, attains a given level of strength earlier than normal concrete does.

concrete, high-strength—concrete that has a specified compressive strength for design of 6000 psi (41 MPa) or greater.

concrete, high-performance—concrete meeting special combinations of performance and uniformity requirements that cannot always be achieved routinely using conventional constituents and normal mixing, placing, and curing practices.

concrete, in-situ—see **concrete, cast-in-place** (preferred term).

concrete, insulating—concrete having low thermal conductivity; used as thermal insulation. (See also **concrete, lightweight** and **concrete, low-density**.)

concrete, lean—concrete of low cementitious material content.

concrete, lightweight—concrete of substantially lower density than that made using aggregates of normal density. (See also **concrete, insulating** and **concrete, low-density**.)

concrete, low-density—concrete having an oven-dry density of less than 50 lb/ft^3 (800 kg/m^3). (See also **concrete, insulating** and **concrete, lightweight**.)

concrete, mass—any volume of concrete with dimensions large enough to require that measures be taken to cope with generation of heat from hydration of the cement and attendant volume change, to minimize cracking.

concrete, monolithic—concrete cast with no joints other than construction joints.

concrete, nailable—concrete, usually made with a suitable low-density aggregate, with or without the addition of sawdust, into which nails can be driven.

concrete, negative-slump—concrete of a consistency such that it not only has zero slump but still has zero slump after adding additional water. (See also **concrete, zero-slump** and **concrete, no-slump**.)

concrete, no-fines—a concrete mixture containing little or no fine aggregate.

concrete, nonair-entrained—concrete in which neither an air-entraining admixture nor air-entraining cement has been used.

concrete, nonslip—
1. a floor, pavement, or walkway of concrete the surface of which has been roughened, before final set, either by sprinkling fine particles of abrasive material thereon and then troweling or by swirling with either a coarse-bristled brush or a trowel; or
2. a concrete surfaced roughened after final set by acid etching, mechanically abrading, or grooving.

concrete, normalweight—concrete having a density of approximately 150 lb/ft^3 (2400 kg/m^3) made with normal-density aggregates.

concrete, normalweight refractory—refractory concrete having a bulk density greater than 100 lb/ft^3 (1600 kg/m^3).

concrete, no-slump—freshly mixed concrete exhibiting a slump of less than 1/4 in. (6 mm). (See also **concrete, zero-slump** and **concrete, negative-slump**.)

concrete, plain—structural concrete with no reinforcement or with less reinforcement than the minimum amount specified in ACI 318 for reinforced concrete; also used loosely to designate concrete containing no admixture and prepared with no special treatment.

concrete, polymer—concrete in which an organic polymer serves as the binder; also known as resin concrete; sometimes erroneously employed to designate hydraulic cement mortars or concretes in which part or all of the mixing water is replaced by an aqueous dispersion of a thermoplastic copolymer. (See also **concrete**.)

concrete, polymer-cement—a mixture of water, hydraulic cement, aggregate, and a monomer or polymer; polymerized in place when a monomer is used.

concrete, popcorn—no-fines concrete containing insufficient cement paste to fill voids among the coarse aggregate so that the particles are bound only at points of contact. (See also **concrete, no-fines**.)

concrete, precast—concrete cast elsewhere than its final position.

concrete, prepacked—see **concrete, preplaced-aggregate**.

concrete, preplaced-aggregate—concrete produced by placing coarse aggregate in a form and later injecting a portland cement-sand grout, usually with admixtures, to fill the voids.

concrete (mortar, grout), preshrunk—
1. concrete that has been mixed for a short period in a stationary mixer before being transferred to a transit mixer, or
2. grout, mortar, or concrete that has been mixed one to three hours before placing to reduce shrinkage during hardening.

concrete, prestressed—concrete in which internal stresses of such magnitude and distribution are introduced that the tensile stresses resulting from the service loads are counteracted to a desired degree; in reinforced con-

concrete

crete the prestress is commonly introduced by tensioning the tendons.

concrete, pumped—concrete which is transported through hose or pipe by means of a pump.

concrete, ready-mixed—concrete manufactured for delivery to a purchaser in a fresh state. (See also **concrete, central-mixed**; **concrete, shrink-mixed**; and **concrete, transit-mixed**.)

concrete, recycled—hardened concrete that has been processed for reuse, usually as aggregate.

concrete, refractory—hardened hydraulic-cement concrete that has refractory properties and that is suitable for use at temperatures between 600 and 2400 F (315 to 1315 C).

concrete, refractory-insulating—refractory concrete having low thermal conductivity.

concrete, reinforced—structural concrete reinforced with no less than the minimum amount of prestressing tendons or nonprestressed reinforcement as specified by ACI 318.

concrete, resin—see **concrete, polymer** (preferred term).

concrete, rich—concrete of high cement content. (See also **concrete, lean**.)

concrete, roller-compacted—concrete compacted by roller compaction; concrete that, in its unhardened state, will support a roller while being compacted.

concrete, rubble—
1. concrete similar to cyclopean concrete except that small stones (such that one person can handle them) are used.
2. concrete made with rubble from demolished structures. (See also **concrete, cyclopean**.)

concrete, sand-lightweight—concrete made with a combination of expanded clay, shale, slag, or slate or sintered fly ash and natural sand; its density is generally between 105 and 120 lb/ft^3 (1680 and 1920 kg/m^3).

concrete, sawdust—concrete in which the aggregate consists mainly of sawdust from wood.

concrete (mortar or grout), self-stressing—expansive-cement concrete (mortar or grout) in which expansion, if restrained, induces persistent compressive stresses in the concrete (mortar or grout); also known as chemically prestressed concrete.

concrete, shielding—concrete, employed as a biological shield to attenuate or absorb nuclear radiation, usually characterized by high density or high hydrogen (water) content or boron content, having specific radiation attenuation effects. (See also **biological shielding**.)

concrete, shrink-mixed—ready-mixed concrete mixed partially in a stationary mixer and then mixed in a truck mixer. (See also **concrete, preshrunk**.)

concrete, shrinkage-compensating—concrete containing expansive components usually based on the formation of calcium sulfoaluminate (ettringite) in a mixture of calcium aluminate and gypsum. (See also **cement, expansive**.)

concrete, siliceous-aggregate—concrete made with normal-density aggregates having constituents composed mainly of silica or silicates.

concrete, sprayed—see **shotcrete** (preferred term).

concrete, spun—see **concrete, centrifugally cast** (preferred term).

concrete, structural—concrete used to carry load.

concrete, structural lightweight—structural concrete made with low-density aggregate; having an air-dry density of not more than 115 lb/ft^3 (1850 kg/m^3) and a 28-day compressive strength of more than 2500 psi (17.2 MPa).

concrete, subaqueous—see **concrete, underwater**.

concrete, terrazzo—marble-aggregate concrete that is cast-in-place or precast and ground smooth for decorative surfacing purposes on floors and walls.

concrete, transit-mixed—concrete, the mixing of which is wholly or principally accomplished in a truck mixer.

concrete, translucent—a combination of glass and concrete used together in precast and prestressed panels.

concrete, truck-mixed—see **concrete, transit-mixed**.

concrete, underwater—concrete placed underwater by tremie or other means.

concrete, unhardened—see **concrete, fresh** (preferred term).

concrete, unreinforced—see **concrete, plain**.

concrete, vacuum—concrete from which excess water and entrapped air are extracted by a vacuum process before hardening occurs.

concrete, vermiculite—concrete in which the aggregate consists of exfoliated vermiculite.

concrete, vibrated—concrete consolidated by vibration during and after placing.

concrete, visual—see **concrete, exposed** and **concrete, architectural**.

concrete, zero-slump—concrete of stiff or extremely dry consistency showing no measurable slump after removal of the slump cone. (See also **slump**; **concrete, no-slump**; and **concrete, negative-slump**.)

concrete block—see **block, concrete**.

concrete breaker—a compressed-air tool specially designed and constructed to break up concrete.

concrete brick—see **brick, concrete**.

concrete cart—see **buggy**.

concrete containment structure—a composite concrete and steel assembly that is designed as an integral part of a pressure retaining barrier, which in an emergency prevents the release of radioactive or hazardous effluents from nuclear power plant equipment enclosed therein.

concrete finishing machine—a machine mounted on flanged wheels that ride on the forms or on specially set tracks, used to finish surfaces such as those of pavements; or a portable power-driven machine for floating and finishing of floors and other slabs.

concrete flatwork—see **flatwork, concrete**.

concrete masonry unit—see **masonry unit, concrete**.

concrete paver—see **paver, concrete**.

concrete pile—see **pile, cast-in-place** and **pile, precast**.

concrete pump—see **pump, concrete**.

concrete reactor vessel—a composite concrete and steel assembly that functions as a component of the principal pressure-containing barrier for the nuclear fuel's primary heat extraction fluid (primary coolant).

concrete spreader—see **spreader, concrete**.

concrete strength—see **strength, compressive**; **strength, fatigue**; **strength, flexural**; **strength, shear**; **strength, splitting tensile**; **strength, tensile**; and **strength, ultimate**.

concrete vibrating machine—a machine that consolidates a layer of freshly mixed concrete by vibration.

condensed silica fume—see **silica fume** (preferred term).

conductance, thermal—time rate of heat flow through a unit area of body induced by a unit temperature difference between the body surfaces; the thermal conductance is the reciprocal of the thermal resistance.

conductivity, thermal—the property (of a homogeneous body) measured by the ratio of the steady-state heat flux (time-rate of heat flow per unit area) to the temperature.

cone—

cone, flow—a device for measurement of grout consistency in which a predetermined volume of grout is permitted to escape through a precisely sized orifice, the time of efflux (flow factor) being used as the indication of consistency; also the mold used to prepare a specimen for the flow test.

cone, pyrometric—a small, slender, three-sided oblique pyramid made of ceramic or refractory material for use in determining the time-temperature effect of heating and in obtaining the pyrometric cone equivalent (PCE) of refractory material.

cone, slump—a mold in the form of the lateral surface of the frustum of a cone with a base diameter of 8 in. (203 mm), top diameter of 4 in. (102 mm), and height of 12 in. (305 mm), used to fabricate a specimen of freshly mixed concrete for the slump test; a cone 6 in. (152 mm) high is used for tests of freshly mixed mortar and stucco.

cone bolt—a type of tie rod for wall forms with cones at each end inside the forms so that a bolt can act as a spreader as well as a tie.

confined concrete—see **concrete, confined**.

confined region—region with transverse reinforcement within beam-column joints.

connection, scarf—a connection made by precasting, beveling, halving, or notching two pieces to fit together; after overlapping, the pieces are secured by bolts or other means.

consistency—the relative mobility or ability of freshly mixed concrete or mortar to flow; the usual measurements are *slump* for concrete, *flow* for mortar or grout, and *penetration resistance* for neat cement paste.

consistency, flowable—the consistency at which a grout will form a nearly level surface when lightly rodded; the consistency of a grout with at least 125% at five drops on the ASTM C 230 flow table and an efflux time through the ASTM C 939 flow cone of more than 30 s.

consistency, fluid—the consistency at which a grout will form a nearly level surface without vibration or rodding; the consistency of a grout that has an efflux time of less than 30 s from the ASTM C 939 flow cone.

consistency, normal—
1. the degree of wetness exhibited by a freshly mixed concrete, mortar, or neat cement grout when the workability of the mixture is considered acceptable for the purpose at hand; or
2. the physical condition of neat cement paste as determined with the Vicat apparatus in accordance with a standard method test (for example, ASTM C 187).

consistency, plastic—
1. condition of freshly mixed cement paste, mortar, or concrete such that deformation when a stress is applied will be sustained continuously in any direction without rupture; or
2. the consistency at which a grout will form a nearly level surface only when rodded or vibrated with a vibrator, the consistency of a grout with a flow between 100 to 125% at five drops on the ASTM C 230 flow table.

consistency, wettest stable—the condition of maximum water content at which cement grout and mortar will adhere to a vertical surface without sloughing.

consistency factor—a measure of grout fluidity, roughly analogous to viscosity, which describes the ease with which grout may be pumped into voids or fissures; usually a laboratory measurement in which consistency is reported in degrees of rotation of a torque viscosimeter in a specimen of grout.

consistometer—an apparatus for measuring the consistency of cement pastes, mortars, grouts, or concretes.

consolidation—the process of inducing a closer arrangement of the solid particles in freshly mixed concrete or mortar during placement by the reduction of voids, usually by vibration, centrifugation, rodding, tamping, or some combination of these actions; also applicable to similar manipulation of other cementitious mixtures, soils, aggregates, or the like. (See also **rodding** and **tamping**.)

construction—

construction, alternate-lane—a method of constructing soil-supported concrete roads, runways, building floors, or other paved areas, in which alternate lanes are placed and allowed to harden before the remaining intermediate lanes are placed.

construction, cellular—a method of constructing concrete elements in which part of the interior concrete is replaced by voids.

construction, composite—a type of construction using members produced by combining different materials (for example, concrete and structural steel); members produced by combining cast-in-place and precast concrete, or cast-in-place concrete elements constructed in separate placements but so interconnected that the combined components act together as a single member and respond to loads as a unit.

construction

construction, shell—construction using thin curved slabs.

construction, structural sandwich—a laminar construction comprising a combination of alternating dissimilar simple or composite materials assembled and intimately fixed in relation to each other so as to use the properties of each to attain specific structural and thermal advantages for the whole assembly.

construction joint—see **joint, construction**.

construction loads—the loads to which a permanent or temporary structure is subjected during construction.

contact ceiling—a ceiling that is secured in direct contact with the construction above without use of furring.

contact pressure—pressure acting at and perpendicular to the contact area between soil and a concrete element.

contact splice—see **splice, contact**.

containment grouting—see **grouting, perimeter**.

continuous beam—see **continuous slab or beam**.

continuous footing—see **footing, continuous**.

continuous grading—see **grading, continuous**.

continuous mixer—see **mixer, continuous**.

continuous sampling—see **sampling, continuous**.

continuous slab or beam—a slab or beam that extends as a unit over three or more supports in a given direction.

continuously reinforced pavement—a pavement with uninterrupted longitudinal steel reinforcement and no intermediate transverse expansion or contraction joints.

contract documents—see **documents, contract**.

contraction—decrease in either length or volume. (See also **expansion**; **shrinkage**; **swelling**; **volume change**; and **volume change, autogenous**.)

contraction, thermal—see **thermal contraction**.

contraction joint—see **joint, contraction**.

contraction-joint grouting—see **grouting, contraction-joint**.

contractor—the person, firm, or corporation with whom the owner enters into an agreement for construction of the work.

control factor—the ratio of the minimum compressive strength to the average compressive strength.

control joint—see **joint, contraction** (preferred term).

control-joint grouting—see **grouting, contraction-joint**.

controlled low-strength cementitious material—material that is intended to result in a compressive strength of 1200 psi (8.3 MPa) or less.

conventional design—design procedure using moments or stresses determined by widely accepted methods.

conveying hose—see **hose, delivery** (preferred term).

conveyor—a device for moving materials; usually a continuous belt, an articulated system of buckets, a confined screw, or a pipe through which material is moved by air or water.

coping—the material or units used to form a cap or finish on top of a wall, pier, pilaster, or chimney.

coquina—a type of limestone formed of sea shells in loose or weakly cemented condition, found along present or former shorelines; used as a calcareous raw material in cement manufacture and other industrial operations.

corbel—a projection from the face of a beam, girder, column, or wall used as a beam seat or a decoration.

core (*n.*)—
1. the soil material enclosed within a tubular pile after driving (it may be replaced with concrete);
2. the mandrel used for driving casings for cast-in-place piles;
3. a structural shape used to internally reinforce a drilled-in-caisson;
4. a cylindrical sample of hardened concrete or rock obtained by means of a core drill;
5. the molded open space in a concrete masonry unit or precast concrete unit (see also **blockout**); or
6. the area enclosed by ties or spiral reinforcement in a concrete column.

core (*v.*)—the act of obtaining cores from concrete structures, rock foundations, or soils.

core test—compression test on a concrete sample cut from hardened concrete by means of a core drill.

cored beam—a beam whose cross section is partially hollow or a beam from which cored samples of concrete have been taken.

coring—the act of obtaining cores from hardened concrete or masonry structures, rock, or soil.

corner reinforcement—see **reinforcement, corner**.

corrosion—destruction of metal by a chemical, electrochemical, or electrolytic reaction within its environment.

corrosion, bacterial—destruction of a material by bacterial processes brought about by the activity of certain bacteria that consume the material and produce substances, such as hydrogen sulfide, ammonia, and sulfuric acid.

corrosion inhibitor—a chemical compound, either liquid or powder, usually intermixed in concrete and sometimes applied to concrete, and that effectively decreases corrosion of steel reinforcement.

cotton mats—see **mats, cotton**.

coupler—
1. a device for connecting reinforcing bars or prestressing tendons end to end;
2. a device for locking together the component parts of a tubular metal scaffold (also known as a clamp); or
3. internal threaded device for joining reinforcing bars with matching threaded ends for the purpose of providing transfer of either axial compression or axial tension or both from one bar to the other. (See also **coupling sleeve**, **end-bearing sleeve**, **mechanical connection**.)

coupling agent—a substance used between the transducer and test surface to permit or improve transmission of ultrasonic energy.

coupling pin—an insert device used to connect lifts or tiers or formwork scaffolding vertically.

coupling sleeve—device fitting over the ends of two reinforcing bars for the eventual purpose of providing transfer of either axial compression or axial tension or both from one bar to the other. (See also **coupler**, **end-bearing sleeve**, **mechanical connection**.)

course—in concrete construction, a horizontal layer of concrete, usually one of several making up a lift; in masonry construction, a horizontal layer of block or brick. (See also **lift**.)

cover—in reinforced concrete, the least distance between the surface of embedded reinforcement and the outer surface of the concrete.

cover block—see **spacer** and **spreader** (preferred terms).

crack—a complete or incomplete separation, of either concrete or masonry, into two or more parts produced by breaking or fracturing. (See also **fracture**.)

crack, diagonal—in a flexural member, an inclined crack caused by shear stress, usually at approximately 45 degrees to the axis; or a crack in a slab, not parallel to either the lateral or longitudinal directions.

crack, longitudinal—a crack that develops parallel to the length of a member.

crack, shrinkage—crack due to restraint of shrinkage.

crack-control reinforcement—see **reinforcement, crack-control**.

cracked section—a section designed or analyzed on the assumption that concrete has no resistance to tensile stress.

cracking—

cracking, diagonal—development of diagonal cracks. (See also **tension, diagonal**.)

cracking, map—
1. intersecting cracks that extend below the surface of hardened concrete; caused by shrinkage of the drying surface concrete that is restrained by concrete at greater depths where either little or no shrinkage occurs; vary in width from fine and barely visible to open and well-defined; or
2. the chief symptom of a chemical reaction between alkalies in cement and mineral constituents in aggregate within hardened concrete; due to differential rate of volume change in different portions of the concrete; cracking is usually random and on a fairly large scale, and in severe instances the cracks may reach a width of 0.50 in. (12.7 mm). (See also **checking** and **crazing**; also known as pattern cracking.)

cracking, pattern—see **cracks** and **cracking, map**.

cracking, plastic—cracking that occurs in the surface of fresh concrete soon after it is placed and while it is still plastic.

cracking, shrinkage—cracking of a structure or member due to failure in tension caused by external or internal restraints as reduction in moisture content develops, carbonation occurs, or both.

cracking, stress-corrosion—a cracking process that requires the simultaneous action of a corrodent and sustained tensile stress. (This excludes corrosion-reduced sections that fail by fast fracture; also excludes intercrystalline or transcrystalline corrosion that can disintegrate an alloy without either applied or residual stress).

cracking, temperature—cracking due to tensile failure, caused by a temperature drop in members subjected to external restraints or by a temperature differential in members subjected to internal restraints.

cracking load—see **load, cracking**.

cracks—

cracks, craze—fine random cracks or fissures in a surface of plaster, cement paste, mortar, or concrete.

cracks, D-line—see **D-cracks** (preferred term.)

cracks, hairline—cracks in an exposed concrete surface having widths so small as to be barely perceptible.

cracks, pattern—see **cracks** and **cracking, map**.

cracks, plastic shrinkage—see **cracking, plastic**.

cracks, transverse—cracks that develop across the long dimension of the member.

craze cracks—see **cracks, craze**.

crazing—the development of craze cracks; the pattern of craze cracks existing in a surface. (See also **checking** and **cracks**.)

creep—time-dependent deformation due to sustained load. (See also **deformation, inelastic**.)

creep, basic—creep that occurs without migration of moisture to or from the concrete. (See also **creep**; and **creep, drying**.)

creep, drying—creep caused by drying. (See **creep**; and **creep, basic**.)

creep, nonrecoverable—the residual or nonreversible deformation remaining in hardened concrete after removal of sustained load.

crimped wire—see **wire, crimped**.

critical saturation—see **saturation, critical**.

cross bracing—crossing members usually designed to act only in tension, often used in scaffolding systems. (See also **sway brace** and **X-brace**.)

cross joint—see **joint, cross**.

cross section—a plane through a body perpendicular to a given axis of the body; a drawing showing such a plane.

cross-tee—a light-gage metal member resembling an upside-down "tee" used to support the abutting ends of formboards in insulating concrete roof constructions.

crush plate—an expendable strip of wood attached to the edge of a form or intersection of fitted forms, to protect the form from damage during prying, pulling, or other stripping operations. (See also **strip, wrecking**.)

crushed gravel—see **gravel, crushed**.

crushed stone—see **stone, crushed**.

crusher—

crusher, primary—a heavy crusher suitable for the first stage in a process of size reduction of rock, slag, or the like.

crusher, secondary—a crusher used for the second stage in a process of size reduction of aggregate and the like. (See also **crusher, primary**.)

crusher-run aggregate—see **aggregate, crusher-run**.

C/S—the molar or mass ratio, whichever is specified, of calcium oxide (CaO) to silicon dioxide (SiO_2), usually of binder materials cured in an autoclave.

cube

cube strength—see **strength, cube**.
cubical piece (of aggregate)—one in which length, breadth, and thickness are approximately equal.
cumulative batching—see **batching, cumulative**.
curb form—a retainer or mold used in conjunction with a curb tool to give the necessary shape and finish to a concrete curb.
curb tool—a tool used to give the desired finish and shape to the exposed surfaces of a concrete curb.
curing—action taken to maintain moisture and temperature conditions in a freshly placed cementitious mixture to allow hydraulic cement hydration and (if applicable) pozzolanic reactions to occur so that the potential properties of the mixture may develop. (See ACI 308.)
curing, adiabatic—the maintenance of adiabatic conditions in concrete or mortar during the curing period.
curing, atmospheric-pressure steam—steam curing of concrete products or cement at atmospheric pressure, usually at maximum ambient temperature between 100 to 200 F (40 to 95 C).
curing, autoclave—curing of concrete products in an autoclave at maximum ambient temperature generally between 340 to 420 F (170 to 215 C).
curing, electrical—a system in which a favorable temperature is maintained in freshly placed concrete by supplying heat generated by electrical resistance.
curing, fog—
  1. storage of concrete in a moist room in which the desired high humidity is achieved by the atomization of water (see also **moist room**); and
  2. application of atomized water to concrete, stucco, mortar, or plaster.
curing, high-pressure steam—see **curing, autoclave** (preferred term).
curing, low-pressure steam—see **curing, atmospheric-pressure steam**.
curing, mass—adiabatic curing in sealed containers.
curing, membrane—a process that involves either liquid sealing compound (for example, bituminous and paraffinic emulsions, coal tar cut-backs, pigmented and non pigmented resin suspensions, or suspension of wax and drying oil) or nonliquid protective coating (for example sheet plastics or "waterproof" paper), both of which types function as a film to restrict evaporation of mixing water from concrete surfaces.
curing, moist-air—curing in air of not less than 95% relative humidity at atmospheric pressure and normally at a temperature approximating 73 F (23 C).
curing, single-stage—autoclave curing process in which precast concrete products are put on metal pallets for autoclaving and remain there until stacked for delivery or yard storage.
curing, standard—exposure of test specimens to specified conditions of moisture and temperature. (See also **fog curing**.)
curing, steam—curing of concrete, mortar, grout, or neat-cement paste in water vapor at atmospheric or higher pressures and at temperatures between about 100 and 420 F (40 and 215 C). (See also **atmospheric-pressure steam curing**, **autoclave curing**, **single-stage curing**, and **two-stage curing**.)
curing, two-stage—a process in which concrete products are cured in low-pressure steam, stacked, and then autoclaved.
curing agent—see **catalyst** and **hardener**.
curing blanket—see **blanket, curing**.
curing compound—see **compound, curing**.
curing cycle—see **cycle, autoclave** and **steam-curing cycle**.
curing delay—see **period, prestreaming** (preferred term).
curing kiln—see **curing, autoclave**.
curing membrane—see **membrane curing** and **curing compound**.
curling—the distortion of an originally essentially linear or planar member into a curved shape, such as the warping of a slab to differences in temperature or moisture content in the zones adjacent to its opposite faces. (See also **warping**.)
curtain grouting—see **grouting, curtain**.
curtain reinforcement—see **reinforcement, curtain**.
curvature friction—friction resulting from bends or curves in the specified prestressing cable profile.
curve, grading—a graphical representation of the proportions of different particle sizes in a granular material; obtained by plotting the cumulative or individual percentages of the material passing through sieves in which the aperture sizes form a given series.
cutting screed—see **screed, cutting**.
cycle, autoclave— the time interval between the start of the temperature-rise period and the end of the blowdown period; also, a schedule of the time and temperature-pressure conditions of periods which make up the cycle.
cyclopean concrete—see **concrete, cyclopean**.
cylinder strength—see **strength, compressive** and **strength, splitting tensile**.
cylinders, field-cured—test cylinders that are left at the jobsite for curing as nearly as practicable in the same manner as the concrete in the structure to indicate when supporting forms may be removed, additional construction loads may be imposed, or the structure may be placed in service.

D

damage, abrasion—wearing away of a surface by rubbing and friction. (See also **damage, cavitation** and **erosion**.)
damage, cavitation—pitting of concrete caused by implosion, that is, the collapse of vapor bubbles in flowing water which form in areas of low pressure and collapse as they enter areas of higher pressure. (See also **damage, abrasion**, and **erosion**.)
damp—either partial saturation or moderate covering of moisture; implies less wetness than that connoted by "wet" and slightly wetter than that connoted by "moist." (See also **moist** and **wet**.)

dampproofing—treatment of concrete or mortar to retard the passage or absorption of water, or water vapor, either by application of a suitable coating to exposed surfaces, by use of a suitable admixture or treated cement, or by use of a preformed film such as polyethylene sheets placed on grade before placing a slab. (See also **vapor barrier**.)

darby—a hand-manipulated straightedge, usually 3 to 8 ft (1 to 2.5 m) long, used in the early stage leveling operations of concrete or plaster, preceding supplemental floating and finishing.

dash-bond coat—see **coat, dash-bond**.

day—for concrete, a time period of 24 consecutive hours.

D-cracks—a series of cracks in concrete near and roughly parallel to joints, edges, and structural cracks.

dead end—in the stressing of a tendon from one end only, the end opposite that to which the load is applied.

dead-end anchorage—see **anchorage, dead-end**.

dead load—see **load, dead**.

deadman—an anchor for a guy line, usually a beam, block, or other heavy item buried in the ground, to which a line is attached.

debonding—procedures whereby specific tendons in pretensioned construction are prevented from becoming bonded to the concrete for a predetermined distance from the ends of flexural members.

decenter—to lower or remove centering or shoring.

deck—the form on which concrete for a slab is placed, also the floor or roof slab itself. (See also **deck, bridge**.)

deck, bridge—the structural concrete slab or other structure that is supported on the bridge superstructure and serves as the roadway or other traveled surface.

decking—sheathing material for a deck or slab form.

deflected tendons—see **tendons, deflected**.

deflection—movement of a point on a structure or structural element, usually measured as a linear displacement or as succession displacements transverse to a reference line or axis.

deflection, dowel—deflection caused by the transverse load imposed on a dowel.

deformation—a change in dimension or shape. (See also **contraction**; **expansion**; **creep**; **length change**; **volume change**; **shrinkage**; **deformation, inelastic**; and **deformation, time-dependent**.)

deformation, anchorage—the loss of elongation or stress in the tendons of prestressed concrete due to the deformation or seating of the anchorage when the prestressing force is transferred from the jack to the anchorage; known also as anchorage loss.

deformation, elastic—elastic deformation proportional to the applied stress. (See also **deformation**.)

deformation, inelastic—nonelastic deformation not proportional to the applied stress. (See also **deformation**; **creep**; **deformation, time-dependent**.)

deformation, nonreversible—see **creep, nonrecoverable**.

deformation, residual—see **creep, nonrecoverable**.

deformation, time-dependent—deformation resulting from effects such as autogenous volume change, thermal contraction or expansion, creep, shrinkage, and swelling, each of which is a function of time.

deformed bar—see **bar, deformed**.

deformed plate—see **plate, deformed**.

deformed reinforcement—see **reinforcement, deformed**.

deformed tie bar—see **bar, tie**.

degree-hour—a measure of strength gain of concrete as a function of the product of temperature multiplied by time for a specific interval. (See also **factor, maturity**.)

dehydration—removal of chemically bound, adsorbed, or absorbed water from a material.

deicer—a chemical, such as sodium or calcium chloride, used to melt ice or snow on slabs and pavements, such melting being due to depression of the freezing point.

delamination—a separation along a plane parallel to a surface, as in the separation of a coating from a substrate or the layers of a coating from each other, or in the case of a concrete slab, a horizontal splitting, cracking, or separation within a slab in a plane roughly parallel to, and generally near, the upper surface; found most frequently in bridge decks and caused by the corrosion of reinforcing steel or freezing and thawing; similar to spalling, scaling, or peeling except that delamination affects large areas and can often only be detected by nondestructive tests, such as tapping or chain dragging.

delay—see **period, presteaming**.

delivery hose—see **hose, delivery**.

demold—to remove molds from concrete test specimens or precast products. (See also **strip**.)

dense concrete—see **concrete, dense**.

dense-graded aggregate—see **aggregate, dense-graded**.

density—mass per unit volume (preferred over deprecated term **unit weight**.)

density, bulk—the mass of a material (including solid particles and any contained water) per unit volume including impermeable and permeable voids in the material. (See also **specific gravity, absolute**.)

density, dry—the mass per unit volume of a dry substance at a stated temperature. (See also **specific gravity, absolute**.)

density, dry-rodded—mass per unit volume of dry aggregate compacted by rodding under standardized conditions; used in measuring density of aggregate.

density, fired—the density of refractory concrete, upon cooling, after having been exposed to a specified firing temperature for a specified time.

density control—control of density of concrete in field construction to ensure that specified values as determined by standard tests are obtained.

depth, effective—depth of a beam or slab section measured from the compression face to the centroid of the tensile reinforcement.

design, elastic—a method of analysis in which the design of a member is based on a linear stress-strain relationship and corresponding limiting elastic properties of the material.

design

design, probabilistic—method of design of structures using the principles of statistics (probability) as a basis for evaluation of structural safety.

design, working-stress—a method of proportioning either structures or members for prescribed service loads at stresses well below the ultimate, and assuming linear distribution of flexural stresses and strains. (See also **design, elastic**.)

design load—see **load, design**.

design strength—see **strength, design**.

deterioration—
1. physical manifestation of failure of a material (for example, cracking, delamination, flaking, pitting, scaling, spalling, and staining) caused by environmental or internal autogenous influences on rock and hardened concrete as well as other materials; or
2. decomposition of material during either testing or exposure to service. (See also **disintegration** and **weathering**.)

detritus—loose material produced by the disintegration of rocks through geological agencies or processes simulating those of nature.

development bond stress—see **bond stress, anchorage**.

development length—see **length, development**.

device, anchorage—see **anchorage** (preferred term).

device, extension—any device, other than an adjustment screw, used to obtain vertical adjustment of shoring towers.

devil's float—see **float, devil's**.

diagonal crack—see **crack, diagonal**.

diagonal cracking—see **cracking, diagonal**.

diagonal tension—see **tension, diagonal**.

diametral compression test—see **splitting tensile test**.

diamond mesh—see **mesh, diamond**.

diatomaceous earth—a friable earthy material composed primarily of nearly pure hydrous amorphous silica (opal) in the form of frustules of the microscopic plants called diatoms.

dicalcium silicate—a compound having the composition $2CaO \cdot SiO_2$, abbreviated C_2S, an impure form of which (belite) occurs in portland-cement clinker. (See also **belite**.)

differential thermal analysis (DTA)—indication of thermal reaction by differential thermocouple recording of temperature changes in a sample under investigation compared with those of a thermally passive control sample, that are heated uniformly and simultaneously.

diffusivity, thermal—thermal conductivity divided by the product of specific heat and density; an index of the facility with which a material undergoes temperature change.

dilation—an expansion of concrete during cooling or freezing generally calculated as the maximum deviation from the normal thermal contraction predicted from the length change-temperature curve or length change-time curve established at temperatures before initial freezing.

diluent—a substance, liquid or solid, mixed with the active constituents of a formulation to increase the bulk or lower the concentration.

direct dumping—discharge of concrete directly into place from crane bucket or mixer.

discoloration—departure of color from that which is normal or desired.

disintegration—reduction into small fragments and subsequently into particles. (See also **deterioration** and **weathering**.)

dispersant—a material that deflocculates or disperses finely ground materials by satisfying the surface energy requirements of the particles; used as a slurry thinner or grinding aid.

dispersant agent—an agent capable of increasing the fluidity of pastes, mortars, or concretes by reduction of interparticle attraction.

displacement, positive—see **positive displacement**.

distortion—see **deformation**.

distress—physical manifestation of cracking and distortion in a concrete structure as the result of stress, chemical action, or both.

distribution-bar reinforcement—see **reinforcement, distribution-bar**.

divider strips—see **strips, divider**.

D-line cracks—see **D-cracks** (preferred term).

documents, contract—documents comprising aspects of the required work and the results and products thereof, including plans, specifications, and project drawings.

dolomite—a mineral having a specific crystal structure and consisting of calcium carbonate and magnesium carbonate in equivalent chemical amounts which are 54.27 and 45.73% by mass, respectively; a rock containing dolomite as the principal constituent.

dolomite, hard-burned—the product of heating dolomitic rock at temperatures high enough to change the magnesium carbonate to magnesium oxide, a constituent that slowly expands on reaction with water.

dome—square prefabricated pan form used in two-way (waffle) concrete joist floor construction.

double-headed nail—a nail with two heads at, or near, one end to permit easy removal; widely used in concrete formwork.

double-tee beam—see **beam, double-tee**.

double-up—a method of plastering characterized by application in successive operations with no setting or drying time between coats.

doughnut (donut)—a large washer of any shape for increasing bearing area of bolts and ties; also a round concrete spacer with a hole in the center to hold bars the desired distance from the forms.

dowel—
1. a steel pin, commonly a plain or coated round steel bar that extends into adjoining portions of a concrete construction, as at an expansion or contraction joint in a pavement slab, so as to transfer shear loads; or
2. a deformed reinforcing bar intended to transmit tension, compression, or shear through a construction joint.

dowel-bar reinforcement—see **dowel**.

dowel deflection—see **deflection, dowel**.

dowel lubricant—see **lubricant, dowel**.
dowel rod—see **rod, dowel**.
drainage—the interception and removal of water from, on, or under an area or roadway; the process of removing surplus ground water or surface water artificially; a general term for gravity flow of liquids in conduits.
drainage fill—
  1. base course of granular material placed between floor slab and sub-grade to impede capillary rise of moisture; or
  2. lightweight concrete placed on floors or roofs to promote drainage.
  draped tendons—see **tendons, deflected** (preferred term).
dried strength—see **strength, dried**.
drier—chemical that promotes oxidation or drying of a paint or adhesive.
drilled pier—see **pier, drilled**.
drip—a transverse groove in the underside of a projecting piece of wood, stone, or concrete to prevent water from flowing back to a wall.
dropchute—a device used to confine or to direct the flow of a falling stream of fresh concrete.
  1. *dropchute, articulated*—a device consisting of a succession of tapered metal cylinders so designed that the lower end of each cylinder fits into the upper end of the one below; or
  2. *dropchute, flexible*—a device consisting of a heavy rubberized canvas or plastic collapsible tube.
  drop-in beam—see **beam, drop-in**.
drop panel—see **panel, drop**.
drop-panel form—see **form, drop-panel**.
dry-batch weight—see **weight, dry-batch**.
dry-cast process—see **process, dry-cast**.
dry mix—see **mix, dry**.
dry-mix concrete—see **concrete, dry mix**.
dry-mix shotcrete—see **shotcrete, dry-mix**.
dry mixing—see **mixing, dry**.
dry pack—see **pack, dry**.
dry-packed concrete—see **concrete, dry-packed**.
dry packing—see **packing, dry**.
dry process—see **process, dry**.
dry-rodded density—see **density, dry-rodded**.
dry-rodded volume—see **volume, dry-rodded**.
dry-rodded weight—deprecated term; see **density, dry-rodded**.
dry rodding—see **rodding, dry**.
dry-shake—a dry mixture of hydraulic cement and fine aggregate (either natural or special metallic) that is distributed evenly over the surface of concrete flatwork and worked into the surface before time of final setting and then floated and troweled to desired finish; the mixture either may or may not contain pigment.
dry-tamp process—see **packing, dry** (preferred term).
dry topping—see **dry-shake** (preferred term).
dry-volume measurement—measurement of the ingredients of grout, mortar, or concrete by their bulk volume.
drying creep—see **creep, drying**.
drying shrinkage—see **shrinkage, drying**.
duct—a hole formed in a concrete member to accommodate a tendon for post-tensioning; a pipe or runway for electric, telephone, or other utilities.
ductility—that property of a material by virtue of which it may undergo large permanent deformation without rupture.
dummy joint—see **joint, construction** and **joint, groove**.
Dunagan analysis—a method of separating the ingredients of freshly mixed concrete or mortar to determine the proportions of the mixture.
durability—the ability of concrete to resist weathering action, chemical attack, abrasion, and other conditions of service.
durability factor—see **factor, durability**.
dust of fracture (in aggregate)—rock dust created during production processing or handling.
dusting—the development of a powdered material at the surface of hardened concrete.
dye, fugitive—see **fugitive dye**.
dynamic analysis—see **analysis, dynamic**.
dynamic load—see **load, dynamic**.
dynamic loading—see **loading, dynamic**.
dynamic modulus of elasticity—see **modulus of elasticity, dynamic**.

E

early ages (of concrete)—the period following the time of final setting during which properties are changing rapidly and heat evolution is important; for concrete made with Type I cement stored moist at 73 F (23 C), it is the first 72 h.
early strength—see **strength, early**.
early stiffening—see **stiffening, early**.
earth pigments—the class of pigments that are produced by physical processing of materials mined directly from the earth; also frequently termed natural or mineral pigments or colors.
eccentric tendon—see **tendon, eccentric**.
edge—
  edge, feather—a wood or metal tool having a beveled edge and used to straighten re-entrant angles in finish plaster coat; also the edge of a concrete or mortar patch or topping that is beveled at an acute angle.
  edge, pressed—edge of a footing along which the greatest soil pressure occurs under conditions of overturning.
edge-bar reinforcement—see **reinforcement, edge-bar**.
edge beam—see **beam, edge**.
edge form—see **form, edge**.
edger—a finishing tool used on the edges of fresh concrete to provide a rounded edge.
edging—the operation of tooling the edges of a fresh concrete slab to provide a rounded corner.
effective area of concrete—area of a concrete section assumed to resist shear or flexural stresses.

effective

effective area of reinforcement—the area obtained by multiplying the right cross-sectional area of the metal reinforcement by the cosine of the angle between its centroidal axis and the direction for which its effectiveness is considered.

effective depth—see **depth, effective**.

effective flange width—see **width, effective flange**.

effective prestress—see **prestress, effective**.

effective span—see **span, effective**.

effective width of slab—that part of the width of a slab taken into account when designing T- or L-beams.

efflorescence—a deposit of salts, usually white, formed on a surface, the substance having emerged in solution from within either concrete or masonry and subsequently been precipitated by reaction, such as carbonation, or evaporation.

elastic deformation—see **deformation, elastic**.

elastic design—see **design, elastic**.

elastic limit—see **limit, elastic**.

elastic loss—see **loss, elastic**.

elastic modulus—see **modulus of elasticity** (preferred term).

elastic shortening—see **shortening, elastic**.

elasticity—that property of a material by virtue of which it tends to recover its original size and shape after deformation.

electrical curing—see **curing, electrical**.

electrolysis—production of chemical changes by the passage of current through an electrolyte.

electrolyte—a conducting medium in which the flow of current is accompanied by movement of matter; usually an aqueous solution.

elephant trunk—an articulated tube or chute used in concrete placement. (See also **dropchute** and **tremie**.)

elongated piece (of aggregate)—particle of aggregate for which the ratio of the length to the width of its circumscribing rectangular prism is greater than a specified value. (See also **flat piece [of aggregate.]**)

elongation—increase in length. (See also **expansion**, **shortening**, and **swelling**.)

embedment length—see **length, embedment**.

embedment-length equivalent—the length of embedded reinforcement which can develop the same stress as that which can be developed by a hook or mechanical anchorage.

emery—a rock consisting essentially of an intercrystalline mixture of corundum and either magnetite or hematite; also manufactured aggregate composed of emery used to produce a wear-and slip-resistant concrete floor surface. (See also **dry-shake**.)

emulsion—a colloidal dispersion of a liquid in another liquid.

encastré—the end fixing of a built-in beam.

enclosure wall—see **wall, enclosure**.

encrustation—see **incrustation** (preferred term).

end anchorage—see **anchorage, end**.

end-bearing sleeve—device fitting over the abutting ends of two reinforcing bars for the purpose of assuring transfer of only axial compression from one bar to the other. (See also **coupler**; **coupling sleeve**; and **mechanical connection**.)

end block—see **block, end**.

endothermic reaction—see **reaction, endothermic**.

engineer-architect—see **architect-engineer**.

entrained air—see **air, entrained**.

entrapped air—see **air, entrapped**.

epoxy—a thermosetting polymer that is the reaction product of epoxy resin and an amino hardener. (See also **epoxy resin**.)

epoxy-coated bar—see **bar, epoxy-coated**.

epoxy concrete—see **concrete, epoxy**.

epoxy grout—see **grout, epoxy**.

epoxy mortar—see **mortar, epoxy**.

epoxy resins—see **resins, epoxy**.

equivalent rectangular stress-distribution—an assumption of uniform stress on the compression side of the neutral axis in the strength method of design to determine flexural capacity.

erosion—progressive disintegration of a solid by abrasion or cavitation of gases, liquids, or solids in motion. (See also **abrasion damage** and **cavitation damage**.)

ettringite—a mineral, high-sulfate calcium sulfoaluminate ($3\ CaO \cdot Al_2O_3 \cdot 3\ CaSO_4 \cdot 30\text{-}32\ H_2O$), occurring in nature or formed by sulfate attack on mortar and concrete; the product of the principal expansion-producing reaction in expansive cements; designated as "cement bacillus" in older literature.

evaporable water—see **water, evaporable**.

evaporation retardant—a long-chain organic material such as cetyl alcohol which when spread on a water film on the surface of concrete retards the evaporation of bleed water. (See also **monomolecular**.)

exfoliation—disintegration occurring by peeling off in successive layers, swelling up and opening into leaves or plates like a partly opened book.

exothermic reaction—see **reaction, exothermic**.

expanded blast-furnace slag—see **blast-furnace slag**.

expanded-metal fabric reinforcement—see **lath, expanded-metal**.

expanded-metal lath—see **lath, expanded-metal**.

expanded shale (clay or slate)—see **shale, expanded**.

expanding cement—see **cement, expansive**.

expansion—increase in either length or volume. (See also **contraction**; **moisture movement**; **shrinkage**; **volume change**; and **volume change, autogenous**.)

expansion, thermal—see **thermal expansion**.

expansion joint—see **joint, expansion**.

expansion sleeve—see **sleeve, expansion**.

expansive cement—see **cement, expansive**.

expansive-cement concrete (mortar or grout)—see **concrete (mortar or grout)** and **expansive cement**.

expansive-cement mortar—see **concrete (mortar or grout)** and **expansive cement**.

expansive component—see **component, expansive**.

exposed-aggregate finish—see **finish, exposed-aggregate**.

exposed concrete—see **concrete, exposed**.

exposed masonry—see **masonry, exposed**.

extender—a finely divided inert mineral added to provide economical bulk in paints, synthetic resins and adhesives, or other products.

extensibility—the maximum tensile strain that hardened cement paste, mortar, or concrete can sustain before cracking occurs.

extension device—see **device, extension**.

exterior panel—see **panel, exterior**.

external vibrator—see **vibrator**.

extreme compression fiber—see **fiber, extreme compression**.

extreme tension fiber—see **fiber, extreme tension**.

exudation—a liquid or viscous gel-like material discharged through a pore, crack, or opening in the surface of concrete.

F

fabric, welded-wire—a series of longitudinal and transverse wires arranged approximately at right angles to each other and welded together at all points of intersection.

fabric, woven-wire—a prefabricated steel reinforcement composed of cold-drawn steel wires mechanically twisted together to form hexagonally shaped openings.

face, pilaster—the form for the front surface of a pilaster parallel to the wall.

factor—

factor, bulking—ratio of the volume of moist sand to the volume of the sand when dry.

factor, coarse-aggregate—the ratio, expressed as a decimal, of the amount (mass or solid volume) of coarse aggregate in a unit volume of well-proportioned concrete to the amount of dry-rodded coarse aggregate compacted into the same volume b/b_0.

factor, durability—
1. a measure of the change in a material property over a period of time as a response to exposure to a treatment that can cause deterioration, usually expressed as a percentage of the value of the property before exposure; or
2. in ASTM C 666, a measure of the effects of freezing and thawing action on concrete specimens, in which resonant frequency of vibration is used as the property measured.

factor, flow—see **cone, flow**.

factor, maturity—a factor that is a function of the age of the concrete (hours or days) multiplied by the difference between the mean temperature of the concrete (degrees) during curing and a datum temperature below which hydration stops. (See also **degree-hour**.)

factor, phi (ϕ)—see **factor, strength-reduction** (preferred term).

factor, Philleo—a distance, used as an index of the extent to which hardened cement paste is protected from the effects of freezing, so selected that only a small portion of the cement paste (usually 10%) lies farther than that distance from the perimeter of the nearest air void. (See also **protected paste volume**.)

factor, Powers' spacing—see **factor, spacing** (preferred term.)

factor, spacing—an index related to the maximum distance of any point in a cement paste or in the cement paste fraction of mortar or concrete from the periphery of an air void; also known as Powers' spacing factor. (See also **factor, Philleo**.)

factor, stiffness—a measure of the stiffness of a structural member; for a prismatic member, it is equal to the ratio of the product of the moment of inertia of the cross section and the modulus of elasticity for the material to the length of the member.

factor, strength reduction—capacity-reduction factor (in structural design); a number less than 1.0 (usually 0.65 to 0.90) by which the strength of a structural member or element (in terms of load, moment, shear, or stress) is required to be multiplied to determine design strength or capacity; the magnitude of the factor is stipulated in applicable codes and construction specifications for respective types of members and cross sections.

factor of safety—the ratio of load, moment, or shear of a structural member at the ultimate to that at the service level.

factored load—see **load, factored**.

failure, fatigue—the phenomenon of rupture of a material, when subjected to repeated loadings, at a stress substantially less than the static strength.

fair-face concrete—see **concrete, fair-face**.

false header—see **header**.

false set—see **set, false**.

falsework—the temporary structure erected to support work in the process of construction; composed of shoring or vertical posting, formwork for beams and slabs, and lateral bracing. (See also **centering**.)

fascia—a flat member or band at the surface of a building or the edge beam of a bridge; also exposed eave of a building.

fastener—a device designed to attach, join, or hold two or more objects, one to another, in juxtaposition; commonly readily removed.

fat concrete—see **concrete, fat**.

fat mortar—see **mortar, fat**.

fatigue—the weakening of a material by repeated or alternating loads.

fatigue failure—see **failure, fatigue**.

fatigue strength—see **strength, fatigue**.

fault—differential displacement of a portion of a structure along a joint or crack.

feather edge—see **edge, feather**.

feed, pneumatic—shotcrete delivery equipment in which material is conveyed by a pressurized air stream.

feed wheel—see **wheel, feed**.

felite—a name used to identify one form of the constituent of portland-cement clinker now known when pure as dicalci-

ferrocement

um silicate (2CaO·SiO$_2$). (See also **alite**; **belite**; and **celite**.)

ferrocement—a composite structural material comprising thin sections consisting of cement mortar reinforced by a number of very closely spaced layers of steel wire mesh.

fiber, extreme compression—farthest fiber from the neutral axis on the compression side of a member subjected to bending.

fiber, extreme tension—farthest fiber from the neutral axis on the tension side of a member subjected to bending

fiber-reinforced concrete—see **concrete, fiber-reinforced**.

fibrous concrete—see **concrete, fiber-reinforced**.

field bending—bending of reinforcing bars on the job rather than in a fabricating shop.

field concrete—see **concrete, field**.

field-cured cylinders—see **cylinders, field-cured**.

field-proportioned grout—see **grout, field-proportioned**.

fill, porous—see **drainage fill**.

filler—
1. finely divided inert material, such as pulverized limestone, silica, or colloidal substances, sometimes added to portland-cement paint or other materials to reduce shrinkage, improve workability, or act as an extender, or
2. material used to fill an opening in a form.

filler, joint—compressible material used to fill a joint to prevent the infiltration of debris and provide support for sealants applied to the exposed surface.

fillet—see **strip, chamfer**.

fin—a narrow linear projection on a formed concrete surface, resulting from mortar flowing into spaces in the formwork; also a type of blade in a concrete mixer drum.

final prestress—see **stress, final**.

final set—see **set, final**.

final setting time—see **time, final setting**.

final stress—see **stress, final**.

fine aggregate—see **aggregate, fine**.

fine-grained soil—see **soil, fine-grained**.

fineness—a measure of particle size.

fineness modulus—see **modulus, fineness**.

finish—the texture of a surface after consolidating and finishing operations have been performed.

   finish, bush-hammer—the finish on concrete surface obtained by means of a bush-hammer.

   finish, broom—the surface texture obtained by stroking a broom over freshly placed concrete. (See also **surface, brushed**.)

   finish, exposed-aggregate—a decorative finish for concrete work achieved by removing, generally before the concrete has fully hardened, the outer skin of mortar and exposing the coarse aggregate.

   finish, float—a rather rough, granular concrete surface texture obtained by finishing with a float.

   finish, granolithic—a surface layer of granolithic concrete which may be laid on a base of either fresh or hardened concrete.

   finish, gun—undisturbed final layer of shotcrete as applied from nozzle, without hand finishing.

   finish, rubbed—a finish obtained by using an abrasive to remove surface irregularities from concrete. (See also **sack rub**.)

   finish, rustic or washed—a type of terrazzo topping in which the matrix is recessed by washing before setting so as to expose the chips without destroying the bond between chip and matrix; a retarder is sometimes applied to the surface to facilitate this operation. (See also **finish, exposed-aggregate**.)

   finish, swirl—a nonskid texture imparted to a concrete surface during final troweling by keeping the trowel flat and using a rotary motion.

   finish, trowel—the smooth or textured finish of an unformed concrete surface obtained by troweling.

finish coat—see **coat, finish**.

finish grinding—see **grinding, finish**.

finish screens—see **screens, finish**.

finishing—leveling, smoothing, consolidating, and otherwise treating surfaces of fresh or recently placed concrete or mortar to produce desired appearance and service. (See also **float** and **trowel**.)

finishing machine—see **machine, finishing**.

fire clay—see **clay, fire**.

fire resistance—see **resistance, fire**.

fired strength—see **strength, fired**.

fired density—see **density, fired**.

fishtail—a wedge-shaped piece of wood used as part of the support form between tapered pans in concrete joist construction.

flange, compression—the widened portion of an I, T, or similar cross-section beam that is shortened or compressed by bending under normal loads, such as the horizontal portion of the cross section of a simple span T-beam.

flame photometer—see **photometer, flame**.

flash coat—see **coat, flash**.

flash set—see **set, flash**.

flashing—a thin impermeable sheet, narrow in comparison with its length, installed as a cover to exclude water from exposed joints, at roof valleys, hips, roof parapets, or intersections of roof and chimney.

flat jack—see **jack, flat**.

flat piece (of aggregate)—one in which the ratio of the width to thickness of its circumscribing rectangular prism is greater than a specified value. (See also **elongated piece [of aggregate.]**)

flat plate—see **plate, flat**.

flat slab—see **slab, flat**.

flatwork, concrete—a general term applicable to concrete floors and slabs that require finishing operations.

flexible joint—see **joint, hinge**; **Mesnager**; and **semiflexible**.

flexible pavement—see **pavement, flexible**.

flexural bond stress—see **bond, flexural stress**.

flexural rigidity—see **rigidity, flexural**.

flexural strength—see **strength, flexural**.

flint—a variety of chert. (See also **chert**.)

float—a tool (not a darby), usually of wood, aluminum, or magnesium, used in finishing operations to impart a relatively even but still open texture to an unformed fresh concrete surface. (See also **darby**.)

  float, angle—a finishing tool having a surface bent to form a right angle; used to finish re-entrant angles.

  float, bull—a tool comprising a large, flat, rectangular piece of wood, aluminum, or magnesium, usually 8 in. (200 mm) wide and 42 to 60 in. (1 to 1.50 m) long, and a handle 4 to 16 ft (1 to 5 m) in length used to smooth unformed surfaces of freshly placed concrete.

  float, devil's—a wooden float with two nails protruding from the toe, used to roughen the surface of a brown plaster coat. (See also **texturing**.)

  float, power—see **float, rotary** (preferred term).

  float, rotary—a motor-driven revolving disc that smooths, flattens, and compacts the surface of concrete floors and floor toppings.

float finish—see **finish, float**.

floating—the operation of finishing a fresh concrete or mortar surface by use of a float, preceding troweling when that is to be the final finish.

flow—
1. time-dependent irrecoverable deformation (see also **creep** and **rheology**); or
2. a measure of the consistency of freshly mixed concrete, mortar, or cement paste expressed in terms of the increase in diameter of a molded truncated cone specimen after jigging a specified number of times.

  flow, capillary—flow of moisture through a capillary pore system, such as in concrete.

flow, plastic—increase in the concrete strain of members subject to constant stress, and decrease in concrete stress of members subject to constant strain; an obsolete term (see **creep** and **stress relaxation**).

flow cone—see **cone, flow**.

flow factor—see **cone, flow**.

flow line—detectable line on a concrete wall or column usually departing somewhat from horizontal, that shows where the concrete in one placement has flowed horizontally before succeeding placement has been made.

flow promoter—see **promoter, flow**.

flow table—see **table, flow**.

flow trough—see **trough, flow**.

flowable consistency—see **consistency, flowable**.

flowing concrete—see **concrete, flowing**.

fluid consistency—see **consistency, fluid**.

fluidifier—an admixture employed in grout to decrease the flow factor without changing water content. (See also **admixture, water-reducing**.)

fluosilicate—magnesium or zinc silico-fluoride used to prepare aqueous solutions sometimes applied to concrete as surface-hardening agents.

flush water—see **wash (or flush) water**.

fly ash—the finely divided residue that results from the combustion of ground or powdered coal and that is transported by flue gases from the combustion zone to the particle removal system.

flying forms—see **forms, flying**.

foam, preformed—foam produced in a foam generator prior to introduction of the foam into a mixer with other ingredients to produce cellular concrete. (See also **concrete, cellular**.)

foamed blast-furnace slag—see **blast-furnace slag (2)**.

foamed concrete—see **concrete, foamed**.

fog curing—see **curing, fog**.

fog room—see **moist room**.

folded plate—see **plate, folded**.

footing—a structural element that transmits loads directly to the soil.

  footing, combined—a structural unit or assembly of units supporting more than one column.

  footing, continuous—a combined footing of prismatic or truncated shape, supporting two or more columns in a row.

  footing, sloped—a footing having sloping top or side faces.

  footing, stepped—a step-like support consisting of prisms of concrete of progressively diminishing lateral dimensions superimposed on each other to distribute the load of a column or wall to the subgrade.

  footing, strip—see **footing, continuous**.

force, jacking—in prestressed concrete, the temporary force exerted by the device which introduces tension into the tendons.

form—a temporary structure or mold for the support of concrete while it is setting and gaining sufficient strength to be self-supporting. (See also **formwork**.)

  form, climbing—a form which is raised vertically for succeeding lifts of concrete in a given structure.

  form, drop-panel—a retainer or mold so erected as to give the necessary shape, support, and finish to a drop panel.

  form, edge—formwork used to limit the horizontal spread of fresh concrete on flat surfaces such as pavements or floors.

  form, paper—a heavy paper mold used for casting concrete columns and other structural shapes.

  form, permanent—any form that remains in place after the concrete has developed its design strength; it may or may not become an integral part of the structure.

  form, sliding—see **slipform**.

  form, top—form required on the upper or outer surface of a sloping slab or thin shell.

  form, vented—a form so constructed as to retain the solid constituents of concrete and permit the escape of water and air.

  form, wall—a retainer or mold so erected as to give the necessary shape, support, and finish to a concrete wall.

form anchor—see **anchor, form**.

form coating—see **coating, form**.

form hanger—see **hanger, form**.
form insulation—see **insulation, form**.
form lining—materials used to line the concreting face of formwork either to impart a smooth or patterned finish to the concrete surface, to absorb moisture from the concrete, or to apply a set-retarding chemical to the formed surface. (See also **sheathing**.)
form oil—see **oil, form**.
form paper—see **paper, form**.
form pressure—see **pressure, form**.
form release agent—see **agent, release**.
form scabbing—inadvertent removal of the surface of concrete because of adhesion to the form.
form sealer—coating applied to the surface of a form to reduce or prevent absorption of water from the concrete.
form spacer—see **spacer**. (See also **spreader**.)
form spreader—see **spreader**.
form tie—see **tie, form**.
forms—
    forms, flying—large prefabricated units of formwork incorporating support, and designed to be moved from place to place.
    forms, ganged—prefabricated panels joined to make a much larger unit (up to 30 by 50 ft [9 by 15 m]) for convenience in erecting, stripping, and reusing; usually braced with wales, strongbacks, or special lifting hardware.
    forms, moving—large prefabricated units of formwork incorporating supports, and designed to be moved horizontally on rollers or similar devices with a minimum amount of dismantling between successive uses.
formwork—total system of support for freshly placed concrete including the mold or sheathing that contacts the concrete as well as supporting members, hardware, and necessary bracing; sometimes called shuttering in the UK (See also **falsework** and **centering**.)
foundation—the structural elements through which the load of a structure is transmitted to the earth.
    foundation, grid—a combined footing formed by intersecting continuous footings, loaded at the intersection points, and covering much of the total area within the outer limits of the assembly.
    foundation, mat—a continuous footing supporting an array of columns in several rows in each direction, having a slab-like shape with or without depressions or openings, covering an area at least 75% of the total area within the outer limits of the assembly. (See also **foundation, raft**.)
    foundation, raft—a continuous slab of concrete, usually reinforced, laid over soft ground or where heavy loads must be supported to form a foundation. (See also **foundation, mat**.)
    foundation, strip—a continuous foundation wherein the length considerably exceeds the breadth.
foundation bolt—see **bolt, anchor** (preferred term).
four-way reinforcement—see **reinforcement, four-way**.
fracture—a crack or break, as of concrete or masonry; the configuration of a broken surface; also the action of cracking or breaking. (See also **crack**.)
frame, rigid—a frame depending on moment in joints for stability.
free fall—descent of freshly mixed concrete into forms without dropchutes or other means of confinement; also the distance through which such descent occurs; also uncontrolled fall of aggregate.
free lime—see **lime, free**.
free moisture—see **moisture, free**.
free water—see **moisture, free**. (See also **moisture, surface**.)
fresh concrete—see **concrete, fresh**.
fresno trowel—a thin steel trowel that is rectangular or rectangular with rounded corners, usually 4 to 10 in. (100 to 250 mm) wide and 20 to 36 in. (420 to 900 mm) long, having 4 to 16 ft (1 to 5 m) long handle, and used to smooth surfaces of nonbleeding concrete and shotcrete.
friction loss—see **loss, friction**.
friction pile—see **pile, friction**.
friction, wobble—in prestressed concrete, the friction caused by the unintended deviation of the prestressing sheath or duct from its specified profile.
frog—a depression in the bed surface of a masonry unit; sometimes called a panel.
fugitive dye—a dye whose color fades in a few days to neutral on exposure, usually to ultraviolet rays in sunlight; used to temporarily color membrane-curing compounds so that coverage of the concrete surface can be observed.
Fuller-Thompson ideal grading curve—see **Fuller's curve** (preferred term).
Fuller's curve—an empirical curve for gradation of aggregates; also known as the Fuller-Thompson ideal gradation curve; the curve is designed by fitting either a parabola or an ellipse to a tangent at the point where the aggregate fraction is one-tenth of the maximum size fraction. (See also **grading curve**.)
furring—strips of wood or metal fastened to a wall or other surface to even it, to form an air space, to give appearance of greater thickness, or for the application of an interior finish such as plaster.

G

ganged forms—see **forms, ganged**.
ganister—a highly refractory siliceous sedimentary rock used for furnace linings.
gap-graded aggregate—see **aggregate, gap-graded**.
gap-graded concrete—see **concrete, gap-graded**.
gas concrete—see **concrete, gas**.
gauge water—see **batched water** (preferred term).
gehlenite—a mineral of the melilite group, $Ca_2Al(AlSi)O_7$. (See also **akermanite**; **melilite**; **merwinite**.)
gel—matter in a colloidal state that does not dissolve, but remains suspended in a solvent from which it fails to precipitate without the intervention of heat or of an electrolyte. (See also **gel, cement**.)

gel, cement— the colloidal material that makes up the major portion of the porous mass of which mature hydrated cement paste is composed.

gel, tobermorite—the binder of concrete cured moist or in atmospheric-pressure steam; a lime-rich gel-like solid containing 1.5 to 1.0 mols of lime per mol of silica.

Gillmore needle—see **needle, Gillmore**.

girder—a large beam, usually horizontal, that serves as a main structural member.

girt—small beam spanning between columns, generally used in industrial buildings to support outside walls. (See also **beam**.)

glass—an inorganic product of fusion that has cooled too a rigid condition without crystallizing, sometimes reactive with alkalies in concrete.

glass-fiber reinforced cement—a composite material consisting essentially of a matrix of hydraulic cement paste or mortar reinforced with glass fibers; typically precast into units less than 1 in. (25 mm) thick.

glass-transition temperature—see **temperature, glass-transition**.

go-devil—a ball of rolled-up burlap or paper or a specially fabricated device put into the pump end of a pipeline and forced through the pipe by water pressure in order to clean the pipeline; also a device used with tremie concrete operations.

grab set—see **set, flash** (preferred term).

gradation—see **grading** (preferred term).

grade—the prepared surface on which a concrete slab is cast; the process of preparing a plane surface of granular material or soil on which to cast a concrete slab.

grade beam—see **beam, grade**.

grade strip—see **strip, grade**.

graded standard sand—see **sand, standard**.

gradient—rate of change in a variable over a distance, as of temperature or moisture.

grading—the distribution of particles of granular material among various sizes; usually expressed in terms of cumulative percentages larger or smaller than each of a series of sizes (sieve openings) or the percentages between certain ranges of sizes (sieve openings).

grading, combined-aggregate—particle-size distribution of a mixture of fine and coarse aggregate.

grading, continuous—a particle size distribution in which intermediate size fractions are present, as opposed to gap-grading. (See also **aggregate, gap-graded**.)

grading curve—see **curve, grading**.

granolithic concrete—see **concrete, granolithic**.

granolithic finish—see **finish, granolithic**.

granulated blast-furnace slag—see **blast-furnace slag**.

gravel—
1. granular material predominantly retained on the 4.75 mm (No. 4) sieve and resulting either from natural disintegration and abrasion of rock or processing of weakly bound conglomerate; and
2. that portion of an aggregate retained on the 4.75 mm (No. 4) sieve and resulting either from natural disintegration and abrasion of rock or processing of weakly bound conglomerate. (See also **aggregate, coarse**.)

gravel, crushed—the product resulting from the artificial crushing of gravel with a specified minimum percentage of fragments having one or more faces resulting from fracture. (See also **aggregate, coarse**.)

gravel, pea— screened gravel, most of the particles of which pass a 9.5 mm (3/8 in.) sieve and are retained on a 4.75 mm (No. 4) sieve.

green concrete—see **concrete, green**.

grid foundation—see **foundation, grid**.

grinding, finish—the final grinding of clinker into cement, with calcium sulfate in the form of gypsum or anhydrite generally being added; the final grinding operation required for a finished concrete surface, for example, bump cutting of pavement, fin removal from structural concrete, and terrazzo floor grinding.

grinding aids—see **aids, grinding**.

grinding medium—see **medium, grinding**.

grizzly—a simple, stationary screen or series of equally spaced parallel bars set at an angle to remove oversized particles in processing aggregate or other material.

grog—burned refractory material; usually calcined clay or crushed brick bats.

groove joint—see **joint, contraction** (preferred term).

groover—a tool used to form grooves or weakened-plane joints in a concrete slab before hardening to control crack location or provide pattern.

gross vehicle load—the mass of a vehicle plus the mass of any load thereon.

gross volume (of concrete mixers)—in the case of a revolving-drum mixer, the total interior volume of the revolving portion of the mixer drum; in the case of an open-top mixer, the total volume of the trough or pan calculated on the basis that no vertical dimension of the container exceeds twice the radius of the circular section below the axis of the central shaft.

ground-granulated slag—see **blast-furnace slag**.

ground wire—see **wire, ground**.

grout—a mixture of cementitious material and water, with or without aggregate, proportioned to produce a pourable consistency without segregation of the constituents; also a mixture of other composition but of similar consistency. (See also **grout, neat cement** and **grout, sanded**.)

grout, colloidal—grout in which a substantial proportion of the solid particles have the size range of a colloid.

grout, epoxy—a grout that is a mixture of ingredients consisting of an epoxy bonding system, aggregate or fillers, and possibly other materials.

grout, expansive-cement—see **concrete (mortar or grout)** and **expansive-cement**.

grout, field-proportioned—a hydraulic-cement grout batched at the jobsite using water and predetermined portions of portland cement, aggregate, and other ingredients.

grout

grout, hydraulic-cement—a grout which is a mixture of hydraulic cement, aggregate, water and possibly admixtures.

grout, machine-base—a grout which is used in the space between plates or machinery and the underlying foundation and which is expected to maintain essentially complete contact with the base and to maintain uniform support.

grout, masonry—a mixture of hydraulic cement, aggregate, water and possibly other materials (ASTM C 476), used for filling designated spaces in masonry construction.

grout, neat cement—a fluid mixture of hydraulic cement and water, with or without other ingredients; also the hardened equivalent of such mixture.

grout, preblended—a hydraulic-cement grout which is a commercially available mixture of hydraulic cement, aggregate, and other ingredients, which requires only the addition of water and mixing at the jobsite; sometime termed premixed grout.

grout, sanded—grout in which fine aggregate is incorporated into the mixture.

grout slope—the natural slope of fluid grout injected into preplaced-aggregate concrete.

grouted-aggregate concrete—see **concrete, preplaced-aggregate**.

grouted masonry—see **masonry, grouted**.

grouting—the process of filling with grout. (See also **grout**.)

grouting, advancing-slope—a method of grouting by which the front of a mass of grout is caused to move horizontally through preplaced aggregate by use of a suitable grout injection sequence.

grouting, closed-circuit—injection of grout into a hole intersecting fissures or voids that are to be filled at such volume and pressure that grout input to the hole is greater than the grout take of the surrounding formation, excess grout being returned to the pumping plant for recirculation.

grouting, containment—see **grouting, perimeter**.

grouting, contraction-joint—injection of grout into contraction joints.

grouting, control-joint—see **grouting, contraction-joint**.

grouting, curtain—injection of grout into a subsurface formation in such a way as to create a zone of grouted material transverse to the direction of anticipated water flow.

grouting, high-lift—a technique in masonry wall construction in which the grouting operation is delayed until the wall has been laid up to a full story height.

grouting, low-lift—a technique of masonry wall construction in which the wall sections are built to a height of not more than 5 ft (1.7 m) before the cells of the masonry units are filled with grout.

grouting, open-circuit—a grouting system with no provision for recirculation of grout to the pump.

grouting, perimeter—injection of grout, usually at relatively low pressure, around the periphery of an area that is subsequently to be grouted at greater pressure; intended to confine subsequent grout injection within the perimeter.

grouting, slush—distribution of grout, with or without fine aggregate, as required over a rock or concrete surface that is subsequently to be covered with concrete, usually by brooming it into place to fill surface voids and fissures.

grouting, staged—sequential grouting of a hole in separate steps or stages in lieu of grouting the entire length at once.

gun—
1. shotcrete material delivery equipment, usually consisting of double chambers under pressure; equipment with a single pressure chamber is used to some extent (see also **gun, cement**); or
2. pressure cylinder used to propel freshly mixed concrete pneumatically.

gun, cement—a machine for pneumatic placement of mortar or small aggregate concrete; in the "Dry Gun," water from a separate hose meets the dry material at the nozzle of the gun; with the "Wet Gun," the delivery hose conveys the premixed mortar or concrete. (See also **shotcrete**.)

gun finish—see **finish, gun**.

Gunite—a proprietary term for shotcrete.

gunman—workman on shotcreting crew who operates delivery equipment.

gunning—act of applying shotcrete; ejection of material from nozzle and impingement on surface to be gunned.

gunning pattern—
1. conical outline of material discharge stream in shotcrete operation; or
2. the sequence of gunning operations to ensure complete filling of the space, total encasement of reinforcing bars, easy removal of rebound, and thickness of shotcrete layers.

gutter tool—see **tool, gutter**.

gypsum—a mineral having the composition calcium sulfate dihydrate ($CaSO_4 \cdot 2H_2O$).

gypsum concrete—see **concrete, gypsum**.

gypsum plaster—plaster made with plaster of paris. (See **plaster** and **plaster of paris**.)

H

hacking—the roughening of a surface by striking with a tool.

hairline cracks—see **cracks, hairline**.

hairpin—the wedge used to tighten some types of form ties; a hairpin-shaped anchor set in place while concrete is unhardened; a light hairpin-shaped reinforcing bar used for shear reinforcement in beams, tie reinforcement in columns, or prefabricated column shear heads.

Hamm tip—flared shotcrete nozzle having a larger diameter at midpoint than at either inlet or outlet; also designated premixing tip.

hammer—

hammer, impact—see **hammer, rebound** (preferred term).

hammer, rebound—an apparatus that provides a relative indication of the strength or hardness of concrete based on the rebound distance of a spring-driven mass after it impacts a rod in contact with the concrete surface.

hammer, Schmidt—see **hammer, rebound** (preferred term).

hammer, Swiss—see **hammer, rebound** (preferred term).

hanger—a device used to suspend one object from another object such as the hardware attached to a building frame to support forms. (See also **beam hanger**.)

hanger, form—device used to support formwork from a structural framework; the dead load of forms, mass of concrete, and construction and impact loads must be supported.

hard-burned dolomite—see **dolomite, hard-burned**.

hard-burned lime—see **lime, hard-burned**.

hardened concrete—see **concrete, hardened**.

hardener—

1. a chemical (including certain fluosilicates or sodium silicate) applied to concrete floors to reduce wear and dusting; or
2. in a two-component adhesive or coating, the chemical component that causes the resin component to cure.

Hardy Cross method—see **moment distribution**.

harped tendons—see **tendons, deflected** (preferred term).

harsh mixture—see **mixture, harsh**.

haunch—a deepened portion of a beam in the vicinity of a support.

haunching—

1. concrete support to the sides of a drain or sewer pipe above the bedding; or
2. work done in strengthening or improving the outer strip of a roadway.

hawk—a tool used by plasterers to hold and carry plaster mortar; generally a flat piece of wood or metal approximately 10 to 12 in. (0.25 to 0.3 m) square, with a wooden handle centered and fixed to the underside. (See also **hod** and **mortar board**.)

header—a masonry unit laid flat with its greatest dimension at a right angle to the face of the wall; when the unit is only the depth of the face wythe it is known as a false header. (See also **bonder** and **wythe [leaf.]**)

header, false—see **header**.

healing, autogenous—a natural process of filling and sealing cracks in concrete or in mortar when kept damp.

heat-deflection temperature—see **temperature, heat-deflection**.

heat of hydration—heat evolved by chemical reactions with water, such as that evolved during the setting and hardening of portland cement, or the difference between the heat of solution of dry cement and that of partially hydrated cement. (See also **heat of solution**.)

heat of solution—heat evolved or absorbed when a substance is dissolved in a solvent.

heat-resistant concrete—see **concrete, heat resistant**.

heating rate—the rate expressed in degrees per hour at which the temperature is raised to the desired maximum temperature.

heavy concrete—see **concrete, high-density** (preferred term).

heavy-edge reinforcement—see **reinforcement, heavy-edge**.

heavy-media separation—see **separation, heavy-media**.

heavyweight aggregate—see **concrete, high-density** (preferred term).

heavyweight concrete—see **concrete, high-density**.

helical reinforcement—see **reinforcement, helical**.

hematite—a mineral, iron oxide (Fe_2O_3), used as aggregate in high density concrete and in finely divided form as a red pigment in colored concrete.

hemihydrate—a hydrate containing one-half molecule of water to one molecule of compound; the most commonly known hemihydrate is partially dehydrated gypsum (also known as plaster of paris), $CaSO_4 \cdot 1/2H_2O$. (See also **bassanite**.)

hesitation set—see **set, false** (preferred term).

Hessian—see **burlap** (preferred term).

high-alumina cement—see **cement, calcium-aluminate** (preferred term).

high-bond bar—see **bar, deformed**.

high-density concrete—see **concrete, high-density**.

high-discharge mixer—see **mixer, inclined-axis** (preferred term).

high-early-strength cement—see **cement, high-early-strength**.

high-fineness cement—see **cement, high-fineness**.

high-early-strength concrete—see **concrete, high-early-strength**.

high-lift grouting—see **grouting, high-lift**.

high-performance concrete—see **concrete, high-performance**.

high-pressure steam curing—see **curing, autoclave** (preferred term).

high-range water-reducing admixture—see **admixture, water-reducing (high-range)**.

high-strength concrete—see **concrete, high-strength**.

high-strength reinforcement—see **steel, high-strength**.

high-strength steel—see **steel, high-strength**.

high-temperature steam curing—see **curing, atmospheric-pressure steam** and **curing, autoclave**.

hinge, Mesnager—a permanent semiarticulation or flexible joint in a reinforced concrete arch, wherein the angles of rotation at the hinge are very small; by crossing steel reinforcing bars within the opening between the concrete structural segments, the resultant articulation presents very small resistance to rotation, resists either axial thrust or

hinge

shearing forces, and is permanently flexible; the center of rotation occurs at the intersection of the reinforcing bars.

hinge, plastic—region where ultimate moment capacity in a member may be developed and maintained with corresponding significant inelastic rotation as main tensile steel elongates beyond yield strain.

hinge joint—see **joint, hinge**.

hod—a V-shaped trough or a tray, supported by a pole handle that is borne on the carrier's shoulder, for carrying small quantities of brick, tile, mortar, or similar load. (See also **hawk** and **mortar board**.)

hold-down bolt—see **bolt, anchor** (preferred term).

holding period—see **period, presteaming** (preferred term).

hollow-unit masonry—see **masonry, hollow-unit**.

honeycomb—voids left in concrete due to failure of the mortar to effectively fill the spaces among coarse-aggregate particles.

hook—a bend in the end of a reinforcing bar.

hooked bar—see **bar, hooked**.

Hooke's law—see **law, Hooke's**.

hoop reinforcement—see **reinforcement, hoop**.

horizontal-axis mixer—see **mixer, horizontal-axis**.

horizontal-shaft mixer—see **mixer, horizontal-shaft**.

horizontal shoring—see **shoring, horizontal**.

hose, delivery—hose through which shotcrete, grout, or pumped concrete or mortar passes; also known as conveying hose or material hose.

hot cement—see **cement, hot**.

hot face—the surface of a refractory section exposed to the source of heat.

hot-load test—see **test, hot-load**.

Hoyer effect—in pretensioned, prestressed concrete, frictional forces that result from the tendency of the tendons to regain the diameter which they had before they were stressed.

hydrate—a chemical combination of water with another compound or element.

hydrate, calcium-silicate—any of the various reaction products of calcium silicate and water. (See also **dicalcium silicate** and **tricalcium silicate**.)

hydrated lime—see **lime, hydrated**.

hydration—formation of a compound by the combining of water with some other substance; in concrete, the chemical reaction between hydraulic cement and water.

hydraulic cement—see **cement, hydraulic**.

hydraulic-cement grout—see **grout, hydraulic-cement**.

hydraulic hydrated lime—see **lime, hydraulic hydrated**.

hydrochloric acid—a mineral acid sometimes used for cleaning or acid etching concrete or removing efflorescence; also known as muriatic acid, which is a 33% HCl solution.

hydrophobic cement—see **cement, hydrophobic**.

hydrous calcium chloride—see **calcium chloride, hydrous**.

I

ignition loss—see **loss on ignition** (preferred term).

ilmenite—a mineral, iron titanate ($FeTiO_3$), which in pure or impure form is commonly used as aggregate in high-density concrete.

impact hammer—see **hammer, rebound** (preferred term).

impending slough—a consistency of a shotcrete mixture containing the maximum amount of water so that the product will not flow or sag after placement.

inclined-axis mixer—see **mixer, inclined-axis**.

incrustation—a crust or coating, generally hard, formed on the surface of concrete or masonry construction or on aggregate particles.

indented strand—see **strand, indented**.

indented wire—see **wire, indented**.

index, plasticity—the range in water content through which a soil remains plastic; numerical difference between the liquid limit and the plastic limit. (See also **limits, Atterberg**.)

index, pozzolanic-activity—an index that measures pozzolanic activity based on the strength of cementitious mixtures containing hydraulic cement with and without the pozzolan; or containing the pozzolan with lime.

industrialized building—the integration of planning, design, programming, manufacturing, site operations, scheduling, financing, and management into a disciplined method of mechanized production of buildings, sometimes called systems building.

inelastic behavior—see **deformation, inelastic** (preferred term).

inelastic deformation—see **deformation, inelastic**.

infrared spectroscopy—see **spectroscopy, infrared**.

initial drying shrinkage—see **shrinkage, initial drying**.

initial prestress—see **prestress, initial**.

initial set—see **set, initial**.

initial setting time—see **time, initial setting**.

initial stresses—see **stresses, initial**.

initial-tangent modulus—see **modulus of elasticity**.

insert—anything other than reinforcing steel that is rigidly positioned within a concrete form for permanent embedment in the hardened concrete.

in-situ concrete—see **concrete, cast-in-place** (preferred term).

insoluble residue—the portion of a cement or aggregate that is not soluble in dilute hydrochloric acid of stated concentration.

insulating concrete—see **concrete, insulating**.

insulation, form—insulating material applied to the outside of forms between studs and over the top in sufficient thickness and air tightness to conserve heat of hydration to maintain concrete at required temperatures in cold weather.

insulation, roof—low-density concrete used for insulating purposes only and placed over a structural roof system.

intermittent sampling—see **sampling, intermittent**.

internal vibration—see **vibration**.

inverted L-beam—a beam having a cross section in the shape of an inverted L. (See also **L-beam**.)

inverted T-beam—a beam having a cross section in the shape of an inverted T. (See also **T-beam**.)

I-section—beam cross section consisting of top and bottom flanges connected by a vertical web.

isolation joint—see **joint, isolation**.

isotropy—the behavior of a medium having the same properties in all directions.

J

jack—a mechanical device used for applying force to prestressing tendons, for adjusting elevation of forms or form supports, and for raising objects small distances.

jack, flat—a hydraulic jack consisting of light gage metal that is folded and welded to a flat shape that expands under internal pressure.

jack shore—telescoping, or otherwise adjustable, single-post metal shore.

jacking device—the device used to stress the tendons for prestressed concrete; also the device for raising a vertical slipform.

jacking force—see **force, jacking**.

jacking stress—see **stress, jacking**.

jaw crusher—a machine having two inclined jaws, one or both being actuated by a reciprocating motion so that the charge is repeatedly nipped between the jaws.

jet, air-water—a high-velocity jet of air and water mixed at the nozzle, used in clean-up of surfaces of rock or concrete, such as horizontal construction joints.

jitterbug—a grate tamper for pushing coarse aggregate slightly below the surface of a slab to facilitate finishing. (See also **tamper**.)

joint—a physical separation in a concrete system, whether precast or cast-in-place, including cracks if intentionally made to occur at specified locations; also the region where structural members intersect, such as a beam-column joint.

joint, butt—a plain square joint between two members.

joint, cold—a joint or discontinuity resulting from a delay in placement of sufficient duration to preclude intermingling and bonding of the material in two successive lifts of concrete, mortar, or the like.

joint, construction—the surface where two successive placements of concrete meet, across which it may be desirable to achieve bond and through which reinforcement may be continuous.

joint, contraction—formed, sawed, or tooled groove in a concrete structure to create a weakened plane to regulate the location of cracking resulting from the dimensional change of different parts of the structure. (See also **joint, isolation**; **joint, expansion**; and **joint, construction**.)

joint, control—see **joint, contraction** (preferred term).

joint, cross—the joint at the end of individual formboards between subpurlins.

joint, expansion—
1. a separation provided between adjoining parts of a structure to allow movement where expansion is likely to exceed contraction; or
2. a separation between pavement slabs on grade, filled with a compressible filler material; or
3. an isolation joint intended to allow independent movement between adjoining parts.

joint, flexible—see **joint; hinge; joint, Mesnager**; and **joint, semiflexible**.

joint, groove—see **joint, contraction** (preferred term).

joint, hinge—any joint which permits rotation with no appreciable moment developed in the members at the joint. (See also **joint, hinge**; **joint, Mesnager**; and **joint, semiflexible**.)

joint, isolation—a separation between adjoining parts of a concrete structure, usually a vertical plane, at a designed location such as to interfere least with performance of the structure, yet such as to allow relative movement in three directions and avoid formation of cracks elsewhere in the concrete and through which all or part of the bonded reinforcement is interrupted. (See also **joint, contraction** and **joint, expansion**.)

joint, lift—surface at which two successive lifts meet.

joint, longitudinal—a joint parallel to the length of a structure or pavement.

joint, raked—a masonry-wall joint that has the mortar raked out to a specified depth while it is only slightly hardened.

joint, sawed—a joint cut in hardened concrete, generally not to the full depth of the member, by means of special equipment.

joint, scarf—see **connection, scarf**.

joint, semiflexible—a connection in which the reinforcement is arranged to permit some rotation of the joint. (See also **joint, hinge** and **Mesnager, joint**.)

joint, separation—see **joint, isolation** (preferred term).

joint, transverse—a joint normal to the longitudinal dimension of a structural element, assembly of elements, slab, or structure.

joint, warping—a joint with the sole function of permitting warping of pavement slabs when moisture and temperature differentials occur between the top and bottom of the slabs, that is, longitudinal or transverse joints with bonded steel or tie bars passing through them.

joint, weakened-plane—see **joint, groove** and **joint, contraction** (preferred term).

joint filler—see **filler, joint**.

joint sealant—see **sealant, joint**.

joint-sealing compound—see **compound, joint-sealing**.

joint spall—a spall adjacent to a joint.

jointer (concrete)—a metal tool approximately 6 in. (150 mm) long and from 2 to 4-1/2 in. (50 to 100 mm) wide and having shallow, medium, or deep bits (cutting edges) ranging from 3/16 to 3/4 in. (5 to 20 mm) or deeper used to cut a joint partly through fresh concrete. (See also **jointing**.)

jointing—the process of producing joints in a concrete slab. (See also **jointer [concrete]**.)

joist—a comparatively narrow beam used in closely spaced arrangements to support floor or roof slabs (that require no reinforcement except that required for temperature and shrinkage stresses); also a horizontal structural member such as that which supports deck form sheathing. (See also **beam**.)

jumbo—traveling support for forms, commonly used in tunnel work.

K

kaolin—a rock, generally white, consisting primarily of clay minerals of the kaolinite group, composed principally of hydrous aluminum silicate of low iron content, used as raw material in the manufacture of white cement.

kaolinite—a common clay mineral having the general formula $Al_2(Si_2O_5)(OH_4)$, the primary constituent of kaolin.

Keene's cement—see **cement, Keene's**.

Kelly ball—an apparatus used for indicating the consistency of fresh concrete, consisting of a cylindrical weight 6 in. (150 mm) in diameter, weighing 30 lb (14 kg) with a hemispherically shaped bottom, a handle consisting of a graduated rod, and a stirrup to guide the handle and serve as a reference for measuring depth of penetration. (See also **test, ball**.)

Kelly ball test—see **test, ball** and **Kelly ball**.

kerb form; kerb tool—see **curb form** and **curb tool** (preferred terms in the U.S.; kerb is used in the UK).

kerf—cut or notch, as a beam, transversely along the underside to curve it; also a cut or notch in a member, such as a rustication strip, to avoid damage from swelling of the wood and permit easier removal.

kern area—the area within a geometric shape in which a compressive force may be applied without tensile stresses resulting in any of the extreme fibers of the section.

kern distance—the distance from the centroid of a section to the farthest point from the centroid at which a resultant force can act without inducing a stress of opposite sign at the extreme fiber on the opposite side of the centroid.

key—see **keyway**.

keyed—fastened or fixed in position in a notch or other recess.

keyway—a recess or groove in one lift or placement of concrete that is filled with concrete of the next lift, giving shear strength to the joint. (See also **tongue and groove**.)

kick strip—see **kicker** (preferred term).

kicker—a wood block or board attached to a formwork member in a building frame or formwork to make the structure more stable; in formwork it acts as a haunch. (See also **wall, stub**.)

kiln—a furnace or oven for drying, charring, hardening, baking, calcining, sintering, or burning various materials. (See also **steam-curing room**.)

kiln, cement—a kiln in which the ground and proportioned raw mixture is dried, calcined, and burned into clinker at a temperature of 2600 to 3000 F (1420 to 1650 C); can be of the rotary, shaft, fluid-bed, or traveling-grate type; fuel may be coal, oil, or gas.

kiln, rotary—a long steel cylinder with a refractory lining, supported on rollers so that it can rotate about its own axis, and erected with a slight inclination from the horizontal so that prepared raw materials fed into the higher end move to the lower end where fuel is blown in by air blast.

kiln, steam—see **steam-curing room** (preferred term).

kip—1000 lb force, equals 4448 N.

knee brace—brace between horizontal and vertical members in a building frame or formwork to make the structure more stable; in formwork it acts as a haunch.

L

lacing—horizontal bracing between shoring members.

lagging—heavy sheathing used as in underground work to withstand earth pressure. (See also **sheathing**.)

laitance—a layer of weak material derived from cementitious material and aggregate fines either: 1) carried by bleeding to the surface or to internal cavities of freshly placed concrete; or 2) separated from the concrete and deposited on the concrete surface or internal cavities during placement of concrete underwater.

lap—the length by which one bar or sheet of fabric reinforcement overlaps another.

lap splice—see **splice, lap**.

lapping (reinforcing steel)—the overlapping of reinforcing steel bars, welded-wire fabric, or expanded metal so that there may be continuity of stress in the reinforcing when the concrete member is subjected to loading.

larnite—a mineral, beta dicalcium silicate (Ca_2SiO_4); occurs naturally at Scawt Hill, Northern Ireland, and artificially in slags and as a major constituent of portland cement.

lateral reinforcement—see **reinforcement, lateral**.

latex—a water emulsion of a high molecular-weight polymer, used especially in coatings, adhesives, leveling compounds, and patching compounds.

lath, expanded-metal—a metal network, often used as reinforcement in concrete or mortar construction, formed by suitably stamping or cutting sheet metal and stretching in to form open meshes, usually of diamond shape. (See also **mesh, diamond**.)

law, Abrams'—a rule stating that, with given concrete materials and conditions of test, the ratio of the amount of water to the amount of the cement in the mixture determines the strength of the concrete provided the mixture is of a workable consistency. (See also **water-cement ratio**.)

law, Hooke's—the law, which holds practically for strains within the elastic limit, that the strain is proportional to the stress producing it. (See also **limit, proportional** and **modulus of elasticity**.)

layer—see **course** and **lift**.

layer, bonding—a layer of mortar, usually 1/8 to 1/2 in. (3 to 13 mm) thick, which is spread on a moist and prepared, hardened concrete surface before placing fresh concrete.

L-beam—a beam having a cross section in the shape of an L; a beam having a ledge on one side only.

L-column—the portion of a precast concrete frame comprising the column, the haunch, and part of the girder.

leaf—see **wythe (leaf)**.

lean concrete—see **concrete, lean**.

lean mixture—see **concrete, lean**.

lean mortar—see **mortar, lean**.

ledger—any member with a protrusion or protrusions that support other structural members. (See also **L-beam** and **inverted T-beam**.)

length—

length, development—the embedment length required to develop the design strength of a reinforcement at a critical section; formerly called bond length.

length, embedment—the length of embedded reinforcement provided beyond a critical section.

length, transfer—the length from the end of the member where the tendon stress is zero, to the point along the tendon where the prestress is fully effective; also called transmission length.

length, transmission—see **length, transfer**.

length change—increase or decrease in length. (See also **volume change** and **deformation**.)

length change, autogenous—length change caused by autogenous volume change. (See **volume change, autogenous**.)

lever arm—in a structural member, the distance from the center of the tensile reinforcement to the center of action of the compression zone; also the perpendicular distance of a transverse force from a point about which moment is taken.

L-head—the top of a shore formed with a braced horizontal member projecting from one side, producing an inverted L-shaped assembly.

lift—the concrete placed between two consecutive horizontal construction joints, usually consisting of several layers or courses.

lift joint—see **joint, lift**.

lift slab—a method of concrete construction in which floor and roof slabs are cast on or at ground level and hoisted into position by jacking; also a slab that is a component of such construction.

lifts (or tiers)—the number of frames of scaffolding erected one above the other.

lightweight aggregate—see **aggregate, lightweight**.

lightweight concrete—see **concrete, lightweight**.

lime—specifically, calcium oxide (CaO); loosely, a general term for the various chemical and physical forms of quicklime, hydrated lime, and hydraulic hydrated lime. (See also **lime, hydrated**; **lime, hydraulic hydrated**; and **quicklime**.)

lime, free—calcium oxide (CaO), as in clinker and cement, which has not combined with SiO_2, Al_2O_3, or Fe_2O_3 during the burning process usually because of underburning, insufficient grinding of the raw mixture, or the presence of traces of inhibitors.

lime, hard-burned—the product of heating limestone to temperatures sufficient to change the calcium carbonate to calcium oxide, which can undergo expansion when it slowly reacts with water.

lime, hydrated—calcium hydroxide, a dry powder obtained by treating quicklime with water.

lime, hydraulic hydrated—the hydrated dry cementitious product obtained by calcining a limestone containing silica and alumina to a temperature short of incipient fusion so as to form sufficient free calcium oxide to permit hydration and at the same time leaving unhydrated sufficient calcium silicates to give the dry powder its hydraulic properties.

lime, spray—a hydrated lime of such fineness that at least 95% of the particles pass a 45 μm (No. 325) sieve.

limestone—a sedimentary rock consisting primarily of calcium carbonate.

limit—

limit, elastic—the limit of stress beyond which the strain is not wholly recoverable.

limit, liquid—water content, expressed as a percentage of the dry weight of the soil at which the soil passes from the plastic to the liquid state under standard test conditions. (See also **limits, Atterberg**.)

limit, plastic—the water content at which a soil will just begin to crumble when rolled into a thread approximately 1/8 in. (3 mm) in diameter. (See also **limits, Atterberg**.)

limit, proportional—the greatest stress that a material is capable of developing without any deviation from proportionality of stress to strain. (See also **law, Hooke's**.)

limit, shrinkage—the maximum water content at which a reduction in water content will not cause a decrease in volume of the soil mass. (See also **limits, Atterberg**.)

limit, vibration—the age at which fresh concrete has hardened sufficiently to prevent its becoming mobile when subjected to vibration.

limits, Atterberg—arbitrary water contents (shrinkage limit, plastic limit, liquid limit) determined by standard tests that define the boundaries between the different states of consistency of plastic soils.

limit design—a method of proportioning reinforced-concrete members based on calculation of their strength. (See also **strength-design method**.)

limonite—an iron ore composed of a mixture of hydrated ferric oxides; occasionally used in heavyweight concrete because of its high density and combined-water content, which contribute to its effectiveness in radiation shielding; a mineral occurring commonly as a constituent of particles of natural aggregate. (See also **oxide, brown**.)

linear prestressing—prestressing applied to linear members, such as beams and columns.

linear transformation—the method of altering the path of the prestressing tendon in any statically indeterminate prestressed structure by changing the location of the tendon at one or more interior supports without altering its

linear

position at the end supports and without changing the basic shape of the path between any supports; linear transformation does not change the location of the path of the pressure line.

linear-traverse method—determination of the volumetric composition of a solid by integrating the distance traversed across areas of each component along a line or along regularly spaced lines in one or more planes intersecting a sample of the solid; frequently employed to determine characteristics of the air-void system in hardened concrete by microscopical examination along a series of traverse lines on finely ground sections of the concrete; sometimes called the Rosiwal method. (See also **point count method** and **point count method [modified]**.)

lining—any sheet, plate, or layer of material attached directly to the inside face of formwork to improve or alter the surface texture and quality of the finished concrete. (See also **form lining**, **tunnel lining**, and **sheathing**.)

lintel—a horizontal supporting member above an opening, such as a window or a door.

liquid limit—see **limit, liquid**.

liquid-volume measurement—measurement of grout on the basis of the total volume of solid and liquid constituents.

lithology—the study of rocks. (See also **petrography** and **petrology**.)

live load—see load, live.

load—

  load, allowable—see **load, service dead** and **load, service live**.

  load, axle—the portion of the gross weight of a vehicle transmitted to a structure or a roadway through wheels supporting a given axle.

  load, balanced—load capacity at simultaneous compressive failure of concrete and yielding of tension steel. (See also **load balancing**.)

  load, cracking—the load that causes tensile stress in a member to exceed the tensile strength of the concrete.

  load, dead—a constant load that in structures is due to the mass of the members, the supported structure, and permanent attachments or accessories.

  load, design—obsolete term for factored load.

  load, dynamic—a load that is variable, that is, not static, such as a moving live load, earthquake, or wind.

  load, factored—load, multiplied by appropriate load factors, used to proportion members by the strength-design method.

  load, live—any load that is not permanently applied to a structure; transitory load.

  load, point—a load whose area of contact with the resisting body is negligible in comparison with the area of the resting body.

  load, safe leg—the load that can safely be directly imposed on the frame leg of a scaffold. (See also **load, service**.)

  load, service—all loads, static or transitory, imposed on a structure, or element thereof, during operation of a facility.

  load, service dead—the dead weight supported by a member.

  load, service live—the live load specified by the general building code or other bridge specification, or the actual nonpermanent load applied in service.

  load, shock—impact of material, such as aggregate or concrete, as it is released or dumped during placement.

  load, static—the mass of a single stationary body or the combined masses of stationary bodies in a structure (such as the load of a stationary vehicle on a roadway); or, during construction, the combined mass of forms, stringers, joists, reinforcing bars, and the actual concrete to be placed. (See also **load, dead**.)

  load, superimposed—the load, other than its own weight, that is resisted by a structural member or system.

  load, ultimate—the maximum load that may be placed on a structure or structural element before its failure.

  load, wheel—the portion of the gross mass of a loaded vehicle transferred to the supporting structure under a given wheel of the vehicle.

  load, working—forces normally imposed on a member in service (obsolete term).

load balancing—a technique used in the design of prestressed-concrete members in which the amount and path of the prestressing is selected so that the forces imposed upon the member or structure by the prestressing counteract or balance a portion of the dead and live loads for which the member or structure must be designed.

load binder—a device used to tighten chains holding loads in place on a truck bed.

load factor—a factor by which a service load is multiplied to determine a factored load used in the strength-design method.

load-bearing wall—see **wall, load-bearing**.

load-transfer assembly—the unit (basket or plate) designed to support or link dowel bars during concreting operations so as to hold them in place while in the desired alignment.

loading, bulk—loading of unbagged cement in containers, specially designed trucks, railroad cars, or ships.

loading, dynamic—loading from units (particularly machinery) that, by virtue of their movement or vibration, impose stresses in excess of those imposed by their dead load.

loading, ribbon—method of batching concrete in which the solid ingredients, and sometimes also the water, enter the mixer simultaneously.

loading hopper—a hopper in which concrete or other free-flowing material is deposited for discharge into buggies or other conveyances used for delivery to the forms or to other place of processing, use, or storage.

locking device—a device used to secure a cross brace in scaffolding to the frame or panel.

long column—see **column, long**.

longitudinal bar—see **reinforcement, longitudinal** (preferred term).

longitudinal crack—see **crack, longitudinal**.

longitudinal joint—see **joint, longitudinal**.

longitudinal reinforcement—see **reinforcement, longitudinal**.

Los Angeles abrasion test—see **test, Los Angeles abrasion**.

loss—

loss, anchorage—see **deformation, anchorage** or **slip**.

loss, elastic—in prestressed concrete, the reduction in prestressing load resulting from the elastic shortening of the member.

loss, friction—the stress loss in a prestressing tendon resulting from friction between the tendon and duct or other device during stressing.

loss, ignition—see **loss on ignition** (preferred term).

loss, plastic—see **creep**.

loss, shrinkage—reduction of stress in prestressing steel resulting from shrinkage of concrete.

loss, slump—the amount by which the slump of freshly mixed concrete changes during a period of time after an initial slump test was made on a sample or samples thereof.

loss of prestress—the reduction in the prestressing force which results from the combined effects of slip at anchorage, relaxation of steel stress, frictional loss due to curvature in the tendons, and the effects of elastic shortening, creep, and shrinkage of the concrete.

loss on ignition—the percentage loss in mass of a sample ignited to constant weight at a specified temperature, usually 1650 to 1830 F (900 to 1000 C).

lot—a defined quantity, usually merchandise.

low-alkali cement—see **cement, low-alkali**.

low-density concrete—see **concrete, low-density** and **concrete, lightweight**.

low-heat cement—see **cement, low heat**.

low-lift grouting—see **grouting, low-lift**.

low-pressure steam curing—see **curing, atmospheric-pressure steam** (preferred term).

low-strength materials—see **controlled low-strength cementitious material** (preferred term).

L-shore—a shore with an L-head. (See also **L-head**.)

lubricant, dowel—a material applied to part of the surface of a dowel to reduce bond with the concrete and permit axial movement.

M

macadam, cement-bound—a road consisting of crushed stone, crushed slag, or gravel and either a grout or mortar filler; formed by rolling a base of stone, slag, or gravel to a compacted mass having an even surface, and then rolling in the cementitious filler.

machine, finishing—a power-operated machine used to produce the desired surface texture on a concrete slab.

machine-base grout—see **grout, machine base**.

macroscopic—visible to the naked eye (preferred term).

magnetite—a mineral, ferrous ferric oxide ($FeO \cdot Fe_2O_3$); the principal constituent of magnetic black iron ore; density approximately 5.2 g/cc and Mohs hardness approximately 6; used as an aggregate in high-density concrete.

manual batcher—see **batcher manual**.

manufactured sand—see **sand**.

map cracking—see **cracking, map**.

marble—a metamorphic rock composed essentially of recrystallized calcite, dolomite, or both.

marl—calcareous clay, usually containing from 35 to 65% calcium carbonate ($CaCO_3$), found in the bottoms of shallow lakes, swamps, or extinct fresh-water basins.

mason—an artisan who builds with concrete masonry units, bricks, stone, and tile; name sometimes given a concrete finisher.

masonry—construction composed of shaped or molded units, usually small enough to be handled by one person and composed of stone, ceramic brick or tile, concrete, glass, adobe, or the like.

masonry, ashlar—masonry composed of bonded blocks of concrete, either rectangular or square, always of two or more sizes; if the pattern is repeated, it is patterned ashlar; if the pattern is not repeated, it is random ashlar.

masonry, bonded hollow-wall—a cavity wall, built of masonry units, in which the inner and outer walls are tied together by bonders.

masonry, exposed—masonry constructed to have no surface finish other than paint.

masonry, grouted—unit masonry composed of either hollow units wherein the cells are filled with grout or multiple wythes where spaces between the wythes are filled with grout.

masonry, hollow-unit—masonry consisting either entirely or partially of hollow masonry units laid in mortar.

masonry, plain—
1. masonry without reinforcement; or
2. masonry reinforced only for shrinkage or thermal change.

masonry, reinforced—unit masonry in which reinforcement is embedded in such a manner that the two materials act together in resisting forces.

masonry, solid-unit—masonry consisting wholly of solid masonry units laid in mortar.

masonry, unit—a structural element consisting of concrete masonry units usually bonded by mortar, grout, or both.

masonry cement—see **cement, masonry**.

masonry filler unit—masonry unit used to fill in between joists or beams to provide a platform for a cast-in-place concrete slab.

masonry grout—see **grout, masonry**.

masonry lift—the height to which masonry is laid between periods of grouting.

masonry unit, concrete—either a hollow or solid unit (block) composed of portland-cement concrete; often referred to by indicating the type of mineral aggregate incorporated (for example, lightweight or sand-gravel block).

masonry

masonry wall, solid—a wall built of blocks or solid masonry units, the mortar completely filling the joints between units.

mason's putty—a pasty substance, composed of water and hydrated lime mixed with portland cement and stone dust; used only for jointing ashlar masonry.

mass—the physical property of matter that causes it to have weight in a gravitational field; the quantity of matter in a body.

mass concrete—see **concrete, mass**.

mass curing—see **curing, mass**.

mass density—see **density**.

mat—see **bar mat**.

mat foundation—see **foundation, mat**.

material hose—see **hose, delivery**.

materials, cementitious—cements and pozzolans used in concrete and masonry construction. (See also **blast-furnace slag**; **cement, hydraulic**; **masonry**; and **mortar**.)

matrix—in the case of mortar, the cement paste in which the fine aggregate particles are embedded; in the case of concrete, the mortar in which the coarse aggregate particles are embedded.

mats, cotton—cotton-filled quilts fabricated for use as a water-retaining covering in curing concrete surfaces.

maturity factor—see **factor, maturity**.

maximum service temperature (refractory concrete)—the temperature above which excessive shrinkage occurs in refractory concrete; usually between 150 F (66 C) and 200 F (93 C) below the temperature at which the refractory concrete softens.

maximum size (of aggregate)—in specifications for and in description of aggregate, the smallest sieve opening through which the entire amount of aggregate is required to pass. (See also **nominal maximum size [of aggregate]**.)

maximum-temperature period—a time interval throughout which the maximum temperature is held constant in an autoclave or steam-curing room.

mean stress—see **stress, mean**.

mechanical analysis—the process of determining particle-size distribution of an aggregate. (See **analysis, sieve**.)

mechanical anchorage—see **anchorage, mechanical**.

mechanical bond—see **bond, mechanical**.

mechanical connection—the complete assembly of an end-bearing sleeve, a coupler, or a coupling sleeve, and possibly additional intervening material or other components to effect connection of reinforcing bars. (See also **bar-end check**; **coupler**; **coupling sleeve**; and **end-bearing sleeve**.)

medium, grinding—a hard, free-moving charge in a ball or tube mill to reduce the particle size of introduced materials by attrition or impact.

megascopic—see **macroscopic** (preferred term).

melilite—a group of minerals ranging from the calcium magnesium silicate (akermanite) to the calcium aluminate silicate (gehlenite) that occur as crystals in blast-furnace slag. (See also **akermanite**; **gehlenite**; and **merwinite**.)

melt—the molten portion of the raw material mass during the burning of cement clinker, firing of lightweight aggregates, or expanding of blast-furnace slags.

member, compression—any member in which the primary stress is longitudinal compression.

member, segmental—a structural member made up of individual elements prestressed together to act as a monolithic unit under service loads.

membrane curing—see **curing, membrane**.

membrane theory—a theory of design for thin shells, based on the premise that a shell cannot resist bending because it deflects; the only stresses that exist in any section, therefore, are shear stress and direct compression or tension.

merwinite—one of the principal crystalline phases found in blast-furnace slags; the chemical formula is $Ca_3Mg(SiO_4)_2$, the crystal system is monoclinic, and the density is 3.15 g/cc. (See also **akermanite**; **gehlenite**; and **melilite**.)

mesh—the number of openings (including fractions thereof) per unit of length in either a screen or sieve in which the openings are 1/4 in. (6 mm) or less.

mesh, diamond—a metallic fabric having rhomboidal openings in a geometric pattern. (See also **lath, expanded-metal**.)

mesh reinforcement—see **fabric, welded-wire** and **reinforcement, welded-wire fabric**.

mesh roller—a finishing tool consisting of a rolling drum attached to a handle, of which the surface of the drum is made of mesh, sometimes used for rolling over the surface of fresh concrete to embed coarse aggregate.

Mesnager hinge—see **hinge, Mesnager**.

meter, air—a device for measuring the air content of concrete and mortar.

method, advancing-slope—a method of placing concrete as in tunnel linings in which the face of the fresh concrete is not vertical and moves forward as concrete is placed.

microconcrete—a mixture of portland cement, water, and suitably graded sand for simulating concrete in small-scale structural models.

microcracks—microscopic cracks within concrete.

micron—an obsolete term designating a unit of length equal to one thousandth of a millimeter (mm) or one millionth of a meter (m); superseded by micrometer (μm).

microsand—fine aggregate, passing the U.S. Standard 150 μm (No. 100) sieve, and essentially free of clay and shale.

microscope, polarizing—a microscope equipped with elements permitting observations and determinations to be made using polarized light. (See also **Nicol prism**.)

microscope, scanning electron (SEM)—an electron microscope in which the image is formed by a beam operating in synchronism with an electron probe scanning the object; the intensity of the image-forming beam is proportional to the scattering or secondary emission of electrons by the specimen where the probe beam strikes it.

microscopic—discernible only with the aid of a microscope.

microsilica—see **silica fume** (preferred term).

middle strip—see **strip, middle**.

mill, ball—horizontal, cylindrical, rotating mill charged with large grinding media. (See also **mill, rod**.)

mill, rod—horizontal, cylindrical, rotating mill charged with steel rods for grinding. (See also **mill, ball**.)

mill scale—the partially adherent layers of oxidation products (heavy oxides) developed on metallic surfaces during either hot fabrication or heat treatment of metals, as on hot-rolled steel reinforcing bars.

mineral aggregate—see **aggregate, mineral**.

mineral filler—a finely divided mineral product at least 65% of which passes the U.S. Standard 75 µm (No. 200) sieve. (See also **silt**.)

mix (*n.*)—see **mixture**.

mix (*v.*)—the act or process of mixing; also, a mixture of materials, such as mortar or concrete.

mix, dry—a concrete, mortar, or plaster mixture, commonly sold in bags, containing all components except water; also a concrete of near zero slump.

mix design—see **mixture proportioning** (preferred term).

mixer—a machine used for blending the constituents of concrete, grout, mortar, cement paste, or other mixture.

    mixer, batch—a machine that mixes batches of either concrete or mortar.

    mixer, central—a stationary concrete mixer from which the freshly mixed concrete is transported to the work.

    mixer, colloidal—a mixer designed to produce colloidal grout.

    mixer, continuous—a mixer into which the ingredients of the mixture are fed without stopping, and from which the mixed product is discharged in a continuous stream.

    mixer, high-discharge—see **mixer, inclined-axis** (preferred term).

    mixer, horizontal-axis—a concrete mixer of the revolving drum type in which the drum rotates about a horizontal axis.

    mixer, horizontal-shaft—a mixer having a stationary cylindrical mixing compartment, with the axis of the cylinder horizontal, and one or more rotating horizontal shafts to which mixing blades or paddle are attached; also called pugmill.

    mixer, inclined-axis—a truck with a revolving drum that rotates about an axis inclined to the bed of the truck chassis.

    mixer, nontilting—a horizontal rotating drum mixer that charges, mixes, and discharges without tilting.

    mixer, open-top—a truck-mounted mixer consisting of a trough or a segment of a cylindrical mixing compartment within which paddles or blades rotate about the horizontal axis of the trough. (See also **mixer, horizontal-shaft** and **mixer, open-top**.)

    mixer, paddle—see **open-top mixer** (preferred term).

    mixer, pan—see **mixer, vertical shaft**.

    mixer, revolving-blade (or paddle)—see **mixer, open-top**.

    mixer, tilting—a revolving-drum mixer that discharges by tilting the drum about a fixed or movable horizontal axis at right angles to the drum axis; the drum axis may be horizontal or inclined while charging and mixing.

    mixer, transit—see **mixer, truck**.

    mixer, trough—see **mixer, open-top** (preferred term).

    mixer, truck—a concrete mixer suitable for mounting on a truck chassis and capable of mixing concrete in transit. (See also **mixer, horizontal-axis**; **mixer, inclined-axis**; **mixer, open-top**; and **agitator**.)

    mixer, tub—see **mixer, open-top** (preferred term).

    mixer, turbine—see **mixer, open-top** (preferred term).

    mixer, vertical-shaft—a cylindrical or annular mixing compartment having an essentially level floor and containing one or more vertical rotating shafts to which blades or paddles are attached; the mixing compartment may be stationary or rotate about a vertical axis.

mixer efficiency—the adequacy of a mixer in rendering a homogeneous product within a stated period; homogeneity is determinable by testing for relative differences in physical properties or composition of samples extracted from different portions of a freshly mixed batch.

mixing cycle—the time taken for a complete cycle in a batch mixer, that is, the time elapsing between successive repetitions of the same operation (for example, successive discharges of the mixer).

mixing, dry—blending of the solid materials for mortar or concrete before adding the mixing water.

mixing plant—see **batch plant** (preferred term).

mixing speed—rotation rate of a mixer drum or of the paddles in an open-top, pan, or trough mixer, when mixing a batch; expressed in revolutions per minute (rpm), or in peripheral feet per minute of a point on the circumference at maximum diameter.

mixing time—the period during which the constituents of a batch of concrete are mixed by a mixer; for a stationary mixer, time is given in minutes from the completion of mixer charging until the beginning of discharge; for a truck mixer, time is given in total minutes at a specified mixing speed or expressed in terms of total revolutions at a specified mixing speed. (See also **amount of mixing**.)

mixing water—see **water, mixing**.

mixture—the assembled, blended, commingled ingredients of mortar, concrete, or the like; or the proportions for their assembly.

    mixture, harsh—a concrete mixture that lacks desired workability and consistency due to a deficiency of mortar or aggregate fines.

    mixture, lean—see **concrete, lean**.

    mixture, rich—see **rich mixture**.

    mixture proportion—the proportions of ingredients that make the most economical use of available materials to produce mortar or concrete of the required properties. (See also **proportion**.)

mobile placer—a small belt conveyor, mounted on wheels or truck-mounted, that can be readily moved to the job

moderate

site for conveying concrete from the ready-mixed concrete truck to the forms or slab.

moderate sulfate-resisting cement—see **cement, moderate sulfate-resisting**.

modified cube—a portion of a rectangular beam of hardened concrete previously broken in flexure; used in determining the compressive strength of the concrete.

modified portland cement—a portland cement having moderate heat of hydration; this term was replaced by Type II cement beginning in 1960. (See also **cement, modified**.)

modular ratio—the ratio of modulus of elasticity of steel E_s to that of concrete E_c; usually denoted by the symbol n.

module—any in a series of standardized units for use together in erecting a structure.

modulus—

  modulus, bulk—the ratio of the change in average stress to the change in unit volume. (See also **modulus of compression**.)

  modulus, chord—see **modulus of elasticity**.

  modulus, elastic—see **modulus of elasticity** (preferred term).

  modulus, fineness—a factor obtained by adding the total percentages of material in the sample that are coarser than each of the following sieves (cumulative percentages retained), and dividing the sum by 100: 150 μm (No. 100); 300 μm (No. 50); 600 μm (No. 30); 1.18 mm (No. 16); 2.36 mm (No. 8); 4.75 mm (No. 4); 9.5 mm (3/8 in.); 19.0 mm (3/4 in.); 37.5 mm (1-1/2 in.); 75 mm (3 in.); and 150 mm (6 in.)

  modulus, initial-tangent—see **modulus of elasticity**.

  modulus, secant—see **modulus of elasticity**.

  modulus, section—a term pertaining to the cross section of a flexural member; the section modulus with respect to either principal axis is the moment of inertia with respect to that axis divided by the distance from that axis to the most remote point of the tension or compression area of the section, as required; the section modulus is used to determine the flexural stress in a beam.

  modulus, shear—see **modulus of rigidity**.

  modulus, sonic—see **modulus of elasticity, dynamic**.

  modulus, subgrade—see **coefficient of subgrade reaction**.

  modulus, tangent—see **modulus of elasticity**.

  modulus, Young's—see **modulus of elasticity** (preferred term).

modulus of compression—the ratio of compressive stress to cubical compression; always positive for physical substances; also known as bulk modulus; related to Young's modulus and Poisson's ratio by the equation $K = E \div 3(1-2\mu)$, where k = bulk modulus; E = Young's modulus; and μ = Poisson's ratio of the material under consideration.

modulus of deformation—

1. a concept of modulus of elasticity expressed as a function of two time variables; strain in loaded concrete as a function of the age at which the load is initially applied and of the length of time the load is sustained; and

2. the ratio of stress to strain for a material that does not deform in accordance with Hooke's law when subjected to applied load. (See also **modulus of elasticity**.)

modulus of elasticity—the ratio of normal stress to corresponding strain for tensile or compressive stress below the proportional limit of the material; also referred to as elastic modulus, Young's modulus, and Young's modulus of elasticity; denoted by the symbol E. (See also **modulus of rigidity**.)

Note: few materials conform to Hooke's law throughout the entire range of stress-strain relations; deviations therefrom are caused by inelastic behavior. If the deviations are significant, the slope of the tangent to the stress-strain curve at the origin, the slope of the tangent to the stress-strain curve at any given stress, the slope of the secant drawn from the origin to any specified point on the stress-strain curve, or the slope of the chord connecting any two specified points on the stress-strain curve, may be considered as the modulus; in such cases, the modulus is designated, respectively, as the initial tangent modulus, the tangent modulus, the secant modulus, or the chord modulus, and the stress stated. The modulus is expressed as force per unit of area (for example, psi or Pa).

  modulus of elasticity, dynamic—the modulus of elasticity computed from the size, weight, shape, and fundamental frequency of vibration of a concrete test specimen, or from pulse velocity. (See also **modulus of elasticity, static** and **velocity, pulse**.)

  modulus of elasticity, static—the value of Young's modulus of elasticity obtained by arbitrary criteria from measured stress-strain relationships derived from other than dynamic loading. (See also **modulus of elasticity**.)

  modulus of elasticity, sustained—term including elastic and inelastic effects in one expression to aid in visualizing net effects of stress-strain up to any given time; computed by dividing the unit sustained stress by the sum of the elastic and inelastic deformations at that time. (See also **modulus of elasticity**.)

  modulus of resilience—see **resilience**.

modulus of rigidity—the ratio of unit shearing stress to the corresponding unit shearing strain; referred to as shear modulus and modulus of elasticity in shear, denoted by the symbol G. (See also **modulus of elasticity**.)

modulus of rupture—a measure of the load-carrying capacity of a beam and sometimes referred to as rupture modulus or rupture strength; it is calculated for apparent tensile stress in the extreme fiber of a transverse test specimen under the load that produces rupture. (See also **strength, flexural**.)

Note: the actual stress in the extreme fiber is less than the apparent stress since the flexure formula employed in the calculation is valid only for stresses within the proportional limit of the material; nevertheless, the nominal rupture strength so obtained is considered the rupture modulus.

modulus of subgrade reaction—see **coefficient of subgrade reaction**.

Mohs scale—arbitrary quantitative units, ranging from 1 through 10, by means of which the scratch hardness of a mineral is determined; each unit of hardness is represented by a mineral that can scratch any other mineral having a lower-ranking number; the minerals are ranked from: talc, or 1 (the softest); gypsum, or 2; calcite, or 3; fluorite, or 4; apatite, or 5; orthoclase, or 6; quartz, or 7; topaz, or 8; corundum, or 9; and diamond, or 10 (the hardest).

moist—slightly damp but not quite dry to the touch; the terms "wet" implies visible free water, "damp" implies less wetness than "wet," and "moist" implies not quite dry. (See also **damp** and **wet**.)

moist-air curing—see **curing, moist-air**.

moist cabinet—see **cabinet, moist**.

moist room—a room in which the atmosphere is maintained at a selected temperature (usually 23.0 ± 2 C or 73.0 ± 3.0 F) and a relative humidity of at least 95%, for the purpose of curing and storing cementitious test specimens; the facilities must be sufficient to maintain free moisture continuously on the exteriors of test specimens; also known as a fog room.

moisture—

    moisture, absorbed—moisture that has entered the permeable voids of a solid and has physical properties not substantially different from ordinary water at the same temperature and pressure. (See also **absorption**.)

    moisture, free—moisture having essentially the properties of pure water in bulk; moisture not absorbed by aggregate. (See also **moisture, surface**.)

    moisture, surface—free water retained on surfaces of aggregate particles and considered to be part of the mixing water in concrete, as distinguished from absorbed moisture.

moisture barrier—see **barrier, moisture**.

moisture content of aggregate—the ratio, expressed as a percentage, of the mass of water in a given granular mass to the dry weight of the mass.

moisture content of concrete masonry unit—the amount of water contained in the hardened concrete at the time of sampling and expressed as a percentage of its capacity for total absorption.

moisture-free—the condition of a material that has been dried in air until there is no further significant change in its mass. (See also **mass** and **ovendry**.)

moisture movement—

    1. the movement of moisture through a porous medium; and
    2. in the UK, the effects of such movement on efflorescence and volume change in hardened cement paste, mortar, concrete, or rock. (See also **shrinkage** and **swelling**.)

mold—

    1. a device containing a cavity into which neat cement, mortar, or concrete test specimens are cast; and
    2. a form used in the fabrication of precast mortar or concrete units (for example, masonry units).

mold, plaster—a mold or form made from gypsum plaster, usually to permit concrete to be formed or cast in intricate shapes or in conspicuous relief. (See also **mold** and **form**.)

mold oil—see **oil, mold**.

moment—the colloquial expression for the more descriptive term, bending moment. (See also **moment, bending**.)

    moment, bending—the bending effect at any section of a structural element; it is equal to the algebraic sum of the moments of the vertical and horizontal forces, with respect to the centroidal axis of a member, acting on a freebody of the member.

    moment, negative—a condition of flexure in which top fibers of a horizontally placed member, or external fibers of a vertically placed exterior member, are subjected to tensile stresses.

    moment, positive—a condition of flexure in which, for a horizontal simply supported member, the deflected shape is normally considered to be concave downward and the top fibers subjected to compression stresses; for other members and other conditions consider positive and negative as relative terms. (See also **moment, negative**.)

    Note: for structural design and analysis, moments may be designated as positive or negative with satisfactory results as long as the sign convention adopted is used consistently.

    moment, secondary—in statically indeterminate structures, the additional moments caused by deformation of the structure due to the applied forces; in statically indeterminate prestressed-concrete structures, the additional moments caused by the use of a nonconcordant prestressing tendon.

    moment, ultimate—obsolete term; see **strength, flexural**.

moment distribution—a method of structural analysis for continuous beams and rigid frames whereby successive converging corrections are made to an assumed set of moments until the desired precision is obtained; also known as the Hardy Cross method.

monolith—a body of plain or reinforced concrete cast or erected as a single integral mass or structure.

monolithic concrete—see **concrete, monolithic**.

monolithic surface treatment—see **dry-shake**.

monolithic terrazzo—the application of a 5/8 in. (15 mm) terrazzo topping directly to a specially prepared concrete substrate, eliminating an underbed.

monolithic topping—see **topping, monolithic**.

monomer—an organic molecule of relatively low molecular weight that creates a solid polymer by reacting with itself or other compounds of low molecular weight or both.

monomolecular—composed of single molecules; specifically, films that are one molecule thick; denotes a thickness equal to one molecule, for example, certain chemical compounds develop a monomolecular film over bleeding water at the surface of freshly placed concrete or mortar as a means of reducing the rate of evaporation. (See also **evaporation retardant**.)

montmorillonite—a swelling clay mineral of the smectite group; main constituent of bentonite. (See also **smectite**.)

mortar—a mixture of cement paste and fine aggregate; in fresh concrete, the material occupying the interstices among particles of coarse aggregate; in masonry construction, joint mortar may contain masonry cement, or may contain hydraulic cement with lime (and possibly other admixtures) to afford greater plasticity and workability than are attainable with standard portland cement mortar. (See also **cement, hydraulic** and **masonry**.)

mortar, air-blown—see **shotcrete** (preferred term).

mortar, expansive-cement—see **concrete (mortar or grout), expansive-cement**.

mortar, epoxy—a mixture of epoxy resin, catalyst, and fine aggregate. (See also **resins, epoxy**.)

mortar, fat—mortar containing a high percentage of fine-grained solid components; sufficiently sticky to adhere to a steel trowel.

mortar, lean—mortar that is harsh and difficult to spread because of either insufficient cement content or the presence of coarse sand.

mortar, plastic—a mortar of plastic consistency.

mortar, resin—see **concrete, polymer**.

mortar, sprayed—see **shotcrete** (preferred term).

mortar, stringing—the procedure of spreading enough mortar on the bed joint to ensure laying several masonry units.

mortar board—a platform or tray for holding freshly mixed mortar. (See also **hawk** and **hod**.)

mortar-flow—see **flow 2**.

mosaic—inlaid exposed surface designs of aggregates or other material.

moving forms—see **forms, moving**.

mud balls—lumps of clay or silt ("mud").

mudjacking—see **slabjacking** (preferred term).

mud pumping—see **pumping (of pavements)**.

mud sill—a timber or timber assembly bedded into the earth at grade to support framed construction.

mud slab—a 2 to 6 in. (50 to 150 mm) layer of concrete beneath a structural concrete floor or footing over soft, wet soil; also called mud mat.

multielement prestressing—prestressing accomplished by stressing an assembly of several individual structural elements as a means of producing one integrated structural member.

multistage stressing—prestressing performed in stages as the construction progresses.

multiwall-bag—a flexible container for transporting a cementitious material and usually consisting of four plies of kraft paper previously treated to ensure resistance to moisture.

muriatic acid—see **hydrochloric acid** (preferred term).

mushroom system of flat-slab construction—a four-way reinforced-concrete girderless floor slab in which the column reinforcing bars are bent down into the slab around the column head in radial directions and additional reinforcing bars are bent into rings laid upon the radials, thus forming a spider web to provide additional reinforcement at the column head and to support the slab steel; mushroom designs of the true flat-slab type do not involve drop panels around the capitals of the columns.

N

nailable concrete—see **concrete, nailable**.

nailer—a strip of wood or other fitting attached to or set in concrete, or attached to steel, to facilitate making nailed connections.

natural air-drying—the process of drying cured concrete masonry units without any special equipment (for example, the drying that occurs in a covered storage area).

natural cement—see **cement, natural**.

natural pozzolan—see **pozzolan, natural**.

natural sand—see **sand, natural**.

neat cement grout—see **grout, neat-cement**.

neat cement paste—see **cement paste, neat**.

neat line—a line defining the proposed or specified limits of an excavation or structure.

neat plaster—see **plaster, neat**.

necking—the localized and permanent reduction of cross-sectional area of a test specimen of metal; due to stretching produced by applied tensile load.

needle, Gillmore—a device used in determining time of setting of hydraulic cement.

needle, Vicat—a weighted needle for determining time of setting of hydraulic cements.

negative catalyst—see **catalyst, negative**.

negative moment—see **moment, negative**.

negative reinforcement—see **reinforcement, negative**.

negative-slump concrete—see **concrete, negative-slump**.

net cross-sectional area (of masonry)—the gross cross-sectional area of a section of masonry minus the area of cavities, cells, or cored spaces.

net mixing water—see **water, mixing**.

neutral axis—see **axis, neutral**.

neutral refractory—see **refractory, neutral**.

Nicol prism—a system of two optically clear crystals of calcite ("Iceland spar") used in producing plane-polarized light.

nip—the seizing of stone between either the jaws or the rolls of a crusher.

no-fines concrete—see **concrete, no-fines**.

nominal flexural strength—see **strength, nominal flexural**.

nominal maximum size (of aggregate)—in specifications for and in descriptions of aggregate, the smallest sieve opening through which the entire amount of the aggregate is permitted to pass. (See also **maximum size [of aggregate]**.)

nominal mixture—the proportions of the constituents of a proposed concrete mixture.

nominal shear strength—see **strength, nominal shear**.

nominal size—see **nominal maximum size (of aggregate)**.

nominal strength—see **strength, nominal**.

nonagitating unit—a truck-mounted container for transporting central-mixed concrete, not equipped to provide agitation (slow mixing) during delivery.
nonair-entrained concrete—see **concrete, nonair-entrained**.
nonbearing wall—see **wall, nonbearing**.
noncombustible—any material that neither ignites nor supports combustion in air when exposed to fire.
nonconcordant tendons—see **tendons, nonconcordant**.
nonevaporable water—see **water, nonevaporable**.
nonferrous—relating to metals other than iron; not containing or including iron.
nonprestressed reinforcement—see **reinforcement, nonprestressed**.
nonrecoverable creep—see **creep, nonrecoverable**.
nonreversible deformation—see **creep, nonrecoverable** (preferred term).
nonsimultaneous prestressing—see **prestressing, nonsimultaneous**.
nonslip concrete—see **concrete, nonslip**.
nonstaining cement—see **cement, nonstaining**.
nonstructural reinforcement—see **reinforcement, temperature**.
nontilting mixer—see **mixer, nontilting**.
normal cement—see **cement, normal**.
normal consistency—see **consistency, normal**.
normal portland cement—see **cement, normal**.
normal stress—see **stress, normal**.
normalweight aggregate—see **aggregate, normalweight**.
normalweight concrete—see **concrete, normalweight**.
normalweight refractory concrete—see **concrete, normalweight refractory**.
no-slump concrete—see **concrete, no-slump**.
nozzle—a metal or rubber tip attached to the discharge end of a heavy thick-walled rubber hose from which a continuous stream of shotcrete is ejected at high velocity.
nozzle liner—a replaceable rubber lining, fitted into the nozzle tip, to prevent abrasion of the interior surface of the nozzle.
nozzle operator—the operator who manipulates the nozzle and controls placement of the shotcrete; in the case of dry-mix shotcrete, the operator also controls the water content of the shotcrete.
nozzle velocity—the rate at which shotcrete is ejected from the nozzle, usually stated in ft per s or m per s.

O

obsidian—a natural volcanic glass of relatively low water content; usually of rhyolite composition. (See also **perlite**.)
offset—an abrupt change in alignment or dimension, either horizontally or vertically; a horizontal ledge occurring along a change in wall thickness of the wall above.
offset bend—an intentional distortion from the normal straightness of a steel reinforcing bar to move the center line of a segment of the bar to a position parallel to the original position of the center line; a mechanical operation commonly applied to vertical bars that reinforce concrete columns.
offset yield strength—see **strength, offset yield**.
oil, form—oil applied to the interior surfaces of forms to promote easy release from the concrete when the forms are removed. (See also **agent, release** and **bond breaker**.)
oil, mold—an oil that is applied to the interior surface of a clean mold, before casting concrete or mortar therein, to facilitate removal of the mold after the concrete or mortar has hardened. (See also **bond breaker**; **oil, form**; and **agent, release**.)
oil-well cement—see **cement, oil-well**.
one-way system—see **system, one-way**.
opal—a mineral composed of amorphous hydrous silica ($SiO_2 \cdot nH_2O$).
opaline chert—chert composed entirely or mainly of opal.
open-circuit crushing—a crushing system in which material passes through the crusher without recycling of oversize particles.
open-circuit grouting—see **grouting, open-circuit**.
open-graded aggregate—see **aggregate, open-graded**.
open-top mixer—see **mixer, open-top**.
ordinary portland cement—see **cement, ordinary portland**.
orthotropic—a contraction of the terms "orthogonal anisotropic" as in the phrase "orthogonal anisotropic plate"; a hypothetical plate consisting of beams and a slab acting together with different flexural rigidities in the longitudinal and transverse directions, as in a composite beam bridge.
ovals—marble chips that have been tumbled until a smooth oval shape has resulted.
oven-dry—the condition resulting from having been dried to essentially constant mass, in an oven, at a temperature that has been fixed, usually between 221 and 239 F (105 and 115 C).
oven dry—the process of drying in an oven at a temperature usually between 221 and 239 F (105 and 115 C) until the mass of the test specimen becomes essentially constant.
overdesign—to require adherence to structural design requirements higher than service demands, as a means of compensating for statistical variation or for anticipated deficiencies or both.
overlay—a layer of concrete or mortar, seldom thinner than 1 in. (25 mm), placed on and usually bonded onto the worn or cracked surface of a concrete slab to either restore or improve the function of the previous surface; also polymeric concrete usually less than 0.4 in. (10 mm) thick.
oversanded—containing more sand than would be necessary to produce adequate workability and a satisfactory condition for finishing.
overstretching—stressing of tendons to a value higher than designed for the initial stress to: (a) overcome frictional losses; (b) temporarily overstress the steel to reduce steel creep that occurs after anchorage, and (c) counteract loss of prestressing force that is caused by subsequent prestressing of other tendons.

overvibration—excessive use of vibrators during placement of freshly mixed concrete, causing segregation, stratification, and excessive bleeding.

owner—the corporation, association, partnerships, individual, or public body or authority with whom the contractor enters into an agreement and for whom the work is provided.

oxide, brown—a brown mineral pigment having an iron oxide content between 28 and 95%. (See also **limonite**.)

P

pack, dry—concrete or mortar mixtures deposited and consolidated by dry packing.

pack, warehouse—see **set, warehouse**.

pack set—see **cement, sticky** and **set, warehouse**.

packaged concrete, mortar, grout—mixtures of dry ingredients in packages, requiring only the addition of water to produce concrete, mortar, or grout.

packer—a device inserted into a hole in which grout is to be injected which acts to prevent return of the grout around the injection pipe; usually an expandable device actuated mechanically, hydraulically, or pneumatically.

packerhead process—see **process, packerhead**.

packing, dry—placing of zero-slump or near zero-slump concrete, mortar, or grout by ramming into a confined space.

paddle mixer—see **mixer, open-top** (preferred term).

paint, cement—a paint consisting generally of white portland cement and water, pigments, hydrated lime, water repellents, or hygroscopic salts.

paint, cold-water—a paint in which the binder or vehicle portion is composed of latex, casein, glue, or some similar material dissolved or dispersed in water.

Palladiana—see **Berliner**.

pan—
1. a prefabricated form unit used in concrete joist floor construction; and
2. a container that receives particles passing the finest sieve during mechanical analysis of granular materials.

pan mixer—see **mixer, vertical shaft** (preferred term).

panel—
1. a section of form sheathing, constructed from boards, plywood, metal sheets, etc., that can be erected and stripped as a unit; and
2. a concrete member, usually precast, rectangular in shape, and relatively thin with respect to other dimensions.

panel, drop—the thickened structural portion of a flat slab in the area surrounding column, column capital, or bracket, to reduce the intensity of stresses.

panel, exterior—in a flat slab, a panel having at least one edge that is not in common with another panel.

panel, ribbed—a panel composed of a thin slab reinforced by a system of ribs in one or two directions, usually orthogonal.

panel, sandwich—a prefabricated panel that is a layered composite, formed by attaching two thin facings to a thicker core, for example, a precast-concrete panel consisting of two layers of concrete separated by a nonstructural insulating core.

panel, solid—a solid slab, usually of constant thickness.

panel strip—see **strip, panel**.

paper form—see **form, paper**.

parallel-wire unit—a post-tensioning tendon composed of a number of wires or strands that are approximately parallel.

parapet—the part of a wall that extends above the roof level; a low wall along the top of a dam.

parge—to coat with plaster, particularly foundation walls and rough masonry. (See also **back plastering**.)

partial prestressing—see **prestressing, partial**.

partial release—see **release, partial**.

particle, colloidal—an electrically charged particle, generally smaller than 0.1 mm, dispersed in a second continuous medium.

particle shape—the form of a particle. (See also **cubical piece [of aggregate]**; **elongated piece [of aggregate]**; and **flat piece [of aggregate]**.)

particle-size distribution—see **grading**.

parting agent—see **agent, release**.

pass—layer of shotcrete placed in one movement over the area of operation.

paste—see **cement paste, neat**.

paste, cement—binder of concrete and mortar consisting essentially of cement, water, hydration products and any admixtures together with very finely divided materials included in the aggregates. (See also **cement paste, neat**.)

paste content—proportional volume of cement paste in concrete, mortar, or the like, expressed as volume percent of the entire mixture. (See also **cement paste, neat**.)

paste volume—see **paste content**.

pat—a specimen of neat cement paste, approximately 3 in. (76 mm) in diameter and 1/2 in. (13 mm) in thickness at the center and tapering to a thin edge, on a flat glass plate for indicating setting time.

path of prestressing force—the locus of points defining the resultant effective prestress force in a concrete member.

pattern cracking—see **cracks, craze** and **cracking, map**.

pattern cracks—see **cracks, craze** and **cracking, map**.

patterned ashlar—see **masonry, ashlar**.

pavement (concrete)—a layer of concrete on such areas as roads, sidewalks, canals, playgrounds, and those used for storage or parking. (See also **pavement, rigid**.)

pavement, flexible—a pavement structure that maintains intimate contact with and distributes loads to the subgrade and depends on aggregate interlock, particle friction, and cohesion for stability; cementing agents, where used, are generally bituminous materials as contrasted to hydraulic cement in the case of rigid pavement. (See also **pavement, rigid**.)

pavement, rigid—pavement that will provide high bending resistance and distribute loads to the foundation over a comparatively large area.

paver, concrete—
1. a concrete mixer, usually mounted on crawler tracks, that mixes and places concrete pavement on the subgrade.
2. precast-concrete paving brick.

paving train—an assemblage of equipment designed to place and finish a concrete pavement.

pea gravel—see **gravel, pea**.

pedestal—an upright compression member whose height does not exceed three times its average least dimension, such as a short pier or plinth used as the base for a column.

pedestal pile—see **pile, pedestal**.

peeling—a process in which thin flakes of mortar are broken away from a concrete surface, such as by deterioration or by adherence of surface mortar to forms as forms are removed.

pencil rod—see **rod, pencil**.

penetration—an opening through which pipe, conduit, or other item passes through a wall or floor.

penetration probe—see **probe, penetration**.

penetration resistance—see **resistance, penetration**.

percent fines—the amount, expressed as a percentage, of material in aggregate finer than a given sieve, usually the 75 μm (No. 200); also the amount of fine aggregate in a concrete mixture expressed as a percent by absolute volume of the total amount of aggregate.

percentage of reinforcement—the ratio of cross-sectional area of reinforcing steel to the effective cross-sectional area of a member, expressed as a percentage.

periclase—a crystalline mineral, magnesia, MgO, the equivalent of which may be present in portland-cement clinker, portland cement, and other materials, such as open-hearth slags and certain basic refractories.

perimeter grouting—see **grouting, perimeter**.

period—
period, precuring—see **period, presteaming** (preferred term).
period, presteaming—in the manufacture of concrete products, the time between molding of a concrete product and start of the temperature-rise period.
period, soaking—in high-pressure and low-pressure steam curing, the time during which the live steam supply to the kiln or autoclave is shut off and the concrete products are exposed to the residual heat and moisture.
period, temperature-rise—the time interval during which the temperature of a concrete product rises at a controlled rate to the desired maximum in autoclave or atmospheric-pressure steam curing.

period at maximum temperature—see **maximum-temperature period**.

perlite—a volcanic glass having a perlitic structure, usually having a higher water content than obsidian; when expanded by heating, used as an insulating material and as a lightweight aggregate in concretes, mortars, and plasters.

perlitic structure—a structure produced in a homogeneous material by contraction during cooling and consisting of a system of irregular convolute and spheroidal cracks; generally confined to natural glass.

permanent form—see **form, permanent**.

permanent set—see **set, permanent**.

permeability to water, coefficient of—the rate of discharge of water under laminar flow conditions through a unit cross-sectional area of a porous medium under a unit hydraulic gradient and standard temperature conditions, usually 20 C.

pessimum—worst; the opposite of optimum.

petrography—the branch of petrology dealing with description and systematic classification of rocks aside from their geologic relations, mainly by laboratory methods, largely chemical and microscopical; also, loosely, petrology or lithology; also the techniques and knowledge of petrography applied to mortar, concrete, and the like.

petrology—the science of rocks, treating their origin, structure, composition, etc., from aspects and in all relations. (See also **petrography**.)

phenolic resin—see **resin, phenolic**.

phi (φ) factor—see **factor, strength-reduction** (preferred term).

Philleo factor—see **factor, Philleo**.

photometer, flame—an instrument used to determine elements (especially sodium and potassium in portland cement) by the color intensity of their unique flame spectra resulting from introducing a solution of a compound of the element into a flame. (Also known as flame spectrophotometer.)

pier—isolated foundation member of either plain or reinforced concrete.

pier, drilled—a concrete pier with or without a casing, cast-in-place in a hole previously bored in soil or rock. (See also **pile, cast-in-place**.)

pigment—a coloring matter, usually in the form of an insoluble fine powder.

pilaster—column built with a wall, usually projecting beyond the wall.

pilaster face—see **face, pilaster**.

pilaster side—see **side, pilaster**.

pile—a timber, concrete, or steel structural element, driven, jetted, or otherwise embedded on end in the ground for the purpose of supporting a load or compacting the soil. (See also **pile, composite**.)
pile, batter—a pile installed at an angle to the vertical; a raking pile or raker pile.
pile, bored—see **pier, drilled**.
pile, caisson—a cast-in-place pile made by driving a tube, excavating it, and filling the cavity with concrete.
pile, cast-in-place—a concrete pile concreted either with or without a casing in its permanent location, as distinguished from a precast pile. (See also **pier, drilled** and **pile, precast**.)
pile, composite—a pile made up of different materials, usually concrete and wood, or steel fastened together end to end, to form a single pile.
pile, concrete—see **pile, cast-in-place** and **pile, precast**.
pile, drilled—see **pier, drilled**.

pile

pile, friction—a load-bearing pile that receives its principal vertical support from skin friction between the surface of the buried pile and the surrounding soil.

pile, pedestal—a cast-in-place concrete pile constructed so that concrete is forced out into a widened bulb or pedestal shape at the foot of the pipe that forms the pile.

pile, pipe—a steel cylinder, usually between 10 and 24 in. (250 and 600 mm) in diameter, generally driven with open ends to firm bearing and then excavated and filled with concrete.

pile, precast—a reinforced pile manufactured in a casting plant or at the site but not in its final position. (See also **pile, cast-in-place**.)

pile, raking—see **pile, batter** (preferred term).

pile, sheet—a pile in the form of a plank driven in close contact or interlocking with others to provide a tight wall to resist the lateral pressure of water, adjacent earth, or other materials; may be tongued and grooved if made of timber or concrete and interlocking if made of metal.

pipe, vent—a small-diameter pipe used in concrete construction to permit escape of air in a structure being concreted or grouted.

pile, wing—a bearing pile, usually of concrete, widened in the upper portion to form part of a sheet pile wall.

pile bent—see **bent, pile**.

pile cap—see **cap, pile**.

pipe column—see **column, pipe**.

pipe pile—see **pile, pipe**.

pitting—development of relatively small cavities in a surface; in concrete, localized disintegration, such as a popout; in steel, localized corrosion evident as minute cavities on the surface.

placeability—see **workability**.

placement—the process of placing and consolidating concrete; a quantity of concrete placed and finished during a continuous operation; inappropriately referred to as pouring.

placing—the deposition, distribution, and consolidation of freshly mixed concrete in the place where it is to harden; inappropriately referred to as pouring.

plain bar—see **bar, plain**.

plain concrete—see **concrete, plain**.

plain masonry—see **masonry, plain**.

plane of weakness—the plane along which a body under stress will tend to fracture; may exist by design, by accident, or because of the nature of the structure and its loading.

plaster—a cementitious material or combination of cementitious material and fine aggregate that, when mixed with a suitable amount of water, forms a plastic mass or paste that when applied to a surface, adheres to it and subsequently hardens, preserving in a rigid state the form or texture imposed during the period of plasticity; also the placed and hardened mixture. (See also **stucco**.)

plaster, neat—plaster devoid of sand.

plaster mold—see **mold, plaster**.

plaster of paris—$CaSO_4 \cdot 1/2H_2O$; gypsum, from which 3/4 of the chemically bound water has been driven off by heating; when wetted it recombines with water and hardens quickly. (See also **hemihydrate**.)

plastic—possessing plasticity, or possessing adequate plasticity. (See also **plasticity**.)

plastic cement—see **cement, plastic**.

plastic centroid—centroid of the resistance to load computed for the assumptions that the concrete is stressed uniformly to 85% of its design strength, and the steel is stressed uniformly to its specified yield point.

plastic consistency—see **consistency, plastic**.

plastic cracking—see **cracking, plastic**.

plastic deformation—see **deformation, inelastic**.

plastic flow—obsolete term for creep and stress relation. (See also **creep**; **flow, plastic**; and **stress relaxation**.)

plastic hinge—see **hinge, plastic**.

plastic limit—see **limit, plastic**.

plastic loss—see **creep**.

plastic mortar—see **mortar, plastic**.

plastic or bond fire clay—a fire clay of sufficient natural plasticity to bond nonplastic material; a fire clay used as a plasticizing agent in mortar.

plastic shrinkage—see **shrinkage, plastic**.

plastic shrinkage cracks—see **cracking, plastic**.

plasticity—a complex property of a material involving a combination of qualities of mobility and magnitude of yield value; the property of freshly mixed cement paste, concrete, or mortar that determines its resistance to deformation or ease of molding.

plasticity index—see **index, plasticity**.

plasticize—to produce plasticity or to render plastic.

plasticizer—a material that increases the plasticity of a fresh cement paste, mortar, or concrete.

plate—
1. in formwork for concrete: a flat, horizontal member either at the top or bottom, or both, of studs or posts; a mud sill if on the ground (see also **mud sill**); and
2. in structural design: a member, the depth of which is substantially less than its length and width. (See also **plate, flat** and **load-transfer assembly**.)

plate, deformed—a flat piece of metal, thicker than 1/4 in. (6 mm), having horizontal deformations or corrugations; used in construction to form a vertical joint and provide a mechanical interlock between adjacent sections.

plate, flat—a flat slab without column capitals or drop panels. (See also **slab, flat**.)

plate, folded—
1. a framing assembly composed of sloping slabs in a hipped or gabled arrangement; and
2. prismatic shell with open polygonal section.

plum—a large random-shaped stone dropped into freshly placed mass concrete to economize on the amount of the other concrete ingredients. (See also **concrete, cyclopean**.)

plumb—vertical or to make vertical.

pneumatic feed—see **feed, pneumatic**.

pneumatically applied mortar—see **shotcrete**.

point count method—method for determination of the volumetric composition of a solid by observation of the frequency with which areas of each component coincide with a regular system of points in one or more planes intersecting a sample of the solid. (See also **linear-traverse method**.)

point count method (modified)—the point count method supplemented by a determination of the frequency with which areas of each component of a solid are intersected by regularly spaced lines in one or more planes intersecting a sample of the solid.

point load—see **load, point**.

point of contraflexure—see **point of inflection** (preferred term).

point of inflection—the point on the length of a structural member subjected to flexure where the curvature changes from concave to convex or conversely and at which the bending moment is zero; also called "point of contraflexure."

Poisson's ratio—see **ratio, Poisson's**.

polarizing microscope—see **microscope, polarizing**.

pole shore—see **shore, post**.

polish or final grind—the final operation in which fine abrasives are used to hone a surface to its desired smoothness and appearance.

polyester—one of a large group of synthetic resins, mainly produced by reaction of dibasic acids with dihydroxy alcohols; commonly prepared for application by mixing with a vinyl-group monomer and free-radical catalysts at ambient temperatures and used as binders for resin mortars and concretes, fiber laminates (mainly glass), adhesives, and the like. (See also **concrete, polymer**.)

polyethylene—a thermoplastic high-molecular-weight organic compound used in formulating protective coatings or, in sheet form, as a protective cover for concrete surfaces during the curing period, or to provide a temporary enclosure for construction operations.

polymer—the product of polymerization; more commonly a rubber or resin consisting of large molecules formed by polymerization.

polymer concrete—see **concrete, polymer**.

polymer-cement concrete—see **concrete, polymer-cement**.

polymerization—the reaction in which two or more molecules of the same substance combine to form a compound containing the same elements and in the same proportions but of higher molecular weight.

polystyrene resin—see **resin, polystyrene**.

polysulfide coating—see **coating, polysulfide**.

polyurethane—reaction product of an isocyanate with any of a wide variety of other compounds containing an active hydrogen group; used to formulate tough, abrasion-resistant coatings.

polyvinyl acetate—colorless, permanently thermoplastic resin; usually supplied as an emulsion or water-dispersible powder characterized by flexibility, stability towards light, transparency to ultraviolet rays, high dielectric strength, toughness, and hardness; the higher the degree of polymerization, the higher the softening temperature; may be used in paints for concrete.

polyvinyl chloride—a synthetic resin prepared by the polymerization of vinyl chloride, used in the manufacture of nonmetallic waterstops for concrete.

ponding—the creation and maintaining of a shallow pond of water on the surface of a concrete slab to assist curing; accidental or incidental occurrence of a shallow pond or ponds on a nominally flat surface of concrete; a condition in which a horizontal slab deforms downward between supports.

popcorn concrete—see **concrete, popcorn**.

popout—the breaking away of small portions of a concrete surface due to localized internal pressure that leaves a shallow, typically conical, depression; small popouts leave holes up to 0.4 in (10 mm) in diameter; medium popouts leave holes 0.4 to 2 in. (10 to 50 mm) in diameter; and large popouts leave holes greater than 2 in. (50 mm) in diameter.

porosity—the ratio, usually expressed as a percentage of the volume of voids in a material to the total volume of the material including the voids.

portland blast-furnace slag cement—see **cement, portland blast-furnace slag**.

portland cement—see **cement, portland**.

portland-cement clinker—see **clinker, portland-cement**.

portland-cement concrete—see **concrete**.

portland-pozzolan cement—see **cement, portland-pozzolan**.

portlandite—the mineral, calcium hydroxide ($Ca(OH)_2$); occurs naturally in Ireland; equivalent to a product of hydration of portland cement.

porous fill—see **drainage fill**.

positive displacement—wet-mix shotcrete delivery equipment in which the material is pushed through the material hose in a solid mass by a piston or auger.

positive moment—see **moment, positive**.

positive reinforcement—see **reinforcement, positive**.

post—vertical formwork member used as a support; also known as shore, prop, or jack.

post shore—see **shore, post**.

post-tensioning—a method of prestressing reinforced concrete in which tendons are tensioned after the concrete has hardened.

post-tensioning, bonded—post-tensioned construction in which the annular spaces around the tendons are grouted after stressing, thereby bonding the tendon to the concrete section.

pot life—time interval after preparation during which a liquid or plastic mixture is to be used.

pouring (of concrete)—see **placement** and **placing**.

power float—see **float, rotary** (preferred term).

Powers' spacing factor—see **factor, Powers' spacing** (preferred term).

pozzolan—a siliceous or siliceous and aluminous material that in itself possesses little or no cementitious value but that will, in finely divided form and in the presence of moisture, chemically react with calcium hydroxide at or-

pozzolan

dinary temperatures to form compounds having cementitious properties; there are both natural and artificial pozzolans.

pozzolan, artificial—materials such as fly ash and silica fume. (See also **fly ash**, and **silica fume**.)

pozzolan, natural—a raw or calcined natural material that has pozzolanic properties (for example, volcanic tuffs or pumicites, opaline cherts and shales, clays, and diatomaceous earths).

pozzolanic—of or pertaining to a pozzolan.

pozzolanic-activity index—see **index, pozzolanic-activity**.

pozzolanic reaction—see **pozzolan**.

preblended grout—see **grout, preblended**.

precast—a concrete member that is cast and cured in other than its final position; the process of placing and finishing precast concrete. (See also **cast-in-place**.)

precast concrete—see **concrete, precast**.

precast pile—see **pile, precast**.

precompressed zone—see **zone, precompressed**.

precuring period—see **period, presteaming** (preferred term).

prefire—to raise the temperature of refractory concrete under controlled conditions before placing it in service.

preformed foam—see **foam, preformed**.

premature stiffening—see **set, false** and **set, flash**.

prepacked concrete—see **concrete, preplaced-aggregate**.

preplaced-aggregate concrete—see **concrete, preplaced-aggregate** and **concrete, colloidal**.

pre-post-tensioning—a method of fabricating prestressed concrete in which some of the tendons are pretensioned and a portion of the tendons are post-tensioned.

preservation—the process of maintaining a structure in its present condition and arresting further deterioration. (See also **rehabilitation**; **repair**; and **restoration**.)

preset period—see **period, presteaming** (preferred term).

preshrunk concrete (mortar, grout)—see **concrete (mortar, grout), preshrunk**.

pressed edge—see **edge, pressed**.

pressure—

pressure, form—lateral pressure acting on vertical or inclined formed surfaces, resulting from the fluid-like behavior of the unhardened concrete confined by the forms.

pressure, lateral—see **pressure, form**.

pressure line—locus of force points within a structure resulting from combined prestressing force and externally applied load.

presteaming period—see **period, presteaming**.

prestress—to place a hardened concrete member or an assembly of units in a state of compression before application of service loads; the stress developed by prestressing, such as by pretensioning or post-tensioning. (See also **concrete, prestressed**; **steel, prestressing**; **pretensioning**; and **post-tensioning**.)

prestress, effective—the prestressing force at a specific location in a prestressed-concrete member under the effects of service dead load or total service load after losses of prestress have occurred.

prestress, final—see **stress, final**.

prestress, initial—the prestressing stress (or force) applied to the concrete at the time of stressing.

prestress, transverse—prestress that is applied at right angles to the longitudinal axis of a member or slab.

prestressed concrete—see **concrete, prestressed**.

prestressing, nonsimultaneous—the post-tensioning of tendons individually rather than simultaneously.

prestressing, partial—prestressing to a stress level such that, under design loads, tensile stresses exist in the precompressed tensile zone of the prestressed member.

prestressing steel—see **steel, prestressing**.

pretensioning—a method of prestressing reinforced concrete in which the tendons are tensioned before the concrete has hardened.

pretensioning bed (or bench)—the casting bed on which pretensioned members are manufactured and which resists the pretensioning force prior to release.

primary crusher—see **crusher, primary**.

primary nuclear vessel—interior container in a nuclear reactor designed for sustained loads and for working conditions.

principal planes—see **stress, principal**.

principal stress—see **stress, principal**.

probabilistic design—see **design, probabilistic**.

probe, penetration—a device for obtaining a measure of the resistance of concrete to penetration; customarily determined by the distance that a steel pin is driven into the concrete from a special gun by a precisely measured explosive charge.

process—

process, centrifugal—a process for producing concrete products, such as pipe, that uses an outer form that is rotated about a horizontal axis and into which concrete is fed by a conveyor; also called spinning process. (See also **concrete, centrifugally cast**; **process, dry-cast**; **packerhead**; **process, tamp**; and **process, wet-cast**.)

process, dry—in the manufacture of cement, the process in which the raw materials are ground, conveyed, blended, and stored in a dry condition. (See also **process, wet**.)

process, dry-cast—a process for producing concrete products, such as pipe, using low-frequency high-amplitude vibration to consolidate dry-mix concrete in the form. (See also **centrifugal process**; **process, packerhead**; **tamp process**; **process, wet-cast**.)

process, dry-tamp—see **packing, dry**.

process, packerhead—a process for producing concrete pipe that uses a rotating device that forms the interior surface of the pipe as concrete is fed into the form from above. (See also **centrifugal process**; **process, dry-cast**; **tamp process**; **process, wet-cast**.)

process, tamp—a process for producing concrete products, such as pipe, that uses direct mechanical action to consolidate the concrete by the action of tampers that rise automatically as the form is rotated and filled with concrete from above. (See also **process, centrifugal**; **process, dry-cast**; **process, packerhead**; and **process, wet-cast**.)

process, wet-cast—a process for producing concrete items, such as pipe, that uses concrete having a measurable slump, generally placed from above, and consolidated by vibration. (See also **centrifugal process**; **process, dry-cast**; **process, packerhead**; and **tamp process**.)

process, wet—in the manufacture of cement, the process in which the raw materials are ground, blended, mixed, and pumped while mixed with water; the wet process is chosen where raw materials are extremely wet and sticky which would make drying before crushing and grinding difficult. (See also **process, dry**.)

promoter—see **catalyst** (preferred term).

promoter, flow—substance added to coating to enhance brushability, flow, and leveling.

proof stress—see **stress, proof**.

prop—see **post** and **shore**.

proportional limit—see **limit, proportional**.

proportion—to select proportions of ingredients to make the most economical use of available materials to produce mortar or concrete of the required properties. (See also **mixture**.)

protected paste volume—the portion of hardened cement paste that is protected from the effects of freezing by proximity to an entrained air void. (See also **factor, Philleo** and **factor, spacing**.)

protection period—the required time during which the concrete is maintained at or above a specific temperature to prevent freezing of the concrete or ensure the necessary strength of development.

proving ring—see **ring, proving**.

psychrometer, sling— a psychrometer containing independently matched dry- and wet-bulb thermometers, suitably mounted for manually swinging through the ambient air, to simultaneously indicate dry- and wet-bulb temperatures.

pugmill—see **mixer, horizontal-shaft** (preferred term).

pulse velocity—see **velocity, pulse**.

pulverized-fuel ash (pfa)—see **fly ash** (preferred term in the U.S.; pulverized-fuel ash is used in the UK).

pumice—a highly porous and vesicular lava usually of relatively high silica content composed largely of glass drawn into approximately parallel or loosely entwined fibers, which themselves contain sealed vesicles.

pumicite—naturally occurring finely divided pumice and glass shards.

pump, concrete—an apparatus that forces concrete to the placing position through a pipeline or hose.

pumped concrete—see **concrete, pumped**.

pumping (of pavements)—the ejection of water, or water and solid materials, such as clay or silt, along transverse or longitudinal joints and cracks, and along pavement edges caused by downward slab movement activated by the passage of loads over the pavement after the accumulation of free water on or in the base course, subgrade, or subbase.

punching shear—failure of a base or slab when a heavily loaded column punches a hole through it.

punching shear stress—shear stress calculated by diving the load on the slab that is transferred to the column by the product of the perimeter and the thickness of the base or cap or by the product of the perimeter taken at 1/2 the slab thickness away from the column and the thickness of the base or cap.

punning—an obsolete term designating a light form of ramming. (See also **ramming** and **tamping**.)

purlin—in roofs, a horizontal member supporting the common rafters. (See also **beam**.)

putty—a plaster composed of quicklime or hydrated lime and water with or without plaster of paris or sand.

pyrite—a mineral, iron disulfide (FeS_2), that, if it occurs in aggregate used in concrete, can cause popouts and dark brown or orange-colored staining.

pycnometer—a vessel for determination of specific gravity of liquids or solids.

pyrometric cone—see **cone, pyrometric**.

pyrometric-cone equivalent (PCE)—the number of that cone whose tip would touch the supporting plaque simultaneously with that of a cone of the refractory material being investigated when tested in accordance with a specified procedure such as ASTM C 24.

Q

quality assurance—actions taken by an owner or representative to provide and document assurance that what is being done and what is being provided are in accordance with the applicable standards of good practice and following the contract documents for the work.

quality control—actions taken by a producer or contractor to provide and document control over what is being done and what is being provided so that the applicable standards of good practice and the contract documents for the work are followed.

quicklime—calcium oxide (CaO).

quick set—see **stiffening, early** (preferred term).

R

R-value—see **resistance, thermal**.

raft foundation—see **foundation, raft**.

rail-steel reinforcement—see **reinforcement, rail-steel**.

rake classifier—machine for separating coarse and fine particles of granular material temporarily suspended in water; the coarse particles settle to the bottom of a vessel and are scraped up an incline by a set of blades, the fine particles remaining in suspension to be carried over the edge of the classifier.

raker—a sloping brace for a shore head.

raked joint—see **joint, raked**.

raker pile—see **pile, batter** (preferred term).

raking

raking pile—see **pile, batter** (preferred term).
ramming—a form of heavy tamping of concrete, grout, or the like by means of a blunt tool forcibly applied. (See also **pack, dry**; **punning**; and **tamping**.)
random ashlar—see **masonry, ashlar** (preferred term).
ranger—see **wale** (preferred term).
ratio, A/F—the molar or mass ratio of aluminum oxide (Al_2O_3) to iron oxide (Fe_2O_3), as in portland cement.
ratio, aggregate-cement—the ratio of cement to total aggregate, either by mass or volume.
ratio, Poisson's—the absolute value of the ratio of transverse (lateral) strain to the corresponding axial (longitudinal) strain resulting from uniformly distributed axial stress below the proportional limit of the material; the value will average approximately 0.2 for concrete and 0.25 for most metals.
raw mix—blend of raw materials, ground to desired fineness, correctly proportioned, and blended ready for burning; such as that used in the manufacture of cement clinker.
Rayleigh wave—an ultrasonic surface wave in which the particle motion is elliptical and effective penetration is approximately one wavelength.
reaction—
  reaction, alkali-aggregate—chemical reaction in either mortar or concrete between alkalies (sodium and potassium) from portland cement or other sources and certain constituents of some aggregates; under certain conditions, deleterious expansion of concrete or mortar may result.
  reaction, alkali-carbonate rock—the reaction between the alkalies (sodium and potassium) in portland cement and certain carbonate rocks, particularly calcitic dolomite and dolomitic limestones, present in some aggregates; the products of the reaction may cause abnormal expansion and cracking of concrete in service.
  reaction, alkali-silica—the reaction between the alkalies (sodium and potassium) in portland cement and certain siliceous rocks or minerals, such as opaline chert, strained quartz, and acidic volcanic glass, present in some aggregates; the products of the reaction may cause abnormal expansion and cracking of concrete in service.
  reaction, endothermic—a chemical reaction that occurs with the absorption of heat.
  reaction, exothermic—a chemical reaction that occurs with the evolution of heat.
  reaction, pozzolanic—see **pozzolan**.
  reaction, subgrade—see **contact pressure** and **coefficient of subgrade reaction**.
reactive aggregate—see **aggregate, reactive**.
reactive silica material—several types of materials that react at high temperatures with portland cement or lime during autoclaving, includes pulverized silica, natural pozzolan, and fly ash.
reactivity (of aggregate), alkali—susceptibility of aggregate to alkali-aggregate reaction.

ready-mixed concrete—see **concrete, ready-mixed**.
rebar—colloquial term for reinforcing bar. (See also **reinforcement**.)
rebound—aggregate and cement, or wet shotcrete, that bounces away from the surface against which shotcrete is being projected.
rebound hammer—see **hammer, rebound**.
recycled concrete—see **concrete, recycled**.
refractories—materials, usually nonmetallic, used to withstand high temperatures.
refractoriness—in refractories, the property of being resistant to softening or deformation at high temperatures.
refractory—resistant to high temperatures.
  refractory, castable—a packaged, dry mixture of hydraulic cement, generally calcium-aluminate cement, and specially selected and proportioned refractory aggregates that, when mixed with water, will produce refractory concrete or mortar.
  refractory, neutral—a refractory that is resistant to chemical attack by either acidic or basic substances.
refractory aggregate—see **aggregate, refractory**.
refractory concrete—see **concrete, refractory**.
refractory-insulating concrete—see **concrete, refractory-insulating**.
reglet—a groove in a wall to receive flashing.
regulated-set cement—see **cement, regulated-set**.
rehabilitation—the process of repairing or modifying a structure to a desired useful condition. (See also **preservation**; **repair**; and **restoration**.)
reinforced concrete—see **concrete, reinforced**.
reinforced masonry—see **masonry, reinforced**.
reinforcement—bars, wires, strands, or other slender members that are embedded in concrete in such a manner that they and the concrete act together in resisting forces.
  reinforcement, auxiliary—in a prestressed member, any reinforcement in addition to that participating in the prestressing function.
  reinforcement, axle-steel—either plain or deformed reinforcing bars rolled from axle steel.
  reinforcement bar—see **reinforcement**.
  reinforcement, cold-drawn wire—steel wire made from rods that have been hot rolled from billets cold-drawn through a die; for concrete reinforcement of a diameter not less than 0.080 in. (2 mm) nor greater than 0.625 in. (16 mm).
  reinforcement, cold-worked steel—steel bars or wires that have been rolled, twisted, or drawn at normal ambient temperatures.
  reinforcement, compression—reinforcement designed to carry compressive stresses. (See also **stress**.)
  reinforcement, corner—metal reinforcement for plaster at reentrant corners to provide continuity between two intersecting planes; or concrete reinforcement used at wall intersections or near corners of square or rectangular openings in walls, slabs, or beams.
  reinforcement, crack-control—reinforcement in concrete construction designed to minimize opening of

cracks, often effective in limiting them to uniformly distributed small cracks.

reinforcement, curtain—a mat of orthogonal reinforcing steel in a member such as a wall; known as a double curtain (of reinforcement) when a mat is at each face.

reinforcement, deformed—metal bars, wire, or fabric with a manufactured pattern of surface ridges that provide a locking anchorage with surrounding concrete.

reinforcement, distribution-bar—small diameter bars, usually at right angles to the main reinforcement, intended to spread a concentrated load on a slab and to prevent cracking.

reinforcement, dowel-bar—see **dowel**.

reinforcement, edge-bar—tension steel sometimes used to strengthen otherwise inadequate edges in a slab without resorting to edge thickening.

reinforcement, expanded-metal fabric—see **lath, expanded-metal**.

reinforcement, four-way—a system of reinforcement in flat-slab construction comprising bands of bars parallel to two adjacent edges and also to both diagonals of a rectangular slab.

reinforcement, heavy-edge—wire-fabric reinforcement for highway pavement slabs having one to four edge wires heavier than the other longitudinal wires.

reinforcement, helical—steel reinforcement of hot-rolled bar or cold-drawn wire fabricated into a helix (more commonly known as spiral reinforcement).

reinforcement, high-strength—see **steel, high-strength**.

reinforcement, hoop—a one-piece closed tie or continuously wound tie not less than No. 3 in size, the ends of which have a standard 135 degree bend with a ten-bar diameter extension, that encloses the longitudinal reinforcement.

reinforcement, lateral—transverse reinforcement, usually applied to ties, hoops, and spirals in columns or column-like members.

reinforcement, longitudinal—reinforcement parallel to the length of a concrete member or pavement.

reinforcement, mesh—see **fabric, welded-wire** and **reinforcement, welded-wire fabric**.

reinforcement, negative—steel reinforcement for negative moment.

reinforcement, nonprestressed—reinforcing steel, not subjected to either pretensioning or post-tensioning.

reinforcement, nonstructural—see **reinforcement, temperature**.

reinforcement, positive—reinforcement for positive moment.

reinforcement, rail-steel—reinforcing bars hot-rolled from standard T-section rails.

reinforcement, shear—reinforcement designed to resist shear or diagonal tension stresses. (See also **dowel**.)

reinforcement, shrinkage—reinforcement designed to resist shrinkage stresses in concrete.

reinforcement, spiral—continuously wound reinforcement in the form of a cylindrical helix. (See also **reinforcement, helical**.)

reinforcement, temperature—reinforcement designed to carry stresses resulting from temperature changes; also the minimum reinforcement for areas of members that are not subjected to primary stresses or necessarily to temperature stresses.

reinforcement, tension—reinforcement designed to carry tensile stresses such as those in the bottom of a simple beam.

reinforcement, transverse—reinforcement at right angles to the longitudinal reinforcement.

reinforcement, twin-twisted bar—two bars of the same nominal diameter twisted together.

reinforcement, two-way—reinforcement arranged in bands of bars at right angles to each other.

reinforcement, web—reinforcement placed in a concrete member to resist shear and diagonal tension.

reinforcement, welded—reinforcement joined together by welding.

reinforcement, welded-wire fabric—welded-wire fabric in either sheets or rolls, used to reinforce concrete.

reinforcement, woven-wire—see **fabric, welded-wire** (preferred term).

reinforcement displacement—movement of reinforcing steel from its specified position in the forms.

reinforcement ratio—ratio of the effective area of the reinforcement to the effective area of the concrete at any section of a structural member. (See also **percentage of reinforcement**.)

relative humidity—the ratio of the quantity of water vapor actually present to the amount present in a saturated atmosphere at a given temperature; expressed as a percentage.

release agent—see **agent, release**.

release, partial—release into a prestressed-concrete member of a portion of the total prestress initially held wholly in the prestressed reinforcement.

remoldability—the readiness with which freshly mixed concrete responds to a remolding effort such as jigging or vibration, causing it to reshape its mass around reinforcement and to conform to the shape of the form. (See also **flow**.)

remolding test—see **test, remoldability**.

render—to apply a coat of mortar by a trowel or float.

repair—to replace or correct deteriorated, damaged, or faulty materials, components, or elements of a structure. (See also **preservation**; **rehabilitation**; and **restoration**.)

repeatability—variability among replicate test results obtained on the same material within a single laboratory by one operator; a quantity that will be exceeded in only about 5% of the repetitions by the difference, taken in absolute value, of two randomly selected test results obtained in the same laboratory on a given material; in use of the term, variable factors should be specified.

repost—see **reshoring**.

reproducibility—variability among replicate test results obtained on the same material in different laboratories; a quantity that will be exceeded in only approximately 5% of the repetitions by the difference, taken in absolute value, of two single test results made on the same material in two different, randomly selected laboratories; in use of the term, variable factors should be specified.

required strength—see **strength, required**.

resetting (of forms)—setting of forms separately for each successive lift of a wall to avoid offsets at construction joints.

reshoring—the construction operation in which the original shoring or posting is removed and replaced in such a manner as to avoid deflection of the shored element or damage to partially cured concrete.

residual deformation—see **creep, nonrecoverable**.

resilience—the work done per unit volume of a material in producing strain.

resin—a natural or synthetic, solid or semisolid, organic material of indefinite and often high molecular weight having a tendency to flow under stress, usually has a softening or melting range, and usually fractures conchoidally.

    resin, acrylic—one of a group of thermoplastic resins formed by polymerizing the esters or amides of acrylic acid used to make polymer-modified concrete and polymer concretes; also used in concrete construction as a bonding agent, surface sealer, or an integral concrete component.

    resin, phenolic—a class of synthetic, oil-soluble resins (plastics) produced as condensation products of phenol, substituted phenols and formaldehyde, or some similar aldehyde that may be used in paints for concrete.

    resin, polystyrene—synthetic resins, varying from colorless to yellow, formed by the polymerization of styrene on heating with or without catalysts, that may be used in paints for concrete, or for making sculptured molds, or as insulation.

resin concrete—see **concrete, polymer** (preferred term).

resin mortar—see **concrete, polymer**.

resins, epoxy—a class of organic chemical bonding systems used in the preparation of special coatings or adhesives for concrete or as binders in epoxy-resin mortars and concretes.

resistance refractory aggregate—see **aggregate, refractory**.

    resistance, abrasion—ability of a surface to resist being worn away by rubbing and friction.

    resistance, fire—the property of a material or assembly to withstand fire or give protection from it; as applied to elements of buildings, it is characterized by the ability to confine a fire or, when exposed to fire, to continue to perform a given structural function, or both.

    resistance, penetration—the resistance, usually expressed in lb/in.^{2} (psi) or megapascals (MPa), of either mortar or cement paste to penetration by a plunger or needle under standard conditions, such as to determine time of setting.

    resistance, skid—a measure of the frictional characteristics of a surface.

    resistance, sulfate—ability of concrete or mortar to withstand sulfate attack. (See also **sulfate attack**.)

    resistance, thermal—the reciprocal of thermal conductance expressed by the symbol R.

restoration—the process of re-establishing the materials, form, and appearance of a structure to those of a particular era of the structure. (See also **preservation**; **rehabilitation**; and **repair**.)

restraint (of concrete)—restriction of free movement of fresh or hardened concrete following completion of placing in formwork or molds or within an otherwise confined space; restraint can be internal or external and may act in one or more directions.

retardation—reduction in the rate of either hardening, setting, or both, that is, an increase in the time required to reach time of initial and final setting or to develop early strength of fresh concrete, mortar, or grout. (See also **retarder**.)

retarder—an admixture that delays the setting of cement paste and mixtures, such as mortar or concrete, containing cement. (See also **admixture, retarding**.)

retarder, surface—a retarder applied to the contact surface of a form or to the surface of newly placed concrete to delay setting of the cement, to facilitate construction joint cleanup, or to facilitate production of exposed-aggregate finish.

retarding admixture—see **admixture, retarding**.

retemper—to add water and remix concrete or mortar to restore workability to a condition in which the mixture is placeable or usable. (See also **temper**.)

reveal (n.)—the vertical surface forming the side of an opening in a wall, as for a window or door; depth of exposure of aggregate in an exposed aggregate finish. (See also **exposed-aggregate finish**.)

revibration—one or more applications of vibration to fresh concrete after completion of placing and initial consolidation but preceding initial setting of the concrete.

revolving-blade (or paddle) mixer—see **mixer, open-top**.

rheology—the science dealing with flow of materials, including studies of deformation of hardened concrete, the handling and placing of freshly mixed concrete, and the behavior of slurries, pastes, and the like.

rib—one of a number of parallel structural members backing sheathing; the portion of a T-beam which projects below the slab; in deformed reinforcing bars, the deformations or the longitudinal parting ridge.

ribbed panel—see **panel, ribbed**.

ribbed slab—see **panel, ribbed**.

ribbon—a narrow strip of wood or other material used in formwork.

ribbon loading—see **loading, ribbon**.

rich concrete—see **concrete, rich**.

rich mixture—a concrete mixture containing a high proportion of cement.

rider cap—see **cap, pile**.
rigid frame—see **frame, rigid**.
rigid pavement—see **pavement, rigid**.
rigidity, flexural—a measure of stiffness of a member, indicated by the product of modulus of elasticity and moment of inertia divided by the length of the member.
ring, air—perforated manifold in nozzle of wet-mix shotcrete equipment through which high pressure air is introduced into the material flow.
ring, proving—a device for calibrating load indicators of testing machines, consisting of a calibrated elastic ring and a mechanism or device for indicating the magnitude of deformation under load.
rock pocket—a porous, mortar-deficient portion of hardened concrete consisting primarily of coarse aggregate and open voids; caused by leakage of mortar from the form, separation (segregation) during placement, or insufficient consolidation. (See also **honeycomb**.)
rod—sharp-edged cutting screed used to trim shotcrete to forms or ground wires. (See also **screed**.)
  rod, dowel—see **dowel** (preferred term).
  rod, pencil—plain metal rod of about 1/4 in. (6 mm) diameter.
  rod, tamping—a straight steel rod of circular cross-section and having one or both ends rounded to a hemispherical tip.
  rod, tie—see **tie, form** and **tieback**.
rodability—the susceptibility of fresh concrete or mortar to consolidation by means of a tamping rod.
rod buster (colloquial)—one who installs reinforcement for concrete.
rodding—consolidation of concrete by means of a tamping rod. (See also **rod**; **rodability**; and **tamping**.)
rodding, dry—in measurement of the mass per unit volume of coarse aggregates, the process of consolidating dry material in a calibrated container by rodding under standardized conditions.
rod mill—see **mill, rod**.
roller-compacted concrete—see **concrete, roller-compacted**.
roller compaction—a process for compacting concrete using a roller, often a vibratory roller.
rolling—the use of heavy metal or stone rollers on terrazzo topping to extract excess matrix.
Roman cement—see **cement, Roman**.
roof, barrel-vault—a thin concrete roof in the form of a part of a cylinder.
roof insulation—see **insulation, roof**.
room, fog—see **moist room** (preferred term).
Rosiwal method—see **linear-traverse method**.
rotary float (also called power float)—see **float, rotary**.
rotary kiln—see **kiln, rotary**.
rough grind—the initial operation in which coarse abrasives are used to reduce the projecting stone chips in hardened terrazzo down to a level surface.
rout—to deepen and widen a crack to prepare it for patching or sealing.
rub brick—see **brick, rubbing** (preferred term).
rubbing brick—see **brick, rubbing**.
rubbed finish—see **finish, rubbed**.
rubber set—see **set, false** (preferred term).
rubble—rough stones of irregular shape and size, broken from larger masses by geological processes or by quarrying; concrete reduced to irregular fragments, as by demolition or natural catastrophe.
rubble concrete—see **concrete, rubble**.
runway—decking over the area of concrete placement, usually of movable panels and supports, on which buggies of concrete travel to points of placement.
rupture modulus—see **modulus of rupture**.
rupture strength—see **modulus of rupture**.
rustic or washed finish—see **finish, rustic or washed**.
rustication—a groove in a concrete surface.
rustication strip—see **strip, rustication**.

S

sack—see **bag (of cement)** (preferred term).
sack rub—a finish for formed concrete surfaces, designed to produce even texture and fill pits and air holes; after dampening the surface, mortar is rubbed over the surface, then, before the surface dries, a mixture of dry cement and sand is rubbed over it with either a wad of burlap or a sponge-rubber float to remove surplus mortar and fill voids. (See also **surface air voids** and **finish, rubbed**.)
safe leg load—see **load, safe leg**.
sagging—see **sloughing** (preferred term).
salamander—a portable source of heat, customarily oil-burning, used to heat an enclosure around or over newly placed concrete to prevent the concrete from freezing.
sample—either a group of units or portion of material taken, respectively, from a larger collection of units or a larger quantity of material, that serves to provide information that can be used as a basis for action on the larger collection or quantity or on the production process; the term is also used in the sense of a sample of observations.
sample, composite—sample obtained by blending two or more individual samples of a material.
sampling, continuous—sampling without interruptions throughout an operation or for a predetermined time.
sampling, intermittent—sampling successively for limited periods of time throughout an operation or for a predetermined period of time; the duration of sampling periods and the intervals are not necessarily regular and are not specified.
sampling plan—
  1. a procedure that specifies the number of units of product from a lot that is to be inspected to establish acceptability of the lot; and
  2. a prearranged program stipulating locations and procedures for securing samples of a material for testing purposes, for example, as concrete in construction or aggregates in a quarry, pit, or stockpile.

sand

sand—
1. granular material passing the 9.5 mm (3/8 in.) sieve and almost entirely passing the 4.75 mm (No. 4) sieve and predominantly retained on the 75 μm (No. 200) sieve, and resulting either from natural disintegration and abrasion of rock or processing of completely friable sandstone; and
2. that portion of an aggregate passing the 4.75 mm (No. 4) sieve and predominantly retained on the 75 μm (No. 200) sieve, and resulting either from natural disintegration and abrasion of rock or processing of completely friable sandstone. (See also **aggregate, fine**.)

Note: the definitions are alternatives to be applied under differing circumstances. Definition 1 is applied to an entire aggregate either in a natural condition or after processing. Definition 2 is applied to a portion of an aggregate. Requirements for properties and grading should be stated in the specifications. Fine aggregate produced by crushing rock, gravel, or slag is commonly known as manufactured sand.

sand, graded standard—see **sand, standard**.

sand, manufactured—see **sand**.

sand, natural—sand resulting from natural disintegration and abrasion of rock. (See also **sand** and **aggregate, fine**.)

sand, sharp—coarse sand consisting of particles of angular shape.

sand, standard—silica sand, composed almost entirely of naturally rounded grains of nearly pure quartz, used for preparing mortars in the testing of hydraulic cements.

Note: standard sand is produced in two gradings.
1. *20-30 sand*—standard sand, predominantly graded to pass a 850 μm (No. 20) sieve and be trained on a 600 μm (No. 30) sieve and the 150 μm (No. 100) sieve.
2. *graded sand*—standard sand, predominantly graded between the 600 μm (No. 30) sieve and the 150 μm (No. 100) sieve.

sand, stone—fine aggregate resulting from the mechanical crushing and processing of rock. (See also **aggregate, fine** and **sand**.)

sandblast—a system of cutting or abrading a surface such as concrete by a stream of sand ejected from a nozzle at high speed by compressed air; often used for cleanup of horizontal construction joints or for exposure of aggregate in architectural concrete.

sand box (or sand jack)—a tight box filled with clean, dry, sand on which rests a tight-fitting timber plunger that supports the bottom of posts used in centering; removal of a plug from a hole near the bottom of the box permits the sand to run out when it is necessary to lower the centering.

sand-coarse aggregate ratio—ratio of fine-to-coarse aggregate in a batch of concrete, by mass or by volume.

sand equivalent—a measure of the relative proportions of detrimental fine dust, claylike material or both in soils or fine aggregate.

sand jack—see **sand box**.

sand-lightweight concrete—see **concrete, sand-lightweight**.

sand-lime brick—see **brick, calcium-silicate** (preferred term).

sand plate—a flat steel plate or strip welded to the legs of bar supports for use on compacted soil.

sand pocket—a zone in concrete or mortar containing fine aggregate with little or no cement.

sand streak—a streak of exposed fine aggregate in the surface of formed concrete, caused by bleeding.

sanded grout—see **grout, sanded**.

sandstone—a cemented or otherwise indurated sedimentary rock composed predominantly of sand grains.

sandwich panel—see **panel, sandwich**.

Santorin earth—a volcanic tuff originating on the Grecian island of Santorin and used as a pozzolan.

saponification—the alkaline hydrolysis of fats forming a soap; more generally, the hydrolysis of an ester by an alkali with the formation of an alcohol and a salt of the acid portion.

saturated surface-dry—condition of an aggregate particle or other porous solid when the permeable voids are filled with water and no water is on the exposed surfaces.

saturated surface-dry (SSD) particle density—the mass of the saturated surface-dry aggregate divided by its displacement volume in water or in concrete.

saturation—
1. in general: the condition of coexistence in stable equilibrium of either a vapor and a liquid or a vapor and solid phase of the same substance at the same temperature; and
2. as applied to aggregate or concrete: the condition such that no more liquid can be held or placed within it.

saturation, critical—a condition describing the degree of filling by freezable water of a pore space in cement paste or aggregate that affects the response of the material to freezing; usually taken to be 91.7% because of the 9% increase in volume of water undergoing the change of state to ice.

saturation, vacuum—a process for increasing the amount of filling of the pores in a porous material, such as lightweight aggregate, with a fluid, such as water, by subjecting the porous material to reduced pressure while immersed in the fluid.

saw cut—a cut in hardened concrete made using abrasive blades or discs.

sawdust concrete—see **concrete, sawdust**.

sawed joint—see **joint, sawed**.

scab—a short piece of wood fastened to two formwork members to secure a butt joint.

scaffolding—a temporary structure for the support of deck forms, cartways, or workers, or a combination of these, such as an elevated platform for supporting workers, tools, and materials; adjustable metal scaffolding is frequently adapted for shoring in concrete work.

scale—the oxide formed on the surface of metal during heating. (See also **scaling**.)

scaling—local flaking or peeling away of the near-surface portion of hardened concrete or mortar; also peeling or flaking of a layer from metal. (See also **mill scale, peeling,** and **spalling**.)

Note: light scaling of concrete does not expose coarse aggregate; medium scaling involves loss of surface mortar to 5 to 10 mm in depth and exposure of coarse aggregate; severe scaling involves loss of surface mortar to 5 to 10 mm in depth with some loss of mortar surrounding aggregate particles 10 to 20 mm in depth; very severe scaling involves loss of coarse aggregate particles as well as mortar generally to a depth greater than 20 mm.

scalper—a sieve for removing oversize particles.

scalping—the removal, by sieving, of particles larger than a specified size.

scanning electron microscope (SEM)—see **microscope, scanning electron (SEM)**.

scarf connection—see **connection, scarf**.

scarf joint—see **scarf connection** (preferred term).

schist—a finely layered metamorphic rock that splits easily and in which the grain is coarse enough to permit identification of the principal minerals.

Schmidt hammer—see **hammer, rebound**.

scoria—vesicular volcanic ejecta of larger size, usually of basic composition and characterized by dark color; the material is relatively heavy and partly glassy, partly crystalline; the vesicles do not generally interconnect. (See also **aggregate, lightweight**.)

scour—erosion of a concrete surface, exposing the aggregate.

scratch coat—see **coat, scratch**.

screed—
1. to strike off concrete lying beyond the desired plane or shape; and
2. a tool for striking off the concrete surface, sometimes referred to as a strikeoff.

screed, cutting—sharp-edged tool used to trim shotcrete to the finished outline. (See also **rod**.)

screed guide—firmly established grade strips or side forms for unformed concrete that guide the strikeoff in producing the desired plane or shape.

screed rails—see **screed guide**.

screed wire—see **wire, ground**.

screeding—the operation of forming a surface by the use of screed guides and a strikeoff. (See also **strikeoff**.)

screen—production equipment for separating granular material according to size, using woven-wire cloth or other similar device with regularly spaced apertures of uniform size.

screens, finish—vibrating screens (preferably horizontal) operated at a batching plant so that excessive amounts of significant undersize material are removed and delivered directly to the appropriate batcher bin without intermediate storage.

screw, adjustment—a leveling device or jack composed of a threaded screw and an adjusting handle; used for the vertical adjustment of shoring and formwork.

sealant—see **sealant, joint**.

sealant, joint—compressible material used to exclude water and solid foreign materials from joints.

sealer—a liquid that is applied to the surface of hardened concrete to either prevent or decrease the penetration of liquid or gaseous media, for example water, aggressive solutions, and carbon dioxide, during service exposer, that is absorbed by the concrete, is colorless, and leaves little or nothing visible on the surface. (See also **coating** and **compound, curing**.)

sealing compound—see **sealer**.

seating—see **deformation, anchorage**.

secant modulus—see **modulus of elasticity**.

secondary crusher—see **crusher, secondary**.

secondary moment—see **moment, secondary**.

secondary nuclear vessel—exterior container or safety container in a nuclear reactor subjected to design load only once in its lifetime, if at all.

section, transformed—a hypothetical section of one material arranged so as to have the same elastic properties as a section of two or more materials.

section modulus—see **modulus, section**.

segmental member—see **member, segmental**.

segregation—the differential concentration of the components of mixed concrete, aggregate, or the like, resulting in nonuniform proportions in the mass. (See also **bleeding** and **separation**.)

seismometer—instrument to detect linear (vertical, horizontal) or rotational displacement, velocity, or acceleration.

self-desiccation—the removal of free water by chemical reaction so as to leave insufficient water to cover the solid surfaces and cause a decrease in the relative humidity of the system; applied to an effect occurring in sealed concretes, mortars, and pastes.

self-furring—metal lath or welded-wire fabric formed in the manufacturing process to include means by which the material is held away from the supporting surface, thus creating a space for "keying" of the insulating concrete, plaster, or stucco.

self-furring nail—nails with flat heads and a washer or a spacer on the shank; for fastening reinforcing wire mesh and spacing it from the nailing member.

self-stressing cement—see **cement, expansive**.

self-stressing concrete (mortar or grout)—see **concrete (mortar or grout), self-stressing**.

selvage—a finished edge of woven-wire screen cloth produced in the weaving process of the finer meshes.

semiautomatic batcher—see **batcher**.

semiflexible joint—see **joint, semiflexible**.

sensor—a device designed to respond to a physical stimulus (as temperature, illumination, and motion) and transmit a resulting signal for interpretation, measurement, or for operating a control.

separation—the tendency, as concrete is caused to pass from the unconfined ends of chutes or conveyor belts or similar arrangements, for coarse aggregate to separate from the concrete and accumulate at one side; the tendency, as

separation

processed aggregate leaves the ends of conveyor belts, chutes, or similar devices with confining sides, for the larger aggregate to separate from the mass and accumulate at one side; or the tendency for the solids to separate from the water by gravitational settlement. (See also **bleeding** and **segregation**.)

separation joint—see **joint, isolation** (preferred term).

separation, heavy-media—a method in which a liquid or suspension of given specific gravity is used to separate particles into a portion lighter than (those that float) and a portion heavier than (those that sink) the medium.

separator, air—an apparatus that separates various size fractions of ground materials pneumatically; fine particles are discharged as product; oversized are returned to the mill as tailing.

sequence-stressing loss—in post-tensioning, the elastic loss in a stressed tendon resulting from the shortening of the member when additional tendons are stressed.

service dead load—see **load, service dead**.

service live load—see **load, service live**.

service load— see **load, service**.

set (*n.*)—the condition reached by a cement paste, mortar, or concrete when it has lost plasticity to an arbitrary degree, usually measured in terms of resistance to penetration or deformation; initial set refers to first stiffening; final set refers to attainment of significant rigidity; also, strain remaining after removal of stress. (See also **set, permanent**.)

set, false—the rapid development of rigidity in a freshly mixed portland cement paste, mortar, or concrete without the evolution of much heat, in which rigidity can be dispelled and plasticity regained by further mixing without addition of water; premature stiffening, hesitation set, early stiffening, and rubber set are terms referring to the same phenomenon, but false set is the preferred designation. (See also **set, flash**.)

set, final—a degree of stiffening of a mixture of cement and water greater than initial set, generally stated as an empirical value indicating the time in hours and minutes required for a cement paste to stiffen sufficiently to resist, to an established degree, the penetration of a weighted test needle; also applicable to concrete and mortar mixtures with use of suitable test procedures. (See also **set, initial**.)

set, flash—the rapid development of rigidity in a freshly mixed portland cement paste, mortar, or concrete, characteristically with the evolution of considerable heat, in which rigidity cannot be dispelled nor can the plasticity be regained by further mixing without the addition of water; also referred to as quick set or grab set. (See also **set, false**.)

set, grab—see **set, flash** (preferred term).

set, hesitation—see **set, false** (preferred term).

set, initial—a degree of stiffening of a mixture of cement and water less than final set, generally stated as an empirical value indicating the time in hours and minutes required for cement paste to stiffen sufficiently to resist to an established degree, the penetration of a weighted test needle; also applicable to concrete or mortar with use of suitable test procedures. (See also **set, final**.)

set, pack—see **cement, sticky** and **set, warehouse**.

set, permanent—inelastic elongation or shortening.

set, rubber—see **set, false** (preferred term).

set, stockhouse—see **cement, sticky** and **set, warehouse**.

set, warehouse—
 1. the partial hydration of cement stored for a time and exposed to atmospheric moisture; and
 2. mechanical compaction occurring during storage. (See also **cement, sticky**.)

set-accelerating admixture—see **accelerator**.

set-control addition—material, composed essentially of calcium sulfate in any hydration state from $CaSO_4$ to $CaSO_4 \cdot 2H_2O$, interground with the clinker during manufacture of cement to modify the setting time of the cement.

set-retarding admixture—see **admixture, retarding** and **retarder**.

setting time—time of setting (preferred term).

setting time, final—the time required for a freshly mixed cement paste, mortar, or concrete to achieve final set. (See also **time, initial setting**.)

setting time, initial—the time required for a freshly mixed cement paste, mortar, or concrete to achieve initial set. (See also **time, final setting**.)

settlement—sinking of solid particles in grout, mortar, or fresh concrete, after placement and before initial set. (See also **bleeding**.)

settlement shrinkage—see **shrinkage, settlement**.

settling—the lowering in elevation of sections of pavement or structures due to their mass, the loads imposed on them, or shrinkage or displacement of the support.

settling velocity—see **velocity, settling**.

shale—a laminated and fissile sedimentary rock, the constituent particles of which are principally in clay and silt sizes; the laminations are bedding planes of the rock.

shale, expanded (clay or slate)—lightweight vesicular aggregate obtained by firing suitable raw materials in a kiln or on a sintering grate under controlled conditions.

sharp sand—see **sand, sharp**.

she bolt—see **bolt, she**.

shear—an internal force tangential to the plane on which it acts.

shearhead—assembled unit in the top of the columns of flat slab or flat plate construction for transmitting loads from slab to column.

shear modulus—see **modulus of rigidity**.

shear reinforcement—see **reinforcement, shear**.

shear strength—see **strength, shear**.

shear stress—see **stress, shear**.

shearwall—a wall portion of a structural frame intended to resist lateral forces, such as earthquake, wind, and blast, acting in the plane of the wall.

sheath—an enclosure in which post-tensioning tendons are encased to prevent bonding during concrete placement. (See also **duct**.)

sheathing—the material forming the contact face of forms; also called lagging or sheeting.

sheet pile—see **pile, sheet**.

sheeting—see **sheathing** (preferred term).

shelf angles—structural angles with holes or slots in one leg for bolting to the structure to support brick work, stone, or terra cotta.

shelf life—the length of time packaged materials can be stored under specified conditions and remain usable.

shell construction—see **construction, shell**.

shelly structure—see **perlitic structure** (preferred term).

shielding concrete—see **concrete, shielding**.

shim—a strip of metal, wood, or other material employed to set base plates or structural members at the proper level for placement of grout, or to maintain the elongation in some types of post-tensioning anchorages.

shiplap—a type of joint in lumber or precast concrete made by using pieces having a portion of the width cut away on both edges, but on opposite sides, so as to make a flush joint with similar pieces.

shock, thermal—the subjection of newly hardened concrete to a rapid change in temperature that may be expected to have a potentially deleterious effect.

shock load—see **load, shock**.

shooting—placing of shotcrete. (See also **gunning**.)

shoot wire—a wire running across the width of the sieve cloth, as woven; also known as fill, filler, weft, or woof wire.

shore—a temporary support for formwork and fresh concrete or for recently built structures that have not developed full design strength; also called prop, tom, post, and strut. (See also **L-head** and **T-head**.)

shore, pole—see **shore, post**.

shore, post—individual vertical member used to support loads; also known as pole shore.
1. *adjustable timber single-post shore*—individual timber used with a fabricated clamp to obtain adjustment; not normally manufactured as a complete unit;
2. *fabricated single-post shore: Type I*—single all-metal post with a fine-adjustment screw or device in combination with pin-and-hole adjustment or clamp; Type II: single or double wooden post members adjustable by a metal clamp or screw and usually manufactured as a complete unit; and
3. *timber single-post shore*—timber used as a structural member for shoring support.

shore head—wood or metal horizontal member placed on and fastened to a vertical shoring member. (See also **raker**.)

shoring—props or posts of timber or other material in compression used for the temporary support of excavations, formwork, or unsafe structures; the process of erecting shores.

shoring, horizontal—metal or wood load-carrying strut, beam, or trussed section used to carry a shoring load from one bearing point, column, frame, post, or wall to another; may be adjustable.

shoring layout—a drawing prepared before erection showing arrangements of equipment for shoring.

short column—see **column, short**.

shorten—to decrease in length. (See also **contraction**; **elongation**; and **shrinkage**.)

shortening, elastic—in prestressed concrete, the shortening of a member that occurs immediately on the application of forces induced by prestressing.

shotcrete—mortar or concrete pneumatically projected at high velocity onto a surface; also known as air-blown mortar, pneumatically applied mortar or concrete, sprayed mortar, and gunned concrete. (See also **feed, pneumatic**; **positive displacement**; **shotcrete, dry-mix**; and **shotcrete, wet-mix**.)

shotcrete, dry-mix—shotcrete in which most of the mixing water is added at the nozzle.

shotcrete, wet-mix—shotcrete in which the ingredients, including water, are mixed before introduction into the delivery hose; accelerator, if used, is normally added at the nozzle.

shoulder—an unintentional offset in a formed concrete surface usually caused by bulging or movement of formwork.

shrink-mixed concrete—see **concrete, shrink-mixed**.

shrinkage—decrease in either length or volume.

Note: may be restricted to the effects of moisture content or chemical changes.

shrinkage, carbonation—shrinkage resulting from carbonation.

shrinkage, drying—shrinkage resulting from loss of moisture.

shrinkage, initial drying—the difference between the length of a specimen (molded and cured under stated conditions) and its length when first dried to constant length, expressed as a percentage of the moist length.

shrinkage, plastic—shrinkage that takes place before cement paste, mortar, grout, or concrete sets.

shrinkage, settlement—a reduction in volume of concrete before the final set of cementitious mixtures, caused by settling of the solids and displacement of fluids. (See also **shrinkage, plastic**, and **volume change, autogenous**.)

shrinkage-compensating—a characteristic of grout, mortar, or concrete made using expansive cement in which volume increases after setting, and if properly elastically restrained, induces compressive stresses that are intended to approximately offset the tendency of drying shrinkage to induce tensile stresses. (See also **cement, expansive**.)

shrinkage-compensating cement—see **cement, expansive**.

shrinkage-compensating concrete—see **concrete, shrinkage-compensating**.

shrinkage crack—see **crack, shrinkage**.

shrinkage cracking—see **cracking, shrinkage**.

shrinkage limit—see **limit, shrinkage**.

shrinkage loss—see **loss, shrinkage**.

shrinkage reinforcement—see **reinforcement, shrinkage**.

shuttering—see **formwork**.

SI (Système International)—the modern metric system. (See ASTM E 380.)

side, pilaster—the form for the side surface of a pilaster perpendicular to the wall.

sieve—a metallic plate or sheet, a woven-wire cloth, or other similar device with regularly spaced apertures of uniform size, mounted in a suitable frame or holder for use in separating granular material according to size.

sieve analysis—see **analysis, sieve**.

sieve correction—correction of a sieve analysis to adjust for deviation of sieve performance from that of standard calibrated sieves.

sieve fraction—that portion of a sample that passes through a standard sieve of specified size and is retained by some finer sieve of specified size.

sieve number—a number used to designate the size of a sieve, usually the approximate number of openings per linear inch; applied to sieves with openings smaller than 6.3 mm (1/4 in.). (See also **mesh**.)

sieve size—nominal size of openings between cross wires of a testing sieve.

significant (statistically significant)—values of a test statistic that lie outside of predetermined limits of test precision and so taken to indicate a difference between populations.

silica—silicon dioxide (SiO_2).

silica flour—very finely divided silica, a siliceous binder component that reacts with lime under autoclave curing conditions; prepared by grinding silica, such as quartz, to a fine powder; also known as silica powder.

silica fume—very fine noncrystalline silica produced in electric arc furnaces as a byproduct of the production of elemental silicon or alloys containing silicon.

silica powder—see **silica flour** (preferred term).

silicate—salt of a silicic acid. (See **alite**; **belite**; **blast-furnace slag**; **bredigite**; **celite**; **brick, calcium-silicate**; **hydrate, calcium-silicate**; **concrete, siliceous-aggregate**; **clay**; **dicalcium silicate**; **clay, fire**; **fluosilicate**; **lime, hydraulic hydrated**; **kaolin**; **larnite**; **melilite**; **smectite**; **Stratling's compound**; **tobermorite**; **tricalcium silicate**; **vermiculite**; and **xonotlite**.)

siliceous-aggregate concrete—see **concrete, siliceous-aggregate**.

silicon carbide—an artificial product (SiC), granules of which may be embedded in concrete surfaces to increase resistance to wear or as a means of reducing skidding or slipping on stair treads or pavements; also used as an abrasive in saws and drills for cutting concrete and masonry, and as abrasive grit in a range of particle sizes.

silicone—a resin, characterized by water-repellent properties, in which the main polymer chain consists of alternating silicon and oxygen atoms with carbon-containing side groups; silicones may be used in caulking or coating compounds or as admixtures for concrete.

sill—see **mud sill**.

silt—a granular material resulting from the disintegration of rock, with grains largely passing a 75 μm (No. 200) sieve; alternatively, such particles in the range from 2 to 50 mm diameter.

simple beam—see **beam, simple**.

single-sized aggregate—see **aggregate, single-sized**.

single-stage curing—see **curing, single-stage**.

sinter—a ceramic material or mixture fired to less than complete fusion, resulting in a coherent mass; also the process involved.

sintering—the formation of a porous mass of material by the agglomeration of fine particles during particle fusion.

sintering grate—a grate on which material is sintered.

size, nominal—see **nominal maximum size (of aggregate)**.

skew back—sloping surface against which the end of an arch rests, such as a concrete thrust block supporting thrust of an arch bridge. (See also **strip, chamfer**.)

skid resistance—see **resistance, skid**.

slab—a molded layer of plain or reinforced concrete, flat, horizontal (or nearly so), usually of uniform but sometimes of variable thickness, either on the ground or supported by beams, columns, walls, or other framework. (See also **slab, flat** and **plate, flat**.)

   slab, flat—a concrete slab reinforced in two or more directions and having drop panels, column capitals or both. (See also **plate, flat**.)

   slab, ribbed—see **panel, ribbed**.

slab bolster—see **bolster, slab**.

slabjacking—the process of either raising concrete pavement slabs or filling voids under them, or both, by injecting a material (cementitious, noncementitious, or asphaltic) under pressure.

slab-on-grade—a slab, continuously supported by ground, whose total loading when uniformly distributed would impart a pressure to the grade or soil that is less than 50% of the allowable bearing capacity thereof; the slab may be of uniform or variable thickness, and it may include stiffening elements such as ribs or beams; the slab may be plain, reinforced, or prestressed concrete; reinforcement or prestressing steel may be provided to accommodate the effects of shrinkage and temperature or structural loading. (Also referred to as slab-on-ground; slab-on-grade is the preferred term.)

slab spacer—see **spacer, slab**.

slab strip—see **strip, middle** (preferred term).

slag—see **blast-furnace slag**.

slag cement—see **cement, slag**.

slate—a fine-grained metamorphic rock possessing a well-developed fissility (slaty cleavage), usually not parallel to the bedding planes of the rock.

sleeve—a pipe or tube passing through formwork for a wall or slab through which pipe, wires, or conduit can be passed after the forms have been stripped.

sleeve, expansion—a tubular metal covering for a dowel bar to allow its free longitudinal movement at a joint.

slender beam—see **beam, slender**.

slender column—see **column, slender**.

slenderness ratio—the effective unsupported length of a uniform column divided by the least radius of gyration of the cross-sectional area.

slick line—end section of a pipeline used in placing concrete by pump which is immersed in the placed concrete and moved as the work progresses.

sliding form—see **slipform** (preferred term).

sling psychrometer—see **psychrometer, sling**.

slip—movement occurring between steel reinforcement and concrete in stressed reinforced concrete, indicating anchorage breakdown.

slip, anchorage—see **deformation, anchorage** or **slip**.

slipform—a form that is pulled or raised as concrete is placed; may move in a generally horizontal direction to lay concrete evenly for highway paving or on slopes and inverts of canals, tunnels, and siphons; or may move vertically to form walls, bins, or silos.

sloped footing—see **footing, sloped**.

sloughing—subsidence of shotcrete, plaster, or the like, due generally to excessive water in the mixture; also called sagging.

slugging—pulsating and intermittent flow of shotcrete material due to improper use of delivery equipment and materials.

slump—a measure of consistency of freshly mixed concrete, mortar, or stucco equal to the subsidence measured to the nearest 1/4 in. (6 mm) of the molded specimen immediately after removal of the slump cone.

slump cone—see **cone, slump**.

slump loss—see **loss, slump**.

slump test—see **test, slump**.

slurry—a mixture of water and any finely divided insoluble material, such as portland cement, slag, or clay in suspension.

slush grouting—see **grouting, slush**.

smectite—a group of clay minerals, including montmorillonite, characterized by a sheet-like internal atomic structure; consisting of extremely finely-divided hydrous aluminum or magnesium silicates that swell on wetting, shrink on drying, and are subject to ion exchange.

snap tie—a proprietary concrete wall-form tie, the end of which can be twisted or snapped off after the forms have been removed.

soaking period—see **period, soaking**.

soffit—the underside of a part or member of a structure, such as a beam, stairway, or arch.

soft particle—an aggregate particle possessing less than an established degree of hardness or strength as determined by a specific testing procedure.

soil—a generic term for unconsolidated natural surface material above bedrock.

soil, fine-grained—soil in which the smaller grain sizes predominate, such as fine sand, silt, and clay.

soil, coarse-grained—soil in which the larger grain sizes, such as sand and gravel, predominate.

soil cement—a mixture of soil and measured amounts of portland cement and water, compacted to a high density.

soil pressure—see **contact pressure**.

soil stabilization—chemical or mechanical treatment designed to either increase or maintain the stability of a mass of soil or otherwise to improve its engineering properties.

soldier—a vertical wale used to strengthen or align formwork or excavations.

solid masonry unit—a unit whose net cross-sectional area in every plane parallel to the bearing surface is 75% or more of its gross cross-sectional area measured in the same plane.

solid masonry wall—see **masonry wall, solid**.

solid panel—see **panel, solid**.

solid-unit masonry—see **masonry, solid-unit**.

solid volume—see **volume, absolute**.

solubility—the amount of one material that will dissolve in another, generally expressed as mass percent, as volume percent, or parts per 100 parts of solvent by mass or volume at a specified temperature.

solution—a liquid consisting of at least two substances, one of which is a liquid solvent in which the other or others, that may be either solid or liquid, are dissolved.

solvent—a liquid in which another substance may be dissolved.

sonic modulus—see **modulus of elasticity, dynamic**.

sounding well—a vertical conduit in the mass of coarse aggregate for preplaced-aggregate concrete, provided with continuous or closely spaced openings to permit entrance of grout; the grout level is determined by means of a float on a measured line.

soundness—the freedom of a solid from cracks, flaws, fissures, or variations from an accepted standard; in the case of a cement, freedom from excessive volume change after setting; in the case of aggregate, the ability to withstand the aggressive action to which concrete containing it might be exposed, particularly that due to weather.

space, capillary—void space in concrete resembling microscopic channels small enough to draw liquid water through them by the molecular attraction of the water adsorbed on their inner surfaces.

spacer—device that maintains reinforcement in proper position; also a device for keeping wall forms apart at a given distance before and during concreting. (See also **spreader**.)

spacer, slab—bar support and spacer for slab reinforcement; similar to slab bolster but without corrugations in top wire; no longer in general use. (See also **bolster, slab**.)

spacing factor—see **factor, spacing**.

spading—consolidation of mortar or concrete by repeated insertion and withdrawal of a flat, spadelike tool.

spall—a fragment, usually in the shape of a flake, detached from a larger mass by a blow, by the action of weather, by pressure, or by expansion within the larger mass; a small spall involves a roughly circular depression not greater than 20 mm in depth and 150 mm in any dimension; a large spall may be roughly circular or oval or in some cases elongate and is more than 20 mm in depth and 150 mm in greatest dimension.

spalling—the development of spalls.

span—distance between the support reactions of members carrying transverse loads.

span-depth ratio—the numerical ratio of total span-to-member depth.

span, effective—the lesser of the two following distances: a) the distance between supports; or b) the clear distance between supports plus the effective depth of the beam or slab.

span length—see **span, effective**.

spandrel—that part of a wall between the head of a window and the sill of the window above it.

spandrel beam—see **beam, spandrel**.

spatterdash—a rich mixture of portland cement and coarse sand; it is thrown onto a background by a trowel, scoop, or other appliance so as to form a thin, coarse-textured, continuous coating; as a preliminary treatment before rendering, it assists bond of the undercoat to the background, improves resistance to rain penetration, and evens out the suction of variable backgrounds. (See also **coat, dash-bond** and **parge**.)

specific gravity, absolute—ratio of the mass (referred to as vacuum) of a given volume of a solid or liquid at a stated temperature to the mass (referred to as vacuum) of an equal volume of gas-free distilled water at a stated temperature.

  specific gravity, apparent—the ratio of the mass of a volume of the impermeable portion of a material at a stated temperature to the mass of an equal volume of distilled water at a stated temperature;

  specific gravity, bulk—the ratio of the mass of a volume of a material (including the permeable and impermeable voids in the material, but not including the voids between particles of the material) at a stated temperature to the mass of an equal volume of distilled water at a stated temperature; and

  specific gravity, bulk (saturated-surface-dry)—the ratio of the mass of a volume of a material (including the mass of water within the voids, but not including the voids between particles) at a stated temperature to the mass of an equal volume of distilled water at a stated temperature. (See also **density**.)

specific gravity factor—the ratio of the mass of aggregates (including moisture), as introduced into the mixer, to the effective volume displaced by the aggregates.

specific heat—the amount of heat required per unit mass to cause a unit rise of temperature, over a small range of temperature.

specific surface—see **surface, specific**.

specification (in ASTM)—an explicit set of requirements to be satisfied by a material, product, system, or service.

specimen—a piece or portion of a sample used to make a test.

spectrophotometer—instrument for measuring the intensity of radiant energy of desired frequencies absorbed by atoms or molecules; substances are analyzed by converting the absorbed energy to electrical signals proportional to the intensity of radiation. (See also **spectroscopy, infrared** and **photometer, flame**.)

spectroscopy, infrared—the use of a spectrophotometer for determination of infrared absorption spectra (2.5 to 18 μm wave lengths) of materials; used for detection, determination, and identification especially of organic materials.

spectroscopy, X-ray emission—see **X-ray fluorescence**.

speed, agitating—the rate of rotation of the drum of a truck mixer or agitator when used for agitating mixed concrete.

spinning—the essential factor of the process of producing spun concrete. (See also **concrete, spun**.)

spiral reinforcement—see **reinforcement, spiral**.

spirally reinforced column—see **column, spirally reinforced**.

splice—connection of one reinforcing bar to another by lapping, welding, mechanical couplers, or other means; connection of welded-wire fabric by lapping; connection of piles by mechanical couplers.

  splice, contact—a means of connecting reinforcing bars in which the bars are lapped and in direct contact. (See also **splice, lap**.)

  splice, lap—a connection of reinforcing steel made by lapping the ends of bars.

  splice, welded-butt—a reinforcing bar splice made by welding the butted ends.

split-batch charging—method of charging a mixer in which the solid ingredients do not enter the mixer together; cement, and sometimes different sizes of aggregate, may be added separately.

split block—see **split-face block**.

split-face block—a concrete masonry unit with one or more faces purposely fractured to provide architectural effects in masonry wall construction.

splitting tensile strength—see **strength, splitting tensile**.

splitting tensile test (diametral compression test)—a test for tensile strength in which a cylindrical specimen is loaded to failure in diametral compression applied along the entire length.

spray drying—a method of evaporating the liquid from a solution or dispersion by spraying it into a heated gas.

spray lime—see **lime, spray**.

sprayed concrete—see **shotcrete** (preferred term).

sprayed mineral fiber—a blend of mineral fibers and inorganic binders to which water is added during the spraying operation.

sprayed mortar—see **shotcrete**.

spread footing—a generally rectangular prism of concrete, larger in lateral dimensions than the column or wall it supports, to distribute the load of a column or wall to the subgrade.

spreader—
1. a piece of lumber, usually about 1 by 2 in. (25 by 50 mm), cut to the thickness of a wall or other formed element and inserted in the form to hold it temporarily at the correct dimension against tension of form ties; wires are usually attached to spreaders so they can be pulled

up out of the forms as the pressure of concrete permits their removal; and

2. a device consisting of reciprocating paddles, a revolving screw, or other mechanism for distributing concrete to required uniform thickness in a paving slab.

spreader, concrete—a machine, usually carried on side forms or on rails parallel thereto, designed to spread concrete from heaps already dumped in front of it, or to receive and spread concrete in a uniform layer.

spreader, form—see **spreader**.

spud vibrator—see **vibrator, spud**.

spun concrete—see **concrete, centrifugally cast** (preferred term).

stabilizer—a substance that makes either a solution or suspension more stable, usually by keeping particles from precipitating.

stacking tube—a slender, free-standing tubular structure used to store granular materials; the material is loaded into the top of the tube and spills out of wall openings to make a conical pile surrounding the tube.

staged grouting—see **grouting, staged**.

stain—discoloration by foreign matter.

stalactite—a downward-pointing deposit formed as an accretion of mineral matter produced by evaporation of dripping water from the surface of rock or of concrete, commonly shaped like an icicle. (See also **stalagmite**.)

stalagmite—an upward-pointing deposit formed as an accretion of mineral matter produced by evaporation of dripping water, projecting from the surface of rock or of concrete, commonly roughly conical in shape. (See also **stalactite**.)

standard curing—see **curing, standard**.

standard deviation—the root mean square deviation of individual values from their average.

standard fire test—the test prescribed by ASTM E 119.

standard hook—a hook at the end of a reinforcing bar made in accordance with a standard.

standard hooked bar—see **bar, standard hooked**.

standard matched—tongue-and-groove lumber with the tongue and groove offset rather than centered as in center matched lumber. (See also **center matched**.)

standard sand—see **sand, standard**.

standard time-temperature curve—the graphic time table for application of temperature to a material or member for the ASTM E 119 fire test.

static load—see **load, static**.

static modulus of elasticity—see **modulus of elasticity, static**.

stationary hopper—a container used to receive and temporarily store freshly mixed concrete.

steam box—enclosure for steam-curing concrete products. (See also **steam-curing room**.)

steam curing—see **curing, steam**.

steam-curing cycle—the time interval between the start of the temperature rise period and the end of the soaking period or the cooling-off period; also a schedule indicating the duration of and the temperature range of the periods that make up the cycle.

steam-curing room—a chamber for steam curing of concrete products at atmospheric pressure.

steam kiln—see **steam-curing room** (preferred term).

stearic acid—a white crystalline fatty acid, obtained by saponifying tallow or other hard fats containing stearin. (See also **butyl stearate**.)

steel—

steel, axle—steel from carbon-steel axles for railroad cars.

steel, billet—steel, either produced directly from ingots or continuously cast, made from properly identified heats of open-hearth, basic oxygen, or electric-furnace steel, or lots of acid Bessemer steel, and conforming to specified limits of chemical composition.

steel, high-strength—steel with a high yield point; in the case of reinforcing bars, 60,000 psi (414 MPa) and greater. (See also **steel, prestressing**.)

steel, prestressing—high-strength steel used to prestress concrete; commonly seven-wire strands, single wires, bars, rods, or groups of wires or strands. (See also **prestress**; **concrete, prestressed**; **pretensioning**, and **post-tensioning**.)

steel sheet—cold-formed sheet or strip steel shaped as a structural member for the purpose of carrying the live and dead loads in lightweight concrete roof construction.

steel temperature—see **reinforcement, temperature**.

steel trowel—see **trowel**.

stem bars—see **bars, stem**.

stepped footing—see **footing, stepped**.

sticky cement—see **cement, sticky**.

stiffback—see **strongback** (preferred term).

stiffening, early—the early development of an abnormal reduction in the working characteristics of a hydraulic-cement paste, mortar, or concrete, which may be further described as false set, quick set, or flash set.

stiffening, premature—see **set, false** and **set, flash** (preferred term).

stiffness—resistance to deformation.

stiffness factor—see **factor, stiffness**.

stirrup—reinforcement used to resist shear and diagonal tension stresses in a structural member; typically a steel bar bent into a U or box shape and installed perpendicular to or at an angle to the longitudinal reinforcement, and properly anchored; lateral reinforcement formed of individual units, open or closed, or of continuously wound reinforcement.

stockhouse set—see **cement, sticky** and **set, warehouse**.

stoichiometric—

1. characterized by or being a proportion of substances or energy in a specific chemical reaction in which there is no excess of any reactant or product; and

2. proportioning based on atomic or molecular weight.

stone, cast—concrete or mortar cast into blocks or small slabs in special molds so as to resemble natural building stone.

stone, crushed—the product resulting from the artificial crushing of rocks, boulders, or large cobblestones, substantially all faces of which possess well-defined edges resulting from the crushing operation. (See also **aggregate, coarse**.)

stone sand—see **sand, stone**.

storage hopper—see **stationary hopper**.

straightedge—
1. a rigid, straight piece of either wood or metal used to strikeoff or screed a concrete surface to proper grade, or to check the planeness of a finished grade (see also **rod**; **screed**; and **strikeoff**); and
2. a highway tool for truing surfaces instead of a bull float.

straight-line theory—an assumption in reinforced concrete analysis according to which the strains and stresses in a member under flexure are assumed to vary in proportion to the distance from the neutral axis.

strain—the change in length, per unit of length, in a linear dimension of a body; a dimensionless quantity that may be measured conveniently in percent, in inches per inch, in millimeters per millimeters, but preferably in millionths.

strain, unit—deformation of a material expressed as the ratio of linear unit deformation to the distance within which that deformation occurs.

strand—a prestressing tendon composed of a number of wires twisted above the center wire or core.

strand, indented—strand having machine-made surface indentations intended to improve bond.

strand grip—a device used to anchor strands.

strand wrapping—application of high tensile strand, wound under tension by machines, around circular concrete or shotcrete walls, domes, or other tension-resisting structural components.

stratification—the separation of overwet or overvibrated concrete into horizontal layers with increasingly lighter material toward the top; water, laitance, mortar, and coarse aggregate tend to occupy successively lower positions in that order; a layered structure in concrete resulting from placing of successive batches that differ in appearance; occurrence in aggregate stockpiles of layers of differing grading or composition; a layered structure in a rock foundation.

Stratling's compound—dicalcium aluminate monosilicate-8-hydrate, a compound that has been found in reacted lime-pozzolan and cement-pozzolan mixtures.

strength—a generic term for the ability of a material to resist strain or rupture induced by external forces. (See also **strength, compressive**; **strength, fatigue**; **strength, flexural**; **strength, shear**; **strength, splitting tensile**; **strength, tensile**; **strength, ultimate**; and **strength, yield**.)

strength, bond—resistance to the separation of mortar and concrete from reinforcing and other materials with which it is in contact; a collective expression for forces such as adhesion, friction due to shrinkage, and longitudinal shear in the concrete engaged by the bar deformations that resist separation.

strength, cold—the compressive or flexural strength of refractory concrete determined before drying or firing.

strength, compressive—the measured maximum resistance of a concrete or mortar specimen to axial compressive loading; expressed as force per unit cross-sectional area; or the specified resistance used in design calculations.

strength, cube—the load per unit area at which a standard cube fails when tested in a specified manner.

strength, cylinder—see **strength, compressive** and **strength, splitting tensile**.

strength, design—nominal strength of a member multiplied by a strength-reduction (ϕ) factor. (See also **strength, nominal** and **factor, phi**.)

strength, dried—the compressive or flexural strength of refractory concrete determined within three hours after first drying in an oven at 220 to 230 F (105 to 110 C) for a specified time.

strength, early—strength of concrete or mortar usually as developed at various times during the first 72 h after placement.

strength, fatigue— the greatest stress that can be sustained for a given number of stress cycles without failure.

strength, fired—the compressive or flexural strength of refractory concrete determined upon cooling after first firing to a specified temperature for a specified time.

strength, flexural—the property of a material or a structural member that indicates its ability to resist failure in bending; in concrete flexural members, the stress at which a section reaches its maximum usable bending capacity; for under-reinforced concrete flexural members, the stress at which the compressive strain in the concrete reaches 0.003; for over-reinforced concrete flexural members, the stress at which the compressive stress reaches 85% of the cylinder strength of the concrete; for unreinforced-concrete members, the stress at which the concrete tensile strength reaches the modulus of rupture. (See also **modulus of rupture**.)

strength, nominal—strength of a member or cross section calculated in accordance with provisions and assumptions of the strength design method before application of any strength-reduction (Φ) factor.

strength, nominal flexural—the flexural strength of a member or cross section calculated in accordance with provisions and assumptions of the strength-design method before application of any strength-reduction (Φ) factor.

strength, nominal shear—the shear strength of a member or cross section calculated in accordance with provisions and assumptions of the strength-design method before application of any strength-reduction (Φ) factor.

strength, offset yield—the stress at which the strain exceeds, by a specified amount, an extension of the initially proportional part of the stress-strain curve;

expressed either as a percentage of the original gage length in conjunction with the strength value (yield strength at... percent offset =...psi) or as force per unit area ([psi] or [MPa].)

strength, required—strength of a member or cross section required to resist factored loads or related internal moments and forces in such combinations as are stipulated in the applicable code or specification.

strength, shear—the maximum shearing stress a flexural member can support at a specific location as controlled by the combined effects of shear forces and bending moment.

strength, splitting tensile—tensile strength of concrete determined by a splitting tensile test.

strength, tensile—maximum unit stress that a material is capable of resisting under axial tensile loading; based on the cross-sectional area of the specimen before loading.

strength, transfer—the concrete strength required before stress is transferred from the stressing mechanism to the concrete.

strength, transverse—see **strength, flexural** and **modulus of rupture**.

strength, ultimate—an obsolete term; see **strength, nominal**.

strength, yield—the engineering stress at which a material exhibits a specific limiting deviation from the proportionality of stress to strain.

strength-design method—a design method that requires service loads to be increased by specified load factors and computed nominal strengths to be reduced by the specified phi (ϕ) factors.

strength-reduction factor—see **factor, strength-reduction**.

stress—intensity of internal force (that is, force per unit area) exerted by either of two adjacent parts of a body on the other across an imagined plane of separation; when the forces are parallel to the plane, the stress is called shear stress; when the forces are normal to the plane, the stress is called normal stress; when the normal stress is directed toward the part on which it acts, it is called compressive stress; when the normal stress is directed away from the part on which it acts, it is called tensile stress.

stress, allowable—maximum permissible stress used in the design of members of a structure and based on a factor of safety against rupture or yielding of any type.

stress, anchorage bond—the bar forces divided by the product of the bar perimeter or perimeters and the embedment length.

stress, bond—the force of adhesion per unit area of contact between two bonded surfaces, such as concrete and reinforcing steel, or any other material, such as foundation rock; shear stress at the surface of a reinforcing bar, preventing relative movement between the bar and the surrounding concrete when the bar carries tensile force.

stress, compressive—see **stress**.

stress, effective—see **prestress, effective**.

stress, final—in prestressed concrete, the stress that exists after substantially all losses have occurred.

stress, jacking—the maximum stress occurring in a prestressed tendon during stressing.

stress, mean—the average of the maximum and minimum stress in one cycle of fluctuating loading (as in a fatigue test); tensile stress is considered positive and compressive stress, negative.

stress, normal—the stress component that is perpendicular to the plane on which the force is applied; designated tensile if the force is directed away from the plane and compressive if the force is directed toward the plane. (See also **stress**.)

stress, principal—maximum and minimum stresses at any point acting at right angles to the mutually perpendicular planes of zero shearing stress, which are designated as the principal planes.

stress, proof—stress applied to materials sufficient to produce a specified permanent strain; a specific stress to which some types of tendons are subjected in the manufacturing process as a means of reducing the deformation of anchorage, reducing the relaxation of steel, or ensuring that the tendon is sufficiently strong.

stress, shear—the stress component acting tangentially to a plane.

stress, temperature—stress in a structure or a member due to changes or differentials in temperature in the structure or member.

stress, temporary—a stress that may be produced in a precast-concrete member or in a component of a precast-concrete member during fabrication or erection, or in cast-in-place concrete structures due to construction or test loadings.

stress, tensile—see **stress**.

stress, thermal—see **stress, temperature**.

stress, torsional—the shear stress on a transverse cross section resulting from a twisting action.

stress, ultimate shear—see **strength, shear**.

stress, working—maximum permissible design stress using working-stress design methods.

stress corrosion—corrosion of a metal either initiated or accelerated by stress.

stress-corrosion cracking—see **cracking, stress-corrosion**.

stress relaxation—the time-dependent decrease in stress in a material held at constant strain. (See also **flow, plastic** and **creep**.)

stress-strain diagram—a diagram in which corresponding values of stress and strain are plotted against each other; values of stress are usually plotted as ordinates (vertically) and values of strain as abscissas (horizontally).

stresses, initial—the stresses occurring in prestressed-concrete members before any losses occur.

stressing end—in prestressed concrete, the end of the tendon at which the load is applied when tendons are stressed from one end only.

stretcher—a masonry unit laid with its length horizontal and parallel with the face of a wall or other masonry member. (See also **header**.)

strike—see **striking**.

strikeoff—to remove concrete in excess of that which is required to fill the form evenly or bring the surface to grade; performed with a straightedged piece of wood or metal by means of a forward sawing movement or by a power operated tool appropriate for this purpose; also the name applied to the tool. (See also **screed** and **screeding**.)

striking—the releasing or lowering of centering or other temporary support.

stringer—a secondary flexural member that is parallel to the longitudinal axis of a bridge or other structure. (See also **beam**.)

stringing mortar—see **mortar, stringing**.

strip—to remove formwork or a mold; also a long thin piece of wood, metal, or other material. (See also **demold** and **stripping**.)

    strip, cant—see **strip, chamfer** (preferred term).

    strip, chamfer—either a triangular or curved insert placed in an inside form corner to produce either a rounded or flat chamfer or to form a rustication; also called cant strip, fillet, dummy joint, and skew back.

    strip, grade—usually a thin strip of wood tacked to the inside surface of forms at the elevation to which the top of the concrete lift is to rise, either at a construction joint or the top of the structure.

    strip, kick—see **kicker**.

    strip, middle—in flat-slab framing, the slab portion that occupies the middle half of the span between columns. (See also **column strip**.)

    strip, panel—a strip extending across the length or width of a flat slab for structural design and construction or for architectural purposes.

    strip, rustication—a strip of wood or other material attached to a form surface to produce a groove or rustication in the concrete.

    strip, slab—see **strip, middle** (preferred term).

    strip, wrecking—small piece or panel fitted into a formwork assembly in such a way that it can be easily removed ahead of main panels or forms, making it easier to strip those major form components.

strip footing—see **footing, continuous**.

strip foundation—see **foundation, strip**.

stripper—a liquid compound formulated to remove coatings by either chemical or solvent action, or both.

stripping—the removal of formwork or a mold. (See also **demold**.)

strips, divider—in terrazzo work, nonferrous metal or plastic strips of different thicknesses, usually embedded from 5/8 to 1-1/4 in. (10 to 40 mm), used to form panels in the topping.

strongback—a frame attached to the back of a form or precast structural member to stiffen or reinforce the form or member during concrete placing operations or handling operations.

structural adhesive—a bonding agent used for transferring required loads between adherents exposed to service environments typical for the structure involved.

structural concrete—see **concrete, structural**.

structural end-point—the acceptance criterion of ASTM E 119, which states that the specimen shall sustain the applied load without collapse.

structural lightweight concrete—see **concrete, structural lightweight**.

structural sandwich construction—see **construction, structural sandwich**.

strut—see **shore**.

stub wall—see **wall, stub**.

stucco—a cement plaster used for coating exterior walls and other exterior surfaces of buildings. (See also **plaster**.)

stud—
1. member of appropriate size and spacing to support sheathing of concrete forms; and
2. a headed steel device used to anchor steel plates or shapes to concrete members.

subaqueous concrete—see **concrete, underwater**.

subbase—a layer in a pavement system between the subgrade and the base course, or between the subgrade and a portland-cement concrete pavement.

subgrade—the soil prepared and compacted to support a structure or a pavement system.

subgrade modulus—see **coefficient of subgrade reaction**.

subgrade reaction—see **contact pressure** and **coefficient of subgrade reaction**.

subpurlin—a light structural section used as a secondary structural member; in lightweight concrete roof construction, used to support the form boards over which the lightweight concrete is placed.

subsample—a sample taken from another sample.

subsieve fraction—particles all of which pass through a U.S. Standard 45 μm (No. 325) sieve.

substructure—all of that part of a structure below grade.

sulfate attack—either a chemical reaction, physical reaction, or both between sulfates usually in soil or ground water and concrete or mortar; the chemical reaction is primarily with calcium aluminate hydrates in the cement-paste matrix, often causing deterioration.

sulfate resistance—see **resistance, sulfate**.

sulfate-resistant cement—see **cement, sulfate-resistant**.

sulfoaluminate cement—see **cement, expansive, Type K**.

superimposed load—see **load, superimposed**.

superplasticizer—see **admixture, water-reducing (high-range)** (preferred term).

superstructure—all of that part of a structure above grade.

supersulfated cement—see **cement, supersulfated**.

surface—
    surface, brushed—a sandy texture obtained by brushing the surface of freshly placed or slightly hardened concrete with a stiff brush for architectural effect or, in pavements, to increase skid resistance. (See also **finish, broom**.)

surface, specific—the surface area of particles or of air voids contained in a unit mass or unit volume of a material; in the case of air voids in hardened concrete, the surface area of the air-void volume expressed as square inches per cubic inch or square millimeters per cubic millimeter.

surface active—having the ability to modify surface energy and to facilitate wetting, penetrating, emulsifying, dispersing, solubilizing, foaming, frothing, etc., of other substances.

surface-active agent—agent, surface-active.

surface air voids—small regular or irregular cavities, usually not exceeding 15 mm in diameter, resulting from entrapment of air bubbles in the surface of formed concrete during placement and consolidation. (See also **sack rub**.)

surface area—see **surface, specific**.

surface bonding (of masonry)—bonding of dry-laid masonry by parging with a thin layer of fiber-reinforced mortar.

surface moisture—see **moisture, surface**.

surface retarder—see **retarder, surface**.

surface tension—an internal molecular force that exists in the surface film of all liquids and tends to prevent the liquid from flowing.

surface texture—degree of roughness or irregularity of the exterior surfaces of aggregate particles and also of hardened concrete.

surface vibrator—see **vibrator, surface**.

surface voids—see **voids, surface**.

surface water—see **moisture, surface** (preferred term).

surfactant—a shortened form of the term "surface-active agent."

surkhi—a pozzolan consisting of burned clay powder principally produced in India.

sustained modulus of elasticity—see **modulus of elasticity, sustained**.

sway brace—a diagonal brace used to resist wind or other lateral forces. (See also **bracing**; **cross bracing**; and **X-brace**.)

swelling—increase in either length or volume. (See also **contraction**; **expansion**; **volume change**; and **volume change, autogenous**.)

swift—a reel or turntable on which prestressing tendons are placed to facilitate handling and placing.

swirl finish—see **finish, swirl**.

Swiss hammer—see **hammer, rebound** (preferred term).

syneresis—the contraction of a gel, usually evidenced by the separation from the gel of small amounts of liquid; a process possibly significant in the bleeding and cracking of fresh hydraulic-cement mixtures.

syngenite—potassium calcium sulfate hydrate, a compound sometimes produced during hydration of portland cement, found in deteriorating portland-cement concrete and said to form in portland cement during storage by reaction of potassium sulfate and gypsum.

system—

    system, one-way—the arrangement of steel reinforcement within a slab that presumably bends in only one direction.

    system, two-way—a system of reinforcement; bars, rods, or wires placed at right angles to each other in a slab and intended to resist stresses due to bending of the slab in two directions.

Système International—see **SI**.

systems building—see **industrialized building**.

T

T & G—see **tongue and groove**.

table, flow—a flat, circular jigging device used in making flow tests for consistency of cement paste, mortar, or concrete. (See also **flow, [2]**).

talc—a mineral with a greasy or soapy feel, very soft, having the composition $Mg_3Si_4O_{10}(OH)_2$. (See also **cement, masonry** and **Mohs scale**.)

tamp process—see **process, tamp**.

tamper—

1. an implement used to consolidate concrete or mortar in molds or forms; and
2. a hand-operated device for consolidating floor topping or other unformed concrete by impact from the dropped device in preparation for strikeoff and finishing; contact surface often consists of a screen or a grid of bars to force coarse aggregates below the surface to prevent interference with floating or troweling. (See also **jitterbug**.)

tamping—the operation of consolidating freshly placed concrete by repeated blows or penetrations with a tamper. (See also **consolidation** and **rodding**.)

tamping rod—see **rod, tamping**.

tangent modulus—see **modulus of elasticity**.

T-beam—a beam composed of a stem and a flange in the form of a T.

telltale—any device designed to indicate movement of formwork or of a point on the longitudinal surface of a pile under load.

temperature—

    temperature, glass-transition—the midpoint of the temperature range over which an amorphous material (such as glass or a high polymer) changes from (or to) a brittle, vitreous state to (or from) a plastic state.

    temperature, heat-deflection—the temperature at which a plastic material has an arbitrary deflection when subjected to an arbitrary load and test condition; this is an indication of the glass-transition temperature.

    temperature, steel—see **reinforcement, temperature**.

temperature cracking—see **cracking, temperature**.

temperature reinforcement—see **reinforcement, temperature**.

temperature rise—the increase of temperature caused by either absorption of heat or internal generation of heat, for example, hydration of cement in concrete.

temperature-rise period—see **period, temperature-rise**.

temperature stress—see **stress, temperature**.

temper—to add water to concrete or mortar as necessary to bring the mixture initially to the desired workability (see also **retempering**.)

template—a thin plate or board frame used as a guide in positioning or spacing form parts, reinforcement, or anchors; also a full-size mold, pattern, or frame, shaped to serve as a guide in forming or testing contour or shape.

temporary stress—see **stress, temporary**.

tendon—a steel element, such as wire, cable, bar, rod, strand, or a bundle of such elements, primarily used in tension to impart compressive stress to concrete.

  tendon, bonded—a prestressing tendon that is bonded to the concrete either directly or through grouting.

  tendon, eccentric—a prestressing tendon that follows a trajectory not coincident with the gravity axis of the concrete member.

  tendon, unbonded—a tendon that is permanently prevented from bonding to the concrete after stressing.

  tendons, concentric— tendons following a line coincident with the gravity axis of the prestressed-concrete member.

  tendons, concordant—tendons, in statically indeterminate structures, that are coincident with the pressure line produced by the tendons; such tendons do not produce secondary moments.

  tendons, deflected—tendons that have a trajectory that is curved or bent with respect to the gravity axis of the concrete member.

  tendons, draped—see **tendons, deflected**.

  tendons, harped—see **tendons, deflected**.

  tendons, nonconcordant—in statically indeterminate structures, tendons, the center of gravity of which is not coincident with the pressure line due to prestressing alone. (See also **cap cables**.)

tendon profile—the path or trajectory of the prestressing tendon.

tensile strength—see **strength, tensile**.

tensile strength, splitting—tensile strength of concrete determined by a splitting tensile test.

tensile stress—see **stress**.

tension, diagonal—the principal tensile stress resulting from the combination of normal and shear stresses acting upon a structural element.

tension reinforcement—see **reinforcement, tension**.

terrazzo concrete—see **concrete, terrazzo**.

tesserae—small pieces of glass or marble tile used in mosaics.

test—a trial, examination, observation, or evaluation used as a means of measuring either a physical or a chemical characteristic of a material, or a physical characteristic of either a structural element or a structure.

  test, air-permeability—a procedure for measuring the fineness of powdered materials such as portland cement.

  test, Atterberg—a method for determining the plasticity of soils.

  test, ball—a test to determine the consistency of freshly mixed concrete by measuring the depth of penetration of a cylindrical metal weight with a hemispherical bottom. (See also **kelly ball**.)

  test, Blaine—a method for determining the fineness of cement or other fine material on the basis of the permeability to air of a sample prepared under specified conditions.

  test, compression—test made on a test specimen of mortar or concrete to detemrine the compressive strength; in the U.S., unless otherwise specified, compression tests of mortars are made on 2 in. (50 mm) cubes and compression tests of concrete are made on cylinders 6 in. (152 mm) in diameter and 12 in. (305 mm) high.

  test, hot-load—a test for determining the resistance to deformation or shear of a refractory material when subjected to a specified compressive load at a specified temperature for a specified time.

  test, Los Angeles abrasion—test for abrasion resistance of concrete aggregates.

  test, remolding—a test to measure remoldability.

  test, slump—the procedure for measuring slump.

testing machine—a device for applying test conditions and accurately measuring results.

tetracalcium aluminoferrite—a compound in the calcium aluminoferrite series, having the composition $4CaO \cdot Al_2O_3 \cdot Fe_2O_3$, abbreviated C_4AF, that is usually assumed to be the aluminoferrite present when compound calculations are made from the results of chemical analysis of portland cement. (See also **brownmillerite**.)

texture—the pattern or configuration apparent in an exposed surface, as in concrete and mortar, including roughness, streaking, striation, or departure from flatness.

texturing—the process of producing a special texture on either unhardened or hardened concrete.

T-head—in precast framing, a segment of girder crossing the top of an interior column; also the top of a shore formed with a braced horizontal member projecting on two sides forming a T-shaped assembly.

thermal conductance—see **conductance, thermal**.

thermal conductivity—see **conductivity, thermal**.

thermal contraction—contraction caused by decrease in temperature.

thermal diffusivity—see **diffusivity, thermal**.

thermal expansion—expansion caused by increase in temperature.

thermal movement—change of dimension of concrete or masonry resulting from change of temperatures. (See also **contraction** and **expansion**.)

thermal resistance—see **resistance, thermal**.

thermal shock—see **shock, thermal**.

thermal stress—see **stress, temperature**.

thermal volume change—see **volume change, thermal**.

thermocouple—two conductors of different metals joined together at both ends, producing a loop in which an electric current will flow when there is a difference in temperature between the two junctions.

thermoplastic—becoming soft when heated and hard when cooled.

thermosetting—becoming rigid by chemical reaction and not remeltable.

thin-shell precast—precast concrete characterized by thin slabs and web sections. (See also **construction, shell**.)

thixotropy—the property of a material that enables it to stiffen in a short period while at rest, but to acquire a lower viscosity when mechanically agitated, the process being reversible; a material having this property is termed thixotropic or shear thinning. (See also **rheology**.)

threaded anchorage—see **anchorage, threaded**.

tie—
1. loop of reinforcing bars encircling the longitudinal steel in columns; and
2. a tensile unit adapted to holding concrete forms secure against the lateral pressure of unhardened concrete.

tie, form—a mechanical connection in tension used to prevent concrete forms from spreading due to the fluid pressure of fresh concrete.

tie bar—see **bar, tie**.

tie bar, deformed—see **bar, tie**.

tie rod—see **tie, form** and **tieback**.

tieback—a rod fastened to a deadman, a rigid foundation, or either a rock or soil anchor to prevent lateral movement of formwork, sheet pile walls, retaining walls, bulkheads, etc.

tied column—see **column, tied**.

tiers—see **lifts** (preferred term).

tilting mixer—see **mixer, tilting**.

tilt-up—a construction technique for casting concrete elements in a horizontal position at the jobsite and then tilting them to their final position in a structure.

time-dependent deformation—see **deformation, time-dependent**.

time, final setting—the time required for a freshly mixed cement paste, mortar, or concrete to achieve final set. (See also **time, initial setting**.)

time, initial setting—the time required for a freshly mixed cement paste, mortar, or concrete to achieve initial set. (See also **time, final setting**.)

time of haul—in production of ready-mixed concrete, the period from first contact between mixing water and cement until completion of discharge of the freshly mixed concrete.

time of set—see **time of setting**.

time of setting—
1. the time required for a freshly mixed cement paste, mortar, or concrete to achieve initial set (see **set, initial**) or;
2. the time required for a freshly mixed cement paste, mortar, or concrete to achieve final set (see **set, final**).

tobermorite—a mineral found in Northern Ireland and elsewhere, having the approximate formula $Ca_5Si_6O_{16}(OH)_2 \cdot 4H_2O$, identified approximately with the artificial product tobermorite (G) of Brunauer, a hydrated calcium silicate having a $CaO:SiO_2$ ratio in the range 1.39 to 1.75 and forming minute layered crystals that constitute the principal cementing medium in portland-cement concrete; a mineral with 5 mols of lime to 6 mols of silica, usually occurring in plate-like crystals, which is easily synthesized at steam pressures of about 100 psi and higher; the binder in several properly autoclaved products.

tobermorite gel—see **gel, tobermorite**.

toenail—
1. an obliquely driven nail; and
2. to drive a nail at an angle.

tolerance—
1. the permitted variation from a given dimension or quantity;
2. the range of variation permitted in maintaining a specified dimension; and
3. a permitted variation from location or alignment.

tom—see **shore** (preferred term).

tongue and groove—a joint in which a protruding rib on the edge of one side fits into a groove in the edge of the other side; abbreviated "T & G." (See also **keyway**.)

tool, arrissing—a tool similar to a float, but having a form suitable for rounding an edge of freshly placed concrete.

tool, gutter—a tool used to give the desired shape and finish to concrete gutters.

tooling—the act of compacting and contouring a material in a joint.

top form—see **form, top**.

topping—
1. a layer of concrete or mortar placed to form a floor surface on a concrete base;
2. a structural, cast-in-place surface for precast floor and roof systems; and
3. the mixture of marble chips and matrix that, when properly processed, produces a terrazzo surface.

topping, dry—see **dry-shake** (preferred term).

topping, monolithic—on flatwork, a higher quality, more serviceable topping course placed promptly after the base course has lost all slump and bleed water.

torque viscometer—see **viscometer, torque**.

torsional stress—see **stress, torsional**.

toughness—the property of matter that resists fracture by impact or shock.

tower—a composite structure of frames, braces, and accessories.

trajectory of prestressing force—see **path of prestressing force**.

transfer—the act of transferring the stress in prestressing tendons from the jacks or pretensioning bed to the concrete member.

transfer bond—see **bond, transfer**.

transfer length—see **length, transfer** (preferred term).

transfer strength—see **strength, transfer**.

transformed section—see **section, transformed**.

transit-mixed concrete—see **concrete, transit-mixed**.

transit-mixer—see **mixer, truck**.

translucent concrete—see **concrete, translucent**.

transmission length—see **length, transfer**.

transverse cracks—see **cracks, transverse**.

transverse joint—see **joint, transverse**.

transverse prestress—see **prestress, transverse**.

transverse reinforcement—see **reinforcement, transverse**.

transverse strength—see **strength, flexural** and **modulus of rupture**.

traprock—any of various fine-grained, dense, dark colored igneous rocks, typically basalt or diabase; also called "trap."

trass—a natural pozzolan of volcanic origin found in Germany, namely, trachytic tuffs that are intensely altered by geologic processes.

traveler—an inverted-U-shaped structure usually mounted on tracks that permit it to move from one location to another to facilitate the construction of an arch, bridge, or building.

travertine—dense to irregularly porous, commonly stratified or banded calcium carbonate, either aragonite or calcite, formed by deposition from hot spring waters.

tremie—a pipe or tube through which concrete is deposited under water, having at its upper end a hopper for filling and a bail for moving the assemblage.

tremie seal—the depth to which the discharge end of the tremie pipe is kept embedded in the fresh concrete that is being placed; a layer of tremie concrete placed in a cofferdam for the purpose of preventing the intrusion of water when the cofferdam is dewatered.

trench form (for cast-in-place concrete pipe)—the vertical sides and semicircular bottom of the trench shaped to provide full, firm, and uniform support for the lower 210 degrees of the pipe.

trial batch—see **batch, trial**.

triaxial compression test—a test in which a specimen is subjected to a confining hydrostatic pressure and then loaded axially to failure.

triaxial test—a test in which a specimen is subjected simultaneously to lateral and axial loads.

tricalcium aluminate—a compound having the composition $3CaO \cdot Al_2O_3$, abbreviated C_3A.

tricalcium silicate—a compound having the composition $3CaO \cdot SiO_2$, abbreviated C_3S, an impure form of which (alite) is a main constituent of portland cement. (See also **alite**.)

trough, flow—a sloping trough used to convey concrete by gravity flow from either a truck mixer or a receiving hopper to the point of placement. (See also **chute**.)

trough mixer—see **mixer, open-top**.

trowel—a flat, broad-blade steel hand tool used in the final stages of finishing operations to impart a relatively smooth surface to concrete floors and other unformed concrete surfaces; also a flat triangular-blade tool used for applying mortar to masonry. (See also **fresno trowel**.)

trowel finish—see **finish, trowel**.

troweling—smoothing and compacting the unformed surface of fresh concrete by strokes of a trowel.

troweling machine—a motor driven device that operates orbiting steel trowels on radial arms from a vertical shaft.

truck, agitating—a vehicle in which freshly mixed concrete can be conveyed from the site of mixing to the site of placement; while being agitated, the truck body can either be stationary and contain an agitator, or it can be a drum rotated continuously so as to agitate the contents; designated "agitating lorry" in the UK.

truck-mixed concrete—see **concrete, transit-mixed**.

truck mixer—see **mixer, truck**.

T-shore—a shore with a T-head.

tub mixer—see **mixer, open-top** (preferred term).

tube-and-coupler shoring—a load-carrying assembly of tubing or pipe which serves as posts, braces, ties, a base supporting the posts, and special couplers that connect the uprights and join the various members.

tunnel lining—a structural system of concrete, steel, or other materials to provide support for a tunnel for exterior loads, to reduce water seepage, or to increase flow capacity.

turbidimeter—a device for measuring the particle-size distribution of a finely divided material by taking successive measurements of the turbidity of a suspension in a fluid.

turbidimeter fineness—the fineness of a material such as portland cement, usually expressed as total surface area in square centimeters per gram, as determined with a turbidimeter. (See also **Wagner fineness**.)

turbine mixer—see **mixer, open-top** (preferred term).

twin-twisted bar reinforcement—see **reinforcement, twin-twisted bar**.

two-stage curing—see **curing, two-stage**.

two-way reinforced footing—a footing having reinforcement in two directions generally perpendicular to each other.

two-way reinforcement—see **reinforcement, two-way**.

two-way system—see **system, two-way**.

Type I cement—see **cement, normal** (preferred term).

Type II cement—see **cement, modified** (preferred term).

Type III cement—see **cement, high-early strength** (preferred term).

Type IV cement—see **cement, low-heat** (preferred term).

Type V cement—see **cement, sulfate-resistant** (preferred term).

U

U-value—overall coefficient of heat transmission; a standard measure of the rate at which heat will flow through a unit area of a material of known thickness.

ultimate-design resisting moment—the moment at which a reinforced-concrete section reaches its usable flexural strength, commonly accepted for under-reinforced concrete flexural members to be the bending moment at which the concrete compressive strain equals 0.003; an obsolete term.

ultimate load—see **load, ultimate**.

ultimate moment—an obsolete term; see **strength, nominal flexural**.

ultimate shear strength—an obsolete term; see **strength, nominal shear**.

ultimate strength—an obsolete term; see **strength, nominal**.

ultimate-strength design—see **strength-design method**.

ultrasonic—pertaining to mechanical vibrations having a frequency greater than approximately 20,000 Hz.

unbonded member—a prestressed concrete member post-tensioned with tendons that are not bonded to the concrete between the end anchorages after stressing.

unbonded post-tensioning—post-tensioning in which the tendons are not grouted after stressing.

unbonded tendon—see **tendon, unbonded**.

unbraced length of column—distance between lateral supports.

underbed—the base mortar, usually horizontal, into which strips are embedded and on which terrazzo topping is applied.

undersanded—concrete containing an insufficient proportion of fine aggregate to produce optimum properties in the fresh mixture, especially workability and finishing characteristics.

undersize—particles of aggregate passing a designated sieve.

underwater concrete—see **concrete, underwater**.

unhardened concrete—see **concrete, fresh** (preferred term).

unreinforced concrete—see **concrete, plain**.

unit masonry—see **masonry, unit**.

unit strain—see **strain, unit**.

unit water content—the quantity of water per unit volume of freshly mixed concrete, often expressed as lb or gal./yd^3; the quantity of water on which the water-cement ratio is based, not including water absorbed by the aggregate.

unit weight—deprecated term; see **density**.

unit weight, fired—see **density, fired**.

unsound—not firmly made, placed, or fixed; subject to deterioration or disintegration during service exposure.

V

vacuum concrete—see **concrete, vacuum**.

vacuum dewatering—see **concrete, vacuum**.

vacuum saturation—see **saturation, vacuum**.

valve bag—paper bag for cement or other material, either glued or sewn, made of four or five plies of kraft paper and completely closed except for a self-sealing paper valve through which the contents are introduced and released.

vapor barrier—see **barrier, vapor**.

vapor pressure—a component of atmospheric pressure; caused by the presence of vapor; expressed in inches, centimeters, or millimeters of height of a column of mercury; or, in SI, in pascals.

variation—see **coefficient of variation and standard deviation**.

vebe apparatus—an apparatus for measuring workability of very low-slump or no-slump concrete, including a vibrating table, a sample container, and other ancillary items, that permits measurement of the time (vebetime) required to be consolidated in a mold.

vehicle—liquid carrier or binder of solids.

velocity, pulse—the velocity at which compressional waves are propagated through a medium.

velocity, settling—the terminal rate of fall of a particle through a fluid as induced by gravity or other external force; the rate at which frictional drag balances the accelerating force (or the external force).

veneer—a masonry facing that is attached to the backup, but not so bonded as to act with it under load.

Venetian—a type of terrazzo topping that incorporates large chips of stone.

vent pipe—see **pipe, vent**.

vented form—see **form, vented**.

vermiculite—a micaceous mineral; also a group name for certain platy minerals, hydrous silicates of aluminum, magnesium, and iron, characterized by marked exfoliation on heating; also a constituent of clays.

vermiculite concrete—see **concrete, vermiculite**.

vertical-shaft mixer—see **mixer, vertical-shaft**.

vibrated concrete—see **concrete, vibrated**.

vibration—energetic agitation of freshly mixed concrete during placement by mechanical devices, either pneumatic or electric, that create vibratory impulses of moderately high frequency to assist in consolidating the concrete in the form or mold.

1. external vibration employs vibrating devices attached at strategic positions on the forms and is particularly applicable to manufacture of precast items and for vibration of tunnel-lining forms; in manufacture of concrete products, external vibration or impact may be applied to a casting table;
2. internal vibration employs one or more vibrating elements that can be inserted into the fresh concrete at selected locations, and is more generally applicable to in-place construction; and
3. surface vibration employs a portable horizontal platform on which a vibrating element is mounted.

vibration limit—see **limit, vibration**.

vibrator—an oscillating machine used to agitate fresh concrete so as to eliminate gross voids, including entrapped air but not entrained air, and to produce intimate contact with form surfaces and embedded materials. (See also **vibration**.)

vibrator, external—see **vibrator**.

vibrator, spud—a vibrator, having a vibrating casing or a vibrating head, used to consolidate freshly placed concrete by insertion into the mass.

vibrator, surface—a vibrator used for consolidating concrete by application to the surface of a mass of freshly mixed concrete; four principal types exist: vibrating screeds, pan vibrators, plate or grid vibratory tampers, and vibratory roller screeds.

Vicat apparatus—a penetration device used in the testing of hydraulic cements and similar materials.

Vicat needle—see **needle, Vicat**.

viscometer—instrument for determining viscosity of slurries, mortars, or concretes.

viscometer, torque—an apparatus used for measuring the consistency of slurries in which the energy required to rotate a device suspended in a rotating cup is proportional to viscosity.

viscosity—the property of a material that resists change in the shape or arrangement of its elements during flow, and the measure thereof.

visual concrete—see **concrete, architectural** and **concrete, exposed**.

void—

void, air—a space in cement paste, mortar, or concrete filled with air; an entrapped air void is characteristically 1 mm or more in size and irregular in shape; an entrained air void is typically between 10 μm and 1 mm in diameter and spherical, or nearly so.

void, water—void along the underside of an aggregate particle or reinforcing steel which formed during the bleeding period; initially filled with bleed water.

void-cement ratio—volumetric ratio of air plus net mixing water to cement in a concrete or mortar mixture.

voids, surface—cavities visible on the surface of a solid. (See also **bug holes**.)

volatile material—material that is subject to release as a gas or vapor; liquid that evaporates readily.

volume—

volume, absolute—in the case of solids, the displacement volume of particles themselves, including their permeable and impermeable voids, but excluding space between particles; in the case of fluids, their volume.

volume, dry-rodded—the bulk volume occupied by a dry aggregate compacted by rodding under standardized conditions; used in measuring density of aggregate.

volume batching—measuring the constituents of mortar or concrete by volume.

volume change—an increase or decrease in volume due to any cause. (See also **deformation** and **deformation, time-dependent**.)

volume change, autogenous—change in volume produced by continued hydration of cement, exclusive of effects of applied load and change in either thermal condition or moisture content.

volume change, thermal—the increase or decrease in volume caused by changes in temperature. (See **thermal contraction** and **thermal expansion**.)

W

waffle—see **dome**.

Wagner fineness—the fineness of portland cement, expressed as total surface area in square centimeters per gram, determined by the Wagner turbidimeter apparatus and procedure.

wale—a long formwork member (usually double) used to gather loads from several studs (or similar members) to allow wider spacing of the restraining ties; when used with prefabricated panel forms, this member is used to maintain alignment; also called waler or ranger.

waler—see **wale**.

wall—a vertical element used primarily to enclose or separate spaces.

wall, enclosure—a non-load-bearing wall intended only to enclose space.

wall, load-bearing—a wall designed and built to carry superimposed vertical or in-plane and shear loads, or both. (See also **wall, nonbearing**.)

wall, nonbearing—a wall that supports no vertical load other than its own weight and no in-plane shear loads. (See also **load-bearing wall**.)

wall, stub—low wall, usually 4 to 8 in. (100 to 200 mm) high, placed monolithically with a concrete floor or other members to provide for control and attachment of wall forms; called kicker in the UK.

wall form—see **form, wall**.

warehouse pack—see **set, warehouse** and **cement, sticky**.

warehouse set—see **set, warehouse**.

warping—a deviation of a slab or wall surface from its original shape, usually caused by either temperature or moisture differentials or both within the slab or wall. (See also **curling**.)

warping joint—see **joint, warping**.

wash (or flush) water—see **water, wash (or flush)**.

water—

water, absorbed—see **moisture, absorbed**.

water, adsorbed—water held on surfaces of a material by electrochemical forces and having physical properties substantially different from those of absorbed water or chemically combined water at the same temperature and pressure. (See also **adsorption**.)

water, evaporable—water in set cement paste present in capillaries or held by surface forces; measured as that removable by drying under specified conditions. (See also **water, nonevaporable**.)

water, flush—see **water, wash (or flush)**.

water, free—see **moisture, free**.

water, gage—see **batched water**.

water, mixing—the water in freshly mixed sand-cement grout, mortar, or concrete, exclusive of any previously absorbed by the aggregate (for example, water considered in the computation of the net water-cement ratio). (See also **batched water** and **moisture, surface**.)

water, nonevaporable—the water that is chemically combined during cement hydration; not removable by specified drying. (See also **water, evaporable**.)

water, wash (or flush)—water carried on a truck mixer in a special tank for flushing the interior of the mixer after discharge of the concrete.

water blast—a system of cutting or abrading a surface such as concrete by a stream of water ejected from a nozzle at high velocity.

water-cement ratio—the ratio of the mass of water, exclusive only of that absorbed by the aggregates, to the mass of portland cement in concrete, mortar, or grout, stated as a decimal and abbreviated as w/c. (See also **water-cementitious material ratio**.)

water-cementitious material ratio—the ratio of the mass of water, exclusive only of that absorbed by the aggregate, to the mass of cementitious material (hydraulic) in concrete, mortar, or grout, stated as a decimal and abbreviated as *w/cm*. (See also **water-cement ratio**.)

water gain—see **bleeding**.

water pocket—see **void, water**.

waterproof—impervious to water in either liquid or vapor state. (See also **dampproof**.) (Note: Because nothing can be completely impervious to water under infinite pressure over infinite time, this term should not be used.)

waterproofed cement—see **water repellent**.

waterproofing—see **dampproofing** (preferred term).

waterproofing compound—see **compound, waterproofing**.

water-reducing admixture—see **admixture, water-reducing**.

water-reducing admixture (high-range)—see **admixture, water-reducing (high-range)**.

water-repellent—property of a surface that resists wetting (by matter in either liquid or vapor state) but permits passage of water when hydrostatic pressure occurs. (See also **watertight**.)

water-resistant—see **water-repellent** (preferred term).

water ring—a device in the nozzle body of dry-mix shotcrete equipment through which water is added to the materials.

waterstop—a thin sheet of metal, rubber, plastic, or other material inserted across a joint to obstruct the seepage of water through the joint.

watertight—impermeable to water except when under hydrostatic pressure sufficient to produce structural discontinuity by rupture.

water void—see **void, water**.

w/c—see **water-cement ratio**.

w/cm—see **water-cementitious material ratio**.

weakened-plane joint—see **joint, groove** and **joint, contraction** (preferred term).

wearing course—a topping or surface treatment to increase the resistance of a concrete pavement or slab to abrasion.

weathering—changes in color, texture, strength, chemical composition or other properties of a natural or artificial material due to the action of the weather.

web bar—see **reinforcement, web** (preferred term).

web reinforcement—see **reinforcement, web**.

wedge—a piece of wood or metal tapering to a thin edge; used to adjust elevation or tighten formwork.

wedge anchorage—see **anchorage, wedge**.

weigh batching—measuring the constituent materials for mortar or concrete by mass.

weight, dry-batch—the mass of the materials, excluding water, used to make a batch of concrete.

weight, dry-rodded—deprecated term; see **density, dry-rodded**.

welded-butt splice—see **splice, welded-butt**.

welded reinforcement—see **reinforcement, welded**.

welded-wire fabric—see **fabric, welded-wire**.

welded-wire fabric reinforcement—see **reinforcement, welded-wire fabric**.

well-graded aggregate—see **aggregate, well-graded**.

wet—covered with visible free moisture; not dry. (See also **damp** and **moist**.)

wet-cast process—see **process, wet-cast**.

wet process—see **process, wet**.

wet screening—screening to remove fresh concrete aggregate particles larger than a certain size.

wet-mix shotcrete—see **shotcrete, wet-mix**.

wet sieving—use of water to ficilitate sieving of a granular material on standard sieves.

wettest stable consistency—see **consistency, wettest stable**.

wetting agent—see **agent, wetting**.

wheel, feed—material distributor or regulator in certain types of shotcrete equipment.

wheel load—see **load, wheel**.

white cement—see **cement, white**.

width, effective flange—width of slab adjoining a beam stem where the slab is assumed to function as the flange element of a T-beam section.

wing pile—see **pile, wing**.

wire—

   wire, alignment—see **wire, ground**.

   wire, cold-drawn—wire made from rods that are hot-rolled from billets and then cold-drawn through dies. (See also **reinforcement, cold-drawn wire**.)

   wire, crimped—wire deformed into a curve that approximates a sine curve as a means of increasing the capacity of the wire to bond to concrete; also welded wire fabric crimped to provide an integral chair. (See also **reinforcement, deformed** and **wire, indented**.)

   wire, ground—small-gage high-strength steel wire used to establish line and grade as in shotcrete work; also called alignment wire and screed wire.

   wire, indented—wire having machine-made surface indentations intended to improve bond; depending on the type of wire, used for either concrete reinforcement or pretensioning tendons.

wire mesh—see **fabric, welded-wire**.

wire wrapping—application of high tensile wire, wound under tension by machines, around circular concrete or shotcrete walls, domes, or other tension-resisting structural components.

wobble coefficient—a coefficient used in determining the friction loss occurring in post-tensioning, which is assumed to account for the secondary curvature of the tendons.

wobble friction—see **friction, wobble**.

wood block—see **block, wood**.

workability—that property of freshly mixed concrete or mortar that determines the ease with which it can be mixed, placed, consolidated, and finished to a homogenous condition.

working load—see **load, working**.

working stress—see **stress, working**.

working-stress design—see **design, working-stress**.

woven-wire fabric—see **fabric, woven-wire**.

woven-wire reinforcement—see **fabric, welded-wire** (preferred term).
wrapping—see **strand wrapping** and **wire wrapping**.
wrecking strip—see **strip, wrecking**.
wythe (leaf)—each continuous vertical section of a wall that is one masonry unit or grouted space in thickness.

X

X-brace—paired set of crossing sway braces. (See also **brace**, **cross bracing**, and **sway brace**.)
xonotlite—calcium silicate monohydrate ($Ca_6Si_6O_{17}(OH)_2$), a natural mineral that is readily synthesized at 302 to 662 F (150 to 350 C) under saturated steam pressure; a constituent of sand-lime masonry units.
X-ray diffraction—the diffraction of X-rays by substances having a regular arrangement of atoms; a phenomenon used to identify substances having such structure.
X-ray emission spectroscopy—see **X-ray fluorescence**.
X-ray fluorescence—characteristic secondary radiation emitted by an element as a result of excitation by X-rays, used to yield chemical analysis of a sample.

Y

yellowing—development of yellow color or cast in white or clear coatings as a consequence of aging.
yield—the volume of freshly mixed concrete produced from a known quantity of ingredients; the total mass of ingredients divided by the density mass of the freshly mixed concrete; also the number of units produced per bag of cement or per batch of concrete.
yield point—the first engineering stress in a test in which stresses and strains are determined for a material that exhibits the phenomenon of discontinuous yielding, of which an increase in strain occurs without an increase in stress.
yield strength—see **strength, yield**.
yoke—a tie or clamping device around column forms or over the top of wall or footing forms to keep them from spreading because of the lateral pressure of fresh concrete; also part of a structural assembly for slipforming which keeps the forms from spreading and transfers form loads to the jacks.
Young's modulus—see **modulus of elasticity** (preferred term).

Z

zero-slump concrete—see **concrete, zero slump**.
zone, anchorage—in post-tensioning, the region adjacent to the anchorage subjected to secondary stresses resulting from the distribution of the prestressing force; in pretensioning, the region in which the transfer bond stresses are developed.
zone, precompressed—the area of a flexural member that is compressed by the prestressing tendons.

ACI 117-90
(Reapproved 2002)

Standard Specifications for Tolerances for Concrete Construction and Materials (ACI 117-90)

Reported by ACI Committee 117

W. Robert Little
Chairman

Russell S. Fling
Chairman, Editorial Subcommittee

S. Allen Face, III
Thomas C. Heist
Richard A. Kaden
Ross Martin
Peter Meza

Andrawos Morcos
Clark B. Morgan, Jr.
Harry M. Palmbaum
William S. Phelan

B. J. Pointer
Dean E. Stephan, Jr.*
Eldon Tipping
Carl S. Togni
Joe V. Williams, Jr.

This specification provides standard tolerances for concrete construction. This document is intended to be used as the reference document for establishing tolerances for concrete construction by specification writers and ACI committees writing Standards.

Keywords: bending (reinforcing steels); building codes; **concrete construction**; concrete piles; concretes; floors; formwork (construction); masonry; mass concrete; piers; precast concrete; prestressed concrete; reinforcing steels; **specifications**; splicing; **standards**; **tolerances (mechanics)**.

FOREWORD

F1. This foreword is included for explanatory purposes only; it is not a part of Standard Specification 117.

F2. Standard Specification 117 is a Reference Standard which the Architect/Engineer may cite in the Project Specifications for any construction project, together with supplementary requirements for the specific project.

This standard is not intended to apply to special structures not cited in the standard such as nuclear reactors and containment vessels, bins and silos, and prestressed circular structures. It is also not intended to apply to the specialized construction procedure of shotcrete.

F3. Standard Specification 117 *addresses each of the Three-Part Section Format of the Construction Specifications Institute, organized by structural elements, structural components and types of structures; the numbering system reflects this organization.* The language is imperative and terse to preclude an alternative.

F4. A Specification Checklist is included as a preface to, but not forming a part of, Standard Specification 117. The purpose of this Specification Checklist is to assist the Architect/Engineer in properly choosing and specifying the necessary mandatory and optional requirements for the Project Specification.

PREFACE TO SPECIFICATION CHECKLIST

P1. Standard Specification 117 is intended to be used in its entirety by reference in the Project Specification. Individual sections, articles, or paragraphs should not be copied into the Project Specifications since taking them out of context may change their meaning.

P2. Building codes establish minimum requirements necessary to protect the public. Some of the requirements in this Standard Specification may be more stringent than the minimum in order to insure the level of quality and performance that the Owner expects the structure to provide. Adjustments to the needs of a particular project should be made by the Architect/Engineer by reviewing each of the items in the Specification Checklist and then including the Architect/Engineer's decision on each item as a mandatory requirement in the Project Specifications.

P3. These mandatory requirements should designate the specific qualities, procedures, materials, and performance criteria for which alternatives are permitted or for which provisions were not made in the Standard Specification. Exceptions to the Standard Specification should be made in the Project Specifications, if required.

P4. A statement such as the following will serve to make Standard Specification ACI 117 an official part of the Project Specifications:

Tolerances for Concrete Construction and Materials shall conform to all requirements of ACI 117, Standard Specifications for Tolerances for Concrete Construction and Materials, published by the American Concrete Institute, Detroit, Michigan, except as modified by the requirements of these Contract Documents.

Adopted as a Standard of the American Concrete Institute in November 1989 in accordance with the Institute's standardization procedures.
Copyright © 1990, American Concrete Institute. All rights reserved, including the making of copies unless permission is obtained from the copyright proprietors.
*Chairman during initial development of this document.

P5. The Specification Checklist that follows is addressed to each item of the Standard Specification where the Architect/Engineer must or may make a choice of alternatives; may add provisions if not indicated; or may take exceptions. The Specification Checklist consists of two columns; the first identifies the sections, parts, and articles of the Standard Specification and the second column contains notes to the Architect/Engineer to indicate the type of action required by the Architect/Engineer.

MANDATORY SPECIFICATION CHECKLIST

Section/Part/Article	Notes to the Architect/Engineer
Section 2—Materials	
2.2—Reinforcement	Tolerances for fabrication, placement, and lap splices for welded wire fabric must be specified by the specifier.
Section 3—Foundations	
3.1.1 Drilled piers	Specify category of caisson. The designer should be aware that the recommended vertical alignment tolerance of 1.5 percent of the shaft length indicated in Category B caissons is based on experience in a wide variety of soil situations combined with a limited amount of theoretical analysis using the beam on elastic foundation theory and minimum assumed horizontal soil restraint.
Section 4—Cast-in-place concrete for buildings	
4.5.4 Form offsets	Designate class of surface (A, B, C, D):
	Class A — For surfaces prominently exposed to public view where appearance is of special importance.
	Class B — Coarse-textured concrete-formed surfaces intended to receive plaster, stucco, or wainscoting.
	Class C — General standard for permanently exposed surfaces where other finishes are not specified.
	Class D — Minimum quality surface where roughness is not objectionable, usually applied where surfaces will be concealed.
4.5.5 Floor finish	Specify floor finish tolerance measurement method (either Section 4.5.6 *or* Section 4.5.7).
4.5.5.1 For Section 4.5.6	Designate floor classification (15/13; 20/15; 30/20; or, 50/30).
4.5.5.2 For Section 4.5.7	Designate maximum gap under a freestanding straightedge (½ in., 5/16 in., 3/16 in., or ⅛ in.).

OPTIONAL SPECIFICATION CHECKLIST

Section 1 — General	
1.1.2 Scope	Tolerance values affect construction cost. Specific use of a toleranced item may warrant less or more stringent tolerances than contained in the specification. Such variances must be individually designated by the specifier in the contract documents.
1.1.2 Scope	Tolerances in this specification are for standard concrete construction and construction procedures. Specialized concrete construction or construction procedures require specifier to include specialized tolerances. ACI committee documents covering specialized construction may provide guidance on specialized tolerances.
	The tolerances in this Specification do not apply to special structures or procedures not cited in the document such as nuclear reactors and containment vessels, bins and silos, circular prestressed concrete tank structures and shotcrete.
1.2.3 Requirements	Where a specific application uses multiply toleranced items that together yield a toleranced result, the specifier must analyze the tolerance envelope with respect to practical limits and design assumptions and specify its value where the standard tolerances values in this specification are inadequate or inappropriate.

OPTIONAL SPECIFICATION CHECKLIST, continued

Section 2 — Materials	
2.2.3 Concrete cover	The tolerance for reduction in cover in reinforcing steel may require a reduction in magnitude where the reinforced concrete is exposed to chlorides or the environment. Where possible excess cover or other protection of the reinforcing steel should be specified in lieu of reduced tolerance because of the accuracy of locating reinforcing steel utilizing standard fabrication accessories and installed procedures.
2.3.2 Embedded items	Tolerance given is for general application. Specific design use of embedded items may require the specifier to designate tolerances of reduced magnitude for various embedded items.
Section 3 — Cast-in-place concrete for foundations	
3.4.1.2 Footings	Plus tolerance for the vertical dimensions is not specified because no limit is imposed. Specifier must designate plus tolerance if desired.
Section 4 — Cast-in-place concrete for buildings	
4.5.5 Floor finish	The procedures for specifying and measuring floor finish tolerances set forth herein are not appropriate for narrow aisle warehouse floors with defined traffic lanes designed for use by specialized wheeled equipment. Consult specific equipment manufacturers for their recommendations.
Section 5 — Precast concrete	The tolerances for precast concrete are intended to apply to all types of precast concrete construction cast onsite *(including tilt-up)* and offsite except as set forth below. Variations to these tolerances may be advisable after consideration of panel size and construction techniques required.
	Tolerances set forth herein are not intended to apply to plant production of patented or copyrighted structural systems and/or elements. Designers, specifiers and contractors should contact the Licensors of such systems and/or products for applicable tolerances.
5.1.4 Camber	For members with a span-to-depth ratio equal to or exceeding 30, the stated camber tolerance may require special production measures and result in cost premiums. Where feasible, a greater tolerance magnitude should be utilized where the span-to-depth ratio is equal to or greater than 30.
5.3 Planer elements	Industrial precast products may not conform to the planar tolerances. Manufacturers should be consulted for appropriate tolerances for their products.

CONTENTS

Section 1 — General, p. 117-4
1.1 — Scope
1.2 — Requirements
1.3 — Definitions

Section 2 — Materials, p. 117-6
2.1 — Reinforcing steel fabrication
2.2 — Reinforcement placement
2.3 — Placement of embedded items
2.4 — Concrete batching
2.5 — Concrete properties

Section 3 — Foundations, p. 117-8
3.1 — Vertical alignment
3.2 — Lateral alignment
3.3 — Level alignment
3.4 — Cross-sectional dimensions
3.5 — Relative alignment

Section 4 — Cast-in-place concrete for buildings, p. 117-9
4.1 — Vertical alignment
4.2 — Lateral alignment
4.3 — Level alignment
4.4 — Cross-sectional dimensions
4.5 — Relative alignment
4.6 — Openings through members

Section 5 — Precast concrete, p. 117-10
5.1 — Fabrication tolerances in linear elements except piles
5.2 — Fabrication tolerances for piles
5.3 — Fabrication tolerances in planar elements
5.4 — Erection tolerances

Section 6 — Masonry, p. 117-11
6.1 — Vertical alignment
6.2 — Lateral alignment
6.3 — Level alignment

6.4 — Cross-sectional dimensions
6.5 — Relative alignment

Section 7 — Cast-in-place, vertically slipformed building elements, p. 117-11
7.1 — Vertical alignment
7.2 — Lateral alignment
7.3 — Cross-sectional dimensions
7.4 — Relative alignment

Section 8 — Mass concrete structures other than buildings, p. 117-11
8.1 — Vertical alignment
8.2 — Lateral alignment
8.3 — Level alignment
8.4 — Relative alignment

Section 9 — Canal lining, p. 117-11
9.1 — Lateral alignment
9.2 — Level alignment
9.3 — Cross-sectional dimensions

Section 10 — Monolithic siphons and culverts, p. 117-11
10.1 — Lateral alignment
10.2 — Level alignment
10.3 — Cross-sectional dimensions

Section 11 — Cast-in-place bridges, p. 117-12
11.1 — Vertical alignment
11.2 — Lateral alignment
11.3 — Level alignment
11.4 — Cross-sectional dimensions
11.5 — Relative alignment

Section 12 — Pavement and sidewalks, p. 117-12
12.1 — Lateral alignment
12.2 — Level alignment

Section 13 — Chimneys and cooling towers, p. 117-12
13.1 — Vertical alignment
13.2 — Diameter
13.3 — Wall thickness

Section 14 — Cast-in-place nonreinforced pipe, p. 117-12
14.1 — Wall thickness
14.2 — Pipe diameter
14.3 — Offsets
14.4 — Surface indentations

SECTION 1 — GENERAL REQUIREMENTS
1.1 — Scope
1.1.1 This specification designates standard tolerances for concrete construction.
1.1.2 The indicated tolerances govern unless otherwise specified.

1.2 — Requirements
1.2.1 Concrete construction shall meet the specified tolerances.
1.2.2 Tolerances shall not extend the structure beyond legal boundaries.

1.2.3 Tolerances are not cumulative. The most restrictive tolerance controls.
1.2.4 Plus (+) tolerance increases the amount or dimension to which it applies, or raises a level alignment. Minus (−) tolerance decreases the amount or dimension to which it applies, or lowers a level alignment. A nonsigned tolerance means + or −. Where only one signed tolerance is specified (+ or −), there is no limit in the other direction.

1.3 — Definitions
Arris — The line, edge, or hip in which two straight or curved surfaces of a body, forming an exterior angle, meet; a sharp ridge, as between adjoining channels of a Doric column.

Bowing — The displacement of the surface of a planar element from a plane passing through any three corners of the element.

Clear distance — In reinforced concrete, the least distance between the surface of the reinforcement and the referenced surface, i.e., the form, adjacent reinforcement, embedment, concrete, or other surface.

Concealed surface — Surface not subject to visual observation during normal use of the element.

Contract documents — The project contract, the project drawings, and the project specifications.

Cover — In reinforced concrete, the least distance between the surface of the reinforcement and the outer surface of the concrete.

Flatness — The degree to which a surface approximates a plane.

Lateral alignment — The location relative to a specified horizontal line or point in a horizontal plane.

Level alignment — The location relative to a specified horizontal plane. When applied to roadways, bridge decks, slabs, ramps, or other nominally horizontal surfaces established by elevations, level alignment is defined as the vertical location of the surface relative to the specified profile grade and specified cross slope.

Levelness — The degree to which a line or surface parallels horizontal.

Precast linear element — Beam, column, or similar unit.

Precast planar element — Wall panel, floor panel, or similar unit.

Project Specifications — The building specifications which employ ACI 117 by reference, and which serve as the instrument for making the mandatory and optional selections available under these and for specifying items not covered herein.

Relative alignment — The distance between two or more elements in any plane, or the distance between adjacent elements, or the distance between an element and a defined point or plane.

Spiral — As used in circular stave silo construction, is defined as the distortion that results when the staves are misaligned so that their edges are inclined while their outer faces are vertical. The resulting assembly appears twisted with the vertical joints becoming long-pitch spirals.

Specified surface, plane, or line — A surface, plane, or line specified by the contract documents; specified planes and lines may slope and specified surfaces may have curvature.

Tolerance —

1. The permitted variation from a given dimension or quantity.

2. The range of variation permitted in maintaining a specified dimension.

3. A permitted variation from location or alignment.

Vertical alignment — The location relative to specified vertical plane or a specified vertical line or from a line or plane reference to a vertical line or plane. When applied to battered walls, abutments or other nearly vertical surfaces, vertical alignment is defined as the

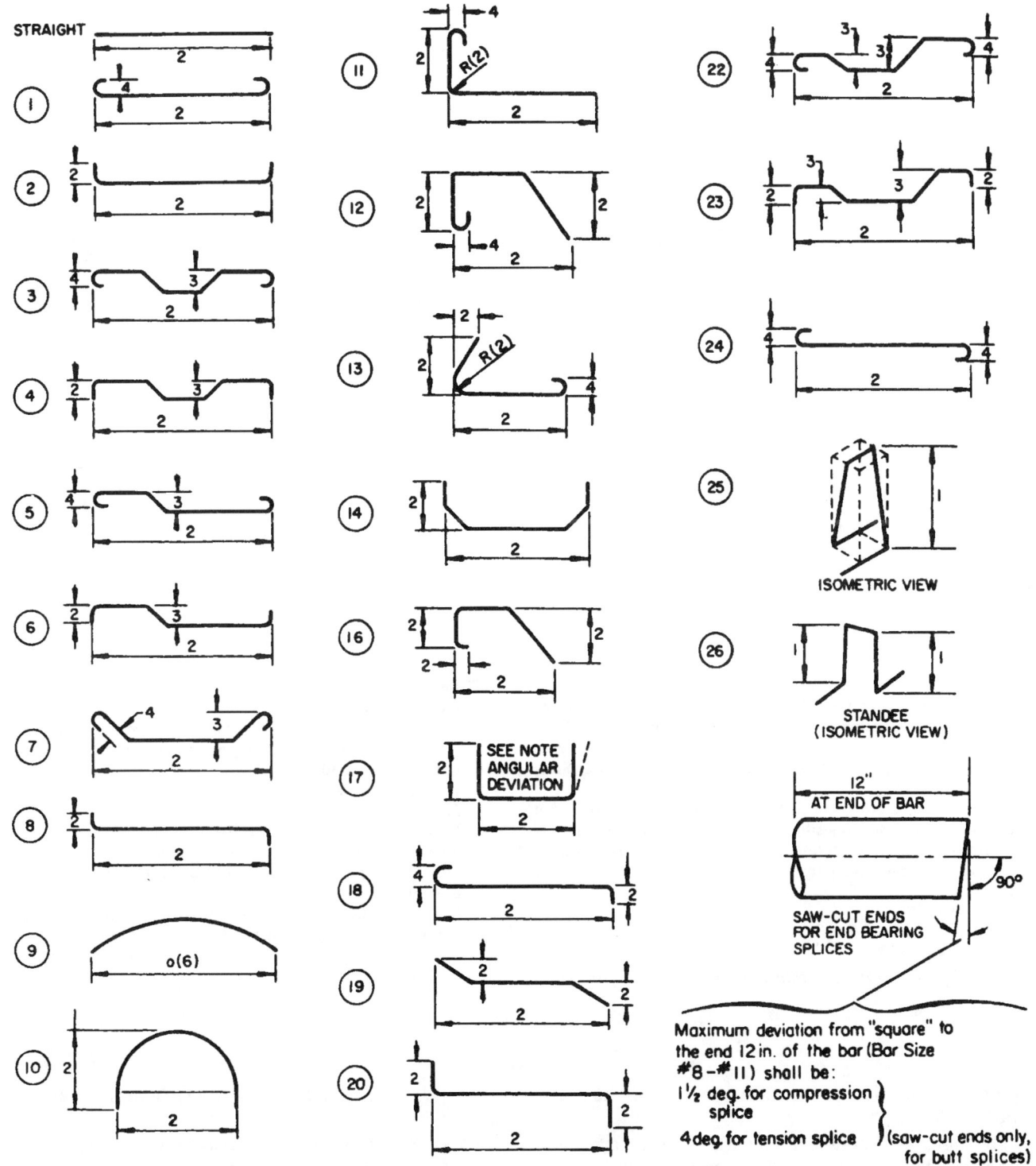

Fig. 2.1(a) — *Standard fabricating tolerances for bar sizes #3 through #11*

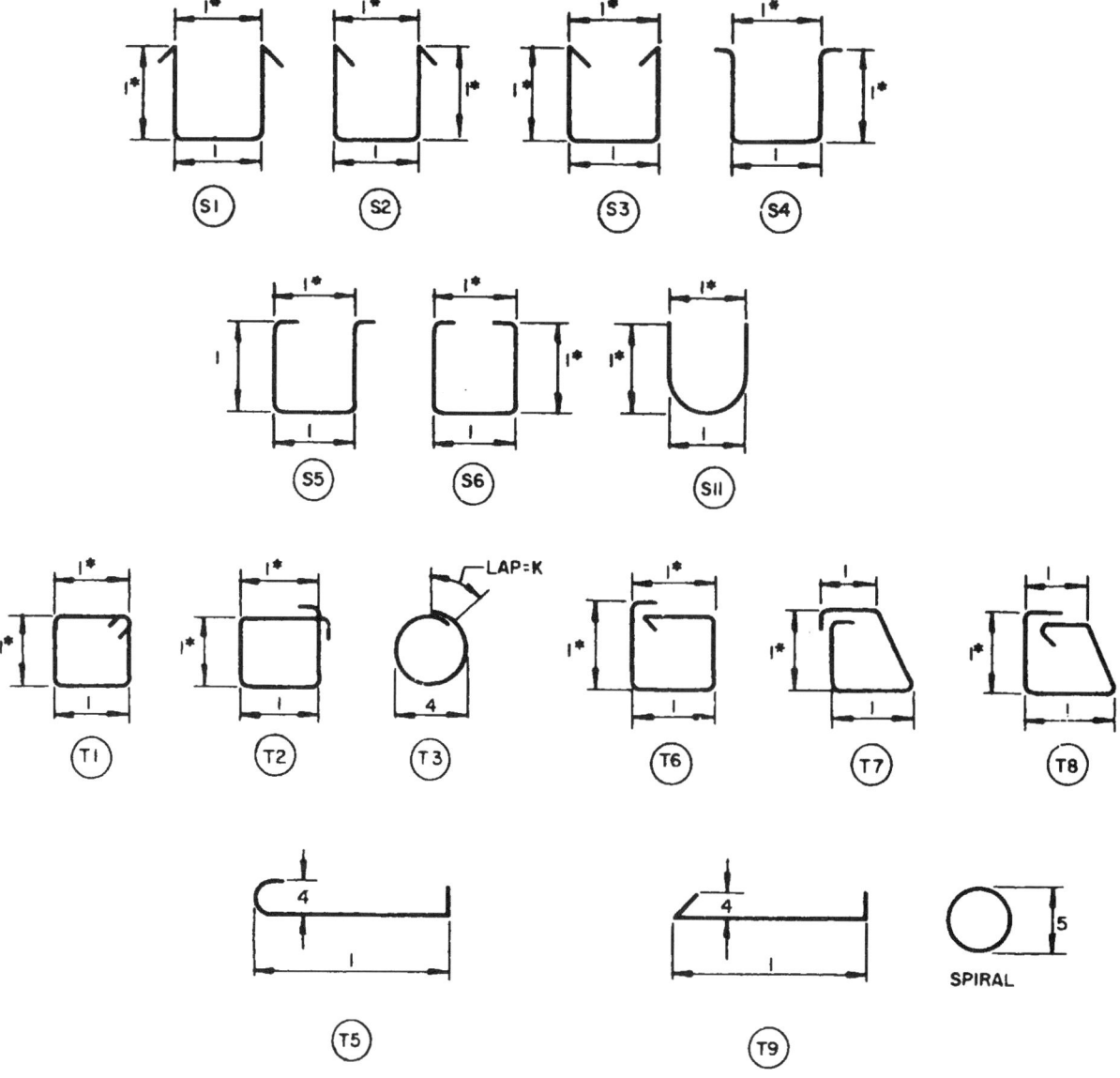

NOTES:
Entire shearing and bending tolerances are customarily absorbed in the extension past the last bend in a bent bar.

All tolerances single plane and as shown. Tolerances for Types S1 through S6, S11, and T1 through T9 apply only the Bar Sizes #3 through #8.

*Dimensions on this line are to be within tolerance shown, but are not to differ from opposite parallel dimension more than ½ in.

Angular deviation—Maximum plus or minus 2½ deg or plus or minus ½ in. per ft, but not less than ½ in., on all 90-deg hooks and bends.

TOLERANCE SYMBOLS:
1. Bar Sizes #3, #4, #5:
   = plus or minus ½ in. when gross bar length < 12 ft
   = plus or minus 1 in. when gross bar length ≥ 12 ft
2. Plus or minus 1 in.
3. Plus 0, minus ½ in.
4. Plus or minus ½ in.
5. Plus or minus ½ in. for diameter ≤ 30 in.
   Plus or minus 1 in. for diameter > 30 in.
6. Plus or minus 1.5 percent of o dimension ≥ plus or minus 2 in. minimum. If application of positive tolerance to Type 9 results in a chord length equal to or greater than the arc or bar length, the bar may be shipped straight.

Fig. 2.1(a) — *Standard fabricating tolerances for bar sizes #3 through #11*

horizontal location of the surface relative to the specified profile.

Warping — The displacement of the surface, portion, or edge of a planar element from a plane passing through any three corners of the element.

SECTION 2 — MATERIALS

2.1 — Reinforcing steel fabrication

For bars #3 and #11 in size, see Fig. 2.1(a).
For bars #14 and #18 in size, see Fig. 2.1(b).

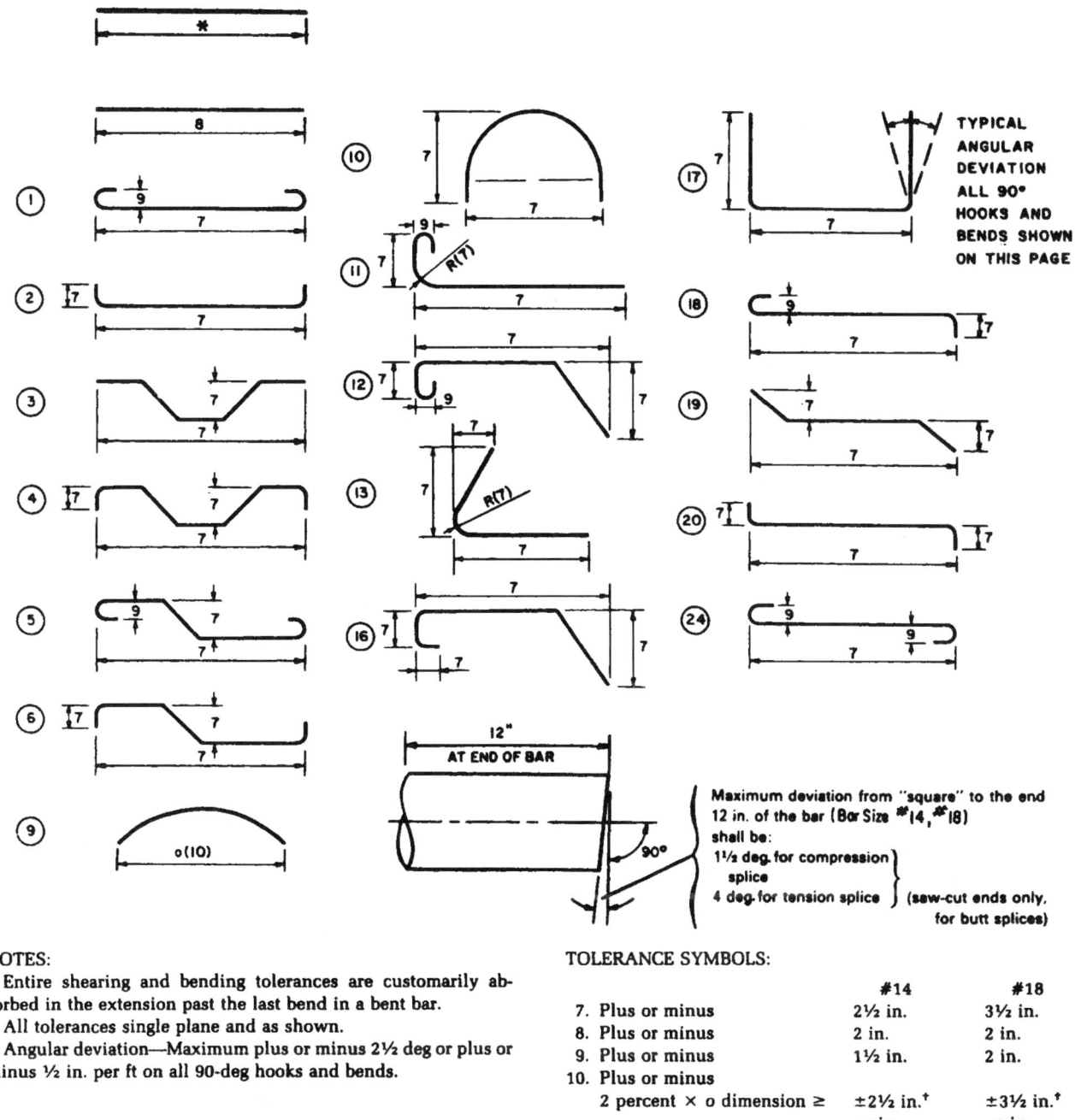

Fig. 2.1(b) — Standard fabricating tolerances for bar sizes #14 and #18

2.2 — Reinforcement placement

2.2.1 *Tolerances shall not permit a reduction in cover except as set forth in Section 2.2.3 hereof.*

2.2.2 *Clear distance to side forms and resulting concrete surfaces and clear distance to formed and resulting concrete soffits in direction of tolerance*

When member size is 4 in. or less .... + ¼ in.
.... − ⅜ in.
When member size is over 4 in. but not over 12 in. .... ⅜ in.
When member size is over 12 in. but not over 2 ft .... ½ in.
When member size is over 2 ft. .... 1 in.

2.2.3 *Concrete cover measured perpendicular to concrete surface in direction of tolerance*

When member size is 12 in. or less ...... − ⅜ in.
When member size is over 12 in. .... − ½ in.
Reduction in cover shall not exceed one-third specified concrete cover.
Reduction in cover to formed soffits shall not exceed .... ¼ in.

2.2.4 *Distance between reinforcement:*

One-quarter specified distance not to exceed .... 1 in.
Providing that distance between reinforcement shall not be less than the greater of the bar di-

ameter or 1 in. for unbundled bars.

For bundled bars, the distance between bundles shall not be less than the greater of 1 in. or 1.4 times the individual bar diameter for 2 bar bundles, 1.7 times the individual bar diameter for 3 bar bundles and 2 times the individual bar diameter for 4 bar bundles.

2.2.5 *Spacing of nonprestressed reinforcement, deviation from specified location*

In slabs and walls other than stirrups and ties .... 3 in.
Stirrups ..... depth of beam in inches/12 × 1 in.
Ties
.... least width of column in inches/12 × 1 in.
However, total number of bars shall not be less than that specified.

2.2.6 *Placement of prestressing reinforcement or prestressing steel ducts*

2.2.6.1 *Lateral placement*
Member depth (or thickness) 24 in. or less .... ½ in.
Member depth (or thickness) over 24 in. .... 1 in.

2.2.6.2 *Vertical placement*
Member depth (or thickness) 8 in. or less .... ¼ in.
Member depth (or thickness) over 8 in. but not over 24 in. .... ⅜ in.
Member depth (or thickness) more than 24 in. .... ½ in.

2.2.7 *Longitudinal location of bends and ends of bars:*

At discontinuous ends of members .... 1 in.
At other locations .... 2 in.

Table 2.4

Material	Tolerance
Cementitious materials	
30% of scale capacity or greater	1% of cumulative weight
Less than 30% of scale capacity	−0% to +4% of the required cumulative weight
Water	
Added water or ice	1% of the total water content which includes added water, ice, and water on aggregates
Total water content	3% of total water content
Aggregates	
a) Cumulative batching	
Over 30% of scale capacity	1% of the required cumulative weight
30% of scale capacity or less	0.3% of scale capacity or 3% of the required cumulative weight, whichever is less
b) Individual material batching	2% of the required weight
Admixtures	3% of the required amount

2.2.8 *Embedded length of bars and length of bar laps:*

#3 through #11 bar sizes .... −1 in.
#14 and #18 bar sizes (embedment only) −2 in.

2.2.9 *Bearing plate for prestressng tendons, deviation from specified plane* .... 1 degree

2.3 — Placement of embedded items
2.3.1 *Clearance to reinforcement the greater of the bar diameter or* .... 1 in.
2.3.2 *Vertical alignment, lateral alignment, and level alignment* .... 1 in.

2.4 — Concrete batching
See Table 2.4.

2.5 — Concrete properties
2.5.1 *Slump, where specified as "maximum" or "not to exceed," for all values* .... +0 in.
Specified slump 3 in. or less .... −1½ in.
Specified slump more than 3 in. .... −2½ in.
Slump, when specified as a single value
Specified slump 4 in. or less .... 1 in.
Specified slump more than 4 in. .... 1½ in.
Where range is specified there is no tolerance.
2.5.2 *Air content, where no range is specified and specified air content by volume is 4 percent or greater* .... 1½ percent
Where range is specified, there is no tolerance.

SECTION 3 — FOUNDATIONS
3.1 — Vertical alignment
3.1.1 *Drilled piers*

3.1.1.1 *Category A* — For unreinforced shafts extending through materials offering no or minimal lateral restraint (i.e., water, normally consolidated organic soils, and soils that might liquefy during an earthquake) — 12.5 percent of shaft diameter.

3.1.1.2 *Category B* — For unreinforced shafts extending through materials offering lateral restraint (soils other than those indicated in Category A) — not more than 1.5 percent of the shaft length.

3.1.1.3 *Category C* — For reinforced concrete shafts — not more than 2.0 percent of the shaft length.

3.2 — Lateral alignment
3.2.1 *Footings*
As cast to the center of gravity as specified; 0.02 times width of footing in direction of misplacement but not more than .... 2 in.
Supporting masonry .... ½ in.
3.2.2 *Drilled piers*
3.2.2.1 1/24 of shaft diameter but not more than .... 3 in.

3.3 — Level alignment
3.3.1 *Footings*
3.3.1.1 *Top of footings supporting masonry* ½ in.
3.3.1.2 *Top of other footings* .... +½ in.
.... −2 in.
3.3.2 *Drilled piers*
3.3.2.1 *Cut-off elevation* .... +1 in.
.... −3 in.

3.4 — Cross-sectional dimensions
3.4.1 Footings
3.4.1.1 *Horizontal dimension of formed members*
.... + 2 in.
.... − ½ in.
3.4.1.2 *Horizontal dimension of unformed members cast against soil*
2 ft. or less...................................... + 3 in.
.... − ½ in.
Greater than 2 ft. but less than 6 ft........ + 6 in.
.... − ½ in.
Over 6 ft.. + 12 in.
.... − ½ in.
3.4.1.3 *Vertical dimension (thickness)* − 5 percent

3.5 — Relative alignment
3.5.1 Footing side and top surfaces may slope with respect to the specified plane at a rate not to exceed the following amounts in 10 ft..............................1 in.

SECTION 4 — CAST-IN-PLACE CONCRETE FOR BUILDINGS

4.1 — Vertical alignment
4.1.1 For heights 100 ft or less
Lines, surfaces, and arrises ....1 in.
Outside corner of exposed corner columns and control joint grooves in concrete exposed to view... ½ in.
4.1.2 For heights greater than 100 ft
Lines, surfaces, and arrises, $1/1000$ times the height but not more than..............................6 in.
Outside corner of exposed corner columns and control joint grooves in concrete, $1/2000$ times the height but not more than........................3 in.

4.2 — Lateral alignment
4.2.1 *Members*.....................................1 in.
4.2.2 *In slabs, centerline location of openings 12 in. or smaller and edge location of larger openings*.. ½ in.
4.2.3 *Sawcuts, joints, and weakened plane embedments in slabs* .... ¾ in.

4.3 — Level alignment
4.3.1 Top of slabs:
4.3.1.1 Elevation of slabs-on-grade .... ¾ in.
4.3.1.2 Elevation of top surfaces of formed slabs before removal of supporting shores
.... ¾ in.
4.3.2 *Elevation of formed surfaces before removal of shores* .... ¾ in.
4.3.3 *Lintels, sills, parapets, horizontal grooves, and other lines exposed to view* .... ½ in.

4.4 — Cross-sectional dimensions
4.4.1 *Members, such as columns, beams, piers, walls (thickness only), and slabs (thickness only)*
12 in. dimension or less................... + ⅜ in.
.... − ¼ in.
More than 12 in. dimension but not over 3 ft dimension .... + ½ in.
.... − ⅜ in.
Over 3 ft dimension .... + 1 in.
.... − ¾ in.

4.5 — Relative alignment
4.5.1 Stairs
Difference in height between adjacent risers
.... ⅛ in.
Difference in width between adjacent trends
.... ¼ in.
4.5.2 Grooves
Specified width 2 in. or less .... ⅛ in.
Specified width more than 2 in. but not more than 12 in. .... ¼ in.
4.5.3 *Formed surfaces may slope with respect to the specified plane at a rate not to exceed the following amounts in 10 ft*
4.5.3.1 Vertical alignment of outside corner of exposed corner columns and control joint grooves in concrete exposed to view
.... ¼ in.
4.5.3.2 All other conditions .... ⅜ in.
4.5.4. *The offset between adjacent pieces of formwork facing material shall not exceed:*
Class of surface:
Class A .... ⅛ in.
Class B .... ¼ in.
Class C .... ½ in.
Class D .... 1 in.
4.5.5 *Floor finish tolerances shall meet the requirements of either Section 4.5.6 or 4.5.7, as set forth by the specifier.*
4.5.6 *Floor finish tolerances as measured in accordance with ASTM E 1155-87 Standard Test Method for Determining Floor Flatness and Levelness Using the F-Number System (Inch-Pound Units)*

Floor profile quality classification	Minimum F_f, F_l number required			
	Test area		Minimum local F number	
	Flatness F_f	Level F_l	Flatness F_f	Level F_l
Conventional Bullfloated Straightedged	15 20	13 15	13 15	10 10
Flat	30	20	15	10
Very flat	50	30	25	15

4.5.6.1 The F_l levelness tolerance shall not apply to slabs placed on unshored form surfaces and/or shored form surfaces after the removal of shores. F_l levelness tolerances shall not apply to cambered or inclined surfaces and shall be measured within 72 hr after slab concrete placement.

4.5.7 Floor finish tolerances as measured by placing a freestanding (unleveled) 10 ft. straightedge anywhere on the slab and allowing it to rest upon two high spots within 72 hr after slab concrete placement. The gap at any point between the straightedge and the floor (and between the highspots) shall not exceed:

Classification:
Conventional
Bullfloated .... ½ in.
Straightedged .... 5/16 in.

Flat .... 3/16 in.
Very flat .... 1/8 in.

4.6 — Openings through members
4.6.1 *Cross-sectional size of opening* .... − 1/4 in.
.... + 1 in.
4.6.2 *Location of centerline of opening* .... 1/2 in.

SECTION 5 — PRECAST CONCRETE
5.1 — Fabrication tolerances in linear elements except piles
5.1.1 *Length of member*
Per 10 ft .... 1/8 in.
Total not more than .... 3/4 in.
5.1.2 *Cross-sectional dimensions*
6 in. or less .... 1/8 in.
Over 6 in. but not over 18 in. .... 3/16 in.
Over 18 in. but not over 36 in. .... 1/4 in.
Over 36 in. .... 3/8 in.
5.1.3 *Lateral alignment (sweep) of noncambered member surfaces relative to centerline of member*
Member length
40 ft and less .... 1/4 in.
Over 40 ft but not over 60 ft .... 3/8 in.
Over 60 ft .... 1/2 in.
5.1.4 *Camber variation from design chamber, at time of erection*
For nonprestressed elements, 1/8 in. per 10 ft of length but not more than .... 1/2 in.
For prestressed elements, 1/4 in. per 10 ft of length but not more than...................... 1 in.
5.1.5 *Surface irregularities, deviation from a 10 ft straightedge*..
For elements which will not receive topping
.... 1/4 in.
For elements to receive topping.............. 1/2 in.
For elements to be used as concrete guideways support and steering surfaces
.... 1/8 in.

5.2 — Fabrication tolerances for piles
5.2.1 *Length*.. + 6 in.
.... − 2 in.
5.2.2 *Cross-sectional dimensions*
Overall .... 3/8 in.
Wall thickness of hollow sections .... + 1/2 in.
.... − 0 in.
5.2.3 *Lateral alignment of pile surfaces relative to pile centerline in length of pile, per 10 ft*............ 1/8 in.
5.2.4 *Location of internal void* .... 3/8 in.
5.2.5 *Pile head*
From the plane perpendicular to the longitudinal axis of pile, 1/4 in. in 12 in. but not more than .... 1/2 in.
5.2.6 *Surface irregularities*
Pile head .... 1/8 in.
Other surfaces, deviation from a 10 ft. straightedge .... 1/4 in.

5.3 — Fabrication tolerances in planar elements
5.3.1 *Length and width*
10 ft or less .... 1/8 in.
Over 10 ft but not over 20 ft. .... + 1/8 in.
.... − 3/16 in.
Over 20 ft but not over 40 ft. .... 1/4 in.
Each additional 10 ft increment in excess of 40 ft. .... 1/16 in.
Difference in length of the two diagonals, of a rectangular member the greater of 1/8 in. per 6 ft of diagonal or .... 1/2 in.
5.3.2 *Cross-sectional dimensions*
thickness .... + 1/4 in.
.... − 1/8 in.
5.3.3 *Openings in panels*
Size of opening .... 1/4 in.
Location of centerline of opening........... 1/4 in.
5.3.4 *Lateral alignment of embedded items*
Reglets for glazing gaskets .... 1/8 in.
Bolts .... 1/4 in.
Flashing reglets .... 1/4 in.
Flashing reglets at panel edge .... 1/8 in.
Electrical outlets and pipe sleeves .... 1/2 in.
Weld plates .... 1 in.
Inserts .... 1/2 in.
5.3.5 *Bowing and warping at time of erection*
Bowing
1/360 *times* the panel diagonal dimension in inches but not more than...................... 1 in.
Warping
1/16 in. per ft. of distance from nearest adjacent corner but not more than .... 1 in.

5.4 — Erection tolerances
5.4.1 *Vertical, lateral, and level alignment*
5.4.1.1 *Building elements*
Same as for cast-in-place concrete in Section 4.0.
5.4.1.2 *Concrete guideways*
Concrete guideway construction misalignment of support or steering surfaces shall not exceed
.... 1/16 in.

5.4.2 *Alignment of exposed wall panels*
5.4.2.1 Width of joints between exposed wall panels .... 1/4 in.
5.4.2.2 Taper (difference in width) of joint between adjacent exposed wall panels, the greater of, 1/40 in. per linear foot of joint, or .... 1/16 in.
Not to exceed .... 3/8 in.
5.4.2.3 Alignment of joints at adjoining corners
.... 1/4 in.
5.4.2.4 Offset in exterior face of adjacent panels .... 1/4 in.
5.4.3 *Offset of top surfaces of adjacent elements in erected positon . . .*
With topping slab .... 3/4 in.
Floor elements without topping slab .... 1/4 in.
Roof elements without topping slab .... 3/4 in.
Guideway elements to be used as riding surface .... 1/16 in.

SECTION 6 — MASONRY

6.1 — Vertical alignment
In surface of wall .... ¾ in.
In alignment of head joints .... ½ in.

6.2 — Lateral alignment
6.2.1 *Vertical members* .... ½ in.

6.3 — Level alignment
6.3.1 *In bed joints and top of wall,*
exposed .... ½ in.
Not exposed .... 1 in.
6.3.2 *Top of wall used for a bearing surface* ... ½ in.
6.3.3 *Top of wall, other than a bearing surface*
.... ¾ in.

6.4 — Cross-sectional dimensions
6.4.1 *Multiwythed walls* .... + ½ in.
.... − ¼ in.
6.4.2 *Other members* .... + ½ in.
.... − ¼ in.
6.4.3 *Joint thickness* .... ⅛ in.

6.5 — Relative alignment
6.5.1 *Masonry surfaces may slope with respect to the specified plane at a rate not to exceed the following amounts in 10 ft*
    6.5.1.1 *Walls and columns* .... ¼ in.
    6.5.1.2 *Bed joints, head joints,*
    and top of wall .... ¼ in.
    6.5.1.3 *Top of wall* .... ¼ in.

SECTION 7 — CAST-IN-PLACE, VERTICALLY SLIPFORMED BUILDING ELEMENTS

7.1 — Vertical alignment
7.1.1 *Translation and rotation from a fixed point at the base of the structure:*
For heights 100 ft. or less .... 2 in.
For heights greater than 100 ft., 1/600 times the height but not more than .... 8 in.

7.2 — Lateral alignment
Between adjacent elements .... 2 in.

7.3 — Cross-sectional dimensions
Walls .... + ¾ in.
.... − ⅜ in.

7.4 — Relative alignment
Formed surfaces may slope with respect to the specified plane at a rate not to exceed the following amount in 10 ft .... ¼ in.

SECTION 8 — MASS CONCRETE STRUCTURES OTHER THAN BUILDINGS

8.1 — Vertical alignment
8.1.1 *Surfaces*
Visible surfaces .... 1¼ in.
Concealed surfaces .... 2½ in.
8.1.2 *Side walls for radial gates and similar watertight joints* .... 3/16 in.

8.2 — Lateral alignment
Visible surfaces .... 1¼ in.
Concealed surfaces .... 2½ in.

8.3 — Level alignment
8.3.1 *General*
Visible flatwork and formed surfaces ...... ½ in.
Concealed flatwork and formed surfaces ... 1 in.
8.3.2 *Sills for radial gates and similar watertight joints* .... 3/16 in.

8.4 — Relative alignment
8.4.1 *Formed surfaces may slope with respect to the specified plane at a rate not to exceed the following amount in 10 ft*
    8.4.1.1 *Slopes in lateral and level alignments*
Visible surfaces .... ¼ in.
Concealed surfaces .... ½ in.
    8.4.1.2 *Slopes in vertical alignment*
Visible surfaces .... ½ in.
Concealed surfaces .... 1 in.

SECTION 9 — CANAL LINING

9.1 — Lateral alignment
9.1.1 *Alignment of tangents* .... 2 in.
9.1.2 *Alignment of curves* .... 4 in.
9.1.3 *Width of section at any height: 0.0025 times specified width* W *plus one in.* .... 0.0025W + 1 in.

9.2 — Level alignment
9.2.1 *Profile grade* .... 1 in.
9.2.2 *Surface of invert* .... ¼ in.
9.2.3 *Surface of side slope* .... ½ in.
9.2.4 *Height of lining: 0.005 times established height* H *plus one in.* .... 0.005H + 1 in.

9.3 — Cross-sectional dimensions
Thickness of lining cross section: 10 percent of specified thickness provided average thickness is maintained as determined by daily batch volumes.

SECTION 10 — MONOLITHIC SIPHONS AND CULVERTS

10.1 — Lateral alignment
10.1.1 *Centerline alignment* .... 1 in.
10.1.2 *Inside dimensions:*
.... 0.005 times inside dimension

10.2 — Level alignment
10.2.1 *Profile grade* .... 1 in.
10.2.2 *Surface of invert* .... ¼ in.
10.2.3 *Surface of side slope* .... ½ in.

10.3 — Cross-sectional dimensions
10.3.1 *Cross section at any point*
Increase thickness: greater of 0.05 times thickness, or .... + ½ in.
Decrease thickness: greater of 0.25 times thickness, or .... − ¼ in.

SECTION 11 — CAST-IN-PLACE BRIDGES

11.1 — Vertical alignment
11.1.1 *Exposed surfaces* .... ¾ in.
11.1.2 *Concealed surfaces* .... 1½ in.

11.2 — Lateral alignment
Centerline alignment .... 1 in.

11.3 — Level alignment
11.3.1 *Profile grade* .... 1 in.
11.3.2 *Top of other concrete surfaces and horizontal grooves*
Exposed .... ¾ in.
Concealed .... 1½ in.
11.3.3 *Mainline pavements in longitudinal direction, the gap below a 10 ft unleveled straightedge resting on highspots shall not exceed* .... ⅛ in.
11.3.4 *Mainline pavements in transverse direction, the gap below a 10 ft unleveled straightedge resting on highspots shall not exceed* .... ¼ in.
11.3.5 *Ramps, sidewalks, and intersections, in any direction, the gap below a 10 ft unleveled straightedge resting on highspots shall not exceed* .... ¼ in.

11.4 — Cross-sectional dimensions
11.4.1 *Bridge slabs vertical dimension (thickness)* .... + ¼ in.
.... − ⅛ in.
11.4.2 *Members such as columns, beams, piers, walls, and other (slabs thickness only)* .... + ½ in.
.... − ¼ in.
11.4.3 *Openings through concrete members* .... ½ in.

11.5 — Relative alignment
11.5.1 *Location of openings through concrete members* .... ½ in.
11.5.2 *Formed surfaces may slope with respect to the specified plane at a rate not to exceed the following amounts in 10 ft*
Watertight joints .... ⅛ in.
Other exposed surfaces .... ½ in.
Concealed surfaces .... 1 in.
11.5.3 *Unformed exposed surfaces, other than pavements and sidewalks, may slope with respect to the specified plane at a rate not to exceed the following amounts*
In 10 ft .... ¼ in.
In 20 ft .... ⅜ in.

SECTION 12 — PAVEMENTS AND SIDEWALKS

12.1 — Lateral alignment
12.1.1 *Placement of dowels* .... 1 in.
12.1.2 *Alignment of dowels, relative to centerline of pavement, 18 in. or less projection* ...... ¼ in.
greater than 18 in. projection
.... *Not established*

12.2 — Level alignment
12.2.1 *Mainline pavements in longitudinal direction, the gap below a 10 ft unleveled straightedge resting on highspots shall not exceed* .... ⅛ in.
12.2.2 *Mainline pavements in transverse direction, the gap below a 10 ft unleveled straightedge resting on highspots shall not exceed* .... ¼ in.
12.2.3 *Ramps, sidewalks, and intersections, in any direction, the gap below a 10 ft unleveled straightedge resting on highspots shall not exceed* .... ¼ in.

SECTION 13 — CHIMNEYS AND COOLING TOWERS

13.1 — Vertical alignment
Translation, rotation or variance form vertical axis the greater of 1/1000 times the height at time of measurement, or 1 in.

In any 10 ft of height the centerpoint shall not change more than 1 in.

13.2 — Diameter
Outside shell diameter 1/100 times the specified diameter plus 1 in.

13.3 — Wall thickness
The average of four wall thickness measurements taken over a 60 deg arc.
Specified wall thickness 10 in. or less .... − ¼ in.
.... + ½ in.
Specified wall thickness greater than 10 in. ... − ½ in.
.... + 1 in.

SECTION 14 — NONREINFORCED CAST-IN-PLACE PIPE

14.1 — Wall thickness
Minimum wall thickness at any point shall be 1/12 times the specified internal diameter of the pipe plus ½ in., but in no case less than .... 2 in.

14.2 — Pipe diameter
The internal diameter at any point shall not be less than 95 percent of the specified diameter, the average of any four measurements taken at 45 deg intervals shall not be less than the specified diameter.

14.3 — Offsets
At formlaps and horizontal edges shall not exceed:
For pipe with an internal diameter not greater than 42 in. .... ½ in.
For pipe with an internal diameter 43 through 72 in.
.... ¾ in.
For pipe with an internal diameter greater than 72 in. .... 1 in.

14.4 — Surface indentations
Maximum allowable .... ½ in.

This standard was submitted to letter ballot of the committee and approved in accordance with the Institute's balloting procedures.

ACI 117R-90
(Reapproved 2002)

Commentary on Standard Specifications for Tolerances for Concrete Construction and Materials (ACI 117-90)

Reported by ACI Committee 117

W. Robert Little
Chairman

Russell S. Fling
Chairman, Editorial Subcommittee

S. Allen Face
Thomas C. Heist
Richard A. Kaden
Ross Martin
Peter Meza

Andrawos Morcos
Clark B. Morgan Jr.
Harry M. Palmbaum
William S. Phelan

B. J. (Duke) Pointer
Dean B. Stephan Jr.*
Eldon Tipping
Carl S. Togni
Joe V. Williams, Jr.

This report is a commentary on the Standard Specifications for Tolerances for Concrete Construction and Materials. It is intended to be used with ACI 117 for clarity of interpretation and insight into the intent of the committee regarding the application of the tolerances set forth therein.

Keywords: bending (reinforcing steels); building codes; **concrete construction**; concrete piles; concretes; floors; formwork (construction); masonry; mass concrete; piers; precast concrete; prestressed concrete; reinforcing steels, **specifications**; splicing; **standards**; **tolerances (mechanics)**.

INTRODUCTION

This commentary pertains to "Standard Specifications for Tolerances for Concrete Construction and Materials (ACI-117)." The purpose of the report is to provide graphic and written interpretations for the specification and its application.

No structure is exactly level, plumb, straight, and true. Fortunately, such perfection is not necessary. Tolerances are a means to establish permissible variation in dimension and location, giving both the designer and the contractor parameters within which the work is to be performed. They are the means by which the designer conveys to the contractor the performance expectations upon which the design is based or the use of the project requires. Such specified tolerances should reflect design assumptions and project needs, being neither overly restrictive nor lenient. Necessity rather than desirability should be the basis of selecting tolerances.

As the title "Standard Specifications for Tolerances for Concrete Construction and Materials (ACI 117)" implies, the tolerances given are standard or usual tolerances that apply to various types and uses of concrete construction. They are based upon normal needs and common construction techniques and practices. Specific tolerances at variance with the standard values can cause both increases and decreases in the cost of construction.

The required degree of accuracy of performance depends on the interrelationship of several factors:

Structural strength and function requirements
The structure must be safe and strong, reflecting the design assumptions, and accurate enough in size and shape to do the job for which it was designed and constructed.

Esthetics
The structure must satisfy the appearance needs or wishes of the owner and the designer.

Economic feasibility
The specified degree of accuracy has a direct impact on the cost of production and the construction method. In general, the higher degree of accuracy required, the higher the cost of obtaining it.

Relationship of all components
The required degree of accuracy of individual parts can be influenced by adjacent units and materials, joint and connection

---

ACI Committee Reports, Guides, Standard Practices, and Commentaries are intended for guidance in planning, designing, executing, and inspecting construction. This document is intended for the use of individuals who are competent to evaluate the significance and limitations of its content and recommendations and who will accept responsibility for the application of the material it contains. The American Concrete Institute disclaims any and all responsibility for the stated principles. The Institute shall not be liable for any loss or damage arising therefrom.

Reference to this document shall not be made in contract documents. If items found in this document are desired by the Architect/Engineer to be a part of the contract documents, they shall be restated in mandatory language for incorporation by the Architect/Engineer.

---

*Chairman during initial development of this document.
Copyright © 1990, American Concrete Institute.
All rights reserved including rights of reproduction and use in any form or by any means, including the making of copies by any photo process, or by electronic or mechanical device, printed, written, or oral, or recording for sound or visual reproduction or for use in any knowledge or retrieval system or device, unless permission in writing is obtained from the copyright proprietors.

details, and the possibility of the accumulation of tolerances in critical dimensions.

Construction techniques
The feasibility of a tolerance depends on available craftsmanship, technology, and materials.

Properties of materials
The specified degree of accuracy for shrinkage and prestressed camber should recognize the degree of difficulty of predetermining deflection due to shrinkage and prestressed camber.

Compatibility
Designers are cautioned to use finish and architectural details that are compatible with the type and anticipated method of construction. Finish and architectural details used should be compatible with the concrete tolerances which are achievable.

Job conditions
Unique job situations and conditions must be considered. The designer must specify and clearly identify those items that require either closer or more lenient tolerances as the needs of the project dictate.

Measurement
Mutually agreed-upon control points and bench marks must be provided as reference points for measurements to establish the degree of accuracy of items produced and for verifying the tolerances of the items produced. Control points and bench marks should be established and maintained in an undisturbed condition until final completion and acceptance of the project.

Project document references
ACI Specification documents—The following American Concrete Institute documents provide mandatory requirements for concrete construction and may be referenced in the Project Documents:

ACI 117	Standard Specifications for Tolerances for Concrete Construction and Materials
ACI 301	Specifications for Structural Concrete for Buildings
ACI 531.1	Specification for Concrete Masonry Construction

ACI informative documents—ACI Committee Reports, Guides, Standard Practices, and Commentaries are intended for guidance in designing, planning, executing, or inspecting construction, and in preparing plans and specifications. Reference to these Reports, Guides, and Standard Practices should not be included in the Project Documents. If the Architect/Engineer desires to include items found in these ACI documents in the Project Documents, they should be rephrased in mandatory language and incorporated into the Project Documents.

The documents of the following American Concrete Institute Committees cover practice, procedures, and state-of-the-art guidance for the categories of construction as listed.

General building	ACI 302, 303, 304, 318, 347
Special structures	ACI 307, 313, 316, 325, 332, 334, 344, 345, 349, 350, 357, 358
Precast construction	ACI 347
Masonry construction	ACI 531
Materials	ACI 211, 223, 302, 304, 315, 318, 531, 543

TABLE OF CONTENTS
Introduction, p. 117R-1

Section 1—General requirements, p. 117R-2

Section 2—Materials, p. 117R-4

Section 3—Foundations, p. 117R-5

Section 4—Cast-in-place concrete for buildings, p. 117R-5

Section 5—Precast concrete, p. 117R-8

Section 6—Masonry, p. 117R-10

Section 7—Cast-in-place, vertically slipformed structures, p. 117R-10

Section 8—Mass concrete structures other than buildings, p. 117R-10

Section 9—Canal lining, p. 117R-10

Section 10—Monolithic siphons and culverts, p. 117R-10

Section 11—Cast-in-place bridges, p. 117R-10

Section 12—Pavement, p. 117R-10

Section 13—Chimneys and cooling towers, p. 117R-11

Section 14—Cast-in-place nonreinforced pipe, p. 117R-11

Section 15—References, p. 117R-11

SECTION 1—GENERAL REQUIREMENTS
1.3—Definitions
Bowing—See Fig. 1.3.1.
Flatness—See Fig. 1.3.2.
Lateral alignment—See Fig. 1.3.3.
Level alignment—See Fig. 1.3.4.
Relative alignment—See Fig. 1.3.5.
Vertical alignment—See Fig. 1.3.6.
Warping—See Fig. 1.3.7.

Level alignment, lateral alignment, and vertical alignment are used to establish a tolerance envelope within which permissible variations can occur. Relative alignment, in addition to designating allowable relative displacements of elements, is used to determine the rate of change of adjacent

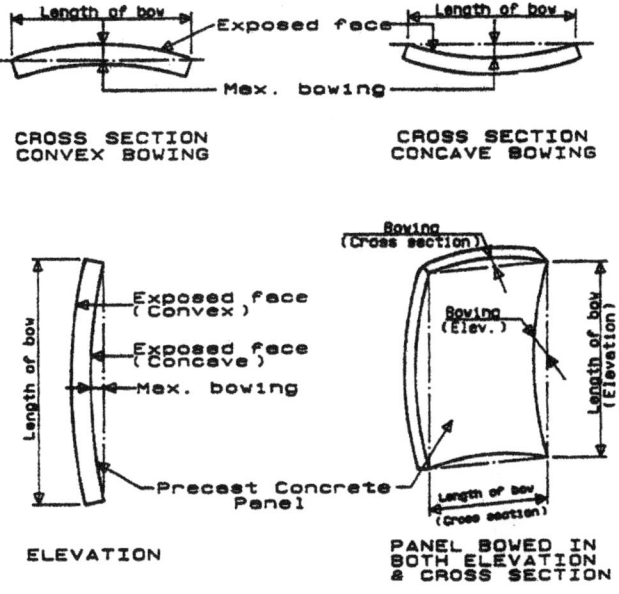

Fig. 1.3.1—Bowing

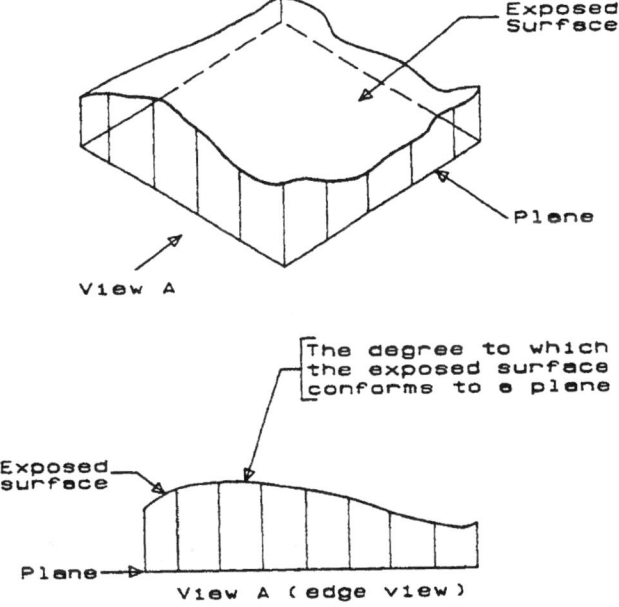

Fig. 1.3.2—Flatness

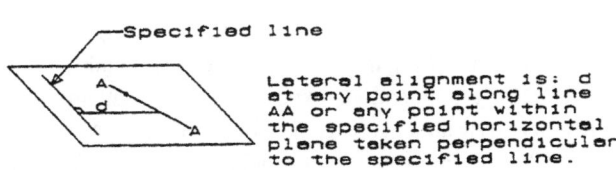

Fig. 1.3.3—Lateral alignment

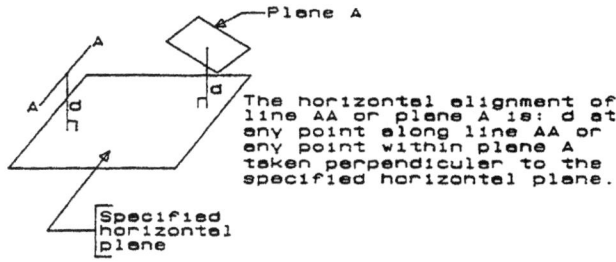

Fig. 1.3.4—Level alignment

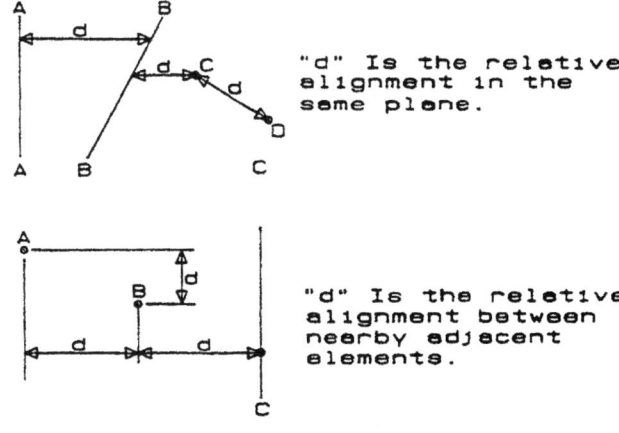

Fig. 1.3.5—Relative alignment

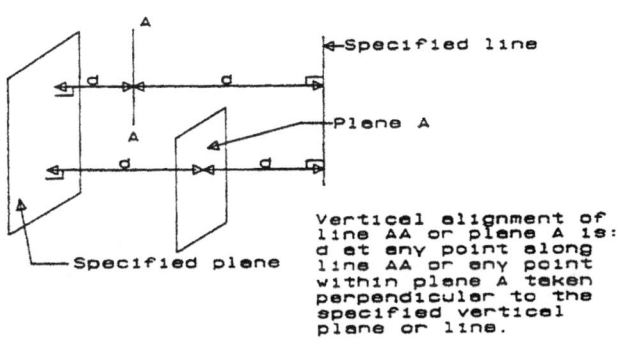

Fig. 1.3.6—Vertical alignment

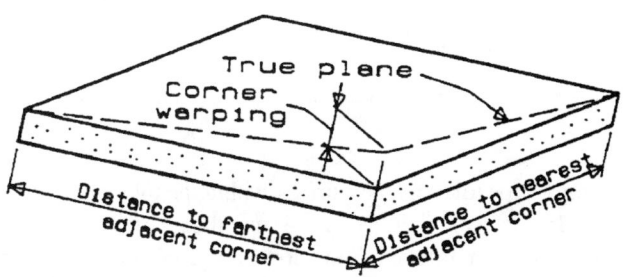

Fig. 1.3.7—Warping

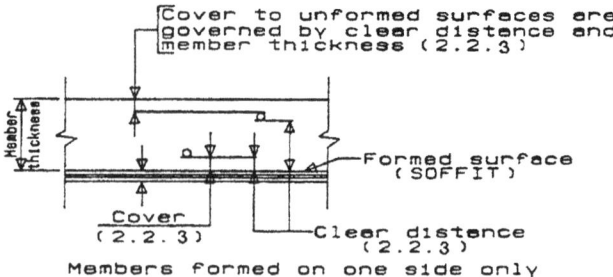

Fig. 2.2.2(a)—Reinforcement placement

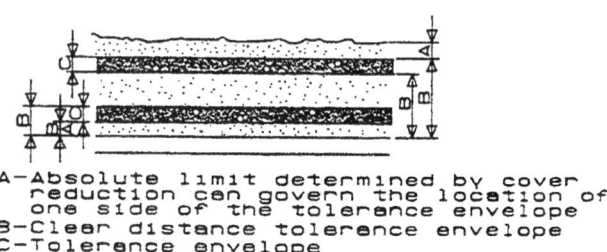

Fig. 2.2.2(b) and 2.2.3(b)—Reinforcement placement

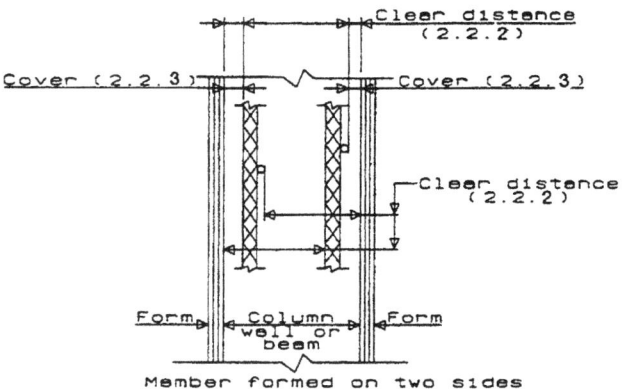

Fig. 2.2.3(a)—Reinforcement placement

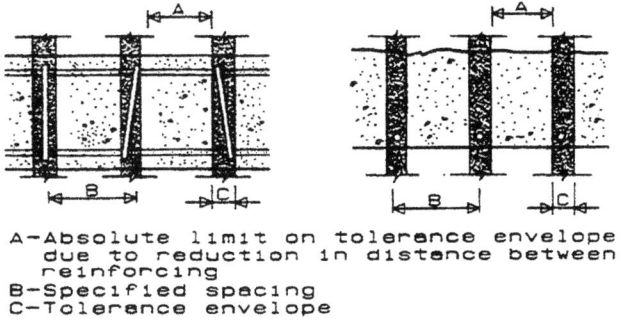

Fig. 2.2.4 and 2.2.5—Reinforcement placement

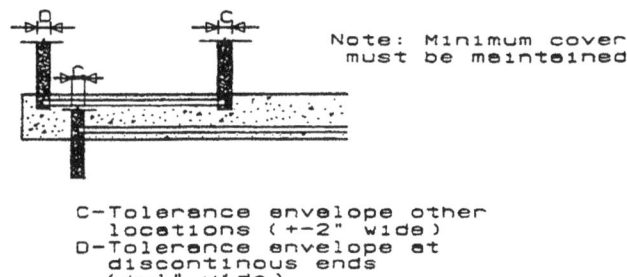

Fig. 2.2.7—Reinforcement placement, longitudinal location

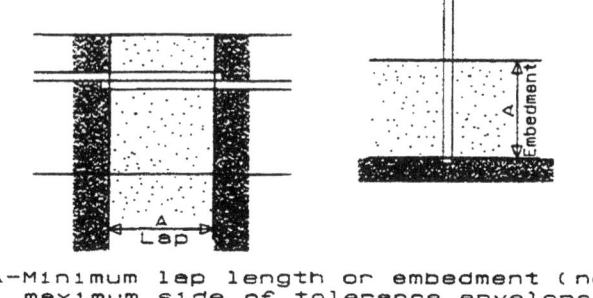

Fig. 2.2.8—Reinforcement placement, embedment and laps

objectionable when exposed to view. The acceptable relative alignment of points on a surface or line is determined by using a slope tolerance.

SECTION 2—MATERIALS
2.2—Reinforcement

In the absence of specific design details shown or specified on the contract documents, CRSI MSP-l, Appendix D, should be followed by estimators, detailers, and placers.

2.2.2 and 2.2.3 The tolerance for placing reinforcing steel is predicated upon measurements of the formed surfaces for quality control during construction and from the resulting surfaces for forensic analysis. It consists of an envelope with an absolute limitation on one side of the envelope determined by the limit on the reduction in cover. See Fig. 2.2.2(a), 2.2.2(b), 2.2.3(a), and 2.2.3(b).

2.2.4 and 2.2.5 The spacing tolerance of reinforcing consists of an envelope with an absolute limitation on one side of the envelope determined by the limit on the reduction in distance between reinforcement. In addition, the allowable tolerance on spacing shall not cause a reduction in the specified number of reinforcing bars utilized. See Fig. 2.2.4 and 2.2.5.

2.2.6 The vertical deviation tolerance should be considered in establishing minimum prestressing tendon covers, particularly in applications exposed to deicer chemicals or salt water environments where use of additional cover is recommended to compensate for placing tolerances. Slab behavior is relatively insensitive to horizontal location of tendons.

points (slope tolerance) occurring within the tolerance envelope. In this fashion the slope and smoothness of surfaces and lines within a tolerance envelope are controlled. Abrupt changes, offsets, sawtoothing, sloping, etc., of lines and surfaces properly located within a tolerance envelope may be

2.2.7 and 2.2.8 The tolerance for the location of the ends of reinforcing steel is determined by these two sections. See Fig. 2.2.7 and 2.2.8.

2.5—Concrete
2.5.1 Where the specification has specified slump as a maximum, the project specifications should provide for the addition of water at the jobsite for slump adjustment. This is because the concrete must be batched at a lesser slump to avoid rejection because of a lack of a plus tolerance for the slump. The water added at the jobsite must be within the water/cement limitations of the specifications or approved mixture proportions.

Flowable concrete achieved by the incorporation of high range water reducers (HRWR) (superplasticizers), are difficult to control within tight tolerances at specified slumps of 7 in. or greater. In addition, it is difficult to accurately measure high slumps. Consideration should be given to eliminating a maximum slump when a HRWR is used to achieve flowable concrete.

When a slump range is specified, caution should be exercised and jobsite conditions should be considered and evaluated to determined if the range is suitable for delivery and placing requirements.

2.5.2 When an air content range is specified, care should be given to address aggregate size and jobsite requirements. The range should be adequately wide to accommodate the preceding.

SECTION 3—FOUNDATIONS
3.2—Lateral alignment
3.2.1 Determines the permissible location of a footing. The magnitude of tolerance for the location of footings is governed by the width (i.e., least dimension in plan view) of the footing with an absolute limit depending on the subsequent construction material supported by the footing. See Fig. 3.2.1.

3.3—Level alignment
Determines the location of any point on the top surface of a footing relative to the specified plane. See Fig. 3.3.1.

3.4—Cross-sectional dimension
Determines the permissible size of a footing. See Fig. 3.4.

3.5—Relative alignment
The relative alignment of points on the surfaces cannot exceed the distance determined by the slope tolerance. Determines the permissible top surface roughness or irregularity of a footing. See Fig. 3.5.

SECTION 4—CAST-IN-PLACE CONCRETE FOR BUILDINGS
4.1, 4.4, and 4.5—Vertical and relative alignment and thickness
Determines the permissible location of surfaces and lines in a vertical plane and the smoothness of those surfaces or straightness of lines and the relative location of adjacent sur-

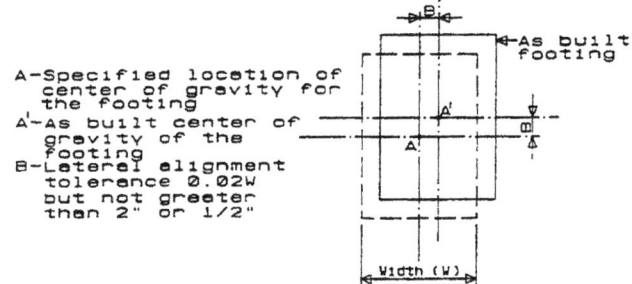

Fig. 3.2.1—Footing lateral alignment

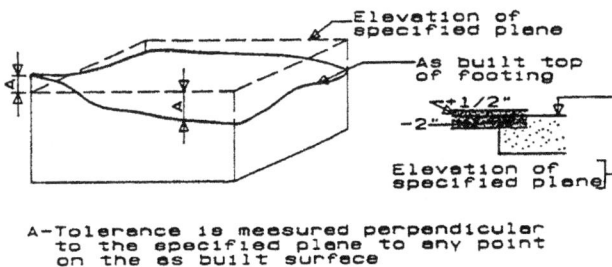

Fig. 3.3.1—Level alignment

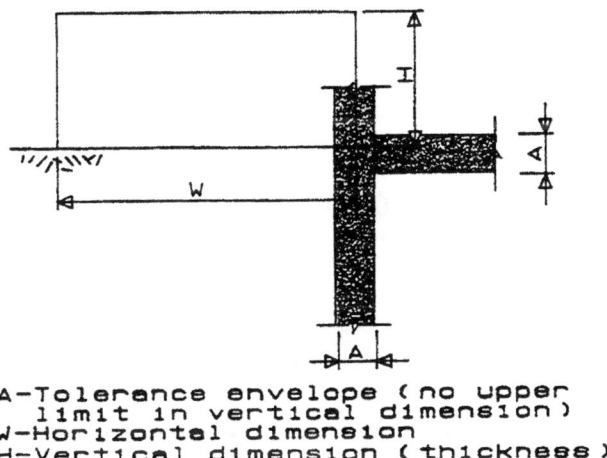

Fig. 3.4—Footing cross-sectional dimension

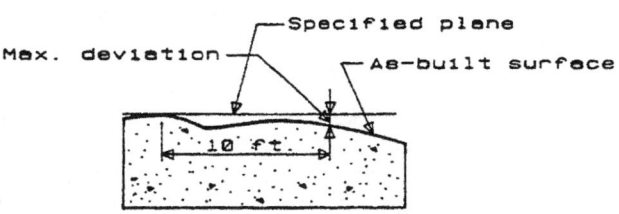

Fig. 3.5—Relative alignment of footing surface

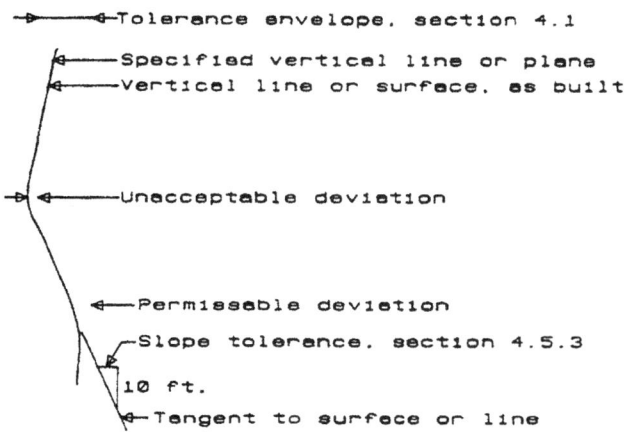

Fig. 4.1(a) and 4.5.3(a)

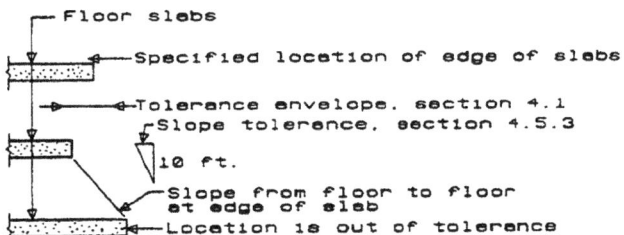

Fig. 4.1(a), (b) and 4.5.3(a), (b)—Vertical and relative alignment

faces in a vertical plane. See Fig. 4.1(a) and (b) and 4.5.3(a) and (b).

4.3, 4.4, and 4.5—Level and vertical alignment and cross-sectional dimensions

If the level and cross-sectional dimension tolerances are given, then a suspended (elevated) slab is fully toleranced.

Example: 12 in. slab—The envelope for the slab element extends 3/4 in. above the specified surface elevation to 3/4 in. below the specified soffit elevation. Thus the slab surface and/or soffit can be 3/4 in. higher or lower than specified. The slab thickness can be 3/8 in. greater or 1/4 in. less than specified; the rate of change in slope of the top surface is toleranced by the F_L, and the soffit is toleranced by the relative alignment and formed surface tolerances. See Fig. 4.3, 4.4, and 4.5.3 (c).

The acceptable elevation envelope of the slab surface and soffit is ± 3/4 in. The rate of change of the adjacent surface elevation points within the acceptable elevation is governed by specification Section 4.5.5.

4.5.5 Floor profile finish quality has traditionally been measured by limiting the gap to be measured under either a freestanding or leveled 10-ft straightedge, according to the specifier's requirements. The technology for measuring floor profiles has rapidly evolved in response to the needs of random vehicular traffic industrial users. This technology provides a welcome alternative and a solution to the general-

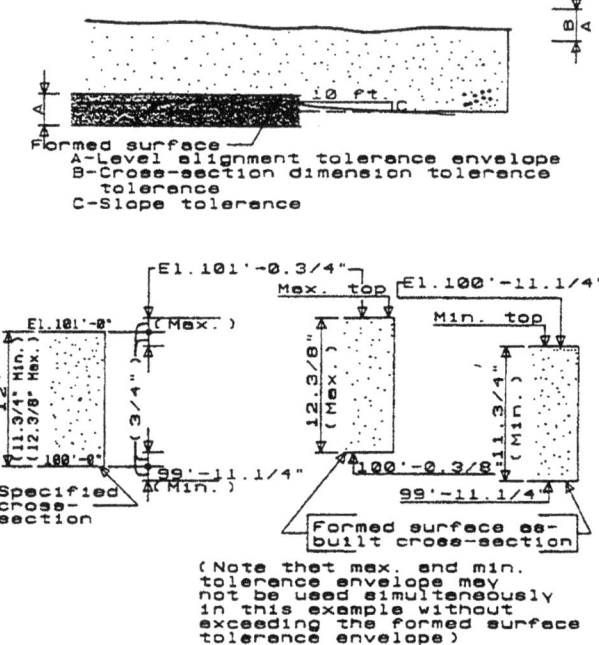

Fig. 4.3, 4.4, and 4.5.3(c)—Level and relative alignment cross-sectional dimension

ly recognized inadequacies of the 10-ft straightedge to describe and define floor surfaces. It is not the intention of the ACI 117 specification to limit floor finish measurement technology to that currently available. As new technology is developed, improved, and perfected, specifiers may consider utilizing alternate techniques for specifying and measuring floor finish tolerances. Random sampling and statistical analysis is particularly appropriate for high-performance floors or portions of floors where irregularities must be rigidly controlled.

The specifying of narrow aisle warehouse floors with defined traffic lanes requires specialized techniques not addressed in this specification.

4.5.6 The F_F-F_L system set forth in Section 4.5.6 of this specification provides the specifier, contractor, and owner with a convenient and precise method of communication, measurement, and determination of compliance of the floor surfaces required and achieved, using the procedures set forth in ASTM E 1155. Floor profile quality has traditionally been specified by limiting the size of the gap to be observed under a freestanding or leveled 10 ft long straightedge. However, recent improvements in floor profile measurement technology have surpassed all variations of this "gap-under-the-straightedge" format.[1]

F-numbers provide a convenient means for specifying the local floor profile in statistical terms. Two distinct profile variables are controlled:

- The 12 in. incremental curvature q measures the local *flatness* of the floor. See Fig. 4.5.6(a).
- The 120 in. elevation difference d measures the local *levelness* of the floor. See Fig. 4.5.6(b).

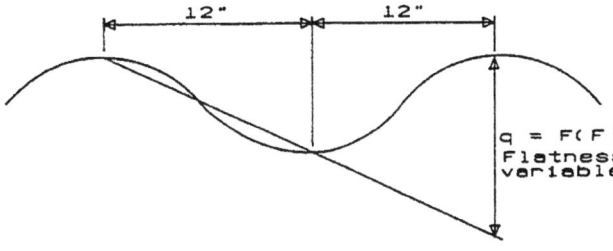

Fig. 4.5.6(a)—Flatness of the floor

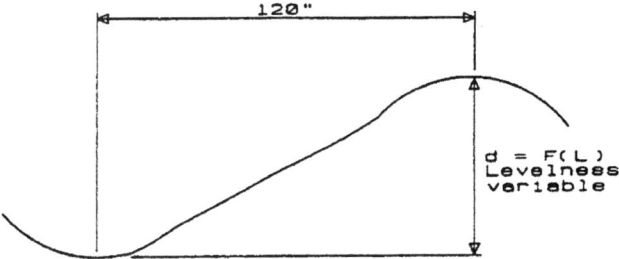

Fig. 4.5.6(b)—Levelness of the floor

The required data may be gathered by several methods, including measurements taken from leveled straightedges, optical levels, and instruments developed for this purpose. Samples of q and d readings are collected from the floor according to the procedures set forth in ASTM E 1155. The means q and d and standard deviations S_q and S_d of these q and d reading samples are calculated, and these statistics are then used to determine the floor's flatness and levelness F-numbers.

Any individual floor section that measures less than either of the specified minimum local F-numbers is rejected. If, after combining all of the individual section results, the entire floor measures less than either of the specified overall F-numbers, then the whole floor is rejected.

To aid in the determination of equitable remedy, the system provides a method for calculating the exact percentage compliance between the floor's specified and estimated F-numbers. To avoid any dispute regarding remedy, the specification should clearly state the specific corrective measures to be applied in the event of an out-of-tolerance result.

Shrinkage, curling, and deflection can all adversely affect floor levelness. Measuring F_L within 72 hr after floor slab installation and before shores and/or forms are removed insures that the floor's "as-built" levelness is accurately assessed. None of the conventional concrete placement techniques in use today can adequately compensate for form or structure deflections that occur during the concrete placement and, for this reason, it is inappropriate to specify levelness tolerances on unshored floor construction.

Since neither deflection nor curling will significantly change a floor's F_F value, there is no time limit on the measurement of this characteristic. Nonetheless, the prudent specifier will provide for the measurement of both F_F and F_L as soon as possible after slab installation to avoid any possible conflict over the acceptability of the floor (and to alert the contractor of the need to modify finishing techniques on subsequent placements if necessary to achieve compliance.)

F-number	Gap under an unleveled 10-ft straightedge
$F_F 12$	1/2 in.
$F_F 20$	5/16 in.
$F_F 25$	1/4 in.
$F_F 32$	3/16 in.
$F_F 50$	1/8 in.

While there is no direct equivalent between F-numbers and straightedge tolerances (see Fig. 4.5.6c), the following table does give a rough correlation between the two systems:

The F-numbers to be obtained using different floor construction methods are given in ACI 302.1R. An increase in flatness from F_F 15 to F_F 20 may generally be achieved by the use of a highway straightedge (or equivalent) rather than a bullfloat following the strike-off. The values listed are for general guidance only. Particular job requirements and conditions can result in F-numbers significantly different from those shown.

To insure user satisfaction, the F_F-F_L values required may be determined by measuring successful installations. of projects with similar uses.

Note that ASTM E 1155 excludes measurements within 2 ft of an imbed or a construction joint. The specifier should provide a limitation on the variation and possible offset potential at these locations appropriate to the use and function of the structure.

Other statistical floor tolerancing systems are being developed and may be used at the option of the specifier providing such methods are shown to give comparable results.

IN GENERAL, TO ACHIEVE HIGHER FLOOR FLATNESS/LEVELNESS VALUES WILL REQUIRE MORE INTENSIVE EFFORT WITH ATTENDANT INCREASES IN LABOR AND CONSTRUCTION COSTS.

4.5.7 Although the 10 ft straightedge procedure has been used for more than 50 years for judging floor irregularities, the procedure has a number of serious deficiencies. These include:
- The difficulty in testing large areas of floors.
- The difficulty of randomly sampling floors.
- The inability to reproduce testing results.
- The inability using normal construction procedures to meet the tolerance limits normally specified, that is, 1/8 in. in 10 ft or 1/4 in. in 10 ft and the widespread lack of conformance and lack of testing for conformance of slab surfaces.
- Failure of the method to predict acceptability of irregularities or roughness in the floor surface. The evaluation of the roughness for a given amplitude should be based upon the frequency of the wave forms.[2]
- The inability of the unleveled straightedge to evaluate levelness of the surface.

The major deficiency of the straightedge measuring system in evaluating floor finishes is demonstrated in Fig. 4.5.6(c).

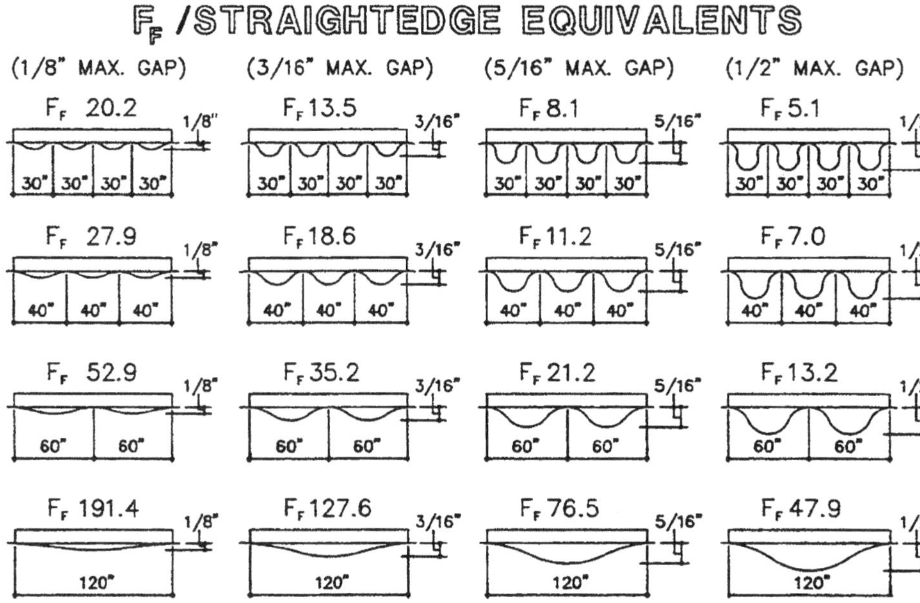

F_F/STRAIGHTEDGE EQUIVALENTS

(1/8" MAX. GAP)	(3/16" MAX. GAP)	(5/16" MAX. GAP)	(1/2" MAX. GAP)
F_F 20.2 (30" spans)	F_F 13.5 (30" spans)	F_F 8.1 (30" spans)	F_F 5.1 (30" spans)
F_F 27.9 (40" spans)	F_F 18.6 (40" spans)	F_F 11.2 (40" spans)	F_F 7.0 (40" spans)
F_F 52.9 (60" spans)	F_F 35.2 (60" spans)	F_F 21.2 (60" spans)	F_F 13.2 (60" spans)
F_F 191.4 (120" span)	F_F 127.6 (120" span)	F_F 76.5 (120" span)	F_F 47.9 (120" span)

Fig. 4.5.6(c)—F-number system is clearly superior to the "gap under a straightedge" approach for distinguishing between the surfaces of obviously different qualities shown in this diagram

The unleveled straightedge measuring system is adversely affected by shrinkage and curling; therefore, measurements are to be taken within 72 hr after floor slab installation and before shores and/or forms are removed.

SECTION 5—PRECAST CONCRETE

5.0
For guidance and recommended tolerances for precast elements not set forth in ACI 117, the specifier should refer to "Tolerances for Precast and Prestressed Concrete," published in *Journal,* Prestressed Concrete Institute, V. 30, No. 1, Jan.-Feb. 1985, pp. 26 to 112.[3]

5.1—Fabrication tolerances
5.1.1 The fabricated length can be longer or shorter than specified by an amount dependent on its design length with an absolute limit of either $3/4$ in. shorter or $3/4$ in. longer. See Fig. 5.1.1.

DESIGNERS ARE CAUTIONED TO PROVIDE LONGER BEARING ELEMENTS TO ACCOMMODATE SHORTER MEMBER LENGTHS AND ROOM FOR OVERLENGTH MEMBERS (WITHIN TOLERANCES.)

5.1.3 The lateral alignment is the displacement of any point on the surface relative to the centerline of the as-built member. The centerline is determined by passing a line through the midpoint of the as-built end. See Fig. 5.1.3 and 5.2.3.

5.1.4 Camber is measured at the midpoint between the as-built ends of the member. The allowable deviation is a function of the length of the member with an absolute limit. Camber tolerances in prestressed members may require reevaluation after initial member castings due to the inaccuracies inherent in initial engineering predications based upon the member design. The specified camber may require adjustment based upon the actual camber that results from the specified design or the design may require modification. See Fig. 5.1.4.

5.1.5 *Surface irregularities*—See Fig. 5.1.5.

5.2—Fabrication tolerances for piles
5.2.3 Tolerance determination is similar to Section 5.1.1. The exception is that there is no absolute limit applied to the tolerance envelope.

5.2.5 The slope across the pile head can vary as a function of the width of the pile head with an absolute limit. The width is the diameter of circular piles and the cross-sectional dimension in the direction of slope measurement of noncircular piles. See Fig. 5.2.5.

5.3—Fabrication tolerances in planar elements
5.3.1 The allowable skew of planar elements is determined by comparing the length of the diagonals. This pre-presumes rectangular units for the application of this fabrication control. For irregularly shaped units the comparison of diagonals may not be possible or meaningful and the concept of skew may not apply. See Fig. 5.3.1.

5.4—Erection tolerances
5.4.2.2 The allowable taper of the joint between exposed panels is a function of the length of the joint with absolute limits on the minimum and maximum width of the tolerance envelope. See Fig. 5.4.2.2.

5.4.3 The control over the offset of top surfaces of adjacent elements applies to members immediately adjacent to each other or separated members that will ultimately be joined in the structure (see Fig 5.4.3). The roofing system must be coordinated with the tolerance for roof elements without topping slabs. Roofing systems that are to be applied directly to

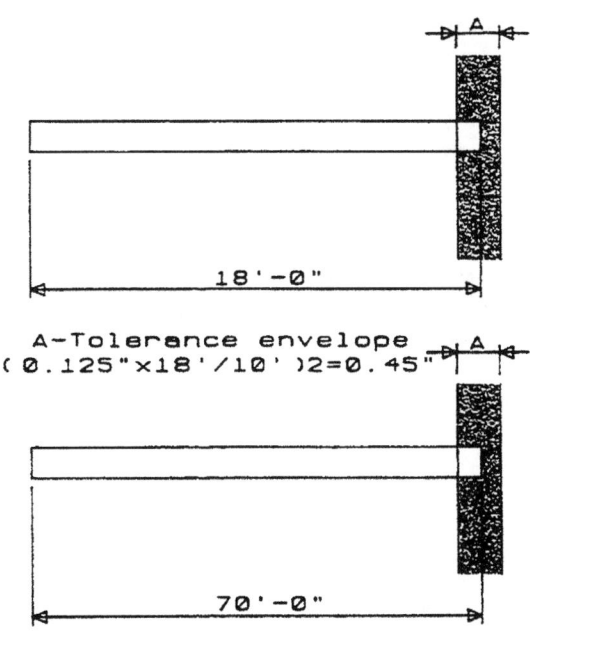

Fig. 5.1.1—Length of member

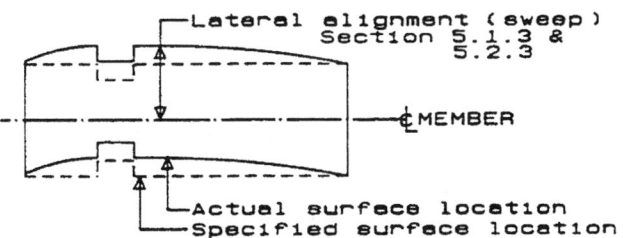

Fig. 5.1.3 and 5.2.3—Lateral alignment

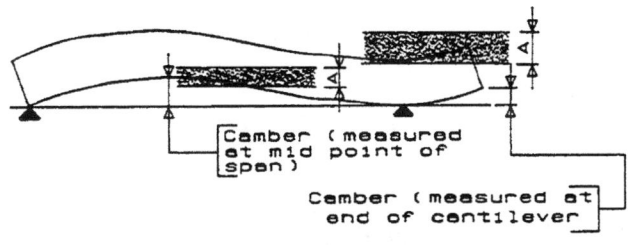

Fig. 5.1.4—Camber

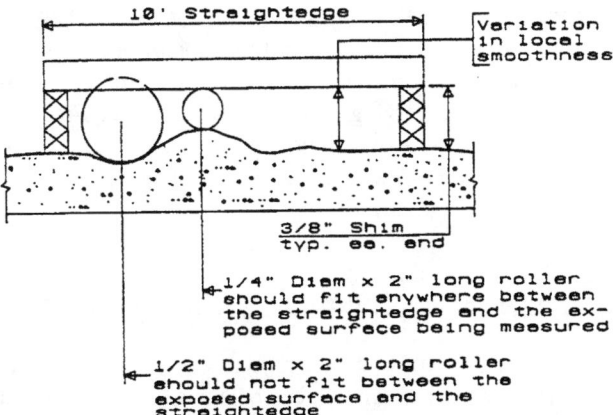

Fig. 5.1.5—Surface irregularities

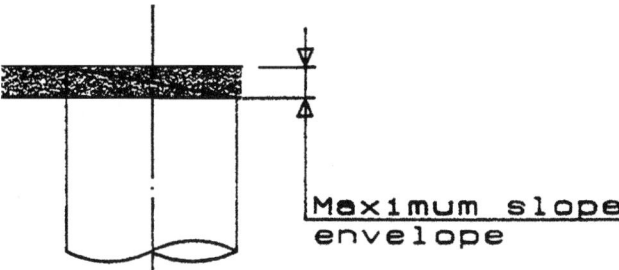

Fig. 5.2.5—Pile head

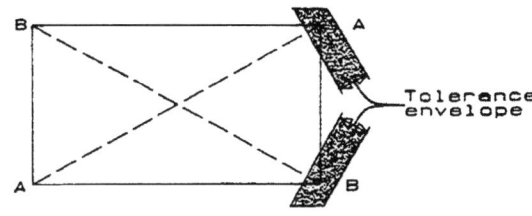

Fig. 5.3.1—Panel length and width

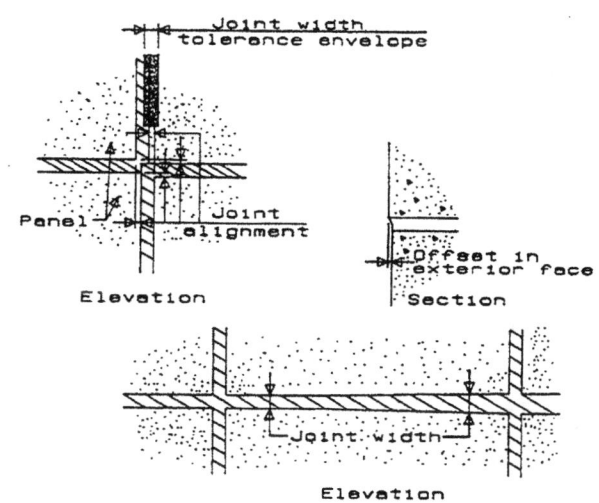

Fig. 5.4.2.2—Alignment of panels

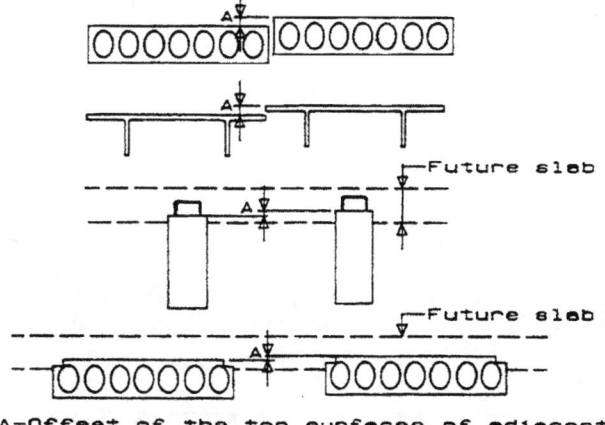

Fig. 5.4.3—Difference in elevation

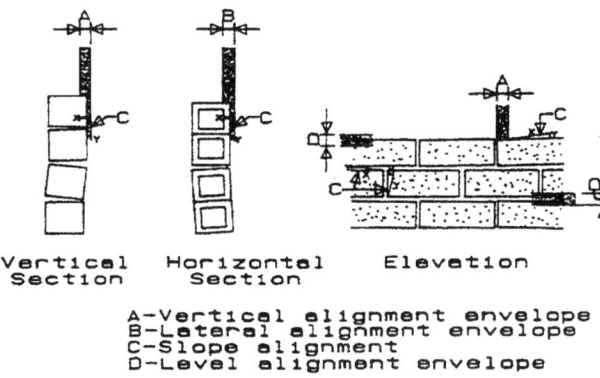

Fig. 6.1, 6.2, 6.3, and 6.5—Masonry alignment

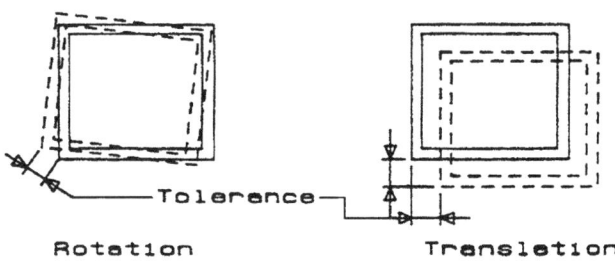

Fig. 7.1—Slipform vertical alignment

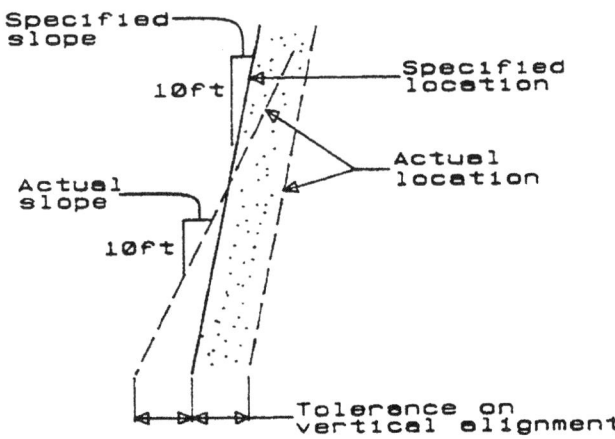

Fig. 11.1 and 11.5.2—Vertical section

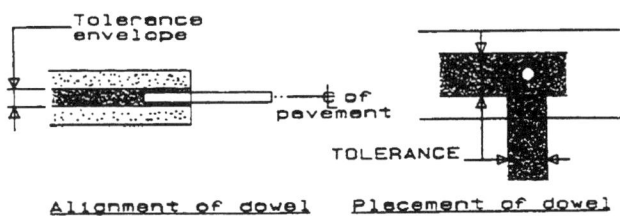

Fig. 12.1—Pavement dowels

the precast surface may require a leveling grout to fill and feather the resulting offset.

SECTION 6—MASONRY
6.1, 6.2, 6.3, and 6.5—Alignments
See Fig. 6.1., 6.2, 6.3, and 6.5.

SECTION 7—CAST-IN-PLACE, VERTICALLY SLIPFORMED BUILDING ELEMENTS
7.1—Vertical alignment
See Fig. 7.1.

7.2, 7.3, and 7.4
Refer to the commentary in Section 4.

SECTION 8—MASS CONCRETE STRUCTURES OTHER THAN BUILDINGS
8.1, 8.2, 8.3, and 8.4
Refer to the commentary in Section 4.

SECTION 9—CANAL LINING
9.1, 9.2, and 9.3
Refer to the commentary in Section 4.

SECTION 10—MONOLITHIC SIPHONS AND CULVERTS
10.1, 10.2, and 10.3
Refer to the commentary in Section 4.

SECTION 11—CAST-IN-PLACE BRIDGES
11.1, 11.2, 11.3, 11.4, and 11.5
Refer to the commentary in Section 4. See Fig. 11.1 and 11.5.2.

SECTION 12—PAVEMENT
12.1—Lateral alignment
   12.1.1 *Placement of dowels*—See Fig. 12.1.

SECTION 13—CHIMNEYS AND COOLING TOWERS
13.1 Tolerances on the size and location of openings and embedments in the concrete shell cannot be uniformly established due to the varying degree of accuracy required depending on the nature of their use. Appropriate tolerances for opening and embedment sizes and locations should be established for each chimney.

SECTION 14—CAST-IN-PLACE NONREINFORCED PIPE
14.1 Cast-in-place concrete pipe tolerances relate to the accuracy of construction that can be achieved with tracked excavators.

SECTION 15—REFERENCES
15.1—Recommended references
The documents of the various standards producing organizations referred to in this document are listed below with their serial designation.

American Concrete Institute

211.1-81 (Revised 1985)	Standard Practice for Selecting Proportions for Normal, Heavyweight and Mass Concrete
223-83	Standard Practice for the Use of Shrinkage-Compensating Concrete
302.1R-80	Guide for Concrete Floor and Slab Construction
303R-74 (Revised 1982)	Guide to Cast-in-Place Architectural Concrete Practice
304R-85	Guide for Measuring, Mixing, Transporting, and Placing Concrete
307-88	Design and Construction of Cast-in-Place Reinforced Concrete Chimneys
313-77 (Revised 1983)	Recommended Practice for Design and Construction of Concrete Bins, Silos, and Bunkers for Storing Granular Materials
315-80	Details and Detailing of Concrete Reinforcement
316R-82	Recommendations for Construction of Concrete Pavements and Concrete Bases
318R-83	Commentary on Building Code Requirements for Reinforced Concrete (318-83)
325.3R-85 (Revised 1987)	Guide for Design of Foundations and Shoulders for Concrete Pavements
332R-84	Guide to Residential Cast-in-Place Concrete Construction
334.1R-64 (Revised 1982) (Reapproved 1986)	Concrete Shell Structures-Practice and Commentary
344R-W	Design and Construction of Circular Wire and Strand Wrapped Prestressed Concrete Structures
344R-T	Design and Construction of Circular Prestressed Concrete Structures with Circumferential Tendons
345-82	Standard Practice for Concrete Highway Bridge Deck Construction
347-78 (Reapproved 1984)	Recommended Practice for Concrete Formwork
349R-85	Commentary on Code Requirements for Nuclear Safety Related Concrete Structures
350R-83	Concrete Sanitary Engineering Structures
357R-84	Guide for the Design and Construction of Fixed Offshore Concrete Structures
358R-80	State-of-the-Art Report on Concrete Guideways
531R-79 (Revised 1983)	Commentary on Building Code Requirements for Concrete Masonry Structures
531.1-76 (Revised 1983)	Specifications for Concrete Masonry Construction
543R-74 (Reapproved 1980)	Recommendations for the Design, Manufacture, and Installation of Concrete Piles

ASTM

E1155-87	Standard Test Method for Determining Floor Flatness and Levelness Using the F-Number System (Inch-Pound Units)

Concrete Reinforcing Steel Institute

MSP-l-86	Manual of Standard Practice (24th Edition)

The preceding publications may be obtained from the following organizations:

American Concrete Institute
P.O. Box 9094
Farmington Hills, MI 48333-9094

ASTM
100 Barr Harbor Drive
West Conshohocken, PA 19428

Concrete Reinforcing Steel Institute
933 North Plum Grove Road
Schaumburg, IL 60173-4758

15.2—Cited references
1. Face, Allen, "Specification and Control of Concrete Floor Flatness," *Concrete International: Design & Construction,* V. 6, No. 2, Feb. 1984, pp. 56-63.
2. Hudson, W. Ronald; Halbach, Dan; Zaniewski, John P.; and Moser, Len, "Root-Mean-Square Vertical Acceleration as a Summary Roughness Statistic," *Measuring Road Roughness and its Effect on User Cost and Comfort,* STP-884, pp. 20-21.
3. PCI Committee on Tolerances, "Tolerances for Precast and Prestressed Concrete," *Journal,* Prestressed Concrete Institute, V. 30, No. 1, Jan.-Feb. 1985, pp. 26-112.

This report was submitted to letter ballot of the committee and was approved in accordance with the Institute's balloting procedures.

ACI 212.3R-04

Chemical Admixtures for Concrete

Reported by ACI Committee 212

Marshall L. Brown[*]
Chair

Joseph J. Hiznay[*]
Secretary

J. Floyd Best	Robert E. Moore	Donald L. Schlegel
Bayard M. Call	William F. Perenchio	Raymond J. Schutz[*]
Lewis J. Cook	William S. Phelan[*]	Billy M. Scott
Edwin A. Decker[†]	Michael F. Pistilli	David B. Stokes
Ross S. Martin	Ken B. Rear[*]	Brad K. Violetta[*]
Bryant Mather[*†]	M. Roger Rixom	David A. Whiting[*†]
Richard C. Mielenz		J. Francis Young[*]

[*]Chair of the task groups that prepared this report.
[†]Deceased.
Note: ACI Committee 212 would like to thank Kari Yuers and Lawrence Roberts for their contributions to the development of this document. The following former members of ACI Committee 212 contributed to the preparation of this document: Barry Butler, Dale Rech, and Charles Taylor.

This report updates the previous 1991 report. Chemical admixtures, which are primarily water-soluble substances, are discussed. Finely divided mineral admixtures are dealt with by ACI Committees 232, Fly Ash and Natural Pozzolans in Concrete, and 234, Silica Fume in Concrete. For the purpose of this report, chemical admixtures are classified into five groups: air-entraining; accelerating; water-reducing and set-controlling; admixtures for flowing concrete; and miscellaneous. Admixtures possessing properties identifiable with more than one group are discussed with the group that describes its most important effect on concrete.

Keywords: adhesives; admixture; air-entraining; alkali-aggregate reaction; bond; calcium chloride; concrete; corrosion inhibitor; flowing concrete; grouting; high-range water-reducing admixture; pigment; plasticizer; pumped concrete; retardon; water-reducing admixture.

ACI Committee Reports, Guides, Standard Practices, and Commentaries are intended for guidance in planning, designing, executing, and inspecting construction. This document is intended for the use of individuals who are competent to evaluate the significance and limitations of its content and recommendations and who will accept responsibility for the application of the material it contains. The American Concrete Institute disclaims any and all responsibility for the stated principles. The Institute shall not be liable for any loss or damage arising therefrom.
Reference to this document shall not be made in contract documents. If items found in this document are desired by the Architect/Engineer to be a part of the contract documents, they shall be restated in mandatory language for incorporation by the Architect/Engineer.

It is the responsibility of the user of this document to establish health and safety practices appropriate to the specific circumstances involved with its use. ACI does not make any representations with regard to health and safety issues and the use of this document. The user must determine the applicability of all regulatory limitations before applying the document and must comply with all applicable laws and regulations, including but not limited to, United States Occupational Safety and Health Administration (OSHA) health and safety standards.

CONTENTS
Chapter 1—General information, p. 212.3R-2
1.1—Introduction
1.2—Reasons for using admixtures
1.3—Specifications for admixtures
1.4—Sampling and testing
1.5—Cost effectiveness
1.6—Considerations for using admixtures
1.7—Preparation and batching

Chapter 2—Air-entraining admixtures, p. 212.3R-6
2.1—Introduction
2.2—Entrained air-void system
2.3—Effects on concrete properties
2.4—Materials for air entrainment
2.5—Applications
2.6—Evaluation and testing
2.7—Batching and storage
2.8—Factors influencing the amount of entrained air
2.9—Controlling air content in concrete

Chapter 3—Accelerating admixtures, p. 212.3R-9
3.1—Introduction
3.2—Types of accelerating admixtures
3.3—Use with special cements
3.4—Effects on hardening concrete
3.5—Effects on concrete durability
3.6—Corrosion of metals
3.7—Discoloration of flatwork
3.8—Dosage levels

ACI 212.3R-04 supersedes ACI 212.3R-91 (Reapproved 1999).
Copyright © 2004, American Concrete Institute.
All rights reserved including rights of reproduction and use in any form or by any means, including the making of copies by any photo process, or by electronic or mechanical device, printed, written, or oral, or recording for sound or visual reproduction or for use in any knowledge or retrieval system or device, unless permission in writing is obtained from the copyright proprietors.

Chapter 4—Water-reducing and set-controlling admixtures, p. 212.3R-12
4.1—Introduction
4.2—Classification and composition
4.3—Applications
4.4—Dosage
4.5—Preparation, batching, and quality control
4.6—Proportioning concrete
4.7—Effects on fresh concrete
4.8—Effects on hardened concrete
4.9—Evaluations

Chapter 5—Admixtures for flowing concrete, p. 212.3R-16
5.1—Introduction
5.2—Materials
5.3—Selection and evaluation
5.4—Applications
5.5—Proportioning concrete
5.6—Effect on fresh and hardening concrete
5.7—Effect on hardened concrete
5.8—Quality assurance

Chapter 6—Miscellaneous admixtures, p. 212.3R-19
6.1—Gas-forming admixtures
6.2—Grouting admixtures
6.3—Extended set-control admixtures
6.4—Bonding admixtures
6.5—Pumping aids
6.6—Pigments
6.7—Flocculating admixtures
6.8—Fungicidal, germicidal, and insecticidal admixtures
6.9—Permeability-reducing admixtures
6.10—Chemical admixtures to reduce alkali-aggregate reaction expansion
6.11—Corrosion-inhibiting admixtures
6.12—Antiwashout admixtures
6.13—Freeze-resistant admixtures

Chapter 7—References, p. 212.3R-24
7.1—Referenced standards and reports
7.2—Cited references

CHAPTER 1—GENERAL INFORMATION
1.1—Introduction
An admixture is defined in ACI 116R and in ASTM C 125 as "a material other than water, aggregates, hydraulic cement, and fiber reinforcement used as an ingredient of concrete or mortar, and added to the batch immediately before or during its mixing." This report deals with commonly used admixtures other than those assigned to other ACI committees. Materials such as admixtures used to produce expansive-cement concrete (ACI Committee 223); fly ash and natural pozzolans (ACI Committee 232); silica fume (ACI Committee 234); admixtures for insulating and cellular concrete (ACI Committee 523); and polymers (ACI Committee 548), are not discussed in this report.

The chemical admixtures are classified generically or with respect to their characteristics. Information to characterize each class is presented along with brief statements of the general purposes and expected effects of using each group of materials. The wide scope of the admixture field, the continued entrance of new or modified materials into this field, and the variations of effects with different concreting materials and conditions preclude a complete listing of all admixtures and their effects on concrete. Summaries of the state-of-the-art of chemical admixtures include Ramachandran (1984), Ramachandran and Mailvaganam (1992), and Mather (1994).

1.2—Reasons for using admixtures
Chemical admixtures are designed to enhance the properties of concrete or mortar in the plastic and hardened states, increase efficiency of cementitious material, and improve the economy of the concrete mixture. The use of an admixture or combination of admixtures may be the only viable strategy to achieve the desired results. In certain instances, the desired objectives may be best achieved by changes to the mixture proportions in addition to using the proper admixture. Chemical admixtures are not substitutes for suitable concrete mixture proportions and acceptable construction practices. Refer to Section 1.6.

1.2.1 *Modification of fresh concrete, mortar, and grout*—Admixtures are used to modify properties of fresh concrete, mortar, and grout to:
- Increase workability without increasing water content, or decrease the water content without changing the workability;
- Retard or accelerate the time of initial setting;
- Reduce or prevent settlement, or create a slight expansion;
- Modify bleeding characteristics;
- Reduce segregation;
- Improve pumpability;
- Reduce the rate of slump loss; and
- Improve finishability.

1.2.2 *Modification of hardened concrete, mortar, and grout*—Admixtures are used to modify properties of hardened concrete, mortar, and grout to:
- Reduce the rate of heat evolution during early cement hydration;
- Accelerate the rate of strength development at early ages;
- Increase strength (compressive, tensile, or flexural);
- Increase resistance to freezing and thawing;
- Reduce scaling caused by deicing salts;
- Decrease permeability;
- Reduce expansion caused by alkali-aggregate reaction;
- Increase bond to steel reinforcement and between existing and new concrete;
- Improve impact resistance and abrasion resistance;
- Inhibit corrosion of embedded metal;
- Produce colored concrete or mortar; and
- Reduce drying shrinkage.

1.3—Specifications for admixtures
The following standard specifications cover the admixture types that make up the bulk of products covered in this report:
- Air-entraining admixtures: ASTM C 260, AASHTO M 154;

- Water-reducing and set-controlling admixtures: ASTM C 494, AASHTO M 194;
- Calcium chloride: ASTM D 98, AASHTO M 144;
- Admixtures for use in producing flowing concrete: ASTM C 1017; and
- Pigments for integrally colored concrete: ASTM C 979.

1.4—Sampling and testing

Admixture samples for testing and evaluation should be obtained by the procedures prescribed for each admixture's specifications using random sampling from plant production, previously unopened packages or containers, or fresh bulk shipments.

Admixtures are tested to determine compliance with specifications; evaluate effects on the properties of concrete made with materials under the anticipated ambient conditions and construction procedures; determine uniformity of the product within or between batches, lots, or containers; or reveal any undesirable effects. The quality-control procedures used by producers of admixtures should ensure product compliance with provisions of ASTM or other applicable specifications, including uniformity. Because a producer's quality-control test methods can be developed around a particular proprietary product, they may not be applicable for general use or use by consumers.

ASTM tests provide screening procedures for selecting admixtures for a particular application. Producing concrete should be preceded by testing that allows observation and measurement of the performance of the admixture under concrete-plant operating conditions in combination with the constituent materials that will be used. Uniformity of results is as important as the average result, with respect to each significant property of the admixture or the concrete.

1.5—Cost effectiveness

Economic evaluation of an admixture should be based on the test results obtained when used with the specified concrete under conditions simulating those expected on the job. The characteristics of the cementitious materials and aggregate, their relative proportions, and the temperature, humidity, and curing conditions influence the test results. When evaluating an admixture, its effect on the volume of a given batch should be taken into account. If the admixture increases the volume of the batch (the yield), the admixture should be considered as a basic ingredient together with the cementitious materials, aggregate, and water. All changes in the composition of a unit volume of concrete should be taken into account when testing the direct effect of the admixture and in estimating its benefits.

The cost effectiveness of an admixture should address the increase in cost required to handle an additional ingredient, as well as the economic effect the admixture may have on the cost of transporting, placing, and finishing the concrete. An admixture can allow the use of less-expensive construction methods, or allow structural designs that offset the added cost due to its use. For example, novel and economical structural designs have resulted from the use of water-reducing and set-retarding admixtures (Schutz 1959). High-range water-reducing admixtures (HRWRAs) are often essential ingredients of cost-effective, high-performance concrete.

Water-reducing and set-retarding admixtures permit placement of large volumes of concrete over extended periods, minimizing the need for forming, placing, and joining separate units. Accelerating admixtures reduce finishing and forming costs. Required physical properties of lightweight concrete may be achieved at a lower density by using air-entraining and water-reducing admixtures.

1.6—Considerations for using admixtures

Careful attention should be given to the instructions and recommendations provided by the manufacturer of the admixture. An admixture's effects should be evaluated whenever possible using the specified materials under site conditions. This is particularly important when:

- The admixture has not been used previously with the particular combination of materials;
- Special types of cementitious material are specified;
- More than one admixture is to be used; or
- Mixing and placing is done at temperatures outside recommended temperature ranges for concrete.

The use of admixtures may require:

- A change in type, source, or amount of cement;
- A change in aggregate type or grading; or
- A change in mixture proportions. The effects of some admixtures are significantly modified by factors such as water and cementitious material content of the mixture, and by the type and length of mixing.

Many admixtures affect more than one property of concrete and can adversely affect desirable properties. Admixtures that modify the properties of fresh concrete can cause problems with early stiffening or excessive retardation, delaying the setting time. The cause of abnormal setting behavior should be determined through studies on how such admixtures affect the cementitious material to be used. Early stiffening is often caused by changes in the reaction rate between the tricalcium aluminate and sulfate ions in solution in the pore fluid. Excessive retardation can be caused by an overdose of admixture or by a lowering of ambient temperature, both of which delay the hydration of the calcium silicates (Hansen 1960).

Another important consideration when using admixtures arises when there is a limit on the amount of chloride ion permitted in the concrete (ACI 318 and ACI 222R). These limits are usually expressed as maximum percent of chloride ion by mass of cement, although the amount of water-soluble chloride ion per mass of cement or concrete is sometime specified. The procedures of ASTM C 1152 and C 1218 can be used to measure acid-soluble and water-soluble chloride, respectively, in mortar or concrete. It is necessary to know the chloride-ion content of an admixture to ensure that it will not jeopardize the concrete conforming to a specified chloride limit. In spite of the use of such terms as "chloride-free," all admixtures sold as solutions will contain small but measurable amounts of chloride ion.

Although specifications deal primarily with the influence of admixtures on specific properties of fresh and hardened concrete, the concrete supplier, contractor, and owner of the construction project may be interested in other features of concrete construction. Of primary concern may be workability, pumping qualities, placing and finishing qualities, early strength development, reuse of forms or molds, or the appearance of formed surfaces. These additional features are important when an admixture is selected and its dosage rate is determined.

Guidance for using different classes of admixtures is given in the relevant chapters of this report. Those responsible for construction of concrete structures should bear in mind that increasing material costs and continuing development of new and improved admixtures warrant the continuous re-evaluation of the benefits of using admixtures.

1.7—Preparation and batching

1.7.1 *Introduction*—The successful use of admixtures depends on the methods of preparation and batching. Neglect in preparation and batching can significantly affect the properties, performance, and uniformity of the concrete. Most admixtures are furnished in a ready-to-use liquid form. These admixtures are introduced into the concrete mixture at the concrete plant or transported in a truck-mounted admixture tank for introduction into the concrete mixture at the job site.

Solid admixtures such as pigments, expansive admixtures, and pumping aids are used in small dosages, and are often batched by hand from premeasured containers. Other hand-added admixtures include accelerators, permeability-reducing admixtures, and bonding admixtures, which often are packaged in single-unit doses. Solids are batched by mass.

1.7.2 *Preparing admixtures for dispensing*—Admixtures supplied as ready-to-use liquids can be much more concentrated than job-mixed solutions. The preparation of admixtures may require diluting the admixture to various concentrations to facilitate accurate batching or dispensing. The manufacturer's recommendations should be followed. For easier dispensing, it can be convenient to prepare standard solutions of admixtures with uniform strength. Some chemical admixtures are supplied as water-soluble solids, which require mixing at the job site. In some cases, it may be necessary to make low-concentration solutions due to difficulties in fully dissolving the solids. If the solutions contain significant amounts of finely divided solids, they should be kept in uniform suspension before actual batching.

1.7.3 *Storage and protection*—If finely divided insoluble matter is present, it needs to be kept in suspension, but continuous agitation usually is not required. Manufacturers can supply information regarding the degree of agitation or recirculation required for their admixtures. To avoid settlement or polymerization, timing devices are commonly used to control recirculation of the storage tank's contents. Some suppliers will provide complete storage and dispensing systems. In freezing climates, the storage tank and its contents should be either heated or placed in a heated environment. The latter is preferred because it protects not only storage tanks, but also pumps, meters, valves, and admixture hoses from freezing and from other problems such as dust, rain, ice, and vandalism. Because the storage temperature in such an environment varies less throughout the year, admixture viscosity is more constant and dispensers require less frequent calibration.

Heating a storage tank with hot water, steam coil, or with electrical heating tapes can overheat the admixture locally, causing partial pyrolysis and perhaps producing explosive gas. The admixture may freeze and damage equipment if connections to heating devices are inadvertently disconnected. When plastic storage tanks or hoses are used, care should be taken to avoid heating these materials to the point of softening and rupture. The price of heating the storage tank and its contents is normally higher than the cost of maintaining above-freezing temperatures in a heated storage room. Storage tanks should be vented, but precautions should be taken so that foreign materials cannot enter the tank through the opening. To avoid contamination, filling nozzles and any other tank openings should be capped when not in use.

1.7.4 *Batching*—Batching and discharging the measured quantity of a liquid admixture into a batch of concrete or into a mixer or truck-mounted tank is generally accomplished by a system of pumps, meters, timers, calibration tubes, and valves. This equipment is called the admixture-dispensing system or the dispenser. To minimize variations in concrete properties, batching and dispensing equipment should meet and maintain required tolerances. ASTM C 94 requires that volumetric measurement of admixtures be accurate to within ±3% of the total volume required, or plus or minus the volume of the dose required for one bag of cement [43 kg (94 lb)], whichever is greater. ASTM C 94 requires that powdered admixtures be measured by mass but permits liquid admixtures to be measured by mass or volume. Accuracy of batching admixtures by mass is required to be within ±3% of the required mass. Before installing a dispenser, the system should be carefully analyzed with the help of the admixture supplier to identify and eliminate possible sources of batching error. The admixture should be discharged from the calibration tube to the concrete batch at the point that achieves the greatest dispersion throughout the concrete. The discharge end of the waterline leading to the mixer generally is the preferred location.

Dispensing admixtures into a concrete batch involves not only accurately controlling the quantity of admixture and the rate of discharge but also the timing of the batching sequence. In some instances, changing the time at which the admixture is added during mixing can modify the effectiveness of the admixture. For example, Bruere (1963) and Dodson and Farkas (1964) report that the retarding effect of water-reducing retarders depends on the time at which the retarder is added to the mixture. If the retarder is added more than a few seconds after the water and cement are mixed, it will retard much more strongly. The water demand of the concrete may also be affected. For any given condition or project, a procedure for controlling the time and rate of the admixture addition to the concrete batch should be established and adhered to closely. The rate of admixture discharge should be adjustable to ensure uniform distribution of the admixture throughout the concrete mixture during the charging cycle when required.

Foster (1966) noted that two or more admixtures often are not compatible in the same solution. For example, a Vinsol® resin-based, air-entraining admixture and a water-reducing admixture containing a lignosulfonate should never come in contact before mixing into the concrete because of their tendency for instantaneous flocculation and loss of efficiency of both admixtures. Intermixing admixtures before introduction into the concrete should be avoided unless tests indicate there will be no adverse effects, or unless the manufacturer's instructions permit it. It is better to introduce admixtures during charging of other materials into the mixer or during mixing.

1.7.5 *Liquid batching systems*—Ordinarily, liquid admixtures are not batched by mass because mass-batching devices are more expensive than volumetric dispensers, although a few concrete plants do use mass-batching systems. In some cases, it is necessary to dilute admixture solutions to obtain a sufficient quantity for accurate measurement by mass. Most methods of batching liquid admixtures require a visual volumetric container, called a calibration tube, to enable the plant operator to verify the accuracy of the admixture dosage. The simplest batching method consists of a visual volumetric container, while others include positive volumetric displacement. Some of these methods can be used readily with manual, semiautomatic, and automatic systems and can be operated easily by a remote control with appropriate interlocking in the batch sequence. Simple manual dispensing systems, which are designed for low-volume concrete plants, depend on the concrete plant operator batching the proper amount of admixture into a calibration tube and discharging it into the batch. More sophisticated systems intended for automated high-volume plants provide automatic fill and discharge of a sight or calibration tube. Adding admixture at the job site can also be accomplished with a tank and a pressurized dispensing system. A calibrated holding tank and a mechanical or electromechanical dispensing device should be part of the system so that the plant operator can verify that the proper amount of admixture has been batched into the concrete mixer or into the truck-mounted tank.

Flow meters and calibration tubes equipped with floats or probes often are combined with pulse-emitting transmitters that give readouts on electromechanical or electronic counters. These are often set by inputting the dosage per unit of cementitious material. The amount of cementitious material input into the batching panel, combined with the dosage rate, sets the dispensing system to batch the proper amount of admixture. Timer-controlled systems involve the timing of flow through an orifice. Considerable error can be introduced by changes in power supply, partial restrictions of the measuring orifice, and changes in viscosity of the solution. Viscosity is affected by temperature and admixture concentration. Timer-controlled systems should be recalibrated in accordance with project documents or manufacturer recommendations. The plant operator should verify the proper admixture dose by observing the calibration tube. Although timer-controlled systems have been used successfully, their use is not recommended, except for dispensing calcium chloride solution.

A number of different methods are used to fill and discharge calibration tubes. A major objective is to ensure that the fill valve will not open until the discharge valve is completely closed and to prevent overbatching in the event of electrical or mechanical malfunction. Power-operated valves are frequently used; a vacuum release may also be provided to prevent overbatching due to venturi action from the concrete plant's waterline. A low-level indicator in the calibration tube is often used to prevent the discharge valve from being closed before all the admixture is dispensed into the batch.

The calibration tube is emptied either by gravity or by air pressure, and the admixture can have a considerable distance to flow through a discharge hose or pipe before it reaches its ultimate destination. In such cases, the dispenser control panel should be equipped with a timer-relay device to ensure that all the admixture has been discharged from the conveying hoses or pipes. If the admixture dispenser system is operated manually, the plant operator should have a valve to prolong the discharge cycle until he or she is certain that all admixture is in the concrete batch. When more than one admixture is intended for the same concrete batch, the dispenser should be designed so that an appropriate delay is built into the system to prevent the admixtures from becoming intermingled, or each being batched separately before entering the mixer. Likewise, in a manual system, the operator should be instructed on methods to prevent intermixing admixtures.

Job-site introduction of certain HRWRA has become common because of the high rate of slump loss associated with some admixtures. Such admixtures can be from truck-mounted admixture tanks or from job-site tanks or drums using a dispensing system similar to that used in concrete plants. If truck-mounted tanks are used, the proper dosage of admixture for the concrete in the truck is measured at the batch plant and is discharged into the truck-mounted tank at a filling station. A series of lights or other signals advises the driver when the admixture batching is complete and the tank contains the proper amount. At the job site, the driver sets the mixer at mixing speed and discharges the entire amount of admixture from the truck-mounted tank into the concrete. Care should be taken that the mixer continues to operate in the mixing mode until the admixture has been thoroughly distributed and mixed throughout the concrete. The condition of the mixer and its blades influences the distribution. To ensure that all the admixture is properly introduced into the mixer, air pressure should be used to force all the admixture through the line into the mixer drum. To shorten the mixing time, while still ensuring thorough mixing, the truck mixer should operate at maximum speed— approximately 18 rpm.

1.7.6 *Maintenance*—Batching systems require routine periodic maintenance to prevent inaccuracies developing from such causes as sticky valves, buildup of foreign matter in meters or in storage and mixing tanks, or worn pumps. Components should be protected from dust and temperature extremes, and be kept readily accessible for visual observation and maintenance. Although admixture batching systems are usually installed and maintained by the admixture producer, plant operators should thoroughly understand the system and be able to adjust it and perform simple maintenance. Plant operators should recalibrate the system on a regular basis, preferably at intervals of not more than 90 days, noting any trends that indicate worn parts needing replacement.

Tanks, conveying lines, and ancillary equipment should be drained and flushed on a regular basis, and calibration tubes should have a water fitting installed to allow the plant operator to flush the tube so that divisions or markings may be clearly seen at all times. Because of the marked effect of admixtures on concrete performance, care and attention to the timing and accuracy of admixture batching are necessary to avoid serious problems.

CHAPTER 2—AIR-ENTRAINING ADMIXTURES
2.1—Introduction

ACI 116R defines an air-entraining admixture as "an admixture that causes the development of a system of microscopic air bubbles in concrete, mortar, or cement paste during mixing." The entrained air-void system is distinct from air voids physically entrapped in concrete during placement and consolidation. Air entrainment should always be required when concrete will be exposed to moisture and repeated cycles of freezing and thawing, particularly where the use of deicing chemicals is anticipated. Highway pavements, garage floors, and sidewalks located in cold climates probably will be exposed to such conditions. If facilities for batching an air-entraining admixture and measuring the air content of fresh concrete are not available, air-entrained concrete can be made using air-entraining portland cement meeting ASTM C 150 Type IA, Type IIA, or Type IIIA. The resistance of concrete to freezing and thawing is affected by placing, finishing, and curing procedures; therefore, acceptable construction practice in these respects should be followed (ACI 201.2R; ACI 304R; ACI 308R).

Extensive laboratory testing and long-term field experience have demonstrated conclusively that portland-cement concrete should contain at least a minimum amount of properly entrained air to resist the action of freezing and thawing (Cordon 1946; Blanks and Cordon 1949; Mather 1990). The process by which air is entrained in concrete and the mechanism by which such air entrainment prevents damage due to freezing and thawing is beyond the scope of this report, but is summarized in various textbooks (Powers 1968; Mindess and Young 1981; Mehta and Monteiro 1993) and in ACI 201.2R. More detailed discussions can be found in research papers (for example, Cordon [1966]; Litvan [1972]; MacInnis and Beaudoin [1974]; Powers [1975]).

2.2—Entrained air-void system

Improvements in resistance to freezing and thawing are due to the presence of minute air bubbles dispersed uniformly through the cement-paste portion of the concrete that provide relief from the pressure of freezing water. Because of the bubble's size, there are literally billions of bubbles in each cubic meter of air-entrained concrete. To provide adequate protection with a relatively low total volume of void space, the bubbles should be small (10 to 100 μm [0.0004 to 0.004 in.] in diameter).

The cement paste in concrete normally is protected against the effects of freezing and thawing if the spacing factor (Powers 1949) does not exceed 0.20 mm (0.008 in.), as determined in accordance with ASTM C 457. Additional requirements are that the surface area of the air voids should be greater than 24 mm^2/mm^3 (600 $in.^2/in.^3$) of air-void volume, and the number of air voids per 25 mm (1 in.) of traverse should be 1-1/2 times greater than the numerical value of the percentage of air in the concrete (Hover 1994). Many investigators (Tynes 1977; Mather 1979; Schutz 1978; Whiting 1979; Litvan 1983) report that the addition of HRWRA to air-entrained concrete increases the spacing factor and decreases the surface areas of the air-void systems beyond the accepted limits. Numerous studies (Kobayashi 1981; Malhotra and Malanka 1979; Philleo 1986), however, indicate that such admixtures do not reduce the freezing-and-thawing resistance of concrete.

The air content and the size distribution of air voids produced in air-entrained concrete are influenced by many factors (Backstrom et al. 1958a; Mielenz et al. 1958a,b), the more important of which are the nature and quantity of the air-entraining admixture, the nature and quantity of the constituents of the concrete, the type and duration of mixing employed, the consistency of the concrete, and the kind and degree of consolidation applied in placing the concrete. These factors are discussed in more detail in Section 2.9. Air content in hardened concrete is determined either by the linear traverse or point-count technique and generally is slightly lower than values obtained from tests of the fresh concrete (Carlson 1967; Reidenour and Howe 1975). Differences are usually less than two percentage points.

Newlon (1971), analyzing field data on cores taken from bridge decks, found that 22 out of 26 samples were within 1.25 percentage points of the air content measured in the fresh concrete. When considerable amounts of entrapped air are present in core specimens, however, air contents, as determined by linear traverse, can be up to three percentage points less than those determined by the pressure meter (Amsler, Eucker, and Chamberlin 1973). Occasionally, measured air contents in hardened concrete can be as much as twice those measured in as-delivered concrete. Explanations of this phenomenon include the incompressibility of very small (<50 μm [0.0002 in.] diameter) air voids (Hover 1989), and the transfer of air between small and large air bubbles (Fagerlund 1990). Attempts to reproduce this phenomenon under controlled laboratory and field conditions have not been successful (Ozyildirim 1991).

2.3—Effects on concrete properties

Air entrainment alters the properties of fresh concrete and should be taken into account when proportioning a mixture (ACI 211.1; ACI 211.2; Powers 1968). An increase in air content normally increases slump. Air-entrained concrete is considerably more workable and cohesive than similar nonair-entrained concrete with the same slump, except at high contents of cementitious materials. At high cementitious material contents, air-entrained concrete becomes sticky and difficult to finish. Air entraining reduces segregation and bleeding, which helps to prevent the formation of pockets of water beneath coarse aggregate particles and embedded items such as reinforcing steel. Air entraining also reduces the accumulation of laitance or weak material at the surface of a lift.

Air entrainment usually reduces the compressive strength of hardened concrete, particularly in concretes with moderate to high cementitious material contents. The reduction is approximately 5% for each percent of entrained air, but the rate of reduction of strength increases with higher amounts of air. Adding entrained air, however, reduces the water content required to maintain a specified slump. The resulting water-cementitious material ratio (w/cm) can partially offset the reduction in strength. This is particularly true of lean mass concretes or concretes that contain a large maximum-size aggregate. In these cases, air entrainment can cause a small decrease in strength, or even an increase. Nevertheless, while a proper air-void system should be provided, excessive amounts of entrained air should be avoided.

2.4—Materials for air entrainment

Many materials can function as air-entraining admixtures, but those used to create cellular concrete, by creating gas bubbles inside the concrete (ACI 523.1R), such as hydrogen peroxide and powdered aluminum metal, are not acceptable air-entraining admixtures.

2.4.1 *Water-soluble compounds*—Water-soluble, air-entraining admixtures are formulated using salts of wood resins, synthetic detergents, salts of petroleum acids, salts of proteinaceous acids, fatty and resinous acids and their salts, and organic salts of sulfonated hydrocarbons. Not every material that fits the preceding description, however, will produce a desirable air-void system. All air-entraining admixtures should meet the requirements of ASTM C 260. Most commercial air-entraining admixtures are in liquid form, although a few are powders, flakes, or semisolids. The proprietary name and the net quantity in kilograms (pounds) or liters (gallons) should be indicated on the containers in which the admixture is delivered.

2.4.2 *Solid materials*—Solid particles that have a high internal porosity and suitable pore size have been added to concrete and seem to act like air voids. These particles can be hollow plastic spheres, crushed brick, expanded clay or shale, or spheres of suitable diatomaceous earths. Research has indicated that when using inorganic particulate materials, the optimum particle size should be 1.18 mm to 300 µm (No. 16 to 50 sieves). The total porosity of the particles should be at least 30% by volume, and the pore-size distribution should be between 0.05 and 3 µm (Gibbons 1978; Sommer 1978). Inclusion of such particulates in the proper proportion has produced concrete with excellent resistance to freezing and thawing in laboratory tests using ASTM C 666 (Litvan and Sereda 1978; Litvan 1985). Particulate air-entraining admixtures have the advantage of stability of the air-void system. Once added to the fresh concrete, changes in mixing procedure or time; changes in temperature, workability, or finishing procedures; or the addition of other admixtures, such as fly ash, or other cementitious materials, such as ground slag, will not change the air content, as may be the case with conventional air-entraining admixtures. Despite these advantages, particulate air-entraining admixtures are not yet commercially available.

2.5—Applications

Air-entrained concrete should be used wherever water-saturated concrete may be exposed to freezing and thawing, especially when deicing chemicals are used. Because air entrainment also improves the workability of concrete, it is particularly effective in lean mixtures and in various kinds of lightweight-aggregate concrete. Air entrainment is used not only in insulating and fill concrete (ACI 523.1R), but also in structural lightweight concrete. Blisters and delaminations can occur when air-entrained normalweight concrete receives a hard-troweled finish (ACI 302.1R; Bimel 1998).

No general agreement exists on the benefits of using air-entraining admixtures in the manufacture of concrete block (Farmer 1945; Kennedy and Brickett 1986; Kuenning and Carlson 1956), and air entrainment is not normally used in these products. Air entraining admixtures, however, are marketed specifically for zero- and low-slump concrete to produce a stable air-void system with proper bubble size and spacing.

Air entrainment is desirable in wet-process shotcrete for the same purposes as in conventional concrete. The process of pumping, spraying, and impinging on a surface limits the air content of in-place shotcrete to approximately 4%, in spite of higher air contents before pumping (Morgan 1991). In dry-process shotcrete, using air-entraining admixtures is questionable because there is no mixing to develop an acceptable air-void system (Segebrecht, Litvan, and Gebler 1989). Nevertheless, air-entrained dry-process shotcrete exhibits excellent durability when exposed to freezing and thawing (Litvin and Shideler 1966; Gebler 1992).

2.6—Evaluation and testing

To improve resistance to freezing and thawing, intentionally entrained air should have certain characteristics as outlined in Section 2.2. An admixture that meets the requirements of ASTM C 260 will produce a desirable air-void system when recommended air contents are achieved. This specification also sets limits on the effects of any given air-entraining admixture on bleeding, time of setting, compressive and flexural strength, resistance to freezing and thawing, and length change on drying of a hardened concrete mixture in comparison with a similar concrete mixture that contains a standard reference air-entraining admixture, such as neutralized Vinsol® resin. Acceptance testing should follow ASTM C 233. ASTM C 457 can be used to determine the actual parameters of the air-void system in hardened concrete to provide greater assurance that satisfactory resistance to freezing and thawing will be obtained.

2.7—Batching and storage

To achieve the greatest uniformity between batches of a concrete mixture, water-soluble, air-entraining admixtures should be added to the mixture in the form of solutions rather than solids. Generally, only small quantities of air-entraining admixtures (about 0.05% of active ingredients by mass of cementitious materials) are required to entrain the desired amount of air. If the admixture is in solid or semisolid forms, a solution should be prepared before use, following the recommendations of the manufacturer.

The dosage required to achieve the desired air content should be determined by experiment, starting from the manufacturer's recommendations or from experience. For any given set of conditions and materials, the amount of air entrained is roughly proportional to the amount of admixture used. The air content, however, can reach an upper limit in low-slump concrete containing a high cement content, finely ground cement, and fine aggregate with a large fraction passing the 75 μm (No. 200) sieve. In such cases, a change in the air-entraining admixture, cement, or fine aggregate may be necessary. Alternatively, an increase in slump may be required to obtain the required air content. These problems are particularly acute during hot weather.

Air-entraining admixtures should be stored in strict accordance with the manufacturer's recommendations. Although most admixtures usually are not damaged by freezing, the manufacturer's instructions should be followed regarding the effects of freezing the product. An admixture that is stored for more than six months, either by the manufacturer, vendor, or contractor, should be retested before use.

2.8—Factors influencing the amount of entrained air

2.8.1 *Effects of materials and proportions*—Many factors can influence the amount of air entrained in concrete (Whiting 1983; Whiting and Nagi 1998). The dosage of an admixture required to obtain a given air content varies depending on the particle shape and grading of the aggregate. Organic impurities in the aggregate usually decrease the air-entraining admixture requirements. Hardness of water generally does not significantly affect the air-entraining admixture requirements (Wuerpel 1946). As concrete temperature increases, higher dosages of air-entraining admixtures are required to maintain the desired air content.

Increasing the cement content, or the fineness of a cement, decreases the amount of air entrained by a given amount of an admixture. Thus, larger amounts of air-entraining admixture are required in concrete containing high early-strength Type III cement or portland-pozzolan cement (Type IP). High-alkali cements require a smaller amount of air-entraining admixture to obtain a given air content than do low-alkali cements. Similarly, increasing the amount of finely divided materials in concrete (by using fly ash or other pozzolans, carbon black or other pigments, or bentonite) also decreases the amount of air entrained by a fixed dosage of an admixture. Air-entraining admixture generally produces more air when calcium chloride is used as an accelerator. This effect is more pronounced with certain water-reducing admixtures where; the amount of air-entraining admixture required to produce a given air content can be reduced by one-third or more (Dodson 1990). Special care should be taken when other admixtures are used in conjunction with air-entraining admixtures to ensure that a satisfactory air-void system is obtained.

The proportioning of air-entrained concrete is similar to that of nonair-entrained concrete. Methods of proportioning air-entrained concrete should follow the procedures of ACI 211.1 or ACI 211.2. These procedures incorporate the reduction in water and fine aggregate permitted by the improved workability of air-entrained concrete. The highest amounts of air per unit of air-entraining admixture develop at 100 to 175 mm (4 to 7 in.) slumps. As the *w/cm* is increased, air contents can increase, but the required characteristics of the air-void system may not be maintained (Dolch 1971).

2.8.2 *Effect of mixing*—The amount of air entrained varies with the type and condition of the mixers. Mixers become less efficient as the blades become worn or when hardened material is allowed to accumulate in the drum and on the blades. The air content may also change if there is a significant variation in batch size for a given mixer, especially if the batch size is different from the rated capacity of the mixer. Adams and Kennedy (1950) found that for various laboratory mixers, air content increased from approximately 4% to as much as 8% as the batch size was increased from 40% to 100% of rated mixer capacity.

Laboratory studies (Bloem and Walker 1946) have shown that the amount of entrained air increases with mixing time up to a maximum value, beyond which it slowly decreases. The air-void system, as characterized by a specific surface and spacing factor, usually is not harmed by prolonged mixing. If more water is added to achieve the desired slump, the air content should be checked because some adjustment may be required. Adding water without complete mixing may result in nonuniform distribution of air and water within the batch. See ACI 304R for further details.

2.8.3 *Effect of transporting and consolidation*—The methods used to transport concrete after mixing, such as pumping, can reduce the air content (Hover 1989; ACI 304.2R). Some admixtures, however, result in an increase in the air content during pumping (Burg 1985). The increase is apparently caused by the increased shearing action imposed by the screw that moves the concrete from the hopper into the pump cylinder.

The type and degree of consolidation used in placing concrete can reduce the air content. Vibration applied to air-entrained concrete removes air as long as the vibration is continued (Backstrom et al. 1958a); however, laboratory tests have shown that concrete's resistance to freezing and thawing is not reduced by moderate amounts of vibration. Stark (1986) has shown that extended vibration, particularly at high frequencies, can significantly reduce this resistance. The air lost by these manipulations primarily consists of larger bubbles of entrapped air that contribute little to the beneficial effects of air entrainment.

2.9—Controlling air content in concrete

Achieving the benefits of entrained air in a consistent manner requires close control of the air content, which should be checked and controlled in accordance with the recommendations of ACI 311.1R and 311.4R. The air content in the concrete after consolidation is the most important. Air losses due to handling, transportation, and consolidation are not detected by air-content tests performed at the mixer (ACI 309R). For control purposes, samples for determining air content should be collected at the point of delivery (ACI 301). When loss of air occurs during handling, such as pumping, it may be necessary to measure air contents at the point of placement. Air-content tests of freshly mixed concrete should be made at regular intervals and whenever there is reason to suspect a change in air content. The properties of the concrete-making materials, the proportioning of the concrete mixture, and all aspects of mixing, handling, and placing, should be maintained as constant as possible to ensure that the air content will be uniform and within the range specified for the work. Too much air may reduce strength without a commensurate improvement in durability, whereas too little air will fail to provide desired durability and workability. Practices that cause excessive air loss should be avoided.

There are three standard ASTM methods for measuring the air content of fresh concrete:
- The gravimetric method, ASTM C 138;
- The volumetric method, ASTM C 173; and
- The pressure method, ASTM C 231.

The pressure method, however, is not applicable to lightweight concrete. The Chace Air Indicator (Grieb 1958), which is an adaptation of the volumetric method, has not been standardized and should not be used to determine compliance with specification limits. The ASTM methods measure only air volume and not the air-void characteristics. While the spacing factor and other significant parameters of the air-void system in hardened concrete have traditionally been determined only by microscopical method such as those described in ASTM C 457, methods are being developed (Whiting 1993) that are claimed to determine air-void parameters of fresh concrete. These methods are still experimental.

CHAPTER 3—ACCELERATING ADMIXTURES
3.1—Introduction

An accelerating admixture is "an admixture that causes an increase in the rate of hydration of the hydraulic cement and thus shortens the time of setting, increases the rate of strength development, or both" (ACI 116R). Accelerating admixtures purchased for use in concrete should meet the requirements for Type C or E of ASTM C 494.

Accelerating admixtures are useful for modifying the properties of concrete, particularly in cold weather, to expedite the start of finishing operations and, where necessary, the application of insulation for protection; reduce the time required for proper curing and protection; and increase the rate of early-strength development to permit earlier removal of forms and earlier opening of construction. Using accelerators in cold-weather concrete usually is not sufficient to counteract effects of low temperature, and should be combined with other recommended practices (refer to ACI 306R). Quick-setting admixtures permit more efficient plugging of leaks against hydrostatic pressure and produce rapid setting of concrete placed by shotcreting.

Accelerating admixtures should not be used to prevent the water in the concrete from freezing, because at normal dosages, accelerating admixtures lower the freezing point of water in concrete by 2 °C (4 °F). Substantial depression of the freezing point of water requires very high dosages, which causes undesirable side effects. Proprietary accelerating admixtures are available (Brook and Ryan 1989) that provide water reduction and accelerate hydration down to –7 °C (20 °F) without any harmful side effects (refer to Section 6.13).

3.2—Types of accelerating admixtures

For convenience, accelerating admixtures can be divided into four groups: soluble inorganic salts, soluble organic compounds, quick-setting admixtures, and miscellaneous solid admixtures.

3.2.1 *Soluble inorganic salts*—Studies (Edwards and Angstadt 1966; Rosskopf, Linton, and Pepper 1975) have shown that a variety of soluble inorganic salts, such as chlorides, bromides, fluorides, carbonates, thiocyanates, nitrites, nitrates, thiosulfates, silicates, aluminates, and alkali hydroxides, decrease the setting time of portland cement. Of these salts, calcium chloride is the most widely used because it is the most cost-effective. Research by numerous investigators has shown that inorganic accelerating admixtures act primarily by accelerating the hydration of tricalcium silicate (Ramachandran 1984). Calcium chloride should meet the requirements of ASTM D 98. Forms of calcium chloride and their equivalent masses are shown in Table 3.1.

3.2.2 *Soluble organic compounds*—The most common organic accelerating admixtures in this class are triethanolamine and calcium formate, which are commonly used to offset the retarding effects of water-reducing admixtures or to provide noncorrosive acceleration. The effectiveness of calcium formate depends on the tricalcium aluminate-to-sulfur trioxide ratio (C_3A/SO_3) of the cement (Gebler 1983). Cements that are undersulfated ($C_3A/SO_3 > 4.0$) provide the best potential for calcium formate to accelerate the early-age strength of concrete. The production of ettringite is greater in mixtures containing calcium formate (Bensted 1978).

Table 3.1—Calcium chloride: amount introduced

Calcium chloride by mass of cement, %	Solid form, %		Liquid form, 29% solution[‡]	Amount of chloride ion added, %
	Dihydrate[*]	Anhydrous[†]	L/100 kg (qt/100 lb)	
0.5	0.7	0.5	0.57 (0.6)	0.3
0.8	1.0	0.8	0.95 (1.0)	0.5
1.0	1.3	1.0	1.14 (1.2)	0.6
1.5	2.0	1.5	1.70 (1.8)	1.0
2.0	2.6	2.0	2.27 (2.4)	1.3

[*]Commercial flake products generally have an assay of 77 to 80% calcium chloride, which is often close to dihydrate.
[†]Commercial anhydrous calcium chloride generally has an assay of 94 to 97% calcium chloride. Remaining solids are usually chlorides of magnesium, sodium, or potassium, or combinations thereof. Thus, the chloride content, assuming the material is 100% calcium chloride, introduces very little error.
[‡]A 29% solution often is the concentration of commercially used liquid forms of calcium chloride and is made of dissolving 0.45 kg (1 lb) dihydrate to make 0.95 L (1 qt) of solution.

Accelerating properties have been reported for calcium salts of carboxylic acid: acetate (Washa and Withey 1953), propionate (Arber and Vivian 1961), and butyrate (RILEM 1968); however, salts of the higher homologs are retarders (RILEM 1968).

Studies (Ramachandran 1973, 1976a) indicate that triethanolamine accelerates the hydration of tricalcium aluminate but retards hydration of tricalcium silicate. Thus, triethanolamine can act as a retarder of cement at high dosages or low temperatures. A number of other organic compounds have been found to accelerate the setting of portland cement when a low *w/cm* is used. Organic compounds reported as accelerating admixtures include urea (RILEM 1968), oxalic acid (Bash and Rakimbaev 1969; Djabarov 1970), lactic acid (Bash and Rakimbaev 1969; Lieber and Richartz 1972), various ring compounds (Lieber and Richartz 1972; Wilson 1927), and condensation compounds of amines and formaldehyde (Rosskopf, Linton, and Pepper 1975; Kossivas 1971). Retardation can occur if high dosages are used, because like triethanolamine, such compounds will retard the hydration of tricalcium silicate.

3.2.3 *Quick-setting admixtures*—Quick-setting admixtures are used to produce quick-setting mortar or concrete suitable for shotcreting and sealing leaks against hydrostatic pressure. These admixtures are believed to act by promoting the flash setting of tricalcium aluminate (Schutz 1977). Ferric salts, sodium fluoride, aluminum chloride, sodium aluminate, and potassium carbonate are reported to produce quick-setting (Mahar, Parker, and Wuellner 1975) mortars, but many proprietary formulations are mixtures of accelerating admixtures.

Quick-setting admixtures for shotcrete, employed extensively in both the dry and wet processes (ACI 506R), are a specific class of quick-setting admixtures, traditionally based on soluble aluminates, carbonates, and silicates. These materials are caustic, hazardous, and require special handling; refer to material safety data sheets from the manufacturer. Newer, neutral-pH, chloride-free proprietary admixtures, based on specific sugar-acid compounds, are available to overcome these deficiencies. Generally, the wet-process shotcrete mixture quickly stiffens and reaches a rapid initial set, with final set occurring. The early stiffening imparted by the accelerator, however, aids in vertical and overhead placement. Using dry-process shotcrete and a compatible cement and accelerator, an initial setting time of less than 1 min and a final setting time of less than 4 min can be attained. The rate of strength gain can be greatly accelerated using quick-setting admixtures in dry-process shotcrete. Strength in excess of 21 MPa (3000 psi) in 8 h is typical for a noncaustic accelerator and 14 MPa (2000 psi) is typical for a conventional caustic accelerator.

3.2.4 *Miscellaneous solid admixtures*—In certain instances, hydraulic cements have been used in place of accelerating admixtures. For example, calcium-aluminate cement can cause flash setting of portland-cement concrete (Robson 1952) depending on the dosage rate. Various silicates have been found to act as accelerating admixtures (Angstadt and Hurley 1967; Kroone 1968). Finely divided silica gels and soluble quaternary ammonium silicates have been found to accelerate strength development (Stein and Stevels 1974; Nelson and Young 1977; Wu and Young 1984) through the acceleration of tricalcium-silicate hydration. Silica fume also accelerates tricalcium-silicate hydration (Kurdowski and Nocum-Wezelik 1983; Lu, Sun, and Young 1993). Adding finely divided magnesium carbonate (Ulfstedt, Wijard, and Watesson 1961) or calcium carbonate (RILEM 1968) decreases setting times.

3.3—Use with special cements

The effectiveness of calcium chloride on blended cements is similar to that on portland cements, the effects being greater for cements containing ground granulated blast-furnace slag than for those with pozzolanic additions (Collepardi, Marcialis, and Solinas 1973). The effects on a specific concrete should be determined by individual tests. Calcium chloride's effectiveness in producing accelerated strength in concrete containing pozzolans or slag is proportional to the amount of cement in the mixture (USBR 1975). The limited and conflicting data available on the effect of accelerating admixtures on the expansion of concrete containing shrinkage-compensating cements (ACI 223) require evaluation on a case-by-case basis. Calcium chloride should not be used with calcium-aluminate cement because it retards the hydration of the aluminates. Similarly, both calcium chloride and potassium carbonate increase the setting time and decrease the early strength development of rapid-hardening cements based on calcium fluoroaluminate ($C_{11}A_7 \cdot CaF_2$). Strengths at 1 day, however, are improved by these admixtures.

3.4—Effects on hardening concrete

3.4.1 *Time of setting*—Initial and final setting times are reduced by an amount dependent on the dosage of accelerator used, the temperature of the concrete, the ambient temperature, and the characteristics of other materials used in the concrete. Calcium chloride is reported (RILEM 1968) to have a greater accelerating effect at 0 to 5 °C (32 to 41 °F) than at 25 °C (77 °F). Excessive amounts of some accelerating admixtures can cause very rapid setting, particularly in hot weather. On the other hand, excessive dosage rates of certain organic compounds may result in extended times of setting. Similarly, at high dosages (6% by mass of cement), calcium nitrate begins to show retarding properties (Murakami and Tanaka 1969), while ferric chloride retards at additions of 2 to 3% by mass, but accelerates at 5% (Rosskopf, Linton, and Pepper 1975). With quick-setting admixtures, setting times as short as 15 to 30 s can be attained. Prepackaged mortar formulations are available that have an initial time of setting of 1 to 4 min and a final setting time of 3 to 10 min. They are used to seal leaks in below-grade structures, for patching, and for emergency repair. The ultimate strength of such mortar is much lower than if no accelerating admixture had been added.

3.4.2 *Air entrainment*—When an accelerating admixture is used, a lower dosage of air-entraining admixture may be required to produce the required air content. In some cases, however, larger bubble sizes and higher spacing factors are obtained, possibly reducing the beneficial effects of entrained air. Concrete containing a specific admixture may be evaluated to ascertain air-void parameters using ASTM C 457 or resistance to freezing and thawing using ASTM C 666.

3.4.3 *Strength development*—When calcium chloride is used, compressive strength may be increased substantially at early ages; later strength may be reduced slightly. Adding 2% $CaCl_2 \cdot 2H_2O$ by mass of cement can provide increases in strength at 1 day in the range of 100 to 200%, depending on the cement used. The percentage increase in flexural strength usually is less than that of the compressive strength (Ramachandran 1976b). The effects of other accelerating admixtures on strength development are not completely known, although other calcium salts behave similarly. Because accelerated strength development depends on accelerated hydration, heat of hydration also develops faster, but there is no appreciable effect on total heat generation. Quick-setting admixtures, such as carbonates, silicates, and aluminates, may decrease concrete strengths as early as one day as well as ultimate strengths (Mailvaganam 1984). Quick-setting mixtures of portland cement and calcium-aluminate cement behave similarly. Organic compounds, such as triethanolamine and calcium formate, appear to be sensitive in their accelerating action to the particular concrete mixture to which they are added, and to the ambient temperature.

3.5—Effects on concrete durability

3.5.1 *Volume change*—Accelerating admixtures can increase the volume changes that occur under moist curing and drying conditions. Calcium chloride can increase creep and drying shrinkage of concrete (Shideler 1942). Mather (1964) offered an alternative hypothesis to the presumed association of the use of calcium chloride with increased drying shrinkage. Bruere, Newbegin, and Wilson (1971) observed that such volume changes depend on the length of curing before beginning measurements, the length of the drying or loading periods, and the composition of the cement used. They also noted that changes in the rate of deformation are greater than changes in the total deformation. Berger, Kung, and Young (1967) suggested that the influence of calcium chloride in drying shrinkage can result from changes in the size distribution of capillary pores due to calcium chloride's effect on hydration of the cement. Drying shrinkage and swelling in water are higher for mixtures containing both portland and calcium-aluminate cements, and their durability may be affected adversely by using an accelerating admixture (Feret and Venuat 1957).

3.5.2 *Freezing and thawing damage*—The resistance to deterioration due to cycles of freezing and thawing and to scaling caused by the use of deicing salts can be increased at early ages by accelerators but may be decreased at later ages (refer to Section 3.4.2).

3.5.3 *Chemical attack*—Resistance to sulfate attack is decreased when conventional portland-cement concrete mixtures contain calcium chloride (USBR 1975), but when used with Type V cement to mitigate the effects of cold weather, it is not harmful (Mather 1992). The expansion produced by alkali-silica reaction is greater when calcium chloride is used (USBR 1975).

3.6—Corrosion of metals

One of the major disadvantages of calcium chloride is that it induces corrosion of metals embedded in concrete when in the presence of sufficient moisture and oxygen. ACI 318 lists the chloride limits for concrete in new construction (Table 3.2) that should be used to determine the maximum permissible water-soluble chloride-ion content for concrete in various types of construction. Gaynor (1985) discusses the calculation of chloride contents for comparison with these limits (see example in Table 3.3). The user should exercise good judgment when applying these limits, keeping in mind that other factors such as moisture and oxygen are always necessary for electrochemical corrosion. The use of calcium chloride as an accelerating admixture will aggravate the effects of poor-quality concrete construction, particularly when the concrete is exposed to chlorides during service. If good construction practices are not followed, the limits listed in Table 3.2 do not guarantee absence of corrosion. Because some accelerating admixtures contain substantial amounts of chlorides, the user should determine whether or not the admixture contains a significant amount of chlorides and, if so, the percent by mass of the cement that its use will introduce into the concrete. The potential for in-service corrosion should then be evaluated accordingly.

Table 3.2—Maximum chloride-ion concentration for corrosion protection (ACI 318-95)

Type of member	Maximum water-soluble chloride ion in concrete, % by mass of cement
Prestressed concrete	0.06
Reinforced concrete exposed to chloride in service	0.15
Reinforced concrete that will be dry or protected from moisture in service	1.00
Other reinforced concrete construction	0.30

Table 3.3—Calculation of total chloride-ion content

1	2	3	4	5
Ingredient	lb/yd^3 (or dosage)	Total Cl^-, % by weight of each material	Calculation	Total Cl^-, lb
Cement	600	0.005	$(0.005)(600)(10^{-2})$	0.03
Sand, SSD	1150	0.01	$(0.01)(1150)(10^{-2})$	0.115
Coarse aggregate, SSD	1800	0.106	$(0.106)(1800)(10^{-2})$	1.908
Water	280	250 ppm	$(205)(280)(10^{-6})$	0.07
Admixture	(5 oz./100 lb)*	800 ppm	$(800)(5)(6)(1/16)(10^{-6})$	0.0015

Total Cl^- in yd^3 = 2.1245 lb
Total chloride, % by weight of cement (2.1245/600)100 = 0.354%

*Per 100 lb of cement.

Admixtures have been sought that emulate the accelerating properties of calcium chloride without having its corrosive potential. Formulations based on calcium formate with a corrosion inhibitor have been patented (Dodson, Farkas, and Rosenberg 1965). Stannous chloride, ferric chloride, and sodium thiosulfate (Arber and Vivian 1961), calcium thiosulfate (Murakami and Tanaka 1969), ferric nitrite (RILEM 1968), and calcium nitrite (Bruere 1971) are reported to retard the corrosion of steel while still accelerating setting and hardening. The fact that an accelerating admixture does not contain significant amounts of chloride, however, does not necessarily render it noncorrosive; for example, Manns and Eichler (1982) report that thiocyanates may promote corrosion. Nmai and Corbo (1989), however, found that the threshold level for initiation of corrosion by sodium thiocyanate lies between 0.75 and 1.0% by mass of cement, and concluded that the use of sodium thiocyanate-based accelerating admixtures is safe for reinforced concrete applications up to these concentrations. Typical dosages of accelerating admixtures containing sodium thiocyanate contribute between 0.05 and 0.1% sodium thiocyanate by mass of cement, and extremely high dosages may contribute as much as 0.2% sodium thiocyanate. Users should request that suppliers of admixtures containing sodium thiocyanates provide test data regarding the corrosion of steel in concrete. The test data should include corrosion results within the intended dosage range.

3.7—Discoloration of flatwork

Discoloration of concrete flatwork has been associated with the use of calcium chloride (Greening and Landgren 1966). Two major types of mottling discoloration can result from the interaction between cement alkalies and calcium chloride. The first type has light spots on a dark background and is characteristic of mixtures in which the ratio of cement alkalies to calcium chloride is relatively low. The second consists of dark spots on a light background, and is characteristic of mixtures in which the ratio of cement alkalies to chlorides is relatively high. The magnitude and permanence of discoloration increase as the calcium chloride concentration increases from 0 to 2% by mass of cement. The discoloration can be aggravated by high rates of evaporation during curing and by improper placement of vapor barriers. Using continuous fog spray or curing compounds can help alleviate this problem.

3.8—Dosage levels

The dosage of an accelerating admixture needed to obtain the desired time of setting and level of early strength depends on local conditions and specific materials used. If adequate information is not available, the effect of a particular admixture on the properties of concrete should be evaluated using site materials with expected site temperatures and construction procedures. In the case of calcium chloride, generally 1 to 2% of the dihydrate is added, based on the mass of cement. The industry practice has been to equate 0.45 kg (1 lb) of the dihydrate form per 45 kg (100 lb) of cement to represent 1% by mass of cement (Calcium Chloride Institute 1959). Various researchers (Abrams 1924; Ramachandran 1976b) have studied the effects of calcium chloride on concrete using this dosage. For convenience and means of reference to various research data, this practice continues and prevails. In actuality, 0.56 kg (1.31 lb) of dihydrate and 0.45 kg (1 lb) of the anhydrous forms of calcium chloride per 45 kg (100 lb) of cement are equivalent to 1% calcium chloride by mass of cement (refer to Table 3.1).

In many locations, anhydrous solid forms or solutions of calcium chloride are more economical. The common dosage rates of each form and the total chloride contributed to the mixture are shown in Table 3.1. This total chloride includes chlorides contributed by normal impurities such as sodium chloride (NaCl), potassium chloride (KCl), and magnesium chloride ($MgCl_2$), in technical-grade products. Accelerating admixtures based on calcium chloride should meet the requirements of ASTM C 494. The amount of water in the solution should be deducted from the water required for the desired w/cm. Batching systems are available and are recommended to ensure accurate and uniform addition of calcium chloride in liquid form.

Calcium chloride should be introduced into the concrete mixture in solution form; the dihydrate and anhydrous solid forms should be dissolved in water before using. Preparing a standard solution from dry calcium chloride requires that the user be aware of the percent calcium chloride printed on the container. When dissolving the dry product, it should be added slowly to the water, rather than vice versa, as the latter may cause a coating to form that is difficult to dissolve. The concentration of the solution may be verified by checking the density, which should be approximately 1.28 g/mL (10.66 lb/gal.) at 23 °C (73 °F), for a 29% solution. The correct density should be obtained from the supplier.

CHAPTER 4—WATER-REDUCING AND SET-CONTROLLING ADMIXTURES
4.1—Introduction

Water-reducing and set-controlling admixtures reduce the water requirements of a concrete mixture for a given slump, modify the time of setting, or both. Some water-reducing and set-controlling admixtures entrain excessive amounts of air. Another common side effect of many water-reducing admixtures is a tendency to retard the setting time of the concrete, mortar, or grout. In such cases, the water reduction is limited to less than 10% because higher dosages cause excessive retardation. Often, water-reducing admixtures that do not retard setting time are produced by combining water-reducing materials with accelerating admixtures. The change in time of setting (retarding, nonretarding ["normal setting"], or accelerating) depends on the relative amounts of each ingredient used in the formulation.

Admixtures formulated using this approach, but including the addition of nonionic surfactants to aid water reduction and workability, may be marketed as midrange water reducers. Although there is no such classification for midrange water reducers, these admixtures meet the requirements of ASTM C 494 Type A water-reducing admixtures. These formulations provide up to 12% water reduction without significantly delaying the setting time of the concrete. Conventional water-reducing admixtures cannot do this because they become too retarding at higher dosages. High-range water-reducing admixtures (HRWRA), commonly referred to as superplasticizers, differ from conventional water-reducing admixtures in that they do not significantly delay the hydration process until much higher dosages are used (up to 0.75% by mass of cement). Therefore, they can provide significantly greater water reduction without excessive air entrainment or retardation. At lower concentrations, they impart the same water reduction and strength benefits as other water-reducing admixtures.

4.2—Classification and composition

4.2.1 *Classification*—Water-reducing and set-controlling admixtures should meet the applicable requirements of ASTM C 494, which defines seven types:
- Type A—Water-reducing admixtures;
- Type B—Retarding admixtures;
- Type C—Accelerating admixtures;
- Type D—Water-reducing and retarding admixtures;
- Type E—Water-reducing and accelerating admixtures;
- Type F—Water-reducing, high-range admixtures; and
- Type G—Water-reducing, high-range, and retarding admixtures.

Types F and G are also covered by ASTM C 1017 as Type I and Type II. ASTM C 494 gives detailed requirements with respect to water requirement, setting time, flexural and compressive strength, drying shrinkage, and resistance to freezing and thawing. The specific effects of water-reducing and set-controlling admixtures, however, vary with different cements, addition sequences, changes in w/cm, mixing temperature, ambient temperature, and other site conditions (ACI 212.4R). Most water-reducing admixtures perform considerably better than the minimum requirements of ASTM C 494.

4.2.2 *Compositions*—Materials generally available for use as water-reducing and set-controlling admixtures fall into one of nine general categories of compounds. Increasingly, though, formulations can include compounds from more than one category:

1. Lignosulfonic acids and their salts and modifications and derivatives of these;

2. Hydroxylated carboxylic acids and their salts and modifications and derivatives of these;

3. Carbohydrate-based compounds such as sugars, sugar acids, and polysaccharides;

4. Nonionic surface-active agents;

5. Salts of sulfonated melamine polycondensation products;

6. Salts of condensation products of naphthalene sulfonic acid;

7. Carboxylic acrylic ester copolymers;

8. Inorganic compounds such as zinc salts, borates, and phosphates; and

9. Other materials such as: amines and their derivatives; organic phosphonates; and certain polymeric compounds, including cellulose-ethers, silicones, and sulfonated hydrocarbon acrylate derivatives.

4.3—Applications

Water-reducing admixtures lower the w/cm and are used to produce higher strength and increased durability, to obtain a specified strength at lower cement content, or to increase slump without an increase in water content or combinations of these effects. They also may improve the properties of concrete containing aggregates that are harsh, poorly graded, or both, or may be used in concrete that will be placed under difficult conditions. The admixtures are useful when placing concrete by pump or tremie.

Set-controlling admixtures are used primarily to offset the accelerating effect of high ambient temperature (hot weather) and to keep concrete workable during the entire placing period, thereby eliminating form-deflection cracks (Schutz 1959). This method is particularly valuable to prevent cracking of concrete beams, bridge decks, or composite construction caused by form deflections. Set retarders are also used to keep concrete workable long enough so that succeeding lifts can be placed without development of cold joints in the structural unit. Their effects on rate of slump loss vary with the particular combinations of materials used.

HRWRA can produce large reductions in the water content of concrete—up to 30% reduction has been achieved. High-strength concrete with a w/cm as low as 0.22 by mass can be produced with slumps exceeding 75 mm (3 in.). Concrete at slumps of 150 to 175 mm (6 to 7 in.) have been produced with water reductions of 10 to 15%. Flowing concrete with slump in excess of 200 mm (8 in.) is typical for many types of concrete (refer to Chapter 5). HRWRA also have been used to reduce cement content. Because the w/cm affects the strength of concrete, the cement content may be reduced with a proportional reduction of the water content for equivalent or higher-strength concrete due to increased cement efficiency, resulting in cost savings. In mass concrete, low cement content is particularly desirable because it lowers the temperature rise of the concrete (refer to Section 4.7.7).

4.4—Dosage

The expected performance of a given brand, class, or type of admixture may be estimated from one or more of the following sources of information:
- Results from jobs where the admixture has been used under good field control, preferably using the same materials and under conditions similar to those anticipated;
- Laboratory tests made to evaluate the admixture; and

Technical literature and information from the manufacturer. The dosage of the admixture should be determined from information provided by one or more of these sources. Varying results can be expected with a given admixture due to differences in dosage, cements, aggregates, other materials, and weather conditions. Water-reducing and set-controlling admixtures are usually more effective with respect to water reduction and strength increase when used with portland cements with lower tricalcium aluminate (C_3A) and alkali content. Differences in setting time also can be expected. In the production of high-strength concrete (above 41 MPa [6000 psi]), it is beneficial to increase the dosage of the admixture. This usually provides extra water reduction and typically a delay in setting time and slow early-strength gain. Concrete with slow-early strength gain generally exhibits higher later strengths.

Concrete containing HRWRA sometimes has rapid slump loss. To overcome this, a second addition of the HRWRA may be used to restore the slump, without any apparent negative effects. Generally, more than two additions are less effective and concrete may lose its workability faster than with a single dose. Redosing may result in a lower air content, on the order of 1 to 2 percentage points for each redose. When redosages are used, the concrete may have a greater potential for bleeding, segregation, and possible retardation of setting time. Therefore, trial mixtures should be conducted to determine the effects of redosing. If slump loss is a concern, the type of HRWRA should be reviewed. The period of required plasticity should be verified by placing the proposed mixture on the project. New products aimed at increasing efficiency, improving cohesiveness, and maintaining workability for longer periods of time are now available. HRWRA can be added at the batch plant rather than at the job site, thereby reducing wear on truck mixers and lessening the need for ancillary equipment such as truck-mounted admixture tanks and dispensers.

4.5—Preparation, batching, and quality control

Water-reducing and set-controlling admixtures should be batched and dispensed as liquids. When supplied as solids, they should be mixed to a suitable solution concentration following the manufacturer's recommendations. The density of admixtures mixed on site or applied as solutions should be determined and compared with the manufacturer's standards. Density can be determined easily and quickly with a hydrometer or volumetric flask. The determinations should be made at a standard temperature and recorded for future reference as part of the site quality-control program. Storage tanks for solutions should be clearly identified, and the solutions should be protected from contamination, dilution, evaporation, and freezing.

It sometimes is necessary or desirable to determine that an admixture is similar to previously tested material, or that successive lots of shipments are similar. Tests used to identify admixtures include solids content, density, infrared spectro-photometry for organic materials, chloride content, and pH. Guidelines for determining the uniformity of chemical admixtures are given in ASTM C 494. Project inspectors may be instructed to sample deliveries of the admixture as part of the project quality control or quality assurance. Admixture users should become familiar with appearances and odors of the admixtures; this knowledge can prevent errors.

4.6—Proportioning concrete

When a concrete mixture that is considered satisfactory in workability and finishing qualities is modified to incorporate a water-reducing and set-controlling admixture, the ratio of mortar to coarse aggregate by volume should remain the same. Changes in water content, cementitious material content, and air content are compensated for by corresponding changes in the content of fine aggregate—all on a solid or absolute volume basis—so that the volume of mortar remains the same. Procedures for proportioning and adjusting concrete mixtures are covered by ACI 211.1. Most chemical admixtures of the water-reducing type are water solutions. The water they contain becomes a part of the mixing water in the concrete and should be considered in the calculation of w/cm. The proportional volume of the solids included in the admixture is so small in relation to the size of the batch that it can be neglected.

Concrete proportioned for high strength using HRWRA usually has a sufficiently high cement content to supply the fines required for good workability. Such concrete can be reproportioned by making up the volume of water reduced by increasing the volume of coarse and fine aggregate equally. If trial mixtures are sticky, the volume of coarse aggregate should be increased and that of the fine aggregate reduced. This usually results in a mixture that is easier to place and finish.

4.7—Effects on fresh concrete

4.7.1 *Water reduction*—Type A water-reducing admixtures decrease the water required for the same slump concrete by at least 5%, and in some cases up to 12%, as is the case with midrange water-reducing admixtures. Concrete containing lignosulfonate or hydroxylated carboxylic acid salts reduce the water content 5 to 10% for a given slump and cement content. HRWRA should reduce the water requirement at least 12%, but there is no upper limit to water reduction. Reductions as high as 30% have been reported. HRWRA can be used to significantly increase slump without increasing water content and may be used to achieve a combination of these two objectives—a slump increase with a reduction in water content.

As the cement content of a concrete mixture increases, the required dosage of a HRWRA, as a percentage by mass of cement, is reduced (Collepardi 1984). The effects of these admixtures also are dependent on the calculated C_3A, C_3S, and alkali contents of the cement. Concrete made with cements meeting requirements for Type II and Type V cements require lower admixture dosages than concrete containing Type I or Type III cements. In some cases, a higher SO_3 content in the cement may be desirable when using HRWRA.

4.7.2 *Air entrainment*—Lignosulfonates entrain air to various degrees ranging from 2 to 6%, although higher amounts have been reported (Tuthill, Adams, and Hemme 1960). The air-entraining properties may be controlled by modifying formulations. Materials in Categories 4, 5, and 6 (Section 4.2.2) usually do not entrain air, but materials in all nine categories may affect the air-entraining capability of both air-entraining cement and air-entraining admixtures. This is particularly true in the case of some HRWRAs. The entrained air can consist of large, unstable bubbles that contribute little resistance to freezing and thawing. Therefore, the entrained air-void characteristics should be evaluated whenever questions about the performance of the concrete arise. Including testing for resistance to freezing and thawing in the evaluation is prudent, as in some instances the concrete may still be durable even though the spacing factors may exceed accepted limits of 0.20 mm (0.008 in.) (Mather 1979).

4.7.3 *Workability*—When comparing a concrete mixture without a water-reducing admixture at the same slump and air content, differences in workability are difficult to detect because no standard test exists. Howard, Griffiths, and Moulton (1960) reported that use of the Kelly Ball (ASTM C 360) detected increases in workability that were missed by the slump test. Concrete containing a water-reducing admixture, however, is less likely to segregate and sometimes has better flowability.

Water-reducing and set-controlling admixtures affect bleeding capacity in varying degrees. For example, unmodified admixtures in Category 2 (Section 4.2.2) tend to increase bleeding, while their modifications and derivatives do not. Admixtures in Category 1 reduce bleeding and segregation in freshly mixed concrete, partly due to the air entrainment. HRWRA derived from Categories 4, 5, and 6 decrease bleeding, except at very high slump.

4.7.4 *Rate of slump loss*—The rate of slump loss may be decreased by adding water-reducing, set-controlling admixtures, especially in concrete made with HRWRA. Because of the slump loss, HRWRA may be added at the job site. Working time can be extended with the careful use of an ASTM C 494 Type B retarding or Type D water-reducing and retarding admixture or with the use of a Type G retarding HRWRA. The working time depends on many factors, including the HRWRA dosage, use of other chemical admixtures, cement characteristics, *w/cm*, concrete temperature, slump, and age of the concrete when the HRWRA is introduced.

4.7.5 *Finishing*—Some water-reducing or midrange water-reducing admixtures improve the concrete's finishing characteristics compared to concrete containing other types of water-reducing admixtures or no admixture. This is beneficial where deficient aggregate properties or gradation result in finishing difficulties. At the high-water content reduction achieved with HRWRA, finishing may become more difficult due to the decrease in bleeding, and surfaces may have a tendency to crust and promote plastic-shrinkage cracking. The surface can be kept from drying by fogging, using an evaporation retarder, or other procedures (ACI 308R). These treatments should be used with caution so that the durability of the surface is not affected adversely.

4.7.6 *Time of setting*—Water-reducing and set-controlling admixtures cause a delay in setting time that will increase with increasing dosages and lower temperatures. Water-reducing admixtures are generally formulated with an accelerating component to produce a setting time within 1-1/2 h of a reference mixture at normal dosage rates. These admixtures will also produce extended setting times when dosages increase beyond normal rates or when temperatures fall. HRWRA can be used at higher dosages without an appreciable increase in setting times; however, increasing dosages beyond normal rates may also result in setting time delays. Accelerators can decrease or eliminate retardation.

Retarders are not recommended for controlling false set; water-reducing retarders have been reported to contribute to premature stiffening. Admixtures based on lignin delay false set, and sugar-type admixtures can cause flash set if the cement has a low sulfate-to-aluminate ratio.

4.7.7 *Heat of hydration*—Within normal *w/cm* ranges, heat of hydration and adiabatic temperature rise are not reduced at equal cement contents with the use of set-controlling admixtures. Acceleration or retardation can alter the rate of heat generation, which can change the early rate of temperature rise under job conditions. If the use of the admixture permits a reduction in cement content, the heat generated is proportionally reduced.

4.8—Effects on hardened concrete

4.8.1 *Strength*—In addition to the strength increase due to reduction of *w/cm*, strength is further increased due to modification of the paste microstructure by the water-reducing admixture. Unless used at unusually high dosages, retarding admixtures can decrease the strength at ages up to 24 h, while the normal-setting and accelerating types increase the very early strength. Later strength with a set-controlling admixture can increase 20% or more at the same cement content. Cement contents can thus be reduced without lowering 28-day strengths. When HRWRAs are used to decrease the *w/cm*, 28-day compressive strength can increase by 25% or more. Increases in flexural strength of concrete containing a water-reducing admixture are not proportionally as great as increases in compressive strength (Collepardi 1984).

4.8.2 *Shrinkage and creep*—Long-term shrinkage may be less than that of concrete not containing a water-reducing and set-controlling admixture. Creep is reduced in proportion to the increase in the strength of the concrete. How much a particular admixture in a given dosage effects shrinkage and creep depends on the composition of the cement (ACI 209R).

4.8.3 *Resistance to freezing and thawing*—Water-reducing admixtures have little effect on resistance to freezing and thawing, including deicer scaling, because the resistance is almost wholly a function of the air-void system in the hardened concrete. An improvement can result from a decrease in *w/cm* because of increased strength and density and reduced permeability, which allow the concrete to remain less than critically saturated in the presence of water while being tested.

4.9—Evaluations

If adequate information is not available, tests should be made to evaluate the effect of the admixture on the properties of concrete made with job materials under the anticipated ambient conditions and construction procedures. Tests of water-reducing admixtures and set-controlling admixtures should indicate their effect on the following properties of concrete, insofar as they are pertinent to the job:

- Water requirement;
- Air content;
- Slump;
- Bleeding and possible loss of air from the fresh concrete;
- Setting time;

- Compressive and flexural strength at 28 days, or the specified age of the concrete;
- Required strength development;
- Resistance to freezing and thawing; and
- Drying shrinkage.

When admixtures are evaluated in laboratory trial batches before use, a series of mixtures should be planned to provide the necessary information. The mixtures need not follow ASTM C 494 procedures, although these procedures may be helpful guidelines. The trial mixtures should be made with the same cementitious materials (particularly cement) and other concrete-making materials that will be used on the project and as close to job conditions as possible. Temperature is particularly important to setting time and early strength development. Air content and setting time of job concrete can differ considerably from that of laboratory concrete with the same materials and mixture proportions. The action of a water-reducing admixture or a HRWRA may differ in a truck mixer from that seen in a laboratory mixer. The admixture dosage will probably have to be adjusted to get the same performance in the truck. In most cases, the HRWRA will work better in the truck and the dosage will come down. All parties should be alert to this possibility at the start of a job and should be ready to adjust the amounts of materials (particularly air-entraining admixtures) to achieve the specified properties of the concrete at the project site.

CHAPTER 5—ADMIXTURES FOR FLOWING CONCRETE

5.1—Introduction

ASTM C 1017 defines flowing concrete as "concrete that is characterized as having a slump greater than 190 mm (7-1/2 in.) while maintaining a cohesive nature...." Flowing concrete can be placed so as to be self-leveling yet remain cohesive without excessive bleeding, segregation, or abnormal retardation. Flowing concrete should be obtained through the use of a plasticizing admixture, because adding water only would result in extremely low-quality concrete. The admixtures are the same as those used as HRWRA, conforming to ASTM C 494 Types F and G (Chapter 4).

As an example, concrete could be delivered to the job site at an initial slump of 50 to 75 mm (2 to 3 in.) and the plasticizing admixture then added to increase the slump to 200 mm (8 in.) or more. Alternatively, the plasticizing admixture could be added at the plant to achieve a high slump so that the concrete arrives at the job site with the slump required for placement. Such concrete loses slump at a more rapid rate than similar concrete without the plasticizing admixture. The dosage required to increase the slump to flowing consistency varies depending on the cement, the initial slump, *w/cm*, temperature, time of addition, and concrete mixture proportions. The dosage required to increase slump from 50 to 200 mm (2 to 8 in.) may be 50% higher than that required if the starting slump is 75 mm (3 in.).

5.2—Materials

Admixtures used to achieve the flowing concrete should meet the requirements of ASTM C 1017, Type I or Type II (retarding). Commonly used materials are:
- Sulfonated naphthalene condensates;
- Sulfonated melamine condensates;
- Modified lignosulfonates;
- Polycarboxylates;
- A combination of the above compounds with either Type A water-reducing admixture, a Type D water-reducing retarding admixture, or a Type E water-reducing accelerating admixture; and
- High dosages of a Type A water-reducing admixture, plus a Type E water-reducing accelerating admixture. This combination requires higher water contents than are required when using a HRWRA.

5.3—Selection and evaluation

When deciding to produce and use flowing concrete, consider the type of admixture to be selected. Factors include:
- Type of construction;
- Restrictions imposed on the chloride-ion content;
- Time interval from the introduction of cementitious materials and water into the mixer;
- Availability of accurate admixture dispensing equipment at the plant, job site, or both; and
- Ambient temperature.

The proposed flowing concrete mixture should be used initially on noncritical work so that proportions and procedures can be verified before the mixture is used in the areas requiring flowing concrete. The proportions of the various concrete ingredients can be adjusted and the dosage or the type of admixture varied to achieve an acceptable final slump, rate of slump loss, and setting characteristics. If the plasticizing admixture is to be added at the job site, an accurate means of introducing the admixture into the concrete mixer should be ensured. Truck mixers should be equipped with tanks designed to introduce the admixture into the concrete mixer so that it can be distributed evenly throughout the batch. Adequate mixing speed and revolutions should be maintained as defined in ASTM C 94. The concrete plant should be equipped to accurately measure the admixture into the truck-mounted tanks. Accurate measurements of the admixture dosage are critical, and proper mixing is essential to realize the full benefit of the admixture. The expected performance of a given brand, class, or type of admixture can be estimated from one or more of the following sources:
- Results from jobs where the admixture has been used under good technical control, preferably using the same materials and under conditions similar to those anticipated;
- Laboratory tests made to evaluate the admixture; and
- Technical literature and information from the manufacturer of the admixture.

ASTM C 1017 provides for compliance with specifications under controlled conditions of temperature, fixed cementitious material content, slump, and air content, using aggregates graded within stipulated limits. This standard requires certain minimum differences in concrete strength, range of setting time, and requirements regarding other aspects of performance such as shrinkage and resistance to freezing and thawing.

5.4—Applications

Flowing concrete is commonly used in areas requiring maximum placement rates, such as slabs, mats and pavements, and in congested locations where the section is unusually shaped or is heavily reinforced. It can be used in areas of limited access or where the maximum horizontal movement of the concrete is desirable. Flowing concrete is useful for pumping because it reduces pumping pressure and increases both the rate and distance that the concrete can be pumped. It is useful for projects that require rapid form cycling with a maximum volume of concrete placed per day. Coupled with a low w/cm, the early strengths required for stripping or post-tensioning can be achieved. An example is flowing concrete with a low w/cm obtaining 20 MPa (3000 psi) compressive strength in 24 h.

Flowing concrete is often desirable for use in mass placements. The cement content can be kept low, which minimizes heat development, and the lower water content reduces drying shrinkage. The plasticizing admixture does not lower the temperature rise in concrete except as a result of reducing cement content. The early temperature-rise characteristics can be modified with the use of the retarding version of the plasticizing admixture (Type II) or in combination with a conventional water-reducing retarding admixture (Type D).

High-performance concrete (HPC) has higher-than-normal compressive strength, reduced permeability, increased durability, or reduced shrinkage. The necessity to reduce the w/cm to 0.40 or lower to produce HPC while achieving a placeable concrete requires the use of an HRWRA (Aïtcin and Neville 1993). Concrete with a low w/cm and a compressive strength higher than 41 MPa (6000 psi) can be produced as flowing concrete. Flowing concrete, being easier to consolidate, also contributes to the proper bond between reinforcing steel and concrete in areas where reinforcement is congested.

Using plasticizing admixtures to increase slump from 50 to 75 mm (2 to 3 in.) to 125 to 150 mm (5 to 6 in.) also reduces the amount of cement required to achieve a particular strength. Because concrete is seldom placed at a slump level of 50 to 75 mm (2 to 3 in.), the additional water required to raise the slump would need to be matched with an increase in the cement content if the strength and w/cm were kept constant. The higher paste content would result in a concrete with a higher shrinkage.

5.5—Proportioning concrete

When admixtures are evaluated in laboratory trial mixtures before job use, the series of mixtures should be planned to provide the necessary project information. Assuming that specification compliance has been established, the tests need not follow ASTM C 1017 test conditions for slump, air content, and cementitious material content. The test conditions, however, should be consistent with the project requirements with respect to target slump, air content, setting time, and strengths at various ages. The trial mixtures should be made with the same materials, particularly cementitious material, that will be used on the job site, and should simulate the site conditions as closely as possible. Temperature is particularly important to setting times and early-strength development. Trial mixtures can be made with a starting slump and air content in the specified range. The dosage of the plasticizing admixture can be varied to achieve different slump increases. If allowed, the initial slump may also be varied. The specified w/cm should be maintained in each case, and a range of slumps can be reviewed. In this manner, the optimum mixture proportions can be selected and the desired results achieved.

A concrete mixture usually needs reproportioning when a plasticizing admixture is added to achieve flowing concrete. Procedures for proportioning and adjusting concrete mixtures are covered in ACI 211.1 and ACI 211.2. The fine aggregate-to-coarse aggregate ratio may require adjustment to ensure that sufficient fines are present to allow a flowable consistency to be achieved without excessive bleeding or segregation. Increasing the cement content or adding other fine materials such as pozzolan or slag may also be necessary. Because 2 L (0.5 gal.) or more of plasticizing admixture is customarily used per cubic meter (cubic yard) of concrete to produce flowing concrete, the water in the admixture should be accounted for when calculating w/cm and the effect on mixture volume. The air content and setting time of job concrete can differ considerably from those of laboratory concrete with the same materials and mixture proportions. Therefore, adjusting the proposed mixture on the job site before its use in the required locations is beneficial.

5.6—Effect on fresh and hardening concrete

5.6.1 *Setting time*—ASTM C 1017 Type I admixtures are required to have negligible effect on initial or final setting times. Therefore, flowing concrete will set as quickly as concrete without the admixture but with the same water content. At increased concrete mixture temperatures, the setting time of concrete containing the Type I admixture is decreased. A Type II admixture can reduce slump loss significantly and retard the initial setting time of the concrete.

5.6.2 *Workability and finishing*—When concrete mixtures are properly proportioned, flowing concrete is extremely workable without bleeding and segregation. The upper slump limit at which flowable concrete still remains cohesive can be determined by testing the mixture before use. Segregation and bleeding can be reduced by increasing the fine-to-coarse aggregate ratio, improving the fine aggregate grading, or adding other fine material. Flowing concrete should be vibrated to achieve proper consolidation. The response of flowing concrete to machine finishing is similar to that of conventional concrete made with the same ingredients. Successful finishing requires proper timing. If a concrete contains too much fine aggregate, the air content is too high, or both, the surface of the concrete may tend to dry before it sets. This condition can cause the concrete to feel rubbery or jelly-like and can cause finishing problems because of its stickiness and rolling. The problem of excessive air entrainment in concrete used in floor slabs is particularly apparent when the initial machine-finishing operations begin.

5.6.3 *Rate of slump loss*—The rate of slump loss in flowing concrete containing a HRWRA will be affected by: the type of HRWRA used; the dosage used; the simultaneous use of an ASTM C 494 Type A, B, or D admixture; the type and brand of cement; and the concrete temperature. These factors are by no means the only ones affecting slump loss, but they can typically be controlled by the user (Ramachandran and Malhotra 1984). Ambient temperature is not as controllable but can have a dramatic effect on the performance of a HRWRA. The higher the HRWRA dosage, the lower the rate of slump loss (Ravina and Mor 1986). Each product has an operating range beyond which other properties of the concrete may be affected; if this range is exceeded to further reduce the rate of slump loss, the results may include changes in initial setting times, segregation, or bleeding.

As a result of advances in HRWRA technology and the numerous products available, it has become advantageous to describe these products not only by the requirements of ASTM standards but also by the method of addition. Although both specifications for HRWRA (ASTM C 494 and C 1017) mention slump loss, neither currently requires tests for slump-loss characteristics. When some HRWRAs are added at the job site, the concrete exhibits moderate to rapid slump loss and normal or retarded initial setting times. HRWRA designed to be added at the batch plant can extend slump retention in concrete (Collepardi and Corradi 1979) along with either normal or retarded initial setting times. The difference in performance does not indicate that one admixture is better than another, but rather that certain admixtures are more appropriate in some construction situations than in others.

5.6.4 *Additional dosages*—An additional dosage of plasticizing admixture should be used when delays occur and the required slump has not been maintained. Two additional dosages have been used with success; more dosages generally are less effective. In general, the compressive strength level is maintained or increased and the air content is decreased. Therefore, if air entrainment is a concern, it should be checked after the concrete has been redosed and returned to its intended slump.

5.6.5 *Heat of hydration and temperature rise*—The total amount of heat produced by hydration is not changed if the cement content is not altered, but the rate of evolution can be altered. If the use of flowing concrete involves the use of a lower cement content, less heat will be evolved.

5.7—Effect on hardened concrete

5.7.1 *Strength*—Because flowing concrete is often batched initially with a water content resulting in a slump of 50 to 100 mm (2 to 4 in.), the *w/cm* is lower than that of conventional concrete with a similar cementitious material content and a 125 mm (5 in.) slump, improving the strength. Flowing concrete often is stronger than conventional concrete at the same *w/cm*, presumably due to increased dispersion resulting in a more efficient cement hydration. The flexural strength of flowing concrete is not significantly changed from that of the initial concrete with the same *w/cm* at a lower slump.

5.7.2 *Drying shrinkage and creep*—The drying shrinkage of low-slump concrete is about the same as that of flowing concrete with the same water content (Gebler 1982). If the cement content is kept constant to produce flowing concrete with a lowered water content, then drying shrinkage can be reduced. If both cement and water contents are reduced, then drying shrinkage can be reduced. When compared with concrete with the same *w/cm*, concrete with HRWRA shows little change in creep characteristics (Brooks, Wainwright, and Neville 1981; ACI 209R).

5.7.3 *Resistance to freezing and thawing*—Conventional and flowing concrete with the same *w/cm* and a comparable air-void system exhibit a similar resistance to freezing and thawing. Compared with conventional concrete, higher dosages of air-entraining admixture usually are required for flowing concrete to maintain proper air content. For a given air content, the air-void system may have larger spacing factors and a decrease in the number of voids per unit length compared with the control concrete; however, satisfactory resistance to freezing and thawing has been achieved in most cases. As with any air-entrained concrete, the air content in the field should be checked so that the air-entraining admixture dosage can be modified to keep the air content in the specified range.

5.7.4 *Permeability*—Flowing concrete with a *w/cm* below 0.40 can be placed easily; therefore, the resultant concrete, if properly cured, can have extremely low permeability and good resistance to the penetration of aggressive solutions. Resistance to chloride penetration is similar to, or slightly better than, that of conventional concrete with the same *w/cm* (Lukas 1981). When the admixture is used to reduce the *w/cm*, the resistance of the concrete to chloride penetration is even greater. Flowing concrete shows better consolidation, reduced bleeding, and increased cement hydration—all of which contribute to lower permeability.

5.7.5 *Bond*—Flowing concrete can improve bond strength to reinforcing steel when compared to similar concrete with a 100 mm (4 in.) slump (Collepardi and Corradi 1979). Brettman, Darwin, and Donahey (1986) found that flowing concrete showed no change in bond strength compared with lower-slump concrete with an equal *w/cm*, providing the concrete was vibrated and set rapidly after consolidation. Bond strength in reinforced concrete beams having equal *w/cm*, however, was decreased if the flowing concrete's setting time was delayed. Proper consolidation around the reinforcement is more easily achieved with flowing concrete, but adequate vibration is required.

5.8—Quality assurance

5.8.1 *Testing admixtures*—Determining that an admixture is similar to that previously tested or that successive lots or shipments are similar is desirable and sometimes necessary. Tests that can be used to identify admixtures include solids content, density, infrared spectrophotometry for organic materials, chloride content, and pH. Admixture manufacturers can recommend which tests are most suitable for their admixtures and the results that should be expected. Guidelines for determining uniformity of chemical admixtures are given in ASTM C 1017.

5.8.2 *Field control of job-site-added HRWRA*—Controlling flowing concrete requires checking the initial slump or water content before adding HRWRA to ensure that the water content and the *w/cm* are within specification. The initial slump check is conducted at the job site for job-site-added HRWRA and at the batch plant for plant-added HRWRA. After the HRWRA is added and thoroughly mixed into the batch, the resulting slump should be in the specified range. For air-entrained concrete, the air content also should be checked. If the flowing concrete is pumped into place, the air content should be measured at the point of discharge into the forms. Rate of slump loss, initial setting time, and early- and final-strength results may require mixture adjustments. Slump loss and setting characteristics may be adjusted by changes in the HRWRA dosage or by the concurrent use of accelerating or retarding admixtures. When the concrete placement is abnormally slow, the temperature is high, or both, the use of a Type II admixture can be desirable. Because variations in cement composition, aggregate grading, or both can cause significant variations in the flowing-concrete characteristics, these changes should be minimized. When pumping, flowing concrete should be placed in accordance with ACI 304.2R and consolidated in accordance with ACI 309R.

CHAPTER 6—MISCELLANEOUS ADMIXTURES
6.1—Gas-forming admixtures

6.1.1 *Introduction*—The gas-void content of concrete can be increased by using admixtures that generate or liberate gas bubbles in the fresh mixture during and immediately following placement and before setting of the cement paste. These materials are normally added to the concrete mixture to counteract settlement and bleeding, causing the concrete to retain its initial volume. They can be used at higher addition rates to produce lightweight concrete. The air voids obtained do not improve resistance to freezing and thawing; any such effect is incidental.

6.1.2 *Materials*—Admixtures that produce these effects are: hydrogen peroxide, which generates oxygen; metallic aluminum powder (Menzel 1943; Shideler 1942), which generates hydrogen; and certain forms of activated carbon that liberate adsorbed air. Only aluminum powder is in common use as an admixture. An unpolished powder usually is preferred; although polished powder may be used when a slower reaction is desired. Under normal conditions, the addition rate varies from 0.006 to 0.02% by mass of cement, although larger quantities can be used to produce low-strength cellular concrete. The rate and duration of gas evolution depends on the fineness and composition of the cement (particularly its alkali content), temperature, *w/cm*, and the fineness and particle shape of the aluminum powder. Because of the very small quantities of aluminum powder generally used (a few grams per 45 kg [100 lb] of cement), and because aluminum powder has a tendency to float on the mixing water, it generally is premixed with fine sand, cement, or pozzolan. It also can be incorporated into commercially available powdered admixtures with water-reducing and set-retarding effects. In cold weather, it may be necessary to speed up the rate of gas generation by adding caustic materials such as sodium hydroxide, hydrated lime, or trisodium phosphate to ensure sufficient gas generation before the mixture has set and hardened. To produce the same amount of expansion, approximately twice the amount of aluminum powder is required at 4 °C (40 °F) than at 21 °C (70 °F).

6.1.3 *Effectiveness*—The effectiveness of the admixture is determined by the duration of mixing, handling, and placing operations relative to the speed of gas generation. The controlled release of gas causes a slight expansion of freshly mixed concrete. When such expansion is restrained, there is an increase in the bond to reinforcing steel without an excessive reduction in compressive strength. For each 1% of added gas volume, a 5% reduction in compressive strength is typical. Too much gas-producing material may produce large voids, seriously weakening the matrix. The effect on compressive strength depends on the degree to which the tendency of the mixture to expand is restrained; therefore, it is important that confining forms be tight and adequately closed. As an example, ASTM C 1107 describes using a restraint in the production of specimens of expansive cement grout. Gas-forming admixtures will not overcome shrinkage of the hardened concrete caused by drying or carbonation.

6.2—Grouting admixtures

6.2.1 *Introduction*—Many of the admixtures in concrete are used as grouting admixtures to impart special properties to the grout. Geotechnical cement grouts (ACI 552) and oil-well cementing grouts should remain pumpable over large distances under high temperatures and pressures. Grout for preplaced-aggregate concrete requires extreme fluidity and resistance to segregation. Nonshrink grout requires a material that will not exhibit a reduction in volume at placement. Bonding and joint-filling grout for tile installation needs to resist fast drying and loss of water through absorption by the substrate and the tile. A wide variety of special-purpose admixtures are used to obtain the special properties required.

6.2.2 *Materials*—Tile grouts and other grouts use materials such as gels, clays, pregelatinized starch, methyl cellulose, and other cellulosics to prevent a rapid loss of water. Grout fluidifiers for preplaced aggregate concrete grouts usually contain water-reducing admixtures along with admixtures to prevent settlement of heavy constituents of the grout. Nonshrink grouts may contain gas-forming or expansion-producing admixtures, or both. Such grouts are required to exhibit no shrinkage under the moist environment stipulated in ASTM C 1107; when dried they typically shrink. Special grout applications can require accelerating or air-entraining admixtures as described in other sections of this document. Tests should be conducted to determine the compatibility of admixtures with the cement to be used.

6.2.3 *Effect*—Extended set-control admixtures can be used to keep grout fluid for 1 h or more at temperatures up to 200 °C (400 °F) and pressures as high as 125 MPa

(18,000 psi). These admixtures are discussed in Section 6.3. Grouting is a highly specialized field, usually requiring material properties not necessary for ordinary concreting operations. The admixture manufacturer's suggestions on addition rates should be followed; however, the grout should be tested to determine if its properties meet the project requirements.

6.3—Extended set-control admixtures

6.3.1 *Introduction*—Extended set-control admixtures are used to stop or severely retard the cement hydration process in unhardened concrete. The extended set-control admixture should be added and thoroughly mixed with the unhardened concrete (either freshly batched or returned) before the concrete approaches its initial setting time. Extended set-control admixtures differ from conventional set-controlling admixtures in that they stop the hydration process of both the silicate and aluminate phases in portland cement. Regular set-controlling admixtures act only on the silicate phases. Using an extended set-control admixture at high dosage rates can cause stiffening in the concrete. The most effective materials are carboxylic acids and phosphorus-containing organic acids and salts (Kinney 1989; Senbetta and Dolch 1991; Senbetta and Scanlon 1991). Such admixtures have gained increased acceptance since their introduction in 1986.

6.3.2 *Applications*—The technology of extended set-control admixtures addresses several problems in the concrete industry:
- The disposal of returned unhardened concrete;
- The disposal of residue from truck mixer drums; and
- The concerns of the Environmental Protection Agency (EPA) pertaining to the alkalinity of returned unhardened concrete, and its disposal.

Using extended set-control admixtures provides flexibility when using concrete materials and eliminates highly alkaline wastewater that otherwise would be disposed of in ponds, but which may be considered hazardous waste by local environmental agencies.

6.3.2.1 *Treating concrete wash water*—This process eliminates the disposal of water used to wash out the inside of truck mixers while still keeping the blades and inner surface of the drum clean. This process is designed for overnight or weekend stabilization of 40 to 270 L (10 to 70 gal.) of wash water using a low dosage of an extended set-control admixture. Following the stabilization period, the water content of the freshly batched concrete is reduced to compensate for the stabilized residue held in the truck mixer. Concrete containing stabilized residue shows setting times equal to concrete without stabilized residue, with performance characteristics equal to or better than conventional concrete.

6.3.2.2 *Stabilization of returned unhardened concrete*—In this application, returned concrete can receive same-day stabilization for 1 to 4 h or longer with a low dosage of an extended set-control admixture. This application reuses returned unhardened concrete during the same production day instead of disposal. Another option is to treat returned unhardened concrete overnight. An accelerating admixture (chloride-bearing or nonchloride) should be added the next morning to reinitiate hydration. Fresh concrete is batched on top of the stabilized or stabilized and activated concrete to complete the normal load. The combination of fresh concrete and stabilized concrete should harden similarly to conventionally batched concrete without an extended set-controlled admixture, provided the manufacturer's guidelines are carefully followed.

6.3.2.3 *Stabilizing of freshly batched concrete for long hauls*—The technology of extended set-control admixtures can be used to stop the hydration process for extended hauls and to reduce or eliminate slump loss and concrete temperature increase during transit. For this application, the extended set-control admixture is added during or immediately after the initial batching process. Once the admixture is thoroughly mixed into the fresh concrete, the mixer drum should turn as slowly as possible. The stabilizer dosage should be adjusted so the setting time of the concrete is extended for the duration of the haul. Ideally, on arrival at the job site, the effects of the admixture should be completed. This will allow the stabilized concrete to set similarly to normal concrete at the site. This application is especially helpful in summer months when hydration is accelerated due to elevated ambient and material temperatures.

6.3.3 *Effect on concrete*—Senbetta and Scanlon (1991) discuss tests results of concrete in which an extended set-control admixture was used and reactivation was carried out after 18 h of storage. When tested for resistance to freezing and thawing, it gave results as good as with untreated concrete and also complied with ASTM C 494. The same extended setting-control materials reported by Senbetta and Scanlon (1991) were studied by Senbetta and Dolch (1991). The effects on the cement paste were evaluated using x-ray diffraction, thermogravimetric analysis, differential thermal analysis, and scanning electron microscopy. The nonevaporable water content, surface area, and pore-size distribution were determined. No significant differences were noted between treated and untreated pastes. The effect of using these admixtures on compressive strength, tensile strength, flexural strength, air void system parameters, durability, and shrinkage are discussed by Ragan and Gay (1995). No detrimental effects were reported.

The extended set-control admixture manufacturer's recommendations on dosage rates should be followed, and tests should be conducted on the concrete to determine if the performance properties meet individual project requirements.

6.4—Bonding admixtures

6.4.1 *Materials*—Admixtures formulated to enhance bonding properties of hydraulic-cement-based mixtures generally consist of an organic polymer dispersed in water (latex) (Goeke 1958; Ohama 1984). In general, the latex forms a film throughout the mixture. A polymer latex for use as a concrete admixture is formulated to be compatible with the alkaline nature of the portland-cement paste and the various ions present. An unstable latex will coagulate in the mixture, rendering it unsuitable for use. When used in the quantities normally recommended by manufacturers (5 to 20% of polymer solids by mass of cement), different polymers can affect the unhardened mixture differently. For example, a film-forming latex can cause skinning upon contact with air. Concrete and mortar modified with polymers are more fully addressed in ACI 548.3R.

6.4.2 *Curing*—Water is still necessary to hydrate the portland cement of the cement-polymer system. The polymer latex carries a portion of the mixing water into the mixture; the water is released to the cement during the hydration process. Removing water causes the latex to coalesce, forming a polymer film. Therefore, after an initial 24 h of moist curing to reduce plastic-shrinkage cracking, additional moist curing is not necessary and actually is undesirable because the latex film needs an opportunity to dry and develop the desired properties. The polymer improves the bond between the various phases and also fills microvoids and bridges microcracks that develop during the shrinkage associated with curing (Isenburg et al. 1971; Whiting 1981). This secondary bonding action preserves some of the potential strength normally lost due to microcracking.

6.4.3 *Effect on concrete properties*—Greater tensile strength and durability are associated with latex mixtures. The surfactants used in producing latex act as water-reducing admixtures, resulting in more fluidity than in mixtures without latex, but with a similar *w/cm*. The compressive strength of moist-cured grouts, mortars, and concrete made with these materials is often less than that of mixtures with the same cementitious material content without the admixture, depending on the admixture used. The increases in bond, tensile, and flexural strengths, however, can outweigh the disadvantage of a compressive-strength reduction. Polymer-modified concrete has better abrasion resistance, better resistance to freezing and thawing, and reduced permeability compared with similar concrete not containing the polymer.

6.4.4 *Limitations*—Surfactants present in latex can entrain air and require that a foam-suppressing agent (defoamer) be used. Air-entraining agents are not recommended for use with polymer-modified concrete. Some polymers, such as vinyl acetate homopolymer, decompose (hydrolyze) and soften in the presence of water and should not be used in concrete that will become moist during service. The result obtained with a bonding admixture is only as good as the surface to which the mixture is applied. The surface should be clean, sound, and free from such foreign matter as paint, grease, and dust.

6.5—Pumping aids

6.5.1 *Materials*—Pumping aids for concrete are lubricants and fine fillers used to overcome friction or segregation that cannot be controlled by changes in the concrete mixture proportions. Many pumping aids are thickeners that increase the cohesiveness of concrete. Australian Standard MP 20.2 identifies five categories of thickening admixtures for concrete and mortar but does not include all of the materials listed by McCutcheon Division (1975). The five categories are:

1. Water-soluble synthetic and natural organic polymers that increase the viscosity of water (thickeners). These include cellulose derivatives (methyl, ethyl, hydroxyethyl, and other cellulose gums); polyethylene oxides; acrylic polymers, including polyacrylamides; carboxyvinyl polymers; natural water-soluble gums; starches; and polyvinyl alcohol;

2. Organic flocculants, such as carboxyl-containing styrene copolymers, other synthetic polyelectrolytes, and natural water-soluble gums;

3. Emulsions of various organic materials, such as paraffin, coal tar, asphalt, and synthetic polymers;

4. High-surface-area inorganic materials, such as bentonites, organic-modified bentonites, and silica fume; and

5. Finely divided inorganic materials that supplement cement in cement paste-fly ash and various raw or calcined pozzolanic materials, hydrated lime, and natural or precipitated calcium carbonates, and various rock dusts.

Classifications can be misleading because the performance of a given admixture can change drastically with a change in dosage rates, cementitious material composition, mixing temperature and time, and other factors. An example is provided by the polyethylene oxides. When used in low dosages of 0.01 to 0.05% by mass of cementitious material, they improve pumpability, while larger amounts produce thickening that may not disappear upon prolonged mixing. Other examples are provided by synthetic polyelectrolytes, which act as flocculants or thickeners depending upon dosage levels. Inducing flocculation and increasing bleeding in pumped concrete would appear to be highly undesirable. Nevertheless, these admixtures are considered effective in pumped concrete because they lower bleeding capacity or total bleeding, despite causing increases in initial rates of bleeding. Natural gums, such as algins, tragacanth, or gum arabic, can function as thickeners or flocculants as well as having dispersing or water-reducing effects. Gum arabic is a powerful water-reducer for calcium sulfate plasters, but can produce a glue-like stickiness in portland-cement pastes. A factor to consider in the use of emulsions (paraffins, polymers) is whether they function in the concrete by remaining stable or by breaking the emulsion. Two types of paraffin emulsion are considered to be useful in Australian concrete technology (Australian Standard MP 20.3).

6.5.2 *Effects on concrete*—A pumping aid may affect properties of fresh concrete other than pumpability and can also change the characteristics of the hardened concrete. Because the main effect of a water thickener is to increase viscosity, substantial thickening can increase the water requirement with the usual consequence of reduced strength, unless a suitable dispersant is used. Many thickening agents cause entrainment of air so that a defoamer such as tributyl phosphate (Stoll 1958) can be needed, especially if high dosages of the pumping aid are used. Many organic thickening agents retard the setting of portland cement. Retardation may be substantial, especially with methyl or hydroxyethyl cellulose. A pumping aid should always be evaluated to determine effects on the fresh and hardened concrete.

6.6—Pigments

6.6.1 *Materials*—Pigments specifically prepared for use in concrete and mortar are available both as natural and synthetic materials. They produce adequate color without materially affecting the desirable physical properties of the mixture in compliance with ASTM C 979. The pigments listed in Table 6.1 may be used to obtain a variety of colors.

The addition rate of any pigment to concrete normally should not exceed 10% by mass of the cementitious material (Wilson 1927; ASTM C 979); however, some fine pigments, such as carbon black, should be used at much lower addition rates. Natural pigments are usually not as finely ground nor as pure as synthetic materials and, consequently, do not produce as intense a color. Brilliant concrete colors are not possible with either natural or synthetic pigments due to their low allowable addition rates and the masking effects of the cement and aggregates. Cleaner colors can be obtained if white rather than gray cement is used.

6.6.2 *Effects on concrete properties*—Except for carbon black, adding less than 6% of pigment has little or no effect on the physical properties of the fresh or hardened concrete. Larger quantities can increase the water requirement of the mixture to such an extent that the strength and other properties, such as abrasion resistance, are adversely affected. Adding unmodified carbon black considerably increases the amount of air-entraining admixture needed to provide resistance of the concrete to freezing and thawing (Taylor 1948). Most carbon blacks available for coloring concrete, however, contain surfactants in sufficient quantity to offset the inhibiting effect of the carbon black.

6.7—Flocculating admixtures

Synthetic polyelectrolytes, such as vinyl acetate-maleic anhydride copolymer, have been used as flocculating admixtures. Published reports (Bruere and McGowan 1958; Vivian 1962) indicate that these materials increase the bleeding rate, decrease the bleeding capacity, reduce flow, increase cohesiveness, and increase early strength. Although the mechanism of this action is not understood fully, it is believed that these compounds, containing highly charged groups in their chains, are absorbed on cement particles, linking them together. The net result is equivalent to an increase in interparticle attraction, which increases the tendency of the paste to behave as one large floc. Bruere and McGowan (1958) discuss uses for these admixtures, including use as an alternative to importing fine sand to correct deficiencies in sand gradings, reducing bleeding and segregation, and improving green strength of molded products.

6.8—Fungicidal, germicidal, and insecticidal admixtures

Certain materials have been suggested as admixtures for concrete or mortar to impart fungicidal, germicidal, and insecticidal properties. The primary purpose of these admixtures is to inhibit and control the growth of bacteria and fungi on concrete floors and walls or joints. The materials that have been found to be most effective are polyhalogenated phenols (Levowitz 1952), dieldrin emulsion (Gay and Wetherly 1959), and copper compounds (Robinson and Austin 1951; Young and Talbot 1945). Addition rates vary from 0.1 to 10% by mass of the cement, depending on the concentration and composition of the chemical. Rates above 3% may reduce the strength of the concrete. The effectiveness of these materials (particularly the copper compounds) is reportedly temporary and probably varies with the type of exposure and cleaning methods used.

6.9—Permeability-reducing admixtures

6.9.1 *Introduction*—The term "waterproofing" means preventing water penetration into dry concrete or stopping water transmission through unsaturated concrete. Because no admixture has been found that produces such effects, the term should not be used. Admixtures are used to make the concrete hydrophobic and capable of repelling water that is not under hydrostatic pressure. Permeability-reducing admixtures delay the ingress of moisture when the concrete is subjected to short or intermittent exposure to rain. Although permeability-reducing admixtures can reduce the rate of penetration of water into concrete and the rate of ingress by aggressive chemicals, they will not indefinitely prevent it. Reducing the rate of water inflow can delay the effects of damage caused by freezing and thawing. Permeability-reducing admixtures are not as reliable or effective as a barrier system in retarding liquid transmission.

6.9.2 *Materials*—Permeability-reducing admixtures include soaps, butyl stearate, and certain petroleum products (Dunagan and Ernst 1934; Uppal and Bahadur 1958; Stoll 1958; Ramachandran 1984). The soaps are composed of salts of fatty acids, usually calcium or ammonium stearate or oleate. The soap content of a permeability-reducing admixture usually is 25% or less. Total soap added should not exceed 0.2% by mass of the concrete. Soaps entrain air during mixing. Butyl stearate reportedly performs better than soap as a water repellent. It does not entrain air and has a negligible effect on strength. It is added as an emulsion with the stearate being 1% by mass of the cement. Petroleum products include mineral oils, wax emulsions, asphalt emulsions, and certain cutback asphalts. Heavy mineral oil is effective for making concrete water-repellent and for reducing its permeability. The petroleum product should be fluid and have a viscosity approximately equal to SAE 6O.

6.9.3 *Effectiveness*—These admixtures may aid in repelling water from the surface of the concrete or concrete block.

Table 6.1—Pigments used to produce various colors

Color	Pigments
Gray to black	Black iron oxide
	Carbon black (indoors)
Blue	Phthalocyanine blue (indoors)
	Cobalt blue
Red	Red iron oxide
Brown	Brown iron oxide
	Raw burnt umber
Green	Chromium oxide
	Phthalocyanine green
Yellow	Yellow iron oxide
Ivory, cream, or buff	Yellow iron oxide (with white cement)
White	Titanium dioxide (with white cement)

Some can reduce rain penetration of the concrete surface or reduce wicking or wetting of the concrete. Test data show that they reduce the rate of moisture penetration into the micropores of concrete (Iob 1993; Amey 1997). Based on data from tests of moisture transmission through unsaturated concrete slabs, however, a special advisory committee to the Building Research Advisory Board (1958) concluded: "The Committee does not find adequate data to demonstrate the effectiveness of any admixture to reduce the transmission of moisture through concrete slabs-on-ground in a manner sufficient to replace either a vapor barrier or granular base, or both, under conditions where such protection would be needed."

6.9.4 *Reduction of permeability*—Permeability refers to the rate at which water is transmitted through a saturated specimen of concrete under an externally maintained hydraulic pressure gradient. Water-reducing admixtures should reduce permeability by reducing the total water content in concrete mixtures. When included in properly proportioned and well-cured concrete mixtures, mineral admixtures, such as fly ash, raw or calcined natural pozzolans, or silica fume (ACI 232.1R, 232.2R, and 234R) or ground granulated blast-furnace slag (ACI 233R), reduce the permeability of those mixtures in which the cement content of the paste is relatively low. Polymer-latex admixtures have been used to reduce permeability of concrete overlays for bridge decks and parking decks (refer to Section 6.4). Accelerating admixtures such as calcium chloride increase the hydration rate, reducing the length of time required to reduce porosity and permeability. Any advantage attained this way, however, is likely to be temporary because if conditions are such that water is being transmitted through the concrete, they also are conducive to continued hydration of the cement.

6.10—Chemical admixtures to reduce alkali-aggregate reaction expansion

Using pozzolans to reduce expansion by alkali-aggregate reaction has been widely studied and reported (Stanton 1950). As early as 1951 (McCoy and Caldwell), soluble salts of lithium or barium, certain air-entraining admixtures, and some water-reducing, set-retarding admixtures have been reported to reduce the expansion of laboratory mortar specimens. Significant reductions have been obtained in the laboratory using 1% additions of the lithium salts and 2 to 7% additions (by mass of cement) of certain barium salts (Stark 1992). Once the molar ratio of lithium to sodium reaches 0.67, expansions due to ASR apparently are totally suppressed (Stark 1993; Ong 1993).

6.11—Corrosion-inhibiting admixtures

6.11.1 *Introduction*—A major contributor to corrosion of reinforcing steel is the presence of chlorides in the concrete. The chlorides may come from exposure of the concrete to saline or brackish waters, from exposure to saline soils from which chlorides can reach the steel by diffusion through the concrete, from deicing solutions entering through cracks or pores in the concrete, or from the use of admixtures or other ingredients containing chloride (Chapter 3). Once started, corrosion aggravated by chlorides in concrete is difficult to control. Some chemical admixtures can inhibit corrosion.

6.11.2 *Materials*—Many chemicals have been evaluated as potentially corrosion-inhibiting admixtures for concrete (Verbeck 1975; Clear and Hay 1973; Griffin 1975; Berke 1989). These include chromates, phosphates, hypophosphites, alkalies, nitrites, and fluorides. Calcium nitrite has been reported as an effective corrosion inhibitor (Rosenberg 1977; Virmani, Clear, and Pasko 1983). Using sodium benzoate at a rate of 2% in the mixing water, as a 10% benzoate-cement slurry painted on reinforcement, or both, have been effective (Lewis, Mason, and Brereton 1956). Analysis showed that the sodium benzoate remained in the concrete after five years of exposure; it also accelerates compressive strength development. Calcium lignosulfonate reduces the tendency for corrosion of steel in concrete containing calcium chloride (Kondo, Takeda, and Hideshima 1959).

Two to 3% sodium nitrite by mass of cement was found to be an efficient inhibitor in autoclaved products (Moskvin and Alekseyev 1958). Those authors suggest that high alkalinity, which normally is present in concrete and serves to passivate the steel, can be reduced considerably by autoclave treatment, especially when siliceous admixtures are present. Two percent sodium nitrite also was effective in delaying corrosion of steel in concrete containing calcium chloride (Sarapin 1958). Low-solubility salts, such as certain phosphates or fluosilicates and fluoaluminates, are beneficial. According to limited reports, dosage should be limited to 1% by mass of cement. Loto (1989) reported that a mixture of potassium dichromate and formaldehyde has a synergistic passivating effect on steel in concrete immersed in seawater.

Nmai, Farrington, and Bobrowski (1992) reported on a commercial admixture using a combination of organic amines and esters in a water medium. The admixture is claimed to both reduce ingress of chlorides and enhance the passivating layer on the steel surface. Other similar amine products are claimed to migrate through concrete in the vapor phase to provide protection to embedded steel.

6.12—Antiwashout admixtures

Antiwashout admixtures (AWAs) increase the cohesiveness of concrete to be placed under water by pumping. The most common AWAs are based on natural or synthetic gums and cellulose-based thickeners. Silica fume also can be beneficial in enhancing cohesion. Loss of cementitious material due to washout is typically reduced by as much as 50% by the inclusion of an AWA (U.S. Army Corps of Engineers 1994). In combination with some water-reducing admixtures, AWA may entrain excessive amounts of air, requiring the use of antifoaming agent. The type of HRWRA used affects the washout characteristics of concrete mixtures, and naphthalene-based HRWRA can cause loss of workability. Most natural gum-based AWAs retard setting time (up to 24 h at high dose rates) and most AWAs eliminate bleeding (Neely 1988).

6.13—Freeze-resistant admixtures

Freeze-resistant admixtures have been used in the former USSR since the 1950s to suppress the freezing point of concrete and permit placement and curing of concrete below the freezing point of water (Brook and Ryan 1989). Typically, nonchloride set accelerators are basic ingredients of freeze-resistant admixtures. When used for this purpose, dosage rates are much higher than those used at temperatures above freezing.

For concrete cured at low temperatures, early strengths are low, even though strengths can be similar to normally cured concrete at later ages. Admixtures made from nonchloride accelerators have shown no tendency to cause corrosion of embedded reinforcing steel (U.S. Army Corps of Engineers 1994). Nitrite salts actually reduce corrosion potential when used in amounts suitable for anti-freeze performance. Proprietary accelerating admixtures that provide water-reduction and accelerate hydration down to –7 °C (20 °F) without harmful side effects are available (Brook and Ryan 1989). Sodium and potassium salts, however, should not be used with potentially reactive aggregates due to an increased risk of alkali-silica reaction, nor should they be used in concrete subject to wetting and drying in an aggressive (marine or sulfate) environment.

CHAPTER 7—REFERENCES
7.1—Referenced standards and reports

The standards and reports listed below were the latest editions at the time this document was prepared. Because these documents are revised frequently, the reader is advised to contact the proper sponsoring group if it is desired to refer to the latest version.

American Association of State Highway and Transportation Officials (AASHTO)

M 144	Calcium Chloride
M 154	Air-Entraining Admixtures for Concrete
M 194	Chemical Admixtures for Concrete

American Concrete Institute

116R	Cement and Concrete Terminology (SP-19)
201.2R	Guide to Durable Concrete
209R	Prediction of Creep, Shrinkage, and Temperature Effects in Concrete Structures
211.1	Standard Practice for Selecting Proportions for Normal, Heavyweight, and Mass Concrete
211.2	Standard Practice for Selecting Proportions for Structural Lightweight Concrete
212.4R	Guide for the Use of High-Range Water-Reducing Admixtures (Superplasticizers) in Concrete
222R	Protection of Metals in Concrete Against Corrosion
223	Standard Practice for Use of Shrinkage-Compensating Concrete
232.1R	Use of Raw or Processed Natural Pozzolans in Concrete
232.2R	Use of Fly Ash in Concrete
233R	Slag Cement in Concrete and Mortar
234R	Guide for the Use of Silica Fume in Concrete
301	Specifications for Structural Concrete
302.1R	Guide to Concrete Floor and Slab Construction
304R	Guide for Measuring, Mixing, Transporting, and Placing Concrete
304.2R	Placing Concrete by Pumping Methods
306R	Cold Weather Concreting
308R	Guide to Curing Concrete
309R	Guide for Consolidation of Concrete
311.1R	ACI Manual of Concrete Inspection (SP-2)
311.4R	Guide for Concrete Inspection
318	Building Code Requirements for Structural Concrete
506R	Guide to Shotcrete
523.1R	Guide for Cast-in-Place Low-Density Concrete
548.3R	Polymer-Modified Concrete
552	Designing, Specifying, and Construction for Durable Concrete

ASTM International

C 94	Specification for Ready-Mixed Concrete
C 125	Terminology Relating to Concrete and Concrete Aggregates
C 138	Test Method for Unit Weight, Yield, and Air Content (Gravimetric) of Concrete
C 150	Specification for Portland Cement
C 173	Test Method for Air Content of Freshly Mixed Concrete by the Volumetric Method
C 231	Test Method for Air Content of Freshly Mixed Concrete by the Pressure Method
C 233	Test Method for Air-Entraining Admixtures for Concrete
C 260	Specification for Air-Entraining Admixtures for Concrete
C 360	Test Method for Ball Penetration in Freshly Mixed Hydraulic Cement Concrete
C 457	Practice for Microscopical Determination of Parameters of the Air-Void System in Hardened Concrete
C 494	Specification for Chemical Admixtures for Concrete
C 666	Test Method for Resistance of Concrete to Rapid Freezing and Thawing
C 979	Specification for Pigments for Integrally Colored Concrete
C 1017	Specification for Chemical Admixtures for Use in Producing Flowing Concrete
C 1152	Test Method for Acid-Soluble Chloride in Mortar and Concrete
C 1107	Specification for Packaged Dry, Hydraulic-Cement Grout (Nonshrink)
C 1218	Test Method for Water-Soluble Chloride in Mortar and Concrete
D 98	Specification for Calcium Chloride

Australian Standards

MP 20.2	Thickening Admixtures for Use in Concrete and Mortar
MP 20.3	Expanding Admixtures for Use in Concrete, Mortar and Grout

These publications may be obtained from the following organizations:

American Association of State Highway and Transportation Officials
444 N. Capitol Street NW, Suite 225
Washington, DC 20001

American Concrete Institute
P.O. Box 9094
Farmington Hills, MI 48333-9094

ASTM International
100 Barr Harbor Drive
West Conshohocken, PA 19428-2959

Australian Standards
P.O. Box 1055
Strathfield NSW 2135
Australia

7.2—Cited references

Abrams, D. A., 1924, "Calcium Chloride as an Admixture in Concrete," *Proceedings*, ASTM International, V. 24, Part 2, pp. 781-834.

Adams, R. F., and Kennedy, J. C., 1950, "Effect of Batch Size and Different Mixers on the Properties of Air-Entrained Concrete," *Laboratory Report* No. C 532, U.S. Bureau of Reclamation, Denver, Colo.

Aïtcin, P. C., and Neville, A., 1993, "High-Performance Concrete Demystified," *Concrete International*, V. 15, No. 1, Jan., pp. 21-26.

Amey, S. L., 1997, "Durability of Concrete Containing an Ester-Amine Admixture Exposed to Sulfate and Sulfuric Acid Solutions," *Proceedings of Superplasticizers and Other Chemical Admixtures in Concrete,* SP-173, V. M. Malhotra, ed., American Concrete Institute, Farmington Hills, Mich., pp. 999-1015.

Amsler, D. E.; Eucker, A. J.; and Chamberlin, W. P., 1973, "Techniques for Measuring Air Void Characteristics of Concrete," *Research Report* No. NYSDOT-ERD-73-RR-11, New York State Department of Transportation, Engineering Research and Development Bureau, Albany, N.Y., 43 pp.

Angstadt, R. L., and Hurley, F. R., 1967, "Spodumene as an Accelerator for Hardening Portland Cement," U.S. Patent No 3,331,695, Washington, D.C.

Arber, M. G., and Vivian, H. E., 1961, "Inhibition of the Corrosion of Steel Imbedded in Mortars," *Australian Journal of Applied Science*, V. 12, No. 12, pp. 339-347.

Backstrom, J. E. et al., 1958a, "Origin, Evolution, and Effects of the Air Void System in Concrete: Part 2—Influence of Type and Amount of Air-Entraining Agent," ACI JOURNAL, *Proceedings*, V. 55, pp. 261-272.

Backstrom, J. E. et al., 1958b, "Origin, Evolution, and Effects of the Air Void System in Concrete: Part 3—Influence of Water-Cement Ratio and Compaction," ACI JOURNAL, *Proceedings*, V. 55, pp. 359-376.

Bash, S. M., and Rakimbaev, S. M., 1969, "Quick-Setting Cement Mortars Containing Organic Additives," *Beton i Zhelezobeton*, V. 15, No. 7, pp. 44-45. (in Russian)

Bensted, J., 1978, "Effect on Accelerator Additives on the Early Hydration of Portland Cement," *Il Cemento*, V. 75, pp. 13-19.

Berke, N. S., 1989, "A Review of Corrosion Inhibitors in Concrete," *Materials Performance*, V.28, No. 10, p. 41.

Berger, R. L.; Kung, J. H.; and Young, J. F., 1967, "Influence of Calcium Chloride in the Drying Shrinkage of Alite Paste," *ASTM Journal of Testing and Evaluation*, V. 4, No. 1, pp. 85-93.

Bimel, C., 1998, "Is Delamination Really a Mystery?" *Concrete International*, V. 20, No. 1, Jan., pp. 29-34.

Blanks, R. F., and Cordon W. A., 1949, "Practices, Experiences, and Tests with Air-Entraining Agents in Making Durable Concrete," ACI JOURNAL, *Proceedings* V. 45, No. 6, pp. 469-488.

Bloem, D. L., and Walker, S., 1946, "Preliminary Report on Effect of Mixing Time on Properties of Concrete Containing Air-Entraining Admixtures," *Preliminary Report*, NRMCA Series 72, National Ready Mix Concrete Association, Silver Spring, Md., 4 pp.

Brettman, B. B.; Darwin, D.; and Donahey, R. C., 1986, "Bond of Reinforcement to Superplasticized Concrete," ACI JOURNAL, *Proceedings* V. 83, No. 1, Jan.-Feb., pp. 98-107.

Brook, J. W., and Ryan, R. J., 1989, "A Year-Round Accelerating Admixture in Superplasticizers and Other Chemical Admixtures in Concrete," *Superplasticizers and Other Chemical Admixtures in Concrete*, SP-119, V. M. Malhotra, ed., American Concrete Institute, Farmington Hills, Mich., pp. 535-555.

Brooks, J. J.; Wainwright, P. J.; and Neville, A. M., 1981, "Time-Dependent Behavior of High-Early-Strength Concrete Containing of Superplasticizer," *Developments in the Use of Superplasticizers*, SP-68, V. M. Malhotra, ed., American Concrete Institute, Farmington Hills, Mich., pp. 81-100.

Bruere, G. M., 1963, "Importance of Mixing Sequence When Using Set-Retarding Agents with Portland Cement," *Nature*, V. 199, pp. 32-33.

Bruere, G. M., 1971, "Air-Entraining Actions of Anionic Surfactants in Portland Cement Pastes," *Journal of Applied Chemistry and Biotechnology*, Oxford, UK, V. 21, No. 3, pp. 61-64.

Bruere, G. M., and McGowan, J. K., 1958, "Synthetic Polyelectrolytes as Concrete Admixtures," *Australian Journal of Applied Science*, V. 9, No. 2, pp. 127-140.

Bruere, G. M.; Newbegin, J. D.; and Wilson, L. M., 1971, "Laboratory Investigation of the Drying Shrinkage of Concrete Containing Various Types of Chemical Admixtures," *Technical Paper* No. 1, Division of Applied Mineralogy, Commonwealth Scientific and Industrial Research Organization, East Melbourne, Australia, 26 pp.

Building Research Advisory Board, 1958, "Effectiveness of Concrete Admixtures in Controlling the Transmission of Moisture Through Slabs-on-Ground," *Publication* No. 596, National Research Council, Washington, D.C.

Burg, G. R., 1985, "You Now Have a Choice of Air-Entraining Admixtures," *Concrete Products*, Oct., 4 pp.

Calcium Chloride Institute, 1959, *Calcium Chloride in Concrete*, 3rd Edition, Washington, D.C., pp. 40-41.

Carlson, C. C., 1967, "Comparison of Air Contents Measured in Fresh and Hardened Concrete," *PCA Research and Developments Laboratories Technical Information Department Internal Report*, Portland Cement Association, Skokie, Ill., 17 pp.

Clear, K. C., and Hay, R. E., 1973, "Time-to-Corrosion of Reinforcing Steel in Concrete Slabs, V. 1: Effect of Mix Design and Construction Parameters," *Report* No. FHWA-RD-73-32, Federal Highway Administration, Washington, D.C., 103 pp.

Collepardi, M., 1984, *Water Reducers/Retarders in Concrete Admixtures Handbook: Properties, Science and Technology*, V. S. Ramachandran, ed., Noyes Publications, Park Ridge, N.J., pp. 176-210.

Collepardi, M., and Corradi, M., 1979, "Influence of Naphthalene-Sulfonated Polymer Based Superplasticizers on the Strength of Ordinary and Lightweight Concrete," *Superplasticizers in Concrete*, SP-62, V. M. Malhotra, ed., American Concrete Institute, Farmington Hills, Mich., pp. 325-366.

Collepardi, M.; Marcialis, A.; and Solinas, V., 1973, "The Influence of $CaCl_2$ on the Properties of Cement Paste," *Il Cemento*, V. 70, pp. 83-92.

Cordon, W. A., 1946, "Entrained Air—A Factor in the Design of Concrete Mixes," ACI JOURNAL, *Proceedings* V. 42., No. 6, pp. 605-620.

Cordon, W. A., 1966, "Freezing and Thawing of Concrete-Mechanisms and Control," *ACI Monograph* No. 3, American Concrete Institute/Iowa State University Press, Farmington Hills, Mich., 99 pp.

Djabarov, N. D., 1970, "Oxalic Acid as an Additive to Cement," *Zement-Kalk-Gips*, V. 23, No. 2, pp. 88-90. (in German)

Dodson, V. H., 1990, *Concrete Admixtures*, Van Nostrand Reinhold, N.Y., 211 pp.

Dodson, V. H., and Farkas, E., 1964, "Delayed Additions of Set Retarding Admixtures to Portland Cement Concrete," *ASTM Proceedings*, V. 64, pp. 816-820.

Dodson, V. H.; Farkas, E.; and Rosenberg, A. M., 1965, "Non-Corrosive Accelerator for Setting of Cement," U.S. Patent No. 3,210,207, Washington, D.C.

Dolch, W. L., 1971, "Air-Entrained Concrete in Admixtures in Concrete," *Special Report* 119, Highway Research Board, NRC, Washington, D.C., pp. 7-13.

Dunagan, W. M., and Ernst, G. C., 1934, "Study of the Permeability of a Few Integrally Water-Proofed Concretes," *Proceedings*, ASTM International, Part 1, V. 34, pp. 383-392.

Edwards, G. C., and Angstadt, R. L., 1966, "Effect of Some Soluble Inorganic Admixtures on the Early Hydration of Portland Cement," *Journal of Applied Chemistry and Biotechnology*, V. 16, No. 5, pp. 166-168.

Fagerlund, G., 1990, "Air-Pore Instability and Its Effect on Concrete Properties," *Publication* No. 9, Nordic Concrete Research, Oslo, Norway, pp. 34-52.

Farmer, H. G., 1945, "Air-Entraining Portland Cement in Concrete Block," *Rock Products*, V. 48, p. 209.

Feret, L., and Venuat, N., 1957, "Effect on Shrinkage and Swelling of Mixing Different Cements to Obtain Rapid Set," *Revue des Materiaux de Construction*, No. 496, pp. 1-10.

Foster, B., 1966, "Chemical Admixtures in Significance of Tests and Properties of Concrete and Concrete Making Materials," *STP-169A*, ASTM International, West Conshohocken, Pa., pp. 556-564, 571.

Gay, F. J., and Wetherly, A. H., 1959, "Termite Proofing of Concrete," *Constructional Review*, V. 32, No. 9, pp. 26-28.

Gaynor, R. D., 1985, "Understanding Chloride Percentages," *Concrete International*, V. 7, No. 9, Sept., pp. 26-27.

Gebler, S., 1983, "Evaluation of Calcium Formate and Sodium Formate as Accelerating Admixtures for Portland Cement Concrete," ACI JOURNAL, *Proceedings* V. 80, No. 5, Sept.-Oct., pp. 439-444.

Gebler, S. H., 1982, "Effects of High-Range Water Reducers on the Properties of Freshly Mixed and Hardened Flowing Concrete," *RD 081-01T*, Portland Cement Association, Skokie, Ill., 12 pp.

Gebler, S. H., 1992, "Durability of Dry-Mix Shotcrete Containing Rapid-Set Accelerators," *ACI Materials Journal*, V. 89, No. 3, May-June, pp. 259-262.

Gibbons, C. S., 1978, "Porous Particulate Materials for Imparting Freezing and Thawing Resistance to Concrete," *Report* No. 78-4, Ontario Research Foundation, Toronto, Ontario, Canada.

Goeke, D. M., 1958, "Bonding of Cementitious Materials," *Concrete Construction*, V. 3, No. 5, May, pp. 18-30.

Greening, N. R., and Landgren, R., 1966, "Surface Discoloration of Concrete Flatwork," *Journal*, PCA Research and Development Laboratories, V. 8, No. 3, pp. 34-50.

Grieb, W. E., 1958, "AE-55 Indicator for Air in Concrete," *Bulletin* No. 176, Highway Research Board, Washington, D.C., pp. 23-27; also discussion, pp. 27-28.

Griffin, D. F., 1975, "Corrosion Inhibitors for Reinforced Concrete," *Corrosion of Metals in Concrete,* SP-49, L. Pepper, R. G. Pike, and J. A. Willett, eds., American Concrete Institute, Farmington Hills, Mich., pp. 95-102.

Hansen, W. C., 1960, "Actions of Calcium Sulfate and Admixtures in Portland Cement Pastes," *Symposium on Effect of Water-Reducing Admixtures and Set-Retarding Admixtures on Properties of Concrete*, STP 266, ASTM International, West Conshohocken, Pa., pp. 3-37.

Hover, K. C., 1989, "Some Recent Problems with Air-Entrained Concrete," *Cement, Concrete, and Aggregates*, V. 11, No. 1, pp. 67-72.

Hover, K. C., 1994, "Air Content and Unit Weight of Hardened Concrete in Significance of Tests and Properties of Concrete and Concrete-Making Materials," P. Klieger and J. F. Lamond, eds., STP169C, ASTM International, West Conshohocken, Pa., pp. 296-319.

Howard, E. L.; Griffiths, K. K.; and Moulton, W. E., 1960, "Field Experience Using Water-Reducers in Ready-Mixed Concrete," *Symposium on the Effect on Water-Reducing Admixtures and Set-Retarding Admixtures on Properties of Concrete*, STP 266, ASTM International, West Conshohocken, Pa., pp. 140-147.

Iob, A.; Saricimen, H.; Narasimhan, S.; and Abbas, N. M., 1993, "Spectroscopic and Microscopic Studies of a Commercial Concrete Water Proofing Material," *Cement and Concrete Research*, V. 23, Pergamon, New York, N.Y., pp. 1085-1094.

Isenberg, J. E., 1971, "Microstructure and Strength of the Bond Between Concrete and Styrene-Butadiene Latex-Modified Mortar," *Highway Research Record* No. 370, Highway Research Board, Washington, D.C., pp. 75-89.

Kennedy, H., and Brickett, E. M., 1986, "Application of Air-Entraining Agents in Concrete and Products," *Pit and Quarry*, V. 38, p. 144.

Kinney, F. D., 1989, "Reuse of Returned Concrete by Hydration Control: Characterization of a New Concept," *Superplasticizers and Other Chemical Admixtures in Concrete*, SP-119, V. M. Malhotra, ed., American Concrete Institute, Farmington Hills, Mich., pp. 19-40.

Kobayashi, M., 1981, "Frost Resistance of Superplasticized Concrete," *Developments in the Use of Superplasticizers*, SP-68, V. M. Malhotra, ed., American Concrete Institute, Farmington Hills, Mich., pp. 269-282.

Kondo, Y.; Takeda, A.; and Hideshima, S., 1959, "Effects of Admixtures on Electrolytic Corrosion of Steel Bars in Reinforced Concrete," ACI JOURNAL, *Proceedings* V. 56, No. 4, pp. 299-312.

Kossivas, T. G., 1971, "Setting Accelerators for Portland Cement," German Patent No. 2,114,081, Bonn, Germany.

Kroone, B., 1968, "Reaction Between Hydrating Portland Cement and Ultramarine Blue," *Chemistry and Industry*, London, pp. 287-288.

Kuenning, W. H., and Carlson, C. C., 1956, "Effect of Variations in Curing and Drying on the Physical Properties of Concrete Masonry Units," *Development Department Bulletin* No. Dl3, Portland Cement Association, Skokie, Ill., 129 pp.

Kurdowski, W., and Nocum-Wezelik, W., 1983, "The Tricalcium Silicate Hydration in the Presence of Active Silica," *Cement and Concrete Research*, V. 13, pp. 341-348.

Levowitz, L. D., 1952, "Anti-Bacterial Cement Gives Longer Lasting Floors," *Food Engineering*, V. 224, pp. 57-60 and 134-135.

Lewis, J. I. M.; Mason, E. E.; and Brereton, D., 1956, "Sodium Benzoate in Concrete," *Civil Engineering and Public Works Review*, V. 51, No. 602, pp. 881-882.

Lieber, W., and Richartz, W., 1972, "Effect of Triethanolamine, Sugar, and Boric Acid on the Setting and Hardening of Cements," *Zement-Kalk-Gips*, V. 25, No. 9, pp. 43-409. (in German)

Litvan, G. G., 1972, "Phase Transitions of Adsorbates: IV. Mechanism of Frost Action in Hardened Cement Paste," *Journal of the American Ceramic Society*, V. 55, No. 1, pp. 38-42.

Litvan, G. G., 1983, "Air Entrainment in the Presence of Superplasticizers," ACI JOURNAL, *Proceedings* V. 80, No. 4, July-Aug., pp. 326-331.

Litvan, G. G., 1985, "Further Study of Particulate Admixtures for Enhanced Freeze-Thaw Resistance of Concrete," ACI JOURNAL, *Proceedings* V. 82, No. 5, Sept.-Oct., pp. 724-730.

Litvan, G. G., and Sereda, P. J., 1978, "Particulate Admixtures for Enhanced Freeze-Thaw Resistance of Concrete," *Cement and Concrete Research*, V. 8, No. 1, pp. 53-60.

Litvin, A., and Shideler, J. J., 1966, "Laboratory Study of Shotcrete," *Shotcreting*, SP-14, T. J. Reading, ed., American Concrete Institute, Farmington Hills, Mich., pp. 165-184.

Loto, C. A., 1989, "Effect of Inhibitors and Admixed Chloride on Electrochemical Corrosion Behavior of Mild Steel Reinforcement in Concrete in Seawater," *Corrosion*, V. 48, No. 9, pp. 759-763.

Lu, P.; Sun, G. K.; and Young, J. F., 1993, "Phase Composition of Hydrated DSP Cement Pastes," *Journal of the American Ceramic Society*, V. 76, No. 4, pp. 1003-1007.

Lukas, W., 1981, "Chloride Penetration in Standard Concrete, Water-Reduced Concrete, and Superplasticized Concrete," *Developments in the Use of Superplasticizers*, SP-68, V. M. Malhotra, ed., American Concrete Institute, Farmington Hills, Mich., pp. 253-257.

MacInnis, C., and Beaudoin, J. J., 1974, "Mechanism of Frost Damage in Hardened Cement Paste," *Cement and Concrete Research*, V. 4, No. 2, pp. 139-148.

Mahar, J. W.; Parker, H. W.; and Wuellner, W. W., 1975, "Shotcrete Practice in Underground Construction," *Report* No. FRA-OR&D 75-90, Department of Transportation, Federal Railroad Administration, Washington, D.C., 505 pp.

Mailvaganam, N. P., 1984, *Miscellaneous Admixtures in Concrete Admixtures Handbook: Properties, Science and Technology*, V. S. Ramachandran, ed., Noyes Publications, Park Ridge, N.J., pp. 480-557.

Malhotra, V. M., and Malanka, D., 1979, "Performance of Superplasticizers in Concrete: Laboratory Investigations—Part I," *Superplasticizers in Concrete*, SP-62, V. M. Malhotra, ed., American Concrete Institute, Farmington Hills, Mich., pp. 209-244.

Manns, W., and Eichler, W. R., 1982, "Corrosion-Promotion Action of Concrete Admixtures Containing Thiocyanate," *Betonwerk+Fertigteil-Technik*, V. 48, No. 3, pp. 154-162.

Mather, B., 1964, "Drying Shrinkage—Second Report," *Highway Research News*, No. 15, Highway Research Board, Washington, D.C., pp. 34-38.

Mather, B., 1979, "Tests of High-Range Water-Reducing Admixtures," *Superplasticizers in Concrete*, SP-62, V. M. Malhotra, ed., American Concrete Institute, Farmington Hills, Mich., pp. 157-166.

Mather, B., 1990, "How to Make Concrete that Will Be Immune to the Effects of Freezing and Thawing," *Paul Klieger Symposium on Performance of Concrete*, SP-122, David Whiting, ed., American Concrete Institute, Farmington Hills, Mich., pp. 1-18.

Mather, B., 1992, "Calcium Chloride in Type V-Cement Concrete," *Durability of Concrete, G. M. Idorn International Symposium*, SP-131, J. Holm and M. Geckel, eds., American Concrete Institute, Farmington Hills, Mich., pp. 169-177.

Mather, B., 1994, "Chemical Admixtures," *Significance of Tests and Properties of Concrete and Concrete-Making Materials*, STP 169-C, P. Klieger and J. F. Lamond, eds., ASTM International, West Conshohocken, Pa., pp. 491-499.

McCoy, W. J., and Caldwell, A. G., 1951, "New Approach to Inhibiting Alkali-Aggregate Expansion," ACI JOURNAL, *Proceedings* V. 47, pp. 693-706.

McCutcheon Division, 1975, *Functional Materials*, McCutcheon Division, MC Publishing Co., Ridgewood, N.J., 124 pp.

Mehta, P. K., and Monteiro, P. J. M., 1993, *Concrete: Microstructure, Properties and Materials*, 2nd Edition, Prentice Hall, 548 pp.

Menzel, C. A., 1943, "Some Factors Influencing the Strength of Concrete Containing Admixtures of Powdered Aluminum," ACI JOURNAL, *Proceedings* V. 39, No. 3, pp. 165-184.

Mielenz, R. C.; V. E. Wolkodoff; J. E. Backstrom; and H. L. Flack, 1958a, "Origin, Evolution, and Effects of the Air-Void System in Concrete—Part 1: Entrained Air in Unhardened Concrete" ACI JOURNAL, *Proceedings* V. 55, pp. 95-122.

Mielenz, R. C.; V. E. Wolkodoff; J. E. Backstrom; and R. W. Burrows, 1958b, "Origin, Evolution, and Effects of the Air-Void System in Concrete—Part 4: The Air-Void System in Job Concrete," ACI JOURNAL, *Proceedings*, V. 55, pp. 507-518.

Mindess, S., and Young, J. F., 1981, *Concrete*, Prentice Hall, Englewood Cliffs, N.J.

Morgan, D. R., 1991, "High Early Strength Blended-Cement Wet-Mix Shotcrete," *Concrete International*, V. 13, No. 5, May, pp. 35-39.

Moskvin, V. M., and Alekseyev, S. N., 1958, "Methods for Increasing the Resistance to Corrosion of Reinforcement in Reinforced Concrete Structural Members," *Beton i. Zhelezobeton*, V. 36, No. 12, pp. 21-23. (in Russian)

Murakami, J., and Tanaka, G., 1969, "Contribution of Calcium Thiosulfate to the Acceleration of the Hydration of Portland Cement and Comparison with Other Soluble Inorganic Salts," *Proceedings, 5th International Symposium on the Chemistry of Cement*, Cement Association of Japan, V. 2, Tokyo, Japan, pp. 422-436.

Neeley, B. D., 1988, "Evaluation of Concrete Mixtures for Use in Underwater Repairs," *Technical Report* REMR-CS-18, AD A193 897, U. S. Army Engineer Waterways Experiment Station, Vicksburg, Miss., 130 pp.

Nelson, J. A., and Young, J. F., 1977, "Addition of Colloidal Silicas and Silicates to Portland Cement Pastes," *Cement and Concrete Research*, V. 7, No. 3, pp. 277-282.

Newlon, H. H., Jr., 1971, "Comparison of Properties of Fresh and Hardened Concrete in Bridge Decks," *Virginia Highway Research Council Report* 70-R56, Virginia Highway Research Council, Charlottesville, Va.

Nmai, C. K., and Corbo, J. M., 1989, "Sodium Thiocyanate and the Corrosion Potential of Steel in Concrete and Mortar," *Concrete International*, V. 11, No. 11, Nov., pp. 59-67.

Nmai, C. K.; Farrington, S. A.; and Bobrowski, G. S., 1992, "Organic-Based Corrosion-Inhibiting Admixture for Reinforced Concrete," *Concrete International*, V. 14, No. 4, Apr., pp. 45-51.

Ohama, Y., 1984, *Polymer Modified Mortars and Concretes in Concrete Admixtures Handbook: Properties, Science, and Technology*, V. S. Ramachandran, ed., Noyes Publications, Park Ridge, N.J., pp. 337-429.

Ong, S., 1993, "Studies on Effects of Steam Curing and Alkali Hydroxide Additions on Pore Solution Chemistry, Microstructure, and Alkali Silica Reactions," PhD thesis, Purdue University, West Lafayette, Ind., 497 pp.

Ozyildirim, C., 1991, "Comparison of Air Contents of Freshly Mixed and Hardened Concretes," *Cement, Concrete, and Aggregates*, V. 13, No. 1, pp. 11-17.

Philleo, R. E., 1986, "Freezing and Thawing Resistance of High Strength Concrete, Synthesis," *NCHRP Synthesis of Highway Practice*, Transportation Research Board, NRC, Washington, D.C., V. 129, 31 pp.

Powers, T. C., 1949, "The Air Requirement of Frost-Resistant Concrete," *Proceedings*, Highway Research Board, Washington, D.C., V. 29, pp. 184-202.

Powers, T. C., 1968, *Properties of Fresh Concrete*, John Wiley & Sons, New York, 664 pp.

Powers, T. C., 1975, "Freezing Effects in Concrete in Durability of Concrete," *Durability of Concrete*, SP-47, American Concrete Institute, Farmington Hills, Mich., pp. 1-11.

Ragan, S. A., and Gay, F. T., 1995, "Evaluation of Applications of DELVO Technology," *Final Report* CPAR-SL-95-2, U.S. Army Engineer Waterways Experiment Station, Vicksburg, Miss., 192 pp.

Ramachandran, V. S., 1973, "Action of Triethanolamine on the Hydration of Tricalcium Aluminate," *Cement and Concrete Research*, V. 3, No. 1, pp. 41-54.

Ramachandran, V. S., 1976a, "Hydration of Cement—Role of Triethanolamine," *Cement and Concrete Research*, V. 6, No. 5, pp. 623-632.

Ramachandran, V. S., 1976b, *Calcium Chloride in Concrete: Science and Technology*, Applied Science Publishers, London, 216 pp.

Ramachandran, V. S., 1984, *Concrete Admixtures Handbook: Properties, Science, and Technology*, Noyes Publications, Park Ridge, N.J., 626 pp.

Ramachandran, V. S., and Mailvaganam, N. P., 1992, "New Developments in Chemical Admixtures in Advances in Concrete Technology," *Energy, Mines, and Resources*, V. M. Malhotra, ed., Ottawa, Canada, pp. 859-898.

Ramachandran, V. S., and Malhotra, V. M., 1984, *Superplasticizers in Concrete Admixtures Handbook*, Noyes Publications, Park Ridge, N.J., pp. 211-268.

Ravina, D., and Mor, A., 1986, "Effects of Superplasticizers," *Concrete International*, V. 8, No. 7, July, pp. 53-55.

Reidenour, D. R., and Howe, R. H., 1975, "Air Content of Plastic and Hardened Concrete," *Pennsylvania Department of Transportation Report* No. 73-1, Pennsylvania Department of Transportation, Harrisburg, Pa.

RILEM, 1968, "RILEM-ABEM International Symposium on Admixtures for Mortar and Concrete (Brussels, 1967)," *Proceedings*, RILEM, 6 volumes. Also, General Reports,

Materials and Structures Research and Testing, V. 1, No. 2, RILEM, Paris, pp. 75-149.

Robinson, R. F., and Austin, C. R., 1951, "Effect of Copper-Bearing Concrete on Molds," *Industrial and Engineering Chemistry*, V. 43, pp. 2077-2082.

Robson, J. D., 1952, "Characteristics and Applications of Mixtures of Portland Cement and High-Alumina Centers," *Chemistry and Industry*, No. 1, pp. 2-7.

Rosenberg, A. M., 1977, "Corrosion Inhibitor Formulated with Calcium Nitrite for Use in Reinforced Concrete," *Chloride Corrosion of Steel in Concrete*, STP-629, ASTM International, West Conshohocken, Pa., pp. 89-99.

Rosskopf, P. A.; Linton, F. J.; and Pepper, R. B., 1975, "Effect of Various Accelerating Chemical Admixtures on Setting and Strength Development of Concrete," *Journal of Testing and Evaluation*, V. 3, No. 4, pp. 322-330.

Sarapin, E. G., 1958, "Corrosion of Wire Reinforcement in Concrete Containing Calcium Chloride," *Promyshlennoye Stroitel'stvo*, V. 36, No. 12, pp. 12-23. (in Russian)

Schutz, R. J., 1959, "Setting Time of Concrete Controlled by the Use of Admixtures," ACI JOURNAL, *Proceedings* V. 55, No. 7, pp. 769-781.

Schutz, R. J., 1977, "Properties of Shotcrete Admixtures," *Shotcrete for Ground Support*, SP-54, American Concrete Institute/American Society of Civil Engineers, Farmington Hills, Mich., pp. 45-58.

Schutz, R. J., 1978, "Superplasticizers in Concrete," Proceedings of an International Symposium, Ottawa, Canada, V. 2, May.

Segebrecht, G. W.; Litvin, A.; and Gebler, S. H., 1989, "Durability of Dry-Mix Shotcrete," *Concrete International*, V. 11, No. 10, Oct., pp. 47-50.

Senbetta, E., and Dolch, W. L., 1991, "The Effects on Cement Paste of Treatment with an Extended Set Control Admixture," *Cement and Concrete Research*, V. 21, pp. 750-756.

Senbetta, E., and Scanlon, J. M., 1991, "Effects of Three New Innovative Chemical Admixtures on Durability of Concrete," Supplementary Papers, *Second CANMET/ACI Industry Conference on Durability of Concrete*, Montreal, Quebec, Canada, pp. 29-48.

Shideler, J. J., 1942, "Use of Aluminum Powder to Produce Non-Settling Concrete," *Report* No. C-192, *Engineering Laboratories*, U.S. Bureau of Reclamation, Denver, Colo., 7 pp.

Sommer, H., 1978, "New Method of Making Concrete Resistant to Frost and Deicing Salts," *Bentonwerk and Fertigteil-Technik*, V. 44, No. 9, Sept., pp. 476-484.

Stanton, T. E., 1950, "Studies of Use of Pozzolans for Counteracting Excessive Concrete Expansion Resulting from Reaction Between Aggregates and the Alkalies in Cement," *Symposium on Use of Pozzolanic Materials in Mortars and Concretes,* STP-99, ASTM International, West Conshohocken, Pa., pp. 178-201.

Stark, D. B., 1993, "Eliminating or Minimizing Alkali-Silica Reactivity," *SHRP-C-343*, National Research Council, Strategic Highway Research Program, Washington, D.C., 266 pp.

Stark, D. C., 1986, "Effect of Vibration on the Air-Void System and Freeze-Thaw Durability of Concrete," *PCA Research and Development Bulletin* RDO92.01T, Portland Cement Association, Skokie, Ill., 10 pp.

Stark, D. C., 1992, "Lithium Salt Admixtures—An Alternative Method to Prevent Alkali-Silica Reactivity," *Proceedings,* 9th International Conference on Alkali-Aggregate Reaction in Concrete, The Concrete Society, V. 2, London, pp. 1017-1025.

Stein, H. N., and Stevels, J. M., 1964, "Influence of Silica on the Hydration of $3CaO \cdot SiO_2$," *Journal of Applied Chemistry*, V. 19, No. 8, pp. 338-346.

Stoll, U. W., 1958, "Hydrophobic Cement," STP No. 205, ASTM International, West Conshohocken, Pa., pp. 7-15.

Taylor, T. G., 1948, "Effect of Carbon Black and Black Iron Oxide on Air Content and Durability of Concrete," ACI JOURNAL, *Proceedings* V. 44, No. 8, pp. 613-624.

Tuthill, L. H.; Adams, R. F.; and Hemme, J. R., 1960, "Observations in Testing and Use of Water-Reducing Admixtures," *Symposium on Effect of Water-Reducing Admixtures and Set-Retarding Admixtures on Properties of Concrete*, STP 266, ASTM International, West Conshohocken, Pa., pp. 97-123.

Tynes, W. O., 1977, "Investigation of Proprietary Admixtures," *Technical Report* No. C-77-1, U.S. Army Engineer Waterways Experiment Station, Vicksburg, Miss., 33 pp.

Ulfstedt, L.; Wijard, E.; and Watesson, A. G., 1961, "Accelerating the Setting of Hydraulic Binders," U.S. Patent No. 2,987,407, Washington, D.C.

Uppal, I. S., and Bahadur, S. R., 1958, "Water-Proofing Cement-Sand Mortar: Part 3—Effect of Pretreatment of Sand on Permeability of Cement, Mortar and Concrete," *Indian Concrete Journal*, V. 32, p. 55.

U.S. Army Corps of Engineers, 1994, "Standard Practice for Concrete for Civil Works Structures," *Technical Engineering and Design Guides as Adapted from the U.S. Army Corps of Engineers, No. 8, ISBN 0-7844-0039*, ASCE, New York, N.Y., 122pp.

United States Bureau of Reclamation (USBR), 1975, *Concrete Manual*, 8th Edition, U.S. Bureau of Reclamation, Denver, Colo., p. 627.

Verbeck, G. J., 1975, "Mechanisms of Corrosion of Steel in Concrete," *Corrosion of Metals in Concrete*, SP-49, L. Pepper, R. G. Pike, and J. A. Willett, eds., American Concrete Institute, Farmington Hills, Mich., pp. 21-38.

Virmani, Y. P.; Clear, K. C.; and Pasko, T. J., Jr., 1983, "Time-to-Corrosion of Reinforcing Steel in Concrete Slabs, V. 5: Calcium Nitrite Admixture or Epoxy-Coated Reinforcing Bars as Corrosion Protection Systems," *Report* No. FHWA-RD-83-012, Federal Highway Administration, Washington, D.C., 71 pp.

Vivian, H. E., 1962, "Some Chemical Additions and Admixtures in Cement Paste and Concrete," *Proceedings, 4th International Symposium on the Chemistry of Cement*, Monograph No. 43, National Bureau of Standards, Washington, D.C., pp. 909-923.

Washa, G. W., and Withey, N. H., 1953, "Strength and Durability of Concrete Containing Chicago Fly Ash," ACI JOURNAL, *Proceedings* V. 49, No. 8, pp. 701-712.

Whiting, D., 1979, "Effects of High-Range Water Reducers on Some Properties of Fresh and Hardened

Concrete," *Research and Development Bulletin* No. RD-61-01T, Portland Cement Association, Skokie, Ill., 15 pp.

Whiting, D., 1981, "Rapid Determination of the Chloride Permeability of Concrete," *Report* No. FHWA/RD-81/119, Federal Highway Administration, Washington, D.C., 174 pp.

Whiting, D., 1983, "Addendum to NCHRP Report 258 Control of Air Content in Concrete: Appendix F. State-of-the-Art Report," National Cooperative Highway Research Program, Washington, D.C., 261 pp.

Whiting, D., 1993, "Synthesis of Current and Projected Highway Technology," SHRP-C-345, Strategic Highway Research Program, Washington, D.C., 286 pp.

Whiting, D., and Nagi, M. A., 1998, "Manual on Control of Air Content in Concrete," *EB 116*, Portland Cement Association, Skokie, Ill., 42 pp.

Wilson, R., 1927, "Tests of Colors for Portland Cement Mortars," ACI JOURNAL, *Proceedings* V. 23, pp. 226-252.

Wu, Z. Q., and Young, J. F., 1984, "The Hydration of Tricalcium Silicate in the Presence of Colloidal Silica," *Journal of Materials Science*, V. 19, pp. 3477-3486.

Wuerpel, C. E., 1946, "Influence of Mixing Water Hardness on Air Entrainment," ACI JOURNAL, *Proceedings* V. 42, pp. 401-402.

Young, R. S., and Talbot, H. L., 1945, "Copper-Containing Cements Inhibiting Footborne Diseases," *South African Mining Engineering Journal*, V. 56, pp. 475- 577.

ACI 213R-03

Guide for Structural Lightweight-Aggregate Concrete

Reported by ACI Committee 213

	John P. Ries Chair		G. Michael Robinson Secretary	
David J. Akers		Ralph D. Gruber		Bruce W. Ramme
Michael J. Boyle		Jiri G. Grygar		Steven K. Rowe
Theodore W. Bremner		Edward S. Kluckowski		Shelley R. Sheetz
Ronald G. Burg		Mervyn J. Kowalsky		Peter G. Snow
David A. Crocker		Michael L. Leming		Jeffrey F. Speck
Calvin L. Dodl		W. Calvin McCall		William X. Sypher
Per Fidjestol		Avi A. Mor		Alexander M. Vaysburd
Dean M. Golden		Dipak T. Parekh		Ming-Hong Zhang

Special thanks goes to the following associate members for their contribution to the revision of this document: Kevin Cavanaugh, Shawn P. Gross, Thomas A. Holm, Henry J. Kolbeck, David A. Marshall, Hesham Marzouk, Karl F. Meyer, Jessica S. Moore, Tarun R. Naik, Robert D. Thomas, Victor H. Villarreal, Jody R. Wall, and Dean J. White, II.

The guide summarizes the present state of technology. It presents and interprets the data on lightweight-aggregate concrete from many laboratory studies, accumulated experience resulting from successful use, and the performance of structural lightweight-aggregate concrete in service.

This guide includes a definition of lightweight-aggregate concrete for structural purposes, and discusses, in condensed fashion, the production methods for and inherent properties of structural lightweight aggregates. Other chapters follow on current practices for proportioning, mixing, transporting, and placing; properties of hardened concrete; and the design of structural concrete with reference to ACI 318.

Keywords: abrasion resistance; aggregate; bond; contact zone; durability; fire resistance; internal curing; lightweight aggregate; lightweight concrete; mixture proportion; shear; shrinkage; specified density concrete; strength; thermal conductivity.

FOREWORD

This guide covers the unique characteristics and performance of structural lightweight-aggregate concrete. General historical information is provided along with detailed information on lightweight aggregates and proportioning, mixing, and placing of concrete containing these aggregates. The physical properties of the structural lightweight aggregate along with design information and applications are also included.

Structural lightweight concrete has many and varied applications, including multistory building frames and floors, curtain walls, shell roofs, folded plates, bridges, prestressed or precast elements of all types, marine structures, and others. In many cases, the architectural expression of form combined with functional design can be achieved more readily with structural lightweight concrete than with any other medium. Many architects, engineers, and contractors recognize the inherent economies and advantages offered by this material, as evidenced by the many impressive lightweight concrete structures found today throughout the world.

ACI Committee Reports, Guides, Standard Practices, and Commentaries are intended for guidance in planning, designing, executing, and inspecting construction. This document is intended for the use of individuals who are competent to evaluate the significance and limitations of its content and recommendations and who will accept responsibility for the application of the material it contains. The American Concrete Institute disclaims any and all responsibility for the stated principles. The Institute shall not be liable for any loss or damage arising therefrom.
Reference to this document shall not be made in contract documents. If items found in this document are desired by the Architect/Engineer to be a part of the contract documents, they shall be restated in mandatory language for incorporation by the Architect/Engineer.

It is the responsibility of the user of this document to establish health and safety practices appropriate to the specific circumstances involved with its use. ACI does not make any representations with regard to health and safety issues and the use of this document. The user must determine the applicability of all regulatory limitations before applying the document and must comply with all applicable laws and regulations, including but not limited to, United States Occupational Safety and Health Administration (OSHA) health and safety standards.

CONTENTS
Chapter 1—Introduction, p. 213R-2
1.1—Objectives

ACI 213R-03 supersedes ACI 213R-87 (Reapproved 1999) and became effective September 26, 2003.
Copyright © 2003, American Concrete Institute.
All rights reserved including rights of reproduction and use in any form or by any means, including the making of copies by any photo process, or by electronic or mechanical device, printed, written, or oral, or recording for sound or visual reproduction or for use in any knowledge or retrieval system or device, unless permission in writing is obtained from the copyright proprietors.

1.2—Historical background
1.3—Terminology
1.4—Economy of lightweight concrete

Chapter 2—Structural lightweight aggregates, p. 213R-5
2.1—Internal structure of lightweight aggregates
2.2—Production of lightweight aggregates
2.3—Aggregate properties

Chapter 3—Proportioning, mixing, and handling, p. 213R-8
3.1—Scope
3.2—Mixture proportioning criteria
3.3—Materials
3.4—Proportioning and adjusting mixtures
3.5—Mixing and delivery
3.6—Placing
3.7—Pumping lightweight concrete
3.8—Laboratory and field control

Chapter 4—Physical and mechanical properties of structural lightweight-aggregate concrete, p. 213R-12
4.1—Scope
4.2—Method of presenting data
4.3—Compressive strength
4.4—Density of lightweight concrete
4.5—Specified-density concrete
4.6—Modulus of elasticity
4.7—Poisson's ratio
4.8—Creep
4.9—Drying shrinkage
4.10—Splitting tensile strength
4.11—Modulus of rupture
4.12—Bond strength
4.13—Ultimate strength factors
4.14—Durability
4.15—Absorption
4.16—Alkali-aggregate reaction
4.17—Thermal expansion
4.18—Heat flow properties
4.19—Fire endurance
4.20—Abrasion resistance

Chapter 5—Design of structural lightweight-aggregate concrete, p. 213R-24
5.1—Scope
5.2—General considerations
5.3—Modulus of elasticity
5.4—Tensile strength
5.5—Shear and diagonal tension
5.6—Development length
5.7—Deflection
5.8—Columns
5.9—Prestressed lightweight concrete
5.10—Thermal design considerations
5.11—Seismic design
5.12—Fatigue
5.13—Specifications

Chapter 6—High-performance lightweight concrete, p. 213R-30
6.1—Scope and historical developments
6.2—Structural efficiency of lightweight concrete
6.3—Applications of high-performance lightweight concrete
6.4—Reduced transportation cost
6.5—Enhanced hydration due to internal curing

Chapter 7—References, p. 213R-35
7.1—Referenced standards and reports
7.2—Cited references
7.3—Other references

CHAPTER 1—INTRODUCTION
1.1—Objectives
The objectives of this guide are to provide information and guidelines for designing and using lightweight concrete. By using such guidelines and construction practices, the structures can be designed and performance predicted with the same confidence and reliability as normalweight concrete and other building materials.

1.2—Historical background
The first known use of lightweight concrete dates back over 2000 years. There are several lightweight concrete structures in the Mediterranean region, but the three most notable structures were built during the early Roman Empire and include the Port of Cosa, the Pantheon Dome, and the Coliseum.

The Port of Cosa, built in about 273 B.C., used lightweight concrete made from natural volcanic materials. These early builders learned that expanded aggregates were better suited for marine facilities than the locally available beach sand and gravel. They went 25 mi. (40 km) to the northeast to quarry volcanic aggregates at the Volcine complex for use in the harbor at Cosa (Bremner, Holm, and Stepanova 1994). This harbor is on the west coast of Italy and consists of a series of four piers (~ 13 ft [4 m] cubes) extending out into the sea. For two millennia they have withstood the forces of nature with only surface abrasion. They became obsolete only because of siltation of the harbor.

The Pantheon, finished in 27 B.C., incorporates concrete varying in density from the bottom to the top of the dome. Roman engineers had sufficient confidence in lightweight concrete to build a dome whose diameter of 142 ft (43.3 m) was not exceeded for almost two millenniums. The structure is in excellent condition and is still being used to this day for spiritual purposes (Bremner, Holm, and Stepanova 1994).

The dome contains intricate recesses formed with wooden formwork to reduce the dead load, and the imprint of the grain of the wood can still be seen. The excellent cast surfaces that are visible to the observer show clearly that these early builders had successfully mastered the art of casting concrete made with lightweight aggregates. Vitruvius took special interest in building construction and commented on what was unusual. The fact that he did not single out lightweight concrete for comment might simply imply that these early builders were fully familiar with this material (Morgan 1960).

The Coliseum, built in 75 to 80 A.D., is a gigantic amphitheater with a seating capacity of 50,000 spectators. The foundations were cast with lightweight concrete using crushed volcanic lava. The walls were made using porous, crushed-brick aggregate. The vaults and spaces between the walls were constructed using porous-tufa cut stone. After the fall of the Roman Empire, lightweight concrete use was limited until the 20th century when a new type of manufactured, expanded shale, lightweight aggregate became available for commercial use.

Stephen J. Hayde, a brick manufacturer and ceramic engineer, invented the rotary kiln process of expanding shale, clay, and slate. When clay bricks are manufactured, it is important to heat the preformed clay slowly so that evolved gases have an opportunity to diffuse out of the clay. If they are heated too rapidly, a "bloater" is formed that does not meet the dimensional uniformity essential for a successfully fired brick. These rejected bricks were recognized by Hayde as an ideal material for making a special concrete. When reduced to appropriate aggregate size and grading, these bloated bricks could be used to produce a lightweight concrete with mechanical properties similar to regular concrete. After almost a decade of experimentation, in 1918 he patented the process of making these aggregates by heating small particles of shale, clay, or slate in a rotary kiln. A particle size was discovered that, with limited crushing, produced an aggregate grading suitable for making lightweight concrete (ESCSI 1971).

Commercial production of expanded slag began in 1928, and in 1948 the first structural-quality, sintered-shale, lightweight aggregate was produced using shale in eastern Pennsylvania.

One of the earliest uses of reinforced lightweight concrete was in the construction of ships and barges around 1918. The U.S. Emergency Fleet Building Corporation found that, for concrete to be effective in ship construction, the concrete would need a maximum density of about 110 lb/ft^3 (1760 kg/m^3) and a compressive strength of approximately 4000 psi (28 MPa). Concrete was obtained with a compressive strength of approximately 5000 psi (34 MPa) and a unit weight of 110 lb/ft^3 (1760 kg/m^3) or less using rotary-kiln-produced expanded shale and clay aggregate.

Considerable impetus was given to the development of lightweight concrete in the late 1940s when a National Housing Agency survey was conducted on the potential use of lightweight concrete for home construction. This led to an extensive study of concrete made with lightweight aggregates. Sponsored by the Housing and Home Finance Agency, parallel studies were conducted simultaneously in the laboratories of the National Bureau of Standards (Kluge, Sparks, and Tuma 1949) and the U.S. Bureau of Reclamation (Price and Cordon 1949) to determine properties of concrete made with a broad range of lightweight aggregate types. These studies and earlier works focused attention on the potential structural use of some lightweight-aggregate concrete and initiated a renewed interest in lightweight members for building frames, bridge decks, and precast products in the early 1950s. Following the collapse of the original Tacoma Narrows Bridge, the replacement suspension structure design used lightweight concrete in the deck to incorporate additional roadway lanes without the necessity of replacing the original piers.

During the 1950s, many multistory structures were designed from the foundations up, taking advantage of reduced dead weight using lightweight concrete. Examples are the 42-story Prudential Life Building in Chicago, which used lightweight concrete floors, and the 18-story Statler Hilton Hotel in Dallas, designed with a lightweight concrete frame and flat plate floors.

These structural applications stimulated more-concentrated research into the properties of lightweight concrete. In energy-related floating structures, great efficiencies are achieved when a lightweight material is used. A reduction of 25% in mass in reinforced normalweight concrete will result in a 50% reduction in load when submerged. Because of this, the oil and gas industry recognized that lightweight concrete could be used to good advantage in its floating structures as well as structures built in a graving dock and then floated to the production site and bottom-founded. To provide the technical data necessary to construct huge offshore concrete structures, a consortium of oil companies and contractors was formed to evaluate lightweight aggregate candidates suitable for making high-strength lightweight concrete that would meet their design requirements. The evaluations started in the early 1980s, with the results made available in 1992. As a result of this research, design information became readily available and has enabled lightweight concrete to be used for new and novel applications where high strength and high durability are desirable (Hoff 1992).

1.3—Terminology

aggregate, insulating—*Nonstructural aggregate meeting the requirements of ASTM C 332.* This includes Group I aggregate, Perlite with a bulk density between 7.5 and 12 lb/ft^3 (120 and 192 kg/m^3), Vermiculite with a bulk density between 5.5 and 10 lb/ft^3 (88 and 160 kg/m^3), and group II aggregate that meets the requirements of ASTM C 330 and ASTM C 331. (See **aggregate, structural-lightweight**, and **aggregate, masonry-lightweight**.)

aggregate, lightweight—See **aggregate, structural lightweight**; **aggregate, masonry lightweight**; or **aggregate, insulating**.

aggregate, masonry-lightweight (MLWA)—*Aggregate meeting the requirements of ASTM C 331 with bulk density less than 70 lb/ft^3 (1120 kg/m^3) for fine aggregate and less than 55 lb/ft^3 (880 kg/m^3) for coarse aggregate.* This includes aggregates prepared by expanding, pelletizing, or sintering products such as blast-furnace slag, clay, diatomite, fly ash, shale, or slate; aggregates prepared by processing natural materials such as pumice, scoria, or tuff; and aggregates derived from and products of coal or coke combustion.

aggregate, structural lightweight (SLA)—*Structural aggregate meeting the requirements of ASTM C 330 with bulk density less than 70 lb/ft^3 (1120 kg/m^3) for fine aggregate and less than 55 lb/ft^3 (880 kg/m^3) for coarse aggregate.* This includes aggregates prepared by expanding, pelletizing, or sintering products such as blast-furnace slag, clay, fly ash, shale or slate, and aggregates prepared by processing natural materials such as pumice, scoria or tuff.

aggregate, low-density—See **aggregate, structural lightweight**.

concrete, all lightweight—Concrete in which both the coarse- and fine-aggregate components are lightweight aggregates. (Deprecated term—use preferred term; **concrete, lightweight**; **concrete, structural lightweight**; or **concrete, specified-density**.)

concrete, high-strength lightweight—Structural lightweight concrete with a 28-day compressive strength of 6000 psi (40 MPa) or greater.

concrete, lightweight—See **concrete, structural lightweight** or **specified density**.

concrete, low-density—See **concrete, lightweight**.

concrete, normalweight—Concrete having a density of 140 to 155 lb/ft^3 (2240 to 2480 kg/m^3) made with ordinary aggregates (sand, gravel, crushed stone).

concrete, sand lightweight—Concrete with coarse lightweight aggregate and normalweight fine aggregate. (Deprecated term—use preferred term; **concrete, structural lightweight**; **concrete, lightweight**; or **concrete, specified-density**.)

concrete, specified density (SDC)—Structural concrete having a specified equilibrium density between 50 to 140 lb/ft^3 (800 to 2240 kg/m^3) or greater than 155 lb/ft^3 (2480 kg/m^3) (see **concrete, normalweight**). SDC may consist as one type of aggregate or of a combination of lightweight or normal-density aggregate. This concrete is project specific and should include a detailed mixture testing program and aggregate supplier involvement before design.

concrete, structural lightweight aggregate—See **concrete, structural lightweight**.

concrete, structural lightweight (SLC)—*Structural lightweight-aggregate concrete made with structural lightweight aggregate as defined in ASTM C 330*. The concrete has a minimum 28-day compressive strength of 2500 psi (17 MPa), an equilibrium density between 70 and 120 lb/ft^3 (1120 and 1920 kg/m^3), and consists entirely of lightweight aggregate or a combination of lightweight and normal-density aggregate.

This definition is not a specification. Project specifications vary. While lightweight concrete with an equilibrium density of 70 to 105 lb/ft^3 (1120 to 1680 kg/m^3) is infrequently used, most lightweight concrete has an equilibrium density of 105 to 120 lb/ft^3 (1680 to 1920 kg/m^3). Because lightweight concrete is often project-specific, contacting the aggregate supplier before project design is advised to ensure an economical mixture and to establish the available range of density and strength.

contact zone—The transitional layer of material connecting aggregate particles with the enveloping continuous mortar matrix.

curing, internal—Internal curing refers to the process by which the hydration of cement continues because of the availability of internal water that is not part of the mixing water. The internal water is made available by the pore system in structural lightweight aggregate that absorbs and releases water.

density, equilibrium—As defined in ASTM 567, it is the density reached by structural lightweight concrete (low density) after exposure to relative humidity of 50 ± 5% and a temperature of 73.5 ± 3.5 °F (23 ± 2 °C) for a period of time sufficient to reach a density that changes less than 0.5% in a period of 28 days.

density, oven-dry—As defined in ASTM C 567, the density reached by structural lightweight concrete after being placed in a drying oven at 230 ± 9 °F (110 ± 5 °C) for a period of time sufficient to reach a density that changes less than 0.5% in a period of 24 h. The oven-dry density test is to be performed at the age specified.

lightweight — The generic name of a group of aggregates having a relative density lower than normal-density aggregates. (See **aggregate, lightweight**). The generic name of concrete or concrete products having lower densities than normal-weight concrete products. (See **concrete, structural lightweight**, and **concrete, lightweight**).

1.4—Economy of lightweight concrete

The use of lightweight concrete is usually predicated on the reduction of project cost, improved functionality, or a combination of both. Estimating the total cost of a project is necessary when considering lightweight concrete because the cost per cubic yard (cubic meter) is usually higher than a comparable unit of ordinary concrete. The following example is a typical comparison of unit cost between lightweight and normalweight concrete on a bridge project.

For example, assume the in-place cost of a typical short-span bridge may vary from 50 to 200 $/ft^2$ (540 to 2150 $/m^2$).

If the average thickness of the deck was 8 in. (200 mm) then one cubic yard (cubic meter) of concrete would yield approximately 40 ft^2/yd^3 (5 m^2/m^3).

The increased cost of using lightweight concrete with a cost of 20 $/yd^3$ (26 $/m^3$) over normalweight concrete would be 20 $/yd^3$/40 ft^2/yd^3 = 0.50 $/ft^2$ (5 $/m^2$), or generally less than a 1% increase.

This increase would easily be offset by any of the following economies, or more importantly, by significant increases in bridge, building, or marine structure functionality:

- The reduction in foundation loads may result in smaller footings, fewer piles, smaller pile caps, and less reinforcing;
- Reduced dead loads may result in smaller supporting members (decks, beams, girder, and piers), resulting in a major reduction in cost;
- Reduced dead load will mean reduced inertial seismic forces;
- In bridge rehabilitation, the new deck may be wider or an additional traffic lane may be added without structural or foundation modification;
- On bridge deck replacements or overlays, the deck may be thicker to allow more cover over reinforcing or to provide better drainage without adding additional dead load to the structure;
- With precast-prestress use, longer or larger elements can be manufactured without increasing overall mass. This may result in fewer columns or pier elements in a

Table 1.1—Analysis of shipping costs of concrete products[*]

	Project Example No. 1	Project Example No. 2
Shipping cost per truck load	$1100	$1339
Number of loads required		
Normalweight	431	87
Lightweight	287	66
Reduction in truck loads:	144	21
Transportation savings		
Shipping cost per load	$1100	$1339
Reduction in truck loads	× 144	× 21
Transportation savings:	$158,400	$28,119
Profit impact		
Transportation savings	$158,400	$28,119
Less: premium cost of lightweight concrete	17,245	3799
Transportation cost savings by using lightweight concrete	$141,155	$24,320

[*]Courtesy of Big River Industries, Inc.

system that is easier to lift or erect, and fewer joints or more elements per load when transporting. There are several documented cases where the savings in shipping costs far exceeded the increased cost of using lightweight concrete. At some precast plants, each element's shipping cost is evaluated by computer to determine the optimum concrete density;

- In marine applications, increased allowable topside loads and the reduced draft resulting from the use of lightweight concrete may permit easier movement out of dry docks and through shallow shipping channels; and
- Due to the greater fire resistance of lightweight concrete, as reported in ACI 216.1, the thickness of slabs may be reduced, resulting in significantly less concrete volumes.

Lightweight concrete is often used to enhance the architectural expression or construction of a structure. In building construction, this usually applies to cantilevered floors, expressive roof design, taller buildings, or additional floors added to existing structures. With bridges, this may allow a wider bridge deck (additional lanes) being placed on existing structural supports. Improved constructibility may result in cantilever bridge construction where lightweight concrete is used on one side of a pier and normalweight concrete used on the other to provide weight balance while accommodating a longer span on the lightweight side of the pier. The use of lightweight concrete may also be necessary when better insulating qualities are needed in thermally sensitive applications like hot water, petroleum storage or building insulation.

1.4.1 Transportation costs—In situations where transportation costs are directly related to the weight of concrete products, there can be significant economies developed through the use of lightweight concrete. The range of products includes large structural members (girders, beams, walls, hollow-core panels, double tees) to smaller consumer products (precast stair steps, fireplace logs, wall board, imitation stone). Two trucking studies conducted at a U.S. precast plant are shown in Table 1.1. These studies demonstrated that the transportation cost savings were seven times more than the additional cost of lightweight aggregate. Savings vary with the size and mass of the product and are most significant for the smaller consumer-type products. For example, one manufacturer of wallboard has shipped products to all 48 mainland states from one manufacturing facility. Less trucks in congested cities is not only environmentally friendly but also generates fewer public complaints. The potential for lower costs is possible when shipping by rail or barge but is most often realized in trucking where highway loadings are posted. The example given in Table 1.1 is a typical analysis of cost for shipping prestressed double-tee members to projects in the late 1990s.

CHAPTER 2—STRUCTURAL LIGHTWEIGHT AGGREGATES
2.1—Internal structure of lightweight aggregates

Lightweight aggregates have a low-particle relative density because of the cellular pore system. The cellular structure within the particles is normally developed by heating certain raw materials to incipient fusion; at this temperature, gases are evolved within the pyroplastic mass, causing expansion, which is retained upon cooling. Strong, durable, lightweight aggregates contain a uniformly distributed system of pores that have a size range of approximately 5 to 300 µm, developed in a continuous, relatively crack-free, high-strength vitreous phase. Pores close to the surface are readily permeable and fill with water within the first few hours of exposure to moisture. Interior pores, however, fill extremely slowly, with many months of submersion required to approach saturation. A small fraction of interior pores are essentially noninterconnected and remain unfilled after years of immersion.

2.2—Production of lightweight aggregates

Structural-grade lightweight aggregates are produced in manufacturing plants from raw materials, including suitable shales, clays, slates, fly ashes, or blast-furnace slags. Naturally occurring lightweight aggregates are mined from volcanic deposits that include pumice and scoria. Pyroprocessing methods include the rotary kiln process (a long, slowly rotating, slightly inclined cylinder lined with refractory materials similar to cement kilns); the sintering process wherein a bed of raw materials, including fuel, is carried by a traveling grate under an ignition hood; and the rapid agitation of molten slag with controlled amounts of air or water. No single description of raw material processing is all-inclusive, and the reader is urged to consult local lightweight aggregate manufacturers for physical and mechanical properties of lightweight aggregates and the concrete made with them.

The increased usage of processed lightweight aggregates is evidence of environmentally sound planning, as these products require less trucking and use of materials that have limited structural applications in their natural state, thus minimizing construction industry demands on finite resources of natural sands, stones, and gravels.

Table 2.1—Bulk-density requirements of ASTM C 330 and C 331 for dry, loose, lightweight aggregates

Aggregate size and group	Maximum density, lb/ft^3 (kg/m^3)
ASTM C 330 and C 331	
-fine aggregate	70 (1120)
-coarse aggregate	55 (880)
-combined fine and coarse aggregate	65 (1040)

2.3—Aggregate properties

Each of the properties of lightweight aggregates may have some bearing on the properties of the fresh and hardened concrete. It should be recognized, however, that properties of lightweight concrete, in common with those of normal-weight concrete, are greatly influenced by the quality of the cementitous matrix. Specific properties of aggregates that may affect the properties of the concrete are listed in Sections 2.3.1 through 2.3.8.

2.3.1 *Particle shape and surface texture*—Lightweight aggregates from different sources, or produced by different methods, may differ considerably in particle shape and texture. Shape may be cubical and reasonably regular, essentially rounded, or angular and irregular. Surface textures may range from relatively smooth with small exposed pores to irregular with small to large exposed pores. Particle shape and surface texture of both fine and coarse aggregates influence proportioning of mixtures in such factors as workability, pumpability, fine-to-coarse aggregate ratio, binder content, and water requirement. These effects are analogous to those obtained with normalweight aggregates with such diverse particle shapes as exhibited by rounded gravel, crushed limestone, traprock, or manufactured sand.

2.3.2 *Relative density*—Due to their cellular structure, the relative density of lightweight-aggregate particles are lower than that of normalweight aggregates. The lightweight particle relative density of lightweight aggregate also varies with particle size, being highest for the fine particles and lowest for the coarse particles, with the magnitude of the differences depending on the processing methods. The practical range of coarse lightweight aggregate relative densities, corrected to the dry condition, are from almost 1/3 to 2/3 that for normalweight aggregates. Particle densities below this range may require more cement to achieve the required strength and may thereby fail to meet the density requirements of the concrete.

2.3.3 *Bulk density*—The bulk density of lightweight aggregate is significantly lower, due to the cellular structure, than that of normalweight aggregates. For the same grading and particle shape, the bulk density of an aggregate is essentially proportional to particle relative densities. Aggregates of the same particle density, however, may have markedly different bulk densities because of different percentages of voids in the dry-loose or dry-rodded volumes of aggregates of different particle shapes. The situation is analogous to that of rounded gravel and crushed stone, where differences may be as much as 10 lb/ft^3 (160 kg/m^3), for the same particle density and grading, in the dry-rodded condition. Rounded and angular lightweight aggregates of the same particle density may differ by 5 lb/ft^3 (80 kg/m^3) or more in the dry-loose condition, but the same mass of either will occupy the same volume in concrete. This should be considered in assessing the workability when using different aggregates. Table 2.1 summarizes the maximum densities for the lightweight aggregates listed in ASTM C 330 and C 331.

2.3.4 *Strength of lightweight aggregates*—The strength of aggregate particles varies with type and source and is measurable only in a qualitative way. Some particles may be strong and hard and others weak and friable. For compressive strengths up to approximately 5000 psi (35 MPa), there is no reliable correlation between aggregate strength and concrete strength.

2.3.4.1 *Strength ceiling*—The concept of "strength ceiling" may be useful in indicating the maximum compressive and tensile strength attainable in concrete made with a given lightweight aggregate using a reasonable quantity of cement. A mixture is near its strength ceiling when similar mixtures containing the same aggregates and with higher cement contents have only slightly higher strengths. It is the point of diminishing returns, beyond which an increase in cement content does not produce a commensurate increase in strength. The strength ceiling for some lightweight aggregates may be quite high, approaching that of some normalweight aggregates.

The strength ceiling is influenced predominantly by the coarse aggregate. The strength ceiling can be increased appreciably by reducing the maximum size of the coarse aggregate for most lightweight aggregates. This effect is more apparent for the weaker and more friable aggregates. In one case, the strength attained in the laboratory for concrete containing 3/4 in. (19 mm) maximum size of a specific lightweight aggregate was 5000 psi (35 MPa); for the same cement content, the strength was increased to 6100 and 7600 psi (42 and 52 MPa) when the maximum size of the aggregate was reduced to 1/2 and 3/8 in. (13 and 10 mm), respectively, whereas concrete unit weights were concurrently increased by 3 and 5 lb/ft^3 (48 and 80 kg/m^3).

Meyer and Kahn (2002) reported that, for a given lightweight aggregate, the tensile strength may not increase in a manner comparable to the increase in compressive strength. Increases in tensile strength occur at a lower rate relative to increases in compressive strength. This becomes more pronounced as compressive strength increases beyond 5000 psi.

2.3.5 *Total porosity*—Proportioning concrete mixtures and making field adjustments of lightweight concrete require a comprehensive understanding of porosity absorbtion and the degree of saturation of lightweight-aggregate particles. The degree of saturation (the fractional part of the pores filled with water) can be evaluated from pychnometer measurements, which determine the relative density at various levels of absorbtion, thus permitting proportioning by the absolute volume procedure. Normally, pores are defined as the air space inside an individual aggregate particle and voids are defined as the interstitial space between aggregate particles. Total porosity (within the particle and between the particles) can be determined from measured values of particle relative density and bulk density.

For example, if measurements on a sample of lightweight coarse aggregate are:
- Bulk density, dry, loose 48 lb/ft^3 (770 kg/m^3), BD = 0.77 (ACI 211.2; ASTM 138);
- Dry-particle relative density 87 lb/ft^3 (1400 kg/m^3) RD = 1.4 (ACI 211.2; ASTM 138); and
- Relative density of the solid particle material without pores 162 lb/ft^3 (2600 kg/m^3) RD = 2.6 (ACI 211.1; ASTM 138).

Note: The particle relative density of the solids (ceramic material without pores) used in this example, 162 lb/ft^3 (2600 kg/m^3), RD = 2.6, was the average value determined by the following procedure: small samples of three different expanded aggregates were ground separately in a jar ball mill for 24 h. After each sample was reduced, it was then tested in accordance with ASTM C 150 to determine the relative density of the ground lightweight aggregate. According to Weber and Reinhardt (1995), the pore structure of expanded aggregates reveals that a small percentage of pores are less than 10 m and exist unbroken within the less than 200 sieve (75 μm) sized particles. The relative densities of the vitreous structure are typically in excess of 162 lb/ft^3 (2600 kg/m^3). The true particle porosity may be slightly greater than that determined by the following calculations. When very small pores are encapsulated by a strong, relatively crack-free vitreous structure, however, the pores are not active in any moisture dynamics.

Using the values given previously, the following results:

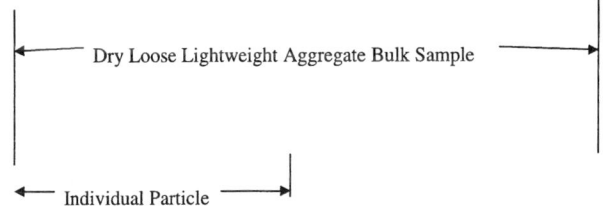

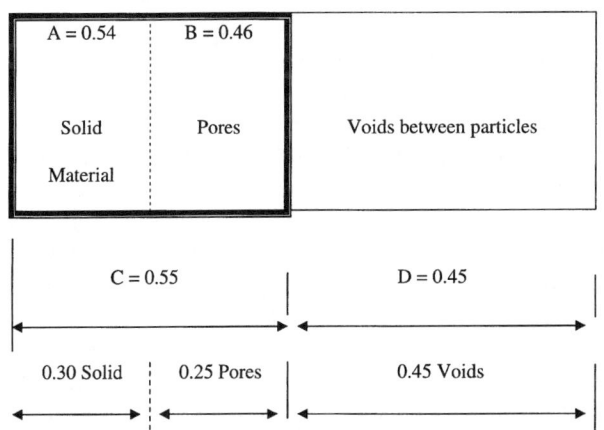

Then the total porosity (pores and voids) equals:

0.45 (voids) + (0.46 (pores) × 0.55 (particles) = 0.70, where A = the fractional solid volume (without pores) of the vitreous material of an individual particle, equals 1.4/2.6 = 0.54; B = the subsequent fractional volume of pore (within the particle), equals 1.00 – 0.54 = 0.46; C = for this example, the fractional volume of particles equals 0.77/1.4 = 0.55; and D = the fractional volume of interstitial voids (between particles) = 1.00 – 0.55 = 0.45.

2.3.6 *Grading*—Grading requirements for lightweight aggregates deviate from those of normalweight aggregates (ASTM C 33) by requiring a larger mass of the lightweight aggregates to pass through the finer sieve sizes. This modification in grading (ASTM C 330) recognizes the increase in density with decreasing particle size of lightweight expanded aggregates. This modification yields the same volumetric distribution of aggregates retained on a series of sieves for both lightweight and normalweight aggregates.

Producers of lightweight aggregate normally stock materials in several standard sizes such as coarse, intermediate, and fine aggregate. By combining size fractions or replacing some or all of the fine fraction with a normalweight sand, a wide range of concrete densities can be obtained. The aggregate producer is the best source of information for the proper aggregate combinations to meet fresh concrete density specifications and equilibrium density for dead-load design considerations.

Normalweight sand replacement will typically increase the equilibrium concrete density from about 5 to 10 lb/ft^3 (80 to 160 kg/m^3). Using increasing amounts of cement to obtain high-strength concrete may increase the density from 2 to 6 lb/ft^3 (32 to 96 kg/m^3). With modern concrete technology, however, it will seldom be necessary to significantly increase cement content to obtain the reduced water-cementitious material ratios *(w/cm)* needed to obtain the specified strength because this can be done using water-reducing or high-range water-reducing admixtures.

2.3.7 *Moisture content and absorption*—Lightweight aggregates, due to their cellular structure, are capable of absorbing more water than normalweight aggregates. Based on a standard ASTM C 127 absorption test expressed at 24 h, lightweight aggregates generally absorb from 5 to 25% by mass of dry aggregate, depending on the aggregate pore system.

In contrast, most normalweight aggregates will absorb less than 2% of moisture. The moisture content in a normalweight aggregate stockpile, however, may be as high as 5 to 10% or more. The important difference is that the moisture content with lightweight aggregates is absorbed into the interior of the particles as well as on the surface, while in normalweight aggregates, it is largely surface moisture. These differences become important as discussed in the following sections on mixture proportioning, batching, and control.

The rate of absorption in lightweight aggregates is a factor that also has a bearing on mixture proportioning, handling, and control of concrete, and depends on the aggregate pore characteristics. The water, that is internally absorbed in the lightweight aggregate, is not immediately available to the cement and should not be counted as mixing water. Nearly all moisture in the natural sand, on the other hand, may be surface moisture and, therefore, part of the mixing water.

2.3.8 *Modulus of elasticity of lightweight aggregate particles*—The modulus of elasticity of concrete is a function of the moduli of its constituents. Concrete may be considered

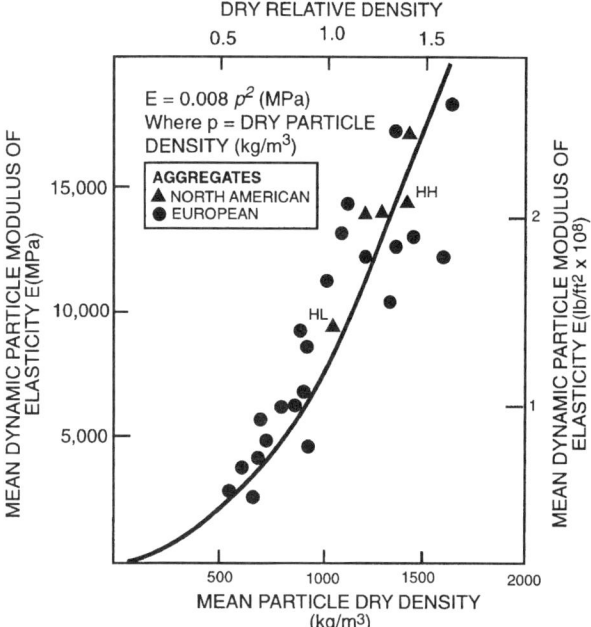

Fig. 2.1—Relationship between mean particle density and the mean dynamic modulus of elasticity for the particles of lightweight aggregates (Bremner and Holm 1986).

as a two-phase material consisting of coarse-aggregate inclusions within a continuous mortar fraction that includes cement, water, entrained air, and fine aggregate. Dynamic measurements made on aggregates alone have shown a relationship corresponding to the function $E = 0.008p^2$, where E is the dynamic modulus of elasticity of the particle, in MPa, and p is the dry mean particle density, in k/m^3 (Fig. 2.1).

Dynamic moduli for typical expanded aggregates have a range of 1.45 to 2.3 × 10^6 psi (10 to 16 GPa), whereas the range for strong normalweight aggregates is approximately 4.35 to 14.5 × 10^6 psi (30 to 100 GPa) (Muller-Rochholz 1979).

CHAPTER 3—PROPORTIONING, MIXING, AND HANDLING

3.1—Scope

The proportioning of lightweight concrete mixtures is determined by economical combinations of the constituents that typically include portland cement; aggregate; water; chemical admixtures, mineral admixtures, or both; in a way that the optimum combination of properties is developed in both the fresh and hardened concrete. A prerequisite to the selection of mixture proportions is a knowledge of the properties of the constituent materials and their compliance with pertinent ASTM specifications.

Based on a knowledge of the properties of the constituents and their interrelated effects on the concrete, lightweight concrete can be proportioned to have the properties specified for the finished structure.

This chapter discusses:
- Criteria on which concrete mixture proportions are based;
- The materials that make up the concrete mixture; and
- The methods by which these are proportioned.

Mixing, delivery, placing, finishing, and curing also will be discussed, particularly where these procedures differ from those associated with normalweight concrete. The chapter concludes with a brief discussion on laboratory and field quality control.

3.2—Mixture proportioning criteria

Chapter 4 indicates a broad range of values for many physical properties of lightweight concrete. Specific values depend on the properties of the particular aggregates being used and on other conditions. In proportioning a lightweight-concrete mixture, the engineer is concerned with obtaining predictable values of specific properties for a particular application.

Specifications for lightweight concrete usually require minimum permissible values for compressive and tensile strength, maximum values for slump, and both minimum and maximum values for air content. For lightweight concrete, a limitation is always placed on the maximum value for fresh and equilibrium density.

From a construction standpoint, the workability of fresh concrete should also be considered. In proportioning lightweight concrete mixtures, these properties may be optimized. Some properties are interdependent, and improvement in one property, such as workability, may affect other properties such as density or strength. The final criterion to be met is overall performance in the structure as specified by the architect/engineer.

3.2.1 *Specified physical properties*

3.2.1.1 *Compressive strength*—Compressive strength is further discussed in Chapter 4. The various types of lightweight aggregates available will not always produce similar compressive strengths for concrete of a given cement content and slump.

Compressive strength of structural concrete is specified according to design requirements of a structure. Normally, strengths specified will range from 3000 to 5000 psi (21 to 35 MPa) and less frequently up to 7000 psi (48 MPa) or higher. Although some lightweight aggregates are capable of producing very high strengths consistently, it should not be expected that concrete made with every lightweight aggregate classified as "structural" can consistently attain the higher strength values.

3.2.1.2 *Density*—From the load-resisting considerations of structural members, reduced density of lightweight concrete can lead to improved economy of structures despite an increased unit cost of concrete. Therefore, density is a very important consideration in the proportioning of lightweight-concrete mixtures. While this property depends primarily on aggregate density and the proportions of lightweight and normalweight aggregate, it is also influenced by the cement, water, and air contents. Within limits, concrete density can be maintained by adjusting proportions of lightweight and normalweight aggregates. For example, if the cement content is increased to provide additional compressive strength, the unit weight of the concrete will be increased. On the other hand, complete replacement of the lightweight-aggregate fines with normalweight sand could increase the concrete density by approximately 10 lb/ft^3 (160 kg/m^3) or more at the same strength level. This should also be considered in the overall economy of lightweight concrete.

If the concrete producer has several different sources of lightweight aggregate available, the optimum balance of cost and concrete performance may require a detailed investigation. Only by comparing concrete of the same compressive strength and of the same equilibrium density can the fundamental differences of concrete made with different aggregates be properly evaluated. In some areas, only a single source of lightweight aggregate is economically available. In this case, the concrete producer needs only to determine the density of concrete that satisfies the economy and specified physical properties of the structure.

3.2.1.3 *Modulus of elasticity*—Although values for E_c are not always specified, this information is usually available for concrete made with specific lightweight aggregates. This property is further discussed in detail in Chapters 4 and 5.

3.2.1.4 *Slump*—Slump should be the lowest value consistent with the ability to satisfactorily place, consolidate, and finish the concrete and should be measured at the point of discharge.

3.2.1.5 *Entrained-air content*—Air entrainment in lightweight concrete, as in normalweight concrete, is required for resistance to freezing and thawing, as shown in ACI 201.2R, Table 1.1. In concrete made with some lightweight aggregates, it is also an effective means of improving workability. Because entrained air reduces the mixing water requirement while maintaining the same slump, as well as reducing bleeding and segregation, it is normal practice to use air entrainment in lightweight concrete regardless of its exposure to freezing and thawing.

Recommended ranges of total air contents for lightweight concrete are shown in Table 3.1

Attempts to use a large proportion of normalweight aggregate in lightweight concrete to reduce costs and then to use a high air content to meet density requirements are counterproductive. Such a practice usually becomes self-defeating because compressive strength is thereby lowered for each increment of air beyond the recommended ranges. The cement content should then be increased to meet strength requirements. Although the percentages of entrained air required for workability and freezing-and-thawing resistance reduce the density of the concrete, it is not recommended that air contents be increased beyond the upper limits given in Table 3.1 simply to meet density requirements. Adjustment of proportions of aggregates, principally by limiting the normal-weight aggregate constituent, is the most reliable, and usually the more economical, way to meet specified density requirements. Nonstructural or insulating concrete may use higher air contents to lower density.

3.2.2 *Workability*—Workability is an important property of freshly mixed lightweight concrete. The slump test is the most widely used method to measure workability. Similar to normalweight concrete, properly proportioned, lightweight concrete mixtures will have acceptable finishing characteristics.

Water-cementitious material ratio—The *w/cm* can be determined for lightweight concrete proportioned using the specific gravity factor as described in ACI 211.2, Method 1.

Table 3.1—Recommended air content for lightweight concrete

Maximum size of aggregate	Air content percent by volume
3/4 in. (19 mm)	4.5 to 7.5
3/8 in. (10 mm)	6 to 9

When lightweight aggregates are adequately prewetted,[*] there will be a minimal amount of water absorbed during mixing and placing. This allows the net *w/cm* to be computed with an accuracy similar to that associated with normal-weight concrete.

3.3—Materials
Lightweight concrete is composed of cement, aggregates, water, and chemical and mineral admixtures similar to normalweight concrete. Admixtures are added to entrain air, reduce mixing water requirements, and modify the setting time or other property of the concrete. Laboratory tests should be conducted on all the ingredients, and trial batches of the concrete mixtures proportions be performed with the actual materials proposed for use.

3.3.1 *Cementitious and pozzolanic material*—These materials should meet ASTM C 150, C 595, C 618, or C 1157.

3.3.2 *Lightweight aggregates*—For structural concrete, lightweight aggregate should meet the requirements of ASTM C 330. Because of differences in particle strength, the cement contents necessary to produce a specific concrete strength will vary with aggregates from different sources. This is particularly significant for concrete strengths above 5000 psi (35 MPa). Mixture proportions recommended by lightweight-aggregate producers generally provide appropriate cement content and other proportions that should be used as a basis for trial batches.

3.3.3 *Normalweight aggregates*—Normalweight aggregates used in lightweight concrete should conform to the provisions of ASTM C 33.

3.3.4 *Admixtures*—Admixtures should conform to appropriate ASTM specifications, and guidance for use of admixtures may be obtained from ACI 212.3R, 232.2R, 233R, and 234R.

3.4—Proportioning and adjusting mixtures
Proportions for concrete should be selected to make the most economical use of available materials to produce concrete of the required physical properties. Basic relationships have been established that provide guidance in developing optimum combinations of materials. Final proportions, however, should be established by laboratory trial mixtures, which are then adjusted to provide practical field batches, in accordance with ACI 211.2.

The principles and procedures for proportioning normal-weight concrete, such as the absolute volume method, may be applied in many cases to lightweight concrete. The local

---
[*]Note: The time required to reach adequate prewetting will vary with each aggregate and the method of wetting used. The thermal and vacuum saturation method may provide adequate prewetting quickly. The sprinkling or soaking method may take several days to reach an adequate prewetted condition from a dry condition. Therefore, it is essential to contact the aggregate supplier on the prewetting method and length of time required. The percent moisture content achieved at an adequate prewetted condition is normally greater than what would be reached after 24 h submersion.

aggregate producers should be consulted for the particular recommended procedures.

3.4.1 *Absolute volume method*—In using the absolute volume method, the volume of fresh concrete produced by any combination of materials is considered equal to the sum of the absolute volumes of cementitious materials, aggregate, net water, and entrained air. Proportioning by this method requires the determination of water absorption and the particle relative density factor of the separate sizes of aggregates in an as-batched moisture condition. The principle involved is that the mortar volume consists of the total of the volumes of cement, fine aggregate, net water, and entrained air. This mortar volume should be sufficient to fill the voids in a volume of rodded coarse aggregate plus sufficient additional volume to provide satisfactory workability. This recommended practice is set forth in ACI 211.1 and represents the most widely used method of proportioning for normalweight concrete mixtures.

The density factor method, trial mixture basis, is described with examples in ACI 211.1. Displaced volumes are calculated for the cement, air, and net water (total water less amount of water absorbed by the aggregate). The remaining volume is then assigned to the coarse and fine aggregates. This factor may be used in calculations as though it were the apparent particle relative density and should be determined at the moisture content of the aggregate being batched.

3.4.2 *Volumetric method*—The volumetric method is described with examples in ACI 211.1. It consists of making a trial mixture using estimated volumes of cementitious materials, coarse and fine aggregates, and sufficient added water to produce the required slump. The resultant mixture is observed for workability and finishability characteristics. Tests are made for slump, air content, and fresh density. Calculations are made for yield (the total batch mass divided by the fresh density) and for actual quantities of materials per unit volume of concrete. Necessary adjustments are calculated and further trial mixtures made until satisfactory proportions are attained. Information on the dry-loose bulk densities of aggregates, the moisture contents of the aggregates, the optimum ratio of coarse-to-fine aggregates, and an estimate of the required cementitious material to provide the strength desired can be provided by the aggregate supplier.

3.5—Mixing and delivery

The fundamental principles of ASTM C 94 apply to lightweight concrete as they do to normalweight concrete. Aggregates with relatively low or high water absorption need to be handled according to the procedures that have been established by the aggregate supplier or the ready-mixed concrete producer. The absorptive nature of the lightweight aggregate requires prewetting to be as uniform a moisture content as possible before adding the other ingredients of the concrete. The proportioned volume of the concrete is then maintained, and slump loss during transport is minimized.

3.6—Placing

There is little or no difference in the techniques required for placing lightweight concrete from those used in properly placing normalweight concrete. ACI 304.5R discusses in detail the proper and improper methods of placing concrete. The most important consideration in handling and placing concrete is to avoid segregation of the coarse aggregate from the mortar matrix. The basic principles required for a good lightweight concrete placement are:
- A workable mixture using a minimum water content;
- Equipment capable of expeditiously handling and placing the concrete;
- Proper consolidation; and
- Good workmanship.

A well-proportioned lightweight concrete mixture can generally be placed, screeded, and floated with less effort than that required for normalweight concrete. Overvibration or overworking of lightweight concrete should be avoided. Overmanipulation only serves to drive the heavier mortar away from the surface where it is required for finishing and to bring the lower-density coarse aggregate to the surface. Upward movement of coarse lightweight aggregate may also occur in mixtures where the slump exceeds the recommendations provided in this chapter.

3.6.1 *Finishing floors*—Satisfactory floor surfaces are achieved with properly proportioned quality materials, skilled supervision, and good workmanship. The quality of the finishing will be in direct proportion to the efforts expended to ensure that proper principles are observed throughout the finishing process. Finishing techniques for lightweight concrete floors are described in ACI 302.1R.

3.6.1.1 *Slump*—Slump is an important factor in achieving a good floor surface with lightweight concrete and generally should be limited to a maximum of 5 in. (125 mm). A lower slump of about 3 in. (75 mm) imparts sufficient workability and also maintains cohesiveness and body, thereby preventing the lower-density coarse particles from working to the surface. This is the reverse of normalweight concrete where segregation results in an excess of mortar at the surface. In addition to surface segregation, a slump in excess of 5 in. (125 mm) may cause unnecessary finishing delays.

3.6.1.2 *Surface preparation*—Surface preparation before troweling is best accomplished with magnesium or aluminum screeds and floats, which minimize surface tearing and pullouts.

3.6.1.3 *Good practice*—A satisfactory finish on lightweight concrete floors can be obtained as follows:

a. Prevent segregation by:

    1. Using a well-proportioned and cohesive mixture;

    2. Requiring a slump as low as possible;

    3. Avoiding over-vibration;

b. Time the placement operations properly;

c. Use magnesium, aluminum, or other satisfactory finishing tools;

d. Perform all finishing operations after free surface bleeding water has disappeared; and

e. Cure the concrete properly.

3.6.2 *Curing*—Upon completion of the finishing operation, curing of the concrete should begin as soon as possible. Ultimate performance of the concrete will be influenced by the extent of curing provided. ACI 302.1R and ACI 308.1 contain information on proper curing of concrete floor slabs.

Unlike traditional curing where moisture is applied to the surface of the concrete, internal curing occurs by the release of water absorbed within the pores of lightweight aggregate. Absorbed water does not enter the *w/cm* that is established at the time of set. As the pore system of the hydrating cement becomes increasingly smaller, water contained within the relatively larger pores of the lightweight aggregate particle is wicked into the matrix, thus providing an extended period of curing. The benefits of internal curing have been known for several decades where ordinary concrete incorporating lightweight aggregate with a high degree of absorbed water has performed extremely well in bridges, parking structures, and other exposed structures. Internal curing is beneficial for high-performance concrete mixtures containing supplementary cementitious materials, especially where the *w/cm* is less than 0.45. These low *w/cm* mixtures are relatively impervious and vulnerable to self-desiccation because external surface curing moisture is unable to penetrate.

3.7—Pumping lightweight concrete

3.7.1 *General considerations*—Unless the lightweight aggregates are satisfactorily prewetted, they may absorb mixing water and subsequently cause difficulty in pumping the concrete. For this reason, it is important to adequately condition the aggregate by fully prewetting before batching the concrete. The conditioning of the lightweight aggregate can be accomplished by any of the following:

- *Atmospheric*—Using a soaker hose or sprinkler system. The length of time required to adequately prewet a lightweight aggregate is dependent on the absorption characteristics of the aggregate. The lightweight aggregate supplier may be able to supply useful information. Uniform prewetting can be accomplished by several methods, including sprinkling, using a soaker hose, and by applying water to aggregate piles at either or both the aggregate plant or batch plants.
- *Thermal*—By immersion of partially cooled aggregate in water. It should be carefully controlled and is feasible only at the aggregate plant.
- *Vacuum*—By introducing dry aggregate into a vessel from which the air can be evacuated. The vessel is then filled with water and returned to atmospheric pressure. This should be performed only at the aggregate plant.

Prewetting minimizes the mixing water being absorbed by the aggregate, therefore minimizing the slump loss during pumping. This additional moisture also increases the density of the lightweight aggregate, which in turn increases the density of the fresh concrete. This increased density due to prewetting will eventually be lost to the atmosphere in drying and provides for extended internal curing.

3.7.2 *Proportioning pump mixtures*—When considering pumping lightweight concrete, some adjustments may be necessary to achieve the desired characteristics. The architect/engineer and contractor should be familiar with any mixture adjustments required before the decision is made as to the method of placement. The ready-mixed concrete producer and aggregate supplier should be consulted so that the best possible pump mixture can be determined. Pumping lightweight concrete is extensively covered in a report by the Expanded Shale, Clay, and Slate Institute (ESCSI) (1996).

When the project requirements call for pumping, the following general rules apply. These are based on the use of lightweight coarse aggregate and normalweight fine aggregate.

- Prewet lightweight aggregate to a moisture content recommended by the aggregate supplier;
- Maintain a 564 lb/yd^3 (335 kg/m^3) minimum cementitous content;
- Use selected liquid and mineral admixtures that will aid in pumping;
- To facilitate pumping, adjustments in the standard mixture proportion may result in a slight reduction in the volume of coarse aggregate, with a corresponding increase in the volume of fine aggregate;
- Cementitious content should be sufficient to accommodate a 4 to 6 in. (100 to 150 mm) slump at the point of placement;
- Use a well-graded natural sand with a good particle shape and a fineness modulus range of 2.2 to 2.7; and
- Use a properly combined coarse- and fine-aggregate grading. The grading should be made by absolute volume rather than by mass to account for differences in relative density of the various particle sizes.

Sometimes it is advisable to plan on various mixture designs as the height of a structure or distance from the pump to the point of discharge changes. Final evaluation of the concrete shall be made at the discharge end of the pumping system (ACI 304.5R).

3.7.3 *Pump and pump system*—Listed as follows are some of the key items pertinent to the pump and pumping system.

- Use the largest size line available, with a recommended minimum of 5 in. (125 mm) diameter without reducers;
- All lines should be clean, the same size, and "buttered" with grout at the start;
- Avoid rapid size reduction from the pump to line; and
- Reduce the operating pressure by:

  1. Slowing down the rate of placement;
  2. Using as much steel line and as little rubber line as possible;
  3. Limiting the number of bends; and
  4. Making sure the lines are gasketed and braced by a thrust block at turns.

A field trial should be conducted using the pump and mixture design intended for the project. Observers present should include representatives of the contractor, ready-mixed concrete producer, architect/engineer, pumping service, testing agency, and aggregate supplier. In the pump trial, the height and length to the delivery point of the concrete to be moved should be taken into account. Because most test locations will not allow the concrete to be pumped vertically as high as it would be during the project, the following rules of thumb can be applied for the horizontal runs with steel lines:

1.0 ft (0.3 m) vertical	=	4.0 ft (1.2 m) horizontal
1.0 ft (0.3 m) rubber hose	=	2.0 ft (0.6 m) of steel
1.0 ft (0.3 m) 90-degree bend	=	10.0 ft (3.0 m) of steel

3.8—Laboratory and field control

Changes in absorbed moisture or relative density of lightweight aggregates, which result from variations in initial moisture content or grading, and variations in entrained-air content suggest that frequent checks of the fresh concrete should be made at the job site to ensure consistent quality (ACI 211.1). Sampling should be in accordance with ASTM C 172. Tests normally required are: density of the fresh concrete (ASTM C 567); standard slump test (ASTM C 143); air content (ASTM C 173); and Standard Practice for Making and Curing Concrete Test Specimens in the Field (ASTM C 31).

At the job start, the fresh properties, density, air content, and slump of the first batch or two should be determined to verify that the concrete conforms to the laboratory mixture. Small adjustments may then be made as necessary. In general, when variations in fresh density exceed 3 lb/ft^3 (50 kg/m^3), an adjustment in batch weights may be required to meet specifications. The air content of lightweight concrete should not vary more than ±1-1/2 percentage points from the specified value to avoid adverse effects on concrete density, compressive strength, workability, and durability.

CHAPTER 4—PHYSICAL AND MECHANICAL PROPERTIES OF STRUCTURAL LIGHTWEIGHT-AGGREGATE CONCRETE

4.1—Scope

This chapter presents a summary of the properties of lightweight concrete. The information is based on many laboratory studies and records of a large number of existing structures that have provided satisfactory service for more than eight decades. The customary requirements for structural concrete are that mixture proportions proposed for the project should be based on laboratory tests or on mixtures with established records of performance.

4.2—Method of presenting data

In the past, properties of lightweight concrete have been compared with those of normalweight concrete, and usually the comparison standard has been a single normalweight material. With several million cubic yards of lightweight concrete being placed each year, such a comparison of properties may no longer be appropriate. The data on various structural properties are presented as reasonable conservative values to be expected in relationship to some fixed property such as compressive strength, density, or in the case of fire resistance, slab thickness.

4.3—Compressive strength

Compressive strength levels commonly required by the construction industry for design strengths of cast-in-place, precast, or prestressed concrete are economically obtained with lightweight concrete (Shideler 1957; Hanson 1964; Holm 1980a). Design strengths of 3000 to 5000 psi (21 to 35 MPa) are common. In precast and prestressing plants, design strengths above 5000 psi (35 MPa) are usual. In several civil structures, such as the Heidrun Platform and Norwegian bridges, concrete cube strengths of 60 MPa (8700 psi) have been specified (fib 2000). As discussed in Chapter 2, all aggregates have strength ceilings, and with lightweight aggregates, the strength ceiling generally can be increased by reducing the maximum size of the coarse aggregate. As with normalweight concrete, water-reducing plasticizing and mineral admixtures are frequently used with lightweight concrete mixtures to increase workability and facilitate placing and finishing.

4.4—Density of lightweight concrete

4.4.1—The fresh density of lightweight concrete is a function of mixture proportions, air contents, water demand, particle relative density, and absorbed moisture content of the lightweight aggregate. Decrease in the density of exposed concrete is due to moisture loss that, in turn, is a function of ambient conditions and surface area/volume ratio of the member. The architect/engineer should specify a maximum fresh density as limits of acceptability should be controlled at time of placement.

Although there are numerous structural applications of lightweight concrete using both coarse and fine lightweight aggregates, usual commercial practice in North America is to design concrete with natural sand fine aggregates. Long-span bridges using concrete with three-way blends (coarse and fine lightweight aggregates and small amounts of natural sand) have provided long-term durability and structural efficiency (density/strength ratios) (Holm and Bremner 1994). Earlier research reports (Kluge, Sparks, and Tuma 1949; Price and Cordon 1949; Reichard 1964; and Shideler 1957) compared all concrete containing both fine and coarse lightweight aggregates with "reference" normalweight concrete. Later studies (Hanson 1964; Pfeifer 1967) supplemented the early findings with data based on lightweight concrete where the fine aggregate was a natural sand.

4.4.2—Self loads used for design should be based on equilibrium density that, for most conditions and members, may be assumed to be achieved after 90 days air-drying. Extensive North American studies demonstrated that despite wide initial variations of aggregate moisture content, equilibrium density was found to be 3 lb/ft^3 (50 kg/m^3) above oven-dry density (Fig. 4.1) for lightweight concrete. European recommendations for in-service density are similar (FIP 1983). Concrete containing high cementitious contents, and particularly those containing efficient pozzolans, will develop densities with a reduced differential between fresh and equilibrium density.

When weights and moisture contents of all the constituents of the concrete are known, a calculated equilibrium density can be determined according to ASTM C 567 from the following equations

$$O = (W_{df} + W_{dc} + 1.2W_{ct})/V \qquad (4\text{-}1)$$

$$E = O + 3 \text{ lb/ft}^3 \, (O + 50 \text{ kg/m}^3) \qquad (4\text{-}2)$$

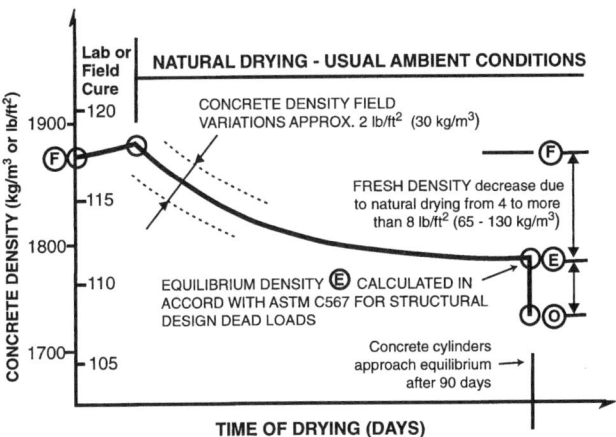

F - FRESH DENSITY: Specified for field control (unit weight bucket). Measurements on 6' x 12' (150 x 300 mm) cylinders will average 2.5 lb/ft³ (40 kg/m³) higher than filed measurements on 0.5 ft³ (.014 m³) unit weight bucket.

E - EQUILIBRIUM DENSITY: Typically 3 lb/ft³ (50 kg/m³) greater than OVEN DRY DENSITY - O

Fig. 4.1—Concrete density versus time of drying for structural lightweight concrete (Holm 1994).

where
O = calculated oven-dry density, lb/ft³ (kg/m³);
W_{df} = mass of dry fine aggregate in batch, lb (kg);
W_{dc} = mass of coarse aggregate in batch, lb (kg);
1.2 = factor to account for water of hydration;
W_{ct} = mass of cement in batch, lb (kg);
V = volume of concrete produced, ft³ (m³); and
E = calculated equilibrium density, lb/ft³ (kg/m³).

4.5—Specified-density concrete

Concrete containing limited amounts of lightweight aggregate that result in equilibrium concrete densities greater than 120 lb/ft³ (1920 kg/m³) but less than concrete composed entirely of normalweight aggregates is defined as specified-density concrete. The increasing usage of specified-density concrete is driven by engineers' decisions to optimize the concrete density to improve structural efficiency (strength-to-density ratio), to reduce concrete product transportation and construction costs, and to enhance the hydration of high cementitous content concrete with very low w/cm.

4.6—Modulus of elasticity

The modulus of elasticity of concrete depends on the relative amounts of paste and aggregate and the modulus of each constituent (LaRue 1946; Pauw 1960). Normalweight concrete has a higher E_c because the moduli of sand, stone, and gravel are greater than the moduli of lightweight aggregates. Figure 4.2 gives the range of modulus of elasticity values for lightweight concrete. Generally, the modulus of elasticity for lightweight concrete is considered to vary between 1/2 to 3/4 that of sand and gravel concrete of the same strength. Variations in lightweight aggregate grading usually have little effect on modulus of elasticity if the relative volumes of cement paste and aggregate remain fairly constant.

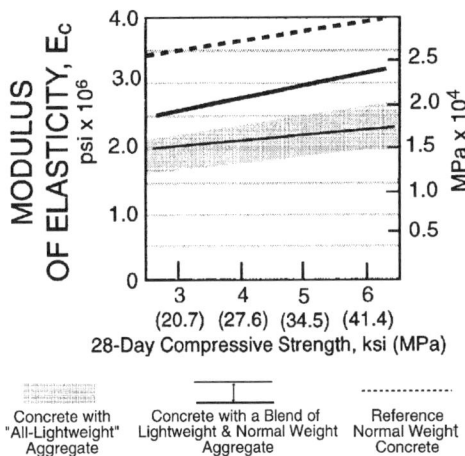

Fig. 4.2—Modulus of elasticity.

The formula for $E_c = w_c^{1.5} 33\sqrt{f_c'}$ ($w_c^{1.5} 0.043\sqrt{f_c'}$) given in ACI 318, may be used for values of w between 90 and 155 lb/ft³ (1440 and 2480 kg/m³) and strength levels of 3000 to 5000 psi (21 to 35 MPa). Further discussion of this formula is given in Section 5.3. Concretes in service may deviate from this formula by up to 20%. When an accurate evaluation of E_c is required for a particular concrete, a laboratory test in accordance with the methods of ASTM C 469 should be carried out.

4.7—Poisson's ratio

Tests to determine Poisson's ratio of lightweight concrete by resonance methods showed that it varied only slightly with age, strength, or aggregate used, and that the values varied between 0.16 and 0.25 with the average being 0.21 (Reichard 1964). Tests to determine Poisson's ratio by the static method for lightweight and normalweight concrete gave values that varied between 0.15 and 0.25 and averaged 0.2.

While this property varies slightly with age, test conditions, and physical properties of the concrete, a value of 0.20 may be usually assumed for practical design purposes. An accurate evaluation can be obtained for a particular concrete by testing according to ASTM C 469.

4.8—Creep

Creep is the increase in strain of concrete under a sustained stress. Creep properties of concrete may be either beneficial or detrimental, depending on the structural conditions. Concentrations of stress, either compressive or tensile, may be reduced by stress transfer through creep, or creep may lead to excessive long-time deflection, prestress loss, or loss of camber. The effects of creep along with those of drying shrinkage should be considered and, if necessary, taken into account in structural designs.

4.8.1 *Factors influencing creep*—Creep and drying shrinkage are closely related phenomena that are affected by many factors, such as type of aggregate, type of cement, grading of aggregate, water content of the mixture, moisture content of aggregate at time of mixture, amount of entrained

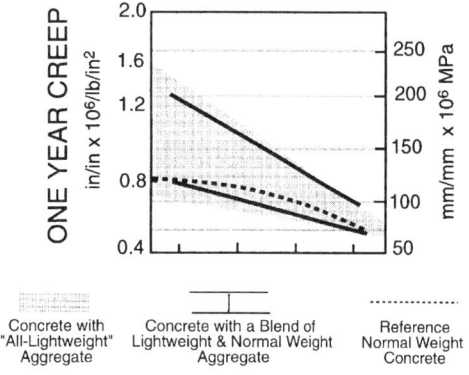

Fig. 4.3—Creep: normally cured concrete.

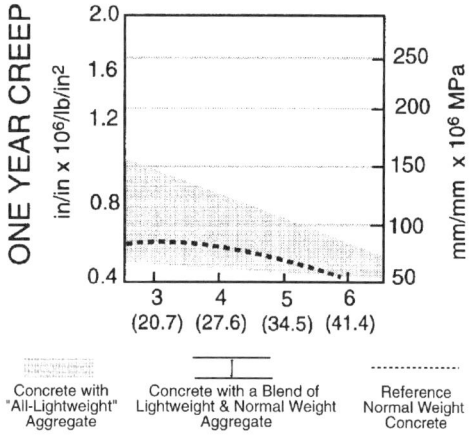

Fig. 4.4—Creep: steam-cured concrete.

air, age at initial loading, magnitude of applied stress, method of curing, size of specimen or structure, relative humidity of surrounding air, and period of sustained loading.

4.8.2 *Normally cured concrete*—Figure 4.3 shows the range in values of specific creep (creep per psi of sustained stress) for normally cured concrete, as measured in the laboratory (ASTM C 512), when under constant loads for 1 year. These diagrams were prepared with the aid of two common assumptions: superposition of creep effects are valid (that is, creep is proportional to stress within working stress ranges); and shrinkage strains, as measured on nonloaded specimens, may be directly separated from creep strains. The band for lightweight concrete containing normalweight sand is considerably narrower than that for the concrete containing both fine and coarse lightweight aggregate. Figure 4.3 suggests that a very effective method of reducing creep of lightweight concrete is to use a higher-strength concrete. A strength increase from 3000 to 5000 psi (21 to 35 MPa) significantly reduces creep.

4.8.3 *Steam-cured concrete*—Several investigations have shown that creep may be significantly reduced by low-pressure curing and very greatly reduced by high-pressure steam curing. Figure 4.4 shows that the reduction for low-pressure steamed concrete may be from 25 to 40% of the creep of similar concrete subjected only to moist curing.

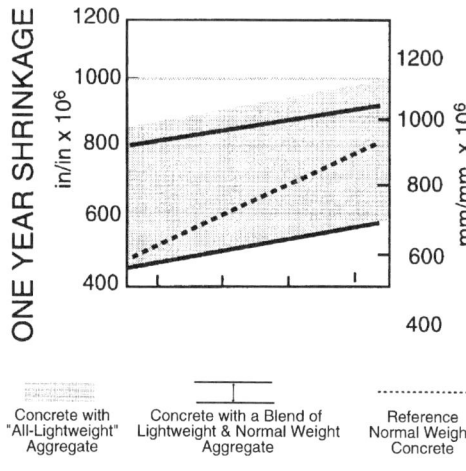

Fig. 4.5—Drying shrinkage: normally cured concrete.

4.9—Drying shrinkage

Drying shrinkage is an important property that can affect the extent of cracking, prestress loss, effective tensile strength, and warping. It should be recognized that large-size concrete members, or those in high ambient relative humidity, might undergo substantially less shrinkage than that exhibited by small laboratory specimens stored at 50% relative humidity.

4.9.1 *Normally cured concrete*—Figure 4.5 indicates wide ranges of shrinkage values after 1 year of drying for lightweight concrete with normalweight sand. Noting the position within these ranges of the reference concrete, it appears that low-strength lightweight concrete generally has greater drying shrinkage than that of the reference concrete. At higher strengths, however, some lightweight concrete exhibits lower shrinkage. Partial or full replacement of the lightweight fine aggregate by natural sand usually reduces shrinkage for concrete made with most lightweight aggregates.

4.9.2 *Atmospheric steam-cured concrete*—Figure 4.6 demonstrates the reduction of drying shrinkage obtained through steam curing. This reduction may vary from 10 to 40%. The lower portion of this range is not greatly different from that for the reference normalweight concrete.

4.10—Splitting tensile strength

The splitting tensile strength of concrete cylinders (ASTM C 496) is an effective method of measuring tensile strength.

4.10.1 *Moist-cured concrete*—Figure 4.7 indicates a narrow range of this property for continuously moist-cured lightweight concrete. The splitting tensile strength of the normalweight reference concrete is nearly intermediate within these ranges.

4.10.2 *Air-dried concrete*—The tensile strength of lightweight concrete that undergoes drying is more relevant in respect to the shear strength behavior of concrete in structures. During drying of the concrete, moisture loss progresses at a slow rate into the interior of concrete members, resulting in the development of tensile stresses at the exterior faces and balancing compressive stresses in the still-moist interior zones. Thus, the tensile resistance to external loading of

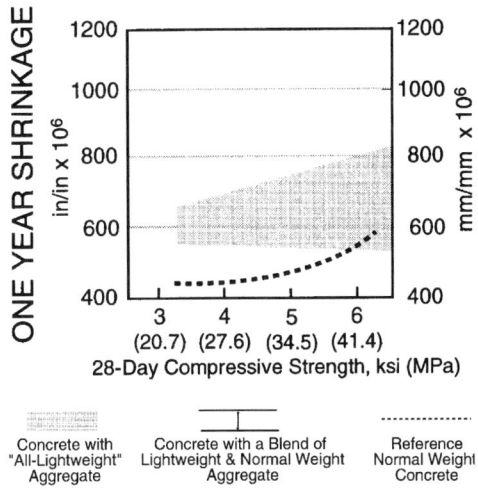

Fig. 4.6—Drying shrinkage: steam-cured concrete.

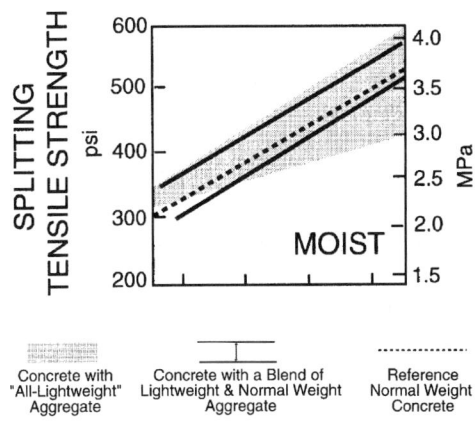

Fig. 4.7—Splitting tensile strength: moist-cured concrete.

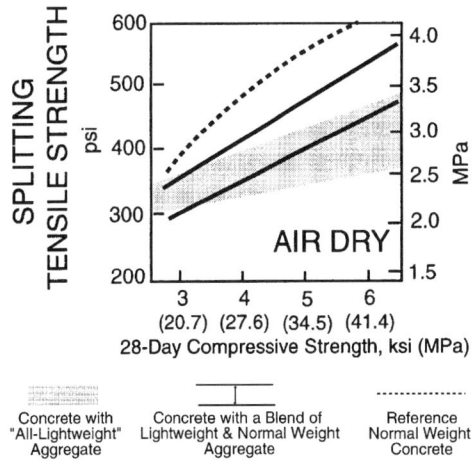

Fig. 4.8—Splitting tensile strength: air-dried concrete.

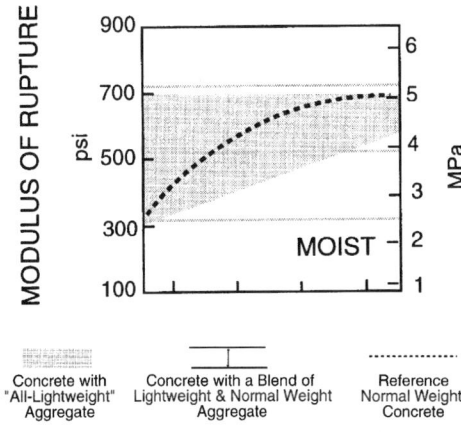

Fig. 4.9—Modulus of rupture: normally cured concrete.

drying lightweight concrete will be reduced from that indicated by continuously moist-cured concrete (Hanson 1961; Pfeifer 1967). Figure 4.8 indicates this reduced strength for concrete that has been moist-cured for 7 days followed by 21 days storage at 50% relative humidity (ASTM C 330). The splitting tensile strength of lightweight concrete varies from approximately 70 to 100% that of the normalweight reference concrete when comparisons are made at equal compressive strength.

The replacement of the lightweight fines by sand generally increases the splitting tensile strength of lightweight concrete subjected to drying (Pfeifer 1967; Ivey and Bluth 1966). In some cases, this increase is nonlinear with respect to the sand content so that, with some aggregates, partial sand replacement is as beneficial as complete replacement.

For lightweight concrete with a compressive strength up to 5000 psi (35 MPa), splitting tensile strength is used for estimating the diagonal tension resistance of lightweight concrete in structures. Tests have shown that the diagonal tension strengths of beams and slabs correlate closely with this property of the concrete (Hanson 1961).

4.11—Modulus of rupture

The modulus of rupture (ASTM C 78) is also a measure of the tensile strength of concrete. Figure 4.9 and 4.10 indicate ranges for normally cured and steam-cured concrete, respectively, when tested in the moist condition. For prism specimens, a nonuniform moisture distribution will reduce the modulus of rupture, but the moisture distribution within the structural member is not known and is unlikely to be completely saturated or completely dry. Studies have indicated that modulus of rupture tests of concrete undergoing drying are extremely sensitive to the transient moisture content and, under these conditions, may not furnish reliable results that are satisfactorily reproducible (Hanson 1961).

The values of the modulus of rupture determined from tests on high-strength lightweight concrete yield inconsistent correlation with code requirements. While Huffington (2000) reported that the tensile splitting and modulus of rupture test results generally met AASHTO requirements for high-strength lightweight concrete, Nassar (2002) found that in his investigation, the modulus of rupture levels were about 60 to 85% of code requirements of $\phi_m \times 7.5\sqrt{f_c'}$ where ϕ_m for sanded lightweight concrete is recommended to be 0.85.

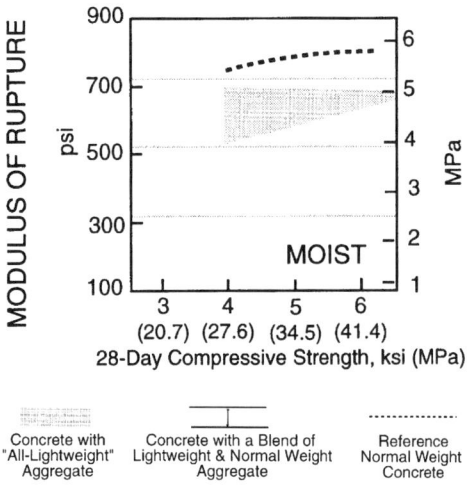

Fig. 4.10—Modulus of rupture: steam-cured concrete.

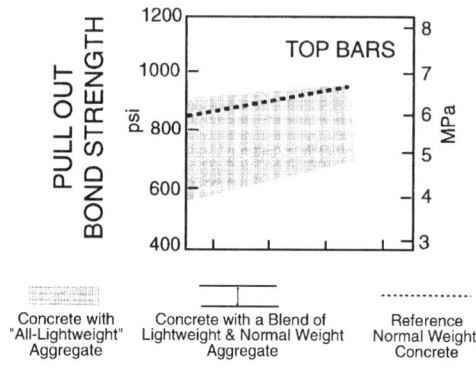

Fig. 4.11—Bond strength: pullout tests.

Nassar recommended that additional testing be conducted to verify the 0.85 factor for high-strength lightweight concrete.

4.12—Bond strength

ACI 318 includes a factor for development length of 1.3 to reflect the lower tensile strength of lightweight-aggregate concrete and allows that factor to be taken as $6.7\sqrt{f_c'}/f_{ct} \geq 1.0$ if the average splitting strength f_{ct} of the lightweight-aggregate concrete is specified. In general, design provisions require longer development lengths for lightweight-aggregate concrete.

Due to the lower strength of the aggregate, lightweight concrete should be expected to have lower tensile strength, fracture energy, and local bearing capacity than normal-weight concrete with the same compressive strength. As a result, the bond strength of bars cast in lightweight concrete, with or without transverse reinforcement, is lower than that in normalweight concrete, with that difference tending to increase at higher strength levels (Fig. 4.11) (Shideler 1957).

Previous reports by Committee 408 (1966, 1970) have emphasized the paucity of experimental data on the bond strength of reinforced concrete elements made with lightweight-aggregate concrete.

The majority of experimental results found in the literature are from different configurations of pullout tests. Early research by Lyse (1934), Petersen (1948), and Shideler (1957) concluded that the bond strength of steel in lightweight-aggregate concrete was comparable to that of normalweight concrete. Petersen tested beams made with expanded shale and expanded slag, and concluded that bond strength of reinforcement in lightweight-aggregate concrete was comparable to that of normalweight concrete. Shideler (1957) conducted pullout tests on 9 in. (230 mm) cube specimens with six different types of aggregates. No. 6 (19 mm) bars were embedded in specimens with compressive strengths of 3000 and 4500 psi (21 and 31 MPa), and No. 9 (29 mm) bars were used in 9000 psi (62 MPa) specimens. Although the bond strength of normalweight concrete specimens was slightly higher than that of lightweight concrete specimens, Shideler stated that the difference was not significant.

Similar behavior has been observed in more recent studies. Based on a series of pullout test, Martin (1982) concluded that there was no significant difference between the bond strength of normalweight and lightweight-aggregate concrete. Berg (1981) obtained similar results from a pullout test; although, in a limited testing program involving beams, he observed lower bond strengths in specimens made with lightweight-aggregate concrete. The observed difference in strength was approximately 10%.

Clarke and Birjandi (1993) used a specimen developed by the British Cement Association (Chana 1990) and tested four lightweight aggregates with various densities available in the United Kingdom. In addition to the type of aggregate, the study investigated the effect of casting position. The fine aggregate in all mixtures was natural sand. Test results indicated that, with the exception of the lightest insulation grade aggregate, all specimens had higher bond strengths than those of specimens made with normalweight aggregate. This was partially attributed by the authors to the fact that natural sand, as opposed to lightweight aggregate, was used as fine aggregate.

In contrast to the studies just described, there are several studies that indicate significant differences between bond strengths in lightweight and normalweight aggregate concrete. In pullout tests, Baldwin (1965) obtained bond strengths for lightweight concrete that were only 65% of those obtained for normalweight concrete. These results contradicted the prevailing assumption at the time that bond strength in lightweight-aggregate concrete was similar to that of normal weight concrete (ACI Committee 408, 1966).

Robins and Standish (1982) conducted a series of pullout tests to investigate the effect of lateral stress on the bond strength of plain and deformed bars in specimens made with lightweight-aggregate concrete. As the lateral pressure applied to the specimens increased, the mode of failure changed from splitting to pullout. Bond strength increased with confining pressure for both normalweight and lightweight concrete. For specimens that failed by splitting, bond strength was 10 to 15% higher for normalweight concrete than for lightweight concrete; however, when the lateral pressure was large enough to prevent a splitting failure, the difference in bond strength was much higher—on the order of 45%.

Mor (1992) tested No. 6 (19 mm) bars embedded in 3 x 3 x 20 in. (76 x 76 x 508 mm) pullout specimens to investigate the effect of condensed silica fume on the bond strength of normalweight and lightweight-aggregate concrete. For his specimens without silica fume, the maximum bond stress for specimens made with lightweight concrete was 88% of that of specimens made with normalweight concrete. For concrete with 13 to 15% condensed silica fume, the ratio was 82%. The specimens made with lightweight concrete developed splitting failures at 75 to 80% of the slip of specimens made with normalweight concrete. The use of silica fume had little effect on bond strength, with an increase of 2% for normal-weight concrete and a decrease of 5% for lightweight concrete.

Overall, the data indicate that the use of lightweight concrete can result in bond strengths that range from nearly equal to 65% of the values obtained with normalweight concrete. For special structures such as long-span bridges with very high strengths and major offshore platforms, a testing program based on the materials selected to the project is recommended.

4.13—Ultimate strength factors

4.13.1 *Ultimate strain*—Figure 4.12 gives a range of values for ultimate compressive strain for concrete containing both coarse and fine lightweight aggregate and for normalweight concrete. These data were obtained from unreinforced specimens eccentrically loaded to simulate the behavior of the compression side of a reinforced beam in flexure (Hognestad, Hanson, and McHenry 1955). The diagram indicates that the ultimate compressive strain of most lightweight concrete (and of the reference normal-weight concrete) may be somewhat greater than the value of 0.003, assumed for design purposes.

4.14—Durability

Numerous accelerated freezing-and-thawing testing programs conducted on lightweight concrete in North America and in Europe researching the influence of entrained-air volume, cement content, aggregate moisture content, specimen drying times, and testing environment have arrived at similar conclusions: air-entrained lightweight concrete proportioned with a high-quality binder provides satisfactory durability results when tested under usual laboratory freezing-and-thawing programs. Observations of the resistance to deterioration in the presence of deicing salts on mature bridges indicate similar performance between lightweight and normalweight concrete. Comprehensive investigations into the long-term weathering performance of bridge decks and marine structures exposed for many years to severe environments support the findings of laboratory investigations and suggest that properly proportioned and placed lightweight concrete performs equal to or better than normalweight concrete (Holm 1994).

Core samples taken from hulls of 80-year-old lightweight concrete ships as well as 40- to 50-year-old lightweight concrete bridges have shown that concrete having a dense

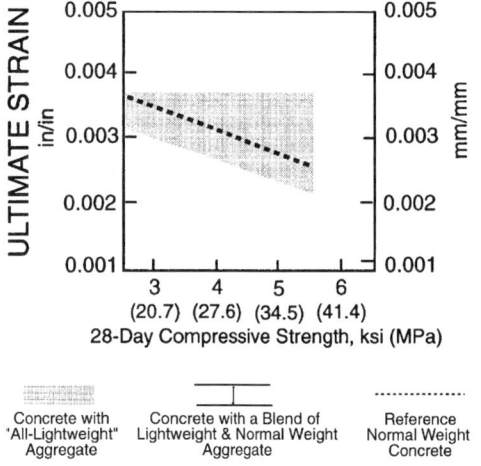

Fig. 4.12—Ultimate strain.

contact zone at the aggregate/matrix interface has low levels of microcracking throughout the mortar matrix. The explanation for this demonstrated record of high resistance to weathering and corrosion is due to several physical and chemical mechanisms, including superior resistance to microcracking developed by significantly higher aggregate/matrix adhesion and the reduction of internal stresses due to elastic matching of coarse aggregate and matrix phases (Holm, Bremner, and Newman 1984). High ultimate strain capacity is also provided by lightweight concrete as it has a high strength/modulus ratio. The strain at which the disruptive dilation of concrete starts is higher for lightweight concrete than for equal-strength normalweight concrete. A well-dispersed pore system provided by the surface of the lightweight fine aggregates may also assist the air-entrainment system and serve an absorption function by reducing concentration levels of deleterious materials in the matrix phase (Holm 1980b).

Long-term pozzolanic action is provided when the silica-rich expanded aggregate combines with calcium hydroxide liberated during cement hydration. This will decrease permeability and minimize leaching of soluble compounds and may also reduce the possibility of sulfate salt disruptive behavior.

It is widely recognized that while the ASTM Test Method for Resistance of Concrete to Rapid Freezing and Thawing (C 666) provides a useful comparative testing procedure, there remains an inadequate correlation between accelerated laboratory test results and the observed behavior of mature concrete exposed to natural freezing and thawing. When freezing-and-thawing tests are conducted, ASTM C 330 requires the following modification to the procedures of ASTM C 666, "Unless otherwise specified, remove the lightweight concrete specimens from moist curing at an age of 14 days and allow to air-dry for another 14 days exposed to a relative humidity of 50 ± 5% and a temperature of 73.5 + 3.5 °F (23 ± 2 °C). Then submerge the specimens in water for 24 h before the freezing and thawing test." Durability characteristics of any concrete, both normalweight and lightweight, are primarily determined by the protective qualities of the cement paste matrix. It is imperative that permeability characteristics of the concrete matrix be of high quality to protect steel reinforcing from corrosion, which is clearly the dominant form of structural deterioration observed in current construction. The matrix protective quality of nonstructural, insulating concrete proportioned for thermal resistance by using high-air and low-cement contents will be significantly reduced. Very low density, nonstructural concrete will not provide resistance to the intrusion of chlorides and carbonation, comparable to the long-term satisfactory performance demonstrated with high-quality, lightweight concrete.

4.14.1—*Carbonation in mature marine structures*

4.14.1.1 *General*—Carbonation in concrete is the reaction of carbon dioxide from the air with calcium hydroxide liberated from the hydration process. This reaction produces calcium carbonate that can neutralize the natural protection of steel reinforcement afforded by the concrete.

The concern for carbonation is predicated on the pH in concrete lowering from approximately 13 to the vicinity of 9, which in turn neutralizes the protective layer over the reinforcing steel, making it vulnerable to corrosion. Two primary mechanisms protect steel from corrosion: the combination of an adequate depth of cover with a sufficiently high-quality cover concrete. This quality is usually related to *w/cm* or strength (relatively easy properties to quantify), but is more closely related to permeability and strain capacity of the cover concrete.

4.14.1.2 *Concrete ships, Cape Charles, Va.*—Holm, Bremner, and Vaysburd (1988) reported the results of carbonation measurements conducted on cores drilled from the hull of several concrete ships built during World War II. The ships were used as breakwaters for a ferryboat dock in the Chesapeake Bay at Cape Charles, Virginia. They were constructed with carefully inspected high-quality concrete made with rotary kiln-produced fine and coarse expanded aggregates and a small volume of natural sand. High-cement contents were used to achieve compressive strengths in excess of 5000 psi (35 MPa) at 28 days with a density of 108 lb/ft^3 (1730 kg/m^3) (McLaughlin 1944). Despite freezing and thawing in a marine environment, the hulls and superstructure of this nonair-entrained concrete are in good condition after more than five decades of exposure. The only less-than-satisfactory performance was observed in some areas of the main decks. These areas experienced a delamination of the 3/4 in. (20 mm) concrete cover protecting four layers of large-sized undeformed (typically 1 in. [25 mm] square) reinforcing bars spaced 4 in. (100 mm) on centers. In retrospect, this failure plane is understandable and would have been avoided by the use of modern prestressing procedures. Cover for hull reinforcing was specified at 7/8 in. (22 mm), with all other reinforcement protected by only 1/2 in. (13 mm).

Without exception, the reinforcing steel bars cut by the 18 cores taken were rust-free. Cores that included reinforcing steel were split along an axis parallel to the plane of the reinforcing in accordance with the procedures of ASTM C 496. Visual inspection revealed negligible corrosion when the bar was removed. After the interface was sprayed with phenolphthalein, the surfaces stained a vivid red, indicating no carbonation at the steel-concrete interface.

Carbonation depth, as revealed by spraying the freshly fractured surface with a standard solution of phenolphthalein, averaged 0.04 in. (1 mm) for specimens taken from the main deck, was between 0.04 and 0.08 in. (1 and 2 mm) for concrete in exposed wing walls, and was virtually nonexistent in the hull and bulkheads. Coring was conducted from the waterline to as much as 16 ft (5 m) above high water. In no instances could carbonation depths greater than 0.08 in. (2 mm) be found. In isolated instances, flexural cracks up to 0.31 in. (8 mm) in depth were encountered, and these had carbonated in the plane of the crack. The carbonation did not appear to progress more than 0.004 in. (0.1 mm) perpendicular to the plane of the crack.

High-quality, low-permeability concrete will inhibit the diffusion of carbon dioxide, and concrete with a high moisture content will reduce the diffusion rate to that of a gas through water rather than that of a gas through air.

4.14.1.3 *Chesapeake Bay Bridge, Annapolis, Md.*—Concrete cores taken from the 35-year-old Chesapeake Bay Bridge revealed carbonation depths of 0.08 to 0.31 in. (2 to 8 mm) from the top of the bridge deck and 0.08 to 0.51 in. (2 to 13 mm) from the underside of the bridge deck. The higher carbonation depth on the underside reflects increased gas diffusion associated with the drier surface of the bridge. The 1.14 in. (36 mm) asphalt-wearing course appears to have inhibited drying and thus reduced carbonation depth on top (Holm 1983; Holm, Bremner, and Newman 1984).

4.14.1.4 *Coxsackie Bridge, New York*—Cores drilled with the cooperation of the New York State Thruway Authority from the 15-year-old exposed deck surface of the Interchange Bridge at Coxsackie revealed 0.20 in. (5 mm) carbonation depths for the top surface and 0.39 in. (10 mm) from the bottom. Despite almost 1000 saltings of the exposed deck, there was no evidence of corrosion in any of the reinforcing bars cut by the six cores taken (Holm, Bremner, and Newman 1984).

4.14.1.5 *Bridges and viaducts in Japan*—The results of measurements of carbonation depths on mature marine structures in North America are paralleled by data reported by Ohuchi et al. (1984). These investigators studied the chloride penetration, depth of carbonation, and incidence of microcracking in both lightweight and normalweight concrete on the same bridges, aqueducts, and caissons after 19 years of exposure. The high-durability performance of those structures (as measured by the carbonation depths, microcracking, and chloride penetration profiles reported by Ohuchi et al. [1984]) is similar to studies conducted on mature lightweight concrete bridges in North America (Holm, Bremner, and Newman 1984).

4.14.2 *Permeability and corrosion protection*—While current technical literature contains numerous reports on the permeability of concrete, only a limited number of papers report experiments in which lightweight and normalweight concrete were tested under the same conditions. Furthermore, almost all studies measuring permeability used test conditions that were static. While this approach is appropriate for dams and water-containing structures, it is not relevant to bridges and parking structures, which are constantly subjected to dynamic stress and strain. Cover concrete is expected to maintain its protective impermeable integrity despite the accumulation of shrinkage, thermal, and structural load-related strains.

Permeability investigations conducted on lightweight and normalweight concrete exposed to the same testing criteria have been reported by Khokrin (1973), Nishi et al. (1980), Keeton (1970), Bamforth (1987) and Bremner, Holm and McInerny (1992). It is of interest that, in every case, despite wide variations in concrete strengths, testing media (water, gas, and oil), and testing techniques (specimen size, media pressure, and equipment), lightweight concrete had equal or lower permeability than its normalweight counterpart. Khokrin (1973) further reported that the lower permeability of lightweight concrete was attributed to the elastic compatibility of the constituents and the enhanced bond between the coarse aggregate and the matrix. In the Onoda Cement Company tests (Nishi et al. 1980), concrete with a *w/cm* of 0.55, moist-cured for 28 days when tested at 128 psi (0.88 MPa) water pressure had a depth of penetration of 1.38 in. (35 mm) for normalweight concrete and 0.95 in. (24 mm) for lightweight concrete. When tested with seawater, penetration was 0.59 and 0.47 in. (15 and 12 mm) for normalweight concrete and lightweight concrete, respectively. The author suggested that the reason for this behavior was "a layer of dense hardened cement paste surrounding the particles of artificial lightweight coarse aggregate." The U.S. Navy-sponsored work by Keeton (1970) reported the lowest permeability with high-strength lightweight concrete. Bamforth (1987) incorporated lightweight concrete as one of the four concretes tested for permeability to nitrogen gas at 145 psi (1 MPa) pressure level. The normalweight concrete specimens included high-strength 13,000 psi (90 MPa) concrete and concrete with a 25% fly ash replacement, by mass or volume. The sanded lightweight concrete (7250 psi [50 MPa]), 6.4% air with a density of 124 lb/ft^3 (1985 kg/m^3) demonstrated the lowest water and air permeability of all mixtures tested.

Fully hydrated portland cement paste of low *w/cm* has the potential to form an essentially impermeable matrix that should render concrete impermeable to the flow of liquids and gases. In practice, however, this is not the case, as microcracks form in concrete during the hardening process, as well as later, due to shrinkage, thermal, and applied stresses. In addition, excess water added to concrete for easier placing will evaporate, leaving pores and conduits in the concrete. This is particularly true in exposed concrete decks where concrete has frequently provided inadequate protection for steel reinforcement.

Mehta (1986) observed that the permeability of a concrete composite is significantly greater than the permeability of either the continuous matrix system or the suspended coarse-aggregate fraction. This difference is primarily related to extensive microcracking caused by mismatched concrete components responding differentially to temperature gradients, service load strains, and volume changes associated with chemical reactions taking place within the concrete. In addition, channels develop in the transition zone surrounding normalweight coarse aggregates, giving rise to unimpeded moisture movements. While separations caused by the evaporation of bleed water adjacent to ordinary aggregates are frequently visible to the naked eye, such defects are essentially unknown in lightweight concrete. The continuous, high-quality matrix fraction surrounding lightweight aggregate is the result of several beneficial processes. Khokrin (1973) reported on several investigations that documented the increased transition zone microhardness due to pozzolanic reaction developed at the surface of the lightweight aggregate. Bremner, Holm, and deSouza (1984) conducted measurements of the diffusion of the silica out of the coarse lightweight-aggregate particles into the cement paste matrix using energy-dispersive x-ray analytical techniques. The results correlated with Khokrin's observations that the superior contact zone in lightweight concrete extended approximately 60 μm from the lightweight aggregate particles into the continuous matrix phase.

The contact zone in lightweight concrete is the interface between two porous media: the lightweight-aggregate particle and the hydrating cement binder. This porous media interface allows for hygrol equilibrium to be reached between the two phases, thus eliminating weak zones caused by water concentration. In contrast, the contact zone of normalweight concrete is an interface between the nonabsorbent surface (wall effect) of the dense aggregate and a water-rich binder. The accumulation of water at that interface is subsequently lost during drying, leaving a porous, low-quality matrix at the interface.

One laboratory report comparing normalweight concrete and lightweight concrete indicated that, in the unstressed state, the permeabilities were similar. At higher levels of stress, however, the lightweight concrete could be loaded to a higher percentage of its ultimate compressive strength before microcracking causes a sharp increase in permeability (Sugiyama, Bremner, and Holm 1996). In laboratory testing programs, the concrete is maintained at constant temperature, there are no significant shrinkage restraints, and field-imposed stresses are absent. Because of the as-batched moisture content of the lightweight aggregate before mixing, this absorbed water provides for extended moist curing. The water tends to wick out from the coarse aggregate pores into the finer capillary pores in the cement paste, thereby extending moist curing. Because the potential pozzolanic surface reaction is developed over a long time, usual laboratory testing that is completed in less than a few months may not adequately take this into account.

4.14.3 *Influence of contact zone on durability*—The contact zone is the transition layer of material connecting the coarse-aggregate particle with the enveloping continuous mortar matrix. Analysis of this linkage layer requires consideration of more than the adhesion developed at the interface (contact zone) and should include the transitional layer that forms between the two phases. Collapse of the structural integrity of a conglomerate may come from the failure of one of the two phases or from a breakdown in the contact zone causing a separation of the still-intact phases. The various mechanisms that act to maintain continuity, or

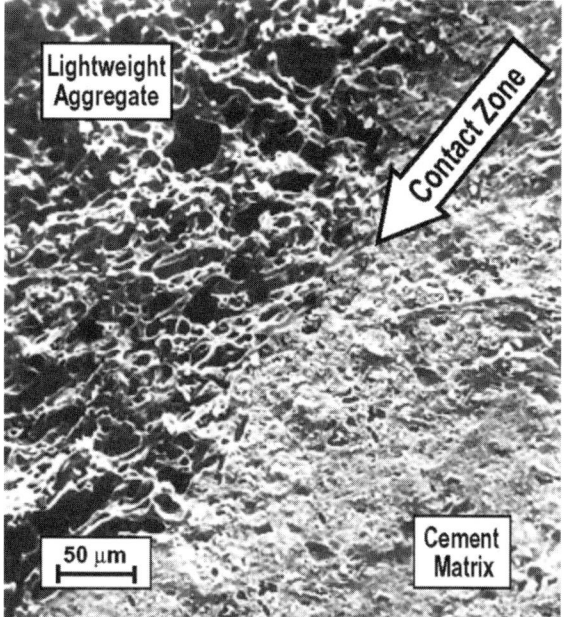

Fig. 4.13—Micrograph of contact zone.

Table 4.1—Microhardness in and beyond the contact zone *c/z* of concrete with differing *w/cm* and various coarse aggregates (Khokrin 1973)

	w/cm					
	0.3		0.4		0.5	
Coarse aggregate type	In c/z	Beyond c/z	In c/z	Beyond c/z	In c/z	Beyond c/z
Lightweight aggregate B	160	92	143	78	136	76
Lightweight aggregate O	167	94	138	73	125	68
Crushed diabase	81	79	—	—	—	—
Crushed limestone	81	81	—	—	—	—

that cause separation, have not received the same attention as has the air-void system necessary to protect the paste. Aggregates are often inappropriately dismissed as being inert fillers and, as a result, they and the associated transition zone have until recently received very modest attention.

For concrete to perform satisfactorily in severe exposure conditions, it is essential that a good bond develops and is maintained between the aggregate and the enveloping continuous mortar matrix. A high incidence of interfacial cracking or aggregate debonding will have a serious effect on durability if these cracks fill with water and subsequently freeze. An equally serious consequence of microcracking is the easy path provided for the ingress of salt water into the mass of the concrete. To provide an insight into the performance of different types of concrete, a number of mature structures that have withstood severe exposure were examined. The morphology and distribution of chemical elements at the interface were studied and reported by Bremner, Holm, and deSouza (1984).

The contact zone in lightweight concrete has been demonstrated to be significantly superior to that of normalweight concrete that does not contain silica fume (Holm, Bremner, and Newman 1984; Khokrin 1973). This profound improvement in the quality, integrity, and microstructure stems from a number of characteristics unique to lightweight concrete, including, but not limited to, the following:

• The alumina- and silicate-rich surface of the fired ceramic aggregate, which is pozzolanic and combines with $Ca(OH)_2$ liberated by hydration of the portland cement;

• Reduced microcracking at the matrix lightweight aggregate interface because of the elastic similarity of the aggregate and the surrounding cementitious matrix; and

• Hygrol equilibrium between two porous materials (lightweight aggregate and a porous cementitious matrix) as opposed to the usual condition with normalweight aggregate where bleed-water lenses around coarse natural aggregates have *w/cm* significantly higher than in the bulk of the matrix. When silica fume is added, the high-quality microstructure of the contact zone of concrete containing lightweight aggregate is moderately enhanced. When used in concrete containing normalweight aggregates, however, this zone of weakness is profoundly improved.

4.14.3.1 *Contact zone of mature concrete subjected to severe exposure*—Micrographs of the contact zone of specimens were prepared for examination in a scanning electron microscope equipped with an energy-dispersive x-ray analyzer. An example is Fig. 4.13, which shows a micrograph from the waterline of a more than 60-year-old concrete ship that was reported by Holm, Bremner, and Newman (1984). This micrograph confirms that a very tight bond develops between the lightweight aggregate and the mortar matrix. Normalweight cores taken from bridges that also contained lightweight decks were also examined and revealed separation between the normalweight aggregate and the matrix, but not at the lightweight-aggregate interface.

Russian studies on the durability of lightweight concrete (Khokrin 1973) included results of scanning electron microscopy that revealed new chemical formations at the contact zone between the matrix and keramzite (rotary kiln-produced expanded clay or shale). These micrographs confirmed earlier tests in which x-ray diffraction of ground keramzite taken before and after immersion in a saturated lime solution attested to the presence of a chemical reaction.

Khokrin (1973) also reported on microhardness tests of the contact zone of lightweight concrete and normalweight concrete, which established the width of the contact zone as approximately 60 μm. These results are shown in Table 4.1.

Virtually all commercial concrete exhibits some degree of bleeding and segregation. This is primarily due to the difference in density of the various ingredients and can be minimized with the use of proper mixture proportioning. The influence of bleeding upon the tensile strength of normalweight concrete was studied by Fenwick and Sue (1982). This report described the effects of the rise of bleed water through the mixture, the entrapment of air pockets below the larger coarse aggregate particles, and the poor paste quality at the interface due to the excessive concentrations of water. Reductions in mechanical properties are inevitable as a result of the interface flaws, as they limit interaction between the two distinctly different phases.

However significant any reduction in compressive and tensile strength due to a poorly developed contact zone, the effect on permeability is even greater. Permeability leads to increased penetration of aggressive agents that accelerate corrosion of embedded reinforcement. The permeability of concrete is then the permeability of each of its two fractions. A plausible explanation could be the effect of the interface flaws linking up with microcracking in the mortar phase of the matrix.

The phenomenon of bleed water collecting and being trapped under coarse particles of lightweight aggregate is considerably diminished, if not essentially eliminated, by the absorption of a small but significant amount of water from the fresh concrete into the interior of the lightweight aggregate. This has been verified in practice by the examination of the contact zone of lightweight concrete split cylinders and by visual examination of sand-blasted vertical surfaces of North American building structures. This observation should not be surprising because with lightweight concrete, the aggregate/matrix interface is a boundary between two porous media, while with normalweight concrete, there is an abrupt transition at the porous/solid phase interface.

Fagerlund (1972, 1978) has presented reports that analyze the contact zone in mortar and concrete. These reports provide equations that describe the influence of the contact zone on strength parameters. Fagerlund supported the analyses with micrographs that clearly identified various degrees of interaction, from almost complete phase separation for normalweight aggregates to cases involving expanded aggregates in which the boundary between the two phases is not possible to discern. The fact that the high-quality contact zones in lightweight concrete have maintained their integrity throughout their service life of the structures provides reassurance of effective long-term interaction of the components of the concrete composite (Holm 1983).

4.14.3.2 Implications of contact zone on failure mechanisms—Exposed concrete must endure the superposition of a dynamic system of forces, including variable live loads, variable temperatures, moisture gradients, and dilation due to chemical changes. These factors cause a predominantly tensile-related failure. Yet, the uniaxial compressive strength is traditionally considered the preeminent single index of quality despite the fact that concrete almost never fails in compression. The simplicity and ease of compression testing has perhaps diverted the focus from a perceptive understanding and development of appropriate measurement techniques that quantify durability characteristics.

In general, weakest-link mechanisms are undetected in uniaxial compression tests due to concrete's forgiving load-sharing characteristics in compression, that is, localized yielding and closing of stress related, temperature, and volume change cracks. Weakest-link mechanisms, however, are important in tensile zones that arise from applied loads and exposure conditions. In many types of concrete, the contact zone may be the weakest link that is decisive in determining the long-term behavior of the contact zone.

4.14.3.3 Accommodation at the aggregate-matrix interface—Additionally, a full understanding has yet to be developed regarding the accommodation mechanism, by which the pores closest to the aggregate-matrix interface provide an accessible space for products of various reactions to form in spaces without causing deleterious expansion. Research has identified ettringite, alkali-silica gel, marine salts, and corrosion products in these near-surface pores. There remains the unfinished work of integrating these findings to explain how these products impact long-term performance (Holm and Bremner 2000).

4.15—Absorption

Lightweight concrete planned for exposed applications will, of necessity, be of high quality. Testing programs have revealed that high-quality lightweight concrete specimens absorbed very little water and thus maintained their low density. As mentioned previously, this was not unexpected. In a series of publications, it was reported that the permeability of lightweight concrete was extremely low and generally equal to or significantly lower than that reported for the normalweight concrete specimens. Similar results and conclusions by Russian, Japanese, and English investigators confirmed these findings. All attributed the low permeability to the profound influence of the high-integrity contact zone possessed by lightweight concrete.

In investigations of high-strength lightweight concrete for the Arctic, Hoff (1992) reported that specimens that had a period of drying followed by water immersion at atmospheric pressure did not refill all the void space caused by drying. Pressurization caused an additional density increase of approximately 2.5 lb/ft^3 (40 kg/m^3). Before the introduction of the test specimens into the seawater, all concrete lost mass during the drying phase of curing, although concrete with a compressive strength of 9000 psi (62 MPa) lost little due to its very dense matrix.

4.16—Alkali-aggregate reaction

ACI 201.1R reports no documented instance of in-service distress caused by alkali reactions with lightweight aggregate. Mielenz (1994) indicates that although the potential exists for alkali-aggregate reaction with some natural lightweight aggregates, the volume change may be accommodated without necessarily causing structural distress. The surface of fine aggregate fractions of expanded shales, clays, and slates are known to be pozzolanic and may also serve to inhibit disruptive expansion (Boyd, Bremner and Holm 2000; Holm and Bremner 2000). No evidence of alkali-lightweight-aggregate reactions were observed in tests conducted on 70-year-old marine structures and several more than 30-year-old lightweight concrete bridge decks (Holm 1994).

Though laboratory studies and field experience have indicated no deleterious expansion resulting from the reaction between cement and silica in the lightweight component of the aggregates, the natural aggregate portion of a sand-lightweight concrete mixture should be evaluated in accordance with applicable ASTM standards.

Many lightweight concrete mixtures designed for an equilibrium density in the range of 110 lb/ft^3 (1760 kg/m^3) and above are produced using either natural sand or a naturally

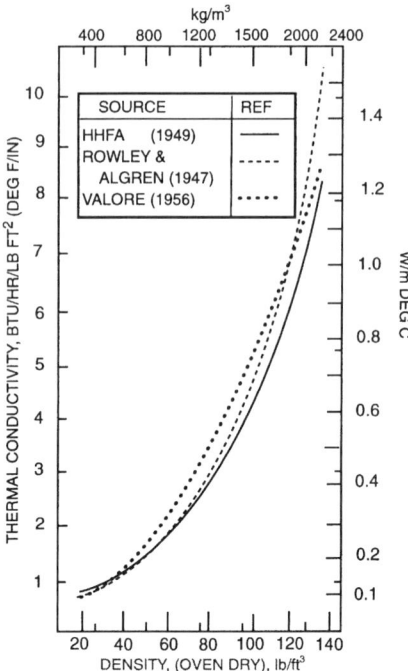

Fig. 4.14—Relation of average thermal conductivity k *values of concrete in oven-dry condition to density (Valore 1980).*

occurring coarse aggregate. In either case, these natural aggregates should be considered a potential source to develop alkali-aggregate reactions until they have been demonstrated by an appropriate ASTM test procedure or by having an established service history to be of negligible effect.

4.17—Thermal expansion

Determinations (Price and Cordon 1949) of linear thermal expansion coefficients made on lightweight concrete indicate values are 4 to 5×10^{-6} in./in./°F (7 to 11×10^{-6} mm/mm/°C), depending on the amount of natural sand used.

4.18—Heat flow properties

4.18.1 *Thermal conductivity*—The value of thermal conductivity k is a specific property of a material rather than of a construction and is a measure of the rate at which heat (energy) passes perpendicularly through a unit area of homogeneous material of unit thickness for a temperature gradient of 1 degree:

U.S. units, k = Btu/h·ft^2 (°F/in.)

(SI units, k = W/m · °C)

Thermal resistivity is the resistance per unit of thickness and is equal to $1/k$.

Thermal conductivity has been determined for concrete ranging in oven-dry density from less than 20 lb/ft^3 to over 200 lb/ft^3 (320 to 3200 kg/m^3). Conductivity values are generally obtained from guarded hot-plate specimens (ASTM C 177) tested in an oven-dry condition.

When conductivity values for concrete having a wide range of densities are plotted against oven-dry density, best-fit curves show a general dependence of k on density, as shown in Fig. 4.14. Also shown is the fact that different investigators have found different relationships. These differences are accounted for by differences in aggregate mineralogical type and microstructure, as well as in grading. Differences due to cement content, as well as matrix density and pore structure also result. Some differences in test methods and specimen sizes also existed.

Valore (1980) plotted over 400 published test results of density w against the logarithm of conductivity k and suggested the following equation

$$k = 0.5e^{0.02w}(k = 0.072e^{0.00125w}) \qquad (4\text{-}3)$$

An accurate k value for a given concrete, based on testing by the method of ASTM C 177, is preferable to a calculated value. For usual construction, however, the formula provides a good base for estimating k for concrete in the oven-dry condition and, in addition, may easily be revised for air-dry conditions.

4.18.2 *Effect of moisture on thermal conductivity of concrete*—Increasing the free moisture content of hardened concrete causes an increase in thermal conductivity. Valore (1980) stated that k increases by 6% for each 1% increment in free or evaporable moisture by weight in relation to oven-dry density. The corrected conductivity may be calculated as follows

$$k(\text{corrected}) = k(\text{oven-dry}) \times \left(1 + 6\frac{(w_m - w_o)}{w_o}\right) \qquad (4\text{-}4)$$

where w_m and w_o are densities in moist and oven-dry conditions, respectively.

4.18.3 *Equilibrium moisture content of concrete*—Concrete in a wall is not in an oven-dry condition; it is in equilibrium with the relative humidity of the environment. Because k values shown are for oven-dry concrete, it is necessary to know the moisture content for concrete in equilibrium with its normal environment in service and then apply a moisture correction factor for estimating k under anticipated service conditions. The relative humidity within masonry units in a wall will vary with type of occupancy, geographical location, exposure, and with the seasons, and it is normally assumed to be 50%. Also, it is normally assumed that exterior surfaces of single-wythe walls are protected by a breathing-type paint, stucco, or surface-bonding fibered-cement plaster. For single-wythe walls, such protection is necessary to minimize rain penetration. For cavity walls, the average moisture content of both wythes, even with the exterior wythe unpainted, will be approximately equal to that of the protected single-wythe wall.

Data from various sources for normalweight and lightweight concrete and for low-density insulation concrete have been summarized by Valore (1956, 1980). Average long-term moisture contents for concrete are in good agreement with data given herein for concrete masonry units.

Under certain conditions, condensation within a wall can cause high moisture content, and this should be considered in selecting an appropriate conductivity value.

4.18.4 *Recommended moisture factor correction for thermal conductivity*—A 6 to 9% increase in k per 1% increase of moisture content, by weight, are recommended for lightweight-aggregate concrete (of all types) and normal-weight concrete, respectively. These factors are for use where exposure conditions or other factors produce moisture contents known to depart appreciably from recommended standard moisture contents of 2% for ordinary concrete and 4% (by volume) for lightweight concrete.

A simple constant factor can be used for masonry unit and concrete under conditions of normal protected exposure. The k values, when corrected for equilibrium moisture in normal protected exposure, are to be increased by 20% over standard values for oven-dry concrete. This results in the Valore equation of Fig. 4.15, which now becomes in that figure

$$k = 0.6e^{0.02w}, \text{Btu/h} \cdot \text{ft}^2 \, {}^\circ\text{F} \; (k = 0.00125w, \text{W/m} \cdot {}^\circ\text{C}) \quad (4\text{-}5)$$

where $e = 2.71828$ and w is the oven-dry density of concrete, in kg/m^3 and lb/ft^3, respectively.

4.18.5 *Cement paste as insulating material*—The oven-dry density of mature portland cement paste ranges from 100 lb/ft^3 (1600 kg/m^3) for a w/cm of 0.4, to 67 lb/ft^3 (1075 kg/m^3) for a w/cm of 0.8. This range for w/cm encompasses structural concrete. Other data on moist-cured neat cement cellular concrete (aerated cement paste) permit the development of k-density relationships for oven-dried, air-dried, and moist pastes (Valore 1980). The latter work shows that neat cement cellular concrete and autoclaved cellular concrete follow a common k-density curve.

4.18.6 *Thermal transmittance*—U-value is thermal transmittance; it is a measure of the rate of heat flow through a building construction, under certain specified conditions. It is expressed in the following units: $U = \text{Btu/h} \cdot \text{ft}^2 \cdot {}^\circ\text{F}$ ($U = \text{W/m}^2 \cdot {}^\circ\text{C}$).

The U-value of a wall or roof consisting of homogeneous slabs of material of uniform thickness is calculated as the reciprocal of the sum of the thermal resistance of individual components of the construction

$$U = \frac{1}{R_1 + R_2 + R_3 + \ldots R_n} \quad (4\text{-}6)$$

where R_1, R_2, etc., are resistances of the individual components and also include standard constant R values for air spaces and interior and exterior surface resistances. R is expressed in the following units: $R = \text{h} \cdot \text{ft}^2 \cdot {}^\circ\text{F/Btu}$ ($R = \text{m}^2 \cdot \text{K/W}$).

Thermal resistances of individual solid layers of a wall are obtained by dividing the thickness of each layer by the thermal conductivity k for the particular material of which the layer consists.

4.19—Fire endurance

Lightweight concrete is more fire resistant than ordinary normalweight concrete because of its lower thermal conductivity, lower coefficient of thermal expansion, and the inherent fire stability of an aggregate already heated to

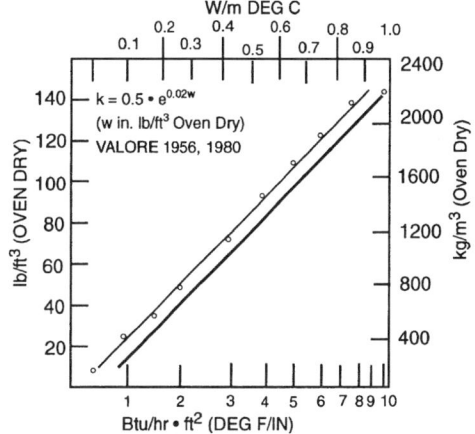

Fig. 4.15—Relation of average k *values of concrete to dry density (Valore 1980).*

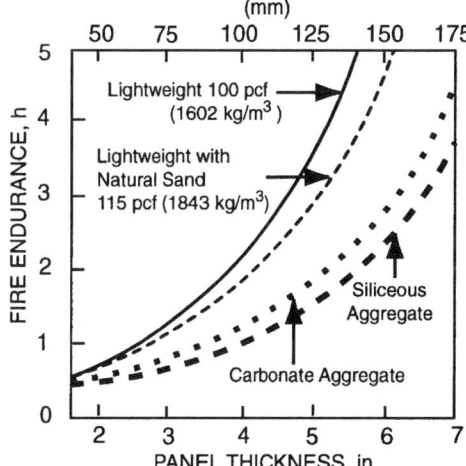

Fig. 4.16—File endurance (heat transmission) of concrete slabs as a function of thickness for naturally dried specimens (Abrams and Gustaferro 1968).

over 2000 °F (1100 °C) (Abrams and Gustaferro 1968; Abrams 1971; Carlson 1962; Selvaggio and Carlson 1964; ESCSI 1980; Prestressed Concrete Institute 1992).

4.19.1 *Heat transmission*—Research on fire endurance comparing lightweight-aggregate concrete with normal-weight concrete is shown in Fig. 4.16.

4.19.2 *Cover requirements*—The thickness of concrete between reinforcing steel (or structural steel) and the nearest fire-exposed surface is called cover. For fire ratings, the reinforcing steel cover requirements for lightweight concrete may be slightly lower than those for normalweight concrete.

4.19.3 *Fire resistance of high-strength lightweight concrete*—While there is more than 50 years of experience and a multitude of fire tests conducted on lightweight concrete of strength levels appropriate for commercial construction—3000 to 5000 psi (21 to 35 MPa)—the availability of data on high-strength lightweight concrete has, until recently, been very limited.

Tests by Bilodeau et al. (1995) and Bilodeau, Malhotra, and Hoff (1998) have reported that, because of the extremely low permeability generally associated with high-strength concrete, there is significantly reduced resistance to damage due to spalling. Because of the higher moisture contents of concrete containing lightweight aggregate with high, as-batched absorbed water contents, there is increased risk of spalling. Because of the use of high-strength lightweight concrete on several offshore platforms where intense hydrocarbon fires could develop, there was an obvious need for finding a remedy for this serious potential problem.

Several reports have documented the beneficial influence of adding small quantities of polypropylene fibers to high-strength concrete as demonstrated by exposure to fire testing that was more intense than the exposure conditions (time-temperature criteria) specified by ASTM E 119. Apparently, the fibers melt and provide conduits for release of the pressure developed by the conversion of moisture to steam. Jensen et al. (1995) reported the results of tests conducted at the Norwegian Fire Research Laboratories. These studies included the determination of mechanical properties at high temperature, the improvement of spalling resistance through material design, and the verification of fire resistance and residual strength of structural elements exposed to fire. The addition of 0.1 to 0.2% polypropylene fibers in the lightweight concrete mixture provided a significant reduction of spalling. Fire tests on beams confirmed previous findings that greater spalling (exposed reinforcement) occurred on reinforced and prestressed lightweight concrete beams than occurred on normalweight concrete beams. Reduced or no spalling, however, occurred on lightweight concrete beams with polypropylene fibers. Also, no spalling was observed on lightweight concrete beams with passive fire protection (a special cement-based mortar with expanded polystyrene balls that did not contain fibers).

4.20—Abrasion resistance

Abrasion resistance of concrete depends on strength, hardness, and toughness characteristics of the cement paste and the aggregates, and the bond between these two phases. Most lightweight aggregates suitable for structural concrete are composed of solidified vitreous material comparable to quartz on the Mohs scale of hardness. Due to its pore system, however, the net resistance to wearing forces may be less than that of a solid particle of most natural aggregates. Lightweight concrete bridge decks that have been subjected to more than 100 million vehicle crossings, including truck traffic, show wearing performance similar to that of normalweight concrete. Limitations are necessary in certain commercial applications where steel-wheeled industrial vehicles are used, but such surfaces generally receive specially prepared surface treatments. Hoff (1992) reported that specially developed testing procedures that measured ice abrasion of concrete exposed to arctic conditions demonstrated essentially similar performance for lightweight and normalweight concrete.

CHAPTER 5—DESIGN OF STRUCTURAL LIGHTWEIGHT-AGGREGATE CONCRETE

5.1—Scope

The availability and proven performance of lightweight aggregates has led to the improved functionality and economical design of buildings, bridges, and marine structures for more than 80 years. During much of this period, designs were based on the usual properties of concrete, properly adjusted by the engineers but without adequate guidance of recommended practices specifically pertaining to lightweight concrete. With the adoption of the 1963 ACI Building Code, lightweight-aggregate concrete received full recognition as an acceptable structural medium. General guidelines for the engineer and for the construction industry were included.

This chapter assists in the interpretation of the ACI 318 requirements for lightweight concrete. It also condenses many practical design aspects pertaining to lightweight concrete and provides the engineer with additional information for design.

A engineer should obtain information on the properties of concrete made with specific lightweight aggregate (or aggregates) available for a given project. These aggregates should fall within the frame of reference presented in this guide, and the specifications should be prepared so that only suitable lightweight aggregates will be used.

5.2—General considerations

Lightweight concrete has been shown by test and performance (refer to Chapter 4) to behave structurally in much the same manner as normalweight concrete, but at the same time, to provide some improved concrete properties—notably reduced weight, better insulation, and improved microstructure. For certain properties of concrete, the differences in performance are those of degree. Generally those properties that are a function of tensile strength (shear, development length, and modulus of elasticity) are sufficiently different from those of normalweight concrete to require design modification.

5.3—Modulus of elasticity

It has been shown that the modulus of elasticity of concrete is a function of density and compressive strength. The formula $E_c = w_c^{1.5} 33 \sqrt{f_c'}$ ($E_c = w_c^{1.5} 0.043 \sqrt{f_c'}$) presented in ACI 318, defines this relationship. Variations of the ACI formula for E_c at the high strength used in prestressed concrete are covered later in this section. Depending on how critically the values E_c will affect the nature of the design, the engineer should decide whether the values determined by the formula are sufficiently accurate or whether to determine E_c values from tests on the specified concrete.

Essentially, a lower E_c value for lightweight concrete means that it is more flexible because stiffness is defined as the product of modulus of elasticity and moment of inertia, EI. Reduced stiffness can be beneficial at times, and the use of lightweight concrete should be considered in these cases instead of normalweight concrete. In cases requiring improved impact or dynamic response, where differential foundation settlement may occur, and in certain types or configurations of shell roofs, the property of reduced stiffness may be desirable.

5.4—Tensile strength

Shear, torsion, anchorage, bond strength, development length, and crack resistance are related to tensile strength that is, in turn, dependent on the tensile strength of the coarse aggregate and mortar phases and the degree to which the two phases are securely bonded. Traditionally, tensile strength has been defined as a function of compressive strength, but this is known to be only a first approximation that does not reflect aggregate particle strength, surface characteristics, or the concrete's moisture content and distribution. The tensile-splitting strength, as determined by ASTM C 496, is used throughout North America as a simple, practical design criteria that is known to be a more reliable indicator of tensile-related properties than beam flexural tests. A minimum tensile-splitting strength of 290 psi (2.0 MPa) is a requirement for structural-grade lightweight aggregates conforming to the requirements of ASTM C 330.

Tests have shown that diagonal tensile strengths of beams and slabs correlate closely with the concrete splitting strengths (Hanson 1958, 1961). As tensile splitting results vary for different combinations of materials, the specifier should consult with the aggregate suppliers for laboratory-developed splitting strength data. Special tensile strength test data should be developed before the start of projects where development of early-age tensile-related handling forces occur such as precast or tilt-up members.

Tensile strength tests on lightweight concrete specimens that undergo some drying correlate well with the behavior of concrete in actual structures. Moisture loss progressing slowly into the interior of concrete members will result in the development of outer envelope tensile stresses that balance the compressive stresses in the still-moist interior zones. ASTM C 496 requires a 7-day moist and 21-day laboratory air drying at 73.4 °F (23 °C) at 50% relative humidity before conducting splitting tests. Lightweight concrete splitting tensile strengths vary from approximately 75 to 100% of normalweight concrete of equal compressive strength. Replacing lightweight fine aggregate with normalweight fine aggregate will normally increase tensile strength. Further, natural drying will increase tensile-splitting strengths.

5.5—Shear and diagonal tension

From a shear and diagonal tension perspective, lightweight concrete members behave in fundamentally the same manner as normalweight concrete members. In both cases, the shear and diagonal tension *contribution* of the concrete member is determined primarily on the tensile capacity of an unreinforced web. Because most concrete in construction is subjected to air-drying, lightweight concrete will generally have lower tensile strength than normalweight concrete of equal compressive strength. ACI 318 provides two alternate approaches by which the permissible shear capacity in a lightweight concrete member may be determined. The permissible shear capacity may be determined by using the splitting-tensile strength f_{ct} for the specific aggregate to be used or by using a fixed percentage of a similar-strength normalweight concrete.

Using the first approach to calculate the permissible shear, the value of $f_{ct}/6.7$ is substituted for $\sqrt{f_c'}$ in the provisions of ACI 318.

Most lightweight aggregate producers have sufficient data available to estimate *realistically* the range of values that *can* be achieved. A realistic value of f_{ct} for design purposes should be established for each desired compressive strength and composition of concrete. The f_{ct} values on which the structural design is based should be incorporated in the concrete specifications for the job. Splitting cylinder strength tests, if required, should be performed on laboratory mixtures similar to those proposed for the project. These tests should be performed in accordance with ASTM C 496. Splitting cylinder strength is a laboratory aggregate evaluation and is not to be conducted on field concrete.

A second, generally conservative approach in calculating the permissible shear may be used when the engineer is unable or is hesitant to specify f_{ct} values. Reduction factors are available that may be used to determine the shear capacity of lightweight concrete as a fixed percentage of normalweight concrete shear. Research on the splitting-tensile strength of lightweight concrete shows an improvement in tensile strength when natural sand is used in place of the lightweight fine aggregate (Pfeifer 1967). Two reduction factors have therefore been established: 75% of normalweight values for concrete containing both fine and coarse lightweight aggregates; and 85% of normalweight values for combinations of natural sand fine aggregates and lightweight coarse aggregates.

Most of the research addressing tensile strength, shear strength, and development lengths of structural lightweight concrete that formed the basis for existing ACI 318 Building Code requirements were limited to concrete with a compressive strength of less than 6000 psi (41 MPa). When concrete strengths of greater than 6000 psi (41 MPa) are *specified*, the determination of the appropriate tension, shear, and development length parameters should be based on a comprehensive testing program that is conducted on the materials selected for the project. For some lightweight aggregates, the tensile strength ceiling may be reached earlier than the compressive strength ceiling.

A comprehensive investigation into the shear strength of higher-strength (41 to 69 MPa [6 to 10 ksi]) reinforced and prestressed lightweight concrete beams has been reported by Ramirez et al. (1999). Measurements during the beam tests and observations of the structural behavior enabled the evaluation of the ACI 318-95, AASHTO Standard (1995) and AASHTO LRFD (1994) shear design methods for the types of beams tested.

Ramirez et al. (1999) reported that for the reinforced concrete specimens:
- Despite the fact that the sand-lightweight concrete beams had higher measured shear capacities than those calculated using code/specification methods considered in their report, the lightweight concrete beams were, on average, 82% of the measured shear capacity of the companion normalweight beams. The 0.85 reduction factor used by the current specifications does not

adequately account for the reduction of shear strength in sand-lightweight concrete beams. The trend observed is important especially for the case of beams with low to minimum amounts of shear reinforcement where the concrete contribution is a larger fraction of the total shear strength;
- While all reinforced (nonprestressed) concrete beams had measured shear capacities that exceeded both the ACI 318-95/AASHTO (simple method) and the AASHTO LRFD (general method), the degree of conservatism was greater for the normalweight concrete then the lightweight concrete beams;
- The degree of conservatism in the calculated capacities decreases for the lightweight concrete beams tested; and
- For the beams tested, the ACI 318-95/AASHTO (simple) method produced estimates of shear strength 6% more conservative than did the AASHTO LRFD (general method).

For the high-strength prestressed and lightweight concrete concrete beams tested, Ramierez et al. found the following:
- The measured shear capacities of the beams using a normal 41 and 69 MPa (6 and 10 ksi) concrete were nearly equal. Therefore, the minimum amount of transverse reinforcement presented by the AASHTO LRFD did not provide the same level of conservatism for the higher strength beams;
- For the high strength prestressed lightweight concrete beams tested both the AASHTO LRFD (general method) and the ACI 318-95 / AASHTO (simple method) provide conservative estimates of the shear strength; and
- For the high-strength prestressed lightweight concrete beams tested, the degree of conservatism afforded by the AASHTO (simple) method were nearly equal.

Based on the results of this comprehensive testing program, Ramirez et al. (1999) recommended more research in the area of high-strength prestressed lightweight concrete beams, especially with regard to the minimum requirements of transverse reinforcement needed.

Because a reduction in self-weight leads to a substantial reduction in total load on lightweight concrete members, shear capacity reduced to as much as 75% of normalweight concrete may not necessarily lead to a decrease in relative structural efficiency.

5.6—Development length

5.6.1 *Deformed reinforcement*—Because of the lower particle strength, lightweight concrete has lower bond-splitting capacities and a lower postelastic strain capacity than normalweight concrete. North American design practice (ACI 318) requires longer embedment lengths for reinforcement bars in lightweight concrete than for normalweight concrete. Unless tensile-splitting strengths are specified, ACI 318 requires the development lengths for lightweight concrete to be increased by a factor of 1.3 over the lengths required for normalweight concrete.

5.6.2 *Prestressed concrete*—Meyer and Kahn (2000) in their paper on development length in high-strength lightweight concrete reported the following:

- An evaluation of code provisions using the results of 12 tests on high-strength prestressed lightweight concrete girders showed the transfer and development length requirements of the current AASHTO and ACI equations to be conservative; and
- Test results showed that shear cracking in the transfer length region across the bottom strands did not induce strand slip if stirrup density was doubled over the current AASHTO specified density in that region.

Thatcher et al. (2002) reported that while the ACI and AASHTO codes provide a conservative estimate of the transfer length of normalweight concrete, their test results showed that transfer length of lightweight concrete was underestimated. Kolozs suggested that the modulus of elasticity was a consistent factor in determining the transfer length for both normal and lightweight concretes and that most models do not accurately predict the behavior of lightweight concrete. On the other hand, Thatcher et al.'s tests indicate that the ACI and AASHTO codes provide a conservative estimate of the development length for both normalweight and lightweight concretes tested in his study.

Nassar's (2002) conclusions differ. Based on the results of tests on large-span high-performance prestressed lightweight concrete beams, he reported:

- That until additional data emerges for transfer length in high-strength lightweight concrete beams, code guidance be raised to $60d_b$ per AASHTO LRFD stipulation and/or $f_{si}d_b/3$ to maintain a more conservative representation; and
- The development length results from his tests were inconclusive, and the ACI and AASHTO code requirements may be marginally acceptable for high-strength prestressed lightweight concrete. Until additional testing is conducted, it is recommended that the equation for the development length be modified by a factor of 1/0.85, resulting in an 18% increase in code requirements.

With closely spaced and larger-diameter prestressing strands that can cause high splitting forces, this increase may no longer be conservative. A conservative design approach or a preproject testing program may be advisable for special structures, larger-diameter strands, short-span decks, or combinations of highly reinforced thin members using high-strength, lightweight concrete. Additional research on development-length requirements and the need for greater amounts of confining reinforcement for prestressing strands in high-strength lightweight concrete and specified-density concrete is clearly warranted.

5.7—Deflection

5.7.1 *Initial deflection*—ACI 318 specifically includes modifications of formulas and minimum thickness requirements that reflect the lower modulus of elasticity, lower tensile strength, and lower modulus of rupture of lightweight concrete.

ACI 318 also lists the minimum thickness of beams for one-way slabs unless deflections are computed and requires a minimum increase of 9% in thickness for lightweight members over normalweight. Thus, using the values in this table, lightweight structural members with increased thickness are not expected to deflect more than normalweight members under the same superimposed load.

5.7.2 *Long-term deflection*—Analytical studies of long-term deflections can be made, taking into account the effects that occur from creep and shrinkage. Final deflection can then be compared with the initial deflection due to elastic strains only. Comparative shrinkage values for concrete vary appreciably with variations in component materials. In typical cases, the shrinkage of lightweight concrete may be somewhat greater than normalweight concrete of the same strength. An analysis of deflection due to elastic strain, creep, and shrinkage leads to the same factor given in ACI 318, and this factor for obtaining long-term deflections should be used for both types of concrete. More refined approaches to estimating deflections are usually not warranted.

5.8—Columns

The design of columns using lightweight concrete is essentially the same as for normalweight concrete. The reduced modulus should be used in the code sections in which slenderness effects are considered.

Extensive tests (Pfeifer 1968; Washa and Fluck 1952) comparing the time-dependent behavior of lightweight and normalweight columns developed the following facts:

- Instantaneous shortening caused by initial loading can be accurately predicted by elastic theory. Such shortening of a lightweight concrete column will be greater than that of a comparable normalweight column due to the lower modulus of elasticity of lightweight concrete;
- Time-dependent shortening of lightweight and normalweight concrete may differ when small unreinforced specimens are compared. These differences, however, are minimized when large reinforced concrete columns are tested as both increasing size and amount of longitudinal reinforcements reduces time-dependent shortening. Measured time-dependent shortening was compared with those predicted by theory, and satisfactory correlations were found; and
- Measured ultimate strengths were compared with theory and good correlations were found. Both concrete type and previous loading had no effect on this correlation.

5.9—Prestressed lightweight concrete

5.9.1 *Applications*—Prestressed lightweight concrete has been widely used for more than 40 years in North America, in nearly every application for which prestressed normalweight concrete has been used. The most beneficial applications are those in which the unique properties of prestressed lightweight concrete are fully utilized.

Prestressed lightweight concrete has been used extensively in roofs, walls, and floors of buildings. Prestressed lightweight concrete has found extensive use in flat plate and beam types of construction. For these uses, the reduced dead weight with its lower structural, seismic, and foundation loads, the better thermal insulation, significantly better fire resistance, and lower transportation cost have usually been the determining factors in the selection of prestressed lightweight concrete.

5.9.2 *Properties*—When lightweight concrete is used with prestressing, it should possess two important properties: the aggregates should be of high quality, and the concrete mixture must have high strength.

The following is a summary of the properties of prestressed lightweight concrete:

Equilibrium density—The range is typically between 100 to 120 lb/ft^3 (1600 to 1920 kg/m^3). Several bridges have incorporated a specified equilibrium density of approximately 130 lb/ft^3 (2080 kg/m^3) (Holm and Ries 2000).

Compressive strength—Typically, higher-strength concrete is used with prestressing. In general, the commercial range of strength is between 5000 to 6000 psi (35 to 41 MPa) or higher.

Modulus of elasticity—An approximate formula for evaluating the modulus of elasticity of lightweight concrete in high-strength prestressed applications can be achieved by a modification of the formula listed in ACI 318. In general, the ACI formula for evaluating E_c tends to overestimate E_c values for high-strength normalweight concrete, and the disparity is even greater with high-strength lightweight concrete.

When accurate values of E_c are required, it is suggested that either a laboratory test or the following formula modified for lightweight concrete be used for a first estimate

$$E_c = w_c^{1.5} C \sqrt{f_c'}$$

where C is a coefficient depending upon the strength of the concrete and the other symbols are the same as those used in ACI 318 (Pauw 1960).

$$C = 31 \text{ when } f_c' = 5000 \text{ psi} \qquad (5\text{-}1)$$

$$(C = 0.040 \text{ when } f_c' = 35 \text{ MPa})$$

$$C = 29 \text{ when } f_c' = 6000 \text{ psi} \qquad (5\text{-}2)$$

$$(C = 0.038 \text{ when } f_c' = 41 \text{ MPa})$$

Combined loss of prestress—The Prestressed Concrete Institute's Design Handbook (1992) provides guidance for estimating the prestress loss due to elastic shortening, creep, shrinkage, and other factors. Estimates for creep strains for lightweight concrete are shown to be greater than for normalweight concrete. No distinction is made between lightweight and normalweight concrete for estimated shrinkage after both moist and accelerated curing. The handbook recommends that total loss of prestress in typical members will range from about 25,000 to 50,000 psi (170 to 340 MPa) for normalweight concrete and from about 30,000 to 55,000 psi (210 to 380 MPa) for members using lightweight coarse aggregate and natural sand.

Thermal insulation—The thermal insulation of lightweight concrete has a significant effect on prestressing applications because of the following factors:

- Greater temperature differential in service between the

side exposed to sun and the inside may cause greater camber. The top member of a stack of precast products should be covered during the initial drying stage;
- Better response to steam curing;
- Greater suitability for winter concreting; and
- Better fire resistance.

Dynamic, shock, vibration, and seismic resistance—Prestressed lightweight concrete appears at least as good as normalweight concrete and may be even better due to its lower modulus of elasticity.

Cover requirements—Where fire requirements dictate the cover requirements, the insulating effects developed by the lower density and the fire stability offered by a preheated-to-1200 °C aggregate may be used advantageously.

5.10—Thermal design considerations

In concrete elements exposed to the environment, the choice of lightweight concrete will provide several distinct advantages over normalweight concrete (Fintel and Khan 1965, 1966, 1968). These physical properties are covered in detail in Chapter 4:
- The lower thermal diffusivity provides a thermal inertia that lengthens the time for exposed members to reach any steady-state temperature;
- Due to this resistance, the effective interior temperature change will be smaller under transient temperature conditions. This time lag will moderate the solar build-up and nightly cooling effects;
- The lower coefficient of linear thermal expansion that is developed in the concrete due to the lower coefficient of thermal expansion of the lightweight aggregate itself is a fundamental design consideration in exposed members; and
- The lower modulus of expansion will develop lower stress changes in members exposed to thermal strains.

A comparative thermal investigation studying the shortening developed by the average temperature of an exposed column restrained by the interior frame demonstrated the fact that the axial shortening effects were about 30% smaller for lightweight concrete, and the stresses due to restrained bowing were about 35% less with lightweight concrete than with normalweight concrete (Fintel and Kahn 1965, 1966, 1968).

For an exact structural analysis, use data on local aggregates obtained from lightweight and normalweight aggregate suppliers.

5.11—Seismic design

Lightweight concrete is particularly adaptable to seismic design and construction because of the significant reduction in inertial forces. A large number of multistory buildings and bridge structures have effectively used lightweight concrete in areas subject to earthquakes.

The lateral or horizontal forces acting on a structure during earthquake motions are directly proportional to the inertia or mass of that structure. These lateral forces may be calculated by recognized formulas and are applied with the other load factors.

5.11.1 *Ductility*—The ductility of concrete structural frames should be analyzed as a composite system—that is, as reinforced concrete. Studies by Ahmad and Batts (1991) and Ahmad and Barker (1991) indicate, for the materials tested, that the ACI rectangular stress block is adequate for strength predictions of high-strength lightweight concrete beams, and the recommendation of 0.003 as the maximum usable concrete strain is an acceptable lower bound for reinforced lightweight concrete members with strength greater than 6000 psi. Moreno (1986) found that while lightweight concrete exhibited a rapidly descending portion of the stress-strain curve, it was possible to obtain a flat descending curve with reinforced members that were provided with a sufficient amount of confining reinforcement slightly greater than that with normalweight concrete. Additional confining steel is recommended to compensate for the lower postelastic strain behavior of lightweight concrete. This report also included results that showed that it was economically feasible to obtain the desired ductility by increasing the amount of steel confinement.

Rabbat et al. (1986) came to similar conclusions when analyzing the seismic behavior of lightweight and normalweight concrete columns. This report focused on how properly detailed reinforced concrete column-beam assemblages could provide ductility and maintain strength when subjected to inelastic deformations from moment reversals. These investigations concluded that properly detailed columns made with lightweight concrete performed as well under moment reversals as normalweight concrete columns. ACI 318 places a compressive strength limit of 5000 psi (35 MPa) for concrete members unless supported by test results for higher strengths.

5.12—Fatigue

The first recorded North American comparison of the fatigue behavior between lightweight and normalweight was reported by Gray, McLaughlin, and Antrim (1961). These investigators concluded that the fatigue properties of lightweight concrete are not significantly different from the fatigue properties of normalweight concrete.

This work was followed by Ramakrishnan, Bremner, and Malhotra (1992), who found that, under wet conditions, the fatigue endurance limit was the same for lightweight and normalweight concrete.

Because of the significance of oscillating stresses that would be developed by wave action on offshore structures, and due to the necessity for these marine structures to use lightweight concrete for buoyancy considerations, a considerable amount of research has been completed determining the fatigue resistance of high-strength lightweight concrete and comparing these results with the characteristics of normalweight concrete. Hoff (1994) reviewed much of the North American and European data and concluded that, despite the lack of a full understanding of failure mechanisms, "under fatigue loading, high-strength lightweight concrete performs as well as high-strength normalweight concrete and, in many instances, provides longer fatigue life." It is, however, the long-term service performance of real structures that provides improved confidence in material behavior

Fig. 5.1—Barge-mounted frame-placed beams. To the right is the old truss bridge. Both will carry U.S. 19 traffic. (Engineering News Record, June 4, 1964.)

Fig. 5.2—Concrete weighing less than 120 lb/ft³ permitted 120 ft spans for Florida bridge. (Engineering News Record, June 4, 1964.)

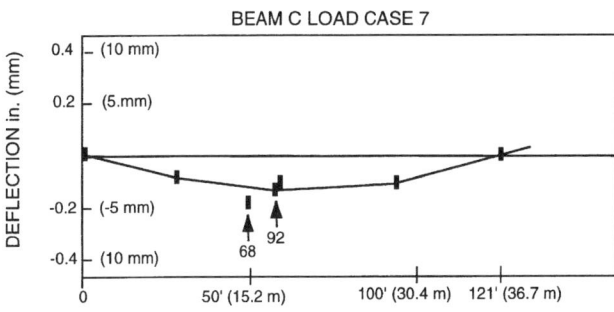

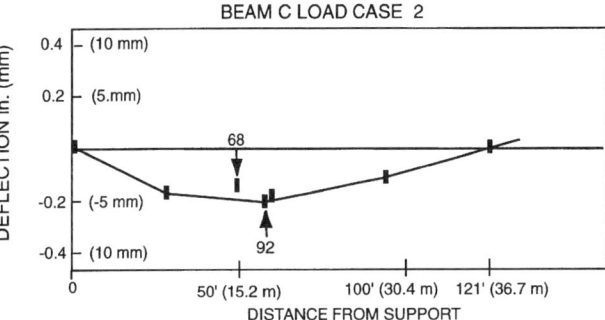

Fig. 5.3—Florida DOT predicted deflections compared with 1968 and 1992 measurements (Brown, Larsen, and Holm 1995).

rather than the extrapolation of conclusions obtained from laboratory investigations.

The long-term field performance of lightweight-concrete bridge members constructed in Florida in 1964 (Fig. 5.1 and Fig. 5.2) was evaluated in an in-depth investigation conducted in 1992. Comprehensive field measurements of service load strains and deflections taken in 1968 and 1992 were compared with the theoretical bridge responses predicted by a finite element model that is part of the Florida Department of Transportation bridge rating system (Brown and Davis 1993). The original 1968 loadings and measurements of the bridge were duplicated in 1992 and compared with calculated deflections, as shown in Fig. 5.3 (Brown, Larsen, and Holm 1995). Maximum deflection for one particular beam due to a midpoint load was 0.28 in. (7.1 mm) measured at 60.5 ft (18.4 m) from the unrestrained end of the span. This compares very well with the original deflection, which was recorded to be 0.26 in. (6.6 mm) measured at 50.5 ft (15.4 m). Rolling load deflections measured in 1968 and 1992 were also comparable, but slightly less in magnitude than the static loads.

Strain measurements across the bridge profile were also duplicated, and these compared very closely for most locations in areas of significant strain. Comparison of the 1992 and 1968 data shows bridge behavior to be essentially similar, with the profiles closely matched.

It appears that dynamic testing of the flexural characteristics of the 31-year-old long-span lightweight-concrete bridge corroborates the conclusions of fatigue investigations conducted on small specimens tested under controlled conditions in several laboratories (Hoff 1994; Gjerde 1982; Gray, McLaughlin, and Antrim 1961). In these investigations, it was generally observed that the lightweight concrete performed as well as and, in most cases, somewhat better than the normalweight control specimens. Several investigators have suggested that the improved performance was due to the elastic compatibility of the lightweight aggregate particles to that of the surrounding cementitious matrix. In lightweight concrete, the elastic modulus of the constituent phases (coarse aggregate and the enveloping mortar phase) are relatively similar, while with normalweight concrete, the elastic modulus of most normalweight concrete may be as much as three to five times greater than its enveloping matrix (Bremner and Holm 1986). With lightweight concrete, the elastic similarity of the two phases of a composite system results in a profound reduction of stress concentrations and a leveling out of the average stress over the cross section of the loaded member. Normalweight concrete having a significant elastic mismatch will inevitably develop stress concentrations that may result in extensive microcracking in the concrete composite.

Additionally, because of the pozzolanic reactivity of the surface of the vesicular aggregate that has been fired at temperatures above 2012 °F (1100 °C) (Khokrin 1973), the quality and integrity of the contact zone of lightweight concrete is considerably improved. As the onset of microcracking is most often initiated at the weak link interface between the dense aggregate and the enveloping matrix, it follows that lightweight concrete will develop a lower incidence of microcracking (Holm, Bremner, and Newman 1984).

5.13—Specifications

Lightweight concrete may be specified and proportioned on the basis of laboratory trial batches or on field experience with the materials to be used. Most lightweight aggregate suppliers have mixture composition information available for their material, and many producers provide field control and technical service to ensure that the specified concrete quality will be used.

The average strength requirements for lightweight concrete do not differ from those for normalweight concrete for the same degree of field control.

Splitting-tensile strength tests should not be used as a basis for field acceptance of lightweight concrete.

The analysis of the load-carrying capacity of a lightweight concrete structure, either by cores or load tests, should be the same as for normalweight concrete.

Equilibrium density should be calculated in accordance with ASTM C 567.

Maximum fresh density should be determined by the designer, ready-mix supplier, and the lightweight aggregate supplier before starting the project.

CHAPTER 6—HIGH-PERFORMANCE LIGHTWEIGHT CONCRETE
6.1—Scope and historical development

While it is clearly understood that high strength and high performance are not synonymous, one may consider the first modern use of high-performance concrete to be when the American Emergency Fleet Corporation built lightweight concrete ships with specified compressive strengths of 5000 psi (35 MPa) during 1917 to 1920. Commercial normalweight concrete strengths of that time were approximately 2500 psi (17 MPa).

Lightweight concrete has achieved high strength levels by incorporating various pozzolans (fly ash, silica fume, metakaolin, calcined clays, and shales) combined with mid-range or high-range water-reducing admixtures, or both. Because of durability concerns, the *w/cm* has, in many cases, (that is, bridges, marine structures) been specified to be less than 0.45, and for severe environments, a significantly lower *w/cm* has been specified. Limiting water content and designing to an air content of 4 to 5% may result in an equilibrium density higher than 120 lb/ft^3 (1920 kg/m^3).

While structural-grade lightweight aggregates are capable of producing concrete with compressive strengths in excess of 5000 psi (35 MPa), several lightweight aggregates have been used in concrete that developed compressive strengths from 7000 to 10,000 psi (48 to more than 69 MPa). In general, an increase in density will be necessary when developing higher compressive strengths. High-strength lightweight concrete with compressive strengths of 6000 psi (41 MPa) are widely available commercially and testing programs on lightweight concrete with a compressive strength approaching 10,000 psi (69 MPa) are ongoing.

6.2—Structural efficiency of lightweight concrete

The entire hull structure of the USS Selma and 18 other concrete ships were constructed with 5000 psi, high-performance lightweight concrete in the ship building

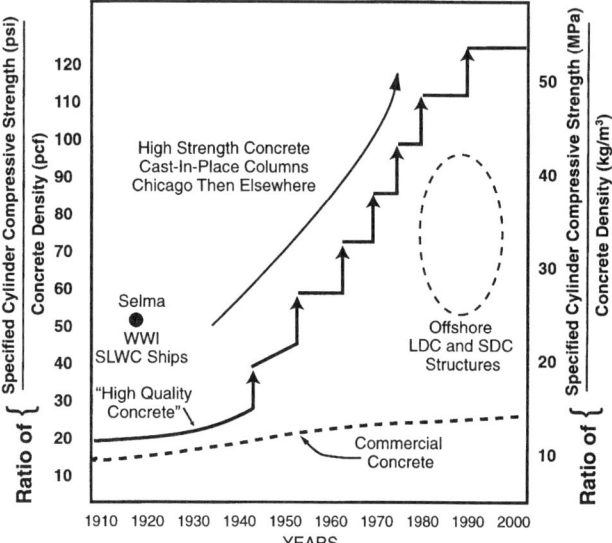

Fig. 6.1—The structural efficiency of concrete. The ratio of specified compressive strength to density (psi/[lb/ft^3]) (Holm and Bremner 1994).

program in Mobile, Alabama starting in 1917. The structural efficiency as defined by the strength/density (*S/D*) ratio of the concrete used in the USS Selma was extraordinary for that time. Improvements in structural efficiency of concrete since that time are shown schematically in Fig. 6.1—an upward trend in the 1950s with the introduction of prestressed concrete, followed by production of high-strength normalweight concrete for columns of very tall cast-in-place concrete-frame commercial buildings. Most increases came as a result of improvements in the cementitious matrix brought about by new generations of admixtures such as high-range water-reducers, and the incorporation of high-quality pozzolans such as silica fume, metakaolin, and fly ash. History suggests, however, that the first major breakthrough came as a result of the lightweight concrete ship-building program in 1917.

6.3—Applications of high-performance lightweight concrete

6.3.1 *Precast structures*—High-strength lightweight concrete with a compressive strength in excess of 5000 psi (35 MPa) has been successfully used for almost four decades by North American precast and prestressed concrete producers. Presently, there are ongoing investigations into longer-span lightweight precast concrete members that may be feasible from a trucking/lifting/logistical point of view.

The 1994 Wabash River Bridge is a good example where a 17% density reduction was realized. The 96 lightweight prestressed post-tensioned bulb-tee girders were 175 ft (53.4 m) long, 7.5 ft (2.3 m) deep, and weighed 96 tons (87.3 metric tons) each. The 5-day strengths exceeded 7000 psi (48 MPa). High-performance concrete was used because it saved the owner $1.7 million, or 18% of the total project cost. (ESCSI 2001).

Parking structure members with 50 to 60 ft (15 to18 m) spans are often constructed with double tees with an equilibrium density of approximately 115 lb/ft^3 (1850 kg/m^3). This mass

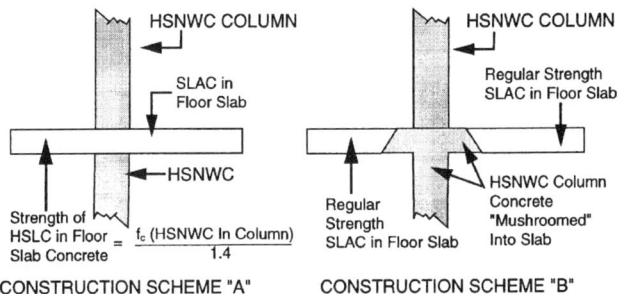

Fig. 6.2—Alternative construction schemes for transfer of high-strength normalweight concrete column loads through floor slabs (Holm and Bremner 1994).

reduction is primarily for lifting efficiencies and lowering transportation costs.

Precast lightweight concrete has frequently been used in long-span roof framing as was the case in the 120 ft (37 m) long single tees used in 1974 in the University of Nebraska sports center.

6.3.2 *Buildings*—Among the thousands of buildings built in North America incorporating high-strength lightweight concrete, the following examples have been selected for their pioneering and unique characteristics.

6.3.2.1 *Federal Post Office and Office Building*—The 450 ft (140 m) multipurpose building constructed in 1967 with five post office floors and 27 office tower floors was the first major New York City building application of post-tensioned floor slabs. Concrete tensioning strengths of 3500 psi (24 MPa) were routinely achieved for 3 days for the 30 x 30 ft (9 x 9 m) floor slabs with a design target strength of 6000 psi (41 MPa) at 28 days. Approximately 30,000 yd^3 (23,000 m^3) of lightweight concrete were incorporated into the floors and the cast-in-place architectural envelope, which serves a structural as well as an aesthetic function. (Holm and Bremner 1994).

6.3.2.2 *The North Pier Apartment Tower, Chicago, 1991*—This project used high-performance lightweight concrete in the floor slabs as an innovative structural solution to avoid construction problems associated with the load transfer from high-strength normalweight concrete columns through the floor slab system. ACI 318 requires a maximum ratio of column compressive strength, which in this project was 9000 psi (62 MPa) and the intervening floor slab concrete to be less than 1.4. By using high-strength lightweight concrete in the slabs with a strength greater than 6430 psi (44 MPa), the floor slabs could be placed using routine placement techniques, thus avoiding scheduling problems associated with the mushroom technique (Fig. 6.2). In the mushroom technique, the high-strength column concrete is overflowed from the column and intermingled with the floor slab concrete. The simple technique of using high-strength floor slab concrete in the North Pier project avoided delicate timing considerations that were necessary to avoid cold joints (Holm and Bremner 1994).

6.3.2.3 *The Bank of America, Charlotte, N.C.*—This concrete structure is the tallest in the southeastern United States with a high-strength concrete floor system consisting

Fig. 6.3—Bank of America, Charlotte, N.C. (from Holm and Bremner 1994, with permission of Edward Arnold Publishers, London).

of 4-5/8 in. (117 mm) thick slabs supported on 18 in. (460 mm) deep post-tensioned, concrete beams centered on 10 ft (3.0 m). The lightweight concrete floor system was selected to minimize the dead weight and to achieve the required 3 h fire rating (Fig. 6.3 and Table 6.1).

6.3.3 *Bridges*—More than 500 bridges have incorporated lightweight concrete into decks, beams, girders, or piers. Transportation engineers generally specify higher concrete strengths primarily to ensure high-quality mortar fractions (high compressive strength combined with high air content) that will minimize maintenance. Several mid-Atlantic state transportation authorities have completed more than 20 bridges using a laboratory target strength of 5200 psi (36 MPa), 6 to 9% air content, and a density of 115 lb/ft^3 (1840 kg/m^3). The following are the principal advantages of using lightweight concrete in bridges and the rehabilitation of existing bridges:

- Increased width or number of traffic lanes;
- Increased load capacity;
- Balanced cantilever construction;
- Reduction in seismic inertial forces;
- Increase cover with equal weight, thicker slabs;
- Improve deck geometry with thicker slabs; and
- Longer spans save pier costs.

6.3.3.1 *Increased number of lanes during bridge rehabilitation*—Thousands of bridges in the United States are functionally obsolete with unacceptably low load capacity or an insufficient number of traffic lanes. To remedy limited

Table 6.1—Mixture proportions and physical properties for concrete pumped on Bank of America project, Charlotte, N.C., 1991

Mixture no.	1	2*	3
Mixture proportions			
Cement, Type III, lb/yd³ (kg/m³)	550 (326)	650 (385)	750 (445)
Fly ash, lb/yd³ (kg/m³)	140 (83)	140 (83)	140 (83)
LWA 20 mm to No. 5, lb/yd³ (kg/m³)	900 (534)	900 (534)	900 (534)
Sand, lb/yd³ (kg/m³)	1370 (813)	1287 (763)	1203 (714)
Water, gal./yd³ (L/m³)	296 (175)	304 (180)	310 (184)
WRA, fl oz./yd³ (L/m³)	27.6 (0.78)	31.6 (0.90)	35.6 (1.01)
HRWRA, fl oz./yd³ (L/m³)	53.2 (1.56)	81.4 (2.31)	80.1 (2.27)
Fresh concrete properties			
Initial slump, in. (mm)	2-1/2 (63)	2 (51)	2-1/4 (57)
Slump after HRWRA, in. (mm)	5-1/8 (130)	7-1/2 (191)	6-3/4 (171)
Air content	2.5	2.5	2.3
Unit weight, lb/ft³ (kg/m³)	117.8 (1887)	118.0 (1890)	118.0 (1890)
Compressive strength, psi (MPa)			
4 days	4290 (29.6)	5110 (35.2)	5710 (39.4)
7 days	4870 (33.6)	5790 (39.9)	6440 (44.4)
28 days (average)	6270 (43.2)	6810 (47.0)	7450 (51.4)
Splitting-tensile strength, psi (MPa)	520 (3.59)	540 (3.72)	565 (3.90)

*Mixture selected and used on project.

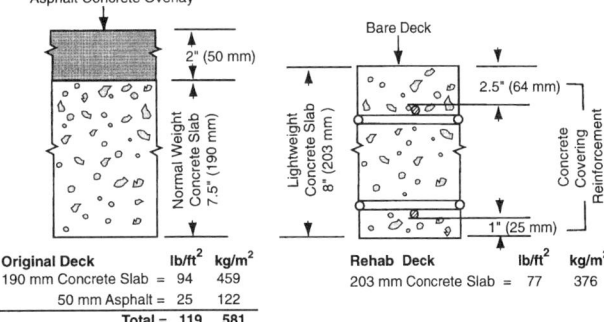

Fig. 6.4—Original and rehabilitated decks for Whitehurst Freeway (Stolldorf and Holm 1996).

lane capacity, Washington, D.C. engineers replaced a four-lane bridge originally constructed with normalweight concrete with five new lanes made with lightweight concrete providing a 50% increase in one-way, rush-hour traffic without replacing the existing structure, piers, or foundations. Similarly, on Interstate 84, crossing the Hudson River at Newburgh, N.Y., two lanes of normalweight concrete were replaced with three lanes of lightweight concrete on a parallel span, allowing three-lane traffic in both east- and west-bound lanes.

6.3.3.2 *Increased load capacity*—The elevated section of the Whitehurst Freeway was upgraded to an HS20 loading criteria during the rehabilitation of the Washington, D.C., corridor system structure with only limited modifications to the steel framing superstructure. An improved load-carrying

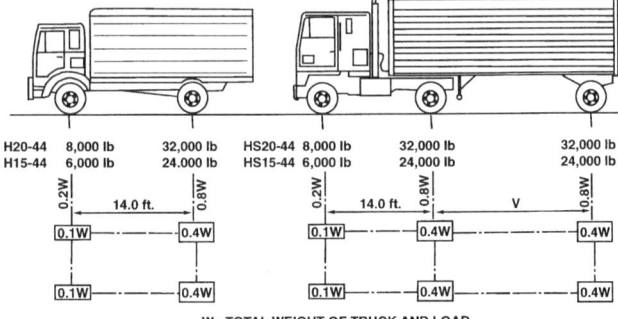

Fig. 6.5—AASHTO H20-44 and HS20-44 loadings (Stolldorf and Holm 1996).

capacity was obtained because of the significant dead load reduction brought about by using lightweight concrete to replace the normalweight concrete and asphalt overlay used in the original deck slab (Fig. 6.4).

The original elevated freeway structure was designed for HS20 live load according to the AASHTO 1941 specifications. With the significantly lighter replacement concrete deck, a minimum of the structural steel framing required strengthening, and little interruption at the street level below was required to upgrade the substructure to an HS20 live load criteria (Fig. 6.5) (Stolldorf and Holm 1996).

6.3.3.3 *Bridges incorporating both lightweight-concrete spans and normalweight concrete spans*—A number of bridges have been constructed where high-performance lightweight concrete has been used to achieve balanced load-free cantilever construction. On the Sandhornoya Bridge, completed in 1989 near the Arctic Circle city of Bodo, Norway, the 350 ft (110 m) sidespans of a three-span bridge were constructed with high-strength lightweight concrete with a cube strength of 8100 psi (55 MPa) that balanced the construction of the center span of 505 ft (154 m) that used normalweight concrete with a cube strength of 6500 psi (45 MPa) (Fergestad 1996).

The Raftsundet Bridge in Norway, also north of the Arctic Circle, with a main span of 978 ft (298 m), was the longest concrete cantilevered span in the world when the cantilevers were joined in June 1998; 722 ft (220 m) of the main span was constructed with high-strength, lightweight-aggregate concrete with a cube strength of 8700 psi (60 MPa). The side spans and piers in normalweight concrete had a cube strength of 9400 psi (65 MPa) (Fig. 6.6) (ESCSI 2001).

6.3.4 *Marine structures*—Because offshore concrete structures may be constructed in shipyards or graving docks located considerable distances from the site where the structure may be, then floated and towed to the project site, there is a special need to reduce mass and improve structural efficiency, especially where shallow-water conditions mandate lower draft structures. The structural efficiency is even more pronounced when lightweight concrete is submerged as shown as follows.

The density ratio

(heavily reinforced normalweight concrete)
(heavily reinforced lightweight concrete)

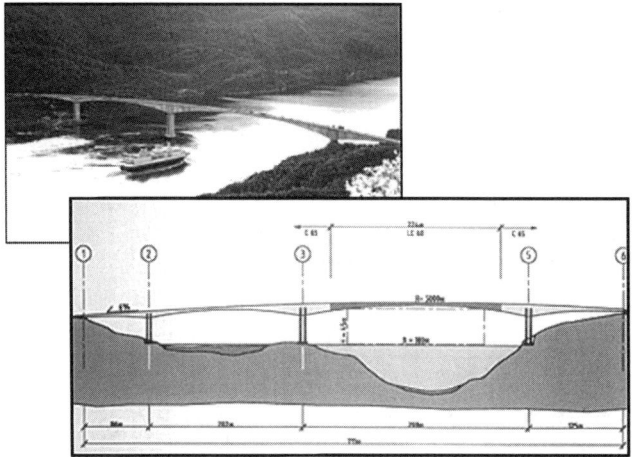

Fig. 6.6—Raftsundet Bridge (ESCSI 2000).

in air is $(2.50[156\ lb/ft^3])/[2.00(125\ lb/ft^3)] = 1.25$; when submerged is $(2.50 - 1.00)/(2.00 - 1.00) = 1.50$.

6.3.4.1 *Tarsiut Caisson Retained Island, 1981*—The first arctic structure using high-performance lightweight concrete was the Tarsiut Caisson retained island built in Vancouver, British Columbia, and barged to the Canadian Beaufort Sea (Fig. 6.7). Four large, prestressed concrete caissons 226 x 50 x 35 ft (69 x 15 x 11 m) high were constructed in a graving dock in Vancouver, towed around Alaska on a submersible barge, and founded on a berm of dredged sand 25 mi (40 km) from shore. The extremely high concentration of reinforcement resulted in a steel-reinforced concrete density of 140 lb/ft^3 (2240 kg/m^3). The use of high-strength lightweight concrete was essential to achieving the desired floating and draft requirements. (ESCSI 2001).

6.3.4.2 *Heidron floating platform, 1996*—Because of the deep water, 1130 ft (345 m), over the Heidron oil fields, an early decision was made to improve buoyancy and construct the first floating platform with high-performance lightweight concrete. The hull of the floating platform, approximately 91,000 yd^3 (70,000 m^3), was constructed entirely of high-strength lightweight concrete with a maximum density of 125 lb/ft^3 (2000 kg/m^3). Heidron was built in Norway and towed to the North Sea. A mean density of 121 lb/ft^3 (1940 kg/m^3), a mean 28-day cube compressive strength of 11460 psi (79 MPa), and a documented cylinder/cube strength ratio of 0.90 to 0.93 are reported in reference (FIB 2000) (ESCSI 2001).

6.3.4.3 *Hibernia oil platform, 1998*—The ExxonMobil Oil Hibernia offshore gravity-based structure is a significant application of specified-density concrete. To improve buoyancy of the largest floating structure built in North America, lightweight aggregate replaced approximately 50% of the normalweight coarse fraction in the high-strength concrete used (Fig. 6.8). The resulting density was 135 lb/ft^3 (2160 kg/m^3). Hibernia was built in a dry dock in Newfoundland, Canada, and then floated out to a deep water harbor area where construction continued. When finished, the more than 1-million ton structure was towed to the Hibernia North Sea oil field site and set in place on the ocean floor. A comprehensive testing program was reported by Hoff et al. (1995).

Fig. 6.7—Tarsuit Caisson Retained Island (from Concrete International *1982).*

Fig. 6.8—Hibernia Offshore Platform (ESCSI 2001).

6.3.5 *Floating bridge pontoons*—High-performance lightweight concrete was used very effectively in both the cable-stayed bridge deck and the separate but adjacent floating concrete pontoons supporting a low-level steel box-girder bridge near the city of Bergen, Norway (Fig 6.9). The pontoons are 138 ft (42 m) long and 67 ft (20.5 m) wide and were cast in compartments separated by watertight bulkheads. The design of the compartments was determined by the concept that the floating bridge would be serviceable despite the loss of two adjacent compartments due to an accident.

6.4—Reduced transportation cost

For more than 20 years, precast manufacturers have evaluated trade-offs between physical properties and transportation costs. In one study, a typically used limestone control concrete was paralleled by other mixtures in which 25, 50, 75, and 100% of the limestone coarse aggregate was

Fig. 6.9—Nordhordland Bridge, Bergen, Norway (Elkem Micro Silica 2000).

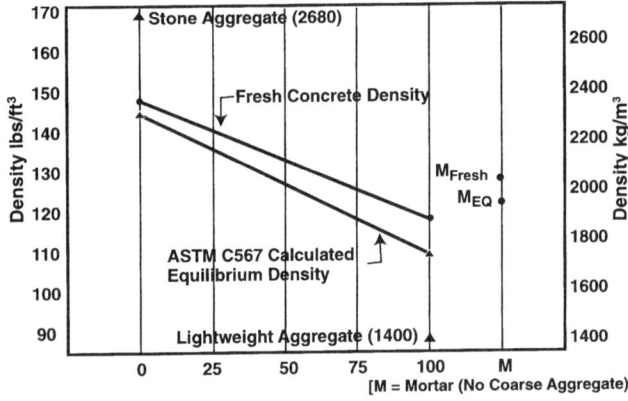

Fig. 6.10—Fresh and ASTM C 567-calculated equilibrium concrete density with varying replacements of limestone coarse aggregate with structural lightweight aggregate (Holm and Ries 2000).

replaced by an equal absolute volume of lightweight aggregate. Results of the testing program that measured compressive, tension, and modulus with density data shown in Fig. 6.10 are reported (Holm and Ries 2000).

Because of weight limits on roads, this precast producer developed lightweight mixtures that reduced the weight of members allowing an increased number of precast elements per truck. By adjusting the density of the concrete, precasters are able to minimize the number of truck deliveries without exceeding highway load limits, while lowering project cost. Opportunities for increased trucking efficiency are greater when transporting smaller concrete products, such as hollow core plank, wallboard, precast steps, and imitation stone.

6.5—Enhanced hydration due to internal curing

Expanded lightweight aggregates with high internal moisture contents may be substituted for normalweight aggregates to provide internal curing in concrete containing a high volume of cementitious materials. High cementitious concrete is vulnerable to self-desiccation and benefits significantly from the added internal moisture. This application is especially helpful for concrete containing high volumes of silica fume that are sensitive to curing procedures. In this application, density reduction is a by-product.

Time-dependent improvement in the quality of concrete containing lightweight aggregate is greater than that with normalweight aggregate. This is due to better hydration of the cementitious fraction provided by moisture available from the slowly released reservoir of water absorbed within the pores of the expanded aggregate. This process of internal curing is made possible when the moisture content of expanded aggregate, at the time of mixing, is in excess of that achieved in 1-day submersion. The fact that absorbed moisture within an expanded aggregate batched with a high degree of saturation (percent of internal pore volume occupied by water) was available for internal curing has been known for several decades and first documented in 1967 (Campbell and Tobin 1967). This comprehensive program compared strengths of cores taken from field-cured exposed slabs with test results obtained from laboratory specimens cured strictly in accordance with ASTM procedures. Their tests confirmed that availability of absorbed moisture within the expanded aggregate produced a more forgiving concrete that was less sensitive to poor field-curing conditions.

It appears that Philleo (1991) was the first to recognize the potential benefits to high-performance normalweight concrete with the addition of expanded lightweight aggregate containing high volumes of absorbed moisture. Weber and Reinhardt (1995) have also conclusively demonstrated reduced sensitivity to poor curing conditions in high-strength normalweight concrete containing an adequate volume of high moisture content expanded aggregates.

The benefits of internal curing are increasingly important when pozzolans (silica fume, fly ash, metokaolin, calcined shales, clays, and lightweight aggregate fines) are included in the mixture. It is well known that the pozzolanic reaction of finely divided alumina-silicates with calcium hydroxide liberated as cement hydrates is contingent upon the availability of moisture. Additionally, internal curing provided by absorbed water minimizes the plastic (early) shrinkage due to rapid drying of concrete exposed to unfavorable drying conditions.

While the improvements in long-term strength gain have been observed, the principal contribution of internal curing rests in the reduction of permeability that develops from a significant extension in the time of curing. Powers, Copeland, and Mann (1959) showed that extending the time of curing increased the volume of cementitious products formed, which caused the capillaries to become segmented and discontinuous. While internal curing is typically provided by an expanded coarse aggregate in high-performance concrete applications, expanded fine aggregate is more effective in distributing available moisture for internal curing. As Hoff (2003) and Bentz and Snyder (1999) have pointed out, a much more efficient spatial distribution could be accomplished by a partial replacement of the sand fraction with expanded fine aggregate.

CHAPTER 7—REFERENCES
7.1—Referenced standards and reports

The standards and reports listed below were the latest editions at the time this document was prepared. Because these documents are revised frequently, the reader is advised to contact the proper sponsoring group if it is desired to refer to the latest version.

American Concrete Institute

ACI 201.1R	Guide for Making a Condition Survey of Concrete in Service
ACI 201.2R	Guide to Durable Concrete
ACI 211.1	Standard Practice for Selecting Proportions for Normal, Heavyweight and Mass Concrete
ACI 211.2	Standard Practice for Selecting Proportions for Structural Lightweight Concrete
ACI 212.3R	Chemical Admixtures for Concrete
ACI 216.1	Standard Method for Determining Fire Resistance of Concrete and Masonry Construction Assemblies
ACI 232.2R	Use of Fly Ash in Concrete
ACI 233R	Slag Cement in Concrete and Mortar
ACI 234R	Guide for the Use of Silica Fume in Concrete
ACI 302.1R	Guide for Concrete Floor and Slab Construction
ACI 304.5R	Batching, Mixing, and Job Control of Lightweight Concrete
ACI 308.1	Standard Specification for Curing Concrete
ACI 318	Building Code Requirements for Structural Concrete and Commentary

ASTM International

ASTM C 31	Practice for Making and Curing Concrete Test Specimens in the Field
ASTM C 33	Standard Specification for Concrete Aggregates
ASTM C 78	Test Method for Flexural Strength of Concrete (Using Simple Beam with Third-Point Loading)
ASTM C 94	Specification for Ready-Mixed Concrete
ASTM C 127	Standard Test Method for Specific Gravity and Absorption of Coarse Aggregate
ASTM C 138	Standard Test Method for Density (Unit Weight), Yield, and Air Content (Gravimetric) of Concrete
ASTM C 143	Test Method for Slump of Hydraulic-Cement Concrete
ASTM C 150	Standard Specification for Portland Cement
ASTM C 172	Standard Practice for Sampling Freshly Mixed Concrete
ASTM C 173	Standard Test Method for Air Content of Freshly Mixed Concrete by the Volumetric Method
ASTM C 177	Standard Test Method for Steady-State Heat Flux Measurements and Thermal Transmission Properties by Means of the Guarded-Hot-Plate Apparatus
ASTM C 330	Standard Specification for Lightweight Aggregates for Structural Concrete
ASTM C 331	Standard Specification for Lightweight Aggregates for Concrete Masonry Units
ASTM C 332	Standard Specification for Lightweight Aggregates for Insulating Concrete
ASTM C 469	Standard Test Method for Static Modulus of Elasticity and Poisson's Ratio of Concrete in Compression
ASTM C 496	Standard Test Method for Splitting Tensile Strength of Cylindrical Concrete Specimens
ASTM C 512	Standard Test Method for Creep of Concrete in Compression
ASTM C 567	Standard Test Method for Density of Structural Lightweight Concrete
ASTM C 595	Specification for Blended Hydraulic Cements
ASTM C 618	Standard Specification for Coal Fly Ash and Raw or Calcined Natural Pozzolan for use as a Mineral Admixture in Concrete
ASTM C 666	Standard Test Method for Resistance of Concrete to Rapid Freezing and Thawing (Procedure A)
ASTM C 1157	Standard Performance Specification for Hydraulic Cement
ASTM E 119	Standard Test Method for Fire Tests for Building Construction and Materials

These publications may be obtained from these organizations:

American Concrete Institute
PO Box 9094
Farmington Hills, Mich. 48333-9094

ASTM International
100 Barr Harbor Dr.
West Conshohocken, Pa. 19428

7.2—Cited references

AASHTO, 1994, *AASHTO LRFD*, American Association of State and Highway Transportation Officials, Washington, D.C.

AASHTO, 1995, *AASHTO Standard*, American Association of State and Highway Transportation Officials, Washington, D.C.

Abrams, M. S., 1971, "Compressive Strength of Concrete at Temperatures to 1600 °F," *Temperature and Concrete*, SP-25, American Concrete Institute, Farmington Hills, Mich., pp. 33-58.

Abrams, M. S., and Gustaferro, A. H., 1968, "Fire Endurance of Concrete Slabs as Influenced by Thickness, Aggregate Type, and Moisture," *Journal PCA Research and Development Laboratories,* V. 10, No. 2, pp. 9-24.

ACI Committee 318, 1995, "Building Code Requirements for Structural Concrete (ACI 318-95) and Commentary (318R-95)," American Concrete Institute, Farmington Hills, Mich., 369 pp.

ACI Committee 408, 1966, "Bond Stress—The State of the Art," ACI JOURNAL, *Proceedings* V. 63, No. 11, Nov., pp. 1161-1190.

ACI Committee 408, 1970, "Bond Stress—The State of the Art (ACI 408-1)," American Concrete Institute, Farmington Hills, Mich., 22 pp.

ACI Committee 408, 2001, "Splice and Development Length of High Relative Rib Area Reinforcing Bars in Tension (ACI 408.3-01) and Commentary (408.3R-01)," American Concrete Institute, Farmington Hills, Mich., 6 pp.

Ahmad, S. H., and Barker, R., 1991, "Flexural Behavior of Reinforced High-Strength Lightweight Concrete Beams," *ACI Structural Journal*, V. 88, No. 1, Jan.-Feb., pp. 69-77.

Ahmad, S. H., and Batts, J., 1991, "Flexural Behavior of Doubly Reinforced High-Strength Lightweight Concrete Beam with Web Reinforcement," *ACI Structural Journal*, V. 88, No. 3, May-June, pp. 351-358.

Baldwin, J. W., Jr., 1965, "Bond of Reinforcement in Lightweight Aggregate Concrete," presented at ACI Convention, University of Missouri, Columbia, Mo., Mar.

Bamforth, P. B., 1987, "The Relationship Between Permeability Coefficients for Concrete Obtained Using Liquid and Gas," *Magazine of Concrete Research*, V. 39, No. 13, pp. 3-11.

Bentz, D. P., and Snyder, K. A., 1999, "Protected Paste Volume in Concrete: Extension to Internal Curing Using Saturated Lightweight Fine Aggregate," *Cement and Concrete Research*, V. 29, pp. 1863-1867.

Berg, O., 1981, "Reinforced Structures In Lightweight-Aggregate Concrete," *Publication* 81.3, Chalmers University of Technology, Goteborg.

Bilodeau, A.; Chevrier, R.; Malhotra, V. M.; and Hoff, G. C., 1995, "Mechanical Properties, Durability and Fire Resistance of High-Strength Lightweight Concrete," *International Symposium on Structural Lightweight-Aggregate Concrete*, Sandefjord, Norway, pp. 432-443.

Bilodeau, A.; Malhotra, V. M; and Hoff, G. C., 1998, "Hydrocarbon Fire Resistance of High Strength Normal Weight and Lightweight Concrete Incorporating Polypropylene Fibers," *Fly Ash, Silica Fume, Slag, and Natural Pozzolans in Concrete*, Sixth CANMET/ACI Conference, Supplementary Volume, V. M. Malhotra, ed., Bangkok, Thailand.

Boyd, S.; Bremner, T. W.; and Holm, T. A., 2000, "Addition of Lightweight Aggregate Reduces Expansion in Concrete Containing Highly Reactive Normalweight Aggregate," *11th ICAAR International Conference*, M. A. Berube, B. Fournier, and B. Durand, eds., Quebec, Canada, pp. 593-602.

Bremner, T. W., and Holm T. A., 1986, "Elastic Compatibility and the Behavior of Concrete," ACI JOURNAL, *Proceedings* V. 83, No. 2, Mar.-Apr., pp. 244-250.

Bremner, T. W.; Holm, T. A.; and deSouza, J., 1984, "Aggregate-Matrix Interaction in Concrete Subject to Severe Exposure," *FIP-CPCI International Symposium on Concrete Sea Structures in Arctic Regions*, Calgary, Canada, 7 pp.

Bremner, T. W.; Holm, T. A.; and McInerney, J. M., 1992, "Influence of Compressive Stress on the Permeability of Concrete," *Structural Lightweight Concrete Performance*, SP-136, T. A. Holm and A. M. Vaysburd, eds., American Concrete Institute, Farmington Hills, Mich., pp. 345-356.

Bremner, T. W.; Holm, T. A.; and Stepanova, V. F., 1994, "Lightweight Concrete—A Proven Material for Two Millennia," *Proceedings of Advances in Cement and Concrete*, S. Sarkar and M. W. Grutzeck, eds., University of New Hampshire, Durham, S.C., pp. 37-41.

Brown, W. R., III; and Davis, C. R., 1993, "A Load Response Investigation of Long Term Performance of a Prestressed Lightweight Concrete Bridge at Fanning Springs, Florida," Florida Department of Transportation, State Materials Office, Gainesville, Fla.

Brown, W. R., III; Larsen, T. J.; and Holm, T. A., 1995, "Long Term Service Performance of Lightweight Concrete Bridge Structures," *International Symposium on Structural Lightweight-Aggregate Concrete*, Sandefjord, Norway.

Campbell, R. H., and Tobin, R. E., 1967, "Core and Cylinder Strengths of Natural and Lightweight Concrete," ACI JOURNAL, *Proceedings* V. 64, No. 4, Apr., pp. 190-195.

Carlson, C. C., 1962, "Fire Resistance of Prestressed Concrete Beam—Study and Influence of Thickness of Concrete Covering Over Prestressing Steel Strands," *Research Dept. Bulletin*, No. 147, Portland Cement Association, Skokie, Ill.

Chana, P. S., 1990, "A Test Method To Establish Realistic Bond Stresses," *Magazine of Concrete Research*, V. 42, No. 151, pp. 83-90.

Clarke, J. L., and Birjandi, F. K., 1993, "Bond Strength Tests For Ribbed Bars in Lightweight-Aggregate Concrete," *Magazine of Concrete Research*, V. 45, No. 163, pp. 79-87.

Expanded Shale, Clay and Slate Institute (ESCSI), 1971, "Lightweight Concrete—History, Application, Economics," Salt Lake City, Utah, 44 pp.

Expanded Shale, Clay and Slate Institute (ESCSI), 1980, "Fire Resistance of Expanded Shale, Clay and Slate Structural Lightweight Concrete," Salt Lake City, Utah., pp. 1-4.

Expanded Shale, Clay and Slate Institute (ESCSI), 1996, "Pumping Structural Lightweight Concrete," Salt Lake City, Utah., 4 pp.

Expanded Shale, Clay and Slate Institute (ESCSI), 2001, "Building Bridges and Marine Structures," Salt Lake City, Utah., 16 pp.

Fagerlund, G., 1972, "Studier av Fasgranser Ballastkorn-Cementpasta I Cementbruk Ogh Betong," *Report* 29, Institutionen for Byognadsteknik Tekntska Hogskolen, Lund. (in Swedish)

Fagerlund, G., 1978, "Frost Resistance of Concrete with Porous Aggregate," Cement Ogh Betong Institue, Stockholm, Sweden.

Fenwick, R. C., and Sue, C. F., 1982, "The Influence of Water Gain Upon the Tensile Strength of Concrete," *Magazine of Concrete Research*, V. 34, No. 120, 10 pp.

Fergestad, S., 1996, "Bridges Built with Lightweight Concrete," *Proceedings of the International Symposium on Lightweight Concrete Bridges*, Sponsored by CALTRANS, Sacramento, Calif.

fib, 2000, "Lightweight-Aggregate Concrete," *Bulleting* 8, Federation Internationale du Beton, Lausanne, Switzerland.

Fintel, M., and Khan, F. R., 1965, "Effects of Column Exposure in Tall Structures—Temperature Variations and their Effects," ACI JOURNAL, *Proceedings* V. 62, No. 12, Dec., pp. 1533-1536.

Fintel, M., and Khan, K. F., 1966, "Analysis of Length Changes of Exposed Columns," ACI JOURNAL, *Proceedings* V. 63, No. 8, Aug., pp. 843-864.

Fintel, M., and Kahn, K. F., 1968, "Design Considerations and Field Observations of Buildings," ACI JOURNAL, *Proceedings* V. 65, No. 2, Feb., pp. 99-110.

FIP (Ferdeation Internationale de la Precontrainte), 1983, *FIP Manual of Lightweight Concrete*, 2nd Edition, John Wiley and Sons, New York.

Gjerde, 1982, "Structural Lightweight-Aggregate Concrete for Marine and Offshore Applications," Norwegian Contractors, Oslo, Norway.

Gray, W. H.; McLaughlin, J. F.; and Antrim, J. O., 1961, "Fatigue Properties of Lightweight-Aggregate Concrete," ACI JOURNAL, *Proceedings* V. 58, No. 6, Aug., pp. 142-62.

Hanson, J. A., 1958, "Shear Strength of Lightweight Reinforced Concrete Beams," ACI JOURNAL, *Proceedings* V. 55, No. 3, pp. 387-404.

Hanson, J. A., 1961, "Tensile Strength and Diagonal Tension Resistance of Structural Lightweight Concrete," ACI JOURNAL, *Proceedings* V. 58, No. 1, pp. 1-40.

Hanson, J. A., 1964, "Replacement of Lightweight Aggregate Fines with Natural Sand in Structural Concrete," ACI JOURNAL, *Proceedings* V. 61, No. 7, pp. 779-793.

HHFA, 1949, "Lightweight Aggregate Concrete," Housing and Home Finance Agency, Washington, D.C., Aug.

Hoff, G. C., 1992, "High Strength Lightweight-Aggregate Concrete for Arctic Applications," *Structural Lightweight Aggregate Concrete Performance*, SP-136, T. A. Holm and A. M. Vaysburd, eds., American Concrete Institute, Farmington Hills, Mich., pp. 1-245.

Hoff, G. C., 1994, "Observations on the Fatigue Behavior of High Strength Lightweight Concrete," *High-Performance Concrete*, Proceedings of the ACI International Conference, SP-149, V. M. Malhotra, ed., American Concrete Institute, Farmington Hills, Mich., pp. 785-822.

Hoff, G. C., 2003, "Internal Curing of Concrete Using Lightweight Aggregate," *Theodore Bremner Symposium on High-Performance Lightweight Concrete*, J. P. Ries and T. A. Holm, eds., Presented at Sixth CANMET/ACI International Conference on Durability of Concrete, pp. 185-204.

Hoff, G. C., et al., 1995, "The Use of Structural Lightweight Aggregates in Offshore Concrete Platforms," *International Symposium on Structural Lightweight-Aggregate Concrete,* Sandefjord, Norway, pp. 349-362.

Hognestad, E.; Hanson, N. W.; and McHenry, D., 1955, "Concrete Stress Distribution in Ultimate Strength Design," ACI JOURNAL, *Proceedings* V. 52, No. 4, Dec., pp. 455-480.

Holm, T. A., 1980a, "Physical Properties of High Strength Lightweight-Aggregate Concrete," *Second International Congress of Lightweight Concrete,* London., 10 pp.

Holm, T. A., 1980b, "Performance of Structural Lightweight Concrete in a Marine Environment," *Performance of Concrete in Marine Environment*, SP-65, V. M. Malhotra, ed., American Concrete Institute, Farmington Hills, Mich., pp. 589-608.

Holm, T. A., 1983, "Three Decades of Durability," *The Military Engineer*, Sept.-Oct., 4 pp.

Holm, T. A., 1994, "Lightweight Concrete and Aggregates," *Tests and Properties of Concrete and Concrete-Making Materials,* STP 169C, 522-32, P. Klieger and J. F. Lamond, eds., ASTM International, West Conshohocken, Pa.

Holm, T. A., and Bremner, T. W., 1994, "High-Strength Lightweight-Aggregate Concrete," *High-Performance Concrete and Applications*, S. P. Shah and S. H. Ahmad, eds., Edward Arnold, London, pp. 341-374.

Holm, T. A., and Bremner, T. W., 2000, "State-of-the-Art Report on High-Strength, High-Durability Structural Low-Density Concrete for Applications in Severe Marine Environments," U.S. Army Corps of Engineers, Engineering Research and Development Center.

Holm, T. A.; Bremner, T. W.; and Newman, J. B., 1984, "Lightweight Aggregate Concrete Subject to Severe Weathering," *Concrete International*, V. 6, No. 6, June, pp. 49-54.

Holm, T. A.; Bremner, T. W.; and Vaysburd, A., 1988, "Carbonation of Marine Structural Lightweight Concretes," *Performance of Concrete in Marine Environment,* Second International Conference, SP-109, V. M. Malhotra, ed., American Concrete Institute, Farmington Hills, Mich., pp. 667-676.

Holm, T. A., and Ries, J. P., 2000, "Specified-Density Concrete—A Transition" *Second International Symposium on Structural Lightweight-Aggregate Concrete*, Kristiansand, Norway.

Huffington, J. A., 2000, "Development of High-Performance Lightweight Concrete Mixes for Prestressed Bridge Girders," University of Texas at Austin, Tex., May.

Ivey, D. L., and Bluth, E., 1966, "Splitting Tension Test of Structural Lightweight Concrete," *Journal of Materials*, ASTM International, V. 1, No. 4, pp. 859-871.

Jensen, J. J.; Hammer, T. A.; Opheim, E.; and Hansen, P. A, 1995, "Fire Resistance of Lightweight-Aggregate Concrete," *International Symposium on Structural Lightweight-Aggregate Concrete,* Ivar Holand, ed., Sandefjord, Norway, Tor Arne Hammer, Finn Fluge, pp. 192-204.

Keeton, P., 1970, "Permeability Studies of Reinforced Thin-Shell Concrete," *Technical Report R692 YF51.001, 01.001,* Naval Engineering Laboratory, Port Hueneme, Calif., 52 pp.

Khokrin, 1973, "The Durability of Lightweight Concrete Structural Members," Kuibyshev, USSR, 114 pp. (in Russian)

Kluge, R. M.; Sparks, M. M.; and Tuma, E. C., 1949, "Lightweight Aggregate Concrete, ACI JOURNAL, *Proceedings* V. 45, No. 5, May, pp. 625-642.

LaRue, H. A., 1946, "Modulus of Elasticity of Aggregates and its Effect on Concrete," *Proceedings* 46, ASTM International, West Conshohocken, pp. 1298-3098.

Lyse, I., 1934, "Lightweight Slag Concrete," ACI JOURNAL, *Proceedings* V. 31, No. 1, pp. 1-20.

Martin, H., 1982, "Bond Performance of Ribbed Bars," *Bond in Concrete—Proceedings of the International Conference on Bond in Concrete*, Paisley, Applied Science Publishers, London, pp. 289-299.

McLaughlin, T., 1944, "Powered Concrete Ships," *Engineering News-Record,* V. 19, Oct., pp. 94-98.

Mehta, P. K., 1986, *Concrete: Structure Properties and Materials,* Prentice Hall, Englewood Cliffs, N.J., 548 pp.

Meyer, K. F., and Kahn, L. F., 2002, "Transfer and Development Length of High Strength Lightweight Concrete,"

Presented at ACI Symposium on High Performance Structural Lightweight Concrete, Phoenix, Ariz., Oct.

Mielenz, R. C., 1994, "Petrographic Evaluation of Concrete Aggregates," Chapter 31, also ASTM C 169, *Significance of Tests and Properties of Concrete and Concrete-Making Materials*, pp. 341-365.

Mor, A., 1992, "Steel-Concrete Bond in High-Strength Lightweight Concrete," *ACI Materials Journals*, V. 89, No. 1, Jan.-Feb., pp. 76-82.

Moreno, J., 1986, "Lightweight Concrete Ductility," *Concrete International*, V. 8, No. 11, Nov., pp. 15-18.

Morgan, M. H., 1960, *Vitruvius, the Ten Books on Architecture Translation*, Dover Publications, New York.

Muller-Rochholz, J., 1979, "Determination of the Elastic Properties of Lightweight Aggregate by Ultrasonic Pulse Velocity Measurements," *International Journal of Lightweight Concrete*, V. 1, No. 2, Lancaster, U.K.

Nassar, A. J., 2002, "Investigation of Transfer Length, Development Length, Flexural Strength and Prestress Loss Trend in Fully Bonded High-Strength Lightweight Prestressed Girders," Virginia Polytechnic Institute and State University, May 15, 136 pp.

Nishi, S.; Oshio, A.; Sone, T.; and Shirokuni, S., 1980, "Watertightness of Concrete Against Sea-Water," Onoda Cement Co., Ltd., Japan.

Ohuchi, T.; Hara, M.; Kubota, N.; Kobayoshi, A.; Nishioka, S.; and Yokoyama, M., 1984, "Some Long-Term Observation Results of Artificial Lightweight Aggregate Concrete for Structural Use in Japan," *International Symposium on Long-Term Observation of Concrete Structures*, Budapest, Hungary, V. II, pp. 274-282.

Pauw, A., 1960, "Static Modulus of Elasticity of Concrete as Affected by Density," ACI JOURNAL, *Proceedings* V. 57, No. 6, pp. 679-688.

Petersen, P. H., 1948, "Properties of Some Lightweight-Aggregate Concretes with and without an Air-Entraining Admixture," *Building Materials and Structures Report* BMS 112, U.S. Department of Commerce, National Bureau of Standards, Aug., 7 pp.

Pfeifer, D. W., 1967, "Sand Replacement in Structural Lightweight Concrete," ACI JOURNAL, *Proceedings* V. 64, No. 7, July, pp. 384-392.

Pfeifer, D. W., 1968, "Sand Replacement in Structural Lightweight Concrete-Creep and Shrinkage Studies," ACI JOURNAL, *Proceedings* V. 65, No. 2, Feb., pp. 131-140.

Philleo, R., 1991, "Concrete Science and Reality," *Materials Science of Concrete II*, J. P. Skalny and S. Mindess, eds., American Ceramic Society, Westerville, Ohio, 8 pp.

Powers, T. E.; Copeland, L. E.; and Mann H. M., 1959, "Capillary Continuity of Discontinuity in Cement Pastes," *Journal Portland Cement Research and Development Labs*, No. 2, May, pp. 38-48.

Prestressed Concrete Institute, 1995, *PCI Design Handbook*, 5th Edition, Chicago, Ill., 614 pp.

Price, W. H., and Cordon, W. A., 1949, "Tests of Lightweight-Aggregate Concrete Designed for Monolithic Construction," ACI JOURNAL, *Proceedings* V. 45, pp. 581-600.

Rabbat, B. G.; Daniel, J. I.; Weinman, T. L.; and Hanson, N. W., 1986, "Seismic Behavior of Lightweight and Normal Weight Concrete Columns," ACI JOURNAL, *Proceedings* V. 83, No. 1, Jan.-Feb., pp. 69-79.

Ramakrishnan, V.; Bremner, T. W.; and Malhotra, V. M., 1992, "Fatigue Strength and Endurance Limit of Lightweight Concrete," *Structural Lightweight Aggregate Concrete Performance*, SP-136, V. M. Malhotra, ed., American Concrete Institute, Farmington Hills, Mich., pp. 397-420.

Ramirez, J.; Olek, J.; Rolle, E.; and Malone, B., 1999, "Performance of Bridge Decks and Girders with Lightweight-Aggregate Concrete," FHWA/IN/JTRP – 98/17, Purdue University, West Lafayette, Ind., May, 161 pp.

Reichard, T. W., 1964, "Creep and Drying Shrinkage of Lightweight and Normalweight Concretes," *Monogram* No. 74, National Bureau of Standards, Washington, D.C., 30 pp.

Robins, P. J., and Standish, I. G., 1982, "Effect of Lateral Pressure on Bond of Reinforcing Bars in Concrete," *Bond in Concrete—Proceedings of the International Conference on Bond in Concrete*, Paisley, Applied Science Publishers, London, pp. 262-272.

Selvaggio, S. L., and Carlson, C. C., 1964, "Fire Resistance of Prestressed Concrete Beams: Study and Influence of Aggregate and Load Intensity," *PCA Research and Development Laboratories Journal*, V. 6, No. 1, Portland Cement Association, Skokie, Ill., pp. 41-64.

Shideler, J. J., 1957, "Lightweight Aggregate Concrete for Structural Use," ACI JOURNAL, *Proceedings* V. 54, pp. 298-328.

Stolldorf, D. W., and Holm, T. A., 1996, "Bridge Rehabilitation Permits Higher Live Loads," *Materials for the New Millennium*, ASCE Materials Conference, Washington, D.C., Nov.

Sugiyama, T.; Bremner, T. W.; and Holm, T. A., 1996, "Effect of Stress on Gas Permeability in Concrete," *ACI Materials Journal*, V. 93, No. 5, Sept.-Oct., pp. 443-450.

Thatcher, D. B.; Heffington, J. A.; Kolozs, G. S.; Sylva, G. S.; Breen, J. E.; and Burns, N. H., 2002, "Structural Lightweight Concrete Prestressed Girders and Panels," Center for Transportation Research, University of Texas at Austin, Jan.

Valore, R. C., Jr., 1956, "Insulating Concretes," ACI JOURNAL, *Proceedings* V. 53, No. 5, pp. 509-532.

Valore, R. C., 1980, "Calculation of U-Values of Hollow Concrete Masonry," *Concrete International*, V. 2, No. 2, Feb., pp. 40-63.

Washa, G. W., and Fluck, P. G., 1952, "Effect of Compressive Reinforcement on the Plastic Flow of Reinforced Concrete Beams," ACI JOURNAL, *Proceedings* V. 49, No. 10, Oct., pp. 89-108.

Weber S., and Reinhardt, 1995, "A Blend of Aggregates to Support Curing of Concrete," *Proceedings of the International Structural Lightweight Concrete*, Sandefjord, Norway.

7.3—Other references

Rowley, F. R., and Algren, A. R., 1947, "Thermal Conductivity of Building Materials," *Bulletin* No. 12, Engineering Experiment Station, University of Minneapolis, Minn., 134 pp.

ACI 214R-02

Evaluation of Strength Test Results of Concrete

Reported by ACI Committee 214

James E. Cook[*]
Chair

Jerry Parnes
Secretary

David J. Akers	Gilbert J. Haddad	Robert E. Neal
M. Arockiasamy	Kal R. Hindo	Terry Patzias
William L. Barringer	William J. Irwin	Venkataswamy Ramakrishnan
F. Michael Bartlett[*]	Alfred L. Kaufman, Jr.[*]	D. V. Reddy
Casimir Bognacki[*]	William F. Kepler	Orrin Riley[*]
Jerrold L. Brown	Peter A. Kopac	James M. Shilstone, Jr.
Ronald L. Dilly	Michael L. Leming[*]	Luke M. Snell
Donald E. Dixon	Colin L. Lobo[*]	Patrick J. Sullivan
Richard D. Gaynor[*]	John J. Luciano[*]	Michael A. Taylor[*]
Steven H. Gebler	Richard E. Miller	J. Derle Thorpe
Alejandro Graf	Avi A. Mor	Roger E. Vaughan
Thomas M. Greene	Tarun R. Naik	Woodward L. Vogt

[*]Committee members who prepared this revision.

Statistical procedures provide tools of considerable value when evaluating the results of strength tests. Information derived from such procedures is also valuable in defining design criteria and specifications. This report discusses variations that occur in the strength of concrete and presents statistical procedures that are useful in the interpretation of these variations with respect to specified testing and criteria.

Keywords: coefficient of variation; quality control; standard deviation; strength.

ACI Committee Reports, Guides, Standard Practices, and Commentaries are intended for guidance in planning, designing, executing, and inspecting construction. This document is intended for the use of individuals who are competent to evaluate the significance and limitations of its content and recommendations and who will accept responsibility for the application of the material it contains. The American Concrete Institute disclaims any and all responsibility for the stated principles. The Institute shall not be liable for any loss or damage arising therefrom.

Reference to this document shall not be made in contract documents. If items found in this document are desired by the Architect/Engineer to be a part of the contract documents, they shall be restated in mandatory language for incorporation by the Architect/Engineer.

CONTENTS
Chapter 1—Introduction, p. 214R-2

Chapter 2—Variations in strength, p. 214R-2
2.1—General
2.2—Properties of concrete
2.3—Testing methods

Chapter 3—Analysis of strength data, p. 214R-3
3.1—Terminology
3.2—General
3.3—Statistical functions
3.4—Strength variations
3.5—Interpretation of statistical parameters
3.6—Standards of control

Chapter 4—Criteria, p. 214R-8
4.1—General
4.2—Data used to establish minimum required average strength

ACI 214R-02 supersedes ACI 214-77 (Reapproved 1997) and became effective June 27, 2002.
Copyright © 2002, American Concrete Institute.
All rights reserved including rights of reproduction and use in any form or by any means, including the making of copies by any photo process, or by electronic or mechanical device, printed, written, or oral, or recording for sound or visual reproduction or for use in any knowledge or retrieval system or device, unless permission in writing is obtained from the copyright proprietors.

4.3—Criteria for strength requirements

Chapter 5—Evaluation of data, p. 214R-12
5.1—General
5.2—Numbers of tests
5.3—Rejection of doubtful specimens
5.4—Additional test requirements
5.5—Basic quality-control charts
5.6—Other evaluation techniques

Chapter 6—References, p. 214R-16
6.1—Referenced standards and reports
6.2—Cited references

Appendix A—Examples of CUSUM technique, p. 214R-17
A.1—Introduction
A.2—Theory
A.3—Calculations
A.4—Analysis and comparison with conventional control charts
A.5—Management considerations of interference
A.6—Establishing limits for interference
A.7—Difficulties with CUSUM chart

CHAPTER 1—INTRODUCTION

This document provides an introduction to the evaluation of concrete strength tests. The procedures described are applicable to the compressive-strength test results required by ACI 301, ACI 318, and other similar specifications and codes. The statistical concepts described are applicable for analysis of other common concrete test results including flexural strength, slump, air content, and density.

Most construction projects in the United States and Canada require routine sampling and fabrication of standard molded cylinders. These cylinders are usually cast from samples of concrete taken from the discharge of a truck or a batch of concrete and molded, cured, and tested under standardized procedures. The results represent the potential strength of the concrete rather than the actual strength of the concrete in the structure.

Inevitably, strength test results vary. Variations in measured strength may originate from any of the following sources:
- Batch-to-batch variations of the proportions and characteristics of the constituent materials in the concrete, the production, delivery, and handling process, and climatic conditions; and
- Variations in the sampling, specimen preparation, curing, and testing procedures (within-test).

Conclusions regarding the strength of concrete can only be derived from a series of tests. The characteristics of concrete strength can be estimated with reasonable accuracy only when an adequate number of tests are conducted, strictly in accordance with standard practices and test methods.

Statistical procedures provide tools of considerable value when evaluating the results of strength tests. Information derived from such procedures is also valuable in refining design criteria and specifications. This report discusses variations that occur in the strength of concrete and presents statistical procedures that are useful in the interpretation of these variations with respect to specified testing and acceptance criteria.

Table 2.1—Principal sources of strength variation

Variations due to the properties of concrete	Variations due to testing methods

•Changes in w/cm caused by:
  -Poor control of water
  -Excessive variation of moisture in aggregate or variable aggregate moisture measurements
  -Retempering
	•Improper sampling procedures
	
	•Variations due to fabrication techniques:
	  -Handling, storing, and curing of newly made cylinders
	  -Poor quality, damaged, or distorted molds

•Variations in water requirement caused by:
  -Changes in aggregate grading, absorption, particle shape
  -Changes in cementitious and admixture properties
  -Changes in air content
  -Delivery time and temperature changes
	•Changes in curing:
	  -Temperature variation
	  -Variable moisture control
	  -Delays in bringing cylinders to the laboratory
	  -Delays in beginning standard curing

•Variations in characteristics and proportions of ingredients:
  -Aggregates
  -Cementitious materials, including pozzolans
  -Admixtures
	•Poor testing procedures:
	  -Specimen preparation
	  -Test procedure
	  -Uncalibrated testing equipment

•Variations in mixing, transporting, placing, and consolidation	
•Variations in concrete temperature and curing	

For the statistical procedures described in this report to be valid, the data should be derived from samples obtained by means of a random sampling plan designed to reduce the possibility that selection will be exercised by the sampler. Random sampling means that each possible sample has an equal chance of being selected. To ensure this condition, the selection should be made by some objective mechanism such as a table of random numbers. If sample batches are selected on the basis of judgement by the sampler, biases are likely to be introduced that will invalidate the analysis using the procedures presented here. Natrella (1963) and ASTM D 3665 provide a discussion of random sampling and a useful short table of random numbers.

This report begins with a discussion of the sources of variability in concrete as produced, mixed, and transported, and the additional variability of samples obtained from the concrete as delivered and tested. The report then describes the statistical tools used to evaluate the variability of concrete and determine compliance with a given specification, including both random variation and variation due to assignable causes. Statistically based specifications are also reviewed.

CHAPTER 2—VARIATIONS IN STRENGTH
2.1—General

The magnitude of variations in the strength of concrete test specimens is a direct result of the degree of control exerted over the constituent materials, the concrete production and transportation process, and the sampling, specimen preparation, curing and testing procedures. Variability in strength can be traced to two fundamentally different sources: variability in strength-producing properties of the concrete mixture and ingredients, including batching and production, and variability in the measured strength caused by variations inherent in the testing process. Table 2.1 summarizes the principal sources of strength variation.

Variation in the measured characteristics may be either random or assignable depending on cause. Random variation is normal for any process; a stable process will show only random variation. Assignable causes represent systematic changes that are typically associated with a shift in some fundamental statistical characteristic, such as mean, standard deviation or coefficient of variation, or other statistical measure.

2.2—Properties of concrete

For a given set of raw materials, strength is governed to a large extent by the water-cementitious materials ratio (w/cm). The first criterion for producing concrete of consistent strength, therefore, is to keep tight control over the w/cm. Because the quantity of cementitious material can be measured reasonably accurately, maintaining a constant w/cm primarily requires strict control of the total quantity of water used.

The water requirement of concrete is strongly influenced by the source and characteristics of the aggregates, cement, and mineral and chemical admixtures used in the concrete, as well as the desired consistency, in the sense of workability and placeability. Water demand also varies with air content and can increase with temperature. Variations in water content can be caused by variations in constituent materials and variations in batching. A common source of variation is from water added on the job site to adjust the slump.

Water can be introduced into concrete in many ways—some of which may be intentional. The amount of water added at the batch plant and job site is relatively easy to record. Water from other sources, such as free moisture on aggregates, water left in the truck, or added but not recorded, can be difficult to determine. For a similar concrete mixture at the same temperature and air content, differences in slump from batch to batch can be attributed to changes in the total mixing water content among other factors.

The AASHTO Standard Test Method for Water Content of Freshly Mixed Concrete Using Microwave Oven Drying (TP 23) is one method of determining water content of fresh concrete. The accuracy of the test method is still under study. The test may be useful in detecting deviations in water content in fresh concrete at the construction site.

Variations in strength are also influenced by air content. The entrained air content influences both water requirement and strength. There is an inverse relationship between strength and air content. The air content of a specific concrete mixture varies depending on variations in constituent materials, extent of mixing, and ambient site conditions. For good concrete control, the entrained air content should be monitored closely at the construction site.

The temperature of fresh concrete affects both the amount of water needed to achieve the proper consistency and the entrained air content. In addition, the concrete temperature during the first 24 hours of curing can have a significant effect on the later-age strengths of the concrete. Concrete cylinders that are not protected from temperatures outside the range specified in ASTM C 31 may not accurately reflect the potential strength of the concrete.

Admixtures can contribute to variability, because each admixture introduces another variable and source of variation. Batching and mixing of admixtures should be carefully controlled. Changes in water demand are also associated with variations in aggregate grading.

Construction practices will cause variations of the in-place strength due to inadequate mixing, improper consolidation, delays in placement, improper curing, and insufficient protection at early ages. These differences will not be reflected in specimens fabricated and stored under standard laboratory conditions. Construction practices can affect the strength results of cores, however, which may be drilled and tested when strength test results do not conform to project specifications.

2.3—Testing methods

Deviations from standard sampling and testing procedures will affect the measured or reported strength. Testing to determine compliance with contract specifications should be conducted strictly according to the methods specified in the appropriate contract documents, for example ASTM C 31 and ASTM C 39. Acceptance tests provide an estimate of the potential strength of the concrete, not necessarily the in-place strength. Deviations from standard moisture and temperature curing is often a reason for lower strength test results. A project can be penalized unnecessarily when variations from this source are excessive. Deviations from standard procedures often result in a lower measured strength. Field sampling, curing, and handling of specimens should be performed by ACI Certified Technicians, or equivalently trained, experienced, and certified personnel, and procedures should be carefully monitored. Provisions for maintaining specified curing conditions should be made. Specimens made from slowly hardening concrete should not be disturbed too soon (ASTM C 31).

The importance of using accurate, properly calibrated testing devices and using proper sample preparation procedures is essential, because test results can be no more accurate than the equipment and procedures used. Less variable test results do not necessarily indicate accurate test results, because a routinely applied, systematic error can provide results that are biased but uniform. Laboratory equipment and procedures should be calibrated and checked periodically; testing personnel should be trained and certified at the appropriate technical level and evaluated routinely.

CHAPTER 3—ANALYSIS OF STRENGTH DATA
3.1—Terminology

3.1.1 *Definitions*—In this chapter, the following terminology is adopted.

concrete sample—a portion of concrete, taken at one time, from a single batch or single truckload of concrete.

single cylinder strength or **individual strength**—the strength of a single cylinder; a single cylinder strength does not constitute a test result.

companion cylinders—cylinders made from the same sample of concrete.

strength test or **strength test result**—the average of two or more single-cylinder strengths of specimens made from the same concrete sample (companion cylinders) and tested at the same age.

range or **within-test range**—the difference between the maximum and minimum strengths of individual concrete specimens comprising one strength test result.

test record—a collection of strength test results of a single concrete mixture. Test records of similar concrete mixtures can be used to calculate the pooled standard deviation. Concrete mixtures are considered to be similar if their nominal strengths are within 6.9 MPa (1000 psi), represent similar materials, and are produced, delivered, and handled under similar conditions (ACI 318).

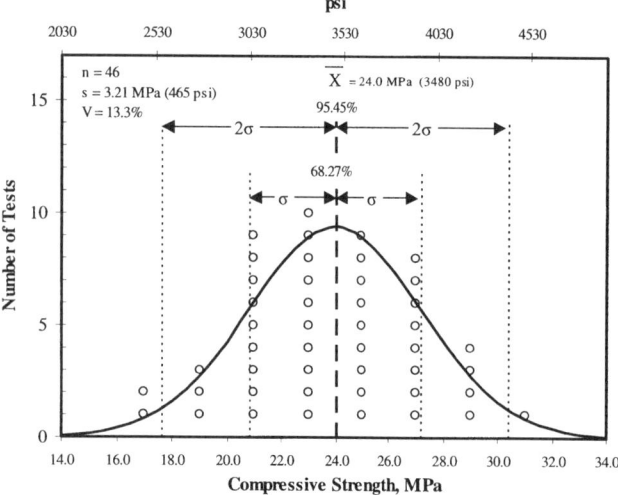

Fig. 3.1—Frequency distribution of strength data and corresponding assumed normal distribution.

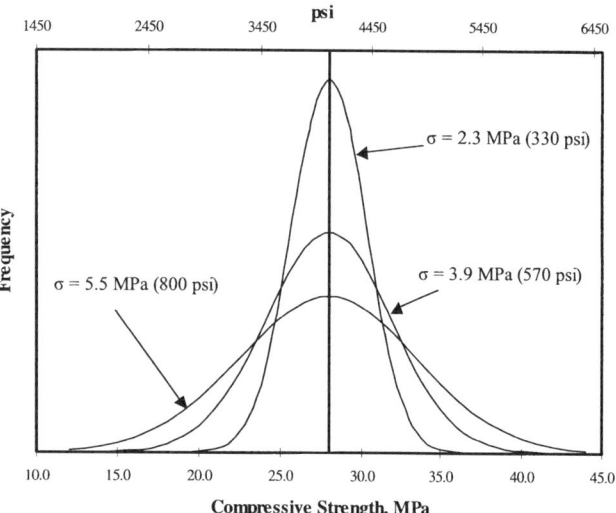

Fig. 3.2—Normal frequency curves for three different distributions with the same mean but different variability.

3.1.2 *Notation*

d_2 = factor for computing within-test standard deviation from average range (See Table 3.1.)
f'_{cr} = required average strength to ensure that no more than the permissible proportion of tests will fall below the specified compressive strength, used as the basis for selection of concrete proportions
f'_c = specified compressive strength
μ = population mean
n = number of tests in a record
R = within-test range
$\overline{R}_m$ = maximum average range, used in certain control charts
$\overline{R}$ = average range
σ = population standard deviation
σ_1 = population within-test standard deviation
σ_2 = population batch-to-batch standard deviation
s = sample standard deviation, an estimate of the population standard deviation, also termed $s_{overall}$
$\overline{s}$ = statistical average standard deviation, or "pooled" standard deviation

Table 3.1—Factors for computing within-test standard deviation from range

No. of specimens	d_2
2	1.128
3	1.693
4	2.059

Note: From Table 49, *ASTM Manual on Presentation of Data and Control Chart Analysis*, MNL 7.

s_1 = sample within-test standard deviation, also termed $s_{within-test}$
s_2 = sample batch-to-batch standard deviation, also termed $s_{producer}$
V = coefficient of variation
V_1 = within-test coefficient of variation
X_i = a strength test result
$\overline{X}$ = average of strength test results
z = a constant multiplier for standard deviation (s) that depends on the number of tests expected to fall below f'_c (See Table 4.3.)

3.2—General

A sufficient number of tests is needed to indicate accurately the variation in the concrete produced and to permit appropriate statistical procedures for interpreting the test results. Statistical procedures provide a sound basis for determining from such results the potential quality and strength of the concrete and for expressing results in the most useful form.

3.3—Statistical functions

A strength test result is defined as the average strength of all specimens of the same age, fabricated from a sample taken from a single batch of concrete. A strength test cannot be based on only one cylinder; a minimum of two cylinders is required for each test. Concrete tests for strength are typically treated as if they fall into a distribution pattern similar to the normal frequency distribution curve illustrated in Fig. 3.1. Cook (1989) reports that a skewed distribution may result for high-strength concrete where the limiting factor is the strength of the aggregate. If the data are not symmetrical about the mean, the data may be skewed. If the distribution is too peaked or too flat, kurtosis exists. Data exhibiting significant skewness or kurtosis are not normally distributed and any analysis presuming a normal distribution may be misleading rather than informative. Available data (Cook 1982) indicate that a normal distribution is appropriate under most cases when the strength of the concrete does not exceed 70 MPa (10,000 psi). Skewness and kurtosis should be considered for statistical evaluation of high-strength concrete. Cook (1989) provides simplified equations that can measure relative skewness and kurtosis for a particular set of data. In this document, strength test results are assumed to follow a normal distribution, unless otherwise noted.

When there is good control, the strength test values will tend to cluster near to the average value, that is, the histogram of test results is tall and narrow. As variation in strength results increases, the spread in the data increases and the normal distribution curve becomes lower and wider (Fig. 3.2). The normal distribution can be fully defined mathematically by two statistical parameters: the mean and standard deviation. These statistical parameters of the strength can be calculated as shown in Sections 3.3.1 and 3.3.2.

3.3.1 *Mean* $\overline{X}$— The average strength tests result $\overline{X}$ is calculated using Eq. (3-1).

$$\overline{X} = \frac{\sum_{i=1}^{n} X_i}{n} = \frac{1}{n}\sum X_i = \frac{1}{n}(X_1 + X_2 + X_3 + \ldots + X_n) \quad (3\text{-}1)$$

where X_i is the *i*-th strength test result, the average of at least two cylinder strength tests. X_2 is the second strength test result in the record, ΣX_i is the sum of all strength test results and n is the number of tests in the record.

3.3.2 *Standard deviation* s—The standard deviation is the most generally recognized measure of dispersion of the individual test data from their average. An estimate of the population standard deviation σ is the sample standard deviation s. The population consists of all possible data, often considered to be an infinite number of data points. The sample is a portion of the population, consisting of a finite amount of data. The sample standard deviation is obtained by Eq. (3-2a), or by its algebraic equivalent, Eq. (3-2b). The latter equation is preferable for computation purposes, because it is simpler and minimizes rounding errors. When using spreadsheet software, it is important to ensure that the sample standard deviation formula is used to calculate s.

$$s = \sqrt{\frac{\sum_{i=1}^{n}(X_i - \overline{X})^2}{n-1}} = \quad (3\text{-}2a)$$

$$\sqrt{\frac{(X_1 - \overline{X})^2 + (X_2 - \overline{X})^2 + \ldots + (X_n - \overline{X})^2}{n-1}}$$

which is equivalent to

$$s = \sqrt{\frac{n\sum_{i=1}^{n} X_i^2 - \left(\sum_{i=1}^{n} X_i\right)^2}{n(n-1)}} = \sqrt{\frac{\sum_{i=1}^{n} X_i^2 - n\overline{X}^2}{n-1}} \quad (3\text{-}2b)$$

where s is the sample standard deviation, n is the number of strength test results in the record, $\overline{X}$ is the mean, or average, strength test result, and ΣX is the sum of the strength test results.

When considering two separate records of concrete mixtures with similar strength test results, it is frequently necessary to determine the statistical average standard deviation, also termed the pooled standard deviation. The statistical average standard deviation of two records is calculated as shown in Eq. (3-3).

$$\overline{s} = \sqrt{\frac{(n_A - 1)(s_A)^2 + (n_B - 1)(s_B)^2}{(n_A + n_B - 2)}} \quad (3\text{-}3)$$

where $\overline{s}$ is the statistical average standard deviation, or pooled standard deviation, determined from two records, s_A and s_B are the standard deviations of Record A and Record B, respectively, and n_A and n_B are the number of tests in Record A and Record B, respectively.

3.3.3 *Other statistical measures*—Several other derivative statistics are commonly used for comparison of different data sets or for estimation of dispersion in the absence of statistically valid sample sizes.

3.3.3.1 *Coefficient of variation* V—The sample standard deviation expressed as a percentage of the average strength is called the coefficient of variation

$$V = \frac{s}{\overline{X}} \times 100 \quad (3\text{-}4)$$

where V is the coefficient of variation, s is the sample standard deviation, and $\overline{X}$ is the average strength test result.

The coefficient of variation is less affected by the magnitude of the strength level (Cook 1989; Anderson 1985), and is therefore more useful than the standard deviation in comparing the degree of control for a wide range of compressive strengths. The coefficient of variation is typically used when comparing the dispersion of strength test results of records with average compressive strengths more than about 7 MPa [1000 psi] different.

3.3.3.2 *Range* R—Range is the statistic found by subtracting the lowest value in a data set from the highest value in that data set. In evaluation of concrete test results, the within-test range R of a strength test result is found by subtracting the lowest single cylinder strength from the highest single cylinder strength of the two or more cylinders used to comprise a strength test result. The average within-test range is used for estimating the within-test standard deviation. It is discussed in more detail in Section 3.4.1.

3.4—Strength variations

As noted in Chapters 1 and 2, variations in strength test results can be traced to two different sources:

1. Variations in testing methods; and
2. Variations in the properties or proportions of the constituent materials in the concrete mixture, variations in the production, delivery or handling procedures, and variations in climatic conditions.

It is possible to compute the variations attributable to each source using analysis of variance (ANOVA) techniques (Box, Hunter, and Hunter 1978) or with simpler techniques.

3.4.1 *Within-test variation*—Variability due to testing is estimated by the within-test variation based on differences in strengths of companion (replicate) cylinders comprising a strength test result. The within-test variation is affected by variations in sampling, molding, consolidating, transporting, curing, capping, and testing specimens. A single strength test result of a concrete mixture, however, does not provide sufficient data for statistical analysis. As with any statistical estimator, the confidence in the estimate is a function of the number of test results.

The within-test standard deviation is estimated from the average range $\overline{R}$ of at least 10, and preferably more, strength test results of a concrete mixture, tested at the same age, and the appropriate values of d_2 in Table 3.1 using Eq. (3-5). In Eq. (3-6), the within-test coefficient of variation, in percent, is determined from the within-test standard deviation and the average strength.

Table 3.2—Standards of concrete control*

Class of operation	Overall variation				
	Standard deviation for different control standards, MPa (psi)				
	Excellent	Very good	Good	Fair	Poor
General construction testing	Below 2.8 (below 400)	2.8 to 3.4 (400 to 500)	3.4 to 4.1 (500 to 600)	4.1 to 4.8 (600 to 700)	Above 4.8 (above 700)
Laboratory trial batches	Below 1.4 (below 200)	1.4 to 1.7 (200 to 250)	1.7 to 2.1 (250 to 300)	2.1 to 2.4 (300 to 350)	Above 2.4 (above 350)
Class of operation	Within-test variation				
	Coefficient of variation for different control standards, %				
	Excellent	Very good	Good	Fair	Poor
Field control testing	Below 3.0	3.0 to 4.0	4.0 to 5.0	5.0 to 6.0	Above 6.0
Laboratory trial batches	Below 2.0	2.0 to 3.0	3.0 to 4.0	4.0 to 5.0	Above 5.0

*$f_c' \leq 34.5$ MPa (5000 psi).

$$s_1 = \frac{1}{d_2} \overline{R} \quad (3\text{-}5)$$

$$V_1 = \frac{s_1}{\overline{X}} \times 100 \quad (3\text{-}6)$$

where s_1 is the sample within-test standard deviation, $\overline{R}$ is the average within-test range of at least 10 tests, d_2 is the factor for computing within-test standard deviation from the average range, V_1 is the sample within-test coefficient of variation, and $\overline{X}$ is the mean, or average, strength test result.

For example, if two cylinders were cast for each of 10 separate strength tests (the minimum number recommended), and the average within-test strength range was 1.75 MPa (254 psi), the estimated within-test standard deviation (d_2 = 1.128 for 2 cylinders) is 1.75/1.128 = 1.55 MPa (254/1.128 = 225 psi). The precision statement in ASTM C 39 indicates the within-test coefficient of variation for cylinder specimens made in the lab to be 2.37% and for cylinders made in the field to be 2.87%.

Consistent errors or bias in testing procedures will not necessarily be detected by comparing test results of cylinders from the same sample of concrete, however. Variations may be small with an improperly conducted test, if performed consistently.

3.4.2 *Batch-to-batch variations*—These variations reflect differences in strength from batch to batch, which can be attributed to variations in:

(a) Characteristics and properties of the ingredients; and
(b) Batching, mixing, and sampling.

Testing effects can inflate the apparent batch-to-batch variation slightly. The effects of testing on batch-to-batch variation are not usually revealed by analyzing test results from companion cylinders tested at the same age, because specimens from the same batch tend to be treated alike. Batch-to-batch variation can be estimated from strength test results of a concrete mixture if each test result represents a separate batch of concrete.

The overall variation, σ (for a population) or s (for a sample), has two component variations, the within-test, σ_1 (population) or s_1 (sample), and batch-to-batch, σ_2 (population) or s_2 (sample) variations. The sample variance—the square

Table 3.3—Standards of concrete control*

Class of operation	Overall variation				
	Coefficient of variation for different control standards, %				
	Excellent	Very good	Good	Fair	Poor
General construction testing	Below 7.0	7.0 to 9.0	9.0 to 11.0	11.0 to 14.0	Above 14.0
Laboratory trial batches	Below 3.5	3.5 to 4.5	4.5 to 5.5	5.5 to 7.0	Above 7.0
Class of operation	Within-test variation				
	Coefficient of variation for different control standards, %				
	Excellent	Very good	Good	Fair	Poor
Field control testing	Below 3.0	3.0 to 4.0	4.0 to 5.0	5.0 to 6.0	Above 6.0
Laboratory trial batches	Below 2.0	2.0 to 3.0	3.0 to 4.0	4.0 to 5.0	Above 5.0

*$f_c' > 34.5$ MPa (5000 psi).

of the sample standard deviation—is the sum of the sample within-test and sample batch-to-batch variances

$$s^2 = s_1^2 + s_2^2 \quad (3\text{-}7)$$

from which the batch-to-batch standard deviation can be computed as

$$s_2 = \sqrt{s^2 - s_1^2} \quad (3\text{-}8)$$

For example, if the overall sample standard deviation s from multiple batches is 3.40 MPa (493 psi), and the estimated within-test sample standard deviation s_1 is 1.91 MPa (277 psi), the batch-to-batch sample standard deviation s_2 can be estimated as 2.81 MPa (408 psi).

The within-test sample standard deviation estimates the variation attributable to sampling, specimen preparation, curing and testing, assuming proper testing methods are used. The batch-to-batch sample standard deviation estimates the variations attributable to constituent material suppliers, and the concrete producer. Values of the overall and the within-test sample standard deviations and coefficients of variation associated with different control standards are provided in Section 3.6 (Table 3.2 and 3.3).

3.5—Interpretation of statistical parameters

Once the statistical parameters have been computed, and with the assumption or verification that the results follow a normal frequency distribution curve, additional analysis of the test results is possible. Figure 3.3 indicates an approximate division of the area under the normal frequency distribution curve. For example, approximately 68% of the area (equivalent to 68% of the results) lies within ±1σ of the average, and 95% lies within ±2σ. This permits an estimate of the portion of the test results expected to fall within given multiples z of σ of the average or of any other specific value.

Agreement between the normal distribution and the actual distribution of the tests tends to increase as the number of tests increases. When only a small number of results are available, they may not fit the standard, bell-shaped pattern. Other causes of differences between the actual and the normal distribution are errors in sampling, testing, and recording.

Table 3.4—Expected percentages of individual tests lower than f_c' *

Average strength μ	Expected percentage of low tests	Average strength μ	Expected percentage of low tests
$f_c' + 0.10\sigma$	46.0	$f_c' + 1.6\sigma$	5.5
$f_c' + 0.20\sigma$	42.1	$f_c' + 1.7\sigma$	4.5
$f_c' + 0.30\sigma$	38.2	$f_c' + 1.8\sigma$	3.6
$f_c' + 0.40\sigma$	34.5	$f_c' + 1.9\sigma$	2.9
$f_c' + 0.50\sigma$	30.9	$f_c' + 2.0\sigma$	2.3
$f_c' + 0.60\sigma$	27.4	$f_c' + 2.1\sigma$	1.8
$f_c' + 0.70\sigma$	24.2	$f_c' + 2.2\sigma$	1.4
$f_c' + 0.80\sigma$	21.2	$f_c' + 2.3\sigma$	1.1
$f_c' + 0.90\sigma$	18.4	$f_c' + 2.4\sigma$	0.8
$f_c' + 1.00\sigma$	15.9	$f_c' + 2.5\sigma$	0.6
$f_c' + 1.10\sigma$	13.6	$f_c' + 2.6\sigma$	0.45
$f_c' + 1.20\sigma$	11.5	$f_c' + 2.7\sigma$	0.35
$f_c' + 1.30\sigma$	9.7	$f_c' + 2.8\sigma$	0.25
$f_c' + 1.40\sigma$	8.1	$f_c' + 2.9\sigma$	0.19
$f_c' + 1.50\sigma$	6.7	$f_c' + 3.0\sigma$	0.13

*where μ exceeds f_c' by amount shown.

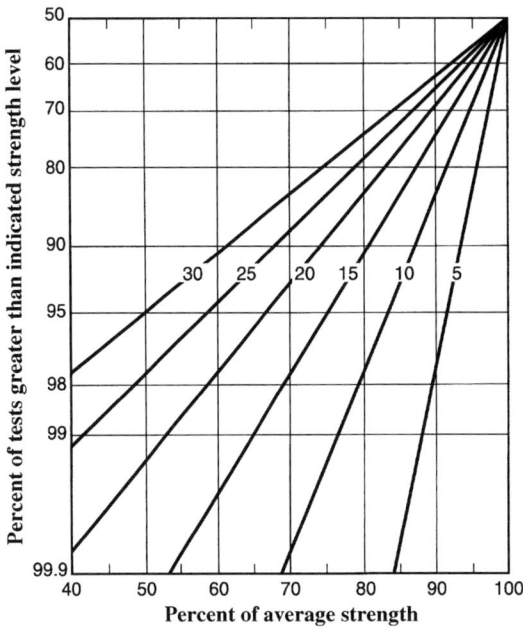

Fig. 3.4—Cumulative distribution curves for different coefficients of variation.

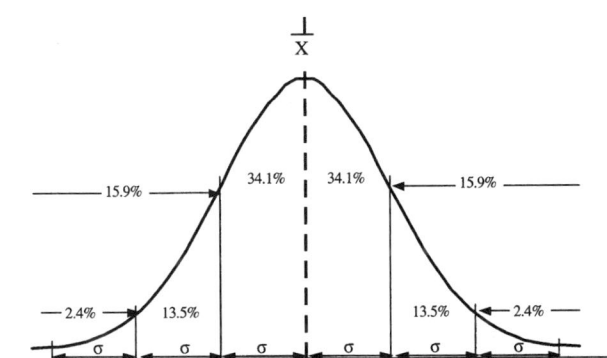

Fig. 3.3—Approximate distribution of area under normal frequency distribution curve.

Failure to sample in a truly random manner, sampling from different populations, or the presence of skew or kurtosis in high-strength concretes (Cook 1989) are examples that would result in substantial differences between the actual and the normal distributions.

Table 3.4 was adapted from the normal cumulative distribution (the normal probability integral) and shows the probability of a fraction of tests falling below f_c' in terms of the average strength of the population of test results when the population average strength μ equals $f_c' + z\sigma$.

Cumulative distribution curves can also be plotted by accumulating the number of tests below any given strength for different coefficients of variation or standard deviations. The below-average half of the normal frequency distribution curve is shown for a variety of coefficients of variation in Fig. 3.4 and a variety of standard deviations in Fig. 3.5. By using the normal probability scale, the curves are plotted as a straight line and can be read in terms of frequencies for which test results will be greater than the indicated percentage of average strength of the population of strength test results (Fig 3.4) or compressive strength below average (Fig. 3.5). When lower coefficients of variation (or standard deviations)

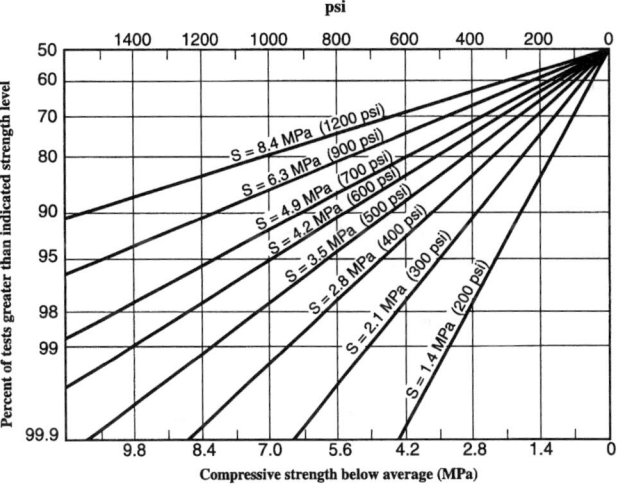

Fig. 3.5—Cumulative distribution curves for different standard deviations.

are attained, the angle formed by the cumulative distribution curve and the 100% ordinate (Fig. 3.4) or 0 standard deviation (Fig. 3.5) decreases; the difference between the lowest and the highest probable strength is reduced, indicating the concrete test results are more consistent. These charts can be used to solve for probabilities graphically. Similar charts can be constructed to compare the performance of different concrete mixtures.

3.6—Standards of control

One of the primary purposes of statistical evaluation of concrete data is to identify sources of variability. This knowledge can then be used to help determine appropriate steps to maintain the desired level of control. Several different techniques can be used to detect variations in concrete production, materials processing and handling, and contractor and testing agency operations. One simple approach is to compare overall variability and within-test variability, using either standard deviation or coefficient of variation, as appropriate, with previous performance.

Whether the standard deviation or the coefficient of variation is the appropriate measure of dispersion to use in any given situation depends upon which of the two measures is more nearly constant over the range of strengths of concern. Present information indicates that the standard deviation remains reasonably constant over a limited range of strengths; however, several studies show that the coefficient of variation is more nearly constant over a wider range of strengths, especially higher strengths (Cook 1982; Cook 1989). Comparison of level of control between compressive and flexural strengths is more easily conducted using the coefficient of variation. The coefficient of variation is also considered to be a more applicable statistic for within-test evaluations (Neville 1959; Metcalf 1970; Murdock 1953; Erntroy 1960; Rüsch 1964; and ASTM C 802). Either the standard deviation or the coefficient of variation can be used to evaluate the level of control of conventional-strength concrete mixtures, but for higher strengths, generally those in excess of 35 MPa (5000 psi), the coefficient of variation is preferred.

The standards of control given in Table 3.2 are appropriate for concrete having specified strengths up to 35 MPa (5000 psi), whereas Table 3.3 gives the appropriate standards of control for specified strengths over 35 MPa (5000 psi). As more high-strength test data become available, these standards of control may be modified. These standards of control were adopted based on examination and analysis of compressive strength data by ACI Committee 214 and ACI Committee 363. The strength tests were conducted using 150 x 300 mm (6 x 12 in.) cylinders, the standard size for acceptance testing in ASTM C 31. The standards of control are therefore applicable to these size specimens, tested at 28 days, and may be considered applicable with minor differences to other cylinder sizes, such as 100 x 200 mm (4 x 8 in.) cylinders, recognized in C 31. They are not applicable to strength tests on cubes or flexural strength test results.

CHAPTER 4—CRITERIA
4.1—General

The strength of concrete in a structure and the strength of test cylinders cast from a sample of that concrete are not necessarily the same. The strength of the cylinders obtained from that sample of concrete and used for contractual acceptance are to be cured and tested under tightly controlled conditions. The strengths of these cylinders are generally the primary evidence of the quality of concrete used in the structure. The engineer specifies the desired strength, the testing frequency, and the permitted tolerance in compressive strength.

Any specified quantity, including strength, should also have a tolerance. It is impractical to specify an absolute minimum strength, because there is always the possibility of even lower strengths simply due to random variation, even when control is good. The cylinders may not provide an accurate representation of the concrete in each portion of the structure. Strength-reduction factors are provided in design methodologies that allow for limited deviations from specified strengths without jeopardizing the safety of the structure. These methodologies evolved using probabilistic methods on the basis of construction practices, design procedures, and quality-control techniques used in the construction industry.

For a given mean strength, if a small percentage of the test results fall below the specified strength, the remaining test results will be greater than the specified strength. If the samples are selected randomly, there is only a small probability that the low strength results correspond to concrete located in a critical area. The consequences of a localized zone of low-strength concrete in a structure depend on many factors, including the probability of early overload; the location and magnitude of the low-quality zone in the structural element; the degree of reliance placed on strength in design; the initial cause of the low strength; and the implications, economic and otherwise, of loss of serviceability or structural failure.

There will always be a certain probability of tests falling below f_c'. ACI 318 and most other building codes and specifications establish tolerances for meeting the specified compressive strength acceptance criteria, analogous to the tolerances for other building materials.

To satisfy statistically based strength-performance requirements, the average strength of the concrete should be in excess of the specified compressive strength f_c'. The required average strength f_{cr}' which is the strength used in mixture proportioning, depends on the expected variability of test results as measured by the coefficient of variation or standard deviation, and on the allowable proportion of tests below the appropriate, specified acceptance criteria.

4.2—Data used to establish the minimum required average strength

To establish the required average strength f_{cr}' an estimate of the variability of the concrete to be supplied for construction is needed. The strength test record used to estimate the standard deviation or coefficient of variation should represent a group of at least 30 consecutive tests. The data used to estimate the variability should represent concrete produced to meet a specified strength close to that specified for the proposed work and similar in composition and production.

The requirement for 30 consecutive strength tests can be satisfied by using a test record of 30 consecutive batches of the same class of concrete or the statistical average of two test records totaling 30 or more tests. If the number of test results available is less than 30, a more conservative approach is needed. Test records with as few as 15 tests can be used to estimate the standard deviation; however, the calculated standard deviation should be increased by as much as 15% to account for the uncertainty in the estimate of the standard deviation. In the absence of sufficient information, a very conservative approach is required and the concrete is proportioned to produce relatively high average strengths.

In general, changes in materials and procedures will have a larger effect on the average strength level than the standard deviation or coefficient of variation. The data used to establish the variability should represent concrete produced to meet a specified strength close to that specified for the proposed work and similar in composition. Significant changes in composition are due to changes in type, brand or source of cementitious materials, admixtures, source of aggregates, and mixture proportions.

If only a small number of test results are available, the estimates of the standard deviation and coefficient of variation become less reliable. When the number of strength test results is between 15 and 30, the calculated standard deviation, multiplied by the appropriate modification factors obtained from Table 4.1, which was taken from ACI 318,

Table 4.1—Modification factors for standard deviation

Number of tests	Modification factors
Less than 15	See Table 4.2
15	1.16
20	1.08
25	1.03
30 or more	1.00

Table 4.2—Minimum required average strength without sufficient historical data

$f'_{cr} = f'_c + 6.9$ MPa (1000 psi)	when $f'_c < 20.7$ MPa (3000 psi)
$f'_{cr} = f'_c + 8.3$ MPa (1200 psi)	when $f'_c \geq 20.7$ MPa (3000 psi) and $f'_c \leq 34.5$ MPa (5000 psi)
$f'_{cr} = 1.10 f'_c + 4.8$ MPa (700 psi)	when $f'_c > 34.5$ MPa (5000 psi)

provides a sufficiently conservative estimate to account for the uncertainty in the calculated standard deviation.

If previous information exists for concrete from the same plant meeting the similar requirements described above, that information can be used to establish a value of standard deviation s to be used in determining f'_{cr}.

Estimating the standard deviation using at least 30 tests is preferable. If it is necessary to use data from two test records to obtain at least 30 strength test results, the records should represent similar concrete mixtures containing similar materials and produced under similar quality control procedures and conditions, with a specified compressive strength f'_c that does not differ by more than 6.9 MPa (1000 psi) from the required strength f'_{cr}. In this case, the pooled standard deviation can be calculated using Eq. (3-3).

When the number of strength test results is less than 15, the calculated standard deviation is not sufficiently reliable. In these cases, the concrete is proportioned to produce relatively high average strengths as required in Table 4.2.

As a project progresses and more strength tests become available, all available strength tests should be analyzed to obtain a more reliable estimate of the standard deviation appropriate for that project. A revised value of f'_{cr}, which is typically lower, may be computed and used.

4.3—Criteria for strength requirements

The minimum required average strength f'_{cr} can be computed using Eq. (4-1a), (4-1b), or, equivalently, (4-2a) or (4-2b), Table 4.2, or Fig. 4.1 or 4.2, depending on whether the coefficient of variation or standard deviation is used. The value of f'_{cr} will be the same for a given set of strength test results regardless of whether the coefficient of variation or standard deviation is used.

$$f'_{cr} = f'_c / (1 - zV) \qquad (4\text{-}1a)$$

$$f'_{cr} = f'_c + zs \qquad (4\text{-}1b)$$

where z is selected to provide a sufficiently high probability of meeting the specified strength, assuming a normal distribution of strength test results. In most cases, f'_c is replaced by a specified acceptance criterion, such as $f'_c - 3.5$ MPa (500 psi) or $0.90 f'_c$.

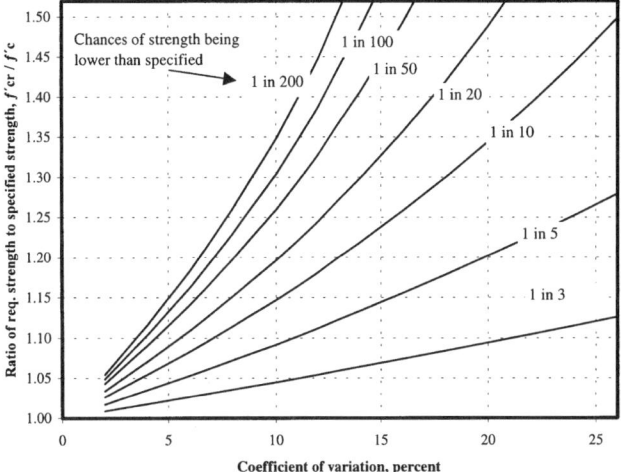

Fig. 4.1—Ratios of required average strength f'_{cr} to specified strength f'_c for various coefficients of variation and chances of falling below specified strength.

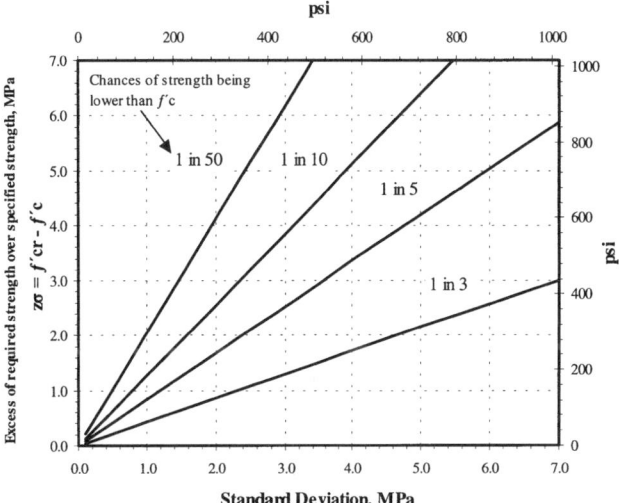

Fig. 4.2—Excess of required average strength f'_{cr} to specified strength f'_c for various standard deviations and chances of falling below specified strength.

When a specification requires computation of the average of some number of tests, such as the average of three consecutive tests, the standard deviation or coefficient of variation of such an average will be lower than that computed using all individual test results. The standard deviation of an average is calculated by dividing the standard deviation of individual test results by the square root of the number of tests (n) in each average. For averages of consecutive tests, Eq. (4-1a) and (4-1b) become:

$$f'_{cr} = f'_c / (1 - zV/\sqrt{n}) \qquad (4\text{-}2a)$$

$$f'_{cr} = f'_c + zs/\sqrt{n} \qquad (4\text{-}2b)$$

The value of n typically specified is 3; this value should not be confused with the number of strength test results used to estimate the mean or standard deviation of the record. Figure 4.3 shows that as the variability increases, f'_{cr} increases and thereby illustrates the economic value of good control.

Table 4.3—Probabilities associated with values of z

Percentages of tests within ± zσ	Chances of falling below lower limit	z
40	3 in 10 (30%)	0.52
50	2.5 in 10 (25%)	0.67
60	2 in 10 (20%)	0.84
68.27	1 in 6.3 (15.9%)	1.00
70	1.5 in 10 (15%)	1.04
80	*1 in 10 (10%)*	*1.28*
90	1 in 20 (5%)	1.65
95	1 in 40 (2.5%)	1.96
95.45	1 in 44 (2.3%)	2.00
98	*1 in 100 (1%)*	*2.33*
99	1 in 200 (0.5%)	2.58
99.73	1 in 741 (0.13%)	3.00

Note: Commonly used values in bold italic.

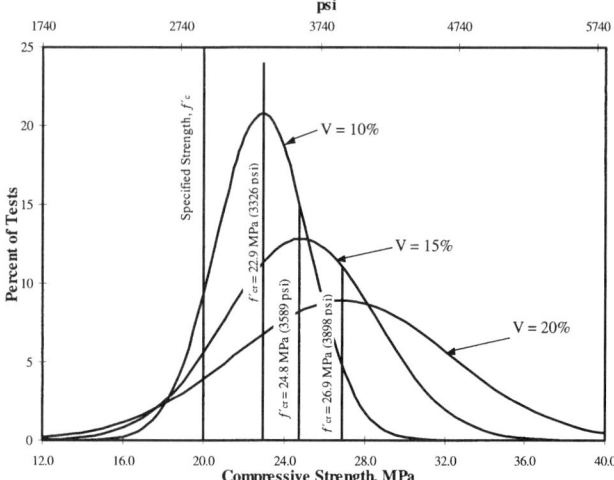

Fig. 4.3—Normal frequency curves for coefficients of variation of 10, 15, and 20%.

Table 4.3 provides values of z for various percentages of tests falling between the mean $+ z\sigma$ and the mean $-z\sigma$.

The amount by which the required average strength f'_{cr} should exceed the specified compressive strength f'_c depends on the acceptance criteria specified for a particular project. The following are criteria examples used to determine the required average strength for various specifications or elements of specifications. The numerical examples are presented in both SI and inch-pound units in a parallel format that have been hard converted and so are not exactly equivalent numerically.

4.3.1 *Criterion no. 1*—The engineer may specify a stated maximum percentage of individual, random strength tests results that will be permitted to fall below the specified compressive strength. This criterion is no longer used in the ACI 318 Building Code, but does occur from time to time in specifications based on allowable strength methods or in situations where the average strength is a fundamental part of the design methodology, such as in some pavement specifications. A typical requirement is to permit no more than 10% of the strength tests to fall below f'_c. The specified strength in these situations will generally be between 21 and 35 MPa (3000 and 5000 psi).

4.3.1.1 *Standard deviation method*—Assume sufficient data exist for which a standard deviation of 3.58 MPa (519 psi) has been calculated for a concrete mixture with a specified strength of 28 MPa. (An example is also given for a mixture with $f'_c = 4000$ psi; these are not equal strengths). From Table 4.3, 10% of the normal probability distribution lies more than 1.28 standard deviations below the mean. Using Eq. (4-1b)

$$f'_{cr} = f'_c + zs$$

$$f'_{cr} = 28 \text{ MPa} + 1.28 \times (3.58) \text{ MPa} = 32.6 \text{ MPa}$$

alternately, $f'_{cr} = 4000 \text{ psi} + 1.28 \times 519 \text{ psi} = 4660 \text{ psi}$

(maintaining appropriate significant figures).

Therefore, for a specified compressive strength of 28 MPa, the concrete mixture should be proportioned for an average strength of not less than 32.6 MPa so that, on average, no more than 10% of the results will fall below f'_c (for a specified strength of 4000 psi, proportioned for not less than 4660 psi).

4.3.1.2 *Coefficient of variation method*—Assume sufficient data exist for which a coefficient of variation of 10.5% has been calculated for a concrete mixture with a specified strength of 28 MPa (or for a mixture with $f'_c = 4000$ psi). From Table 4.3, 10% of the normal probability distribution lies more than 1.28 standard deviations below the mean. Using Eq. (4-1a)

$$f'_{cr} = f'_c / (1 - zV)$$

$$f'_{cr} = 28 \text{ MPa} / [1 - (1.28 \times 10.5/100)] = 32.3 \text{ MPa}$$

alternately, $f'_{cr} = 4000 \text{ psi}/[1 - (1.28 \times 0.105)] = 4620 \text{ psi}$

(maintaining appropriate significant figures).

Therefore, for a specified compressive strength of 28 MPa, the concrete mixture should be proportioned for an average strength of not less than 32.3 MPa so that, on average, no more than 10% of the results will fall below f'_c (for a specified strength of 4000 psi, proportioned for not less than 4620 psi).

4.3.2 *Criterion no. 2*—The engineer can specify a probability that an average of *n* consecutive strength tests will be below the specified compressive strength. For example, one of the acceptance criteria in ACI 318 stipulates that the average of any three consecutive strength test results should equal or exceed f'_c. The required average strength should be established such that nonconformance is anticipated no more often than 1 in 100 times (0.01).

4.3.2.1 *Standard deviation method*—Assume sufficient data exist for which a standard deviation of 3.58 MPa (519 psi) has been calculated for a concrete mixture with a specified strength of 28 MPa (or for a mixture with $f'_c = 4000$ psi). From Table 4.3, 1% of the normal probability distribution lies more than 2.33 standard deviations below the mean. Using Eq. (4-2b)

$$f'_{cr} = f'_c + zs/\sqrt{n}$$

$$f'_{cr} = 28 \text{ MPa} + [(2.33 \times 3.58 \text{ MPa})/\sqrt{3}] = 32.8 \text{ MPa}$$

alternately, $f'_{cr} = 4000 \text{ psi} + [(2.33 \times 519 \text{ psi})/\sqrt{3}] = 4700 \text{ psi}$

(maintaining appropriate significant figures).

Therefore, for a specified compressive strength of 28 MPa, the concrete mixture should be proportioned for an average strength of not less than 32.8 MPa so that, on average, no more than 1% of the moving average of three strength-test results will fall below f'_c (for a specified strength of 4000 psi, proportioned for not less than 4700 psi).

In ACI 318, Eq. (4-2b) is presented in slightly different form. The value 1.34 in ACI 318 is equivalent to the term $z/\sqrt{n} = 2.33/\sqrt{3} = 1.34$, because both z and n are already specified.

4.3.2.2 *Coefficient of variation method*—Assume sufficient data exist for which a coefficient of variation of 10.5% has been calculated for a concrete mixture with a specified strength of 28 MPa (or for a mixture with $f'_c = 4000$ psi). From Table 4.3, 1% of the normal probability distribution lies more than 2.33 standard deviations below the mean. Using Eq. (4-2a)

$$f'_{cr} = f'_c / [1 - (zV/\sqrt{n})]$$

$$f'_{cr} = 28 \text{ MPa}/[1 - (2.33 \times 10.5/100/\sqrt{3})] = 32.6 \text{ MPa}$$

alternately, $f'_{cr} = 4000 \text{ psi}/[1 - (2.33 \times 0.105/\sqrt{3})] = 4660 \text{ psi}$

(maintaining appropriate significant figures).

Therefore, for a specified compressive strength of 28 MPa, the concrete mixture should be proportioned for an average strength of not less than 32.6 MPa so that, on average, no more than 1% of the moving average of three consecutive strength-test results will fall below f'_c (for a specified strength of 4000 psi, proportioned for not less than 4660 psi).

4.3.3 *Criterion no. 3*—The engineer may specify a certain probability that a random individual strength test result will be no more than a certain amount below the specified compressive strength. For example, this criterion is used in ACI 318 by stipulating that no individual strength test result (where a test result is the average of at least two cylinders fabricated from the same batch) falls below f'_c by more than 3.5 MPa (500 psi). An alternative criterion is more appropriate for high-strength concrete. The acceptance criterion for high-strength concrete, 34.5 MPa ($f'_c > 5000$ psi), requires that no individual strength test result falls below 90% of f'_c. These two criteria are equivalent at 34.5 MPa (5000 psi). The minimum required average strength is established so that nonconformance of an individual, random test is anticipated no more often than 1 in 100 times in either case.

4.3.3.1 *Standard deviation method*, $f'_c \leq 34.5$ MPa *(5000 psi)*—Assume sufficient data exist for which a standard deviation of 3.58 MPa (519 psi) has been calculated for a concrete mixture with a specified strength of 28 MPa (or for a mixture with $f'_c = 4000$ psi). From Table 4.3, 1% of the normal probability distribution lies more than 2.33 standard deviations below the mean. Using a modified form of Eq. (4-1b)

$$f'_{cr} = (f'_c - 3.5) + zs$$

$$f'_{cr} = (28 \text{ MPa} - 3.5 \text{ MPa}) + (2.33 \times 3.58 \text{ MPa}) = 32.8 \text{ MPa}$$

alternately, $f'_{cr} = (4000 \text{ psi} - 500 \text{ psi}) + (2.33 \times 519 \text{ psi}) = 4710 \text{ psi}$

(maintaining appropriate significant figures).

Therefore, for a specified compressive strength of 28 MPa, the concrete mixture should be proportioned for an average strength of not less than 32.8 MPa so that, on average, no more than 1% of the individual strength-test results will fall below $f'_c - 3.5$ MPa (for a specified strength of 4000 psi strength, proportioned for not less than 4710 psi).

4.3.3.2 *Standard deviation method*, $f'_c > 34.5$ MPa *(5000 psi)*—Assume sufficient data exist for which a standard deviation of 5.61 MPa (814 psi) has been calculated for a concrete mixture with a specified strength of 60 MPa (or for a mixture with $f'_c = 9000$ psi). From Table 4.3, 1% of the normal probability distribution lies more than 2.33 standard deviations below the mean. Using a modified form of Eq. (4-1b)

$$f'_{cr} = 0.90 \times f'_c + zs$$

$$f'_{cr} = (0.90 \times 60 \text{ MPa}) + (2.33 \times 5.61 \text{ MPa}) = 67.1 \text{ MPa}$$

alternately, $f'_{cr} = 0.90 \times 9000 \text{ psi} + 2.33 \times 814 \text{ psi} = 10,000 \text{ psi}$

(maintaining appropriate significant figures).

Therefore, for a specified compressive strength of 60 MPa the concrete mixture should be proportioned for an average strength of not less than 67.1 MPa so that, on average, no more than 1% of the individual strength-test results will fall below $0.90 f'_c$ (for a specified 9000 psi strength, proportioned for not less than 10,000 psi).

4.3.3.3 *Coefficient of variation method*, $f'_c \leq 34.5$ MPa *(5000 psi)*—Assume sufficient data exist for which a coefficient of variation of 10.5% has been calculated for a concrete mixture with a specified strength of 28 MPa (or for a mixture with $f'_c = 4000$ psi). From Table 4.3, 1% of the normal probability distribution lies more than 2.33 standard deviations below the mean. Using a modified form of Eq. (4-1a):

$$f'_{cr} = (f'_c - 3.5)/(1 - zV)$$

$$f'_{cr} = (28 \text{ MPa} - 3.5 \text{ MPa})/[1 - (2.33 \times 10.5/100)] = 32.4 \text{ MPa}$$

alternately, $f'_{cr} = (4000 \text{ psi} - 500 \text{ psi})/[1 - (2.33 \times 0.105)] = 4630 \text{ psi}$

(maintaining appropriate significant figures).

Therefore, for a specified compressive strength of 28 MPa, the concrete mixture should be proportioned for an average strength of not less than 32.4 MPa so that, on average, no more than 1% of the individual strength-test results will fall below $f'_c - 3.5$ MPa (for a specified strength of 4000 psi, proportioned for not less than 4630 psi).

4.3.3.4 *Coefficient of variation method*, $f'_c > 34.5$ MPa *(5000 psi)*—Assume sufficient data exist for which a coefficient of variation of 8.2% has been calculated for a concrete mixture with a specified strength of 60 MPa (or for a mixture with $f'_c = 9000$ psi). From Table 4.3, 1% of the normal prob-

Table 5.1—Probability of at least one event in *n* tests for selected single-event probabilities

n	Single event probability = 1.5%	Single event probability = 10%
1	1.5%	10.0%
5	7.3%	41.0%
7	10.0%	54.3%
10	14.0%	65.1%
20	26.1%	87.8%
50	53.0%	99.5%
100	77.9%	Approximately 100%

ability distribution lies more than 2.33 standard deviations below the mean. Using a modified form of Eq. (4-1a)

$$f'_{cr} = 0.90 \times f'_c/(1 - zV)$$

$$f'_{cr} = (0.90 \times 60 \text{ MPa})/[1 - (2.33 \times 8.2/100)] = 66.8 \text{ MPa}$$

alternately, $f'_{cr} = (0.90 \times 9000 \text{ psi})/[1 - (2.33 \times 0.082)] = 10{,}010 \text{ psi}$

(maintaining appropriate significant figures).

Therefore, for a specified compressive strength of 60 MPa, the concrete mixture should be proportioned for an average strength of not less than 66.8 MPa so that, on average, no more than 1% of the individual strength test results will fall below $0.90 f'_c$ (for a specified 9000 psi strength, proportioned for not less than 10,010 psi).

4.3.4 *Multiple criteria*—In many instances, multiple criteria will be specified. ACI 318 and 318M require that concrete strengths conform to both individual test criteria and the moving average of three test criteria. Because both criteria are in effect, the required average compressive strength f'_{cr} should meet or exceed all requirements; that is, f'_{cr} should be the largest strength calculated using all relevant criteria. For example, assume sufficient data exist for which a coefficient of variation 8.2% has been calculated for a concrete mixture with a specified strength of 60 MPa (8700 psi). The required average strength for this concrete mixture should meet both of the following criteria:

1. Individual criterion (see 4.3.3.4): $f'_{cr} = 0.90 \times f'_c/(1 - 2.33V) = 66.8$ MPa (9690 psi).
2. Moving average criterion (see 4.3.2.2): $f'_{cr} = f'_c/(1 - 2.33V/\sqrt{3}) = 67.4$ MPa (9780 psi).

The moving average criterion governs, because 67.4 MPa > 66.8 MPa (9780 psi > 9690 psi) and f'_{cr} should be the largest strength calculated using all relevant criteria.

CHAPTER 5—EVALUATION OF DATA
5.1—General

Evaluation of strength data is required in many situations. Three commonly required applications are:
- Evaluation for mixture submittal purposes;
- Evaluation of level of control (typically called quality control); and
- Evaluation to determine compliance with specifications.

A major purpose of these evaluations is to identify departures from desired target values and, where possible, to assist with the formulation of an appropriate response. In all cases, the usefulness of the evaluation will be a function of the amount of test data and the statistical rigor of the analysis. Applications for routine quality control and compliance overlap considerably. Many of the evaluation tools or techniques used in one application are appropriate for use in the other.

Techniques appropriate for concrete mixture submittal evaluation were reviewed in Chapter 4. Techniques for routine quality control and compliance applications are provided and discussed in this chapter. Criteria for rejecting doubtful results, determination of an appropriate testing frequency, and guidelines for additional test procedures are also discussed.

It is informative to determine the likelihood of various outcomes when there is at most a 1% probability of a test less than $f'_c - 3.5$ MPa (500 psi) and, at most, a 1% probability that the moving average of three consecutive tests will be less than f'_c. The maximum probability that at least one event will occur in *n* independent tests may be estimated using Eq. (5-1) (Leming 1999)

$$Pr\{\text{at least 1 event} \mid n \text{ tests}\} = 1 - (1 - p)^n \quad (5\text{-}1)$$

where *p* is the probability of a single event.

One value of interest for *p* is the single event probability of noncompliance with the strength criteria in ACI 318. Because *p* includes both possible cases ($f'_c - (3.5$ MPa [$f'_c - 500$ psi] and the moving average of three consecutive tests less than f'_c), *p* lies between 1.0 and 2.0%. In the absence of more details, the probability of a single test failing to meet the strength criteria of ACI 318 may be assumed to be 1.5%. Table 5.1 gives the probabilities of at least one occurrence of an event given various numbers of independent tests *n* when the single event probability *p* equals 1.5% (a test does not meet ACI 318 strength criteria) and 10% (a test falls below f'_c).

The probability is not trivial even for relatively small projects. For example, there is approximately a 10% probability of having at least one noncompliant test and slightly greater than 50% probability of having at least one test fall below f'_c for a project with only seven tests. There is a very high probability of such an occurrence on most projects, and a virtual certainty on large projects, even if the variation is due exclusively to random effects, and the minimum average strength was determined accurately using statistically valid methods. The probabilities are reduced somewhat for larger projects due to the effects of interference; however, the probabilities are still appreciable (Leming 1999).

5.2—Numbers of tests

For a particular project, a sufficient number of tests should be made to ensure accurate representation of the concrete. A test is defined as the average strength of at least two specimens of the same age fabricated from a sample taken from a single batch of concrete. The frequency of concrete tests can be established on the basis of time elapsed or volume placed. The engineer should establish the number of tests needed based on job conditions.

A project where all concrete operations are supervised by one engineer provides an excellent opportunity for control and for accurate estimates of the mean and standard deviation with a minimum of tests. Once operations are progressing smoothly, tests taken each day or shift, depending on the volume of concrete produced, can be sufficient to obtain data that reflect the variations of the concrete as delivered. The engineer can reduce the number of specimens required by the project specifications as the levels of control of the producer, the laboratory, and the contractor are established. To avoid bias, all sampling for acceptance testing should be conducted using randomly selected batches of concrete.

For routine building construction, ACI 318 requires at least one test per day; one test every 115 m^3 (150 yd^3) or one test for every 460 m^2 (5000 ft^2) of the surface area of slabs and walls, but permits the engineer to waive testing on quantities less than 40 m^3 (50 yd^3). Testing should be conducted so that each of these criteria are satisfied. These testing frequencies generally result in testing concrete in one out of 10 to 20 trucks.

Testing more frequently than this can slow the construction process and should be specified only for compelling reasons. For example, more frequent testing is recommended for specialized or critical members or applications. For members where the structural performance is particularly sensitive to compressive strength, a testing frequency of one test for every 80 m^3 (100 yd^3) may be appropriate; one test for every 40 m^3 (50 yd^3) would be appropriate only for critical applications. Testing each load of concrete delivered for potential strength is rarely required.

In general, make a sufficient number of tests so that each different class of concrete placed during any one day will be represented by at least one test; a minimum of five tests should be conducted for each class of concrete on a given project. Guidelines for routine testing requirements can also be found in ACI 301, ACI 318, and ASTM C 94.

5.3—Rejection of doubtful specimens

The practice of arbitrary rejection of strength test results that appear too far out of line is not recommended because the normal distribution anticipates the possibility of such results. Discarding test results indiscriminately can seriously distort the strength distribution, making analysis of results less reliable. Occasionally, the strength of one cylinder from a group made from a sample deviates so far from the others as to be highly improbable. If questionable variations have been observed during fabrication, curing, or testing of a specimen, the specimen should be rejected on that basis alone.

ASTM E 178 provides criteria for rejecting the test result for one specimen in a set of specimens. In general, the result from a single specimen in a set of three or more specimens can be discarded if its deviation from a test mean is greater than three times the previously established within-test standard deviation (see Chapter 3), and should be accepted with suspicion if its deviation is greater than two times the within-test standard deviation. The test average should be computed from the remaining specimens. A test, that is, the average of all specimens of a single sample tested at the same age, should not be rejected unless it is very likely that the specimens are faulty. The test represents the best available estimate for the sample.

5.4—Additional test requirements

The potential compressive strength and variability of concrete is normally based on test results using a standard cylinder which has been sampled, molded, and cured initially in accordance with ASTM C 31, then moist cured at a controlled temperature (23 ± 2 C [73 ± 3 F]) until the specified test age, normally 28 days. When the nominal size of the coarse aggregate in the mixture exceeds 50 mm (2 in.), a larger test specimen is used, or the larger aggregate is removed by wet sieving. Analysis of concrete strength variability is based on these standard-sized specimens. Specimens of concrete made or cured under other than standard conditions provide additional information but should be analyzed and reported separately. Specimens that have not been produced, cured, or tested under standard conditions may or may not accurately reflect the potential concrete strength. Discrepancies and deviations from standard testing conditions should be noted on strength test reports.

The strength of concrete at later ages, such as 56, 91, or 182 days may be more relevant than the 28-day strength, particularly where a pozzolan or cement of slow strength gain is used or heat of hydration is a concern. Some elements or structures will not be loaded until the concrete has been allowed to mature for longer periods and advantage can be taken of strength gain after 28 days. Some concretes have been found to produce strengths at 28 days, which are less than 50% of their ultimate strength. Others, made with finely ground, Type III portland cements, may not gain appreciable strength after 28 days.

If design is based on strength at later ages, it may be necessary to correlate these later strengths with strength at 28 days because it is not always practical to use later-age specimens for concrete acceptance. This correlation should be established by field or laboratory tests before construction starts. If concrete batching plants are located in one place for long enough periods, establish this correlation for reference even though later-age concrete may not be immediately involved.

Many times, particularly in the early stages of a job, it is necessary to estimate the strength of concrete being produced before the 28-day strength results are available. Concrete cylinders should be made and tested from the same batch at seven days and, in some instances, at three days. Testing at very early ages using accelerated test procedures, such as found in ASTM C 684, can also be adopted. The 28-day strength can be estimated on the basis of a previously established correlation for the specific mixture using the method described in ASTM C 918. These early tests provide only an indication of acceptable performance; tests for the purposes of acceptance are still typically conducted at 28 days and are often the legal standard. A minimum of two cylinders are required for a valid test and more are sometimes specified.

Curing concrete test specimens at the construction site and under job conditions, that is, field or job-cured specimens, is sometimes recommended or required in such applications as fast-track construction or post-tensioning, because an acceptable in-place strength has to be attained, particularly at early ages, before the member can be safely loaded or stressed. Tests of job-cured specimens are highly desirable or necessary when determining the time of form removal, particularly in cold weather, and when establishing the strength of steam-cured concrete or concrete pipe and block. In addition, the adequacy of curing by the contractor can be evaluated only by monitoring strength gain on the job site. Do not confuse nor replace these special test requirements with the required standard control tests.

5.5—Basic quality-control charts

Quality-control charts have been used by manufacturing industries for many years as aids in reducing variability, increasing production efficiency, and identifying trends as early as practicable. Well-established methods for setting up such charts are outlined in convenient form in the ASTM

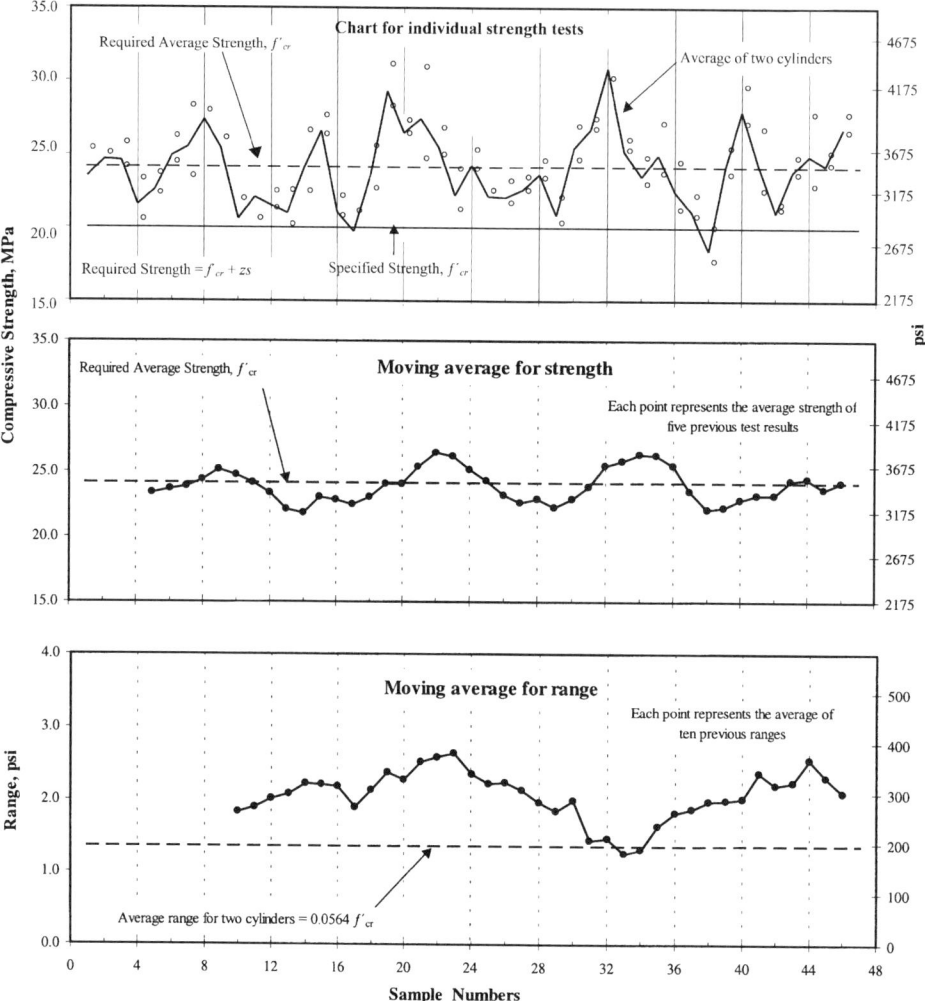

Fig. 5.1—Three simplified quality control charts (individual strength tests, moving average of five strength tests, and range of two cylinders in each test and moving average for range).

Manual on Presentation of Data and Control Chart Analysis, MNL 7. Trends become more readily apparent based on the pattern of previous results and limits established from ASTM MNL 7. Data that fall outside established limits indicate that something has affected the control of the process, and some type of action or interference with the existing process is typically required. In general, these action or process interference limit values are established using the guidelines published in this document, based on contract specifications or other values at which action should be taken. Frequently, the action or interference limits are equal to the acceptance criteria specified for a particular project.

Figure 5.1 illustrates three simplified charts prepared specifically for concrete control and are combined into one diagram. This technique permits evaluation of all charts simultaneously, which can ease analysis. While these charts may not contain all the features of formal control charts, they are useful to the engineer, architect, contractor, and supplier. Control charts are strongly recommended for concrete in continuous production over considerable periods.

5.5.1 *Simple strength chart*—The top chart in Fig. 5.1 shows the results of all strength tests plotted in succession based on casting date. The target for the average strength is established as indicated by Eq. (4-1a), Eq. (4-1b), or Table 4.2. The chart often includes the specified strength and may include the acceptance criteria for individual tests. This chart is useful because it shows all of the available data but it can be difficult to detect meaningful shifts in a timely fashion.

5.5.2 *Moving average strength*—The middle chart in Fig. 5.1 shows the moving average of consecutive tests. This type of chart reduces the noise and scatter in the individual test chart. Trends in performance are more easily identified and will show the influence of effects, such as seasonal changes and changes in materials, more effectively. The chart often includes the acceptance criteria f_c' when the moving average of three tests is plotted.

The larger the number of tests averaged, the more powerful the chart is in helping identify trends. There is an obvious trade-off with timeliness, however. A trend should be identified as soon as possible so that appropriate corrective actions may be taken. Because the moving average of three consecutive strength tests is one of the compliance criteria of ACI 318, this parameter is frequently tracked in a control chart. Because tracking the moving average of three tests may not provide sufficient analytical power, the moving average of five consecutive strength tests is also frequently used. The number of tests averaged for this control chart and the appropriate interference limit can be varied to suit each job.

The concrete supplier with a large number of tests for a particular mixture can elect to track the moving average of 10 or 15 tests. A target value can be established based on f'_{cr}. While requiring significant amounts of data, any trends detected with this approach will necessarily be strong and shifts in average strength can be readily detected. The averages of 10 and 15 tests can also be used in mixture submittal documentation.

5.5.3 *Testing variability*

5.5.3.1 *Purpose*—The lower chart in Fig. 5.1 shows the moving average of the range, the maximum difference between companion cylinders comprising a single strength test, which is used to monitor the repeatability of testing. The laboratory has the responsibility of making accurate tests, and concrete will be penalized unnecessarily if tests show greater variations or lower average strength levels than actually exist. Because the range in strength between companion specimens from the same sample can be assumed to be the responsibility of the laboratory, a control chart for ranges should be maintained by the laboratory as a check on the uniformity of its operations. These changes will not reveal day-to-day differences in testing, curing, capping, and testing procedures or testing procedures that affect measured strength levels over long periods.

The average range of the previous 10 consecutive tests (sets of companion cylinders as discussed in Section 3.4.1) is typically plotted. The interference limits for this control chart are based on average strength and desired level of control.

5.5.3.2 *Calculation of acceptable testing variation*—Calculation of the acceptable range between companion cylinders of a test depends on the number of specimens in the group and the within-test variation, as discussed in Chapter 3. The following process can be used to establish interference limits for the moving average range control chart.

The expected value of the average range $\overline{R}_m$ can be determined by reformulating Eq. (3-5) as shown in Eq. (5-2).

$$\overline{R}_m = f'_{cr} V_1 d_2 \qquad (5\text{-}2)$$

The within-test coefficient of variation V_1 should not be greater than 5% for good control (Table 3.3). Therefore, the estimate of the corresponding average range will be

$$\overline{R}_m = (0.05 \times 1.128) f'_{cr} = 0.05640 f'_{cr} \qquad (5\text{-}3a)$$

for groups of two companion cylinders, or

$$\overline{R}_m = (0.05 \times 1.693) f'_{cr} = 0.08465 f'_{cr} \qquad (5\text{-}3b)$$

for groups of three companion cylinders. These interference limits are effective only after the average range, computed from companion cylinder strengths from at least 10 strength tests, has been calculated.

To be fully effective, maintain control charts on each project for the duration of the project. The testing laboratory should, as a minimum, maintain a control chart for average range and may also offer other control charts as a service to the engineer or architect. Concrete suppliers can track the moving average range on a mixture by mixture basis, because a single mixture can be used on multiple projects. Many suppliers track individual projects to obtain data for their own use.

5.6—Other evaluation techniques

A number of other techniques exist for evaluating series of data for quality-control purposes. As with basic control charts, these techniques were developed for general industrial applications but can be adapted for use with concrete properties. A complete description of these techniques is beyond the scope of this document, but the general outline of the cumulative sum (CUSUM) procedure, along with guidance on interpretation as applied to concrete properties, particularly compressive strength, is provided. A much more detailed description of analytical techniques and interpretation of the CUSUM technique can be found in Day (1991) and Dewar (1995); a simple example of this technique is provided in Appendix A.

5.6.1 *Overall variability and concrete supplier's variability*—In conventional practice, the mean compressive strength is estimated with as few as 10 tests, while at least 15 tests are needed to estimate the standard deviation. Changes in the mixture materials or proportions will have a larger effect on the average strength level than on the standard deviation. For these reasons, most control charts are based on averages of compressive strength. Monitoring the overall standard deviation can also provide insight into changes in the level of control or variability of production or raw materials for the concrete supplier.

An estimate of variation due to testing, the within-test standard deviation, can be obtained from the average range chart or by direct computation. As discussed in Chapter 3, the combined variation due to variation in raw materials and production, which can be termed the concrete supplier's or producer's variability, can be determined knowing the overall standard deviation and the within-test standard deviation. The producer's variability, as measured by the standard deviation, is the square root of the difference of the square of the overall standard deviation and the within-test standard deviation, as shown in Eq. (3-8), provided in slightly different form as Eq. (5-4).

$$s_{Producer} = \sqrt{s^2_{Overall} - s^2_{Within\ test}} \qquad (5\text{-}4)$$

The concrete supplier can directly track the variability of the production process. If the within-test standard deviation is reasonably consistent, as it is in a well-run testing program, the supplier can simply track overall standard deviation, which is easier. For a constant within-test variation, changes in the overall standard deviation can indicate changes in either the raw materials or the production of concrete and are, therefore, of value to the concrete supplier.

Control charts should incorporate a moving standard deviation of at least 10 and preferably 15 tests. With modern, computer-based spreadsheets this type of control chart is not difficult to implement. Due to the large number of tests required, the usefulness of this control chart to rapidly identify changes in the process is limited, however. Another technique (CUSUM), described Section 5.6.2, typically provides rapid identification of changes in various measured properties of concrete.

5.6.2 *CUSUM*—In both quality control and problem resolution there is a need to identify assignable causes in average strength level or in variability of strength. Early detection of changes in the average strength level is useful so that causes may be identified and steps taken to avoid future problems or reduce costs. This requires being able to distinguish between random variations and variations due to assignable causes.

The cumulative sum (CUSUM) chart provides a method for detecting relatively small but real changes in average concrete strength or some other aspect of concrete performance. It can also help identify approximately when those shifts began and the approximate size of the shift. CUSUM will generally provide greater sensitivity in detecting a small, systemic change in average strength than the basic control charts discussed in this chapter and will detect these changes faster (Box, Hunter, and Hunter 1978; Day 1991; Dewar 1995; Day 1995).

There are limitations in using a CUSUM chart, particularly when data are highly variable, but the technique is only slightly more complicated than conventional strength analysis and is easily implemented either manually or using a spreadsheet or commercially available computer program. As with any single technique, the conclusions reached using a CUSUM chart should be confirmed by additional analysis or investigation before making critical decisions.

Although probably most commonly used to monitor compressive strength, it can be used with any number of variables. Day (1995) reports successfully using CUSUM charts to monitor a variety of concrete properties. He also notes that by monitoring multiple CUSUM charts and tracking a variety of properties simultaneously, the probability that a change will be missed or misdiagnosed is reduced. A review of the theory of the CUSUM technique and an example are provided in Appendix A.

CHAPTER 6—REFERENCES
6.1—Referenced standards and reports

The standards and reports listed below were the latest editions at the time this document was prepared. Because these documents are revised frequently, the reader is advised to contact the sponsoring group if it is desired to refer to the latest version.

American Concrete Institute
301 Specifications for Structural Concrete
318 Building Code Requirements for Structural Concrete and Commentary

ASTM
MNL7 Manual 7 on Presentation Data and Control Chart Analysis, 6th Edition
C 31 Practice for Making and Curing Concrete Test Specimens in the Field
C 39 Standard Test Method for Compressive Strength of Cylindrical Concrete Specimens
C 94 Specification for ready-Mixed Concrete
C 684 Standard Test Method for Making, Accelerated Curing, and Testing Concrete Compression Test Specimens
C 802 Practice for Conducting an Interlaboratory Test Program to Determine the Precision of Test Methods for Construction Materials
C 918 Standard Test Method for Developing Early-Age Compression Test Values and Projecting Later-Age Strengths
D 3665 Standard Practice for Random Sampling of Construction Materials
E 178 Practice for Dealing with Outlying Observations

American Association of State Highway & Transportation Officials
TP 23 Edition 1A—Standard Test Method for Water Content of Freshly Mixed Concrete Using Microwave Oven Drying

British Standards Institution
BS 5703-3 Guide to data analysis and quality control using CUSUM techniques. CUSUM methods for process/quality control by measurement

These publications may be obtained from the following organizations:

AASHTO
444 N. Capitol St. NW Ste 249
Washington, D.C. 20001
www.aashto.org

American Concrete Institute
P.O. Box 9094
Farmington Hills, Mich. 48333-9094
www.concrete.org

ASTM International
100 Barr Harbor Dr.
West Conshohocken, Pa. 19428
www.astm.org

British Standards Institution
389 Chiswick High Rd.
London W4 4AL UK
www.bsi.or.uk

6.2—Cited references

Anderson, F. D., 1985, "Statistical Controls for High Strength Concrete," *High Strength Concrete,* SP-87, American Concrete Institute, Farmington Hills, Mich., pp. 71-82.

Box, G. E. P.; Hunter, W. G.; and Hunter, J. S., 1978, *Statistics for Experiments*, Wiley & Sons, New York, 653 pp.

Brown, B. V., 1984, "Monitoring Concrete by the CUSUM System," *Concrete Digest No. 6*, The Concrete Society, London, 8 pp.

Cook, J. E., 1982, "Research and Application of High Strength Concrete Using Class C Fly Ash," *Concrete International*, V. 4, No. 7, July, pp. 72-80.

Cook, J. E., 1989, "10,000 psi Concrete," *Concrete International*, V. 11, No. 10, Oct., pp. 67-75.

Day, K. W., 1995, *Concrete Mix Design, Quality Control and Specification*, E&FN Spon, 461 pp.

Day, K. W., 1991, "Computerized Concrete QC Using Spreadsheets and CUSUM Graphs," *ACI Compilation No. 13*, American Concrete Institute, Farmington Hills, Mich.

Dewar, J. D., 1995, "Developments in CUSUM Control Systems for Concrete Strength," *Proceedings of the 11th ERMCO Congress, Istanbul*, Dec.

Erntroy, H. C., 1960, "The Variation of Works Test Cubes," *Research Report* No. 10, Cement and Concrete Association, London, 28 pp.

Leming, M. L., 1999, "Probabilities of Low Strength Events in Concrete," *ACI Structures Journal*, V. 96, No. 3, May-June, pp. 369-376.

Metclaf, J. B., 1970, "The Specification of Concrete Strength, Part II, The Distribution of Strength of Concrete for Structures in Current Practice," *RRL Report* No. LR 300, Road Research Laboratory, Crawthorne, Berkshire, pp. 22.

Murdock, C. J., 1953, "The Control of Concrete Quality," *Proceedings,* Institution of Civil Engineers (London), V. 2, Part 1, July, pp. 426-453.

Natrella, M. G., 1963, "Experimental Statistics" *Handbook* No. 91 (reprinted 1966 with corrections), U.S. Department of Commerce, National Bureau of Standards (now National Institute of Standards and Technology, NIST, Gaithersburg, Md.).

Neville, A. M., 1959, "The Relation Between Standard Deviation and Mean Strength of Concrete Test Cubes," *Magazine of Concrete Research* (London), V. 11, No. 32, July, pp. 75-84.

Rüsch, H., 1964, "Statistical Quality Control of Concrete," *Materialprufung* (Dusseldorf), V.6, No. 11, Nov., pp. 387-394.

APPENDIX A—EXAMPLES OF CUSUM TECHNIQUE
A.1—Introduction

The cumulative sum, or CUSUM, chart can be used to help detect relatively small changes or shifts in average concrete strength or other concrete characteristics relatively quickly. This Appendix reviews the theory and demonstrates the application of the technique, which is only slightly more complicated than conventional strength analysis and is easily implemented either manually or using computer-based spreadsheets. As with any technique, the conclusions reached using a CUSUM chart should be confirmed by additional analysis or investigation before making critical decisions.

A.2—Theory

Deviations of individual test results from the mean have a normal distribution even if the parent distribution is not normal. Because the distribution of concrete compressive strength frequently approximates a normal distribution, the distribution of deviations from the mean strength is normal to a very good approximation. The average deviation from the mean is approximately zero for a stable process. Therefore, if ε_i is the difference between the average compressive strength and the i-th compressive strength test, or

$$\varepsilon_i = \bar{X} - X_i \qquad \text{(A-1)}$$

where $\bar{X}$ is the average compressive strength (established over a suitable time period), and X_i is the i-th compressive strength test, then:

$$\sum_{i=1}^{N} \varepsilon_i = \sum_{i=1}^{N} (\bar{X} - X_i) \approx 0 \qquad \text{(A-2)}$$

as long as the average strength does not change and the number of tests (Nt) is sufficiently large.

If a change occurs in some element of the concrete materials, production, handling, testing, in seasonal variation, or any other assignable cause, variation deviations of test results about the mean are no longer random and ε_i will no longer average 0. If the assignable cause is constant, the sum of ε_i will change in a linear fashion

$$\sum_{i=1}^{N} \varepsilon_i = \sum_{i=1}^{N} ((\bar{X} + \delta) - X_i) \approx (N - m)\delta \qquad \text{(A-3)}$$

where δ is the value of the change in the average strength, and m is the test in the sequence at which the change occurs. A positive δ means that there has been an increase in the average strength and the cumulative sum of the differences between the original average strength and the individual tests increases approximately linearly. If δ is negative, the average strength has decreased and the cumulative sum will decrease approximately linearly.

A shift in the average compressive strength can be detected by plotting the cumulative sum of the ε_i in sequence. A change in the slope of the CUSUM plot indicates a difference in the average strength from the assumed value. Once a trend is detected, further analysis of both the CUSUM chart and the concrete testing, handling, materials, production, or environment should be conducted to determine the likely source of the change.

A.3—Calculations

Previous data for a certain concrete mixture, produced to provide an f_c' of 30 MPa (4350 psi), indicate an average strength of 35.8 MPa (5190 psi). During a project, sequential compressive strength data become available. The CUSUM chart may be constructed from the data as shown in Table A.1. Sample calculations are shown for the first few entries.

The moving average of three tests (MA3) is also provided, because it is a commonly monitored quality-control variable. All data (compressive strengths, CUSUM, averages, and standard deviation) are reported to three significant figures.

Using these 19 test results only, the average compressive strength is 34.8 MPa (5050 psi) and the sample standard deviation is 2.41 MPa (350 psi). Based on only these 19 strength test results, the required average strength f_{cr}' is 33.5 MPa ($f_{cr}' = 30$ MPa plus the larger of (1.34 × 2.64) or (2.33 × 2.64 – 3.5), where 2.64 is the product of the standard deviation (2.41 MPa) and the interpolated modification factor from Table 4.1). It may be seen that:

1. The low standard deviation indicates apparent excellent control;

2. The average strength is greater than f_c', f_{cr}' but 1.0 MPa (150 psi) less than the average strength determined from the previous data;

3. There are no instances where a test is less than $f_c' - 3.5$ MPa (500 psi); and

Table A.1—Data for CUSUM example

No.	Test result, MPa	Difference, MPa	CUSUM, MPa	MA3, MPa
1	(37.1 + 36.9)/2 = 37.0 (average of two cylinders)	37.0 – 35.8 = 1.2	1.2	—
2	34.7	34.7 – 35.8 = –1.1	1.2 + (–1.1) = 0.1	—
3	32.8	32.8 – 35.8 = –3.0	0.1 + (–3.0) = –2.9	34.8
4	37.8	37.8 – 35.8 = 2.0	–2.9 + 2.0 = –0.9	35.1
5	35.2	–0.6	–1.5	35.3
6	36.5	0.7	–0.8	36.5
7	39.6	3.8	3.0	37.1
8	37.6	1.8	4.8	37.9
9	33.6	–2.2	2.6	36.9
10	33.6	–2.2	0.4	34.9
11	35.1	–0.7	–0.3	34.1
12	31.8	–4.0	–4.3	33.5
13	36.4	0.6	–3.7	34.4
14	32.5	–3.3	–7.0	33.6
15	31.0	–4.8	–11.8	33.3
16	31.7	–4.1	–15.9	31.7
17	37.0	1.2	–14.7	33.2
18	34.5	–1.3	–16.0	34.4
19	32.9	–2.9	–18.9	34.8

Notes: No. is the sequence number. Test is the average compressive strength, MPa; in this case, of at least cylinder strengths. Difference is the difference, MPa, between the compressive strength test result and the previously determined average strength. CUSUM is the cumulative sum, MPa, of the differences. MA3 is the moving average of three consecutive compressive test results, MPa.

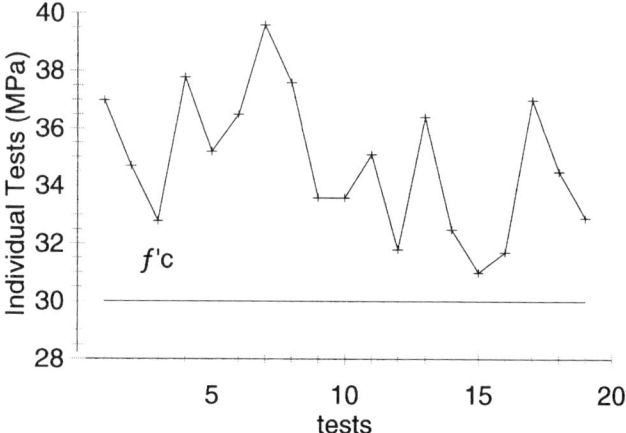

Fig. A.1—Individual test QC chart.

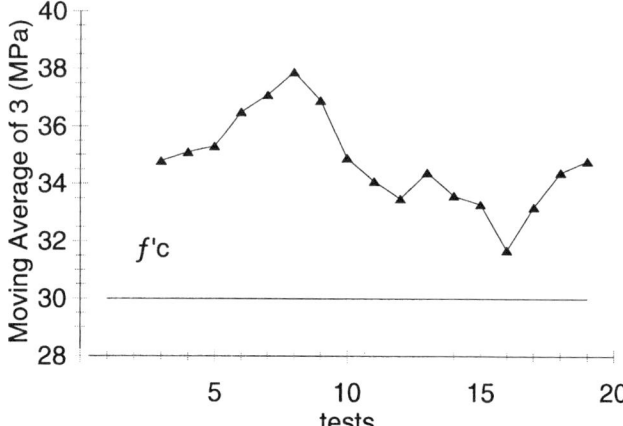

Fig. A.2—Moving average of three QC chart.

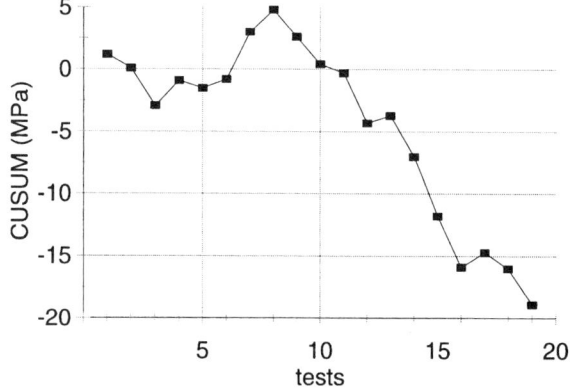

Fig. A.3—CUSUM QC chart for compressive strength.

4. There are no instances where a moving average of a three result is less than f_c'.

All of these indicate satisfactory performance contractually and a process apparently in control.

Simple control charts (Fig. A.1 and A.2) do not indicate any significant problems, although the moving average does trend slightly lower for a period of time. The CUSUM chart, however, (Fig. A.3) clearly indicates that a shift has occurred. A decrease in the average strength level apparently originates no later than the 10th strength test.

A simple estimate of the decrease in strength level that occurred can be made from the slope of the CUSUM chart. The slope from Test No. 10 to Test No. 19 can be estimated as -18.9 (the cumulative sum of the differences at Test No. 19) divided by 9 (19-10 tests), or about 2.1 MPa (300 psi).

A.4—Analysis and comparison with conventional control charts

The preceding example demonstrates several of the potential advantages of the CUSUM chart method. No obvious indication of a change in the data is found in a simple plot of the strength data itself (Fig. A.1), because there is too much scatter due to random variation to easily detect trends or small changes.

Detection of trends or changes in variables can be provided by moving average charts, which improve trend detection by reducing the effect of random variation. Increasing the number of data points averaged increases the ease with which the trend is detected and improves the reliability of the trend identification, that is, the likelihood that the trend indicates a real change. Improvement comes at the price of having to wait for more data points. While the moving average of three provides some improvement in trend detection, averaging over only three tests is not a strong indicator.

The moving average of three charts (Fig. A.2) does show a slightly lower trend in the data for a period of time. It, however, is not immediately apparent from Fig. A.2 that a significant change has occurred, or that if it has occurred, what the size of the change is or whether the trend has, in fact, been reversed near the end of the available data. Additional statistical evaluation might be initiated, but it is not clear that any corrective action is warranted and, in practice, none would probably be undertaken based on this analysis alone.

The primary advantages of the CUSUM chart are that small changes may be detected sooner than with the other methods described, and the timing and size of these changes may be estimated directly. As with any analytical tool, the CUSUM method has some limitations.

A.5—Management considerations of interference

A perfect technique to identify shifts in average strength will identify all real shifts without falsely classifying a random variation as a shift. Practical techniques balance the two types of errors:
- Type I (rejecting a true shift); and
- Type II (accepting a false shift) errors.

The probability of an error increases when analysis is based on fewer data points, or shorter runs, but analysis based on shorter runs is frequently preferable so that deviations can be corrected as soon as possible. Both types of errors have associated costs.

An unexpected decrease in average strength will typically prompt both corrective action to increase the average reported strength and an investigation to determine the source(s) of the decrease. There is a management cost associated with this investigation, and there may be a cost associated with at least a temporary increase in cost of the concrete as the average strength of the mixture is increased. These costs can be offset by the reduction in risk associated with a low strength on the job. If no real problem is detected or subsequent analysis indicates the original analysis was incorrect, the losses are real but may be small compared with the reduction in potential costs. Overreaction and overcorrection can also cause problems, however.

Interference with the process in the absence of an assignable cause can lead to several difficulties. Once a change in the average strength has been implemented, the change in the average affects the CUSUM chart as would any other assignable cause. Both the chart and the process should be "rezeroed" to the new average to account for the interference with the process. Multiple changes over a relatively short time can shift the true average sufficiently from the presumed average that the chart provides considerably less useful information. Over-correction should also be avoided due to the additional costs of unnecessary changes and to inducing more variation in the data than would have occurred in the absence of the interference.

The relative costs of these two errors versus the delay in identifying a real change usually mean that interference will occur more often than actually needed, but analyze each situation separately. Determining when to change the average strength is not always obvious.

A.6—Establishing limits for interference

Interference as early as possible is usually desirable. Analysis of the CUSUM chart can provide an estimate of δ using the slope and the approximate time when the change occurred using the intercept with 0. In general, the steeper the slope, the stronger the trend and the larger the change; the longer the trend, the more certain the change. In a fast-moving project, nine or 10 test results can easily accrue in a short period of time and early interference can be impractical. Where results between test number 10 and test number 19, in the example, accrue over several days, the strength of the trend makes it unlikely that a change in the process would be delayed past, perhaps, test number 15, or only after about five data points.

In the example data used, the change is readily apparent due to the relatively low standard deviation and relatively large strength change (about 2.1 MPa [300 psi]). Smaller changes in average strength result in flatter slopes that take a longer time to identify. Identification in these situations is not as critical as larger changes resulting in steeper slopes that are easier to identify conclusively. When small changes in average strength occur combined with high variability, identification in a timely manner can be difficult. Both qualitative and statistically rigorous guidelines for interference depend on variability and the required level of certainty in the analysis. In practical problems, the costs of changes relative to the effect of the changes are be considered.

To determine the cause or causes of a change, the time at which the change occurred should be estimated relatively accurately. In the example data, it is not clear from Fig. A.3 if the change occurred at the 10th test, the 8th test, or some other test. Methods for determining when a change in the average strength has occurred and when a run is simply due to random variation based on statistically rigorous analysis are desirable; however, they are neither self evident nor as powerful as might be desired.

Day (1995) and Brown (1984) report the use of a truncated V-mask as described in BS 5703 to identify when a statistically significant change has occurred. The V-mask, which is a linear function of the process standard deviation, was developed for any CUSUM analysis. Day (1995), however, notes that the V-mask for concrete strength CUSUM analysis often requires many data points and that simple examination of the CUSUM chart is frequently adequate for an experienced concrete analyst, particularly if multiple measures of concrete mixture behavior are plotted simultaneously. Day reports that a real change in the concrete mixture can frequently be identified with only a few points on CUSUM charts, which plot several different measures of concrete mixture behavior.

Judgement is required in visual trend identification. When developing a spreadsheet analysis, it is possible to graph the data such that random changes appear to be significant. In Fig. A.4, data from only the first 12 tests are shown. There appears to be an upward trend in the data from Test no. 3 to Test no. 8, and a downward trend from Test no. 8 to Test no. 12. These trends are not statistically significant, however. The scale of the graph should be established appropriately.

A.7—Difficulties with CUSUM chart

There are several situations that can produce difficulties with CUSUM analysis. Some of the more common problems arising with interpretation are listed below:

1. The CUSUM graph is sensitive to the average strength value used in calculating the cumulative sum. An error in determining this value or the use of a target strength instead of the population average strength will result in a non-zero average for the cumulative sum. In Fig. A.5, the CUSUM graph is shown with three different initial estimates of the average compressive strength. One curve represents using 35.8 MPa (5190 psi) as the initial estimate of the average strength (except for the scale, this is the same as Fig. A.3). The other two curves represent the effects of errors of ± 2.0 MPa (290 psi) in the estimate. Small errors in estimating the

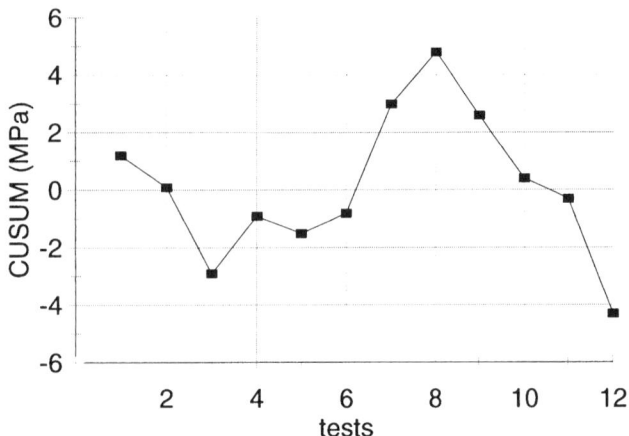

Fig. A.4—CUSUM chart with partial data.

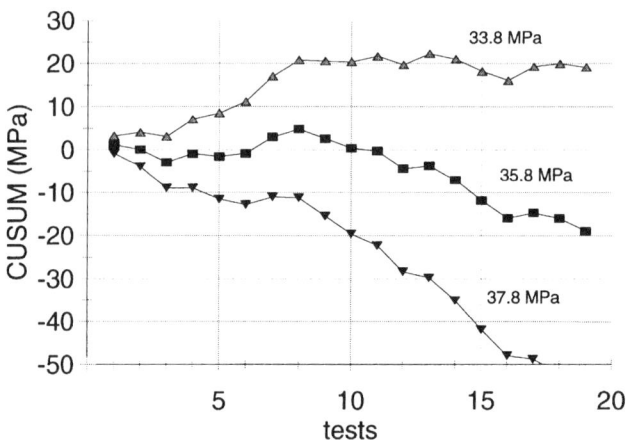

Fig. A.5—CUSUM QC chart for compressive strength with three different initial average strengths.

average strength can compound rapidly creating misleading results.

2. A single aberration in the data can create what appears to be an offset in a trend. If a trend occurs with a single offset, it may often be ignored in the analysis.

3. A single concrete mixture will typically have a slightly different average strength for each different project. Different contractors can exercise different levels of control and different testing agencies will invariably provide test data with slightly different averages. A series of test results from different jobs and testing agencies that are intermixed may show random variation. If the plotted data consist of runs of test results from different jobs and testing agencies, the differences in average of each set of data may produce a trend in the CUSUM chart. When a statistically significant trend has been found with multiple projects, the sample standard deviation should be calculated based on multiple data sets rather than one set. This will provide a more accurate, and typically smaller, estimate of the true standard deviation.

4. An estimate of the point at which the change in average strength occurred can be obtained from regression analysis of the CUSUM chart. The extra precision presumably obtained in such an analysis is rarely of practical value.

ACI 214.4R-03

Guide for Obtaining Cores and Interpreting Compressive Strength Results

Reported by ACI Committee 214

	James E. Cook Chair	Jerry Parnes Secretary	
David J. Akers	Steven H. Gebler	Michael L. Leming	D. V. Reddy
M. Arockiasamy	Alejandro Graf	Colin L. Lobo	Orrin Riley
William L. Barringer	Thomas M. Greene	John J. Luciano	James M. Shilstone, Jr.
F. Michael Bartlett[*]	Gilbert J. Haddad	Richard E. Miller	Luke M. Snell
Casimir Bognacki	Kal R. Hindo	Avi A. Mor	Patrick J. E. Sullivan
Jerrold L. Brown	Robert S. Jenkins	Tarun R. Naik	Michael A. Taylor
Ronald L. Dilly[*]	Alfred L. Kaufman, Jr.[*]	Robert E. Neal	Derle J. Thorpe
Donald E. Dixon	William F. Kepler	Terry Patzias	Roger E. Vaughan
Richard D. Gaynor	Peter A. Kopac	V. Ramakrishnan	Woodward L. Vogt[*]

[*]Task force that prepared this document.

Core testing is the most direct method to determine the compressive strength of concrete in a structure. Generally, cores are obtained either to assess whether suspect concrete in a new structure complies with strength-based acceptance criteria or to evaluate the structural capacity of an existing structure based on the actual in-place concrete strength. In either case, the process of obtaining core specimens and interpreting the strength test results is often confounded by various factors that affect either the in-place strength of the concrete or the measured strength of the test specimen. The scatter in strength test data, which is unavoidable given the inherent randomness of in-place concrete strengths and the additional uncertainty attributable to the preparation and testing of the specimen, may further complicate compliance and evaluation decisions.

This guide summarizes current practices for obtaining cores and interpreting core compressive strength test results. Factors that affect the in-place concrete strength are reviewed so locations for sampling can be selected that are consistent with the objectives of the investigation. Strength correction factors are presented for converting the measured strength of non-standard core-test specimens to the strength of equivalent specimens with standard diameters, length-to-diameter ratios, and moisture conditioning. This guide also provides guidance for checking strength compliance of concrete in a structure under construction and methods for determining an equivalent specified strength to assess the capacity of an existing structure.

Keywords: compressive strength; core; hardened concrete; sampling; test.

ACI Committee Reports, Guides, Standard Practices, and Commentaries are intended for guidance in planning, designing, executing, and inspecting construction. This document is intended for the use of individuals who are competent to evaluate the significance and limitations of its content and recommendations and who will accept responsibility for the application of the material it contains. The American Concrete Institute disclaims any and all responsibility for the stated principles. The Institute shall not be liable for any loss or damage arising therefrom.

Reference to this document shall not be made in contract documents. If items found in this document are desired by the Architect/Engineer to be a part of the contract documents, they shall be restated in mandatory language for incorporation by the Architect/Engineer.

It is the responsibility of the user of this document to establish health and safety practices appropriate to the specific circumstances involved with its use. ACI does not make any representations with regard to health and safety issues and the use of this document. The user must determine the applicability of all regulatory limitations before applying the document and must comply with all applicable laws and regulations, including but not limited to, United States Occupational Safety and Health Administration (OSHA) health and safety standards.

CONTENTS
Chapter 1—Introduction, p. 214.4R-2

Chapter 2—Variation of in-place concrete strength in structures, p. 214.4R-2
2.1—Bleeding
2.2—Consolidation
2.3—Curing
2.4—Microcracking
2.5—Overall variability of in-place strengths

Chapter 3—Planning the testing program, p. 214.4R-4
3.1—Checking concrete in a new structure using strength-based acceptance criteria

---
ACI 214.4R-03 became effective September 25, 2003.
Copyright © 2003, American Concrete Institute.
All rights reserved including rights of reproduction and use in any form or by any means, including the making of copies by any photo process, or by electronic or mechanical device, printed, written, or oral, or recording for sound or visual reproduction or for use in any knowledge or retrieval system or device, unless permission in writing is obtained from the copyright proprietors.

3.2—Evaluating the capacity of an existing structure using in-place strengths

Chapter 4—Obtaining specimens for testing, p. 214.4R-5

Chapter 5—Testing the cores, p. 214.4R-6

Chapter 6—Analyzing strength test data, p. 214.4R-6
6.1—ASTM C 42/C 42M precision statements
6.2—Review of core strength correction factors
6.3—Statistical analysis techniques

Chapter 7—Investigation of low-strength test results in new construction using ACI 318, p. 214.4R-9

Chapter 8—Determining an equivalent f_c' value for evaluating the structural capacity of an existing structure, p. 214.4R-9
8.1—Conversion of core strengths to equivalent in-place strengths
8.2—Uncertainty of estimated in-place strengths
8.3—Percentage of in-place strengths less than f_c'
8.4—Methods to estimate the equivalent specified strength

Chapter 9—Summary, p. 214.4R-12

Chapter 10—References, p. 214.4R-13
10.1—Referenced standards and reports
10.2—Cited references
10.3—Other references

Appendix—Example calculations, p. 214.4R-15
A1—Outlier identification in accordance with ASTM E 178 criteria
A2—Student's t test for significance of difference between observed average values
A3—Equivalent specified strength by tolerance factor approach
A4—Equivalent specified strength by alternate approach

CHAPTER 1—INTRODUCTION

Core testing is the most direct method to determine the in-place compressive strength of concrete in a structure. Generally, cores are obtained to:
a) Assess whether suspect concrete in a new structure complies with strength-based acceptance criteria; or
b) Determine in-place concrete strengths in an existing structure for the evaluation of structural capacity.

In new construction, cylinder strength tests that fail to meet strength-based acceptance criteria may be investigated using the provisions given in ACI 318. This guide presents procedures for obtaining and testing the cores and interpreting the results in accordance with ACI 318 criteria.

If strength records are unavailable, the in-place strength of concrete in an existing structure can be evaluated using cores. This process is simplified when the in-place strength data are converted into an equivalent value of the specified compressive strength f_c' that can be directly substituted into conventional strength equations with customary strength reduction factors. This guide presents procedures for carrying out this conversion in a manner that is consistent with the assumptions used to derive strength reduction factors for structural design.

The analysis of core test data can be difficult, leading to uncertain interpretations and conclusions. Strength interpretations should always be made by, or with the assistance of, an investigator experienced in concrete technology. The factors that contribute to the scatter of core strength test results include:
a) Systematic variation of in-place strength along a member or throughout the structure;
b) Random variation of concrete strength, both within one batch and among batches;
c) Low test results attributable to flawed test specimens or improper test procedures;
d) Effects of the size, aspect ratio, and moisture condition of the test specimen on the measured strengths; and
e) Additional uncertainty attributable to the testing that is present even for tests carried out in strict accordance with standardized testing procedures.

This guide summarizes past and current research findings concerning some of these factors and provides guidance for the interpretation of core strength test results. The presentation of these topics follows the logical sequence of tasks in a core-testing program. Chapter 2 reviews factors that affect the in-place concrete strength so that sampling locations consistent with the objectives of the investigation can be identified. Chapters 3, 4, and 5 present guidelines for planning the test program, obtaining the cores, and conducting the tests. Chapter 6 discusses the causes and magnitudes of the scatter usually observed in core test strengths and provides statistical methods for data analysis. Chapter 7 summarizes criteria given in ACI 318 for investigating low-strength tests in new construction. Chapter 8 presents methods for determining an equivalent f_c' for use in evaluating the capacity of an existing structure. Various example calculations appear in the Appendix.

CHAPTER 2—VARIATION OF IN-PLACE CONCRETE STRENGTH IN STRUCTURES

This chapter discusses the variation of in-place concrete strength in structures so that the investigator can anticipate the relevant factors in the early stages of planning the testing program. Selecting locations from which cores will be extracted and analyzing and interpreting the data obtained are simplified and streamlined when the pertinent factors are identified beforehand.

The quality of "as-delivered" concrete depends on the quality and relative proportions of the constituent materials and on the care and control exercised during batching, mixing, and handling. The final in-place quality depends on placing, consolidation, and curing practices. Recognizing that the delivery of quality concrete does not ensure quality in-place concrete, some project specifications require minimum core compressive strength results for concrete acceptance (Ontario Ministry of Transportation and Communications 1985). If excess mixing water was added at the site, or poor placing, consolidation, or curing practices were followed, core test results may not represent the quality of concrete as delivered to the site.

Generally, the in-place strength of concrete at the top of a member as cast is less than the strength at the bottom (Bloem 1965; Bungey 1989; Dilly and Vogt 1993).

2.1—Bleeding

Shallow voids under coarse aggregate caused by bleeding can reduce the compressive strength transverse to the direction of casting and consolidation (Johnson 1973). The strength of cores with axes parallel to the direction of casting can therefore be greater than that of cores with axes perpendicular to the direction of casting. The experimental findings, however, are contradictory because some investigators observed appreciable differences between the strengths of horizontally and vertically drilled cores (Sanga and Dhir 1976; Takahata, Iwashimizu, and Ishibashi 1991) while others did not (Bloem 1965). Although the extent of bleeding varies greatly with mixture proportions and constituent materials, the available core strength data do not demonstrate a relationship between bleeding and the top-to-bottom concrete strength differences.

For concrete cast against earth, such as slabs and pavements, the absorptive properties of the subgrade also affect core strength. Cores from concrete cast on subgrades that absorb water from the concrete will generally be stronger than cores from concrete cast against metal, wood, polyethylene, concrete, or wet, saturated clay.

2.2—Consolidation

Concrete is usually consolidated by vibration to expel entrapped air after placement. The strength is reduced by about 7% for each percent by volume of entrapped air remaining after insufficient consolidation (Popovics 1969; Concrete Society 1987; ACI 309.1R). The investigator may need to assess the extent to which poor consolidation exists in the concrete in question by using the nondestructive techniques reported in ACI 228.2R.

Consolidation of plastic concrete in the lower portion of a column or wall is enhanced by the static pressure of the plastic concrete in the upper portion. These consolidation pressures can cause an increase of strength (Ramakrishnan and Li 1970; Toossi and Houde 1981), so the lower portions of cast vertical members may have relatively greater strengths.

2.3—Curing

Proper curing procedures, which control the temperature and moisture environment, are essential for quality concrete. Low initial curing temperatures reduce the initial strength development rate but may result in higher long-term strength. Conversely, high initial-curing temperatures increase the initial strength development but reduce the long-term strength.

High initial temperatures generated by hydration can significantly reduce the strength of the interior regions of massive elements. For example, the results shown in Fig 2.1 indicate that the strength of cores obtained from the middle of mock 760 x 760 mm (30 x 30 in.) columns is consistently

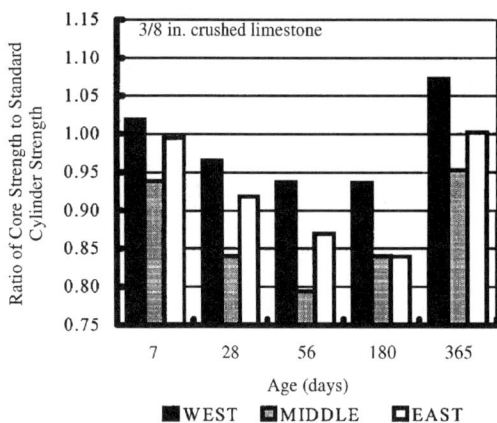

Fig. 2.1—Relationships between compressive strengths of column core samples and standard-cured specimens cast with high-strength concrete (Cook 1989).

less than the strength of cores obtained from the exterior faces (Cook 1989). The mock columns were cast using a high-strength concrete with an average 28-day standard cylinder strength in excess of 77 MPa (11,200 psi). Similarly, analysis of data from large specimens reported by Yuan et al. (1991), Mak et al. (1990, 1993), Burg and Ost (1992), and Miao et al. (1993) indicate a strength loss of roughly 6% of the average strength in the specimen for every 10 °C (3% per 10 °F) increase of the average maximum temperature sustained during early hydration (Bartlett and MacGregor 1996a). The maximum temperatures recorded in these specimens varied between 45 and 95 °C (110 and 200 °F).

In massive concrete elements, hydration causes thermal gradients between the interior, which becomes hot, and the surfaces of the element, which remain relatively cool. In this case, the surfaces are restrained from contracting by the interior of the element, which can cause microcracking that reduces the strength at the surface. This phenomenon has been clearly observed in some investigations (Mak et al. 1990) but not in others (Cook et al. 1992).

The in-place strength of slabs or beams is more sensitive to the presence of adequate moisture than the in-place strength of walls or columns because the unformed top surface is a relatively large fraction of the total surface area. Data from four studies (Bloem 1965; Bloem 1968; Meynick and Samarin 1979; and Szypula and Grossman 1990) indicate that the strength of cores from poorly cured shallow elements averages 77% of the strength of companion cores from properly cured elements for concrete ages of 28, 56, 91, and 365 days (Bartlett and MacGregor 1996b). Data from two studies investigating walls and columns (Bloem 1965; Gaynor 1970) indicate that the strength loss at 91 days attributable to poor curing averages approximately 10% (Bartlett and MacGregor 1996b).

2.4—Microcracking

Microcracks in a core reduce the strength (Szypula and Grossman 1990), and their presence has been used to explain why the average strengths of cores from two ends of a beam cast from a single batch of concrete with a cylinder strength

Table 2.1—Coefficient of variation due to in-place strength variation within structure V_{WS}

Structure composition	One member	Many members
One batch of concrete	7%	8%
Many batches of concrete		
Cast-in-place	12%	13%
Precast	9%	10%

of 54.1 MPa (7850 psi) differed by 11% of their average (Bartlett and MacGregor 1994a). Microcracks can be present if the core is drilled from a region of the structure that has been subjected to stress resulting from either applied loads or restraint of imposed deformations. Rough handling of the core specimen can also cause microcracking.

2.5—Overall variability of in-place strengths

Estimates of the overall variability of in-place concrete strengths reported by Bartlett and MacGregor (1995) are presented in Table 2.1. The variability is expressed in terms of the coefficient of variation V_{WS}, which is the ratio of the standard deviation of the in-place strength to the average in-place strength. The overall variability depends on the number of members in the structure, the number of concrete batches present, and whether the construction is precast or cast-in-place. The values shown are for concrete produced, placed, and protected in accordance with normal industry practice and may not pertain to concrete produced to either high or low standards of quality control.

CHAPTER 3—PLANNING THE TESTING PROGRAM

The procedure for planning a core-testing program depends on the objective of the investigation. Section 3.1 presents procedures for checking whether concrete in a new structure complies with strength-based acceptance criteria, while Section 3.2 presents those procedures for evaluating the strength capacity of an existing structure using in-place strengths.

As noted in Chapter 2, the strength of concrete in a placement usually increases with depth. In single-story columns, cores should be obtained from the central portion, where the strength is relatively constant, and not in the top 450 to 600 mm (18 to 24 in.), where it may decrease by 15%, or in the bottom 300 mm (12 in.), where it may increase by 10% (Bloem 1965).

3.1—Checking concrete in a new structure using strength-based acceptance criteria

To investigate low-strength test results in accordance with ACI 318, three cores are required from that part of the structure cast from the concrete represented by the low-strength test result. The investigator should only sample those areas where the suspect concrete was placed.

In some situations, such as a thin composite deck or a heavily reinforced section, it is difficult or impossible to obtain cores that meet all of the length and diameter requirements of ASTM C 42/C 42M. Nevertheless, cores can allow a relative comparison of two or more portions of a structure representing different concrete batches. For example, consider two sets of columns placed with the same concrete mixture proportion: one that is acceptable based on standard strength tests and one that is questionable because of low strength test results. Nondestructive testing methods (ACI 228.1R) may indicate that the quality of concrete in the suspect columns exceeds that in the acceptable columns. Alternatively, it is appropriate to take 50 mm (2 in.) diameter cores from the columns where 25 mm (1 in.) maximum size aggregate was used. After trimming the cores, however, the ℓ/d will be less than 1.0 if the cover is only 50 mm (2 in.) and reinforcing bars cannot be cut. Acknowledging that strength tests of the "short" cores may not produce strength test results that accurately reflect the strength of the concrete in the columns, a relative comparison of the two concrete placements may be sufficient to determine if the strength of the concrete in question is comparable to the other placement or if more investigation is warranted.

3.2—Evaluating the capacity of an existing structure using in-place strengths

To establish in-place strength values for existing structures, the sample size and locations from which the cores will be extracted need to be carefully selected using procedures such as those described in ASTM E 122 and ASTM C 823.

As the sample size increases, the accuracy of the result improves; the likelihood of detecting a spurious value in the data set also improves, but greater costs are incurred and the risk of weakening the structure increases. ASTM E 122 recommends sample sizes be computed using Eq. (3-1) to achieve a 1-in-20 chance that the difference between the measured average of the sample and the average of the population, expressed as a percentage of the average of the population, will be less than some predetermined error.

$$n = \left(\frac{2V}{e}\right)^2 \qquad (3\text{-}1)$$

where
n = the recommended sample size;
e = the predetermined maximum error expressed as a percentage of the population average; and
V = the estimated coefficient of variation of the population, in percent, and may be estimated from the values shown in Table 2.1 or from other available information.

For example, if the estimated coefficient of variation of the in-place strength is 15%, and it is desired that the measured average strength should be within 10% of the true average strength approximately 19 times out of 20, Eq. (3-1) indicates that (for $V = 0.15$ and $e = 0.10$) a total of nine cores should be obtained. If a higher confidence level is desired, or if a smaller percentage error is necessary, then a larger sample size is required. Statistical tests for determining whether extreme values should be rejected, such as those in ASTM E 178, become more effective as the sample size increases. As indicated by the relationships between the percentage error and the recommended number of specimens shown in Fig. 3.1, however, the benefits of larger sample sizes tend to diminish. ASTM C 823 recommends that a

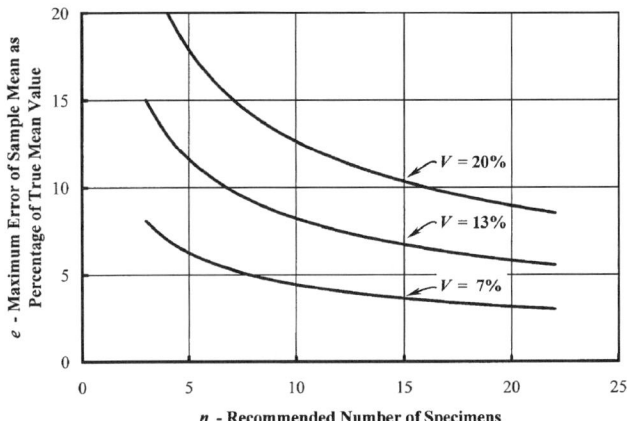

Fig. 3.1—Maximum error of sample mean for various recommended number of specimens.

minimum of five core test specimens be obtained for each category of concrete with a unique condition or specified quality, specified mixture proportion, or specified material property. ASTM C 823 also provides guidance for repeating the sampling sequence for large structures.

The investigator should select locations from which the cores will be extracted based on the overall objective of the investigation, not the ease of obtaining samples. To characterize the overall in-place strength of an existing structure for general evaluation purposes, cores should be drilled from randomly selected locations throughout the structure using a written sampling plan. If the in-place strength for a specific component or group of components is sought, the investigator should extract the cores at randomly selected locations from those specific components.

When determining sample locations, the investigator should consider whether different strength categories of concrete may be present in the structure. For example, the in-place strengths of walls and slabs cast from a single batch of concrete may differ (Meininger 1968) or concrete with different required strengths may have been used for the footings, columns, and floor slabs in a building. If the concrete volume under investigation contains two or more categories of concrete, the investigator should objectively select sample locations so as not to unfairly bias the outcome. Alternatively, he or she should randomly select a sufficient number of sampling locations for each category of concrete with unique composition or properties. The investigator can use nondestructive testing methods (ACI 228.1R) to perform a preliminary survey to identify regions in a structure that have different concrete properties.

ACI 311.1R (SP-2) and ASTM C 823 contain further guidance concerning sampling techniques.

CHAPTER 4—OBTAINING SPECIMENS FOR TESTING

Coring techniques should result in high-quality, undamaged, representative test specimens. The investigator should delay coring until the concrete being cored has sufficient strength and hardness so that the bond between the mortar and aggregate will not be disturbed. ASTM C 42/C 42M suggests that the concrete should not be cored before it is 14 days old, unless other information indicates that the concrete can withstand the coring process without damage. ASTM C 42/C 42M further suggests that in-place nondestructive tests (ACI 228.1R) may be performed to estimate the level of strength development of the concrete before coring is attempted.

Core specimens for compression tests should preferably not contain reinforcing bars. These can be located before drilling the core using a pachometer or cover meter. Also, avoid cutting sections containing conduit, ductwork, or prestressing tendons.

As described in Chapter 6, the strength of the specimen is affected by the core diameter and the ratio of length-to-diameter, ℓ/d, of the specimen. Strength correction factors for these effects are derived empirically from test results (Bartlett and MacGregor 1994b) and so are not universally accurate. Therefore, it is preferable to obtain specimens with diameters of 100 to 150 mm (4 to 6 in.) and ℓ/d ratios between 1.5 and 2 to minimize error introduced by the strength correction factors (Neville 2001).

The drilling of the core should be carried out by an experienced operator using a diamond-impregnated bit attached to the core barrel. The drilling apparatus should be rigidly anchored to the member to avoid bit wobble, which results in a specimen with variable cross section and the introduction of large strains in the core. The drill bit should be lubricated with water and should be resurfaced or replaced when it becomes worn. The operator should be informed beforehand that the cores are for strength testing and, therefore, require proper handling and storage.

Core specimens in transit require protection from freezing and damage because a damaged specimen will not accurately represent the in-place concrete strength.

A core drilled with a water-cooled bit results in a moisture gradient between the exterior and interior of the core that adversely affects its compressive strength (Fiorato, Burg, Gaynor 2000; Bartlett and MacGregor 1994c). ASTM C 42/C 42M presents moisture protection and scheduling requirements that are intended to achieve a moisture distribution in core specimens that better represent the moisture distribution in the concrete before the concrete was wetted during drilling. The restriction concerning the commencement of core testing provides a minimum time for the moisture gradient to dissipate.

The investigator, or a representative of the investigator, should witness and document the core drilling. Samples should be numbered and their orientation in the structure indicated by permanent markings on the core itself. The investigator should record the location in the structure from which each core is extracted and any features that may affect the strength, such as cracks or honeycombs. Similar features observed by careful inspection of the surrounding concrete should also be documented. Given the likelihood of questionable low-strength values, any information that may later identify reasons for the low values will be valuable.

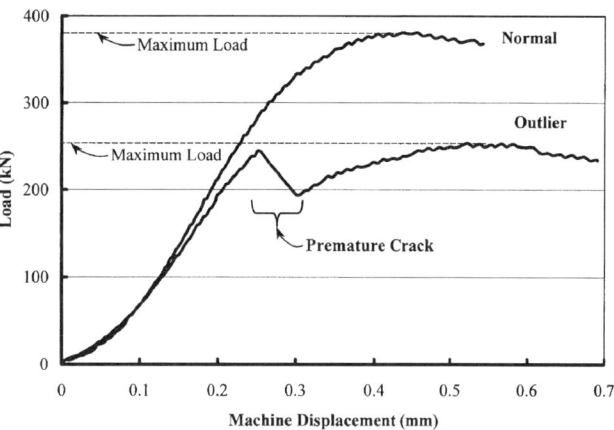

Fig. 5.1—Use of load-machine displacement curves to identify outlier due to flawed specimen (Bartlett and MacGregor 1994a).

CHAPTER 5—TESTING THE CORES

ASTM C 42/C 42M presents standard methods for conditioning the specimen, preparing the ends before testing, and correcting the test result for the core length-to-diameter ratio. Other standards for measuring the length of the specimen and performing the compression test are referenced and information required in the test report is described.

Core densities, which can indicate the uniformity of consolidation, are often useful to assess low core test results. Before capping, the density of a core can be computed by dividing its mass by its volume, calculated from its average diameter and length.

When testing cores with small diameters, careful alignment of the specimen in the testing machine is necessary. If the diameter of the suspended spherically seated bearing block exceeds the diameter of the specimen, the spherical seat may not rotate into proper alignment, causing nonuniform contact against the specimen. ASTM C 39 limits the diameter of the upper bearing face to avoid an excessively large upper spherical bearing block.

A load-machine displacement response graph can be a useful indicator of abnormal behavior resulting from testing a flawed specimen. For example, the two curves in Fig. 5.1 are for 100 x 100 mm (4 x 4 in.) cores, obtained from one beam, that were given identical moisture treatments. The lower curve is abnormal because the load drops markedly before reaching its maximum value. This curve is consistent with a premature splitting failure and may be attributed to imperfect preparation of the ends of the specimen. Thus, the low result can be attributed to a credible physical cause and should be excluded from the data set.

Sullivan (1991) describes the use of nondestructive tests to check for abnormalities in cores before the compressive strength tests are conducted.

If the investigator cannot find a physical reason to explain why a particular result is unusually low or unusually high, then statistical tests given in ASTM E 178 can be used to determine whether the observation is an "outlier." When the sample size is less than six, however, these tests do not consistently classify values as outliers that should be so classified (Bartlett and MacGregor 1995). An example calculation using ASTM E 178 criteria to check whether a low value is an outlier is presented in the Appendix. If an outlier can be attributed to an error in preparing or testing the specimen, it should be excluded from the data set. If an observation is an outlier according to ASTM E 178 criteria but the reason for the outlier cannot be determined, then the investigator should report the suspect values and indicate whether they have been used in subsequent analyses.

CHAPTER 6—ANALYZING STRENGTH TEST DATA

The analysis and interpretation of core strength data are complicated by the large scatter usually observed in the test results. This chapter describes the expected scatter of properly conducted tests of cores from a sample of homogeneous material, discusses other possible reasons for strength variation that require consideration, and briefly reviews statistical techniques for identifying sources of variability in a specific data set. Detailed descriptions of these statistical techniques can be found in most statistical references, such as Ang and Tang (1975) or Benjamin and Cornell (1970).

6.1—ASTM C 42/C 42M precision statements

ASTM C 42/C 42M provides precision statements that quantify the inherent error associated with testing cores from a homogeneous material tested in accordance with the standardized procedures. The single operator coefficient of variation is 3.2%, and the multilaboratory coefficient of variation is 4.7%. In the interlaboratory study used to derive these values, the measured values of the single operator coefficient of variation varied from 3.1 to 3.4% for cores from the three different slabs, and measured values of the multilaboratory coefficient of variation varied between 3.7 and 5.3% (Bollin 1993).

These precision statements are a useful basis for preliminary checks of core strength data if the associated assumptions and limitations are fully appreciated. Observed strength differences can exceed the limits stated in ASTM C 42/C 42M due to one or more of the following reasons:

a) The limits stated in ASTM C 42/C 42M are "difference 2 sigma" (d2s) limits so the probability that they are exceeded is 5%. Therefore, there is a 1-in-20 chance that the strength of single cores from the same material tested by one operator will differ by more than 9% of their average, and also a 1-in-20 chance that the average strength of cores from the same material tested by different laboratories will differ by more than 13% of their average;

b) The variability of the in-place concrete properties can exceed that in the slabs investigated for the multilaboratory study reported by Bollin (1993); and

c) The testing accuracy can be less rigorous than that achieved by the laboratories that participated in the study reported by Bollin (1993).

The single-operator coefficient of variation is a measure of the repeatability of the core test when performed in accordance with ASTM C 42/C 42M. A practical use of this measure is to check whether the difference between strength test results of two individual cores obtained from the same sample of material does not differ by more than 9% of their average. The difference between consecutive tests (or any two randomly selected tests) is usually much less than the overall range between the largest and least values, which tends to increase as the sample size increases. The expected range and the range that has a 1-in-20 chance of being exceeded, expressed as a fraction of the average value, can be determined for different sample sizes using results originally obtained by Pearson (1941-42). Table 6.1 shows values corresponding to the ASTM C 42/C 42M single-operator coefficient of variation of 3.2%, which indicate, for example, in a set of five cores from the same sample of material, the expected range is 7.2% of the average value and there is a 1-in-20 chance the range will exceed 12.4% of the average value. Table 1 of ASTM C 670 gives multipliers that, when applied to the single-operator coefficient of variation, also estimate the range that has a 1-in-20 chance of being exceeded.

The multilaboratory coefficient of variation is a measure of the reproducibility of the core test, as performed in accordance with ASTM C 42/C 42M. Although the reported values are derived for tests defined as the average strength of two specimens, they can be assumed to be identical to those from tests defined as the average strength of three specimens. Thus, this measure indicates that, for example, if two independent laboratories test cores from the same sample of material in accordance with criteria given in ACI 318, and each laboratory tests three specimens in conformance with ASTM C 42/C 42M, there remains a 1-in-20 chance that the reported average strengths will differ by more than 13% of their average.

6.2—Review of core strength correction factors

The measured strength of a core depends partly on factors that include the ratio of length to diameter of the specimen, the diameter, the moisture condition at the time of testing, the presence of reinforcement or other inclusions, and the direction of coring. Considerable research has been carried out concerning these factors, and strength correction factors have been proposed to account for their effects. The research findings, however, have often been contradictory. Also, published strength correction factors are not necessarily exact and may not be universally applicable because they have been derived empirically from specific sets of data. To indicate the degree of uncertainty associated with these factors, this section summarizes some of the relevant research findings. Chapter 8 presents specific strength correction factor values.

6.2.1 *Length-to-diameter ratio*—The length-to-diameter ratio ℓ/d was identified in the 1927 edition of ASTM C 42/C 42M as a factor that influences the measured compressive strength of a core, and minor variations of the original ℓ/d strength correction factors have been recommended in subsequent editions. Specimens with small ℓ/d fail at greater loads because the steel loading platens of the testing machine restrain lateral expansion throughout the length of the specimen more effectively and so provide confinement (Newman and Lachance 1964; Ottosen 1984). The end effect is largely eliminated in standard concrete compression test specimens, which have a length to diameter ratio of two.

Table 6.1—Probable range of core strengths due to single-operator error

Number of cores	Expected range of core strength as % of average core strength	Range with 5% chance of being exceeded as % of average core strength
3	5.4	10.6
4	6.6	11.6
5	7.2	12.4
6	8.1	12.9
7	8.6	13.3
8	9.1	13.7
9	9.5	14.1
10	9.8	14.3

Table 6.2—Strength correction factors for length-to-diameter ratio

ℓ/d	ASTM C 42/C 42M	BS 1881
2.00	1.00	1.00
1.75	0.98	0.97
1.50	0.96	0.92
1.25	0.93	0.87
1.00	0.87	0.80

Table 6.2 shows values of strength correction factors recommended in ASTM C 42/C 42M and British Standard BS 1881 (1983) for cores with ℓ/d between 1 and 2. Neither standard permits testing cores with ℓ/d less than 1. The recommended values diverge as ℓ/d approaches 1. The ASTM factors are average values that pertain to dry or soaked specimens with strengths between 14 and 40 MPa (2000 and 6000 psi). ASTM C 42/C 42M states that actual ℓ/d correction factors depend on the strength and elastic modulus of the specimen.

Bartlett and MacGregor (1994b) report that the necessary strength correction is slightly less for high-strength concrete and soaked cores, but they recommend strength correction factor values that are similar to those in ASTM C 42/C 42M. They also observed that the strength correction factors are less accurate as the magnitude of the necessary correction increases for cores with smaller ℓ/d. Thus, corrected core strength values do not have the same degree of certainty as strength obtained from specimens having ℓ/d of 2.

6.2.2 *Diameter*—There is conflicting experimental evidence concerning the strength of cores with different diameters. While there is a consensus that differences between 100 and 150 mm (4 and 6 in.) diameter specimens are negligible (Concrete Society 1987), there is less agreement concerning 50 mm (2 in.) diameter specimens. In one study involving cores from 12 different concrete mixtures, the ratio of the average strength of five 50 mm (2 in.) diameter cores to the average strength of three 100 mm (4 in.) diameter cores ranged from 0.63 to 1.53 (Yip and Tam 1988). An analysis of strength data from 1080 cores tested by various investigators indicated that the strength of a 50 mm (2 in.) diameter core was on average 6% less than the strength of a 100 mm (4 in.) diameter core (Bartlett and MacGregor 1994d).

The scatter in the strengths of 50 mm (2 in.) diameter cores often exceeds that observed for 100 or 150 mm (4 or 6 in.) diameter cores. The variability of the in-place strength within the element being cored, however, also inflates the variability of the strength of small-volume specimens. Cores drilled vertically through the thickness of slabs can be particularly susceptible to this effect (Lewis 1976).

In practice it is often difficult to obtain a 50 mm (2 in.) diameter specimen that is not affected by the drilling process or does not contain a small defect that will markedly affect the result. If correction factors are required to convert the strength of 50 mm (2 in.) diameter cores to the strength of equivalent 100 or 150 mm (4 or 6 in.) diameter cores, the investigator should derive them directly using a few cores of each diameter obtained from the structure in question.

6.2.3 *Moisture condition*—Different moisture-conditioning treatments have a considerable effect on the measured strengths. Air-dried cores are on average 10 to 14% (Neville 1981; Bartlett and MacGregor 1994a) stronger than soaked cores, although the actual ratio for cores from a specific concrete can differ considerably from these average values. Soaking causes the concrete at the surface of the specimen to swell, and restraint of this swelling by the interior region causes self-equilibrated stresses that reduce the measured compressive strength (Popovics 1986). Conversely, drying the surface causes shrinkage that, when restrained, creates a favorable residual stress distribution that increases the measured strengths. In both cases the changes in moisture condition are initially very rapid (Bartlett and MacGregor 1994c, based on data reported by Bloem 1965). If cores are not given standardized moisture conditioning before testing, or if the duration of the period between the end of the moisture treatment and the performance of the test varies significantly, then additional variability of the measured strengths can be introduced.

The percentage of strength loss caused by soaking the core depends on several factors. Concrete that is less permeable exhibits a smaller strength loss. Bartlett and MacGregor (1994a) observed a more severe strength loss in 50 mm (2 in.) diameter cores compared with 100 mm (4 in.) diameter cores from the same element. Extending the soaking period beyond 40 h duration can cause further reduction of the core strength. The difference between strengths of soaked and air-dried cores may be smaller for structural lightweight aggregate concrete (Bloem 1965).

6.2.4 *Presence of reinforcing bars or other inclusions*— The investigator should avoid specimens containing embedded reinforcement because it may influence the measured compressive strength. Previous editions of ASTM C 42 have recommended trimming the core to eliminate the reinforcement provided, l/d, of at least 1.0 can be maintained.

6.2.5 *Coring direction*—Cores drilled in the direction of placement and compaction (which would be loaded in a direction perpendicular to the horizontal plane of concrete as placed, according to ASTM C 42/C 42M) can be stronger than cores drilled normal to this direction because bleed water can collect underneath coarse aggregate, as described in Chapter 2. In practice, it is often easier to drill horizontally into a column, wall, or beam in a direction perpendicular to the direction of placement and compaction. The influence of coring direction can be more pronounced near the upper surface of members where bleed water is concentrated. To determine whether the in-place strength is affected by the direction of drilling, the investigator should assess this directly using specimens drilled in different directions from the structure in question, if possible.

6.3—Statistical analysis techniques

Statistical analysis techniques can determine if the data are random or can be grouped into unique sets. For example, statistical tests can verify that the strengths in the uppermost parts of columns are significantly less than the strengths elsewhere, and so the investigation is focused accordingly.

Statistical tests are particularly useful for analyzing preliminary hypotheses developed during an initial review of the data, which are logically consistent with the circumstances of the investigation and are credible in light of past experience. While it is possible to conduct "fishing expeditions" using statistical techniques to look for correlations and trends in data in an exploratory manner, it is rarely efficient to do so. Flawed conclusions are undetectable if statistical analyses are conducted without a clear understanding of the essential physical and behavioral characteristics represented in the data. Instead, it is preferable to first identify the possible factors that affect the strength in a particular instance and then use statistical analyses to verify whether these factors are in fact significant.

Perhaps the most useful analysis method is the Student's t test, which is used to decide whether the difference between two average values is sufficiently large to imply that the true mean values of the underlying populations, from which the samples are drawn, are different. ASTM C 823 recommends the use of the Student's t test to investigate whether the average strength of cores obtained from concrete of questionable quality differs from the average strength of cores obtained from concrete of good quality. Details of the Student's t test can be found in most statistical references (Benjamin and Cornell 1970; Ang and Tang 1975), and a numerical example illustrating its use is presented in the Appendix.

There are two types of error associated with any statistical test. A Type I error occurs when a hypothesis (such as: "the true mean values of two groups are equal") is rejected when, in fact, it is true, and a Type II error occurs when a hypothesis is accepted when, in fact, it is false. In the practice of quality control, these are referred to as the producer's and the consumer's risk, respectively, because the producer's concern is that a satisfactory product will be rejected, and the consumer's concern is that an unsatisfactory product will be accepted. It is not possible to reduce the likelihood of a Type I error without increasing the likelihood of a Type II error, or vice versa, unless the sample size is increased. When decisions are made on the basis of a small number of tests (and so the likelihood of an error is large), the investigator should recognize that most statistical tests, including the Student's t test, are designed to limit the likelihood of a Type I error. If an

Table 8.1—Magnitude and accuracy of strength correction factors for converting core strengths into equivalent in-place strengths*

Factor	Mean value	Coefficient of variation V, %
$F_{\ell/d}$: ℓ/d ratio†		
As-received‡	$1 - \{0.130 - \alpha f_{core}\}\left(2 - \dfrac{\ell}{d}\right)^2$	$2.5\left(2 - \dfrac{\ell}{d}\right)^2$
Soaked 48 h	$1 - \{0.117 - \alpha f_{core}\}\left(2 - \dfrac{\ell}{d}\right)^2$	$2.5\left(2 - \dfrac{\ell}{d}\right)^2$
Air dried‡	$1 - \{0.144 - \alpha f_{core}\}\left(2 - \dfrac{\ell}{d}\right)^2$	$2.5\left(2 - \dfrac{\ell}{d}\right)^2$
F_{dia}: core diameter		
50 mm (2 in.)	1.06	11.8
100 mm (4 in.)	1.00	0.0
150 mm (6 in.)	0.98	1.8
F_{mc}: core moisture content		
As-received‡	1.00	2.5
Soaked 48 h	1.09	2.5
Air dried‡	0.96	2.5
F_d: damage due to drilling	1.06	2.5

*To obtain equivalent in-place concrete strength, multiply the measured core strength by appropriate factor(s) in accordance with Eq. (8-1).
†Constant α equals $3(10^{-6})$ 1/psi for f_{core} in psi, or $4.3(10^{-4})$ 1/MPa for f_{core} in MPa.
‡Standard treatment specified in ASTM C 42/C 42M.

observed difference obtained from a small sample seems large but is not statistically significant, then a true difference may exist and can be substantiated if additional cores are obtained to increase the sample size.

CHAPTER 7—INVESTIGATION OF LOW-STRENGTH TEST RESULTS IN NEW CONSTRUCTION USING ACI 318

In new construction, low cylinder strength tests are investigated in accordance with the provisions of ACI 318. The suspect concrete is considered structurally adequate if the average strength of the three cores, corrected for ℓ/d in accordance with ASTM C 42/C 42M, exceeds $0.85f'_c$, and no individual strength is less than $0.75f'_c$. Generally, these criteria have served producers and consumers of concrete well. ACI 318 recognizes that the strengths of cores are potentially lower than the strengths of cast specimens representing the quality of concrete delivered to the project. This relationship is corroborated by observations that the strengths of 56-day-old soaked cores averaged 93% of the strength of standard 28-day cylinders and 86% of the strength of standard-cured 56-day cylinders (Bollin 1993).

ACI 318 permits additional testing of cores extracted from locations represented by erratic strength results. ACI 318 does not define "erratic," but this might reasonably be interpreted as a result that clearly differs from the rest that can be substantiated by a valid physical reason that has no bearing on the structural adequacy of the concrete in question.

For structural adequacy, the ACI 318 strength requirements for cores need only be met at the age when the structure will be subject to design loads.

CHAPTER 8—DETERMINING AN EQUIVALENT f'_c VALUE FOR EVALUATING THE STRUCTURAL CAPACITY OF AN EXISTING STRUCTURE

This chapter presents procedures to determine an equivalent design strength for structural evaluation for direct substitution into conventional strength equations that include customary strength reduction factors. This equivalent design strength is the lower tenth percentile of the in-place strength and is consistent with the statistical description of the specified strength of concrete f'_c. This chapter presents two methods for estimating the lower tenth-percentile value from core test data.

The procedures described in this chapter are only appropriate for the case where the determination of an equivalent f'_c is necessary for the strength evaluation of an existing structure and should not be used to investigate low cylinder strength test results.

8.1—Conversion of core strengths to equivalent in-place strengths

The in-place strength of the concrete at the location from which a core test specimen was extracted can be computed using the equation

$$f_c = F_{\ell/d} F_{dia} F_{mc} F_d f_{core} \qquad (8\text{-}1)$$

where f_c is the equivalent in-place strength; f_{core} is the core strength; and strength correction factors $F_{\ell/d}$, F_{dia}, and F_{mc} account for the effects of the length-to-diameter ratio, diameter, and moisture condition of the core, respectively. Factor F_d accounts for the effect of damage sustained during drilling including microcracking and undulations at the drilled surface and cutting through coarse-aggregate particles that may subsequently pop out during testing (Bartlett and MacGregor 1994d). Table 8.1 shows the mean values of the strength correction factors reported by Bartlett and MacGregor (1995) based on data for normalweight concrete with strengths between 14 and 92 MPa (2000 and 13,400 psi). The right-hand column shows coefficients of variation V that indicate the uncertainty of the mean value. It follows that a 100 mm (4 in.) diameter core with $\ell/d = 2$ that has been soaked 48 h before testing has $f_c = 1.0 \times 1.0 \times 1.09 \times 1.06 f_{core} = 1.16 f_{core}$.

8.2—Uncertainty of estimated in-place strengths

After the core strengths have been converted to equivalent in-place strengths, the sample statistics can be calculated. The sample mean in-place strength $\bar{f}_c$ is obtained from the following equation

$$\bar{f}_c = \frac{1}{n} \sum_{i=1}^{n} f_{ci} \qquad (8\text{-}2)$$

where n is the number of cores, and f_{ci} is the equivalent in-place strength of an individual core specimen, calculated using Eq. (8-1). The sample standard deviation of the in-place strength s_c is obtained from the following equation

$$s_c = \sqrt{\sum_{i=1}^{n} \frac{(f_{ci} - \bar{f}_c)^2}{(n-1)}} \qquad (8\text{-}3)$$

The sample mean and the sample standard deviation are estimates of the true mean and true standard deviation, respectively, of the entire population. The accuracy of these estimates, which improves as the sample size increases, can be investigated using the classical statistical approach to parameter estimation (Ang and Tang 1975).

The accuracy of the estimated in-place strengths also depends on the accuracy of the various strength correction factors used in Eq. (8-1). The standard deviation of the in-place strength due to the empirical nature of the strength correction factors s_a can be obtained from the following equation

$$s_a = \bar{f}_c \sqrt{V_{\ell/d}^2 + V_{dia}^2 + V_{mc}^2 + V_d^2} \qquad (8\text{-}4)$$

The right column of Table 8.1 shows the values of $V_{\ell/d}$, V_{dia}, V_{mc}, and V_d, the coefficients of variation associated with strength correction factors $F_{\ell/d}$, F_{dia}, F_{mc}, and F_d, respectively. The coefficient of variation due to a particular strength correction factor need only be included in Eq. (8-4) if the corresponding factor used in Eq. (8-1) to obtain the in-place strength differs from 1.0. If the test specimens have different ℓ/d, it is appropriate and slightly conservative to use the $V_{\ell/d}$ value for the core with the smallest ℓ/d. For cores from concrete produced with similar proportions of similar aggregates, cement, and admixtures, the errors due to the strength correction factors remain constant irrespective of the number of specimens obtained.

The overall uncertainty of the estimated in-place strengths is a combination of the sampling uncertainty and the uncertainty caused by the strength correction factors. These two sources of uncertainty are statistically independent, and so the overall standard deviation s_o is determined using the following equation

$$s_o = \sqrt{s_c^2 + s_a^2} \qquad (8\text{-}5)$$

8.3—Percentage of in-place strengths less than f'_c

The criteria in ACI 318 for proportioning concrete mixtures require that the target strength exceeds f'_c to achieve approximately a 1-in-100 chance that the average of three consecutive tests will fall below f'_c, and approximately a 1-in-100 chance that no individual test will fall more than 3.5 MPa (500 psi) below f'_c if the specified strength is less than 35 MPa (5000 psi), or below $0.90 f'_c$ if the specified strength exceeds 35 MPa (5000 psi). These criteria imply that f'_c represents approximately the 10% fractile, or the lower tenth-percentile value, of the strength obtained from a standard test of 28-day cylinders. In other words, one standard strength test in 10 will be less than f'_c if the target strength criteria required by ACI 318 are followed. Various methods for converting in-place strengths obtained by nondestructive testing into an equivalent f'_c are therefore based on estimating the 10% fractile of the in-place strength (Bickley 1982; Hindo and Bergstrom 1985; Stone, Carino, and Reeve 1986).

This practice was corroborated by a study that showed f'_c represents roughly the 13% fractile of the 28-day in-place strength in walls and columns and roughly the 23% fractile of the 28 day in-place strength in beams and slabs (Bartlett and MacGregor 1996b). The value for columns is more appropriate for defining an equivalent specified strength because the nominal strength of a column is more sensitive to the concrete compressive strength than a beam or slab. Therefore, a procedure that assumes that the specified strength is equal to the 13% fractile of the in-place strength is appropriate, and one that assumes that f'_c is equivalent to the 10% fractile of the in-place strength is slightly conservative.

8.4—Methods to estimate the equivalent specified strength

There is no universally accepted method for determining the 10% fractile of the in-place strength, which, as described in Section 8.3, is roughly equivalent to f'_c. In general, the following considerations should be addressed:

a) Factors that bias the core test result, which can be accounted for using the strength correction factors discussed in Chapter 6;

b) Uncertainty of each strength correction factor used to estimate the in-place strength;

c) Errors of the measured average value and measured standard deviation that are attributable to sampling and therefore decrease as the sample size increases;

Table 8.2—K-factors for one-sided tolerance limits on 10% fractile (Natrella 1963)

n	Confidence level		
	75%	90%	95%
3	2.50	4.26	6.16
4	2.13	3.19	4.16
5	1.96	2.74	3.41
6	1.86	2.49	3.01
8	1.74	2.22	2.58
10	1.67	2.06	2.36
12	1.62	1.97	2.21
15	1.58	1.87	2.07
18	1.54	1.80	1.97
21	1.52	1.75	1.90
24	1.50	1.71	1.85
27	1.49	1.68	1.81
30	1.48	1.66	1.78
35	1.46	1.62	1.73
40	1.44	1.60	1.70

Note: n = number of specimens tested.

Table 8.3—Z-factors for use in Eq. (8-7) and (8-8) (Natrella 1963)

Confidence level, %	Z
75	0.67
90	1.28
95	1.64

Table 8.4—One-sided T-factors for use in Eq. (8-8) (Natrella 1963)

n	Confidence level		
	75%	90%	95%
3	0.82	1.89	2.92
4	0.76	1.64	2.35
5	0.74	1.53	2.13
6	0.73	1.48	2.02
8	0.71	1.41	1.90
10	0.70	1.38	1.83
12	0.70	1.36	1.80
15	0.69	1.34	1.76
18	0.69	1.33	1.74
21	0.69	1.33	1.72
24	0.69	1.32	1.71
30	0.68	1.32	1.70

Note: n = number of specimens tested.

d) Variability attributable to acceptable deviations from standardized testing procedures that can cause the measured standard deviation of strength tests to exceed the true in-place strength variation; and

e) Desired confidence level, which represents the likelihood that the fractile value calculated using the sample data will be less than the true fractile value of the underlying population from which the sample is drawn.

This section presents two methods for estimating the 10% fractile of the in-place strength. To use either method, it is necessary to assume a type of probability distribution for the in-place strengths and to determine the desired confidence level.

There is a general consensus that concrete strengths are normally distributed if control is excellent or follow a lognormal distribution if control is poor (Mirza, Hatzinikolas, and MacGregor 1979). The assumption of a normal distribution always gives a lower estimate of the 10% fractile; although, if the coefficient of variation of the in-place strength is less than 20%, any difference is of little practical significance. It is convenient to adopt the normal distribution because this permits the use of many other statistical tools and techniques that have been derived on the basis of normality. If a lognormal distribution is adopted, however, these tools can be used by working with the natural logarithms of the estimated in-place strengths.

There is less available guidance concerning the appropriate confidence level. Hindo and Bergstrom (1985) suggest that the 75% confidence level should be adopted for ordinary structures, 90% for very important buildings, and 95% for crucial components in nuclear power plants. ACI 228.1R reports that a confidence level of 75% is widely used in practice when assessing the in-place strength of concrete during construction. Tables 8.2, 8.3, and 8.4 give parameters, based on a normal distribution of strengths, to facilitate the use of one of these three confidence levels in calculating the equivalent specified strength.

8.4.1 *Tolerance factor approach*—The conventional approach to estimate a fractile value is to use a tolerance factor K that accommodates the uncertainties of both the sample mean and the sample standard deviation caused by smaller sample sizes (Philleo 1981). If the samples are drawn from a normal population, values of K are based on a noncentral t distribution (Madsen, Krenk, and Lind 1986) and are tabulated for various sample sizes, confidence levels, and fractile values in Natrella (1963). The tolerance factor approach is presented in detail in ACI 228.1R as a relatively simple statistically based method for estimating the tenth percentile of the strength. Neglecting errors due to the use of empirically derived strength correction factors, the lower tolerance limit on the 10% fractile of the in-place strength data $f_{0.10}$ is obtained from the following equation

$$f_{0.10} = \bar{f}_c - K s_c \qquad (8\text{-}6)$$

where $\bar{f}_c$ and s_c are obtained from Eq. (8-2) and (8-3), respectively. The value of K for one-sided tolerance limits on the 10% fractile value, shown in Table 8.2, decreases markedly as the sample size n increases.

The estimate of the lower tenth-percentile of the in-place strength obtained from Eq. (8-6) does not account for the uncertainty introduced by the use of the strength correction factors. This uncertainty, which does not diminish as the number of specimens increases, can be accounted for using a factor Z shown in Table 8.3, which is derived from the standard normal distribution. Thus, the equivalent design strength $f'_{c,eq}$, following the tolerance factor approach, is obtained from the equation

$$f'_{c,eq} = \bar{f}_c - \sqrt{(Ks_c)^2 + (Zs_a)^2} \qquad (8\text{-}7)$$

An example calculation using the tolerance factor approach is given in the Appendix.

8.4.2 *Alternate approach*—Bartlett and MacGregor (1995) suggest that the tolerance factor approach may be unduly conservative in practice because core tests tend to overestimate the true variability of the in-place strengths. Therefore, the resulting value of $f'_{c,eq}$ is too low because the value of s_c used in Eq. (8-7) is too high. Also, the precision inherent in the tolerance factor approach is significantly higher than that associated with current design, specification, and acceptance practices.

A study of a large number of cores from members from different structures indicated that the variability of the average in-place strength between structures dominates the overall variability of the in-place strength (Bartlett and MacGregor 1996b). Thus, core data can be used to estimate the average in-place strength and a lower bound on this average strength for a particular structure. Assuming that the actual within-structure strength variation is accurately represented by the generic values shown in Table 2.1, the approximate 10% fractile of the in-place strength can then be obtained. Thus, the variability of the measured core strengths, which can exceed the true in-place strength variability due to testing factors that are hard to quantify, affects only the estimate of the lower bound on the mean strength.

In this approach, the equivalent specified strength is estimated using a two-step calculation. First, a lower bound estimate on the average in-place strength is determined from the core data. Then the 10% fractile of the in-place strength, which is equivalent to the specified strength, is obtained.

The lower-bound estimate of the mean in-place strength $(\bar{f}_c)_{CL}$ can be determined for some desired confidence level CL using the following equation

$$(\bar{f}_c)_{CL} = \bar{f}_c - \sqrt{\frac{(Ts_c)^2}{n} + (Zs_a)^2} \qquad (8\text{-}8)$$

The first term under the square root represents the effect of the sample size on the uncertainty of the mean in-place strength. The factor T is obtained from a Student's t distribution with $(n-1)$ degrees of freedom (Natrella 1963), which depends on the desired confidence level. The second term under the square root reflects the uncertainty attributable to the strength correction factors. As in the tolerance factor approach, it depends on a factor Z obtained from the standard normal distribution for the desired confidence level. Tables 8.3 and 8.4 show values of Z and T for the 75, 90, and 95% one-sided confidence levels, respectively. Bartlett and MacGregor (1995) suggest that a 90% confidence level is probably conservative for general use, but a greater confidence level may be appropriate if the reliability is particularly sensitive to the in-place concrete strength.

The estimated equivalent specified strength is defined using $(\bar{f}_c)_{CL}$ from the following expression

Table 8.5—C-factors for use in Eq. (8-9)

Structure composed of:	One member	Many members
One batch of concrete	0.91	0.89
Many batches of concrete		
Cast-in-place	0.85	0.83
Precast	0.88	0.87

$$f'_{c,eq} = C(\bar{f}_c)_{CL} \qquad (8\text{-}9)$$

Assuming the in-place strengths to be normally distributed, the desired 10% strength fractile is obtained using the constant C equal to $(1-1.28V_{WS})$, where V_{WS} is the within-structure coefficient of variation of the strengths shown in Table 2.1. Therefore, values of C depend on the number of batches, number of members, and type of construction, as shown in Table 8.5. To estimate the 13% fractile of the in-place concrete strength, Bartlett and MacGregor (1995) recommend values of C equal to 0.85 for cast-in-place construction consisting of many batches of concrete, or 0.90 for precast construction or cast-in-place members cast using a single batch of concrete. An example illustrating this approach is presented in the Appendix.

CHAPTER 9—SUMMARY

This guide summarizes current practices for obtaining cores and interpreting core compressive strength test results in light of past and current research findings. Parallel procedures are presented for the cases where cores are obtained to assess whether concrete strength in a new structure complies with strength-based acceptance criteria, and to determine a value based on the actual in-place concrete strength that is equivalent to the specified compressive strength f'_c and so can be directly substituted into conventional strength equations with customary strength reduction factors for the strength evaluation of an existing structure. It is inappropriate to use the procedures for determining an equivalent specified concrete strength to assess whether concrete strength in a new structure complies with strength-based acceptance criteria.

The order of contents parallels the logical sequence of activities in a typical core-test investigation. Chapter 2 describes how bleeding, consolidation, curing, and microcracking affect the in-place concrete strength in structures so that the investigator can account for this strength variation when planning the testing program. Chapter 3 identifies preferred sample locations and provides guidance on the number of specimens that should be obtained. Chapter 4 summarizes coring techniques that should result in high-quality, undamaged, representative test specimens. It is recommended that specimens with diameters of 100 to 150 mm (4 to 6 in.) and length-to-diameter ratios between 1.5 and 2 be obtained wherever possible to minimize any errors introduced by the strength correction factors for nonstandard specimens.

Chapter 5 describes procedures for testing the cores and detecting "outliers" by inspection of load-machine displacement curves or using statistical tests from ASTM E 178. Chapter 6 summarizes the subsequent analysis of strength test data including the use of ASTM C 42/42 M precision statements that quantify the expected variability of properly conducted tests for a sample of homogeneous material, research findings concerning the accuracy of empirically derived core strength correction factors, and statistical analysis techniques that can determine if the data can be grouped into unique categories.

Chapter 7 briefly elaborates on criteria presented in ACI 318 for using core test results to investigate low-strength cylinder test results in new construction.

Chapter 8 presents two methods for estimating the lower tenth-percentile value of the in-place concrete strength using core test data to quantify the in-place strength. This value is equivalent to the specified concrete strength f'_c and so can be directly substituted into conventional strength equations with customary strength reduction factors for the strength evaluation of an existing structure.

Example calculations are presented in an appendix for: outlier identification in accordance with ASTM E 178 criteria; determining whether a difference in mean strengths of cores from beams and columns is statistically significant; and computing the equivalent specified strength using the two approaches presented in Chapter 8.

CHAPTER 10—REFERENCES
10.1—Referenced standards and reports

The standards and reports listed were the latest editions at the time this document was prepared. Because these documents are revised frequently, the reader is advised to contact the proper sponsoring group if it is desired to refer to the latest version.

American Concrete Institute

228.1R	In-Place Methods for Determination of Strength of Concrete
228.2R	Nondestructive Test Methods for the Evaluation of Concrete in Structures
309.1R	Behavior of Fresh Concrete During Vibration
311.1R	ACI Manual of Concrete Inspection, SP-2
318	Building Code Requirements for Reinforced Concrete and Commentary

ASTM International

C 39	Standard Test Method for Compressive Strength of Cylindrical Concrete Specimens
C 42/ C 42M	Standard Method for Obtaining and Testing Drilled Cores and Sawed Beams of Concrete
C 670	Standard Practice for Preparing Precision and Bias Statements for Test Methods for Construction Materials
C 823	Standard Practice for Examination and Sampling of Hardened Concrete in Constructions
E 122	Standard Practice for Choice of Sample Size to Estimate the Average Quality of a Lot or Process
E 178	Standard Practice for Dealing with Outlying Observations

10.2—Cited references

Ang, A. H.-S., and Tang, W. H., 1975, *Probability Concepts in Engineering Planning and Design*, V. 1, Basic Principles, John Wiley and Sons, Inc., New York, 409 pp.

Bartlett, F. M., and MacGregor, J. G., 1994a, "Cores from High Performance Concrete Beams," *ACI Materials Journal*, V. 91, No. 6, Nov.-Dec., pp. 567-576.

Bartlett, F. M., and MacGregor, J. G., 1994b, "Effect of Core Length-to-Diameter Ratio on Concrete Core Strengths," *ACI Materials Journal*, V. 91, No. 4, July-Aug., pp. 339-348.

Bartlett, F. M., and MacGregor, J. G., 1994c, "Effect of Moisture Condition on Concrete Core Strengths," *ACI Materials Journal*, V. 91, No. 3, May-June, pp. 227-236.

Bartlett, F. M., and MacGregor, J. G., 1994d, "Effect of Core Diameter on Concrete Core Strengths," *ACI Materials Journal*, V. 91, No. 5, Sept.-Oct., pp. 460-470.

Bartlett, F. M., and MacGregor, J. G., 1995, "Equivalent Specified Concrete Strength from Core Test Data," *Concrete International*, V. 17, No. 3, Mar., pp. 52-58.

Bartlett, F. M., and MacGregor, J. G., 1996a, "In-Place Strength of High-Performance Concretes," *High Strength Concrete: An International Perspective*, SP-167, J. A. Bickley, ed., American Concrete Institute, Farmington Hills, Mich., pp. 211-228.

Bartlett, F. M., and MacGregor, J. G., 1996b, "Statistical Analysis of the Compressive Strength of Concrete in Structures," *ACI Materials Journal*, V. 93, No. 2, Mar.-Apr., pp. 158-168.

Benjamin, J. R., and Cornell, C. A., 1970, *Probability, Statistics, and Decision for Civil Engineers*, McGraw-Hill Book Co., New York, 684 pp.

Bickley, J. A., 1982, "Variability of Pullout Tests and In-Place Concrete Strength," *Concrete International*, V. 4. No. 4, Apr., pp. 44-51.

Bloem, D. L., 1965, "Concrete Strength Measurements—Cores versus Cylinders," *Proceedings*, V. 65, ASTM International, West Conshohocken, Pa., pp. 668-696.

Bloem, D. L., 1968, "Concrete Strength in Structures," ACI JOURNAL, *Proceedings* V. 65, No. 3, Mar., pp. 176-187.

Bollin, G. E., 1993, "Development of Precision and Bias Statements for Testing Drilled Cores in Accordance with ASTM C 42," *Cement, Concrete and Aggregates*, CCAGDP, V. 15, ASTM International, West Conshohocken, Pa., No. 1, pp. 85-88.

British Standards Institution, 1983, "BS 1881: Part 120, Method for Determination of the Compressive Strength of Concrete Cores," London, 6 pp.

Bungey, J. H., 1989, *Testing of Concrete in Structures*, 2nd Edition, Surrey University Press, Blackie & Son Ltd., 228 pp.

Burg, R. G., and Ost, B. W., 1992, "Engineering Properties of Commercially Available High-Strength Concretes," *Research and Development Bulletin* RD 104T, Portland Cement Association, Skokie, Ill., 55 pp.

Concrete Society, 1987, "Concrete Core Testing for Strength," *Technical Report* No. 11, The Concrete Society, London, 44 pp.

Cook, J. E., 1989, "10,000 psi Concrete," *Concrete International*, V. 11, No. 10, Oct., pp. 67-75.

Cook, W. D.; Miao, B.; Aïtcin, P.-C.; and Mitchell, D., 1992, "Thermal Stresses in Large High-Strength Concrete Columns," *ACI Materials Journal*, V. 89, No. 1, Jan.-Feb., pp. 61-68.

Dilly, R. L., and Vogt, W. L., 1993, "Statistical Methods for Evaluating Core Strength Results," *New Concrete Technology: Robert E. Philleo Symposium,* SP-141, T. C. Liu and G. C. Hoff, eds., American Concrete Institute, Farmington Hills, Mich., pp. 65-101.

Fiorato, A. E.; Burg, R. G.; and Gaynor, R. D., 2000, "Effects of Conditioning on Measured Compressive Strength of Concrete Cores," CTOO3, *Concrete Technology Today,* V. 21, No. 3, Portland Cement Association, Skokie, Ill, pp. 1-5.

Gaynor, R. D., 1970, "In-Place Strength: A Comparison of Two Test Systems," *Cement, Lime and Gravel,* V. 45, No. 3, pp. 55-60.

Hindo, K. R., and Bergstrom, W. R., 1985, "Statistical Evaluation of the In-Place Compressive Strength of Concrete," *Concrete International,* V. 7, No. 2, Feb., pp. 44-48.

Johnson, C. D., 1973, "Anisotropy of Concrete and Its Practical Implications," *Highway Research Record* No. 423, pp. 11-16.

Lewis, R. K., 1976, "Effect of Core Diameter on the Observed Strength of Concrete Cores," *Research Report* No. 50, CSIRO Division of Building Research, Melbourne, 13 pp.

Madsen, H. O.; Krenk, S.; and Lind, N. C., 1986, *Methods of Structural Safety,* Prentice-Hall Inc., Englewood Cliffs, N.J., 403 pp.

Mak, S. L.; Attard, M. M.; Ho, D. W. S.; and Darvall, P., 1990, "In-Situ Strength of High Strength Concrete," *Civil Engineering Research* Report No. 4/90, Monash University, Australia, 120 pp.

Mak, S. L.; Attard, M. M.; Ho, D. W. S.; and Darvall, P., 1993, "Effective In-Situ Strength of High Strength Columns," *Australian Civil Engineering Transactions,* V. CE35, No, 2, pp. 87-94.

Meininger, R. C., 1968, "Effect of Core Diameter on Measured Concrete Strength," *Journal of Materials,* JMLSA, V. 3, No. 2, pp. 320-326.

Meynick, P., and Samarin, A., 1979, "Assessment of Compressive Strength of Concrete by Cylinders, Cores, and Nondestructive Tests," *Controle de Qualite des Structures en Beton,* Proceedings of the RILEM Conference, V. 1, Stockholm, Sweden, pp. 127-134.

Miao, B.; Aïtcin, P.-C.; Cook, W. D.; and Mitchell, D., 1993, "Influence of Concrete Strength on In-Situ Properties of Large Columns," *ACI Materials Journal,* V. 90, No. 3, May-June, pp. 214-219.

Mirza, S. A.; Hatzinikolas, M.; and MacGregor, J. G., 1979, "Statistical Descriptions of Strength of Concrete," *Journal of the Structural Division, Proceedings,* ASCE, V. 105, No. ST6, pp. 1021-1037.

Natrella, M., 1963, "Experimental Statistics," *Handbook* No. 9, National Bureau of Standards, United States Government Printing Office, Washington.

Neville, A. M., 1981, *Properties of Concrete,* 3rd Edition, Pitman Publishing Ltd., London, 779 pp.

Neville, A. M., 2001, "Core Tests: Easy to Perform, Not Easy to Interpret," *Concrete International,* V. 23, No. 11, Nov., pp. 59-68.

Newman, K., and Lachance, L., 1964, "The Testing of Brittle Materials under Uniform Uniaxial Compressive Stresses," *Proceedings,* ASTM International, V. 64, pp. 1044-1067.

Ontario Ministry of Transportation and Communications, 1985, "Development of Special Provisions for the Acceptance of Lean Concrete, Base, Concrete Base and Concrete Pavement," *Report* No. MI-76, Ontario MTC, Downsview, Ontario, Mar.

Ottosen, N. S., 1984, "Evaluation of Concrete Cylinder Tests Using Finite Elements," *Journal of Engineering Mechanics,* ASCE, V. 110, No. 3, pp. 465-481.

Pearson, E. S., 1941-42, "The Probability Integral of the Range in Samples of n Observations from a Normal Population," *Biometrika,* pp. 301-308.

Philleo, R. E., 1981, "Increasing the Usefulness of ACI 214: Use of Standard Deviation and a Technique for Small Sample Sizes," *Concrete International,* V. 3, No. 9, Sept., pp. 71-74.

Popovics, S., 1969, "Effect of Porosity on the Strength of Concrete," *Journal of Materials,* JMLSA, V. 4, No. 2, pp. 356-371.

Popovics, S., 1986, "Effect of Curing Method and Final Moisture Condition on Compressive Strength of Concrete," ACI JOURNAL, *Proceedings* V. 83, No. 4, July-Aug., pp. 650-657.

Ramakrishnan, V., and Li, Shy-t'ien, 1970, "Maturity Strength Relationship of Concrete under Different Curing Conditions," *Proceedings of the 2nd Inter-American Conference on Materials Technology,* ASCE, New York, pp. 1-8.

Sanga, C. M., and Dhir, R. K., 1976, "Core-Cube Relationships of Plain Concrete," *Advances in Ready Mixed Concrete Technology,* R. K. Dhir, ed., Pergamon Press, Oxford, pp. 193-292.

Stone, W. C.; Carino, N. J.; and Reeve, C. P., 1986, "Statistical Methods for In-Place Strength Predictions by the Pullout Test," ACI JOURNAL, *Proceedings* V. 83, No. 5, Sept.-Oct., pp. 745-756.

Sullivan, P. J. E., 1991, "Testing and Evaluating Strength in Structures," *ACI Materials Journal,* V. 88, No. 5, Sept.-Oct., pp. 530-535.

Szypula, A., and Grossman, J. S., 1990, "Cylinder vesus Core Strength," *Concrete International,* V. 12, No. 2, Feb., pp. 55-61.

Takahata, A.; Iwashimizu, T.; and Ishibashi, U., 1991, "Construction of a High-Rise Reinforced Concrete Residence Using High-Strength Concrete," *High-Strength Concrete,* SP-121, W. T. Hester, ed., American Concrete Institute, Farmington Hills, Mich., pp. 741-755.

Toossi, M., and Houde, J., 1981, "Evaluation of Strength Variation Due to Height of Concrete Members," *Cement and Concrete Research,* V. 11, pp. 519-529.

Yip, W. K., and Tam, C. T., 1988, "Concrete Strength Evaluation Through the Use of Small Diameter Cores," *Magazine of Concrete Research,* V. 40, No. 143, pp. 99-105.

Yuan, R. L.; Ragab, M.; Hill, R. E.; and Cook, J. E., 1991, "Evaluation of Core Strength in High-Strength Concrete," *Concrete International,* V. 13, No. 5, May, pp. 30-34.

10.3—Other references
ACI Committee 214, 1977, "Recommended Practice for Evaluation of Strength Test Results of Concrete (ACI 214-77)," American Concrete Institute, Farmington Hills, Mich., 14 pp.

ACI Committee 446, 1999, "Fracture Mechanics of Concrete: Concepts, Models, and Determination of Material Properties (ACI 446.1R-91 (Reapproved 1999))," American Concrete Institute, Farmington Hills, Mich., 146 pp.

APPENDIX—EXAMPLE CALCULATIONS
A1—Outlier identification in accordance with ASTM E 178 criteria

Six cores are obtained from a single element. All have the same diameter ℓ/d and are given identical conditioning treatments in accordance with ASTM C 42/C 42M before testing. The measured strengths are 22.1, 29.4, 30.2, 30.8, 31.0, and 31.7 MPa (3200, 4270, 4380, 4470, 4500, and 4600 psi). The average strength is 29.2 MPa (4240 psi), and the standard deviation is 3.56 MPa (520 psi). If the smallest strength value is an outlier and so can be removed from the data set, the average strength will increase by almost 5% and the standard deviation will be markedly reduced.

The test statistic for checking if the smallest measured strength is an outlier according to ASTM E 178 criteria is the difference between the average and minimum values divided by the sample standard deviation. In this case it equals SI: (29.2 MPa – 22.1 MPa)/3.56 MPa = 1.99 [(4240 psi – 3200 psi)/520 psi = 2.00]. From Table 1 of ASTM E 178-80, the critical value for the two-sided test is 1.973 at the 1.0% significance level for a set of six observations. Thus, an observation this different from the mean value would be expected to occur by chance less than once every 100 times, and because this is unlikely, the low value of 22.1 MPa (3200 psi) is an outlier and can be removed from the data set. This decision conforms to the ASTM E 178 recommendation that a low significance level, such as 1%, be used as the critical value to test outlying observations.

If, in this example, the smallest core strength was 26.9 MPa (3900 psi) instead of 22.1 MPa (3200 psi), the average of the six strengths would be 30.0 MPa (4350 psi) with a standard deviation of 1.71 MPa (250 psi). The low value is (30.0 MPa – 26.9 MPa)/1.71 MPa = 1.81 standard deviations below the mean value [(4350 psi – 3900 psi)/250 psi = 1.80], which is less than the critical value of 1.822 given in Table 1 of ASTM E 178-80 for the two-sided test at the 10% significance level. The low test result would be expected to occur by chance at least once every 10 times, and because this is likely the 26.9 MPa (3900 psi) value is not an outlier according to ASTM E 178 and should not be removed from the data set.

A2—Student's *t* test for significance of difference between observed average values

It is not always obvious that any difference between average concrete strengths observed for cores from different structural components indicate a true difference of concrete quality between the components. For example, assume four cores obtained from four beams have measured strengths of 27.3, 29.0, 29.6, and 29.4 MPa (3960, 4210, 4300, and 4270 psi), which average 28.8 MPa (4180 psi) with a standard deviation of 1.05 MPa (155 psi). Five cores obtained from five columns have measured strengths of 31.2, 31.8, 30.9, 31.4, and 31.9 MPa (4520, 4610, 4480, 4560, and 4630 psi), which average 31.4 MPa (4560 psi) and have a standard deviation of 0.42 MPa (62 psi). Clearly the column cores are stronger, but is the difference large enough, given the small sample sizes, to consider the two data sets separately instead of combining them into a single set of nine observations for subsequent analysis?

To check whether the observed 2.6 MPa (380 psi) difference between the average strengths is statistically significant and not simply a value that might often be exceeded by chance given the scatter of the data, a test based on the Student's *t* distribution (Benjamin and Cornell 1970; Ang and Tang 1975) can be performed. The test statistic *t* for testing the hypothesis that the mean values of the underlying populations are equal is

$$t = \frac{|\bar{x}_2 - \bar{x}_1|}{S_P\sqrt{\frac{1}{n_1} + \frac{1}{n_2}}} \quad \text{(A-1)}$$

where the standard deviation of the pooled sample S_p is

$$S_p = \sqrt{\frac{(n_1 - 1)s_1^2 + (n_2 - 1)s_2^2}{(n_1 + n_2 - 2)}} \quad \text{(A-2)}$$

In these equations, $\bar{x}$ is the sample mean, s is the sample standard deviation, n is the number of observations, and subscripts 1 and 2 are used to distinguish between the two populations. The test is only valid when the true variances of the two populations σ^2 are equal, which can be verified using an F test (Benjamin and Cornell 1970; Ang and Tang 1975).

The rejection region is defined at a significance level α with degrees of freedom $df = v_1 + v_2 - 2$. Should the observed *t* value exceed the critical value, $t_{1-\alpha/2}$, which is tabulated in most statistical references (Benjamin and Cornell 1970; Ang and Tang 1975), then the probability that a difference at least as large as that observed will occur by chance is α. Most engineers and statisticians would not consider a difference to be statistically significant if the associated significance level is greater than 5%. As noted in the first example, more stringent significance levels are recommended for outlier detection.

Thus, for the example data:

$$S_p = \sqrt{\frac{(4-1)(1.05\text{ MPa})^2 + (5-1)(0.42\text{ MPa})^2}{(4+5-2)}} \quad \text{(A-3)}$$

$$= 0.76 \text{ MPa}$$

$$\left(S_p = \sqrt{\frac{(4-1)(155\text{ psi})^2 + (5-1)(62\text{ psi})^2}{(4+5-2)}} = 112 \text{ psi}\right)$$

so

$$t = \frac{|31.4 \text{ MPa} - 28.8 \text{ MPa}|}{0.76 \text{ MPa}\sqrt{\left(\frac{1}{4} + \frac{1}{5}\right)}} = 5.1 \quad (A\text{-}4)$$

$$\left(t = \frac{|4560 \text{ psi} - 4185 \text{ psi}|}{112 \text{ psi}\sqrt{\left(\frac{1}{4} + \frac{1}{5}\right)}} = 5.0\right)$$

For this case with seven degrees of freedom, the critical values for the two-sided test are 2.37 at the 95% significance level, 3.50 at the 99% significance level, and 4.78 at the 99.9% significance level (Ang and Tang 1975). Because the observed t statistic is slightly larger than the critical value at the 99.9% significance level, the value $1 - \alpha/2$ exceeds 99.9%, and so α is less than 0.2%. Thus, the probability of a difference of this magnitude occurring by chance is less than 1-in-500, and it can be concluded that the average strengths of the cores from the beams and the columns are significantly different. The data sets should not be combined, and distinct strength values should be computed separately for the columns and for the beams.

A3—Equivalent specified strength by tolerance factor approach

An equivalent specified strength is to be computed using the tolerance factor approach for five 100 x 200 mm (4 x 8 in.) cores that have been air-dried in accordance with ASTM C 42/C 42M before testing. The test strengths are 27.1, 29.8, 32.7, 34.8, and 39.6 MPa (3930, 4320, 4740, 5040, and 5740 psi). Only strength corrections for the effects of the moisture condition and the damage due to drilling are necessary to obtain the equivalent in-place strengths. Thus, using Eq. (8-1) and the factors from Table 8.4, $f_c = 1.02 \, f_{core}$, and the corresponding in-place strengths, rounded to the nearest 0.1 MPa (10 psi) in accordance with ASTM practice, are 27.6, 30.4, 33.3, 35.5, and 40.4 MPa (4010, 4410, 4830, 5140, and 5850 psi). The mean in-place strength $\bar{f}_c$ is 33.4 MPa (4850 psi), and the sample standard deviation of the in-place strength values s_c is 4.9 MPa (700 psi). If the uncertainty associated with the use of the strength correction factors is neglected, then the 75% confidence limit on the 10% fractile of the in-place strength is obtained using Eq. (8-6) with, from Table 8.2, $K = 1.96$

$$f_{0.10} = 33.4 \text{ MPa} - 1.96 \times 4.90 \text{ MPa} = 23.8 \text{ MPa} \quad (A\text{-}5)$$

$$(f_{0.10} = 4850 \text{ psi} - 1.96 \times 700 \text{ psi} = 3480 \text{ psi})$$

The uncertainty introduced by strength correction factors F_d and F_{mc} is determined using Eq. (8-4)

$$s_a = 33.4 \text{ MPa}\sqrt{0^2 + 0^2 + 0.025^2 + 0.025^2} = 1.18 \text{ MPa} \quad (A\text{-}6)$$

$$(s_a = 4850 \text{ psi}\sqrt{0^2 + 0^2 + 0.025^2 + 0.025^2} = 171 \text{ psi})$$

Thus, from Eq. (8-7), the 75% confidence limit on the 10% fractile of the in-place strength, determined using $Z = 0.67$ from Table 8.3 is

$$f'_{c,eq} = 33.4 \text{ MPa} - \sqrt{(1.96 \times 4.9 \text{ MPa})^2 + (0.67 \times 1.18 \text{ MPa})^2} \quad (A\text{-}7)$$

$$= 23.8 \text{ MPa}$$

$$(f'_{c,eq} = 4850 \text{ psi} - \sqrt{(1.96 \times 700 \text{ psi})^2 + (0.67 \times 171 \text{ psi})^2}$$

$$= 3470 \text{ psi})$$

In this example, the uncertainty due to the strength correction factors does not greatly influence the result because the 10% fractile of the in-place strength, Eq. (A-5), is essentially identical to the equivalent specified strength, Eq. (A-7). The equivalent specified strength is 23.8 MPa (3470 psi).

A4—Equivalent specified strength by alternate approach

For the core test results from the previous example, the equivalent specified strength is to be determined using the alternate approach. The 90% one-sided confidence interval on the mean in-place strength is, using Eq. (8-8) with $Z = 1.28$ from Table 8.3 and $T = 1.53$ from Table 8.4,

$$(\bar{f}_c)_{90} = 33.4 \text{ MPa} - \sqrt{\frac{(1.53 \times 4.9 \text{ MPa})^2}{5} + (1.28 \times 1.18 \text{ MPa})^2} \quad (A\text{-}8)$$

$$= 29.7 \text{ MPa}$$

$$\left((\bar{f}_c)_{90} = 4850 \text{ psi} - \sqrt{\frac{(1.53 \times 700 \text{ psi})^2}{5} + (1.28 \times 171 \text{ psi})^2}\right.$$

$$= 4320 \text{ psi})$$

Hence, from Eq. (8-9) with $C = 0.83$ for a cast-in-place structure composed of many members cast from many batches

$$f'_{c,eq} = 0.83 \times 29.7 \text{ MPa} = 24.7 \text{ MPa} \quad (A\text{-}9)$$

$$(f'_{c,eq} = 0.83 \times 4320 \text{ psi} = 3580 \text{ psi})$$

The equivalent specified strength is therefore 24.7 MPa (3580 psi) using the alternate approach. It is slightly greater than that computed using the tolerance factor method because, as described in Section 8.4.2, core test data tend to overestimate the true variability of the in-place strengths.

ACI 224.1R-93
(Reapproved 1998)

Causes, Evaluation and Repair of Cracks in Concrete Structures

Reported by ACI Committee 224

Grant T. Halvorsen*†
Chairman

Randall W. Poston
Secretary

Peter Barlow†
Florian Barth†
Alfred G. Bishara*
Howard L. Boggs
Merle E. Brander†
David Darwin‡
Fouad H. Fouad

David W. Fowler§
Peter Gergely*
Will Hansen
M. Nadim Hassoun
Tony C. Liu‡
Edward G. Nawy
Harry M. Palmbaum

Keith A. Pashina
Andrew Scanlon‡
Ernest K. Schrader
Wimal Suaris
Lewis H. Tuthill
Zenon A. Zielinski

* Contributing Author
† Member of Task Group which prepared these revisions
‡ Principal Author
§ Chairman of Task Group which prepared these revisions
Note: Associate members Masayatsu Ohtsu, Robert L. Yuan, and Consulting Member LeRoy Lutz contributed to the revision of this document.

The causes of cracks in concrete structures are summarized. The procedures used to evaluate cracking in concrete and the principal techniques for the repair of cracks are presented. The key methods of crack repair are discussed and guidance is provided for their proper application.

Keywords: autogenous healing; beams (supports); cement-aggregate reactions; concrete construction; concrete pavements; concrete slabs; concretes; consolidation; corrosion; cracking (fracturing); drilling; drying shrinkage; epoxy resins; evaluation; failure; grouting; heat of hydration; mass concrete; methacrylates; mix proportioning; plastics; polymers and resins; precast concrete; prestressed concrete; reinforced concrete; repairs; resurfacing; sealing; settlement (structural); shrinkage; specifications; structural design; tension; thermal expansion; volume change.

CONTENTS

Preface, pg. 224.1R-1

Chapter 1—Causes and control of cracking, pg. 224.1R-2
  1.1—Introduction
  1.2—Cracking of plastic concrete
  1.3—Cracking of hardened concrete

Chapter 2—Evaluation of cracking, pg. 224.1R-9
  2.1—Introduction
  2.2—Determination of location and extent of concrete cracking
  2.3—Selection of repair procedures

Chapter 3—Methods of crack repair, pg. 224.1R-13
  3.1—Introduction
  3.2—Epoxy injection
  3.3—Routing and sealing
  3.4—Stitching
  3.5—Additional reinforcement
  3.6—Drilling and plugging
  3.7—Gravity filling
  3.8—Grouting
  3.9—Drypacking
  3.10—Crack arrest
  3.11—Polymer impregnation
  3.12—Overlay and surface treatments
  3.13—Autogenous healing

ACI Committee Reports, Guides, Standard Practices, and Commentaries are intended for guidance in planning, designing, executing, and inspecting construction. This document is intended for the use of individuals who are competent to evaluate the significance and limitations of its content and recommendations and who will accept responsibility for the application of the material it contains. The American Concrete Institute disclaims any and all responsibility for the stated principles. The Institute shall not be liable for any loss or damage arising therefrom.
  Reference to this document shall not be made in contract documents. If items found in this document are desired by the Architect/Engineer to be a part of the contract documents, they shall be restated in mandatory language for incorporation by the Architect/Engineer.

ACI 224.1R-93 supersedes ACI 224.1R-90 and became effective September 1, 1993.
Copyright © 1993, American Concrete Institute.
All rights reserved including rights of reproduction and use in any form or by any means, including the making of copies by any photo process, or by electronic or mechanical device, printed, written, or oral, or recording for sound or visual reproduction or for use in any knowledge or retrieval system or device, unless permission in writing is obtained from the copyright proprietors.

Chapter 4—Summary, pg. 224.1R-19

Acknowledgment, pg. 224.1R-19

Chapter 5—References, pg. 224.1R-20
5.1—Recommended references
5.2—Cited references

PREFACE

Cracks in concrete have many causes. They may affect appearance only, or they may indicate significant structural distress or a lack of durability. Cracks may represent the total extent of the damage, or they may point to problems of greater magnitude. Their significance depends on the type of structure, as well as the nature of the cracking. For example, cracks that are acceptable for buildings may not be acceptable in water-retaining structures.

The proper repair of cracks depends on knowing the causes and selecting the repair procedures that take these causes into account; otherwise, the repair may only be temporary. Successful long-term repair procedures must attack the causes of the cracks as well as the cracks themselves.

To aid the practitioner in pinpointing the best solution to a cracking problem, this report discusses the causes, evaluation procedures, and methods of repair of cracks in concrete. Chapter 1 presents a summary of the causes of cracks and is designed to provide background for the evaluation of cracks. Chapter 2 describes evaluation techniques and criteria. Chapter 3 describes the methods of crack repair and includes a discussion of a number of techniques that are available. Many situations will require a combination of methods to fully correct the problem.

Preface to the 1991 Revision

Following the initial publication of ACI 224.1R in 1985, the Committee processed two minor revisions. One revision, published as ACI 224.1R-89 simply updated the format of recommended references. A second minor revision contained minor technical revisions and editorial corrections in the document, and added a new section to Chapter 3, regarding the use of high-molecular-weight methacrylates as sealer/healers.

During 1990 a Committee 224 Task Group reviewed the document and recommended the revisions contained herein. Chapter 1 has been altered in only minor detail. The introduction to Chapter 2 has been revised extensively, and additional minor revisions have been made to the rest of the Chapter. In Chapter 3, the section on routing and sealing has been rewritten to include flexible sealing and overbanding of cracks, and it is updated to reflect current materials and construction practices. Section 3.2 on epoxy injection has been revised to be somewhat more general and reflect current practice. The former section on high-molecular-weight methacrylates has been moved to Section 3.7 and retitled "Gravity Filling." This recognizes the point that "high-molecular-weight methacrylate" is a material, and not a method. References are presented in Chapter 5; citations throughout the text have been revised to employ the author/date format. Several new references have been added.

Additional revision of the report is ongoing. Committee 224 invites comment from the readers and users of this report on new developments, or alternate viewpoints on the *Causes, Evaluation, and Repair of Cracks in Concrete Structures*.

CHAPTER 1—CAUSES AND CONTROL OF CRACKING

1.1—Introduction

This chapter presents a brief summary of the causes of cracks and means for their control. Cracks are categorized as occurring either in plastic concrete or hardened concrete (Kelly 1981; Price 1982). In addition to the information provided here, further details are presented in ACI 224R and articles by Carlson et al. (1979), Kelly (1981), Price (1982), and Abdun-Nur (1983). Additional references are cited throughout the chapter.

1.2—Cracking of plastic concrete

1.2.1 *Plastic shrinkage cracking*—"Plastic shrinkage cracking (Fig. 1.1) occurs . . . when subjected to a very rapid loss of moisture caused by a combination of factors which include air and concrete temperatures, relative humidity, and wind velocity at the surface of the concrete. These factors can combine to cause high rates of surface evaporation in either hot or cold weather."

When moisture evaporates from the surface of freshly placed concrete faster than it is replaced by bleed water, the surface concrete shrinks. Due to the restraint provided by the concrete below the drying surface layer, tensile stresses develop in the weak, stiffening plastic concrete, resulting in shallow cracks of varying depth which

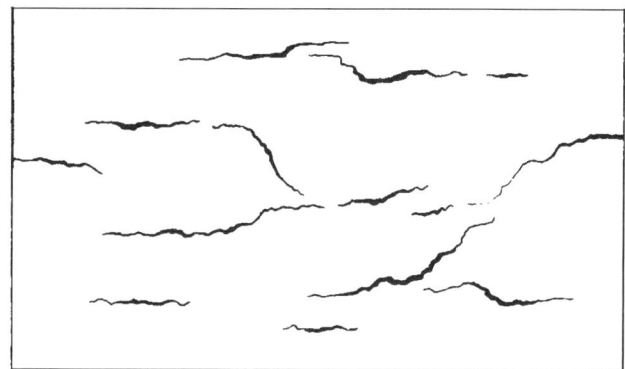

Fig. 1.1—Typical plastic shrinkage cracking (Price 1982)

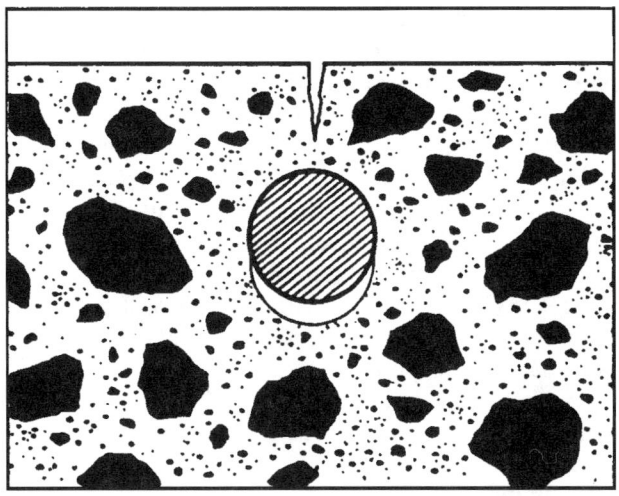

Fig. 1.2—Crack formed due to obstructed settlement (Price 1982)

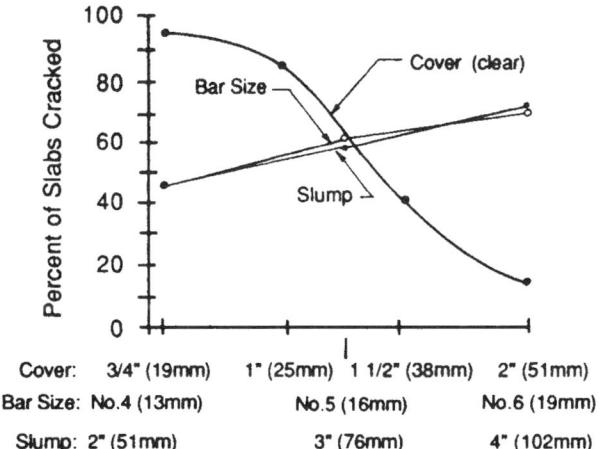

Fig. 1.3—Settlement cracking as a function of bar size, slump and cover (Dakhil et al. 1975)

may form a random, polygonal pattern, or may appear as essentially parallel to one another. These cracks are often fairly wide at the surface. They range from a few inches to many feet in length and are spaced from a few inches to as much as 10 ft (3 m) apart. Plastic shrinkage cracks begin as shallow cracks but can become full-depth cracks.

Since plastic shrinkage cracking is due to a differential volume change in the plastic concrete, successful control measures require a reduction in the relative volume change between the surface and other portions of the concrete.

Steps can be taken to prevent a rapid moisture loss due to hot weather and dry winds (ACI 224R, ACI 302.1R, ACI 305R). These measures include the use of fog nozzles to saturate the air above the surface and the use of plastic sheeting to cover the surface between finishing operations. Windbreaks to reduce the wind velocity and sunshades to reduce the surface temperature are also helpful, and it is good practice to schedule flat work after the windbreaks have been erected.

1.2.2 *Settlement cracking* — After initial placement, vibration, and finishing, concrete has a tendency to continue to consolidate. During this period, the plastic concrete may be locally restrained by reinforcing steel, a prior concrete placement, or formwork. This local restraint may result in voids and/or cracks adjacent to the restraining element (Fig. 1.2). When associated with reinforcing steel, settlement cracking increases with increasing bar size, increasing slump, and decreasing cover (Dakhil et al. 1975). This is shown in Fig. 1.3 for a limited range of these variables. The degree of settlement cracking may be intensified by insufficient vibration or by the use of leaking or highly flexible forms.

Form design (ACI 347R) and vibration (and revibration), provision of a time interval between the placement of concrete in columns or deep beams and the placement of concrete in slabs and beams (ACI 309.2R), the use of the lowest possible slump, and an increase in concrete cover will reduce settlement cracking.

1.3—Cracking of hardened concrete

1.3.1 *Drying shrinkage*—A common cause of cracking in concrete is restrained drying shrinkage. Drying shrinking is caused by the loss of moisture from the cement paste constituent, which can shrink by as much as 1 percent. Fortunately, aggregate provides internal restraint that reduces the magnitude of this volume change to about 0.06 percent. On wetting, concrete tends to expand.

These moisture-induced volume changes are a characteristic of concrete. If the shrinkage of concrete could take place without restraint, the concrete would not crack. It is the combination of shrinkage and restraint (usually provided by another part of the structure or by the subgrade) that causes tensile stresses to develop. When the tensile strength of concrete is exceeded, it will crack. Cracks may propagate at much lower stresses than are required to cause crack initiation.

In massive concrete elements, tensile stresses are caused by differential shrinkage between the surface and the interior concrete. The larger shrinkage at the surface causes cracks to develop that may, with time, penetrate deeper into the concrete.

The magnitude of the tensile stresses induced by volume change is influenced by a combination of factors, including the amount of shrinkage, the degree of restraint, the modulus of elasticity, and the amount of creep. The amount of drying shrinkage is influenced mainly by the amount and type of aggregate and the water content of the mix. The greater the amount of aggregate, the smaller the amount of shrinkage (Pickett 1956). The higher the stiffness of the aggregate, the more effective it is in reducing the shrinkage of the concrete (*i.e.*, the shrinkage of concrete containing sandstone aggregate may be more than twice that of concrete with granite, basalt, or limestone (Carlson 1938)). The higher the water content, the greater the amount of drying shrinkage (U.S. Bureau of Reclamation 1975).

Surface crazing (alligator pattern) on walls and slabs is an example of drying shrinkage on a small scale. Crazing usually occurs when the surface layer of the concrete has a higher water content than the interior concrete. The result is a series of shallow, closely spaced, fine cracks.

Drying shrinkage can be reduced by increasing the amount of aggregate and reducing the water content. A procedure that will help reduce settlement cracking, as well as drying shrinkage in walls, is reducing the water content of the concrete as the wall is placed from the bottom to the top. Using this procedure, bleed water from the lower portions of the wall will tend to equalize the water content within the wall. To be successful, this procedure needs careful control of the concrete and proper consolidation.

Shrinkage cracking can be controlled by using contraction joints and steel detailing. Shrinkage cracking may also be reduced by using shrinkage-compensating cement. The reduction or elimination of subslab restraint can also be effective in reducing shrinkage cracking in slabs-on-grade (Wimsatt et al. 1987). In cases where crack control is particularly important, the minimum requirements of ACI 318 are not always adequate. These points are discussed in greater detail in ACI 224R, which describes additional construction practices designed to help control the drying shrinkage cracking that does occur, and in ACI 224.3R, which describes the use and function of joints in concrete construction.

1.3.2 *Thermal stresses*—Temperature differences within a concrete structure may be caused by portions of the structure losing heat of hydration at different rates or by the weather conditions cooling or heating one portion of the structure to a different degree or at a different rate than another portion. These temperature differences result in differential volume changes. When the tensile stresses due to the differential volume changes exceed the tensile stress capacity, concrete will crack. The effects of temperature differentials due to different rates of heat dissipation of the heat of hydration of cement are normally associated with mass concrete (which can include large columns, piers, beams, and footings, as well as dams), while temperature differentials due to changes in the ambient temperature can affect any structure.

Cracking in mass concrete can result from a greater temperature on the interior than on the exterior. The temperature gradient may be caused by either the center of the concrete heating up more than the outside due to the liberation of heat during cement hydration or more rapid cooling of the exterior relative to the interior. Both cases result in tensile stresses on the exterior and, if the tensile strength is exceeded, cracking will occur. The tensile stresses are proportional to the temperature differential, the coefficient of thermal expansion, the effective modulus of elasticity (which is reduced by creep), and the degree of restraint (Dusinberre 1945; Houghton 1972, 1976). The more massive the structure, the greater the potential for temperature differential and restraint.

Procedures to help reduce thermally-induced cracking include reducing the maximum internal temperature, delaying the onset of cooling, controlling the rate at which the concrete cools, and increasing the tensile strength of the concrete. These and other methods used to reduce cracking in massive concrete are presented in ACI 207.1R, ACI 207.2R, ACI 207.4R, and ACI 224R.

Hardened concrete has a coefficient of thermal expansion that may range from 4 to 9 x 10^{-6} F (7 to 11 x 10^{-6} C), with a typical value of 5.5 x 10^{-6} F (10 x 10^{-6} C). When one portion of a structure is subjected to a temperature-induced volume change, the potential for thermally-induced cracking exists. Designers should give special consideration to structures in which some portions are exposed to temperature changes, while other portions of the structure are either partially or completely protected. A drop in temperature may result in cracking in the exposed element, while increases in temperature may cause cracking in the protected portion of the structure. Temperature gradients cause deflection and rotation in structural members; if restrained, serious stresses can result (Priestly 1978; Hoffman et al. 1983; ACI 343R). Allowing for movement by using properly designed contraction joints and correct detailing will help alleviate these problems.

1.3.3 *Chemical reaction*—Deleterious chemical reactions may cause cracking of concrete. These reactions may be due to materials used to make the concrete or materials that come into contact with the concrete after it has hardened.

Some general concepts for reducing adverse chemical reactions are presented here, but only pretesting of the mixture or extended field experience will determine the effectiveness of a specific measure.

Concrete may crack with time as the result of slowly developing expansive reactions between aggregate containing active silica and alkalies derived from cement hydration, admixtures, or external sources (*e.g.*, curing water, ground water, alkaline solutions stored or used in the finished structure.)

The alkali-silica reaction results in the formation of a swelling gel, which tends to draw water from other portions of the concrete. This causes local expansion and accompanying tensile stresses, and may eventually result in the complete deterioration of the structure. Control measures include proper selection of aggregates, use of low alkali cement, and use of pozzolans, which themselves contain very fine, highly active silicas. The first measure may preclude the problem from occurring, while the later two measures have the effect of decreasing the alkali to reactive silica ratio, resulting in the formation of a nonexpanding calcium alkali silicate.

Certain carbonate rocks participate in reactions with alkalies which, in some instances, produce detrimental expansion and cracking. These detrimental alkali-carbonate reactions are usually associated with argillaceous dolomitic limestones which have a very fine grained (cryptocrystalline) structure (ACI 201.2R). The affected concrete is characterized by a network pattern of cracks. The reaction is distinguished from the alkali-silica reaction by the general absence of silica gel surface deposits at the crack. The problem may be minimized by avoiding reactive aggregates, dilution with nonreactive aggregates, use of a smaller maximum size aggregate, and use of low-alkali cement (ACI 201.2R).

Sulfate-bearing waters are a special durability problem for concrete. When sulfate penetrates hydrated cement paste, it comes in contact with hydrated calcium aluminate. Calcium sulfoaluminate is formed, with a subsequently large increase in volume, resulting in high local tensile stresses that lead to cracking which causes development of closely spaced cracking and deterioration. ASTM C 150 Types II and V portland cement, which are low in tricalcium aluminate, will reduce the severity of the problem. The blended cements specified in ASTM C 595 are also useful in this regard. In severe cases, some pozzolans, known to impart additional resistance to sulfate attack, could be used after adequate testing.

Detrimental conditions may also occur from the application of deicing salts to the surface of hardened concrete. Concrete subjected to water soluble salts should be amply air entrained, have adequate cover of the reinforcing steel, and be made of high-quality, low permeability concrete.

The effects of these and other problems relating to the durability of concrete are discussed in greater detail in ACI 201.2R.

The calcium hydroxide in hydrated cement paste will combine with carbon dioxide in the air to form calcium carbonate. Since calcium carbonate has a smaller volume than the calcium hydroxide, shrinkage will occur (commonly known as carbonation shrinkage). This situation may result in significant surface crazing and may be especially serious on freshly placed surfaces during the first 24 hours when improperly vented combustion heaters are used to keep concrete warm during the winter months.

With the exception of surface carbonation, very little can be done to protect or repair concrete that has been subjected to the types of chemical attack described above (ACI 201.2R).

1.3.4 *Weathering*—The weathering processes that can cause cracking include freezing and thawing, wetting and drying, and heating and cooling. Cracking of concrete due to natural weathering is usually conspicuous, and it may give the impression that the concrete is on the verge of disintegration, even though the deterioration may not have progressed much below the surface.

Damage from freezing and thawing is the most common weather-related physical deterioration. Concrete may be damaged by freezing of water in the paste, in the aggregate, or in both (Powers 1975).

Damage in hardened cement paste from freezing is caused by the movement of water to freezing sites and by hydraulic pressure generated by the growth of ice crystals (Powers 1975).

Aggregate particles are surrounded by cement paste which prevents the rapid escape of water. When the aggregate particles are above a critical degree of saturation, the expansion of the absorbed water during freezing may crack the surrounding cement paste or damage the aggregate itself (Callan 1952; Snowdon and Edwards 1962).

Concrete is best protected against freezing and thawing through the use of the lowest practical water-cement ratio and total water content, durable aggregate, and adequate air entrainment. Adequate curing prior to exposure to freezing conditions is also important. Allowing the structure to dry after curing will enhance its freezing and thawing durability.

Other weathering processes that may cause cracking in concrete are alternate wetting and drying, and heating and cooling. Both processes produce volume changes that may cause cracking. If the volume changes are excessive, cracks may occur, as discussed in Sections 1.3.1 and 1.3.2.

1.3.5 *Corrosion of reinforcement*—Corrosion of a metal is an electro-chemical process that requires an oxidizing agent, moisture, and electron flow within the metal; a series of chemical reactions takes place on and adjacent to the surface of the metal (ACI 201.2R).

The key to protecting metal from corrosion is to stop or reverse the chemical reactions. This may be done by cutting off the supplies of oxygen or moisture or by supplying excess electrons at the anodes to prevent the formation of the metal ions (cathodic protection).

Reinforcing steel usually does not corrode in concrete because a tightly adhering protective oxide coating forms in the highly alkaline environment. This is known as passive protection.

Reinforcing steel may corrode, however, if the alkalinity of the concrete is reduced through carbonation or if the passivity of this steel is destroyed by aggressive ions (usually chlorides). Corrosion of the steel produces iron oxides and hydroxides, which have a volume much greater than the volume of the original metallic iron (Verbeck 1975). This increase in volume causes high radial bursting stresses around reinforcing bars and results in local radial cracks. These splitting cracks can propagate along the bar, resulting in the formation of longitudinal cracks (*i.e.*, parallel to the bar) or spalling of the concrete. A broad crack may also form at a plane of bars parallel to a concrete surface, resulting in delamination, a well-known problem in bridge decks.

Cracks provide easy access for oxygen, moisture, and chlorides, and thus, minor splitting cracks can create a condition in which corrosion and cracking are accelerated.

Cracks transverse to reinforcement usually do not cause continuing corrosion of the reinforcement if the concrete has low permeability. This is due to the fact that the exposed portion of a bar at a crack acts as an anode.

At early ages, the wider the crack, the greater the corrosion, simply because a greater portion of the bar has lost its passive protection. However, for continued corrosion to occur, oxygen and moisture must be supplied to other portions of the same bar or bars that are electrically connected by direct contract or through hardware such as chair supports. If the combination of density and cover thickness is adequate to restrict the flow of oxygen and moisture, then the corrosion process is self sealing (Verbeck 1975).

Corrosion can continue if a longitudinal crack forms parallel to the reinforcement, because passivity is lost at many locations, and oxygen and moisture are readily available along the full length of the crack.

Other causes of longitudinal cracking, such as high bond stresses, transverse tension (for example, along stirrups or along slabs with two-way tension), shrinkage, and settlement, can initiate corrosion.

For general concrete construction, the best protection against corrosion-induced splitting is the use of concrete with low permeability and adequate cover. Increased concrete cover over the reinforcing is effective in delaying the corrosion process and also in resisting the splitting and spalling caused by corrosion or transverse tension (Gergely 1981; Beeby 1983). In the case of large bars and thick covers, it may be necessary to add small transverse reinforcement (while maintaining the minimum cover requirements) to limit splitting and to reduce the surface crack width (ACI 345R).

In very severe exposure conditions, additional protective measures may be required. A number of options are available, such as coated reinforcement, sealers or overlays on the concrete, corrosion-inhibiting admixtures, and cathodic protection (NCHRP Synthesis 57). Any procedure that effectively prevents access of oxygen and moisture to the steel surface or reverses the electron flow at the anode will protect the steel. In most cases, concrete must be allowed to breathe, that is any concrete surface treatment must allow water to evaporate from the concrete.

1.3.6 *Poor construction practices*—A wide variety of poor construction practices can result in cracking in concrete structures. Foremost among these is the common practice of adding water to concrete to improve workability. Added water has the effect of reducing strength, increasing settlement, and increasing drying shrinkage. When accompanied by a higher cement content to help offset the decrease in strength, an increase in water content will also mean an increase in the temperature differential between the interior and exterior portions of the structure, resulting in increased thermal stresses and possible cracking. By adding cement, even if the water-cement ratio remains constant, more shrinkage will occur since the relative paste volume is increased.

Lack of curing will increase the degree of cracking within a concrete structure. The early termination of curing will allow for increased shrinkage at a time when the concrete has low strength. The lack of hydration of the cement, due to drying, will result not only in decreased long-term strength, but also in the reduced durability of the structure.

Other construction problems that may cause cracking are inadequate formwork supports, inadequate consolidation, and placement of construction joints at points of high stress. Lack of support for forms or inadequate consolidation can result in settlement and cracking of the concrete before it has developed sufficient strength to support its own weight, while the improper location of construction joints can result in the joints opening at these points of high stress.

Methods to prevent cracking due to these and other poor construction procedures are well known (see ACI 224R, ACI 302.1R, ACI 304R, ACI 305R, ACI 308, ACI 309R, ACI 345R, and ACI 347R), but require special attention during construction to insure their proper execution.

1.3.7 *Construction overloads*—Loads induced during construction can often be far more severe than those experienced in service. Unfortunately, these conditions may occur at early ages when the concrete is most susceptible to damage and they often result in permanent cracks.

Precast members, such as beams and panels, are most frequently subject to this abuse, but cast-in-place concrete can also be affected. A common error occurs when precast members are not properly supported during transport and erection. The use of arbitrary or convenient lifting points may cause severe damage. Lifting eyes, pins, and other attachments should be detailed or approved by the designer. When lifting pins are impractical, access to the bottom of a member must be provided so that a strap may be used. The PCI Committee on Quality Control Performance Criteria (1985, 1987) provides additional information on the causes, prevention and repair of cracking related to fabrication and shipment of precast or prestressed beams, columns, hollow core slabs and double tees.

Operators of lifting devices must exercise caution and be aware that damage may be caused even when the proper lifting accessories are used. A large beam or panel lowered too fast, and stopped suddenly, results in an impact load that may be several times the dead weight of the member. Another common construction error that should be avoided is prying up one corner of a panel to lift it off its bed or "break it loose."

When considering the support of a member for shipment, the designer must be aware of loads that may be induced during transportation. Some examples that occur during shipment of large precast members via tractor and trailer are jumping curbs or tight highway corners, torsion due to differing roadway superelevations between the trailer and the tractor, and differential acceleration of the trailer and the tractor.

Pretensioned beams can present unique cracking problems at the time of stress release—usually when the beams are less than one day old. Multiple strands must be detensioned following a specific pattern, so as not to place unacceptable eccentric loads on the member. If all of the strands on one side of the beam are released while the strands on the other side are still stressed, cracking may occur on the side with the unreleased strands. These cracks are undesirable, but should close with the release of the balance of the strands.

In the case of a T-beam with a heavily reinforced flange and a highly prestressed thin web, cracks may develop at the web-flange junction.

Another practice that can result in cracks near beam ends is tack welding embedded bearing plates to the casting bed to hold them in place during concrete placement. The tack welds are broken only after enough prestress is induced during stress transfer to break them. Until then, the bottom of the beam is restrained while the rest of the beam is compressed. Cracks will form near the bearing plates if the welds are too strong.

Thermal shock can cause cracking of steam-cured concrete if it is treated improperly. The maximum rate of cooling frequently used is 70 F (40 C) per hour (ACI 517.2R; Verbeck 1958; Shideler and Toennies 1963; Kirkbride 1971b). When brittle aggregate is used and the strain capacity is low, the rate of cooling should be decreased. Even following this practice, thermally induced cracking often occurs. Temperature restrictions should apply to the entire beam, not just locations where temperatures are monitored. If the protective tarps used to contain the heat are pulled back for access to the beam ends when cutting the strands, and if the ambient temperatures are low, thermal shock may occur. Temperature recorders are seldom located in these critical areas.

Similar conditions and cracking potential exist with precast blocks, curbs, and window panels when a rapid surface temperature drop occurs.

It is believed by many (ACI 517.2R; Mansfield 1948; Nurse 1949; Higginson 1961; Jastrzebski 1961; Butt et al. 1969; Kirkbride 1971a; Concrete Institute of Australia 1972; PCI Energy Committee 1981) that rapid cooling may cause cracking only in the surface layers of very thick units and that rapid cooling is not detrimental to the strength or durability of standard precast products (PCI Energy Committee 1981). One exception is transverse cracking observed in pretensioned beams subjected to cooling prior to detensioning. For this reason, pretensioned members should be detensioned immediately after the steam-curing has been discontinued (PCI Energy Committee 1981).

Cast-in-place concrete can be unknowingly subjected to construction loads in cold climates when heaters are used to provide an elevated working temperature within a structure. Typically, tarps are used to cover windows and door openings, and high volume heaters are operated inside the enclosed area. If the heaters are located near exterior concrete members, especially thin walls, an unacceptably high thermal gradient can result within the members. The interior of the wall will expand in relation to the exterior. Heaters should be kept away from the exterior walls to minimize this effect. Good practice also requires that this be done to avoid localized drying shrinkage and carbonation cracking.

Storage of materials and the operation of equipment can easily result in loading conditions during construction far more severe than any load for which the structure was designed. Tight control must be maintained to avoid overloading conditions. Damage from unintentional construction overloads can be prevented only if designers provide information on load limitations for the structure and if construction personnel heed these limitations.

1.3.8 Errors in design and detailing—The effects of improper design and/or detailing range from poor appearance to lack of serviceability to catastrophic failure. These problems can be minimized only by a thorough understanding of structural behavior (meant here in the broadest sense).

Errors in design and detailing that may result in unacceptable cracking include use of poorly detailed reentrant corners in walls, precast members and slabs, improper selection and/or detailing of reinforcement, restraint of members subjected to volume changes caused by variations in temperature and moisture, lack of adequate contraction joints, and improper design of foundations, resulting in differential movement within the structure. Examples of these problems are presented by Kaminetzky (1981) and Price (1982).

Reentrant corners provided a location for the concentration of stress and, therefore, are prime locations for the initiation of cracks. Whether the high stresses result from volume changes, in-plane loads, or bending, the designer must recognize that stresses are always high near reentrant corners. Well-known examples are window and door openings in concrete walls and dapped end beams, as shown in Fig. 1.4 and 1.5. Additional properly anchored diagonal reinforcement is required to keep the inevitable cracks narrow and prevent them from propagating.

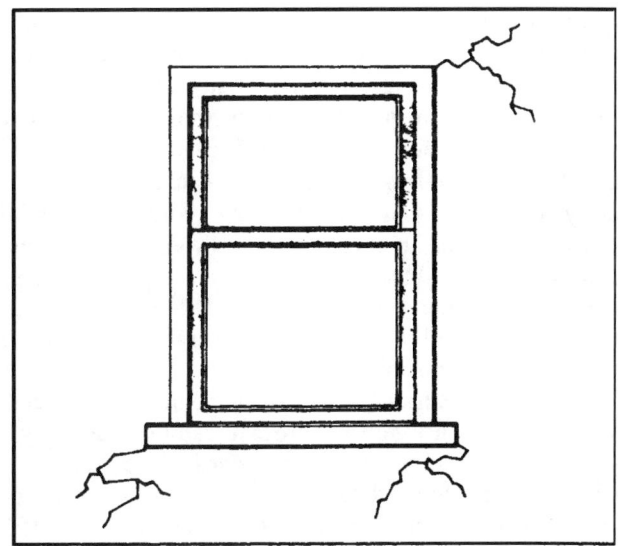

Fig. 1.4—*Typical crack patterns at reentrant corners (Price 1982)*

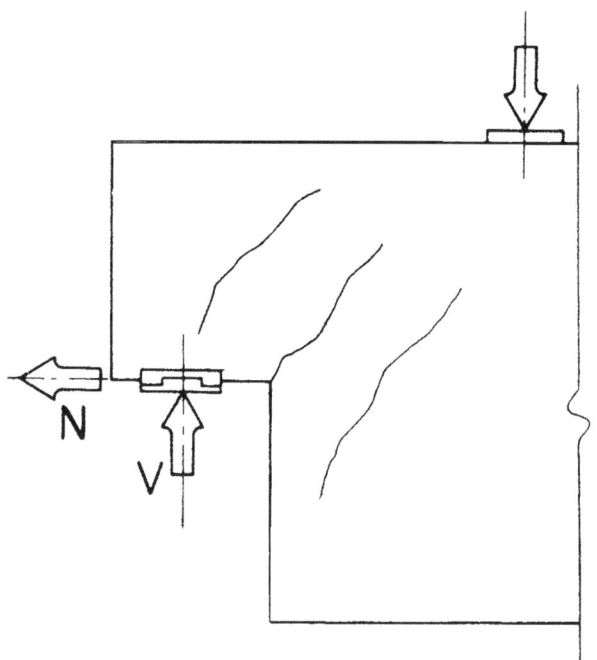

Fig. 1.5—*Typical cracking pattern of dapped end at service load* *

The use of an inadequate amount of reinforcing may result in excessive cracking. A typical mistake is to lightly reinforce a member because it is a "nonstructural member." However, the member (such as a wall) may be tied to the rest of the structure in such a manner that it is required to carry a major portion of the load once the structure begins to deform. The "nonstructural element" then begins to carry loads in proportion to its stiffness. Since this member is not detailed to act structurally, unsightly cracking may result even though the safety of the structure is not in question.

The restraint of members subjected to volume changes results frequently in cracks. Stresses that can occur in concrete due to restrained creep, temperature differential, and drying shrinkage can be many times the stresses that occur due to loading. A slab, wall, or a beam restrained against shortening, even if prestressed, can easily develop tensile stresses sufficient to cause cracking. Properly designed walls should have contraction joints spaced from one to three times the wall height. Beams should be allowed to move. Cast-in-place post-tensioned construction that does not permit shortening of the prestressed member is susceptible to cracking in both the member and the supporting structure (Libby 1977). The problem with restraint of structural members is especially serious in pretensioned and precast members that may be welded to the supports at both ends. When combined with other problem details (such as reentrant corners), results may be catastrophic (Kaminetzky 1981; Mast 1981).

Improper foundation design may result in excessive differential movement within a structure. If the differential movement is relatively small, the cracking problems may be only visual in nature. However, if there is a major differential settlement, the structure may not be able to redistribute the loads rapidly enough, and a failure may occur. One of the advantages of reinforced concrete is that, if the movement takes place over a long enough period of time, creep will allow at least some load redistribution to take place.

The importance of proper design and detailing will depend on the particular structure and loading involved. Special care must be taken in the design and detailing of structures in which cracking may cause a major serviceability problem. These structures also require continuous inspection during all phases of construction to supplement the careful design and detailing.

1.3.9 *Externally applied loads*—It is well known that load-induced tensile stresses result in cracks in concrete members. This point is readily acknowledged and accepted in concrete design. Current design procedures (ACI 318 and AASHTO Standard Specifications for Highway Bridges) use reinforcing steel, not only to carry the tensile forces, but to obtain both an adequate distribution of cracks and a reasonable limit on crack width.

Current knowledge of flexural members provides the basis for the following general conclusions about the variables that control cracking: Crack width increases with increasing steel stress, cover thickness and area of concrete surrounding each reinforcing bar. Of these, steel stress is the most important variable. The bar diameter is not a major consideration. The width of a bottom crack increases with an increasing strain gradient between the steel and the tension face of the beam.

The equation considered to best predict the most probable maximum surface crack width in bending was developed by Gergely and Lutz (1968). A simplified version of this equation is:

$$w = 0.076 \beta f_s (d_c A)^{0.33} \times 10^{-3} \qquad (1.1)$$

in which w = most probable maximum crack width, in.; β = ratio of distance between neutral axis and tension face to distance between neutral axis and centroid of reinforcing steel (taken as approximately 1.20 for typical beams in buildings); f_s = reinforcing steel stress, ksi; d_c = thickness of cover from tension fiber to center of bar closest thereto, in.; and A = area of concrete symmetric with reinforcing steel divided by number of bars, in.^{2}

A modification of this equation is used in ACI 318, which effectively limits crack widths to 0.016 in. (0.41 mm) for interior exposure and 0.013 in. (0.33 mm) for exterior exposure. However, considering the information presented in Section 1.3.5 which indicates little correlation between surface crack width for cracks transverse to bars and the corrosion of reinforcing, these limits do not appear to be justified on the basis of corrosion control.

---

* From Alan H. Mattock and Timothy C. Chan (1979), "Design and Behavior of Dapped-end Beams," *Journal*, Prestressed Concrete Institute, V. 24, No. 6, Nov.-Dec., pp. 28-45.

There have been a number of equations developed for prestressed concrete members (ACI 224R), but no single method has achieved general acceptance.

The maximum crack width in tension members is larger than that predicted by the expression for flexural members (Broms 1965; Broms and Lutz 1965). Absence of a strain gradient and compression zone in tension members is the probable reason for the larger crack widths.

On the basis of limited data, the following expression has been suggested to estimate the maximum crack width in direct tension (ACI 224R):

$$w = 0.10 f_s (d_c A)^{0.33} \times 10^{-3} \qquad (1.2)$$

Additional information on cracking of concrete in direct tension is provided in ACI 224.2R.

Flexural and tensile crack widths can be expected to increase with time for members subjected to either sustained or repetitive loading. Although a large degree of scatter is evident in the available data, a doubling of crack width with time can be expected (Abeles et al. 1968; Bennett and Dave 1969; Illston and Stevens 1972; Holmberg 1973; Rehm and Eligehausen 1977).

Although work remains to be done, the basic principles of crack control for load-induced cracks are well understood. Well-distributed reinforcing offers the best protection against undesirable cracking. Reduced steel stress, obtained through the use of a larger amount of steel, will also reduce the amount of cracking. While reduced cover will reduce the surface crack width, designers must keep in mind, as pointed out in Section 1.3.5, that cracks (and therefore, crack widths) perpendicular to reinforcing steel do not have a major effect on the corrosion of the steel, while a reduction in cover will be detrimental to the corrosion protection of the reinforcing.

CHAPTER 2—EVALUATION OF CRACKING

2.1—Introduction

When anticipating repair of cracks in concrete, it is important to first identify the location and extent of cracking. It should be determined whether the observed cracks are indicative of current or future structural problems, taking into consideration the present and anticipated future loading conditions. The cause of the cracking should be established before repairs are specified. Drawings, specifications, and construction and maintenance records should be reviewed. If these documents, along with field observations, do not provide the needed information, a field investigation and structural analysis should be completed before proceeding with repairs.

The causes of cracks are discussed in Chapter 1. A detailed evaluation of observed cracking can determine which of those causes applies in a particular situation.

Cracks need to be repaired if they reduce the strength, stiffness, or durability of the structure to an unacceptable level, or if the function of the structure is seriously impaired. In some cases, such as cracking in water-retaining structures, the function of the structure will dictate the need for repair, even if strength, stiffness, or appearance are not significantly affected. Cracks in pavements and slabs-on-grade may require repair to prevent edge spalls, migration of water to the subgrade, or to transmit loads. In addition, repairs that improve the appearance of the surface of a concrete structure may be desired.

2.2—Determination of location and extent of concrete cracking

Location and extent of cracking, as well as information on the general condition of concrete in a structure, can be determined by both direct and indirect observations, nondestructive and destructive testing, and tests of cores taken from the structure. Information may also be obtained from drawings and construction and maintenance records.

2.2.1 *Direct and indirect observation*—The locations and widths of cracks should be noted on a sketch of the structure. A grid marked on the surface of the structure can be useful to accurately locate cracks on the sketch.

Crack widths can be measured to an accuracy of about 0.001 in. (0.025 mm) using a crack comparator, which is a small, hand-held microscope with a scale on the lens closest to the surface being viewed (Fig. 2.1). Crack widths may also be estimated using a clear comparator card having lines of specified width marked on the card. Observations such as spalling, exposed reinforcement, surface deterioration, and rust staining should be noted on the sketch. Internal conditions at specific crack locations can be observed with the use of flexible shaft fiberscopes or rigid borescopes.

Crack movement can be monitored with mechanical movement indicators of the types shown in Fig. 2.2. The indicator, or crack monitor, shown in Fig. 2.2 (a) gives a direct reading of crack displacement and rotation. The indicator in Fig. 2.2 (b) (Stratton et al. 1978) amplifies the crack movement (in this case, 50 times) and indicates the maximum range of movement during the measurement period. Mechanical indicators have the advantage

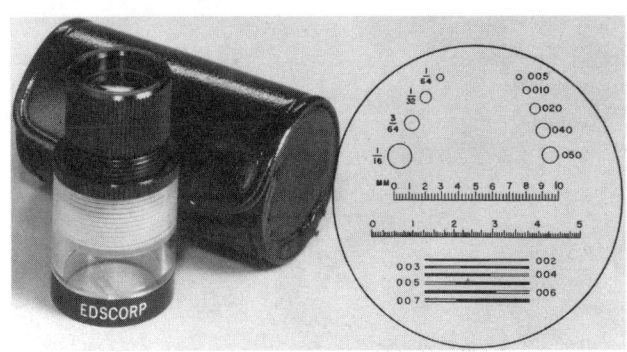

Fig. 2.1—Comparator for measuring crack widths (courtesy of Edmound Scientific Co.)

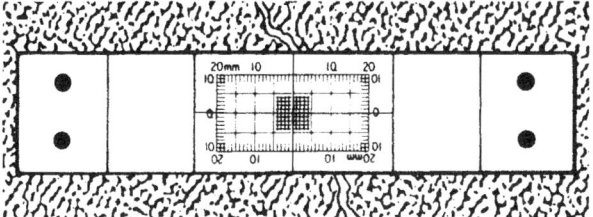

Newly Mounted Monitor

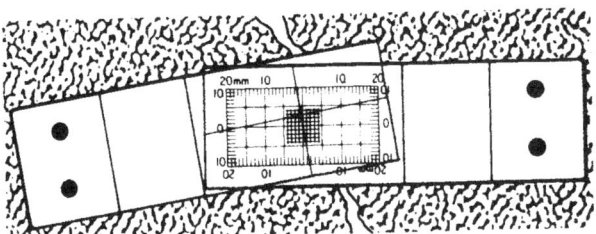

Monitor After Crack Movement

(a)—Crack monitor (courtesy of Avongard)

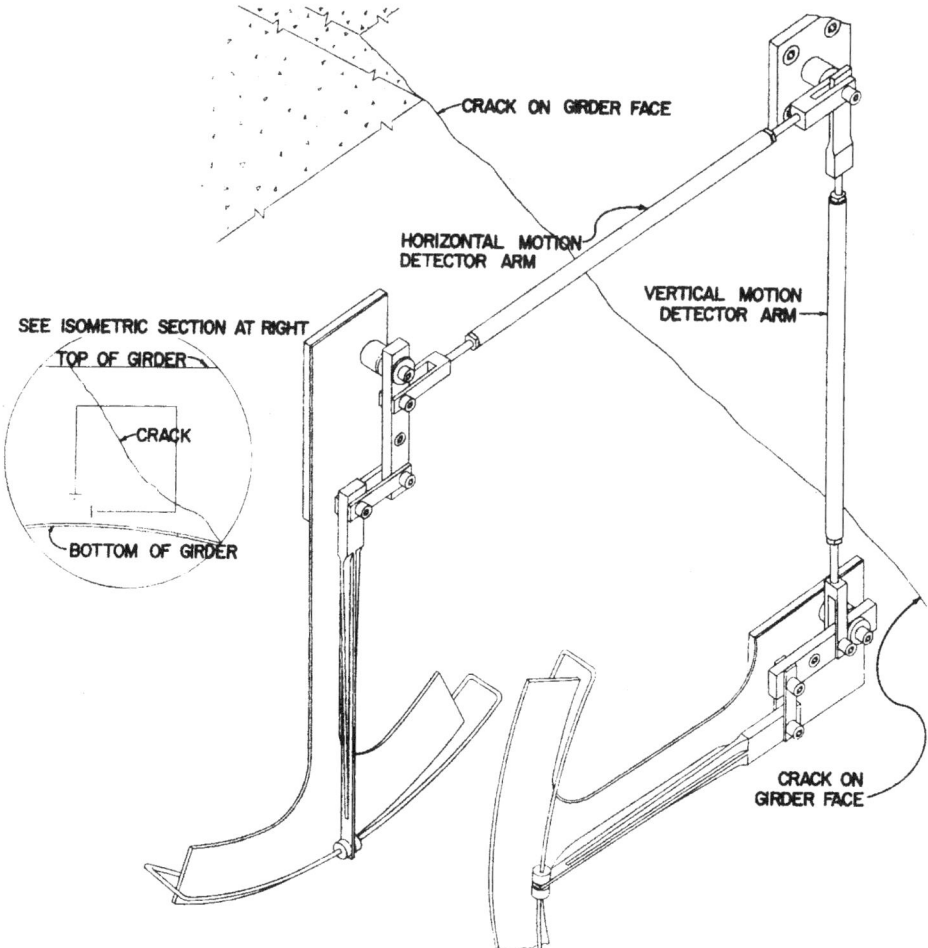

(b)—Crack movement indicator (Stratton et al. 1978)

Figure 2.2

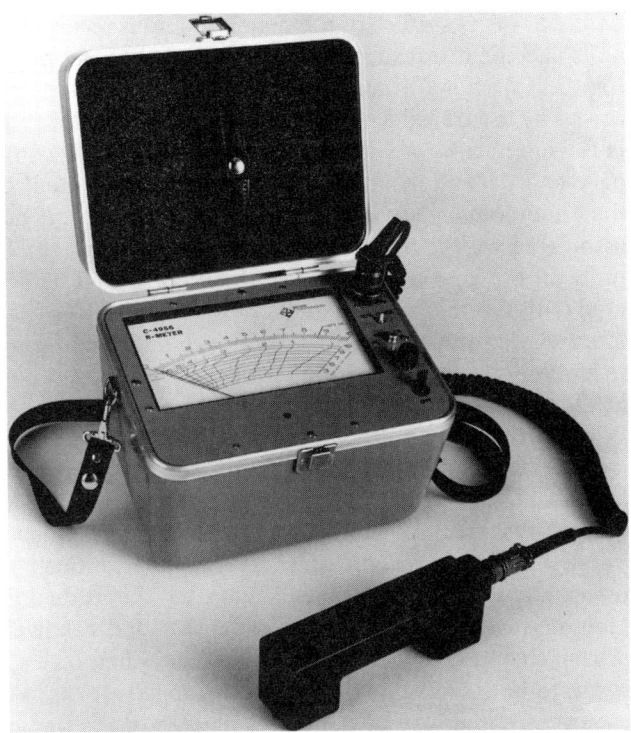

Fig. 2.3—*Pachometer (reinforcing bar locator) (courtesy of James Instruments)*

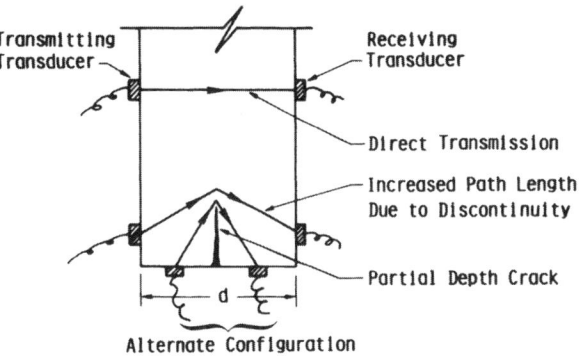

a) Pulse transmitted through member

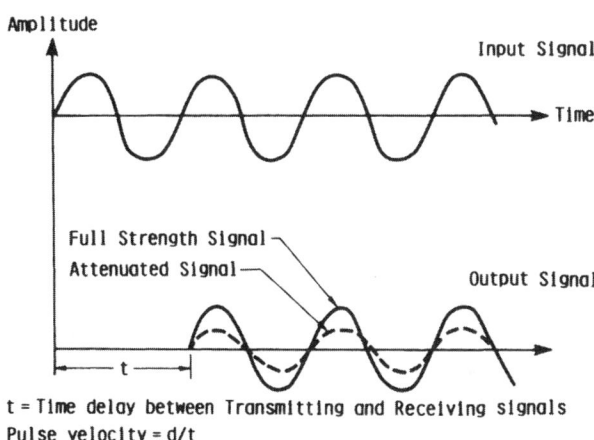

t = Time delay between Transmitting and Receiving signals
Pulse velocity = d/t

b) Oscilloscope Signal

Fig. 2.4—*Ultrasonic testing, through-transmission technique*

that they do not require moisture protection. If more detailed time histories are desired, a wide range of transducers (most notably linear variable differential transformers or LVDT's) and data acquisition systems (ranging from strip chart recorders to computer-based systems) are available.

Sketches can be supplemented by photographs documenting the condition of the structure at the time of investigation. Guidance for making a condition survey of concrete in service is given in ACI 201.1R, ACI 201.3R, ACI 207.3R, ACI 345.1R, and ACI 546.1R.

2.2.2 *Nondestructive testing*—Nondestructive tests can be made to determine the presence of internal cracks and voids and the depth of penetration of cracks visible at the surface.

Tapping the surface with a hammer or using a chain drag are simple techniques to identify laminar cracking near the surface. A hollow sound indicates one or more cracks below and parallel to the surface.

The presence of reinforcement can be determined using a pachometer (Fig. 2.3) (Malhotra 1976). A number of pachometers are available that range in capability from merely indicating the presence of steel to those that may be calibrated to allow the experienced user a closer determination of depth and the size of reinforcing steel. In some cases, however, it may be necessary to remove the concrete cover (often by drilling or chipping) to identify the bar sizes or to calibrate cover measurements, especially in areas of congested reinforcement.

If corrosion is a suspected cause of cracking, the easiest approach to investigate for corrosion entails the removal of a portion of the concrete to directly observe the steel. Corrosion potential can be detected by electrical potential measurements using a suitable reference half cell. The most commonly used is a copper-copper sulfate half cell (ASTM C 876; Clear and Hay 1973); its use also requires access to a portion of the reinforcing steel.

With properly trained personnel and careful evaluation, it is possible to detect cracks using ultrasonic nondestructive test equipment (ASTM C 597). The most common technique is through-transmission testing using commercially available equipment (Malhotra and Carino 1991; Knab et al. 1983). A mechanical pulse is transmitted to one face of the concrete member and received at the opposite face, as shown Fig. 2.4. The time taken for the pulse to pass through the member is measured electronically. If the distance between the transmitting and receiving transducers is known, the pulse velocity can be calculated.

When access is not available to opposite faces, transducers may be located on the same face [Fig. 2.4(a)]. While this technique is possible, the interpretation of results is not straightforward.

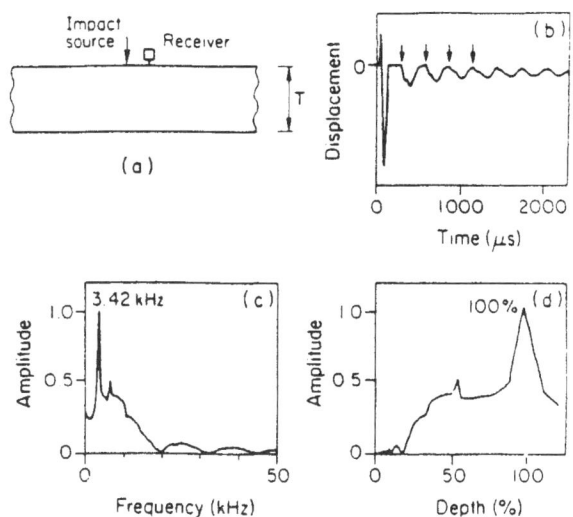

Fig. 2.5—Impact-echo response of a solid plate: a) schematic of test configuration; b) displacement waveform; c) amplitude spectrum; and d) normalized amplitude spectrum

A significant change in measured pulse velocity can occur if an internal discontinuity results in an increase in path length for the signal. Generally, the higher the pulse velocity, the higher the quality of the concrete. The interpretation of pulse velocity test results is significantly improved with the use of an oscilloscope that provides a visual representation of the received signal [Fig. 2.4(b)].

Some equipment provides only a digital readout of the pulse travel time, with no oscilloscope display. If no signal arrives at the receiving transducer, a significant internal discontinuity, such as a crack or void, is indicated. An indication of the extent of the discontinuity can be obtained by taking readings at a series of positions on the member.

Ultrasonic equipment should be operated by a trained person, and the results should be evaluated cautiously by an experienced person, because moisture, reinforcing steel, and embedded items may affect the results. For example, with fully saturated cracks, ultrasonic testing will generally be ineffective. In some cases, it is difficult to discern between a group of close cracks and a single large crack.

An alternative to through-transmission testing is the pulse-echo technique in which a simple transducer is used to send and receive ultrasonic waves. It has been difficult to develop a practical pulse-echo test for concrete. Pitch-catch systems have been developed which use separate transmitting and receiving transducers (Alexander 1980). More detailed information on pulse-echo and other wave propagation methods is provided by Malhotra and Carino (1991).

Significant advances in use of wave propagation techniques for flaw detection in concrete by the impact-echo technique have been made by Sansalone and Carino (1988, 1989). A mechanical pulse is generated by impact on one face of the member as illustrated in Fig. 2.5. The wave propagates through the member, reflects from a defect or other surface of the member, and is received by a displacement transducer placed near the impact point. Fig. 2.5(b) shows a surface time-domain waveform received by the transducer. A resonance condition is set up in the member between the member boundaries or boundary and defect. By analyzing the frequency content of the time-domain waveform [Fig. 2.5(c)] the frequency associated with the resonance appears as a peak amplitude. In the case of Fig. 2.5(a), the peak is that associated with the thickness frequency [see Fig. 2.5(d)]. If an internal flaw exists, then a significant amplitude peak from the reflections from the flaw depth will be observed at the associated flaw depth frequency.

Radiography can also be used to detect internal discontinuities. Both x-ray and gamma-ray equipment are available (Malhotra and Carino 1991; Bungey 1990). The procedures are best suited for detecting crack planes parallel to the direction of radiation; it is difficult to discern crack planes perpendicular to the radiation. Gamma-ray equipment is less expensive and relatively portable compared to x-ray equipment and therefore appears to be more suitable for field testing.

An important use of nondestructive testing is finding those portions of the structure that require a more detailed investigation, which may include core tests.

2.2.3 *Tests on concrete cores*—Significant information can be obtained from cores taken from selected locations within the structure. Cores and core holes afford the opportunity to accurately measure the width and depth of cracks. In addition, an indication of concrete quality can be obtained from compressive strength tests; however, cores that contain cracks should not be used to determine concrete strength.

Petrographic examinations of cracked concrete can identify material causes of cracking, such as alkali reactivities, cyclic freezing damage, "D" cracking, expansive aggregate particles, fire-related damage, shrinkage, and corrosion. Petrography can also identify other factors that may be related to cracking such as the water-to-cement ratio, relative paste volume, and distribution of concrete components. Petrography can frequently determine the relative age of cracks and can identify secondary deposits on fracture surfaces, which have an influence on repair schemes.

Chemical tests for the presence of excessive chlorides indicate the potential for corrosion of embedded reinforcement.

2.2.4 *Review of drawings and construction data*—The original structural design and reinforcement placing or other shop drawings should be reviewed to confirm that the concrete thickness and quality, along with installed reinforcing, meets or exceeds strength and serviceability requirements noted in the governing building code(s). A detailed review of actual applied loading compared to design loads should get special consideration. Concrete configurations, restraint conditions, and the presence of construction and other joints should be considered in calculating the tensile stresses induced by concrete deformation (creep, shrinkage, temperature, etc.). Special consideration should be given to cracks that develop parallel to one-way reinforced slabs primarily supported on beams, but also bear on the girders that support those beams.

2.3—Selection of repair procedures

Based on the careful evaluation of the extent and cause of cracking, procedures can be selected to accomplish one or more of the following objectives:

1. Restore and increase strength;
2. Restore and increase stiffness;
3. Improve functional performance;
4. Provide watertightness;
5. Improve appearance of the concrete surface;
6. Improve durability; and/or
7. Prevent development of corrosive environment at reinforcement.

Depending on the nature of the damage, one or more repair methods may be selected. For example, tensile strength may be restored across a crack by injecting it with epoxy or other high strength bonding agent. However, it may be necessary to provide additional strength by adding reinforcement or using post-tensioning. Epoxy injection alone can be used to restore flexural stiffness if further cracking is not anticipated (ACI 503R).

Cracks causing leaks in water-retaining or other storage structures should be repaired unless the leakage is considered minor or there is an indication that the crack is being sealed by autogenous healing (See section 3.14). Repairs to stop leaks may be complicated by a need to make the repairs while the structures are in service.

Cosmetic considerations may require the repair of cracks in concrete. However, the crack locations may still be visible, and it is likely that some form of coating over the entire surface may be required.

To minimize future deterioration due to the corrosion of reinforcement, cracks exposed to a moist or corrosive environment should be sealed.

The key methods of crack repair available to accomplish the objectives outlined are described in Chapter 3.

CHAPTER 3—METHODS OF CRACK REPAIR

3.1—Introduction

Following the evaluation of the cracked structure, a suitable repair procedure can be selected. Successful repair procedures take into account the cause(s) of the cracking. For example, if the cracking was primarily due to drying shrinkage, then it is likely that after a period of time the cracks will stabilize. On the other hand, if the cracks are due to a continuing foundation settlement, repair will be of no use until the settlement problem is corrected.

This chapter provides a survey of crack repair methods, including a summary of the characteristics of the cracks that may be repaired with each procedure, the types of structures that have been repaired, and a summary of the procedures that are used. Readers are also directed to ACI 546.1R and ACI Compilation No. 5 (1980), which specifically address the subject of concrete repair.

3.2—Epoxy injection

Cracks as narrow as 0.002 in. (0.05 mm) can be bonded by the injection of epoxy. The technique generally consists of establishing entry and venting ports at close intervals along the cracks, sealing the crack on exposed surfaces, and injecting the epoxy under pressure.

Epoxy injection has been successfully used in the repair of cracks in buildings, bridges, dams, and other types of concrete structures (ACI 503R). However, unless the cause of the cracking has been corrected, it will probably recur near the original crack. If the cause of the cracks cannot be removed, then two options are available. One is to rout and seal the crack, thus treating it as a joint, or, establish a joint that will accommodate the movement and then inject the crack with epoxy or other suitable material. Epoxy materials used for structural repairs should conform to ASTM C 881 (Type IV). ACI 504R describes practices for sealing joints, including joint design, available materials, and methods of application.

With the exception of certain moisture tolerant epoxies, this technique is not applicable if the cracks are actively leaking and cannot be dried out. Wet cracks can be injected using moisture tolerant materials, but contaminants in the cracks (including silt and water) can reduce the effectiveness of the epoxy to structurally repair the cracks.

The use of a low-modulus, flexible adhesive in a crack will not allow significant movement of the concrete structure. The effective modulus of elasticity of a flexible adhesive in a crack is substantially the same as that of a rigid adhesive (Adams et al. 1984) because of the thin layer of material and high lateral restraint imposed by the surrounding concrete.

Epoxy injection requires a high degree of skill for satisfactory execution, and application of the technique may be limited by the ambient temperature. The general procedures involved in epoxy injection are as follows (ACI 503R):

- Clean the cracks. The first step is to clean the cracks that have been contaminated, to the extent this is possible and practical. Contaminants such as oil, grease, dirt, or fine particles of concrete prevent epoxy penetration and bonding, and reduce the effectiveness of repairs. Preferably, contamination should be removed by vacuuming or flushing with water or other specially effective cleaning solutions. The solution is then flushed out using compressed air and a neutralizing agent or adequate time is provided for air drying. It is important, however, to recognize the practical limitations of accomplishing complete crack cleaning. A reasonable evaluation should be made of the extent, and necessity, of cleaning. Trial cleaning may be required.

- Seal the surfaces. Surface cracks should be sealed to keep the epoxy from leaking out before it has gelled. Where the crack face cannot be reached, but where there is backfill, or where a slab-on-grade is being repaired, the backfill material or subbase material is sometimes an adequate seal; however, such a condition can rarely be determined in advance, and uncontrolled injection can cause damage such as plugging a drainage system. Extreme caution must therefore be exercised when injecting cracks that are not visible on all surfaces. A surface can be sealed by applying an epoxy, polyester, or other appropriate sealing material to the surface of the crack and allowing it to harden. If a permanent glossy appearance along the crack is objectionable and if high injection pressure is not required, a strippable plastic surface sealer may be applied along the face of the crack. When the job is completed, the surface sealer can be stripped away to expose the gloss-free surface. Cementitious seals can also be used where appearance of the completed work is important. If extremely high injection pressures are needed, the crack can be cut out to a depth of ½ in. (13 mm) and width of about ¾ in. (20 mm) in a V-shape, filled with an epoxy, and struck off flush with the surface.
- Install the entry and venting ports. Three methods are in general use:
  a. Fittings inserted into drilled holes. This method was the first to be used, and is often used in conjunction with V-grooving of the cracks. The method entails drilling a hole into the crack, approximately ¾ in. (20 mm) in diameter and ½ to 1 in. (13 to 25 mm) below the apex of the V-grooved section, into which a fitting such as a pipe nipple or tire valve stem is usually bonded with an epoxy adhesive. A vacuum chuck and bit, or a water-cooled corebit, is useful in preventing the cracks from being plugged with drilling dust.
  b. Bonded flush fitting. When the cracks are not V-grooved, a method frequently used to provide an entry port is to bond a fitting flush with the concrete face over the crack. The flush fitting has an opening at the top for the adhesive to enter and a flange at the bottom that is bonded to the concrete.
  c. Interruption in seal. Another system of providing entry is to omit the seal from a portion of the crack. This method can be used when special gasket devices are available that cover the unsealed portion of the crack and allow injection of the adhesive directly into the crack without leaking.
- Mix the epoxy. This is done either by batch or continuous methods. In batch mixing, the adhesive components are premixed according to the manufacturer's instructions, usually with the use of a mechanical stirrer, like a paint mixing paddle. Care must be taken to mix only the amount of adhesive that can be used prior to commencement of gelling of the material. When the adhesive material begins to gel, its flow characteristics begin to change, and pressure injection becomes more and more difficult. In the continuous mixing system, the two liquid adhesive components pass through metering and driving pumps prior to passing through an automatic mixing head. The continuous mixing system allows the use of fast setting adhesives that have a short working life.
- Inject the epoxy. Hydraulic pumps, paint pressure pots, or air-actuated caulking guns may be used. The pressure used for injection must be selected carefully. Increased pressure often does little to accelerate the rate of injection. In fact, the use of excessive pressure can propagate the existing cracks, causing additional damage.

  If the crack is vertical or inclined, the injection process should begin by pumping epoxy into the entry port at the lowest elevation until the epoxy level reaches the entry port above. The lower injection port is then capped, and the process is repeated until the crack has been completely filled and all ports have been capped.

  For horizontal cracks, the injection should proceed from one end of the crack to the other in the same manner. The crack is full if the pressure can be maintained. If the pressure can not be maintained, the epoxy is still flowing into unfilled portions or leaking out of the crack.
- Remove the surface seal. After the injected epoxy has cured, the surface seal should be removed by grinding or other means as appropriate.
- Alternative procedure. For massive structures, an alternate procedure consists of drilling a series of holes [usually ⅞ to 4-in. (20 to 100-mm) diameter] that intercepts the crack at a number of locations. Typically, holes are spaced at 5-ft (1.5-m) intervals.

Another method recently being used is a vacuum or vacuum assist method. There are two techniques: one is to entirely enclose the cracked member with a bag and introduce the liquid adhesive at the bottom and to apply a vacuum at the top. The other technique is to inject the cracks from one side and pull a vacuum from the other. Typically, epoxies are used; however, acrylics and polyesters have proven successful.

Stratton and McCollum (1974) describe the use of epoxy injection as an effective intermediate-term repair procedure for delaminated bridge decks. As reported by Stratton and McCollum the first, second, and sixth steps are omitted and the process is terminated at a specific location when epoxy exits from the crack at some distance from the injection ports. This procedure does not arrest ongoing corrosion. The procedure can also be attempted for other applications, and is available as an option, although not accepted universally. Success of the repair depends on the absence of bond-inhibiting contaminants from the crack plane. Epoxy resins and injection procedures should be carefully selected when attempting to inject delaminations. Unless there is sufficient depth or anchorage to surrounding concrete the injection process can be unsuccessful or increase the extent of delamination. Smith (1992) provides information on bridge decks observed for up to seven years after injection. Smithson and Whiting describe epoxy injection as a method to rebond delaminated bridge deck overlays. Committee 224 is developing additional information on this application for inclusion in a future revision of this Report.

3.3—Routing and sealing

Routing and sealing of cracks can be used in conditions requiring remedial repair and where structural repair is not necessary. This method involves enlarging the crack along its exposed face and filling and sealing it with a suitable joint sealant (Fig. 3.1). This is a common technique for crack treatment and is relatively simple in comparison to the procedures and the training required for epoxy injection. The procedure is most applicable to approximately flat horizontal surfaces such as floors and pavements. However, routing and sealing can be accomplished on vertical surfaces (with a non-sag sealant) as well as on curved surfaces (pipes, piles and pole).

Routing and sealing is used to treat both fine pattern cracks and larger, isolated cracks. A common and effective use is for waterproofing by sealing cracks on the concrete surface where water stands, or where hydrostatic pressure is applied. This treatment reduces the ability of moisture to reach the reinforcing steel or pass through the concrete, causing surface stains or other problems.

The sealants may be any of several materials, including epoxies, urethanes, silicones, polysulfides, asphaltic materials, or polymer mortars. Cement grouts should be avoided due to the likelihood of cracking. For floors, the sealant should be sufficiently rigid to support the anticipated traffic. Satisfactory sealants should be able to withstand cyclic deformations and should not be brittle.

The procedure consists of preparing a groove at the surface ranging in depth, typically, from ¼ to 1 in. (6 to 25 mm). A concrete saw, hand tools or pneumatic tools may be used. The groove is then cleaned by air blasting, sandblasting, or waterblasting, and dried. A sealant is placed into the dry groove and allowed to cure.

A bond breaker may be provided at the bottom of the groove to allow the sealant to change shape, without a concentration of stress on the bottom (Fig. 3.2). The bond breaker may be a polyethylene strip or tape which will not bond to the sealant.

Careful attention should be applied when detailing the joint so that its width to depth aspect ratio will accommodate anticipated movement (ACI 504R).

In some cases overbanding (strip coating) is used in-

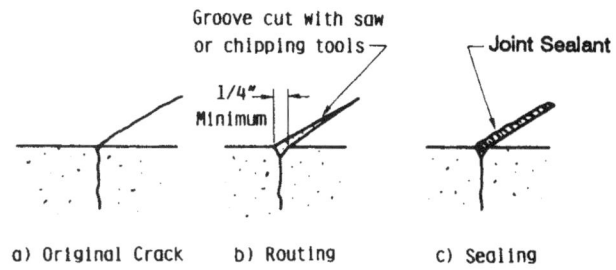

Fig. 3.1—Repair of crack by routing and sealing (Johnson 1965)

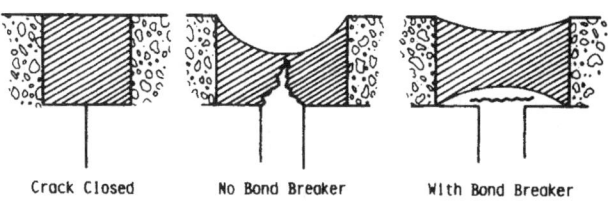

Fig. 3.2—Effect of bond breaker

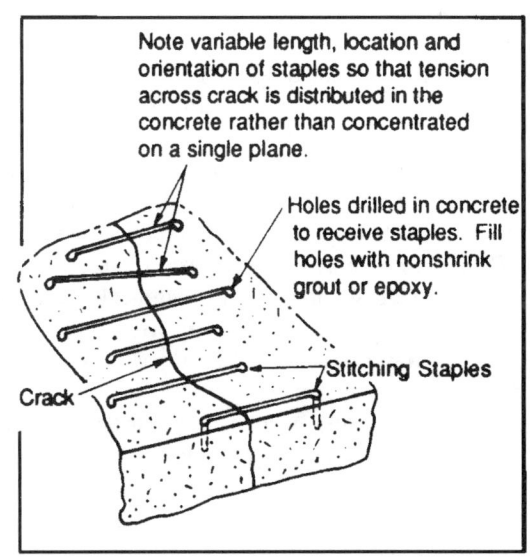

Fig. 3.3—Repair of crack by stitching (Johnson 1965)

dependently of or in conjunction with routing and sealing. This method is used to enhance protection from edge spalling and for aesthetic reasons to create a more uniform appearing treatment. A typical procedure for overbanding is to prepare an area approximately 1 to 3 in. (25 to 75 mm) on each side of the crack by sandblast-

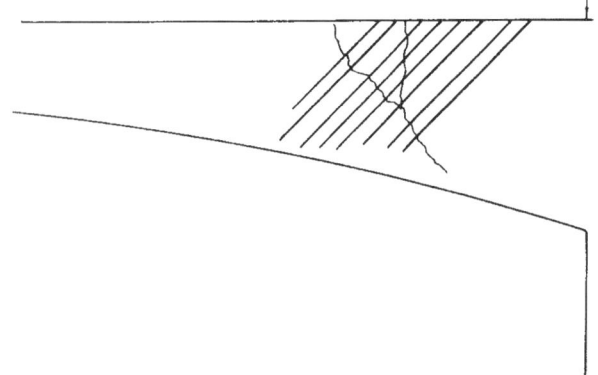

Fig. 3.4—Reinforcing bar orientation used to effect the repair (Stratton et al. 1978)

ing or other means of surface preparation, and apply a coating (such as urethane) 0.04 to 0.08 in. (1 to 2 mm) thick in a band over the crack. Before overbanding in non-traffic areas a bond breaker is sometimes used over a crack that has not been routed, or over a crack previously routed and sealed. In traffic areas a bond breaker is not recommended. Cracks subject to minimal movement may be overbanded, but if significant movement can take place, routing and sealing must be used in conjunction with overbanding to ensure a waterproof repair.

3.4—Stitching

Stitching involves drilling holes on both sides of the crack and grouting in U-shaped metal units with short legs (staples or stitching dogs) that span the crack as shown in Fig 3.3 (Johnson 1965). Stitching may be used when tensile strength must be reestablished across major cracks (Hoskins 1991). Stitching a crack tends to stiffen the structure, and the stiffening may increase the overall structural restraint, causing the concrete to crack elsewhere. Therefore, it may be necessary to strengthen the adjacent section or sections using technically corrected reinforcing methods. Because stresses are often concentrated, using this method in conjunction with other methods may be necessary.

The stitching procedure consists of drilling holes on both sides of the crack, cleaning the holes, and anchoring the legs of the staples in the holes, with either a non-shrink grout or an epoxy resin-based bonding system. The staples should be variable in length, orientation, or both, and they should be located so that the tension transmitted across the crack is not applied to a single plane within the section but is spread over an area.

3.5—Additional reinforcement

3.5.1 *Conventional reinforcement*—Cracked reinforced concrete bridge girders have been successfully repaired by inserting reinforcing bars and bonding them in place with epoxy (Stratton et al. 1978, 1982; Stratton 1980). This technique consists of sealing the crack, drilling holes that intersect the crack plane at approximately 90 deg (Fig. 3.4), filling the hole and crack with injected epoxy and placing a reinforcing bar into the drilled hole. Typically, No. 4 or 5 (10 M or 15 M) bars are used, extending at least 18 in. (0.5 m) each side of the crack. The reinforcing bars can be spaced to suit the needs of the repair. They can be placed in any desired pattern, depending on the design criteria and the location of the in-place reinforcement. The epoxy bonds the bar to the walls of the hole, fills the crack plane, bonds the cracked concrete surfaces back together in one monolithic form, and thus reinforces the section. The epoxy used to rebond the crack should have a very low viscosity and conform to ASTM C 881 Type IV.

3.5.2 *Prestressing steel*—Post-tensioning is often the desirable solution when a major portion of a member must be strengthened or when the cracks that have formed must be closed (Fig. 3.5). This technique uses prestressing strands or bars to apply a compressive force. Adequate anchorage must be provided for the prestressing steel, and care is needed so that the problem will not merely migrate to another part of the structure. The effects of the tensioning force (including eccentricity) on the stress within the structure should be carefully analyzed. For indeterminate structures post-tensioned using this procedure, the effects of secondary moments and induced reactions should be considered (Nilson 1987; Lin and Burns 1981).

3.6—Drilling and plugging

Drilling and plugging a crack consists of drilling down the length of the crack and grouting it to form a key (Fig. 3.6).

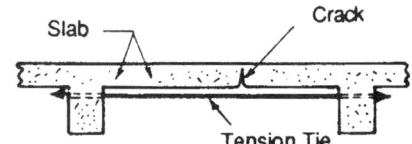

a) To Correct Cracking of Slab

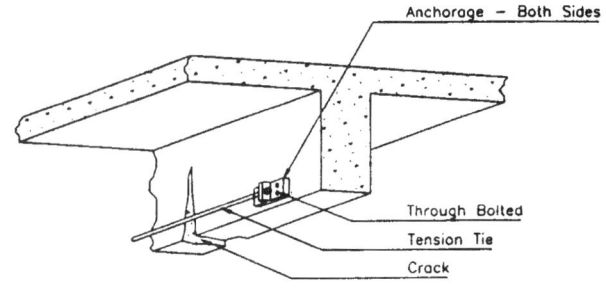

b) To Correct Cracking of Beam

Fig. 3.5—Examples of external prestressing (Johnson 1965)

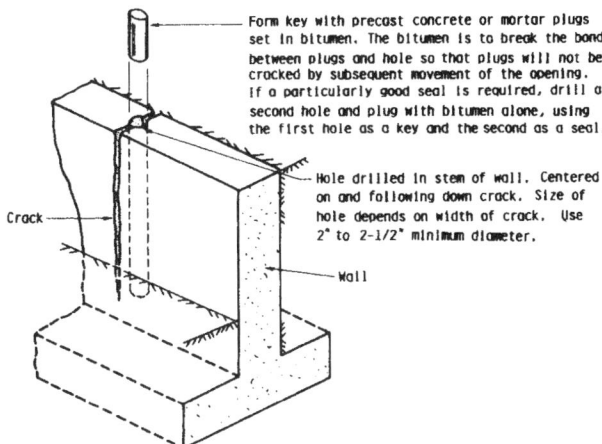

Fig. 3.6—Repair of crack by drilling and plugging

This technique is only applicable when cracks run in reasonable straight lines and are accessible at one end. This method is most often used to repair vertical cracks in retaining walls.

A hole [typically 2 to 3 in. (50 to 75 mm) in diameter] should be drilled, centered on and following the crack. The hole must be large enough to intersect the crack along its full length and provide enough repair material to structurally take the loads exerted on the key. The drilled hole should then be cleaned, made tight, and filled with grout. The grout key prevents transverse movements of the sections of concrete adjacent to the crack. The key will also reduce heavy leakage through the crack and loss of soil from behind a leaking wall.

If water-tightness is essential and structural load transfer is not, the drilled hole should be filled with a resilient material of low modulus in lieu of grout. If the keying effect is essential, the resilient material can be placed in a second hole, the first being grouted.

3.7—Gravity Filling

Low viscosity monomers and resins can be used to seal cracks with surface widths of 0.001 to 0.08 in. (0.03 to 2 mm) by gravity filling (Rodler, et al. 1989). High-molecular-weight methacrylates, urethanes, and some low viscosity epoxies have been used successfully. The lower the viscosity, the finer the cracks that can be filled.

The typical procedure is to clean the surface by air blasting and/or waterblasting. Wet surfaces should be permitted to dry several days to obtain the best crack filling. The monomer or resin can be poured onto the surface and spread with brooms, rollers, or squeegees. he material should be worked back and forth over the cracks to obtain maximum filling since the monomer or resin recedes slowly into the cracks. Excess material should be broomed off the surface to prevent slick, shining areas after curing. If surface friction is important, sand should be broadcast over the surface before the monomer or resin cures.

If the cracks contain significant amounts of silt, moisture or other contaminants, the sealant cannot fill them. Water blasting followed by a drying time may be effective in cleaning and preparing these cracks.

Cores taken at cracks can be used to evaluate the effectiveness of the crack filling. The depth of penetration of the sealant can be measured. Shear (or tension) tests can be performed with the load applied in a direction parallel to the repaired cracks (as long as reinforcing steel is not present in the core in or near the failure area). For some polymers the failure crack will occur outside the repaired crack.

3.8—Grouting

3.8.1 *Portland cement grouting*—Wide cracks, particularly in gravity dams and thick concrete walls, may be repaired by filling with portland cement grout. This method is effective in stopping water leaks, but it will not structurally bond cracked sections. The procedure consists of cleaning the concrete along the crack; installing built-up seats (grout nipples) at intervals astride the crack (to provide a pressure tight connection with the injection apparatus); sealing the crack between the seats with a cement paint, sealant, or grout; flushing the crack to clean it and test the seal; and then grouting the whole area. Grout mixtures may contain cement and water or cement plus sand and water, depending on the width of the crack. However, the water-cement ratio should be kept as low as practical to maximize the strength and minimize shrinkage. Water reducers or other admixtures may be used to improve the properties of the grout. For small volumes, a manual injection gun may be used; for larger volumes, a pump should be used. After the crack is filled, the pressure should be maintained for several minutes to insure good penetration.

3.8.2 *Chemical grouting*—Chemical grouts consist of solutions of two or more chemicals (such as urethanes, sodium silicates, and acrylomides) that combine to form a gel, a solid precipitate, or a foam, as opposed to cement grouts that consist of suspensions of solid particles in a fluid. Cracks in concrete as narrow as 0.002 in. (0.05 mm) have been filled with chemical grout.

The advantages of chemical grouts include applicability in moist environments (excess moisture available), wide limits of control of gel time, and their ability to be applied in very fine fractures. Disadvantages are the high degree of skill needed for satisfactory use and their lack of strength.

3.9—Drypacking

Drypacking is the hand placement of a low water content mortar followed by tamping or ramming of the mortar into place, producing intimate contact between the mortar and the existing concrete (U.S. Bureau of Reclamation 1978). Because of the low water-cement ratio of the material, there is little shrinkage, and the patch remains tight and can have good quality with respect to durability, strength, and watertightness.

Drypack can be used for filling narrow slots cut for the

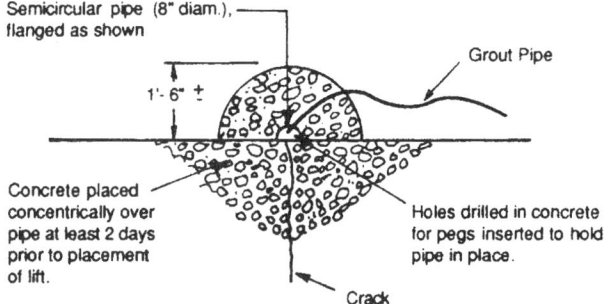

Fig. 3.7—Crack arrest method of crack repair

repair of dormant cracks. The use of drypack is not advisable for filling or repairing active cracks.

Before a crack is repaired by drypacking, the portion adjacent to the surface should be widened to a slot about 1 in. (25 mm) wide and 1 in. (25 mm) deep. The slot should be undercut so that the base width is slightly greater than the surface width.

After the slot is thoroughly cleaned and dried, a bond coat, consisting of cement slurry or equal quantities of cement and fine sand mixed with water to a fluid paste consistency, or an appropriate latex bonding compound (ASTM C 1059), should be applied. Placing of the dry pack mortar should begin immediately. The mortar consists of one part cement, one to three parts sand passing a No. 16 (1.18 mm) sieve, and just enough water so that the mortar will stick together when molded into a ball by hand.

If the patch must match the color of the surrounding concrete, a blend of grey portland cement and white portland cement may be used. Normally, about one-third white cement is adequate, but the precise proportions can be determined only by trial.

To minimize shrinkage in place, the mortar should stand for ½ hour after mixing and then should be remixed prior to use. The mortar should be placed in layers about ⅜ in. (10 mm) thick. Each layer should be thoroughly compacted over the surface using a blunt stick or hammer, and each underlying layer should be scratched to facilitate bonding with the next layer. There need be no time delays between layers.

The mortar may be finished by laying the flat side of a hardwood piece against it and striking it several times with a hammer. Surface appearance may be improved by a few light strokes with a rag or sponge float. The repair should be cured by using either water or a curing compound. The simplest method of moist curing is to support a strip of folded wet burlap along the length of the crack.

3.10—Crack arrest

During construction of massive concrete structures, cracks due to surface cooling or other causes may develop and propagate into new concrete as construction progresses. Such cracks may be arrested by blocking the crack and spreading the tensile stress over a larger area (U.S. Army Corps of Engineers 1945).

A piece of bond-breaking membrane or a grid of steel mat may be placed over the crack as concreting continues. A semicircular pipe placed over the crack may also be used (Fig. 3.7). A description of installation procedures for semicircular pipes used during the construction of a massive concrete structure follows: (1) The semicircular pipe is made by splitting an 8-in. (200-mm), 16-gage pipe and bending it to a semicircular section with about a 3-in. (75-mm) flange on each side; (2) the area in the vicinity of the crack is cleaned; (3) the pipe is placed in sections so as to remain centered on the crack; (4) the sections are then welded together; (5) holes are cut in the top of the pipe to receive grout pipes; and (6) after setting the grout pipes, the installation is covered with concrete placed concentrically over the pipe by hand. The installed grout pipes are used for grouting the crack at a later date, thereby restoring all or a portion of the structural continuity.

3.11—Polymer impregnation

Monomer systems can be used for effective repair of some cracks. A monomer system is a liquid consisting of monomers which will polymerize into a solid. Suitable monomers have varying degrees of volatility, toxicity and flammability, and they do not mix with water. They are very low in viscosity and will soak into dry concrete, filling the cracks, much as water does. The most common monomer used for this purpose is methyl methacrylate.

Monomer systems used for impregnation contain a catalyst or initiator plus the basic monomer (or combination of monomers). They may also contain a cross-linking agent. When heated, the monomers join together, or polymerize, creating a tough, strong, durable plastic that greatly enhances a number of concrete properties.

If a cracked concrete surface is dried, flooded with the monomer, and polymerized in place, some of the cracks will be filled and structurally repaired. However, if the cracks contain moisture, the monomer will not soak into the concrete at each crack face, and consequently, the repair will be unsatisfactory. If a volatile monomer evaporates before polymerization, it will be ineffective. Polymer impregnation has not been used successfully to repair fine cracks. Polymer impregnation has primarily been used to provide more durable, impermeable surfaces (Webster et al. 1978; Hallin 1978).

Badly fractured beams have been repaired using polymer impregnation. The procedure consists of drying the fracture, temporarily encasing it in a watertight (monomer proof) band of sheet metal, soaking the fractures with monomer, and polymerizing the monomer. Large voids or broken areas in compression zones can be filled with fine and coarse aggregate before being flooded with monomer, providing a polymer concrete repair. A more detailed discussion of polymers is given in ACI 548R.

3.12—Overlay and surface treatments

Fine surface cracks in structural slabs and pavements may be repaired using either a bonded overlay or surface treatment if there will not be further significant movement across the cracks. Unbonded overlays may be used to cover, but not necessarily repair a slab. Overlays and surface treatments can be appropriate for cracks caused by one-time occurrences and which do not completely penetrate the slab. These techniques are not appropriate for repair of progressive cracking, such as that induced by reactive aggregates, and D-cracking.

Slabs-on-grade in freezing climates should not be repaired by an overlay or surface treatment that is a vapor barrier. An impervious barrier will cause condensation of moisture passing from the subgrade, leading to critical saturation of the concrete and rapid disintegration during cycles of freezing and thawing.

3.12.1 *Surface treatments*—Low solids and low-viscosity resin-based systems have been used to seal the concrete surfaces, including treatment of very fine cracks. They are most suited for surfaces not subject to significant wear.

Bridge decks and parking structure slabs, as well as other interior slabs may be coated effectively after cracks are treated by injecting with epoxy or by routing and sealing. Materials such as urethanes, epoxies, polyesters, and acrylics have been applied in thickness of 0.04 to 2.0 in. (1 to 50 mm), depending on the material and purpose of the treatment. Skid-resistant aggregates are often mixed into the material or broadcast onto the surface to improve traction.

3.12.2 *Overlays*—Slabs containing find dormant cracks can be repaired by applying an overlay, such as polymer-modified portland cement mortar or concrete, or by silica fume concrete. Slabs with working cracks can be overlaid if joints are placed in the overlay directly over the working cracks. In highway bridge applications, an overlay thickness as low as 1¼ in. (30 mm) has been used successfully (NCHRP Synthesis 57). Suitable polymers include styrene butadiene or acrylic latexes. The resin solids should be at least 15 percent by weight of the portland cement, with 20 percent usually being optimum (Clear and Chollar 1978).

The surface to be overlaid should be cleaned to remove laitance, carbonated or otherwise weak material, or contaminants, such as grease or oil. A bond coat consisting of the mortar fraction broom-applied, or an epoxy adhesive should be applied immediately before placing the overlay. Since polymer-modified concretes normally solidify rapidly, continuous batching and mixing equipment should be used. Polymer-modified overlays should be mixed, placed and finished rapidly (within 15 min in warm weather). A 24-hr moist curing is typical for these overlays.

3.13—Autogenous healing

A natural process of crack repair known as "autogenous healing" can occur in concrete in the presence of moisture and the absence of tensile stress (Lauer and Slate 1956). It has practical application for closing dormant cracks in a moist environment, such as may be found in mass concrete structures.

Healing occurs through the continued cement hydration and the carbonation of calcium hydroxide in the cement paste by carbon dioxide, which is present in the surrounding air and water. Calcium carbonate and calcium hydroxide crystals precipitate, accumulate, and grow within the cracks. The crystals interlace and twine, producing a mechanical bonding effect, which is supplemented by a chemical bonding between adjacent crystals and between the crystals and the surfaces of the paste and the aggregate. As a result, some of the tensile strength of the concrete is restored across the cracked section, and the crack may become sealed.

Healing will not occur if the crack is active and is subjected to movement during the healing period. Healing will also not occur if there is a positive flow of water through the crack, which dissolves and washes away the lime deposit, unless the flow of water is so slow that complete evaporation occurs at the exposed face causing redeposition of the dissolved salts.

Saturation of the crack and the adjacent concrete with water during the healing process is essential for developing any substantial strength. Submergence of the cracked section is desirable. Alternatively, water may be ponded on the concrete surface so that the crack is saturated. The saturation must be continuous for the entire period of healing. A single cycle of drying and reimmersion will produce a drastic reduction in the amount of healing strength. Healing should be commenced as soon as possible after the crack appears. Delayed healing results in less restoration of strength than does immediate correction.

CHAPTER 4—SUMMARY

This report is intended to serve as a tool in the process of crack evaluation and repair of concrete structures.

The causes of cracks in concrete are summarized along with the principal procedures used for crack control. Both plastic and hardened concrete are considered. The importance of design, detailing, construction procedures, concrete proportioning, and material properties are discussed.

The techniques and methodology for crack evaluation are described. Both analytical and field requirements are discussed. The need to determine the causes of cracking as a necessary prerequisite to repair is emphasized. The selection of successful repair techniques should consider the causes of cracking, whether the cracks are active or dormant, and the need for repairs. Criteria for the selection of crack repair procedures are based on the desired outcome of the repairs.

Twelve methods of crack repair are presented, including the techniques, advantages and disadvantages, and areas of application of each.

ACKNOWLEDGMENT

ACI Committee 224 — Cracking, gratefully acknowledges the assistance of Robert Gaul, Paul Krauss, and James Warner, non-members of the Committee, for their suggestions and review of the revisions to this document. The Committee would also like to recognize the contributions of Raymond J. Schutz, former Committee Member Paul H. Karr and deceased Committee Members Donald L. Houghton and Robert E. Philleo who were Contributing Authors of the original version of ACI 224.1R.

CHAPTER 5—REFERENCES

5.1 Recommended references

The Documents of the various standards-producing organizations referred to in this document are listed below with their serial designation.

American Association of State Highway and Transportation Officials
   Standard Specification for Highway Bridges

American Concrete Institute
201.1R Guide for Making a Condition Survey of Concrete in Service
201.2R Guide to Durable Concrete
207.1R Mass Concrete
207.2R Effect of Restraint, Volume Change, and Reinforcement on Cracking of Mass Concrete
207.4R Cooling and Insulating Systems for Mass Concrete
224R   Control of Cracking in Concrete Structures
224.2R Cracking of Concrete Members in Direct Tension
224.3R Joints in Concrete Construction
302.1R Guide to Concrete Floor and Slab Construction
304R   Guide for Measuring, Mixing Transporting and Placing Concrete
305R   Hot Weather Concreting
308   Standard Practice for Curing Concrete
309R   Guide for Consolidation of Concrete
309.2R Identification and Control of Consolidation-Related Surface Defects in Formed Concrete
318   Building Code Requirements for Reinforced Concrete
343R   Analysis and Design of Reinforced Concrete Bridge Structures
345R   Guide for Concrete Highway Bridge Deck Construction
347R   Guide to Concrete Formwork
350R   Environmental Engineering of Concrete Structures
503R   Use of Epoxy Compounds with Concrete
504R   Guide to Sealing Joints in Concrete Structures
517.2R Accelerated Curing of Concrete at Atmospheric Pressure—State of the Art
546.1R Guide for Repair of Concrete Bridge Superstructures
548R   Polymers in Concrete

American Society for Testing and Materials
C 150   Standard Specification for Portland Cement
C 595   Standard Specification for Blended Hydraulic Cements
C 597   Standard Test Method for Pulse Velocity through Concrete
C 876   Standard Test Method for Half Cell Potentials of Reinforcing Steel in Concrete
C 881   Standard Specification for Epoxy-Resin-Base Bonding Systems for Concrete
C 1059 Standard Specification for Latex Agents for Bonding Fresh to Hardened Concrete

The above publications may be obtained from the following organizations:

American Association of State Highway and Transportation Officials
444 N Capitol Street NW
Suite 224
Washington, D.C. 20001

American Concrete Institute
P.O. Box 9094
Farmington Hills, Mich. 48333-9094

ASTM
100 Barr Harbor Dr.
West Conshohocken, Pa. 19428-2959

5.2—Cited references

Abdun-Nur, Edward A. (1983), "Cracking of Concrete — Who Cares?," *Concrete International: Design and Construction*, V. 5, No. 7, July, pp. 27-30.

Abeles, Paul W.; Brown, Earl L., II, and Morrow, Joe W. (1968), "Development and Distribution of Cracks in Rectangular Prestressed Beams During Static and Fatigue Loading,"*Journal*, Prestressed Concrete Institute, V. 13, No. 5, Oct., pp. 36-51.

ACI Compilation No. 5 (1980), "Concrete Repair and Restoration," American Concrete Institute, Detroit 118 pp. (compiled from *Concrete International: Design and Construction*, V. 2, No. 9, Sept.).

Adams, Robert D., and Wake, William C. (1984), *Structural Adhesive Joints in Engineering*, Elsevier Applied Science Publishers, Ltd., Essex, England, pp. 121-125.

Alberta Ministry of Highways (1990), "Alberta Test Procedures for Evaluation of Concrete Sealer," BT001, 5 pp.

Alexander, A. Michel (1980), "Development of Procedures for Nondestructive Testing of Concrete Structures: Report 2, Feasibility of Sonic Pulse-Echo Technique," *Miscellaneous Paper* No. C-77-11, U.S. Army Engineer Waterways Experiment Station, Vicksburg, 25 pp.

Beeby, A.W. (1983), "Cracking, Cover, and Corrosion of Reinforcement," *Concrete International: Design and Construction*, V. 5, No. 2, Feb., pp. 35-40.

Bennett, E.W., and Dave, N.J. (1969), "Test Performances and Design of Beams with Limited Prestress," *The Structural Engineer* (London), V. 47, No. 12, Dec., pp. 487-496.

Broms, Bengt B. (1965), "Crack Width and Spacing in Reinforced Concrete Members," ACI JOURNAL, *Proceedings*, V. 62, No. 10, Oct., pp. 1237-1256.

Broms, Bengt B., and Lutz, LeRoy A. (1965), "Effects of Arrangement of Reinforcement on Crack Width and Spacing of Reinforced Concrete Members," ACI JOURNAL, *Proceedings*, V. 62, No. 11, Nov., pp. 1395-1410.

Bungey, J.H. (1990), *Testing of Concrete in Structures*, 2nd ed., Chapman and Hall, NY.

Butt, Yu M.; Kolbasov, V.M.; and Timashev, V.V. (1969), "High Temperature Curing of Concrete Under Atmospheric Pressure," *Proceedings*, 5th International Symposium on the Chemistry of Cement (Tokyo, 1968), Cement Association of Japan, pp.437-476.

Callan, Edwin J. (1952), "Thermal Expansion of Aggregates and Concrete Durability," ACI JOURNAL, *Proceedings*, V. 48, No. 6, Feb., pp. 485-504.

Carlson, Roy W. (1938), "Drying Shrinkage of Concrete as Affected by Many Factors," *Proceedings* ASTM, V. 38, Part 2, pp. 419-437.

Carlson, Roy W.; Houghton, Donald L.; and Polivka, Milos (1979), "Causes and Control of Cracking in Unreinforced Mass Concrete," ACI JOURNAL, *Proceedings*, V. 76, No. 7, July, pp. 821-837.

Clear, Kenneth C., and Chollar, Brian H. (1978), "Styrene-Butadiene Latex Modifiers for Bridge Deck Overlay Concrete," *Report* No. FHWA-RD-78-35, Federal Highway Administration, Washington, D.C., 124 pp.

Clear, K.C., and Hay, R.E. (1973), "Time-to-Corrosion of Reinforcing Steel in Concrete Slabs. V. 1, Effects of Mix Design and Construction Parameters," *Report* No. FHWA-RD-73-32, Federal Highway Administration, Washington, D.C., 103 pp.

Concrete Institute of Australia (1972), *Third Progress Report* of the Low Pressure Steam-Curing of Concrete, North Sydney, 26 pp.

Dakhil, Fadh H.; Cady, Philip D.; and Carrier, Roger E. (1975), "Cracking of Fresh Concrete as Related to Reinforcement," ACI JOURNAL, *Proceedings*, V. 72, No. 8, Aug., pp. 421-428.

Dusinberre, D.M. (1945), "Numerical Methods for Transient Heat Flow," *Transactions*, American Society of Mechanical Engineers, V. 67, Nov., pp. 703-712.

Gergely, Peter (1981), "Role of Cover and Bar Spacing in Reinforced Concrete," *Significant Developments in Engineering Practice and Research*, SP-72, American Concrete Institute, Detroit, pp. 133-147.

Gergely, Peter, and Lutz, LeRoy A. (1968), "Maximum Crack Width in Reinforced Concrete Members," *Causes, Mechanism, and Control of Cracking in Concrete*, SP-20, American Concrete Institute, pp. 87-117.

Higginson, Elmo C. (1961), "Effect of Steam Curing on the Important Properties of Concrete," ACI JOURNAL, *Proceedings*, V. 58, No. 3, Sept., pp. 281-298.

Hallin, J.P., "Field Evaluation of Polymer Impregnation of New Bridge Deck Surfaces," *Polymers in Concrete*, SP-58, American Concrete Institute, Detroit, 1978, pp. 267-280.

Hoffman, P.C.; McClure, R.M; and West, H.H. (1983), "Temperature Study of an Experimental Segmental Concrete Bridge," *Journal*, Prestressed Concrete Institute, V. 28, No. 2, Mar.-Apr., pp. 78-97.

Holmberg, Ake (1973), "Crack Width Prediction and Minimum Reinforcement for Crack Control," *Dansk Slesab for Bygningsstatitik* (Copenhagen), V. 44, No. 2, June, pp. 41-50.

Hoskins, B.E.; Fowler, D.W.; and McCullough, B.F. (1991), "An Evaluation of Repair Techniques Used for Uncontrolled Longitudinal Cracking and Failed Longitudinal Joints," *Research Report* 920-4, Center for Transportation Research, The University of Texas at Austin, 21 pp.

Houghton, Donald L. (1972), "Concrete Strain Capacity Tests — Their Economic Implications," *Economical Construction of Concrete Dams*, American Society of Civil Engineers, New York, pp. 75-99.

Houghton, Donald L. (1976), "Determining Tensile Strain Capacity of Mass Concrete," ACI JOURNAL, *Proceedings*, V. 73, No. 12, Dec., pp. 691-700.

Illston, J.M, and Stevens, R.F. (1972), "Long-Term Cracking in Reinforced Concrete Beams," *Proceedings*, Institution of Civil Engineers (London), Part 2, V. 53, pp. 445-459.

Jastrzebski, Zbigniew D. (1961), *Nature and Properties of Engineering Materials*, John Wiley and Sons, New York, 571 pp.

Johnson, Sydney M. (1965), *Deterioration, Maintenance and Repair of Structures*, McGraw-Hill Book Co., New York, 373 pp.

Kaminetzky, Dov (1981), "Failures During and After Construction," *Concrete Construction*, V. 26, No, 8, Aug., pp. 641-649.

Kelly, Joe W. (1981), "Cracks in Concrete: Part 1, Part 2," *Concrete Construction*, V. 26, No. 9, Sept., pp. 725-734.

Kirkbride, T. (1971a), "Review of Accelerated Curing Procedures," *Precast Concrete* (London), V. 1, No. 2, Feb., pp. 87-90.

Kirkbride, T.W. (1971b), "Burner Curing," *Precast Concrete* (London), V. 1, No. 11, Nov., pp. 644-646.

Knab, L.I.; Blessing, G.V.; and Clifton, J.R. (1983), "Laboratory Evaluation of Ultrasonics for Crack Detection in Concrete," ACI JOURNAL, *Proceedings*, V. 80, No. 1, Jan.-Feb., pp. 17-27.

Lauer, Kenneth R., and Slate, Floyd O. (1956), "Autogenous Healing of Cement Paste," ACI JOURNAL, *Proceedings*, V. 27, No. 10, June, pp. 1083-1098.

Libby, James R. (1977), *Modern Prestressed Concrete*, 2nd Edition, Van Nostrand Reinhold, New York, pp. 388-390.

Lin, T.Y, and Burns, Ned H. (1981), *Design of Prestressed Concrete Structures*, 3rd edition, John Wiley &

Sons, New York, 646 pp.

Malhotra, V.M. and Carino, N.J. (Editors) (1991), *Handbook on Nondestructive Testing of Concrete*, CRC Press, Boca Raton, Fl. 343 p.

Mansfield, G.A. (1948), "Curing — A Problem in Thermodynamics," *Rock Products*, V. 51, No. 8, Aug., p. 212.

Mast, Robert F. (1981), "Roof Collapse at Antioch High School," *Journal*, Prestressed Concrete Institute, V. 26, No. 3, May-June, pp. 29-53.

NCHRP Synthesis of Highway Practice No. 57 (1979), "Durability of Concrete Bridge Decks," Transportation Research Board, Washington, D.C., May, 61 pp.

Nilson, Arthur H. (1987), *Design of Prestressed Concrete*, 2nd Ed., John Wiley and Sons, New York, 526 pp.

Nurse, R.W. (1949), "Steam Curing of Concrete," *Magazine of Concrete Research* (London), V. 1, No. 2, June, pp. 79-88.

PCI Committee on Quality Control Performance Criteria (1985), "Fabrication and Shipment Cracks in Precast or Prestressed Beams and Columns," *PCI Journal*, May-June, pp. 24-49.

PCI Committee on Quality Control Performance Criteria (1983), "Fabrication and Shipment Cracks in Prestressed Hollow-Core Slabs and Double Tees," *PCI Journal*, V. 28, No. 1, Jan.-Feb., pp. 18-39.

PCI Energy Committee (1981), Discussion of "Accelerated Curing of Concrete at Atmospheric Pressure — State of the Art," ACI JOURNAL, *Proceedings*, Jul.-Aug., pp. 320-324.

Pickett, Gerald (1956), "Effect of Aggregate on Shrinkage of Concrete," ACI JOURNAL, *Proceedings*, V. 52, No. 5, Jan., pp. 581-590.

Powers, T.C. (1975), "Freezing Effects in Concrete," *Durability of Concrete*, SP-47, American Concrete Institute, Detroit, pp. 1-11.

Price, Walter H. (1982), "Control of Cracking During Construction," *Concrete International: Design and Construction*, V. 4, No. 1, Jan., pp. 40-43.

Priestly, M.J. Nigel (1978), "Design of Concrete Bridges for Temperature Gradients," ACI JOURNAL, *Proceedings*, V. 75, No. 5, May, pp. 209-217.

Rehm, Gallus, and Eligehausen, Rolf (1977), "Lapped Splices of Deformed Bars Under Repeated Loadings (Ubergreifungstosse von Rippenstahlen unter nicht ruhender Belastung)," *Beton und Stahlbeton* (Berlin), No. 7, pp. 170-174.

Rodler, D.J.; D.P. Whitney; D.W. Fowler; and D.L. Wheat, "Repair of Cracked Concrete with High Molecular Weight Methacrylates," *Polymers in Concrete: Advances and Applications*, SP-116, American Concrete Institute, Detroit, 1989, pp. 113-127.

Sansalone, M. and Carino, N.J. (1988), "Laboratory and Field Studies of the Impact-Echo Method for Flaw Detection of Concrete," SP-112, American Concrete Institute, Detroit, pp. 1-20.

Sansalone, M. and Carino, N.J. (1989), "Detecting Delaminations in Concrete Slabs with and without Overlays Using the Impact-Echo Method," *ACI Materials Journal*, V. 86, No. 2, Mar-Apr., pp. 175-184.

Shideler, Joseph J., and Toennies, Henry T. (1963), "Plant Drying and Carbonation of Concrete Block — NCMA-PCA Cooperative Program," ACI JOURNAL, *Proceedings*, V. 60, No. 5, May 1963, pp. 617-634. Also, *Development Department Bulletin* No. D64, Portland Cement Association.

Smith, Barbara J. (1992), "Epoxy Injection of Bridge Deck Delaminations," *Transportation Research Record 1533*, Transportation Research Board, National Research Council, pp. 10-18.

Smithson, Leland D., and Whiting, John E. (1992), "Rebonding Delaminated Bridge Deck Overlays," *Concrete Repair Digest*, V. 3, No. 3, June/July, pp. 100-101.

Snowdon, L.C., and Edwards, A.G. (1962), "The Moisture Movement of Natural Aggregate and Its Effect on Concrete," *Magazine of Concrete Research* (London), V. 14, No. 41, July, pp. 109-116.

Stratton, F. Wayne (1980), "Custom Concrete Drill Helps Repair Shear Cracks in Bridge Girders," *Concrete International: Design and Construction*, V. 2, No. 9, Sept., pp. 118-119.

Stratton, F. Wayne, and McCollom, Bruce F. (1974), "Repair of Hollow or Softened Areas in Bridge Decks by Rebonding with Injected Epoxy Resin or Other Polymers," *Report* No. K-F-72-5, State Highway Commission of Kansas, July, 104 pp.

Stratton, F. Wayne; Alexander, Roger; and Nolting, William (1978), "Cracked Structural Concrete Repair through Epoxy Injection and Rebar Insertion," *Report* No. FHWA-KS-RD.78-3, Kansas Department of Transportation, Topeka, Nov., 56 pp.

U.S. Army Corps of Engineers (1945), "Concrete Operation with Relation to Cracking at Norfolk Dam," Little Rock District, Arkansas, Oct.

U.S. Bureau of Reclamation (1975), *Concrete Manual*, 8th edition, Denver, 627 pp.

Verbeck, George G. (1958), "Carbonation of Hydrated Portland Cement," *Cement and Concrete*, STP-205, American Society for Testing and Materials, Philadelphia, pp. 17-36. Also, *Research Department Bulletin* No. 87, Portland Cement Association.

Verbeck, George G. (1975), "Mechanisms of Corrosion of Steel in Concrete," *Corrosion of Metals in Concrete*, SP-49, American Concrete Institute, Detroit, pp. 21-38.

Webster, R.P.; Fowler, D.W.; and Paul D.R. (1978), "Bridge Deck Impregnation in Texas," *Polymers in Concrete*, SP-58, American Concrete Institute, Detroit, pp. 249-265.

Wimsatt, A.W.; McCullough, B.F.; and Burns, N.H (1987) "Methods of Analyzing and Factors Influencing Frictional Effects of Subbases," *Research Report* 459-2F, Center for Transportation Research, The University of Texas at Austin, November, 77 pp.

ACI 224.1R-93 was submitted to letter ballot of the committee and processed in accordance with ACI standardization procedures.

ACI 229R-99
(Reapproved 2005)

Controlled Low-Strength Materials

Reported by ACI Committee 229

Bruce W. Ramme
Chairman

Wayne S. Adaska	Morris Huffman	Frances A. McNeal	Charles F. Scholer
Richard L. Boone	Bradley M. Klute	Donald E. Milks	Glenn O. Schumacher
Christopher Crouch	Henry J. Kolbeck	Narasimhan Rajendran	Victor Smith
Kurt R. Grabow	Ronald L. Larsen	Kenneth B. Rear	Richard Sullivan
Daniel J. Green	Leo A. Legatski	Paul E. Reinhart	Samuel S. Tyson
Richard R. Halverson	William MacDonald	Harry C. Roof	Harold Umansky
William Hook	Oscar Manz	Edward H. Rubin	Orville R. Werner

Controlled low-strength material (CLSM) is a self-compacted, cementitious material used primarily as a backfill in place of compacted fill. Many terms are currently used to describe this material, including flowable fill, unshrinkable fill, controlled density fill, flowable mortar, flowable fly ash, fly ash slurry, plastic soil-cement, soil-cement slurry and other various names. This report contains information on applications, material properties, mix proportioning, construction, and quality-control procedures. The intent of this report is to provide basic information on CLSM technology, with emphasis on CLSM material characteristics and advantages over conventional compacted fill.

Keywords: aggregates; backfill; compacted fill; controlled density fill; controlled low-strength material; flowable fill; flowable mortar; fly ash; foundation stabilization; low-density material; pipe bedding; plastic soil-cement; preformed foam; soil-cement slurry; trench backfill; unshrinkable fill; void filling.

ACI Committee Reports, Guides, Standard Practices, and Commentaries are intended for guidance in planning, designing, executing, and inspecting construction. This document is intended for the use of individuals who are competent to evaluate the significance and limitations of its content and recommendations and who will accept responsibility for the application of the material it contains. The American Concrete Institute disclaims any and all responsibility for the stated principles. The Institute shall not be liable for any loss or damage arising therefrom.

Reference to this document shall not be made in contract documents. If items found in this document are desired by the Architect/Engineer to be a part of the contract documents, they shall be restated in mandatory language for incorporation by the Architect/Engineer.

CONTENTS
Chapter 1—Introduction, p. 229R-2

Chapter 2—Applications, p. 229R-2
2.1—General
2.2—Backfills
2.3—Structural fills
2.4—Insulating and isolation fills
2.5—Pavement bases
2.6—Conduit bedding
2.7—Erosion control
2.8—Void filling
2.9—Nuclear facilities
2.10—Bridge reclamation

Chapter 3—Materials, p. 229R-5
3.1—General
3.2—Cement
3.3—Fly ash
3.4—Admixtures
3.5—Other additives
3.6—Water
3.7—Aggregates
3.8—Nonstandard materials
3.9—Ponded ash or basin ash

Chapter 4—Properties, p. 229R-6
4.1—Introduction

ACI 229R-99 became effective April 26, 1999.
Copyright © 1999, American Concrete Institute.
All rights reserved including rights of reproduction and use in any form or by any means, including the making of copies by any photo process, or by electronic or mechanical device, printed, written, or oral, or recording for sound or visual reproduction or for use in any knowledge or retrieval system or device, unless permission in writing is obtained from the copyright proprietors.

4.2—Plastic properties
4.3—In-service properties

Chapter 5—Mixture proportioning, p. 229R-9

Chapter 6—Mixing, transporting, and placing, p. 229R-9
6.1—General
6.2—Mixing
6.3—Transporting
6.4—Placing
6.5—Cautions

Chapter 7—Quality control, p. 229R-11
7.1—General
7.2—Sampling
7.3—Consistency and unit weight
7.4—Strength tests

Chapter 8—Low-density CLSM using preformed foam, p. 229R-13
8.1—General
8.2—Applications
8.3—Materials
8.4—Properties
8.5—Proportioning
8.6—Construction

Chapter 9—References, p. 229R-14
9.1—Specified references
9.2—Cited references

CHAPTER 1—INTRODUCTION

Controlled low-strength material (CLSM) is a self-compacted, cementitious material used primarily as a backfill as an alternative to compacted fill. Several terms are currently used to describe this material, including flowable fill, unshrinkable fill, controlled density fill, flowable mortar, plastic soil-cement, soil-cement slurry, and other various names.

Controlled low-strength materials are defined by ACI 116R as materials that result in a compressive strength of 8.3 MPa (1200 psi) or less. Most current CLSM applications require unconfined compressive strengths of 2.1 MPa (300 psi) or less. This lower-strength requirement is necessary to allow for future excavation of CLSM.

The term CLSM can be used to describe a family of mixtures for a variety of applications. For example, the upper limit of 8.3 MPa (1200 psi) allows use of this material for applications where future excavation is unlikely, such as structural fill under buildings. Chapter 8 of this report describes low-density (LD) CLSM produced using preformed foam as part of the mixture proportioning. The use of preformed foam in LD-CLSM mixtures allow these materials to be produced having unit weights lower than those of typical CLSM. The distinctive properties and mixing procedures for LD-CLSM are discussed in the chapter. Future CLSM mixtures can be developed as anticorrosion fills, thermal fills, and durable pavement bases.

CLSM should not be considered as a type of low-strength concrete, but rather a self-compacted backfill material that is used in place of compacted fill. Generally, CLSM mixtures are not designed to resist freezing and thawing, abrasive or erosive forces, or aggressive chemicals. Nonstandard materials can be used to produce CLSM as long as the materials have been tested and found to satisfy the intended application.

Also, CLSM should not be confused with compacted soil-cement, as reported in ACI 230.IR. CLSM typically requires no compaction (consolidation) or curing to achieve the desired strength. Long-term compressive strengths for compacted soil-cement often exceed the 8.3 MPa (1200 psi) maximum limit established for CLSM.

Long-term compressive strengths of 0.3 to 2.1 MPa (50 to 300 psi) are low when compared with concrete. In terms of allowable bearing pressure, however, which is a common criterion for measuring the capacity of a soil to support a load, 0.3 to 0.7 MPa (50 to 100 psi) strength is equivalent to a well-compacted fill.

Although CLSM generally costs more per yd^3 than most soil or granular backfill materials, its many advantages often result in lower in-place costs. In fact, for some applications, CLSM is the only reasonable backfill method available.[1-3] Table 1 lists a number of advantages to using CLSM.[4]

CHAPTER 2—APPLICATIONS
2.1—General

As stated earlier, the primary application of CLSM is as a structural fill or backfill in lieu of compacted soil. Because CLSM needs no compaction and can be designed to be fluid, it is ideal for use in tight or restricted-access areas where placing and compacting fill is difficult. If future excavation is anticipated, the maximum long-term compressive strength should generally not exceed 2.1 MPa (300 psi). The following applications are intended to present a range of uses for CLSM.[5]

2.2—Backfills

CLSM can be readily placed into a trench, hole or other cavity (Fig. 2.1 and 2.2). Compaction is not required; hence, the trench width or size of excavation can be reduced. Granular or site-excavated backfill, even if compacted properly in the required layer thickness, can not achieve the uniformity and density of CLSM.[5]

When backfilling against retaining walls, consideration should be given to the lateral pressures exerted on the wall by flowable CLSM. Where the lateral fluid pressure is a concern, CLSM can be placed in layers, allowing each layer to harden prior to placing the next layer.

Following severe settlement problems of soil backfill in utility trenches, the city of Peoria, Ill., in 1988, tried CLSM as an alternative backfill material. The CLSM was placed in trenches up to 2.7 m (9 ft) deep. Although fluid at time of placement, the CLSM hardened to the extent that a person's weight could be supported within 2 to 3 hr. Very few shrinkage cracks were observed. Further tests were conducted on patching the overlying pavement within 3 to 4 hr. In one test, a pavement patch was successfully placed over a sewer

Table 1—Cited advantages of controlled low-strength materials[4]

Readily available	Using locally available materials, ready-mixed concrete suppliers can produce CLSM to meet most project specifications.
Easy to deliver	Truck mixers can deliver specified quantities of CLSM to job site whenever material is needed.
Easy to place	Depending on type and location of void to be filled, CLSM can be placed by chute, conveyor, pump, or bucket. Because CLSM is self-leveling, it needs little or no spreading or compacting. This speeds construction and reduces labor requirements.
Versatile	CLSM mixtures can be adjusted to meet specific fill requirements. Mixes can be adjusted to improve flowability. More cement or fly ash can be added to increase strength. Admixtures can be added to adjust setting times and other performance characteristics. Adding foaming agents to CLSM produces lightweight, insulating fill.
Strong and durable	Load-carrying capacities of CLSM are typically higher than those of compacted soil or granular fill. CLSM is also less permeable, thus more resistant to erosion. For use as permanent structural fill, CLSM can be designed to achieve 28-day compressive strength as high as 8.3 MPa (1200 psi).
Allows fast return to traffic	Because many CLSMs can be placed quickly and support traffic loads within several hours, downtime for pavement repairs is minimal.
Will not settle	CLSM does not form voids during placement and will not settle or rut under loading. This advantage is especially significant if backfill is to be covered by pavement patch. Soil or granular fill, if not consolidated properly, may settle after a pavement patch is placed and forms cracks or dips in the road.
Reduces excavation costs	CLSM allows narrower trenches because it eliminates having to widen trenches to accommodate compaction equipment.
Improves worker safety	Workers can place CLSM in a trench without entering the trench, reducing their exposure to possible cave-ins.
Allows all-weather construction	CLSM will typically displace any standing water left in a trench from rain or melting snow, reducing need for dewatering pumps. To place CLSM in cold weather, materials can be heated using same methods for heating ready-mixed concrete.
Can be excavated	CLSM having compressive strengths of 0.3 to 0.7 MPa (50 to 100 psi) is easily excavated with conventional digging equipment, yet is strong enough for most backfilling needs.
Requires less inspection	During placement, soil backfill must be tested after each lift for sufficient compaction. CLSM self-compacts consistently and does not need this extensive field testing.
Reduces equipment needs	Unlike soil or granular backfill, CLSM can be placed without loaders, rollers, or tampers.
Requires no storage	Because ready-mixed concrete trucks deliver CLSM to job site in quantities needed, storing fill materials on site is unnecessary. Also, there is no leftover fill to haul away.
Makes use of coal combustion product	Fly ash is by-product produced by power plants that burn coal to generate electricity. CLSM containing fly ash benefits environment by making use of this industrial product material.

trench immediately after backfilling with CLSM. As a result of these initial tests, the city of Peoria has changed its backfilling procedure to require the use of CLSM on all street openings.[4]

Some agencies backfill with a CLSM that has a setting time of 20 to 35 min. (after which time a person can walk on it). After approximately 1 hr, the wearing surface consisting of either a rapid-setting concrete or asphalt pavement is placed, resulting in a total traffic-bearing repair in about 4 hr.[6]

2.3—Structural fills

Depending upon the strength requirements, CLSM can be used for foundation support. Compressive strengths can vary from 0.7 to 8.3 MPa (100 to 1200 psi) depending upon application. In the case of weak soils, it can distribute the structure's load over a greater area. For uneven or nonuniform subgrades under foundation footings and slabs, CLSM can provide a uniform and level surface. Compressive strengths will vary depending upon project requirements. Because of its strength, CLSM may reduce the required thickness or strength requirements of the slab. Near Boone, Iowa, 2141 m^3 (2800 yd^3) of CLSM was used to provide proper bearing capacity for the footing of a grain elevator.[7]

2.4—Insulating and isolation fills

LD-CLSM material is generally used for these applications. Chapter 8 addresses LD-CLSM material using preformed foam.

2.5—Pavement bases

CLSM mixtures can be used for pavement bases, subbases, and subgrades. The mixture would be placed directly from the mixer onto the subgrade between existing curbs. For base course design under flexible pavements, structural coefficients differ depending upon the strength of the CLSM. Based on structural coefficient values for cement-treated bases derived from data obtained in several states, the structural coefficient of a CLSM layer can be estimated to range from 0.16 to 0.28 for compressive strengths from 2.8 to 8.3 MPa (400 to 1200 psi).[8]

Good drainage, including curb and gutter, storm sewers, and proper pavement grades, is required when using CLSM mixtures in pavement construction. Freezing and thawing damage could result in poor durability if the base material is frozen when saturated with water.

A wearing surface is required over CLSM because it has relatively poor wear-resistance properties. Further information regarding pavement base materials is found in ACI 325.3R.

Fig. 2.1—Using CLSM to backfill adjacent to building foundation wall.

2.6—Conduit bedding

CLSM provides an excellent bedding material for pipe, electrical, telephone, and other types of conduits. The flowable characteristic of the material allows the CLSM to fill voids beneath the conduit and provide a uniform support.

The U.S. Bureau of Reclamation (USBR) began using CLSM in 1964 as a bedding material for 380 to 2400 mm (15 to 96 in.) diameter concrete pipe along the entire Canadian River Aqueduct Project, which stretches 518 km (322 miles) from Amarillo to Lubbock, Tex. Soil-cement slurry pipe bedding, as referred to by the USBR, was produced in central portable batching plants that were moved every 16 km (10 miles) along the route. Ready-mixed concrete trucks then delivered the soil-cement slurry to the placement site. The soil was obtained from local blow sand deposits. It was estimated that the soil-cement slurry reduced bedding costs 40%. Production increased from 120 to 300 m (400 to 1000 linear ft) of pipe placed per shift.[9]

CLSM can be designed to provide erosion resistance beneath the conduit. Since the mid-1970s, some county agencies in Iowa have been placing culverts on a CLSM bedding. This not only provides a solid, uniform pipe bedding, but prevents water from getting between the pipe and bedding, eroding the support.[10]

Encasing the entire conduit in CLSM also serves to protect the conduit from future damage. If the area around the conduit is being excavated at a later date, the obvious material change in CLSM versus the surrounding soil or conventional granular backfill would be recognized by the excavating crew, alerting them to the existence of the conduit. Coloring agents have also been used in mixtures to help identify the presence of CLSM.

2.7—Erosion control

Laboratory studies, as well as field performance, have shown that CLSM resists erosion better than many other fill materials. Tests comparing CLSM with various sand and clay fill materials showed that CLSM, when exposed to a water velocity of 0.52 m/sec (1.7 ft/sec), was superior to the other materials, both in the amount of material loss and suspended solids from the material.[11]

Fig. 2.2—Backfilling utility cut with CLSM.

CLSM is often used in riprap for embankment protection and in spilling basins below dam spillways, to hold rock pieces in place and resist erosion. CLSM is used to fill flexible fabric mattresses placed along embankments for erosion protection, thereby increasing their strength and weight. In addition to providing an erosion resistance under culverts, CLSM is used to fill voids under pavements, sidewalks, bridges and other structures where natural soil or noncohesive granular fill has eroded away.

2.8—Void filling

2.8.1 *Tunnel shafts and sewers*—When filling abandoned tunnels and sewers, it is important to use a flowable mixture. A constant supply of CLSM will help keep the material flowing and make it flow greater distances. CLSM was used to fill an abandoned tunnel that passed under the Menomonee River in downtown Milwaukee, Wis. The self-leveling material flowed over 71.6 m (235 ft). On another Milwaukee project, 635 m^3 (831 yd^3) were used to fill an abandoned sewer. The CLSM reportedly flowed up to 90 m (300 linear ft).[12]

Before constructing the Mount Baker Ridge Tunnel in Seattle, Wash., an exploratory shaft 37 m (120 ft) deep, 3.7 m (12 ft) in diameter with 9.1 m (30 ft) long branch tunnels was excavated. After exploration, the shaft had to be filled before construction of the tunnel. Only 4 hr were needed to fill the shaft with 601 m^3 (786 yd^3) of CLSM.[13]

2.8.2 *Basements and underground structures*—Abandoned basements are often filled in with CLSM by pumping or conveying the mixture through an open window or doorway. An industrial renovation project in LaSalle, Ill., required the filling of an existing basement to accommodate expansion plans. Granular fill was considered, but access problems made CLSM a more attractive alternative. About 300 m^3 (400 yd^3) of material were poured in one day. A 200 mm (8 in.) concrete floor was then placed directly on top of the CLSM mixture.[14]

In Seattle, buses were to be routed off busy streets into a tunnel with pedestrian stations.[13] The tunnel was built by a conventional method, but the stations had to be excavated from the surface to the station floor. After the station was built, there was a 19,000 m^3 (25,000 yd^3) void over each station to the street. So as not to disrupt traffic with construction equipment and materials, the voids were filled with CLSM, which required no layered placement or compaction.

CLSM has been used to fill abandoned underground storage tanks (USTs). Federal and State regulations have been developed that address closure requirements for underground fuel and chemical tanks. USTs taken out of service permanently must either be removed from the ground or filled with an inert solid material. The Iowa Department of Natural Resources has developed a guidance document for storage tank closures, which specifically mentions flowable fill.

2.8.3 *Mines*—Abandoned mines have been filled with CLSM to eliminate access, prevent subsidence, bottle up hazardous gases, cut off the oxygen supply for fires, and reduce or eliminate acid drainage. It is important that a flowable mixture be placed with a constant supply to facilitate the spread and minimize the quantity of injection/placement points. The western U.S. alone contains approximately 250,000 abandoned mines with various hazards.[15] CLSM can be used to fill mine voids completely, or in areas of particular concern, to prevent subsidence, block trespasser entry, and eliminate or reduce acid or other harmful drainage. Abandoned underground coal mines in the eastern U.S. have been filled using CLSM that was manufactured from various coal combustion products for this purpose.[6,15-17]

2.9—Nuclear facilities

CLSM is used in nuclear facilities for conventional applications such as those described previously. It provides a significant advantage over conventional granular backfill in that remote placement decreases personnel exposure to radiation. CLSM can also be used in unique applications at nuclear facilities, such as waste stabilization, encapsulation of decommissioned pipelines and tanks, encapsulation of waste-disposal sites, and new landfill construction. CLSM can be used to address a wide range of chemical and radionuclide-stabilization requirements.[18-20]

2.10—Bridge reclamation

CLSM has been used in several states as part of a cost-effective process for bridge rehabilitation. The process requires putting enough culverts under the bridge to handle the hydrology requirements. A dam is placed over both ends of the culvert(s) and the culvert(s) are covered with fabric to keep the CLSM from flowing into the joints. These culvert(s) are set on granular backfill. The CLSM is then placed until it is 150 mm (6 in.) from the lower surface of the deck. A period of at least 72 hr is required before the CLSM is brought up to the bottom of the deck through holes cored in the deck. Later, the railing is removed and the deck is widened. The same procedure is then completed on the opposite side of the bridge. The work is done under traffic conditions. The camber of the roadway over the culvert(s) is the only clue that a bridge had ever been present. Iowa DOT officials estimate that the cost of four reclamations is equivalent to one replacement when this technology can be employed.[10,21,22]

CHAPTER 3—MATERIALS
3.1—General

Conventional CLSM mixtures usually consist of water, portland cement, fly ash or other similar products, and fine or coarse aggregates or both. Some mixtures consist of water, portland cement, and fly ash only. Special low-density CLSM (LD-CLSM) mixtures, as described in Chapter 8 of this report, consist of portland cement, water, and preformed foam.

Although materials used in CLSM mixtures meet ASTM or other standard requirements, the use of standardized materials is not always necessary. Selection of materials should be based on availability, cost, specific application, and the necessary characteristics of the mixture, including flowability, strength, excavatability, and density.

3.2—Cement

Cement provides the cohesion and strength for CLSM mixtures. For most applications, Type I or Type II portland cement conforming to ASTM C 150 is normally used. Other types of cement, including blended cements conforming to ASTM C 595, can be used if prior testing indicates acceptable results.

3.3—Fly ash

Coal-combustion fly ash is sometimes used to improve flowability. Its use can also increase strength and reduce bleeding, shrinkage, and permeability. High fly ash-content mixtures result in lower-density CLSM when compared with mixtures with high aggregate contents. Fly ashes used in CLSM mixtures do not need to conform to either Class F or C as described in ASTM C 618. Trial mixtures should be prepared to determine whether the mixture will meet the specified requirements. Refer to ACI 232.2R for further information.[23,24]

3.4—Admixtures

Air-entraining admixtures and foaming agents can be valuable constituents for the manufacture of CLSM. The inclusion of air in CLSM can help provide improved workability, reduced shrinkage, little or no bleeding, minimal segregation, lower unit weights, and control of ultimate strength development. Higher air contents can also help enhance CLSM's thermal insulation and freeze-thaw properties. Water content can be reduced as much as 50% when using air-entraining admixtures. The use of these materials may require modifications to typical CLSM mixtures. To prevent segregation when utilizing high air contents, the mixtures need to be proportioned with sufficient fines to promote cohesion. Most air-entrained CLSM mixtures are pumpable but can require higher pump pressures when piston pumps are used. To prevent extended setting times, extra cement or the use of an accelerating admixture may be required. In all cases, pretesting should be performed to determine acceptability.[6,25,26]

3.5—Other additives

In specialized applications such as waste stabilization, CLSM mixtures can be formulated to include chemical and/or mineral additives that serve purposes beyond that of simple backfilling. Some examples include the use of swelling clays such as bentonite to achieve CLSM with low permeability. The inclusion of zeolites, such as analcime or chabazite, can be used to absorb selected ions where water or sludge treatment is required. Magnetite or hematite fines can be added to CLSM to provide radiation shielding in applications at nuclear facilities.[18-20]

3.6—Water

Water that is acceptable for concrete mixtures is acceptable for CLSM mixtures. ASTM C 94 provides additional information on water-quality requirements.

3.7—Aggregates

Aggregates are often the major constituent of a CLSM mixture. The type, grading, and shape of aggregates can affect the physical properties, such as flowability and compressive strength. Aggregates complying with ASTM C 33 are generally used because concrete producers have these materials in stock.

Granular excavation materials with somewhat lower-quality properties than concrete aggregate are a potential source of CLSM materials, and should be considered. Variations of the physical properties of the mixture components, however, will have a significant effect on the mixture's performance. Silty sands with up to 20% fines passing through a 75 μm (No. 200) sieve have proven satisfactory. Also, soils with wide variations in grading have shown to be effective. Soils with clay fines, however, have exhibited problems with incomplete mixing, stickiness of the mixtures, excess water demand, shrinkage, and variable strength. These types of soils are not usually considered for CLSM applications. Aggregates that have been used successfully include:[27]

- ASTM C 33 specification aggregates within specified gradations;
- Pea gravel with sand;
- 19 mm (3/4 in.) minus aggregate with sand;
- Native sandy soils, with more than 10% passing a 75 μm (No. 200) sieve;
- Quarry waste products, generally 10 mm (3/8 in.) minus aggregates.

3.8—Nonstandard materials

Nonstandard materials, which can be available and more economical, can also be used in CLSM mixtures, depending upon project requirements. These materials, however, should be tested prior to use to determine their acceptability in CLSM mixtures.

Examples of nonstandard materials that can be substituted as aggregates for CLSM include various coal combustion products, discarded foundry sand, glass cullet, and reclaimed crushed concrete.[28-30]

Aggregates or mixtures that might swell in service due to expansive reactions or other mechanisms should be avoided. Also, wood chips, wood ash, or other organic materials may not be suitable for CLSM. Fly ashes with carbon contents up to 22% have been successfully used for CLSM.[31]

In all cases, the characteristics of the nonstandard material should be determined, and the suitability of the material should be tested in a CLSM mixture to determine whether it meets specified requirements. In certain cases, environmental regulations could require prequalification of the raw material or CLSM mixture, or both, prior to use.

3.9—Ponded ash or basin ash

Ponded ash, typically a mixture of fly ash and bottom ash slurried into a storage/disposal basin, can also be used in CLSM. The proportioning of the ponded ash in the resulting mixtures depends on its particle size distribution. Typically, it can be substituted for all of the fly ash and a portion of the fine aggregate and water. Unless dried prior to mixing, ponded ash requires special mixing because it is usually wet. Basin ash is similar to ponded ash except it is not slurried and can be disposed of in dry basins or stockpiles.[18-20]

CHAPTER 4—PROPERTIES

4.1—Introduction

The properties of CLSM cross the boundaries between soils and concrete. CLSM is manufactured from materials similar to those used to produce concrete, and is placed from equipment in a fashion similar to that of concrete. In-service CLSM, however, exhibits characteristic properties of soils. The properties of CLSM are affected by the constituents of the mixture and the proportions of the ingredients in the mixture. Because of the many factors that can affect CLSM, a wide range of values may exist for the various properties discussed in following sections.[32]

4.2—Plastic properties

4.2.1 *Flowability*—Flowability is the property that distinguishes CLSM from other fill materials. It enables the materials to be self-leveling; to flow into and readily fill a void; and be self-compacting without the need for conventional placing and compacting equipment. This property represents a major advantage of CLSM compared with conventional fill materials that must be mechanically placed and compacted. Because plastic CLSM is similar to plastic concrete and grout, its flowability is best viewed in terms of concrete and grout technology.

A major consideration in using highly flowable CLSM is the hydrostatic pressure it exerts. Where fluid pressure is a concern, CLSM can be placed in lifts, with each lift being allowed to harden before placement of the next lift. Examples where multiple lifts can be used are in the case of limited-strength forms that are used to contain the material, or where buoyant items, such as pipes, are encapsulated in the CLSM.

Flowability can be varied from stiff to fluid, depending upon requirements. Methods of expressing flowability include the use of a 75 x 150 mm (3 x 6 in.) open-ended cylinder modified flow test (ASTM D 6103), the standard concrete slump cone (ASTM C 143), and flow cone (ASTM C 939).

Good flowability, using the ASTM D 6103 method, is achieved where there is no noticeable segregation and the CLSM material spread is at least 200 mm (8 in.) in diameter. Flowability ranges associated with the slump cone can be expressed as follows:[33]

- Low flowability: less than 150 mm (6 in.);
- Normal flowability: 150 to 200 mm (6 to 8 in.);
- High flowability: greater than 200 mm (8 in.)

ASTM C 939, for determining flow of grout, has been used successfully with fluid mixtures containing aggregates not greater than 6 mm (1/4 in.) The method is briefly described in Chapter 7 on Quality Control. The Florida and Indiana Departments of Transportation (DOT) require an efflux time of 30 ± 5 sec, as measured by this method.

4.2.2 *Segregation*—Separation of constituents in the mixture can occur at high levels of flowability when the flowability is primarily produced by the addition of water. This situation is similar to segregation experienced with some high-slump concrete mixtures. With proper mixture proportioning and materials, a high degree of flowability can be attained without segregation. For highly flowable CLSM without segregation, adequate fines are required to provide suitable cohesiveness. Fly ash generally accounts for these fines, although silty or other noncohesive fines up to 20% of total aggregate have been used. The use of plastic fines, such as clay, should be avoided because they can produce deleterious results, such as increased shrinkage. In flowable mixtures, satisfactory performance of CLSM has been obtained with Class F fly ash contents as high as 415 kg/m^3 (700 lb/yd^3) in combination with cement, sand, and water. Some CLSM mixtures have been designed without sand or gravel, using only fly ash as filler material. These mixtures require much higher water content, but produce no noticeable segregation.

4.2.3 *Subsidence*—Subsidence deals with the reduction in volume of CLSM as it releases its water and entrapped air through consolidation of the mixture. Water used for flowability in excess of that needed for hydration is generally absorbed by the surrounding soil or released to the surface as bleed water. Most of the subsidence occurs during placement and the degree of subsidence is dependent upon the quantity of free water released. Typically, subsidence of 3 to 6 mm (1/8 to 1/4 in.) per ft of depth has been reported.[34] This amount is generally found with mixtures of high water content. Mixtures of lower water content undergo little or no subsidence, and cylinder specimens taken for strength evaluation exhibited no measurable change in height from the time of filling the cylinders to the time of testing.

4.2.4 *Hardening time*—Hardening time is the approximate period of time required for CLSM to go from the plastic state to a hardened state with sufficient strength to support the weight of a person. This time is greatly influenced by the amount and rate of bleed water released. When this excess water leaves the mixture, solid particles realign into intimate contact and the mixture becomes rigid. Hardening time is greatly dependent on the type and quantity of cementitious material in the CLSM.

Normal factors affecting the hardening time are:
- Type and quantity of cementitious material;
- Permeability and degree of saturation of surrounding soil that is in contact with CLSM;
- Moisture content of CLSM;
- Proportioning of CLSM;
- Mixture and ambient temperature;
- Humidity; and
- Depth of fill.

Hardening time can be as short as 1 hr, but generally takes 3 to 5 hr under normal conditions.[4,25,34] A penetration-resistance test according to ASTM C 403 can be used to measure the hardening time or approximate bearing capacity of CLSM. Depending upon the application, penetration numbers of 500 to 1500 are normally required to assure adequate bearing capacity.[35]

4.2.5 *Pumping*—CLSM can be successfully delivered by conventional concrete pumping equipment. As with concrete, proportioning of the mixture is critical. Voids must be adequately filled with solid particles to provide adequate cohesiveness for transport through the pump line under pressure without segregation. Inadequate void filling results in mixtures that can segregate in the pump and cause line blockage. Also, it is important to maintain a continuous flow through the pump line. Interrupted flow can cause segregation, which also could restrict flow and could result in line blockage.

In one example, CLSM using unwashed aggregate with a high fines content was pumped through a 127 mm (5 in.) pump system at a rate of 46 m^3/hr (60 yd^3/hr).[36] In another example, CLSM with a slump as low as 51 mm (2 in.) was successfully delivered by concrete pump without the need for added consolidation effort.[37]

CLSM with high entrained-air contents can be pumped, although care should be taken to keep pump pressures low. Increased pump pressures can cause a loss in air content and reduce pumpability.

Pumpability can be enhanced by careful proportioning to provide adequate void filling in the mixture. Fly ash can aid pumpability by acting as microaggregate for void filling. Cement can also be added for this purpose. Whenever cementitious materials are added, however, care must be taken to limit the maximum strength levels if later excavation is a consideration.

Fig. 4.1—*Excavating CLSM with backhoe.*

4.3—In-service properties

4.3.1 *Strength (bearing capacity)*—Unconfined compressive strength is a measure of the load-carrying ability of CLSM. A CLSM compressive strength of 0.3 to 0.7 MPa (50 to 100 psi) equates to an allowable bearing capacity of a well-compacted soil.

Maintaining strengths at a low level is a major objective for projects where later excavation is required. Some mixtures that are acceptable at early ages continue to gain strength with time, making future excavation difficult. Section 4.3.7 provides additional information on excavatability.

4.3.2 *Density*—Wet density of normal CLSM in place is in the range of 1840 to 2320 kg/m^3 (115 to 145 lb/ft^3), which is greater than most compacted materials. A CLSM mixture with only fly ash, cement, and water should have a density between 1440 to 1600 kg/m^3 (90 to 100 lb/ft^3).[12] Ponded ash or basin ash CLSM mixture densities are typically in the range of 1360 to 1760 kg/m^3 (85 to 110 lb/ft^3).[19] Dry density of CLSM can be expected to be substantially less than that of the wet density due to water loss. Lower unit weights can be achieved by using lightweight aggregates, high entrained-air contents, and foamed mixtures, which are discussed in detail in Chapter 8.

4.3.3 *Settlement*—Compacted fills can settle even when compaction requirements have been met. In contrast, CLSM does not settle after hardening. Measurements taken months after placement of a large CLSM fill showed no measurable shrinkage or settlement.[13] For a project in Seattle, Wash., 601 m^3 (786 yd^3) were used to fill a 37 m (120 ft) deep shaft. The placement took 4 hr and the total settlement was reported to be about 3 mm (1/8 in.).[37]

4.3.4 *Thermal insulation/conductivity*—Conventional CLSM mixtures are not considered good insulating materials. Air-entrained conventional mixtures reduce the density and increase the insulating value. Lightweight aggregates, including bottom ash, can be used to reduce density. Foamed or cellular mixtures as described in Chapter 8 have low densities and exhibit good insulating properties.

Where high thermal conductivity is desired, such as in backfill for underground power cables, high density and low porosity (maximum surface contact area between solid particles) are desirable. As the moisture content and dry density increase, so does the thermal conductivity. Other parameters to consider (but of lesser importance) include mineral composition, particle shape and size, gradation characteristics, organic content and specific gravity.[31,38-40]

4.3.5 *Permeability*—Permeability of most excavatable CLSM is similar to compacted granular fills. Typical values are in the range of 10^{-4} to 10^{-5} cm/sec. Mixtures of CLSM with higher strength and higher fines-content can achieve permeabilities as low as 10^{-7} cm/sec. Permeability is increased as cementitious materials are reduced and aggregate contents are increased.[4] However, materials normally used for reducing permeability, such as bentonite clay and diatomaceous soil, can affect other properties and should be tested prior to use.

4.3.6 *Shrinkage (cracking)*—Shrinkage and shrinkage cracks do not affect the performance of CLSM. Several reports have indicated that minute shrinkage occurs with CLSM. Ultimate linear shrinkage is in the range of 0.02 to 0.05%.[12,27,34]

4.3.7 *Excavatability*—The ability to excavate CLSM is an important consideration on many projects. In general, CLSM with a compressive strength of 0.3 MPa (50 psi) or less can be excavated manually. Mechanical equipment, such as backhoes, are used for compressive strengths of 0.7 to 1.4 MPa (100 to 200 psi) (Fig. 4.1). The limits for excavatability are somewhat arbitrary, depending upon the CLSM mixture. Mixtures using high quantities of coarse aggregate can be difficult to remove by hand, even at low strengths. Mixtures using fine sand or only fly ash as the aggregate filler have been excavated with a backhoe up to strengths of 2.1 MPa (300 psi).[11]

When the re-excavatability of the CLSM is of concern, the type and quantity of cementitious materials is important. Acceptable long-term performance has been achieved with cement contents from 24 to 59 kg/m^3 (40 to 100 lb/yd^3) and Class F fly ash contents up to 208 kg/m^3 (350 lb/yd^3). Lime (CaO) contents of fly ash that exceed 10% by weight can be a concern where long-term strength increases are not desired.[27]

Because CLSM will typically continue to gain strength beyond the conventional 28-day testing period, it is suggested, especially for high cementitious-content CLSM, that long-term strength tests be conducted to estimate the potential for re-excavatability.

In addition to limiting the cementitious content, entrained air can be used to keep compressive strengths low.

4.3.8 *Shear modulus*—The shear modulus, which is the ratio of unit shearing stress to unit shearing strain, of normal density CLSM is typically in the range of 160 to 380 MPa (3400 to 7900 ksf).[7,18,20] The shear modulus is used to evaluate the expected shear strength and deformation of CLSM material.

4.3.9 *Potential for corrosion*—The potential for corrosion on metals encased in CLSM has been quantified by a variety of methods specific to the material that is in contact with CLSM. Electrical resistivity tests can be performed on CLSM in the same manner that natural soils are compared for their corrosion potential on corrugated metal culvert pipes (California Test 643). The moisture content of the sample is an important parameter for the resistivity of a sample, and the samples should be tested at their expected long-term field moisture content.

The Ductile Iron Pipe Research Association has a method for evaluating the corrosion potential of backfill materials. The evaluation procedure is based upon information drawn from five tests and observations: soil resistivity; pH; oxidation-reduction (redox) potential; sulfides; and moisture. For a given sample, each parameter is evaluated and assigned points according to its contribution to corrosivity.[41-43]

These procedures are intended as guides in determining a soil's potential corrosivity to ductile iron pipe and should be used only by qualified engineers and technicians experienced in soil analysis and evaluation.

One cause of galvanic corrosion is the differences in potential from backfill soils of varying composition. The uniformity of CLSM reduces the chance for corrosion caused by the use of dissimilar backfill materials and their varying moisture contents.

4.3.10 *Compatibility with plastics*—High-, medium-, and low-density polyethylene materials are commonly used as protection for underground utilities or as the conduits themselves. CLSM is compatible with these materials. As with any backfill, care must be exercised to avoid damaging the protective coating of buried utility lines. The fine gradation of many CLSMs can aid in minimizing scratching and nicking these polyethylene surfaces.[31]

CHAPTER 5—MIXTURE PROPORTIONING

Proportioning for CLSM has been done largely by trial and error until mixtures with suitable properties are achieved. Most specifications require proportioning of ingredients; some specifications call for performance features and leave proportioning up to the supplier. ACI 211 has been used; however, much work remains to be done in establishing consistent reliability when using this method.[37]

Where proportions are not specified, trial mixtures are evaluated to determine how well they meet certain goals for strength, flowability, and density. Adjustments are then made to achieve the desired properties.

Table 5.1 presents a number of mixture proportions that have been used by state DOTs and others; however, requirements and available materials can vary considerably from project to project. Therefore, the information in Table 5.1 is provided as a guide and should not be used for design purposes without first testing with locally available materials.

The following summary can be made regarding the materials used to manufacture CLSM:

Cement—Cement contents generally range from 30 to 120 kg/m^3 (50 to 200 lb/yd^3), depending upon strength and hardening-time requirements. Increasing cement content while maintaining all other factors equal (that is, water, fly ash, aggregate, and ambient temperature) will normally increase strength and reduce hardening time.

Fly ash—Class F fly ash contents range from none to as high as 1200 kg/m^3 (2000 lb/yd^3) where fly ash serves as the aggregate filler. Class C fly ash is used in quantities of up to 210 kg/m^3 (350 lb/yd^3). The quantity of fly ash used will be determined by availability and flowability needs of the project.

Ponded ash/basin ash—Ponded ash/basin ash contents range from 300 to 500 kg/m^3 (500 to 950 lb/yd^3), depending upon the fineness of ash.[18-20]

Aggregate—The majority of specifications call for the use of fine aggregate. The amount of fine aggregate varies with the quantity needed to fill the volume of the CLSM after considering cement, fly ash, water, and air contents. In general, the quantities range from 1500 to 1800 kg/m^3 (2600 to 3100 lb/yd^3). Coarse aggregate is generally not used in CLSM mixtures as often as fine aggregates. When used, however, the coarse aggregate content is approximately equal to the fine aggregate content.

Water—More water is used in CLSM than in concrete. Water provides high fluidity and promotes consolidation of the materials. Water contents typically range from 193 to 344 kg/m^3 (325 to 580 lb/yd^3) for most CLSM mixtures containing aggregate. Water content for Class F fly ash and cement-only mixtures can be as high as 590 kg/m^3 (1000 lb/yd^3) to achieve good flowability. This wide range is due primarily to the characteristics of the materials used in CLSM and the degree of flowability desired. Water contents will be higher with mixtures using finer aggregates.

Admixtures—High doses of air-entraining admixtures and specifically formulated or packaged air-entraining admixtures, or both, can be used to lower the density or unit weight of CLSM. Accelerating admixtures can be used to accelerate the hardening of CLSM. When these products are used, the manufacturer's recommendations for use with CLSM should be followed.

Other additives—Additives such as zeolites, heavy minerals, and clays can be added to typical CLSM mixes in the range of 2 to 10% of the total mixture. Fly ash and cement can be adjusted accordingly while maintaining all other factors.[18-20]

CHAPTER 6—MIXING, TRANSPORTING, AND PLACING

6.1—General

The mixing, transporting, and placing of CLSM generally follows methods and procedures given in ACI 304R. Other methods can be acceptable, however, if prior experience and performance data are available. Whatever methods and procedures are used, the main criteria is that the CLSM be homogeneous, consistent, and satisfy the requirements for the purpose intended.

6.2—Mixing

CLSM can be mixed by several methods, including central-mixed concrete plants, ready-mixed concrete trucks, pugmills, and volumetric mobile concrete mixers. For high fly ash mixtures where fly ash is delivered to the mixer from existing silos, batching operations can be slow.

Table 5.1—Examples of CLSM mixture proportions*

Source	CO DOT	IA DOT	FL DOT	IL DOT	IN DOT Mix 1	IN DOT Mix 2[4]	OK DOT	MI DOT Mix 1	MI DOT Mix 2[4]	OH DOT Mix 1	OH DOT Mix 2
Cement content, kg/m³	30 (50)	60 (100)	30 to 60 (50 to 100)	30 (50)	36 (60)	110 (185)	30 (50) min	60 (100)	30 (50)	60 (100)	30 (50)
Fly ash, kg/m³ (lb/yd³)	—	178 (300)	0 to 356 (0 to 600)[2]	178 (300) Class F or 119 (200) Class C	196 (330)	—	148 (250)	1187 (2000) Class F	326 (550) Class F	148 (250)	148 (250)
Coarse aggregate, kg/m³ (lb/yd³)	1010 (1700)[1]	—	—	—	—	—	—	—	Footnote no. 5	—	—
Fine aggregate, kg/m³ (lb/yd³)	1096 (1845)	1543 (2600)	1632 (2750)[3]	1720 (2900)	1697 (2860)	1587 (2675)	1727 (2910)	—	Footnote no. 5	1691 (2850)	1727 (2910)
Approximate water content, kg/m³ (lb/yd³)	193 (325)	347 (585)	297 (500) maximum	222 to 320 (375 to 540)	303 (510)	297 (500)	297 (500) maximum	395 (665)	196 (330)	297 (500)	297 (500)
Compressive strength at 28 days, MPa (psi)	0.4 (60)	—	0.3 to 1.0 (50 to 150)	—	—	—	—	—	—	—	—

Table 5.1(continued)—Examples of CLSM mixture proportions*

Source	SC DOT	DOE-SR[16]	Unshrinkable fill[6]	Pond ash/basin ash mix[17] Mix AF	Pond ash/basin ash mix[17] Mix D	Coarse aggregate CLSM[8] Non-air entrainment[9]	Coarse aggregate CLSM[8] Air entrainment[11]	Flowable fly ash slurry[12] Mix S-2[13]	Flowable fly ash slurry[12] Mix S-3[14]	Flowable fly ash slurry[12] Mix S-4[15]
Cement content, kg/m³	30 (50)	30 (50)	36 (60)	98 (165)	60 (100)	30 (50)	30 (50)	58 (98)	94 (158)	85 (144)
Fly ash, kg/m³ (lb/yd³)	356 (600)	356 (600) Class F	—	481 (810)[18]	326 (550)[19]	148 (250)	148 (250)	810 (1366) Class F	749 (1262) Class F	685 (1155) Class F
Coarse aggregate, kg/m³ (lb/yd³)	—	—	1012 (1705) (3/4-in. maximum)	1300 (2190)	1492 (2515)	1127 (1900) (1-in. maximum)	1127 (1900) (1-in. maximum)	—	—	—
Fine aggregate, kg/m³ (lb/yd³)	1483 (2500)	1492 (2515)	1173 (1977)	—	—	863 (1454)	795 (1340)	—	—	—
Approximate water content, kg/m³ (lb/yd³)	273 to 320 (460 to 540)	397 to 326 (500 to 550)	152 (257)[7]	415 (700)	301 (507)	160 (270)[10]	151 (255)[10]	634 (1068)	624 (1052)	680 (1146)
Compressive strength at 28 days, MPa (psi)	0.6 (80)	0.2 to 1.0 (30 to 150)	0.1 (17) at 1 day	0.4 (65)	0.4 (65)	0.7 (100)	—	0.3 (40) (40 at 56 days)	0.4 (60) [0.5 (75) at 56 days]	0.3 (50) [0.5 (70) at 56 days]

*Table examples are based on experience and test results using local materials. Yields will vary from 0.76 m³ (27 ft³). This table is given as a guide and should not be used for design purposes without first testing with locally available materials.
[1]Quantity of cement can be increased above these limits only when early strength is required and future removal is unlikely.
[2]Granulated blast-furnace slag can be used in place of fly ash.
[3]Adjust to yield 1 yd³ of CLSM.
[4]5 to 6 fl oz of air-entraining admixture produces 7 to 12% air contents.
[5]Total granular material of 1690 kg/m³ (2850 lb/yd³) with 19 mm (3/4 in.) maximum aggregate size.
[6]Reference 44.
[7]Produces 150 mm (6 in.) slump.
[8]Reference 37.
[9]Produces approximately 1.5% air content.
[10]Produces 150 to 200 mm (6 to 8 in.) slump.
[11]Produces 5% air content.
[12]Reference 6.
[13]Produces modified flow of 210 mm (8-1/4 in.) diameter (Table 7.1); air content of 0.8%; slurry density of 1500 kg/m³ (93.7 lb/ft³).
[14]Produces modified flow of 270 mm (10-1/2 in.) diameter; air content of 1.1%; slurry density of 1470 kg/m³ (91.5 lb/ft³).
[15]Produces modified flow of 430 mm (16-3/4 in.) diameter; air content of 0.6%; slurry density of 1450 kg/m³ (90.6 lb/ft³).
[16]Department of Energy (DOE) Savannah River Site CLSM mix.
[17]DOE Savannah River Site CLSM mix using pond/basin ash.
[18]Basin ash mix.
[19]Pond ash mix.

Truck mixers are commonly used by ready-mixed concrete producers to mix CLSM; however, in-plant central mixers can be used as well. In truck-mixing operations, the following is one procedure that can be used for charging truck mixers with batch materials.

Load truck mixer at standard charging speed in the following sequence:

- Add 70 to 80% of water required.
- Add 50% of the aggregate filler.
- Add all cement and fly ash required.
- Add balance of aggregate filler.
- Add balance of water.

For CLSM mixtures consisting of fly ash, cement, water, and no aggregate filler, an effective mixing method consists of initially charging the truck mixer with cement then water. After thoroughly mixing these materials, the fly ash is added. Additional mixing for a minimum of 15 min was required in one case to produce a homogeneous slurry.[12]

Pugmill mixing works efficiently for both high and low fly ash mixtures and other high fines-content mixtures. For high fly ash mixtures, the fly ash is fed into a hopper with a front-end loader, which supplies a belt conveyor under the hopper. This method of feeding the mixer is much faster than silo feed. To prevent bridging within the fly ash, a mechanical agitator or vibrator is used in the hopper. Cement is usually added to the mixer by conveyor from silo storage. If bagged cement is used, it is added directly into the mixer. The measurement for payment of CLSM mixed through a pugmill is generally based on weight rather than volume, which is typically used for concrete.

6.3—Transporting

Most CLSM mixtures are transported in truck mixers. Agitation of CLSM is required during transportation and waiting time to keep the material in suspension. Under certain on-site circumstances, CLSM has been transported in nonagitating equipment such as dump trucks. Agitator trucks, although providing some mixing action, may not provide enough action to prevent the solid materials from settling out.

CLSM has been transported effectively by pumps and other types of conveying equipment. In pumping CLSM, the fly ash serves as a lubricant to reduce the friction in the pipeline. However, the fine texture of the fly ash requires that the pump be in excellent condition and properly cleaned and maintained.

CLSM has also been transported effectively by volumetric-measuring and continuous-mixing concrete equipment (VMCM) (ACI 304.6R), particularly if it is desired to reduce waiting time. The major advantage of this equipment is its ability to mix at the job site and vary the water content to attain desired flowability. This is particularly true for fast-setting CLSM mixtures. VMCMs are equipped with separate bins for water, cementitious materials, and selected aggregates. The materials are transported to the job site where continuous mixing of water and dry materials make a good, easily regulated CLSM.

6.4—Placing

CLSM can be placed by chutes, conveyors, buckets, or pumps, depending upon the application and its accessibility. Internal vibration or compaction is not required because the CLSM consolidates under its own weight. Although it can be placed year round, CLSM should be protected from freezing until it has hardened. Curing methods specified for concrete are not considered essential for CLSM.[27]

For trench backfill, CLSM is usually placed continuously. To contain CLSM when filling long, open trenches in stages or open-ended structures such as tunnels, the end points can be bulkheaded with sandbags, earth dams, or stiffer mixtures of CLSM.

For pipe bedding, CLSM can be placed in lifts to prevent floating the pipe. Each lift should be allowed to harden before continued placement. Other methods of preventing flotation include sand bags placed over the pipe, straps around the pipe anchored into the soil, or use of faster-setting CLSM placed at strategic locations over the pipe.

In the plastic state, CLSM is not self-supporting and places a load on the pipe. For large, flexible wall pipes, CLSM should be placed in lifts so that lateral support can develop along the side of the pipe before fresh CLSM is placed over the pipe.[4] Backfilling retaining walls can also require the CLSM be placed in lifts to prevent overstressing the wall.

CLSM has been effectively placed by tremie under water[11] without significant segregation. In confined areas, the CLSM displaces the water to the surface where it can easily be removed. Because of its very fluid consistency, CLSM can flow long distances to fill voids and cavities located in hard-to-reach places. Voids need not be cleaned, as the slurry will fill in irregularities and encapsulate any loose materials.

6.5—Cautions

6.5.1 *Hydrostatic pressure*—CLSM is often placed in a practically liquid condition and thus will exert a hydrostatic pressure against basement walls and other structures until it hardens. On deep fills, it is often necessary to place the CLSM in multiple lifts.

6.5.2 *Quick condition*—Liquid CLSM in deep excavations is essentially a quick-sand hazard and therefore should be covered until hardening occurs.

6.5.3 *Floating tanks, pipes, and cables*—Underground utilities and tanks must be secured against floating during CLSM placement.[45]

CHAPTER 7—QUALITY CONTROL

7.1—General

The extent of a quality-control (QC) program for CLSM can vary depending upon previous experience, application, raw materials used, and level of quality desired. A QC program can be as simple as a visual check of the completed work where standard, pretested mixtures are being used. Where the application is critical, the materials are nonstandard, or where product uniformity is questionable, regular tests for consistency and strength may be appropriate.

Both as-mixed and in-service properties can be measured to evaluate the mixture consistency and performance. For most projects, CLSM is pretested using the actual raw materials to develop a mixture having certain plastic (flowability, consistency, unit weight) and hardened (strength, durability, permeability) characteristics. Following the initial testing program, field testing can consist of simple visual checks, or can include consistency measurements or compressive strength tests.

As stated above, the QC program can be simple or detailed. It is the responsibility of the specifier to determine an appropriate QC program that will assure that the product will be adequate for its intended use. The following procedures and test methods have been used to evaluate CLSM mixtures.

Table 7.1—Test procedures for determining consistency and unit weight of CLSM mixtures

Consistency	
Fluid mixtures	
ASTM D 6103	"Standard Test Method for Flow Consistency of Controlled Low Strength Material." Procedure consists of placing 75 mm diameter x 150 mm long (3 in. diameter x 6 in. long) open-ended cylinder vertically on level surface and filling cylinder to top with CLSM. Cylinder is then lifted vertically to allow material to flow out onto level surface. Good flowability is achieved where there is no noticeable segregation and material spread is at least 200 mm (8 in.) in diameter.
ASTM C 939	"Flow of Grout for Preplaced-Aggregate Concrete." Florida Department of Transportation and Indiana Department of Transportation specifications require efflux time of 30 sec ±5 sec. Procedure is not recommended for CLSM mixtures containing aggregates greater than 6 mm (1/4 in.).
Plastic mixtures	
ASTM C 143	"Slump of Portland Cement Concrete."
Unit weight	
ASTM D 6023	"Standard Test Method for Unit Weight, Yield and Air Content (Gravimetric) of Controlled Low Strength Material." Ohio Ready Mixed Concrete Association has similar test method [FF3(94)].
ASTM C 1152	"Acid Soluble Chloride in Mortar and Concrete."
ASTM D 4380	"Density of Bentonitic Slurries." Not recommended for CLSM containing aggregate greater than 1/4 in.
ASTM D 1556	"Density of Soil In-Place by Sand-Cone Method."
ASTM D 2922	"Density of Soil and Soil Aggregate In-Place by Nuclear Method (Shallow Depth)."

Table 7.2—Test procedures for determining in-place density and strength of CLSM mixtures

ASTM D 6024	"Standard Test Method for Ball Drop on Controlled Low Strength Material to Determine Suitability for Load Application." This specification covers determination of ability of CLSM to withstand loading by repeatedly dropping metal weight onto in-place material.
ASTM C 403	"Time of Setting of Concrete Mixtures by Penetration Resistance." This test measures degree of hardness of CLSM. California Department of Transportation requires penetration number of 650 before allowing pavement surface to be placed.
ASTM D 4832	"Preparation and Testing of Soil-Cement Slurry Test Cylinders." This test is used for molding cylinders and determining compressive strength of hardened CLSM.
ASTM D 1196	"Nonrepetitive Static Plate Load Tests of Soils and Flexible Pavement Components for Use in Evaluation and Design of Airport and Highway Pavements." This test is used to determine modulus of subgrade reaction (K values).
ASTM D 4429	"Bearing Ratio of Soils in Place." This test is used to determine relative strength of CLSM in place.

7.2 Sampling

Sampling CLSM that has been delivered to the project site should be performed in accordance with ASTM D 5971.

7.3—Consistency and unit weight

Depending upon application and placement requirements, flow characteristics can be important. CLSM consistency can vary considerably from plastic to fluid; therefore, several methods of measurement are available. Most CLSM mixtures perform well with various flow and unit weight properties. Table 7.1 describes methods that can be used to measure consistency and unit weight.

7.4—Strength tests

CLSM is used in a variety of applications requiring different load-carrying characteristics. The maximum loads to be imposed on the CLSM should be identified to determine the minimum strength requirements. In many cases, however, CLSM needs to be limited in its maximum strength. This is especially true where removal of the material at a later date is anticipated.

The strength of CLSM can be measured by several methods (Table 7.2). Unconfined compressive strength tests are the most common; however, other methods, such as penetrometer devices or plate load tests, can also be used. Compressive-strength specimens can vary in size from 50 x 50 mm (2 x 2 in.) cubes to 150 x 300 mm (6 x 12 in.) cylinders. Special care may be needed removing very low-strength CLSM mixtures from test molds. Additional care in the handling, transporting, capping, and testing procedures shall be taken because the specimens are often very fragile. Mold stripping techniques have included: placement of a hole on the center of the bottom of standard watertight cylinder molds by drilling or use of a hot probe, and addition of a dry polyester fleece pad on the inside bottom of the cylinder; for easy release of the specimen with or without air compression, splitting of the molds with a hot knife, and presplitting the molds and reattachment with duct tape for easy removal later. The use of grout molds has also been employed for testing CLSM. In this method, four 150 x 150 x 200 mm (6 x 6 x 8 in.) high concrete masonry units are arranged to provide a nominal 100 mm (4 in.) square space in the center. The four sides and bottom of the inside of the molds are lined with blotting paper to serve as a bond breacher for easy removal.

CHAPTER 8—LOW-DENSITY CLSM USING PREFORMED FOAM

8.1—General

This chapter is limited to low-density CLSM mixtures (LD-CLSM) produced using preformed foam as part of the mixture proportioning. Preformed foam is made up of air cells generated from foam concentrates or gas-forming chemicals. The use of preformed foam in LD-CLSM mixtures allows mixture proportionings to be developed having lower unit weights than those typical of standard CLSM mixtures. Preformed foam is used in LD-CLSM proportions to attain stable air void or cell structures within the paste of the mix. LD-CLSM mixtures can be batched at ready-mix plants or in specially designed job site batch plants. The preformed foam can be added to LD-CLSM mixtures during batching at the ready-mix plant, into the mixers of transit-mix trucks at the job site, or directly into the mixer during the batching operations of specially designed job site batch plants.

8.2—Applications

LD-CLSM mixtures can be alternatively considered in situations where standard CLSM mixtures have been determined applicable. LD-CLSMs are typically designed by unit weight. The ability to proportion mixtures having low unit weights is especially advantageous where weak soil conditions are encountered and the weight of the fill must be minimized. LD-CLSM is also effective as an insulating and isolation fill. The air void or cell structure inherent in LD-CLSM mixtures provides thermal insulation and can add some shock mitigation properties to the fill material.

8.3—Materials

Portland cement is a typical binder component used to produce most LD-CLSM mixtures. Neat cement paste LD-CLSMs can be produced by adding preformed foam to the paste during mixing. The encapsulated air within the preformed foam is often the primary volume-producing component in the LD-CLSM mixtures. LD-CLSMs can also be designed to include mineral fillers such as fly ash or sand. When considering the use of nonstandard binders or mineral filler materials in LD-CLSM mixture proportioning, pretesting is recommended.

Generally all preformed foams are pregenerated by the use of devices known as foam generators. These foam-generating devices, however, can be configured specifically to be used with a particular foaming agent. The manufacturer of the foaming agent to be used should be consulted to obtain specific foam-generating recommendations.

Foaming agents used to produce the preformed foam must have a chemical composition capable of producing stable air cells that resist the physical and chemical forces imposed during the mixing, placing, and setting of the LD-CLSM mixture. If the air void or cellular structure within the mixture is not stable, a nonuniform increase in density will re-

Table 8.1—Typical strength properties of low-density CLSM based on density

Class	In-service density, kg/m^3 (lb/ft^3)	Minimum compressive strength, MPa (psi)
I	290 to 380 (18 to 24)	0.1 (10)
II	380 to 480 (24 to 30)	0.3 (40)
III	480 to 580 (30 to 36)	0.6 (80)
IV	580 to 670 (36 to 42)	0.8 (120)
V	670 to 800 (42 to 50)	1.1 (160)
VI	800 to 1300 (50 to 80)	2.2 (320)
VII	1300 to 1900 (80 to 120)	3.4 (500)

sult. Procedures for the evaluation of foaming agents are specified in ASTM C 796 and ASTM C 869. Additional information can be found in ACI 523.1R.

8.4—Properties

The properties of LD-CLSM are primarily density-related. When batched using standard component materials, LD-CLSM can be produced having properties that fall within ranges described by the manufacturer of the foaming agent. When nonstandard component materials are used, trial batches should be produced and tested to confirm theoretical predictions.

The most significant property of LD-CLSM is the in-service density. Table 8.1 divides the in-service density into convenient ranges relating density with typical minimum compressive-strength values. Classes VI and VII may be subdivided into smaller ranges for specific applications.

8.5—Proportioning

Mixture proportioning of LD-CLSM typically begins with the designation of the desired in-place dry density and minimum compressive strength. Within these parameters, the mixture constituents are designed on a rational basis. Basic LD-CLSM mixtures consist of portland cement as a binder, water, and preformed foam. In addition to this base proportioning, fly ash can be included as a pozzolan or a densifying mineral filler. Sand aggregate is also often used to achieve density in mixture proportionings having unit weights more than 800 kg/m^3 (50 lb/ft^3). The manufacturer of the foam concentrate is generally responsible for the mixture proportioning, which is based on desired physical properties (density, compressive strength, etc.) of the in-place material.

8.6—Construction

8.6.1 *Batching*—The batching sequence used to produce most LD-CLSM mixtures begins by metering the required water into a mechanical mixer. The portland cement binder, fly ash, or aggregates (if used) are individually weighed before entering the mixer. After the components are mixed to a uniform consistency, the required amount of preformed foam is added. The preformed foam is measured into the mixture through calibrated nozzle or by filling and weighing a mixing vessel of known volume. The accuracy of the foam-generating device and the batching apparatus is critical to the final mixture's density and its subsequent reproducibility.

8.6.2 *Mixing*—All LD-CLSM component materials should be mechanically mixed to a uniform consistency prior to the addition of the preformed foam. To properly combine the mixture ingredients (including the foam) sufficient mixing action and speeds are required. When producing neat cement or cement/fly ash pastes for LD-CLSM mixtures, mixers that provide vigorous mixing action, such as high-speed paddle mixers, are preferred. Truck mixers readily blend LD-CLSM mixtures to the consistency required for the addition of preformed foam. When truck mixers are used to produce neat cement or cement/fly ash paste mixtures, slightly longer mixing times are required. Other mixing processes, such as volumetric mixing, that produce uniformly consistent mixtures are also acceptable. The manufacturer of the foaming agent to be used should be consulted for specific recommendations on mixing procedures and approved mixing equipment.

8.6.3 *Placing*—LD-CLSM can be placed by chutes, buckets, or pumps. The method of placement must not cause a change in density by loss of air content beyond predictable ranges. Often, site-produced LD-CLSMs are delivered to the point of placement through pumplines. Progressing cavity pumps can be used, which provide nonpulsating and constant flow, minimizing air volume losses between the mixer and the point of deposit. By this method, LD-CLSMs can be pumped over 300 m (1000 ft).

CONVERSION FACTORS

1 ft = 0.305 m
1 in. = 25.4 mm
1 lb = 0.454 kg
1 yd^3 = 0.7646 m^3
1 psi = 6.895 kPa
1 lb/ft^3 = 16.02 kg/m^3
1 lb/yd^3 = 0.5933 kg/m^3
1 ft/sec = 0.305 m/sec

CHAPTER 9—REFERENCES
9.1—Specified references

The documents of the various standard-producing organizations referred to in this document are listed below with their serial designation.

American Concrete Institute

116R	Cement and Concrete Terminology
211.1	Standard Practice for Selecting Proportions for Normal, Heavyweight and Mass Concrete
230.1R	State-of-the-Art Report on Soil Cement
232.2R	Use of Fly Ash in Concrete
304.6R	Guide for Measuring, Mixing, Transporting and Placing Concrete
325.3R	Guide for Design of Foundations and Shoulders for Concrete Pavements
523.1R	Guide for Cast-in-Place Low Density Concrete

ASTM International

C 33	Specification for Concrete Aggregates
C 94	Specifications for Ready-Mixed Concrete
C 138	Test Method for Unit Weight, Yield and Air Content (Gravimetric) of Concrete
C 143	Test Method for Slump of Hydraulic Cement Concrete
C 150	Specification for Portland Cement
C 403	Test Method for Time of Setting of Concrete Mixtures by Penetration Resistance
C 595	Specification for Blended Hydraulic Cements
C 618	Specification for Fly Ash and Raw or Calcined Natural Pozzolan for Use as a Mineral Admixture in Portland Cement Concrete
C 796	Test Method of Testing Foaming Agents for Use in Producing Cellular Concrete Using Preformed Foam
C 869	Specification for Foaming Agents Used in Making Preformed Foam for Cellular Concrete
C 939	Test Method for Flow of Grout for Preplaced-Aggregate Concrete
C 1152	Acid-Soluble Chloride in Mortar and Concrete
C 1556	Density of Soil in-place by Sand-cone Method
C 2922	Density of Soil and Soil Aggregate in-place by Nuclear Method (Shallow Depth)
D 1196	Test Methods for Nonrepetitive Static Plate Load Tests of Soils and Flexible Pavement Components for Use in Evaluation and Design of Airport and Highway Pavements
D 4380	Test Method for Density of Bentonitic Slurries
D 4429	Test Method for Bearing Ratio of Soils in Place
D 4832	Test Method for Preparation and Testing of Soil-Cement Slurry Test Cylinders
D 5971	Practice for Sampling Freshly Mixed Controlled Low Strength Material
D 6023	Test Method for Unit Weight, Yield and Air Content (Gravimetric) of Controlled Low Strength Material
D 6024	Test Method of Ball Drop on Controlled Low Strength Material to Determine Suitability for Load Application
D 6103	Test Method for Flow Consistency of Controlled Low Strength Material

The above publications may be obtained from the following organizations:

American Concrete Institute
P.O. Box 9094
Farmington Hills, MI 48333-9094

ASTM International
100 Barr Harbor Drive
West Conshohocken, PA 19428-2959

9.2—Cited references

1. Adaska, W. S., ed., *Controlled Low-Strength Materials*, SP-150, American Concrete Institute, Farmington Hills, Mich., 1994, 113 pp.
2. Ramme, B. W., "Progress in CLSM: Continuing Innovation," *Concrete International*, V. 19, No. 5, May 1997, pp. 32-33.
3. Adaska, W. S., "Controlled Low-Strength Materials," *Concrete International*, V. 19, No. 4, Apr. 1997, pp. 41-43.
4. Smith, A., "Controlled Low-Strength Material," *Concrete Construction*, May 1991.
5. Sullivan, R. W., "Boston Harbor Tunnel Project Utilizes CLSM," *Concrete International*, V. 19, No. 5, May 1997, pp. 40-43.
6. Howard, A. K., and Hitch, J. L., eds., *The Design and Application of Controlled Low-Strength Materials (Flowable Fill)*, ASTM STP 1331, Symposium on the Design and Application of CLSM (Flowable Fill), St. Louis, Mo., June 19-20, 1997.

7. Larsen, R. L., "Use of Controlled Low-Strength Materials in Iowa," *Concrete International*, V. 10, No. 7, July 1988, pp. 22-23.

8. "AASHTO Guide for Design of Pavement Structures," American Association of State Highway and Transportation Officials, Washington, D.C., 1986.

9. Lowitz, C. A., and Defroot, G., "Soil-Cement Pipe Bedding, Canadian River Aqueduct," *Journal of the Construction Division*, ASCE, V. 94, No. C01, Jan. 1968.

10. Larsen, R. L., "Sound Uses of CLSM in the Environment," *Concrete International*, V. 12, No. 7, July 1990, pp. 26-29.

11. Krell, W. C., "Flowable Fly Ash," *Concrete International*, V. 11, No. 11, Nov. 1989, pp. 54-58.

12. Naik, T. R.; Ramme, B. W.; and Kolbeck, H. J., "Filling Abandoned Underground Facilities with CLSM Fly Ash Slurry," *Concrete International*, V. 12, No. 7, July 1990, pp. 19-25.

13. Flechsig, J. L., "Downtown Seattle Transit Project," *International Symposium on Unique Underground Structures*, Denver, June 1990.

14. "Flowable Fill," *Illinois Ready Mixed Concrete Association Newsletter*, July 1991.

15. Celis III, W., "Mines Long Abandoned to Dark Bringing New Dangers to Light," *USA Today*, Mar. 24, 1997.

16. Petzrick, P. A., "Ash Utilization for Elimination of Acid Mine Drainage," *Proceedings of the American Power Conference*, V. 59-II, 1997, pp. 834-836.

17. Dolance, R. C., and Giovannitti, E. F., "Utilization of Coal Ash/Coal Combustion Products for Mine Reclamation," *Proceedings of the American Power Conference*, V. 59-II, 1997, pp. 837-840.

18. Rajendran, N., and Venkata, R., "Strengthening of CMU Wall through Grouting," *Concrete International*, V. 19, No. 5, May 1997, pp. 48-49.

19. Langton, C. A., and Rajendran, N., "Utilization of SRS Pond Ash in Controlled Low-Strength Material(U)," U.S. Department of Energy *Report* WSRC-RP-95-1026, Dec. 1995.

20. Langton, C. A., and Rajendran, N., "Chemically Reactive CLSM for Radionuclides Stabilization," ACI Fall Convention, Montreal, Canada, Nov. 1995.

21. Buss, W. E., "Iowa Flowable Mortar Saves Bridges and Culverts," *Transportation Research Record* 1234, Concrete and Construction New Developments in Management TRB National Research Council, Washington, D.C., 1989.

22. Golbaum, J.; Hook, W.; and Clem, D. A., "Modification of Bridges with CLSM," *Concrete International*, V. 19, No. 5, May 1997, pp. 44-47.

23. Naik, T. R.; Ramme, B. W.; and Kolbeck, H. J., "Controlled Low-Strength Material (CLSM) Produced with High-Lime Fly Ash," *Proceedings*, Shanghai 1991 Ash Utilization Conference, Electric Power Research Institute Publication GS-7388, V. 3, 1991, pp. 110–1 through 110–11.

24. Landwermeyer, J. S., and Rice, E. K., "Comparing Quick-Set and Regular CLSM," *Concrete International*, V. 19, No. 5, May 1997, pp. 34-39.

25. Hoopes, R. J., "Engineering Properties of Air-Modified Controlled Low-Strength Material," *The Design and Application of Controlled Low-Strength Materials (Flowable Fill)*, ASTM STP 1331, A. K. Howard and J. L. Hitch, eds., ASTM, 1997.

26. Nmai, C. K.; McNeal, F.; and Martin, D., "New Foaming Agent for CLSM Applications," *Concrete International*, V. 19, No. 4, Apr. 1997, pp. 44-47.

27. Tansley, R., and Bernard, R., "Specification for Lean Mix Backfill," U.S. Department of Housing and Urban Development, *Contract* No. H-5208, Oct. 1981.

28. Naik, T. R.; Singh, S.; Taraniyil, M.; and Wendorf, R., "Application of Foundry Byproduct Materials in Manufacture of Concrete and Masonry Products," *ACI Materials Journal*, V. 93, No. 1, Jan.-Feb. 1996, pp. 41-50.

29. Naik, T. R., and Singh, S., "Flowable Slurry Containing Foundry Sands," *AE Materials Journal*, May 1997.

30. Naik, T. R., and Singh, S., "Permeability of Flowable Slurry Materials Containing Foundry Sand and Fly Ash," *ASCE Journal of Geotechnical Engineering*, May 1997.

31. Ramme, B. W.; Naik, T. R.; and Kolbeck, H. J., "Construction Experience with CLSM Fly Ash Slurry for Underground Facilities," *Fly Ash, Slag, Silica Fume, and Other Natural Pozzolans—Proceedings, Fifth International Conference*, SP-153, American Concrete Institute, Farmington Hills, Mich., 1995, pp. 403–416.

32. Glogowski, P. E., and Kelly, J. M., "Laboratory Testing of Fly Ash Slurry," Electric Power Research Institute, EPRI CS-6100, *Project* 2422-2, Dec. 1988.

33. "Suggested Specifications for Controlled Density Fill," Washington Aggregates and Concrete Association, Seattle, Wash., 1992.

34. McLaren, R. J., and Balsamo, N. J., "Fly Ash Design Manual for Road and Site Applications; Volume 2: Slurried Placement," Electric Power Research Institute, *Research Report* No. CS-4419, Oct. 1986.

35. "What, Why and How? Flowable Fill Materials," National Ready Mixed Concrete Association, CIP17, 1989.

36. "Soil-Cement Pumped for Unique Siphon Project," *Rocky Mountain Construction*, Feb. 17, 1986.

37. Fox, T. A., "Use of Coarse Aggregate in Controlled Low-Strength Materials," *Transportation Research Board* 1234, 1989.

38. Steinmanis, J. E., "Underground Cable Thermal Backfill," *Proceedings of the Symposium on Underground Cable Thermal Backfill*, Toronto, Canada, Sept. 1981.

39. Parmar, D., "Current Practices for Underground Cable Thermal Backfill," UTTF Meeting, Montreal, Canada, Sept. 1991.

40. Parmar, D., "Optimizing the Use of Controlled Backfill to Achieve High Ampacities on Transmission Cable," *Proceedings of Power Engineering Society Insulated Conductors Committee*, 1992.

41. Straud, Troy F., "Corrosion Control Measures for Ductile Iron Pipe," *Proceedings of Corrosion 89*, Paper 585, Apr. 1989, 38 pp.

42. "American National Standard for Polyethylene Encasement for Ductile Iron Piping for Water and Other Liquids," ANSI/AWWA C105/A21.5-88, AWWA, Denver, 1988, p. 7.

43. Hill, J. C., and Sommers, J., "Production and Marketing of Flowable Fill Utilizing Coal Combustion Byproducts," *Proceedings: 12th International Symposium on Coal Combustion Byproduct (CCB) Management and Use*, EPRI TR-107055-V2 3176, Jan. 1997.

44. Emery, J., and Johnston, T., "Unshrinkable Fill for Utility Cut Restorations," *Concrete in Transportation*, SP-93, American Concrete Institute, Farmington Hills, Mich., 1986, pp. 187-212.

45. Ramme, B. W., and Naik, T. R., "Controlled Low-Strength Materials (CLSM) State-of-the-Art New Innovations," *Supplementary Proceedings of Third CANMET/ACI International Symposium on Advances in Concrete Technology*, Auckland, New Zealand, Aug. 24-27, 1997.

ACI 302.1R-04

Guide for Concrete Floor and Slab Construction

Reported by ACI Committee 302

Eldon Tipping
Chair

Dennis Ahal
Secretary

Robert B. Anderson	C. Rick Felder	John P. Munday
Charles M. Ault	Edward B. Finkel	Joseph P. Neuber, Jr.
Charles M. Ayers	Jerome H. Ford	Russell E. Neudeck
Kenneth L. Beaudoin	Barry E. Foreman	Scott E. Niemitalo
Carl Bimel	Terry J. Fricks	Mark E. Patton
Michael G. Callas	Robert J. Gulyas	William S. Phelan
Douglas W. Deno	Patrick J. Harrison	Dennis W. Phillips
Gregory Dobson	Eugene D. Hill, Jr.	John W. Rohrer
Alphonse E. Engleman	Jerry A. Holland	Philip A. Smith
Robert A. Epifano	Arthur W. McKinney	Bruce A. Suprenant
Samuel A. Face, III	Steven N. Metzger	R. Gregory Taylor

FOREWORD

The quality of a concrete floor or slab is highly dependent on achieving a hard and durable surface that is flat, relatively free of cracks, and at the proper grade and elevation. Properties of the surface are determined by the mixture proportions and the quality of the concreting and jointing operations. The timing of concreting operations—especially finishing, jointing, and curing—is critical. Failure to address this issue can contribute to undesirable characteristics in the wearing surface such as cracking, low resistance to wear, dusting, scaling, high or low spots, poor drainage, and increasing the potential for curling.

Concrete floor slabs employing portland cement, regardless of slump, will start to experience a reduction in volume as soon as they are placed. This phenomenon will continue as long as any water, heat, or both, is being released to the surroundings. Moreover, because the drying and cooling rates at the top and bottom of the slab will never be the same, the shrinkage will vary throughout the depth, causing the as-cast shape to be distorted and reduced in volume.

This guide contains recommendations for controlling random cracking and edge curling caused by the concrete's normal volume change. Application of present technology permits only a reduction in cracking and curling, not elimination. Even with the best floor designs and proper construction, it is unrealistic to expect crack-free and curl-free floors. Consequently, every owner should be advised by both the designer and contractor that it is normal to expect some amount of cracking and curling on every project, and that such occurrence does not necessarily reflect adversely on either the adequacy of the floor's design or the quality of its construction (Ytterberg 1987; Campbell et al. 1976).

Refer to the latest edition of ACI 360R for a detailed discussion of shrinkage and curling in slabs-on-ground. Refer to the latest edition of ACI 224R for a detailed discussion of cracking in reinforced and nonreinforced concrete slabs.

This guide describes how to produce high-quality concrete slabs-on-ground and suspended floors for various classes of service. It emphasizes aspects of construction such as site preparation, concreting materials, concrete mixture proportions, concreting workmanship, joint construction, load transfer across joints, form stripping procedures, finishing methods, and curing. Flatness/levelness requirements and measurements are outlined. A thorough preconstruction meeting is critical to facilitate communication among key participants and to clearly establish expectations and procedures that will be employed during construction to achieve the floor qualities required by the project specifications. Adequate supervision and inspection are required for job operations, particularly those of finishing.

ACI Committee Reports, Guides, Standard Practices, and Commentaries are intended for guidance in planning, designing, executing, and inspecting construction. This document is intended for the use of individuals who are competent to evaluate the significance and limitations of its content and recommendations and who will accept responsibility for the application of the material it contains. The American Concrete Institute disclaims any and all responsibility for the stated principles. The Institute shall not be liable for any loss or damage arising therefrom.

Reference to this document shall not be made in contract documents. If items found in this document are desired by the Architect/Engineer to be a part of the contract documents, they shall be restated in mandatory language for incorporation by the Architect/Engineer.

It is the responsibility of the user of this document to establish health and safety practices appropriate to the specific circumstances involved with its use. ACI does not make any representations with regard to health and safety issues and the use of this document. The user must determine the applicability of all regulatory limitations before applying the document and must comply with all applicable laws and regulations, including but not limited to, United States Occupational Safety and Health Administration (OSHA) health and safety standards.

Keywords: admixture; aggregate; concrete; consolidation; contract documents; curing; curling; deflection; durability; form; fracture; joint; mixture proportioning; mortar; paste; placing; quality control; slab-on-ground; slabs; slump test; specification.

CONTENTS
Chapter 1—Introduction, p. 302.1R-2
 1.1—Purpose and scope
 1.2—Terminology
 1.3—Related work of other committees

ACI 302.1R-04 supersedes ACI 302.1R-96 and became effective March 23, 2004.
Copyright © 2004, American Concrete Institute.
All rights reserved including rights of reproduction and use in any form or by any means, including the making of copies by any photo process, or by electronic or mechanical device, printed, written, or oral, or recording for sound or visual reproduction or for use in any knowledge or retrieval system or device, unless permission in writing is obtained from the copyright proprietors.

Chapter 2—Classes of floors, p. 302.1R-5
  2.1—Classification of floors
  2.2—Single-course monolithic floors: Classes 1, 2, 4, 5, and 6
  2.3—Two-course floors: Classes 3, 7, and 8
  2.4—Class 9 floors
  2.5—Special finish floors

Chapter 3—Design considerations, p. 302.1R-6
  3.1—Scope
  3.2—Slabs-on-ground
  3.3—Suspended slabs
  3.4—Miscellaneous details

Chapter 4—Site preparation and placing environment, p. 302.1R-17
  4.1—Soil-support system preparation
  4.2—Suspended slabs
  4.3—Bulkheads
  4.4—Setting screed guides
  4.5—Installation of auxiliary materials
  4.6—Concrete placement conditions

Chapter 5—Materials, p. 302.1R-20
  5.1—Introduction
  5.2—Concrete
  5.3—Portland cement
  5.4—Aggregates
  5.5—Water
  5.6—Curing materials
  5.7—Admixtures
  5.8—Liquid surface treatments
  5.9—Reinforcement
  5.10—Evaporation reducers
  5.11—Gloss-imparting waxes
  5.12—Joint materials
  5.13—Volatile organic compounds (VOC)

Chapter 6—Concrete properties and consistency, p. 302.1R-27
  6.1—Concrete properties
  6.2—Recommended concrete mixture
  6.3—Concrete mixture analysis

Chapter 7—Batching, mixing, and transporting, p. 302.1R-34
  7.1—Batching
  7.2—Mixing
  7.3—Transporting

Chapter 8—Placing, consolidating, and finishing, p. 302.1R-35
  8.1—Placing operations
  8.2—Tools for spreading, consolidating, and finishing
  8.3—Spreading, consolidating, and finishing operations
  8.4—Finishing Class 1, 2, and 3 floors
  8.5—Finishing Class 4 and 5 floors
  8.6—Finishing Class 6 floors and monolithic-surface treatments for wear resistance
  8.7—Finishing Class 7 floors
  8.8—Finishing Class 8 floors (two-course unbonded)
  8.9—Finishing Class 9 floors
  8.10—Toppings for precast floors
  8.11—Finishing lightweight concrete
  8.12—Nonslip floors
  8.13—Decorative and nonslip treatments
  8.14—Grinding as a repair procedure
  8.15—Floor flatness and levelness
  8.16—Treatment when bleeding is a problem
  8.17—Delays in cold-weather finishing

Chapter 9—Curing, protection, and joint filling, p. 302.1R-59
  9.1—Purpose of curing
  9.2—Methods of curing
  9.3—Curing at joints
  9.4—Curing special concrete
  9.5—Length of curing
  9.6—Preventing plastic-shrinkage cracking
  9.7—Curing after grinding
  9.8—Protection of slab during construction
  9.9—Temperature drawdown in cold storage and freezer rooms
  9.10—Joint filling and sealing

Chapter 10—Quality control checklist, p. 302.1R-61
  10.1—Introduction
  10.2—Partial list of important items to be observed

Chapter 11—Causes of floor and slab surface imperfections, p. 302.1R-62
  11.1—Introduction
  11.2—Cracking
  11.3—Low wear resistance
  11.4—Dusting
  11.5—Scaling
  11.6—Popouts
  11.7—Blisters and delamination
  11.8—Spalling
  11.9—Discoloration
  11.10—Low spots and poor drainage
  11.11—Curling
  11.12—Analysis of surface imperfections

Chapter 12—References, p. 302.1R-71
  12.1—Referenced standards and reports
  12.2—Cited references
  12.3—Other references

CHAPTER 1—INTRODUCTION
1.1—Purpose and scope

This guide presents state-of-the-art information relative to the construction of slab-on-ground and suspended-slab floors for industrial, commercial, and institutional buildings. It is applicable to the construction of normalweight and structural lightweight concrete floors and slabs made with conventional portland and blended cements. Slabs specifically intended for the containment of liquids are beyond the scope of this document.

The design of slabs-on-ground should conform to the recommendations of ACI 360R. Refer to ACI 223 for procedures for the design and construction of shrinkage-compensating concrete slabs-on-ground. The design of suspended floors should conform to requirements of ACI 318 and ACI 421.1R. See Section 1.2 for relevant work by these and other committees.

This guide identifies the various classes of floors as to
- Use;
- Design details as they apply to construction;
- Necessary site preparation; and
- Type of concrete and related materials.

In general, the characteristics of the concrete slab surface and the performance of joints have a powerful impact on the serviceability of floors and other slabs. Because the eventual success of a concrete floor installation depends on the mixture proportions and floor finishing techniques used, considerable attention is given to critical aspects of achieving the desired finishes and the required floor surface tolerances. This guide emphasizes choosing and proportioning of materials, design details, proper construction methods, and workmanship.

1.1.1 *Prebid meeting*—While this guide does provide a reasonable overview of concrete floor construction, it should be emphasized that every project is unique; circumstances can dictate departures from the recommendations contained herein. Accordingly, contractors and suppliers are urged to make a thorough review of contract documents before bid preparation.

The best forum for such a review is the prebid meeting. This meeting offers bidders an opportunity to ask questions and clarify their understanding of contract documents before submitting their bids. A prebid meeting also provides the owner and the owner's designer an opportunity to clarify intent where documents are unclear and to respond to last-minute questions in a manner that provides bidders an opportunity to be equally responsive to the contract documents.

1.1.2 *Preconstruction meeting*—Construction of any slab-on-ground or suspended floor or slab involves the coordinated efforts of many subcontractors and material suppliers. It is strongly recommended that the designer require a preconstruction meeting to be held to establish and to coordinate procedures that will enable key participants to produce the best possible product under the anticipated field conditions. This meeting should be attended by responsible representatives of organizations and material suppliers directly involved with either the design or construction of floors.

The preconstruction meeting should confirm and document the responsibilities and anticipated interaction of key participants involved in floor slab construction. Following is a list of agenda items appropriate for such a meeting; many of the items are those for which responsibility should be clearly established in the contract documents. The following list is not necessarily all-inclusive:

1. Site preparation;
2. Grades for drainage, if any;
3. Work associated with installation of auxiliary materials, such as vapor barriers, vapor retarders, edge insulation, electrical conduit, mechanical sleeves, drains, and embedded plates;
4. Class of floor;
5. Floor thickness;
6. Reinforcement, when required;
7. Construction tolerances: base (rough and fine grading), forms, slab thickness, surface configuration, and floor flatness and levelness requirements (including how and when measured);
8. Joints and load-transfer mechanism;
9. Materials: cements, fine aggregate, coarse aggregate, water, and admixtures (usually by reference to applicable ASTM standards);
10. Special aggregates, admixtures, or monolithic surface treatments, where applicable;
11. Concrete specifications, to include the following:
    a. Compressive strength, flexural strength, or both, and finishability (Section 6.2);
    b. Minimum cementitious material content, if applicable (Table 6.2);
    c. Maximum size, grading, and type of coarse aggregate;
    d. Grading and type of fine aggregate;
    e. Combined aggregate grading;
    f. Air content of concrete, if applicable (Section 6.2.7);
    g. Slump of concrete (Section 6.2.5);
    h. Water-cement ratio (w/c) or water-cementitious material ratio (w/cm); and
    i. Preplacement soaking requirement for lightweight aggregates.
12. Measuring, mixing, and placing procedures (usually by reference to specifications or recommended practices);
13. Strikeoff method;
14. Recommended finishing methods and tools, where required;
15. Coordination of floor finish requirements with those required for floor coverings such as vinyl, ceramic tile, or wood that are to be applied directly to the floor;
16. Curing procedures, length of curing, necessary protection, and time before opening slabs for traffic (ACI 308R);
17. Testing and inspection requirements; and
18. Acceptance criteria and remedial measures to be used, if required.

Additional issues specific to suspended slab construction are as follows:

1. Form tolerances and preplacement quality assurance survey procedures for cast-in-place construction;
2. Erection tolerances and preplacement quality assurance survey procedures for composite slab construction (see ANSI/ASCE 3 and ANSI/ASCE 9 [Section 12.1]);
3. Form stripping procedures, if applicable; and
4. Items listed in Section 3.3 that are appropriate to the structural system(s) used for the project.

1.1.3 *Quality assurance*—Adequate provisions should be made to ensure that the constructed product meets or exceeds the requirements of the project documents. Toward this end, quality control procedures should be established and maintained throughout the entire construction process.

The quality of a completed concrete slab depends on the skill of individuals who place, finish, and test the material. As an aid to ensuring a high-quality finished product, the specifier or owner should consider requiring the use of prequalified concrete contractors, concrete suppliers, accredited testing laboratories, and concrete finishers who have had their proficiency and experience evaluated through an independent third-party certification program. ACI has developed programs to train and certify concrete flatwork finishers and concrete inspectors and testing technicians throughout the United States, Mexico, and Canada.

1.2—Terminology

adjusted mix optimization indicator (MOI-Adj)—intersection of the coarseness factor value and the adjusted workability factor on the coarseness factor chart.

adjusted workability factor (W-Adj)—the workability factor adjusted for cementitious content. For each 94 lb (43 kg) of total cementitious material above 564 lb/yd^3 (335 kg/m^3), increase the workability factor by 2.5%. For each 94 lb (43 kg) of total cementitious material below 564 lb/yd^3 (335 kg/m^3), decrease the workability factor by 2.5%. (Example for a workability factor of 33% and 600 lb/yd^3 [356 kg/m^3] of cementitious material: 600 lb/yd^3 [356 kg/m^3] – 564 lb/yd^3 [335 kg/m^3] = 36 lb/yd^3 [21 kg/m^3]; 36 lb [16 kg]/94 lb [43 kg] = 0.38; 0.38 × 2.5% = 0.95%; W-Adj = 33% workability factor + 0.95% = 33.95%).

coarseness factor—the percentage of combined aggregate that is larger than the 3/8 in. (9.5 mm) sieve, divided by the percentage of combined aggregate that is larger than the No. 8 (2.36 mm) sieve, expressed as a percent. (Example: 33% retained on the 3/8 in. [9.5 mm] sieve/45% retained on the No. 8 [2.36 mm] sieve = 73.3%).

differential set time—the difference in timing of initial power floating of sequential truck loads of concrete as they are delivered to the jobsite.

dry shake—metallic or mineral hardener mixed with cement and applied dry to the surface of concrete during finishing operations.

floating—a term used to describe smoothing and subsequent compaction and consolidation of the unformed concrete surface.

mix optimization indicator (MOI)—intersection of the coarseness factor value and the workability factor on the coarseness factor chart.

pumping—the vertical displacement and rebound of the soil support system in response to applied wheel loads.

rutting—the creation of troughs in the soil support system in response to applied wheel loads.

score—the creation of lines or notches in the surface of a concrete slab.

water slump—the magnitude of slump, measured in accordance with ASTM C 143, which is directly attributed to the amount of water in the concrete mixture.

window of finishability—the time period available for finishing operations after the concrete has been placed, consolidated, and struck-off, and before final troweling.

workability factor—the percentage of combined aggregate that passes the No. 8 (2.36 mm) sieve.

1.3—Related work of other committees

1.3.1 *ACI committees*

117—Prepares and updates tolerance requirements for concrete construction.

201—Reviews research and recommendations on durability of concrete and reports recommendations for appropriate materials and methods.

211—Develops recommendations for proportioning concrete mixtures.

223—Develops and reports on the use of shrinkage-compensating concrete.

224—Studies and formulates recommendations for the prevention or control of cracking in concrete construction.

301—Develops and maintains reference specifications for structural concrete for buildings.

308—Prepares guidelines for type and amount of curing required to develop the desired properties in concrete.

309—Studies and reports on research and development in consolidation of concrete.

311—Develops guides and procedures for inspection and testing.

318—Develops and updates building code requirements for reinforced concrete and structural plain concrete, including suspended slabs.

325—Reports on the structural design, construction, maintenance, and rehabilitation of concrete pavements.

330—Reports on the design, construction, and maintenance of concrete parking lots.

332—Gathers and reports on the use of concrete in residential construction.

347—Gathers, correlates, and reports information, and prepares recommendations for formwork for concrete.

350—Develops and updates code requirements for concrete in environmental structures.

360—Develops and reports on criteria for design of slabs-on-ground, except highway and airport pavements.

421—Develops and reports on criteria for suspended slab design.

423—Develops and reports on technical status, research, innovations, and recommendations for prestressed concrete.

435—Provides recommendations for deflection control in concrete slabs.

503—Studies and reports information and recommendations on the use of adhesives for structurally joining concrete, providing a wearing surface, and other uses.

504—Studies and reports on materials, methods, and systems used for sealing joints and cracks in concrete structures.

515—Prepares recommendations for selection and application of protective systems for concrete surfaces.

544—Studies and reports information and recommendations on the use of fiber-reinforced concrete.

640—Develops, maintains, and updates programs for use in certification of concrete construction workers.

1.3.2 *The American Society of Civil Engineers*—ASCE publishes documents that can be helpful for floor and slab construction. Two publications that deal with suspended slab construction are ASCE Standard for the Structural Design of Composite Slabs (ANSI/ASCE 3) and ASCE Standard Prac-

Table 2.1—Classes of floors on the basis of intended use and the suggested final finish technique

Class	Anticipated type of traffic	Use	Special considerations	Final finish
1. Single course	Exposed surface—foot traffic	Offices, churches, commercial, institutional, multi-unit residential	Uniform finish, nonslip aggregate in specific areas, curing	Normal steel-troweled finish, nonslip finish where required
		Decorative	Colored mineral aggregate, color pigment or exposed aggregate, stamped or inlaid patterns, artistic joint layout, curing	As required
2. Single course	Covered surface—foot traffic	Offices, churches, commercial, multi-unit residential, institutional with floor coverings	Flat and level slabs suitable for applied coverings, curing. Coordinate joints with applied coverings	Light steel-troweled finish
3. Two course	Exposed or covered surface—foot traffic	Unbonded or bonded topping over base slab for commercial or non-industrial buildings where construction type or schedule dictates	*Base slab*—good uniform level surface tolerance, curing *Unbonded topping*—bondbreaker on base slab, minimum thickness 3 in. (75 mm), reinforced, curing *Bonded topping*—properly sized aggregate, 3/4 in. (19 mm) minimum thickness curing	*Base slab*—troweled finish under unbonded topping; clean, textured surface under bonded topping *Topping*—for exposed surface, normal steel-troweled finish. For covered surface, light steel-troweled finish
4. Single course	Exposed or covered surface—foot and light vehicular traffic	Institutional or commercial	Level and flat slab suitable for applied coverings, nonslip aggregate for specific areas, curing. Coordinate joints with applied coverings	Normal steel-troweled finish
5. Single course	Exposed surface—industrial vehicular traffic, that is, pneumatic wheels and moderately soft solid wheels	Industrial floors for manufacturing, processing, and warehousing	Good uniform subgrade, joint layout, abrasion resistance, curing	Hard steel-troweled finish
6. Single course	Exposed surface—heavy-duty industrial vehicular traffic, that is, hard wheels and heavy wheel loads	Industrial floors subject to heavy traffic; may be subject to impact loads	Good uniform subgrade, joint layout, load transfer, abrasion resistance, curing	Special metallic or mineral aggregate surface hardener; repeated hard steel-troweling
7. Two course	Exposed surface—heavy-duty industrial vehicular traffic, that is, hard wheels and heavy wheel loads	Bonded two-course floors subject to heavy traffic and impact	*Base slab*—good uniform subgrade, reinforcement, joint layout, level surface, curing *Topping*—composed of well-graded all-mineral or all-metallic aggregate. Minimum thickness 3/4 in. (19 mm). Mineral or metallic aggregate surface hardener applied to high-strength plain topping to toughen, curing	Clean, textured base slab surface suitable for subsequent bonded topping. Special power floats for topping are optional, hard steel-troweled finish
8. Two course	As in Classes 4, 5, or 6	Unbonded topping—on new or old floors where construction sequence or schedule dictates	Bondbreaker on base slab, minimum thickness 4 in. (100 mm), abrasion resistance, curing	As in Classes 4, 5, or 6
9. Single course or topping	Exposed surface—superflat or critical surface tolerance required. Special materials-handling vehicles or robotics requiring specific tolerances	Narrow-aisle, high-bay warehouses; television studios, ice rinks, or gymnasiums. Refer to ACI 360R for design guidance	Varying concrete quality requirements. Special application procedures and strict attention to detail are recommended when shake-on hardeners are used. F_F 50 to F_F 125 ("superflat" floor). Curing	Strictly following techniques as indicated in Section 8.9

tice for Construction and Inspection of Composite Slabs (ANSI/ASCE 9).

CHAPTER 2—CLASSES OF FLOORS
2.1—Classification of floors

Table 2.1 classifies floors on the basis of intended use, discusses special considerations, and suggests finishing techniques for each class of floor. Intended use requirements should be considered when selecting concrete properties (Section 6.2), and the step-by-step placing, consolidating, and finishing procedures in Chapter 8 should be closely followed for different classes and types of floors.

Wear resistance and impact resistance should also be considered. Currently, there are no standard criteria for evaluating the wear resistance of a floor, and it is not possible to specify concrete quality in terms of ability to resist wear. Wear resistance is directly related to the concrete-mixture proportions, types of aggregates, finishing, curing, and other construction techniques used.

2.2—Single-course monolithic floors: Classes 1, 2, 4, 5, and 6

Five classes of floors are constructed with monolithic concrete; each involves some variation in strength and final finishing techniques. If abrasion from grit or other materials is anticipated, a higher-quality floor surface may be required for satisfactory service (ASTM 1994). Under these conditions, a higher-class floor, a special mineral or metallic aggregate monolithic surface treatment, or a higher-strength concrete is recommended.

2.3—Two-course floors: Classes 3, 7, and 8

2.3.1 *Unbonded topping over base slab*—The base courses of Class 3 (unbonded, two course) floors and Class 8 floors can be either slabs-on-ground or suspended slabs, with the finish to be coordinated with the type of topping. For Class 3 floors, the concrete topping material is similar to the base slab concrete. The top courses for Class 8 floors require a hard-steel troweling and usually have a higher compressive strength than the base course. Class 8 floors can also make use of an embedded hard aggregate, a premixed (dry-shake) mineral aggregate, or metallic hardener for addition to the surface (Section 5.4.5).

Class 3 (with unbonded topping) and Class 8 floors are used when it is preferable to not bond the topping to the base course, so that the two courses can move independently (for example, with precast members as a base), or so that the top courses can be more easily replaced at a later period. Two-course floors can be used when mechanical and electrical equipment require special bases and when their use permits more expeditious construction procedures. Two-course unbonded floors can also be used to resurface worn or damaged floors when contamination prevents complete bond or when it is desirable to avoid scarifying and chipping the base course and the resultant higher floor elevation is compatible with adjoining floors. Class 3 floors are used primarily for commercial or nonindustrial applications, whereas Class 8 floors are primarily for industrial applications.

Plastic sheeting, roofing felt, or a bond-breaking compound is used to prevent bond to the base slab. Reinforcement, such as deformed bars, welded wire fabric, bar mats, or fibers, may be placed in the topping to reduce the width of shrinkage cracks. Unbonded toppings should have a minimum thickness of 3 in. (75 mm). The concrete should be proportioned to meet the requirements of Chapter 6. Joint spacing in the topping should be coordinated with joint spacing in the base slab.

Additional joints should be considered if the topping slab thickness mandates a closer spacing than the base slab to limit uncontrolled cracking and slab curl. Curl or warping will be more probable due to the effects of drying from the top surface only.

2.3.2 *Bonded topping over base slab*—Class 3 (bonded topping) and Class 7 floors use a topping bonded to the base slab. Class 3 (bonded topping) floors are used primarily for commercial or nonindustrial applications; Class 7 floors are used for heavy-duty, industrial applications subject to heavy traffic and impact. The base slabs can either be a conventional portland cement concrete mixture or shrinkage-compensating concrete. The surface of the base slab should have a rough, open pore finish and be free of any substances that would interfere with the bond of the topping to the base slab.

The topping can be either a same-day installation (before hardening of the base slab) or a deferred installation (after the base slab has hardened). The topping for a Class 3 floor is a concrete mixture similar to that used in Class 1 or 2 floors. The topping for a Class 7 floor requires a multiple-pass, hard-steel-trowel finish, and it usually has a higher strength than the base course. A bonded topping can also make use of an embedded hard aggregate or a premixed (dry-shake) mineral aggregate or metallic hardener for addition to the surface (Section 5.4.5). Bonded concrete toppings should have a minimum thickness of 3/4 in. (19 mm). Proprietary products should be applied per manufacturers' recommendations. Joint spacing in the topping should be coordinated with construction and contraction joint spacing in the base slab. Saw-cut contraction joints should penetrate into the base slab a minimum of 1 in. (25 mm).

If the topping is placed on a base slab before the joints are cut, joints in the topping should extend into the base slab and depth should be appropriate for the total thickness of the combined slab. If the topping is installed on a previously placed slab where joints have activated, additional joints in the topping are unnecessary as shrinkage relief cannot occur between the slab joints in the bonded topping. When topping slabs are placed on shrinkage-compensating base slabs, the joints in the base slab can only be reflected in the bonded topping slab if the bonded topping slab is installed shortly after the maximum expansion occurs. Maximum expansion usually occurs within seven to 14 days.

2.4—Class 9 floors

Certain materials-handling facilities (for example, high-bay, narrow-aisle warehouses) require extraordinarily level and flat floors. The construction of such superflat floors (Class 9) is discussed in Chapter 8. A superflat floor could be constructed as a single-course floor or it could be constructed as a two-course floor with a topping, either bonded (similar to a Class 7 topping) or unbonded (similar to a Class 8 topping).

2.5—Special finish floors

Floors with decorative finishes and those requiring skid resistance or electrical conductivity are covered in appropriate sections of Chapter 8.

Floors exposed to mild acids, sulfates, or other chemicals require special preparation or protection. ACI 201.2R reports on means of increasing the resistance of concrete to chemical attack. Where attack will be severe, wear-resistant protection suitable for the exposure should be used. Such environments and the methods of protecting floors against them are discussed in ACI 515.1R.

In certain chemical and food processing plants, such as slaughterhouses, exposed concrete floors are subject to slow disintegration due to organic acids. In many instances, it is preferable to protect the floor with other materials such as acid-resistant brick, tile, or resinous mortars (ACI 515.1R).

CHAPTER 3—DESIGN CONSIDERATIONS
3.1—Scope

This chapter addresses the design of concrete floors as it relates to their constructibility. Specific design requirements for concrete floor construction are found in other documents: ACI 360R for slabs-on-ground, ACI 223 for shrinkage-compensating concrete floors, ACI 421.1R for suspended floors, ANSI/ASCE 3 for structural design of composite slabs, and ANSI/ASCE 9 for construction and inspection of composite slabs. Refer to ACI 318 for requirements relating to the building code.

3.2—Slabs-on-ground

3.2.1 *Required design elements*—The following items should be specified in the contract documents prepared by the designer:
- Base and subbase materials, preparation requirements, and vapor retarder, if required;
- Concrete thickness;

- Concrete compressive strength, flexural strength, or both;
- Concrete mixture proportion requirements;
- Joint locations and details;
- Reinforcement (type, size, and location), if required;
- Surface treatment, if required;
- Surface finish;
- Tolerances (base, subbase, slab thickness, and surface);
- Concrete curing;
- Joint filling material and installation;
- Special embedments; and
- Preconstruction meeting, quality assurance, and quality control.

3.2.2 *Soil-support system*—The performance of a slab-on-ground depends on the integrity of both the soil-support system and the slab; therefore, specific attention should be given to the site preparation requirements, including proof-rolling, discussed in Section 4.1.1. In most cases, proof-rolling results are far more indicative of the ability of the soil-support system to withstand loading than the results from in-place tests of moisture content or density are. A thin layer of graded, granular, compactible material is normally used as fine grading material to better control the thickness of the concrete and to minimize friction between the base material and the slab. For detailed information on soil-support systems, refer to ACI 360R.

3.2.3 *Moisture protection*—Proper moisture protection is essential for any slab-on-ground where the floor will be covered by moisture-sensitive flooring materials such as vinyl, linoleum, wood, carpet, rubber, rubber-backed carpet tile, impermeable floor coatings, adhesives, or where moisture-sensitive equipment, products, or environments exist, such as humidity-controlled or refrigerated rooms.

A vapor retarder is a material that is intended to minimize the transmission of moisture upward through the slab from sources below. The performance requirements for plastic vapor retarder materials in contact with soil or granular fill under concrete slabs are listed in ASTM E 1745. It is generally recognized that a vapor retarder should have a permeance (water vapor transmission rate) of less than 0.3 perms, as determined by ASTM E 96.

The selection of a vapor retarder or barrier material should be made on the basis of protective requirements and the moisture-related sensitivity of the materials to be applied to the floor surface. Although conventional polyethylene film with a thickness of as little as 6 mils (0.15 mm) has been used, the committee strongly recommends that the material be in compliance with ASTM E 1745 and that the thickness be no less than 10 mils (0.25 mm). The increased thickness offers increased resistance to moisture transmission while providing greater durability during and after installation.

A number of vapor retarder materials have been incorrectly referred to and used by designers as vapor barriers. True vapor barriers are products that have a permeance (water-vapor transmission rating) of 0.00 perms when tested in accordance with ASTM E 96. The laps or seams in either a vapor retarder or barrier should be overlapped 6 in. (150 mm) (ASTM E 1643) or as instructed by the manufacturer. The joints and penetrations should be sealed with the manufacturer's recommended adhesive, pressure-sensitive tape, or both.

The decision whether to locate the vapor retarder or barrier in direct contact with the slab or beneath a layer of granular fill should be made on a case-by-case basis (Suprenant and Malisch 1998b). For moisture-sensitive flooring materials and environments, placing concrete in direct contact with the vapor retarder or barrier eliminates the potential for water from sources such as rain, saw-cutting, curing, cleaning, or compaction to become trapped within the fill course. Wet or saturated fill above the vapor retarder can significantly increase the time required for a slab to dry to levels required by the manufacturers of floor coverings, adhesives, and coatings.

Placing concrete in direct contact with the vapor retarder or barrier, however, requires additional consideration if potential slab-related problems are to be avoided. When compared with identical concrete cast on a draining base, concrete placed in direct contact with a vapor retarder or barrier has been shown to exhibit significantly larger length change in the first hour after casting, during drying shrinkage, and when subject to environmental change; there is also more settlement (Suprenant 1997). Care should be taken in design detailing to minimize restraint to such movement (Anderson and Roper 1977). Where reinforcing steel is present, settlement cracking over the steel is more likely because of the increased settlement resulting from a longer bleeding period. The potential for a greater measure of slab curl is also increased.

Concrete that does not lose water to the base does not stiffen as rapidly as concrete that does lose part of its excess water to the base. If rapid, surface drying conditions are present, the surface of concrete placed directly on a vapor retarder will have a tendency to dry and crust over while the concrete below the top fraction of an inch remains relatively less stiff or unhardened. When this occurs, it may be necessary to begin machine operations on the concrete surface before the concrete below the top surface is sufficiently set. Under such conditions, a reduction in surface flatness and some blistering or delamination can occur as air, water, or both become trapped below the finish surface.

The committee recommends that each proposed installation be independently evaluated as to the moisture sensitivity of subsequent floor finishes, anticipated project conditions, and the potential effects of slab curling, crusting, and cracking. The anticipated benefits and risks associated with the specified location of the vapor retarder should be reviewed with all appropriate parties before construction. Figure 3.1 can be used to assist this evaluation process.

3.2.4 *Reinforcement for crack-width control*—Reinforcement restrains movement resulting from slab shrinkage and can actually increase the number of random cracks experienced, particularly at wider joint spacing (Section 3.2.5.3). Reinforcement in nonstructural slab-on-ground installations is provided primarily to control the width of cracks that occur (Dakhil, Cady, and Carrier 1975; CRSI 1990). This reinforcement is normally furnished in the form of deformed steel bars, welded wire reinforcing, steel fibers, or post-

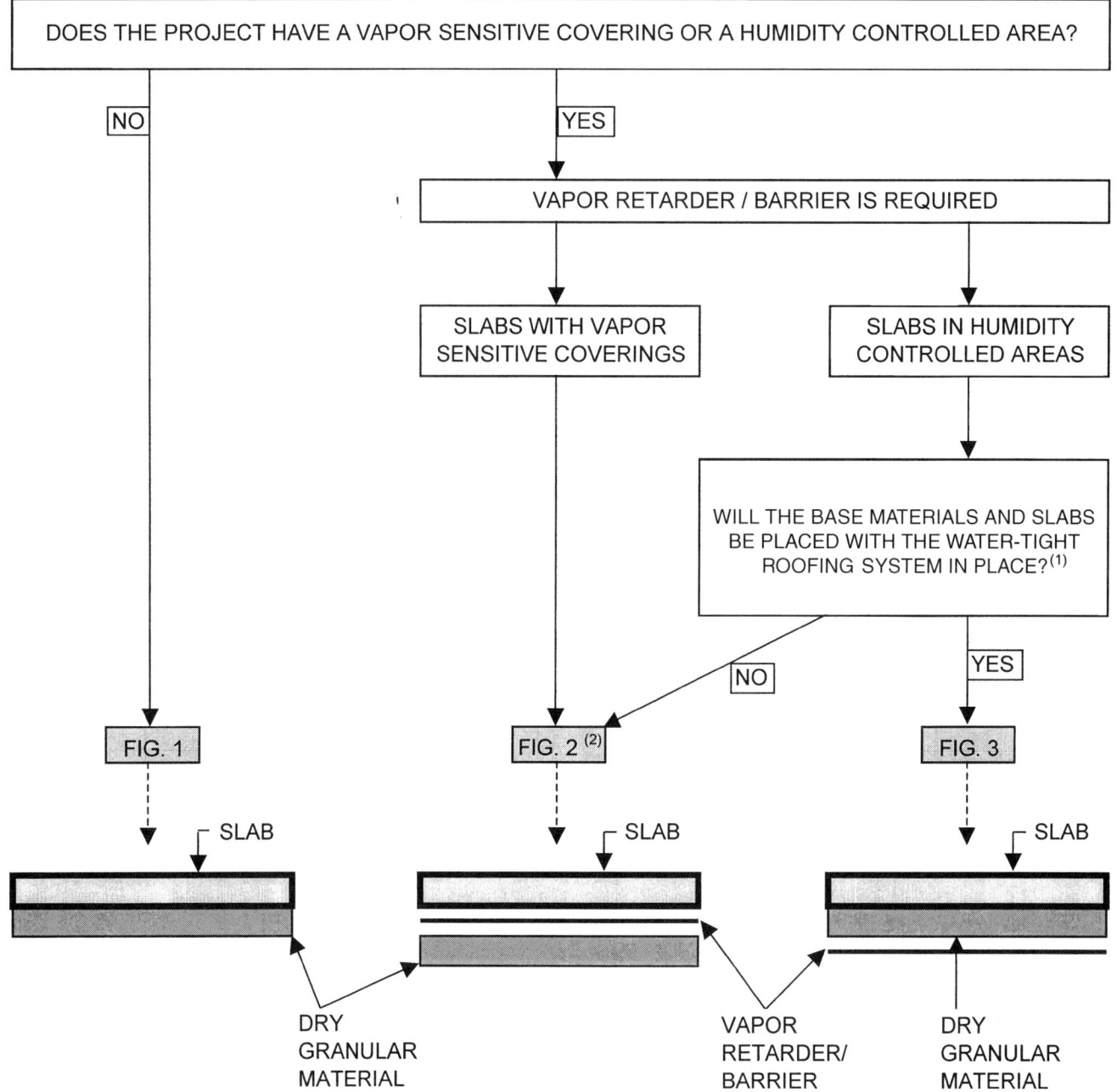

Fig. 3.1—Decision flow chart to determine if a vapor retarder/barrier is required and where it is to be placed.

tensioning tendons. Combinations of various forms of reinforcement have proved successful.

Normally, the amount of reinforcement used in nonstructural slabs is too small to have a significant influence on restraining movement resulting from volume changes. Refer to Section 3.2.5 for a detailed discussion of the relationship between joint spacing and amount of reinforcement.

Temperature and shrinkage cracks in unreinforced slabs-on-ground originate at the surface of the slab and are wider at the surface, narrowing with depth. For maximum effectiveness, temperature and shrinkage reinforcement in slabs-on-ground should be positioned in the upper third of the slab thickness. The Wire Reinforcement Institute recommends that welded wire reinforcement be placed 2 in. (50 mm) below the slab surface or within the upper third of slab thickness, whichever is closer to the surface (CRSI 2001; Snell 1997). Reinforcement should extend to within 2 in. (50 mm) of the slab side edge.

Deformed reinforcing steel or post-tensioning tendons should be supported and tied together sufficiently to minimize movement during concrete placing and finishing operations. Chairs with sand plates or precast-concrete bar supports are generally considered to be the most effective method of providing the required support. When precast-concrete bar supports are used, they should be at least 4 in. (100 mm) square at the base, have a compressive strength at least equal to the specified compressive strength of the concrete being placed, and be thick enough to support reinforcing steel or post-tensioning tendons at the proper elevation while maintaining minimum concrete cover requirements.

When welded wire reinforcement is used, its larger flexibility dictates that the contractor pay close attention to establishing and maintaining adequate support of the reinforcement during the concrete placing operations. Welded wire reinforcement should not be placed on the ground and pulled up after placement of the concrete, nor should the mats be walked in after placing the concrete. Proper support spacing is necessary to maintain welded wire reinforcement at the proper elevation; supports should be close enough that the welded wire reinforcement cannot be forced out of location by construction foot traffic. Support spacing can be increased when heavier gage wires or a double mat of small gage wires is used.

Reinforcing bars or welded wire reinforcement should be discontinued at any joints where the intent of the designer is to let the joint open and reduce the possibility of shrinkage and temperature cracks in an adjacent panel. Where the reinforcement is continued through the joint, cracks are likely to occur in adjacent panels because of restraint at the joint (WRI/CRSI 1991). When used in sufficient quantity, reinforcement will hold out-of-joint cracks tightly closed. Some designers prefer partial discontinuance of the reinforcement at contraction joints to obtain some load-transfer capacity without the use of dowel baskets. Refer to Section 3.2.7.

3.2.4.1 *Steel fibers*—In some installations, steel fibers specifically designed for such use can be used with or without conventional mild steel shrinkage and temperature reinforcement in slab-on-ground floors. As in the case of conventional reinforcement, steel fibers will not prevent cracking of the concrete. Use of steel fibers through the contraction joints reduces the width of joint openings and that increases the likelihood of cracking occurring between joints. The crack width, however, should remain narrow and, in most cases, there are nondetectible microcracks providing sufficient quantities of fibers used for the given slab joint spacing and thickness, and subgrade conditions and concrete material shrinkage properties are taken into consideration.

3.2.4.2 *Synthetic fibers*—Polypropylene, polyethylene, nylon, and other synthetic fibers can help reduce segregation of the concrete mixture and formation of shrinkage cracks while the concrete is in the plastic state and during the first few hours of curing. As the modulus of elasticity of concrete increases with hardening of concrete, however, most synthetic fibers at typical dosage rates recommended by the fiber manufacturers will not provide sufficient restraint to inhibit cracking.

3.2.4.3 *Post-tensioning reinforcement*—The use of high-strength steel tendons as reinforcement instead of conventional mild steel temperature and shrinkage reinforcement allows the contractor to introduce a relatively high compressive stress in the concrete by means of post-tensioning. This compressive stress provides a balance for the crack-producing tensile stresses that develop as the concrete shrinks during the curing process. Stage stressing, or partial tensioning, of the slab on the day following placement can result in a significant reduction of shrinkage cracks. Construction loads on the concrete should be minimized until the slabs are fully stressed (PTI 1990; PTI 1996). For guidelines on installation details, contact a concrete floor specialty contractor who is thoroughly experienced with this type of installation.

3.2.4.4 *Causes of cracking over reinforcement*—Plastic settlement cracking over reinforcement is caused by inadequate consolidation of concrete, inadequate concrete cover over the reinforcement, use of large diameter bars (Dakhil, Cady, and Carrier 1975), higher temperature of reinforcing bars exposed to direct sunlight, higher-than-required slump in concrete, revibration of the concrete, inadequate curing of the concrete, or a combination of these items.

3.2.5 *Joint design*—Joints are used in slab-on-ground construction to limit the frequency and width of random cracks caused by volume changes and to reduce the magnitude of slab curling. Generally, if limiting the number of joints by increasing the joint spacing can be accomplished without increasing the number of random cracks, floor maintenance will be reduced. The layout of joints and joint details should be provided by the designer. If the joint layout is not provided, the contractor should submit a detailed joint layout and placing sequence for approval of the designer before proceeding.

As stated in ACI 360R, every effort should be made to isolate the slab from restraint that might be provided by any other element of the structure. Restraint from any source, whether internal or external, will increase the potential for random cracking.

Three types of joints are commonly used in concrete slabs-on-ground: isolation joints, contraction joints, and construction joints. Appropriate locations for isolation joints and contraction joints are shown in Fig. 3.2. With the designer's approval, construction joint and contraction joint details can be interchanged. Refer to ACI 360R for a detailed discussion of joints. Joints in topping slabs should be located directly over joints in the base slab.

3.2.5.1 *Isolation joints*—Isolation joints should be used wherever complete freedom of vertical and horizontal movement is required between the floor and adjoining building members. Isolation joints should be used at junctions with walls (not requiring lateral restraint from the slab), columns, equipment foundations, footings, or other points of restraint such as drains, manholes, sumps, and stairways. Isolation joints are formed by inserting preformed joint filler

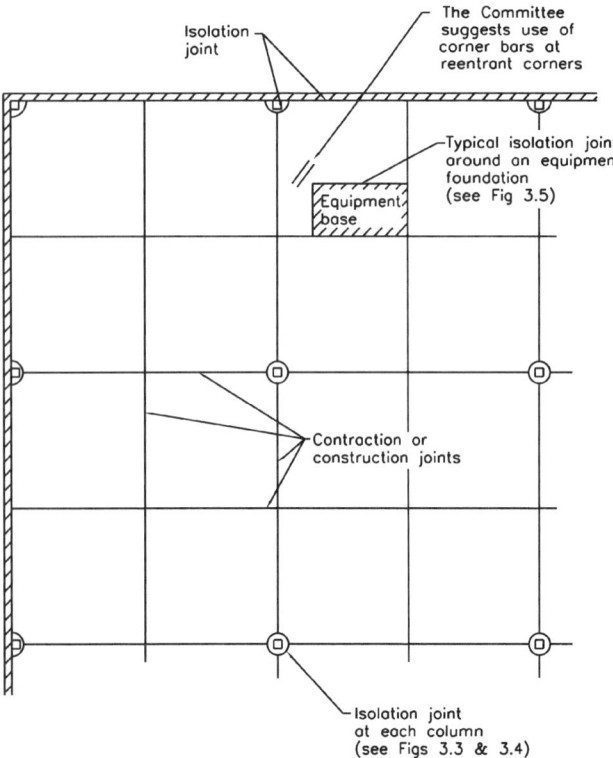

Fig. 3.2—Appropriate locations for joints.

between the floor and the adjacent member. The joint material should extend the full depth of the slab and not protrude above it. The joint filler will be objectionably visible where there are wet conditions, or hygienic or dust-control requirements. Two methods of producing a relatively uniform depth of joint sealant are as follows:

1) Score both sides of the preformed filler at the depth to be removed by using a saw. Insert the scored filler in the proper location and remove the top section after the concrete hardens by using a screwdriver or similar tool.

2) Cut a strip of wood equal to the desired depth of the joint sealant. Nail the wood strip to the preformed filler and install the assembly in the proper location. Remove the wood strip after the concrete has hardened.

Alternatively, a premolded joint filler with a removable top portion can be used. Refer to Fig. 3.3 and 3.4 for typical isolation joints around columns. Figure 3.5 shows an isolation joint at an equipment foundation.

Isolation joints for slabs using shrinkage-compensating concrete should be dealt with as recommended in ACI 223.

3.2.5.2 *Construction joints*—Construction joints are placed in a slab to define the extent of the individual concrete placements, generally in conformity with a predetermined joint layout. If concreting is ever interrupted long enough for the placed concrete to harden, a construction joint should be used. If possible, construction joints should be located 5 ft (1.5 m) or more from any other joint to which they are parallel.

In areas not subjected to traffic, a butt joint is usually adequate. In areas subjected to hard-wheeled traffic, heavy loadings, or both, joints with dowels are recommended (Fig. 3.6). Refer to Section 3.2.7 for a detailed discussion on

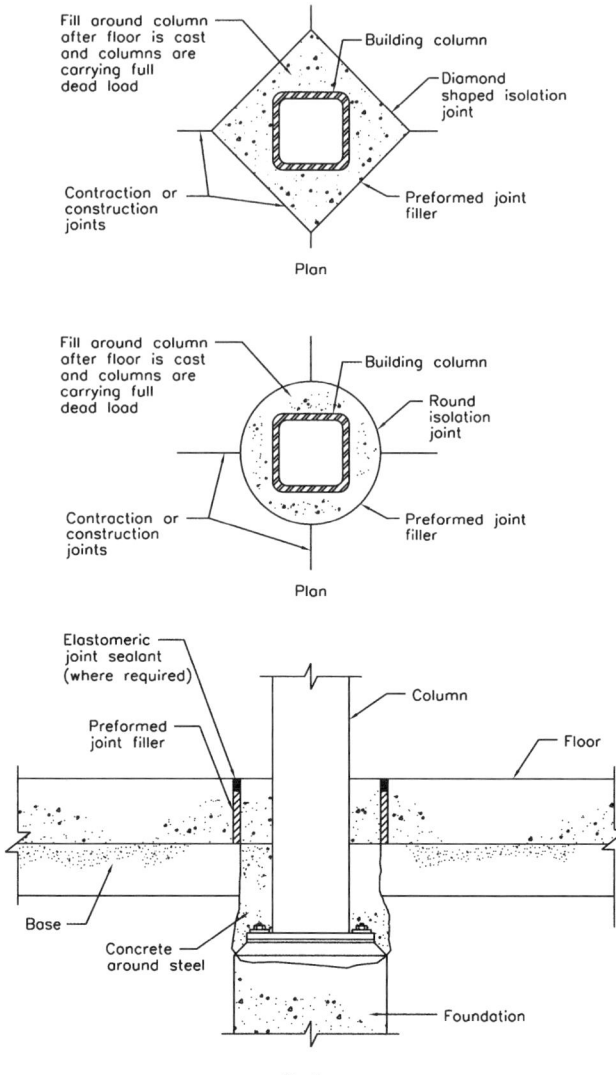

Fig. 3.3—Typical isolation joints at tube columns.

dowel joints. Keyed joints are not recommended where load transfer is required because the two sides of the keyway lose contact when the joint opens due to drying shrinkage (Section 3.2.7).

3.2.5.3 *Contraction joints*—Contraction joints are usually located on column lines with intermediate joints located at equal spaces between column lines as shown in Fig. 3.2. The following factors are normally considered when selecting spacing of contraction joints:
- Method of slab design (ACI 360R);
- Thickness of slab;
- Type, amount, and location of reinforcement;
- Shrinkage potential of the concrete (cement type and quantity; aggregate size, quantity, and quality; w/cm; type of admixtures; and concrete temperature);
- Base friction;
- Floor slab restraints;
- Layout of foundations, racks, pits, equipment pads, trenches, and similar floor discontinuities;
- Environmental factors such as temperature, wind, and humidity; and

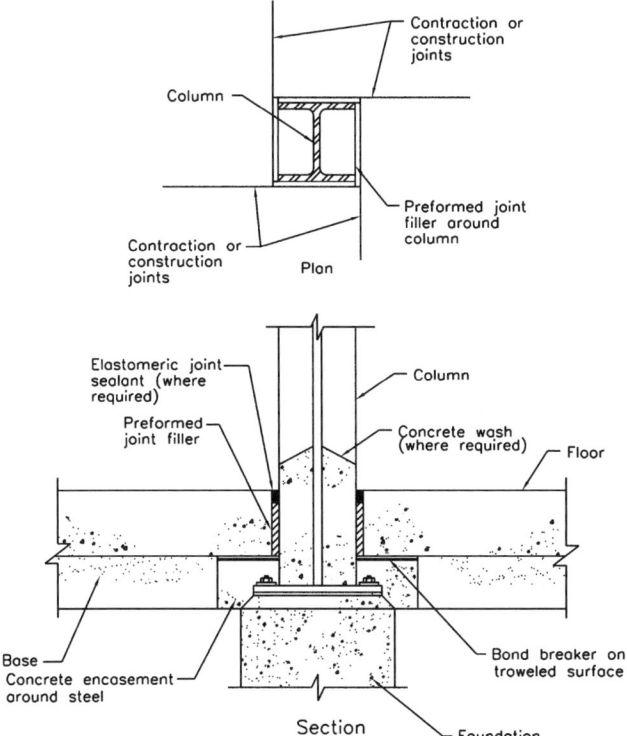

Fig. 3.4—*Typical isolation joint at wide flange column.*

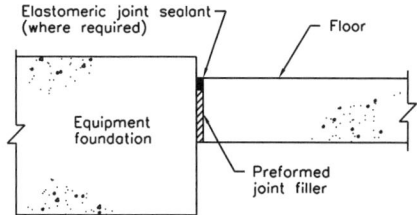

Fig. 3.5—*Typical isolation joint around an equipment foundation.*

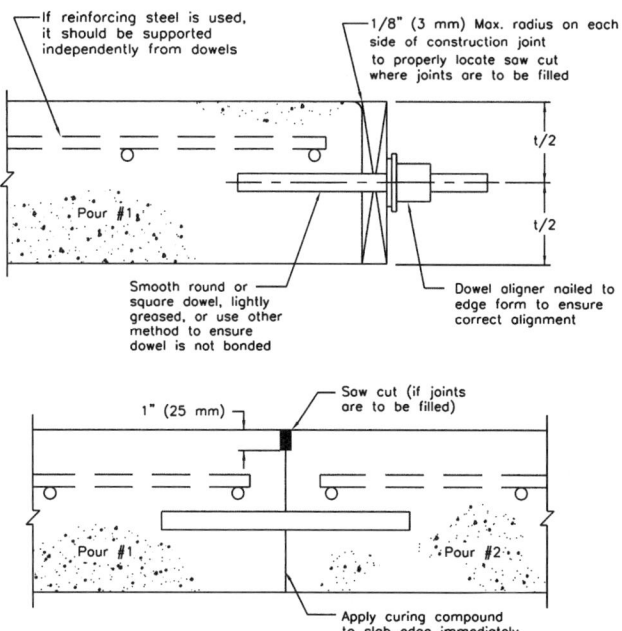

Fig. 3.6—*Typical doweled construction joint.*

- Methods and quality of concrete curing.

As previously indicated, establishing slab joint spacing, thickness, and reinforcement requirements is the responsibility of the designer. The specified joint spacing will be a principal factor dictating both the amount and the character of random cracking to be experienced, so joint spacing should always be carefully selected.

Curling of the floor surface at joints is a normal consequence of volume change resulting from differential moisture loss from concrete slab to the surrounding environment. This distortion can result in conflict with respect to installation of some floor coverings in the months after concrete placement. Current national standards for ceramic tile and wood flooring, such as gymnasium floors, are two instances that require the concrete slab surface to comply with stringent surface tolerances that cannot be met under typical slab curling behavior. The designer should consider the requirements for successful installation of floor coverings contained in Division 9 of the project specifications when performing the concrete slab design (ACI 360R).

For unreinforced, plain concrete slabs, joint spacings of 24 to 36 times the slab thickness, up to a maximum spacing of 18 ft (5.5 m), have produced acceptable results. Some random cracking should be expected; a reasonable level might be random visible cracks to occur in 0 to 3% of the floor slab panels formed by saw-cutting, construction joints, or a combination of both. If slab curl is of greater concern than usual, joint spacing, mixture proportion, and joint details should be carefully analyzed.

Joint spacing in nominally reinforced slabs (approximately 0.2% steel placed within 2 in. [50 mm] of the top of the slab) can be increased somewhat beyond that recommended for unreinforced, plain concrete slabs, but the incidence of random cracking and curling will increase. Reinforcement will not prevent cracking. If the reinforcement is properly sized and located, cracks that do occur should remain tightly closed.

Contraction joints can be reduced or eliminated in slabs reinforced with at least 0.5% continuous reinforcing steel placed within 2 in. (50 mm) of the top of the slab or upper one-third of slab thickness, whichever is closer to the slab top surface. This will typically produce a larger number of closely spaced fine cracks throughout the slab.

Joints in either direction can be reduced or eliminated by post-tensioning that introduces a net compressive force in the slab after all tensioning losses.

The number of joints can also be reduced with the use of shrinkage-compensating concrete; however, the recommendations of ACI 223 should be carefully followed.

Contraction joints should be continuous, not staggered or offset. The aspect ratio of slab panels that are unreinforced, reinforced only for shrinkage and temperature, or made with shrinkage-compensating concrete should be a maximum of 1.5 to 1; however, a ratio of 1 to 1 is preferred. L- and T-shaped panels should be avoided. Figure 3.7 shows various types of contraction joints. Floors around loading docks have a tendency to crack due to their configuration and restraints.

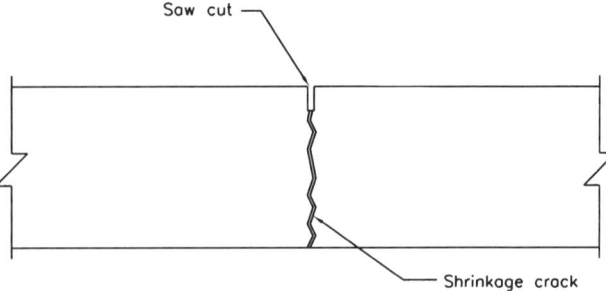

Fig. 3.7—Saw-cut contraction joint.

Figure 3.8 shows two methods that can be used to minimize slab cracking at reentrant corners of loading docks.

Plastic or metal inserts are not recommended for constructing or forming a contraction joint in any exposed floor surface that will be subjected to wheeled traffic.

3.2.5.4 *Saw cutting joints*—Contraction joints in industrial and commercial floors are usually formed by sawing a continuous slot in the slab to result in a weakened plane, below which a crack will form (Fig. 3.7). Further details on saw cutting of joints are given in Section 8.3.12

3.2.6 *Joint filling*—Contraction and construction joints in floor areas subject to the hard wheels of material handling vehicle traffic should be filled with a semirigid filler to minimize wear and damage to joint edges. Construction joints should be saw-cut 1 in. (25 mm) deep before filling. Joints should be as narrow as possible to minimize damage due to wheels loads while still being wide enough to be properly filled.

Where wet conditions or hygienic requirements exist, joints should be sealed with an elastomeric liquid sealant or a preformed elastomeric device. If there is also industrial vehicular traffic in these areas, consideration should be given to strengthening the edge of the joint through alternative means.

Refer to Section 5.12 for a discussion of joint materials, Section 9.10 for installation of joint fillers, and ACI 504R for joint sealants.

3.2.7 *Load-transfer mechanisms*—Doweled construction and contraction joints (Fig. 3.6 and 3.9) are recommended when load transfer is required, unless a sufficient post-tensioning force is provided across the joint to transfer the shear. Dowels force the concrete sections on both sides of a joint to undergo approximately equal vertical displacements subjected to a load and help prevent damage to an exposed edge when the joint is subjected to vehicles with hard-wheels such as forklifts. Table 3.1 provides recommended dowel sizes and spacing for round, square, and rectangular dowels. For dowels to be effective, they should be smooth, aligned, and supported so they will remain parallel in both the horizontal and the vertical planes during the placing and finishing operation. All dowels should be sawn and not sheared. Properly aligned, smooth dowels allow the joint to open as concrete shrinks. Dowel baskets (Fig. 3.9 to 3.11) should be used to maintain alignment of dowels in contraction joints, and alignment devices similar to the one shown in Fig. 3.6 should be used when detailing the doweled construction joints. Dowels should be placed no closer than 12 in. (300 mm) from the intersection of any joints.

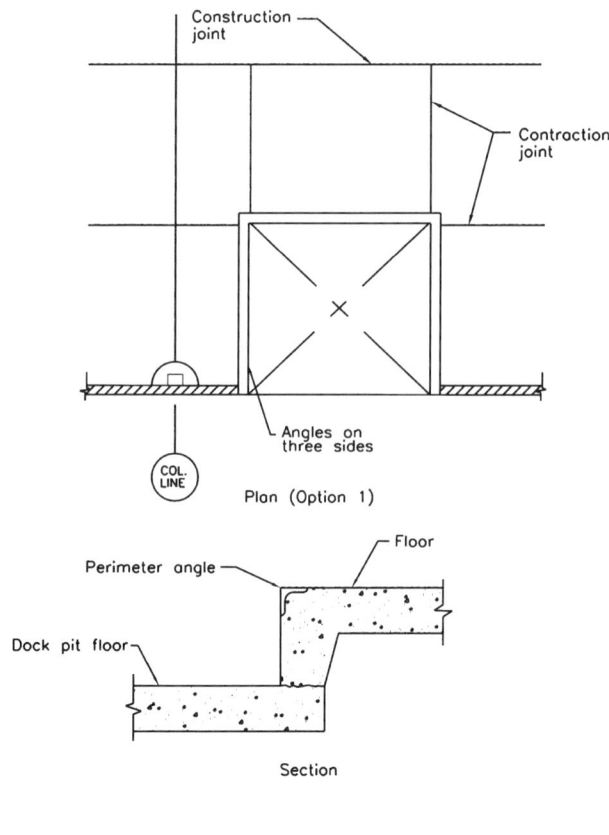

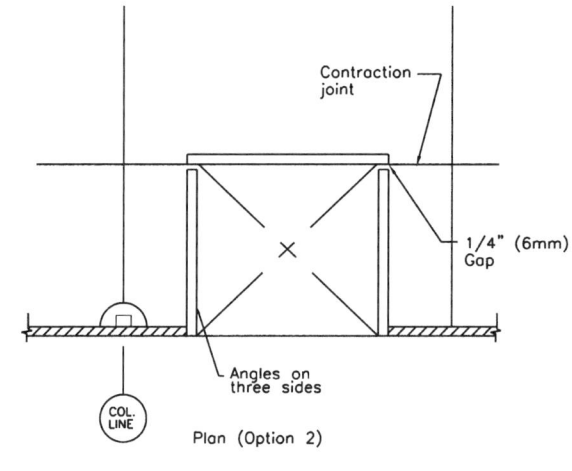

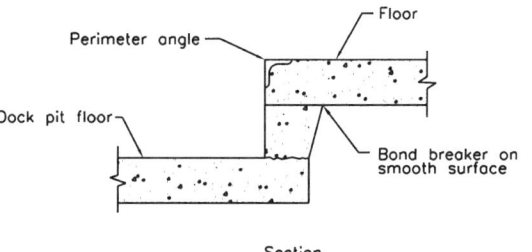

Fig. 3.8—Joint details at loading dock.

Diamond-shaped load plates (a square plate turned so that two corners line up with the joint, Fig. 3.12) can be used to replace dowels in construction joints (Walker and Holland 1998). The diamond shape allows the slab to move horizontally without restraint when the slab shrinkage opens the joint (Fig. 3.13). Table 3.2 provides the recommended size and

Table 3.1—Dowel size and spacing for round, square, and rectangular dowels (ACI Committee 325 1956)

Slab depth, in. (mm)	Dowel dimensions*, in. (mm)			Dowel spacing center-to-center, in. (mm)		
	Round	Square	Rectangular†	Round	Square	Rectangular
5 to 6 (125 to 150)	3/4 x 14 (19 x 350)	3/4 x 14 (19 x 350)	3/8 x 2 x 12 (10 x 50 x 300)	12 (300)	14 (350)	19 (475)
7 to 8 (175 to 200)	1 x 16 (25 x 400)	1 x 16 (25 x 400)	1/2 x 2-1/2 x 12 (12 x 60 x 300)	12 (300)	14 (350)	18 (450)
9 to 11 (225 to 275)	1-1/4 x 18 (30 x 450)	1-1/4 x 18 (30 x 450)	3/4 x 2-1/2 x 12 (19 x 60 x 300)	12 (300)	12 (300)	18 (450)

*Total dowel length includes allowance made for joint opening and minor errors in positioning dowels.
†Rectangular plates are typically used in contraction joints.
Notes: Table values based on a maximum joint opening of 0.20 in. (5 mm). Dowels must be carefully aligned and supported during concrete operations. Misaligned dowels cause cracking.

Table 3.2—Dowel size and spacing for diamond-shaped load plates

Slab depth, in. (mm)	Diamond load plate dimensions, in. (mm)	Diamond load plate spacing center-to-center, in. (mm)
5 to 6 (125 to 150)	1/4 x 4-1/2 x 4-1/2 (6 x 115 x 115)	18 (450)
7 to 8 (175 to 200)	3/8 x 4-1/2 x 4-1/2 (10 x 115 x 115)	18 (450)
9 to 11 (225 to 275)	3/4 x 4-1/2 x 4-1/2 (19 x 115 x 115)	20 (500)

Notes: Table values based on a maximum joint opening of 0.20 in. (5 mm). The construction tolerances required make it impractical to use diamond-shaped load plates in contraction joints.

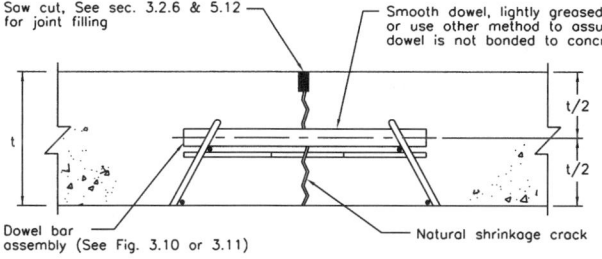

Fig. 3.9—Typical doweled contraction joint.

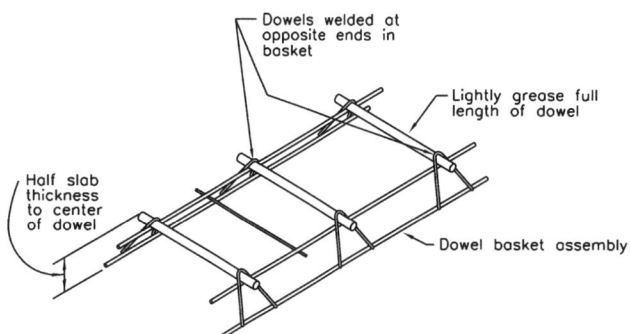

Fig. 3.10—Dowel basket assembly.

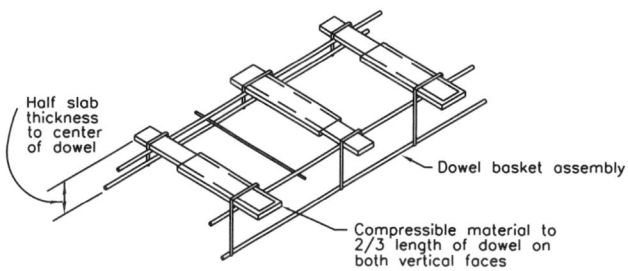

Fig. 3.11—Rectangular load plate basket assembly.

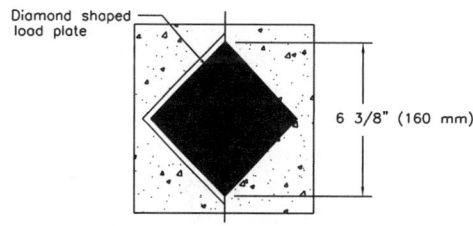

Fig. 3.12—Diamond-shaped load plate at construction joint.

spacing of diamond-shaped load plates. This type of load-transfer device can be placed within 6 in. (150 mm) of an intersection (Fig. 3.13). Square and rectangular dowels cushioned on the vertical sides by a compressible material also permit movement parallel and perpendicular to the joint (Fig. 3.14). These types of load-transfer devices are useful in other slab types where the joint should have load-transfer capability while allowing some differential movement in the direction of the joint, such as might be necessary in post-tensioned and shrinkage-compensating concrete slabs or in slabs with two-directional doweling (Schrader 1987). In saw-cut contraction joints, aggregate interlock should not be relied upon for effective load transfer for wheeled traffic if the expected joint width exceeds 0.035 in. (0.9 mm) (Colley and Humphrey 1967).

Deformed reinforcing bars should not be used across contraction joints or construction joints because they restrain joints from opening as the slab shrinks during drying. Continuation of a part of the slab reinforcing through contraction joints can provide some load-transfer capability without using dowels but significantly increases the probability of out-of-joint random cracking.

Round, square, and rectangular dowels for slab-on-ground installation should meet ASTM A 36. The diameter or cross-sectional area, length, shape, and specific location of dowels as well as the method of support should be specified by the designer. Refer to Table 3.1 and Fig. 3.9 to 3.14.

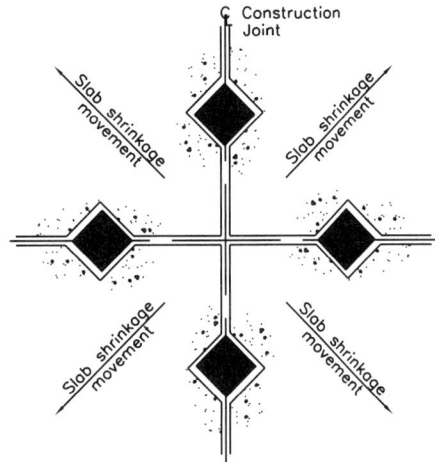

Fig. 3.13—Diamond-shaped load plates at slab corner.

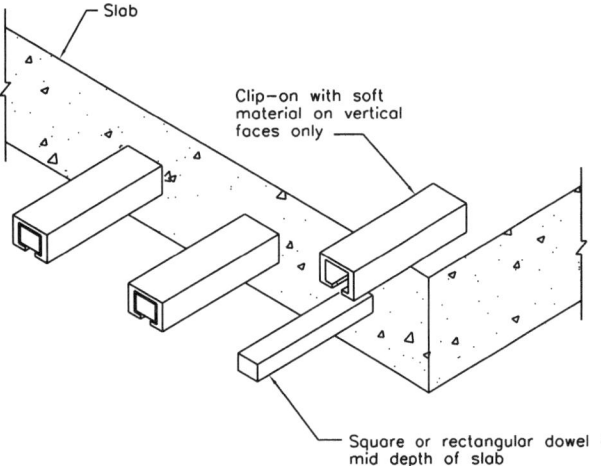

Fig. 3.14—Doweled joint detail for movement parallel and perpendicular to the joint.

Keyed joints are not recommended for load transfer in slabs-on-ground where heavy-wheeled traffic load is anticipated, because they do not provide effective load transfer. When the concrete shrinks, the keys and keyways do not retain contact and do not share the load between panels; this can eventually cause a breakdown of the concrete joint edges. For long post-tensioned floor strips and floors using shrinkage-compensating concrete with long joint spacing, care should be taken to accommodate significant slab movements. In most instances, post-tensioned slab joints are associated with a jacking gap. The filling of jacking gaps should be delayed as long as possible to accommodate shrinkage and creep (PTI 1990; PTI 2000). Where significant slab movement is expected, steel plating of the joint edges is recommended; for strengthening the edges (Fig. 3.15 and 3.16).

A doweled joint detail at a jacking gap in a post-tensioned slab (ASTM 1994; Spears and Panarese 1992) is shown in Fig. 3.16.

3.3—Suspended slabs

3.3.1 *Required design elements*—In addition to many of the items listed in Section 1.1.2, the following items specifically impact the construction of suspended slabs and should

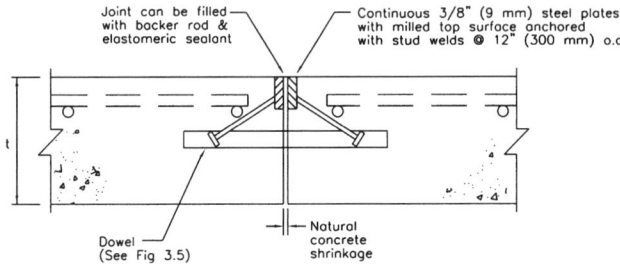

Fig. 3.15—Typical armored construction joint detail.

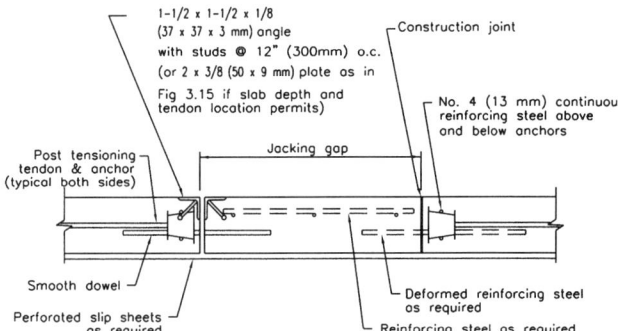

Fig. 3.16—Typical doweled joint detail for post-tensioned slab.

be included in the contract documents prepared by the designer:
- Frame geometry (member size and spacing);
- Reinforcement (type, size, location, and method of support);
- Shear connectors, if required;
- Construction joint location;
- Metal deck (type, depth, and gage), if required;
- Shoring, if required; and
- Tolerances (forms, structural steel, reinforcement, and concrete).

3.3.2 *Suspended slab types*—In general, suspended floor systems fall into four main categories:
1. Cast-in-place suspended floors;
2. Slabs with removable forms;
3. Slabs on metal decking; and
4. Topping slabs on precast concrete.

Design requirements for cast-in-place concrete suspended floor systems are covered by ACI 318 and ACI 421.1R. Refer to these documents to obtain design parameters for various cast-in-place systems. Slabs on metal decking and topping slabs on precast concrete are hybrid systems that involve design requirements established by ANSI, ASCE, The American Institute of Steel Construction, Precast/Prestressed Concrete Institute, and tolerances of ACI 117.

The levelness of suspended slabs depends on the accuracy of formwork and strikeoff but is further influenced (especially in the case of slabs on metal decking) by the behavior of the structural frame during and after completion of construction. Each type of structural frame behaves somewhat differently; it is important for the contractor to recognize these differences and plan accordingly.

The presence of camber in some floor members and the ACI 117 limitation on tolerances in slab thickness dictate that concrete be placed at a uniform thickness over the supporting steel. When placing slabs on metal decking, the contractor is cautioned that deflections of the structural steel members can vary from those anticipated by the designer. Achieving a level deflected surface can require increasing the slab thickness more than 3/8 in. (9.5 mm) in local areas. The committee recommends that concrete placement procedures and the basis for acceptance of the levelness of a completed concrete floor surface be established and agreed upon by key parties before beginning suspended floor construction (Tipping 1992).

3.3.3 *Slabs with removable forms*—Cast-in-place concrete construction can be either post-tensioned or conventionally reinforced. Both of these systems are supported during initial concrete placement, and they will deflect when supporting shores are removed.

Post-tensioned systems are normally used when larger spans are necessary or when the structural system is relatively shallow for the spans considered. Post-tensioned systems use high-strength steel tendons that are tensioned using a hydraulic jack designed for that purpose. The magnitude of floor slab deflection after supports are removed is less than that of comparable floors reinforced with conventional deformed reinforcing steel. At times, dead load deflection is entirely eliminated by the use of post-tensioning.

The magnitude of deflection in a conventionally mild steel reinforced floor system is dependent on a number of variables such as span, depth of structure, age at the time forms are stripped, concrete strength, and amount of reinforcement. In locations where the anticipated dead load deflection of a member is deemed excessive by the designer, an initial required camber, generally 1/2 in. (13 mm) or more, can be required. The amount of camber is determined by the designer based on an assessment of the loading conditions discussed. Ideally, the cambered floor system will deflect down to a level position after removal of the supporting shores.

3.3.4 *Slabs on carton forms*—Slabs on carton forms are a special application of slabs with removable forms (Tipping and North 1998). These slabs are necessary when slabs at ground level should remain independent of soil movement. Slabs on carton forms are most commonly used when soils at the building site are expansive clays subject to significant movement as a result of moisture variation. They provide a more economical construction solution than conventional framing systems, which require a crawl space to remove forms. The cardboard carton forms deteriorate in the months following construction, eventually leaving the desired void space below the slab and forcing the slab to span between supporting foundation elements.

Experience has shown that certain types of wet cardboard carton forms can fail locally under the weight of concrete and construction activities, with a resultant loss of part or all of the desired void space in the vicinity of the form failure. This failure can be instantaneous but can also occur 30 or 45 min after strikeoff. The latter type of failure, in addition to reducing desired void space, can result in a loss of local slab levelness. Forms that have been damaged by rain should be replaced or allowed to dry thoroughly, with their capacity verified, before placement of concrete.

3.3.5 *Slabs on metal deck*—Construction of slabs on metal deck involves the use of a concrete slab and a supporting platform consisting of structural steel and metal deck. The structural steel can be shored or unshored at the time of concrete placement, and the metal deck serves as a stay-in-place form for the concrete slab. This construction can be composite or noncomposite.

The supporting steel platform for slabs on metal deck is seldom level. Variation in elevations at which steel beams connect to columns and the presence of camber in some floor members combine to create variations in the initial elevation of steel members. Regardless of the initial levelness of the steel frame, unshored frames will deflect during concrete placement. These factors make the use of a laser or similar instrument impractical for the purpose of establishing a uniform elevation for strikeoff of the concrete surface of a slab on metal deck, unless the frame is preloaded to allow deflection to take place before strikeoff, and slab thickness is allowed to vary outside norms dictated by ACI 117. The presence of camber in some floor members and the ACI 117 limitation on variation in slab thickness generally dictates that concrete be placed to a uniform thickness over the supporting steel.

3.3.5.1 *Composite slabs on metal deck*—In composite construction, the composite section (concrete slab and steel beams) will work together to support any loads placed on the floor surface after the concrete has hardened. Composite behavior is normally developed through the use of shear connectors welded to the structural steel beam. These shear connectors physically connect the concrete slab to the beam and engage the concrete slab within a few feet of the steel beam; the resulting load-carrying element is configured much like a capital T. The steel beam forms the stem of the T, and the floor slab forms the cross-bar. Construction joints that are parallel to structural steel beams should be located far enough away to eliminate their impact on composite behavior. Questions about the location of construction joints should be referred to the designer on the project (Ryan 1997).

Unshored composite construction is the more common method used by designers because it is less expensive than shored construction. In unshored construction, the structural steel beams are sometimes cambered slightly during the fabrication process. This camber is intended to offset the anticipated deflection of that member under the weight of concrete. Ideally, after concrete has been placed and the system has deflected, the resulting floor surface will be level (Tipping 2002).

Shored composite concrete slabs on metal deck are similar to slabs with removable forms in that both are supported until the concrete has been placed and reaches the required strength. Structural steel floor framing members for shored composite slabs on metal deck are usually lighter and have less camber than those used for unshored construction with similar column spacings and floor loadings. One major concern with shored composite construction is the tendency for cracks wider than 1/8 in. (3 mm) to form in the concrete slab when the supporting shores are removed. These cracks do not normally impair the structural capacity of the floor but can become a severe aesthetic problem. The contractor is cautioned that this issue and any measures taken by the designer to avoid the formation of this type of crack should be addressed to the satisfaction of key parties before beginning suspended floor construction.

3.3.5.2 *Noncomposite slabs on metal deck*—In noncomposite construction, the slab and supporting structural steel work independently to support loads imposed after hardening of the concrete slab.

3.3.6 *Topping slabs on precast concrete*—A cast-in-place concrete topping on precast-prestressed concrete units involves the use of precast elements as a combination form and load-carrying element for the floor system. The cast-in-place portion of the system consists of a topping of some specified thickness placed on top of the precast units. The topping can be composite or noncomposite. In either case, added deflection of precast units under the weight of the topping slab is normally minor, so the finished surface will tend to follow the surface topography established by the supporting precast units. The camber in precast members, if they are prestressed, can change with time as a result of concrete creep. Depending on the length of time between casting of precast units and erection, this potential variation in camber of similar members can create significant challenges for the contractor. Care should be taken in the scheduling of such operations to minimize the potential impact of these variations. Precast members are less flexible and adaptable to changes or modifications that can be required on the jobsite than are the previously discussed systems.

3.3.7 *Reinforcement*—For cast-in-place concrete suspended slabs, reinforcing steel location varies as dictated by the contract documents. Post-tensioning reinforcement, when used, is enclosed in a plastic or metal sleeve and is tensioned by a hydraulic jack after the concrete reaches sufficient compressive strength. Elongation and subsequent anchoring of the ends of post-tensioning tendons results in the transfer of compressive force to the concrete (PTI 1990).

For slabs on metal deck, reinforcement is normally provided by deformed reinforcing steel, welded wire reinforcement, or a combination thereof.

3.3.8 *Construction joints*—The designer should provide criteria for location of construction joints in suspended slabs. The following is a general discussion of criteria that can influence these decisions.

3.3.8.1 *Slabs on removable forms*—Construction joints can introduce weak vertical planes in an otherwise monolithic concrete member, so they should be located where shear stresses are low. Under most gravity load conditions, shear stresses in flexural members are low in the middle of the span. ACI 318 requires that construction joints in floors be located within the middle third of spans of slabs, beams, and primary beams. Joints in girders should be offset a minimum distance of two times the width of any intersecting beams.

3.3.8.2 *Composite slabs on metal deck*—An important consideration when deciding on the location of construction joints in composite slabs on metal deck is that the joint location can influence deflection of the floor framing near the joint. A composite member (steel beam and hardened concrete slab working together) is stiffer and, therefore, deflects less than a noncomposite member (steel beam acting alone). Most composite slabs on metal deck are placed on an unshored structural steel floor frame. Often, structural steel members have initial camber to offset anticipated noncomposite deflection resulting from concrete placement. After hardening of the concrete, however, the composite member deflects much less than a comparable noncomposite beam or primary beam.

Following are general guidelines for deciding on the location of construction joints in composite slabs on metal deck:

1. Construction joints that parallel secondary structural steel beams should normally be placed near the midspan of the slab between beams;

2. Construction joints that parallel primary structural steel beams and cross secondary structural steel beams should be placed near the primary beam. The primary structural steel beam should not be included in the initial placement. It is important to place the construction joint far enough away from the primary beam to allow sufficient distance for development of the primary beam flange width. Placing the construction joint a distance of 4 ft (1.2 m) from the primary beam is usually sufficient for this purpose. This construction joint location allows nearly the full dead load from concrete placement to be applied to secondary beams that are included in the initial concrete placement. The primary beam should generally be included in the second placement at the construction joint. This will allow the primary beam to deflect completely before concrete at the primary beam hardens; and

3. Construction joints that cross primary structural steel beams should be placed near a support at one end of the primary beam. This will allow the beam to deflect completely before concrete at the beam hardens.

3.3.8.3 *Noncomposite slabs on metal deck*—The placement of construction joints in noncomposite slabs on metal deck should follow the same general guidelines discussed for slabs on removable forms in Section 3.3.8.1.

3.3.8.4 *Topping slabs on precast concrete*—Construction joints in topping slabs on precast concrete should be placed over joints in the supporting precast concrete.

3.3.9 *Cracks in slabs on metal deck*—Cracks often develop in slabs on metal deck. These cracks can result from drying shrinkage and thermal contraction or variations in flexibility of the supporting structural steel and metal deck. In a composite floor framing system, primary beams are the stiffest elements and generally deflect less than secondary beams. The most flexible part of the floor framing assembly is the metal deck, which is often designed for strength rather than for flexibility consideration.

Vibration as a result of power floating and power troweling operations can produce cracking over the structural steel beams during concrete finishing operations if the metal deck is flexible. As the concrete cures and shrinks, these cracks will open wide if not restrained by reinforcing steel, usually welded wire reinforcement, located near the top surface of the slab.

3.4—Miscellaneous details

3.4.1 *Heating ducts*—Heating ducts embedded in a concrete slab can be of metal, rigid plastic, or wax-impregnated cardboard. Ducts with waterproof joints are recommended. When metal ducts are used, calcium chloride should not be used in the concrete. Refer to Section 5.7.3 for a discussion on chlorides in concrete and Section 4.5.2 for installation of heating ducts.

3.4.2 *Edge insulation*—Edge insulation for slabs-on-ground is desirable in most heated buildings. The insulation should be in accordance with ASHRAE 90.1. It should not absorb moisture and should be resistant to fungus, rot, and insect damage; it should not be easily compressed.

Insulation should preferably be placed vertically on the inside of the foundation. It can also be placed in an L-shape configuration adjacent to the inside of the foundation and under the edge of the slab. If the L-shape configuration is used, the installation should extend horizontally under the slab a total distance of 24 in. (600 mm).

3.4.3 *Radiant heating: piped liquids*—Slabs can be heated by circulating heated liquids through embedded piping. Ferrous, copper, or plastic pipe is generally used with approximately 2 in. (50 mm) of concrete cover (not less than 1 in. [25 mm]) under the pipe and with 2 to 3 in. (50 to 75 mm) of concrete cover over the pipe. The slab is usually monolithic and the concrete is placed around the piping, which is fixed in place. Two-course slab construction has also been used, wherein the pipe is laid, connected, and pressure tested for tightness on a hardened concrete base course. Too often, however, the resulting cold joint is a source of distress during the service life.

Insulating concrete made with vermiculite or perlite aggregate or cellular foam concrete can be used as a subfloor. The piping should not rest directly on this or any other base material. Supports for piping during concreting should be inorganic and nonabsorbent; precast concrete bar supports (Section 3.2.4) are preferred to random lengths of pipe for use as supports and spacers. Wood, brick, or fragments of concrete or concrete masonry should not be used.

Sloping of the slab, where possible, can simplify sloping of the pipe. Reinforcement, such as welded wire reinforcement, should be used in the concrete over the piping. Where pipe passes through a contraction or construction joint, a provision should be made for possible movement across the joint. The piping should also be protected from possible corrosion induced by chemicals entering the joint. The piping should be pressure-tested before placing concrete, and air pressure (not water pressure) should be maintained in the pipe during concreting operations. After concreting, the slab should not be heated until curing is complete. The building owner should be warned to warm the slabs gradually using lukewarm liquid in the system to prevent cracking of the cold concrete.

3.4.4 *Radiant heating: electrical*—In some electrical radiant heating systems, insulated electrical cables are laid singly in place within the concrete or fastened together on transverse straps to form a mat. One system employs cable fastened to galvanized wire sheets or hardware cloth. The cables are embedded 1 to 3 in. (25 to 75 mm) below the concrete surface, depending on their size and operating temperature. In most systems the wires, cables, or mats are laid over a bottom course of unhardened concrete, and the top course is placed immediately over this assemblage with little lapse of time, thus avoiding the creation of a horizontal cold joint (ASHVE 1955).

Calcium chloride should not be used where copper wiring is embedded in the concrete; damage to insulation and subsequent contact between the exposed wiring and reinforcing steel will cause corrosion. If admixtures are used, their chloride contents should comply with the limits recommended by ACI 222R.

3.4.5 *Snow-melting*—Systems for melting snow and ice can be used in loading platforms or floor areas subjected to snow and ice. The concrete should be air-entrained for freezing-and-thawing resistance. Concrete surfaces should have a slope not less than 1/4 in./ft (20 mm/m) to prevent puddles from collecting. Piping systems should contain a suitable liquid heat-transfer medium that does not freeze at the lowest temperature anticipated. Calcium chloride should not be used (Section 5.7.3). Experience has shown that these snow-melting piping systems demand high energy consumption while displaying a high potential for failure and thermal cracking. The most successful applications appear to have been at parking garage entrances.

Some electrical systems are in use. These internally heated snow-melting systems have not been totally satisfactory.

3.4.6 *Pipe and conduit*—Water pipe and electrical conduit, if embedded in the floor, should have at least 1-1/2 in. (38 mm) of concrete cover on both the top and bottom.

3.4.7 *Slab embedments in harsh environments*—Care should be exercised in using heating, snow-melting, water, or electrical systems embedded in slabs exposed to harsh environments such as parking garages in northern climates and marine structures. If not properly embedded, systems can accelerate deterioration by increasing seepage of saltwater through the slab or by forming electrical corrosion circuits with reinforcing steel. If concrete deterioration occurs, the continuity and effective functioning of embedded systems are invariably disrupted.

CHAPTER 4—SITE PREPARATION AND PLACING ENVIRONMENT

4.1—Soil-support system preparation

The soil-support system should be well drained and provide adequate and uniform load-bearing support.

The ability of a slab to resist loads depends on the integrity of both the slab and full soil-support system. As a result, it is essential that the full soil-support system be tested or thoroughly evaluated before the slab is placed upon it (Ringo 1958).

The in-place density of the subgrade, subbase (if used), and base should be at least the minimum required by the

Fig. 4.1—Proof-rolling by loaded ready mix truck.

specifications, and the base should be free of frost before concrete placing begins and able to support construction traffic such as loaded truck mixers (Fig. 4.1).

The base should normally be dry at the time of concreting. If protection from the sun and wind cannot be provided as mentioned in Section 4.6 or if the concrete is placed in hot, dry conditions, the base should be lightly dampened with water in advance of concreting. There should be no free water standing on the base, nor should there be any muddy or soft spots when the concrete is placed (Section 4.1.5).

4.1.1 *Proof-rolling*—Proof-rolling is one of the most effective ways to determine if the full soil-support system is adequate to provide a uniformly stable and adequate bearing support during and after construction. If applicable, this process should be done after completion of the rough grading and, if required, can be repeated before the placement of the slab (Fig. 4.1).

Proof-rolling, observed and evaluated by the designer or the designer's representative, should be accomplished by a loaded tandem-axle dump truck, a loaded truck mixer, roller, or equivalent. In any case, multiple passes should be made using a pre-established grid pattern.

If rutting or pumping is evident at any time during the preparation of the subgrade, subbase, or base rolling, corrective action should be taken.

Rutting normally occurs when the surface of the base or subbase is wet and the underlying soils (subgrade) are firm. Pumping normally occurs when the surface of the base or subbase is dry and the underlying soils are wet. Any depression in the surface deeper than 1/2 in. (13 mm) should be repaired. Repair should include, but not be limited to, raking smooth or consolidating with suitable compaction equipment.

4.1.2 *Subgrade tolerance*—Industry practice is to plan and execute grading operations so that the final soil elevation is at the theoretical bottom of the slab-on-ground immediately before commencing concreting operations. Variations in grading equipment, subgrade material, and construction methods will result in inevitable local departures from this theoretical elevation. Studies have shown that these departures commonly result in slabs that vary in thickness as much as 1-1/2 in. (38 mm) from that shown on the contract documents (Gustaferro 1989; Gustaferro and Tipping 2000).

The designer should explicitly address the slab thickness issue in mandatory language in the project specifications on occasions where a result other than that which would commonly result from industry practice is desired. Further, the issue of minimum allowable slab-on-ground thickness should be addressed in a preconstruction meeting to ensure that all parties are aware of the designer's expectations. The committee recommends the following rough-grade and fine-grade tolerances as a necessary component of this process.

The necessary grading of the subgrade, often referred to as rough grading, should conform to a tolerance of +0 in./ –1-1/2 in. (+0 mm/–38 mm). Compliance should be confirmed before removal of excavation equipment. A rod and level survey should be performed by a surveyor. Measurements should be taken at 20 ft (6 m) intervals in each of two perpendicular directions.

4.1.3 *Base tolerance*—Base tolerances, often referred to as fine grading, should conform to a tolerance of +0 in./–1 in. (+0 mm/–25 mm) for floor Classes 1 to 3 and +0 in./–3/4 in. (+0 mm/–19 mm) for floor Classes 4 to 9, when measured from bottom of slab elevation. Compliance with these fine-grade values should be based on the measurements of individual floor sections or placements. A rod and level survey should be performed; measurements should be taken at 20 ft (6 m) intervals in each direction.

4.1.4 *Base material*—Use of the proper materials is essential to achieve the tolerances suggested in Section 4.1.3 (Suprenant and Malisch 1999b). The base material should be a compactible, easy to trim, granular fill that will remain stable and support construction traffic. The tire of a loaded concrete truck mixer should not penetrate the surface more than 1/2 in. (13 mm) when driven across the base. The use of so-called cushion sand or clean sand with uniform particle size, such as concrete sand, meeting requirements of ASTM C 33, will not be adequate. This type of sand will be difficult, if not impossible, to compact and maintain until concrete placement is completed.

A clean, fine-graded material with at least 10 to 30% of particles passing a No. 100 (150 μm) sieve but not contaminated with clay, silt, or organic material is recommended. Manufactured sand from a rock-crushing operation works well; the irregular surfaces tend to interlock and stabilize the material when compacted. The material should have a uniform distribution of particle sizes ranging from No. 4 (4.75 mm) to the No. 200 (75 μm) sieve. Refer to ASTM C 33, Table 1, for limitation of deleterious material finer than No. 200 (75 μm) sieve. Unwashed size No. 10 (2 mm) per ASTM D 448 works well.

4.1.5 *Vapor barrier/vapor retarder*—If a vapor barrier or retarder is required to reduce the impact of moisture transmission from below the slab on moisture-sensitive floor finishes, adhesives, coatings, equipment, or environments, the decision whether to locate the material in direct contact with the slab or beneath a fill course should be made on a case-by-case basis. Each proposed installation should be independently evaluated as to the moisture-related sensitivity of subsequent floor finishes, project conditions, schedule, and the potential effects of slab curling and cracking.

When a fill course is used over the vapor barrier/retarder, it should be a minimum of 4 in. (100 mm) of trimmable, compactible, granular fill (not sand), a so-called crusher-run material. Usually graded from 1-1/2 to 2 in. (38 to 50 mm) down to rock dust is suitable. Following compaction, the surface can be choked off with a fine-grade material (Section 4.1.4) to reduce friction between the base material and the slab.

If it is not practical to install a crusher-run material, the vapor barrier/retarder should be covered with at least 3 in. (75 mm) of fine-graded material such as crusher fines or manufactured sand (Section 4.1.4). The granular fill and fine-graded material should have sufficient moisture content to be compactible but still be dry enough at the time of concrete placement to act as a blotter (Section 4.1).

If a fill course is used, it should be protected from taking on additional water from sources such as rain, curing, cutting, or cleaning. Wet fill courses have been directly linked to a significant lengthening of the time required for a slab to reach an acceptable level of dryness for floor covering applications. If a vapor barrier/retarder is to be placed over a rough granular fill, a thin layer of approximately 1/2 in. (13 mm) of fine-graded material should be rolled or compacted over the fill before installation of the vapor barrier/retarder to reduce the possibility of puncture (Section 4.1.4). Vapor barriers/retarders should be overlapped 6 in. (150 mm) at the joints and carefully fitted around service openings. See Section 3.2.3 for more information on vapor barriers/retarders for slabs-on-ground (Suprenant and Malisch 1998A).

4.2—Suspended slabs

Before concrete placement, bottom-of-slab elevation and the elevation of reinforcing steel and any embedments should be confirmed. Forms that are too high often result in reinforcements being above the desired elevation for the slab surface. Screed rails or guides should be set at elevations that will accommodate initial movement of the forms during concreting. Screed rails may also be set at elevations that will offset downward deflection of the structure following concrete placement (Section 3.3).

4.3—Bulkheads

Bulkheads can be wood or metal; they should be placed at the proper elevation with stakes and necessary support required to keep the bulkheads straight, true, and firm during the entire placing and finishing procedure. Keyways are not recommended. If specified, however, small wood or metal keys should be attached to the inside of the form.

When it is necessary to set bulkheads on insulation material, such as in cold storage or freezer rooms, extra attention should be given to keeping the forms secure during the placing and finishing process. The insulation material should not be punctured by stakes or pins. It may be necessary to place sand bags on top of form supports to ensure stability during concrete placement.

Circular or square forms can be used to isolate the columns. Square forms should be rotated 45 degrees (Fig. 3.3) or installed in a pinwheel configuration as indicated in Fig. 3.4. Walls, footings, and other elements of the structure should be isolated from the floors. Asphalt-impregnated sheet or other suitable preformed compressible joint material (ACI 504R) should be used. These joint materials should never be used as freestanding forms at construction joints or column block outs but should be installed after the original forms have been removed. After removal of forms around columns, preformed joint materials should be placed at the joint to the level of the floor surface and the intervening area concreted and finished. These preformed joint materials can be placed at the proper elevation to serve as screed guides during the concreting operations. The preformed joint material should be of the type specified and should conform to one of the following specifications, depending on the conditions of its use: ASTM D 994, D 1751, or D 1752.

4.4—Setting screed guides

The screed guides can be 2 in.-thick (50 mm) lumber, pieces of pipe, T-bars, or rails, the tops of which are set to the finished concrete grade without changing the design elevation of the reinforcing steel. Each type should have a tight-radius edge. If the wet-screed approach is used to establish concrete grade, the finished floor elevation for a slab-on-ground may be laid out by driving removable grade stakes into the subgrade at predetermined intervals that are appropriate for the width of placement strips being installed. The tops of these stakes should be set to the required concrete grade.

4.4.1 *Establishing grades for adequate drainage on the slab surface*—When positive drainage is desired, the forms and screed guides should be set to provide for a minimum slope of 1/4 in./ft (20 mm/m) to prevent ponding. Positive drainage should always be provided for exterior slabs and can be desirable for some interior slabs.

4.5—Installation of auxiliary materials

4.5.1 *Edge insulation*—Insulation (Section 3.4.2) should preferably be placed vertically on the inside of the foundation. It can also be placed in an inverted L-shape configuration adjacent to the foundation and under the edge of the slab.

4.5.2 *Heating ducts*—Metal, rigid plastic, or wax-impregnated cardboard ducts with watertight joints are recommended; they can be set on a sand-leveling bed and back-filled with sand to the underside of the slab. Precautions should be taken to ensure that the position of the ducts is not disturbed during concreting and that they are adequately protected from corrosion or deterioration.

If the ducts to be used are not waterproof, they should be completely encased in at least 2 in. (50 mm) of concrete to prevent the entrance of moisture.

4.6—Concrete placement conditions

When slabs are placed on ground, there should be no more than 30 °F (17 °C)—ideally 20 °F (11 °C)—difference between the temperature of the base and concrete at the time of placement.

Floor slab installations should be undertaken in a controlled environment where possible. Protection from the sun and wind is crucial to the placing and finishing process. The roof of the structure should be waterproof, and the walls should be completely up. The site should provide easy access for concrete trucks and other necessary materials and suppliers. The site should be have adequate light and ventilation. Temperatures inside the building should be maintained above 50 °F (10 °C) during placing, finishing, and curing the concrete. If heaters are required, they should be vented to the outside (Kauer and Freeman 1955). Salamanders or other open flame heaters that might cause carbonation of the concrete surface should not be used. When installation procedures are carried out each day under the same conditions, the resulting floors are significantly superior to those floors installed under varying or poor environmental conditions. Also, refer to Sections 9.5.1 and 9.5.2 for cold- and hot-weather considerations.

CHAPTER 5—MATERIALS
5.1—Introduction
Concrete produced in accordance with ASTM C 94 varies and produces concrete with different setting and finishing characteristics. These standards offer a wide window of acceptance (Bimel 1993). Therefore, the specific concrete mixture should be investigated before the preparation of mixture proportions for floors and slabs.

5.2—Concrete
Because minimizing shrinkage is of prime importance, special attention should be given to selecting the best possible concrete mixture proportions. The shrinkage characteristics of a concrete mixture can be determined by ASTM C 157. Should it be necessary to determine if a proposed concrete mixture has other than normal shrinkage (ACI 209R), the proposed concrete mixture should be compared to the specified or a reference concrete mixture using ASTM C 157. It is essential that the concrete used in these tests be made with the same materials that will be used in the actual construction.

In addition to meeting the specified compressive strength based on standard laboratory samples, a concrete mixture proportion for use in a floor slab should, if specified, also meet the flexural-strength requirements and the limits on *w/cm* for durability, if applicable (Section 6.2.3). The portland cement content and the content of other cementitious products, if used, should be sufficient to permit satisfactory finishability under the anticipated field conditions. The setting characteristics of the concrete should be relatively predictable. The concrete should not experience excessive retardation, differential set time, or surface crusting difficulties under the conditions of temperature and humidity expected on the project. Some admixture-cement combinations can cause these difficulties, particularly when multiple admixtures are used. Because there is not a generally recognized procedure for establishing these performance characteristics, the committee recommends placement of a sample floor slab as indicated in Section 6.2.4. Floor concrete requirements differ from those of other concrete used in the structure. Project requirements should be reviewed thoroughly before mixture proportioning. If possible, the concrete contractor should have the opportunity to review the proposed mixture proportions and to prepare a sample placement to verify the workability, finishability, and setting time for the proposed usage.

5.3—Portland cement
5.3.1—Concrete floors can incorporate a variety of portland cements that meet ASTM C 150, C 595, C 845, and C 1157.

Of the four cements used in floors and slabs described in ASTM C 150, Type I is the most common, and it is used when the special properties of another type are not required. Type II is also for general use, especially when moderate sulfate resistance or moderate heat of hydration is desired. Type III is used when high early strength is desired. Type V is used when high sulfate resistance is required. When the aggregate to be used on the project is possibly susceptible to alkali-aggregate reaction, it is recommended that the maximum equivalent alkali limits of ASTM C 150 (Table 2 Optional Chemical Requirements) be specified if supplementary cementitious materials demonstrated to control alkali-silica reactivity, or alkali silica reaction-inhibiting admixtures, are not available. Refer to the appendix of ASTM C 33 for further information.

If air-entrained concrete is required, air-entrainment should be obtained with an admixture, rather than by using an air-entraining cement, allowing for better control of air content.

5.3.2 *Blended hydraulic cements*—Blended hydraulic cements are produced by intimately and uniformly blending two or more types of fine materials, such as portland cement, ground-granulated blast-furnace slag, fly ash and other pozzolans, hydrated lime, and preblended cement combinations of these materials.

There are six recognized classes of blended cements that conform with ASTM C 595: Type IS portland blast-furnace slag cement; Type IP and P portland-pozzolan cements; Type I (PM) pozzolan-modified portland cement; Type S slag cement; and Type I (SM) slag-modified portland cement. Types P and S, however, are normally not available for use in general concrete construction. The manufacturers of these cements should be contacted for information regarding the specific product and the effect its use will have on setting time, strength, water demand, and shrinkage of concrete proposed for the project under anticipated field conditions. Conformance to the requirements of ASTM C 595 does not impose sufficient restrictions on the cement to be used. If the 28-day design strength is achieved but shrinkage is excessive and retardation is significant, the cement may not be suitable for the project.

ASTM C 1157 is a performance specification that establishes requirements for six types of cement mirroring the attributes of ASTM C 150 and ASTM C 595 cement types.

For information on pozzolans used as cement replacements or cementitious additions, refer to Section 5.7.5.

5.3.3 *Expansive cements*—Types K, M, and S are expansive cements meeting ASTM C 845 specifications that are used in shrinkage-compensating concrete floors. Refer to ACI 223 for specific details on shrinkage-compensating concrete floors. Shrinkage-compensating concrete can also be made by adding an expansive component as discussed in Section 5.7.4. When a component is used, it is essential that the component manufacturers work with the concrete producer and testing laboratory to determine the rate and level of expansion that can be expected under anticipated job conditions.

5.4—Aggregates

Aggregates should conform to ASTM C 33 or to ASTM C 330. These specifications are satisfactory for most Class 1, 2, 3, 4, 5, and 6 floors. Additional limitations on grading and quality can be required for the surface courses of heavy-duty Class 7 and 8 floors.

Although these ASTM standards set guidelines for source materials, they do not establish combined gradation requirements for the aggregate used in concrete floors. Compliance with the aggregate gradations discussed in Section 5.4.3 will produce a desirable matrix while reducing water demand of the concrete mixture and reducing the amount of cement paste required to coat the aggregate (Shilstone 1990). ASTM C 33 limits coal and lignite to no more than 0.5% in fine or coarse aggregate and limits low specific gravity chert to no more than 5.0% in coarse aggregate. Although the concrete used may comply with this standard, some popouts are always possible. In fact, concrete containing as little as 0.2% or less coal, lignite, or low-density deleterious material may not be acceptable, as this quantity of those products can affect both the overall durability and appearance of the finished floor.

5.4.1 *Fine aggregate grading*—Although ASTM C 33 and C 330 are acceptable specifications, Table 5.1 contains preferred grading specifications for the toppings for Class 7 floors. The amount of material passing through the No. 50 and 100 sieves (300 and 150 µm) should be limited as indicated for heavy-duty floor toppings for Class 7. When fine aggregates contain minimum percentages of material passing the No. 50 and 100 sieves (300 and 150 µm), however, the likelihood of excessive bleeding is increased and limitations on water content of the mixture become increasingly important. Natural sand is preferred to manufactured sand; the gradation indicated in Table 5.1 will minimize water demand.

5.4.2 *Coarse aggregate grading*—The maximum size of coarse aggregate should not exceed 3/4 the minimum clear spacing of the reinforcing bars in structural floors, or 1/3 the thickness of nonreinforced slabs (ACI 318-02, Section 3.3.2). In general, natural aggregate larger than 1-1/2 in. (38 mm) or lightweight aggregate larger than 1 in. (25 mm) is not used. Although the use of large aggregate is generally desired for lower water demand and shrinkage reduction, it is important to recognize the overall gradation of all the aggregate (Section 5.1). When aggregate sizes larger than 1 in. (25 mm) are used, the coarse aggregate can be batched as two sizes to prevent segregation. Drying shrinkage can be minimized by the use of the largest practical-size coarse aggregate. If flexural strength is of primary concern, however, the use of smaller-size coarse aggregate can help achieve better uniformity in strength.

Table 5.1—Preferred grading of fine aggregates for floors

Sieve designations		Percent passing		
Standard	Alternative	Normalweight aggregate	Lightweight aggregate	Heavy-duty toppings, Class 7 floors
9.5 mm	3/8 in.	100	100	100
4.75 mm	No. 4	85 to 100	85 to 100	95 to 100
2.36 mm	No. 8	80 to 90	—	65 to 80
1.18 mm	No. 16	50 to 75	40 to 80	45 to 65
600 µm	No. 30	30 to 50	30 to 65	25 to 45
300 µm	No. 50	10 to 20	10 to 35	5 to 15
150 µm	No. 100	2 to 5	5 to 20	0 to 5

5.4.3 *Combined aggregate grading*—Gradations requiring between 8 and 18% for large top size aggregates such as 1-1/2 in. (38 mm) or 8 and 22% for smaller maximum-size aggregates such as 1 or 3/4 in. (25 or 19 mm) retained on each sieve below the top size and above the No. 100 (150 µm) sieve have proven satisfactory in reducing water demand while providing good workability. The ideal range for No. 30 and 50 (600 and 300 µm) sieves is 8 to 15% retained on each. Often, a third aggregate is required to achieve this gradation (Shilstone 1990). Typically, 0 to 4% retained on the top size sieve and 1.5 to 5.0% on the No. 100 (150 µm) sieve will be a well-graded mixture. This particle-size distribution is appropriate for round or cubically shaped particles in the No. 4 to 16 (4.75 to 1.18 mm) sieve sizes. If the available aggregates for these sizes are slivered, sharp, or elongated, 4 to 8% retained on any single sieve is a reasonable compromise. Mixture proportions should be adjusted whenever individual aggregate grading varies during the course of the work. Refer to Sections 6.3.2 and 6.3.4 for additional information.

Limitations in locally available material may require some deviations from the aforementioned optimum recommendations. The following limitations should always be imposed:

1. Do not permit the percent retained on two adjacent sieve sizes to fall below 5%;

2. Do not allow the percent retained on three adjacent sieve sizes to fall below 8%; and

3. When the percent retained on each of two adjacent sieve sizes is less than 8%, the total percent retained on either of these sieves and the adjacent outside sieve should be at least 13% (for example, if both the No. 4 and No. 8 [4.75 and 2.36 mm] sieves have 6% retained on each, then: 1) the total retained on the 3/8 in. and No. 4 [9.5 and 4.75 mm] sieves should be at least 13%, and 2) the total retained on the No. 8 and No. 16 [2.36 and 1.18 mm] sieves should be at least 13%.)

5.4.4 *Aggregate quality*—Compliance with ASTM C 33 and C 330 generally ensures aggregate of adequate quality, except where either chemical attack or abrasion in Class 7 and 8 floors is severe. See ACI 201.2R for a more complete discussion of precautions under these conditions. Sections 5.4.6 and 5.4.8 discuss special abrasion-resistant and nonslip aggregates, respectively. The guidelines of ACI 201.2R and

ASTM C 33 and its appendix should be followed where there is concern about the possibility of alkali-aggregate reaction.

5.4.5 *Special-purpose aggregates*—Decorative and nondecorative mineral aggregate and metallic hardeners are used to improve the properties of the slab surface. These materials, applied as dry shakes on top of the concrete, are floated and troweled into the floor surface to improve the abrasion resistance, impact resistance, achieve nonslip surfaces, or to obtain a decorative finish. In this document, the term dry-shake is applied to premixed materials, which may be mineral aggregate, metallic, or colored. The term embedded is a more generic term used where the material can be furnished in either premixed or bulk form. Trap rock and emery are two examples of materials that can be furnished in bulk form. These bulk materials should be blended with locally available portland cement and meet the requirements of ASTM C 150 or C 1157 before being introduced to the concrete surface.

5.4.6 *Wear-resistant aggregates*—Hard, wear-resistant aggregates, such as quartz, emery, and traprock, as well as malleable metallic hardeners, are frequently used as surface treatments (ASTM 1994). They are applied as dry shakes and finished into the surface of the floor to improve its abrasion and wear resistance.

Nonmetallic surface hardeners should be used on floors subjected to heavy frequent forklift or hard-wheeled traffic (Table 2.1). Metallic hardeners in sufficient quantity should be considered for use when heavy steel wheel or intense point impact loading is anticipated. Chloride-bearing admixtures should not be used in conjunction with a metallic floor hardener.

Mineral aggregate and metallic surface hardeners are factory premixed with specially selected portland cement and plasticizers. Some mineral aggregates can be supplied in bulk and mixed with cement on site. These aggregates, in properly graded sizes, can also be used in topping mixtures.

5.4.7 *Surface treatment for electrically conductive floors*—Concrete floors can be made electrically conductive by using specially prepared metallic hardeners (dry shakes). Electrically conductive floors are also required to be spark-resistant under abrasion or impact. For protection against abrasion sparks, care should be taken in the choice of aggregates. Because construction techniques for these floors are rather specialized, specific recommendations of the product manufacturer and designer should be followed (Boone et al. 1958).

The electrical resistance of such floors can be determined by reference to the appropriate specification of the Naval Facilities Engineering Command (NFEC 1984). A typical test for spark resistance under abrasion or impact is given in the aforementioned specification and the National Fire Protection Association, NFPA 99, specification. A factory premixed metallic surface hardener containing a conductive binder is commonly used for these floors. This hardener is floated and troweled into the surface of freshly placed concrete (Section 8.6).

Special conductive curing compounds should be used to cure these floors. Conductive floors should not be used in areas expected to be continuously moist.

5.4.8 *Slip-resistant aggregates*—Slip-resistant aggregates should be hard and nonpolishing. Fine aggregates are usually emery or a manufactured abrasive. The slip resistance of some aggregates can be improved by replacing the fines with those of a more slip-resistant aggregate. To improve slip resistance, extremely soft aggregates like vermiculite can be troweled into the surface of freshly placed concrete and then removed later by scrubbing after the concrete has hardened.

5.4.9 *Decorative aggregates*—Decorative aggregates can be of many minerals and colors. They should be sound, clean, nonreactive, and of consistent quality. The most common are quartz, marble, granite, and some ceramics. Rocks, shells, brass turnings or other brass pieces, and ball bearings have also been used. Shapes resembling spheres and cubes are preferable to flat or highly irregularly shaped pieces, which can become dislodged easily. It is usually preferable to have aggregate of only one sieve size.

5.5—Water

Mixing water should be potable. Nonpotable water can be used if 7- and 28-day strengths of 2 in. (50 mm) mortar cubes made with it are equal to at least 90% of the strengths of cubes made from similar mixtures using distilled water and tested in accordance with ASTM C 109 (Section 3.4, ACI 318-02). ACI 301 discusses mixing water, as do Steinour (1960) and others (Kosmatka, Kerkhoff, and Panarese 2002a). Also refer to AASHTO T 26.

5.6—Curing materials

ACI 308R lists many coverings and membrane-forming liquids that are acceptable for curing concrete floors. Because curing is so vital to good flatwork, the characteristics of curing materials suitable for flatwork are set forth here in great detail. Also refer to Chapter 9 for the purpose, methods, and length of curing.

5.6.1 *Wet burlap*—Wet burlap, plastic film, waterproof paper, and combination polyethylene/burlap sheets may all be generically referred to as reusable wet cure covers or blankets. All should meet requirements of AASHTO M 182. When water is used in the curing procedure, the difference in temperature between water and the concrete surface should not exceed 20 °F (11 °C).

If kept continually moist, burlap is an effective material for curing concrete surfaces. Old burlap from which the sizing has disappeared or has been removed is easier to wet than new burlap.

Care should be taken so the burlap used does not stain the concrete or come from sacks that once contained sugar; sugar retards the hardening of concrete and its presence could result in a soft surface. The requirements for burlap are described in AASHTO M 182. White, polyethylene-coated burlap is available; the polyethylene is helpful in keeping the burlap moist longer, but it makes rewetting more difficult. Refer to ASTM C 171.

5.6.2 *Plastic film, waterproof paper, or combination polyethylene/burlap sheets*—Plastic film, waterproof paper, or polyethylene/burlap sheets for curing should allow a moisture loss of no more than 1.8 oz./ft^2 (0.55 kg/m^2) in 72 h when tested according to ASTM C 156. Polyethylene plastic film with the same thickness and permeance used for vapor retarders below slabs-on-ground (Section 3.4) should be satisfactory. Waterproof paper should meet the requirements of ASTM C 171. These products should not be used on colored floors.

5.6.3 *Membrane-forming curing compounds*—Liquid membrane-forming curing compounds should meet the provisions of ASTM C 309, which describes the requirements for both clear and pigmented types. White or gray compounds are used for their good light reflection. Colored curing compounds are available for colored concrete. Dissipating or strippable resin-based materials can be used on slabs receiving applied finishes or subsequent liquid surface treatments. ASTM C 309 allows moisture loss of 1.8 oz./ft^2 (0.55 kg/m^2) in 72 h at a curing compound coverage of 200 ft^2/gal. (5.0 m^2/L) when applied in compliance with ASTM C 156. Special conductive curing compounds should be used to cure electrically conductive and spark-resistant floors. It is important to determine if a dissipating or nondissipating product should be used. The use of a nondissipating compound can be incompatible with the installation or application of future floor coverings.

For floors designed for high wear resistance, optimum top surface strength development, and minimal cracking, it is desirable to use curing compounds that offer high water retention. When a mineral aggregate or metallic surface hardener is used, the curing procedure and specific product used for curing should be approved by the manufacturer of the hardener. A high-solids-type curing compound can limit maximum moisture loss to 0.008 lb/ft^2 (0.04 kg/m^2) at a coverage of 300 ft^2/gal. (7.50 m^2/L)—less than 50% of that allowed by ASTM C 309 (ACI 308R).

More stringent criteria can be appropriate for some projects. Manufacturers' written instructions should be followed for both the number of coats and the coverage rate needed to meet the appropriate ASTM or project requirements. Periodic field testing to evaluate actual performance is recommended. One practical test for concrete surfaces to receive a moisture-sensitive covering is to apply a 18 x 18 in. (460 x 460 mm) transparent polyethylene sheet, sealed to the slab with tape at the edges. Visible liquid water should not be present when the sheet is removed after a minimum test period of 16 h. Three tests should be conducted for areas up to 500 ft^2 (46.5 m^2) and one for each additional 500 ft^2 (46.5 m^2). Refer to ASTM E 1907 and D 4263 for additional information.

When a mineral aggregate or metallic surface hardener is used, the curing method should be compatible with recommendations of the hardener manufacturer.

5.7—Admixtures

Admixtures should be used when they will effect a specific desired change in the properties of the freshly mixed or hardened concrete. They should be used in accordance with the instruction and principles given in ACI 212.1R and 212.2R and the guidelines for chloride limits given in Section 5.7.3. If more than one type of admixture is used in the same concrete, each should be batched separately. A second admixture can significantly affect the required dosage of both admixtures; therefore, preliminary tests are recommended to ensure compatibility. Sample slabs made under the anticipated job conditions of temperature and humidity can also be used to help evaluate admixture performance, and to allow necessary adjustments affecting workability, finishability, and setting time before the start of the slab installation. Some admixtures are not compatible with shrinkage-compensating concrete because they adversely affect expansion, bond to steel, and shrinkage (ACI 223).

5.7.1 *Air-entraining admixtures*—Concrete for use in areas that will be exposed to freezing temperatures while moist should contain entrained air (Section 6.2.7). Entrained air is not recommended for concrete to be given a smooth, dense, hard-troweled finish because blistering and delamination may occur (Suprenant and Malisch 1999a). For nontroweled finishes, smaller percentages of entrained air than those normally used for exposure to freezing-and-thawing cycles may reduce bleeding and segregation. Air-entraining admixtures, when used in the concrete as recommended in Chapter 6, should meet the requirements of ASTM C 260. Consistent control of air entrainment is necessary.

In most cases, concrete for trowel-finished interior concrete floors made with normalweight aggregates should not include an air-entraining admixture; the maximum total air content for this concrete should normally not exceed 3% at the point of placement. Air contents in excess of 3% make the surface difficult to finish and can lead to surface blistering and peeling during finishing. Troweled concrete with intentionally added air will typically not retain the proper bubble size required to provide scale resistance and freezing-and-thawing durability for most applications. Troweling can also reduce the ability of the concrete mortar at the surface to have adequate protection for resistance to freezing and thawing.

The committee recommends that the total air content of the concrete initially delivered to the jobsite be tested at the point of placement for air in accordance with ASTM C 173 or C 231. Air content can then later be checked as may be required by comparing density of concrete as defined in ASTM C 138.

5.7.2 *Chemical admixtures*—Chemical admixtures should meet the requirements of ASTM C 494 for whichever of the following types are to be used:
- Type A water-reducing;
- Type B retarding;
- Type C accelerating;
- Type D water-reducing and retarding;
- Type E water-reducing and accelerating;
- Type F high-range water-reducing; and
- Type G high-range water-reducing and retarding.

The high-range water reducers (Types F and G) should also meet the requirements of ASTM C 1017. Water-reducing and combination admixtures should provide the additional advantage of increased compressive and flexural strength at ages less than six months. The retarding admixtures can be useful in delaying initial set and possibly extending time available for final finishing in hot weather; however, excessive retardation can cause surface crusting or plastic-shrinkage cracking.

Accelerating admixtures increase the rate of strength gain at early ages and can be useful in cold weather.

High-range water-reducing admixtures meeting ASTM C 494, Types F or G, and midrange water-reducing admixtures meeting ASTM C 494, Type A, can be used to either reduce the water content required for a given slump or increase the slump of a given concrete while maintaining the same total water content. Water-reducing and high-range water-reducing admixtures used in industrial floor construction are most effective when the initial slump of the concrete, before introducing admixtures, is between 2 and 4 in. (50 and 100 mm). The admixture's impact on the workability and setting characteristics of the concrete for floor construction appear to be optimized when they are used in this manner.

The use of mid- or high-range water-reducing admixtures will not necessarily reduce the total water content of a concrete mixture as compared to that required for a Type A low-range water-reduced admixture. Even though these products have the capacity to reduce water content to a level below that which would correspond to pre-admixture slumps of 1 in. (25 mm) or less, the water content should not be reduced to less than that which would produce a minimum slump of 2 to 3 in. (50 to 75 mm). Water-reducing admixtures should not be relied on to reduce concrete shrinkage. Whiting and Dziedzic (1992) suggest that certain water-reducing admixtures can increase concrete shrinkage.

If the goal is to reduce concrete shrinkage by reducing the total water content in the concrete mixture, the designer should consider improving the characteristics of aggregate used to produce the concrete. Careful selection of characteristics, such as density, particle shape and texture, maximum size of the aggregate, and combined aggregate grading have a profound impact on reducing total water content, cementitious paste content, and long-term shrinkage (Section 6.3).

Considerations influencing a reduction in cement content should include the amount necessary to properly cut, trim, finish, and compact the floor slab surface. Reducing the total cement content may provide more than proportional reduction in shrinkage (Kosmatka and Panarese 2002b).

Although an initial slump of 2 to 4 in. (50 to 100 mm) is often recommended before the introduction of water-reducing admixtures, design water slump can be increased to 3 to 4 in. (75 to 100 mm) for lightweight concrete or when an embedded aggregate type hardener will be applied.

When using a high-range water-reducing admixture, the target slump at the point of placement can be increased to 6 to 8 in. (150 to 200 mm) without increasing the water content of the original concrete mixture. A mid-range water-reducing admixture can be used to increase the target slump at the point of placement to 4 to 6 in. (100 to 150 mm) without increasing the water content.

High-slump concrete may require less effort to place, consolidate, and finish compared with lower-slump concrete. If high slumps are used, excessive internal and external vibration can promote segregation of the concrete, excessive fines at the surface, or both, resulting in reduced abrasion resistance, especially for nonoptimized combined aggregate gradings. Caution should be exercised to avoid starting or continuing the finishing process before the concrete has achieved a sufficient degree of stiffness to support the type of finishing process and equipment used.

In special applications, such as parking garage surfaces, where low permeability is desirable, the reduction of water to produce w/c less than 0.42 in high-strength concrete can increase the chances of autogenous shrinkage when an external supply of water is not provided during curing. This shrinkage is caused by internal consumption of water during hydration (McGovern 2002).

The committee recommends that a representative test slab be cast at the jobsite so that the workability, finishability, and setting time of the proposed mixture can be evaluated by the project team (ACI 212.3R and ACI 212.4R).

5.7.3 *Chlorides*—Chlorides are significant contributors to corrosion of steel in concrete. The problem is particularly severe when dissimilar metals are embedded in concrete or when reinforced concrete is placed over galvanized decking. Corrosion products can cause expansion, cracking, and spalling.

Limits on chloride in fresh concrete mixtures are based on the recommendations of ACI 222R.

The following concrete should not include any intentionally added calcium chloride:
- Prestressed concrete;
- Floors over prestressed concrete or galvanized deck;
- Floors containing two kinds of embedded metals;
- Conventionally reinforced concrete in a moist environment and exposed to deicing salts or saltwater mist;
- Parking garage floors in areas where freezing and thawing should be considered;
- Structures near bodies of saltwater;
- Floors or slabs containing snow-melting electrical radiant heating systems; and
- Floors finished with metallic dry shakes.

Noncorrosive, nonchloride accelerators are available for use in cold weather. The admixture manufacturer should be able to provide long-term data (at least a year's duration) demonstrating noncorrosivity using an acceptable accelerated corrosion test method, such as one using electrical potential measurements. Data from an independent laboratory are preferable.

If accelerated set or high early strength is desired, either a noncorrosive nonchloride accelerator or high early-strength (Type III) cement can be used; alternatively, 100 to 150 lb/yd^3 (60 to 90 kg/m^3) of additional Type I or Type II cement can be used in the mixture. A significant decrease in setting time may not be realized with the increased cement content. The increased cement and water demand can increase shrinkage and curling.

Heated concrete may be required for cold-weather construction (ACI 306R). Chloride-based accelerators should only be used in nonreinforced concrete and when specifically permitted by project specifications.

When used, calcium chloride should be added as a water solution in amounts of not more than 1 to 2% by weight of cement. It will accelerate the rate of strength development and decrease setting time. Calcium chloride, in dosages as high as 1 to 2%, does not significantly lower the temperature at which the concrete will freeze. It accelerates the rate of strength development and thereby decreases the length of time during which protection against freezing should be provided. Setting time is decreased, thereby reducing finishing time.

Calcium chloride tends to darken the color of concrete and can cause variations in color of the hardened concrete. The difference in color is most noticeable when slabs with calcium chloride are adjacent to those without (Fig. 5.1). If concrete containing calcium chloride is not adequately cured, the surface can show light and dark spots. Calcium chloride should not be dispensed dry from bags. Dry-flake material frequently absorbs moisture and becomes lumpy. Pellet-type calcium chloride should be completely dissolved before addition to concrete or pop-outs will result from any undissolved pellets.

5.7.4 *Expansive cementitious admixtures*—Specifically formulated dry-powder admixtures can be blended with portland cement at the batch plant to produce shrinkage-compensating concrete. Concrete incorporating the same materials that will be used for the anticipated project should be tested for expansion by ASTM C 878 (refer to ACI 223 for full details). The compatibility of the expansive cementitious admixture and portland cement should be checked by the use of ASTM C 806.

The anticipated rate and quantity of expansion that can be obtained in the field should be established before beginning construction. This can be accomplished by conducting a series of tests using identical materials to those proposed for the project. These tests should be conducted by a testing laboratory that is familiar with ASTM C 878 procedures.

5.7.5 *Pozzolans*—A number of natural materials, such as metakaolin, diatomaceous earth, opaline cherts, clays, shales, volcanic tuffs, and pumicites, are used as pozzolans. ASTM C 618 pozzolans also include fly ash and silica fume and slag. Information on the use of these materials can be found in documents developed by ACI Committee 226. When these materials, excepting silica fume, are used in concrete, the time of set is frequently extended. The color of concrete can be different from that produced when portland cement is the only cementitious component.

ASTM C 618 fly ash, Class F or Class C, is frequently incorporated in concrete. Fly ash can affect the setting time, and it is often helpful in hot weather by delaying set time or as an aid in pumping concrete (ACI 226.3R). In floors and slabs, fly ash is often substituted for portland cement in quantities up to about 20% fly ash by mass of cementitious materials.

In cool weather, fly ash will usually delay the setting and finishing of the concrete unless measures are taken to compensate for the low temperatures, such as increasing the concrete temperature or using an accelerator.

Silica fume is used as a portland cement replacement or as a cementitious addition when using an accelerator to compensate for low temperatures. The amount of silica fume in a mixture typically varies between 5 and 10% by mass of the total cementitious material. Silica fume can increase both the impermeability and the strength of the concrete. Special attention should be given to avoiding plastic-shrinkage cracking during placing and finishing by using evaporation-

Fig. 5.1—Concrete slab discoloration due to the use of calcium chloride mixture. Concrete in upper part of the photo did not contain the admixture.

retardant chemicals sprayed onto the plastic concrete surface or by using fog sprays in the air above the concrete. Early and thorough curing of the slab is also very important to minimize cracking.

5.7.6 *Coloring admixtures*—Pigments for colored floors should be either natural or synthetic mineral oxides or colloidal carbon. Synthetic mineral oxides can offer more intensity in color, but they are normally more expensive. Pigments can be purchased alone or interground with a water-reducing admixture for mixing into the batched concrete to produce integrally colored concrete. Colored aggregate-type surface hardeners containing pigments can also be used. These pigmented mineral aggregates or metallic hardeners contain mineral oxide pigment, portland cement, a well-graded mineral aggregate, or metallic hardener and plasticizers. Pigments for integrally colored concrete should conform to ASTM C 979 and have uniform color. Carbon-black pigments, especially manufactured for this purpose, will appear lighter in color at an early age. The prepared mixtures should not contain pigments that are not mineral oxides. Job-proportioning or job-mixing of material for monolithic colored surfaces is not recommended. The use of these materials is described in Section 8.6. Coloring admixtures should be lime-proof and contain no calcium chloride. Curing compounds for these slabs should be the same as those used on the approved sample panels (Chapter 8).

5.8—Liquid surface treatments

Improperly constructed floor slabs can have relatively pervious and soft surfaces that wear or dust rapidly. Though the life of such surfaces can be short, it can be extended by using surface treatments containing certain chemicals, including sodium silicate and magnesium fluosilicate (Smith 1956; Vail 1952; Bhatty and Greening 1978). When these compounds penetrate the floor surface, they react chemically with calcium hydroxide (a product of cement hydration).

This broad classification covers a diverse group of products that are intended to penetrate the concrete and form a hard, glassy substance (calcium silicate) within the pores of the concrete. Effective use of the products generally results in reduced dusting and improved density of the concrete surface. Depth of penetration into concrete by these products varies with the porosity of the concrete surface and concrete moisture content at the time of application.

Products in this group are not specifically formulated for curing applications and do not meet the requirements of either ASTM C 309 or ASTM C 1315 for liquid membrane-forming compounds. While their use may offer some desirable benefits when applied after curing, they should not be applied on fresh concrete.

If these surface treatments are to be applied to new concrete floors, the floor should be moist cured for at least seven days and allowed to air dry in accordance with the product manufacturer's recommendations before application. Liquid membrane-forming curing compounds should generally be removed before application of surface treatments because they prevent penetration of the liquid. The lone exception to this requirement would be when compatible curing and sealing products from a single manufacturer are used.

Pozzalanic materials, including fly ash and silica fume, are siliceous or siliceous and aluminous materials that react with calcium hydroxide to form hydration products similar to those produced by portland cement and water, which, in turn, can contribute to the strength development and a reduced permeability of the concrete. This is the same calcium hydroxide needed to react with the liquid surface treatment that eventually hardens at the floor slab surface. Thus, less calcium hydroxide may be available to the floor treatment for concrete mixtures incorporating pozzolanic materials, slag cement, or both.

When a liquid surface treatment is applied to concrete mixtures incorporating pozzolanic materials or slag cement, the application should be delayed 28 days to ensure that the strength of the concrete has developed adequately before the application of the surface treatment.

Alternatively, the quantity of pozzolanic material should also be kept, as proposed by one manufacturer, to a maximum of 15% by mass of portland cement. If the percentage is not limited, calcium hydroxide may be depleted below the level necessary for the proper performance of the liquid surface treatment. Contact the manufacturer of the surface treatment for limits specific to the product to be used.

In view of the aforementioned, a number of specific items should be considered when liquid surface treatments are to be used. Specific considerations include, but are not limited to, the following:
- Anticipated application conditions;
- Timing of application for maximum benefit;
- Eventual appearance of the treated surface;
- Resistance of treated surface to wear, dusting, and tire marks;
- Coefficient of friction of treated surface; and
- Anticipated results if applied on a surface that has become carbonated.

Liquid surface treatments react with materials found in cement paste but not aggregate; they are not capable of providing abrasion resistance equal to that obtained by use of an embedded aggregate-type hardener.

5.9—Reinforcement

5.9.1 *Reinforcing steel, mats, or welded wire reinforcement*—Deformed bars, bar mats, or welded wire reinforcement are usually required in suspended structural floors as part of the structural design. They can also be called for in the specifications for slabs-on-ground as discussed in Section 3.2.4. Deformed bars should conform to the requirements of ASTM A 615, A 616, or A 617. Bar mats conforming to ASTM A 184 can also be used. Welded wire reinforcing should conform to ASTM A 185 or A 497. The use of widely spaced deformed reinforcing fabric conforming to ASTM A 497 will typically permit easier placement.

5.9.2 *Post-tensioning*—Post-tensioning can be used in slabs-on-ground and suspended slabs to address specific design requirements. Prestressing steel for use in floors and slabs should conform to the requirements of ASTM A 416. The post-tensioning tendons can be bonded or unbonded. Unbonded tendons should meet or exceed specifications published by the Post-Tensioning Institute (PTI 2000).

5.9.3 *Synthetic fibers*—Synthetic fibers for use in concrete floors increase the cohesiveness of concrete and should meet the requirements outlined in ASTM C 1116. The most widely used synthetic fibers are polypropylene and nylon, although other types are available. Polypropylene fibers are available in both fibrillated and monofilament form; nylon fibers are only available in monofilament form.

Synthetic fibers are added to the concrete mixer in quantities generally less than 0.2% by volume of the concrete. They are generally used in floors and slabs in quantities of from 0.75 to 1.5 lb/yd^3 (0.45 to 0.90 kg/m^3). Synthetic fibers have a tendency to reduce the formation of plastic-shrinkage and settlement cracks at the surface by increasing the tensile strain capacity of the plastic concrete. These fibers should not be used to replace temperature and shrinkage reinforcement because they have little impact on the behavior of concrete after it hardens.

5.9.4 *Steel fibers*—Steel fibers for use in floors and slabs should conform to the requirements of ASTM A 820. Steel fibers made from wire, slit sheet, milled steel, and melt extract are available and are normally deformed or hooked to improve bond to the hardened matrix. Steel fibers are added to the concrete mixer in quantities ranging from 0.0625 to 1% by volume of the concrete (8 to 132 lb/yd^3 [5 to 78 kg/m^3]), although quantities from 0.25 to 0.50% by volume of the concrete (34 to 68 lb/yd^3 [20 to 40 kg/m^3]) are more common.

Steel fibers are used in floors to minimize visible cracking, increase shear strength, increase the flexural fatigue endurance and impact resistance, and increase post-crack flexural toughness. The increases in mechanical properties achieved depend primarily on the type and amount of fiber used, and can result in reduced floor thickness and increased contraction joint spacing (Tatnall and Kuitenbrouwer 1992).

5.9.5 *Fiber characteristics*—Crack reduction, material properties, and mixture proportions are thoroughly discussed by Balaguru and Shah (1992). Additional information is available in ACI 544.1R, 544.2R, 544.3R, and 544.4R.

5.9.6 *Dowels and load-transfer devices*—Dowels required for load transfer can be round, square, or rectangular. Square and rectangular dowels cushioned on the vertical sides by a compressible material allow for some horizontal movement. Diamond-shaped load plates (a square plate turned so that two corners line up with the joint) can be used to replace dowels in construction joints. The diamond shape also allows the slab to move horizontally without restraint when slab shrinkage opens the joint. All dowels and load-transfer devices should meet requirements of ASTM A 36.

The diameter or cross-sectional area, length, shape, and the specific location of load transfer device and the method of support should be specified by the designer. Refer to Section 3.2.7 for more information on load-transfer mechanisms for slabs-on-ground.

5.10—Evaporation reducers

Evaporation-reducing chemicals can be sprayed on the plastic concrete one or more times during the finishing operation to minimize plastic-shrinkage cracking when the evaporation rate is high. These products should be used in strict accordance with the manufacturers' directions; they should never be used during the final troweling operations because they discolor the concrete surface.

5.11—Gloss-imparting waxes

Concrete waxes to impart gloss to concrete surfaces are available from various manufacturers. Some are curing compounds; for such use, they should meet or exceed the water-retention requirements of ASTM C 309.

5.12—Joint materials

Certain two-component semirigid epoxy resins and polyureas can be used to fill joints where the joint edges need support to withstand the action of small, hard-wheeled traffic. These are the only materials known to the committee that can provide sufficient shoulder support to the edges of the concrete and prevent joint breakdown. Two-component fillers are desirable because their curing is independent of jobsite conditions. Such joint materials should be 100% solids and have a minimum Shore A hardness of 80 when measured in accordance with ASTM D 2240. Refer to Section 9.10 for more details on joint filling and sealing.

Preformed elastomeric sealants are useful for some applications. They should not be used where subjected to the traffic of small, hard wheels. They can be quickly installed, they require no curing, and, if properly chosen, they can maintain a tight seal in joints that are subject to opening and closing. Refer to ACI 504R for more information on preformed elastomeric sealants.

Preformed asphalt impregnated or plain fiber materials or compressible foam are used in expansion and isolation joints, depending on the anticipated movement. These materials and their appropriate use are described in detail in ACI 504R.

5.13—Volatile organic compounds (VOC)

Many users and some states require materials to meet VOC limits. Liquid materials are of the greatest concern because they are often solvent-based. Certification of compliance with the applicable VOC limits should be required before the products are used.

Many curing compounds that comply with limits on VOC are water-based. They should not be permitted to freeze. In many cases, they cannot be reconstituted after freezing.

CHAPTER 6—CONCRETE PROPERTIES AND CONSISTENCY

6.1—Concrete properties

A concrete mixture for floor and slab construction should incorporate an optimized combination of locally available materials that will consistently produce concrete with the required workability and finishability properties in the plastic state during placement. After the concrete mixture hardens, it will be required to develop engineering properties that include surface abrasion, wear resistance, impact resistance, adequate flexural strength to accommodate anticipated loads, and shrinkage characteristics that minimize potential cracking from restrained shrinkage and curling stresses.

Concrete with good placing and finishing characteristics, that also meets the required engineering properties, can best be achieved by developing an understanding of the concrete mixture. Concrete mixtures are commonly defined by the proportions of individual materials that do not identify the qualities of the blended mixture delivered and placed on the site. To produce the best results, all parties involved in the design and construction of the slab should understand the characteristics of the combined materials in the mixture delivered to the project.

In most flatwork, the placeability of the concrete and finishability of the surface are at least as important as the abrasion resistance, durability, and strength. The former qualities will have a significant effect on the integrity of the top 1/16 or 1/8 in. (1.5 or 3 mm) of the concrete surface. Unfortunately, placeability and finishability are not easily measured. There is a tendency for specifiers to emphasize more easily determined properties, such as slump and compressive strength.

Other parameters being equal, a given concrete's shrinkage potential will decrease as its paste content is decreased and its paste quality is optimized. The quality of the concrete paste is reflected in the total water content necessary to produce a workable mixture while maintaining a reasonable *w/cm* that minimizes the amount of cementitious materials. The *w/cm* specified should not require arbitrary increases in the cementitious material or chemical admixture content, exceeding what is necessary to achieve the specified strength or required setting and finishing needs of the concrete materials. Acceptance of a mixture based on compliance with *w/cm* alone seldom produces desired results without first minimizing the total water content, which generally can be accomplished through adjustments in the combined aggregate size, uniformity of distribution, or material source. Therefore, the use of the minimum amount of water necessary to produce the required slump and workability is very important.

For steel-trowelled, slab-on-ground concrete, a minimum amount of water is required to produce a workable, finishable mixture with predictable, uniform setting characteristics. Currently available water-reducing admixtures perform best when they are mixed with concrete that has enough water to produce a slump of 2 to 3 in. (50 to 75 mm) if no admixture was added. If this "water slump" is not achievable without the admixture, setting time and finishability can vary when the concrete is subjected to normal variations of ambient temperature and time between batching and discharge (Harrison 2004).

The particular cementitious materials, aggregates, and admixtures used can significantly affect the strength, setting characteristics, workability, finishability, and shrinkage of the concrete at a given w/cm (Tremper and Spellman 1963; Kosmatka, Kerkhoff, and Panarese 2002b). Furthermore, the amount of water required to produce a given slump depends on the maximum size of coarse aggregate, the uniformity of the combined aggregate gradation, particle shape and surface texture of both fine and coarse aggregates, air content, admixtures used, and the temperature and humidity at time of placement. Using larger maximum-size aggregate and improving the overall aggregate gradation reduces the mixing-water requirement.

Air-entraining admixtures produce a system of small air bubbles that reduce the mixing water requirement. Concrete containing entrained air is proportioned to have the same amount of coarse aggregate as similar nonair-entrained concrete. Air-entrained concrete mixtures use less mixing water and less fine aggregate; however, in mixtures with higher cement content, this may not offset the strength reduction that can result from intentional entrainment of air. Air-entraining admixtures should not be used in floors that are to have a dense, smooth, hard-troweled surface.

The optimum quality and content of fine aggregate in concrete for floors should be related to the slump of the concrete and the abrasive exposure to which the floor will be subjected. Concrete should be sufficiently plastic and cohesive to avoid segregation and bleeding (Kosmatka, Kerkhoff, and Panarese 2002b). Less fine aggregate should be used in concrete with low slump—less than 1 in. (25 mm)—because this concrete does not normally bleed or segregate. Decreased fine aggregate contents can improve resistance to abrasion if the concrete exhibits little bleeding and segregation.

Previous field experience or laboratory trial batches should be used to establish the initial proportions of ingredients. Test placements can then be used to optimize the mixture proportions. The laboratory trial batches can be omitted if concrete mixtures have been used successfully under similar conditions in other jobs.

Current records of gradations of fine and coarse aggregates from concrete mixtures should be retained to produce a combined gradation analysis. This gradation analysis should evaluate the amount of aggregate retained on each of the following sieve sizes (percent of total mass): 1-1/2, 1, 3/4, 1/2, and 3/8 in.; 4, 8, 16, 30, 50, 100, and No. 200 (38.1, 25, 19, 12.5, 9.5, and 4.75 mm; 2.36 mm, 1.18 mm, and 600, 300, 150, and 75 μm, respectively).

Trial batch proportions should generally be in accordance with ACI 211.1 or 211.2. Adjustments of fine aggregate content, however, may be necessary to obtain the best workability (Martin 1983). The amount of the total combined aggregate passing the No. 8 (2.36 mm) sieve for a uniformly graded mixture should be between 32 and 42% of the total combined aggregate. This index is referred to as the workability factor and should be evaluated in relation to the coarseness factor of the larger aggregate particles as illustrated later in this chapter (Shilstone 1990). Adjustments in the fine aggregate content directly influence the workability factor.

6.2—Recommended concrete mixture

6.2.1 *Required compressive strength and slump*—Two approaches for selecting mixture proportions are discussed in Section 6.2.4. Regardless of the approach, the specified compressive strength f_c' shown in Table 6.1 should be used for the various classes of concrete floors.

The designer should be consulted as to the strength to be achieved by concrete before subjecting the slab to early construction loads. To obtain this strength quickly, it may be necessary to use more cementitious materials than the recommended amounts shown in Table 6.2, or to proportion the concrete for a 28-day strength higher than that shown in Table 6.1. The designer should take into account that increased concrete strengths achieved through higher cementitious material contents alone may adversely affect the shrinkage characteristics and stiffness (modulus of elasticity) of the concrete. Drying shrinkage potential and stiffness properties can greatly influence slab curling stresses and ultimate load capacity of the slab-on-ground (Walker and Holland 1999). Compressive strengths should be used to monitor batched material consistency.

The slump indicated for each floor class shown in Table 6.1 is the recommended maximum at the point of placement to prevent segregation and yet provide adequate workability of the concrete. A one-time jobsite slump adjustment should be permitted as outlined in the "Tempering and Control of Mixing Water" provisions of ACI 301 or the "Mixing and Delivery" provisions of ASTM C 94. Validation of total water content should be conducted periodically at the point of concrete placement, concurrent with other specified site testing.

6.2.2 *Required finishability*—Concrete for floors should have other desirable characteristics in addition to strength. There should be sufficient mortar content to allow the finisher to completely close the surface and to achieve the required surface tolerance, hardness, and durability (Martin 1983). The mortar fraction, the volume percentage of all materials in the mixture (cementitious materials, aggregate, water, and air) that pass the No. 8 (2.36 mm) sieve, should be balanced between the desired properties of both fresh and hardened concrete. During construction, sufficient mortar is desirable for pumping, placing, and finishing. Excess mortar, however, can increase shrinkage characteristics. Typically, a mortar fraction of 55 to 57% is sufficient for a 3/4 or 1 in. (19 or 25 mm) maximum-size aggregate slab-on-ground concrete placed directly from the mixer truck. Larger aggregates, improved uniform distribution of the combined aggregate particle sizes, or both will decrease the mortar content needed. Smaller 3/8 to 1/2 in. (9.5 to 12.5 mm) maximum-size aggregates can increase the mortar content by as much as 63%. Refer to Section 6.3.3 for other typical mortar contents.

6.2.3 *Required durability*—The procedures for producing durable concrete outlined in ACI 201.2R apply to floors and slabs. Concrete floors exposed to freezing and thawing while in a moist condition should have a *w/cm* not greater than the values given in the following paragraph. These *w/cm* requirements can be lower than those required for strength alone. Additionally, this type of concrete should have adequately entrained air.

Requirements based only on durability may yield concrete compressive strengths much higher than normally required for structural concerns. Concrete floors and slabs subjected to moderate and severe exposures to freezing and thawing, as defined in ACI 201.2R, should have a *w/cm* no greater than 0.50. Concrete subjected to deicing chemicals should have a *w/cm* no greater than 0.45. Reinforced concrete exposed to brackish water, seawater, deicing chemicals, or other aggressive materials should have a *w/cm* no greater than 0.40. The committee recognizes that there is no direct correlation between compressive strengths and *w/cm* and suggests that the two not be combined in a specification.

Entrained air is necessary in concrete subjected to freezing and thawing when moist or when subjected to deicing chemicals. Recommended air contents for hardened concrete for various exposure conditions, aggregate types, and maximum-size aggregates are given in ACI 201.2R. Properly air-entrained concrete should achieve a compressive strength of 4000 psi (28 MPa) before being subjected to freezing and thawing in a moist condition or to deicing chemicals. The use of deicing chemicals is not recommended in the first year of slab service in any event.

Air contents, within the limits recommended, will cause significant strength reductions in richer concrete, but the effect will be less in leaner concrete. Air contents in excess of the recommended quantities will reduce strength in richer mixtures, approximately 3 to 5% per 1% increase in air content, and will reduce abrasion resistance correspondingly. Air-entrained concrete should not be hard-troweled finished (Sections 6.2.7, 8.3.11, and 8.6).

6.2.4 *Concrete mixture*—In addition to meeting structural and drying-shrinkage requirements, concrete for floors should provide adequate workability and setting characteristics necessary to obtain the required finish and floor surface profile. Total water content can have a major impact on the bleeding characteristics of the concrete and the potential for shrinkage, so use of the lowest practical quantity of water in the concrete mixture is recommended. The amount of water needed to produce a workable mixture is generally determined by the characteristics of the combined aggregate materials used in the mixture and is not effectively controlled by specifying *w/cm*. If *w/cm* is specified, *w/cm* in the range of 0.47 to 0.55 are common for most interior floors of Classes 4 to 9. Floors that are required to be impermeable, resistant to freezing-and-thawing and deicing chemicals, or meet the requirements of ACI 211.2, 223, or 318 should conform to more stringent criteria.

The committee recommends that the concrete mixture be accepted on the basis of a satisfactory mixture design submittal and a successful test slab placement (if appropriate for the project). This submittal should include a combined aggregate distribution analysis derived from current, certified reports of gradations of the individual aggregates. The test placement should determine if the proposed concrete mixture is capable of producing a floor of acceptable finish and appearance and meet the project requirements.

If a history of finishing properties is not available for a concrete mixture, a test slab should be placed under job conditions to evaluate the workability, finishability, setting time, slump loss, and appearance of the concrete proposed for use. Materials, including all admixtures, equipment, and personnel proposed for the project, should be used. The test panel should be as large as possible and at least 20 x 20 ft (6 x 6 m), placed at the specified project thickness. A floor slab area in a noncritical section is often chosen as the test panel. The concrete contractor should review the proposed mixture proportions before the preconstruction meeting and placement of the test slab. If a pump will be used for the placement of concrete materials, the test slab should be placed with the same pump equipment.

6.2.5 *Consistency and placeability*—The maximum slump recommended for each class of floor is given in Table 6.1. These slumps should produce concrete of sufficient workability to be properly consolidated in the work without excessive bleeding or segregation during placing and finishing. Excessive bleeding and segregation can contribute to poor performance in concrete floors. If the finished floor is to be uniform in appearance and grade, successive batches placed in the floor should have very nearly the same slump and setting characteristics. See Sections 6.1, 6.2.1, and 7.3.2 regarding

Table 6.1—Recommended strength and maximum slump at point of placement for concrete floors

Floor class[*]	Specified compressive strength f_c' on 28-day tests, psi (MPa)	Maximum slump at placement[†], in. (mm)
1, 2, and 3	3000 (21)	5 (125)
4, 5, and 6	3500 (24)	5 (125)
7 base	3500 (24)	5 (125)
7 bonded topping[‡]	5000 (35)	3 (75)
8 unbonded topping[§]	4000 (28)	3 (75)
9 superflat	4000 (28)	5 (125)

[*]Refer to Table 2.1 for floor class definitions.
[†]Maximum slump is assumed to be achieved using a Type A water-reducing admixture.
[‡]The strength specified will depend on the severity of usage.
[§]Maximum aggregate size not greater than 1/3 the thickness of unbonded topping.

Table 6.2—Recommended cementitious material contents for concrete floors

Nominal maximum-size aggregate,[*] in. (mm)	Cementitious material content, lb/yd^3 (kg/m^3)
1-1/2 (37.5)	470 to 560 (280 to 330)
1 (25)	520 to 610 (310 to 360)
3/4 (19)	540 to 630 (320 to 375)
1/2 (12.5)	590 to 680 (350 to 405)
3/8 (9.5)	610 to 700 (360 to 415)

[*]For normalweight aggregates.
Note: See ACI 318 for minimum portland cement requirements for structural applications.

jobsite slump adjustment. Workability of a concrete mixture is not directly proportional to the slump. Properly proportioned concrete with slumps less than that shown in Table 6.1 can respond very well to vibration and other consolidation procedures. Increased slump alone does not ensure satisfactory workability characteristics.

Recommended slump values in Table 6.1 are for concrete made with both normalweight and structural lightweight aggregate and assume the use of a normal water-reducing admixture, if required. Slumps in excess of those shown in the table, not to exceed 7 in. (175 mm), are acceptable when mid-range or high-range water-reducing admixtures are used. If structural lightweight-aggregate concrete is placed at slumps higher than that shown in Table 6.1, however, the coarse lightweight-aggregate particles can rise to the surface and the concrete can bleed excessively, particularly if the concrete does not contain an adequate amount of entrained air.

6.2.6 *Nominal maximum size of coarse aggregate*—The nominal maximum aggregate sizes in Table 6.2 are for normalweight aggregates. The largest practical-size aggregate should be used if economically available, and if it will satisfy the requirements that maximum size not exceed 3/4 of the minimum clear spacing of reinforcing bars or 1/3 of the depth of the section. Structural lightweight aggregates generally are not furnished in sizes larger than 3/4 or 1 in. (19 or 25 mm); however, some lightweight aggregates provide maximum strength with relatively fine gradings.

6.2.7 *Air content*—Moderate amounts of entrained air for purposes other than durability, as described in Section 6.2.3, can be used to improve workability, particularly with leaner and more harsh concrete mixtures or with poorly graded aggregates. Concrete made with structural lightweight aggregates should contain some entrained air. Minimum air content should be 4%, and specific recommendations for air content secured from the concrete supplier, the manufacturer of the lightweight aggregate, or both.

An air-entraining agent should not be specified or used for concrete to be given a smooth, dense, hard-troweled finish because blistering or delamination may occur. These troublesome finishing problems can develop any time the total air content is in excess of 3%. This is particularly true when embedded hardeners are applied.

Some variation in the air content of air-entrained concrete is common, and this can result in problems associated with finishing of concrete surface. Exposure conditions that dictate the need for air-entrainment should be discussed with the designer before the placement of concrete floors.

6.2.8 *Required yield and concrete mixture adjustment to correct yield*—A concrete mixture should be proportioned to yield a full 27 ft^3/yd^3 (1 m^3/m^3). The yield of the mixture proposed by a testing agency or concrete supplier is the total of the absolute volume of the mixture ingredients plus the volume of air. This proposed mixture should have been tested in accordance with the requirements of ASTM C 138 to determine its density (or unit weight) and yield, and the weights of the ingredients subsequently adjusted as necessary. The concrete mixture should be sampled and tested one or more times at the jobsite during placement of concrete floor for the purpose of confirming yield. These jobsite samples should be obtained from a mixer truck chute (the point of delivery), and the tests performed by a certified field technician as required by ASTM C 94. Concrete samples should be obtained after any necessary jobsite slump adjustment. The concrete supplier should adjust the concrete mixture, as necessary, to produce the proper yield. Adjustments in concrete mixture proportions should be made in accordance with the recommendations of ACI 211.1.

6.3—Concrete mixture analysis

6.3.1 *Evaluation of the concrete mixture*—Due to the many variables involved in the production of concrete, the ultimate evaluation of the concrete materials is when the concrete is mixed, placed, and finished under the anticipated conditions of the jobsite (Section 6.2.4). There are evaluation methods, however, that can be used to identify potential problem areas of defined proportioned materials before mixing and placing concrete.

For proper analysis, major emphasis is placed on the combined aggregates and mortar contents. The optimization of the combined aggregate materials not only improves the long-term strength and durability characteristics of the concrete, but it can also dramatically improve placing characteristics during construction (Shilstone 1990). The ideal mortar content is one that finds the balance between adequate mortar for placing and finishing of fresh concrete while minimizing the shrinkage and curling properties of the hardened material.

6.3.2 *Aggregate blending*—To maximize the uniform gradation distribution of the combined aggregates, blending of three or more individual aggregates may be necessary. Generally, this includes one coarse aggregate, one fine aggregate, and the addition of an intermediate-size aggregate, typically to compensate for deficiencies in particles' sizes retained on the 3/8 in. (9.5 mm) through No. 8 (2.36 mm) size sieves. There are times when the addition of a second fine aggregate source is necessary to supplement deficiencies in the finer aggregate particle sizes.

Many methods are used to blend both coarse and fine aggregate materials to produce an optimized proportioning from the largest to smallest particles. The importance of combined aggregate grading is not a new concept. It was recognized as early as 1918 by Abrams. Other recent publications include Shilstone (1990) and Weymouth (1933).

ASTM C 33—Since 1993, ASTM C 33 has provided a means to optimize combined aggregate grading by providing the following under Part 1, "Scope:" "...Those responsible for selecting the proportions for the concrete mixture shall have the responsibility for selecting the proportions of fine and coarse aggregate and the addition of blending aggregate sizes if required or approved." ASTM D 448 includes sizes 89 and 9 to provide the opportunity to blend these sizes with other classifications to obtain improved particle distribution. Sizes 89 and 9 are abundant in No. 4 (4.75 mm) and No. 8 (2.36 mm) size particles. These size and gradation designations were developed to supplement the intermediate-sized aggregate that is often missing in a standard single coarse plus single fine aggregate combination.

Blending aggregates to meet criteria for a combined grading is another proportioning method that can be used. The different procedures that have been used to determine proportions and potential concrete characteristics due to the gradations of the combined aggregates include:

1. Percentage of the combined aggregate retained on each of the standard sieves;
2. Coarseness factor chart; and
3. 0.45 power chart.

When one of the above or other similar methods is used, the specific combined grading to which aggregate is to be blended, along with tolerances for control, should be included with the mixture proportion submittal. The details of the above-mentioned procedures are described in the following.

1. Percent of the combined aggregate retained on each of the standard sieves—This procedure provides a tolerance of acceptable uniformity of distribution of the total combined aggregate particles found in the mixture. Recommended tolerance limits are defined in Section 5.4.3.

A deficiency in particles retained on the No. 8, 16, and 30 (2.36 mm, 1.18 mm, and 600 μm) sieves and an excess of particles retained on the No. 50 and 100 (300 and 150 μm) sieves occur in many areas of the U.S., leading to problems associated with cracking, curling, blistering, and spalling of concrete.

While the gradation specifications discussed in previous paragraphs set limits on the percent of aggregate retained on each sieve, this should only be a guide as it may not be easily attainable using available local resources. In 1933, Weymouth described the importance of clusters versus individual sieve sizes. If there is a deficiency on one sieve but excess on an adjacent sieve, the two sizes are a cluster and they balance one another. When there is a deficiency in particles on each of two adjacent sieve sizes but abundance on the sieves adjacent to each, the adjacent sizes tend to balance the two-point valley. If there are three adjacent deficient sizes, there is a problem that must be corrected.

2. Coarseness factor chart—Figure 6.1 illustrates an alternative method of analyzing the size and uniformity of the combined aggregate particle distribution, balanced with the fine aggregate content of the mixture. The x-axis, labeled as coarseness factor (CF), defines the relationship between the coarse and intermediate particles. It is the percent of the combined aggregate that is retained on the No. 8 (2.36 mm) sieve that is also retained on the 3/8 in. (9.5 mm) sieve.

The y-axis represents the percent of the combined aggregate that passes the No. 8 (2.36 mm) sieve. A correction based on cementitious material content should be made. This chart was developed for a cementitious material content of 564 lb/yd³ (335 kg/m³) of concrete. When a mixture contains 564 lb/yd³ (335 kg/m³) of cementitious materials, there is no correction factor. With respect to the workability factor, the impact of 94 lb (43 kg) of portland cement is approximately equal to a similar adjustment of 2.5% in the amount of fine aggregate. As cementitious materials are increased, the fine aggregate

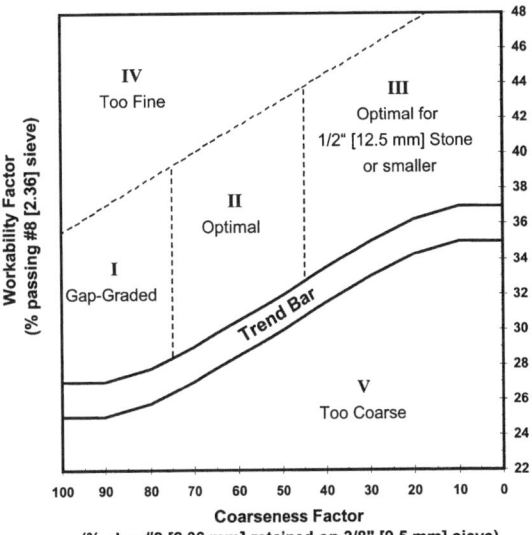

Fig. 6.1—Coarseness factor chart for evaluating the potential performance of a mixture.

content should be reduced to maintain the same workability factor W and vice versa. An increase or decrease in either the cementitious material content or fine aggregate content without a compensating adjustment in the other of these two components will impact the workability of the mixture.

The diagonal trend bar defines a region where combined rounded or cube-shaped crushed stone and well-graded natural sand are in near perfect balance to fill voids with aggregate. Variations in shape and texture of the coarse and fine aggregates that allow the combined mixture to fall within this region reflect maximum packing of aggregate within the concrete volume. Mixtures with aggregate combinations that fall in or near this region should be placed by bottom drop buckets or by paving machines.

Five zones are used to identify regions above the diagonal trend bar where variation in combined aggregate grading is indicative of certain general characteristics based upon the following field experience (Shilstone and Shilstone 2002).

Zone I—A mixture is seriously gap-graded and will have a high potential for segregation during placement or consolidation due to a deficiency in intermediate particles. These mixtures are readily identified in the field when placed by chute. They are not cohesive, so a clear separation between the coarse particles and the mortar can be seen as the concrete is deposited from the chute. Segregation is a major problem, especially for floor slabs and paving. Slab mixtures plotting in this zone and three points above the trend bar described in previous paragraphs segregated at a 1 in. (25 mm) slump. These mixtures lead to blistering, spalling, and scaling. As mixtures approach Zone IV, they can experience additional problems as described as follows.

Zone II—This is the optimum zone for mixtures with nominal maximum aggregate size from 1-1/2 to 3/4 in. (37.5 to 19 mm). Mixtures in this zone generally produce consistent, high-quality concrete. Field observations with multiple materials and construction types have produced outstanding results

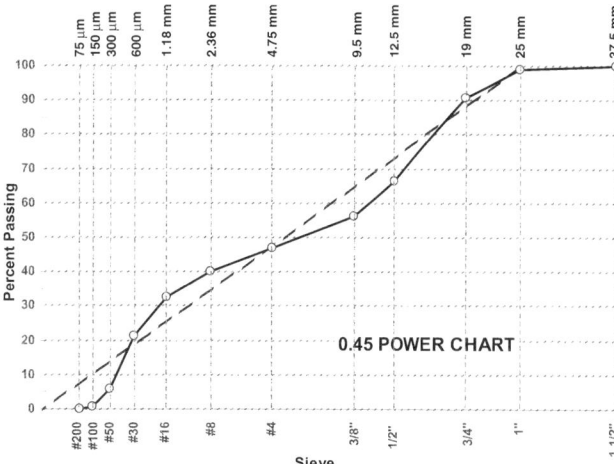

Fig. 6.2—The 0.45 power chart for determining the best combined gradation of aggregates.

when the coarseness factor is approximately 60 and the adjusted workability factor W is approximately 35. Mixtures that plot close to the trend bar require close control of the aggregate. Variations in grading can lead to problems caused by an excess of coarse particles. Mixtures that plot close to Zone IV should be placed with special care or they can experience problems found in that zone. Mixtures with slivered or flat intermediate aggregate require more fine sizes as their shape creates mobility problems.

Zone III—An extension of Zone II for maximum aggregate size equal to or smaller than 1/2 in. (12.5 mm).

Zone IV—Excessive fines lead to a high potential for segregation during consolidation and finishing. Such mixtures will produce variable strength, have high permeability, and exhibit shrinkage, which generally contributes to the development of cracking, curling, spalling, and scaling. They are undesirable.

Zone V—They are too coarse, that is, nonplastic; therefore, an increase in fines content is necessary.

3. 0.45 power chart—The 0.45 power chart (Fig. 6.2) is similar to a semi-log graph, except the x-axis is the sieve opening in microns to the 0.45 power. It has been widely used by the asphalt industry to determine the best combined grading to reduce voids and the amount of asphalt in a mixture. A straight line on this chart defines the densest grading for aggregate for asphalt. Because asphalt includes fine mineral filler while concrete includes cementitious materials, fewer fine particles passing the No. 8 (2.36 mm) sieve are necessary for concrete mixtures.

This chart, historically used to develop uniform gradations in the asphalt industry, can also be adapted for use with concrete materials. The mixture example at the end of this chapter shows acceptable and improved gradations. To create a 0.45 power curve, plot the mathematically combined percent passing for each sieve on a chart having percent passing on the y-axis and sieve sizes raised to the 0.45 power on the x-axis. Plot the maximum density line from the origin of the chart to the sieve one size larger than the first sieve to have 90% or less passing.

The deviations from the optimum line help identify the location of grading problems. Gradings that zigzag across the line are undesirable. A gap-graded aggregate combination will form an S-shaped curve deviating above and below the maximum density line.

6.3.3 *Mortar fraction*—Mortar fraction is an extension of the coarseness factor chart. The mortar fraction consists of all materials passing the No. 8 (2.36 mm) sieve (fine aggregate and paste) and is often at the center of conflicting interests. With reasonably sound and properly distributed aggregate, it is the mortar fraction of the concrete mixture that has a major affect on the designer's interest in strength, drying shrinkage, durability, and creep. The mortar fraction also provides the contractor with necessary workability, pumpability, placeability, and finishability. Neither should dominate. A mixture that is optimized for strength and shrinkage but can't be properly placed and consolidated will perform poorly regardless of the *w/cm* value.

The mortar factor needed for various construction types varies. A mat foundation with the concrete placed by chute requires less mortar than the same strength concrete cast in a thin slab to be trowel finished. Unless aggregate proportions are adjusted to compensate for differing needs, changes in slump to increase mortar content through the addition of water is the only option open to the contractor. Construction requirements that affect the amount and quality of mortar necessary to properly place and finish the concrete materials should be considered when optimizing a mixture. Mortar fractions are influenced by aggregate particle shape, texture, and distribution, and will vary with each mixture. Approximate mortar fractions for ten construction classifications are shown as follows (Shilstone 1990):

Mortar fractions for various construction methods

Construction classification	Placing and construction method	Approximate mortar fractions (% volume of concrete)
1	Steep sided bottom-dropped bucket, conveyor, or paving machine	48 to 50
2	Bottom-drop bucket or chute in open vertical construction	50 to 52
3	Chute, buggy, or conveyor in an 8 in. (200 mm) or deeper slab	51 to 53
4	1-1/2 in. (37.5 mm) maximum size aggregate mixture receiving high tolerance finish 5 in. (125 mm) or larger pump for use in vertical construction, thick flat slabs and larger walls, beams, and similar elements	52 to 54
5	3/4 to 1 in. (19 to 25 mm) maximum size aggregate mixture receiving high tolerance finish 5 in. (125 mm) pump for pan joist slabs, thin or small castings, and high reinforcing steel density	53 to 55
6	4 in. (100 mm) pump	55 to 57
7	Long cast-in-place piling shells	56 to 58
8	Pump smaller than 4 in. (100 mm)	58 to 60
9	Less than 4 in.-thick (100 mm) topping	60 to 62
10	Flowing fill	63 to 66

Table 6.3—Mixture design example

Mixture components	Density		Original mixture (gap graded)					Adjusted mixture (optimized)				
			Mass		Volume			Mass		Volume		
	lb/ft³	kg/m³	lb	kg	ft³	m³	% by volume	lb	kg	ft³	m³	% by volume
Portland cement	196.6	3150	480	217.7	2.44	0.07	9.0	440	199.6	2.24	0.06	8.3
Pozzolan	146.7	2350	84	38.1	0.57	0.02	2.1	77	34.9	0.52	0.01	1.9
1 in. (25 mm) aggregate	164.2	2630	1706	773.8	10.39	0.29	38.5	1223	554.7	7.45	0.21	27.6
3/8 in. (9.5 mm) aggregate	164.2	2630	0	0.0	0.00	0.00	0.0	658	298.5	4.01	0.11	14.8
Fine aggregate	162.3	2600	1380	626.0	8.50	0.24	31.5	1292	586.0	7.96	0.23	29.5
Water	62.4	1000	292	132.4	4.68	0.13	17.3	275	124.7	4.41	0.12	16.3
Air	0.0	0	0	0.0	0.42	0.01	1.6	0	0.0	0.42	0.01	1.6
Totals			3942	1788.1	27.00	0.76	100.0	3965	1798.5	27.00	0.76	100.00
Combined fineness modulus					5.16					5.09		
Paste + air fraction					30%					27%		
Mortar fraction					58%					55%		
Coarseness factor					84%					60%		
Workability factor					40%					36%		

Table 6.4—Aggregate gradation example

Sieve size	Individual percent retained			Combined percent retained	
	1 in. (25 mm)	3/8 in. (9.5 mm)	Fine	Original	Adjusted
1-1/2 in. (37.5 mm)	—	—	—	—	—
1 in. (25.0 mm)	2.0	—	—	1.1	0.8
3/4 in. (19.0 mm)	18.0	—	—	9.9	6.9
1/2 in. (12.5 mm)	52.0	—	—	28.6	19.9
3/8 in. (9.5 mm)	20.0	12.0	—	11.0	10.1
No. 4 (4.75 mm)	6.0	72.0	0.9	3.7	17.5
No. 8 (2.36 mm)	0.7	13.0	11.7	5.7	7.7
No. 16 (1.18 mm)	0.4	1.7	16.3	7.6	7.2
No. 30 (600 μm)	0.9	1.3	24.1	11.3	10.5
No. 50 (300 μm)	—	—	34.2	15.4	14.0
No. 100 (150 μm)	—	—	11.6	5.2	4.8
No. 200 (75 μm)	—	—	1.2	0.5	0.5
Combined fineness modulus				**5.16**	**5.09**

6.3.4 *Example slab mixture analysis*—Table 6.3 shows a concrete slab mixture design example that compares an original gap-graded mixture that is high in fine content with an adjusted mixture that uses an intermediate-size aggregate material to improve the uniformity of the combined aggregate particle size distribution. The total combined aggregates in the original mixture consist of blending a 1 in. (25 mm) maximum-sized stone source with a single sand source. The adjusted, or optimized, mixture adds a 3/8 in. (9.5 mm) intermediate aggregate to supplement the particle sizes that were previously lacking in the combined gradations of the original mixture (Table 6.4).

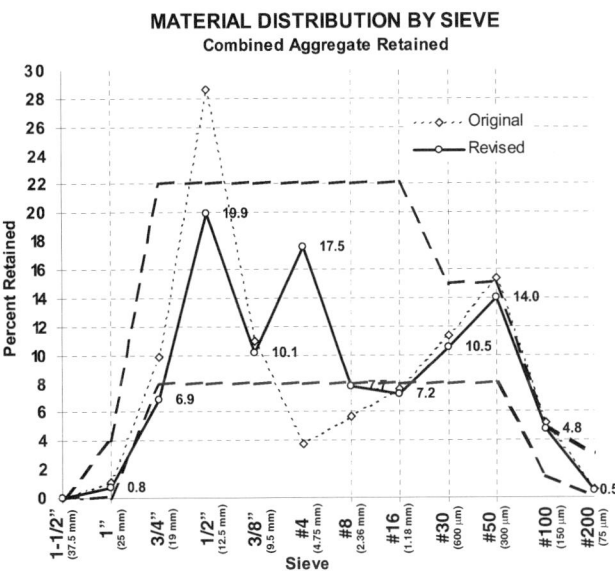

Fig. 6.3—Material sieve analysis showing change in aggregate distribution with blending of 3/8 in. (9.5 mm) aggregate (Section 6.3.4).

As shown in the following, improving the uniformity of aggregate particle size distribution (Fig. 6.3 and 6.4) will many times reduce the amount of paste necessary to coat the particles, also resulting in a reduction in the amount of water necessary to produce the same workability and finishability in the mixture. A reduction in paste also reduces the cost of producing the concrete mixture while maintaining equivalent strength and improved durability of the concrete. The potential for cracking and curling is reduced because shrinkage is reduced.

Affected changes in Coarseness Factor Chart—

1. The blending of the 3/8 in. (9.5 mm) aggregate changes the coarseness factor (Fig. 6.5) from 84 to 60%, moving the mixture optimization indicator (MOI) to the right. At this point, the total cementitious materials content is equal to 564 lb/yd³ (335 kg/m³). Therefore, both the MOI and the

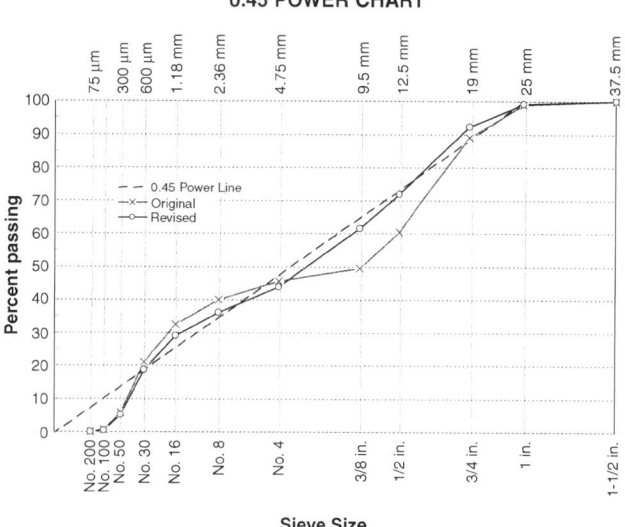

Fig. 6.4—The 0.45 power chart showing change in combined gradation of aggregates after blending of 3/8 in. (9.5 mm) aggregate (Section 6.3.4).

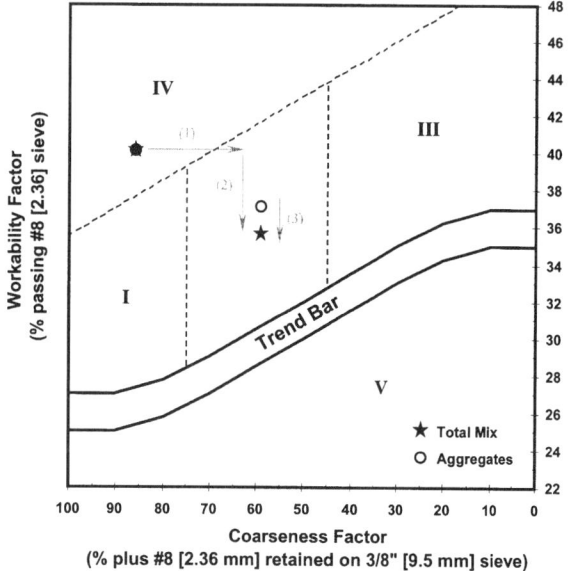

Fig. 6.5—Coarseness factor chart showing the impact of blending 3/8 in. (9.5 mm) aggregate (Section 6.3.4).

MOI-Adj (influenced by the adjusted workability factor W-Adj) are the same.

2. The fine aggregate content is reduced, changing the workability factor (WF) from 40 to 36% optimized for the new coarseness factor of 60%.

3. Due to the improved uniformity of the gradations, it is determined that less cementitious material is necessary to produce the same flexural strength without degrading the finishing characteristics of the mixture. This is due to a decrease in void space between the combined aggregates. The amount of fine paste materials necessary to fill these void spaces is reduced, thus reducing the amount of paste necessary for the mixture. The reduction in cementitious materials reduces the W-Adj value, thus moving MOI-Adj down the y-axis.

CHAPTER 7—BATCHING, MIXING, AND TRANSPORTING

Detailed provisions relating to batching, mixing, and transporting concrete are available in ASTM C 94, ASTM C 1116, and ASTM C 685.

7.1—Batching

Whether the concrete is centrally mixed on-site or in a ready-mixed concrete operation, the materials should be batched within the following ranges of the target batch weights:

Cement	±1%
Added water	±1%
Fine and coarse aggregate	±2% (if cumulative; 1% if individual)
Admixtures and pigments	±3%

Except for site mixing on small jobs, cement should be weighed on a scale separate from that used for weighing aggregates. If batching is by the bag, no fractional bags should be used.

Aggregate should be batched by weight (mass). Batching by volume should not be permitted, except with volumetric batching and continuous-mixing equipment (Section 7.2.2). Batch weights should be adjusted to compensate for absorbed and surface moisture. When the concrete mixture contains special aggregates, particular care should be exercised to prevent segregation or contamination.

Water can be batched by weight (mass) or volume. The measuring device used should have readily adjustable positive cutoff and should have provisions for calibration.

Accurate batching of admixtures and colored pigments is critical because they are used in relatively small quantities. Admixtures should be accurately batched at the concrete plant. Admixtures that are designed to be added to the concrete at the jobsite should be incorporated in accordance with the manufacturers' recommendations. When more than one admixture is batched, each should be batched separately and in such a way that the concentrated admixtures do not come into contact with each other. Care should be taken to avoid the freezing of admixtures in cold weather, as this can damage some of them. It is preferable to purchase pigments or colored admixtures prepackaged in batch-sized quantities. Powdered admixtures should be batched by weight, and paste or liquid admixtures by weight or volume. The volume of admixture batched should not be controlled by timing devices. Liquid admixtures are preferred but can require agitation to prevent the settling of solids.

7.2—Mixing

7.2.1 *Ready-mixed concrete*—Mixing should be in accordance with ASTM C 94 or ASTM C 1116 and should produce the required slump and air content without exceeding the specified *w/cm*. Close attention should be given to the moisture content of the aggregate and to the necessary adjustments to batched weights. Truck mixers should be in compliance with requirements of the project specification. To ensure consistent slump at the point of placement, a small quantity of adjustment water should be held out at the batch plant. The amount of withheld water should be indicated on the ticket; the truck should then leave the plant with a full water tank to allow the addition of any water on-site to be monitored and controlled.

7.2.2 *Site mixing*—Mixers that produce a volume of concrete requiring less than one bag of cement should not be used. For small quantities of concrete, packaged products meeting ASTM C 387 are more convenient and can be more accurately proportioned.

Mixing time should be sufficient to produce uniformly mixed concrete with the required slump and air content. Site mixers less than 1 yd^3 (0.75 m^3) in capacity should mix for no less than 3 min; ordinarily, 15 s should be added for each additional cubic yard (0.75 m^3) of capacity or fraction thereof, unless a turbine mixer is used. A longer mixing time is required for concrete with a slump of less than 3 in. (75 mm).

Equipment for volumetric batching and continuous mixing at the jobsite is available. Concrete produced in this manner should comply with ASTM C 685.

7.2.3 *Architectural concrete*—When special architectural concrete is produced using special aggregates, white cement, special cements, or pigments, mixer drums and equipment should be kept clean and any wash water should be disposed of before a new batch is introduced. Identical ingredients and quantities of materials should be used, using no less than 1/3 of the capacity of the mixing drum, a minimum of 3 yd in a 9 yd drum, and should always be in full yard increments. See ACI 303R for additional details.

7.2.4 *Shrinkage-compensating concrete*—When expansive cement or an expansive-component type admixture specifically designed for producing shrinkage-compensating concrete is required, refer to ACI 223 for details.

7.3—Transporting

7.3.1 *Discharge time*—Concrete mixed or delivered in a truck mixer should be completely discharged while the concrete still has sufficient workability to respond properly during the placing and finishing operations. The period after arrival at the jobsite during which the concrete can be properly worked will generally vary from less than 45 min to more than 90 min, depending on the weather, the concrete mixture proportions, and travel time from the batch plant to the jobsite. Prolonged mixing accelerates the rate of stiffening and can greatly complicate placing and timing of finishing operations.

7.3.2 *Jobsite slump control*—When concrete arrives at the point of placement with a slump below what will result in the specified slump at the point of placement and is unsuitable for placing at that slump, the slump may be adjusted to the required value by adding water up to the amount allowed in the accepted mixture proportions. The addition of water should be in accordance with ASTM C 94. The specified *w/cm* or slump should not be exceeded. Water should not be added to concrete delivered in a transit mix truck or similar equipment acceptable for mixing. Test samples for compressive strength, slump, air content, and temperature should be taken after any necessary adjustment. Refer to ACI 301 for further details.

7.3.3 *Delivery to point of discharge*—Concrete for floor and slab placement can be delivered to the forms directly from a truck mixer chute or by pump, belt conveyor, buggy, crane and bucket, or a combination of these methods. Delivery of concrete should be at a consistent rate, appropriate to the size of the placement, and should be deposited as close as possible to its final location. Concrete should not be moved horizontally by vibration, as this contributes to segregation. See ACI 304R for recommended procedures.

7.3.4 *Required yield and mixture adjustment to correct yield*—Refer to Section 6.2.

CHAPTER 8—PLACING, CONSOLIDATING, AND FINISHING

Most of this chapter applies to both normalweight and lightweight concrete. The proper procedures for finishing lightweight concrete floors differ somewhat from finishing normalweight concrete. Procedures specific to finishing lightweight concrete are discussed separately in Section 8.11.

Various finishing procedures should be executed sequentially and within the proper time period, neither too early nor too late in the concrete-hardening process. This time period is called the window of finishability. It refers to the time available for operations taking place after the concrete has been placed, consolidated, and struck off. Surface finish, surface treatment, and flatness/levelness requirements dictate the type and number of finishing operations. All should take place within the proper time period. If the floor slab is placed during a time period of rapid hardening, this window becomes so small that it can present considerable difficulties to the floor contractor. The preconstruction meeting should include discussion of the measures necessary to ensure a satisfactory window of finishability. The NRMCA/ASCC preconstruction checklist provides an outline of topics that might be considered for inclusion in the meeting agenda. ACI 311.4R also contains guidance that may be valuable.

8.1—Placing operations

8.1.1 *Caution*—All concrete handling operations should minimize segregation, because it is difficult to remix concrete after it has been placed.

8.1.1.1 *Placing sequence*—In many cases, the most efficient way to place concrete in large areas is in long alternating strips, as illustrated in Fig. 8.1. Strip placements allow superior access to the sections being placed. Intermediate contraction joints are installed at specified intervals transverse to the length of the strips. Wide strip placements can require installation of longitudinal contraction joints.

Large block placements with interior contraction joints are an acceptable alternative to strip placements if the contraction joints are installed at specified intervals in a timely manner. The use of shrinkage-compensating concrete (because of the decrease in jointing requirements) or laser screeds (because they provide accurate strikeoff without the use of edge forms) is compatible with large block placements.

A checkerboard sequence of placement with side dimensions of 50 ft (15 m) or less, as shown in Fig. 8.1, has been used in the past in an effort to permit earlier placements to shrink and to obtain minimum joint width. Experience has shown, however, that shrinkage of the earlier placements

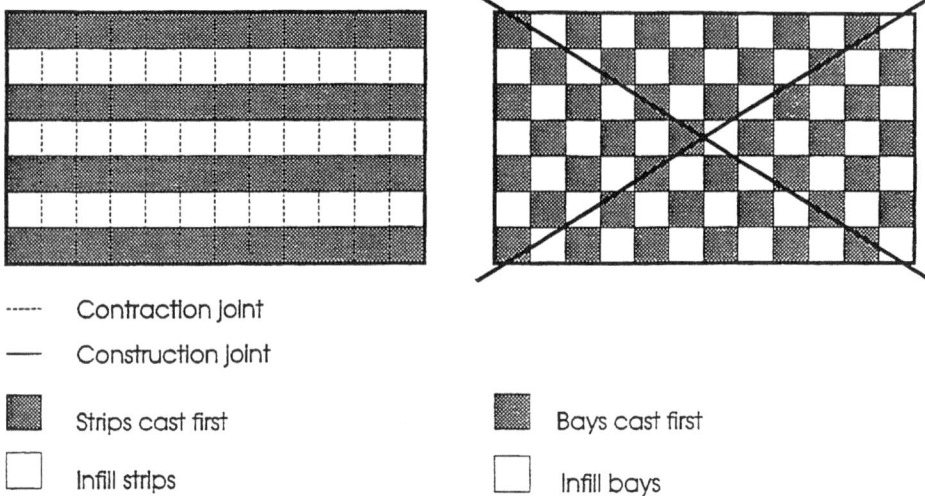

Fig. 8.1—Placing sequence: long-strip construction (left) is recommended; checkerboard construction (right) is not recommended.

occurs too slowly for this method to be effective. Access is more difficult and expensive, and joints may not be as smooth. The committee recommends that the checkerboard sequence of placement not be used.

8.1.1.2 *Placing sequence for shrinkage-compensating concrete*—Neither the strip method nor the checkerboard method described in Section 8.1.1.1 should be used with shrinkage-compensating concrete. Refer to ACI 223 for specific recommendations concerning placement configuration and sequence.

8.1.2 *Discharge rate of concrete*—The rate of discharge of concrete from a truck mixer can be controlled by varying the drum speed.

8.1.3 *Jobsite transfer*—Chutes should have rounded bottoms and be constructed of metal or be metal-lined. The chute slope should be constant and steep enough to permit concrete of the slump required to flow continuously down the chute without segregation. Long flat chutes should be avoided because they encourage the use of high-slump concrete. A baffle at the end of the chute helps to prevent segregation. The discharge end of the chute should be near the surface of previously deposited concrete. When concrete is being discharged directly onto the base, the chute should be moved at a rate sufficient to prevent accumulation of large piles of concrete. Allowing an excessively steep slope on chutes can result in high concrete velocity and segregation.

Regardless of the method of transportation and discharge, the concrete should be deposited as near as possible to its final position and toward previously placed concrete. Advance planning should include access to and around the site, suitable runways, and the use of other devices to avoid the use of concrete with a high w/cm or excessive delays.

8.1.4 *Placing on base*—Mixing and placing should be carefully coordinated with finishing operations. Concrete should not be placed on the base at a faster rate than it can be spread, bull floated or darbied, and restraightened because these latter operations should be performed before bleeding water has an opportunity to collect on the surface.

Proper sizing of finishing crews, with due regard for the effects of concrete temperature and atmospheric conditions on the rate of hardening of the concrete, will assist the contractor in obtaining good surfaces and avoiding cold joints. If construction joints become necessary, they should be produced using suitably placed bulkheads, with provisions made to provide load transfer between current and future work (Section 3.2.7).

8.2—Tools for spreading, consolidating, and finishing

The sequence of steps commonly used in finishing unformed concrete floor surfaces is illustrated in Fig. 8.2. Production of high-quality work requires that proper tools be available for the placing and finishing operations. Following is a list and description of typical tools that are commonly available. Refer to Section 8.3 for suggestions and cautions concerning uses of these tools. Definitions for many of these tools can be found in ACI 116R.

8.2.1 *Spreading*—Spreading is the act of extending or distributing concrete or embedding hardeners—often referred to as shake-on or dry-shake—or other special purpose aggregate over a desired area.

8.2.1.1 *Spreading concrete*—The goal of spreading operations for concrete is to avoid segregation.

8.2.1.1.1 *Hand spreading*—Short-handled, square-ended shovels, or come-alongs—hoe-like tools with blades about 4 in. (100 mm) high, 20 in. (500 mm) wide, and curved from top to bottom—should be used for the purpose of spreading concrete after it has been discharged.

8.2.1.2 *Spreading dry-shake hardeners, colored dry-shake hardeners, or other special-purpose material*—The goal of spreading operations for these materials is to provide an even distribution of product over the desired area. Generally, hand application should be used for distribution of these materials in areas where a mechanical spreader cannot be used.

8.2.1.2.1 *Mechanical spreaders*—Mechanical spreaders are the best method of uniformly applying dry-shake hardeners, colored dry-shake hardeners, or other special purpose materials to concrete during the finishing process. These devices generally consist of a bin or hopper to hold the material, a vibrator or motorized auger to assist in distribution of the material, and a supporting framework that allows the hopper to move smoothly over the concrete surface while distributing the material (Fig. 8.3).

8.2.2 *Tools for consolidating*—Consolidation is the process of removing entrapped air from freshly placed concrete, usually by vibration. Internal vibration and surface vibration are the most common methods of consolidating concrete in supported slabs and slabs-on-ground. Refer to ACI 309R for additional discussion of topics related to the consolidation of concrete.

8.2.2.1 *Internal vibration*—This method employs one or more vibrating elements that can be inserted into the fresh concrete at selected locations. Internal vibration is generally most applicable to supported cast-in-place construction.

8.2.2.2 *Surface vibration*—This process employs a portable horizontal platform on which a vibrating element is mounted. Surface vibration is commonly used in slab-on-ground, strip-type placements with edge forms. Refer to 8.2.3.2 for additional discussion.

8.2.3 *Tools for screeding*—Screeding is the act of striking off concrete lying above the desired plane or shape to a predetermined grade. Screeding can be accomplished by hand, using a straightedge consisting of a rigid, straight piece of wood or metal, or by using a mechanical screed.

8.2.3.1 *Hand screeding*—Hollow magnesium or solid wood straightedges are commonly used for hand-screeding of concrete. The length of these straightedges varies up to approximately 20 ft (6 m). Straightedge cross-sectional dimensions are generally 1 to 2 in. (25 to 50 mm) wide by 4 to 6 in. (100 to 150 mm) deep. Tools specifically made for screeding, such as hollow magnesium straightedges, should be used instead of randomly selected lumber.

8.2.3.2 *Mechanical screeding*—Various types of surface vibrators, including vibrating screeds, vibratory tampers, and vibratory roller screeds, are used mainly for screeding slab-on-ground construction. They consolidate concrete from the top down while performing the screeding function. Refer to ACI 309R for a detailed discussion of equipment and parameters for proper usage.

Vibrating screeds generally consist of either hand-drawn or power-drawn single-beam, double-beam, or truss assemblies. They are best suited for horizontal or nearly horizontal surfaces. Vibrating screeds should be of the low-frequency—3000 to 6000 vibrations per min (50 to 100 Hz)—high-amplitude type, to minimize wear on the machine and provide adequate depth of consolidation without creating an objectionable layer of fines at the surface. Frequency and amplitude should be coordinated with the concrete mixture designs being used.

Laser-controlled variations of this equipment can be used to produce finished slabs-on-ground with improved levelness over that which might otherwise be achieved. Laser-controlled screeds can ride on supporting forms or they can operate from a vehicle using a telescopic boom (Fig. 8.4).

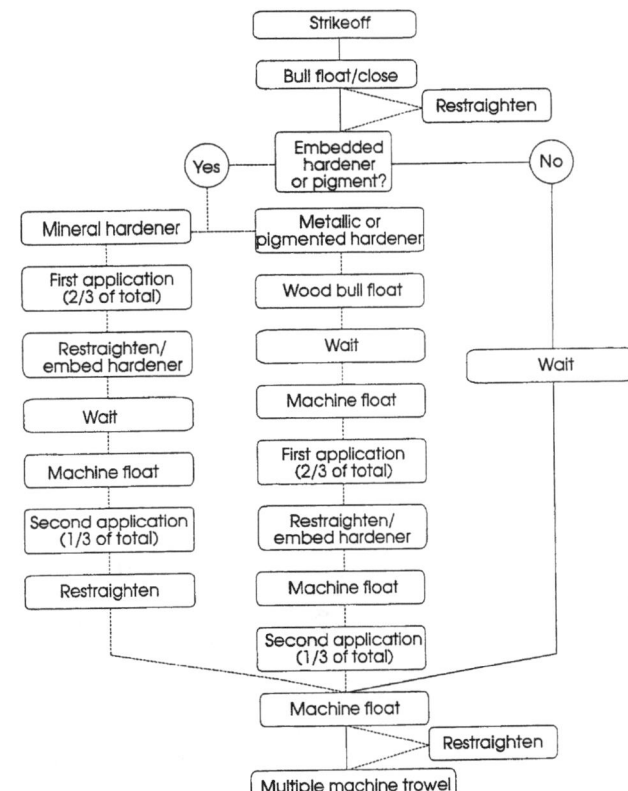

NOTE: Dashed lines indicate optional procedures that may be included as a part of the finishing operation. Restraightening is encouraged as a means of improving surface flatness.

Fig. 8.2—Typical finishing procedures (subject to numerous conditions and variables).

Fig. 8.3—Mechanical spreader.

Plate-tamper screeds are vibratory screeds that are adjusted to a lower frequency and amplitude. Tamper screeds work best on very stiff concrete. These screeds are generally used to embed metallic or mineral aggregate hardeners. The contractor is cautioned that improper use of this screed could embed the hardener too deeply and negate the intended benefit.

Vibratory-roller screeds knock down, strike off, and provide mild vibration. They can rotate at varying rates up to several hundred revolutions per minute, as required by the

Fig. 8.4—Laser-controlled screed.

Fig. 8.5—Double-riding trowel with clip-on pans.

consistency of the concrete mixture. The direction of rotation of the rollers on the screed is opposite to the screed's direction of movement. These screeds are most suitable for concrete mixtures with higher slumps.

8.2.4 *Tools for floating*—Floating is the act of consolidating and compacting the unformed concrete surface in preparation for subsequent finishing operations. Initial floating of a concrete floor surface takes place after screeding and before bleed water comes to the surface and imparts a relatively even but still open texture to the fresh concrete surface. After evaporation of bleed water, additional floating operations prepare the surface for troweling.

8.2.4.1 *Bull floats (long-handled)*—Bull floats are used to consolidate and compact unformed surfaces of freshly placed concrete immediately after screeding operations, while imparting an open texture to the surface. They are usually composed of a large, flat, rectangular piece of wood or magnesium and a handle. The float part of the tool is usually 4 to 8 in. (100 to 200 mm) wide and 3.5 to 10 ft (1 to 3 m) long. The handle is usually 4 to 20 ft (1.2 to 6 m) long. The handle is attached to the float by means of an adjustable head that allows the angle between the two pieces to change during operation.

8.2.4.2 *Darby*—A darby is a hand-manipulated float, usually 3-1/2 in. (90 mm) wide and 3 to 8 ft (1 to 2.4 m) long. It is used in early-stage-floating operations near the edge of concrete placements.

8.2.4.3 *Hand floats*—Hand tools for basic floating operations are available in wood, magnesium, and composition materials. Hand float surfaces are generally about 3-1/2 in. (90 mm) wide and vary from 12 to 20 in. (300 to 500 mm) in length.

8.2.4.4 *Power floats*—Also known as rotary floats, power floats are engine-driven tools used to smooth and to compact the surface of concrete floors after evaporation of the bleed water. Two common types are heavy, revolving, single-disk-compactor types that often incorporate some vibration, and troweling machines equipped with float blades. Most troweling machines have four or more blades mounted to the base and a ring diameter that can vary from 36 to 46 in. (1 to 1.2 m); mass generally varies from about 150 to 250 lb (70 to 110 kg).

Two types of blades can be used for the floating operation. Float blades are designed to slip over trowel blades; they are generally 10 in. (250 mm) wide and 14 to 18 in. (350 to 450 mm) long. Both the leading edge and the trailing edge of float blades are turned up slightly. Combination blades are usually 8 in. (200 mm) wide and vary in length from 14 to 18 in. (350 to 450 mm). The leading edges of combination blades are turned up slightly. The use of float blades is recommended (Section 8.3.10).

Another attachment that is available to assist in power float operations is a pan with small brackets that slide over the trowel blades. These pans are normally used on double- or triple-platform ride-on machines and are very effective on concrete surfaces requiring an embedded hardener or coloring agent. The use of mechanical pan floating (Fig. 8.5) can also materially improve flatness of the finished floor.

8.2.5 *Tools for restraightening*—Straightedges are used to create and to maintain a flat surface during the finishing process. Straightedges vary in length from 8 to 12 ft (2.4 to 3.7 m) and are generally rectangular in cross section (though designs differ among manufacturers). When attached to a handle with an adjustable head (that is, a bull-float handle and head), these tools are frequently referred to as modified highway straightedges (Fig. 8.6).

8.2.6 *Tools for edging*—Edgers are finishing tools used on the edges of fresh concrete to provide a rounded edge. They are usually made of stainless steel and should be thin-lipped. Edgers for floors should have a lip radius of 1/8 in. (3 mm).

8.2.7 *Tools for troweling*—Trowels are used in the final stages of finishing operations to impart a relatively hard and dense surface to concrete floors and other unformed concrete surfaces.

8.2.7.1 *Hand trowels*—Hand trowels generally vary from 3 to 5 in. (75 to 125 mm) in width and from 10 to 20 in. (250 to 500 mm) in length. Larger sizes are used for the first troweling to spread the troweling force over a large area. After the surface has become harder, subsequent trowelings use smaller trowels to increase the pressure transmitted to the surface of the concrete.

8.2.7.2 *Fresno trowels*—A fresno is a long-handled trowel that is used in the same manner as a hand trowel. Fresnos are useful for troweling slabs that do not require a hard-troweled surface. These tools are generally 5 in. (125 mm) wide and vary in length from 24 to 48 in. (0.6 to 1.2 m).

8.2.7.3 *Power trowels*—Power trowels are gasoline engine-driven tools used to smooth and compact the surface of concrete floors after completion of the floating operation. Ring diameters on these machines generally vary from 36 to 46 in. (0.9 to 1.2 m); their mass generally varies from approximately 150 to 250 lb (70 to 110 kg). Trowel blades are usually 6 in. (150 mm) wide and vary in length from 14 to 18 in. (350 to 450 mm). Neither the leading nor the trailing edge of trowel blades is turned up. Power trowels can be walk-behind machines with one set of three or four or more blades or ride-on machines with two or three sets of four blades.

8.2.8 *Tools for jointing*—These tools are used for the purpose of creating contraction joints in slabs. Contraction joints can be created by using groovers, also called jointers, or by saw-cutting.

8.2.8.1 *Groovers*—Groovers can be handheld or walk-behind. Stainless steel is the most common material. Handheld groovers are generally from 2 to 4-3/4 in. (50 to 120 mm) wide and from 6 to 7-1/2 in. (150 to 190 mm) long. Groove depth varies from 3/16 to 1-1/2 in. (5 to 38 mm). Walk-behind groovers usually have a base with dimensions from 3-1/2 to 8 in. (90 to 200 mm) in width and from 6 to 10 in. (150 to 250 mm) in length. Groove depth for these tools varies from 1/2 to 1 in. (13 to 25 mm).

8.2.8.2 *Saw-cutting*—The following three types of tools can be used for saw-cutting joints: conventional wet-cut (water-injection) saws; conventional dry-cut saws; and early-entry dry-cut saws. Timing of the sawing operations will vary with manufacturer and equipment. The goal of saw-cutting is to create a weakened plane to influence the location of shrinkage crack formation as soon as the joint can be cut, preferably without creating spalling at the joint.

Both types of dry-cut tools can use either electrical or gasoline power. They provide the benefit of being generally lighter than wet-cut equipment. Early-entry dry-cut saws do not provide as deep a cut—generally 1-1/4 in. (32 mm) maximum—as can be achieved by conventional wet-cut and dry-cut saws.

Early-entry dry-cut saws use diamond-impregnated blades and a skid plate that helps prevent spalling. Timely changing of skid plates is necessary to effectively control spalling. It is best to change skid plates in accordance with manufacturers' recommendations.

Conventional wet-cut saws are gasoline powered and, with the proper blades, are capable of cutting joints with depths of up to 12 in. (300 mm) or more.

8.3—Spreading, consolidating, and finishing operations

This section describes the manner in which various placing and finishing operations can be completed successfully. The finishing sequence to be used after completion of the initial screeding operation depends on a number of variables

Fig. 8.6—"Modified highway" straightedge.

related to project requirements or to the concrete finishing environment.

Project variables are generally controlled by requirements of the owner and are specified by the designer. Some examples are the choice of admixtures used in concrete, the requirement for an embedded hardener, and the final finish desired.

Variables subject to the environment include such items as setting time of the concrete, ambient temperature, timeliness of concrete delivery, consistency of concrete at the point of deposit, and site accessibility. Figure 8.2 is a flowchart that illustrates the normal sequence of steps in the finishing process.

8.3.1 *Spreading and consolidating*—Concrete, whether from a truck mixer chute, wheelbarrow, buggy, bucket, belt conveyor, pump, or a combination of these methods, should be delivered without segregation of the concrete components (Section 8.1). Spreading, the first operation in producing a plane surface (not necessarily a level surface because, in many cases, it can be sloped for surface drainage) should be performed with a come-along or a short-handled, square-ended shovel (Section 8.2.1.1).

Long-handled shovels, round-ended shovels, or garden-type rakes with widely spaced tines should not be used to spread concrete. Proper leverage, of prime importance for manipulating normalweight concrete, is lost with a long-handled shovel. Round-ended shovels do not permit proper leveling of the concrete. The tines of garden-type rakes can promote segregation and should not be used in any concrete.

Initial consolidation of concrete in floors, with the exception of heavily reinforced slabs, is usually accomplished in the first operations of spreading, vibrating, screeding, darbying or bull floating, and restraightening. The use of grate tampers or mesh rollers is usually neither desirable nor necessary, unless the concrete slump is less than 3 in. (75 mm). If grate tampers are used on lightweight-concrete floors, only one pass over the surface with a very light impact should be permitted. Spreading by vibration should be minimized. Refer to ACI 309R for detailed discussion.

8.3.1.1 *Structural floors*—Structural floors, both suspended and on ground, can be reinforced with relatively large deformed reinforcing bars or with post-tensioning tendons and typically contain other embedded items such as piping and conduit. Proper consolidation around reinforcing steel, post-tensioning anchorages, and embedded elements requires internal vibration, but care should be taken not to use the vibrator for spreading the concrete, especially in deeper sections where over-vibration can easily cause segregation. Restraint of concrete movement by embedded items such as piping and conduits can result in crack formation as the concrete shrinks.

The vibrator head should be completely immersed during vibration. Where slab thickness permits, it is proper to insert the vibrator vertically. On thin slabs, the use of short 5 in. (125 mm) vibrators permits vertical insertion. Where the slab is too thin to allow vertical insertion, the vibrator should be inserted at an angle or horizontally. The vibrator should not be permitted to contact the base because this might contaminate the concrete with foreign materials.

8.3.2 *Screeding*—Screeding is the act of striking off the surface of the concrete to a predetermined grade, usually set by the edge forms. This should be done immediately after placement. When hand strikeoff is used, a slump of 5 in. (125 mm) or higher should be used to facilitate strikeoff and consolidation of concrete without mechanical methods. Refer to Section 8.2.3 for tools used for screeding.

Of all the floor-placing and finishing operations, form setting and screeding have the greatest effect on achieving the specified grade. Accuracy of the screeding operation is directly impacted by the stability and the degree of levelness of the edge forms or screed guides selected by the contractor. Consequently, care should be taken to match the forming system and the screeding method to the levelness tolerance specified.

Edge forms for slab-on-ground and suspended-slab placements are normally constructed of wood or metal. Some edge forms are constructed of concrete. The spacing between edge forms, and the support provided for them, will influence the accuracy of the screeding operation. Where edge-form spacing exceeds the width of the screed strip, intermediate screed guides can improve the accuracy of the screeding operation. The width of these screed strips will generally vary between 10 and 16 ft (3 and 5 m) and will be influenced by column spacings. Generally, screed strips should be equal in width and should have edges that fall on column lines.

In general, slab-on-ground placements are either block placements or strip placements. Block placements generally have edge dimensions that exceed 50 ft (15 m). Strip placements are generally 50 ft (15 m) or less in width and vary in length up to several hundred feet. Suspended-slab placements are usually block placements. Where wood is used for edge forms, the use of dressed lumber is recommended. The base should be carefully fine-graded to ensure proper slab thickness.

Selection of the type of screed guide to be used for screeding operations is somewhat dependent on placement configuration. The maximum practical width of screed strips for hand screeding is approximately 20 ft (6 m). Where strict elevation tolerances apply, it is wise to limit the width of screed strips. The length of the hand screeding device should not be longer than 16 ft (5 m) and should overlap previously placed strips of wet concrete a minimum of 2 ft (600 mm).

Screeding of strip placements for slabs-on-ground is generally completed using some type of a vibrating screed supported by edge forms. Screeding of block placements for slabs-on-ground is usually accomplished using wet-screed guides, dry-screed guides, a combination of these two, or some type of laser-guided screed. For slabs-on-ground, an elevation change no greater than 3/8 in. (9.5 mm) in 10 ft (3 m), approximately $F_L 35$, can be achieved routinely through use of laser-guided screeds (refer to Section 8.15 for discussion of floor flatness and levelness). Screeding of block placements for suspended slabs is usually accomplished using either wet-screed guides, dry-screed guides, or a combination of the two.

Wet-screed guides, when used between points or grade stakes, are established immediately after placement and spreading; refer to Section 4.4 for setting of dry-screed guides. At the time of floor placement, before any excess moisture or bleed water is present on the surface, a narrow strip of concrete not less than 2 ft (600 mm) wide should be placed from one stake or other fixed marker to another, and straightedged to the top of the stakes or markers; then another parallel strip of concrete should be placed between the stakes or markers on the opposite side of the placement strip. These two strips of concrete, called wet-screed guides, establish grade for the concrete located between the guides. Immediately after wet-screed guides have been established, concrete should be placed in the area between, then spread and straightedged to conform to the surface of the wet-screed guides. The contractor should confirm that proper grade has been achieved following strikeoff. High spots and low spots should be identified and immediately corrected. Low spots left behind should be filled by placing additional concrete in them with a shovel, carefully avoiding segregation. Nonconforming areas should then be rescreeded. Difficulty in maintaining the correct grade of the floor while working to wet-screed guides is an indication that the concrete mixture does not have the proper consistency or that vibration is causing the guides to move.

Elevation stakes placed at regular intervals are one method of establishing grade for wet-screed guides in slab-on-ground construction. As screeding progresses, the stakes can be driven down flush with the base if expendable or pulled out one at a time to avoid walking back into the screeded concrete. This early removal of stakes is one of the big advantages in the use of wet-screeds; in addition, grade stakes are much easier and faster to set than dry-screeds. Screeding should be completed before any excess moisture or bleed water is present on the surface.

The benefits of using wet-screed guides include economical and rapid placement of the concrete. The successful use of wet-screed guides, however, requires careful workmanship by craftspeople who strike off the concrete because vibration can change the elevation of the wet-screed. Wet-screed guides are difficult to use when varying surface slopes are required and can produce inconsistent results when variations in slab thickness are required to compensate for deflection of a suspended slab. Special care is necessary to avoid poor consolidation or cold joints adjacent to wet-screed guides.

Wet-screed guides should not be used in suspended-slab construction, unless the finished floor surface is level and formwork is shored at the time of strikeoff. During construction activity, vibration of reinforcing steel and the supporting platform may result in an incorrect finished grade when wet-screed guides are used. It is imperative, therefore, that grade be confirmed after strikeoff and that errors be corrected at that time by restriking the area.

Wet-screed guides should be used only for surfaces where floor levelness is not critical. For slabs on grade where floor levelness requirements are important, dry-screed guides should be used instead of wet-screed guides. In general, surfaces produced using wet-screed guides will exhibit maximum elevation changes of at least 5/8 in. (16 mm) in 10 ft (3 m). This corresponds to an F_L20 floor.

Elevation variation of surfaces produced using dry-screed guides is dependent on placement-strip width and the accuracy the guides are installed with. Generally, the maximum elevation changes that can be anticipated will be reduced as the dry-screed guides are moved closer together.

For suspended-slab construction, the desirability of using dry-screed guides on both sides of each placement strip is diminished by the damage done when the contractor retrieves the guide system. For this reason, a combination of dry-screed guide and wet-screed guide techniques should be employed on suspended slabs.

The first placement strip should always start against a bulkhead or edge of the building. Strikeoff on the interior side of the strip should be controlled through the use of moveable dry-screed guides, which will provide positive control over the surface elevation along that line. The concrete edge along the moveable guide should be kept near vertical and straight. As concrete is placed and struck off, these guides are removed. When the next strip is placed, preferably in the same direction as the initial strip, the prior strip will normally have been in place for 30 min or more. The contractor can extend the straightedge 2 ft (600 mm) or more over the previous partially set placement to control grade of strikeoff on that side of the strip and use moveable dry-screed guides to control grade on the side of the strip not adjacent to previously placed concrete.

For suspended-slab construction, the procedure described in the previous paragraph has several advantages over unmodified wet-screed techniques or those techniques that use dry-screed guides on both sides of each placement strip.

1. Where previously placed concrete is used as a guide for strikeoff, it provides a relatively stable guide because it will have been in place for some time before it is used.

2. Retrieval of the dry-screed guide from areas surrounded by previously placed concrete is unnecessary because dry-rigid guides are not used in these locations.

Moveable dry-screed guides should be used to establish grade on any suspended slabs that are not level and shored at the time of strikeoff, and for any suspended slab where increases in local slab thickness might be used to compensate for anticipated or identified differential deflection of the structure. When an increase in local slab thickness is used to compensate for differential floor deflection, it is likely that the resulting slab will be more than 3/8 in. (9.5 mm) thicker than design thickness. The contractor should secure permission from the designer to exceed the plus tolerance for slab thickness before beginning construction. Refer to Section 3.3 for a discussion of suspended slab deflection and suggested construction techniques.

For construction of slabs-on-ground, the use of vibrating screeds—where edge forms or screed-guide rails can be used—will facilitate strike-off operations. By using a vibrating screed, crews can place concrete at a lower slump than might be practical if screeding were done by hand. Suspended slabs are seldom both level and supported at the time of construction. Vibrating screeds and roller screeds similar to those used for slab-on-ground strip placements are generally not appropriate for use in suspended-slab construction because of the probability that their use will result in slabs that are too thin in localized areas. It is essential that minimum slab thickness be maintained at all locations on suspended slabs because of compliance with contract documents and fire safety requirements.

Slumps up to 5 in. (125 mm) are often recommended for concrete consolidated by vibrating screeds. If slumps in excess of 4 in. (100 mm) are used, the amplitude of vibration should be decreased in accordance with the consistency of the concrete so that the concrete does not have an accumulation of excess mortar on the finished surface after vibration.

Vibrating screeds strikeoff and straightedge the concrete in addition to providing consolidation. To perform significant consolidation, the leading edge of the blade should be at an angle to the surface, and the proper surcharge (height of unconsolidated concrete required to produce a finished surface at the proper elevation) should be carried in front of the leading edge.

Vibrating screeds should be moved forward as rapidly as proper consolidation allows. If not used in this manner, too much mortar will be brought to the surface in normalweight concrete; conversely, too much coarse aggregate will be brought to the surface in lightweight concrete.

8.3.3 *Floating*—Floating is used to describe compaction and consolidation of the unformed concrete surface. Floating operations take place at two separate times during the concrete finishing process.

The first floating, generally called bull floating, is by hand and takes place immediately after screeding. Initial floating should be completed before any excess moisture or bleeding water is present on the surface. Any finishing operation performed while there is excess moisture or bleed water on the surface will cause dusting or scaling. This basic rule of concrete finishing cannot be over-emphasized. The first floating operation is performed using a wide bull float, darby, or modified highway straightedge. The second floating operation takes place after evaporation of most of the bleed water and is usually performed using a power trowel with float blades or a pan attached. The second floating operation is described in Section 8.3.10.

8.3.3.1 *Bull floating*—One of the bull float's purposes is to eliminate ridges and fill in voids left by screeding operations. Bull floating should embed the coarse aggregate only slightly. This process prepares the surface for subsequent edging, jointing, floating, and troweling.

When the specified finished floor flatness, using the F-number system, restricts the difference between successive 1 ft (300 mm) slopes to a maximum of 1/4 in. (6 mm), approximately $F_F 20$ (Section 8.15), a traditional-width bull float of 4 to 5 ft (1.2 to 1.5 m) can be used to smooth and consolidate the concrete surface after screeding. The use of this width bull float, however, can adversely affect floor flatness and make it extremely difficult to achieve flatness F-numbers higher than 20. When the magnitude of difference between successive 1 ft (300 mm) slopes is limited to less than 1/4 in. (6 mm)—floor flatness greater than $F_F 20$ (Section 8.15)—an 8 to 10 ft wide (2.4 to 3 m) bull float can be very useful in removing surface irregularities early in the finishing process. This is particularly true for suspended-slab construction, where local irregularities caused by form- or metal-deck deflection and concrete leakage can be significant.

Many contractors use an 8 to 10 ft wide (2.4 to 3 m) bull float or modified highway straightedge after initial strikeoff to restraighten any local irregularities that can be present. Use of a traditional 4 to 5 ft wide (1.2 to 1.5 m) bull float will provide little assistance to the finisher in correcting these irregularities. Using the wider bull float or modified highway straightedge allows the finisher to recognize and correct irregularities at a time when significant amounts of material can be moved with relatively little effort. This simple substitution of tools can routinely produce up to a 50% increase in floor flatness.

In block placements for slabs-on-ground, and for suspended-slab placements, a wide bull float or modified highway straightedge can also be used. Applied at an angle of approximately 45 degrees to the axis of the placement strip and extending across the joint between the current strip and the strip previously placed, these tools can remove many irregularities that would otherwise remain if they were used only in a direction perpendicular to the axis of the placement strip.

A magnesium bull float can be used for lightweight concrete and sticky mixtures or where a partially closed surface is desired until it is time to float. The magnesium face of the bull float slides along the fines at the surface and thus requires less effort and is much less likely to tear the surface.

When an embedded hardener or other special purpose aggregate is required and rapid stiffening is expected, the use of a bull float, preferably wooden, can be helpful in initially smoothing the surface after the aggregate is applied and before the modified highway straightedge is used in the initial cutting and filling operation. Inevitable variations in the uniformity of coverage when an embedded hardener or other special purpose aggregate is applied will create slight irregularities in the slab surface. Restraightening operations necessary to remove these irregularities will remove embedded material in some locations while adding to the thickness of embedded material in other locations. Experience has shown that some variations in the uniformity of embedded material coverage does not adversely impact the floor's function.

Wooden bull floats are preferable for use on normalweight concrete that receives an embedded hardener. The wood's coarse texture enables it to evenly spread the mortar mixture of cement and fine aggregate across the surface, leaving the surface of the concrete open and promoting uniform bleeding. If a magnesium bull float is used for normalweight concrete, the embedded hardener should first be forced into the concrete using a wooden float. This brings moisture to the surface and ensures proper bond of the hardener to the base slab. This is particularly important where dry shakes will be applied for color, increased wear resistance, or both.

8.3.3.2 *Darbying*—Darbying serves the same purpose as bull floating, and the same rules apply. Because bull floating and darbying have the same effect on the surface of fresh concrete, the two operations should never be performed on the same surface. Because of its long handle, the bull float is easy to use on a large scale, but the great length of the handle detracts from the attainable leverage, so high tolerances are more difficult to achieve. A darby is advantageous on narrow slabs and in restricted spaces. Long-handled darbies should be used for better leverage and control of level. Metal darbies are usually unsatisfactory for producing surfaces meeting high-tolerance requirements. The same principles regarding the use of wooden or magnesium bull floats (Section 8.3.3.1) apply to darbies because both darbies and bull floats are used for the same purpose following screeding.

8.3.3.3 *Hand floating*—Wooden hand floats encourage proper workmanship and timing. If used too early on any type of concrete, they stick, dig in, or can tear the surface. Used too late, they roll the coarser particles of fine aggregate out of the surface, at which time use of a magnesium float held in a flat position would be preferable. Wooden floats more easily fill in low spots with mortar; they should also be used in areas where embedded hardeners or other special purpose aggregates will be applied, floated, and finished by hand only. The use of wooden hand floats has declined largely due to the need for periodic replacement because of wear or breakage, and the greater effort and care in timing required in using them. Used at the proper time, their floating action is unequaled by other hand tools.

Magnesium hand floats require less effort. Like magnesium bull floats, they slide along largely on fines. They can be used on concrete from the time of placement to beyond the point of stiffening when a wooden float cannot be used. Magnesium floats are best used in the initial smoothing of the surface near screeds, walls, columns, or other projections, and during placing, screeding, and bull floating, when a wooden float would dig in or tear the surface. Magnesium floats can also be used on air-entrained concrete that is not to receive a troweled finish, or following wooden or power floating to produce a more uniform swirl finish not quite as roughly textured. Well-worn magnesium floats develop an edge almost as sharp as a steel trowel, so care should be exercised to use them flat to avoid closing the surface too early or causing blisters.

Composition hand floats using resin-impregnated canvas surfaces are smoother than wooden floats and only slightly rougher than magnesium floats. They are similar to magnesium hand floats and should be used in the same manner.

8.3.4 *Highway-type straightedging*—The use of a modified highway straightedge for restraightening of the surface varies with the type of slab being installed. Experienced finishers can use this tool early in the finishing process instead of an 8 to 10 ft wide (2.4 to 3 m) bull float. Care is needed, however, because the straightedge tends to dig into the concrete if it is used improperly. Initial restraightening with the modified highway straightedge should immediately follow screeding. Restraightening should be completed before any excess moisture or bleed water is present on the surface. When specified differences between successive 1 ft (300 mm) slopes are 3/16 in. (5 mm) or less—flatnesses higher than $F_F 20$ (Section 8.15)—a modified highway straightedge is recommended to smooth and restraighten the surface after power floating or any floating operation that generates significant amounts of mortar. A weighted modified highway straightedge can also be used after power-trowel operations to scrape the surface, reducing local high spots. Filling of low spots is generally not appropriate after scraping with a weighted modified highway straightedge.

The flatness exhibited by any concrete floor will be determined almost exclusively by the effectiveness of corrective straightedging employed after each successive strikeoff, floating, and troweling step. Without restraightening, each step performed in a conventional concrete floor installation tends to make the surface less flat. Straightedges are capable of restraightening or reflattening the plastic concrete because they alone contain a reference line the resulting floor profile can be compared against. Restraightening operations are most effective when new passes with the modified highway straightedge overlap previous passes by approximately 50% of the straightedge width. In contrast, traditional 4 to 5 ft wide (1.2 to 1.5 m) bull floats, power floats, and power trowels are wave-inducing devices. To the extent that further restraightedgings can only reduce floor-wave amplitudes and enlarge floor-wave lengths, floor surface flatness can be further improved until Class 9 floor surface quality is obtained.

The modified highway straightedge is used in a cutting and filling operation to achieve surface flatness. When using this or any restraightening tool, it is desirable to overlap previous work with the tool by at least 50% of the tool width. The tool should be used in at least two directions, preferably in perpendicular directions to each other. For strip placements, this can be accomplished by using the straightedge at a 45 degree angle to the axis of the strip and toward the end of the strip, followed by use of the straightedge at a 45 degree angle toward the beginning of the strip. The cutting and filling operation taking place in these two directions from the edge of a placement strip will enable the straightedge passes to cross at right angles and to produce a flatter, smoother floor. Straightedging in a direction parallel to the strip-cast operation and to the construction joints is possible but less desirable because this would require the finisher to stand in the plastic concrete or on a bridge spanning the strip. This cut-and-fill process can also be performed after power-floating operations (Section 8.3.10) to further improve the floor's flatness.

For slabs-on-ground with an embedded metallic or mineral hardener, coloring agents, or other special-purpose material, the use of a modified highway straightedge plays an important part in reestablishing surface flatness after application of the material. These products are generally applied after initial screeding or strikeoff, and even the best of applications will create minor irregularities in the surface. After the hardener or special-purpose material has been worked into the surface of the concrete using a wooden bull float, a follow-up pass using the modified highway straightedge is desirable to restraighten the surface after the embedded metallic, mineral, and special-purpose material or its coating has absorbed sufficient moisture.

Some embedded metallic dry-shake hardeners and colored dry-shake hardeners are applied immediately after the initial power float pass (Section 8.3.10). When these materials are relatively fine, it is necessary to wait until this point in the finishing operation to begin their application. When applied too early in the finishing process, they tend to be forced below the surface by finishing operations. The use of a modified highway straightedge to embed these materials and to restraighten the surface after their application is a critical component of the finishing process.

Mechanical spreaders should be used for use in the application of metallic or mineral hardeners, colored dry-shake hardeners, or other special-purpose materials. Hand spreading often results in an inadequate and uneven application of the material, unless applied by highly skilled craftsmen.

8.3.5 *Waiting*—After initial floating and restraightening have been completed, a slight stiffening of concrete is necessary before proceeding with the finishing process. Depending on job conditions, it may be necessary to wait for this stiffening to occur. Waiting time can be reduced or eliminated by the use of dewatering techniques. No subsequent operation should be done until the concrete will sustain foot pressure with only approximately 1/4 in. (6 mm) indentation (Section 8.3.10).

8.3.6 *Dewatering techniques*—Completing the initial power float operation using a walk-behind machine and clip-on float blades can be very beneficial, particularly when steel fiber reinforcing is used. Mechanical pan floating should not begin until the surface has stiffened sufficiently so that footprints are barely perceived on the concrete surface.

For slabs-on-ground, the use of dewatering techniques as an alternative to waiting should be thoroughly discussed by key parties before implementation by the contractor. Either vacuum mats, or a blotter of cement on top of damp burlap, applied to the surface of freshly placed concrete, can be used to remove significant amounts of water. While this process quickly prepares the surface for final floating and troweling, it should only be undertaken by those with successful experience in the use of these techniques. The application of dry cement directly to the surface of freshly placed concrete should be avoided; this practice promotes dusting of the floor surface and can result in reduced abrasion resistance.

Vacuum mats, or a blotter of cement on top of damp burlap, are applied after the concrete has been placed, compacted, and floated. If vacuum mats are used, vacuum is applied for about 3 to 5 min per 1 in. (25 mm) thickness of slab (Wenander 1975; Wenander, Danielsson and Sendker 1975; Malinowski and Wenander 1975).

Vacuum dewatering has been used extensively in Europe. More detailed information is presented in several sources (Martin and Phelan 1995; ACI Committee 226 1987; USBR).

8.3.7 *Edging*—Edging is not required or recommended on most floors. Edgers should be used only when specifically required by the project documents. Where edging is required, use of walk-behind edgers is discouraged because their use can yield inconsistent results. If the floor is to be covered with tile, an edger should not be used. If edging is required by the project documents, a 1/8 in. (3 mm) or smaller radius edge should be used for construction joints subjected to regular vehicular traffic, although saw-cutting is the preferred method for this type of surface.

The edger is used to form a radius at the edge of the slab (Section 8.2.6). Edging, or stoning when the placement is finished flush with the edge forms, will also allow construction joints to be readily visible for accurate location of sawing, when used. The second placement at a construction joint will often bond to the first placement. Sawing this joint encourages development of a clean, straight crack at the construction joint. Edging is most commonly used on sidewalks, driveways, and steps; it produces a neater looking edge that is less vulnerable to chipping. Edging should not commence until most bleed water and excess moisture have left or been removed from the surface. Instead of being edged, construction joints of most floor work can be finished flush with the edge forms, and then lightly stoned to remove burrs after the bulkheads or edge forms are stripped and before the adjacent slab is placed.

8.3.8 *Hand-tooled joints*—Slabs-on-ground are jointed immediately following edging, or at the same time, unless the floor is to be covered with hard or soft tile. If the floor is to be covered with tile, jointing is unnecessary because random cracks are preferable to tooled joints under tile. For floors to be covered with quarry tile, ceramic tile, terrazzo pavers, or cast-in-place terrazzo, the joints in slabs-on-ground should be aligned with joints in the rigid coverings.

The cutting edge or bit of the jointing tool creates grooves in the slab, called contraction joints (Section 3.2.5.3). For contraction joints, the jointing tool should have a bit deep enough to cut grooves that are 1/4 the thickness of the slab. This forms a plane of weakness along which the slab will crack when it contracts. Jointers with worn-out or shallow bits should not be used except for forming decorative, nonfunctional groves in the concrete surface. The jointer should have a 1/8 in. (3 mm) radius for floors. Because of limitations on bit length, hand-tooled joints are not practical for slabs greater than 5 in. (125 mm) thick where the groove depth is 1/4 of the slab thickness.

It is good practice to use a straight 1 x 8 or 1 x 10 in. (25 x 200 or 25 x 250 mm) board as a guide when making the joint or groove in a concrete slab. If the board is not straight, it should be planed true. The same care should be taken in running joints as in edging because a hand-tooled joint can either add to or detract from the appearance of the finished slab.

8.3.9 *Preformed joints*—Preformed plastic and metal strips are also available as an alternative to the use of jointers or saw cuts for making contraction joints. If used, they are inserted in the fresh concrete at the time hand-tooled jointing would take place. Proper performance of these strips is extremely sensitive to installation. Plastic or metal inserts are not recommended in any floor surface subjected to wheeled traffic (Section 3.2.5.3).

8.3.10 *Power floating*—After edging and hand-jointing operations (if used), slab finishing operations should continue with use of either the hand float or the power float. Power floating is the normal method selected. The purposes of power floating are threefold:

1. To embed the large aggregate just beneath the surface of a mortar composed of cement and fine aggregate from the concrete;

2. To remove slight imperfections, humps, and voids; and

3. To compact the concrete and consolidate mortar at the surface in preparation for other finishing operations.

In the event that multiple floating passes are required, each floating operation should be made perpendicular to the direction of the immediately previous pass.

Nonvibratory, 24 to 36 in. diameter (600 to 900 mm) steel disk-type floats are usually employed to float low-slump or zero-slump concrete or toppings. They can also be used for additional consolidating or floating following normal floating operations when the surface has stiffened to a point where it can support the weight of the machine without disturbing the flatness of the concrete.

Troweling machines equipped with float blades or pans slipped over the trowel blades can be used for floating. Float blades are beneficial when steel fiber reinforcing, surface hardeners, or both are used. Troweling machines with combination blades could be used but are not recommended. Floating with a troweling machine equipped with normal trowel blades should not be permitted. Contract documents should also prohibit the use of any floating or troweling machine that has a water attachment for wetting the concrete surface during finishing of a floor. Application of water by brush or machine during finishing promotes dusting of the floor surface and should be done only to overcome adverse conditions. This should be discussed in the context of the placement environment at a pre-placement meeting.

Many variables—concrete temperature, air temperature, relative humidity, and wind—make it difficult to set a definite time to begin floating. The concrete is generally ready for hand floating when the water sheen has disappeared or has been removed, and the concrete will support a finisher on kneeboards without more than approximately a 1/8 in. (3 mm) indentation. The slab surface is ready for machine floating with the lightest machine available when the concrete will support a finisher on foot without more than approximately a 1/4 in. (6 mm) indentation, and the machine will neither dig in nor disrupt the levelness of the surface. Mechanical pan floating should not begin until the surface has stiffened sufficiently so that foot prints are barely perceived on the concrete surface.

Normally, concrete will be ready for power floating in the same order it was placed in. On a given placement, however, certain areas can become ready for power floating before others. The areas that should be floated first generally include surfaces adjacent to screed guides, edge forms, blockouts, walls, and columns. Areas exposed to sun tend to set more quickly than those protected by shade; surfaces exposed to wind also require attention before those protected from the wind. Generally, one or more finishers should be assigned to look after those areas that will set faster than the overall placement.

As a general rule, and under slow-setting conditions when flatness tolerances are not high, power floating should be started as late as possible; this is indicated by minimum machine indentation or when a footprint is barely perceptible. Under fast-setting conditions or when high-flatness tolerances are required, and with the understanding that abrasion resistance of the slab can be reduced, floating should be started as soon as possible; the maximum practical indentation is approximately 1/4 in. (6 mm). When higher-flatness quality is required, the floating operation should generate sufficient mortar to assist in restraightening operations with the modified highway straightedge. Flatness/levelness tolerances can require restraightening of the surface before and after the floating operation.

The marks left by the edger and jointer should be removed by floating, unless such marks are desired for decoration, in which case the edger or jointer should be rerun after the floating operation.

Generally, when the floating operation produces sufficient mortar, restraightening after the floating operation is very beneficial. After the initial power-float pass, and while the surface mortar is still fresh, the modified highway straightedge can be used to restraighten the slab surface by removing the troughs and ridges generated by the power float with float blades or combination blades attached. This is accomplished by cutting down the ridges and using that mortar to fill the troughs. These operations should be completed during the window of finishability.

The use of the power float tends to create troughs under the center of the machine in the direction of travel, with ridges of mortar occurring just outside the perimeter of the blades. Around projections such as columns and sleeves, the power float tends to push mortar up against the projection. If this mortar buildup is not removed by the hand finisher, it will remain when the concrete hardens and the surface will be at a higher elevation than desired. The use of float pan attachments on riding machines reduces the tendency to create troughs.

One method that allows proper grade to be maintained at projections is to place a bench mark a specified distance above design grade on the projection for subsequent use by the finisher. While completing hand work around the column or sleeve, the finisher can use a template to confirm that proper grade has been maintained. Excess material can then be removed as required.

8.3.11 *Troweling*—The purpose of troweling is to produce a dense, smooth, hard surface. Troweling is done immediately following floating; no troweling should ever be done on a surface that has not been floated by power or by hand. Use of a bull float or darby without following by hand or machine floating is not sufficient.

If troweling is done by hand, it is customary for the concrete finisher to float and then steel trowel an area before moving kneeboards. If necessary, tooled joints and edges should be rerun before and after troweling to maintain uniformity and true lines.

Hand trowels that are short, narrow, or of inferior construction should not be used for first troweling. Mechanical troweling machines can be used. The mechanical trowel can be fitted with either combination blades or with those intended specifically for the troweling operation.

For the first troweling, whether by power or by hand, the trowel blade should be kept as flat against the surface as possible; in the case of power troweling, use a slow speed. If the trowel blade is tilted or pitched at too great an angle, an objectionable washboard or chatter surface will result. A trowel that has been properly broken in can be worked quite flat without the edges digging into the concrete. Each subsequent troweling should be made perpendicular to the previous pass. Smoothness of the surface can be improved by restraightening operations with the modified highway straightedge and by timely additional trowelings. There should be a time lapse between successive trowelings to permit concrete to become harder. As the surface stiffens, each successive troweling should be made with smaller trowel blades or with blades tipped at a progressively higher angle to enable the concrete finisher to apply sufficient pressure for proper finishing. Additional troweling increases the compaction of fines at the surface and decreases the *w/cm* of concrete near the slab surface where the trowel blades agitate surface paste and hasten the evaporation rate of water within the paste; this process results in increased surface density and improved wear resistance. Extensive steel-troweling of surfaces receiving a colored dry-shake hardener can have a negative impact on the uniformity of color. Refer to Section 8.6.2 for a detailed discussion.

The formation of blisters in the surface of the concrete during troweling can be the result of entrained air or excessive fines in the concrete mixture, of early troweling, or of an excessive angle of the trowel blades. Air-entrained concrete should never be used in any normalweight concrete floor slab that is to receive a hard-troweled finish (Section 6.2.7). By hindering the passage of bleed water to the surface, such purposeful air entrainment can compel the finisher to start the troweling process too quickly, leading to the entrapment of a liquid water layer immediately beneath the prematurely closed surface. Unfortunately, the concrete will appear to behave normally in the initial troweling stages, so there is no way for the finisher to know that the slab is being damaged.

If the air content is acceptable, then blister formation is an immediate indication that the angle of the trowel blade is too great for the surface in that area at that particular time for the concrete and job conditions involved.

Extensive steel-troweling leaves the concrete surface with a very high sheen. Such surfaces become quite slippery when wet and should be slightly roughened to produce a nonslip surface if they are to be exposed to the weather. A smooth-textured swirl finish can be produced by using a steel trowel in a swirling motion (also known as a sweat finish) or by brooming the freshly troweled surface.

A fine-broomed surface is created by drawing a soft-bristled broom over a freshly troweled surface. When coarser textures are desired, a stiffer bristled broom can be used after the floating operation. A coarse-textured swirl finish can be created after completion of the power float pass and subsequent restraightening using a modified highway straightedge. A coarse swirl pattern is normally created using a hand-held wood or magnesium float (Section 8.13.4).

During periods of hot, dry, and windy weather, troweling should be kept to the minimum necessary to obtain the desired finish. When ambient conditions create high water loss due to slab evaporation, fog spraying above the concrete or use of an evaporation retardant is necessary. After finishing, any delay in protecting the slab with curing compounds or other water-retaining materials can result in an increase in plastic-shrinkage cracking, crazing, low surface strength, dusting, and early deterioration.

8.3.12 *Saw-cut joints*—On large flat concrete surfaces, rather than hand-tooling joints, it can be more convenient to cut joints with an electric or gasoline-driven power saw fitted with an abrasive or diamond blade, and using one of the following three types of saws: conventional wet-cut, conventional dry-cut, or early-entry dry-cut.

The early-entry dry-cut process is normally used when early sawing is desired. Early-entry dry-cut joints are formed using diamond-impregnated blades. The saw cuts resulting from this process are not as deep as those produced using the conventional wet-cut process—typically no more than 1-1/4 in. (32 mm). The timing of the early-entry process, however, allows joints to be in place before development of significant tensile stresses in the concrete; this increases the probability of cracks forming at the joint when sufficient stresses are developed in the concrete. Care should be taken to make sure the early-entry saw does not ride up over hard or large coarse aggregate. The highest coarse aggregate should be notched by the saw to ensure the proper function of the contraction joint. State-of-the-art early-entry saws have an indicator that shows the operator if the saw cut becomes too shallow.

Typically, joints produced using conventional processes are made within 4 to 12 h after the slab has been finished in an area—4 h in hot weather to 12 h in cold weather. For early-entry dry-cut saws, the waiting period will typically vary from 1 h in hot weather to 4 h in cold weather after completing the finishing of the slab in that joint location. Longer waiting periods can be necessary for all three types of sawing for floors with steel-fiber reinforcement or embedded-mineral-aggregate hardeners with long-slivered particles such as traprock.

The depth of saw-cut using a conventional saw should be at least 1/4 of the slab depth or a minimum of 1 in. (25 mm), whichever is greater. The depth of a saw-cut using an early-entry dry-cut saw should be 1 in. (25 mm) minimum for slab depths up to 9 in. (225 mm). This recommendation assumes that the early-entry dry-cut saw is used within the time constraints noted previously. For steel fiber-reinforced slabs, the saw cut using the conventional saw should be 1/3 of the slab depth. Typically, when timely cutting is done with an early-entry saw, the depth can be the same as for concrete without steel fibers.

Regardless of the process chosen, saw-cutting should be performed before concrete starts to cool, as soon as the concrete surface is firm enough not to be torn or damaged by the blade, and before random drying-shrinkage cracks can form in the concrete slab. Shrinkage stresses start building up in the concrete as it sets and cools. If sawing is unduly delayed, the concrete can crack randomly before it is sawed. Additionally, delay can generate cracks that run off from the saw blade toward the edge of the slab at an obtuse or skewed angle to the saw cut.

Under hot, dry, or windy conditions, especially when placing exterior slabs, initial cracking can occur before final troweling. These random cracks can also appear hours or days after saw-cutting. The tendency for these cracks to form can be reduced by fogging the air over the concrete, using a monomolecular film, and starting the placement at night to minimize the impact of temperature, wind, and exposure to direct sunlight.

When these conditions occur, it may be prudent to stop floor placement until a time when conditions are more favorable. Project delay may be more desirable than random out-of-joint cracking.

8.4—Finishing Class 1, 2, and 3 floors

Class 1, 2, and 3 floors (Table 2.1) include tile covered, offices, churches, schools, hospitals, and garages. The placing and finishing operations described under Section 8.3 should be followed. Multiple restraightening operations and two hand or machine trowelings are recommended, particularly if a floor is to be covered with thin-set flooring or resilient tile; this will give closer surface tolerances and a better surface for application of the floor covering.

The use of silica fume concrete for parking garage construction lends itself to a one-pass finishing approach. After initial strikeoff and bull floating have been completed, the concrete placement strips can be textured using a broom. Normally, a light broom with widely spaced, stiff bristles will be satisfactory for this purpose.

Because silica fume concrete exhibits virtually no bleeding, it is necessary to keep the surface moist during concrete finishing operations to prevent plastic-shrinkage cracking. This normally requires use of an evaporation retarder or a pressure fogger with a reach capable of covering the entire surface. Fogging should be performed continuously between finishing operations until the surface has been textured. The goal of the fogging operation should be to keep the concrete surface moist but not wet. Curing operations should commence as quickly as possible after texturing has been completed (ACI Committee 226 1987).

If decorative or nonslip finishes are desired, refer to the procedures described in Section 8.13.

8.5—Finishing Class 4 and 5 floors

Class 4 and 5 floors (Table 2.1) may be light-duty industrial or commercial. The placing and finishing operations described in Section 8.3 should be followed. Three machine trowelings can be specified for increased wear resistance.

8.6—Finishing Class 6 floors and monolithic-surface treatments for wear resistance

Industrial floors using embedded mineral or metallic hardeners are usually intended for moderate or heavy traffic and, in some cases, to resist impact. These hardeners should be properly embedded near the top surface of the slab to provide the required surface hardness, toughness, and impact resistance.

The total air content of normalweight concrete should exceed 3% only if the concrete is subject to freezing-and-thawing cycles under service conditions and the concrete floor slab is not to receive a hard-troweled finish. As with any commercial or industrial floor subjected to wheeled traffic, special care should be exercised to obtain flat and level surfaces and joints. Metallic hardeners should not be placed over concrete with intentionally added chloride. The proposed mixture proportions should be used in the installation of any test panel or test placement. If adjustments to the concrete mixture are required, they can be made at that time.

8.6.1 *Embedded mineral-aggregate hardener*—The application and finishing of embedded mineral-aggregate hardeners should follow the basic procedures outlined below. Concrete installations are subject to numerous conditions and variables. Experience is necessary to determine proper timing for the required procedures. These procedures should be discussed and agreed upon at the preconstruction meeting:

1. Place, consolidate, and strike off concrete to the proper grade.

2. Compact and consolidate the concrete surface using a bull float.

3. Restraighten the surface using a modified highway straightedge. Occasionally, compacting, consolidating, and restraightening are accomplished in one step by using a wide bull float or a modified highway straightedge with the straightedge rotated so its wide dimension is in contact with the surface.

4. Evenly distribute approximately 2/3 of the specified amount of mineral-aggregate hardener immediately following strike-off, and before the appearance of bleed water on the slab surface. The first application generally consists of a larger, coarser material than will be used in the final application. Distribution of the hardener by mechanical spreader is the preferred method. The concrete mixture should have proportions such that excessive bleed water does not appear on the surface after application of the hardener.

5. As soon as the hardener darkens slightly from absorbed moisture, a modified highway straightedge should be used to embed the hardener and remove any irregularities in the surface.

6. Wait until the concrete sets up sufficiently to support the weight of a power trowel with float blades or a pan attached. Combination blades should not be used. The float breaks the surface and agitates concrete paste at the surface of the slab. The first power-float passes should be across the placement strip in the short direction. This will ensure that irregularities resulting from the power floating can be easily identified and corrected in subsequent operations.

7. Apply the remaining one-third of the specified mineral aggregate, preferably at right angles to the first application. This material generally consists of finer-size aggregate and may be broadcast evenly over the surface of the slab by hand.

8. Restraighten the surface using a modified highway straightedge. Remove irregularities and move excess material to low spots.

9. Embed the mineral-aggregate fines using a power trowel with float blades or a pan attached.

10. Restraighten the surface following the power-floating operation using a weighted modified highway straightedge if its use is seen to be effective or necessary to achieve required surface tolerances. One method of increasing the weight of a modified highway straightedge is to wedge a No. 11 bar (35 mm) inside the rectangular section of the straightedge.

11. Continue finishing with multiple power trowelings as required to produce a smooth, dense, wear-resistant surface (Section 8.3.11). Provide a burnished (hard) troweled surface where required by specification.

12. Cure immediately after finishing by following the curing material manufacturer's recommendations. Curing methods should be in accordance with those used and approved in construction of any test panel.

8.6.2 *Metallic dry-shake hardener and colored dry-shake hardeners*—Metallic dry-shake hardeners and colored dry-shake hardeners can be finer in texture than uncolored mineral-aggregate dry-shake hardeners. This difference, along with the fact that the metallic dry-shake hardener has a higher specific gravity, dictates that the material normally be embedded in the concrete later in the setting process than is common for uncolored mineral-aggregate dry-shake hardeners. Some metallic and colored dry-shake hardeners are designed by their manufacturers to allow application of all the hardener at one time. When such procedures are used, however, caution should be exercised to ensure that manufacturer's recommendations are followed, and that the material is thoroughly wetted-out because a one-time application significantly increases the possibility of surface delamination or related finishing problems. Typical installation techniques for metallic dry-shake hardeners and colored dry-shake hardeners are similar to those described in Section 8.6.1, but the following sequence is recommended (refer to Section 8.13.1):

1. Place, consolidate, and strike off concrete to the proper grade.

2. Compact and consolidate the concrete surface using a bull float.

3. Restraighten the surface using a modified highway straightedge. A wide bull float or a modified highway straightedge can be used to accomplish both steps in one operation.

4. Open the surface to promote movement of bleed water to the top of the slab by using a wooden bull float. Steps 3 and 4 can be accomplished in one operation if the wide bull float or modified highway straightedge is made of wood.

5. Wait until the concrete sets up sufficiently to support the weight of a power trowel.

6. Break the surface using a power trowel with float blades or a pan attached.

7. Evenly distribute approximately 2/3 of the specified amount of metallic dry-shake hardener or colored dry-shake hardener. Application of the material by mechanical spreader is the preferred method.

8. Restraighten the surface after application of the metallic dry-shake hardener or colored dry-shake hardener to remove irregularities. Some contractors find that embedding the materials and restraightening can be accomplished in one step using a modified highway straightedge.

9. Complete initial embedment and prepare the surface for additional material by using a power trowel with float blades or a pan attached.

10. Apply the remaining 1/3 of the specified amount of metallic dry-shake hardener or colored dry-shake hardener, preferably at right angles to the first application.

11. Embed metallic dry-shake hardener or colored dry-shake hardener using a power trowel with float blades or a pan attached. Thorough embedment and integration of the metallic dry-shake hardener or colored dry-shake hardener with the concrete by floating is very important. Failure to accomplish this goal can result in blistering or delamination of the slab.

12. Restraighten the surface following the power-floating operation using a weighted modified highway straightedge, if effective.

13. Continue finishing with multiple power trowelings as required to produce a smooth, dense, wear-resistant surface (Section 8.3.11). Proper and uniform troweling is essential. Colored surfaces should not be burnished (hard-troweled); the result would be uneven color and a darkening of the surface.

14. Cure immediately after finishing by following the curing material manufacturer's recommendations. Curing methods should be in accordance with those used and approved in construction of any test panel. Colored floors should not be cured with plastic sheeting, curing paper, damp sand, or wet burlap. These materials promote uneven color, staining, or efflorescence.

8.7—Finishing Class 7 floors

The topping course of heavy-duty industrial floors should have a minimum thickness of 3/4 in. (19 mm). The concrete topping used should have a maximum slump of 3 in. (75 mm), unless a water-reducing admixture or high-range water-reducing admixture is used to increase the slump, or unless dewatering techniques are used. Because of the relatively small amount of concrete in the topping course and the low slump required, concrete for the topping could be job-mixed.

Embedded metallic dry-shake hardeners, mineral-aggregate dry shakes, and colored dry-shakes can be applied to produce the desired combination of increased wear resistance or color as described in Sections 8.6.1 and 8.6.2, respectively.

The base course should be screeded and bull floated; close maintenance of the elevation tolerance for the base course surface is important. Class 7 floors (Table 2.1) can be constructed in two ways: (1) the topping installation can be bonded monolithically to the base slab before the base slab has completely set, or (2) the topping can be deferred for several days.

For suspended slabs, the deferred bonded approach should be used. This will allow the structure to deflect under its own weight before application of the topping. The additional weight of the topping will have little impact on subsequent deflection of the slab.

8.7.1 *Bonded monolithic two-course floors*—In most cases, wet curing is recommended for the bonded topping. Special precautions should be taken to prevent premature drying of the edges because curling of the topping and delamination from the base slab can result.

For these floors, the topping course is placed before the base course has completely set. Any excess moisture or laitance should be removed from the surface of the base course, and the surface floated before the top course is placed. When the topping is being placed, the concrete in the base slab should be sufficiently hard that footprints are barely perceptible. The use of a disk-type power float can be necessary to bring sufficient paste to the surface to allow restraightening to take place. The power-floating operation should be followed by a minimum of two power trowelings. This method of topping application is generally not appropriate for a suspended slab.

8.7.2 *Deferred bonded two-course floors*—Bonding of two-course floors is a highly critical operation requiring the most meticulous attention to the procedure described. Even with such care, such bonding has not always been successful. As a result, contractors using this type of construction for heavy-duty industrial applications should be experienced and familiar with the challenges presented.

Locations of joints in the base course should be marked so that joints in the topping course can be placed directly over them.

After the base course has partially set, the surface should be brushed with a coarse-wire broom. This removes laitance and scores the surface to improve bond of the topping course.

Concrete base courses should be wet-cured a minimum of 3 days (Sections 9.2.1 and 9.2.2). Shrinkage-compensating concrete base courses should be wet-cured a minimum of 7 to 10 days, and preferably until the topping is applied. Refer to ACI 223 for additional information.

If the topping is to be applied immediately after the minimum 3-day curing time has elapsed, the curing cover or water should be removed from the slab and any collected dirt and debris washed or hosed off. After most free water has evaporated or has been removed from the surface, a bonding grout should be scrubbed in. The bonding grout should be composed, by volume, of one part cement, 1.5 parts fine sand passing the No. 8 sieve (2.36 mm), and sufficient water to achieve the consistency of thick paint. The grout should be applied to the floor in segments, keeping only a short distance ahead of the concrete topping placing operations that follow it.

While the bonding grout is still tacky, the topping course should be spread and screeded. The use of a disk-type power float is suggested, followed by a minimum of two power trowelings.

If 3 to 7 days are to elapse between placing the base and the topping course, the surface of the base course should be protected from dirt, grease, plaster, paint, or other substances that would interfere with the bond. Immediately before placing the topping, the base course should be thoroughly cleaned by scrubbing with a brush and clean water. Most excess water should be removed and a thin scrub-coat of grout applied. While this grout is still tacky, the topping course should be spread and screeded.

If the floor is to be subjected to construction activities after curing and before application of the topping, more thorough cleaning may be necessary. One method of cleaning the base slab is to scrub the surface with water containing detergent. If oil or grease has been spilled on the floor, a mixture of sodium metasilicate and resin soap is useful. If this method is used, the floor should then be rinsed thoroughly with water. Shot-blasting, sandblasting, or mechanical scarification by scabbling can also be employed instead of cleaning with detergent to achieve a bondable surface.

In some circumstances, it can be convenient or desirable to bond the topping with an epoxy adhesive appropriate for the particular application. Methods are described in ACI 503R, and a standard specification is given in ACI 503.2.

Joints in the topping above the joints in the base slab should be saw-cut to a depth equal to twice the thickness of the topping and should match the location of joints in the base slab, where applicable.

8.8—Finishing Class 8 floors (two-course unbonded)

The unbonded topping for two-course unbonded floors should be a minimum of 4 in. (100 mm) thick. An unbonded topping thickness of 3 in. (75 mm) has been used with some success for Class 3 floors, but thickness for strength and control of curling is less important for a Class 3 slab because of its duty, loading, and because it may also be covered. A Class 8 floor (Table 2.1) is intended for industrial applications where strength and control of curling is more important. The base course, whether old or new, should be covered with plastic sheeting, felt, a sand cushion, or other approved bond-breaker, and spread as wrinkle-free as possible.

The topping slab should contain sufficient steel reinforcement to limit the width of shrinkage cracks in the topping and the displacement of the topping concrete on either side of any cracks that might form. Steel fiber and high-volume synthetic fibers in proper quantities may be used effectively to minimize crack opening widths. Although reinforcing steel is normally discontinued at joints, engineering considerations can make it desirable to carry reinforcement through construction joints in specified locations in a topping. Reinforcement that is continuous through contraction and construction joints will cause restraint against movement that will inevitably result in cracks in the concrete.

Concrete for the top course should comply with the requirements of Table 6.1.

Power floats and power trowels are recommended and usually required. The practice of completing troweling by hand is counterproductive because hand troweling is less effective than power troweling in consolidating the surface.

Embedded mineral-aggregate hardeners for increased wear resistance can be applied as described in Section 8.6.1.

Embedded metallic dry-shake hardeners and colored dry-shake hardeners can be applied as described in Section 8.6.2.

8.9—Finishing Class 9 floors

Class 9 floors (Table 2.1) may be superflat or require critical surface tolerance. Floor surfaces of this quality can be subdivided by function into two separate groups. Refer to Section 8.9.1 for special considerations dealing with construction of Class 9 floor surfaces.

The more common group of these floor surfaces should support vehicular traffic along paths that are defined before construction and that do not change during the life of the floor surface (that is, defined traffic). A typical example of a defined-traffic floor would be a distribution center that uses very narrow aisles and high-bay racking systems. In this type of facility, tolerances across aisles and the joints that parallel them are less critical than those along the axis of the aisle. This type of floor surface is often referred to as superflat.

Floor surfaces in the second group are less common but should support traffic in all directions (that is, random traffic). A typical example of a random-traffic floor would be a gymnasium, ice rink, television studio, or movie studio. The random nature of traffic in these facilities requires that tolerances across placement strips and their joints should match those achieved parallel to the axis of the strip.

Finishing procedures required to produce Class 9 floors represent the most rigorous and demanding floor installation technology now being performed. If discipline and preplanning are a part of the overall process, however, installation of Class 9 floors is neither complex nor especially difficult. Proper timing and execution of various procedures will usually ensure that the floor produced is of a predictable quality.

Class 9 defined-traffic floor construction requires that:

1. Slabs be constructed in long strips less than 20 ft (6 m) in width;

2. Concrete slump be adjusted on-site to within ±1/2 in. (±13 mm) of the target slump;

3. Slump at point of deposit be sufficient to permit use of the modified highway straightedge to close the floor surface without difficulty after the initial strikeoff;

4. Window of finishability be sufficient for the concrete contractor to perform the necessary finishing operations; and

5. Concrete supplier use enough trucks to ensure an uninterrupted concrete supply.

In addition, because environmental factors can significantly alter the setting rate of concrete, an effort is usually made to construct Class 9 floors out of the weather.

On Class 9 defined-traffic floors, construction joints between placement strips are located out of the traffic pattern where racks abut each other. These surfaces are evaluated by taking measurements only in locations matching the wheel-paths of the vehicles that will eventually use the floor. The part of the floor surface falling under racks is not tested.

While the same construction techniques can be used to produce Class 9 random-traffic floors—television studios or similar surfaces—the entire floor surface should be evaluated because the entire surface will be subjected to traffic. The contractor is cautioned that grinding of the entire length of the joints will be necessary to produce Class 9 quality across the width of concrete placement strips.

On most projects with Class 9 defined-traffic floors, surfaces are measured for flatness and levelness immediately following the final troweling of each placement; placements are frequently scheduled for consecutive days. Where Class 9 random-traffic quality is required across multiple placement strips, initial testing should take place as each strip is placed, but final testing should be deferred until the installation is complete.

Nonetheless, it is imperative that surface-profile testing and defect identification be accomplished on each new slab as soon as possible. To maintain satisfactory results, the contractor requires continuous feedback to gage the effectiveness of construction techniques against ever-changing job conditions (Section 8.9.1). Refer to Figure 8.7 for additional information.

Achieving Class 9 quality levels on suspended slabs is impractical in a one-course placement. Deflection of the surface between supports occurs after removal of supporting shores. If the surface were to meet Class 9 requirements in a shored condition, it is likely that the deflected surface after shores are removed would be less level than is required to meet Class 9 requirements. Two-course placements using methods similar to those discussed for Class 7 and 8 floors provide the best opportunity for achieving Class 9 quality levels on suspended floors.

8.9.1 *Special considerations for construction of Class 9 floor surfaces*—Certain specialized operations—narrow-aisle warehouses, gymnasiums, ice rinks, television studios, and air-pallet systems—require extraordinarily flat and level floors for proper equipment performance. Such superflat floors generally exhibit F_F numbers and F_L numbers above 50 in the direction of travel for the particular application. Refer to Section 8.15 for additional discussion.

The floor-finish tolerance employed in the contract specification should meet the equipment supplier's published requirements, unless there is reason to doubt the validity of such requirements. In any case, written approval of the contract floor tolerance should be obtained from the appropriate equipment supplier before finalizing the bid package. In this way, equipment warranties will not be jeopardized, and the special superflat nature of the project will be identified to key parties from the outset.

Superflat floors have very specific design requirements. Chief among these is the limit imposed on placement width. In general, superflat floors cannot be produced if the placement strip width exceeds 20 ft (6 m). Because hand-finishing procedures and curling effects are known to make floors in the vicinity of construction joints less flat than in the middle of the slab, joints should be located out of the main traffic areas, or provisions should be made for their correction. Contraction joints oriented transverse to the longitudinal axis of a Class 9 placement strip can curl and reduce surface flatness along aisles. Limited placement width, consequent increased forming requirements, and reduced daily floor production are primary factors that increase the cost of Class 9 floors.

The prebid meeting is an essential component of any superflat project. Because floor flatness/levelness is one of the primary construction requirements, a thorough prebid review of the design, specification, and method of compliance testing is required. This will enable the prospective contractor to price the project realistically (thereby avoiding costly misunderstandings and change orders), and will greatly increase the chances of obtaining the desired results at the lowest possible cost.

To further reduce the risk of problems, the installation of test slabs has become a standard part of superflat floor construction. If the contractor is inexperienced with superflat construction or with the concrete to be used, at least two test slabs should be installed and approved before the contractor is permitted to proceed with the balance of the superflat floor construction.

Superflat floor tolerances should be inspected within 24 h after slab installation. This eliminates the possibility of large areas being placed before any tolerance problem is discovered. In narrow-aisle warehouses, tolerances are measured using a continuous recording floor profilograph or other device. In these facilities, floor tolerances are based on the lift-truck wheel dimensions, and compliance measurements and corrections are required only in the future wheel tracks.

In television studios and other similar random-traffic installations, the use of F_F and F_L to specify the floor-surface tolerances is appropriate. Measurements for compliance should be made in accordance with ASTM E 1155 (Section 8.15), except that measurements should extend to the joints.

8.10—Toppings for precast floors
Many types of precast floors require toppings. These include double tees, hollow-core slabs, and other kinds of precast floor elements. When these floors are to be covered with bonded toppings, the procedures in Section 8.7.2 or 8.8 should be followed, as appropriate. High-strength concrete is often used for precast floor elements; roughening of the surface of such members can be difficult if delayed too long.

8.11—Finishing lightweight concrete
This section concerns finishing lightweight concrete floors. Finishing lightweight insulating-type concrete, having fresh density of 60 lb/ft^3 (960 kg/m^3) or less, that is sometimes used below slabs, generally involves little more than screeding.

Lightweight concrete for floors usually contains expanded shale, clay, slate, or slag coarse aggregate—expanded shale is most common. The fine aggregate can consist of manufactured lightweight sand, natural sand, or a combination of the two, but natural sand is most common. The finishing procedures differ somewhat from those used for a normalweight concrete. In lightweight concrete, the density of the coarse aggregate is generally less than that of the sand and cement. Working the concrete has a tendency to bring coarse aggregate rather than

FLATNESS/LEVELNESS TYPICAL USE GUIDE

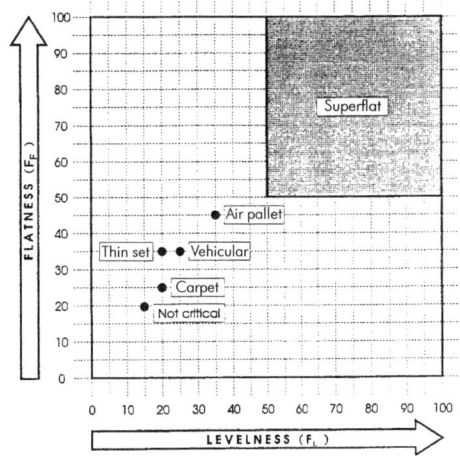

SLABS ON GROUND

Composite Overall Flatness (F_F)	Composite Overall Levelness (F_L)	Typical Use	Typical Class
20	15	Noncritical: mechanical rooms, non-public areas, surfaces to have raised computer flooring, surfaces to have thick-set tile, and parking structure slabs	1 or 2
25	20	Carpeted areas of commercial office buildings or lightly-trafficked office/industrial buildings	2
35	25	Thin-set flooring or warehouse floor with moderate or heavy traffic	2, 3, 4, 5, 6, 7, or 8
45	35	Warehouse with air-pallet use, ice or roller rinks, gymnasium floors[4]	9
>50	>50	Movie or television studios	3 or 9

SUSPENDED SLABS

Composite Overall Flatness (F_F)	Composite Overall Levelness (F_L)	Typical Use	Typical Class
20	15[2] or N/A	Noncritical: mechanical rooms, non-public areas, surfaces to have raised computer flooring, surfaces to have thick-set tile, and parking structure slabs	1 or 2
25	20[1] or N/A	Carpeted areas of commercial office buildings or lightly-trafficked office/industrial buildings	2
35	20[2] or N/A	Surfaces to receive thin-set flooring	2, 3, or 4
45	35[3]	Ice or roller rinks, gymnasium floors[4]	3
>50	>50[1,3]	Movie or television studios	3 or 9

NOTES
1. Multi-directional quality of this level requires grinding of joints.
2. Levelness F-number only applies to level slabs shored at time of testing.
3. This levelness quality on a suspended slab requires a two-course placement.
4. All elevation samples should fall inside a 1/2 in. deep envelope.

Fig. 8.7—Typical use guide for flatness and levelness.

mortar to the surface. This should be taken into account in the finishing operations (ESCSI 1958).

Observing the following rules will control this tendency so that lightweight concrete can be finished as easily as normal-weight concrete, provided the mixture has been properly proportioned:

1. The mixture should not be oversanded in an effort to bring more mortar to the surface for finishing. This usually will aggravate rather than eliminate finishing difficulties.

2. The mixture should not be undersanded in an attempt to meet the unit weight requirements. Neither mixing to the recommended slump nor entrainment of air will effectively control segregation in such a mixture.

3. The lightweight concrete mixture should be proportioned to provide proper workability, pumpability, finishing characteristics, and required setting time, and to minimize segregation or the tendency for coarse aggregate particles to rise above the heavier mortar.

4. Some lightweight aggregates can require further control of segregation, bleeding, or both. For this purpose, use no less than 4% entrained air in accordance with ACI 211.2. Note: unlike its effect on normalweight concrete, the use of an air-entraining admixture in lightweight concrete that is to be hard-troweled does not lead to blistering.

5. Presaturate lightweight aggregates for use in concrete that will be pumped, in accordance with the manufacturer's recommendations.

6. Overworking or overvibrating lightweight concrete should be avoided. A well-proportioned mixture can generally be placed, screeded, and bull floated with approximately half the effort considered good practice for normalweight concrete. Excess darbying or bull floating are often principal causes of finishing problems because they only serve to drive down the heavier mortar that is required for finishing and to bring an excess of the coarse aggregate to the surface.

7. A magnesium darby or bull float should be used in preference to wood. Metal will slide over coarse aggregate and embed it rather than tear or dislodge it.

8. The surface should be floated and flat troweled while the concrete is still plastic, taking care to ensure bleeding is complete. Premature sealing of the surface can trap bleed water and result in blisters and delamination. If floating is being done by hand, use a magnesium float. If evaporation is not taking place soon enough (while concrete is still plastic), other measures should be taken. Water and excess moisture should be removed from the surface with as little disturbance as possible. A simple but reliable method is to drag a loop of heavy rubber garden hose over the surface.

8.12—Nonslip floors

Nonslip surfaces are produced by using the following finishing procedures: swirl or broom finish (Section 8.13.4), or nonslip special-purpose aggregate (Section 8.13.2). The nonslip special-purpose aggregate is recommended for heavy foot traffic.

Brungraber (1976, 1977) describes methods of measuring and evaluating the relative skid resistance of floors.

8.13—Decorative and nonslip treatments

8.13.1 *Colored dry-shake hardener surface treatment*—The installation of a colored surface treatment is particularly sensitive to the finishing and curing techniques employed by the contractor. Sample panels should be constructed before beginning actual placement on the project to confirm that the proposed procedures are adequate and that the uniformity of color is acceptable. Any sample panel should be larger than 200 ft^2 (18 m^2) and should be prepared using the concrete mixture and finishing and curing techniques planned for the project (Kosmatka 1991).

Coloring agents are normally included with an embedded hardener when a hardener is applied and color is desired. Finishing procedures should follow the steps described in Section 8.6.2.

8.13.2 *Nonslip monolithic surface treatment*—Before being applied to the surface, the slip-resistant material (Section 5.4.7) should be mixed with dry portland cement, if not already so formulated. Volumetric proportions usually range from 1:1 to one part slip-resistant material:two parts portland cement, but the manufacturer's directions should be followed. The treatment procedure is the same as that outlined for the colored treatment (Section 8.6.2). A swirl finish produced using natural or colored embedded mineral or metallic hardeners provides increased wear resistance and also produces a long-lasting, nonslip finish (Section 8.13.1).

8.13.3 *Exposed aggregate surface treatment*—Exposed aggregate surfaces are commonly used to create decorative effects. Both the selection of the aggregates and the techniques employed for exposing them are important to the effect obtained; test panels should be made before the job is started. Colorful, uniform-sized gravel or crushed aggregate is recommended.

Aggregates should not be reactive with cement (ACI 201.2R). Aggregates can be tested by using ASTM C 227 or by petrographic examination (ASTM C 295). If information or a service record is lacking, the aggregates and the cement aggregate combinations should be evaluated using the guideline in the appendix to ASTM C 33.

Flat particles, sliver-shaped particles, and particles smaller than 1/4 in. (6 mm) do not bond well. As a result, they can easily become dislodged during the operation of exposing the aggregate. The use of aggregate ordinarily used in concrete is not satisfactory, unless the aggregate is sufficiently uniform in size, bright in color, and can be closely packed, and uniformly distributed.

Immediately after the slab has been screeded and darbied or bull floated, the selected aggregate should be broadcast and evenly distributed so that the entire surface is completely covered with one layer of the select aggregate. The aggregate should be free of dust to promote good bond with the base slab. Initial embedding of the aggregate is usually done by patting with a darby or the broad side of a short piece of 2 x 4 in. (50 x 100 mm) lumber. After the aggregate has been thoroughly embedded and as soon as the concrete will support the weight of a finisher on kneeboards, the surface should be floated using a magnesium hand float, darby, or bull float, until the aggregate is entirely embedded and slightly covered with mortar. This operation should leave no holes in the surface.

Shortly after floating, a reliable surface set retarder can be sprayed over the surface in accordance with the manufacturer's recommendations. Retarders may not be necessary on small jobs, but they are generally used on large jobs to ensure better control of the exposing operations. Use of a surface set retarder ordinarily permits several hours to elapse before brushing and hosing the surface with water to expose the aggregate. The proper timing for exposing the aggregate is critical, whether or not a retarder has been used, and this timing is very dependent upon the temperature and other weather conditions. Recommendations of the retarder manufacturer should be followed closely.

Operations to expose the aggregate should begin as soon as the surface can be brushed and washed without overexposing or dislodging the aggregate. If it becomes necessary for finishers to move about on the newly exposed surface, kneeboards should be used, gently brought into contact with the surface, and neither slid nor twisted on it. If possible, however, finishers should stay off the surface entirely because of the risk of breaking the aggregate bond.

If a smooth surface is desired, as might be the case in an interior area, no retarder should be used. The aggregate is not exposed until the surface has hardened. Exposure is accomplished after hardening entirely by grinding. If grinding is followed by polishing, a terrazzo-like surface can be produced.

Alternative methods of placement are available. A top course, 1 in. (25 mm) or more thick, that contains the select aggregate can be applied, or the monolithic method can be used. The monolithic method does not use aggregate seeding; the select aggregate to be exposed is mixed throughout the concrete during batching.

Tooled joints are not practical in exposed-aggregate concrete, because the aggregate completely covers the surface. Decorative or working joints are best produced by wet-cut sawing (Section 8.3.12). Another method of providing joints is to install permanent strips of wood (redwood, cypress, or cedar) before placing concrete (Fig. 3.7).

Exposed-aggregate slabs should be cured thoroughly. The method of curing should not stain the surface. Straw, earth, and any type of sheet membrane, such as polyethylene or building paper, can cause discoloration (Section 9.2.1).

8.13.4 *Geometric designs, patterns, and textures*—Concrete surfaces are frequently scored or tooled with a jointer to produce various decorative patterns. For random geometric designs, the concrete should be scored after it has been screeded, bull floated, or darbied, and excess moisture has left the surface. Scoring can be done using a jointer, groover, or piece of pipe bent to resemble an S-shaped jointer tool. The tool is made of 1/2 or 3/4 in. (13 or 19 mm) pipe, approximately 18 in. (450 mm) long. Cobblestone, brick, tile, and many other patterns can be impressed deeply into partially set concrete slabs with special imprinting tools (Kosmatka 1991).

A swirl-float finish or swirl design can be produced using a magnesium or wooden hand float or a steel finishing trowel. After the concrete surface has received the first power-float pass and subsequent restraightening using a modified highway straightedge, a float should be worked flat on the surface in a semicircular or fan-like motion using pressure. A finer-textured swirl design can be obtained with the same motion by using a steel finishing trowel held flat. An alternative method is to draw a soft-bristled broom across the slab in a wavy motion.

After the concrete has set sufficiently that these surface textures or patterns will not be marred, the slab should be moist-cured. Plastic membranes or waterproof curing paper should not be used on colored concrete (Sections 9.2.1 and 9.2.2).

8.14—Grinding as a repair procedure

Grinding can be used to repair certain surface defects. Grinding has been used successfully to repair the following kinds of problems:

1. Unacceptable flatness and levelness;
2. Curled joints;
3. Surface irregularities that might show through thin floor coverings, such as resilient tile;
4. Poor resistance to wear, when this is due to a weak floor surface with sound concrete underneath; and
5. Rain damage. Corrective grinding upon completion should return the surface to the specified texture. Treatment of the newly ground area with a liquid chemical surface treatment (Section 5.8) is also recommended.

8.14.1 *Cautions*—Grinding does not always produce the desired effect, and it sometimes makes the floor look worse. If improperly executed, grinding can adversely affect the floor's resistance to wear, particularly in industrial applications where the surface is subject to heavy traffic and abuse. For these reasons, it is usually wise to make a small trial section before starting full-scale repairs. Only wet grinding should be used, primarily to minimize dust, and also because diamond-disk grinders are more effective when used with water.

8.14.2 *Types of grinders*—Many types of grinders are available. The two types most often used on floor slabs are diamond-disk grinders and stone grinders.

8.14.2.1 *Diamond-disk grinder*—This grinder uses one or more diamond-impregnated steel disks. Each disk is mounted horizontally and is driven by a vertical shaft. The most common type of diamond-disk grinder has a single 10 in. (250 mm) grinding disk powered by a gasoline engine or electric motor of 5 to 10 hp (3.7 to 7.5 kW). Bigger, more powerful machines are available for floors that need extensive grinding. Diamond-disk grinders are much faster than stone grinders and are usually the better choice to correct the aforementioned problems 1 and 2.

8.14.2.2 *Stone grinder*—This grinder uses multiple abrasive blocks, called stones, mounted on one or more steel disks. The abrasive material is usually silicon carbide. The most widely used type of stone grinder has two disks with three stones on each disk. One-disk and four-disk machines are also available. Stone grinders can be effective on aforementioned problems 3 to 5, particularly where the floor surface is soft or where the amount of material to be removed is small.

8.15—Floor flatness and levelness

8.15.1 *Floor flatness/levelness tolerances*—Tolerances for various floor uses should conform to the requirements set forth in ACI 117. A discussion of floor flatness/levelness is given in the commentary to ACI 117.

ACI 117 specifies that overall conformance to design grade shall be within 3/4 in. (19 mm) of design elevation. For suspended cast-in-place concrete slabs, this tolerance is to be achieved before removal of any supporting shores. For suspended slabs on metal deck, this tolerance for overall conformance to design grade does not apply because tolerances for erected steel frames are not consistent with those for formwork in cast-in-place concrete frames.

8.15.1.1 F-*number system*—Both flatness and levelness requirements should be described by Face Floor Profile Numbers (Face 1987). Two separate F-numbers are required to define the required flatness and levelness of the constructed floor surface. Refer to the commentary in ACI 117 for additional discussion of this method.

The flatness F-number (F_F) controls local surface bumpiness by limiting the magnitude of successive 1 ft (300 mm) slope changes when measured along sample measurement lines in accordance with ASTM E 1155.

The levelness F-number (F_L) controls local conformance to design grade by limiting differences in departures from design grade over distances of 10 ft (3 m), when measured along sample measurement lines in accordance with ASTM E 1155.

The *F*-number pair is always written in the order F_F/F_L. In theory, the range of flatness and levelness F-numbers extends from zero to infinity. In practice, F_F and F_L values generally fall between 12 and 45. The scale is linear, so the relative flatness/levelness of two different floors will be in proportion to the ratio of their F-numbers. For example, an F_F30/F_L24 floor is twice as flat and twice as level as an F_F15/F_L12 floor.

On random-traffic floors—those with varied and unpredictable traffic patterns—two tiers of specified F_F/F_L values should be indicated: one for the composite values to be achieved (specified overall value), and one for the minimum quality level that will be accepted without repair (minimum local value).

Compliance with the specified overall value is based on the composite of all measured values. For any given floor, the composite F_F/F_L values are derived in accordance with ASTM E 1155.

Minimum local values represent the minimum acceptable flatness and levelness to be exhibited by any individual floor section. Minimum local values are generally set at 67% of the specified overall values and are not normally set lower than 50% of the specified overall F_F/F_L requirements. Minimum local values should never be less than F_F13/F_L10, because these values represent the minimum local results achievable by any concrete floor construction method.

Remedial measures can be required:

- If the composite value of the entire floor installation (when completed) measures less than either of the specified overall F-numbers; or
- If any individual section measures less than either of the specified minimum local F_F/F_L numbers. Sectional boundaries are usually set at the column and half-column lines on suspended slabs, or at the construction and contraction joints for slabs-on-ground. They should be no closer together than 1/2 bay.

Remedial measures for slabs-on-ground might include grinding, planing, surface repair, retopping, or removal and replacement. For suspended slabs, remedial measures are generally limited to grinding or use of an underlayment or topping material. Contract documents should clearly identify the acceptable corrective method(s) to be used.

The selection of proper F_F/F_L tolerances for a project is best made by measurement of a similar satisfactory floor. This measurement is then used as the basis for the F_F/F_L tolerance specification for the new project. If this method is used, the slab-on-ground floor surfaces change after construction as a result of shrinkage and curling, and the surfaces of suspended slabs change as a result of deflection. Because of these post-construction changes, it is likely that measurements of an existing project will yield results of a lower quality than can be achieved by the contractor because all of the post-construction changes create slightly diminished F_F/F_L measurement results. When measurement of a similar satisfactory floor is not possible or practical, the flatness/levelness quality levels provided in Fig. 8.7 have been found to be reasonable for the stated applications.

8.15.1.2 *The 10 ft (3 m) straightedge method*—The older method of using a 10 ft (3 m) straightedge can also be used to measure floor flatness, but it is much less satisfactory than the F-number system. There is no nationally accepted method for taking measurements or for establishing compliance of a test surface using this tolerance approach. This lack of an accepted standard test procedure often leads to conflict and litigation. The straightedge-tolerance method also has a number of other serious deficiencies. Refer to the Commentary on ACI 117 for additional discussion.

When straightedge tolerances are specified, 100% compliance with 10 ft (3 m) straightedge tolerances is unrealistic. Compliance with four of five consecutive measurements is more realistic, with a provision that obvious faults be corrected.

8.15.1.3 *Other measurement methods*—Measurement methods are not limited to the F-number (ASTM E 1155) or the 10 ft (3 m) straightedge systems. Alternative tolerancing systems that adequately control critical floor surface characteristics can be used.

8.15.2 *Precautions*—Floor tolerance specification and measurement procedures are currently undergoing technological change. Much remains to be learned about which tolerances can be reasonably expected from a given construction method. On those projects where floor flatness/levelness constitutes a potential issue, the following precautions are suggested:

- The exact meaning of the flatness/levelness requirement, and the exact method and time of measurement to determine compliance, should be established before beginning construction;
- The contractor should confirm an ability to satisfy the floor tolerance requirement by profiling previous installations;
- Where feasible, test slabs should be installed to verify

the effectiveness of proposed installation procedures under actual job conditions. If necessary, methods and procedures should then be modified for the actual job installation based on these results. The acceptance of the test slab by the owner as to tolerances and surface finish should clarify requirements for the project slab; and
- The exact remedy to be applied to every possible floor tolerance deficiency should be confirmed with the designer.

8.15.3 *Factors influencing floor flatness and levelness*—The flatness and levelness exhibited by a newly installed concrete slab-on-ground will depend upon the effectiveness of the specific placement and finishing procedures employed during its construction. In general, the forming, placement, and initial strikeoff phases of the installation will establish the floor's relative levelness, while subsequent finishing operations (floating, restraightening, and troweling) will determine the floor's relative flatness. Any factor that complicates placing or finishing operations will have an adverse effect upon the flatness/levelness produced.

The flatness and levelness F-numbers normally obtained using a given floor construction procedure are summarized in Table 8.1 and 8.2.

These are the floor-finish tolerances expected to be achieved by competent, knowledgeable finishers under standard job conditions. Difficult job environments could result in significantly lower values. Both specifiers and contractors should approach each new concrete floor project using the guidelines set forth in Section 8.15.2.

8.15.3.1 *Flatness*—On those projects where flatness is an important consideration, precautions should be taken to provide an adequate construction environment. Of particular concern for both slabs-on-ground and suspended slabs are:
- Workability, finishability, and setting times of concrete to be used;
- The window of finishability, which should be sufficient for the contractor to perform the required finishing operations;
- Sun, wind, rain, temperature, and other exposure conditions and their effects on personnel and concrete;
- Amount and angle of light;
- Timeliness of concrete delivery;
- Consistency of delivered slump;
- Consistency of final setting time; and
- Site accessibility.

8.15.3.2 *Levelness*—For slabs-on-ground, accuracy of formwork and initial strikeoff establish the overall levelness of the surface. Form spacing, therefore, is an important consideration when developing a construction program intended to produce a certain quality. The use of block-placement techniques with wet-screed strikeoff provides the least accurate control of grade. Block placements with moveable rigid-screed guides provides an improvement in the levelness quality that can be achieved. Further improvement in levelness generally requires the use of either strip placements and vibrating screeds or self-propelled laser-guided strike-off equipment. Strip widths up to 50 ft (15 m) have provided levelness quality comparable to that which can be achieved using moveable dry-screeds in a block placement. Reducing the width of strips improves the ability of the contractor to produce level surfaces because there is less tendency for the vibrating screed to oscillate or deflect, and the controlling edge form elevations are closer together. The highest quality of levelness can be achieved using strip widths between 10 and 20 ft (3 and 6 m). This width allows the contractor to follow the vibrating screed with hand straightedging operations to remove any imperfections in the surface left by the vibrating screed.

Levelness of suspended slabs is dependent on accuracy of formwork and strikeoff, but is further influenced by behavior of the structural frame during and after completion of construction. Each type of structural frame behaves somewhat differently; the contractor should recognize those differences and plan accordingly. Refer to Chapter 3 for a more detailed discussion of behavior of different types of structural systems.

The F_L tolerance should only be applied to slabs-on-ground that are level and suspended slabs that are both level and shored at the time data are taken. The F_L levelness tolerance should not apply to slabs placed on unsupported form surfaces. It should not be applied to cambered or inclined slab surfaces. Concrete slabs placed over unshored structural steel and metal deck surfaces can exhibit significant deflection in the hardened state. The resulting slab surfaces have occasionally required extensive repair to achieve a product satisfactory for applied finishes or partitions.

8.15.4 *Timeliness of tolerance measurement*—To establish the flooring contractor's compliance with specified floor tolerances, the contract documents should stipulate that floor tolerance compliance tests be performed and defective areas identified. This should be completed by the owner's agent as soon as possible, preferably within 24 h after placement, and be reported to key parties as soon as possible, but not later than 72 h after installation. For suspended cast-in-place slabs, tests for acceptance should be conducted before forms and shoring have been removed. In this way, the effects of deflection and shrinkage on the tolerance data can be minimized.

As a practical matter, measurements for suspended-slab construction should usually be made within a few hours of slab placement. In vertical construction, the only available surface for staging materials is often the slab that has just been placed and finished. Failure to take advantage of this very short window of availability following completion of finishing operations will hamper, if not preclude, the tolerance data collection.

Early measurement also relates directly to the contractor's performance. If methods and procedures require modification, changes can be made after the initial placement, minimizing the amount of unsatisfactory floor surface and repair required. At times, later measurements will be needed to see whether other influences have impacted flatness or levelness. For example, slabs-on-ground are subject to edge curling in the weeks following construction; cast-in-place suspended slabs deflect from their supported position when shores are removed. These possible later changes are affected by various design choices and the implementation of these choices by the contractor. For slabs-on-ground, such

Table 8.1—Slab on ground flatness/levelness construction guide

Notes:
1. These descriptions illustrate typical tolerance levels and construction procedures for floor surfaces in which direction and location of traffic may vary (random-traffic pattern). Most surfaces must accommodate random-traffic patterns.
2. The use of F-numbers to specify tolerances allows the specifier and contractor independent control of surface waviness and levelness. The Flatness F-number (F_F) controls waviness; the Levelness F-number (F_L) controls local levelness. Levelness quality is mainly dependent on accuracy of formwork and initial strikeoff.
3. The tolerance examples illustrate average to high floor tolerances; specified quality levels should be dictated by facility use.
4. Descriptions of placing and finishing methods are intended to assist the contractor in evaluation and "fine-tuning" of relative costs associated with producing the various levels of quality in flatness and levelness.
5. Finishing sequences described in this table require a slight modification when a metallic hardener, mineral-aggregate hardener, pigmented hardener, or pigment is to be applied. Refer to Section 8.6 for detailed discussion of suggested techniques. Proposed techniques for application of hardener and finishing concrete should be confirmed with a successful panel installation.

FLATNESS

Typical specification requirements	Typical finishing requirements
Specified overall value—20 Minimum local value—15	1. Smooth surface using 4 to 5 ft wide bull float. 2. Wait until bleed water sheen has disappeared. 3. Float surface with one or more passes using a power float (float-shoe blades or pans). 4. Make multiple passes with a power trowel (trowel blades).
Specified overall value—25 Minimum local value—17	1. Smooth and restraighten surface using 8 to 10 ft wide bull float. 2. Wait until bleed water sheen has disappeared. 3. Float surface with one or more passes using a power float (float-shoe blades or pans). 4. Restraighten surface following paste-generating float passes using 10 ft wide highway straightedge. 5. Make multiple passes with a power trowel (trowel blades).
Specified overall value—35 Minimum local value—24	1. Smooth and restraighten surface using 8 to 10 ft wide bull float. Apply in two directions at 45 degree angle to strip. 2. Wait until bleed water sheen has disappeared. 3. Float surface with one or more passes using a power float (float-shoe blades or pans). 4. Restraighten surface following paste-generating float passes using 10 ft wide highway straightedge. Use in two directions at 45 degree angle to strip. Use supplementary material to fill low spots. 5. Multiple passes with a power trowel (trowel blades).
Specified overall value—50 Minimum local value—35	1. Smooth and restraighten surface using 8 to 10 ft wide bull float or highway straightedge. Apply in two directions at 45 degree angle to strip. 2. Wait until bleed water sheen has disappeared. 3. Float surface with one or more passes using a power float (float-shoe blades or pans). First float pass should be across width of strip. 4. Restraighten surface following paste-generating float passes using 10 ft wide highway straightedge. Use in two directions at 45 degree angle to strip. Use supplementary material to fill low spots. 5. Multiple passes with a power trowel (trowel blades). 6. Restraighten surface after trowel passes using multiple passes with weighted highway straightedge to scrape the high points. No filling of the low spots is done at this stage.

LEVELNESS

Typical specification requirements	Typical forming and strikeoff requirements
Specified overall value—15 Minimum local value—10	1. Set perimeter forms (optical or laser instruments). 2. Use block placements of varying dimensions. Use wet screed strikeoff techniques to establish initial grade.
Specified overall value—20 Minimum local value—15	1. Set perimeter forms (optical or laser instruments). 2. Use block placements of varying dimensions. Use wet screed strikeoff techniques to establish initial grade. 3. Check grade after strikeoff. Repeat strikeoff as necessary.
Specified overall value—25 Minimum local value—17	1. Set edge forms using optical or laser instruments. Optical instruments provide more accurate elevation control. 2. Use strip placements with maximum widths of 50 ft. Use edge forms to establish initial grade. 3. Use vibratory screed for initial strikeoff.
Specified overall value—30 Minimum local value—20	1. Set edge forms using optical or laser instruments. Optical instruments provide more accurate elevation control. 2. Use strip placements with maximum widths of 30 ft. Use edge forms to establish initial grade. 3. Use vibratory screed for initial strikeoff. 4. Check grade after strikeoff. Repeat strikeoff as necessary. 5. Use a laser screed instead of rigid strikeoff guides and vibratory screed to produce this same quality.
Specified overall value—50 Minimum local value—35	1. Set edge forms using optical instrument to ±1/16 in. in accuracy. Use straightedge to identify form high spots; place top surface to fit inside 1/16 in. envelope. 2. Use strip placements with maximum widths of 20 ft. Use edge forms to establish initial grade. 3. Use vibratory screed for initial strikeoff. 4. Check grade after strikeoff. Repeat strikeoff as necessary. 5. Follow vibratory screed pass with two or three hand straightedge passes along the axis of the strip.

Table 8.2—Suspended slab flatness/levelness construction guide

FLATNESS

Typical specification requirements	Typical finishing requirements

Level and shored until after testing:
  Specified overall value—20
  Minimum local value—15
Unshored:
  Specified overall value—20
  Minimum local value—15
	1. Smooth surface using 4 to 5 ft wide bull float.
	2. Wait until bleed water sheen has disappeared.
	3. Float surface with one or more passes using a power float (float-shoe blades or pans).
	4. Make multiple passes with a power trowel (trowel blades).

Level and shored until after testing:
  Specified overall value—25
  Minimum local value—17
Unshored:
  Specified overall value—25
  Minimum local value—17
	1. Smooth and restraighten surface using 8 to 10 ft wide bull float.
	2. Wait until bleed water sheen has disappeared.
	3. Float surface with one or more passes using a power float (float-shoe blades or pans).
	4. Restraighten surface following paste-generating float passes using 10 ft wide highway straightedge.
	5. Make multiple passes with a power trowel (trowel blades).

Level and shored until after testing:
  Specified overall value—30
  Minimum local value—24
Unshored:
  Specified overall value—30
  Minimum local value—24
	1. Smooth and restraighten surface using 8 to 10 ft wide bull float. Apply in two directions at 45 degree angle to strip.
	2. Wait until bleed water sheen has disappeared.
	3. Float surface with one or more passes using a power float (float-shoe blades or pans).
	4. Restraighten surface following paste-generating float passes using 10 ft wide highway straightedge. Use in two directions at 45 degree angle to strip. Use supplementary material to fill low spots.
	5. Make multiple passes with a power trowel (trowel blades are preferable).

Level and shored until after testing:
  Specified overall value—50
  Minimum local value—35
Unshored:
  Specified overall value—50
  Minimum local value—35
	1. Smooth and restraighten surface using 8 to 10 ft wide bull float or highway straightedge. Apply in two directions at 45 degree angle to strip.
	2. Wait until bleed water sheen has disappeared.
	3. Float surface with one or more (float-shoe blades or pans). First float pass should be across width of strip.
	4. Restraighten surface following paste-generating float passes using 10 ft wide highway straightedge. Use in two directions at 45 degree angle to strip. Use supplementary material to fill low spots.
	5. Make multiple passes with a power trowel (trowel blades are preferable).
	6. Restraighten surface after trowel passes using multiple passes with weighted highway straightedge to scrape the high spots. No filling of low spots is done at this stage.

LEVELNESS

Typical specification requirements	Typical forming and strikeoff requirements

Level and shored until after testing:
  Specified overall value—15
  Minimum local value—10
Unshored:
  Specified overall value—N/A
  Minimum local value—N/A
	1. Set perimeter forms (optical or laser instruments).
	2. Use block placements of varying dimensions. Use wet screed strikeoff techniques to establish initial grade.

Level and shored until after testing:
  Specified overall value—20
  Minimum local value—15
Unshored:
  Specified overall value—N/A
  Minimum local value—N/A
	1. Set perimeter forms (optical or laser instruments).
	2. Use block placements of varying dimensions. Use wet screed strikeoff techniques to establish initial grade.
	3. Check grade after strikeoff. Repeat strikeoff as necessary.

Level and shored until after testing:
  Specified overall value—N/A
  Minimum local value—N/A
Unshored:
  Specified overall value—50
  Minimum local value—30
	1. Use a two-course placement to achieve this levelness quality. Topping slab must be placed using slab on grade techniques after shoring has been removed.
	2. Set edge forms using optical instrument to ±1/16 in. accuracy. Use straightedge to identify form high spots; place top surface to fit inside 1/16 in. in envelope.
	3. Use strip placements with maximum widths of 20 ft. Use edge forms to establish initial grade.
	4. Use vibratory screed for initial strikeoff.
	5. Check grade after strikeoff. Repeat strikeoff as necessary.
	6. Follow vibratory screed pass with two or three hand straightedge passes along the axis of the strip.

Metric Equivalents

1/16 in.	=	1.5 mm
4 ft	=	1.2 m
5 ft	=	1.5 m
8 ft	=	2.4 m
10 ft	=	3 m
20 ft	=	6.1 m
30 ft	=	9.1 m
50 ft	=	15.2 m

design choices include slab thickness, joint spacing, use of reinforcing steel, and vapor retarders. Inadequate curing can also accelerate curling of slabs-on-ground. For cast-in-place suspended slabs, deflection can be influenced by a number of variables, including depth of the structure, quantity of reinforcing steel, form-stripping procedures, and concrete strength when shoring is removed.

Because curling of slabs-on-ground will adversely affect flatness/levelness in service, methods to limit curling (Section 11.10) should be identified in the contract documents. Concrete with the lowest practical water content and low-shrinkage characteristics should be required. In addition, base conditions should not be such that the concrete underside remains wet while the top dries out. Joint spacings, load-transfer device, and reinforcement should be designated to minimize curling. Proper curing measures are essential and should be started as soon as possible after final finishing. These requirements should be clearly defined in the contract documents and adhered to during the concreting operations.

8.16—Treatment when bleeding is a problem

Prolonged bleeding can occur with poorly proportioned mixtures, poorly graded aggregates, excessive slump, or under conditions of low temperature, high humidity, or no air circulation. Bleed water may not evaporate, and the surface may not be sufficiently dry for floating and troweling.

One method to remedy the problem is to use fans or blower heaters of adequate size and in sufficient numbers to evaporate the excess moisture while the concrete is still plastic. Avoid using nonvented heaters, particularly those impinging on the surface of new concrete. They will cause carbonation of the surface, which can create a soft, dusty, chalky surface (Section 11.3).

If the concrete is firm enough for floating but the surface is still wet, the following methods can be used to obtain a drier surface:

1. Drag a rubber hose slowly over the entire surface; the concrete should be stiff enough so that only water is removed. In limited small areas that are difficult to reach with a hose, a single pass of a trowel tipped on edge can be used to remove water; however, slowly dragging a hose is much less likely to damage the surface, and this method should be used for the problem whenever possible.

2. Where required after removal of bleed water, apply additional concrete to fill low spots. This can be accomplished by discharging a small amount of concrete in a container during placing operations. The material in the container should have setting characteristics similar to those of the in-place concrete.

In general, the bleeding tendencies of concrete can be reduced significantly by the following actions. Every reasonable effort should be made to take such measures when bleeding is a problem:

1. Correct any aggregate gradation deficiency problem where materials of the required size gradations are economically available. The use of gap-graded aggregates results in increased bleeding. Ideally, combined gradation of all aggregates should yield a percent retained on each sieve below the largest and above the No. 100 (150 μm) of somewhere between 8 and 18%. The most common deficiency is in the 3/8 in., No. 4, 8, or 16 (9.5, 4.75, 2.36, or 1.18 mm) sieve sizes.

2. Use more cement if paste content is low, or lower the water content of the mixture.

3. Use pozzolan to replace part of cement or as an addition to the cement. (Note: Pozzolan should be finer than cement, and if the pozzolan is fly ash, it should conform to the requirements of ASTM C 618.)

4. Use the maximum allowable amount of entrained air. However, the use of air-entrained concrete containing in excess of 3% air for hard-troweled surfaces can promote development of blisters, delamination, and surface peeling.

5. Increase the amount of fine aggregate passing the No. 50, 100, and 200 (300, 150, and 75 μm) sieves to near the maximum allowable amount. More water (and possibly more cement) may be needed due to more paste being required; more shrinkage could result.

6. Use the lowest practicable water content.

7. Avoid admixtures that augment bleeding.

8. Use an accelerating admixture (see Section 5.7.3 for potentially deleterious effects).

9. Use concrete approaching the highest as-placed temperature permitted by the contract documents. (Note: Except for bleeding, there are benefits to be derived from placing concrete at the lowest permissible temperature.)

10. Use dewatering techniques (Section 8.3.6).

8.17—Delays in cold-weather finishing

Because concrete sets more slowly in cold weather and can be damaged by freezing, measures should be taken to keep the concrete temperature above 50 °F (10 °C). Appropriate curing procedures (Section 9.5.1) should be provided to prevent moisture loss and to keep every portion of the slab, including the edges, above freezing temperature. Any of the concrete's tendency toward bleeding will be considerably aggravated by the slower setting, and more work will be required to take care of it properly (Section 8.16). Many extra hours of finishers' time will be required, unless acceptable means can be found to shorten the setting time. Often, some extra expense to speed up the operation is justified by whichever of the following methods are most appropriate and least costly for a particular situation. Before adopting any method, tests should be made with job materials at job temperature conditions to confirm that acceptable results will be obtained.

Where it can be used without violating the precautions of Sections 5.7.3 and 8.6, a 1 to 2% addition of calcium chloride by weight of cement will accelerate setting significantly. When used, it should be added as a water solution.

Where the use of calcium chloride is prohibited:

1. A change to high early-strength cement (Type III), or use of a larger amount of Type I or II cement than usually required, can provide sufficient acceleration. Increases in cement content above approximately 600 to 625 lb/yd^3 (355 to 370 kg/m^3) can cause additional drying shrinkage and cracking in the hardened concrete.

2. Noncorrosive, nonchloride accelerating admixtures are available. The dosage rate can be varied to provide the optimum acceleration.

3. An increase in concrete temperate to 70 °F (21 °C) will noticeably reduce the setting time, although a low as-placed concrete temperature has many benefits (ACI 306R).

4. Early access for floating can be achieved by the application of dewatering techniques after first strikeoff and bull floating (Section 8.3.6).

The *w/cm* should be reduced and the minimum slump selected that can be easily handled and placed. Overworking the concrete should be avoided during the strikeoff and bull-floating operations.

CHAPTER 9—CURING, PROTECTION, AND JOINT FILLING

9.1—Purpose of curing

After proper placement and finishing of suitable quality concrete, curing is the single most important factor in achieving a high quality slab. The primary purpose of curing is to slow the loss of moisture from the slab and reduce early carbonation of the surface. A longer period of moisture retention permits more complete hydration of the cement, resulting in greater strength. Refer to ACI 308R for details on recommended curing time and minimum recommended temperatures.

9.2—Methods of curing

Moisture retention can be enhanced by several methods including moisture addition, moisture-retaining covers, and liquid membrane-forming curing compound. The characteristics of curing materials are set forth in detail in Section 5.6.

9.2.1 *Water curing*—Water curing by ponding, sprinkling, or fogging is practical only for slab areas without joints or where the water is positively confined by dams to prevent flooding the base course or saturating the subbase/subgrade. This is necessary to limit potential slab curling due to moisture gradients and to preserve compaction of the soil-support system. Water used for curing should be within 20 °F (7 °C) of the concrete temperature to avoid thermal shock. Continuous wetting should be maintained to avoid isolated dry spots. Water curing or wet covering should be used for shrinkage-compensating concrete slabs (ACI 223). Wet covering is generally the more practical and satisfactory method of water curing.

9.2.2 *Wet covering*—When properly applied and maintained, burlap and other wet coverings provide a continuous supply of moisture uniformly distributed on the slab surface. Burlap has been the most commonly used wet covering; wet burlap tends to reduce the temperature of the hydrating concrete slabs. Moist hay, straw, earth, or sand have been used, but their use is usually too labor-intensive for large projects and can discolor the surface. If sand or earth is used, it should be applied at least 1 in. (25 mm) deep and kept continuously wet during the curing period. Wet coverings should be laid over the concrete as soon as finishing operations are complete and surface marring can be avoided. Exposed concrete edges should be carefully covered. The coverings should be kept wet so that a film of moisture remains continuously in contact with the concrete throughout the curing period. Burlaps are available that resist rot and fire or that reflect light—reducing heat absorption from sunlight—or a combination thereof. Coverings with burlap on one side and polyethylene on the other are also available; the polyethylene is helpful in keeping the burlap moist longer, but it makes rewetting more difficult. Other polyethylene-backed fabrics are also available. These fabrics do not stain concrete like some burlaps and are often lighter and more durable than the burlap-backed product.

9.2.3 *Moisture retaining coverings*—Although not usually as effective as water curing and wet coverings, moisture-retaining coverings are widely used due to their convenience.

9.2.3.1 *Polyethylene (plastic) film* — Polyethylene film and other plastic sheet materials are available in clear, white, or black and are easily handled; the white is especially good for covering fresh concrete subject to sunlight. These films avoid leaving a residue that can prevent the bond of new concrete to hardened concrete or the bond of resilient floor coverings to concrete. Plastic films are particularly effective for curing the base slab of two-course floors. They can, however, leave blotchy spots on the slab and should not be used for colored concrete or where appearance of the slab surface is important. The sheets should be spread as soon as possible after finishing operations without marring the surface finish. Edges of sheets should be lapped a sufficient distance to prevent moisture loss and sealed with tape, mastic or glue, or held in place with wood planks or sand. Construction traffic should be restricted because the film can be extremely slippery.

9.2.3.2 *Waterproof paper*—Waterproof paper has the same advantages and disadvantages as plastic film, except that discoloration is less likely. It should be light in color; the edges should be lapped and sealed and left in place for the duration of the curing period. Tears caused by construction traffic should be repaired to maintain proper moisture retention.

9.2.4 *Liquid membrane-forming curing compounds*—Application of liquid membrane-forming curing compounds is the most widely used method for curing concrete. Advantages are relatively low in-place cost, early access to the floor, elimination of need to monitor the curing process, and the opportunity for longer uninterrupted cure. The membrane should be protected from damage due to construction traffic. Disadvantages include the potential for insufficient and uneven coverage, conflict with regulations on the release of volatile organic compounds, interference with bond of surfacing materials, and variability of quality and solids content. Liquid membrane-forming curing compounds should be applied as soon as finishing operations are complete while the surface is still damp but without free water. Machine spraying is preferable, but manual spraying is acceptable if accomplished with sufficient care to ensure uniform and complete coverage. Manual application should be accomplished by either spraying or rolling and by back-rolling with a wide short-nap paint roller. This can ensure full coverage without ponding of the curing compound in low spots. White-pigmented or fugitive-dye compounds help ensure even coverage and can be considered to reflect light and heat for floors exposed to sunlight. Generally, the curing compound should meet or exceed the minimum moisture retention requirements of ASTM C 309 (Section 5.6.3) or ASTM C 1315 (Section 5.6.3).

Curing compounds leave a film that can interfere with the adhesion of other materials to the treated surface; they should not be used on the base slab of a bonded two-course floor. Their use should also be avoided on surfaces that will later be covered with resilient floor coverings, protective coatings, sealers, or other special treatments. Where applicable, a letter of compatibility should be obtained from the manufacturer before the use of a curing compound on a floor receiving a subsequent finish. Curing compounds can also aggravate tire marking problems from forklift traffic; special nonmarking tires can be effective in minimizing these problems.

9.3—Curing at joints

Edges of joints should be cured to ensure maximum concrete strength to increase the durability of joint edges subject to solid wheeled traffic and to further reduce the potential for curling. Joints are cured adequately when wet coverings or moisture-retaining coverings are used. If a liquid membrane-forming curing compound is used, it should be applied to the inner joint walls. The curing compound may later require removal by sawing if a joint filler is installed. Alternatively, joints can be temporarily filled with wet sand or compressed backer rod during the curing period. If sand is used, it should be rewetted periodically.

9.4—Curing special concrete

Colored concrete and metallic-hardened floors require special curing techniques. Refer to Section 8.6 and recommendations of the material's manufacturer.

9.5—Length of curing

Regardless of the method used, the curing process should begin as soon as finishing operations are completed. If concrete begins to dry excessively before completion of finishing operations, the surface should be protected by fogging or use of a monomolecular film. The duration of curing will vary with the method, ambient temperature, humidity, and type of cement. With any type of cement, in temperatures above 40 °F (5 °C) 7 days of uninterrupted curing is normally recommended for water curing or moisture-retaining-cover curing. This time period can be reduced to 3 days when high early-strength concrete is used and temperatures are 73 °F (23 °C) or higher.

9.5.1 *Cold-weather considerations*—Slabs should not be placed on a frozen base. Cold-weather protective measures should maintain a concrete temperature above 50 °F (10 °C), and appropriate curing procedures should be provided to minimize moisture loss. Insulating blankets placed over the top of a curing membrane will retain heat, provided there is no danger of serious loss of heat from below the slab—for example, an upper floor of an open, unheated building. Particular care should be given to the corners and edges of the slab, which are more vulnerable to rapid heat loss. The amount of insulation required can be calculated from tables furnished in ACI 306R. When there is a danger of freezing—particularly when mean daily temperatures are lower than 40 °F (4 °C)—insulation is frequently not sufficient to protect thin slabs of concrete used for floors, and auxiliary heat is required. The area should be enclosed with tarpaulins or plastic sheeting and heated with live steam or vented heaters. The use of salamander-type heaters or other equipment that exhaust carbon dioxide gases into the area above the concrete floor should be avoided because of the danger of carbonation of the fresh concrete; carbonation will result in a soft, dusty surface (Kauer and Freeman 1955).

Where freezing is anticipated during or within a few days following the curing period, consideration should be given to protection of the concrete. Concrete saturated with water is vulnerable to freezing damage; the use of thermal blankets or other protective measures may be necessary. The curing method and procedure should cure the concrete satisfactorily and allow appropriate drying of the concrete before freezing. Refer to Section 4.6 for other concrete placement conditions, and refer to ACI 306R for more information on cold-weather concreting procedures.

9.5.2 *Hot-weather considerations*—In hot weather, curing procedures should begin as soon as an area of slab is finished to prevent surface drying. Continuous moist-curing methods—water curing and wet coverings—are the most effective because they provide adequate moisture and tend to prevent excessive heat build-up. Moisture-retaining coverings limit evaporation; conditions creating temperature gradients in the slab should be avoided. Curing compounds used for exterior work should be white-pigmented. Refer to Section 4.6 for other concrete placement conditions, and refer to ACI 305R for more information on hot-weather concreting procedures.

9.6—Preventing plastic-shrinkage cracking

Plastic-shrinkage cracking in newly floated or troweled slabs results when the rate of drying at the surface is more rapid than the upward movement of bleed water. Plastic-shrinkage cracking occurs in the presence of such factors as moderate to high winds, low relative humidity, and high concrete and air temperatures. The use of latex-modified concrete, high-range water reducers, and silica fume tends to increase plastic-shrinkage cracking potential because these materials usually reduce the bleed rate of concrete.

Vapor retarders/barriers immediately under the concrete may aggravate plastic and drying-shrinkage cracking and slab curling because the bottom of the slab loses little or no moisture, while the top dries and shrinks at a faster rate (Anderson and Roper 1977, Nicholson 1981, Turenne 1978).

If the rate of water evaporation from the concrete exceeds 0.2 lb/ft^2/h (1.0 kg/m^2/h), precautions should be taken to reduce evaporation (ACI 305R). Measures helpful in preventing or reducing plastic-shrinkage cracking are given in Section 11.2.2.1; additional information is presented in ACI 305R.

9.7—Curing after grinding

If grinding is required, it should be initiated as soon as the floor is hard enough to avoid tearing out aggregate particles. Curing should be maintained both before and after early grinding.

9.8—Protection of slab during construction
Protection should be provided against:
a. Heavy construction traffic;
b. Hard-wheeled traffic;
c. Impact and abrasion;
d. Imposed loads (cranes, concrete trucks);
e. Stains (grease, oil, chemicals, paints, plaster, clay soil);
f. Rubber tire marks;
g. Deicers; and
h. Freezing.

9.9—Temperature drawdown in cold storage and freezer rooms
The temperature reduction in freezer and cold storage rooms should be gradual to control cracking caused by differential thermal contraction and to allow drying to remove excess moisture from the slab after curing. A typical drawdown schedule might be as follows:

Temperature	Time
Ambient to 35 °F (2 °C)	10 °F (5.5 °C) Per 24 h
Hold at 35 °F (2 °C)	2 to 5 days
35 °F (2 °C) to final	10 °F (5.5 °C) per 24 h

9.10—Joint filling and sealing
Materials for joint fillers and sealants are discussed in Section 5.12 Contraction joints are normally sawn using the narrowest blade practical. Formed construction joints should be similarly sawn but to a depth of only 1 in. (25 mm). Sawcuts at the construction joints should not be introduced until a crack is perceptible at the cold joint between adjacent placements. Compressible backer rods should not be used in joints that will be exposed to heavy traffic. Isolation joints can be formed with preformed fiberboard, polyethylene foam, or similar materials before concrete placement begins. This is described in Section 3.2.5.1 and detailed in ACI 504R. Isolation joints are sometimes sealed with an elastomeric sealant to prevent accumulation of moisture, dirt, or debris. Asphalt-impregnated or similar materials should not be used in isolation joints that will be sealed.

9.10.1 *Time of filling and sealing*—Concrete slabs-on-ground continue to shrink for years; most shrinkage takes place within the first 4 years. The most significant shrinkage takes place within the first year, especially the first 60 to 90 days. It is advisable to defer joint filling and sealing as long as possible to minimize the effects of shrinkage-related joint opening on the filler or sealant. This is especially important where semirigid fillers are used in traffic-bearing joints; such fillers have minimal extensibility. If the joint should be filled before most of the shrinkage has occurred, separation should be expected between the joint edge and the joint filler or within the joint filler itself. These slight openings can subsequently be filled with a low-viscosity filler recommended by the same manufacturer as the original filler. If construction traffic dictates that joints be filled early, provisions should be made to require that the contractor return at a preestablished date to complete the necessary work using the same manufacturer's products. Earlier filling will result in greater separation and will lead to the need for more substantial correction; this separation does not indicate a failure of the filler. For cold storage and freezer room floors, the joint filler should be installed only after the room has been held at its final operating temperature for a minimum of 48 h. For rooms with operating temperatures below 0 °F (–18 °C), the operating temperature should be maintained for at least 14 days before starting joint filling.

9.10.2 *Installation*—Elastomeric sealants should be installed over a backer rod or other bondbreaker as described in ACI 504R. The use of elastomeric sealants is not recommended in joints exposed to solid-wheel traffic. Semirigid epoxy and polyurea fillers should be installed full-depth in saw-cut joints. Joints should be suitably cleaned to provide optimum contact between the filler or sealant and bare concrete. Vacuuming is recommended rather than blowing the joint out with compressed air. Dirt, debris, saw-cuttings, curing compounds, and sealers should be removed. Cured semirigid fillers should be finished flush with the floor surface to protect the joint edges and to re-create an interruption-free floor surface. Specific installation instructions should be requested of the filler/sealant manufacturer if the floor is to receive a nonbreathing covering such as vinyl, epoxy, or a similar finish.

CHAPTER 10—QUALITY CONTROL CHECKLIST
10.1—Introduction
Details on a quality control program should be included in the contract documents. To ensure that the program will be fully complied with for the duration of the project, procedures should be presented to the involved parties in the prebid meeting and reviewed in detail at the preconstruction meeting. ACI 311.4R is a good source to use in the development of the program. Because the eventual success of any project is the result of a team effort, there should be a complete understanding and agreement regarding the provisions of the program before any concrete construction is started.

Many items involved with quality control will be covered in the preconstruction meeting, but some questions or concerns will invariably come up on site that are not covered in the bid documents or at the meeting. Therefore, it is essential to have a person on site who has the experience and background necessary to use the best possible judgment. Personnel with ACI certification can contribute greatly toward resolving these concerns and ensuring quality construction in the field. The NRMCA/ASCC Preconstruction Checklist is a good source of items that should be addressed before construction.

10.2—Partial list of important items to be observed
Additional background information regarding important items, such as concrete reinforcement, surface hardeners, and joint sealants, can be applicable during the actual construction phase.

10.2.1 *Slump control and testing*—The addition of trim water to the concrete at the jobsite (Section 7.3.2) can be required to ensure consistent placeability, workability, and finishability; no more water should be added than is necessary to meet the overall project requirements. The committee recommends that an agreed-on amount of trim water—part of the design mixture water—be withheld at the plant to permit this on-site adjustment. Two practices that help ensure adequate control of slump at the jobsite are: 1) be sure truck mixers come to the site with full water tanks, and 2) designate one specific person to authorize adding water at the site. An ACI Certified Concrete Field Technician should be present during the entire placement to perform the required slump and other tests.

Testing, including provisions for handling and storing cylinders or cores, should be completed in accordance with ASTM procedures by a testing agency meeting accreditation requirements of ASTM E 329. This is particularly significant when air-entrained concrete is used; the actual air content is subject to change and requires repeated testing. When entrained air is prohibited or less than 3% total air is desired, the air content should be checked on the first truck and occasionally thereafter. ACI 311.5R contains guidance on plant inspection and field testing.

10.2.2 *Avoid delays*—Anything that would result in slump loss should be avoided—delays in delivery of concrete, delays in placing or finishing operations, and interruptions by other trades. Although the mixture proportions may have been approved, some minor adjustments could be required due to locally available materials or jobsite conditions.

10.2.3 *Forms, reinforcement, dowels, and joints*—Forms, reinforcement, and dowels should be secured and remain straight and true during the entire placing and finishing operation. Unless otherwise stated in the contract documents, reinforcement should be discontinued at joints. If the contract documents indicate that reinforcement is to continue through joints, the possibility that some out-of-joint random cracking could occur should be discussed during the preconstruction meeting with the designer and owner. The alignment of reinforcement along joints should permit a straight saw-cut to be effective and allow joints to open. Smooth dowels should be used in joints where load transfer is required. Dowels in contraction joints should be positively supported and aligned. Any conditions that create restraint to the normal shrinkage process should be noted—for example, the condition of the base on which the concrete is placed. Although the practice of cutting every other bar or wire has been used with some success, there is always the possibility of some cracks forming in the intermediate panels due to partial restraint at the joint (Section 3.2.4).

10.2.4 *Finishing*—The finishing process should be discussed with the finishing supervisor because no specification can be sufficiently accurate as to the actual timing of most finishing operations. Slab edges should be given special attention, beginning with the initial floating step and continuing through the entire finishing process. ACI Certified Finishers should be used whenever possible.

If an aggregate- or metallic-surface hardener is used, the hardener should be completely moistened so no dry material will be floated into the surface before machine floating.

10.2.5 *Curing, saw-cutting, joint filling, and tolerances*—The proposed method of curing, the necessary timing for sawing joints, the protection of joint edges until the joints are filled, the timing of joint filling, and the protection required of the completed floor should be reviewed in detail. There should be a complete understanding regarding the order in which the curing, sawing, and floor tolerance testing are to be performed.

10.2.6 *On-site meeting*—After initial placement, additional on-site meetings may be necessary to review actual results and discuss any required adjustments in the overall plan. Also, backup procedures for equipment breakdowns should be discussed with the concrete superintendent—for example, pumps, troweling machines, spreaders, and saws.

CHAPTER 11—CAUSES OF FLOOR AND SLAB SURFACE IMPERFECTIONS

11.1—Introduction

Concrete is a forgiving material; however, concrete quality can be adversely affected by conditions over which the designer or the contractor has little control. This chapter lists the conditions and circumstances that can cause imperfections in concrete floor and slab surfaces. Concrete is capable of providing a highly durable, serviceable, and attractive surface. When it does not do so, there are always reasons. By keeping the causes of certain imperfections in mind, it is possible to reduce the likelihood of unsatisfactory results; these causes will be described briefly in this chapter. When the corrective action to eliminate a particular cause is not obvious, the most promising suitable procedure described in preceding chapters will be referenced.

In reviewing the causes of floor and slab surface imperfections, the reader should keep in mind the inherent characteristics of portland cement concrete, such as drying-shrinkage cracking. Some curling and cracking can be expected on every project. Also, it will be evident that the most common imperfections stem from failure to follow the basic rules of concrete finishing given in Section 8.3.3, such as "Any finishing operation performed while there is excess moisture or bleed water on the surface will cause dusting or scaling" (and also cause crazing and reduced resistance to wear), and as is stated in Section 8.3.5, "No subsequent operation (after bull floating and restraightening) should be done until the concrete will sustain foot pressure with only about 1/4 in. (6 mm) indentation," that is, no premature finishing.

Another common cause of floor and slab surface imperfections is the lack of prompt curing. The keyword is prompt, and the degree to which this can be accomplished, especially in dry or windy weather, will improve the quality of floor and slab surfaces tremendously. Moist curing is best, provided the slab is kept continuously moist (Section 9.2).

Rarely will there be a single cause for a given imperfection; usually some combination will be responsible. The influence of any cause will vary with the degree of its departure from best practice, with the properties of the materials used, and with the ambient temperature and other weather conditions present during the work. Satisfactory results are more likely to be obtained if the causes mentioned for the various kinds of imperfections are carefully avoided.

11.2—Cracking

Cracking of concrete (Fig. 11.1) is a frequent complaint. Cracking is caused by restraint (internal or external) of volume change, commonly brought about by a combination of factors such as drying shrinkage, thermal contraction, curling, settlement of the soil-support system, and applied loads. Cracking can be significantly reduced when the causes are understood and preventive steps are taken. For example, joints that are properly designed, detailed, and installed at the proper spacing and time during construction will cause cracks to occur in the joints where they remain inconspicuous, instead of random locations.

Contractors are not necessarily responsible for all cracks. Many floor or slab design features and concrete mixture proportions are responsible for, or contribute to, cracking of concrete construction. If a contractor believes there are problems with slab design, mixture proportions, or other problems, they should be pointed out before installation; the prebid and preconstruction meetings should be used for this purpose. Designers should pay careful attention to the causes of cracking, and contractors need to understand floor and slab design and concrete mixture proportioning to avoid problems. Designers also should understand slab construction to avoid "building in" problems for the contractor. For more information on control, causes, evaluation, and repair of cracks in concrete structures, refer to ACI 224R and 224.1R.

11.2.1 *Restraint*—Because cracking is caused by restraint of volume changes, normal volume changes would be of little consequence if concrete were free of any restraint. Concrete in service, however, is usually restrained by foundations, subgrade, reinforcement, or connecting members, significant stresses can develop—particularly tensile stresses.

The amount of drying shrinkage will be reduced somewhat by taking practical measures to place the concrete with the lowest possible water content. Water reduction through use of admixtures—water-reducing admixtures meeting ASTM C 494, Types A and D, and air-entraining admixtures—has little effect on drying shrinkage (Ytterberg 1987; Martin and Phelan 1995; Whiting and Dziedzic 1992). Refer to Section 5.4.2 for additional information concerning use of these admixtures. Thus, drying shrinkage of concrete containing water reducers can still cause unsightly cracking, unless the following good practices are employed:

1. Contraction joints not spaced too far apart (Section 3.2.5.3);
2. Contraction joints deep enough;
3. Contraction joints sawn early enough;
4. Slabs not strongly restrained at their perimeters by bond of floor or slab concrete to foundation walls or other construction, or by tying-in reinforcement to foundations, docks, and tilt-up walls (Section 3.2.5);
5. Isolation joints provided around columns (Fig. 3.3 and 3.4);

Fig. 11.1—Drying-shrinkage cracks such as these are a frequent cause of complaint (PCA A5271).

6. Joint or extra reinforcing steel placed diagonally to reentrant corners;
7. Concrete mixtures of necessary strength with the proper amount of cement and water, Also, mixtures that do not include any ingredient, such as aggregates or admixtures, with high-shrinkage characteristics;
8. Proper curing;
9. Slabs not restrained by a rutted or uneven base and changes in slab thickness;
10. Discontinued reinforcement at joints, thus encouraging joints to open; and
11. Slabs cast upon a base which has a low coefficient of friction, such as a fine-graded crushed stone. This will provide a smooth surface on which the slab can slide (Section 4.1).

11.2.2 *Early cracking*—Some cracking can occur before the concrete has hardened. This can complicate the finishing operations considerably. Some examples are:

1. Plastic-shrinkage cracking (Sections 9.6 and 11.2.2.1);
2. Cracking from settlement of concrete around reinforcing bars or other embedments (Sections 5.8 and 6.2.5);
3. Cracking along edges where forms are not rigid;
4. Early thermal cracking; and
5. Damage from form removal.

11.2.2.1 *Plastic-shrinkage cracking*—Plastic-shrinkage cracks (Fig. 11.2) are relatively short, shallow, random (but sometimes parallel) cracks that can occur before final finishing on days when wind, low humidity, and high concrete and ambient temperatures occur. Surface moisture evaporates faster than it can be replaced by rising bleed water, causing the surface to shrink more than the interior concrete. As the interior concrete restrains shrinkage of the surface concrete, stresses that exceed the concrete's tensile strength develop, resulting in surface cracks. These cracks range approximately from 4 in. to 3 ft (100 mm to 1 m) or more in length. They can be roughly parallel to one another and spaced about from a 4 in. to 2 ft (100 to 600 mm) apart, but usually occur in a random, irregular pattern. Crack formation begins at the surface and continues downward for some distance, rapidly becoming narrower with depth. Though usually only 1 to 3 in. (25 to 75 mm) deep, they can go completely through the slab. Plastic-shrinkage cracks in the still-unhardened concrete can sometimes be closed by

Fig. 11.2—Plastic-shrinkage cracks are caused by rapid loss of mixing water from the surface while the concrete is still plastic (PCA 1311).

Fig. 11.3—Crazing is a network of very fine superficial surface cracks (PCA 4099).

tamping and beating the surface with a hand float. While this should be done, the more effective protective measures listed below should also be undertaken immediately to remove the causes of plastic-shrinkage cracking in the remaining work:

1. Dampen the base when no vapor retarder is used;
2. Erect windbreaks;
3. Erect sunshades;
4. Cool aggregates and mixing water before mixing; and
5. Prevent rapid drying by one of the following:
    a. Protect concrete with moisture-retaining coverings (Section 9.2.3) during any delay between placing and finishing.
    b. Cover with damp burlap or with white polyethylene sheeting (Section 9.2.2) immediately after screeding and bull-floating. Keep burlap moist until the concrete is ready for finishing. Uncover only a small area at one time, just ahead of the finishers. Begin curing as soon as possible.
    c. Use monomolecular films to reduce evaporation between the various placing and finishing operations.
    d. Use a fog spray located upwind of the freshly placed concrete. The spray device should use metered heads and discharge spray into the air above the concrete.
6. Postpone each step of finishing (and its inherent reworking of the surface) as long as possible without endangering results; and
7. Avoid the use of a vapor retarder where not needed.

11.2.2.2 *Crazing*—Crazing, a pattern of fine cracks that do not penetrate much below the surface, is caused by minor surface shrinkage (Fig. 11.3). Crazing cracks are very fine and barely visible, except when the concrete is drying after the surface has been wet. They are similar to mud cracking in shape and in generation. The cracks encompass small concrete areas less than approximately 2 in. (50 mm) in dimension, forming a chicken-wire-like pattern. The term map cracking is often used to refer to cracks that are similar to crazing cracks only more visible and involving larger areas of concrete. Although crazing cracks can be unsightly and can collect dirt, crazing is not structurally serious and does not necessarily indicate the start of future deterioration in interior slabs.

When concrete is just beginning to gain strength, climatic conditions, particularly the relative humidity during the drying period in a wetting-and-drying cycle, are an important cause of crazing. Low humidity, high air and concrete temperatures, hot sun, or drying wind, either separately or in any combination, can cause rapid surface drying that encourages crazing. The conditions that contribute to dusting, as described in Section 11.4, also will increase the tendency to craze.

To prevent crazing, curing procedures should begin immediately, within minutes after final finishing, particularly after hard troweling. This is especially important when weather conditions are adverse. When the temperature is high and the sun is shining with high winds and low humidity, some method of moist curing should be used to stop rapid drying. The concrete should be protected against rapid changes in temperature and moisture wherever feasible. Other conditions to be avoided that can cause craze cracking are:

1. Curing with water that is more than 20 °F (11 °C) cooler than the concrete;
2. Alternate wetting and drying of the concrete surface at early ages;
3. Overuse of jitterbugs, vibrating screeds, and bull floats (Section 8.3.2);
4. Overworking and overtroweling, especially when the surface is too wet (Sections 8.3.10 and 8.3.11);
5. Premature floating and troweling (Section 8.3.3);
6. Dusting dry cement onto a surface to hasten drying before finishing;
7. Too much clay and dirt in aggregates; and
8. Sprinkling water onto the surface of a slab during finishing.

11.2.3 *Other causes*—Cracking over the long term can result from causes other than shrinkage. Prominent causes are:

1. Uneven support by a poorly prepared subgrade, subbase, or base; poor drainage; or uneven support due to curling of slab edges (Section 11.11);
2. Expansive clay in the subgrade;
3. Sulfates in subgrade soil or groundwater;

4. Placing concrete over preformed joint filler (when placing adjacent concrete);

5. Improper jointing and sealing (Sections 3.2.5, 5.11, 9.10, and ACI 504R);

6. Structural overloading, especially following the floor construction phase of a building project;

7. Impact loads;

8. Disruption from expansive alkali-silica reaction;

9. Disruption from corrosion of reinforcing steel;

10. Disruption from freezing and thawing along edges and at corners;

11. Earth movements from contiguous construction, for example, blasting or pile driving;

12. Thermal contraction, such as a sharp drop in ambient temperature shortly after casting a floor or slab;

13. Early or excessive construction traffic; and

14. Improper design (for example, selection of an inadequate safety factor), resulting in a slab of inadequate thickness for service conditions.

11.3—Low wear resistance

Low wear resistance is due primarily to low-strength concrete, particularly at the surface. Such low strengths result from:

1. Too much mixing water;

2. Use of concrete with too high a w/cm;

3. Excessive slump, which promotes bleeding and carries softer, lighter-weight material (laitance) to the surface. (After considering the unavoidable causes for slump loss, use the lowest practical water content and slump (Table 6.1));

4. Overworking overly wet concrete. (This does not mean that it is acceptable to use overly wet concrete under any conditions. It should be rejected and removed from the site. Whatever is placed must be worked, but if concrete is overly wet, the ready-mix plant should be called to make sure no more wet batches are delivered. In addition to producing a surface with low wear resistance, overworking of overly wet concrete also will cause segregation; fluid mortar will flow into low areas, settle, and leave low spots.);

5. Premature floating and troweling, which works bleed water into the surface (see Section 8.16 if bleeding is a problem);

6. Excessive use of water by finishers (Section 8.3.3);

7. Excessive entrained air in the surface mortar, although occurrence of this is not common;

8. Deficient curing (Chapter 9);

9. Surface carbonation from unvented heaters used for cold-weather protection (ACI 306R);

10. Impairment of surface strength potential by early-age freezing (ACI 306R);

11. Opening slab to abrasive traffic before sufficient strength has developed; and

12. Poor finishing techniques and improper timing during and between finishing operations (Section 8.3).

11.4—Dusting

Dusting (Fig. 11.4) is another aspect of weak concrete at the surface of a floor or slab. Dusting (the development of a fine,

Fig. 11.4—Dusting is evident when a fine, powdery material can be easily rubbed off the surface of a slab (PCA 1297).

powdery material that easily rubs off the surface of hardened concrete) can occur either indoors or outdoors but is more likely to be a problem when it occurs indoors. Dusting is the result of a thin, weak surface layer, called laitance, which is composed of water, cement, and fine particles.

Fresh concrete is a fairly cohesive mass, with the aggregates, cement, and water uniformly distributed throughout. A certain amount of time must elapse before the cement and water react sufficiently to stiffen and develop hardened concrete. During this period, the cement and aggregate particles are partly suspended in the water. Because the cement and aggregates are heavier than water, they tend to sink. As they move downward, the displaced water and fines move upward and appear at the surface, resulting in more water and fines near and at the surface than in the lower portion of the concrete. This laitance—the weakest, most permeable, and least wear-resistant material—is at the top surface, exactly where the strongest, most impermeable, and most wear-resistant concrete is needed. Floating and troweling concrete with bleed water on the surface mixes the excess water back into the surface, further reducing the strength and wear resistance at the surface and giving rise to dusting (Section 8.3.3). Dusting can also be caused by:

1. Overly wet mixtures with poor finishing characteristics;

2. Insufficient cement (Table 6.2);

3. Excessive clay, dirt, and organic materials in the aggregate;

4. Use of dry cement as a blotter to speed up finishing;

5. Water applied to the surface to facilitate finishing;

6. Carbonation of the surface during winter concreting, caused by unvented heaters (ACI 306R);

7. Inadequate curing, allowing rapid drying of the surface, especially in hot, dry, and windy weather; and

8. Freezing of the surface (ACI 306R).

11.5—Scaling

Scaling is the loss of surface mortar and mortar surrounding the coarse-aggregate particles (Fig. 11.5). The aggregate is usually clearly exposed and often stands out from the concrete. Scaling is primarily a physical action caused by hydraulic pressure from water freezing within the concrete; it is not usually caused by chemical corrosive

Fig. 11.5—Scaling is the loss of surface mortar, usually exposing the coarse aggregate (PCA A5273).

Fig. 11.6—Mortar flaking over coarse aggregate particles is another form of scaling that resembles a surface with popouts (PCA 52225).

action. When pressure exceeds the tensile strength of concrete, scaling can result if entrained air voids are not present in the surface concrete to act as internal pressure-relief valves. The presence of a deicing solution in water-soaked concrete during freezing causes an additional buildup of internal pressure. Properly designed and placed air-entrained concrete, however, will withstand deicers for many years.

Deicers, such as sodium chloride, urea, and weak solutions of calcium chloride, do not chemically attack concrete; however, deicers containing ammonium sulfate or ammonium nitrate will rapidly disintegrate concrete and should not be used. Several deicers, particularly those containing chloride ions, can accelerate corrosion of embedded steel.

Prominent among the causes of scaling are:
1. Permeable and poor quality concrete due to:
   a. High w/cm (over 0.50);
   b. Excessive slump for prevailing job conditions (Table 6.1);
   c. Overworking of wet concrete (Section 8.3.10);
   d. Premature finishing operations (Section 8.3.3);
   e. Inadequate curing (Chapter 9); and
   f. Low compressive strength at the surface (Section 6.2.3);

2. In concrete to be exposed to freezing and thawing in service, little or no entrained air due to:
   a. Failure to use an air-entraining agent;
   b. Air worked out by overworking overly wet concrete in premature finishing operations (Sections 8.3.10 and 8.3.3);
   c. Air content too low due to: mixing too long, concrete temperature too high for a given dosage of air-entraining agent, or improper dispensing of air-entraining agent;

3. Air content too low to resist the effect of chemicals used for snow and ice removal (Section 6.2.3);

4. Inadequate thermal protection, allowing freezing of the surface at an early age;

5. Exposure of new concrete to freezing and thawing before it has been adequately cured, achieving a compressive strength of 4000 psi (28 MPa), and allowed to air dry. Application of deicing chemicals at this early age greatly increases the likelihood of scaling;

6. Blistering (Section 11.7), which increases vulnerability to scaling; and

7. Inadequate slope to properly drain water away from the slab; saturated concrete is more susceptible to damage from freezing and thawing than drier concrete.

11.5.1 *Mortar flaking*—Mortar flaking over coarse-aggregate particles (Fig. 11.6) is another form of scaling. Aggregate particles with flat surfaces are more susceptible to this type of imperfection than round particles. Mortar flaking occasionally precedes more widespread surface scaling, but its presence is not necessarily an indication of an onslaught of more extensive scaling.

Mortar flaking over coarse-aggregate particles is caused essentially by the same actions that cause regular scaling and often results from placing concrete on hot, windy days. Excessive and early drying out of the surface mortar can alone aggravate scaling; however, the moisture loss is accentuated over aggregate particles near the surface as bleed water beneath the aggregates cannot readily migrate to the surface to replenish the evaporated water. This combination of bleed-water blockage, high rate of evaporation, and lack of moisture necessary for cement hydration results in a dry-mortar layer of low strength, poor durability, high shrinkage, and poor aggregate bond. Upon freezing in a saturated condition, this thin, weakened mortar layer breaks away from the aggregate. Poor finishing practices can also aggravate mortar flaking.

11.6—Popouts

Popouts are roughly cone-shaped pits left in the surface of flatwork after a small piece of concrete has broken away by internal pressure (Fig. 11.7). This pressure is generated by the expansion of a piece of chert, soft fine-grained limestone, shale, hard-burned lime, hard-burned dolomite, pyrite, or coal. The first two are natural constituents of some aggregates; the others sometimes find their way into aggregates as impurities. In some materials, the expansion is caused by freezing or absorption of moisture; in others, it is caused by a chemical change. For example, popouts can occur from the chemical reaction between alkalies in concrete and reactive siliceous aggregates. Popouts range in size from about 1/4 to 2 in. (6 to 50 mm) or more in diameter.

Because popouts usually do not significantly diminish the integrity of concrete flatwork, they are sometimes tolerated. Nevertheless, they are usually unsightly and interfere with the performance of any slab required to be smooth. On floors with hard-wheeled traffic, popouts can degenerate into larger imperfections. Early repair should minimize further problems in high-traffic areas.

The occurrence of impurities in the concrete can be beyond the control of the floor constructor, because it usually occurs inadvertently in the production and handling of ready-mixed concrete or its constituents. The presence of naturally occurring chert or soft fine-grained limestone, however, can be a continuing problem in some locales. Measures that can be taken to alleviate the problem are:

1. Switching to a nonoffending source of aggregate for floors and slabs, if possible;

2. Using two-course construction with selected or imported aggregate without popout potential for the topping course;

3. Using aggregates from which the offending particles have been removed by heavy-media separation, if available and economically feasible;

4. Using wet-curing methods such as continuous fogging or covering with wet burlap immediately after final finishing. Wet-cure for a minimum of 7 days, as wet curing can greatly reduce or eliminate popouts caused by alkali-aggregate reactivity (Landgren and Hadley 2002). Avoid plastic film, curing paper, and especially curing compounds, as they allow an accumulation of alkalies at the surface. Impervious floor coverings or membranes, such as wax, epoxy, or other coatings, should be avoided as they can aggravate popout development; and

5. Using the lowest practical slump possible to prevent potential popout-causing particles from floating to the surface.

In some areas and situations, these measures may not be practical. Specific local practices have been developed that have been helpful in minimizing popouts. For example, in some regions ready-mix producers can supply popout-free concrete.

11.7—Blisters and delamination

The appearance of blisters (Fig. 11.8) on the surface of a concrete slab during finishing operations is annoying and an imperfection that can leave portions of the top surface vulnerable to delamination once the concrete hardens. Blisters are "bumps" that can range in size from 1/4 to 4 in. (6 to 100 mm) in diameter and approximately 1/8 in. (3 mm) deep. They appear when bubbles of entrapped air or water rise through the plastic concrete and are trapped under an already sealed, airtight surface. This early closing of the surface frequently happens when the top of a slab stiffens, dries, or sets faster than the underlying concrete. Several factors are attributed to blistering.

An excessive amount of entrapped or entrained air held within the concrete by excessive fines—material passing the No. 30, 50, and 100 sieves (600, 300, and 150 μm)—resulting in a sticky mixture that can become more easily

Fig. 11.7—A popout is a small fragment of concrete broken away from the surface of a slab due to internal pressure, leaving a shallow, typically conical, depression (PCA 0113).

Fig. 11.8—Blisters (courtesy of NRMCA).

sealed during the raised troweling and closing operations. Sticky mixtures have a tendency to crust under drying winds, while the remainder of the concrete stays plastic. Usually, what is needed to relieve this condition is to reduce the amount of sand in the mixture by 100 to 200 lb/yd^3 (60 to 120 kg/m^3) and to replace the removed sand with a like amount of the smallest-size coarse aggregate available (PCA 2001). The resulting slightly harsher mixture should release most of the entrapped air using normal vibration. On days when surface crusting occurs, slightly modified finishing techniques may be needed, such as the use of wooden floats to keep the surface open and flat troweling to avoid enfolding air into the surface under the blade action.

Insufficient vibration during compaction may not adequately release entrapped air or overuse of vibration may leave the surface with excessive fines, inviting crusting and early finishing.

Power finishing operations should continue before the initial set of the slab is over its full depth. Any tool used to compact or finish the surface will tend to force the entrapped air toward the surface. Blisters may not appear after the first finishing pass, but may appear, as the work progresses to the second or third troweling. At this stage in finishing, the trowel blade is tilted to increase surface density; air and water just under the surface are forced ahead of the blade until enough is concentrated (usually near a piece of large aggregate) to form a blister. Blisters, which can be full of air, water, or both, when punctured, can also appear at any time during finishing operations and without apparent cause. Floating the concrete a second time helps to reduce blistering. Delayed troweling will depress the blisters even though it may not reestablish complete bond.

Project specifications that require combination blades, float pans, or trowel blades should be used on concrete with intentionally entrained air (Refer to Sections 5.7.1 and 6.2.7).

To avoid blisters, the following should be considered:

1. Avoid the use of concrete with excessively high slump, water content, air content, or fines;

2. Use appropriate cement contents (Table 6.2);

3. Warm the base before placing concrete during cool weather. During hot, dry, windy weather, reduce evaporation over the slab by using an evaporation retardant (monomolecular film), a fog spray, or a slab cover (polyethylene film or wet burlap);

4. Avoid placing a slab directly on polyethylene film or any other vapor retarder. Use a minimum 4 in. thick (100 mm) layer of trimmable, compactible granular fill (not sand) to separate the vapor retarder from the concrete (Section 4.1.5);

5. Avoid overworking the concrete, especially with vibrating screeds, jitterbugs, or bull floats. Overworking causes coarse aggregate to settle, and bleed water and excess fines to rise to the surface. Properly vibrate concrete to release entrapped air;

6. Do not attempt to seal (finish) the surface too soon. Hand floating should be started when a worker standing on a slab makes a 1/4 in. (6 mm) footprint. For machine floating, the footprint should be only about 1/8 in. (3 mm) deep. If moisture is deficient, a magnesium float should be used;

7. Use a wooden bull float on non-air-entrained concrete to avoid early sealing. Magnesium or aluminum tools should be used on air-entrained concrete. Slabs that incorporate a surface hardener are more prone to blister if not properly finished (Sections 8.6.1 and 8.6.2); and

8. Use proper finishing techniques and proper timing during and between finishing operations (Section 8.3). The formation of blisters is an immediate indication that the angle of the trowel is too great for the surface in that area at that particular time with the concrete and job conditions involved. The position of the trowel should be flattened, and the blistered area retroweled immediately to eliminate and rebond the blisters. If frequent blistering occurs despite reasonable care in the timing and technique employed in the finished troweling, attention should be directed to the job and climatic conditions and to the concrete mixture as discussed as follows.

Most skilled finishers know when a concrete surface is ready for the raised and final troweling and closing of the surface, and how to accomplish this operation; however, circumstances are often beyond their control. For instance, if there are too few finishers for the climatic conditions, finishers may have to close some portions of a floor too early to get it troweled before it has set too much. Similarly, if supervisors insist that a floor be finished by a certain time, whether it is ready or not, blisters, trowel marks, and poor surfaces can result.

11.8—Spalling

Unlike scaling and blistering, spalling is a deeper surface imperfection, often extending to the top layers of reinforcing steel or to the horizontal joint between the base and topping in two-course construction. Spalls can be 6 in. (150 mm) or more in diameter and 1 in. (25 mm) or more in depth; although, smaller spalls also occur (Fig. 11.9 and 11.10). Spalls are caused by pressure or expansion within the concrete, bond failure in two-course construction, impact loads, fire, or weathering. Joint spalls are often caused by improperly constructed joints. Spalls can occur over corroding reinforcing steel because the corrosion products (rust) occupy more volume than the original steel, and the resultant pressure spalls the concrete.

In addition to its poor appearance, spalling can seriously impair the strength or serviceability of a floor or slab. Indoor spalling is more likely to result from improper joint design or installation or bond failure in two-course floor construction, but obviously this can happen outdoors as well. Causes for the various kinds of spalling include:

1. Insufficient depth of cover over reinforcement;

2. Inferior concrete in the cover over reinforcing steel. Such concrete can fail to protect the steel from disruptive corrosion because of its high permeability due to:

   a. Overworking overly wet concrete during finishing (Sections 8.3.10 and 8.3.3);

   b. Serious loss of entrained air during such wet-finishing operations;

   c. Problems with excessive bleeding during finishing, especially in cold weather (Sections 8.16 and 8.17);

   d. Inadequate or delayed curing;

   e. Severe cracking that permits water and salts to attack the steel;

   f. Loss of bond between concrete and reinforcing steel bars, caused by placement of concrete on top of excessively hot steel during hot-weather concreting;

3. Joint edge spalls caused by small hard-wheeled vehicles traveling across improperly installed or filled joints (Sections 3.6, 5.11, and 9.10), and spalls on the upper flange of the female side of keyed-construction joints;

4. Poor bonding of topping to base course in two-course floors (Sections 8.7.1 and 8.7.2) due to:

   a. Inferior quality of surface concrete in the base course;

   b. Unremoved contamination in, or poor preparation of, the surface of the base course;

   c. Differences in shrinkage between topping and base courses;

   d. Drying of the bonding grout before the topping concrete is placed;

   e. Excessive pressure developed at joints, where preformed joint material was topped by continuous concrete; and

   f. Restraint of movement of deck slabs on supporting walls and piers due to inadequate provision for such movement.

11.9—Discoloration

Surface discoloration of concrete flatwork can appear as gross color changes in large areas of concrete, as spotted or mottled light or dark blotches on the surface, or as early light patches of efflorescence. Laboratory studies to determine the effects of various concrete materials and concreting procedures show that no single factor is responsible for discoloration (Landgren and Hadley 2002). Factors found to influence discoloration are calcium chloride admixtures, concrete alkalies, hard-troweled surfaces, inadequate or inappropriate curing, variations in w/cm at the surface, and changes in the concrete mixture. Like many other surface imperfections, discoloration is generally a cosmetic nuisance rather than a structural or serviceability problem.

Dark areas do not necessarily denote inferior serviceability, unless there is evidence that dry cement has been troweled into the surface to absorb excess bleed water (Section 8.16). The following are causes of dark areas:

1. The use of calcium chloride in concrete can discolor the surface (Fig. 5.1). Calcium chloride accelerates the overall hydration process but has a retarding effect on the hydration of the ferrite compounds in portland cement. The ferrite phases normally become lighter with hydration; however, in the presence of calcium chloride, the retarded, unhydrated ferrite phases remain dark in color;

2. Low spots where water stands longer before evaporating can cause dark areas;

3. Curing with waterproof paper and plastic sheets can cause a lighter color where the sheet is in contact with the surface and a darker color where the sheet is not in contact with the surface. This type of discoloration is aggravated when concrete contains calcium chloride;

4. Changes in the w/cm of concrete mixtures can significantly affect color. Such a change can result from localized changes in construction practices, from a batch-to-batch variation in the concrete's water or cementitious material content, or from steel troweling. A high w/cm will usually produce a light-colored concrete, a low ratio a darker color. Repeated hard-steel troweling in areas of advanced setting reduces the w/cm at the surface, darkening its color;

5. Changes in source or type of cement. Individual brands and types of cement can differ in color; therefore, changing brand or type of cement in the middle of a job can noticeably change the color of concrete;

6. Another cause is the uneven application of dry-shake materials, such as mineral-aggregate or metallic hardeners; and

7. Changes in the amount, source, and chemistry of a mineral admixture also affect discoloration. The extent of the discoloration will depend upon the color and the amount of admixture used. Some mineral admixtures resemble portland cement and have no effects on concrete color. Silica fume can give concrete a dark gray tint. Dark gray fly ashes can also give concrete a darker color, whereas tan- or beige-colored fly ashes, if used in large quantities, can produce a tan color in concrete.

Light-colored areas can simply be the result of contrast to adjacent dark areas; these would not normally impair serviceability. If light-colored areas are caused by local overworking

Fig. 11.9—Spalled joint (courtesy of Eldon Tipping).

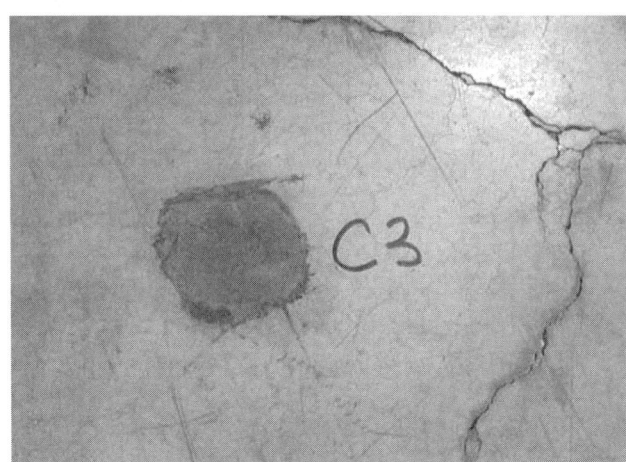

Fig. 11.10—Spalled crack (courtesy of Eldon Tipping).

of excessively wet concrete, however, the surface will be weaker and serviceability may be impaired. This can be caused by high concrete water content or finishing while there is excess moisture or bleed water on the surface.

Light-colored areas also can be caused by efflorescence (a crystalline deposit—usually white in color—that occasionally develops on the surface of concrete slabs after construction is completed). Moisture present in hardened concrete dissolves soluble salts. These salts in solution migrate to the surface by evaporation or hydraulic pressure where the water evaporates and leaves a deposit of salt at the surface. If the water, the evaporation, or the salts are not present, efflorescence will not occur.

11.10—Low spots and poor drainage

Puddles or bird baths on an outdoor concrete slab after a rain, or on a floor after hosing, characterize poor slab or floor surface drainage or serviceability (Fig. 11.11). Among the primary causes:

1. Inadequate slope. Positive drainage requires a slope of 1/4 in./ft (20 mm/m) for an exterior slab; for an interior floor slab, 1/16 in./ft (5 mm/m) minimum is adequate for drainage, but 1/8 in./ft (10 mm/m) is preferred;

2. Inaccuracy in setting grades for forms and screeds;

Fig. 11.11—Low spots in a slab after rain shower (courtesy of Eldon Tipping).

Fig. 11.12—Curling at joint in a 2 in. (50 mm) unbonded topping (courtesy of Eldon Tipping).

3. Damage to grade settings of forms and screeds during construction;

4. Strikeoff operation in which low spots are filled in with extra-wet concrete. The wetter concrete settles more than the surrounding areas during the interval between strikeoff and floating operations;

5. Fresh concrete that is too wet or variably wet. A little working of such concrete results in areas with excessive mortar at the surface, which settles more than the surrounding areas;

6. Failure to frequently check grades, levels, and slopes with long straightedges (Sections 8.2.5 and 8.3.4), and to properly build up low spots in areas detected;

7. By tooling joint grooves without removing the small amount of mortar displaced, the ridge of mortar formed in this way can act as a dam;

8. Failure to check the finished grade following strikeoff when wet-screeds are used (Section 8.3.2);

9. Poor lighting during placing and finishing; and

10. Deflection of suspended slabs between supports after removal of supporting shores.

11.11—Curling

Curling is the distortion (rising up) of a slab's corners and edges due to differences in moisture content or temperature between the top and bottom of a slab (Fig. 11.12). The top dries out or cools and contracts more than the wetter or warmer bottom. If the curled section of a slab is loaded beyond the flexural strength of the concrete, cracks will develop parallel to the joints at which curling occurs.

Slabs also can be dished in the center because the centers were finished lower than the screeds. This is readily apparent from straightedging after finishing. There are a number of ways to reduce slab curling:

1. Equalize moisture content and temperature between the top and bottom of a slab;

2. Use a concrete mixture with low-shrinkage characteristics, that is, a stony concrete mixture with large maximum-size coarse aggregate at the highest quantity consistent with the required workability. Such mixtures minimize water content;

3. Use a permeable (porous) dry—or almost dry—base;

4. Use shrinkage-compensating concrete;

5. Place a generous amount of reinforcement in the top third of the slab. One percent reinforcement could be justified in the direction perpendicular to the slab edge or construction joint, and for approximately 10 ft (3 m) in from the slab edge or construction joint; and

6. Use post-tensioning.

Some of the measures that can reduce moisture differentials between the top and the bottom of a slab are:

1. Cure the slab well, particularly during early ages. Use of a continuous moist cure or a high-solids curing compound (Sections 5.9.3 and 9.2.4)—especially during the first few days—can greatly reduce the rate of water lost from the concrete and help reduce moisture differentials;

2. After proper curing, further reduce moisture loss from the top of slabs by using coatings, sealers, and waxes. These also reduce carbonation, which adds to surface shrinkage; and

3. If a vapor retarder is necessary, use a minimum 4 in. thick (100 mm) layer of trimmable, compactible granular fill (not sand) between the vapor retarder and concrete slab (Section 4.1.5). Material conforming to ASTM D 448, No. 10, with plenty of rock fines, has been used successfully. If the fill is dry, or almost dry, this will permit some moisture loss from the slab bottom. The fill should be designed so that it does not retain water.

Measures to reduce the shrinkage potential of a concrete mixture include:

1. Reduce total water content of concrete by:
   a. Maintaining the proper slump (Table 6.1);
   b. Reducing the as-mixed temperature of the concrete;
   c. Avoiding delays in placement that require large quantities of retempering water;
   d. Selecting hard aggregates that are well graded for good workability at minimum water contents, and contain a minimum of fines; aggregates should be generally rounded or cubical in shape, with a minimum of flat or elongated particles;
   e. Increasing the maximum size of coarse aggregate and using coarser sand;
   f. Reducing the sand content to the lowest level consistent with adequate workability and mixing water requirements; and

g. Using a high-range water-reducing admixture with good shrinkage-reduction history and tests;

2. Avoid aggregates known to have high-shrinkage potential, such as sandstone, slate, hornblende, and some types of basalt. Hard, rigid aggregates that are difficult to compress provide more restraint to shrinkage of cement paste in concrete than softer aggregates. Quartz, granite, feldspar, limestone, dolomite, and some basalt aggregates generally produce concrete with low drying shrinkage (ACI 224R);

3. Minimize aggregate gap-grading;

4. Avoid admixtures or concrete constituents that increase drying shrinkage. Use of a water-reducing admixture, or other admixture conforming to ASTM C 494 and intended for reducing the water demand of concrete, will not necessarily decrease the drying shrinkage of concrete. Unless concrete contains very low levels of calcium chloride or triethanolamine, drying shrinkage generally will be increased. Chlorides can get into concrete from admixtures, water, aggregates, or cement; and

5. Dewatering techniques (Section 8.3.6) of fresh concrete slab surfaces can significantly reduce water content, and thus help reduce slab curling. Because vacuum mats do not extend fully to the edges of the forms and screeds, however, it is possible for the joints at the forms and screeds to end up slightly higher than the overall slab surface after vacuum dewatering is completed. Where wheeled traffic, especially automated guided vehicles, will be involved, this should be taken into consideration during screeding, leveling, and bull floating to the forms and screeds.

Placing concrete at lower temperatures can reduce thermal contraction from cooling. Curling magnitude can diminish with age as moisture and temperature equalize throughout the slab thickness. In addition, creep probably reduces curling over a period of months.

Concrete strength should be only as high as necessary for the floor or slab to fulfill its function (Table 2.1 and 6.1). Excessively high strengths reduce creep, and this can accentuate curling. High–strength, quality concrete slabs, however, have less cracking due to higher early flexural and tensile strengths.

11.12—Analysis of surface imperfections

The cause of most surface imperfections can be determined by petrographic (microscopic) analysis on samples of the concrete. A petrographic analysis of concrete is performed in accordance with ASTM C 856.

Samples for the analysis are usually 4 in. diameter (100 mm) drilled cores or saw-cut sections. Broken sections can be used, but cores or saw-cut sections are preferred because they are less apt to be disturbed. Samples should represent concrete from both the problem and the nonproblem areas. The petrographer should be provided with a description and photographs of the problem, in addition to information on the concrete mixture proportions, construction practices used, and environmental conditions. A field review by a petrographer, designer, or concrete technologist is also helpful in analyzing the imperfection.

The petrographic report often includes the probable cause of the problem, extent of distress, the general quality of the concrete, and expected durability and performance of the concrete. Corrective action, if necessary, would be based to a great extent on the petrographic report.

CHAPTER 12—REFERENCES
12.1—Referenced standards and reports

The standards and reports listed below were the latest editions at the time this document was prepared. Because these documents are revised frequently, the reader is advised to contact the proper sponsoring group if it is desired to refer to the latest version.

American Association of State Highway and Transportation Officials (AASHTO)

M 182	Standard Specification for Burlap Cloth Made from Jute or Kenaf
T 26	Standard Method of Test for Quality of Water to be Used in Concrete

American Concrete Institute (ACI)

116R	Cement and Concrete Terminology
117	Standard Specifications for Tolerances for Concrete Construction and Materials
201.2R	Guide to Durable Concrete
209R	Prediction of Creep, Shrinkage, and Temperature Effects in Concrete Structures
211.1	Standard Practice for Selecting Proportions for Normal, Heavyweight, and Mass Concrete
211.2	Standard Practice for Selecting Proportions for Structural Lightweight Concrete
212.1R/ 212.2R	Admixtures for Concrete and Guide for Use of Admixtures in Concrete
212.3R	Chemical Admixtures for Concrete
212.4R	Guide for the Use of High-Range Water-Reducing Admixtures (Superplasticizers) in Concrete
222R	Corrosion of Metals in Concrete
223	Standard Practice for the Use of Shrinkage-Compensating Concrete
224R	Control of Cracking in Concrete Structures
224.1R	Causes, Evaluation, and Repair of Cracks in Concrete Structures
224.3R	Joints in Concrete Construction
226.1R	Ground Granulated Blast-Furnace Slag as a Cementitious Constituent in Concrete
226.3R	Use of Fly Ash in Concrete
301	Specifications for Structural Concrete
303R	Guide to Cast-in-Place Architectural Concrete Practice
304R	Guide for Measuring, Mixing, Transporting, and Placing Concrete
305R	Hot Weather Concreting
306R	Cold Weather Concreting
306.1	Standard Specification for Cold Weather Concreting
308R	Guide for Curing Concrete

309R	Guide for Consolidation of Concrete
311.4R	Guide for Concrete Inspection
311.5R	Guide for Concrete Plant Inspection and Field Testing of Ready-Mixed Concrete
318/318R	Building Code Requirements for Structural Concrete and Commentary
360R	Design of Slabs on Grade
421.1R	Shear Reinforcement for Slabs
435	Control of Deflection in Concrete Structures
503R	Use of Epoxy Compounds with Concrete
503.2	Standard Specifications for Bonding Plastic Concrete to Hardened Concrete with a Multi-Component Epoxy Adhesive
504R	Guide to Sealing Joints in Concrete Structures
515.1R	A Guide to Use of Waterproofing, Dampproofing, Protective and Decorative Barrier Systems for Concrete
544.1R	State-of-the-Art Report on Fiber Reinforced Concrete
544.2R	Measurement of Properties of Fiber Reinforced Concrete
544.3R	Guide for Specifying, Proportioning, Mixing, Placing, and Finishing Steel Fiber Reinforced Concrete
544.4R	Design Considerations for Steel Fiber Reinforced Concrete

American Society of Civil Engineers (ASCE)

ANSI/ASCE 3	Standard for the Structural Design of Composite Slabs
ANSI/ASCE 9	Standard Practice for Construction and Inspection of Composite Slabs

American Society of Heating, Refrigerating, and Air-Conditioning Engineers (ASHRAE)

90.1	Energy Conservation in New Building Design (Sections 1 through 9)

ASTM International

A 36	Specification for Structural Steel
A 184	Specification for Fabricated Deformed Steel Bar Mats for Concrete Reinforcement
A 185	Specification for Steel Welded Wire Fabric, Plain, for Concrete Reinforcement
A 416	Specification for Steel Strand, Uncoated Seven-Wire for Prestressed Concrete
A 497	Specification for Steel Welded Wire Fabric, Deformed, for Concrete Reinforcement
A 615	Specification for Deformed and Plain Billet-Steel Bars for Concrete Reinforcement
A 616	Specification for Rail-Steel Deformed and Plain Bars for Concrete Reinforcement
A 617	Specification for Axle-Steel Deformed and Plain Bars for Concrete Reinforcement
A 820	Specification for Steel Fibers for Use in Fiber Reinforced Concrete
C 33	Specification for Concrete Aggregates
C 94	Specification for Ready-Mixed Concrete
C 109	Test Method for Compressive Strength of Hydraulic Cement Mortars (Using 2-in. or 50-mm Cube Specimens)
C 138	Standard Test Method for Density (Unit Weight), Yield, and Air Content (Gravimetric) of Concrete
C 143	Standard Test Method for Slump of Hydraulic Cement Concrete
C 150	Specification for Portland Cement
C 156	Test Method for Water Retention by Concrete Curing Materials
C 157	Test Method for Length Change of Hardened Hydraulic-Cement Mortar and Concrete
C 171	Specification for Sheet Materials for Curing Concrete
C 173	Standard Test Method for Air Content of Freshly Mixed Concrete by the Volumetric Method
C 227	Test Method for Potential Alkali Reactivity of Cement-Aggregate Combinations (Mortar-Bar Method)
C 231	Standard Test Method for Air Content of Freshly Mixed Concrete by the Pressure Method
C 260	Specification for Air-Entraining Admixtures for Concrete
C 295	Practice for Petrographic Examination of Aggregates for Concrete
C 309	Specification for Liquid Membrane-Forming Compounds for Curing Concrete
C 330	Specification for Lightweight Aggregates for Structural Concrete
C 387	Specification for Packaged, Dry, Combined Materials for Mortar and Concrete
C 494	Specification for Chemical Admixtures for Concrete
C 595	Specification for Blended Hydraulic Cements
C 618	Specification for Fly Ash and Raw or Calcined Natural Pozzolan for Use as a Mineral Admixture in Portland Cement Concrete
C 685	Specification for Concrete Made by Volumetric Batching and Continuous Mixing
C 806	Test Method for Restrained Expansion of Expansive Cement Mortar
C 845	Specification for Expansive Hydraulic Cement
C 856	Standard Practice for Petrographic Examination of Hardened Concrete
C 878	Test Method for Restrained Expansion of Shrinkage-Compensating Concrete
C 979	Specification for Pigments for Integrally Colored Concrete
C 1017	Specification for Chemical Admixtures for Use in Producing Flowing Concrete
C 1116	Specification for Fiber-Reinforced Concrete and Shotcrete
C 1151	Test Method for Evaluating the Effectiveness of Materials for Curing Concrete

C 1157	Performance Specification for Blended Hydraulic Cement
C 1315	Standard Specification for Liquid Membrane-Forming Compounds Having Special Properties for Curing and Sealing Concrete
D 448	Classification for Sizes of Aggregate for Road and Bridge Construction
D 994	Specification for Preformed Expansion Joint Filler for Concrete (Bituminous Type)
D 1751	Specification for Preformed Expansion Joint Filler for Concrete Paving and Structural Construction (Nonextruding and Resilient Bituminous Types)
D 1752	Specification for Preformed Sponge Rubber and Cork Expansion Joint Fillers for Concrete Paving and Structural Construction
D 2240	Test Method for Rubber Property—Durometer Hardness
D 4263	Test Method for Indicating Moisture in Concrete by the Plastic Sheet Method
E 96	Test Method for Water Vapor Transmission of Materials
E 329	Standard Specification for Agencies Engaged in the Testing and/or Inspection of Materials Used in Construction
E 1077	Standard Practice for Laboratories Testing Concrete and Concrete Aggregates for Use in Construction and Criteria for Laboratory Evaluation
E 1155	Test Method for Determining Floor Flatness and Levelness Using the F-Number System
E 1643	Standard Practice for Installation of Water Vapor Retarders Used in Contact with Earth or Granular Fill Under Concrete Slabs
E 1745	Standard Specification for Water Vapor Retarders Used in Contact with Soil or Granular Fill under Concrete Slabs
E 1907	Standard Practices for Determining Moisture-Related Acceptability of Concrete Floors to Receive Moisture-Sensitive Finishes

The aforementioned publications may be obtained from the following organizations:

American Association of State Highway and
  Transportation Officials
333 West Capitol Street, NW, Suite 225
Washington, DC 20001

American Concrete Institute
P.O. Box 9094
Farmington Hills, MI 48333-9094

American Society of Civil Engineers
345 East 47th Street
New York, NY 10017-2398

American Society of Heating, Refrigerating, and Air-
  Conditioning Engineers
1791 Tullie Circle, NE
Atlanta, GA 30329

American Society of Concrete Contractors
2025 S. Brentwood Blvd.
St. Louis, MO 63144

ASTM International
100 Barr Harbor Drive
West Conshohocken, PA 19428-2959

National Ready Mixed Concrete Association
900 Spring Street
Silver Spring, MD 20910

National Fire Protection Association
1 Batterymarch Park
P.O. Box 1901
Quincy, MA 02269-9101

12.2—Cited references

Abrams, D. A., 1918, "Design of Concrete Mixtures," *Bulletin 1*, Structural Materials Research Laboratory, Lewis Institute, Chicago.

ACI Committee 226, 1987, "Silica Fume in Concrete," *ACI Materials Journal*, V. 84, No. 2, Mar.-Apr., pp. 158-166.

ACI Committee 325, 1956, "Structural Design Considerations for Pavement Joints," ACI JOURNAL, V. 53, July, pp. 1-29.

Anderson, T., and Roper, H., 1977, "Influence of an Impervious Membrane Beneath Concrete Slabs on Grade," *Symposium, Concrete for Engineering*, Institute of Engineers, Brisbane, Australia, Aug., pp. 51-56.

ASHVE, 1955, *Panel Heating*, Heating, Ventilating, Air-Conditioning Guide, V. 36, American Society of Heating and Ventilating Engineers, Atlanta, Ga., pp. 605-644.

ASTM, 1994, *Significance of Tests and Properties of Concrete and Concrete-Making Materials*, STP 169-C, ASTM International, West Conshohocken, Pa. (Note especially Chapter 19, "Abrasion Resistance," pp. 182-191.)

Balaguru, P. N., and Shah, S. P., 1992, *Fiber Reinforced Cement Composites*, McGraw-Hill, New York, 530 pp.

Bhatty, M. S. Y., and Greening, N. R., 1978, "Interaction of Alkalies with Hydrating and Hydrated Calcium Silicates," *Proceedings of the 4th International Conference on the Effects of Alkalies in Cement and Concrete*, Publication No. CE-MAT-1-78, School of Civil Engineering, Purdue University, West Lafayette, Ind., pp. 87-112.

Bimel, C., 1993, "ASTM Specifications are a Start, But," *Concrete International*, V. 15, No. 12, Dec., p. 55.

Boone, T. H. et al., 1958, "Conductive Flooring for Hospital Operating Rooms," *Journal of Research*, V. 630, No. 2, U.S. Department of Commerce, National Bureau of Standards, Washington, D.C., Oct.-Dec., pp. 125-140.

Brungraber, R. J., 1976, "Overview of Floor Slip-Resistance Research with Annotated Bibliography," *NBS Technical Note*

No. 895, U.S. Department of Commerce, National Bureau of Standards, Washington, D.C., Jan., 108 pp.

Brungraber, R., J., 1977, "New Portable Tester for the Evaluation of the Slip-Resistance of Walkway Surfaces," *NBS Technical Note* No. 953, U.S. Department of Commerce, National Bureau of Standards, Washington, D.C., July, 43 pp.

Colley, B. E., and Humphrey, H. A., 1967, "Aggregate Interlock at Joints in Concrete Pavements," *Bulletin* DX124, Portland Cement Association, Skokie, Ill., 23 pp.

CRSI, 2001, "Manual of Standard Practice," *MSP-1-01*, 27th Edition, Concrete Reinforcing Steel Institute, Schaumburg, Ill., 101 pp.

Dakhil, F. H.; Cady, P. D.; and Carrier, R. E., 1975, "Cracking of Fresh Concrete as Related to Reinforcement," ACI JOURNAL, *Proceedings* V. 72, No. 8, Aug., pp. 421-428.

ESCSI 1958, *Floor Finishing*, Lightweight Concrete Information Sheet No. 7, Expanded Shale, Clay and Slate Institute, Bethesda, Md., 4 pp.

Face, A., 1987, "Floor Flatness and Levelness—The F-Number System," *Construction Specifier*, V. 40, No. 4, Apr., pp. 24-32.

Gustaferro, A. H., 1989, "Are Thickness Tolerances for Concrete Floors on Grade Realistic?" *Concrete Construction*, Apr., pp. 389-391.

Gustaferro, A. H., and Tipping, E., 2000, "Slab Thickness Tolerances: Are They Realistic?" *Concrete Construction*, June, pp. 66-67.

Harrison, P. J., 2004, "For the Ideal Slab-on-Ground Mixture," *Concrete International*, V. 26, No. 3, Mar, pp. 49-55.

Kauer, J. A., and Freeman, R. L., 1955, "Effect of Carbon Dioxide on Fresh Concrete," ACI JOURNAL, *Proceedings* V. 52, Dec., pp. 447-454.

Kosmatka, S. H., 1991, "Finishing Concrete Slabs With Color and Texture," *PA124H*, Portland Cement Association, Skokie, Ill., 40 pp.

Kosmatka, S H., Kerkhoff, B., and Panarese, W. C., 2002a, "Mixing Water for Concrete," Chapter 4, *Design and Control of Concrete Mixtures*, 14th Edition, Portland Cement Association, Skokie, Ill., pp. 73-77.

Kosmatka, S. H., Kerkhoff, B., and Panarese, W. C, 2002b, "Designing and Proportioning Normal Concrete Mixtures," Chapter 9, *Design and Control of Concrete Mixtures*, 14th Edition, Portland Cement Association, Skokie, Ill., pp. 149-177.

Landgren, R. and Hadley, D. W., 2003, *Surface Popouts Caused by Alkali-Aggregate Reaction*, RD121, Portland Cement Association, 20 pp.

Martin, R., and Phelan, W. S., 1995, "How Do Admixtures Influence Shrinkage?" *Concrete Construction*, V. 17, No. 7, pp. 611-617.

Martin, R., 1983, Discussion of "Proposed Revisions to Specifications for Structural Concrete for Buildings (ACI 301-72) (Revised 1981)," ACI JOURNAL, *Proceedings* V. 80, No. 6, Nov.-Dec., p. 548.

Malinowski, R., and Wenander, H., 1975, "Factors Determining Characteristics and Composition of Vacuum-Dewatered Concrete," ACI JOURNAL, *Proceedings* V. 72, No. 3, Mar., pp. 98-101.

McGovern, M., 2002, "Wanted: Cause of Early Cracking," *Concrete Technology Today*, CT021, Portland Cement Association, Skokie, Ill., Mar., pp. 3-4.

Nicholson, L. P., 1981, "How to Minimize Cracking and Increase Strength of Slabs on Grade," *Concrete Construction*, Sept., pp. 739-742.

NFEC, 1984, *Metallic Type Conductive and Spark-Resistant Concrete Floor Finish*, Guide Specification No. NFGS-09785, Naval Facilities Engineering Command, Apr., 14 pp.

NFPA 99, 2002, *Standard for Health Care Facilities*.

PCA, 2001, "Concrete Slab Surface Defects: Causes, Prevention, Repair," *Bulletin* IS177, Portland Cement Association, Skokie, Ill., 16 pp.

PTI, 1990, *Post-Tensioning Manual*, 5th Edition, Post-Tensioning Institute, Phoenix, Ariz., 406 pp.

PTI, 1996, "Design and Construction of Post-Tensioned Slabs on Ground," 2nd Edition, Post-Tensioning Institute, Phoenix, Ariz., 90 pp.

PTI, 2000, "Specifications for Unbonded Single Strand Tendons," 2nd Edition, Post-Tensioning Institute, Phoenix, Ariz., 16 pp.

Ringo, B., 1958, "Basics of Subgrade Preparation for Industrial Floors," *Concrete Construction*, Feb., pp. 137-140.

Ryan, T. J., 1997, "Controlling Deflection of Composite Deck Slabs," *Concrete Construction*, Sept., pp. 734-739.

Schrader, E. K., 1987, "A Proposed Solution to Cracking by Dowels," *Concrete Construction*, Dec., pp. 1051-1053.

Shilstone, J. M., Sr., 1990, "Concrete Mixture Optimization," *Concrete International*, V. 12, No. 6, June, pp. 33-39.

Shilstone, J. M., Sr., and Shilstone, J. M., Jr., 2002, "Performance-Based Concrete Mixtures and Specifications for Today," *Concrete International*, V. 24, No. 2, Feb., pp. 80-83.

Smith, F. L., 1956, "Effect of Various Surface Treatments using Magnesium and Zinc Fluosilicate Crystals on Abrasion Resistance of Concrete Surfaces," *Concrete Laboratory Report* No. C-819, U.S. Bureau of Reclamation, Denver, Colo.

Snell, L. N., 1997, "Cover of Welded Wire Fabric in Slabs and Pavements," *Concrete Construction*, July, pp. 580-584.

Spears, R., and Panarese, W. C., 1992, "Concrete Floors on Ground," *Bulletin* EB075D, Portland Cement Association, Skokie, Ill., 23 pp.

Steinour, H. H., 1960, "Concrete Mix Water — How Impure Can It Be?" *Bulletin* RX119, Portland Cement Association, Skokie, Ill., Sept., 23 pp.

Suprenant, B. A., 1997, "Troubleshooting Crusted Concrete," *Concrete Construction*, Apr., pp. 375-378.

Suprenant, B. A., and Malisch, W. R., 1998a, "Don't Puncture the Vapor Retarder," *Concrete Construction*, Dec., pp. 1071-1075.

Suprenant, B. A., and Malisch, W. R., 1998b, "Where to Place the Vapor Retarder," *Concrete Construction*, May, pp. 427-433.

Suprenant, B. A., and Malisch, W. R., 1999a, "Don't Use Loose Sand Under Concrete Slabs," *Concrete Construction*, Mar., pp. 23-31.

Suprenant, B. A., and Malisch, W. R., 1999b, "Beware of Troweling Air-Entrained Concrete," *Concrete Construction*, Feb., pp. 35-37.

Tatnall, P. C., and Kuitenbrouwer, L., 1992, "Steel Fiber Reinforced Concrete in Industrial Floors," *Concrete International*, V. 14, No. 12., Dec. pp. 43-47.

Tipping, E., 1992, "Building Superior Quality Elevated Floors," *Concrete Construction*, Apr., pp. 285-288.

Tipping, E., 2002, "Keys to Constructing Level Suspended Floors," *L&M Concrete News*, V. 3, No. 1, Spring, pp. 6-7.

Tipping, E. and North, J., 1998, "No Texas Tall Tale," *Concrete Construction*, Jan., pp. 21-25.

Tremper, B., and Spellman, D. C., 1963, "Shrinkage of Concrete—Comparison of Laboratory and Field Performance," *Highway Research Record* No. 3, Highway Research Board, Washington, D.C., pp. 30-61.

Turenne, R. G., 1978, "The Use of Vapor Barriers Under Concrete Slabs on Ground," Building Practice Note No. 8, Division of Building Research, National Research Council of Canada, Aug., p. 3.

USBR, "The Effect of Various Surface Treatments Using Magnesium and Zinc Fluosilicate Crystals on Abrasion Resistance of Concrete Surfaces," *Concrete Laboratory Report* No. C-819, U.S. Bureau of Reclamation, Denver.

Vail, J. G., 1952, *Soluble Silicates: Their Properties and Uses, Volume 2: Technology*, New York, Reinhold, pp. 315-319.

Walker, W. W., and Holland, J. A., 1998, "Dowels for the 21st Century—Plate Dowels for Slabs on Ground," *Concrete International*, V. 20, No. 7, July, pp. 32-38.

Walker, W. W., and Holland, J. A., 1999, "The First Commandment for Floor Slabs: Thou Shalt Not Curl nor Crack... (Hopefully)," *Concrete International*, V. 21, No. 1, Jan., pp. 47-53.

Wenander, H., 1975, "Vacuum Dewatering Is Back," *Concrete Construction*, V. 20, No. 2, Feb., pp. 40-42.

Wenander, H.; Danielsson, J. L.; and Sendker, F. T., 1975, "Floor Construction by Vacuum Dewatering," *Concrete Construction*, V. 20, No. 2, Feb., pp, 43-46.

Weymouth, C. A. G., 1933, "Effects of Particle Interference in Mortars and Concretes," *Rock Products*, Feb.

Whiting, D., and Dziedzic, W., 1992, "Effects of Conventional and High-Range Water Reducers on Concrete Properties," *Research and Development Bulletin* RD107, Portland Cement Association, Skokie, Ill., 28 pp.

WRI/CRSI, 1991, "Reinforcing Steel in Slabs-on-Grade," *WRI/CRSI Engineering Data Report* No. 37, Concrete Reinforcing Steel Institute, Schaumburg, Ill.

Ytterberg, R. F., 1987, "Shrinkage and Curling of Slabs on Grade," *Concrete International*, V. 9, No. 4, 5, and 6; Apr., pp. 22-31; May, pp. 54-61; and Jun., pp. 72-81.

12.3—Other references

ACI Committee 211, 2002, "Guide for Selecting Proportions for No-Slump Concrete (ACI 211.3R-02)," American Concrete Institute, Farmington Hills, Mich., 26 pp.

ACI Committee 311, 1999, "ACI Manual of Concrete Inspection (311.1R-99 [SP-2])," American Concrete Institute, Farmington Hills, Mich., 209 pp.

ACI Committee 311, 2000, "Guide for Concrete Inspection (ACI 311.4R-00)," American Concrete Institute, Farmington Hills, Mich., 12 pp.

ACI Committee 325, 1991, "Recommendations for Construction of Concrete Pavements and Concrete Bases (ACI 325.9R-91)," American Concrete Institute, Farmington Hills, Mich., 27 pp.

ACI Committee 330, 2001, "Guide for Design and Construction of Concrete Parking Lots (ACI 330R-01)," American Concrete Institute, Farmington Hills, Mich., 32 pp.

ACI Committee 332, 1962, "Guide for Construction of Concrete Floors on Grade," ACI JOURNAL, *Proceedings* V. 59, No. 10, Oct., pp. 1377-1390.

ACI Committee 332, 1984, "Guide to Residential Cast-in-Place Concrete Construction (ACI 332R-84)," American Concrete Institute, Farmington Hills, Mich., 38 pp.

ACI Committee 347, 2001, "Guide to Formwork for Concrete (ACI 347-01)," American Concrete Institute, Farmington Hills, Mich., 32 pp.

ACI Committee 423, 1996, "Recommendations for Concrete Members Prestressed with Unbonded Tendons (ACI 423.3R-96)," American Concrete Institute, Farmington Hills, Mich., 19 pp.

American Concrete Institute, *Concrete Slabs on Grade: Design, Specification, Construction, and Problem Solving*, 1992, American Concrete Institute, Farmington Hills, Mich., 429 pp.

Anderson, R. B., 1992, "Soil Information Needed for Slab Design," *Concrete Construction*, Apr., pp. 289-290.

ASTM C 989-00, 2000, "Specification for Ground Granulated Blast-Furnace Slag for Use in Concrete and Mortars," ASTM International, West Conshohocken, Pa., 5 pp.

Bimel, C., 1988, "Trap Rock Aggregates for Floor Construction," *Concrete Construction*, Oct., pp. 946 and 948.

Bimel, C., 1993, "Concrete Contractors Don't Create All Cracks," *Concrete International*, V. 15, No. 1, Jan., pp. 46-47.

Campbell, R. H.; Harding, W.; Misenhimer, E.; Nicholson, L. P.; and Sisk, J., 1976, "Job Conditions Affect Cracking and Strength of Concrete In-Place," ACI JOURNAL, *Proceedings* V. 73, No. 1, Jan., pp. 10-13.

Concrete Manual, 8th Edition, U.S. Bureau of Reclamation, Denver, 1975, pp. 457-467.

"Concrete Screed Rails used for Concrete Placed on Metal Decks," *Concrete Construction*, Apr. 1991, pp. 341-342.

"Contractor's Guide to Air-Entraining and Chemical Admixtures," 1990, *Concrete Construction*, Mar., pp. 279-286.

"Finishing $$ Ahead with Surface Vibration," 1982, *Concrete Technology Today*, V. 3, No. 2, Portland Cement Association, Skokie, Ill., June, pp. 1-2.

Fitzpatrick, R., 1996, "Designing Durable Industrial Floor Slabs," *Concrete International*, V. 18, No. 1, Jan., pp., 38-39.

Fling, R. S., 1987, "A Screeding Machine That's More Than a Strike-Off," *Concrete Construction*, Apr., pp. 351-353.

"Fly Ash in Concrete," 1982, *Concrete Construction*, May, pp. 417-427.

Fricks, T. J., 1994, "Misunderstandings and Abuses in Flatwork Specifications," *Concrete Construction*, June, pp. 492-497.

Garber, G., 1983, "Post-Tensioning for Crack-Free Superflat Floors," *Concrete Construction*, May, pp. 396-400.

Goeb, E. O., 1989, "Do Plastic Fibers Replace Wire Mesh in a Slab on Grade?" *Concrete Technology Today*, V. 10, No. 1, Portland Cement Association, Skokie, Ill., Apr., p. 2.

Gray, J. E., 1962, "Report on Skid Resistance of Portland Cement Mortar Surfaces," *Projects 61-34-36-36*, National Crushed Stone Association, Washington, D.C., Mar., p. 22.

Gulyas, R. J., 1984, "Dry Shake for Floors," *Concrete Construction*, Mar., pp. 285-289.

Hays, C. R, 1995, "Achieving Quality in Concrete Construction," *Concrete International*, V. 17, No. 11, Nov., pp. S-2 to S-3.

Hester, W. T., 1979, "Superplasticizers in Ready Mixed Concrete (A Practical Treatment of Everyday Operations)," *NRMC Publication* No. 158, National Ready Mixed Concrete Association, Silver Spring, Md., Jan.

Hoff, P. L., 1986, "Industrial Floors—Before You Build," *Concrete Technology Today*, V. 7, No. 3, Portland Cement Association, Skokie, Ill., Sept., p. 1-3.

Hover, K., 1995, "Investigating Effects of Concrete Handling on Air Content," *Concrete Construction*, Sept., pp. 745-750.

Kosmatka, S. H., 1985a, "Floor Covering Materials and Moisture in Concrete," *Concrete Technology Today*, V. 6, No. 2, Portland Cement Association, Skokie, Ill., Sept., pp. 2-3.

Kosmatka, S. H., 1985b, "Repair With Thin-Bonded Overlay," *Concrete Technology Today*, V. 6, No. 2, Portland Cement Association, Skokie, Ill., Mar., pp. 3-5.

Kosmatka, S. H., 1986a, "Discoloration of Concrete—Causes and Remedies," *Concrete Technology Today*, V. 7, No. 1, Portland Cement Association, Skokie, Ill., Apr., pp. 2-3.

Kosmatka, S. H., 1986b, "Petrographic Analysis of Concrete," *Concrete Technology Today*, V. 7, No. 2, Portland Cement Association, Skokie, Ill., July.

Kosmatka, S. H.; Kerkhoff, B.; and Panarese, W. C., 2002a, "Curing Concrete," Chapter 12, "Hot-Weather Concreting," Chapter 13, and "Cold-Weather Concreting," Chapter 14, *Design and Control of Concrete Mixtures*, EB001, 14th Edition, Portland Cement Association, Skokie, Ill., Revised 2002, pp. 219-238, 229-238, and 239-255, respectively.

Lien, R., 1995, "Pan Floats Help Make Nestle's Floors Sweet," *Concrete Construction*, May, pp. 439-444.

Moens, J., and Nemegeer, D., 1991, "Designing Fiber Reinforced Concrete Based on Toughness Characteristics," *Concrete International*, V. 13, No. 1, Jan., pp. 38-43.

Metzger, S. N., 1988, "Better Industrial Floors Through Better Joints," *Concrete Construction*, Aug., pp. 749-754.

Metzger, S. N., 1989, "Repairing Joints in Industrial Floors," *Concrete Construction*, June, pp. 548-551.

NRMCA, 1962, "Control of Quality of Ready-Mixed Concrete," *Publication* No. 44, 5th Edition, National Ready Mixed Concrete Association, Silver Spring, Maryland, Oct., p. 51.

Nussbaum, P. J., 1992, "Reflections on Reinforcing Steel in Slabs on Grade," *Concrete Technology Today*, V. 13, No. 2, Portland Cement Association, Skokie, Ill., July, pp. 4-6.

Packard, R. G., 1976, "Slab Thickness Design for Industrial Concrete Floors on Grade," *Concrete Information*, IS195D, Portland Cement Association, Skokie, Ill., p. 16.

Panarese, W. C., 1995, "Cement Mason's Guide," PA122H, 6th Edition, Portland Cement Association, Skokie, Ill., p. 20.

PCA, 1981, "Load Transfer Across Joints in Floors," *Concrete Technology Today*, V. 2, No. 4, Portland Cement Association, Skokie, Ill., Dec., pp. 3-4.

PCA, 1982a, "Concrete Myths: Vapor Barriers are Always Required Under Slab-on-Grade Floors," *Concrete Technology Today*, Portland Cement Association, Skokie, Ill., V. 3, No. 3, Sept. 1982, p. 5.

PCA, 1982b, "Proper Curing—Preventive Medicine for Concrete," *Concrete Technology Today*, V. 3, No. 3, Portland Cement Association, Skokie, Ill., Sept., pp. 2-4.

PCA, 1982c, "Slab Curling is Not a Game Played on Ice," *Concrete Technology Today*, V. 3, No. 2, Portland Cement Association, Skokie, Ill., June, p. 5.

PCA, 1983, "Fly Ash—Its Effect on Concrete Performance," *Concrete Technology Today*, PL833B, Portland Cement Association, Skokie, Ill., Sept.

PCA, 1989, "Polymeric Fiber Reinforced Concrete," *Concrete Technology Today*, V. 10, No. 3, Portland Cement Association, Skokie, Ill., Nov., pp. 1-5.

PCA, 1991, "Subgrades and Subbases for Concrete Pavements," *Concrete Information*, IS029P, Portland Cement Association, Skokie, Ill., p. 24.

PCA, 1992, "Reinforcing Steel in Slabs on Grade," *Concrete Technology Today*, V. 13, No. 4, Portland Cement Association, Skokie, Ill., Mar., pp. 4-6.

PCA, 1994, "Concrete Specifications: Read and Write Them Carefully," *Concrete Technology Today*, V. 15, No. 1, Portland Cement Association, Skokie, Ill., Mar.

PCA, 1995a, "Early Sawing to Control Slab Cracking," 1995, *Concrete Technology Today*, V. 16, No. 1, Portland Cement Association, Skokie, Ill., Nov.

PCA, 1995b, "Popouts: Causes, Prevention, Repair," *PCA's Concrete Technology Today*, June.

PCA, 2001, "Concrete Slab Surface Defects: Causes, Prevention, Repair," *Bulletin* IS177, Portland Cement Association, Skokie, Ill., 16 pp.

Phelan, W., 1989, "Floors that Pass the Test," *Concrete Construction*, Jan., pp. 5-11.

Reed, R., and Schmidt, G., 1994, "Long-Strip Concrete Placement," *Concrete Construction*, Jan. pp. 46-50.

Ringo, B., 1992, "Effect of Design Variables on Floor Thickness Requirements," *Concrete Construction*, Jan., pp. 13-14.

Ringo, B. C., and Anderson, R. B., 1994, "Choosing Design Methods for Industrial Floor Slabs," *Concrete Construction*, Apr., pp., 346-352.

Robinson, C.; Colasanti, A.; and Boyd, G., 1991, "Steel Fibers Reinforce Auto Assembly Plant Floor," *Concrete International*, V. 13, No. 4, pp. 30-35.

Rocole, L., 1993, "Silica-Fume Concrete Proves to be an Economical Alternative," *Concrete Construction*, June, pp. 441-442.

Rose, J. G., 1986, "Yield of Concrete," *Concrete Construction*, Mar., pp. 313-316.

Schmidt, N. O., and Riggs, C. O., 1985, "Methods for Achieving and Measuring Soil Compaction," *Concrete Construction*, Aug., pp. 681-689.

Shilstone, J. M., 1982, "Concrete Strength Loss and Slump Loss in Summer," *Concrete Construction*, May, pp. 429-432.

Shilstone, J. M., 1983, "Quality Assurance and Quality Control," *Concrete Construction*, Nov., pp. 813-816.

Snell, L. M., 1989, "A Proposed Method for Determining Compliance with Floor Thickness Specifications," *Concrete Construction,* Jan., pp. 13-16.

Suprenant, B. A., 1994, "Adjusting Slump in the Field," *Concrete Construction*, Jan., pp. 38-44.

Suprenant, B., 1990, "Construction of Elevated Concrete Slabs—Understanding the Effect of Structural Systems," *Concrete Construction*, V. 35, No. 11, Nov., pp. 910-917.

Suprenant, B. A., 1992, "Finishing Non-Bleeding Concrete," *Concrete Construction*, May, pp. 386-389.

Suprenant, B. A., 2002, "Specified Tolerances versus As-Built Data," *Concrete International*, V. 24, No. 5, May, pp. 49-52.

Symposium on Use of Pozzolanic Materials in Mortars and Concretes, 1950, STP-99, ASTM International, West Conshohocken, Pa., Aug., 203 pp.

Teller, L. W., and Cashell, H. D., 1958, "Performance of Doweled Joints Under Repetitive Loading", *Highway Research Board Bulletin* 217, Transportation Research Board, National Research Council, Washington, D.C., pp. 8-43, discussion by B. F. Friberg, pp. 44-49.

Tipping, E., 1992, "Bidding and Building to F-Number Floor Specs," *Concrete Construction*, V. 14, No. 1, Jan., pp. 18-19.

Tipping, E., 1992, "Controlling the Quality of Suspended Slab Construction," *Concrete International*, V. 14, No. 8, Aug., pp. 38-40.

Tipping, E., 1992, "Tolerance Conflicts and Omissions in Suspended Slab Construction," *Concrete International*, V.14, No. 8, Aug., pp. 33-37.

Tipping, E., 1996, "Using the F-Number System to Manage Floor Installations (Part 1 of a 2-Part Series)," *Concrete Construction*, V. 41, No. 1, Jan., pp. 28-34.

Tipping, E., 1996, "Using the F-Number System to Manage Floor Installations (Part 2 of a 2-Part Series)," *Concrete Construction*, V. 41, No. 2, Feb., pp. 176-178.

Tipping, E., and Suprenant, B., 1991 "Construction of Elevated Concrete Slabs—Measuring and Evaluating Quality," *Concrete Construction*, V. 36, No. 3, Mar., pp. 260-268.

Tipping, E., and Suprenant, B., 1991, "Construction of Elevated Concrete Slabs—Practice and Procedures," *Concrete Construction*, V. 36, No. 1, Jan., pp. 32-42.

Tobin, R. E., 1985, "How to Double the Value of Your Concrete Dollar," *Concrete Technology Today*, V. 3, No. 2, Portland Cement Association, Skokie, Ill., June, pp. 1-2.

Transportation Research Board, 1981, "Concrete Sealers for Protection of Bridge Structures," *NCHRP Report* No. 244, Transportation Research Board, Washington, D.C., 138 pp.

"Working with Steel Fiber Reinforced Concrete," 1985, *Concrete Construction*, Jan., pp. 5-11.

Ytterberg, C. F., 1961, "Good Industrial Floors: What It Takes to Get Them and Why—Part 1: Monolithic Floors; Part 2: Concrete Toppings," *Civil Engineering*, V. 31, No. 2, Feb., pp. 55-58, and No. 4, Apr., pp. 60-63.

Zollo, R. F., and Hays, C. D., 1991, "Fibers vs. WWF as Non-Structural Slab Reinforcement," *Concrete International*, V. 13, No. 11, Nov., pp. 50-55.

ACI 303R-04

Guide to Cast-in-Place Architectural Concrete Practice

Reported by ACI Committee 303

Daniel P. Dorfmueller
Chair

Keith Ahal	Jerome H. Ford	Robert W. Nussmeier
Doug Bannister	Chris A. Forster	James M. Shilstone, Jr.
George F. Baty	Thomas J. Grisinger	Michael S. Smith
Eugene H. Boeke, Jr.	Robert P. Harris	David M. Suchorski
Muriel Burns	G. P. Jum Horst	Claude B. Trusty, Jr.
Joseph A. Dobrowolski	Robert D. Kirk	Gregory R. Wagner
Greg Dobson		

Consulting member Louis Tallarico (deceased 2004), a former committee member, was a major contributor to this guide.

This guide presents recommendations for producing cast-in-place architectural concrete. The importance of specified materials, forming, concrete placement, curing, additional treatment, and inspection, and their effect on the appearance of the finished product are discussed. Architectural concrete requires special construction techniques and materials, and each project will have special requirements. Specific recommendations and information presented in this guide should be used accordingly.

Keywords: admixture; aggregate; architectural concrete; beam; bushhammer; cement; coating; column; consolidation; cracking; curing; deflection; exposed-aggregate finish; finish; form lining; formwork; joint; joint sealant; mixture proportion; pigment; placing; quality control; release agent; repair; retarder; sealant; texture; wall.

ACI Committee Reports, Guides, Standard Practices, and Commentaries are intended for guidance in planning, designing, executing, and inspecting construction. This document is intended for the use of individuals who are competent to evaluate the significance and limitations of its content and recommendations and who will accept responsibility for the application of the material it contains. The American Concrete Institute disclaims any and all responsibility for the stated principles. The Institute shall not be liable for any loss or damage arising therefrom.
Reference to this document shall not be made in contract documents. If items found in this document are desired by the Architect/Engineer to be a part of the contract documents, they shall be restated in mandatory language for incorporation by the Architect/Engineer.

It is the responsibility of the user of this document to establish health and safety practices appropriate to the specific circumstances involved with its use. ACI does not make any representations with regard to health and safety issues and the use of this document. The user must determine the applicability of all regulatory limitations before applying the document and must comply with all applicable laws and regulations, including but not limited to, United States Occupational Safety and Health Administration (OSHA) health and safety standards.

CONTENTS
Chapter 1—Introduction, p. 303R-2

Chapter 2—Architectural considerations, p. 303R-2
2.1—Architectural features
2.2—Architectural design
2.3—Coatings and sealers
2.4—Joint sealants
2.5—Specifications

Chapter 3—Structural considerations, p. 303R-7
3.1—Spalling
3.2—Deflections
3.3—Cracking
3.4—Joints
3.5—Beams and slabs
3.6—Columns
3.7—Walls

Chapter 4—Forms, p. 303R-9
4.1—General
4.2—Materials
4.3—Economics
4.4—Formwork accuracy
4.5—Form joints
4.6—Textures and patterns
4.7—Formwork accessories
4.8—Form coatings and sealers

ACI 303R-04 supersedes ACI 303R-91 and became effective August 27, 2004.
Copyright © 2004, American Concrete Institute.
All rights reserved including rights of reproduction and use in any form or by any means, including the making of copies by any photo process, or by electronic or mechanical device, printed, written, or oral, or recording for sound or visual reproduction or for use in any knowledge or retrieval system or device, unless permission in writing is obtained from the copyright proprietors.

4.9—Form release agents
4.10—Form removal

Chapter 5—Reinforcement, p. 303R-17
5.1—General
5.2—Clear space
5.3—Reinforcement supports and spacers
5.4—Tie wire
5.5—Zinc-coated (galvanized) steel reinforcement
5.6—Epoxy-coated reinforcing bars

Chapter 6—Concrete materials and mixture proportioning, p. 303R-19
6.1—General
6.2—Materials
6.3—Proportioning, mixing, and temperature control

Chapter 7—Placing and consolidation, p. 303R-21
7.1—Conveying and placing
7.2—Consolidation

Chapter 8—Curing, p. 303R-22
8.1—General
8.2—Curing in forms
8.3—Moist curing
8.4—Membrane curing
8.5—Hot-weather curing

Chapter 9—Treated architectural surfaces, p. 303R-23
9.1—Surface retarders
9.2—High-pressure water jet
9.3—Acid wash
9.4—Abrasive blasting
9.5—Tooling or other mechanical treatments

Chapter 10—Finishing and final cleanup, p. 303R-25
10.1—General
10.2—Tie holes
10.3—Blemish repair
10.4—Stain removal
10.5—Sealers and coatings

Chapter 11—References, p. 303R-26
11.1—Referenced standards and reports
11.2—Cited references

Appendix A—Architectural concrete photos, p. 303R-29

CHAPTER 1—INTRODUCTION

Architectural concrete is concrete that will be permanently exposed to view and that requires special care in the selection of the concrete materials, forming, placing, and finishing to obtain the desired architectural appearance. This guide presents recommendations for cast-in-place architectural concrete based on the information available to the committee. Various procedures are recommended for determining initial requirements of the architect, contractor, concrete producer, and inspector. Critical areas are indicated for special attention, and means for prevention or correction of defects are discussed. Specific surface treatments and special forming techniques are presented. Applicable codes, specifications, and recommendations are cited throughout the text and listed in Chapter 11. General information is found in References 1, 2, and 3.

The information presented in this guide is very broad and covers many special conditions for specific architectural concrete. Information that may be applicable for use in producing a specific result may not be applicable to another. The user should also be aware that recommendations in this guide are subjective to the means and methods used for accomplishing a specific task for a specific level of architectural effect and should be tested before use to ensure it will produce the required result. Further research is needed to provide additional information on bugholes and other construction problems. This guide does not address all the problems associated with architectural concrete.

CHAPTER 2—ARCHITECTURAL CONSIDERATIONS

2.1—Architectural features

2.1.1 *General acceptance criteria*—Architecturally acceptable concrete surfaces should be aesthetically compatible with minimal color and texture variations and minimal surface defects when viewed at a distance of approximately 20 ft (6 m) or more as agreed upon by the architect, owner, and contractor, or as otherwise specified.

2.1.2 *Measurement*—It is beyond the scope of this guide to establish precise or definitive rules of measurement. Within any discrete building element or series of like elements, however, a high degree of visual uniformity is generally expected and required. The preconstruction mockup panel would normally be used to establish acceptance criteria. Refer to Section 2.5.4, Preconstruction mockup.

2.1.3 *Variations in color and shading*—These are minimized by:

- Quality control of ingredients, concrete mixtures, and consistency (Sections 6.2, and 6.3);
- Uniform concrete delivery schedules (Section 7.1.1);
- Uniformity of form surface, form release agent, application rate, and formwork reusage, erection, and stripping (Sections 4.8 through 4.10.4);
- Uniform rates of placement and consistent methods of placement and consolidation of concrete (Chapter 7);
- Placement schedules to minimize extreme variations of ambient conditions (Section 7.1.1);
- Uniform curing procedure and material (Chapter 8); and
- Properly timed or executed finishing operations (Chapter 10).

2.1.4 *Finishes*—Surface textures are grouped into two general classes:

- Untreated surfaces where the mortar is the principal visible constituent, and the texture is that which is imparted by the formwork sheathing or form liner; and
- Surfaces that are mechanically treated in place by removal of surface mortar to expose the underlying aggregate, thus wholly or partially obscuring the texture of the form sheathing or form liner.

High-build polymer coatings and cementitious or polymer-modified coatings that obscure both color and texture are not included in this guide.

2.2—Architectural design

2.2.1 *General criteria*—Architectural design criteria for readily obtainable and acceptable results should include:

- Isolation or division of concrete surfaces to allow reuse of formwork modules by the incorporation of rustication or joint patterns, or by the employment of a paneled effect;
- Systematic planning of construction joints that allows a reusable formwork module and conforms to structural requirements;
- Use of textured form sheathing (face sheets) or form liners, mechanically or chemically textured concrete finishes, or other relief features; and
- Limitations on the size of panels bounded by rustication or joint patterns. (Large, smooth, uninterrupted expanses of concrete surfaces should be avoided.)

2.2.2 *Details of architectural design*

2.2.2.1 *Unchamfered corners*—Acute- and right-angle corners can be obtained. Proper consolidation of the concrete is essential and self-compacting concrete may be required to achieve a crisp edge (Fig. 2.1). Forms should be designed and sealed to resist concrete placing pressures, and fabricated such that they can be stripped without damaging the concrete. Extended stripping time will probably be necessary to prevent damage to sharp corners which, depending on the size of the project, may require additional forms. Extended stripping time could also affect color because variation in form stripping times will increase the color differential.

2.2.2.2 *Chamfered corners*—Where chamfers are part of the architectural feature, the chamfer form strip should be continuously tight to all form surfaces that they are in contact with. Plastic or metal chamfer strips are available that have integral means of sealing to the form contact surfaces. Wood chamfer strips are difficult to maintain.

2.2.2.3 *Joints*—Panel area joints may be recessed into the concrete surface by applying rustication strips on the formed surfaces. These may also be used for construction joints as needed by the contractor. Waterstops, keyways, and joint sealants may be included where required. Refer to Section 3.4 for recommended joint depths.

Formwork rustication strips should have a draft of not less that 15 degrees to facilitate removal. Draft is defined as a small angle or taper in the formwork for reentrant formed surface that facilitates release when the form is stripped, as shown in Fig. 2.2. Also, wooden strips should be deeply sawcut (kerfed) on the back side to prevent binding due to expansion from absorbed water.

All rustication strips or other inserts should be installed tight to the form contact surface. Back-screwing the strip so it can be released before stripping may help obtain a tight seal. Nailing the strips to the form face for some architectural treatments may be acceptable but will not attain an absolutely tight seal. Insufficient nailing of the form strip will usually

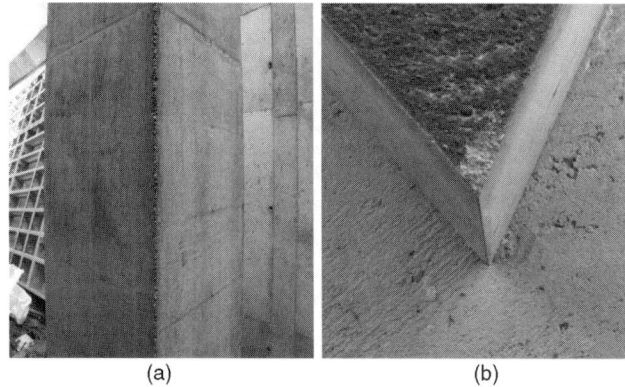

Fig. 2.1—(a) Difficult-to-achieve consolidation at acute corners; and (b) the use of self-compacting concrete allows for proper consolidation at the acute corner.

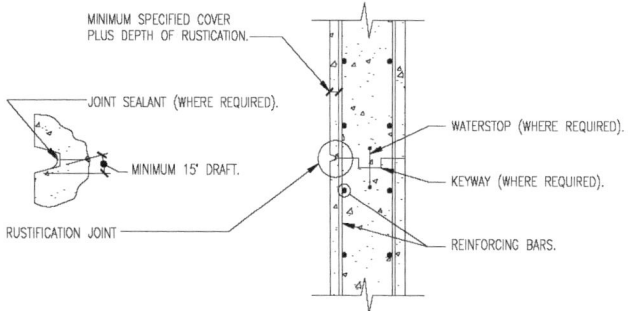

Fig. 2.2—Typical construction joint/detail.

result in some leakage and leave the reveal void ragged and discolored.

Rustication strips should be uniform in dimensions, nonwater absorbent, and of sufficient stiffness to maintain alignment during concrete placement operations. In areas of possible deflection of the sheathing, a method of treatment to prevent mortar leakage should be used.

Metal chamfer or rustication strips and other materials of similar stiffness should have a minimum width of 3/4 in. (20 mm). Widths of wooden rustication strips should be at least equal to their depths. Joints smaller than recommended above can be attained with special form detailing using steel insert strips installed between the forms.

Intersections of chamfer or rustication strips should be mitered or coped to fit snugly. Chamfer, rustication, or isolation strips may be placed so as to cover form joints.

2.2.2.4 *Soffits*—A drip should be installed in soffits near vertical surfaces where there is a potential for downward movement of rainwater on the vertical concrete face (Fig. 2.3). The drip molds should be placed as near to the external vertical face as practicable, but not closer than 1 in. (25 mm) from the finished concrete surface. Note that the drip in Fig. 2.3 is interrupted at either end of the underside horizontal surface to encourage water to fall before it reaches the inside vertical face.

2.2.2.5 *Sloped surfaces*—Accumulation of airborne solids on horizontal surfaces can be minimized by sloping

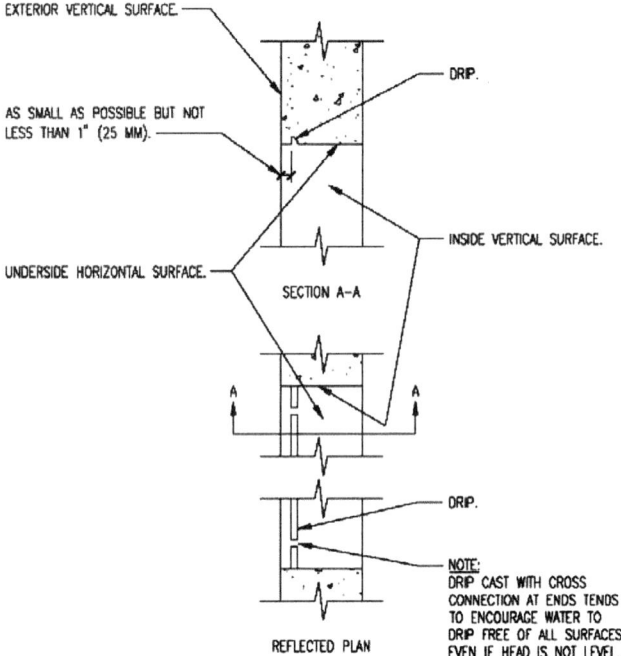

Fig. 2.3—Drip detail.

such surfaces. Sills should have a slight downward slope, and upper surfaces of recesses should have an upward slope from the horizontal relative to the inside face of the recess. Slopes may vary from 1:12 for smooth surfaces to 1:1 for textured surfaces (Fig. 2.4(a) and (b)).

On parapets, the slope should be away from the face. Unsloped horizontal offsets in vertical recesses and the use of textures on horizontal surfaces should be avoided. If a horizontal recess is formed without a drip near the exterior face, a slight upward slope relative to the exterior face should be provided (Fig. 2.4(c)).

Sloped and horizontal surfaces cast against a form will trap air bubbles as the concrete is consolidated. Special techniques and materials may be required to reduce trapped air, such as an absorbent form face. Refer to ACI 309R for more information.

2.2.3 *Combination with precast concrete*—Cast-in-place architectural concrete and precast elements may be successfully combined in either of the two following ways, both of which require detailed effort on the parts of the owner, architect, contractor, and inspector (Fig. 2.5):

- Color and texture may be reasonably matched by on-site precasting at the same time cast-in-place work is done, using the same concrete mixture, materials, formwork techniques, and curing for both types of concrete.
- Provide for contrasting colors and textures in the design between off-site precast concrete and cast-in-place architectural elements. In some cases, color and texture can be closely matched using dissimilar materials. A mockup panel can be used to demonstrate the match for approval by the architect. Trying to achieve an exact match of cast-in-place concrete with precast units cast off-site is extremely difficult and may not be achievable in a cost-effective manner.

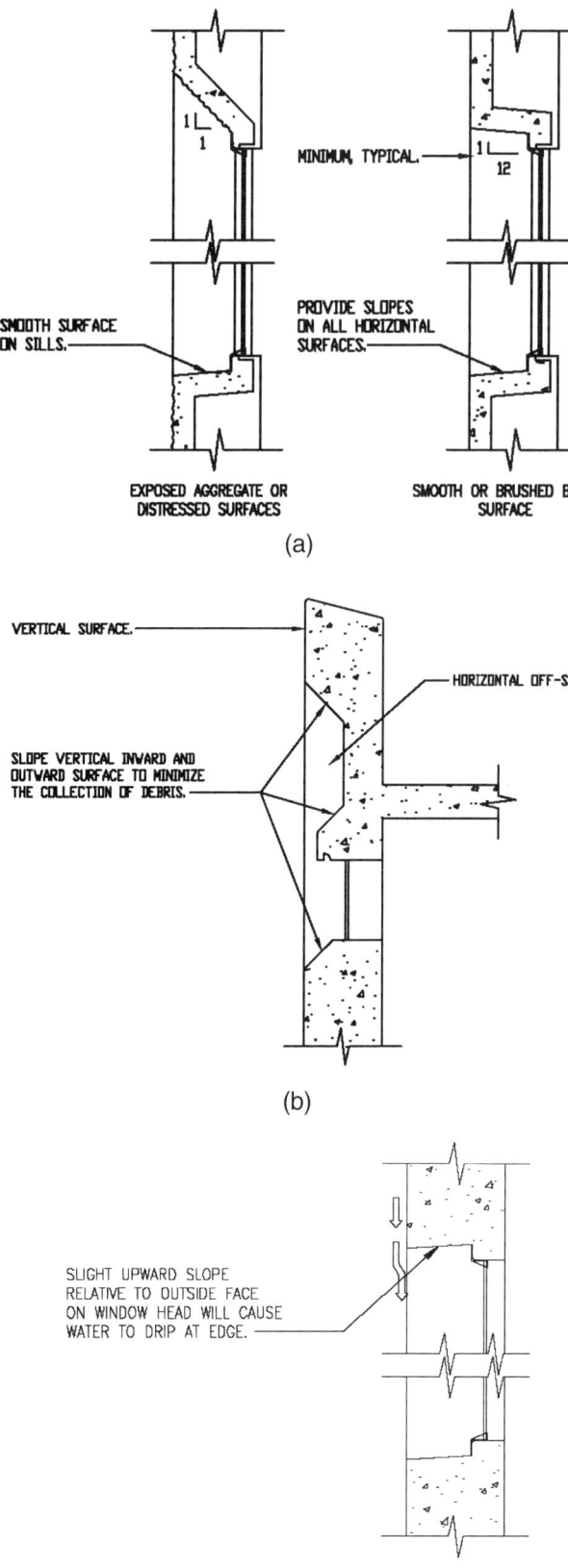

Fig. 2.4—(a) Suggestions for detail that will promote self-cleaning by encouraging water to continue its downward flow; (b) deleterious airborne solids should not be afforded any horizontal surfaces on which to collect. A sloped surface should be provided; and (c) a slight upward slope will encourage rainwater to drip free if no drip is used.

2.3—Coatings and sealers

2.3.1 *Purpose*—Retention of original color and texture may be prolonged by surface application of clear, penetrating (vapor-transmitting) sealers that reduce surface moisture absorption and consequent weather staining. Recommendations for use can be found in Reference 4 and ACI 515.1R.

Stains may be applied where it is desired to alter the natural color of the concrete and to retain its texture. Pigmented coatings drastically reduce vapor transmission, are not penetrating, and often alter both color and texture. They should be warranted by the manufacturer against changing color after exposure.

2.4—Joint sealants

Passage of moisture through construction and control joints should be prevented by filling the joints with sealants as recommended in ACI 504R. For architectural reasons, the sealant should be harmonious in color and shape with the adjacent concrete surface, nonstaining, and compatible with clear penetrating sealers.

2.5—Specifications

2.5.1 *General*—Specifications are customarily prepared as one of the following types or a combination of both.[5]
- **Performance specification**—The quality of the end product is specified. In this case, full responsibility is placed on the contractor. Recommended methods may be suggested.
- **Prescriptive specification**—Detailed methods, materials, and procedures are specified.

Most exposed aggregate projects use a combination of performance and prescriptive specifications, where aggregate proportions, cement source, and minimum vibration frequency are specified to ensure success without specifying actual methods of construction.

2.5.2 *Design reference sample*—Surface quality and appearance requirements should be referenced for bidding purposes to an actual sample or samples exhibiting the desired surfaces, color, and texture.

The sample should be prepared under the architect's direction and labeled as the design reference sample. A minimum size of 18 x 18 x 2 in. (460 x 460 x 50 mm) provides sufficient area for display and thickness for surface tooling and yet can easily be handled. The sample should be cast vertically or horizontally similar to the position in which the final concrete will be cast.

Design reference samples of walls or other representative building elements should be available for inspection and examination by prospective bidders. Mixture proportions, placement method, and sources of materials should also be provided. The samples should be confirmed in writing by both the owner and architect/engineer so as to have equal legal status with the contract documents.

In special cases, such as building additions or additional structures within a pre-existing group or complex, it may be acceptable to use an existing building for reference that contains elements of the desired quality and appearance.

Fig. 2.5—Architectural cast-in-place wall with precast window spandrels to match.

Fig. 2.6—Preconstruction mockup for the Getty Villas, Malibu, Calif.

2.5.3 *Prebid conference*—A prebid conference should be held between the architect/engineer and the prospective bidding contractors. At this conference, the special expectations and requirements should be explained and clarified. The acceptable architectural finish should be presented as established by the contract documents. At the same time, the contractors will have an opportunity to point out any aspects of the specifications that make it difficult or impossible to achieve the desired effect.

2.5.4 *Preconstruction mockup*—The preconstruction mockup (Fig. 2.6) is a full-scale sample of architectural concrete constructed on site by the contractor with proposed equipment, materials, and construction procedures. The contractor should obtain written approval of the finished

Fig. 2.7—Full-scale preconstruction mockup used for the Contemporary Arts Center, Cincinnati, Ohio.

product from the specifying agency before constructing the main structure.

The preconstruction mockup is normally constructed on the job site by the successful contractor before commencing the architectural portion of the major work. On certain projects, an extra mockup may be constructed under a special contract to determine the feasibility of various materials, treatments, and procedures that are to be included in the architectural specifications for the project (Fig. 2.7).

The overall height and width of the full-scale mockup should allow the demonstration of floor, column, and wall construction. The mockup should incorporate both horizontal and vertical form or form liner joints, placing method and equipment, and all of the specified reinforcement, accessories, and curing materials and equipment. All construction materials for the mockup should be the same as those used during the actual construction of the project. Ideally, the main vibrator operator for the concrete in the final structure should also operate the vibrator during construction of the mockup.

The mockup should also include a repaired area to determine ahead of time an acceptable color and texture match for use if remedial work is needed. In evaluating this experimentally repaired area, the repair should be aged at least 1 month to give a true indication of its color. To expedite construction of the project, a minimum of five variations of mixture color should be made for selection of the best match. Perfecting a repair procedure can save both time and money in the final outcome of the project.

Where feasible, architectural concrete placement, treatment, and procedures may be evaluated on portions of the structure eventually hidden from view, such as basement walls.

2.5.5 *Shop drawings*—Shop-drawing submittals are effective tools for reviewing the contractor's plans for construction. The shop drawings should include:
- Details of the formwork aspects of the project, including form butt joints, rustications or reveals, and construction joints. The details should convey to the architect the means and methods the contractor will be using to fabricate and erect the formwork and to show how the forms will be maintained to ensure tightness against leakage and alignment during concrete placement;
- Cover of reinforcing steel; and
- Description of the placing operation, including the method of conveying concrete to the forms, wall placement lengths and lift heights, soffit deposit sequences, and vibration techniques.

2.5.6 *Inspection and quality control*

2.5.6.1 *General*—The architect, contractor, project inspector, suppliers, and other parties involved with the production of the architectural concrete may interpret the results of the finished product differently. These interpretations may vary during the progress of the work when expectations become or seem impossible to achieve with the current practice of construction and materials. The inspector should review the specifications ahead of time and meet with the architect/engineer and contractor after the mockup has been constructed to establish and agree on the criteria that is to be used in evaluating the final product. This section will concern only those points not covered in Reference 6.

2.5.6.2 *Qualification of the inspector*—The inspector should have previous experience in the inspection of architectural concrete of equivalent complexity and scope. The extent of his or her responsibility is determined by the architect/engineer. Confirmation of this responsibility is developed at the prebid and postbid conferences with the contractors. The architect and inspector should hold periodic conferences to discuss the progress and quality of the work. On major projects where different operations are proceeding concurrently, more than one inspector may be required to adequately inspect the work.

2.5.6.3 *Duties of the inspector*—The duties and responsibilities of the inspector should include:
- Reviewing project documents in advance of the prebid conference and advising the architect/engineer of changes that may be required to produce the desired results;
- Inspecting and personally observing the manufacture of the prebid samples. This will provide the inspector with invaluable knowledge and experience regarding the mixture, slump, color, and placing characteristics of the concrete. During the contract period that follows, any variation would immediately become apparent;
- Attending the prebid conference to assist the architect/engineer in clarifying the intent of the specifications;
- Being present during construction of the preconstruction mockup to observe and evaluate the materials and techniques used to produce and repair the panel. The inspector, after completion of these efforts, will be ready to assist the architect/engineer in evaluating the quality of the completed architectural concrete work;
- Providing inspection of materials used for concrete mixture, forming or texturing, form release agents and their application, form alignment, tightness of the form joints, rustication fastening, placement of reinforcing steel relative to the exposed face, curing before and subsequent to the form stripping, required repairs, and final clean down. Each of these contributes to the appearance of the finished surface of the architectural concrete; and

- Observing and recording the condition of the architectural forms after each reuse and consulting with the contractor and architect when the forms need to be replaced or refurbished to achieve uniformity of the architectural surface.

2.5.6.4 *Final acceptance*—If the procedures determined by the approved on-site mockup are continued throughout the project, final acceptance should not be a problem. Due to the inevitable nonuniformity of construction practices, some repairs will normally be required. Their final acceptability will depend on the contractor's blending technique and skill. Periodic review by the inspector and the design architect to allow partial acceptance creates goodwill and confidence with all concerned. After final acceptance, the inspector's records should be completed and filed. If later additions are made or adjoining buildings constructed, these records will be helpful for construction.

CHAPTER 3—STRUCTURAL CONSIDERATIONS

The structural and architectural design should function in harmony to produce a structure capable of withstanding service loads and stresses without excessive cracking, spalling, or deflection that may detract from the architectural appearance of the structure.

3.1—Spalling

In their design, the architect and engineer should pay special attention to applied loads at joints where movement can occur and cause spalling.

3.2—Deflections

The use of strength design requires strict investigation for undesirable deflections in beams or spandrels exposed as architectural concrete. The desired camber should be specified to compensate for the deflection of the completed structural member. Consideration should also be given to additional long-term deflection due to creep of concrete under permanent loads. Architectural requirements may dictate that the engineer design for deflections less than normally acceptable. They may also require additional camber to avoid the impression of sagging in long spans. Excessive camber can, in some cases, be as objectionable as sag.

3.3—Cracking

Cracking in architectural concrete can be very unsightly. Practices to reduce and control cracking to minimize the effect on appearance and durability should be used.

Factors that influence concrete cracking are:
- Gravity or lateral loads producing tension, shear, or torsion in members;
- Restraint of drying shrinkage;
- Creep;
- Abrupt change in geometry, such as concrete thickness;
- Axial or bending stresses due to thermal effects; and
- Foundation settlement.

Cracking may be reduced by:
- Post-tensioning. The resulting axial shortening and perhaps rotation may require attention at bearings or slip joints, and added reinforcing, closure strips, and other methods may be required;
- Providing joints or crack-inducing devices to relieve stress. Cracking can often be minimized by attention to joint placement, which relieves the shrinkage stresses and directs the cracking to joints that are integral with the designed surface presentation. This is often achieved through rustication or reveals on the concrete surface with internal devices—both weaken the concrete at predetermined lines to induce the concrete to crack at that location. An induced crack should be sealed when exposed to weather to prevent water penetration. This is done by applying sealant in the surface rustication or reveal void or by placing a waterstop-type internal crack inducing device in the concrete;
- Limiting the flexural tension stress in reinforcement;
- Appropriate distribution of flexural reinforcement (ACI 318);
- Using special materials, such as shrinkage-reducing admixtures, shrinkage-compensating cements, and aggregates with low shrinkage characteristics, or selected proportions of these materials (ACI 223);
- Minimizing the water content, not just the water-cementitious materials ratio (w/cm) of the concrete mixture by optimizing the aggregate grading. Some water-reducing admixtures may increase shrinkage despite a reduction in water;
- Minimizing the change in member-to-member geometry or mass;
- Evaluate trial batches with various combinations of aggregates, cements, and admixtures, and checking for shrinkage using ASTM C 157;
- Increasing the shrinkage and temperature reinforcement above the minimum requirements in ACI 318; and
- Minimizing thermal stresses by controlling heat of hydration and cooling.

3.4—Joints

Construction joints divide the structure into segments that can be constructed in a logical and efficient manner. Details should be shown in the project drawings.

Contraction joints are used in walls and slabs-on-ground to control cracking.

Isolation joints are used in slabs-on-ground to separate them from walls, columns, or other structural elements.

Rustication strips (Fig. 3.1) provide the simplest method of architecturally treating joints where surfaces in the same plane are joined. They reduce both the effective size of the members and the cover over the reinforcing steel. Additional compensating cover or a protective coating on the reinforcement should be provided (Fig. 2.2).

Recommended joint depths follow:
- Small rustication or pattern grooves: 3/4 in. (20 mm);
- Joint depths to initiate cracking: Approximately 20% of the member thickness, that is, 1-1/2 in. (38 mm) deep for an 8 in. (200 mm) thick wall or slab; and
- Joints between panel divisions should be at least 1-1/2 in. (38 mm) deep.

Fig. 3.1—Rustication strips used to break up a large expanse of concrete.

A large structure containing drastic changes in section size or exposure should be designed with either expansion or delayed pour joints to control cracking and allow for movement. All joints and rustications should be shown on the contract drawings. Refer to ACI 224.3R for detailed information regarding joints in concrete construction.

3.5—Beams and slabs

The structural and architectural design of beams and slabs requires careful consideration because all of the factors that produce tension in concrete may be present in portions of the member.

Rustication depths should be kept to the minimum recommended size at the top and bottom of beams in regions of flexural tension. When the typical concrete cover on the steel is increased excessively by the depth of the rustication strip, any cracks that occur in the flexural tension zone will also be increased in width.

Sandblasting accentuates the width of such cracks due to rounding of the crack edge, especially at soffits of beams and slabs. Connections that are intended to allow substantial rotation or displacement should be designed and detailed to prevent spalling and to prevent leakage if exposed to weather.

3.6—Columns

Concrete columns normally do not have the high tensile stresses present in beams. The lateral dimensions are usually small, and there is less tendency for vertical cracking. The small lateral dimensions also permit rapid concrete placement that tend to eliminate some of the problems associated with slow concrete placement. Columns associated with long span beams may have a tendency to exhibit horizontal cracking as those beams become loaded. Thermal movements and movement due to concrete shrinkage can also cause horizontal cracking in columns. The engineer should take these added effects into consideration when designing the column reinforcing.

3.7—Walls

The typical wall is relatively long compared with its thickness and may be tall. It normally is in compression vertically, and horizontal cracking is usually not a problem. The most common cracking in walls is generally vertical or nearly vertical. Horizontal cracking can occur when the concrete further consolidates subsequent to vibration of thick walls. These cracks can be prevented by ensuring that each lift is integrated with the previous lift by using a good vibrator spacing and insertion time. Because walls are lightly loaded, the required reinforcement percentage is low in each direction. The prime cause of vertical cracks in walls is axial tension due to restrained drying shrinkage, temperature stresses, or both.

Construction joints are normally provided at the bottom and top of slabs at the intersection of the walls and slabs. Other joints should be provided at locations necessary for construction and to control cracking. These joints should be concealed, rusticated, or emphasized. Those required for construction should be coordinated between the contractor and the architect/engineer.

Vertical construction joints in long walls may be necessary at midspan or at some multiple of the bay length. An architectural approach is to provide for vertical rustications at a uniform spacing in all bays. This provides the contractor greater flexibility in planning the construction operation. Vertical construction joints with rustication strips can be detailed as joints to accommodate volume change movements and to reduce the horizontal extent of the casting operation, or can be detailed as crack control joints. This permits the acceleration of the vertical rate of casting which, particularly in hot weather, will eliminate or make manageable the problems associated with bugholes, form spatter, cold joints, and lift lines. (Form spatter is mortar splashed on forms and allowed to dry before covering with concrete, which causes a nonuniform surface and affects the architectural appearance.) The architect/engineer should consider the formwork panel size in planning the joint locations because form-facing material joints may be visible on the concrete surface even though subsequent mechanical finishing is performed on the hardened concrete surfaces.

An effective method for the control of vertical cracking is to provide contraction joints at a uniform spacing of not more than 20 ft (6 m) on center by placing deep (1.5 times the maximum aggregate size), narrow rustication strips on both wall faces to induce cracking. Any of the contraction joint locations can also be used for construction joints.

Consideration should be given to reducing the reinforcement crossing contraction and construction joints that are used for crack control. Suggested reinforcement crossing either type of joint should be not greater than 1/2 of the horizontal reinforcement elsewhere in the wall. The minimum horizontal reinforcement required in walls by ACI 318 may be increased to minimize the widths of cracks that may occur between joints.

Consideration may also be given to the use of fiber reinforcing as a means of reducing the width of the visible crack. Fibers may be seen at the surface and should be approved in the mockup.

Openings in walls induce cracks from the corners of the openings. When horizontal steel is interrupted by openings, additional steel equivalent to that interrupted should be placed

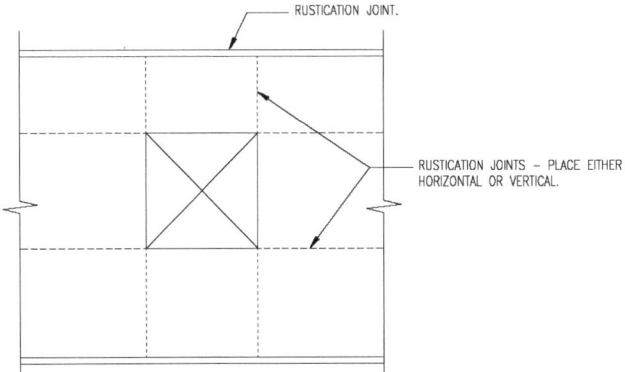

Fig. 3.2—Either vertical or horizontal rustication joints can control cracking at openings in a wall.

1/2 above and below the opening, with a minimum steel extension 1-1/2 times the development length beyond the opening. Diagonal bars placed in the corners of the wall openings are also effective in preventing or minimizing crack width at these locations. The wall should have sufficient thickness to accommodate these bars and the normal horizontal and vertical reinforcement. Vertical or horizontal joints placed at the jambs or at the top and bottom of the opening, respectively, can also effectively be used to control cracks emanating from corners of openings (Fig. 3.2).

CHAPTER 4—FORMS
4.1—General

Through general drawings, specifications, samples, and mockups, the architect defines the structure and the desired appearance of the architectural concrete. These design conditions impose limitations on the selection of forms and forming materials. Satisfactory results are more likely when the architect understands the capabilities and limitations of the forming materials because the forming system is vitally important to the success of the project (Fig. 4.1). It is also vital to the success of the project that the contractor understands the means and methods required to produce the required architectural product. Formwork shop drawings, including a form layout drawing if more than one placement of formwork is to be used, and mockups should be produced and submitted for approval to establish that the materials, logistics, and procedures will meet the specified results.

Formwork and formwork labor can account for more than half of the overall cost of the concrete structure[7] and often more in an architectural concrete structure. Communication and understanding between the architect, contractor, and owner are the foundation of a cost-effective, trouble-free project.

Another necessary ingredient is the ingenuity, know-how, and ability of the contractor to interpret the specifications and select the system that will best accomplish the desired result within the owner's budget.

In an elevation view, reusage of formwork has three variations: identical, similar, and nontypical. Identical reusage means that both the horizontal and the vertical configuration of the forming module never change. The distance between construction joints as a horizontal determinant and the wall height as the vertical determinant

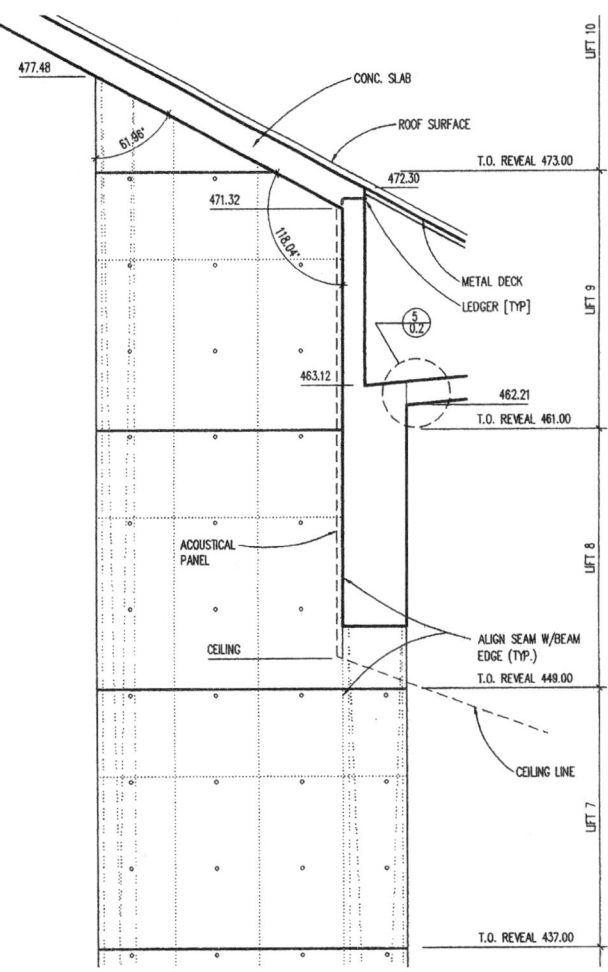

Fig. 4.1—Form layout drawings showing seams, rustication strips, and dimensions for the Cathedral of Our Lady of the Angels (Los Angeles, Calif.).

usually determines the forming module. To be identical, it is necessary to have symmetry on either side of both the vertical and horizontal axes of the forming module. Similar reusage means that only one of the forming module configurations is identical and the other varies. Nontypical means both forming module configurations vary.

Opposite hand means that the configuration must be rotated 180 degrees to be identical. Opposite hand may necessitate assembly and disassembly of formwork for each placement of concrete. Where possible, gang forming should be designed to allow for opposite hand configurations.

Unless special coloring and texture effects are desired, B/B plywood is not recommended for the architectural portions of the structure. Even low-density overlay (LDO) and medium-density overlay (MDO) plywoods impart unexpected color variations, unless limited to one use only. High-density overlay (HDO) plywoods provide the most protection from unwanted color and texture variations.

References 7 and 8 contain more detailed information on forms.

4.2—Materials

A great variety of materials have been used for forms, form liners, and sheathing. The list includes lumber;

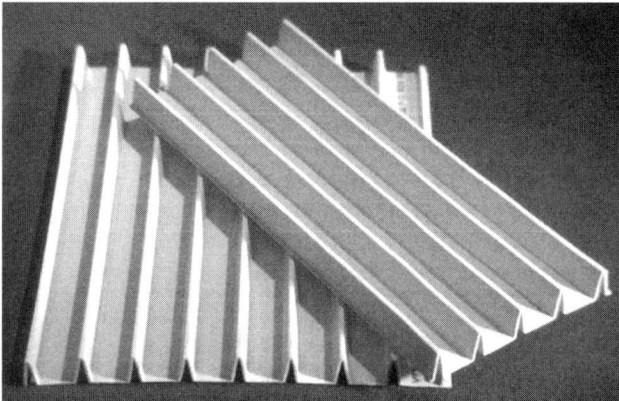

Fig. 4.2—Fiberglass form liner.

Fig. 4.3—Use of rough-sawn lumber to produce a broad form texture, The Getty Villas, Malibu, Calif.

Fig. 4.4—Color variation from the first use of forms to the next.

plywood; coated and plastic overlaid plywood; metals such as steel, aluminum, and magnesium; reinforced and nonreinforced plastics; plaster waste molds; thermosetting plastics; and elastomeric liners of both rigid and flexible plastics such as acrylonitrile butadiene styrene (ABS), fiberglass (glass-reinforced polyester) (Fig. 4.2), and both filled and pure polyurethane elastomers. Each of the materials has advantages and limitations. Special precautions should be taken with the use of aluminum, magnesium, and zinc-coated forms. Refer to Section 4.2.3.

The architect/engineer should give consideration to the effect of impervious and absorptive form surfaces. Each leaves its own particular characteristics. The impervious form surface will usually result in a more uniform appearance. Examples of impervious form surfaces are metal, plastics, high-density overlaid plywood, and other materials with applied coatings. Forms and liners that have a moisture content below their saturation point will absorb water from the fresh concrete, resulting in a darker concrete color. The color will vary with the absorptive capacity of the form. Form release agents will not solve this problem. The most effective method of preventing problems of this type, if objectionable, is to seal the surface of the absorptive form surfaces (Section 4.8).

4.2.1 *Lumber*—Lumber is a readily available forming material. It may have a smooth surface or be rough-sawn (Fig. 4.3) or sandblasted to transfer distinctive textures to the concrete surface.

The reusability and durability characteristics of lumber depend on the wood species, time and exposure condition while in storage, release agents, and other factors.

Lumber forms can affect the color of the concrete surface. A mottled effect is achieved through variations in water absorption of different densities in the grain of the board surfaces. The softer grains of the wood will absorb more water from the surface of fresh concrete, lowering the w/cm of the concrete, which causes a darker surface color. Organic substances in the wood can result in a discolored concrete surface, and wood sugars can cause dusting. Release agents cannot prevent either of these conditions. Care in selecting wood is recommended as wood splits and warps during use, causing changes in surface presentation.

With each reuse of the form, the darkening effect of the lumber on the concrete surface becomes less (Fig. 4.4). When forms are reused several times, considerable variation in concrete surface color and texture may be expected from the first use to the last unless the wood is treated.

Dusting, caused by wood sugar, is significant only in the first use. It may be desirable to simulate a first use by coating a new form face with a cement slurry, washing it off, and reapplying the form release agent. For a uniform surface color, all form lumber should be obtained from the same source, and a form coating or sealer should be used (Section 4.8.2). If controlled variations in color are desired, this may be achieved by lumber from different sources.

4.2.2 *Plywood*—Plywoods can be purchased with surface treatments that provide a nearly impervious and smooth surface. Mixing of different brands or differing surface treatments should be avoided as this will result in variations in color due to different amounts of water absorption (Fig. 4.5). If a raised grain is desirable for reverse transfer to the concrete, the impervious coatings should be avoided. Additional relief can be obtained by sandblasting the plywood surface to expose the grain texture. This type of rough surface does not allow for many reuses.

4.2.3 *Metals*—Metal surfaces are impervious and provide uniform color to the concrete if cleaned of all reactive or potential staining materials before use, and if the face is maintained free of rust pitting, weld heat areas, and dirt. The metal skin should be thick enough to support the load between its support members to keep deflections within acceptable limits.

To prevent staining, steel skins for architectural concrete should be made of cold-rolled steel so that there is no mill scale. Bluing, a coating used in steel making, should be added to the cold-rolling process to decrease the potential for water marking or staining in general. Bluing can be field applied over welded sections and has been found beneficial to avoid staining. Release agents that contain a rust inhibitor to reduce the possibility of staining are recommended.

Aluminum and magnesium alloys may be used successfully if compatible with the concrete. There is no standard method of testing to determine compatibility. Past history of use with the same concrete mixture, forms, and curing conditions is the best-known indicator.

Aluminum, magnesium, and zinc all react with alkaline material in concrete, liberating hydrogen gas that causes sticking and bubbling on forms containing these metals. Extreme care and attention is needed to minimize these effects. Therefore, these materials should not be used to produce architectural concrete.

4.2.4 *Plastics*—Plastics, both reinforced and unreinforced, have an important role in architectural concrete forming because they have an impervious surface and the ability to be molded into any pattern or texture. They do not cause discoloration that is common with many absorptive types of forming materials. The use of smooth forming materials may lead to the nonuniform coloring known as aggregate transparency. Some plastic forming materials may produce a glossy concrete surface that should be used with caution as such surfaces exposed to the weather will soon lose some of their gloss due to the effects of wetting or drying and freezing and thawing. Repairs may be difficult to match when the as-cast surface is glossy.

Some reinforced plastics contain glass fibers in various forms to increase the flexural strength of the resin materials. Such plastics have found considerable acceptance in custom forms for architectural concrete. The appropriate resin (gel coat) should be used on the surface to overlay the glass fiber mat and to ensure good performance through a reasonable number of uses. Unless alkali-resistant glass fibers are used, deterioration can be expected when in contact with the concrete. Maintenance of the resin cover is mandatory for surface uniformity. It can be accomplished by careful cleaning, use of form release agents, or occasional touch-up of the surface.

Unreinforced plastics can be obtained in sheet form with smooth or textured surfaces. Lightly textured patterns transfer to the concrete and change the characteristics of a smooth surface. Sheet plastics need appropriate backup to resist the concrete pressure. Unreinforced plastics are normally used as liners with a forming system designed to meet all of the structural requirements of concrete pressure

Fig. 4.5—Absorption of medium-density plywood (MDO) on left compared to regular plywood on right.

containment. Unreinforced plastics are used only to change the characteristic of the concrete surface. Thermoplastic coatings and form liners may expand or contract and change dimensions due to direct sunlight or elevated ambient heat.

Preformed foamed polystyrene can be used as forms for recesses, textures, and designs. The preformed foam planks are easily cut to size, easily attached to the form, and are inexpensive enough not to require salvaging; however, there are release agents that promote multiple reuses and easier removal. Foamed polystyrene is also used in backing for deep relief, vacuum-formed, plastic form liners where the concrete pressure would cause deformation. Most-oil based release agents will dissolve foamed polystyrene.

4.2.5 *Plaster waste molds*—Highly detailed forms can be made of plaster. The concrete is cast against these molds, and the plaster is then broken away from the finished concrete. Single-use forms are often used for nonrepetitive forming or where intricate shapes are difficult to form by more conventional methods. An effective membrane-forming bond breaker should be used with plaster waste molds. Use of solvent-based release agents is not recommended because most of these products may soak into the plaster, resulting in a defective release.

4.3—Economics

In an analysis of formwork cost, these factors should be considered:
- Crane or hoist equipment available for moving formwork;
- Materials, fabrication, and rental expense;
- Erection labor;
- Stripping and reconditioning;
- Reuse capability;
- Effect of forming method, stripping and curing duration on construction speed, influencing overall project cost;
- Salvage value at completion of use; and
- Governmental safety and environmental regulations.

The number of reuses of formwork is a significant factor in the total cost of formwork. Normally, formwork systems are increasingly cost effective after multiple uses.

Fig. 4.6—Form joints sealed to prevent leakage.

4.4—Formwork accuracy

In general, formwork for architectural use should be designed, constructed, and maintained in accordance with the recommendations of ACI 347R and the additional requirements outlined below. As placing and consolidating requirements are more demanding than for structural concrete, architectural concrete requires particular care in formwork design to eliminate deflection, deformations, pillowing, offsets, and mortar leakage. Conflicts between reveal strips and reinforcing steel should be resolved. Specified clearances between reinforcing steel and formwork should be maintained.

4.4.1 *Bracing and walers*—In most cases, form sheathing deflections will govern design. The form face should be designed as a stable envelope to contain the plastic concrete. Extra walers may be required to satisfy face sheet deflection requirements. Additional anchors and bracing may also be required to maintain alignment if the forms are externally vibrated. These conditions could require more ties and walers than required by ACI 347R. External vibration also requires anchorage of the formwork to the previous placement or footing, because it produces extremely high stresses in the structure of the formwork.

Deflections of sheathing, studs, and walers no greater than 1/400 times the span are generally satisfactory for architectural concrete formwork. Where architectural considerations, adjacent work, or special effects are critical, lesser form deflections may be required. As form deflections may increase with each use, deflection criteria may govern the number of allowable reuses. Where deflections are to be limited, locations and deflection criteria should be included in the project specifications or noted on the contract drawings so that the contractor knows in advance what is required.

4.4.2 *Tolerances*—The dimension and position tolerances required in ACI 117 are generally satisfactory for architectural concrete and should be maintained unless the architect/engineer specifically calls for closer tolerances for particular work items. In these cases, tolerances of 1/2 those called for in ACI 117 are the most reasonably restrictive that are possible to obtain in the field using extreme care in placement and form design.

4.5—Form joints

4.5.1 *Prevention of leakage*—A surface defect will result when the formwork experiences fluid loss. The resulting defect is characterized by a color change and an aggregate-rich surface, inconsistent with the normal, dense, adjacent surfaces. There may be streaking, mottling, or a darker appearance as a result of less water available for hydration. This aggregate-rich condition penetrates the concrete mass to a considerable depth, and the discoloration may still be noticeable after additional surface treatment. Leakage should be minimized where uniform color and texture are critical. Low-slump concrete—less than 5 in. (127 mm)—will reduce the tendency for fluid to escape through fine openings in the formwork. Low-slump concrete is more difficult to vibrate and consolidate and may cause entrapped air on the concrete face. When establishing the slump of the concrete, consideration should be given to the difficulty of placement, placement method, and reinforcing steel congestion. More precautions against leakage will be needed if fluidized concrete is used. Formwork leakage may be minimized by:

- Lining forms with a separate facing material and staggering the joints with those of the structural form;
- Using pressure-sensitive compressible gaskets or sealants within form interfacing joints;
- Face caulking with lumber batten backing; and
- Avoiding horizontal movement of the concrete with vibrator.

To minimize leakage, specifications should require form joints to be sealed (Fig. 4.6). Chamfer and rustication strips should be sealed at the edges to prevent leakage behind the strips.

Pressure-sensitive tape may be used on the form sheathing when significant paste removal, such as medium sandblast of the surface, is planned. Care should be taken to prevent displacement of the tape or gaskets during concrete operations, as this results in blemishes that are difficult to remove. Brush-applied gum adhesive over the tape can successfully stabilized it against movement. Taped joints should be inspected before casting to be sure the tape has not moved. A mockup panel should be used to verify the effectiveness of achieving the desired results with taped joints.

4.5.2 *Fins*—Fins are thin projections of hardened concrete extending from the wall face or soffit, most commonly due to leakage at joints. Although they can be knocked off and stoned smooth, they are generally considered undesirable because this results in a nonuniform appearance of the finished surface and possible staining due to lowering of the *w/cm* at the point of leakage. In some cases, fins are desired by the designer for a specific effect. This effect can be accomplished by lining the form with planks, boards, or plywood kept at specific distances apart or randomly placed. Such fins can be left as stripped or broken back. Provisions should be made, however, to prevent mortar leakage between the liner and structural form to prevent color and texture variation.

4.6—Textures and patterns

4.6.1 *Form marks*—All forms will have some characteristics that may be transferred as texture, pattern, or blemishes to the finished surface, including:
- Size of the unit of forming material or prefabricated panel;
- Plank widths;
- Variations in absorptive characteristics of the face that change the *w/cm* on the surface of the concrete and consequently change the consistency of its color;
- Special perimeter configurations found in proprietary type panels;
- Wood grain;
- Wood grain rise due to moisture;
- Number and size markings;
- Plywood boat patches that may be evident on the concrete surface even if the plywood is overlaid with plastic;
- Hairline checks or pierced holes in plastic overlaid plywood allowing moisture intrusion into the soft grains of the plywood. The moisture migrates along the soft grain, expands, and produces "tiger stripping" or "blisters";
- Fasteners such as nail and screw heads. To avoid this, fasteners should be placed from the backside and go through the form to hold rustication strips to the form; and
- Tie holes.

4.6.2 *Form liners*—Textures and patterns can be obtained by specific design through the use of form liners. The use of liners is a practical approach to many desired results in the finished wall because the facing can be designed separately and allow a choice of a backup forming system. The method of attaching form liners should be studied for its resulting visual effect.

Wood liners can be used to feature planks, grain, rustication strips, or used in a checkerboard fashion by changing the direction of the grain or planks in adjacent panels. Striated liners of various materials may also be used.

Foamed polystyrene liners provide a wide choice of surface textures and designs with smooth, grained, or a variety of fractured finishes.

Thermoplastics can be heat stressed into a wide variety of designs. Plastic liners should be rigidly secured to backup forms. Wide sections of deep relief liners used for deep indentations in the concrete surface should be completely supported between the backup form and the liner.

Elastomeric liners may be considered for relatively shallow textures and deep relief (Fig. 4.7). Some elastomers may deteriorate when exposed to the higher temperatures associated with mid-summer curing conditions or when heat is otherwise used to hasten the cure. PVC elastomers should be checked for resistance to deterioration by oils commonly used as release agents and they should be rigidly glued to formwork to resist wrinkling. The elastomers should also be checked for the possibility of staining the concrete.

Polyurethane elastomer is made in foamed and nonfoamed versions. Foamed polyurethane is either closed or open celled. Because open-cell foams may absorb release agents, tests are recommended to determine form liner-release agent compatibility.

Fig. 4.7—Elastometric form liner used to create a block pattern finish at the University of San Diego Science Center.

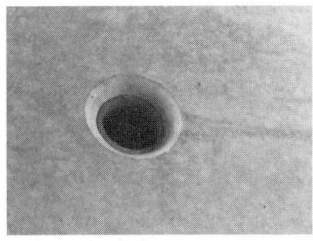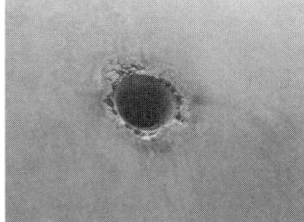

Fig. 4.8—Care should be taken to properly seal tie holes. Concrete will seep between the snap-tie cone and formwork if cones are not tight against form.

Metal liners are available in various textures and ribbed patterns that can be joined with different types of fasteners to achieve an architectural effect. Liner joints should be placed at rustication strips or form corners because leakage is difficult to prevent at butt joints. An investigation should be made to determine whether staining may occur from the liner material or its fastenings.

Thin, 0.50 to 0.60 in. (12.5 to 15.0 mm) vinyl plastic sheathing has been used successfully for lining gang forms. These vinyl liners are manufactured in rolls 3 to 6 ft (0.9 to 1.8 m) wide and 50 to 60 ft (15.2 to 18.3 m) long, and are fastened to the form backing, which is coated fully with rubber cement. The liner joints will self-seal as the form liner is applied to the backing.

4.7—Formwork accessories

4.7.1 *Ties*—Early stripping and finishing requirements may dictate the system of form ties (Fig. 4.8). Recommended form ties should leave no corrosive metal closer than 1-1/2 in. (40 mm) to the finished surface, and fall generally into one of several groups:
- Continuous single-piece proprietary ties for specific wall thicknesses are available in different lengths and capacities with positive break-back characteristics, with or without cones;
- Snap-ties are available in a variety of sizes and strengths with cones. Snap-ties with washers are usually not acceptable for architectural concrete. Care should be taken when snap ties with cones are used to ensure that the cone is maintained tight to the contact form face

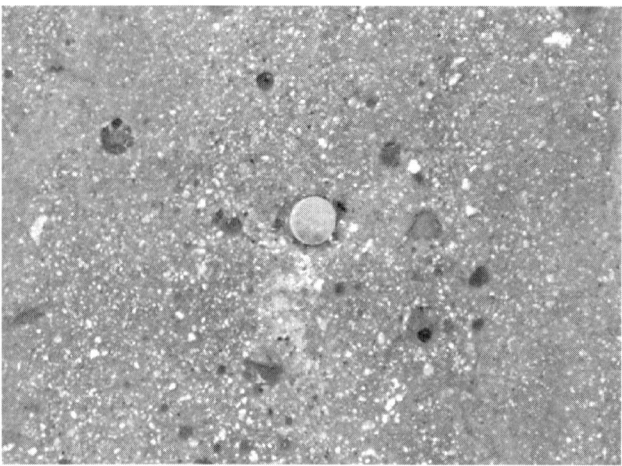

Fig. 4.9—Integral color fiberglass tie left exposed on sandblasted wall finish.

Fig. 4.10—Use of tie-hole pattern for architectural appearance at the Salk Institute Addition, La Jolla, Calif.

under placing pressures. Backup members may contract and allow fluid loss;
- She-bolts have a male-threaded inner unit left in the wall and a female-threaded outer unit that is removed and reused;
- He-bolts are male-threaded devices that are reusable with an expendable female-threaded unit left in the wall;
- Taper-ties are available in a variety of sizes and capacities;
- Fiberglass ties are available in many strengths, sizes, and colors. Fiberglass ties leave a cut end of round rod on the concrete surface (Fig. 4.9). Fiberglass ties are nonoxidizing and can be exposed on the face of the concrete;
- Sleeve and rod ties with or without cones are available from 1-1/2 in. (38 mm) cone sizes and larger; and
- All-thread rods with plastic sleeves.

Ties can be removed early (generally within 24 h) provided a form release agent was used on the embedded section of the ties and they are removed in a torsional motion.

Spacing of ties will normally be dictated by the strength of the ties, the strength of the forming members, the concrete placing rates, allowable deflection, amount of vibration, and architectural requirements.

Each type of tie leaves a characteristic hole, except fiberglass ties that leave a round plastic surface on the face of the concrete. Wire snap ties leave small holes, about 1/4 in. (6 mm) in diameter with a nominal depth of cover of 1 in. (25 mm). Wood or tapered plastic cones or sleeves are often provided for architectural expression or when deeper break-backs, up to 2 in. (50 mm), are required. The cones increase the size of the hole to about 1 in. (25 mm) diameter and are used to reduce grout leakage where the tie passes through the form. Maintaining tightness is essential. The characteristic hole of tapered she-bolts depends on the strength category of the ties, which have diameters in the range of 9/16 to 1-1/2 in. (15 to 40 mm). He-bolts are available with cones and with tapered studs. Cones are available from 1 to 2 in. (25 to 50 mm) in diameter, and tapered studs from 1/2 to 1/-1/2 in. in diameter (10 to 40 mm). Pull, or completely removable, ties may require plastic sleeves and can be from 1/2 to 1-1/2 in. (10 to 40 mm) in diameter, leaving a hole of similar size to the rod diameter passing completely through the wall. All the aforementioned ties leave round and relatively clean holes that may be subsequently patched or plugged flush or left with a slight recess for an architectural shadow effect (Fig. 4.10). Plastic or precast premolded plugs are available with most systems and can be inserted or bonded in the hole. The use of plastic plugs prevents mortar stains on the concrete surface that may be objectionable on some surfaces. Snap ties (without cones or other special seals) may be unacceptable for architectural concrete unless a rustic, crude look is desired.

Leakage at ties is difficult to prevent, especially at taper and sleeved ties. Various methods to minimize leakage at tie locations should be addressed with the mockup panel.

4.7.2 *Tie removal*—Most ties can and should be removed before removal of formwork. Break-back ties should be removed as soon as possible after the formwork has been removed. After forms are removed, uncoated ties or ties that possess staining tendencies should be properly broken off as soon as practical and the ends treated to prevent rust stains. Stainless steel ties present the least trouble with staining and are broken off at least 1 in. (25 mm) back of the exposed surface. Stainless steel ties are softer and are often difficult to break back. They are not a stock item with any manufacturer and therefore require longer lead times for shipments. They are more expensive and always of lesser capacity than regular steel ties. Twisted wire ties should not be used for architectural work.

To reduce spalling, removal of cones should be delayed until the concrete has adequate strength. When the cones are removed, the bond should be broken with a torsional motion. This may require a special tool. The part of the tie remaining in the concrete should be immediately coated with dry-packed mortar or sealed with premolded plugs.

4.8—Form coatings and sealers

4.8.1 *Function*—Form coatings are nonmoisture-transmitting and form sealers are semimoisture-transmitting. Both are usually applied in liquid form to the form sheathing either during manufacture or in the field to serve one or more purposes:

- To protect and prolong the useful life of the form material;
- To prevent color variations and dusting of the concrete surface caused by wood sugar transfer;
- To alter the texture of the contact surfaces, such as preventing transfer of undesirable grain patterns. Multiple coats may be required;
- To facilitate release from concrete during stripping. Despite careful application of a release agent, some of it may be removed accidentally before or during concrete placement and consolidation;
- To aid in obtaining a uniform depth of surface retardation when surface retarder is used; and
- To prevent corrosion on steel-faced forms.

4.8.2 *Types of coatings and sealers*—The selection of a form coating will depend on the form material, concrete surface characteristics required, number of form reuses, and the environment of use. Prior experience is the most valuable standard for evaluation and selection. Pretesting is used to develop guidelines for specifying materials and procedures when a form coating is used. If water tightness is required, sealing of the joints in the form sheathing will be necessary (Section 4.5.1).

4.8.2.1 *Mill-sealed form panels*—HDO plywood has a paper impregnated with phenol-formaldehyde resin bonded to the plywood by high temperature and pressure. The resulting surface hides the timber grain of the plywood and requires only a light application of release agents between uses. The plywood manufacturer's directions for treating the form should be followed for best results. During use, the color of the overlay may turn to a reddish mahogany, which is occasionally transferred to the concrete surface during the first few form uses. This discoloration is called concrete pinking and is more apparent on white concrete. Any alkali-resistant film (cement or lime slurry) between the concrete and the overlaid surface will significantly reduce or eliminate pinking. Very few, if any, form release agents will prevent pinking. Pinking can be removed from concrete with an oxidizing bleach solution.

Proprietary coatings or treatments are available such as glass fiber-reinforced polymer bonded to plywood and epoxy resin formulations that exude oil. Coated forms of the same quality and from the same manufacturer should be used throughout to prevent a difference in concrete color and possible buckling due to different coefficients of thermal expansion.

4.8.2.2 *Field-applied coatings*—Lacquers, shellacs, spar varnish, oil-based paints, and some enamels are not recommended because they degrade in the presence of alkalis in concrete, ultraviolet light, and because of a tendency to chip and peel. Catalyzed low-modulus polymer systems should be of types that cure to a hard surface but retain a degree of flexibility and will resist a pH of 12 to 13. Polyurethane coatings are the most common field-applied coating.

4.9—Form release agents

4.9.1 *General*—Release agents are materials applied to the form sheathing to prevent the bonding of concrete to the sheathing, keep the formwork clean, and assist the successful production of high-quality architectural surfaces.

4.9.2 *Selection*—Release agents help produce the concrete surfaces specified in the design reference sample, contract documents, and mockup. Additionally, the following should be considered:
- Compatibility of the release agent with the form or form liner, admixtures in the concrete mixture and, if used, the form sealer or coating;
- Possible interference with the adhesion of other materials such as sealants, architectural coatings, and curing compounds to the hardened concrete surface;
- Allowable amount of any discoloration or staining and the permissible number and size of bugholes on the concrete surface;
- Effect on stripping time, ease of stripping, and cementitious buildup on the form;
- Effect of seasonal temperature extremes on application procedures when the concrete placing portions of the project overlap more than one season, which may affect both concrete color and bughole blemishes on the surface;
- Effect with accelerated curing procedures (especially steam) on stripping and the appearance of the concrete surface;
- Uniformity of appearance: the same release agent should be used for all the architectural concrete surfaces;
- Local and federal environmental regulations, especially on volatile organic compounds (VOCs);
- Dew point of water-borne materials; and
- Entrapped air migration in the consolidation process.

The safest approach to evaluate several different release agents is under actual use conditions on a test panel, mockup, or nonarchitectural portion of the project concrete. Information should also be obtained from the release agent manufacturer as to the kind of form surface for which the product is intended and the proper method of application to produce the desired surface appearance because the thickness of the application may affect the quality of a finished surface and air voids.

4.9.3 *Types of release agents*—Release agents fall into two main classes: barrier and chemically active. Barrier types are water-insoluble materials that include oils without additives (neat oils), diesel oil, paraffin wax, and silicone oils. The EPA prohibits use of uncut or straight diesel oil as a release agent. Barrier-type release agents are not recommended for architectural effects. They tend toward more stains, bugholes, and difficulty with releasing in both very cold and very hot weather, and they can cause problems with adhesion of coatings and other construction materials to the hardened concrete.

Chemically active release agents are the most common for architectural concrete surfaces. Fatty acids chemically react with the basic materials in concrete and produce soap. Soap is a better lubricant than oil for the removal of entrapped air in fresh concrete.

The formation of the soap film from the ingredients in the cement paste and the chemically active release agent prevents the concrete from bonding to the formwork. Applied at the rate recommended by the manufacturer, the chemical reaction only consumes a very small quantity of the free lime from the fresh concrete. During consolidation, the soap film on the form face is an excellent channel for the migration of the entrapped air out of the fresh concrete.

In a vertical casting, undesirable striping effects are sometimes produced when an immersion vibrator is improperly placed very close to the release agent. It is caused by over application of the release agent. The excess release agent is consumed by the basic materials in the concrete raising the *w/cm* at the points of tangency as the vibration stimulates the reaction. At the secant points, there is not sufficient stimulation of the vibration to change the *w/cm*; consequently, a striping effect is created. This striping effect will not bleach out. For this reason, control of vibrator insertions is critical to the overall appearance. Other unrelated causes of striping effects exist, such as shadows of reinforcement, porous form facings, and overly wet concrete mixtures.

Each brand of release agent exhibits its own fingerprint of final surface color, although vibration and form surface texture also have a pronounced effect. Using the same release agent throughout a project is recommended for achieving uniform color.

The two common categories of chemically active release agents include both buffered reactive (partially reactive) and fully reactive types.

Buffered form release agents tend to produce an improved soap film that not only helps remove entrapped air but may promote better flow of a thin skin of cement paste at the very surface of the form. This may help explain why, in vertical castings, these release agents tend to minimize or eliminate the striped effect from vibrator insertions.

Fully reactive form release agents can provide a good basic soap film that, depending on brand, works well in most cases. Because buffered and fully reactive release agents are similar and proprietary, specifying absolute differences between them is difficult. Generally, the buffered release agents produce a slightly different type of soap film that, with some brands, assists in improving the visual impact.

Properly formulated, both oil-based and water-based form-release agents can meet the Federal Volatile Organic Content regulations of 450 g/L and even the more restrictive value of 250 g/L required in some areas.

4.9.4 *Influence of form materials*—Release agent performance is influenced by the quality of the form face. Nonporous sheathings tend to produce less discoloration caused by moisture absorption.

Nonporous forms and form liners include polymers, such as PVC and ABS, glass fiber-reinforced polymers, high-density overlaid plywood, elastomeric polymers, steel, rubber, and others. Several layers (at least four) of urethane or epoxy coatings on wood and plywood can produce a nonporous coating.

Although aluminum is nonporous, it often reacts chemically with fresh concrete to produce hydrogen gas, resulting in a possible bughole problem.

Nonporous forms and form liners, including many polymers, elastomers, and steel, help produce the best visual impact surfaces. Water-based release agents should form a continuous film and not bead up on new or oily forms. They should also contain a rust inhibitor when used on steel forms. Some PVC elastomer form liners exude aluminum stearate, which acts as a release agent. The amount of aluminum stearate in a PVC elastomer (hot melt) is limited. The projected amount can be obtained from the manufacturer to determine the effective number of uses, if any. If exceeding this number is anticipated, a conventional release agent should be used to produce consistent color, beginning with the first use. Some expanded polystyrene (foam) form liners are soluble in solvent-based release agents. Natural rubber form liners absorb petroleum oils, which may cause softening and expansion. Water/oil emulsion-type release agents that do not affect the foam or rubber are available. Many urethane-elastomer form liners are not adversely affected by applying release agents in thin films. Testing is suggested. Many polyurethane elastomer form liners use mold release agents in processing. This factory mold release should be removed before the first placement, which can be done by scrubbing with a form release agent. Care should be taken to prevent damaging the liner.

4.9.5 *Site storage*—When stored in accordance with the manufacturer's recommendations, release agents should have a reasonably long and stable storage life without being susceptible to damage from extreme temperature changes or from repeated or rough handling. If solids settle out, periodic stirring may be necessary to maintain uniformity. When stored outside, drums should be stored on their sides so moisture will not leak into drum bungholes.

4.9.6 *Application of release agents*—If the treated form surface is protected from precipitation, dust, debris, and prolonged exposure to sunlight, most release agents may be applied up to 4 days before placing the concrete, and some may even be applied up to 2 weeks before. When nonporous (nonabsorbent) form sheathing or liners are used and the form is in a vertical position, certain brands of release agents should be applied the day of concrete placement for best results. A check should be made with manufacturer of each form release agent.

Generally speaking, the thinner the film of release agent applied to the form, the fewer bugholes and stains on the hardened concrete. The performance of some release agents, however, is not affected by film thickness. Testing before use is recommended.

Release agent application should be in accordance with the manufacturer's recommendations on rate of coverage and application method to achieve the desired concrete surface appearance. Best results are obtained with a sprayer having a flat fan-type spray tip. Optimum coverage depends on the type of release agent, form texture, and desired concrete surface appearance. The agents should be applied only to form surfaces thoroughly cleaned before erection. The form surface should dry thoroughly before reinforcement and concrete placement. Some release agents are adversely affected by prolonged exposure to sunlight or precipitation. Figure 4.11 shows a release agent being applied to a form.

Both the form type and release agent should be chosen early to allow sufficient testing.

Fig. 4.11—Release agent applied after form erection.

4.10—Form removal

4.10.1 *General*—Formwork should be removed without damage or shock to the concrete. Prying against the face of any concrete for any reason, including the release of formwork, should be avoided.

4.10.2 *Protection of concrete*—Once formwork is removed, concrete should be protected to prevent damage from any means, including the normal construction operations. Sharp edge lines and corners require special care in form removal as they are vulnerable to chipping at early ages.

Care should be exercised during form removal to prevent sudden drops of concrete temperature or thermal shock (Fig. 4.12). This is especially true when surface retarders have been used on large sections and cool water under pressure is used to expose aggregates.

When concrete is being protected from extremely low temperatures, the rate of cooling should be gradual and should not exceed 40 °F (22 °C) for the 24 h period following the termination of heat application (ACI 306R). Loosening forms slightly, without complete removal, aids in gradual cooling and will minimize the occurrence of map cracking caused by thermal shock.

4.10.3 *Procedures for form removal*—Procedures for formwork removal should follow ACI 347R. Two surfaces of the same age may have different-color hues where adjacent formwork is removed at different times. Uniformity in all operations is required for best visual results.

Early stripping, after the concrete has attained its specified stripping strength, of formwork or form liners is recommended as release agents generally do not continue to break the bond between formwork and hardened concrete after extended periods of time, such as 48 h. Sticking can occur if forms are left in place much longer. This, of course, mandates a nonstaining curing process to be initiated immediately upon form removal. Early stripping-type ties can be removed earlier than 24 h if the tie is stripped in a torsional motion, provided a form release agent was used on the embedded section of the tie.

Fig. 4.12—Map cracking due to thermal shock; sandblasting emphasizes the cracks.

4.10.4 *Protection and care of forms*—Careful cleaning and maintenance of forms is necessary to attain uniform architectural concrete. After multiple uses, complete refurbishing of forms will be required to maintain uniformity of surfaces. Resealing of the form surfaces and application of form release agents should be uniform in quantity and type to ensure a uniform appearance on the final surface. To avoid warping or damage to the face, temporary storage of face forms should be in a clean area, in a near vertical position, and away from traffic. Store plastic-coated forms and plastic form liners away from direct sunlight to prevent deterioration of the form surface.

CHAPTER 5—REINFORCEMENT
5.1—General

Conventional uncoated reinforcement should conform to ASTM A 615 and the requirements specified in the ACI 301 specifications for structural concrete. The use of zinc or epoxy reinforcing bar coatings should be considered for architectural concrete exposed to chloride or other corrosive environments.

Reinforcement should be detailed to provide adequate space between horizontal layers of steel and between adjacent bars to permit satisfactory placement and consolidation of the concrete. It also should be accurately tied and located to prevent displacement and to provide proper cover from the final architectural surface.

Whenever possible, architectural concrete face forms should be erected after installation of reinforcement. Special care should be taken to prevent formwork damage during erection and, when reinforcement is welded after form erection, from weld spatter and smoke.

ACI 318 provisions for bar spacing and cover should apply except as modified in Section 5.2. Where structural design requirements create steel congestion and desired clearances are not possible, mixtures with small coarse

Fig. 5.1—Reinforcing steel shadowing may occur when clear space is not adequate.

Fig. 5.2—Improperly installed reinforcing steel chair.

aggregate may be used; however, every effort should be made to minimize congestion of steel. Where form ties and rustication strips are accurately placed for architectural reasons, slight shifts of the reinforcement may be made for clearance. Reinforcing steel alignment also should be more closely controlled to allow sufficient clearance for later treatments, such as bush hammering, and to avoid staining of the finished surface.

5.2—Clear space
In walls and columns, a 5 in. (125 mm) minimum space between vertical mats of reinforcement is recommended to allow concrete placement and consolidation. At least a 4 in. (100 mm) space should be provided between one form face and the reinforcement in a wall containing a single mat of reinforcement. When practical, the single mat of reinforcement should be located 2 in. (50 mm) from the architectural face so that the concrete may be vibrated between the reinforcement and the back form. If clear space is not adequate, shadowing may occur (Fig. 5.1).

5.2.1 *Clear spacing between bars*—To facilitate placement of concrete and lessen the possibility of rust stains, the minimum clear distance between bars and the minimum cover for beams as permitted by ACI 318 should be increased to the following values:
- The horizontal clear distance between bars should be 2 in. (50 mm), 1.25 times the bar diameter, or 1.75 times the maximum aggregate size, whichever is largest; and
- The horizontal clear distance between bars and the form should be 2 in. (50 mm), 1.25 times the bar size, or 1.5 times the maximum aggregate size, whichever is largest. Where rustication strips are used, ACI 318 minimum cover is in addition to the depth of the strip (Fig. 2.2). If part of the surface is to be removed by further treatment after form removal, additional cover should be provided (Section 9.5.1).

5.3—Reinforcement supports and spacers
Supporting chairs, spacers, side-form spacers, or bolsters in contact with or near exposed surfaces should be all polymer, epoxy-coated, concrete, or stainless steel to ensure absence of surface rust staining, and should be properly installed. (Figure 5.2 shows an improperly installed reinforcing steel chair.) Plastic and epoxy-coated products should not be used on surfaces that are to be abrasive-blasted. The project specifications concerning reinforcing steel should indicate the need to increase the number of chairs to compensate for loads that cannot be tolerated either by the plastic tip or the form materials. Any plastic coating should be investigated for durability when exposed to weather or sunlight.

Concrete blocks are not recommended for spacers between the reinforcement and architectural face due to differences in texture and color.

5.4—Tie wire
Wire for tying reinforcement should preferably be comprised of soft stainless steel to avoid staining exposed surfaces. All tie wires should be bent back away from formed surfaces. Tie wire clippings should be removed from horizontal surfaces to be exposed to view, such as beam soffits, particularly if the concrete is to be sandblasted, mechanically treated, or etched to any degree or is to be exposed to the weather. Tie wire for epoxy-coated reinforcement should be coated with an epoxy or other polymer.

5.5—Zinc-coated (galvanized) steel reinforcement
Zinc-coated reinforcing bars should conform to ASTM A 767. When steel forms are used, the coatings should be passivated to prevent reaction between zinc and alkaline fresh concrete. Consideration should be given to the following methods of passivating the galvanized coating:
- After galvanizing, the reinforcing steel may be stored outdoors, allowing it to get wet, until it oxidizes and is covered with a white oxide coating; and
- A chromate additive may be introduced into the concrete mixture. (This may increase the potential for dermatitis to humans.)

5.6—Epoxy-coated reinforcing bars
Epoxy-coated reinforcement should conform with ASTM A 775 or A 934.

CHAPTER 6—CONCRETE MATERIALS AND MIXTURE PROPORTIONING

6.1—General

The properties of materials used in architectural concrete should not be determined only by strength and proportioning requirements, but also by color or texture, workability, durability, and shrinkage as outlined in this chapter.

6.2—Materials

6.2.1 *Portland cements and special cements*—The portland cements or special cements used for architectural concrete should meet the requirements of the specified type in ASTM C 150, ASTM C 595, ASTM C 845, or ASTM C 1157.

Cements have different color characteristics that affect the desired concrete color. To minimize color variations, the same type and brand of cement from the same mill and the same raw materials should be used for all of the concrete on a given structure. These precautions alone, however, will not automatically ensure color uniformity. Variables in concrete manufacture and handling and delivery equipment may also have marked effects on color. Ordinary gray-colored cement may produce color variations (even if from one manufacturer).

White cement is often used in architectural concrete, both precast and cast-in-place. It is readily available and may be considered as a standard concrete material. White cement is manufactured to conform to the specification of ASTM C 150 white cements and are made of selected raw materials containing negligible amounts of iron and manganese oxides. Variation in uniformity of shade for each brand of white cement is small, but assume there will be differences between brands or mill sources. White cement used with mineral pigments will provide good color intensity and uniformity.

Buff, tan, or light brown cements also can provide uniform color. There is no specification for these cements. They should meet ASTM C 150 or ASTM C 595.

Shrinkage-compensating cements meeting ASTM C 845 are available from some producers. Detailed information on these cements is available in ACI 223. The availability of special cements for the completion of the project should be established before completion of the job specifications and design for the architectural concrete. Cement and concrete samples provided to architects should be marked to show the type, brand, and source of cement. The reinforcing steel provided in concrete with shrinkage-compensating cements should comply with ACI 223, especially in tall, lightly reinforced columns.

6.2.2 *Aggregates*—The gradation of aggregates is a major factor influencing the appearance of architectural concrete. Gap-graded mixtures for specific finishes should be carefully selected and tested to verify constructibility. Well-graded combined mixtures can usually be produced by a combination of coarse and fine aggregates complying with ASTM C 33. Graded aggregates should be proportioned in equal quantities, by mass, according to sieve size to ensure a good *w/cm* and dimensional stability. If flowability is critical, a sample batch of concrete should be cast and evaluated for workability.

Normal-density or light-density aggregates meeting ASTM C 33 or ASTM C 330 may be used in exposed aggregate surfaces to provide countless combinations of color and texture. The maximum allowable percentages of deleterious substances and impurities allowed by these specifications, such as chert, iron, soft particles, and clay lumps, should be significantly lowered or eliminated in architectural surfaces exposed to the weather. Acceptable aggregates include natural gravel, crushed gravel, and crushed stone aggregates of many different colors. Artificial aggregates include expanded shales, clays, slates, blast-furnace slags, nonreactive glass, and ceramic materials. All facing and concrete aggregates for a given structure should come from the same source to provide quality and color similar to the approved sample. Any combination may be used for contrast, provided specified levels of strength, durability, and workability are met. Materials, colors, gradations, size of aggregate, and depths of reveals compatible with aggregate size may be varied for architectural effect. All facing aggregates should have proven service records or satisfactory results from laboratory testing.

The choice of aggregates is more critical with concrete containing white cements. Dark aggregates tend to create shadows where thinner sections of white mortar cannot completely mask the aggregate. Intensity of color may be diminished when dirty aggregate is used from contaminated stockpiles. Fine aggregates have a major effect on the color, especially for light-colored concrete and exposed aggregate surfaces, and can be used to vary the particular color desired.

Special requirements such as gap grading (Reference 9) or a single size of aggregate may be established for exposed aggregate finishes to provide an optimum surface exposure (Fig. 6.1). Maximum aggregate sizes vary from 1/8 to 1-1/2 in. (3 to 40 mm), depending on the desired architectural effect. Gap gradings include coarse aggregate in a limited size range, such as 1-1/2 to 3/4 in., 1 to 1/2 in., and 3/4 to 3/8 in. (40 to 20 mm, 25 to 10 mm, and 20 to 9 mm) or others. Settlement or subsidence can take place in gap-graded mixtures unless there is rigid observance of low slump. Refer to Reference 10. Care should be taken in selecting aggregates with a high degree of contrast between the coarse aggregate and the mortar if the aggregate is to be exposed. High contrast will make variations in finishing and aggregate density much more apparent than low-contrast combinations.

6.2.3 *Admixtures*—Chemical and mineral admixtures can have adverse effects on color when used with certain cements, especially when using white or light-colored cements. Visual tests are needed to determine the effect of admixtures on color. Color compatibility may be established by casting job-site panels before actual use in construction, or by using nonexposed parts of the structure as test areas. Detailed information on admixtures is given in ACI 212.3R.

6.2.3.1 *Air-entraining agents*—Air entrainment, as presented in Table 4.2.2.4 of ACI 301-99,[11] is generally recommended for architectural concrete when the concrete can readily be water-saturated in severe weather zones. Entrained air tends to make concrete sticky so that entrapped air bubbles at the form face become difficult to bring to the surface by vibration. The lower limit of entrained air found in Table 4.2.2.4 (ACI 301-99) is recommended when durability

Fig. 6.1—A gap-graded, cast-in-place concrete mixture after heavy sandblasting. Note the preponderance of one-size material.

is mandated. In mild climates, small percentages of entrained air may be required for workability in special harsh mixtures.

6.2.3.2 *Accelerating admixtures*—Where accelerators are necessary, the use of nonchloride accelerators is recommended. Other newly developed accelerators may be used if they do not adversely affect color and placement properties. Accelerators often interfere with air-entraining agents, so dosages of air entrainment are increased to accommodate this, resulting in extremely sticky mixtures that produce visually unacceptable results. Additional attention to curing is required when accelerators are used.

6.2.3.3 *Water-reducing and set-retarding admixtures*—These admixtures are normally used in architectural concrete to reduce the amount of mixing water or to increase the workability of the concrete, particularly with harsh concrete mixtures.

High-range and medium-range water-reducing admixtures can be used to reduce the water content and increase the workability without adversely affecting the setting time. The admixture should be checked for adverse side effects on color, concrete properties, and formwork deflections. Changes in water content will affect the color.

Retarding admixtures may be used in architectural concrete to delay the initial set of the concrete, minimizing cold joints. High dosages may cause setting, cracking, or discoloration problems, particularly with white or buff cements. When the setting of concrete is retarded, the effect of additional form pressure may need to be considered in the design of formwork.

6.2.3.4 *Pozzolans and slag*—Mineral admixtures or pozzolans meeting ASTM C 618 can be used to increase durability and workability. Before their use, trial batches should be prepared to determine any detrimental changes in the architectural appearance. The use of fly ash, silica fume, slag, or metakaolin in a concrete mixture may darken or lighten the color or produce erratic color variation. They also can slow the set time of the concrete and the rate of concrete placement.

6.2.4 *Pigments and color admixtures*—To augment color tone of architectural concrete, pigments or color admixtures meeting ASTM C 979 may be used. Color admixtures commonly used for this purpose are finely ground natural or synthetic mineral oxides. Synthetic oxides are usually more uniform in color for longer periods of time. All color additives, however, may react chemically with other products used on the surface, such as surface retarders or muriatic acid, and should be tested before use. Various iron oxides produce shades of yellow, buff, tan, brown, maroon, red, and black. Chromium oxide produces shades of green, and cobalt oxide is used to obtain shades of blue. The color shade depends on the amount of these materials used. The quantity of pigment is expressed as a percentage of the cement content by weight. Amounts of pigment in excess of 5% seldom produce greater color intensity whereas amounts more than 10% may be harmful to concrete quality. Pigments will produce more intense colors when used with light cements.

Organic phthalocyanine dyes have been used successfully to produce light to dark shades of blue and green in concrete. The dyes are used in quantities of less than 1% by weight of cement and can be dispersed in the mixture water, eliminating the need for preblending. Ultraviolet light and ozone affect color stability.

For any coloring agent, it is important to have tests or performance records that indicate color stability in concrete. Carbon black is difficult to handle and may cause various shades of black due to leaching of the surface. Common lampblack has a detrimental effect on entrained air, though some air-entraining agents can be dosed to overcome this effect. Trial batches should be cast to determine if air content is affected. Concrete colored by pigments may show varying degrees of weathering. The effect on air content and water requirements should be determined.

A wet pat method can be used to check color uniformity. This test provides a quick method of approximating color uniformity of the concrete being discharged from the ready-mixed concrete truck. References 12 and 13 contain information on designing colored architectural concrete.

6.2.5 *Water*—Potable water and water conforming to ASTM C 94 are acceptable for use in architectural concrete. Water containing iron or rust may cause staining in light or white concrete.

6.3—Proportioning, mixing, and temperature control

Mixture proportions for architectural concrete should provide a mixture of proper workability and strength.

The slump at the point of discharge into the form should be consistent with the particular type of concrete and the methods of placing and consolidation. In general, the water content for any placement should be constant from batch to batch to provide uniformity of color in the end product, even during significant seasonal temperature fluctuations. There is a tendency for a lighter color and an increase in bugholes to occur in the concrete near the top of placement lifts due to decreased form pressures, inadequate vibration, and an increase in the *w/cm* at these locations. Attention should be given to properly consolidate the concrete in the upper layers of placement lifts to improve appearance. Gradually using less water to provide dryer concrete for lifts toward the top could effect a more uniform color; however, this procedure is impractical for most projects. If used, experimentation on nonarchitectural walls with the architectural concrete mixture would determine actual proportions and procedures to be used.

6.3.1 *Gap grading*—Gradation for gap-graded mixtures varies widely. The use of a one-sieve size or a narrow size range of coarse aggregate, with a small percentage of concrete or masonry sand for workability, will help to produce a uniform distribution of exposed aggregate.

Masonry sand is normally used when a high concentration of coarse aggregate is desired for uniform color and texture at the face. Gap-graded mixtures should be placed at as low of a slump as possible to minimize segregation.

6.3.2 *Temperature*—The concrete temperature should be kept consistent; concrete temperatures between 65 and 85 °F (18 and 29 °C) will normally produce concrete uniform in color. Concrete temperatures higher than 80 °F (27 °C) may result in a faster setting rate, visible flow lines, and cold joints if proper scheduling of the concrete placement is not closely coordinated with the concrete producer.

Holding the temperature constant is especially difficult when the project extends over more than one season. If color uniformity is critical, testing for variations of color at different temperatures should be considered.

CHAPTER 7—PLACING AND CONSOLIDATION
7.1—Conveying and placing

7.1.1 *General*—Methods of conveying and placing architectural concrete should conform with the requirements of ACI 304R, except as modified herein. A description of the methods and sequence of placement to be used for the structure should be submitted in writing to the architect/engineer and inspector for review. These should be the same as used in fabricating the approved preconstruction field mockup.

Concrete trucks should be scheduled and dispatched to arrive at the job site just before the concrete is needed. This will avoid excessive mixing while waiting or during delays in placement, which may cause nonuniform color and possibly cause cold joints.

7.1.2 *Conveying*—Care should be taken with any conveying equipment to prevent contamination of architectural concrete by other mixtures. If methods of conveyance are varied during placement of architectural concrete, the uniformity of color may be affected.

7.1.3 *Depositing in the form*—With proper proportioning, and depending on the width of the forms and the amount of reinforcement, lifts can be up to 36 in. (900 mm) deep. Deeper lifts, accompanied by additional careful vibration, can be used with high-density forming to eliminate excess bugholes. The surface of each layer should be fairly level so that the vibrator does not move the concrete laterally, as this might cause segregation. This can be accomplished by depositing the concrete on the leading edge of the prior placement. Placing the concrete at a new location and then transporting it back to the leading edge of the prior placement by using a vibrator should be avoided.

Spattering the form face with high-cement-content mortar that stiffens before being covered by the concrete may cause mortar streaks in exposed aggregate finishes. Metal or polyethylene sheets placed against the form face and raised with the height of concrete will protect the form face against spatter. Also, insertion of the placing pump hose into the concrete while discharging aids in reducing spatter. The use of elephant trunks (tremies) used to contain spatter is another way to eliminate the problems of dried spatter.

7.2—Consolidation

7.2.1 *General*—Good consolidation to eliminate entrapped air throughout the mass and to minimize surface voids is particularly important to architectural concrete. Today most concrete is consolidated by immersion vibration, which is especially required by stiffer consistencies (Fig. 7.1).

An explanation of consolidation, and detailed recommendations on the selection of vibrators and vibration procedures, is given in ACI 309R. Elimination of surface voids is discussed further in ACI 309.2R.

7.2.2 *Internal (immersion) vibration*—Internal vibration is recommended for all vertical, cast-in-place, architectural concrete. The vibrator should be inserted vertically at 1.5 times the radius of influence (ACI 309R), uniformly spaced over the entire area. The distance between insertions should depend on the properties of the mixture, maximum aggregate size, slump, and the amount of power getting to the vibrator. The area visibly affected by the vibrator should overlap the adjacent just-vibrated area. There should be a row of insertions within 6 in. (150 mm) of the form.

The vibrators should operate at frequencies above 9000 vpm. The diameter of the vibrator head should be at least 2 in. (50 mm), except when reinforcing steel congestion or clear space limits vibrator head size. Where space permits, a high-cycle, motor-in-head vibrator should be used. These vibrators require a special generator that produces 180 Hz electrical power (ACI 309R).

The vibrator should penetrate rapidly to the bottom of the layer and at least 6 in. (150 mm) into the preceding layer. The vibrator should be manipulated in an up-and-down motion, generally for 5 to 15 s, to knit the two layers together. The vibrator should then be withdrawn gradually with a series of up-and-down motions. The down motion should be a rapid drop to apply a force to the concrete which, in turn, increases internal pressure in the freshly placed mixture.

Rapidly extract the vibrator from the concrete when the head becomes only partially immersed in the concrete. The concrete should move back into the space vacated by the vibrator. For dry mixtures where the hole does not close during the withdrawal, sometimes reinserting the vibrator a few inches (several centimeters) away will solve the problem; if this is not effective, the mixture or vibrator should be changed.

Where air voids in formed surfaces are excessive, the distance between vibrator insertions should be reduced 20 to 30% from the normal 1.5 times the radius of influence and the time of each withdrawal increased. Sometimes it is practical to insert a small vibrator between the reinforcement

Fig. 7.1—Proper vibration of architectural concrete is vital to a successful and consistent product.

Fig. 7.2—External vibration.

and the form. In general, this practice is not recommended. In some cases, the vibrator should be rubber-tipped; even so, any contact with the form should be avoided if at all possible because this might mar and disfigure the form sheathing. Insertions closer than 3 in. (75 mm) to the architectural formed surface may result in a darker color of the surface opposite these locations. Vibrating too close to the formed surface also effects the aggregate quantities for exposed aggregate concrete. Walls less than 6 in. (150 mm) thick will require special vibration considerations.

Sometimes, very harsh mixtures, such as those with gap grading, are used to produce special architectural effects. They generally require more powerful vibrators and longer vibration times. The vibration should be terminated when the mortar level reaches the top of the aggregate to prevent mortar lines between layers.

If rigid, nonabsorptive forms, such as steel, are to be used, vibration should be minimized to the least amount necessary to consolidate the concrete and remove objectionable air voids. Excessive vibration near a rigid, nonabsorptive surface can cause a color variation at the concrete-form interface.

7.2.3 *Form vibration*—Form vibration is recommended only in areas inaccessible to internal vibration, provided the formwork was specifically designed for external vibration and it is anchored to the footing to prevent movement (Fig. 7.2). Forms for external vibration should stand up under the repeated reversing stresses induced by vibrators attached to the forms. Furthermore, they should be capable of transmitting the vibration more or less uniformly over a considerable area. The form should have adequate skin thickness and suitable stiffeners. The vibrators should be rigidly attached to the form. Special attention should be given to water tightness to prevent leakage. Form vibration produces very dark concrete as it locks in surface moisture when a high-density form face sheet is used. When allowed to dry over a period of time, the surface should lighten. Trials should be made with form vibrators before large-scale use. These trials should simulate the forming conditions to be encountered on the structure.

7.2.4 *Revibration*—After bleeding is substantially complete but before initial set, revibration can sometimes be used to further densify the concrete and reduce air and water pockets against the form. Revibration is of particular benefit for the top few feet of a placement where air and water pockets are most prevalent. Revibration should not be used where harsh, gap-graded mixtures are used to produce exposed aggregate surfaces due to a possible danger of dispersing the aggregate at the surface.

Revibration more than a few feet below the top of the placement may damage well-consolidated concrete.[14]

7.2.5 *Spading*—Spading may be used in conjunction with internal vibration to improve formed surfaces. A flat, spade-like tool that will not damage the form, such as a long, flat ruler or sail batten, is repeatedly inserted and withdrawn adjacent to the form. This forces the coarse particles away from the form and assists the air bubbles in their upward movement toward the top surface. This method is only used when the amount of air bubbles at the form face becomes unacceptable.

CHAPTER 8—CURING
8.1—General
To produce a uniform color, the method and period of curing should provide a consistent concrete temperature, regardless of the ambient temperature. Proposed methods should be tried on the site-cast mockup to determine any possible adverse effects. Standard curing procedures are described in ACI 308.1.

8.2—Curing in forms
All concrete in beam soffits and other supported formwork soffits can be cured by leaving the formwork and shoring in place. All vertical formwork should be removed in a consistent time interval or when the concrete has reached a 1000 psi (6.9 MPa) compressive strength, with curing to follow immediately upon form removal. To prevent staining caused by the type of form material, the forms should be sealed with a nonvapor-transmitting coating, following the manufacturer's instructions.

In-the-form curing increases the possibilities of color variations. Whether these are from absorption of water by the form, wood staining, new versus old forms, form leakage, or temperature variations, proposals to minimize color variations need to be addressed.

Color variations due to rapid surface drying and extreme thermal changes should be minimized by following the recommendations of ACI 305R and ACI 306R and in Sections 4.8 and 4.10.2.

8.3—Moist curing
Extreme care should be taken to ensure that the material used to cover the concrete surface does not cause texture or color variations. Water curing should only be used when the concrete has a *w/cm* below 0.4. When used, the temperature of the water should not vary from the surface temperature of the concrete by more than ±20 °F (±11 °C). Plastic sheets may be useful for covering complex shapes, but texture and color differences will occur where the plastic sheet contacts the concrete.[15] Curing water should be nonstaining.

8.4—Membrane curing
Liquid-membrane curing compounds should be applied to a moist surface. Otherwise they may cause discoloration or staining and prevent bonding of subsequent repairs, architectural coatings, and sealants. Manufacturers should be consulted concerning characteristics of their products and warranties. Curing practices and materials should be thoroughly evaluated on the preconstruction field mockup sample.

8.5—Hot-weather curing
Freshly placed architectural concrete can be adversely affected by high temperature, low humidity, and high winds. To prevent variations in color due to nonuniform drying and to prevent plastic shrinkage cracking, curing should commence as soon as practical, perhaps even before completion of concrete placement (see also ACI 305R). In especially hot and or windy (desert-type) environments, special curing compounds may be required.

CHAPTER 9—TREATED ARCHITECTURAL SURFACES
Architectural surfaces, including horizontal surfaces, can be treated after casting and form removal to expose fine or coarse aggregate in the finished product by brushing and washing at an early age, surface retardation, high-pressure water jet, acid wash, abrasive blasting, bush-hammering, or other mechanical tooling. These methods can impose additional requirements for aggregates in shape, size, texture, or color. As more of the aggregate is exposed, the effect of the cement color diminishes. The total area and the expected distance to the viewer will usually determine the size of the aggregate. Because treated surfaces are more susceptible to atmospheric pollution and weathering, consideration should be given to aggregate shapes that may change due to weathering, which alters the visual effect. Round aggregates have less tendency to collect airborne dust on the matrix portion than rough aggregates. In areas subject to air pollution, a matrix darker than the exposed aggregate may be preferable.

Hand broadcasting of architectural aggregate slab surfaces allows economical use of the costly aggregate and helps ensure uniform coverage. For slabs, brushing and washing with water at an early age is commonly used to expose the aggregate.

9.1—Surface retarders
Personnel using surface retarders should become familiar with their characteristics before use. Surface retarders are used to delay the set of the surface cement paste so that the aggregate can be exposed easily. The use of accelerators or heating during cold-weather concreting may shorten the delay of set at the surface. Prolonged exposure of the forms coated with retarder before placing concrete may also affect the action of the retarder. A sample panel should be made to determine any adverse effects from the form or concrete materials. Further experimentation to determine the effect of heights of placement, form-stripping times, and method of exposure should be done in areas of minor importance, such as basement walls. The recommended minimum concrete strength before removal of the retarded surface is from 1000 to 1500 psi (6.9 to 10.3 MPa). Ensuring uniform results on vertical surfaces requires preplanning and more supervision than for structural concrete. Due to the numerous factors affecting the action of surface retarders applied to form faces for vertical casting, their use should be carefully evaluated for each project.

9.2—High-pressure water jet
High-pressure water jets are used in combination with air to expose aggregate. Proper time of application should be determined for each concrete and its curing conditions to obtain the desired amount of reveal without loosening the aggregate. The minimum compressive strength of the concrete for high-pressure water jetting should be 1500 psi (10.3 MPa). This method can be used with or without surface retarders, and requires an operator pretrained on a test area. Aggregate exposure should be started immediately after forms are stripped when retarders are not used. Reference 16 describes typical equipment.

9.3—Acid wash
Washing with solutions containing acid (of varying content) can be used to etch the surface of the concrete, giving the surface a light matte finish or to bring out the full color of an exposed aggregate surface. When exposing aggregate, the aggregate should be one such as quartz or granite that is acid resistant. Limestones, dolomites, and marbles will discolor or be dissolved by muriatic acid due to their high calcium content. Such treatment should be initiated at consistent concrete strengths. The depth of paste removal and its effect on appearance and color will vary with concrete age. Treating cast-in-place surfaces at a consistent age is difficult to achieve in the field because other factors, such as weather, weekends, and delays, will determine when the work will be performed. At higher concrete strengths, age differences will have less effect on appearance and color. Therefore, it is recommended that acid wash treatments be initiated when the concrete is not less than 14 days old and

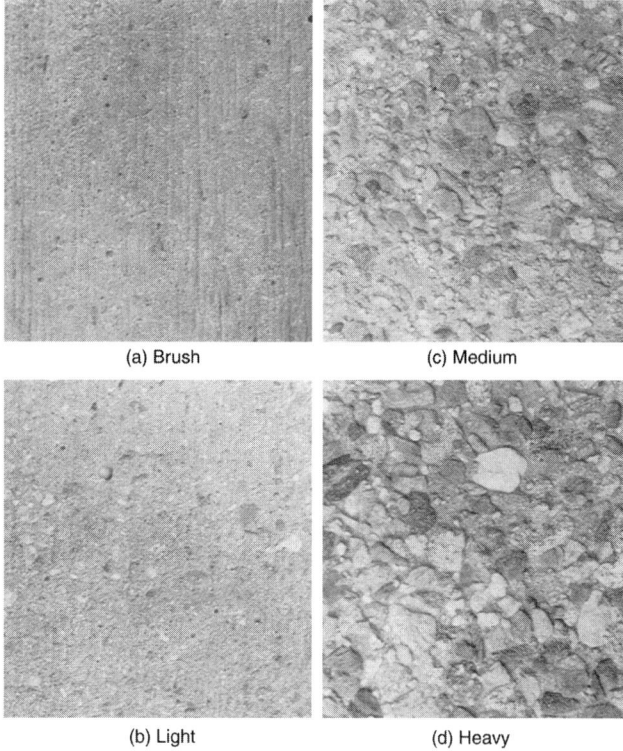

Fig. 9.1—Degrees of abrasive-blast or surface-retarded finish.

after it reaches a compressive strength of at least 3000 psi (20.1 MPa).

Acid washing is not recommended for vertical cast-in-place concrete due to the hazards of such application. All personnel should have protective clothing and covering to prevent injury from spattering. Uniform application of the acid solution is difficult due to runoff. Complete neutralization of the runoff is also difficult, which may create problems at the ground level.

9.4—Abrasive blasting

Abrasive blasting, or sandblasting, is used to dull the surface glaze, make the color uniform, or expose the aggregate of architectural concrete. The depth of treatment is usually defined by the designed reference sample in consultation with the architect, contractor, and inspector, and confirmed by the preconstruction mockup panel. Uniformity is difficult to obtain where concrete mixtures of different strengths are located side by side. Form-joint tightness is particularly important when the surface is to receive an abrasion treatment. Degrees of sandblasted or surface retarded finish (Fig. 9.1) may be described and defined as:

- *Brush*—Sufficient to remove the surface sheen, but may not make the color uniform; this finish will not expose the coarse aggregate from the matrix after application;
- *Light*—Sufficient to expose fine aggregate with occasional exposure of coarse aggregate and to make the color uniform; maximum aggregate exposure 1/16 in. (1.5 mm);
- *Medium*—Sufficient to generally expose coarse aggregate with slight reveal; maximum aggregate exposure 1/4 in. (6 mm); and
- *Heavy*—Sufficient to generally expose and reveal the coarse aggregate to a maximum projection of 1/3 of the diameter; aggregate exposure 1/4 to 1/2 in. (6 to 12 mm); the surface is rugged and uneven.

The lighter the abrasive-blast, the more critical the skill of the operator, because defects from forming and placing, such as bugholes, leakage lines, and lift lines, tend to be accentuated by such treatment. Additional thickness of concrete should be provided to maintain proper cover for reinforcement.

The time for abrasive-blasting is determined by scheduling, economics, visual appearance desired, and hardness of the aggregate. Softer aggregates tend to abrade more when concrete strengths are high. Surface retarders (Section 9.1) can be used in conjunction with medium and heavy texture to reduce blasting time and lessen the abrasion on softer aggregate. If a more pronounced reveal is desired, abrasive-blasting is usually done during the first 24 to 72 h, and after concrete strengths have reached a minimum of 2000 psi (13.8 MPa). Once the time is determined through testing on the preconstruction mockup, all subsequent blasting should be done at approximately the same concrete age for uniformity of appearance.

Complicating the timing of abrasive-blasting are changes in ambient temperatures that can have a large effect on early concrete strengths and, consequently, the visual appearance.

Materials used for abrasive-blasting include silica sand, aluminum carbide, black-slag particles, and walnut shells. The type and grading of the abrasive determines the surface treatment and should remain the same throughout the entire project. Because some aggregates change color after exposure by sandblasting, trials of different abrasive materials with sample panels are desirable to define the textured design reference sample. White-cement concretes require abrasive materials that are stainproof. When wet abrasive-blasting is required by air pollution standards, abraded mortar should be continually washed from previously abrasive-blasted areas to prevent staining. Course abrasives tend to cause more of a matte texture, which minimizes variations due to concrete hardness.

A variety of blasting equipment and techniques are available and are detailed in Reference 17.

9.5—Tooling or other mechanical treatments

Tooling and other mechanical treatments for the exposure of aggregates or other surface modification may be done by any of several processes, including chipping and spalling of the surface by a method called bush-hammering, grinding to produce a smooth exposed aggregate surface, or breaking off the projections of fluted surfaces to produce alternate rough and smooth areas. Orientation of equipment for tooling, blasting, or fracturing should be kept uniform throughout the architectural work. To maintain uniformity, the same individual should perform this work throughout the project or at least on adjacent portions of a structure.[18]

9.5.1 *Bush-hammering*—Bush-hammered surfaces are produced by pneumatic tools fitted with a bush-hammer, comb, chisel, or multiple pointed attachments. The type of tool will be determined by the surface effect desired. Because most bush-hammering will remove approximately 3/16 in. (5 mm) of material, additional concrete cover should be provided. To prevent loosening of the aggregate, a concrete strength of 4500 psi (31.3 MPa) in compression and a minimum age of 14 days is required. In many cases, better uniformity is found when the concrete is allowed to age for 21 days and the surface to dry. Bush-hammering at corners tends to cause jagged edges. If sharp corners are desired, bush-hammering is held back from the corner a distance of 1 to 2 in. (25 to 50 mm). Aggregates should be carefully selected and tested in the mockup panel to avoid internal cracks from occurring in the concrete as a result of the bush-hammering operation.

9.5.2 *Grinding*—Grinding of concrete surfaces is more laborious than the other treatments, especially on vertical and overhead surfaces. Final costs are determined by the hardness of the aggregate and the desired exposure. To define the final product, treatment is accomplished on the preconstruction mockup or a trial area of minor importance. Small samples may not be capable of field reproduction. This type of treatment produces a result similar to terrazzo work.

9.5.3 *Manual treatment*—Vertical surfaces of structures may be formed to produce projections of concrete of triangular or rectangular shape. These may be broken off either by removing the form or by hand at a concrete age sufficient to also fracture the aggregate. Prior testing should be done on the preconstruction mockup to determine the optimum procedures.

CHAPTER 10—FINISHING AND FINAL CLEANUP
10.1—General
Even with good workmanship and positive effort to produce excellent architectural concrete, an occasional blemish will need to be repaired and variations in color and texture can occur. Limits to their acceptance should be determined during review of the field mockup sample.

10.2—Tie holes
Tie holes should be plugged or the exposed metal portion of the tie sealed with sealant to prevent corrosion and possible staining of the surface, except where stainless steel form ties or fiberglass ties are used. The holes left in the surface of the concrete as the result of the form tie may be small or large, depending on the type used. In a rough textured surface, small holes can be plugged flush with the surface and concealed. With smooth-surface concrete, the tie holes will be more apparent, and it is better to only partially fill the holes, leaving the holes as a part of the planned appearance. The color of the tie-hole patches in the mockup panel should be approved by the designer.

Care should be exercised to avoid smearing the fill material on the surface of the concrete. Materials used for plugging tie holes include portland-cement mortar, epoxy mortar, plastic plugs, precast mortar plugs, and lead plugs. They should be carefully selected from among those that have shown no staining or discoloration tendencies in use. Some epoxy mortars change to brilliant orange or yellow after exposure to sunlight. Mortar materials of a dry pack consistency and densely tamped into the hole will be less likely to smear on to the surface than those of a wet consistency. When portland-cement mortar is used, the tie hole should first be prewetted with clean water, and a neat cement slurry bond coat should be applied to the hole surfaces before filling with mortar. If epoxy mortar is used, it should be applied in accordance with the manufacturer's instructions, and a caulking gun used to inject it into the tie hole to prevent smearing it on the surface. Cleaning is difficult and will usually leave a stain on the surface. Plastic inserts are provided by cone tie manufacturers and should be wedged into the tie hole, leaving a standard predetermined recess. Alternatively, lead plugs can be wedged into the hole by hammering. Sometimes the removable cone becomes embedded in the concrete due to form movement or leakage around the cone. It can be removed to produce a neat appearance by drilling out the cone with a diamond bit tool conforming to the hole size produced by the cone. It may be economical to remove all cones in this manner to ensure neat, uniform holes.

10.3—Blemish repair
Blemishes that are beyond the limits of variations as established by the quality of the preconstruction mockup should be repaired. Generally, the repair work should proceed as soon as possible after form removal and surface finishing using the materials and methods already accepted on the approved mockup. Then the repair and the surrounding concrete will age together, and the chance of color variation will be minimized. The importance of establishing a repair method before the need arises cannot be over-stressed. Once proven acceptable on the mockup, repairs can be made without delay and with confidence of the final outcome. Where adjacent abrasive-blasting or bush-hammering treatments have to be matched, prior experimentation should be performed on unimportant areas. Ingenuity may sometimes be used to establish methods and techniques that are as satisfactory as those in standard use. Light abrasive-blasting sometimes greatly emphasizes cracks as well as blemishes or bugholes caused by forming or placing. When their appearance is unacceptable, it may be desirable to decide to accept a heavily blasted surface because additional blasting may diminish the effect of the cracks and other defects after they have been repaired. Filling the cracks with epoxy before blasting has prevented rounding of the crack edges during blasting. After blasting, the resulting epoxy fin is broken off manually at the face of the concrete, leaving a fine line that is less noticeable to the eye at normal viewing distances.

Where the surface is abrasive-blasted to expose substantial aggregate, a needle gun treatment can successfully diminish the contrast of dark leakage lines at form joints and tie cone holes and can remove ragged lift lines. These lines are stronger than the adjacent concrete matrix and cannot be abrasive-blasted away without greatly eroding the surrounding softer mortar. This treatment is accomplished by lightly applying a needle gun (Fig. 10.1) containing 16 to 20 chisel-pointed rods to lighten the dark, hard, and contrasting cement-rich lines. The chisel tips should be maintained to

Fig. 10.1—Needle gun.

prevent a bush-hammered appearance. To prevent shadows from oblique sun rays, the removal should not be deeper than the adjacent mortar surface.

10.4—Stain removal

Rust is the most common stain on architectural concrete surfaces. It is usually caused by water washing the rust from reinforcing bars (extending out of an element for connection to the next concrete element), by ferrous materials (nails, formwork hardware, or other reinforcing steel accessories) carelessly left on top of a surface, or both. Loose ferrous materials should be picked up. Reinforcing steel exposed and likely to cause rust stains should be coated with a neat slurry of portland cement and water to temporarily protect it from rusting. Many proprietary coating products are available for this purpose also.

Stains from various causes may be removed by commercial stain removers, but some alteration of the concrete surface may occur. Reference 19 suggests methods for stain removal.

Objectionable efflorescence and surface deposits can be removed by commercially available efflorescence removers or by use of low weak acids (5 to 10% solutions). The concrete may be sprayed with water first to minimize etching. Following application of the acid, thorough flushing with water is required to prevent formation of scum. Further treatment with detergents or a light abrasive-blast may be useful if the acid is not effective.

10.5—Sealers and coatings

Sealers are defined as vapor-retarding penetrants that are absorbed when applied. Some sealers change the profile of the concrete surface and others do not. Coatings are nonvapor-transmitting, penetrate slightly, and leave a visible film on the surface. The film can be clear or pigmented.

Sealers and coatings are not recommended unless needed for protection from atmospheric or other contaminants. These may be used for the following purposes:
- To reduce attack of the concrete surface by industrial airborne chemicals;
- To inhibit soiling of the surface; however, some sealers have an affinity for airborne contaminants;
- To facilitate cleaning of the surface or to resist impregnation of graffiti;
- To avoid darkening of the surface when wetted; and
- To reduce the effects of carbonation.

Commercial sealers and coatings vary in chemical composition and in effectiveness as reported in Reference 4. The methyl methacrylate and other acrylic sealers, silane or siloxane, and polyurethane coatings offer the best protection for architectural concrete surfaces.

Some sealers or coatings based on polyurethanes, epoxies, polyesters, and their combinations have a glossy appearance and some may tend to yellow or darken the surface of the concrete. The discoloration often takes months to manifest itself. Where waterborne dirt stains vertical surfaces, a clear sealer or coating will prevent the dirt from penetrating the surface, and make cleaning easier or unnecessary. Sealers or coatings can also protect concrete surfaces that would otherwise become stained by the initial runoff of rust from intentionally exposed special steel, which forms its own protective, oxidized, rust-colored coating. Some sealers or coatings have an affinity for airborne contaminants, including hydrocarbon contaminants, as well as incompatibility for caulking, sealants, and paints. Some silicone sealers or coatings have an affinity for hydrocarbon contaminants. Antigraffiti products are either permanent or sacrificial and may alter the color of the concrete in areas on which they are applied. The market for these products is currently changing as a result of changes in the federal regulations governing their production. These new regulations have made many of the present products obsolete. They should be carefully considered before use.

Joints should be caulked before application of sealers or coatings so as not to affect the bond of the caulking compound. The caulking compound should not smear the exposed face and prevent adhesion of the sealers or coatings.

CHAPTER 11—REFERENCES
11.1—Referenced standards and reports

The standards and reports listed below were the latest editions at the time this document was prepared. Because these documents are revised frequently, the reader is advised to contact the proper sponsoring group if it is desired to refer to the latest version.

American Concrete Institute

117	Standard Specifications for Tolerances for Concrete Construction and Materials
212.3R	Chemical Admixtures for Concrete
223	Standard Practice for the Use of Shrinkage-Compensating Concrete
224.3R	Joints in Concrete Construction
301	Specifications for Structural Concrete
304R	Guide for Measuring, Mixing, Transporting, and Placing Concrete
305R	Hot Weather Concreting
306R	Cold Weather Concreting
308.1	Standard Specification for Curing Concrete
309R	Guide for Consolidation of Concrete
309.2R	Identification and Control of Visible Effects of Consolidation on Formed Concrete Surfaces

318	Building Code Requirements for Structural Concrete
347R	Guide to Formwork for Concrete
504R	Guide for Sealing Joints in Concrete Structures
515.1R	Guide to the Use of Waterproofing, Dampproofing, Protective, and Decorative Barrier Systems for Concrete

ASTM International

A 615	Standard Specification for Deformed and Plain Billet-Steel Bars for Concrete Reinforcement
A 767	Specification for Zinc-Coated (Galvanized) Steel Bars for Concrete Reinforcement
A 775	Specification for Epoxy-Coated Reinforcing Steel Bars
A 934	Standard Specification for Epoxy-Coated Prefabricated Steel Reinforcing Bars
C 33	Standard Specification for Concrete Aggregates
C 94	Standard Specification for Ready-Mixed Concrete
C 150	Specification for Portland Cement
C 157	Standard Test Method for Length Change of Hardened Hydraulic-Cement, Mortar, and Concrete
C 330	Standard Specification for Lightweight Aggregates for Structural Concrete
C 595	Standard Specification for Blended Hydraulic Cements
C 618	Specification for Fly Ash and Raw or Calcined Natural Pozzolan for Use as a Mineral Admixture in Portland Cement Concrete
C 845	Specification for Expansive Hydraulic Cement
C 979	Standard Specification for Pigments for Integrally Colored Concrete
C 1157	Standard Performance Specification for Hydraulic Cement

These publications may be obtained from the following organizations:

American Concrete Institute
P.O. Box 9094
Farmington Hills, MI 48333-9094

ASTM International
100 Barr Harbor Dr.
West Conshohocken, PA 19428

11.2—Cited references
1. Nine papers on Architectural Concrete, *Concrete International*, V. 6, No.1, Jan. 1984, pp. 21-66:
 a. "The Twentieth Century Stone," T. G. de Leon
 b. "Colored Architectural Concrete," C. M. Dabney
 c. "Textured Architectural Concrete," J. A. Dobrowolski
 d. "ACI and Architectural Concrete," F. A. Nassaux
 e. "From the ACI Library: Architectural Concrete Guide"
 f. "Problems and Surface Blemishes in Architectural Cast-in-Place Concrete," A. R. Kenney
 g. "Cast-in-Place Concrete Creates Campus Focal Point," J. E. Thyer and M. E. Worsham
 h. "Architects, Artists, and Concrete," ACI Staff
 i. "Architectural Concrete: Defects Demand Discretion," J. R. Smith
2. Eight papers on Concrete and Architecture, *Concrete International*, V. 10, No. 9, Sept. 1988, pp. 18-54.
 a. "Concrete In Context," G. Perkin
 b. "Truth Is Concrete," B. Goldberg
 c. "Exposed Aggregate Concrete Chimney"
 d. "Concrete is Beautiful—From the Pantheon to the d'Orsay Museum," L. Farkas
 e. "Sculpting with Portland Cement," K. M. Page
 f. "Works of Art in Concrete," R. W. Steiger
 g. "Artist Carves his Niche in Concrete," M. M. Miller
 h. "Picasso's Adventures in Concrete," R. C. Heun
3. Kenny, A. R., and Freedman, S., "Chapter 20, Architectural Concrete," *Concrete Construction Engineering Handbook,* CRC Press, 1997.
4. Litvin, A., "Clear Coatings for Exposed Architectural Concrete," *Journal*, PCA Research and Development Laboratories, V. 10, No. 2, May 1968, pp. 49-57. Also, Development Department Bulletin DX137, Portland Cement Association.
5. Bell, L. W., "Writing Specifications for Architectural Concrete," *Concrete International,* V. 18, No. 6, June 1996, pp. 63-66.
6. ACI Committee 311, "ACI Manual of Concrete Inspection (SP-2)," American Concrete Institute, Farmington Hills, MI, 9th Edition, 1999, 209 pp.
7. Hurd, M. K., *Formwork for Concrete (SP-4)*, American Concrete Institute, Farmington Hills, Mich., 6th Edition, 1995, 500 pp.
8. *Forming Economical Concrete Buildings*, Proceedings of Second International Conference, SP-90, American Concrete Institute, Farmington Hills, Mich., 1986, 264 pp.
9. Litvin, A., and Pfeifer, D. W., "Gap-Graded Mixes for Cast-in-Place Exposed Aggregate Concrete," ACI JOURNAL, *Proceedings* V. 62, No. 5, May 1965, pp. 521-538. Also, Development Department Bulletin DX90, Portland Cement Association.
10. *Design-Control of Concrete Mixes*, 13th Edition, Portland Cement Association, Skokie, Ill.
11. ACI Committee 301, "Specifications for Structural Concrete (ACI 301-99)," American Concrete Institute, Farmington Hills, Mich., 1999, 49 pp.
12. Ball, C., and Decandia, M., "Designing with Colored Architectural Concrete," *Concrete International,* V. 24, No. 6, June 2002, pp. 22-26.
13. Dabney, C. M., "Impact of Color on Concrete," *Concrete International*, V. 12, No. 12, Dec. 1990, pp. 20-23.
14. Brettman, B. B.; Darwin, D.; and Donahey, R. C., "Bond of Reinforcement to Superplasticized Concrete," ACI JOURNAL, *Proceedings* V. 83, No. 1, Jan.-Feb. 1986, pp. 98-107.
15. Greening, N. R., and Landgren, R., "Surface Discoloration of Concrete Flatwork," *Journal*, PCA Research and Development Laboratories, V.8, No. 3, Sept. 1966, pp. 34-50. Also, Research Bulletin 203, Portland Cement Association.

16. *Concrete Manual*, 8th Edition, U.S. Bureau of Reclamation, Denver, Colo., 1981, 627 pp.

17. Panarese, W., and Freedman, S., "Exposed Aggregate Concrete," *Modern Concrete*, V. 33, No. 7, Nov. 1969; No. 8, Dec. 1969; and No. 9, Jan. 1970, 16 pp.

18. "Bushhammering of Concrete Surfaces," *IS051*, Portland Cement Association, Skokie, Ill.,1987, 4 pp.

19. "Removing Stains and Cleaning Concrete Surfaces," *IS214*, Portland Cement Association, Skokie, Ill., 1988, 16 pp.

APPENDIX A—ARCHITECTURAL CONCRETE PHOTOS

Fig. A.1—Cathedral of Our Lady of the Angels, Los Angeles, Calif. (Morley Construction Co., Santa Monica, Calif.).

Fig. A.4—Disneyland Parking Structure.

Fig. A.2—Cathedral of Our Lady of the Angels, Los Angeles, Calif.

Fig. A.5—Water features (Shaw & Sons, Costa Mesa, Calif.).

Fig. A.3—Disneyland Parking Structure, Anaheim, Calif. (McCarthy, Newport Beach, Calif.).

Fig. A.6—The Government Building, Boston, Mass.

Fig. A.7—Christian Science Plaza; Colonnade Building featuring sun shade columns.

Fig. A.9—Torre Dataflux, Monterrey, Mexico.

Fig. A.8—Kaiser Permanente Regional Office Building, Oakland, Calif.

Fig. A.10—Rock & Roll Hall of Fame, Cleveland, Ohio. I. M. Pei, Architect.

Fig. A.11—University of San Diego Science Center, Calif. (Morley Construction, Santa Monica, Calif.).

Fig. A.12—Form liner finish and tapered openings.

Fig. A.13—Outward reveals at the Amgen Building 29, Thousand Oaks, Calif. (Morley Construction, Santa Monica, Calif.).

Fig. A.14—Amgen Building 29.

Fig. A.15—J.F.K. Flight Control Tower, New York, N.Y.

Fig. A.16—Stained flatwork (Shaw & Sons, Costa Mesa, Calif.).

Fig. A.17—Planter walls (Shaw & Sons, Costa Mesa, Calif.).

Fig. A.18—Planter walls (completed) (Shaw & Sons, Costa Mesa, Calif.).

Fig. A.19—Bondo and sanding to prepare formwork for use.

Fig. A.20—Salk Institute for Biological Studies, La Jolla, Calif.

Fig. A.21—Decorative flatwork at plaza. (Shaw & Sons, Costa Mesa, Calif.).

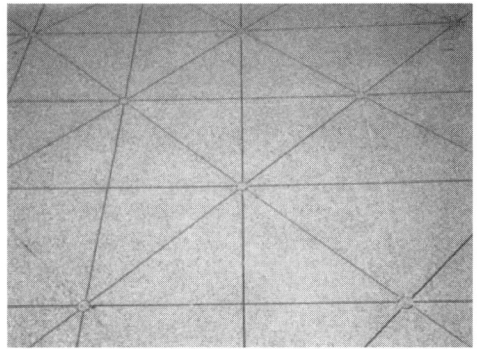

Fig. A.22—Architectural sawcut (Shaw & Sons, Costa Mesa, Calif.).

ACI 304R-00

Guide for Measuring, Mixing, Transporting, and Placing Concrete

Reported by ACI Committee 304

Neil R. Guptill
Chairman

David J. Akers	John C. King	Kenneth L. Saucier
Casimir Bognacki	Gary R. Mass	James M. Shilstone, Jr.
James L. Cope	Patrick L. McDowell	Ronald J. Stickel
Michael R. Gardner	Dipak T. Parekh	William X. Sypher
Daniel J. Green	Roger J. Phares	J.A. Tony Tinker
Brian Hanlin	James S. Pierce	Robert E. Tobin
Terence C. Holland	Paul E. Reinhart	Joel B. Tucker
Thomas A. Johnson	Royce J. Rhoads	Kevin Wolf

This guide presents information on the handling, measuring, and batching of all the materials used in making normalweight, lightweight structural, and heavyweight concrete. It covers both weight and volumetric measuring; mixing in central mixture plants and truck mixers; and concrete placement using buckets, buggies, pumps, and conveyors. Underwater concrete placement and preplaced aggregate concrete are also covered in this guide, as well as procedures for achieving good quality concrete in completed structures.

Keywords: batching; continuous mixing; conveying; heavyweight concretes; lightweight concretes; materials handling; mixing; placing; preplaced aggregate concrete; pumped concrete; tremie concrete; volumetric measuring.

CONTENTS

Chapter 1—Introduction, p. 304R-2
1.1—Scope
1.2—Objective
1.3—Other considerations

Chapter 2—Control, handling, and storage of materials, p. 304R-3
2.1—General considerations
2.2—Aggregates
2.3—Cement
2.4—Ground slag and pozzolans
2.5—Admixtures
2.6—Water and ice
2.7—Fiber reinforcement

Chapter 3—Measurement and batching, p. 304R-6
3.1—General requirements
3.2—Bins and weigh batchers
3.3—Plant type
3.4—Cementitious materials
3.5—Water and ice measurement
3.6—Measurement of admixtures
3.7—Measurement of materials for small jobs
3.8—Other considerations

Chapter 4—Mixing and transporting, p. 304R-9
4.1—General requirements
4.2—Mixing equipment
4.3—Central-mixed concrete
4.4—Truck-mixed concrete
4.5—Charging and mixing
4.6—Mixture temperature
4.7—Discharging
4.8—Mixer performance
4.9—Maintenance
4.10—General considerations for transporting concrete
4.11—Returned concrete

Chapter 5—Placing concrete, p. 304R-13
5.1—General considerations
5.2—Planning

ACI Committee Reports, Guides, Standard Practices, and Commentaries are intended for guidance in planning, designing, executing, and inspecting construction. This document is intended for the use of individuals who are competent to evaluate the significance and limitations of its content and recommendations and who will accept responsibility for the application of the material it contains. The American Concrete Institute disclaims any and all responsibility for the stated principles. The Institute shall not be liable for any loss or damage arising therefrom.

Reference to this document shall not be made in contract documents. If items found in this document are desired by the Architect/Engineer to be a part of the contract documents, they shall be restated in mandatory language for incorporation by the Architect/Engineer.

ACI 304R-00 supersedes ACI 304R-89 and became effective January 10, 2000.
Copyright © 2000, American Concrete Institute.
All rights reserved including rights of reproduction and use in any form or by any means, including the making of copies by any photo process, or by electronic or mechanical device, printed, written, or oral, or recording for sound or visual reproduction or for use in any knowledge or retrieval system or device, unless permission in writing is obtained from the copyright proprietors.

5.3—Reinforcement and embedded items
5.4—Placing
5.5—Consolidation
5.6—Mass concreting

Chapter 6—Forms, joint preparation, and finishing, p. 304R-19
6.1—Forms
6.2—Joint preparation
6.3—Finishing unformed surfaces

Chapter 7—Preplaced-aggregate concrete, p. 304R-21
7.1—General considerations
7.2—Materials
7.3—Grout proportioning
7.4—Temperature control
7.5—Forms
7.6—Grout pipe systems
7.7—Coarse aggregate placement
7.8—Grout mixing and pumping
7.9—Joint construction
7.10—Finishing
7.11—Quality control

Chapter 8—Concrete placed under water, p. 304R-24
8.1—General considerations
8.2—Materials
8.3—Mixture proportioning
8.4—Concrete production and testing
8.5—Tremie equipment and placement procedure
8.6—Direct pumping
8.7—Concrete characteristics
8.8—Precautions
8.9—Special applications
8.10—Antiwashout admixtures

Chapter 9—Pumping concrete, p. 304R-28
9.1—General considerations
9.2—Pumping equipment
9.3—Pipeline and accessories
9.4—Proportioning pumpable concrete
9.5—Field practice
9.6—Field control

Chapter 10—Conveying concrete, p. 304R-30
10.1—General considerations
10.2—Conveyor operation
10.3—Conveyor design
10.4—Types of concrete conveyors
10.5—Field practice

Chapter 11—Heavyweight and radiation-shielding concrete, p. 304R-33
11.1—General considerations
11.2—Materials
11.3—Concrete characteristics
11.4—Mixing equipment
11.5—Formwork
11.6—Placement
11.7—Quality control

Chapter 12—Lightweight structural concrete, p. 304R-36
12.1—General considerations
12.2—Measuring and batching
12.3—Mixing
12.4—Job controls

Chapter 13—Volumetric-measuring and continuous-mixing concrete equipment, p. 304R-38
13.1—General considerations
13.2—Operations
13.3—Fresh concrete properties

Chapter 14—References, p. 304R-39
14.1—Referenced standards and reports
14.2—Cited references

CHAPTER 1—INTRODUCTION
1.1—Scope
This guide outlines procedures for achieving good results in measuring and mixing ingredients for concrete, transporting it to the site, and placing it. The first six chapters are general and apply to all types of projects and concrete. The following four chapters deal with preplaced-aggregate concrete, underwater placing, pumping, and conveying on belts. The concluding three chapters deal with heavyweight, radiation-shielding concrete, lightweight concrete, and volumetric-measuring and continuous-mixing concrete equipment.

1.2—Objective
When preparing this guide, ACI Committee 304 followed this philosophy:
- Progress in improvement of concrete construction is better served by the presentation of high standards rather than common practices;
- In many, if not most, cases, practices resulting in the production and placement of high-quality concrete can be performed as economically as those resulting in poor concrete. Many of the practices recommended in this document improve concrete uniformity as well as quality, yielding a smoother operation and higher production rates, both of which offset potential additional cost; and
- Anyone planning to use this guide should have a basic knowledge of the general practices involved in concrete work. If more specific information on measuring, mixing, transporting, and placing concrete is desired, the reader should refer to the list of references given at the end of this document, and particularly to the work of the U.S. Bureau of Reclamation (1981), the U.S. Department of Commerce (1966), the Corps of Engineers (1994a), ASTM C 94, ACI 311.1R, and ACI 318. To portray more clearly certain principles involved in achieving maximum uniformity, homogeneity, and quality of concrete in place, figures that illustrate good and poor practices are also included in this guide.

1.3—Other considerations
All who are involved with concrete work should know the importance of maintaining the unit water content as low as possible and still consistent with placing requirements (Mielenz 1994; Lovern 1966). If the water-cementitious materials ratio (*w/cm*) is kept constant, an increase in unit water content increases the potential for drying-shrinkage cracking, and with this cracking, the concrete can lose a portion of its durability and other favorable characteristics, such as monolithic properties and low permeability. Indiscriminate addition of water that increases the *w/cm* adversely affects both strength and durability.

The more a form is filled with the right combination of solids and the less it is filled with water, the better the resulting concrete will be. Use only as much cement as is required to achieve adequate strength, durability, placeability, workability, and other specified properties. Minimizing the cement content is particularly important in massive sections subject to restraint, as the temperature rise associated with the hydration of cement can result in cracking because of the change in volume (ACI 207.1R and 207.2R). Use only as much water and fine aggregate as is required to achieve suitable workability for proper placement and consolidation by means of vibration.

CHAPTER 2—CONTROL, HANDLING, AND STORAGE OF MATERIALS
2.1—General considerations
Coarse and fine aggregates, cement, pozzolans, and chemical admixtures should be properly stored, batched, and handled to maintain the quality of the resulting concrete.

2.2—Aggregates
Fine and coarse aggregates should be of good quality, uncontaminated, and uniform in grading and moisture content. Unless this is accomplished through appropriate specifications (ASTM C 33) and effective selection, preparation, and handling of aggregates (Fig. 2.1), the production of uniform concrete will be difficult (Mielenz 1994; ACI 221R).

2.2.1 *Coarse aggregate*—The coarse aggregate should be controlled to minimize segregation and undersized material. The following sections deal with prevention of segregation and control of undersized material.

2.2.1.1 Sizes—A practical method of minimizing coarse aggregate segregation is to separate the material into several size fractions and batch these fractions separately. As the range of sizes in each fraction is decreased and the number of size separations is increased, segregation is further reduced. Effective control of segregation and undersized materials is most easily accomplished when the ratio of maximum-to-minimum size in each fraction is held to not more than four for aggregates smaller than 1 in. (25 mm) and to two for larger sizes. Examples of some appropriate aggregate fraction groupings follow:

Example 1
Sieve designations
No. 8 to 3/8 in. (2.36 to 9.5 mm)
No. 4 to 1 in. (4.75 to 25.0 mm)
3/4 to 1-1/2 in. (19.0 to 37.5 mm)

Example 2
Sieve designations
No. 4 to 3/4 in. (4.75 to 19.0 mm)
3/4 to 1-1/2 in. (19.0 to 37.5 mm)
1-1/2 to 3 in. (37.5 to 75 mm)
3 to 6 in. (75 to 150 mm)

2.2.1.2 Control of undersized material—Undersized material for a given aggregate fraction is defined as material that will pass a sieve having an opening 5/6 of the nominal minimum size of each aggregate fraction (U.S. Bureau of Reclamation 1981). In Example 2 in Section 2.2.1.1, it would be material passing the following sieves: No. 5 (4.0 mm), 5/8 in. (16.0 mm), 1-1/4 in. (31.5 mm), and 2-1/2 in. (63 mm). For effective control of gradation, handling operations that do not increase the undersized materials in aggregates significantly before their use in concrete are essential (Fig. 2.1 and 2.2). The gradation of aggregate as it enters the concrete mixer should be uniform and within specification limits. Sieve analyses of coarse aggregate should be made with sufficient frequency to ensure that grading requirements are met. When two or more aggregate sizes are used, changes may be necessary in the proportions of the sizes to maintain the overall grading of the combined aggregate. When specification limits for grading cannot be met consistently, special handling methods should be instituted. Materials tend to segregate during transportation, so reblending may be necessary. Rescreening the coarse aggregate as it is charged to the bins at the batch plant to remove undersized materials will effectively eliminate undesirable fines when usual storage and handling methods are not satisfactory. Undersized materials in the smaller coarse aggregate fractions can be consistently reduced to as low as 2% by rescreening (Fig. 2.2). Although rescreening is effective in removing undersized particles, it will not regrade segregated aggregates.

2.2.2 *Fine aggregate (sand)*—Fine aggregate should be controlled to minimize variations in gradation, giving special attention to keeping finer fractions uniform and exercising care to avoid excessive removal of fines during processing.

If the ratio of fine-to-coarse aggregate is adjusted in accordance with ACI 211.1 recommendations for mixture proportioning, a wide range of fine aggregate gradings can be used (Tynes 1962). Variations in grading during production of concrete should be minimized, however, and the ASTM C 33 requirement that the fineness modulus of the fine aggregate be maintained within 0.20 of the design value should be met.

Give special attention to the amount and nature of material finer than the No. 200 screen (75 μm sieve). As stated in ASTM C 33, if this material is dust of fracture, essentially free of clay or shale, greater percentages of materials finer than the No. 200 screen (75 μm sieve) are permissible. If the reverse is true, however, permissible quantities should be significantly reduced. The California sand equivalent test is sometimes used to determine quantitatively the type, amount, and activity of this fine material (Mielenz 1994; ASTM D 2419). Excessive quantities of material finer than the No. 200 screen (75 μm sieve) increase the mixing-water requirement, rate of slump loss, and drying shrinkage, and therefore decrease strength.

Avoid blending two sizes of fine aggregate by placing alternate amounts in bins or stockpiles or when loading cars or trucks. Satisfactory results are achieved when different size fractions are blended as they flow into a stream from regulating gates or feeders. A more reliable method of control for a wide range of plant and job conditions, however, is to separate storage, handling, and batching of the coarse and fine fractions.

2.2.3 *Storage*—Stockpiling of coarse aggregate should be kept to a minimum because fines tend to settle and accumulate. When stockpiling is necessary, however, use of correct methods minimizes problems with fines, segregation, aggregate breakage, excessive variation in gradation, and contamination. Stockpiles should be built up in horizontal or gently sloping layers, not by end-dumping. Trucks, loaders, and dozers, or other equipment should not be operated on the stockpiles because, in addition to breaking the aggregate, they frequently track dirt onto the piles (Fig. 2.1).

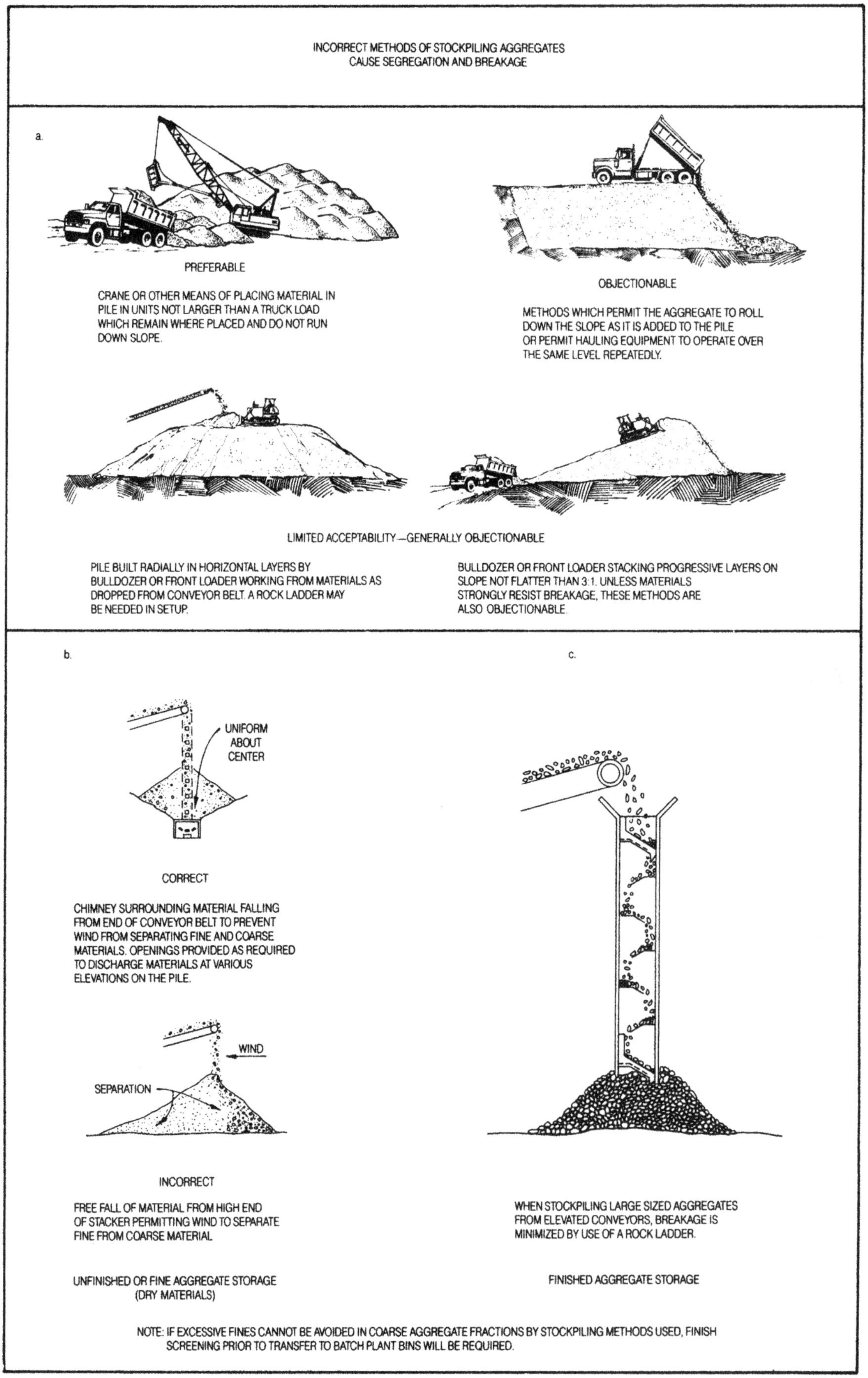

INCORRECT METHODS OF STOCKPILING AGGREGATES CAUSE SEGREGATION AND BREAKAGE

a.

PREFERABLE
CRANE OR OTHER MEANS OF PLACING MATERIAL IN PILE IN UNITS NOT LARGER THAN A TRUCK LOAD WHICH REMAIN WHERE PLACED AND DO NOT RUN DOWN SLOPE.

OBJECTIONABLE
METHODS WHICH PERMIT THE AGGREGATE TO ROLL DOWN THE SLOPE AS IT IS ADDED TO THE PILE OR PERMIT HAULING EQUIPMENT TO OPERATE OVER THE SAME LEVEL REPEATEDLY.

LIMITED ACCEPTABILITY—GENERALLY OBJECTIONABLE

PILE BUILT RADIALLY IN HORIZONTAL LAYERS BY BULLDOZER OR FRONT LOADER WORKING FROM MATERIALS AS DROPPED FROM CONVEYOR BELT. A ROCK LADDER MAY BE NEEDED IN SETUP.

BULLDOZER OR FRONT LOADER STACKING PROGRESSIVE LAYERS ON SLOPE NOT FLATTER THAN 3:1. UNLESS MATERIALS STRONGLY RESIST BREAKAGE, THESE METHODS ARE ALSO OBJECTIONABLE.

b.

CORRECT
UNIFORM ABOUT CENTER

CHIMNEY SURROUNDING MATERIAL FALLING FROM END OF CONVEYOR BELT TO PREVENT WIND FROM SEPARATING FINE AND COARSE MATERIALS. OPENINGS PROVIDED AS REQUIRED TO DISCHARGE MATERIALS AT VARIOUS ELEVATIONS ON THE PILE.

INCORRECT
FREE FALL OF MATERIAL FROM HIGH END OF STACKER PERMITTING WIND TO SEPARATE FINE FROM COARSE MATERIAL.

UNFINISHED OR FINE AGGREGATE STORAGE (DRY MATERIALS)

c.

WHEN STOCKPILING LARGE SIZED AGGREGATES FROM ELEVATED CONVEYORS, BREAKAGE IS MINIMIZED BY USE OF A ROCK LADDER.

FINISHED AGGREGATE STORAGE

NOTE: IF EXCESSIVE FINES CANNOT BE AVOIDED IN COARSE AGGREGATE FRACTIONS BY STOCKPILING METHODS USED, FINISH SCREENING PRIOR TO TRANSFER TO BATCH PLANT BINS WILL BE REQUIRED.

Fig. 2.1—Correct and incorrect methods of handling and storing aggregates.

Provide a hard base with good drainage to prevent contamination from underlying material. Prevent overlap of the different sizes by suitable walls or ample spacing between piles. Protect dry, fine aggregate from being separated by the wind by using tarps or windbreaks. Do not contaminate stockpiles by swinging aggregate-filled buckets or clam-shovels over the other piles of aggregate sizes. In addition, fine aggregate that is transported over wet, unimproved haul roads can become contaminated with clay lumps. The source of this contamination is usually accumulation of mud between the tires and on mud flaps that is dislodged during dumping of the transporting unit. Bottom-dump trailers are particularly susceptible to causing contamination when they drive through discharged piles. Clay lumps or clay balls can usually be removed from the fine aggregate by placing a scalping screen over the batch plant bin.

Keep storage bins as full as practical to minimize breakage and changes in grading as materials are withdrawn. Deposit materials into the bins vertically and directly over the bin outlet (Fig. 3.1b). Pay particular attention to the storage of special concrete aggregates, including lightweight, high-density, and architectural-finish aggregates. Contamination of these materials has compounding effects on other properties of the concrete in which they are to be used (Chapters 11 and 12).

2.2.4 *Moisture control*—Ensure, as practically as possible, a uniform and stable moisture content in the aggregate as batched. The use of aggregates with varying amounts of free water is one of the most frequent causes for loss of control of concrete consistency (slump). In some cases, wetting the coarse aggregate in the stockpiles or on the delivery belts may be necessary to compensate for high absorption or to provide cooling. When this is done, the coarse aggregates should be dewatered to prevent transfer of excessive free water to the bins.

Provide adequate time for drainage of free water from fine aggregate before transferring it to the batch plant bins. The storage time required depends primarily on the grading and particle shape of the aggregate. Experience has shown that a free-moisture content of as high as 6%, and occasionally as high as 8%, can be stable in fine aggregate. Tighter controls, however, may be required for certain jobs. The use of moisture meters to indicate variations in the moisture of the fine aggregate as batched, and the use of moisture compensators for rapid batch weight adjustments, can minimize the influence of moisture variations in the fine aggregate (Van Alstine 1955, Lovern 1966).

2.2.5 *Samples for test*—Samples representing the various aggregate sizes batched should be obtained as closely as possible to the point of their introduction into the concrete. The difficulty in obtaining representative samples increases with the size of the aggregate. Therefore, sampling devices require careful design to ensure meaningful test results. Methods of sampling aggregates are outlined in detail in ASTM D 75.

Maintaining a running average of the results of the five to 10 previous gradation tests, dropping the results of the oldest and adding the most recent to the total on which the average is calculated, is good practice. This average gradation can then be used for both quality control and for proportioning purposes.

2.3—Cement
All cement should be stored in weathertight, properly ventilated structures to prevent absorption of moisture.

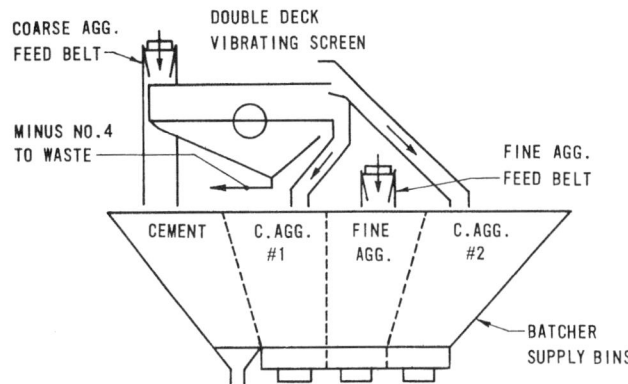

Fig. 2.2—Batching plant rescreen arrangement.

Storage facilities for bulk cement should include separate compartments for each type of cement used. The interior of a cement silo should be smooth, with a minimum bottom slope of 50 degrees from the horizontal for a circular silo and 55 to 60 degrees for a rectangular silo. Silos should be equipped with nonclogging air-diffuser flow pads through which small quantities of dry, oil-free, low-pressure air can be introduced intermittently at approximately 3 to 5 psi (20 to 35 kPa) to loosen cement that has settled tightly in the silos. Storage silos should be drawn down frequently, preferably once per month, to prevent cement caking.

Each bin compartment from which cement is batched should include a separate gate, screw conveyor, air slide, rotary feeder, or other conveyance that effectively allows both constant flow and precise cutoff to obtain accurate batching of cement.

Make sure cement is transferred to the correct silo by closely monitoring procedures and equipment. Fugitive dust should be controlled during loading and transferring.

Bags of cement should be stacked on pallets or similar platforms to permit proper circulation of air. For a storage period of less than 60 days, stack the bags no higher than 14 layers, and for longer periods, no higher than seven layers. As an additional precaution the oldest cement should be used first.

2.4—Ground slag and pozzolans
Fly ash, ground slag, or other pozzolans should be handled, conveyed, and stored in the same manner as cement. The bins, however, should be completely separate from cement bins without common walls that could allow the material to leak into the cement bin. Ensure that none of these materials is loaded into a cement bin on delivery.

2.5—Admixtures
Most chemical admixtures are delivered in liquid form and should be protected against freezing. If liquid admixtures are frozen, they should be properly reblended before they are used in concrete. Manufacturers' recommendations should be followed.

Long-term storage of liquid admixtures in vented tanks should be avoided. Evaporation of the liquid could adversely affect the performance of the admixture (ACI 212.3R).

2.6—Water and ice
Water for concrete production can be supplied from city or municipal systems, wells, truck wash-out systems, or from

Table 3.1.2—Typical batching tolerances

Ingredient	Batch weights greater than 30% of scale capacity		Batch weights less than 30% of scale capacity	
	Individual batching	Cumulative batching	Individual batching	Cumulative batching
Cement and other cementitious materials	±1% of required mass or ±0.3% of scale capacity, whichever is greater		Not less than required weight or 4% more than required weight	
Water (by volume or weight), %	±1	Not recommended	±1	Not recommended
Aggregates, %	±2	±1	±2	±0.3% of scale capacity or ±3% of required cumulative weight, whichever is less
Admixtures (by volume or weight), %	±3	Not recommended	±3	Not recommended

any other source determined to be suitable. If questionable, the quality of the water should be tested for conformance with the requirements given in ASTM C 94. Concrete made with recycled wash water can show variations in strength, setting time, and response to air-entraining and chemical admixtures. Recycled wash water may be required to meet chemical requirements of ASTM C 94. Compensation may be necessary for the solids in recycled water to maintain yield and total water content in the concrete.

The water batcher and the water pipes should be leak-free.

If ice is used, the ice facilities, including the equipment for batching and transporting to the mixer, should be properly insulated to prevent the ice from melting before it is in the mixer.

2.7—Fiber reinforcement

Synthetic fiber reinforcement is available in one cubic yard (one cubic meter) or multicubic yard (cubic meter) increments from most manufacturers. These prepackaged units should be readily accessible so they can be added directly to the mixer during the batching process.

Steel fibers are packaged in various sizes; the most common are 50 or 100 lb (23 or 45 kg) increments. Appropriate equipment should be used to disperse the fibers into the mixer to minimize the potential for the development of fiber balls. Steel fibers should be stored so that they are not exposed to moisture or other foreign matter. For more information on working with steel fibers, see ACI 544.3R.

CHAPTER 3—MEASUREMENT AND BATCHING
3.1—General requirements

3.1.1 *Objectives*—An important objective in producing concrete is to achieve uniformity and homogeneity, as indicated by physical properties such as unit weight, slump, air content, strength, and air-free unit weight of mortar in individual batches and successive batches of the same mixture proportions (U.S. Department of Reclamation 1981, U.S. Department of Commerce 1966, Bozarth 1967, ASTM C 94, Corps of Engineers 1994b). During measurement operations, aggregates should be handled so that the desired grading is maintained, and all materials should be measured within the tolerances acceptable for desired reproducibility of the selected concrete mixture. Another important objective of successful batching is the proper sequencing and blending of the ingredients (U.S. Department of Commerce 1966, Bozarth 1967). Visual observation of each material being batched is helpful in achieving this objective.

3.1.2 *Tolerances*—Most engineering organizations, both public and private, issue specifications containing detailed requirements for manual, semiautomatic, partially automatic, and automatic batching equipment for concrete (U.S. Bureau of Reclamation 1981, Corps of Engineers 1994b, ASTM C 94, AASHTO 1993). Batching equipment currently marketed will operate within the usual specified batch-weight tolerances when the equipment is maintained in good mechanical condition. The "Concrete Plant Standards of the Concrete Plant Manufacturers Bureau" (Concrete Plant Manufacturers Bureau 1996a) and the "Recommended Guide Specifications for Batching Equipment and Control Systems in Concrete Batch Plants" (Concrete Plant Manufacturers Bureau 1996b) are frequently used for specifying batching and scale accuracy. Batching tolerances commonly used are given in Table 3.1.2.

Other commonly used requirements include: beam or scale divisions of 0.1% of total capacity and batching interlock of 0.3% of total capacity at zero balance (Concrete Plant Manufacturers Bureau 1996a); quantity of admixture weighed never to be so small that 0.4% of full scale capacity exceeds 3% of the required weight; isolation of batching equipment from plant vibration; protection of automatic controls from dust and weather; and frequent checking and cleaning of scale and beam pivot points. With good inspection and plant operation, batching equipment can be expected to perform consistently within the required tolerances.

3.2—Bins and weigh batchers

Batch plant bins and components should be of adequate size to accommodate the productive capacity of the plant. Compartments in bins should separate the various concrete materials, and the shape and arrangement of aggregate bins should be conducive to the prevention of aggregate segregation and breakage. The aggregate bins should be designed so that material cannot hang up in the bins or spill from one compartment to another.

Weigh batchers should be charged with easily operated clamshell or undercut radial-type bin gates. Gates used to charge semiautomatic and fully automatic batchers should be power-operated and equipped with a suitable dribble control to allow the desired weighing accuracy. Weigh batchers should be accessible for obtaining representative samples, and they should be arranged to obtain the proper sequencing and blending of aggregates during charging of the mixer.

Illustrations showing proper and improper design and arrangement of batch plant bins and weigh batchers are given in Fig. 3.1.

3.3—Plant type

Factors affecting the choice of the batching systems are: 1) size of job; 2) required production rate; and 3) required standards of batching performance. The production capacity of a batch plant is determined by a combination of the ma-

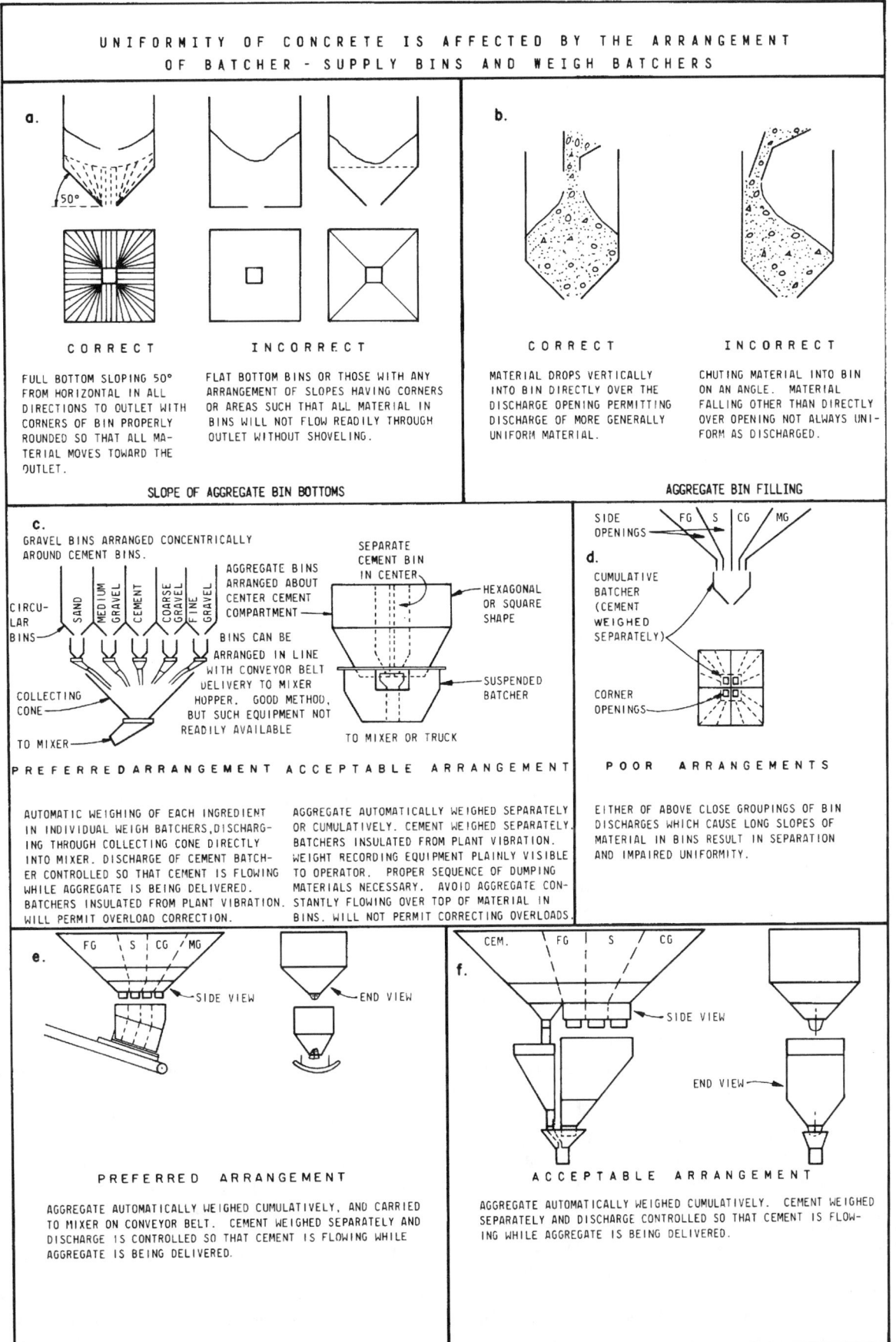

Fig. 3.1—Correct and incorrect methods of batching.

terials handling system, bin size, batcher size, and mixer size and number.

Available weigh batch equipment falls into four general categories: manual; partially automatic; semiautomatic; and fully automatic (Concrete Plant Manufacturers Bureau 1996a).

3.3.1 *Manual weigh batching*—As the name implies, all operations of weighing and batching of the concrete ingredients are controlled manually. Manual plants are acceptable for small jobs having low batching-rate requirements. As the job size increases, automation of batching operations is rapidly justified. Attempts to increase the capacity of manual plants by rapid batching can result in excessive weighing inaccuracies.

3.3.2 *Partially automatic weigh batching*—A partially automatic system consists of a combination of batching controls where at least one of the controls for weighing either cement or aggregates is either semiautomatic or automatic as described as follows. Weighing of the remaining materials is manually controlled and interlocking of the batching system to any degree is optional. This system can also lack accuracy when rapid batching is required.

3.3.3 *Semiautomatic weigh batching*—In this system, aggregate-bin gates for charging are opened by manually operated buttons or switches. Gates are closed automatically when the designated weight of material has been delivered. With satisfactory plant maintenance, the batching accuracy should meet the tolerances given in Section 3.1.2. The system should contain interlocks that prevent batcher charging and discharging from occurring simultaneously. In other words, when the batcher is being charged, it cannot be discharged, and when it is being discharged, it cannot be charged. Visual confirmation of the scale reading for each material being weighed is essential.

3.3.4 *Automatic weigh batching*—Automatic weigh batching of all materials is activated by a single starter switch. Interlocks, however, interrupt the batching cycle when the scale does not return to 0.3% of zero balance or when preset weighing tolerances detailed in Section 3.1.2 are exceeded.

3.3.4.1 *Cumulative automatic weigh batching*—Interlocked sequential controls are required for this type of batching. Weighing will not begin, and it will be automatically interrupted when preset tolerances in any of the successive weighings exceed values such as those given in Section 3.1.2. The charging cycle will not begin when the batcher discharge gate is open, and the batcher discharge cycle will not begin when batcher charging gates are open or when any of the indicated material weights is not within applicable tolerances. Presetting of desired batch weights is completed by such devices as punched cards, digital switches, or rotating dials and computers. Setting of weights, starting the batch cycle, and discharging the batch are all manually controlled. Mixture and batch-size selectors, aggregate moisture meters, manually controlled fine aggregate moisture compensators, and graphic or digital devices for recording the batch weight of each material are required for good plant control (Van Alstine 1955; Lovern 1966). This type of batching system provides greater accuracy for high-speed production than either the manual or semiautomatic systems.

A digital recorder can have a single measuring device for each scale or a series of measuring devices can record on the same tape or ticket. This type of recorder should reproduce the reading of the scale within 0.1% of the scale capacity or one increment of any volumetric batching device. A digital batch-documentation recorder should record information on each material in the mixture along with the concrete mixture identification, size of batch, and production facility identification. Required information can be preprinted, written, or stamped on the document. The recorder should identify the load by a batch-count number or a ticket serial number. The recorder, if interlocked to an automatic batching system, should show a single indication of all batching systems meeting zero or empty balance interlocks. All recorders should produce two or more tickets containing the information stated previously and also leave space for the identification of the job or project, location of placement, sand moisture content, delivery vehicle, driver's signature, purchaser's representative's signature, and the amount of water added at the project site.

3.3.4.2 *Individual automatic weigh batching*—This system provides separate scales and batchers for each aggregate size and for every other material batched. The weighing cycle is started by a single start switch, and individual batchers are charged simultaneously. Interlocks for interrupting weighing and discharge cycles when tolerances are exceeded, mixture selectors, aggregate moisture meters and compensators, and recorders differ only slightly from those described for cumulative automatic batching systems.

3.3.5 *Volumetric batching*—When aggregates or cementitious materials are batched by volume, it is normally a continuous operation coupled with continuous mixing. Volumetric batching and continuous mixing are covered in Chapter 13.

3.4—Cementitious materials

3.4.1 *Batching*—For high-volume production requiring rapid and accurate batching, bulk cementitious materials should be weighed with automatic, rather than semiautomatic or manual, equipment. All equipment should provide access for inspection and permit sampling at any time. The bins and weigh batchers should be equipped with aeration devices, vibrators, or both to aid in the smooth and complete discharge of the batch. Return to zero and weighing tolerance interlocks described in Section 3.1.2 should be used. Cement should be batched separately and kept separate from all ingredients before discharging. When both cement and pozzolan or slag are to be batched, separate silos should be used. They can be batched cumulatively, however, if the cement is weighed first.

3.4.2 *Discharging*—Effective precautions should be taken to prevent loss of cementitious materials during mixer charging. At multiple-stop plants where materials are charged separately, losses can be minimized by discharging the cementitious materials through a rubber drop chute. At one-stop plants, cement and pozzolan can be successfully charged along with the aggregate through rubber telescopic dropchutes. For plant mixers, a pipe should be used to discharge the cementitious materials to a point near the center of the mixer after the water and aggregates have started to enter the mixer. Proper and consistent sequencing and blending of the various ingredients into the mixer during the charging operation will contribute significantly toward the maintenance of batch-to-batch uniformity and, perhaps, reduced mixing time when confirmed by mixer performance tests (U.S. Department of Commerce 1966, Gaynor and Mullarky 1975, ASTM C 94).

3.5—Water and ice measurement

3.5.1 *Batching equipment*—On large jobs and in central batching and mixing plants where high-volume production is required, accurate water and ice measurement can only be obtained by the use of automatic weigh batchers or meters. Equipment and methods used should, under all operating conditions, be capable of routine measurement within the 1% tolerance specified in Section 3.1.2. Tanks or vertical cylinders with a center-siphon discharge can be permitted as an auxiliary part of the weighing, but should not be used as the direct means of measuring water. For accurate measurement, a digital gallon (liter) meter should be used. All equipment for water measurement should be designed for easy calibration so that accuracy can be quickly verified. Ice-batching equipment should be insulated to avoid melting the ice.

3.5.2 *Aggregate moisture determination and compensation*—Measurement of the correct total mixing water depends on knowing the quantity and variation of moisture in the aggregate (particularly in the fine aggregate) as it is batched. Aggregate that is not saturated surface dry will absorb mixture water from the concrete. Fine aggregate moisture meters are frequently used in plants and when properly maintained do satisfactorily indicate changes in fine aggregate moisture content. Use of moisture meters in fine sizes of coarse aggregate is also recommended if these materials vary in moisture content. Moisture meters should be calibrated to oven-dried samples for optimum consistency of readings. Moisture meters should be recalibrated monthly or whenever the slump of the concrete produced is inconsistent.

Moisture-compensating equipment can also be used that can reproportion water and fine aggregate weights for a change in aggregate moisture content, with a single setting adjustment. Compensators are usually used on the fine aggregate, but occasionally are also used on the small coarse aggregate size fractions. The moisture setting on the compensators is made manually with calibration dials, buttons, or levers. The use of moisture compensators is recommended when used in conjunction with calibrated moisture meters or regularly performed conventional moisture-control tests. Under these conditions, compensators can be useful tools for maintaining satisfactory control of the fine aggregate and the mixing water content.

Most computer-controlled batching systems now have software that interlocks moisture meters or compensating equipment with the measuring of fine aggregate and water. Readings are taken automatically and incorporated into the batching of these ingredients. Some systems work with an individual reading, whereas others can continuously record moisture as the fine aggregate is batched. Regardless of the system used, the software should impose user-defined upper and lower moisture limits and alert the operator when moisture values are outside those limits. Proper maintenance and calibration of equipment is essential to satisfactory performance and consistent production of concrete.

3.5.3 *Total mixing water*—In addition to the accurate weighing of added water, uniformity in the measurement of total mixing water involves control of such additional water sources as mixer wash water, ice, and free moisture in aggregates. One specified tolerance (ASTM C 94) for accuracy in measurement of total mixing water from all sources is ± 3%.

The operating mechanism in the water measuring devices should be such that leakage (dribbling or water trail) will not occur when the valve is closed. Water tanks on truck mixers or other portable mixers should be constructed so that the indicating device will register, within the specified accuracy, the quantity of water discharged, regardless of the inclination of the mixer.

3.6—Measurement of admixtures

Batching tolerances (Section 3.1.2) and charging and discharge interlocks described previously for other mixture ingredients should also be provided for admixtures. Batching and dispensing equipment should be readily capable of calibration. When timer-activated dispensers are used for large-volume admixtures such as calcium chloride, a container with a sight tube calibrated to show admixture quantity (usually referred to as a "calibration tube") should be used to allow visual confirmation of the volume being batched. In practice, calibration tubes are usually installed for all liquid admixtures.

Refer to ACI 212.3R for additional information on recommended practices in the use and dispensing of admixtures in concrete.

3.7—Measurement of materials for small jobs

If the concrete volume on a job is small, establishing and maintaining a batch plant and mixer at the construction site may not be practical. In such cases, using ready-mixed concrete or mobile volumetric batching and continuous mixing equipment may be preferable. If neither is available, precautions should be taken to properly measure and batch concrete materials mixed on the job site. Bags of cementitious materials should be protected from moisture and fractional bags should not be used unless they are weighed. The water-measuring device should be accurate and dependable, and the mixer capacity should not be exceeded.

3.8—Other considerations

In addition to accurate measurement of materials, correct operating procedures should also be used if concrete uniformity is to be maintained. Ensure that the batched materials are properly sequenced and blended so that they are charged uniformly into the mixture (U.S. Department of Commerce 1966; Bozarth 1967). Arrange the batching plant control room, if possible, with the plant operator's station located in a position where the operator can closely and clearly see the scales and measuring devices during batching of the concrete, as well as the charging, mixing, and discharging of the mixtures without leaving the operating console. Some common batching deficiencies to be avoided are: overlapping of batches; loss of materials; loss or hanging up of a portion of one batch, or its inclusion with another.

CHAPTER 4—MIXING AND TRANSPORTING
4.1—General requirements

Thorough mixing is essential for the production of uniform, quality concrete. Therefore, equipment and methods should be capable of effectively mixing concrete materials containing the largest specified aggregate to produce uniform mixtures of the lowest slump practical for the work. Recommendations on maximum aggregate size and slump to be used for various types of construction are given in ACI 211.1 for concretes made with ASTM C 150 and C 595M cements, and in ACI 223R for concretes made with ASTM C 845 expansive hydraulic cements. Sufficient mixing, transporting, and placing capacity should be provided so that unfinished concrete lifts can be maintained plastic and free of cold joints.

4.2—Mixing equipment

Mixers can be stationary parts of central mixture plants or of portable plants. Mixers can also be truck mounted. Satisfactorily designed mixers have a blade or fin arrangement and drum shape that ensure an end-to-end exchange of materials parallel to the axis of rotation or a rolling, folding, and spreading movement of the batch over itself as it is being mixed. For additional descriptions of some of the various mixer types, refer to the publications of the Concrete Plant Manufacturers Bureau (1996c) and of the Truck Mixer Manufacturers Bureau (1996).

The more common types of mixing equipment are:

4.2.1 *Tilting drum mixer*—This is a revolving drum mixer that discharges by tilting the axis of the drum. In the mixing mode, the drum axis can be either horizontal or at an angle.

4.2.2 *Nontilting drum mixer*—This is a revolving drum mixer that charges, mixes, and discharges with the axis of the drum horizontal.

4.2.3 *Vertical shaft mixer*—This is often called a turbine or pan-type mixer. Mixing is accomplished with rotating blades or paddles mounted on a vertical shaft in either a stationary pan or one rotating in the opposite direction to the blades. The batch can be easily observed and rapidly adjusted, if necessary. Rapid mixing and low overall profile are other significant advantages. This type of mixer does an excellent job of mixing relatively dry concretes and is often used for laboratory mixing and by manufacturers of concrete products.

4.2.4 *Pugmill mixers*—These mixers are defined in ACI 116R as "a mixer having a stationary cylindrical mixing compartment, with the axis of the cylinder horizontal, and one or more rotating horizontal shafts to which mixing blades or paddles are attached." Although this is an accurate definition, there are many types, styles, and configurations. Pugmills can have single or double shafts. They can have a curved blade configuration or a paddle configuration that is vertical to the shaft. In either case, they are designed to fold and move the concrete from one end of the pugmill to the other.

These mixers are suitable for harsh, stiff concrete mixtures. They have primarily been used in the production of concrete block units, cement-treated bases, and roller compacted concrete. Newer versions of these mixers are used in the production of normal- and high-strength concrete, with slumps of up to 8 in. (200 mm).

4.2.5 *Truck mixers*—There are two types of revolving drum truck mixers currently in use—rear discharge and front discharge. The rear-discharge, inclined-axis mixer predominates. In both, fins attached to the drum mix concrete in the mixing mode and also discharge the concrete when drum rotation is reversed.

4.2.6 *Continuous mixing equipment*—Two types of continuous mixing equipment are available. In the first type, all materials come together at the base of the mixing trough. Mixing is accomplished by a spiral blade rotated at a relatively high speed inside the enclosed trough, which is inclined at 15 to 25 degrees from the horizontal. These can be mobile, mounted either on a truck chassis or a trailer, or stationary. The second type is a continuous-feed pugmill mixer generally used for roller-compacted concrete and cement-treated base. Aggregates, cement, and fly ash are measured by weight or volume and fed into the charging end of the pugmill by variable-speed belts. Water is metered either from an attached tank or an outside source. Mixing is accomplished by paddles attached to one or two rotating horizontal shafts. The mixture is lifted and folded as it is moved from the charging end to the discharging end of the pugmill, where the completed mixture is discharged onto an elevated conveyor belt for easy loading into trucks. These types of continuous-feed mixers can be used for normal concretes as well. These would be considered semimobile plants as they are mounted on wheels and can be broken down for transport. Refer to Chapter 13 for additional information on continuous mixing equipment.

4.2.7 *Separate paste mixing*—Experimental work has shown that the mixing of cement and water into a paste before combining these materials with aggregates can increase the compressive strength of the resulting concrete (Mass 1989). The paste is generally mixed in a high-speed, shear-type mixer at a *w/cm* of 0.30 to 0.45 by mass. The premixed paste is then blended with aggregates and any remaining batch water, and final mixing is completed in conventional concrete mixing equipment.

4.3—Central-mixed concrete

Central-mixed concrete is mixed completely in a stationary mixer and then transferred to another piece of equipment for delivery. This transporting equipment can be a ready-mixed truck operating as an agitator, or an open-top truck body with or without an agitator. The tendency of concrete to segregate limits the distance it can be hauled in transporters not equipped with an agitator. If a truck mixer or a truck body with an agitator is used for central-mixed concrete, ASTM C 94 limits the volume of concrete charged into the truck to 80% of the drum or truck volume.

Sometimes the central mixer will partially mix the concrete with the final mixing and transporting being done in a revolving-drum truck mixer. This process is often called "shrink mixing" as it reduces the volume of the as-charged mixture. When using shrink mixing, ASTM C 94 limits the volume of concrete charged into the truck to 63% of the drum volume.

4.4—Truck-mixed concrete

Truck mixing is a process by which previously proportioned concrete materials from a batch plant are charged into a ready-mixed truck for mixing and delivery to the construction project. To achieve thorough mixing, total absolute volume of all ingredients batched in a revolving drum truck mixer should not exceed 63% of the drum volume (Truck Mixer Manufacturers Bureau 1996; ASTM C 94).

4.5—Charging and mixing

The method and sequence of charging mixers is of great importance in determining whether the concrete will be properly mixed. For central plant mixers, obtaining a preblending or ribboning effect by charging cement and aggregates simultaneously as the stream of materials flow into the mixer is essential (U.S. Department of Commerce 1966; Bozarth 1967; Gaynor and Mullarky 1975).

In truck mixers, all loading procedures should be designed to avoid packing of the material, particularly sand and cement, in the head of the drum during charging. The probability of packing is decreased by placing approximately 10% of the coarse aggregate and water in the mixer drum before the sand and cement.

Generally, approximately 1/4 to 1/3 of the water should be added to the discharge end of the drum after all other ingredients have been charged. Water-charging pipes should be of proper design and of sufficient size so that water enters at a point well inside the mixer and charging is complete within the first 25% of the mixing time (Gaynor and Mullarky 1975). Refer to Section 4.5.3.1 for additional discussion of mixing water.

The effectiveness of chemical admixtures will vary depending upon when they are added during the mixing sequence. Follow the recommendations of the admixture supplier regarding when to add a particular product. Once the appropriate time in the sequence is determined, chemical admixtures should be charged to the mixer at the same point in the mixing sequence for every batch. Liquid admixtures should be charged with the water or on damp sand, and powdered admixtures should be ribboned into the mixer with other dry ingredients. When more than one admixture is used, each should be batched separately unless premixing is allowed by the manufacturer.

Synthetic fiber reinforcement can be added any time during the mixing process as long as at least 5 min of mixing occurs after the addition of the synthetic fibers.

4.5.1 *Central mixing*—Procedures for charging central mixers are less restrictive than those necessary for truck mixers because a revolving-drum central mixer is not charged as full as a truck mixer and the blades and mixing action are quite different. In a truck mixer, there is little folding action compared with that in a stationary mixer. Batch size, however, should not exceed the manufacturer's rated capacity as marked on the mixer name plate.

The mixing time required should be based on the ability of the mixer to produce uniform concrete throughout the batch and from batch to batch. Manufacturers' recommendations and other typical recommendations, such as 1 min for 1 yd^3 (3/4 m^3) plus 1/4 min for each additional cubic yard (cubic meter) of capacity can be used as satisfactory guides for establishing initial mixing time. Final mixing times, however, should be based on the results of mixer performance tests made at frequent intervals throughout the duration of the job (U.S. Bureau of Reclamation 1981; U.S. Department of Commerce 1966; ASTM C 94; CRD-C 55). The mixing time should be measured from the time all ingredients are in the mixer. Batch timers with audible indicators used in combination with interlocks that prevent under- or over-mixing of the batch and discharge before completion of a preset mixing time are provided on automatic plants and are recommended on manual plants. The mixer should be designed for starting and stopping under full-load conditions.

4.5.2 *Truck mixing*—Generally, 70 to 100 revolutions at mixing speed are specified for truck mixing. ASTM C 94 limits the total number of revolutions to a maximum of 300. This limits the grinding of soft aggregates, loss of slump, wear on the mixer, and other undesirable effects that can occur in hot weather. Final mixing can be done at the producer's yard, or, more commonly, at the project site.

If additional time elapses after mixing and before discharge, the drum speed is reduced to the agitation speed or stopped. Then, before discharging, the mixer should be operated at mixing speed for approximately 30 revolutions to enhance uniformity.

Mixer charging, mixing, and agitating speeds vary with each truck and mixer-drum manufacturer. ASTM C 94 requires that these speeds and the mixing and agitating capacity of each drum be shown on a plate attached to the unit.

Maximum transportation time can be extended by several different procedures. These procedures are often called dry batching and evolved to accommodate long hauls and unavoidable delays in placing by attempting to postpone the mixing of cement with water. When cement and damp aggregate come in contact with each other, however, free moisture on the aggregate results in some cement hydration. Therefore, materials cannot be held in this manner indefinitely.

In one method, the dry materials are batched into the ready-mixed truck and transported to the job site where all of the mixing water is added. Water should be added under pressure, preferably at both the front and rear of the drum with it revolving at mixing speed, and then mixing is completed with the usual 70 to 100 revolutions. The total volume of concrete that can be transported in truck mixers by this method is the same as for regular truck mixing, approximately 63% of the drum volume (Truck Mixer Manufacturers Bureau 1996, ASTM C 94).

Another approach to accommodate long hauls is to use extended-set admixtures. The concrete is mixed and treated with the admixture before leaving the plant. The admixture dosage is typically selected to wear off shortly after the concrete arrives at the placement site, allowing the concrete to set normally. In some instances, an accelerator is added to activate the concrete once it arrives at the placement site. Concrete has been transported over 200 miles (320 km) using this technique.

4.5.3 *Water*

4.5.3.1 *Mixing water*—The water required for proper concrete consistency (slump) is affected by variables such as amount and rate of mixing, length of haul, time of unloading, and ambient temperature conditions. In cool weather, or for short hauls and prompt delivery, problems such as loss or variation in slump, excessive mixing water requirements, and discharging, handling, and placing problems rarely occur. The reverse is true, however, when rate of delivery is slow or irregular, haul distances are long, and weather is warm. Loss of workability during warm weather can be minimized by expediting delivery and placement and by controlling the concrete temperature. Good communication between the batching plant and the placement site is essential for coordination of delivery. It may be necessary to use a retarder to prolong the time the concrete will respond to vibration after it is placed. When feasible, all mixing water should be added at the central or batch plant. In hot weather, however, it is better to withhold some of the mixing water until the mixer arrives at the job. With the remaining water added, an additional 30 revolutions at mixing speed is required to adequately incorporate the additional water into the mixture. When loss of slump or workability cannot be offset by these measures, the procedures described in Section 4.5.2. should be considered.

4.5.3.2 *Addition of water on the job*—The maximum specified or approved *w/cm* should never be exceeded.

If all the water allowed by the specification or approved mixture proportions has not been added at the start of mixing, it may be permissible, depending upon project specifications, to add the remaining allowable water at the point of delivery. Once part of a batch has been unloaded, however, it becomes impractical to determine what w/cm is produced by additional water.

The production of concrete of excessive slump or adding water in excess of the proportioned w/cm to compensate for slump loss resulting from delays in delivery or placement should be prohibited. Persistent requests for the addition of water should be investigated.

Where permitted, a high-range water-reducing admixture (superplasticizer) can be added to the concrete to increase slump while maintaining a low w/cm (Cement and Concrete Association 1976; Prestressed Concrete Institute 1981). Addition of the admixture can be made by the concrete supplier or the contractor by a variety of techniques. When this admixture is used, vibration for consolidation is reduced. In walls and sloping formed concrete, however, some vibration is necessary to remove air trapped in the form. Use of this admixture can also increase form pressure.

4.5.3.3 *Wash water*—Most producers find it necessary to rinse off the rear fins of the mixer between loads and wash and discharge the entire mixer only at the end of the day. Hot weather and unusual mixture proportions can require washing and discharge of wash water after every load. Rinse water should not remain in the mixer unless it can be accurately compensated for in the succeeding batch. Rinse water can be removed from the mixer by reversing the drum for 5 to 10 revolutions at medium speed. Pollution-control regulations make it increasingly difficult to wash out after every load and have created an interest in systems to reclaim and reuse both wash water and returned concrete aggregates.

ASTM C 94 describes the reuse of wash water based on prescribed tests. Particular attention is necessary when admixtures are being used because the required dosages can change dramatically. When wash water is used, admixtures should be batched into a limited quantity of clean water or onto damp sand.

Wash water can also be treated using extended-set admixtures. In this case, a limited amount of wash water is added to a drum after all solid materials are discharged. Typically 50 gal. (200 L) instead of the normal 500 gal. (2000 L) are used. The admixture is added to the drum and the drum is rotated to ensure that all surfaces are coated. This treated wash water can be left in the truck overnight or over a weekend. The next morning or after the weekend, concrete can be batched using the treated wash water as part of the mixing water. Given the small amount of the admixture used for this application, use of an activating admixture is not usually required.

4.6—Mixture temperature

Batch-to-batch uniformity of concrete from a mixer, particularly with regard to slump, water requirement, and air content, also depends on the uniformity of the concrete temperature. Controlling the maximum and minimum concrete temperatures throughout all seasons of the year is important.

Concrete can be cooled using ice, chilled mixing water, chilled aggregates, or liquid nitrogen. In-place concrete temperatures as low as 40 F (4 C) are not unusual.

Liquid nitrogen at a temperature of –320 F (–196 C) can be used to chill mixture water, aggregates, or concrete (Anon. 1977). Liquid nitrogen has been injected directly into central mixers, truck mixers, or both to achieve required concrete temperatures (Anon. 1988). Concrete can be warmed by using heated water, aggregates, or both. Recommendations for control of concrete temperatures are discussed in detail in ACI 305R and 306R.

4.7—Discharging

Mixers should be capable of discharging concrete of the lowest slump suitable for the structure being constructed, without segregation (separation of coarse aggregate from the mortar). Before discharge of concrete transported in truck mixers, the drum should again be rotated at mixing speed for about 30 revolutions to reblend possible stagnant spots near the discharge end into the batch.

4.8—Mixer performance

The performance of mixers is usually determined by a series of uniformity tests made on samples taken from two or three locations within the concrete batch after it has been mixed for a given time period (U.S. Bureau of Reclamation 1981, ASTM C 94 and CRD-C 55). Mixer performance requirements are based on allowable differences in test results of samples from any two locations or a comparison of individual locations with the average of all locations. The procedures published by Gaynor and Mullarky (1975) are an excellent reference.

Among the many tests used to check mixer performance, the following are the most common: air content; slump; unit weight of air-free mortar; coarse aggregate content; and compressive strength.

Another important aspect of mixer performance is batch-to-batch uniformity of the concrete, which is also affected by the uniformity of materials and their measurement as well as by the efficiency of the mixer. Visual observation of the concrete during mixing and discharge from the mixer is an important aid in maintaining a uniform mixture, particularly with a uniform consistency. Some consistency-recording meters, such as those operating from the amperage draw on the electric motor drives for revolving-drum mixers, have also proven to be useful. The most positive control method for maintaining batch-to-batch uniformity, however, is a regularly scheduled program of tests of the fresh concrete, including unit weight, air content, slump, and temperature. All plants should have facilities and equipment for conveniently obtaining representative samples of concrete for routine control tests in accordance with ASTM C 172. Although strength tests provide an excellent measure of the efficiency of the quality control procedures that are employed, the strength-test results are available too late to be of practical use in controlling day-to-day production.

4.9—Maintenance

Mixers should be properly maintained to prevent mortar and dry material leakage. Inner mixer surfaces should be kept clean and worn blades should be replaced. Mixers not meeting the performance tests referenced in Section 4.8 should be taken out of service until necessary maintenance and repair corrects their deficient performance.

4.10—General considerations for transporting concrete

4.10.1 *General*—Concrete can be transported by a variety of methods and equipment, such as pipeline, hose, conveyor belts, truck mixers, open-top truck bodies with and without agitators, or buckets hauled by truck or railroad car. The method of transportation should efficiently deliver the concrete to the point of placement without losing mortar or significantly altering the concrete's desired properties associated with w/cm, slump, air content, and homogeneity. Various conditions should be considered when selecting a method of transportation, such as: mixture ingredients and proportions; type and accessibility of placement; required delivery capacity; location of batch plant; and weather conditions. These conditions can dictate the type of transportation best suited for economically obtaining quality in-place concrete.

4.10.2 *Revolving drum*—In this method, the truck mixer (Section 4.2.5) serves as an agitating transportation unit. The drum is rotated at charging speed during loading and is reduced to agitating speed or stopped after loading is complete. The elapsed time before discharging the concrete can be the same as for truck mixing and the volume carried can be increased to 80% of the drum capacity (ASTM C 94).

4.10.3 *Truck body with and without an agitator*—Units used in this form of transportation usually consist of an open-top body mounted on a truck, although bottom-dump trucks have been used successfully. The metal body should have smooth, streamlined contact surfaces and is usually designed for discharge of the concrete at the rear when the body is tilted. A discharge gate and vibrators mounted on the body should be provided at the point of discharge for control of flow. An agitator, if the truck body is equipped with one, aids in the discharge and ribbon-blends the concrete as it is unloaded. Water should never be added to concrete in the truck body because no mixing is performed by the agitator.

Use of protective covers for truck bodies during periods of inclement weather, proper cleaning of all contact surfaces, and smooth haul roads contribute significantly to the quality and operational efficiency of this form of transportation. The maximum delivery time specified is usually 30 to 45 min, although weather conditions can require shorter or permit longer times.

Trucks that have to operate on muddy haul roads should not be allowed to discharge directly on the grade or drive through the discharged pile of concrete.

4.10.4 *Concrete buckets on trucks or railroad cars*—This is a common method of transporting concrete from the batch plant to a location close to the placement area of a mass concrete placement. A crane then lifts the bucket to the final point of placement. Occasionally, transfer cars operating on railroad tracks are used to transport the concrete from the batch plant to buckets operating from cableways. Discharge of the concrete from the transfer cars into the bucket, which can be from the bottom or by some form of tilting, should be closely controlled to prevent segregation. Delivery time for bucket transportation is the same as for other nonagitating units—usually 30 to 45 min.

4.10.5 *Other methods*—Transporting of concrete by pumping methods and by belt conveyors are discussed in Chapters 9 and 10, respectively. Helicopter deliveries have been used in difficult-to-reach areas where other transporting equipment could not be used. This system usually employs one of the methods described previously to transport the concrete to the helicopter, which then lifts the concrete in a lightweight bucket to the placement area.

4.11—Returned concrete

Disposal of returned concrete is becoming more and more difficult for some producers. Two approaches for alleviating this problem are currently being used:

4.11.1 *Admixtures*—Extended-set admixtures were developed to address the need to hold returned concrete overnight. These admixtures are also used to hold concrete during the day for reuse on the same day.

The appropriate dosage of admixture is determined by the mixture characteristics, the quantity of concrete to be stabilized or held, and the length of time that the concrete is to be held. Depending on the length of time that the concrete is held, an accelerating admixture may be required. The stabilized concrete is usually blended with freshly batched concrete before being sold.

Various methods have been developed by concrete producers to handle and determine the volume of returned concrete. In some cases, all returned concrete is transferred at the end of a day to a single mixer for treatment and holding. Other producers have elected to handle the concrete on a truck-by-truck basis.

4.11.2 *Mechanical methods*—Equipment has been developed to process plastic, unused concrete returned to a plant. This equipment typically involves washing the concrete to separate it into two or more components. Some or all of the components are then reused in concrete production. The components can include coarse and fine aggregate, combined aggregate, and a slurry of cement and water, sometimes called gray water.

Although the processed components can often be reused in new concrete, a concrete producer should take care to ensure that these materials will not adversely affect the new concrete. Variations in aggregate grading can occur due to degradation of the previously used aggregate during mixing or reclaiming. Use of the slurry can affect strength and setting time. Conduct appropriate testing to verify that the concrete meets project requirements.

CHAPTER 5—PLACING CONCRETE
5.1—General considerations

This chapter presents guidelines for transferring concrete from the transporting equipment to its final position in the structure.

Placement of concrete is accomplished with buckets, hoppers, manual or motor-propelled buggies, chutes and drop pipes, conveyor belts, pumps, tremies, and paving equipment. Figure 5.1 and 5.2 show a number of handling and placing methods discussed in this chapter and give examples of both satisfactory and unsatisfactory construction procedures.

Placement of concrete by the preplaced aggregate method and by pumps and conveyors is discussed in Chapters 7, 9, and 10, respectively. In addition, placing methods specific to underwater, heavyweight, and lightweight concreting are noted in Chapters 8, 11, and 12, respectively. Another effective placement technique for both mortar and concrete is the shotcrete process. Thin layers are applied pneumatically to areas where forming is inconvenient or impractical, access or location provides difficulties, or normal casting techniques cannot be employed (ACI 506R).

Placing of concrete by the roller-compacted method is not covered in this guide. Refer to ACI 207.5R.

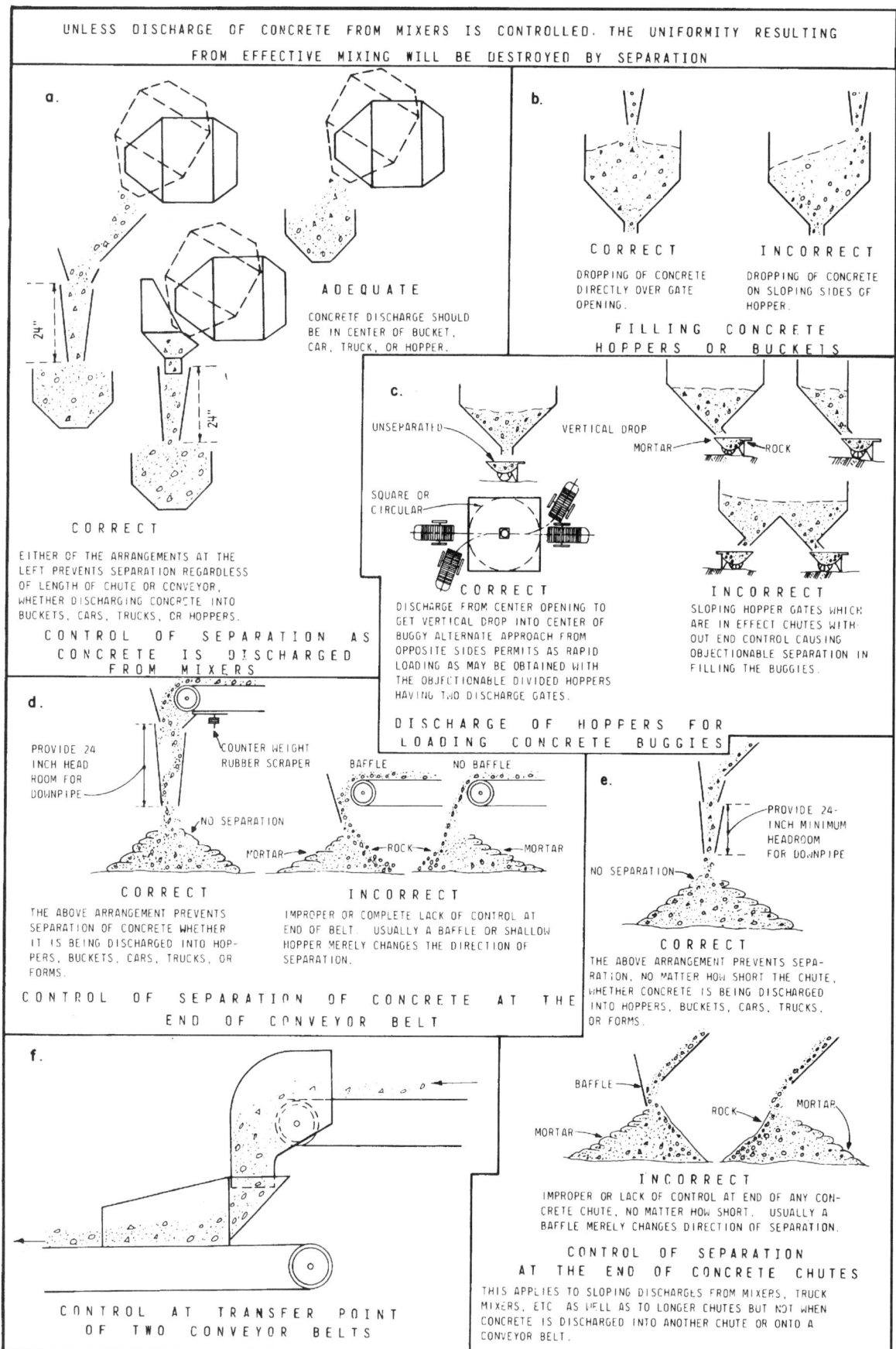

Fig. 5.1—Correct and incorrect methods of handling concrete.

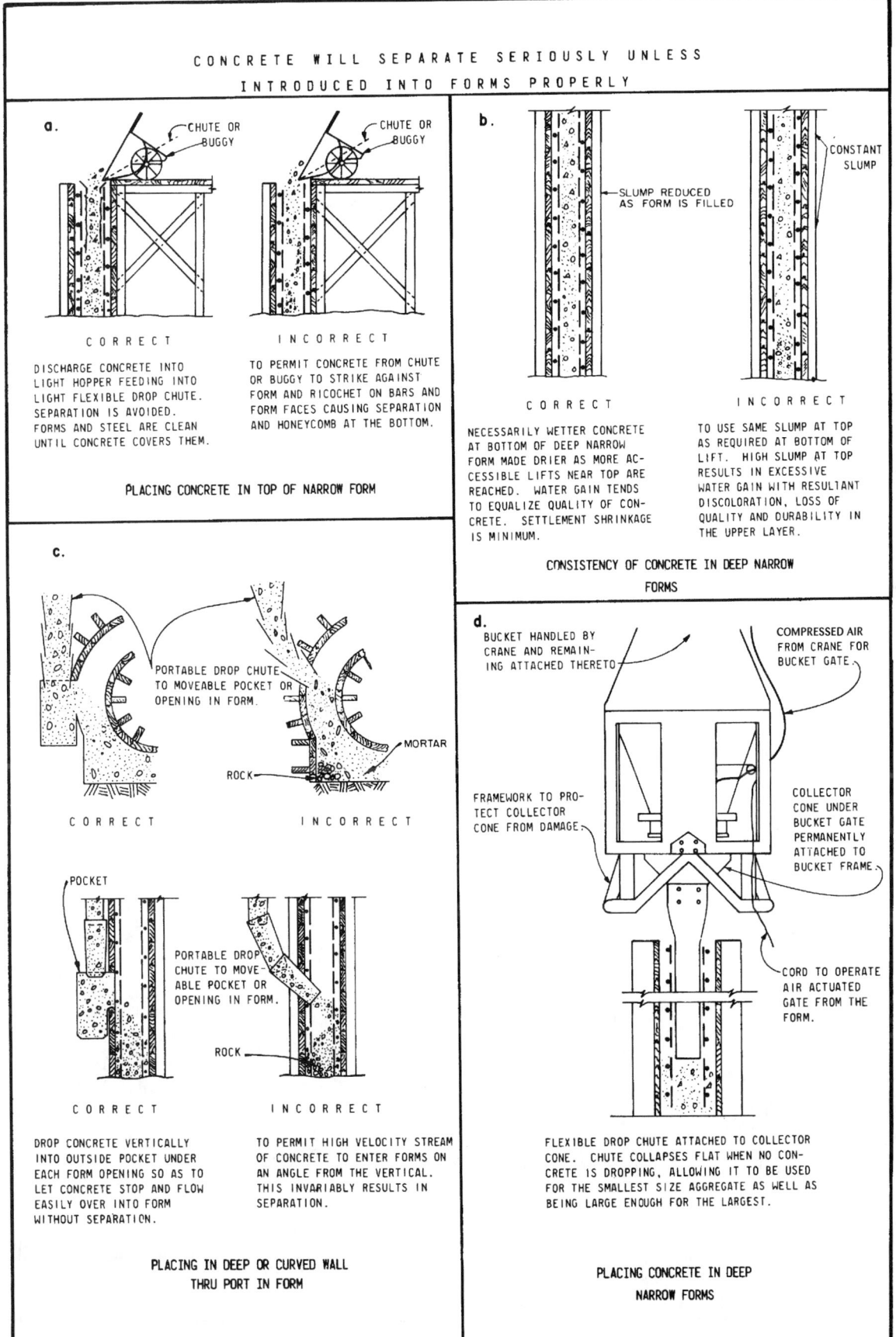

Fig. 5.2(a) to (d)—Correct and incorrect methods of placing concrete.

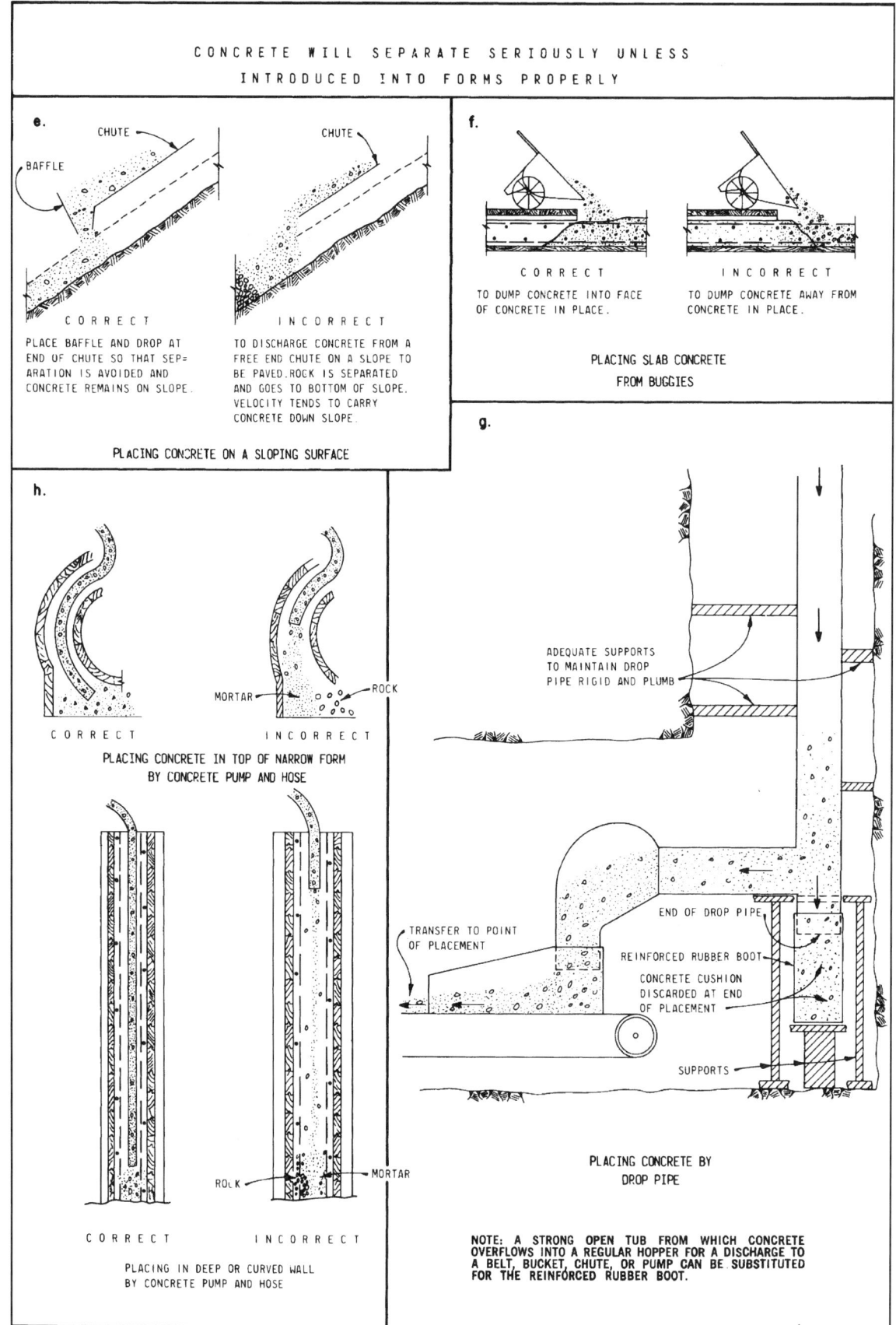

Fig. 5.2(e) to (h)—Correct and incorrect methods of placing concrete.

5.2—Planning

A basic requirement in all concrete handling is that both quality and uniformity of the concrete, in terms of *w/cm*, slump, air content, and homogeneity, have to be preserved. The selection of handling equipment should be based on its capability to efficiently handle concrete of proportions most advantageous for being readily consolidated in place with vibrators. Equipment requiring adjustment of mixture proportions beyond ranges recommended by ACI 211.1 should not be used.

Advance planning should ensure an adequate and consistent supply of concrete. Sufficient placement capacity should be provided so that the concrete can be kept plastic and free of cold joints while it is being placed. All placement equipment should be clean and in proper repair. The placement equipment should be arranged to deliver the concrete to its final position without significant segregation. The equipment should be adequately and properly arranged so that placing can proceed without undue delays and manpower should be sufficient to ensure the proper placing, consolidating, and finishing of the concrete. If the concrete is to be placed at night, the lighting system should be sufficient to illuminate the inside of the forms and to provide a safe work area.

Concrete placement should not commence when there is a chance of freezing temperatures occurring, unless adequate facilities for cold-weather protection have been provided (ACI 306R). Curing measures should be ready for use at the proper time (ACI 308). Where practical, it is advantageous to have radio or telephone communications between the site of major placements and the batching and mixing plant to better control delivery schedules and prevent excessive delays and waste of concrete.

The concrete should be delivered to the site at a uniform rate compatible with the manpower and equipment being used in the placing and finishing processes. If an interruption in the concreting process is a potential problem, consideration should be given to the provision of backup equipment.

A final detailed inspection of the foundation, construction joints, forms, water stops, reinforcement, and any other embedments in the placement should be made immediately before the concrete is placed. A method of documenting the inspection should be developed and approved by all parties before the start of work. All of these features should be carefully examined to make sure they are in accordance with the drawings, specifications, and good practice.

5.3—Reinforcement and embedded items

At the time of concrete placement, reinforcing steel and embedded items should be clean and free from mud, oil, and other materials that can adversely affect the steel's bonding capacity. Most reinforcing steel is covered with either mill scale or rust and such coatings are considered acceptable provided that loose rust and mill scale are removed and that the minimum dimensions of the steel are not less than those required in ACI 318.

Care should be taken to ensure that all reinforcing steel is of the proper size and length and that it is placed in the correct position and spliced in accordance with the plans. Adequate concrete cover of the reinforcing steel has to be maintained.

Mortar coating on embedded items within a lift to be completed within a few hours need not be removed, but loose dried mortar on embedded items projecting into future lifts should be removed prior to placing those lifts.

The method of holding a waterstop in the forms should ensure that it cannot bend to form cavities during concreting.

Bars and embedded items should be held securely in the proper position by suitable supports and ties to prevent displacement during concreting. Concrete blocks are sometimes used for support of the steel. Metal bar chairs with or without plastic protected ends or plastic bar chairs are more commonly used. Whatever system is used, there should be assurance that the supports will be adequate to carry expected loads before and during placement and will not stain exposed concrete surfaces, displace excessive quantities of concrete, or allow bars to move from their proper positions (Concrete Reinforcing Steel Institute 1982).

In some cases when reinforced concrete is being placed, it is useful to have a competent person in attendance to adjust and correct the position of any reinforcement that may be displaced. Structural engineers should identify critical areas where such additional supervision would be advantageous.

5.4—Placing

5.4.1 *Precautions*—Arrange equipment so that the concrete has an unrestricted vertical drop to the point of placement or into the container receiving it. The stream of concrete should not be separated by falling freely over rods, spacers, reinforcement, or other embedded materials. If forms are sufficiently open and clear so that the concrete is not disturbed in a vertical fall into place, direct discharge without the use of hoppers, trunks or chutes is favorable. Concrete should be deposited at or near its final position because it tends to segregate when it has to be flowed laterally into place.

If a project involves monolithic placement of a deep beam, wall, or column with a slab or soffit above, delay placing the slab or soffit concrete until the deep concrete settles. The time allotted for this settling depends on the temperature and setting characteristics of the concrete placed, but is usually about 1 h. Concreting should begin again soon enough to integrate the new layer thoroughly with the old by vibration.

5.4.2 *Equipment*—When choosing placement equipment, consider the ability of the equipment to place the concrete in the correct location economically without compromising its quality.

Equipment selection is influenced by the method of concrete production. Certain types of equipment, such as buckets, hoppers, and buggies will suit batch production; whereas other equipment, such as belt conveyors and pumps, are more appropriate for continuous production.

5.4.2.1 *Buckets and hoppers*—The use of properly designed bottom-dump buckets permits placement of concrete at the lowest practical slump consistent with consolidation by vibration. The bucket should be self-cleaning upon discharge, and concrete flow should start when the discharge gate is opened. Discharge gates should have a clear opening equal to at least five times the maximum aggregate size being used. Side slopes should be at least 60 degrees from the horizontal.

Control the bucket and its gate opening to ensure a steady stream of concrete is discharged against previously placed concrete where possible. Stacking concrete by discharging the bucket too close to the lift surface or discharging buckets while traveling, commonly causes segregation.

To prevent contamination, do not shovel spilled concrete back into buckets or hoppers for subsequent use or swing buckets directly over freshly finished concrete.

To expedite the placement schedule, the use of two or more buckets per crane is recommended.

5.4.2.2 *Manual or motor-propelled buggies*—Buggies should run on smooth, rigid runways independently supported, and set well above reinforcing steel. Concrete being transferred by buggies tends to segregate during motion; therefore, the planking on which the buggies travel should be butted rather than lapped to maintain the smoothest possible surface and subsequently reduce separation of concrete materials in transit.

The recommended maximum horizontal delivery distance to transfer concrete by manual buggies is 200 ft (60 m), and for power buggies, 1000 ft (300 m). Manual buggies range in capacity from 6 to 8 ft^3 (0.2 to 0.3 m^3) with placing capacities averaging from 3 to 5 yd^3 (3 to 5 m^3) per h. Power buggies are available in sizes from 9 to 12 ft^3 (0.3 to 0.4 m^3) with placing capacities ranging from 15 to 20 yd^3 (14 to 18 m^3) per h, depending on the distance traveled.

5.4.2.3 *Chutes and drop chutes*—Chutes are frequently used for transferring concrete from higher to lower elevations. They should have rounded corners, be constructed of steel or be steel-lined, and should have sufficient capacity to avoid overflow. The slope should be constant and steep enough to permit concrete of the required slump to flow continuously down the chute without segregation.

Drop chutes are circular pipes used for transferring concrete vertically from higher to lower elevations. The pipe should have a diameter of at least eight times the maximum aggregate size at the top 6 to 8 ft (2 to 3 m) of the chute, but can be tapered to approximately six times the maximum aggregate size below. It should be plumb, secure, and positioned so that the concrete will drop vertically. The committee is aware of instances in which concrete has been dropped several thousand feet in this manner without adverse effects.

The flow of the concrete at the end of a chute should be controlled to prevent segregation. Plastic or rubber drop chutes or tremies can be used and shortened by cutting them rather than raising them as placement progresses. When using plastic drop chutes, ensure that the chutes do not fold over or kink.

5.4.2.4 *Paving equipment*—The use of large mixers, high-capacity spreaders, and slipform pavers has made it possible to place large volumes of concrete pavement at a rapid rate. Most of the same principles of quality control are required for successful paving as for other forms of concrete placement. The rapid rate at which concrete pavement is placed necessitates routine inspection procedures to detect any deviations from acceptable quality that should be corrected.

Some of the more frequent problems that can detrimentally affect the quality of the concrete in paving are also common in other types of placement, namely, poor batch-to-batch mixing uniformity, variation in slump and air content, and nonuniform distribution of the paste through the aggregates.

Placing concrete with paving equipment is covered in ACI 325.9R.

5.4.2.5 *Slipforming*—This method entails placing concrete in prefabricated forms that are slipped to the next point of placement as soon as the concrete has gained enough dimensional stability and rigidity to retain its design shape.

Careful, consistent concrete control with suitable mixture adjustments for changing ambient temperatures is required.

5.5—Consolidation

Internal vibration is the most effective method of consolidating plastic concrete for most applications. The effectiveness of an internal vibrator depends mainly on the head diameter, frequency, and amplitude of the vibrators. Detailed recommendations for equipment and procedures for consolidation are given in ACI 309R.

Vibrators should not be used to move concrete laterally. They should be inserted and withdrawn vertically, so that they quickly penetrate the layer and are withdrawn slowly to remove entrapped air. Vibrate at close intervals using a systematic pattern to ensure that all concrete is adequately consolidated (Fig. 5.3).

As long as a running vibrator will sink into the concrete by means of its own weight, it is not too late for the concrete to benefit from revibration, which improves compressive and bond strengths. There is no evidence of detrimental effects either to embedded reinforcement or concrete in partially cured lifts that are revibrated by consolidation efforts on fresh concrete above.

In difficult and obstructed placements, supplemental form vibration can be used. In these circumstances, avoid excessive operation of the vibrators, which can cause the paste to weaken at the formed surface.

On vertical surfaces where air-void holes need to be reduced, use additional vibration. Extra vibration, spading, or mechanical manipulation of concrete, however, are not always reliable methods for removing air-void holes from surfaces molded under sloping forms. Conduct trial placements to determine what works best with a particular concrete mixture.

The use of experienced and competent vibrator operators working with well-maintained vibrators and a sufficient supply of standby units is essential to successful consolidation of fresh concrete.

5.6—Mass concreting

The equipment and method used for placing mass concrete should minimize separation of coarse aggregate from the concrete. Although scattered pieces of coarse aggregate are not objectionable, clusters and pockets of coarse aggregate are and should be scattered before placing concrete over them. Segregated aggregate will not be eliminated by subsequent placing and consolidation operations.

Concrete should be placed in horizontal layers not exceeding 2 ft (610 mm) in depth and inclined layers and cold joints should be avoided. For monolithic construction, each concrete layer should be placed while the underlying layer is still responsive to vibration, and layers should be sufficiently shallow to permit the two layers to be integrated by proper vibration.

The step method of placement should be used in massive structures where large areas are involved to minimize the occurrence of cold joints. In this method, the lift is built up in a series of horizontal, stepped layers 12 to 18 in. (300 to 450 mm) thick. Concrete placement on each layer extends for the full width of the block, and the placement operations progress from one end of the lift toward the other, exposing only small areas of concrete at a time. As the placement progresses, part

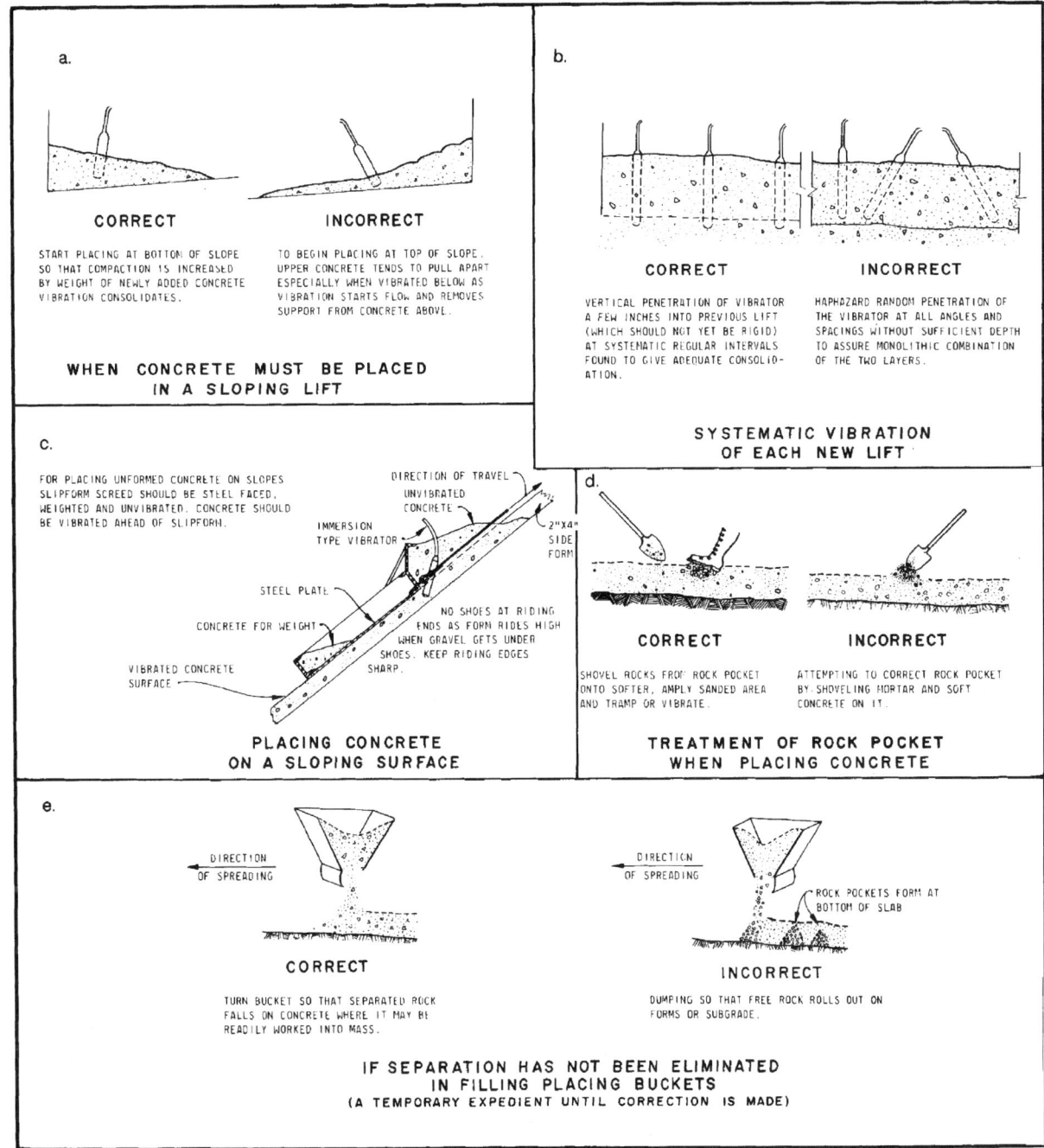

Fig. 5.3—Correct and incorrect methods of consolidation.

of the lift will be completed while concreting continues on the remainder.

For a more complete discussion of mass concrete and the necessary thermal considerations, see ACI 207.1R.

CHAPTER 6—FORMS, JOINT PREPARATION, AND FINISHING

6.1—Forms

Forms are the molds into which concrete is placed and falsework is the structural support and the necessary bracing required for temporary support during construction. Formwork is the total system of support for freshly placed concrete, including forms and falsework. Formwork design should be established before erection, and shop drawings containing construction details, sequence of concrete placing, and loading values used in the design should be approved before construction begins. Shop drawings should be available on site during formwork erection and when placing the concrete.

Design and construction of concrete forms should comply with ACI 347R. The design and construction of concrete formwork should be reviewed to minimize costs without sacrificing either safety or quality. Because workmanship in concrete construction is frequently judged by the appearance of the concrete after removal of the forms, proper performance of formwork while bearing the plastic concrete weight and live construction loading is of vital importance.

Forms should be built with sufficient strength and rigidity to carry the mass and fluid pressure of the actual concrete as well as all materials, equipment, or runways that are to be placed upon them. Fluid pressure on forms should be correlated to the capacity and type of placement equipment, planned rate of placing concrete, slump, temperature, and stiffening characteristics of the concrete.

Form-panel joints, corners, connections, and seams should be mortar-tight. Consolidation will liquefy the mortar in concrete, allowing it to leak from any openings in the formwork, leaving voids, sand streaks, or rock pockets. When forms are set for succeeding lifts, avoid bulges and offsets at horizontal joints by resetting forms with only 1 in. (25 mm) of sheathing overlapping the concrete below the line made by the grade strip from the previous lift and securely tying and bolting the forms close to the joint. The form ties used should result in the minimum practical hole size and their design should permit removal without spalling surrounding concrete. Leakage of mortar around ties should be prevented, and filling of cone holes or other holes left by form ties should be done in a manner that results in a secure, sound, nonshrinking, and inconspicuous patch (ACI 311.1R). Before concreting, forms should be protected from deterioration, weather, and shrinkage by proper oiling or by effective wetting. Form surfaces should be clean and of uniform texture. When reuse is permitted, they should be carefully cleaned, oiled, and reconditioned if necessary.

Steel forms should be thoroughly cleaned and promptly oiled to prevent rust staining. If peeling of concrete is encountered when using steel forms, leaving the cleaned, oiled forms in the sun for a day, vigorously rubbing the affected areas with liquid paraffin, or applying a thin coating of lacquer will usually remedy the problem. Sometimes peeling is the result of abrasion of certain form areas from impact during placement. Abrasion can be reduced by temporarily protecting form areas subject to abrasion with plywood or metal sheets.

Form faces should be treated with a releasing agent to prevent concrete from sticking to the forms and thereby aid in stripping. The releasing agent can also act as a sealer or protective coating for the forms to prevent absorption of water from the concrete into the formwork. Form coatings should be carefully chosen for compatibility with the contact surfaces of the forms being used and with subsequent coatings to be applied to the concrete surfaces. Form coatings that are satisfactory on wood are not always suitable for steel forms; for example, steel forms would require a coating that acts primarily as a releasing agent, whereas plywood requires a coating that also seals the forms against moisture penetration.

Ample access should be provided within the forms for proper cleanup, placement, consolidation, and inspection of the concrete.

For the sake of appearance, proper attention should be paid to the mark made by a construction joint on exposed formed surfaces of concrete. Irregular construction joints should not be permitted. A straight line, preferably horizontal, should be obtained by filling forms to a grade strip. Rustication strips, either a v-shaped or a beveled rectangular strip, can be used as a grade strip and to form a groove at the construction joint when appropriate.

6.2—Joint preparation

Construction joints occur wherever concreting is stopped or delayed so that fresh concrete subsequently placed against hardened concrete cannot be integrated into the previous placement by vibrating. Horizontal construction joints will occur at the levels between lifts, whereas vertical joints occur where the structure is of such length that it is not feasible to place the entire length in one continuous operation. In general, the preparation of a vertical construction joint for acceptable performance and appearance is the same as for horizontal joints.

The surfaces of all construction joints should be cleaned and properly prepared to ensure adequate bond with concrete placed on or adjacent to them and to obtain required water-tightness (U.S. Bureau of Reclamation 1981; Tynes 1959, 1963). Several methods of cleanup are available depending on the size of the area to be cleaned, age of the concrete, skill of workers, and availability of equipment. Creating a satisfactory joint when high-quality concrete has been properly placed is not difficult. When large quantities of bleed water and fines rise to the construction-joint surface, concrete at the surface is so inferior that adequate cleanup becomes difficult. Under normal circumstances, it is necessary only to remove laitance and expose the sand and sound surface mortar by sandblasting or high-pressure water jetting.

Sandblasting is performed to prepare the surface of the construction joint after the concrete has hardened and preferably just before forms are erected for the next placement (U.S. Bureau of Reclamation 1981; Tynes 1959, 1963). Wet sandblasting is usually preferred due to the objectionable dust associated with the dry process. Wet sandblasting produces excellent results on horizontal joint surfaces, particularly on those placed with 2 in. (50 mm) or less slump concrete using internal vibrators.

Another method for cleaning construction joints entails the use of a water jet under a minimum pressure of 6000 psi (40 MPa). As with the sandblasting method, cleanup is delayed until the concrete is sufficiently hard so that only the surface skin of mortar is removed and no undercutting of coarse aggregate particles occurs.

Cloudy pools of water will leave a film on the joint surface when they dry and should be removed by thorough washing after the main cleanup operation is completed. Cleaned joint surfaces should be continuously moist-cured until the next concrete placement or until the specified curing time has elapsed. Before placing new concrete at the joint, the surface should be restored to the clean condition that exists immediately after initial cleanup. If the surface has been properly cured, little final cleaning will be necessary prior to placement.

Hand tools such as wire brushes, wire brooms, hand picks, or bush hammers can be used to remove dirt, laitance, and soft mortar, but are only practical for small areas.

Retarding admixtures can be used, if allowed by the project specifications, to treat concrete surfaces after the finishing operations and before the concrete has set. Manufacturer's instructions for application and coverage rate should be followed. Subsequent removal of the unhardened surface mortar is completed with other cleanup methods such as water jets, air-water jets, or hand tools. Concrete surfaces treated with retarding admixtures should be cleaned as soon as practical after initial set; a longer delay results in less of the retarded surface layer being removed.

The clean concrete joint surface should be saturated, surface dry at the time new concrete is placed on it. Surface moisture weakens the joint by increasing the *w/cm* of the newly placed concrete. Ensure that the first layer of concrete on the construction joint is adequately consolidated to achieve good bond with the previously hardened concrete.

6.3—Finishing unformed surfaces

To obtain a durable surface on unformed concrete, proper procedures should be carefully followed. The concrete used should be of the lowest practical slump that can be properly consolidated, preferably by means of internal vibration. Following consolidation, the operations of screeding, floating, and first troweling should be performed in such a manner that the concrete will be worked and manipulated as little as possible to produce the desired result.

Overmanipulation of the concrete brings excessive fines and water to the surface, which lessens the quality of the finished surface, causing checking, crazing, and dusting. For the same reason, each step in the finishing operation, from the first floating to the final floating or troweling, should be delayed as long as possible while still working toward the desired grade and surface smoothness. Free water is not as likely to appear and accumulate between finishing operations if proper mixture proportions and consistency are used. If free water does accumulate, however, it should be removed by blotting with mats, draining, or pulling off with a loop of hose so that the surface loses its water sheen before the next finishing operation is performed. Under no circumstances should any finishing tool be used in an area before accumulated water has been removed, nor should neat cement or mixtures of sand and cement be worked into the surface to dry such areas.

Satisfactory results can be achieved from a correctly designed mortar topping placed on, and worked into, base concrete before the base concrete sets. The mortar consistency, consolidation, and finishing should be as described previously. A concrete of correct proportions, consistency, and texture placed and finished monolithically with the base concrete, however, is preferable to a mortar topping. See ACI 302.1R for a detailed discussion and recommendations on concrete floor and slab finishing.

Several special floor finishes, such as terrazzo, that are installed over cured concrete surfaces require special techniques and are not covered in this guide.

CHAPTER 7—PREPLACED-AGGREGATE CONCRETE

7.1—General considerations

In this method of construction, forms are first filled with clean, coarse aggregate. The voids in this coarse aggregate are then filled with structural quality grout to produce preplaced-aggregate (PA) concrete. This type of concrete is particularly useful where concrete is to be placed under water, where structures are heavily reinforced for seismic or other reasons, where structural concrete or masonry is to be repaired, or where concrete of low volume change is required (U.S. Bureau of Reclamation 1981; Davis and Haltenhoff 1956; Davis et al. 1955; Anon. 1954; King 1971; Davis 1958; Corps of Engineers 1994a).

PA concrete differs from conventionally placed concrete in that it contains a higher percentage of coarse aggregate; consequently, the properties of the coarse aggregate have a greater effect on the properties of the concrete. For example, the modulus of elasticity is slightly higher than that of conventional concrete. Also, because of point-to-point contact of the coarse aggregate, drying shrinkage is approximately 1/2 the magnitude of that in conventionally placed concrete (Davis 1958, Davis 1960). Structural design for PA concrete, however, is the same as for conventionally placed concrete (U.S. Bureau of Reclamation 1981, Corps of Engineers 1994a).

Structural formwork for PA concrete is usually more expensive than that required for conventionally placed concrete because greater care is needed to prevent grout leaks. In underwater construction, higher placing rates at lower cost have been achieved by this method than by conventional placement methods.

Because PA concrete construction is specialized in nature, the work should be undertaken by qualified personnel experienced in this method of construction. Detailed information on all aspects of PA concrete is given in ACI 304.1R.

7.2—Materials

7.2.1 *Cement*—Grout can be made with any one of the nonair-entraining types of cement that complies with ASTM C 150 or ASTM C 595M. Use of air-entrained cements combined with gas-forming fluidifiers could result in excessive quantities of entrained air in the grout, resulting in reduced strengths. When air entrainment is required to a higher extent than that provided by the gas-forming fluidifier, air-entraining agent should be added separately.

7.2.2 *Coarse aggregate*—Coarse aggregate should be washed, free of surface dust and fines, and in conformance with the requirements of ASTM C 33, except as to grading.

The void content of the aggregate should be as low as possible and is usually attained when the coarse aggregate is graded uniformly from the smallest allowable particle size to the largest (King 1971).

Grading 1 or 2 (Table 7.1) is recommended for general use. Where reinforcement is crowded or the placement is in relatively shallow patches, Grading 1 should be used. Where special circumstances dictate the use of coarser sand, Grading 3 is acceptable.

7.2.3 *Fine aggregate*—Sand should conform to ASTM C 33, except that grading should be as shown in Table 7.1. Fine aggregate that does not fall within these grading limits is usable provided results fall within the requirements of Section 7.3.

7.2.4 *Pozzolan*—Pozzolans conforming to ASTM C 618, Class N or F, can be used in PA concrete. Class F has been used in the great majority of installations as it improves pumpability of the fluid grout and extends grout handling time. Class C fly ash and blast-furnace slag have been used to a limited extent, but extensive data on grout mixture proportions and properties are not currently available.

7.2.5 *Admixtures*

7.2.5.1 *Grout fluidifier*—This admixture is commonly used to offset the effects of bleeding, reduce the *w/cm* for a given fluidity, and retard stiffening. The usual dosage of grout fluidifier is 1% by weight of the total cementitious material in the grout mixture.

7.2.5.2 *Calcium chloride*—A small quantity of calcium chloride may be desirable to promote early strength development. Calcium chloride in excess of 1% by weight of cementitious materials, however, will diminish the expansive

Table 7.1—Gradation limits for coarse and fine aggregates for preplaced-aggregate concrete

	Percentage passing		
	Coarse aggregate		
Sieve size	Grading 1	Grading 2	Grading 3
	For 1/2 in. (1.25 mm) minimum size coarse aggregate	For 3/4 in. (19 mm) minimum size coarse aggregate	For 1-1/2 in. (38 mm) minimum size coarse aggregate
1-1/2 in. (37.5 mm)	95 to 100	—	0 to 5
1 in. (25.0 mm)	40 to 80	—	—
3/4 in. (19.0 mm)	20 to 45	0 to 10	—
1/2 in. (12.5 mm)	0 to 10	0 to 2	—
3/8 in. (9.5 mm)	0 to 2	0 to 1	—
	Fine aggregate (sand)		
No. 4 (4.75 mm)	—		100
No. 8 (2.36 mm)	100		90 to 100
No. 16 (1.18 mm)	95 to 100		80 to 90
No. 30 (600 microns)	55 to 80		55 to 70
No. 50 (300 microns)	30 to 55		25 to 50
No. 100 (150 microns)	10 to 30		5 to 30
No. 200 (75 microns)	0 to 10		0 to 10
Fineness modulus	1.30 to 2.10		1.60 to 2.45

action of the aluminum powder, if present, in the grout fluidifier because the acceleration will reduce the time available for expansion to take place. Pretesting for expansion, bleeding, rate of curing, and strength in PA concrete cylinders is recommended (refer to ASTM C 953).

7.3—Grout proportioning

7.3.1 *Cementitious materials*—Usually, the proportion of portland cement-to-pozzolan is in the range of 2.5:1 to 3.5:1 by mass. Ratios as low as 1.3:1 (equal bulk volumes) for lean mass concrete and as high as 12:1 for high-strength concrete have been used. The w/cm usually ranges from 0.42 to 0.50.

7.3.2 *Fine aggregate*—Compressive strength, pumpability (Anon. 1954; King 1971), and void-penetration requirements control the amount of fine aggregate that can be used in the grout. For structural grade PA concrete, the ratio of cementitious material-to-fine aggregate will usually be 1:1 by mass. For massive placements where the minimum size of coarse aggregate is 3/4 in. (19 mm), the ratio may be increased to 1:1.5. With Grading 3 (Table 7.1), the ratio may be further increased to approximately 1:3.

7.3.3 *Proportioning requirements*—Materials should be proportioned in accordance with ASTM C 938 to produce a grout of required consistency that will provide the specified strength of PA concrete. For best results, bleeding should be less than the total measured expansion. Strength, bleeding, and expansion should be tested according to ASTM C 943.

7.3.4 *Consistency of grout*—For most work, such as walls and structural repairs, a 22 ± 2 s flow (ASTM C 939) is usually satisfactory. For massive sections and underwater work, the flow can be as low as 20 ± 2 s or as high as 24 ± 2 s.

Where special care can be taken in the execution of work and higher strengths are required, flows as high as 35 to 40 s can be used.

7.4—Temperature control

For mass concrete placements, temperature rise in PA concrete can be limited by one or more of the following procedures: chilling coarse aggregate before placement; chilling coarse aggregate in place; chilling the grout with chilled mixing water; and reducing the cement content to the minimum for obtaining the desired properties. Refer to ACI 207.2R and ACI 224R for more detail.

7.5—Forms

Forming materials for PA concrete are similar to those for conventionally placed concrete. The forms, however, should be tight enough to prevent grout leakage and resist high lateral pressures (refer to ACI 347R). After the forms are erected, shored, properly braced, and set to line and grade, all small openings should be caulked. All joints between adjacent panels should be sealed on the inside of the form with tape. Specifications may require that a layer of water 1 to 2 ft (0.3 to 0.6 m) deep be maintained above the rising grout surface to ensure saturation of the coarse aggregate particles. In these cases, the forms should be essentially watertight.

7.6—Grout pipe systems

7.6.1 *Delivery pipes*—The most reliable grout delivery system consists of a single line. To provide for continuous grout flow, a y-shaped fitting can be incorporated. The grout should be injected through only one leg of the y at a time.

The delivery line should be of sufficient diameter to allow grout velocity at the planned operating rate to range between 2 and 4 ft/s (0.6 and 1.2 m/s).

High-pressure grout hose, 400 psi (3 MPa) or higher, is commonly used for delivery lines. A hose diameter of 1-1/4 or 1-1/2 in. (30 or 40 mm) is preferred for distances up to 500 ft (150 m). For longer distances, up to approximately 1000 ft (300 m), 2 in. (50 mm) diameter is preferred.

7.6.2 *Grout insertion pipes*—Insertion pipes are used to inject the grout into the aggregate mass and are normally schedule 40 pipe, 3/4 to 1-1/4 in. (20 to 30 mm) diameter for normal structural concrete and up to 1-1/2 in. (40 mm) for mass concrete. The grout insertion pipes should extend vertically to within 6 in. (150 mm) of the bottom of the aggregate mass, or they can extend horizontally through the formwork at different elevations. When insert pipes are required in depths of aggregate exceeding approximately 50 ft (15 m), flush-coupled schedule 120 pipe or flush-coupled casing is recommended. For deep placements, such as caissons in deep water, telescoping-insertion pipes can be required.

7.6.3 *Vent pipes*—Vent pipes should be used where water or air can be entrapped by the rising grout surface, such as beneath a blockout or under some embedments. Grout is usually injected through insert pipes until it returns through these vent pipes.

7.7—Coarse aggregate placement

7.7.1 *Preparation for placement*—Coarse aggregate should be washed and screened immediately before placing in forms. Coarse aggregate should not be flushed with water after placement in the forms (Anon. 1954; King 1971). This will cause fines to accumulate in the lower strata of aggregate. When it is necessary to flood the coarse aggregate to obtain saturation or precooling (King 1971), the water should be injected through the insert pipes so that the water rises gently through the coarse aggregate.

For underwater placement, all loose, fine material should be removed from the foundation area before placement of aggregate to prevent subsequent coating of the aggregate or filling of voids with stirred-up sediment. Where the concrete will bear on piles, it is only necessary to remove soft material a sufficient depth below pipe encasement depth to provide for a filter cloth on the mud. Additionally, a layer of aggregate is carefully dropped on top of the cloth to stabilize it and form a base for the bulk of the coarse aggregate to follow.

7.7.2 *Aggregate placement*—For structural concrete work, aggregate is commonly delivered to the forms in concrete buckets and placed through a flexible elephant trunk to prevent segregation and breakage of the aggregate. A pipe having a diameter of at least four times the maximum aggregate size has been used for lowering aggregate preplaced under water to depths ranging from 50 to 1000 ft (15 to 300 m) (Davis, Johnson, and Wendell 1955). The pipe is normally lowered to bottom contact, then gradually filled. Discharge is then controlled by raising the pipe only enough to permit discharge at a controllable rate. Where coarse aggregate is being placed through water, it can be discharged directly into the water from bottom-dump barges or self-unloading ships (Davis and Haltenhoff 1956).

Coarse aggregate can also be blown into place around tunnel liners by using 6 in. (150 mm) or larger pipe and large volumes of low-pressure air (Davis, Johnson, and Wendell 1955).

In most placements, there is little to be gained from attempts to consolidate the coarse aggregate in place by rodding or vibration. Rodding or compressed-air lances can be used, however, to achieve placement into heavily reinforced areas and in the construction of overhead repairs.

Around closely spaced piping, reinforcement, and penetrations, such as in some nuclear shielding situations where uniform high density and homogeneity are desired, hand placement in shallow lifts may be required.

7.7.3 *Contamination*—In underwater construction where organic contamination is known or suspected to exist, sample and test the water to estimate the rate of sludge build-up on immersed aggregate and its possible influence on the quality of the concrete.

7.8—Grout mixing and pumping

7.8.1 *Mixers*—Vertical-spindle, paddle-type, and double-tub mixers are commonly used for mixing grout. One tub serves as a mixer while the second, from which grout is being withdrawn, serves as an agitator. Horizontal shaft mixers are used for large-volume work. A separate agitator is used to provide continuous operation.

Pan or turbine mixers are well-suited for mixing grout, although maintenance of a tight seal at the discharge gate can be difficult. Conventional revolving-drum concrete mixers are suitable if the mixing is sufficiently prolonged to ensure thorough mixing. The colloidal, or shear mixer, provides extremely high-speed, first-stage mixing of cement and water in a close-tolerance centrifugal pump followed by mixing of the cement slurry with sand with an open-impeller pump. This type of mixer provides a relatively bleed-free mixture, but because of high-energy input, mixing time should be limited to avoid heating the grout.

7.8.2 *Pumps*—The pump should be a positive-displacement pump such as the piston or progressive cavity type. The pump should be equipped with a bypass line connecting the discharge with the pump inlet or the agitator. On large jobs, providing standby equipment so that continuous discharge can be provided is prudent. A pressure gauge should be installed on the pump line discharge in clear view of the pump operator to indicate incipient line blockage.

7.8.3 *Grout injection*—There are essentially two basic patterns of grout injection: the horizontal layer and advancing slope techniques. With both systems, grout should start from the lowest point within the forms.

In the horizontal layer technique, grout is injected through each insert pipe to raise the grout a short distance at the point of injection, and by sequential injection through adjacent insert pipes, a layer of coarse aggregate is grouted before proceeding to the next horizontal layer above. When injecting through vertical-insert pipes, the injection pipes are withdrawn after each injection, leaving the lower end of the insert pipe embedded a minimum of 1 ft (0.3 m) below the grout surface. When injecting through ports in the forms or horizontal insert pipes, grouting should be continuous through the injection point until grout flows from the next higher point. For the next lift of grout, injection should be into the insert point next above that just completed.

When the horizontal surface procedure is not practical, as when plan dimensions are relatively large compared to the depth, the advancing slope method is used. Intrusion is started at one end of the narrowest dimension of the form and pumping is continued through the first row of insert pipes until the grout appears at the surface. The surface of the grout within the submerged aggregate will assume a generally vertical-to-horizontal slope ranging from 1:5 to 1:10. The slope is advanced by pumping grout through successive rows of insert pipes until the entire slab has been grouted.

Normal injection rates through a given insert pipe vary from less than 1 ft^3/min (0.03 m^3/min) to over 4 ft^3/min (0.11 m^3/min). For a particular application, the injection rate will depend on form configuration, aggregate voids, and grout fluidity.

7.8.4 *Grout surface determination*—The location of the grout surface within the aggregate mass should be known at all times. When grout is injected horizontally through the side of the formwork, grout location can be readily determined by flow from adjacent grouting points, the location of seepage through the forms, or with the aid of closable inspection holes through the forms. Where grout is injected through vertical-insert pipes, sounding wells should be provided. These sounding wells usually consist of 2 in. (50 mm) diam-

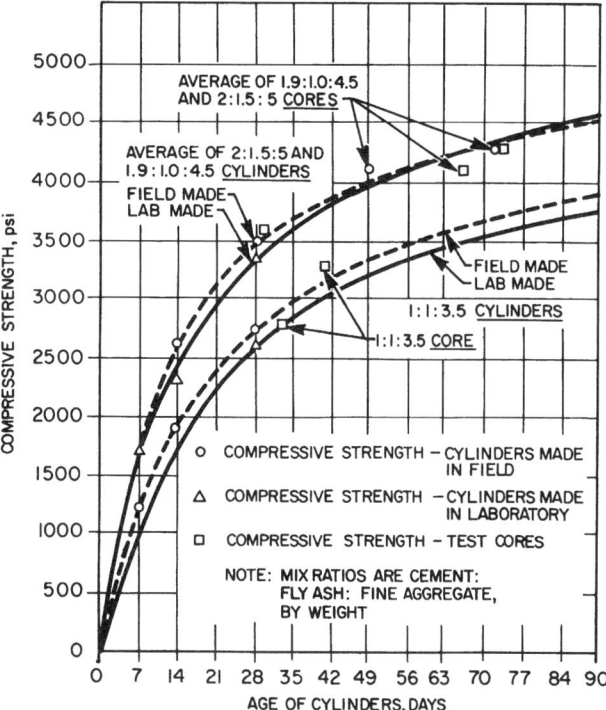

Fig. 7.1—Comparison of results, field- and lab-made cylinders versus cores.

eter thin-wall pipe with 1/2 in. (12 mm) milled (not burned) slots at frequent intervals. Partially rolled, unwelded tubing providing a continuous slot can also be used. The sounding line is equipped with a 1 in. (25 mm) diameter float weighted to sink in water, yet float on the grout surface, within the slotted pipe. Sounding wells are usually left in place and become a permanent part of the structure.

7.9—Joint construction

7.9.1 *Cold joints*—Cold joints are formed within the mass of preplaced aggregate concrete when pumping is stopped for longer than the grout remains plastic. When this occurs, the insert pipes should be pulled to just above the grout surface before the grout stiffens and rodded clear. To resume pumping, the pipes should be worked back to near contact with the hardened grout surface and pumping resumed, slowly for a few minutes, to create a mound of grout around the end of the pipe.

7.9.2 *Construction joints*—Construction joints can be formed in the same manner as cold joints by stopping the grout rise approximately 12 in. (300 mm) below the aggregate surface. Dirt and debris should be prevented from filtering down to the grout surface.

If construction joints are made by bringing the grout up to the surface of the coarse aggregate, the surface should be green-cut, chipped, or sandblasted to present a clean, rough surface for the new grout in the next lift.

7.10—Finishing

Exercise care when topping out to control the grout injection rate and avoid lifting or dislodging the surface aggregate (Anon. 1954). Coarse aggregate at or near the surface can be held in place by wire screening, which is removed before finishing.

Low-frequency, high-amplitude external vibration of forms at or just below the grout surface will permit grout to cover aggregate-form contacts, thereby providing an excellent, smooth surface appearance. Excessive form vibration will cause bleeding, the usual result being sand streaking from the upward movement of the bleed water. Internal vibration should only be used in short bursts to level the grout between insert pipes for topping out purposes. When a screeded or troweled finish is required, the grout should be brought up to flood the aggregate surface and any diluted surface grout should be removed by brooming. A thin layer of pea gravel is then worked down into the surface by raking followed by tamping. When the surface is sufficiently hardened to permit working, a screeded, floated, or troweled finish is then applied.

7.11—Quality control

Job site control of fresh grout characteristics is maintained by following the appropriate ASTM methods. Compressive strength of PA concrete should be determined in accordance with procedures given in ASTM C 943. The strength of grout alone, when determined in cubes or cylinders, may bear little relation to the strength of PA concrete made with the same grout because these units do not duplicate the weakening effect of excessive bleeding of the grout in place. Properly made PA concrete cylinders, however, bear a close relationship to cores taken from the concrete in place. A typical comparison of lab-made and field-made cylinders with cores taken from a major installation is given in Fig. 7.1.

CHAPTER 8—CONCRETE PLACED UNDER WATER
8.1—General considerations

Typical underwater concrete placements include nonstructural elements such as cofferdams or caisson seals, and structural elements such as bridge piers, dry-dock walls and floors, and water intakes. Concrete placed under water has also been used to add weight to sink precast tunnel sections, to join tunnel sections once in place, and to repair erosion or cavitation damage to major hydraulic structures (Gerwick 1964; Gerwick, Holland, and Kommendant 1981).

8.1.1 *Scope*—The recommendations given in this chapter are directed toward relatively large-volume placements of concrete under water, but these recommendations are also generally applicable to small-volume underwater placements, such as thin overlays or deep confined placements. The reader is cautioned to consider the specific problems associated with these placements and how they differ from typical placements.

8.1.2 *Methods available*—The tremie is currently the most frequently used technique to place concrete under water, but use of direct pumping is increasing. These two methods are similar and are described in this chapter.

8.1.3 *Basic technique*—Successful placement of concrete under water requires preventing flow of water across or through the placement site. Once flow is controlled, either tremie or pump placement consists of the following three steps:

1. The first concrete placed is physically separated from the water by using a go-devil or pig in the pipe, or by having the pipe mouth sealed and the pipe dewatered;

2. Once filled with concrete, the pipe is raised slightly to allow the go-devil to escape or to break the end seal. Concrete will then flow out and develop a mound around the mouth of the pipe. This is termed establishing a seal; and

3. Once the seal is established, fresh concrete is injected into the mass of existing concrete. The exact flow mechanism that takes place is not precisely known, but the majority of the concrete apparently is not exposed to direct contact with the water (Gerwick, Holland, and Kommendant 1981).

8.2—Materials

8.2.1 *General requirements*—Concrete materials should meet all appropriate specifications. In addition, materials should be selected for their contribution toward improved concrete flow characteristics.

8.2.2 *Aggregates*— The maximum size of aggregates used in reinforced placements under water is usually 3/4 in. (19 mm). Larger aggregates (1 in. [25 mm]) can be used depending on availability, reinforcing spacing, and maintenance of the workability of the concrete. The maximum size of aggregates for nonreinforced placements should be 1-1/2 in. (38 mm).

8.2.3 *Admixtures*— Admixtures to improve the characteristics of fresh concrete, especially flowability, are frequently used in concrete placed under water (Williams 1959). For example, an air-entraining admixture can be beneficial because of the increased workability that can be achieved with its use.

Water-reducing or water-reducing and retarding admixtures are particularly beneficial in reducing water content to provide a cohesive yet high-slump concrete. Retarding admixtures are beneficial in a large monolithic placement. Because of the extreme importance of maintaining as high a slump as possible for as long as possible, the use of a high-range water-reducing admixture (HRWR) for massive placements is not recommended, unless slump-loss testing has shown no detrimental results. The use of HRWR for smaller volume placements in which flow distances are not as critical may be acceptable.

Admixtures are also available to prevent washout of cementitious materials and fines from concrete placed under water. These antiwashout admixtures are discussed in Section 8.10.

8.3—Mixture proportioning

8.3.1 *Basic proportions*—Pozzolans (approximately 15% by mass of cementitious materials) are generally used because they improve flow characteristics. Relatively rich mixtures, 600 lb/yd^3 (356 kg/m^3) cementitious materials, or more, or a maximum *w/cm* of 0.45 are recommended. Fine aggregate contents of 45 to 55% by volume of total aggregate and air contents of up to approximately 5% are generally used. Refer to 8.8.5 for thermal cracking considerations.

A slump of 6 to 9 in. (150 to 230 mm) is generally necessary, and occasionally a slightly higher range is needed when embedded items obstruct the flow or when relatively long horizontal flow is required.

8.3.2 *Final selection*—If possible, the final selection of a concrete mixture should be based on test placements made under water in a placement box or in a pit that can be dewatered after the placement. Test placements should be examined for concrete surface flatness, amount of laitance present, quality of concrete at the extreme flow distance of the test, and flow around embedded items, if appropriate.

8.4—Concrete production and testing

8.4.1 *Production sampling and testing*—Sampling should be done as near to the tremie hopper as possible to ensure that concrete with the proper characteristics is arriving at the tremies. Once a concrete mixture has been approved, slump, air content, unit weight, and compressive strength testing should be adequate for production control. Because of the importance of the flowability of the concrete to the success of the placement, slump and air content tests should be performed more frequently than is usually done for concrete not placed under water.

Compressive strength specimens should be available for testing at early ages to determine when the concrete has gained enough strength to allow dewatering of the structure.

8.4.2 *Concrete temperature*—The concrete temperature should be kept as low as practical to improve placement and structural qualities. Depending on the volume of the placement and the anticipated thermal conditions within the placement, maximum temperatures in the range of 60 to 90 F (16 to 32 C) are normally specified. While concrete placed under water obviously cannot freeze, a minimum concrete temperature of 40 F (5 C) should be maintained. Because heating either water or aggregates can cause erratic slump-loss behavior, extreme care should be taken when such procedures are used to raise the concrete temperature.

8.5—Tremie equipment and placement procedure

8.5.1 *Tremie pipes*—The tremie should be fabricated of heavy-gage steel pipe to withstand all anticipated handling stresses. In deep placements, buoyancy of the pipe can be a problem if an end plate is used to gain the initial tremie seal. Use of pipe with thicker walls or weighted pipe can overcome buoyancy problems.

Tremie pipes should have a diameter large enough to ensure that aggregate-induced blockages will not occur. Pipes in the range of 8 to 12 in. (200 to 300 mm) diameter are adequate for the range of aggregates recommended herein. For deep placements, the tremie should be fabricated in sections with joints that allow the upper sections to be removed as the placement progresses. Sections can be jointed by flanged, bolted connections, (with gaskets) or screwed together. Whatever joint technique is selected, joints between tremie sections should be watertight and should be tested for watertightness before beginning placement. The tremie pipe should be marked to allow quick determination of the distance from the surface of the water to the mouth of the tremie.

The tremie should have a suitably sized funnel or hopper to facilitate transfer of concrete from the delivery device to the tremie. A stable platform should be provided to support the tremie during placement. Floating platforms are generally not suitable. The platform should be capable of supporting the tremie while sections are being removed from the upper end of the tremie.

8.5.2 *Placement procedures*—All areas in which there is to be bond between steel, wood, or cured concrete and fresh concrete should be thoroughly cleaned immediately before concrete placement.

8.5.2.1 *Pipe spacing*—Pipe spacing should be on the order of one pipe for every 300 ft^2 (28 m²) of surface area or pipes on approximately 15 ft (4.5 m) centers. These spacings are recommended, but concrete has been placed that flowed as far as 70 ft (21 m) with excellent results. For most large placements, it will not be practical to achieve a pipe spacing as close as 15 ft (5 m) on centers simply because it would be impractical to supply concrete to the number of tremies or pumps involved.

Actual pipe spacing should be established on the basis of the thickness of the placement, congestion due to piles or reinforcing steel, available concrete production capacity, and available capacity to transfer concrete to the tremies. The placement method selected should also be considered.

8.5.2.2 *Starting placements*—Tremies started using the end-plate, dry-pipe technique should be filled with concrete before being raised off the bottom. The tremie should then be raised a maximum of 6 in. (150 mm) to initiate flow. These tremies should not be lifted further until a mound is established around the mouth of the tremie pipe. Initial lifting of the tremie should be done slowly to minimize disturbance of material surrounding the mouth of the tremie.

Tremies started using a go-devil should be lifted a maximum of 6 in. (150 mm) to allow water to escape. Concrete should be added to the tremie slowly to force the go-devil downward. Once the go-devil reaches the mouth of the tremie, the tremie should be lifted enough to allow the go-devil to escape. After that, a tremie should not be lifted again until a sufficient mound is established around the mouth of the tremie.

Tremies should be embedded in the fresh concrete 3 to 5 ft (1.0 to 1.5 m) deep. Exact embedment depths will depend on placement rates and setting time of the concrete. All vertical movements of the tremie pipe should be done slowly and carefully to prevent loss of seal. If loss of seal occurs in a tremie, placement through that tremie should be halted immediately. The tremie should be removed, the end plate should be replaced, and flow should be restarted as described above. To prevent washing of concrete already in place, a go-devil should not be used to restart a tremie after loss of seal.

8.5.2.3 *Placing*—Concrete placement should be as continuous as possible through each tremie. Excessive delays in placement can cause the concrete to stiffen and resist flow when placement resumes.

Placement interruptions of up to approximately 30 min should allow restarting without any special procedures. Interruptions of between 30 min and the initial setting time of the concrete should be treated by removing, resealing, and restarting the tremie. Interruptions of a duration greater than the initial setting time of the concrete should be treated as a construction joint. If a break in placement results in a planned (or unplanned) horizontal construction joint, the concrete surface should be green-cut after it sets. Green-cutting by a diver is difficult but can be accomplished where there is no practical alternative for cleaning. The concrete surface should be water-jetted immediately before resuming concrete placement.

Recommendations on the rate of concrete rise are generally in the range of 1 to 10 ft/h (0.3 to 3 m/h). Calculation of a projected rate is somewhat difficult because the exact flow pattern of the concrete will not be known. The most logical approach is to compare concrete production with the entire area that is being supplied. As with pipe spacing, achieving the recommended values can be difficult. Concrete has been successfully placed under water at rates of approximately 0.5 ft (150 mm) of rise per h (Gerwick, Holland, and Kommendant 1981).

The volume of concrete in place should be monitored throughout the placement. Underruns (using less concrete than anticipated) are indicative of loss of tremie seal, because the washed and segregated aggregates will occupy a greater volume. Overruns (using more concrete than anticipated) are therefore also indicative of loss of concrete from the forms.

Once the placement scheme has been developed, flow distances and rates of rise can be calculated. If flow distances seem excessive or if the rate of concrete rise is too low, make a judgment as to the suitability of the available plant or the necessity for breaking the placement into smaller segments.

Tremie blockages that occur during placement should be cleared extremely carefully to prevent loss of seal. If a blockage occurs, the tremie should be quickly raised 6 in. to 2 ft (150 to 610 mm) and then lowered in an attempt to dislodge the blockage. The depth of pipe embedment should be closely monitored during all such attempts. If the blockage cannot be cleared readily, the tremie should be removed, cleared, resealed, and restarted.

8.5.2.4 *Horizontal distribution of concrete*—The pipe delivering concrete should remain fixed horizontally while concrete is flowing. Horizontal movement of the pipe will damage the surface of the concrete in place, create additional laitance, and lead to loss of seal. Horizontal distribution of the concrete is accomplished by flow of the concrete after exiting the pipe or by halting placement, moving the pipe, reestablishing the seal, and resuming placement.

Two methods are typically used to achieve horizontal concrete distribution in large placements: the layer method or the advancing slope method. In the horizontal layer method, the entire area of the placement is concreted simultaneously using a number of tremies. With the advancing slope method, one portion of the placement is brought to finished grade and then the tremies are moved to bring adjacent low areas to grade. Work normally progresses from one end of a large placement to the other. Concrete slopes from nearly flat to 1:6 (vertical to horizontal) can be expected.

8.5.3 *Postplacement evaluation*—To evaluate the underwater placement, the following techniques can be used:
- Coring in areas of maximum concrete flow or in areas of questionable concrete quality;
- After dewatering, accurately surveying the concrete surface to evaluate the adequacy of the concrete mixture and the placement plan; and
- After removal of forms or sheet piling, inspecting the exterior surface of the concrete with divers for evidence of cracking, voids, or honeycomb.

8.6—Direct pumping

Tremie placement techniques are generally applicable to direct pump placement under water. The following minor differences, however, are worth noting:
- The mechanism causing concrete flow through the pipeline is pump pressure rather than gravity;
- The concrete should be proportioned for flow after leaving the pipe rather than simply for pumping;
- Pipes are typically smaller than those used for tremies. Rigid sections should always be used for the portion actually embedded in the concrete;
- The pump action can cause some lateral movement of the pipe where it is embedded in the fresh concrete; this movement can contribute to laitance formation by drawing fines to the pipe-concrete interface; and
- A relief valve (air vent) can be required near the highest point in the pipeline to prevent development of a vacuum blockage.

8.7—Concrete characteristics

Concrete placed under water can be expected to be of excellent quality. Curing conditions are excellent and drying shrinkage is minimal. Compressive strengths of the rich mixtures used will often be from 4000 to 8000 psi (28 to 55 MPa). There is no evidence that other structural properties differ from those of similar concretes placed in the dry. In-place unit weight, often critical in massive placements to offset hydrostatic uplift, will be close to that measured for the fresh concrete before placement. If laitance is entrapped in the concrete, however, unit weight can be significantly below that of the fresh concrete.

Although there have been recent attempts to ascertain the quality and homogeneity of concrete placed under water using nondestructive techniques (Laine et al. 1980), coring is still the recommended technique for evaluation of questionable areas.

8.8—Precautions

The precautions in this section are applicable to either tremie or pump placement.

8.8.1 *Inspection*—Inspection of concrete placements under water is difficult. The water itself will become increasingly murky as the placement progresses and the surface of the fresh concrete will not support the weight of a diver. Therefore, preplacement inspection becomes extremely important and should concentrate on reviews of the proposed procedures and equipment and the proposed concrete mixture. Inspection during the placement will be limited to observing all phases of the concrete production, transportation, and placement procedures. Because the success of an underwater placement depends largely on the concrete itself, sampling and testing during the placement are critical to ensure compliance with approved mixtures and required concrete characteristics (slump, air content, temperature).

An inspection plan detailing locations and frequency of soundings should be developed. Soundings should be taken over the entire area of the placement on a regular basis, such as every hour or every 200 yd^3 (75 m^3). Locations for taking soundings should be marked on the structure to ensure that all soundings are made at the same location. Additionally, soundings should be required on a more frequent basis adjacent to each tremie to monitor pipe embedment. Data obtained from soundings should be plotted immediately to monitor the progress of the placement.

8.8.2 *Loss of seal*—The most common cause of loss of seal is excessive vertical movement of the pipe to clear a blockage or to remove a pipe section. With either placement method, the loss of seal likely will result in washing and segregation. A related and similar problem is the failure to establish a satisfactory seal at the beginning of a placement.

8.8.3 *Go-devils*—The use of go-devils has traditionally been advocated as a technique for sealing tremies or pump lines. Although the technique is effective, the water that is forced out of the pipe ahead of the go-devil can wash and scour the material underlying the placement area. This condition can be alleviated by the placement of a layer of properly graded rock before the start of concreting.

When a pipe is relocated during a placement, the water forced out of the pipe will wash previously placed concrete, resulting in extreme segregation, laitance formation, and possibly entrapped zones of uncemented aggregates. Therefore, the use of a go-devil at the beginning of a placement is acceptable, but not to restart a tremie or pumping line during a placement.

8.8.4 *Laitance*—Because it is physically impossible to separate the concrete and the water completely, a certain amount of laitance will be formed. If the seal is lost, or if the concrete is disturbed in any way, additional laitance will be formed when starting or restarting pipes. The laitance will flow to and accumulate in any low areas on the surface of the concrete. Such accumulations can prevent sound concrete from filling an area and can become entrapped by subsequent concrete flows. In either case, the zones of laitance will be more permeable and lower in strength. Problems with laitance can be avoided by using pumps or air-lifts during the placement to remove unsuitable material as it accumulates. Another way of reducing laitance problems is to discard several inches of concrete from the form. This can only be done where the top of the form coincides with the top of the placement.

8.8.5 *Cracking*—Problems associated with heat development and subsequent cracking in massive underwater placements have generally not been resolved. The following characteristics, however, of underwater placements should be considered.

8.8.5.1 *Cement content*—Underwater concrete mixtures have traditionally used high cement contents (650 lb/yd^3 [385 kg/m^3] or more) to compensate for cement washout and to provide the necessary flow characteristics to the concrete. Measurements made on one large placement indicated maximum internal concrete temperature in excess of 95 F (35 C) above the placement temperature of 60 F (16 C) (Gerwick, Holland, and Kommendant 1981).

8.8.5.2 *Placement environment*—Tremie concrete is usually placed in locations that act as excellent heat sinks. The temperature of the water surrounding the concrete will normally vary little; thus, the outside of the concrete mass cools quickly, developing steep temperature gradients. In the placement mentioned previously, the concrete temperature varied from 150 F (66 C) to river temperature 55 F (13 C) in only 40 in. (1 m).

8.8.5.3 *Volume*—To eliminate construction joint preparation under water, placements tend to be large monoliths placed over short periods of time.

8.8.5.4 *Restraint*—Underwater placements are frequently made on rock or contain many piles with the concrete acting as a pile cap. In either case, a high degree of restraint can be present.

Of the methods recommended for controlling cracking in mass concrete, modifying the materials or mixture proportions appears to have the greatest potential for application in underwater placements. In particular, use of lower-heat cements, replacement of 15 to 30% of the cement with a suitable pozzolan, and cooled aggregates and water are recommended. It is conceivable, but as yet untried, to provide internal cooling using the water available at the site or to include insulation in the fabrication of forms used in structural placements. The reader is referred to the work of Carlson, Houghton, and Polivka (1979), Gerwick and Holland (1983), and ACI 224R for additional information on cracking.

8.8.6 *Detailing*—Concrete placed under water moves to its final position in the structure by gravity, without vibration and inspection. Therefore, all formwork, reinforcing steel, and precast elements to be filled with concrete should be de-

tailed with underwater placement in mind and incorporate the following:
- Reinforcing steel should be sized and placed to allow the maximum possible openings between bars so that concrete flow will not be impeded;
- Forms should be adequately sealed to prevent loss of concrete or mortar; and
- Forms and reinforcing steel should not trap laitance in areas intended to be filled with concrete.

8.8.7 *Preplacement planning*—Underwater concrete placements are infrequent and cannot be treated as just another concrete operation. Planning for an underwater placement should begin as soon as the decision to do the project has been made. Items that have a long lead time include detailing of reinforcing steel (if any), detailing of forms, consideration of overexcavating the placement area to avoid concrete removal if concrete placed under water is above design grade, and consideration of incorporating members required to support the tremie platforms into the internal bracing scheme of a cofferdam, if appropriate.

Consideration of the above items should result in the development of a placement plan that includes pipe spacings and locations throughout the duration of the placement. The plan should also include the locations to be used for relocating pipes as placement progresses.

8.8.8 *Personnel*—Because underwater placements are infrequent and errors can lead to problems that are extremely difficult and expensive to correct, all underwater placements should be done under the direct supervision of qualified, experienced personnel. An experienced individual should be available to interpret soundings and make necessary decisions concerning relocation of placement pipes and air lifts, and to observe overall placement procedures.

8.9—Special applications

8.9.1 *Fabric forming*—Fabric forming offers some unique advantages for specialized types of underwater concrete placement (Lamberton 1980; Koener and Welsh 1980). Normally, a sand-cement mortar, sometimes with the addition of pea gravel, is pumped into a fabric container tailored to the required shape. The fabric acts as a separator between surrounding water and the concrete as it flows into the container preventing segregation.

A high-strength, water-permeable fabric is preferred. This fabric is usually woven of nylon or polyester yarns of industrial tire cord weight at approximately 20 yarns per in. (780 yarns per m). The use of textured multifilament yarns produces a more stable fabric and is also more effective as a filter, permitting the release of excess mixing water from the concrete and thereby increasing the rate of stiffening and long-term strength and durability (Lamberton 1980).

Fabric forming is used in construction of erosion-control revetments produced by injecting mortar into a double-layer fabric envelope and in the construction of concrete jackets used to rehabilitate deteriorated marine piles. Large fabric containers have been used to cast blocks of concrete weighing up to 15 tons (14 Mg) for construction of breakwaters. Specially designed fabric assemblies have been used to cast saddles and weights for underwater pipelines.

8.9.2 *Diaphragm-wall construction*—In diaphragm- or slurry-wall construction (Xanthakos 1979; Nash 1974; Holland and Turner 1980), concrete is placed under water or under a bentonite slurry in trenches to form walls. These placements can serve as retaining walls for open excavations (when suitably braced or tied back) or as cutoff walls to stop flow through or under existing structures, such as earthfill dams or levees.

Because these walls are confined placements, the rate of concrete rise will be high, necessitating frequent removal of tremie sections to maintain flow.

8.10—Antiwashout admixtures

Chemical admixtures intended for use in concrete placed under water have been developed (Saucier and Neeley 1987; Khayat, Gerwick, and Hester 1990). These antiwashout admixtures make the concrete more cohesive and thus less prone to washout of cement or fines from the concrete during placement.

These admixtures were developed for use in situations where freshly placed concrete, may be exposed to flowing water during or after placement where concrete placement is not thick enough to permit the required tremie pipe embedment, or where the wash-out of cement may cause an environmental problem. A Corps of Engineers test method (CRD-C 61) has been developed to evaluate the effectiveness of these admixtures (Neeley 1988). Because of the thixotropic nature of the concrete treated with these admixtures, they should be used with caution for massive placements in which the concrete is expected to flow for long distances once it exits the tremie pipe. Trial placements should be conducted to verify that the concrete proportioned with the antiwashout admixture can maintain adequate slump life and can flow for the required distance.

Applications of these antiwashout admixtures include underwater paving of a canal (Kepler 1990; Klemens 1991) and underwater repair of a dam (Neeley and Wickersham 1989).

CHAPTER 9—PUMPING CONCRETE

9.1—General considerations

This chapter gives an overview of concrete pumping. For a more detailed discussion, refer to ACI 304.2R.

ACI defines pumped concrete as concrete that is transported through rigid or flexible pipeline by means of a pump. Pumping can be used for most concrete construction, but is especially useful where space for construction equipment is limited. A steady supply of pumpable concrete is necessary for satisfactory pumping. A pumpable concrete, like conventional concrete, requires good quality control: that is, uniform, properly graded aggregate, and uniform batching and thorough mixing of all materials.

Pumped concrete moves as a cylinder riding on a thin lubricant film of grout or mortar on the inside diameter of the pipeline.

Maximum volume output and maximum pressure on the concrete cannot be achieved simultaneously from most concrete pumps because this combination requires too much power. Three to four times more pressure is required per foot of vertical rise than is necessary per foot of horizontal movement.

9.2—Pumping equipment

The most common concrete pumps consist of a receiving hopper, two concrete pumping cylinders, and a valving system to alternately direct the flow of concrete into the pumping cylinders, and from them, to the pipeline. One concrete cylinder receives concrete from the receiving hopper while the other discharges into the pipeline to provide a relatively constant flow of concrete through the pipeline to the placement area. The price of concrete pumps varies greatly with maximum pumping capacity and maximum pressure that can be applied to the concrete. Pumps should be selected to provide the desired output, volume, and pressure on the concrete in the pipeline.

The most versatile concrete pumps use hydraulically operated concrete valves that have the ability to crush or displace aggregate that becomes trapped in the valve area. Most of these pumps have an outlet port 5 in. (125 mm) or larger in diameter and use reducers to reach smaller pipeline sizes, if necessary.

Other pumps use steel balls and mating seats to control the flow of concrete from the hopper into the pumping cylinder and out of the pumping cylinder into the pipeline. These units are limited to pumping concrete with smaller than 1/2 in. (13 mm) maximum-sized aggregate. ACI 304.2R describes general-purpose, medium-duty, and special-application pumps in detail. These can be trailer- or truck-mounted units. Truck-mounted pumps can also be equipped with placement booms that support a 5 in. (125 mm) diameter pipeline that receives the discharge from a concrete pump and places it in the forms. Most booms have three or more articulating sections and are mounted on a turret that rotates to enable the discharge of the pipeline to be located where needed. Booms are generally rated according to their vertical reach and range in size from about 72 ft to 175 ft (22 to 53 m).

Concrete pumps are powerful machines that use high hydraulic oil pressures, concrete under high pressure, and compressed air for cleanup. Safe operating practices are necessary for the protection of the pump operator, ready-mixed concrete drivers, and the workers placing and finishing the pumped concrete.

9.3—Pipeline and accessories

9.3.1 *General*—Most concrete transported to the placement area by pumping methods is pumped through rigid steel tubing or heavy-duty flexible hose, both of which are called pipeline. The flexibility of the hose allows workers to place concrete exactly where it is needed. For placements on grade, rubber hose is frequently used at the end of a steel tubing pipeline. Large or elevated placements generally are done by placement booms.

Pipeline surface irregularity or roughness, diameter variations, and directional changes disturb the smooth flow of pumped concrete. This results in increased pressure required to push concrete through the pipeline and increased wear rate throughout the pump and pipeline.

All components of the pipeline should be able to handle, with an adequate safety factor, the maximum internal pressure that the concrete pump being used is capable of producing. The safety factor decreases as the pipeline wears due to the abrasiveness of the coarse and fine aggregate used in the concrete. The rate of wear can vary greatly.

Straight sections of pipeline are made of welded or seamless steel tubing, most commonly 10 ft (3 m) in length. The most common diameters are 4 and 5 in. (100 and 125 mm) with most systems in the 5 in. (125 mm) size. Aluminum pipeline should not be used in concrete pumping (Fowler and Holmgren 1971).

9.3.2 *Pipeline components*—Concrete pipeline components can be assembled in virtually any order, then disassembled and reconfigured in a different manner. To achieve this flexibility, each delivery line component requires the use of connecting ends or collars, a coupling, and a gasket. The coupling connections require a gasket sealing ring to hold the required pressure and prevent grout leakage. The most common connecting ends use a raised section profile to make a joint that can withstand pressures in excess of 2000 psi (14 MPa). They can also withstand considerable stress from external bending forces. Grooved-end connections should not be used on pipeline with diameter greater than 3 in. (75 mm).

Concrete pumping hose is divided into two classifications: hose intended for use at the end of a placement line (discharge hose) and hose used on a placement boom (boom hose). A discharge hose has a lower pressure rating. A boom hose typically connects rigid boom sections and should withstand high pressures. Approximately three times more pressure is required to pump concrete through a given length of hose than is needed to pump through the same length of steel line. Pumping pressure can cause a curved or bent hose to straighten. Injuries have resulted from such movement, and sharp bends should be avoided.

To help achieve maximum component life, safe and thorough cleaning of the pipeline is necessary at the end of each placement or at any time a lengthy delay in pumping operation occurs. The pipeline is cleaned by propelling a sponge ball or rubber go-devil through the line with air or water pressure. Arrangements for disposal of this residual concrete should be made before pumping begins.

9.4—Proportioning pumpable concrete

9.4.1 *Basic considerations*—Concrete pumping is so established in most areas that most ready-mixed concrete producers can supply a concrete mixture that will pump readily if they are informed of the concrete pump volume and pressure capability, pipeline diameter, and horizontal and vertical distance to be pumped.

The shape of the coarse aggregate, whether angular or rounded, has an influence on the required mixture proportions, although both shapes can be pumped satisfactorily. The angular pieces have a greater surface area per unit volume as compared with rounded pieces and thus require more mortar to coat the surface for pumpability.

9.4.2 *Coarse aggregate*—The maximum size of angular or crushed coarse aggregate is limited to 1/3 of the smallest inside diameter of the pump or pipeline. For well-rounded aggregate, the maximum size should be limited to 2/5 of these diameters. The principles of proportioning are covered in ACI 211.1 and ACI 211.2.

Whereas the grading of sizes of coarse aggregate should meet the requirements of ASTM C 33, it is important to recognize that the range between the upper and lower limits of this standard is broader than ACI Committee 304 recommends to produce a pumpable concrete.

9.4.3 *Fine aggregate*—The properties of the fine aggregate have a much more prominent role in the proportioning of pumpable mixtures than do those of the coarse aggregate. Together with the cement and water, the fine aggregate provides the mortar or fluid that conveys the coarse aggregates in suspension, thus rendering a mixture pumpable.

Particular attention should be given to those portions passing the finer screen sizes (Anderson 1977). At least 15 to 30% should pass the No. 50 (300 µm) screen and 5 to 10% should pass the No. 100 (150 µm) screen. ACI 211.1 states that for more workable concrete, which is sometimes required when placement is by pump, it may be desirable to reduce the estimated coarse aggregate content by up to 10%. Exercise caution to ensure that the resulting slump, w/cm, and strength properties of the concrete meet applicable project specification requirements.

9.4.4 *Combined normalweight aggregates*—The combined coarse and fine aggregates occupy about 67 to 77% of the mixture volume. For gradation purposes, the fine and coarse aggregates should be considered as one even though they are usually proportioned separately.

ACI 304.2R includes an analysis worksheet for evaluating the pumpability of a concrete mixture by combining the fine and coarse aggregate with nominal maximum-sized aggregate from 3/4 to 1-1/2 in. (19 to 38 mm). The worksheet makes provision for additional coarse and fine aggregate that can be added to a mixture to improve the overall gradation and recognizes possible overlap of some coarse and fine aggregate components. If a mixture is known to be pumpable is evaluated and graphed first, the curve representing its proportions provides a useful reference for determining the pumpability of a questionable mixture. If that mixture has a curve running in a zigzag fashion, or has one or more values falling below the boundary line, the mixture is questionable for pumping and may not be pumpable by all types of concrete pumps. Those pumps with powered valves, higher pressure on the concrete, and the most gradual and smallest reduction from concrete tube diameter can pump the most difficult mixtures.

Concrete containing lightweight fine and coarse aggregate can be pumped if the aggregate is properly saturated. Refer to ACI 304.2R for more detailed information and procedures.

9.4.5 *Water*—Water requirements and slump control for pumpable normalweight concrete mixtures are interrelated and extremely important considerations. The amount of water used in a mixture will influence the strength and durability (for a given amount of cement) and will also affect the slump or workability.

Mixing water requirements vary for different maximum sizes of aggregate as well as for different slumps.

To establish the optimum slump resulting from water content for a pump mixture and to maintain control of that particular slump through the course of a job are both extremely important factors. Slumps from 2 to 6 in. (50 to 150 mm) are most suitable for pumping. In mixtures with higher slump, the coarse aggregate can separate from the mortar and paste and can cause pipeline blockage. Slumps obtained through the use of superplasticizers, however, are usually pumped without difficulty.

There are several reasons why the slump of concrete can change between initial mixing and final placement. If the slump at the end of the discharge hose can be maintained within specification limitations, it may be satisfactory for the concrete to enter the pump at a higher slump to compensate for slump loss, if the change is due simply to aggregate absorption.

9.4.6 *Cementitious materials*—The determination of the cementitious materials content follows the same basic principles used for any concrete.

In establishing the cement content, remember the need for overstrength proportioning in the laboratory to allow for field variations.

The use of extra quantities of cementitious materials as the only means to correct pumping difficulties is shortsighted and uneconomical. Correcting any deficiencies in the aggregate gradation is more important.

9.4.7 *Admixtures*—Any admixture that increases workability in both normalweight and lightweight concretes will usually improve pumpability. Admixtures used to improve pumpability include regular and high-range, water-reducing admixtures, air-entraining admixtures, and finely divided mineral admixtures.

Increased awareness of the need to incorporate entrained air in concrete to minimize freezing and thawing damage to structures has coincided with increased use of concrete pumps, as well as the development of longer placement booms. This has resulted in considerable research and testing, which has established that the effectiveness of the air-entraining agent (AEA) in producing a beneficial air-void system depends on many factors. The more important factors are:
- The compatibility of the AEA and other admixtures as well as the order in which they are introduced into the batch;
- The mixture proportions and aggregate gradation;
- Mixing equipment and procedures;
- Mixture temperatures; and
- Slump.

AEA effectiveness and the resulting dosage of AEA also depend on the cement fineness, cement factor, and water content, and the chemistry of cement and water, as well as that of other chemical and mineral admixtures used in the concrete. Refer to ACI 304.2R for more detailed information on air content and admixtures.

9.5—Field practice

Preplanning for concrete pumping is essential for successful placements, with increasing detail and coordination required as the size of the placement and the project increases. This planning should provide for the correct amount and type of concrete for the pump being used, provision for necessary pipeline, and agreement as to which personnel will provide the labor necessary to the complete placement operation.

Any trailer- or truck-mounted concrete pump can be used for pipeline concrete placement. The limiting factor in this method is the ability to spread the concrete as needed at the end of the pipeline. Generally, this is done by laborers using a rubber hose at the end of a rigid placement line.

The discharge of powered placement booms can be positioned at almost any point within the radius of the boom and at elevations achieved with the boom from near vertical (up or down) to horizontal. Their use generally reduces the manpower required for a given placement.

9.6—Field Control

Pumped concrete does not require any compromise in quality. A high level of quality control, however, should be maintained to ensure concrete uniformity.

Concrete has been pumped successfully during both hot and cold weather. Precautions may be necessary to provide adequate protection during extreme conditions. Refer to ACI 305R and ACI 306R for guidance.

CHAPTER 10—CONVEYING CONCRETE
10.1—General considerations

This chapter gives an overview of conveying concrete. For a more detailed discussion, refer to ACI 304.4R.

Belt conveyors for handling concrete are unique in that they transport plastic concrete that is approximately 48% heavier than aggregate or other commonly conveyed materials. They transport plastic concrete from a supply source, such as a truck mixer or a batching and mixing plant, to the point of placement or to other equipment that is used to place the concrete. Maximum success for conveyor placement requires a constant supply of properly mixed concrete for charging the belt conveyor and a provision for moving the discharge point during placement so that the plastic concrete is deposited over the entire placement area without the need for rehandling or excessive vibration. Concrete belt conveyors are classified into three types: 1) portable or self-contained; 2) feeder or series; and 3) spreader-radial or side-discharge.

All concrete conveyors require charge and discharge hoppers, belt wipers, and proper combinations of belt support idlers and belt speed to prevent segregation of the concrete. Any normalweight or lightweight aggregate concrete that can be discharged by a truck mixer can be placed by a concrete belt conveyor.

10.2—Conveyor operation

Concrete conveyors running at the correct belt speed and with properly functioning charging hoppers, transfer devices, and belt wipers have only a minor effect on the strength, slump, or air content of the concrete that they carry.

The characteristics of the ribbon of concrete on a conveyor belt are determined by the angle of surcharge of the concrete, the required minimum edge distance, and the load cross section. The angle of surcharge is the angle to the horizontal that the surface of the same concrete assumes while it is being carried on a moving (horizontal) belt conveyor. The angle of surcharge for most concrete falls in a range from 0 to 10 degrees (Anon. 1979). The angle of surcharge determines the cross section of the concrete ribbon that can be efficiently carried on the belt and the maximum slope (ascending or descending) at which concrete can be handled by a belt conveyor.

Concrete cannot be carried across the entire face of a belt. The ribbon of concrete should be centered on the belt with equal widths of clear belt or edge distance between it and each edge of the belt. Failure to observe the minimum edge distance requirement will result in excessive spilling and loss of large aggregate off the edges of the belt.

All concrete belt conveyors use idlers that trough or cup the belt, enabling it to carry a deeper ribbon of concrete than would be possible on a flat belt. As the angle of the belt (ascending or descending) is increased, the ribbon of concrete on the belt becomes shallower. As concrete is loaded on a belt conveyor, any difference between its velocity in the direction of belt travel and the speed of the belt will be equalized by acceleration or deceleration of the concrete, which results in turbulence. Properly designed charging hoppers use this turbulence to produce a remixing of the concrete as it flows onto the belt. A concrete belt conveyor should be equipped with a charging hopper that levels out surges of concrete flow and delivers a uniform ribbon of concrete onto the belt with proper edge distance.

Plastic concrete is traveling at the same speed as the belt when it is discharged from a belt conveyor. The plastic concrete generally leaves the belt as a cohesive mass except that some of the larger pieces of coarse aggregate can segregate from the stream and some mortar clings to the belt.

Using properly designed discharge hoppers, chutes, dropchutes, or elephant trunks will eliminate concrete segregation problems. Equipping every end-discharge conveyor belt with a belt wiper or scraper will limit mortar loss.

10.3—Conveyor design

Concrete conveyor belts are quite flexible because they operate at high speeds over relatively small-diameter head and tail pulleys. Almost all conveyor belting is made in long lengths and is cut to fit the conveyor on which it is installed. The ends of the belt are spliced to make the belt continuous.

Most concrete belt conveyors are moved frequently and it is impossible to ensure that the supporting structure and belt idlers will always be level in the plane at a right angle to the center line of the belt. Whenever a belt conveyor is not level, gravity will cause the belt to drift to the low side. This problem is usually solved with specially designed belt-support idlers or with guide rollers that are in contact with the belt edge.

The single most important factor in determining load cross section is belt width. A relatively small increase in conveyor belt width greatly increases capacity. For example, increasing belt width from 16 to 24 in. (400 to 600 mm) more than doubles the capacity of the conveyor system at the same belt speed.

A convenient method of estimating concrete belt-conveyor capacity is to use conveyor capacity tables published by the conveyor manufacturer. These tables usually assume continuous horizontal operating conditions, average angle of surcharge, and a conventional three-roll idler configuration. These tables are intended to cover average conditions and are usually accurate enough for most purposes. There is a direct relationship between capacity and belt speed so that capacities can be interpolated for belt speeds not shown.

Keeping the weight of concrete on the belt to a minimum will allow the belt conveyor to run at optimum belt speed. Generally, this speed is in the range of 300 to 750 ft/min (90 to 230 m/min) depending on the type of concrete belt conveyor involved, the angle of surcharge of the concrete, and the angle of the conveyor.

The proper combination of idler spacing and belt tension allows concrete belt conveyors to stop and hold concrete on the belt without spillage. Increasing idler spacing decreases the overall weight of the concrete conveyor but increases the belt tension required for successful operation.

Operating conditions for concrete belt conveyors require the use of watertight or waterproof electrical components, sealed bearings, and closed hydraulic circuits. Consequently, there is no equipment-related reason to protect the conveyors from weather and environmental conditions. There is rarely a need to enclose or protect the concrete on portable conveyors or on other types of conveyors up to 200 to 300 ft (60 to 90 m) long. The concrete is conveyed at high speed and is exposed to ambient conditions for only a short time.

If extreme ambient conditions are anticipated when using longer conveyor systems, some form of enclosure may be necessary to maintain the workability of the concrete or to protect it from freezing. Whether such an enclosure will be required should be determined on a project-by-project basis.

All structural concrete can be handled satisfactorily by a concrete belt conveyor. Extremes of slump, either below 1 in. (25 mm) or above 7 in. (180 mm), reduce the placement capacity of a belt conveyor significantly.

Saucier (1974) reported that in tests of concrete conveyed over 3000 ft (900 m), cement hydration, water evaporation, or aggregate absorption resulted in loss of slump for concrete conveyed these long distances. Strength tests indicated a definite increase in strength corresponding to the decrease in slump. The loss of entrained air was less than 0.5% for concrete originally containing approximately 5% air.

No single factor of conveyor design is of such overriding importance that it alone will produce satisfactory or unsatisfactory operation.

10.4—Types of concrete conveyors

Different project requirements have resulted in the development of portable, feeder, and spreading conveyors for concrete placement. Each type can be used alone or combined with others to form a conveyor system.

10.4.1 *Portable conveyors*—Short-lift or short-reach concrete-placement applications require the use of a portable belt conveyor. The most important characteristic is that each unit is self-contained and can be readily moved about the project. Belt widths of 16 or 18 in. (400 or 460 mm) are most common. The weight and mobility trade-off of the portable belt conveyor restricts its overall length to approximately 60 ft (18 m). This, in turn, establishes the maximum discharge height at approximately 35 ft (11 m).

Portable belt conveyors are generally powered with diesel or gasoline engines and use hydraulic drive systems to power the load-carrying belt. A self-propelled conveyor with an overall belt length of 56 ft (17 m), a 30 hp (22 MW) engine, and power steering can place at a rate as high as 100 yd^3/h (76 m^3/h).

10.4.2 *Feeder conveyors*—Long-reach concrete-placement applications require the use of transporting or feeder-type belt conveyors that operate in series with end-discharge transfer points.

Feeder-belt conveyors are normally powered with alternating current electric motors so that the load-carrying belt speed will be controlled by the power supply. Controls and cables should meet the normal electrical code requirements and be safe for use in a wet environment. The motors should be protected against both overload conditions and low-voltages. It is important that the conveyors automatically start in sequence and that the system ensure that each flight or unit of the system is operating at the proper belt speed before concrete is discharged onto the belt.

Feeder-belt conveyors can be operated over a rail or track that allows the feeder train to be extended or retracted without interrupting concrete placement. On large projects, relatively permanent feeder-belt conveyors can be installed. Under these conditions, much longer conveyor units can be used.

Spreading of the concrete at the discharge end of the train requires particular attention because feeder-belt conveyors move such a large volume of concrete. Usually, feeder conveyors discharge into equipment especially designed for spreading concrete.

10.4.3 *Spreading conveyors*

10.4.3.1 *Radial spreaders*—Radial spreaders are mounted on the placement conveyor or on a cantilevered support that swings the discharge end through an arc. They also have provision for extending and retracting the placement conveyor a substantial distance. Cantilevered radial spreaders normally rely on outrigger legs supported by the forms or the base on which the concrete is being cast to resist the overturning moment created by the loaded belt. Radial spreaders are also supported by rigid posts mounted in or near the placement area.

The limitations of reach and weight of radial spreading units have been largely overcome through the use of two- or three-section telescoping conveyors mounted on tracks or the telescoping boom of a hydraulic crane. Radial spreaders have the advantages of relatively quick setup time and the capability of reaching past obstructions. They also cause a minimum obstruction or congestion in the placement area itself.

For wide placements, the most efficient method of equipment use and the best placement pattern for finishing with mechanical equipment are achieved by side-discharge conveyors or straight-line spreaders (Cope 1972).

10.4.3.2 *Side-discharge conveyors*—Side-discharge conveyors span completely across the placement area. By discharging concrete over the side of the belt with a traveling plow or diverter, they place a straight ribbon of concrete that is ideal for mechanical finishing. Side-discharge conveyors normally operate horizontally, so the belt can be loaded heavily. Those equipped with 16 in. (400 mm) wide belts have a capacity of approximately 100 yd^3/h (75 m^3/h), 20 in. (500 mm) wide belt capacity is 200 yd^3 (153 m^3) per h, and 24 in. (600 mm) wide belt capacity is approximately 300 yd^3 (230 m^3) per h.

A crane using a bucket to bring concrete to the relatively stationary and usually visible hopper of a side-discharge conveyor is significantly more efficient than the same crane swinging blind to place concrete for an elevated slab. Side-discharge conveyors have made pumps more practical for wide slabs or decks by eliminating the labor needed to constantly move the discharge end of the pipeline back and forth in front of the commonly used straight-line finishing equipment.

The diverter that removes concrete from the belt and discharges it over the side of the conveyor uses a wiper blade to remove the concrete from the belt. The operation and adjustment of the wiper blade is critical on an end-discharge conveyor. Provisions should be made for adjusting the belt wiper or scraper on side-discharge conveyors while concrete is being placed. Some wear on the wiping strip is normal, and a small amount of grout can be carried past the diverter.

10.4.3.3 *Conveyor combinations*—Each type of conveyor has some limited ability to reach, lift, carry, or spread. On complex or large projects, economics will normally favor using each type of machine for the function it performs best. As long as belt speeds and widths are compatible, it is practical to combine various types of equipment.

10.5—Field practice

It is generally not practical to custom design belt conveyors for each project or application. Normal practice is to select standard, commercially available equipment that has adequate capacity and reach, and to organize and plan its use to meet the general construction sequences required to properly perform the work.

The actual field placement rate of a concrete conveyor will rarely equal the theoretical capacity from charts. This is primarily attributable to the inevitable delays that occur in batching, mixing, and transporting concrete to the belt conveyor at the placement area. Other delays involve consolidation and finishing of the concrete and moving of the conveyor. There is no way that a belt conveyor can place a surge of concrete in excess of design capacity because excess concrete placed on the belt will usually be spilled off the sides.

Hourly production on an efficient project will usually average about 70% of the capacity of the belt conveyor. This adjustment of the theoretical capacity provides the safety factor that most jobs require for successful completion within scheduled times.

As the placement progresses, fresh concrete should always be discharged onto or against concrete of plastic consistency that is already in place so there will be some blending of concrete through vibration and there will be no opportunity for objectionable rock pockets to be formed. Vibration at the delivery point and immediately behind the advancing edge of the concrete will cause the concrete to surround reinforcing steel without significant segregation.

Some conveyor maintenance can be necessary during concrete placement on large-volume projects because conveyor belting will stretch to some degree during concrete placing. Concrete conveyors should have provision for increasing belt tension.

Any spilled concrete should be cleaned off the conveyor before it can harden.

Concrete belt conveyors are an open system where almost all of the concrete being placed can be visually inspected. The ribbon of concrete on the conveyor belt should be observed at the start and frequently through the placement. The main emphasis of inspection should be on the proper discharging of concrete from the conveyors and consolidation of the concrete. Concrete discharged from a conveyor should not free-fall far enough to cause segregation.

Concrete belt conveyor systems should be tested under job conditions before any significant placement is attempted if there is any doubt about the ability of the system to successfully place the concrete. Fortunately, handling of only a few cubic yards of concrete over any belt conveyor system will validate the conveyor design and identify problem areas.

Tests of the plastic concrete and samples for strength determination taken at the discharge from the mixing or transporting equipment and at the concrete belt conveyor discharge point should provide adequate assurance of satisfactory operation. The quality of concrete being placed in the structure can be measured only at the point of placement in the structure. Once a satisfactory correlation between samples taken at the point of placement and the point of discharge of the mixer has been established, sampling at the more convenient point should be satisfactory, provided placement conditions remain unchanged.

CHAPTER 11—HEAVYWEIGHT AND RADIATION-SHIELDING CONCRETE

11.1—General considerations

The procedures for measuring, mixing, transporting, and placing heavyweight and radiation- shielding concrete are similar to those used in conventional concrete construction. Special expertise and thorough planning are necessary for the successful completion of this type of concrete work (Pihlajayaara 1972). For a detailed discussion on heavyweight and radiation-shielding concrete, refer to ACI 304.3R.

Normalweight concrete is generally specified for radiation shielding when space is available. When space is limited, however, the thickness of these shields can be reduced by using both natural and synthetic heavyweight aggregates. Natural mineral aggregates and synthetic aggregate can produce concrete having a typical density as high as 240 lb/ft^3 (3840 kg/m^3) and 340 lb/ft^3 (5450 kg/m^3), respectively. Heavyweight concrete not only has a higher density, but also more desirable attenuation properties.

When heavyweight concrete is used to absorb gamma rays, the density is of prime importance (Pihlajayaara 1972). When the concrete is to attenuate neutrons, material of light atomic weight containing hydrogen should be included in the concrete mixture (Davis 1972a). Some aggregates are used because of their ability to retain chemically bound water at elevated temperatures (above 185 F [85 C]), which ensures a source of hydrogen.

Colemanite, a mineral containing boron, and manufactured boron additives, such as boron frit, ferroboron, and boron carbide have been used in conjunction with normalweight and heavyweight concrete. Their use enhances absorption of thermal neutrons, limits hard gamma radiation, and limits buildup of long-lived activity. Caution should be exercised because of the possibility of retardation due to the presence of soluble borates (Volkman 1994).

11.2—Materials

11.2.1 *General*—Cements, admixtures, and water used in heavyweight concrete should conform to the standards generally required for normalweight concrete, only the aggregate is different and may require special consideration during handling, batching, mixing, transporting and placing.

11.2.2 *Aggregate*—Thorough examination and evaluation of heavyweight aggregate sources are necessary to obtain material suitable for the type of shielding required (Browne and Blundell 1972).

Composition of aggregates for use in radiation-shielding concrete is described in ASTM C 638, and aggregates should meet requirements of ASTM C 637. Some typical properties for shielding aggregates are shown in Table 11.1.

Some aggregates (ferrophosphorous and barite) and some iron ores are brittle and highly crystalline in structure and tend to break up into smaller pieces while being handled. These factors should not preclude the use of the material, provided it is demonstrated that the concrete manufactured has properties meeting the specification requirements.

Fine metallic aggregate should consist of commercial chilled-iron, steel shot, or ground iron meeting the specifications of the Society of Automotive Engineers (1993).

Heavyweight PA concrete usually precludes the use of aggregate larger than 1-1/2 in. (40 mm) due to form configuration and embedment limitations. Coarse aggregate should be uniformly graded from 1/2 to 1-1/2 in. (10 to 40 mm) and conform to Grading 1 in Table 7.1 (Chapter 7). Fine aggregate grading should be within the limits shown in Table 7.1 so that the smaller particles show less tendency to segregate.

Aggregate should be shipped, handled, and stored in a manner that will ensure little loss of fines, no contamination by foreign material, or significant aggregate breakage or segregation.

11.2.3 *Proprietary premixed mortar*—Heavyweight iron mortar and lightweight organic and inorganic mortar concretes produced commercially by manufacturers for biological shielding should be tested before use for radiation-shielding properties. Inspection at the point of manufacture should be as stringent as for natural and synthetic heavyweight aggregates and shielding concretes.

Table 11.1—Typical radiation-shielding aggregates*

Natural mineral					Synthetic mineral			
			Percent by weight				Percent by weight	
Aggregate	Source‡	Specific gravity†	Iron	Fixed water	Aggregate	Specific gravity‡	Iron	Fixed water
Hydrous ore					Crushed aggregate			
Bauxite	—	1.8 to 2.3	0	15 to 25	Heavy slags	5.0	0	0
Geothite	Utah, Michigan	3.4 to 3.8	0	8 to 12	Ferrophosphorous§	5.8 to 6.3	0	0
Limonite	Utah, Michigan	3.4 to 3.8	55	8 to 12	Ferrosilicon	6.5 to 7.0	70	0
Heavy ore					Metallic iron products			
Barite	Nevada, Tennessee	4.0 to 4.4	1 to 10	0	Sheared reinforcing bars	7.7 to 7.8	99	0
Magnetite	Nevada, Wyoming, Montana	4.2 to 4.8	60	1.0 to 2.5	Steel punchings	7.7 to 7.8	99	0
Ilmentite	Quebec	4.2 to 4.8	40	0	Iron and steel shot	7.5 to 7.6	99	0
Hematite	South America, Africa	4.2 to 4.8	70	—	Boron Products			
Boron additives					Boron frit	2.4 to 2.6	0	0
Boro calcite	Turkey	2.3 to 2.4	0	0	Ferroboron	5.0	85	0
Colemanite	California	2.3 to 2.4	0	0	Borated Diatomaceous earth	1.0	0	0
					Boron carbine	2.5 to 2.6	0	0

*Source: Society of Automotive Engineers (1993), Davis (1967), and Anon (1955).
†Material water-saturated with its surface dry.
‡Other sources may be available.
§Ferrophosphorous when used in portland cement will generate flammable and possible toxic gases that can develop high pressure if confined.

11.3—Concrete characteristics

11.3.1 *Physical properties*—High modulus of elasticity, low coefficient of thermal expansion, and low elastic and creep deformation are ideal properties for heavyweight concrete. High compressive strengths may be required if heavyweight concrete is to be subjected to high stresses. Heavyweight concrete with a high cement content and a low w/cm can exhibit increased creep and shrinkage, and in a massive concrete placement could generate high temperatures at early ages, causing undesirable localized cracking from the thermally induced stresses. When structural considerations require this cracking potential to be eliminated, it is necessary to use appropriate temperature control measures, which could include precooling or postcooling the concrete, or both, as described in ACI 207.2R and ACI 224R.

11.3.2 *Mixture proportioning*—Procedures outlined in ACI 211.1 should be used for concrete proportioning. Conventionally placed heavyweight concrete should be proportioned to provide the desired compressive strength and density as well as adequate workability. Also, the chemical constituents and fixed water content of the resulting mixture should provide satisfactory shielding properties (Davis 1972b). Typical proportions for heavyweight, conventionally placed concrete, PA concrete, and grout mixtures are shown in Table 11.2.

11.4—Mixing equipment

Standard mixing equipment is generally used to mix heavyweight concrete, but care should be taken not to overload the equipment. In general, the amount of heavyweight concrete mixed should be equivalent to the mixture weight of normalweight concrete rather than the volume capacity of the mixing equipment. Heavyweight concrete should be agitated when transported from the mixing plant to the point of placement to prevent segregation, consolidation, and packing.

11.5—Formwork

Formwork should follow the practices set forth in ACI 347R. Formwork for conventionally placed heavyweight concrete needs to be stronger than comparable formwork for ordinary concrete because of the increased concrete density.

Typical radiation-shielding structures require a complex shape, and they can contain many penetrations through the formwork. The strutting and bracing system should be carefully designed to avoid unintentionally placing a load on penetrating members and to ensure precise alignment of external fixtures corresponding to these penetrations. Consider the use of permanent steel forms.

Steel penetrations are often precisely machined and fabricated assemblies that can be subject to delays in delivery. It is prudent to allow for such delays by providing for blockouts to receive these penetrations. Blockouts should be provided with normal bends or a stepped configuration to reduce the possibility of radiation streaming or leakage. The basic structure then can be completed around the blockouts. After the items to be embedded are placed, the blockouts are filled with heavyweight grout. Precautions should be taken to ensure that penetrations and blockouts are tightly grouted with a nonshrink grout of appropriate density.

11.6—Placement

11.6.1 *Conventional method*—Placement of conventionally mixed heavyweight concrete is subject to the same considerations of quality control as normalweight concrete, except that it is far more susceptible to variations in quality due to segregation caused by improper handling.

The placement of heavyweight concrete dictates the strictest observance of good placement practice. Regular concrete placement techniques can be used, including pumping. Heavyweight concrete should be placed as close as possible to

Table 11.2—Typical proportions for radiation-shielding concrete conventionally mixed and placed

Mixture no.	Fine aggregate W/R*	Fine aggregate Material	Coarse aggregate W/R*	Coarse aggregate Material	Admixture	W/CM	Slump in.	Slump mm	Wet density lb/ft³	Wet density kg/m³
A	3.40	Ilmenite	—	—	Specify	0.43	0	0	190	3040
B	3.35	Magnetite	—	—	Specify	0.47	0	0	190	3040
C	2.39	Serpentine	5.07	Magnetite	Specify	0.62	2	50	190	3040
D	4.46	Barite	5.44	Barite	Specify	0.60	2	50	220	3520
E	3.66	Ilmenite	4.62	Ilmenite	Specify	0.45	3	75	225	3630
F	3.61	Magnetite	4.58	Magnetite	Specify	0.49	2	50	225	3630
Q	2.95	Ferrophosphorous	2.95	Ferrophosphorous	Specify	0.54	2	50	260	4170
	1.48	Barite	2.11	Barite	Specify	0.54	2	50	260	4170
H	3.01	Magnetite	1.76	Magnetite	Specify	0.49	2	50	270	4330
	—	—	2.69	S330/390 iron shot	—	—	—	—	—	—
	—	—	2.65	S1110/1320 iron shot	—	—	—	—	—	—
I	2.98	Magnetite	2.60	S330/390 iron shot	Specify	0.51	2	50	300	4800
	—	—	5.76	S1110/1320 iron shot	—	—	—	—	—	—
J	3.21	Ilmenite	2.60	S330/390 iron shot	Specify	0.49	2	50	300	4800
	—	—	5.54	S1110/1320 iron shot	—	—	—	—	—	—
K	3.82	Ferrophosphorous	7.10	Ferrophosphorous	Specify	0.53	2	50	300	4800
L	1.00	Magnetite	5.96	S330/390 iron shot	Specify	0.46	2	50	330	5290
	—	—	5.89	S1110/1320 iron shot	—	—	—	—	—	—
Preplaced-aggregate grout										
M	1.15	Serpentine	—	—	—	0.50	—	—	128	2050
N	1.00	Conventional	—	—	—	0.42	—	—	129	2060
O	1.28	Limonite	—	—	—	0.55	—	—	146	2390
P	1.49	Barite	—	—	—	0.54	—	—	155	2480
Q	2.12	Magnetite	—	—	—	0.55	—	—	170	2720
Preplaced-aggregate concrete										
Mixture no.	Grout no.	Material	W/R*	Material	Admixture	W/CM	in.	mm	lb/ft³	kg/m³
R	O	Limonite	1.23	Limonite	Specify	0.55	—	—	215	3440
		—	5.37	Magnetite	—	—	—	—	—	—
S	Q	Magnetite	10.29	Magnetite	Specify	0.55	—	—	240	3850
T	M	Serpentine	2.46	Serpentine	Specify	0.50	—	—	240	3850
—	—	—	7.44	Steel punchings	—	—	—	—	—	—
U	Q	Magnetite	6.16	Magnetite	Specify	0.55	—	—	260	4170
		—	3.38	Steel punchings	—	—	—	—	—	—
V	O	Limonite	2.70	Limonite	Specify	0.55	—	—	260	4170
		—	6.31	Steel punchings	—	—	—	—	—	—
W	Q	Magnetite	8.08	Steel punchings	Specify	0.55	—	—	300	4810
X	Q	Magnetite	13.11	Steel punchings	Specify	0.55	—	—	340	5450

*W/R = weight ratio aggregate/cement.

its final position in the forms with a minimum of vibration to prevent segregation. The use of long, rigid chutes or drop pipes should be avoided. Where concrete is placed in narrow forms or through restricted areas, a short, flexible-type drop chute that tends to collapse and restrict the fall should be employed. Lifts should be limited to a maximum 12 in. (300 mm) thickness.

Consolidation procedures should conform to ACI 309R. In heavyweight concrete, vibrators have a smaller effective area or radius of action; therefore, greater care should be exercised to ensure that the concrete is properly consolidated.

Vibration and revibration for removing entrapped air and to establish aggregate-to-aggregate contact can cause an excessive amount of grout to collect on the top of lift surfaces (Davis 1972c). This grout should be removed from the lift surface while the concrete is still in a fresh state.

11.6.2 *Preplaced-aggregate method*—Precautions for placement of heavyweight PA concretes are given in Chapter 7. Placement of grout for heavyweight PA concrete requires extreme care because of a greater tendency for segregation and line blockages. Therefore, ample preparations should be made for rapid clearing of grout hoses and pipes. Standby equipment should be provided, and a trial run is recommended before operation.

11.7—Quality control

11.7.1 *Samples and testing*—Heavyweight and radiation-shielding concrete materials should be sampled and tested before and during construction to ensure conformance with applicable standards and specifications. Guidance presented in ACI standards and reports, as well as previous experience with the same materials, will determine the required frequency of testing.

The complexity of structures in which heavyweight concrete is placed usually precludes the possibility of taking test cores. It is, therefore, of the utmost importance that a thorough quality control program be established before the start of construction and maintained throughout construction.

11.7.2 *Control tests*—The quality of the concrete produced and of its constituent materials should be controlled by an established program of sampling and testing in accordance with appropriate ASTM test methods. The limits of rejection for heavyweight concrete should be established in the construction specifications and conform to the design parameters of the structures involved. Prior to wasting expensive heavyweight concrete, the engineer should be notified so that the severity of any nonconformance can be evaluated.

Heavyweight PA concrete is adaptable to the use of sophisticated and exacting quality control tests. The extent of control exercised depends on the complexity and importance of the project.

Tests of materials, grouts, and compressive strength of heavyweight PA concrete should be the same as those discussed in Chapter 7.

11.7.3 *Inspection*—The inspection of heavyweight concrete should be in accordance with applicable standards and project specifications.

Other than special modifications discussed in this chapter, those inspection items emphasized as important in ACI 311.4R should be followed for heavyweight concrete as well.

CHAPTER 12—LIGHTWEIGHT STRUCTURAL CONCRETE
12.1—General considerations

This chapter gives an overview of lightweight structural concrete. For a more detailed discussion, refer to ACI 304.5R.

Procedures for measuring, mixing, transporting, and placing lightweight concrete are similar in many respects to comparable procedures for normalweight concrete. There are certain differences, however, especially in proportioning and batching procedures, that should be considered to produce a finished product of comparable quality. The weight and absorptive properties of lightweight aggregates are different and should be considered. This chapter deals primarily with batching methods for coarse lightweight aggregates to correct for changes in weight and moisture content to ensure proper yield (Tobin 1971). It also covers batching of lightweight fine aggregates using a modification of the method used for coarse lightweight aggregates (Expanded Shale, Clay and Slate Institute 1958a, Portland Cement Association 1988).

These proportioning and batching methods have been coordinated with the basic principles set forth in ACI 211.2 and ACI 304.5R. It is necessary for the user to refer to those documents for detailed discussions of the methods available for batching lightweight aggregate, as that material is not duplicated herein.

12.2—Measuring and batching

12.2.1 *Free water and absorbed water*—One of the first considerations in batching lightweight concrete mixtures is a proper understanding of the water used in the mixture (Tobin 1967). The total water used per unit volume is divided into two components. One part is the water absorbed by the aggregates, whereas the other is similar to that in normalweight aggregate concrete and is classified as free water. Free water controls the slump, and when mixed with a given quantity of cement, establishes the strength of the paste, as for any concrete mixture. The amount and weight of absorbed water will vary with different lightweight materials, presoaking, and mixing times (Reilly 1972; Shideler 1957). Absorbed water does not change the volume of the aggregates or the concrete because it is inside the aggregate. Most important, absorbed water does not affect the *w/cm* or the slump of the concrete.

12.2.2 *Unit-weight variations*—The unit weight of lightweight aggregate varies depending on the raw materials used and the size of the aggregate. Smaller particles usually have higher unit weights than larger particles. Unit weights also vary due to changes in absorption or moisture content. If the lightweight aggregates are batched by weight without adjusting for these variations in unit weight, problems of over- or under yield of the concrete can result.

The dry, loose unit weight of aggregate depends primarily on its specific gravity and on the grading and shape of the particles. Angular crushed aggregates have more voids or unfilled spaces between the aggregate particles than rounded or spherically shaped pieces (Tobin 1978; Wills 1974). Poorly graded aggregate (that is, all one size) generally has more voids than a uniformly graded material that has enough smaller pieces to fit into the voids between the larger particles.

Numerous routine tests of both natural and lightweight aggregates show an amazingly close correlation of the void content for specific products being produced by a given plant over a long period.[*] Each production facility has its own characteristic void content values for each size of aggregate being produced and this information can usually be obtained from the source.

The absolute volume of a specific lightweight coarse aggregate is the volume of material remaining after the volume of voids has been subtracted. The absolute volume or the displaced volume in the concrete for a given lightweight material remains the same even though its density or its moisture content changes.

The proper usage of these basic principles makes it possible to batch and deliver lightweight concrete at the proper slump and yield for any job.

12.2.3 *Volume-weight batching of coarse aggregate*—To avoid problems with yield of concrete, it is necessary to maintain the same absolute volumes of lightweight aggregates in each batch of concrete by adjusting the batch weights to compensate for changes in unit weights. Standard unit weight tests on the lightweight aggregates, made frequently during batching operations can be used to adjust batch weights to reflect any changes that can occur in unit weights. This practice is rather time-consuming in a busy production facility, and a volume-weight batching system has been developed and used in some areas as an alternate method. Either method produces satisfactory results. The principal difference between the systems reported herein and that reported in ACI 211.2 is that the volume-weight method provides automatic yield adjustments for each batch of

---

[*]Unpublished data provided by committee member Robert E. Tobin.

lightweight concrete without the need for determining specific gravity factors of structural lightweight aggregate.

12.2.3.1 *Calibrating the weighing hopper*—The volume-weight system can be set up for virtually any batching system that employs a hopper or bin for weighing materials. The first operation is to determine the volume of this weighing hopper.

When the discharge gate in the overhead bin containing the lightweight coarse aggregate is opened, the material will flow into the weighing hopper until it builds up to the level of the discharge gate. Some plants may be slightly different than others but suitable modifications can be made in the overhead bins, the weighing hopper, or both to allow the weighing hopper to be filled to a prescribed level each time.

The volume of lightweight aggregate in this filled weighing hopper can be calibrated for most batching plants in the following manner. The total weight of the material (either dry or containing absorbed water) in the filled hopper can be read directly from the scales. The hopper is then discharged into a dump truck and the unit weight of three or four samples of loose material is determined in a suitable container. The total hopper weight divided by the average unit weight will give the total volume of the material in the weighing hopper in ft^3 or in m^3. For example, if the net weight of the filled hopper is 4650 lb (2100 kg) and the average unit weight of the material in it is 48.2 lb/ft^3 (772 kg/m^3), the volume is simply $4650/48.2 = 96.5$ ft^3 ($2210/772 = 2.73$ m^3). This calibration procedure should be performed about three times to ensure valid measurements. A new calibration can be necessary if the source of lightweight aggregate is changed, because the new material can have a different angle of repose that could alter the overall volume in the weighing hopper. If no major changes occur in the lightweight aggregates, one calibration will suffice for several months or until the materials are changed significantly.

The calibrated weighing hopper can be used as a container to determine the unit weight of the lightweight coarse aggregate for each batch of concrete. A batching chart can be prepared for any specified mixture proportions based on a full range of unit weights of aggregate as measured in the weighing hopper. This procedure is explained in detail in ACI 304.5R.

12.2.4 *Batching lightweight fine aggregate*—It is not practical to batch the lightweight fine aggregate by volumetric methods because their volume changes due to variable bulking with different amounts of surface water (Portland Cement Association 1944). For this reason, the lightweight fines are batched by mass in much the same manner as natural sand with allowances made for total moisture content.

Because the moisture in lightweight fines can be partly absorbed water as well as surface or free water, the moisture meters used in batch plant storage bins for natural sand are not satisfactory for lightweight sand. Satisfactory batching results have been achieved by drying a small sample [approximately 1 lb (500 g)] of the lightweight sand being used in a suitable container to a constant weight at the temperature of 212 to 230 F (100 to 110 C). The total moisture (absorbed plus free moisture) is calculated by comparing the moist weight of the sample with its dry weight. Moisture tests should be conducted at least once per day or whenever a fresh supply of lightweight sand is introduced.

To adjust for the proper amount of lightweight fines, the oven-dry unit weight of the material being used is determined as indicated previously. If this dry unit weight differs from that shown on the laboratory mixture proportion, then the dry batch weight is changed by multiplying the loose volume by the new dry unit weight just determined. This dry batch weight is increased by the moisture content as determined previously to give the actual scale weight to be used.

12.3—Mixing

The absorptive properties of lightweight aggregates require consideration during mixing. The time rate of absorption as well as the maximum total absorption have to be properly integrated into the mixing cycle to control consistency properly (Expanded Shale, Clay and Slate Institute 1958b; Tobin 1971).

12.3.1 *Charging mixers*—The sequence of introducing the ingredients for lightweight concrete into a mixer can vary from one plant to another. Once acceptable procedures for both wetting and batching have been established, it is important to repeat these as closely as possible at all times to produce uniformity. Weather conditions, such as ambient temperature and humidity, can exert significant influences on lightweight concrete production and should be properly considered.

12.3.1.1 *Stationary mixers*—Stationary plant mixers are commonly used in precasting or prestressing operations and occasionally on building sites where concrete is not moved a great distance. They can also be used at a ready-mixed concrete production plant.

Coarse aggregates should be placed in the mixer first, followed by the fine aggregates. Then add in sequence the required water, cement, and any specified additives.

After all of the ingredients have been fed into the plant mixer, it should be operated at mixing speed to produce a complete mixture that will meet the evaluation tests as described in ASTM C 94. When stationary mixers are used for the purpose of partial or shrink mixing, they are only required to blend the materials together as mixing is completed in the truck mixer. If the lightweight aggregate has not reached its full saturation, further absorption during and after mixing can cause the mixture to stiffen.

12.3.1.2 *Truck mixers*—Charging or loading a truck mixer follows the same general practice used with stationary mixers. Larger volumes of lightweight concrete can sometimes be hauled in truck mixers without exceeding the legal weight or axle load limits. The volume of concrete in the drum should not exceed 63% of the drum volume when used as a mixer or 80% of the drum volume when used as an agitator (Gaynor and Mullarky 1975).

12.3.2 *Mixer operation*—Delivery time has an important role in slump control and can require changes in the amount of water needed to produce the desired slump. Construction jobs at different distances from the batch plant require longer or shorter haul periods, and it is not uncommon to have a delay in unloading. These factors make it difficult to determine the total time that a mixture will be in the drum for any particular load. Most lightweight aggregates continue to absorb water with time, even though they are prewetted. Prewetting slows the rate of absorption but does not necessarily eliminate absorption. It can be desirable to hold back 2 or 3 $gal./yd^3$ (10 to 15 L/m^3) of water to make certain that the batch is not too wet upon arrival. It is often necessary and permissible to add water to a lightweight concrete mixture on the job to replace free water that has been absorbed by the lightweight aggregate to bring the concrete back to the desired slump. Mixing is done as described in Section 4.5.2.

12.4—Job controls

Field control of the yield of lightweight concrete is most important. Overyield produces a larger volume of concrete than intended, whereas underyield produces less. Overyield is nearly always associated with a loss in strength due to a reduction in the net cement content. Underyield results in less concrete being delivered than was expected or ordered. ASTM C 127, C 138, C 173, and C 231 give methods of establishing field control.

The unit weight of the fresh concrete is used to measure the yield of a mixture. The total weight of all the ingredients that are placed in a mixer drum as given on the delivery ticket is computed, or the entire truck can be weighed before and after discharging. The weight of all of the ingredients divided by the unit weight of the concrete will give the total volume of concrete in the mixer drum. When the calculated volume is more than 2% above or below the volume shown on the delivery ticket, an adjustment is required.

If the change in yield is due to entrained air content, then an adjustment in the amount of air-entraining admixture can correct this condition.

If the unit weight measured in the field is greater than the wet unit weight shown for the mixture proportioning, this indicates an underyield; conversely, if the weight is less, an overyield can occur. When there have been no appreciable changes in the weights of the lightweight aggregates themselves, in all probability, the differences in yield can be attributed to an incorrect amount or an incorrect absolute volume of lightweight aggregates. In this case, steps should be taken at the batch plant to correct the absolute volume of lightweight aggregates used in the concrete as it is being batched.

CHAPTER 13—VOLUMETRIC-MEASURING AND CONTINUOUS-MIXING CONCRETE EQUIPMENT
13.1—General Considerations

This chapter gives an overview of volumetric-measuring and continuous-mixing concrete equipment (VMCM). For a more detailed discussion, refer to ACI 304.6R.

When aggregates or cementitious materials are batched by volume, the method of batching is considered volumetric. It is normally a continuous operation coupled with continuous mixing. Accurate volumetric batching is achieved by passing material through a calibrated rotary vane feeder, conveying material through a calibrated gate opening, or by any other method that would provide a known volume in a calibrated unit time.

Volumetric batching is suitable for the production of most concrete, provided the equipment is operated in accordance with ASTM C 685 and with the same attention to detail as that required for weigh batching. The available equipment is highly mobile, requires little or no setup time, and often serves as its own material transport.

VMCM units carry enough materials to produce 6 to 10 yd^3 (5 to 8 m^3) of concrete (Fig. 13.1). This limitation is based on axle loading limitations. Production of larger volumes of concrete or high rates of production will require special provisions for recharging the material storage compartments.

The portability of the equipment makes it practical to bring the VMCM unit to the point of use, which can be an advantage in many applications. Having the unit at the placement site also allows close control of concrete quality at the site.

VMCM equipment lends itself to many different applications. Many of these applications involve relatively low-volume production of concrete, but large jobs have also been done with this equipment. In addition to producing conventional concrete, VMCM equipment is well suited for a variety of special applications, such as:
- Mixtures with short working times;
- Low-slump mixtures;
- Long unloading times;
- Concrete at remote sites;
- Making small deliveries;
- Precast operations;
- Hot weather concreting;
- Mining applications;
- Grouting and pile filling;
- Colored concretes;
- Emergency applications;
- Variable slumps within same load; and
- Flowable-fill mixtures.

13.2—Operations

Quality control—The production of concrete by volumetric measurement and continuous mixing is subject to the same rules of quality control as any other concrete production method.

Calibration—To ensure production of quality concrete, calibrate each volumetric-measuring unit for each respective concrete ingredient, following the manufacturer's recommendations and ASTM C 685.

Operational precautions—The VMCM should be in good condition. All shields and covers should be in place. All controls should operate smoothly and be connected according to the manufacturer's recommendations. All material-feed operations should start and stop simultaneously. The cement-measuring device should be inspected and cleaned regularly. Indicating meters and dials should be checked for proper

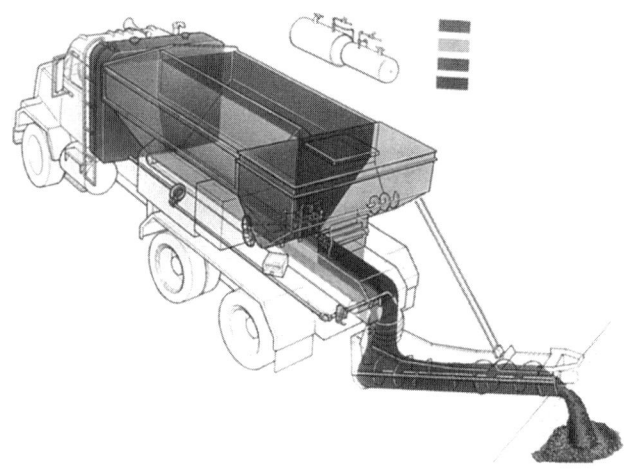

Fig. 13.1—Typical system.

flow and operation. All filters should be clean and allow full flow of water. Aggregate feed systems should be free of any blockage. Checks of the various feeding systems should be carried out according to the manufacturer's recommendations and as job experience indicates.

13.3—Fresh concrete properties

Fresh concrete produced by VMCM behaves slightly differently than ready-mixed concrete. Elapsed hydration time at discharge is measured in seconds rather than in minutes. This means that, although the actual setting time (from start of hydration) is the same, the apparent setting time (from time in place) can seem longer. Finally, the apparent slump at discharge is often higher than the measured slump 3 to 5 min after discharge. Finishers and inspectors should be made aware of these differences.

CHAPTER 14—REFERENCES
14.1—Referenced standards and reports

The documents of the various standards-producing organizations referred to in this document are listed below with their serial designation.

American Concrete Institute

116R	Cement and Concrete Terminology
207.1R	Mass Concrete
207.2R	Effect of Restraint, Volume Change, and Reinforcement on Cracking of Mass Concrete
207.5R	Roller Compacted Mass Concrete
211.1	Standard Practice for Selecting Proportions for Normal, Heavyweight, and Mass Concrete
211.2	Standard Practice for Selecting Proportions for Structural Lightweight Concrete
212.3R	Chemical Admixtures for Concrete
221R	Guide for Use of Normalweight and Heavyweight Aggregates in Concrete
223	Standard Practice for the Use of Shrinkage Compensating Concrete
224R	Control of Cracking in Concrete Structures
302.1R	Guide for Concrete Floor and Slab Construction
304.1R	Guide for the Use of Preplaced Aggregate Concrete for Structural and Mass Concrete Applications
304.2R	Placing Concrete by Pumping Methods
304.3R	Heavyweight Concrete: Measuring, Mixing, Transporting and Placing
304.4R	Placing Concrete with Belt Conveyors
304.5R	Batching, Mixing, and Job Control of Lightweight Concrete
304.6R	Guide for the Use of Volumetric-Measuring and Continuous Mixing Concrete Equipment
305R	Hot Weather Concreting
306R	Cold Weather Concreting
308	Standard Practice for Curing Concrete
309R	Guide for Consolidation of Concrete
311.1R	ACI Manual of Concrete Inspection (SP-2)
311.4R	Guide for Concrete Inspection
318	Building Code Requirements for Structural Concrete
325.9R	Guide for Construction of Concrete Pavements and Concrete Bases
347R	Guide to Formwork for Concrete
506R	Guide for Shotcrete
544.3R	Guide for Specifying, Proportioning, Mixing, Placing, and Finishing Steel Fiber-Reinforced Concrete

ASTM International

C 33	Specification for Concrete Aggregates
C 94	Specification for Ready-Mixed Concrete
C 127	Test Method for Specific Gravity and Absorption of Coarse Aggregate
C 138	Test Method for Unit Weight, Yield and Air Content (Gravimetric) of Concrete
C 150	Specification for Portland Cement
C 172	Practice for Sampling Freshly Mixed Concrete
C 173	Test Method for Air Content of Freshly Mixed Concrete by the Volumetric Method
C 231	Test Method for Air Content of Freshly Mixed Concrete by the Pressure Method
C 595M	Specification for Blended Hydraulic Cements (Metric)
C 618	Specification for Coal Fly Ash and Raw or Calcined Natural Pozzolan for Use as a Mineral Admixture in Portland Cement Concrete
C 637	Specification for Aggregates for Radiation-Shielding Concrete
C 638	Descriptive Nomenclature of Constituents of Aggregates for Radiation-Shielding Concrete
C 685	Specification for Concrete Made by Volumetric Batching and Continuous Mixing
C 845	Specification for Expansive Hydraulic Cement
C 938	Practice for Proportioning Grout Mixtures for Preplaced-Aggregate Concrete
C 939	Test Method for Flow of Grout for Preplaced-Aggregate Concrete (Flow Cone Method)
C 943	Practice for Making Test Cylinders and Prisms for Determining Strength and Density of Preplaced-Aggregate Concrete in the Laboratory
C 953	Test Method for Time of Setting of Grouts for Preplaced-Aggregate Concrete in the Laboratory
D 75	Practice for Sampling Aggregates
D 2419	Test Method for Sand Equivalent Value of Soils and Fine Aggregate

U.S. Army Corps of Engineers

CRD-C 55 Test Method for Within-Batch Uniformity of Freshly Mixed Concrete

CRD-C 61 Test Method for Determining the Resistance of Freshly Mixed Concrete to Washing Out in Water

The above publications may be obtained from the following organizations:

American Concrete Institute
P.O. Box 9094
Farmington Hills, MI 48333-9094

ASTM International
100 Barr Harbor Drive
West Conshohocken, PA 19428

U. S. Army Corps of Engineers Waterways Experiment Station
3909 Halls Ferry Road
Vicksburg, MS 39180

14.2—Cited references

AASHTO, 1993, *Guide Specifications for Highway Construction*, American Association of State Highway and Transportation Officials, Washington, D.C., 296 pp.

Anderson, W. G., 1977, "Analyzing Concrete Mixtures for Pumpability," ACI JOURNAL, *Proceedings* V. 74, No. 9, Sept., pp. 447-451.

Anon., 1954, "Investigation of the Suitability of Prepacked for Mass and Reinforced Concrete Structure," *Technical Memorandum No. 6-330*, U.S. Army Engineer Waterways Experiment Station, Vicksburg, Miss., 44 pp.

Anon., 1955, *Nucleonics*, McGraw-Hill Publishing Co., New York, pp. 60-65.

Anon., 1977, "Cooling Concrete Mixes with Liquid Nitrogen," *Concrete Construction*, V. 22, No. 5, pp. 257-258.

Anon., 1979, *Belt Conveyors for Bulk Materials*, 2nd Edition, Conveyor Equipment Manufacturers Association Engineering Conference, Washington, D. C.

Anon., 1988, "Cooled Concrete Controls Cracking for Base Mat Pour," *Concrete Construction*, V. 33, No. 11, pp.1032-1033.

Bozarth, F. M., 1967, "Case Study of Influences of Imbalances in Charging of Cement and Water on Mixing Performance of an Eight Cubic Yard Central Plant Mixer," U.S. Bureau of Public Roads, Washington, D.C.

Browne, R. D., and Blundell, R., 1972, "Relevance of Concrete Property Research to Pressure Vessel Design," *Concrete for Nuclear Reactors*, SP-34, C. E. Kesler, ed., American Concrete Institute, Farmington Hills, Mich., pp. 69-102.

Carlson, R. W.; Houghton, D. L.; and Polivka, M., 1979, "Causes and Control of Cracking in Unreinforced Mass Concrete," ACI JOURNAL, *Proceedings* V. 76, No. 7, July, pp. 821-837.

Cement and Concrete Association, 1976, "Superplasticizing Admixtures in Concrete," *Publication* No. 45.030, Wexham Springs, 32 pp.

Concrete Plant Manufacturers Bureau, 1996a, "Concrete Plant Standards of the Concrete Plant Manufacturers Bureau," *CMPB-101*, Silver Spring, 13 pp.

Concrete Plant Manufacturers Bureau, 1996b, "Recommended Guide Specifications for Batching Equipment and Control Systems in Concrete Batch Plants," *CPMB-102*, Silver Spring, 9 pp.

Concrete Plant Manufacturers Bureau, 1996c, "Concrete Plant Mixer Standards of the Plant Mixer Manufacturers Division, Concrete Plant Manufacturers Bureau," *PMMD-100*, Silver Spring, 4 pp.

Concrete Reinforcing Steel Institute, 1982, *CRSI Handbook*, 5th Edition, Schaumburg, Ill.

Cope, J., L., 1972, "Conveying Concrete to Lower Dam Construction Costs," *Economical Construction of Concrete Dams*, American Society of Civil Engineers, New York, pp. 252-255.

Corps of Engineers, 1994a, "Standard Practice for Concrete for Civil Works Structures," *EM-1110-2-2000*, Washington, D. C., 119 pp.

Corps of Engineers, 1994b, "Cast-in-Place Structural Concrete," *Civil Works Guide Specification 03301*, Washington, D. C., 57 pp.

Davis, H. S., 1958, "High Density Concrete for Shielding Atomic Energy Plants," ACI JOURNAL, *Proceedings* V. 54, No. 11, May, pp. 965-978.

Davis, H. S., 1967, "Aggregates for Radiation Shielding Concrete," *Materials Research and Standards*, V. 7, No. 11, pp. 494-501.

Davis, H. S., 1972a, "Concrete for Radiation Shielding—In Perspective," *Concrete for Nuclear Reactors*, SP-34, C. E. Kesler, ed., American Concrete Institute, Farmington Hills, Mich., pp. 3-13.

Davis, H. S., 1972b, "Iron-Serpentine Concrete," *Concrete for Nuclear Reactors*, SP-34, C. E. Kesler, ed., American Concrete Institute, Farmington Hills, Mich., pp. 1195-1224.

Davis, H. S., 1972c, "N-Reactor Shielding," *Concrete for Nuclear Reactors*, SP-34, C. E. Kesler, ed., American Concrete Institute, Farmington Hills, Mich., pp. 1109-1161.

Davis, R. E., 1960, "Prepacked Method of Concrete Repair," ACI JOURNAL, *Proceedings* V. 57, No. 2, Aug., pp. 155-172.

Davis, R. E., Jr., and Haltenhoff, C. E., 1956, "Mackinac Bridge Pier Construction," ACI JOURNAL, *Proceedings* V. 53, No. 6, Dec., pp. 581-596.

Davis, R. E., Jr.; Johnson, G. D.; and Wendell, G. E., 1955, "Kemano Penstock Tunnel Liner Backfilled with Prepacked Concrete," ACI JOURNAL, *Proceedings* V. 52, No. 3, Sept., pp. 287-308.

Expanded Shale, Clay, and Slate Institute, 1958a, "Workability is Easy," *Lightweight Concrete Information Sheet* No. 1, Bethesda, Md., 3 pp.

Expanded Shale, Clay, and Slate Institute, 1958b, "Suggested Mix Design for Job Mixed Structural Lightweight Concrete," *Lightweight Concrete Information Sheet* No. 3, 2 pp.

Fowler, E. L., and Holmgren, E. F., 1971, "Expansion of Concrete Pumped Through Aluminum Pipeline," ACI JOURNAL, *Proceedings* V. 68, No. 12, Dec., pp. 950-958.

Gaynor, R. D., and Mullarky, J. I, 1975, "Mixing Concrete in a Truck Mixer," *NRMCA Publication* No. 148, National Ready Mixed Concrete Association, Silver Spring, 24 pp.

Gerwick, B. C., 1964, "Placement of Tremie Concrete," *Symposium on Concrete in Aqueous Environments*, SP-8, American Concrete Institute, Farmington Hills, Mich., pp. 9-20.

Gerwick, B. C., and Holland, T. C, 1983, "Cracking of Mass Concrete Placed Under Water," *Concrete International: Design and Construction*, V. 5, No. 4, Apr., pp. 29-36.

Gerwick, B. C.; Holland, T. C.; and Kommendant, G. J., 1981, "Tremie Concrete for Bridge Piers and Other Massive Underwater Placements," *Report* No. FHWA/RD-81/153, Federal Highway Administration, Washington, D.C., 203 pp.

Holland, T. C., and Turner, J. R., 1980, "Construction of Tremie Concrete Cutoff Wall, Wolf Creek Dam, Kentucky," *Miscellaneous Paper* No. SL-80-10, U.S. Army Engineer Waterways Experiment Station, Vicksburg, Miss., 85 pp.

Kepler, W. F., 1990, "Underwater Placement of a Canal Lining," *Concrete International*, V. 12, No. 6, June, pp. 54-59.

Khayat, K. H.; Gerwick, B. C.; and Hester, W. T., 1990, "High-Quality Tremie Concretes for Underwater Repairs," *Proceedings, Paul Klieger Symposium on Performance of Concrete*, SP-122, D. Whiting, ed., American Concrete Institute, Farmington Hills, Mich., pp. 125-138.

King, J. C., 1971, "Special Concretes and Mortars," *Handbook of Heavy Construction*, 2nd Edition, McGraw-Hill Book Co., New York, pp. 22-1 to 22-30.

Klemens, T. L., 1991, "Who Says You Can't Pave Underwater?" *Highway & Heavy Construction*, V. 134, No. 10, pp. 64-66.

Koerner, R. M., and Welsh, J. P., 1980, "Fabric Forms Conform to Any Shape," *Concrete Construction*, V. 25, No. 5, pp. 401-409.

Laine, E. F.; Dines, K. A.; Okada, J. T.; and Lytle, R. J., 1980, "Probing Concrete with Radio Waves," *Proceedings* ASCE, V. 106, GT7, pp. 759-766.

Lamberton, B. A., 1980, "Fabric Forms for Erosion Control and Pile Jacketing," *Concrete Construction*, V. 25, No. 5, pp. 395-399.

Lovern, J. D., 1966, "Important Variables Affecting Moisture Control," *Modern Concrete*, V. 30, No. 4, pp. 44-46.

Mass, G. R., 1989, "Premixed Cement Paste," *Concrete International*, V. 11, No. 11, Nov., pp. 82-86.

Mielenz, R.C., 1994, "Petrographic Evaluation of Concrete Aggregates," *Significance of Tests and Properties of Concrete and Concrete-Making Materials*, STP-169, ASTM, Philadelphia, Pa., pp. 341-364.

Nash, K. L., 1974, "Diaphragm Wall Construction Techniques," *Proceedings* ASCE, V. 100, C04, pp. 605-620.

Neeley, B. D., 1988, "Evaluation of Concrete Mixtures for Use in Underwater Repairs," *Technical Report* No. REMR-CS-18, U. S. Army Engineers Waterways Experiment Station, Vicksburg, Miss., 124 pp.

Neeley, B. D., and Wickersham, J., 1989, "Repair of Red Rock Dam," *Concrete International*, V. 11, No. 10, Oct., pp. 36-39.

Pihlajavaara, S. E., 1972, "Preliminary Recommendation for Design, Making and Control of Radiation Shielding Structures," *Concrete for Nuclear Reactors,* SP-34, C. E. Kesler, ed., American Concrete Institute, Farmington Hills, Mich., pp. 57-67.

Portland Cement Association, 1944, "Bulking of Sand Due to Moisture," *Concrete Information Sheet* No. ST20, Skokie, Ill., 2 pp.

Portland Cement Association, 1988, *Design and Control of Concrete Mixtures*, 13th Edition, Skokie, Ill., 205 pp.

Prestressed Concrete Institute, 1981, "Recommended Practice for Use of High-Range Water Reducing Admixtures in Precast Prestressed Concrete Operations," *Journal*, V. 26, No. 5, Sept.-Oct. pp. 28-48.

Reilly, W. E., 1972, "Hydrothermal and Vacuum Saturated Lightweight Aggregate for Pumped Structural Concrete," ACI JOURNAL, *Proceedings* V. 69, No. 7, July, pp. 428-432.

Saucier, K. L., 1974, "Use of Belt Conveyors to Transport Mass Concrete," *Technical Report* No. C-74-4, U. S. Army Engineer Waterways Experiment Station, Vicksburg, Miss.

Saucier, K. L., and Neeley, B. D., 1987, "Antiwashout Admixtures in Underwater Concrete," *Concrete International*, V. 9, No. 5, May, pp. 42-47.

Shideler, J. J., 1957, "Lightweight Aggregate Concrete for Structural Use," ACI JOURNAL, *Proceedings* V. 54, No. 4, Sept., pp. 299-328.

Society of Automotive Engineers, 1993, "Recommended Practice for Cast Shot and Grit Size Specification for Peening and Cleaning," *SAE J 444, SAE Handbook*, New York, 5 pp.

Tobin, R.E., 1967, "Lightweight Ready Mix—A New Approach," *Concrete Products*, 5 pp.

Tobin, R.E., 1971, "Handling Lightweight Concrete on the Job," *Lightweight Concrete*, SP-29, D. P. Jenny and A. Litvin, eds., American Concrete Institute, Farmington Hills, Mich., pp. 63-70.

Tobin, R.E., 1978, "Flow Cone Sand Tests," ACI JOURNAL, *Proceedings* V. 75, No. 1, Jan., pp. 1-12.

Truck Mixer Manufacturers Bureau, 1996, "Truck Mixer, Agitator, and Front Discharge Concrete Carrier Standards," *TMMB-100*, Silver Spring, 5 pp.

Tynes, W. O., 1959, "Investigation of Methods of Preparing Horizontal Construction Joints in Concrete," *Technical Report* No. 6-518, U.S. Army Engineer Waterways Experiment Station, Vicksburg, Miss., 19 pp.

Tynes, W. O., 1962, "Influence of Fine Aggregate Grading on Properties of Concrete," *Technical Report* No. 6-544, U.S. Army Engineer Waterways Experiment Station, Vicksburg, Miss., 25 pp.

Tynes, W. O., 1963, "Investigation of Methods of Preparing Horizontal Construction Joints in Concrete—Report 2, Tests of Joints of Large Blocks," *Technical Report* No. 6-518, U.S. Army Engineer Waterways Experiment Station, Vicksburg, Miss., 19 pp.

U.S. Bureau of Reclamation, 1981, *Concrete Manual*, 8th Edition, Denver, 627 pp.

U.S. Department of Commerce, 1966, "A Study of Mixing Performance of Large Central Plant Concrete Mixers," Bureau of Public Roads, Washington, D.C.

Van Alstine, C. B., 1955, "Mixing Water Control by Use of a Moisture Meter," ACI JOURNAL, *Proceedings* V. 52, No. 3, Nov. pp. 341-348. Also, Discussion, Part 2, Dec. 1956, p. 1209.

Volkman, D. E., 1994, "Concrete for Radiation Shielding," *Significance of Tests and Properties of Concrete and Concrete-Making Materials*, ASTM STP 169C, P. Klieger and J. Lamond, eds., ASTM, pp. 540-546.

Williams, J. Wayman, Jr., 1959, "Tremie Concrete Controlled with Admixtures," ACI JOURNAL, *Proceedings,* V. 55, No. 8, Feb., pp. 839-850.

Wills, M. H., Jr., 1974, "Lightweight Aggregate Particle Shape Effect on Structural Concrete," ACI JOURNAL, *Proceedings* V. 71, No. 3, Mar., pp. 134-142.

Xanthakos, Petros P.,1979, *Slurry Walls*, McGraw-Hill Book Co., New York, 622 pp.

ACI 305R-99

Hot Weather Concreting

Reported by ACI Committee 305

Robert J. Ryan		Kenneth B. Rear
Chairman		Secretary

Muwafaq A. Abu-Zaid	D. Gene Daniel	Alexander Leschinsky
Bijan Ahmadi	Richard D. Gaynor	William C. Moore
J. Howard Allred	John G. Gendrich	Dan Ravina
Zawde Berhane	G. Terry Harris, Sr.	John M. Scanlon
Karl P. Brandt	Barry L. Houseal	Victor H. Smith
Terence M. Browne	Frank A. Kozeliski	George V. Teodoru
Joseph G. Cabrera	Mark E. Leeman	Habib M. Zein Al-Abidien
James N. Cornell, II		

Concrete mixed, transported, and placed under conditions of high ambient temperature, low humidity, solar radiation, or wind, requires an understanding of the effects these environmental factors have on concrete properties and construction operations. Measures can be taken to eliminate or minimize undesirable effects of these environmental factors. Experience in hot weather with the types of construction involved will reduce the potential for serious problems.

This committee report defines hot weather, lists possible potential problems, and presents practices intended to minimize them. Among these practices are such important measures as selecting materials and proportions, precooling ingredients, special batching, length of haul, consideration of concrete temperature as placed, facilities for handling concrete at the site, and during the early curing period, placing, and curing techniques, and appropriate testing and inspecting procedures in hot weather conditions. A selected bibliography is included.

These revisions involve an editorial revision of the document. The revisions focus in particular on the effects of hot weather on concrete properties, and the use of midrange water-reducing admixtures and extended set-control admixtures in hot weather.

ACI Committee Reports, Guides, Standard Practices, and Commentaries are intended for guidance in planning, designing, executing, and inspecting construction. This document is intended for the use of individuals who are competent to evaluate the significance and limitations of its content and recommendations and who will accept responsibility for the application of the material it contains. The American Concrete Institute disclaims any and all responsibility for the stated principles. The Institute shall not be liable for any loss or damage arising therefrom.
 Reference to this document shall not be made in contract documents. If items found in this document are desired by the Architect/Engineer to be a part of the contract documents, they shall be restated in mandatory language for incorporation by the Architect/Engineer.

Keywords: air entrainment; cooling; curing; evaporation; high temperature; hot weather construction; plastic shrinkage; production methods; retempering; slump tests; water content.

CONTENTS
Chapter 1—Introduction, p. 305R-2
1.1—General
1.2—Definition of hot weather
1.3—Potential problems in hot weather
1.4—Potential problems related to other factors
1.5—Practices for hot weather concreting

Chapter 2—Effects of hot weather on concrete properties, p. 305R-3
2.1—General
2.2—Temperature of concrete
2.3—Ambient conditions
2.4—Water requirements
2.5—Effect of cement
2.6—Supplementary cementitious materials
2.7—Chemical admixtures
2.8—Aggregates
2.9—Proportioning

Chapter 3—Production and delivery, p. 305R-11
3.1—General
3.2—Temperature control of concrete
3.3—Batching and mixing

ACI 305R-99 supersedes ACI 305R-91 and became effective October 27, 1999.
Copyright © 2000, American Concrete Institute.
All rights reserved including rights of reproduction and use in any form or by any means, including the making of copies by any photo process, or by electronic or mechanical device, printed, whitten, or oral, or recording for sound or visual reproduction or for use in any knowledge or retrieval system or device, unless permission in writing is obtained from the copyright proprietors.

3.4—Delivery
3.5—Slump adjustment
3.6—Properties of concrete mixtures
3.7—Retempering

Chapter 4—Placing and curing, p. 305R-13
4.1—General
4.2—Preparations for placing and curing
4.3—Placement and finishing
4.4—Curing and protection

Chapter 5—Testing and inspection, p. 305R-16
5.1—Testing
5.2—Inspection

Chapter 6—References, p. 305R-17
6.1—Referenced standards and reports
6.2—Cited references

Appendix A—Estimating concrete temperature, p. 305R-19

Appendix B—Methods for cooling fresh concrete, p. 305R-19

CHAPTER 1—INTRODUCTION
1.1—General
Hot weather may create problems in mixing, placing, and curing hydraulic cement concrete. These problems can adversely affect the properties and serviceability of the concrete. Most of these problems relate to the increased rate of cement hydration at higher temperature and increased evaporation rate of moisture from the freshly mixed concrete. The rate of cement hydration is dependent on concrete temperature, cement composition and fineness, and admixtures used.

This report will identify problems created by hot weather concreting and describe practices that will alleviate these potential adverse effects. These practices include suggested preparations and procedures for use in general types of hot weather construction, such as pavements, bridges, and buildings. Temperature, volume changes, and cracking problems associated with mass concrete are treated more thoroughly in ACI 207.1R and ACI 224R.

A maximum "as placed" concrete temperature is often used in an effort to control strength, durability, plastic-shrinkage cracking, thermal cracking, and drying shrinkage. The placement of concrete in hot weather, however, is too complex to be dealt with by setting a maximum "as placed" or "as delivered" concrete temperature. Concrete durability is a general term that is difficult to quantify, but it is perceived to mean resistance of the concrete to weathering (ACI 201.2R). Generally, if concrete strengths are satisfactory and curing practices are sufficient to avoid undesirable drying of surfaces, durability of hot weather concrete will not differ greatly from similar concrete placed at normal temperatures. The presence of a desirable air-void system is needed if the concrete is going to be exposed to freezing cycles.

If an acceptable record of field tests is not available, concrete proportions may be determined by trial batches (ACI 301 and ACI 211.1). Trial batches should be made at temperatures anticipated in the work and mixed following one of the procedures described in Section 2.9, Proportioning. The concrete supplier and contractor are generally responsible for determining concrete proportions to produce the required quality of concrete unless specified otherwise.

According to ASTM C 31/C 31M, concrete test specimens made in the field that are used for checking adequacy of laboratory mixture proportions for strength or as a basis for acceptance or quality control should be cured initially at 60 to 80 F (16 to 27 C). If the initial 24 h curing is at 100 F (38 C), the 28-day compressive strength of the test specimens may be 10 to 15% lower than if cured at the required ASTM C 31/C 31M curing temperature (Gaynor et al 1985). If the cylinders are allowed to dry at early ages, strengths will be reduced even further (Cebeci 1987). Therefore, proper fabrication, curing, and testing of the test specimens during hot weather is critical, and steps should be taken to ensure that the specified procedures are followed.

1.2—Definition of hot weather
1.2.1 For the purpose of this report, hot weather is any combination of the following conditions that tends to impair the quality of freshly mixed or hardened concrete by accelerating the rate of moisture loss and rate of cement hydration, or otherwise causing detrimental results:
- High ambient temperature;
- High concrete temperature;
- Low relative humidity;
- Wind speed; and
- Solar radiation.

1.2.2 The effects of high air temperature, solar radiation, and low relative humidity may be more pronounced with increases in wind speed (Fig. 2.1.5). The potential problems of hot weather concreting may occur at any time of the year in warm tropical or arid climates, and generally occur during the summer season in other climates. Early cracking due to thermal shrinkage is generally more severe in the spring and fall. This is because the temperature differential for each 24 h period is greater during these times of the year. Precautionary measures required on a windy, sunny day will be more strict than those required on a calm, humid day, even if air temperatures are identical.

1.3—Potential problems in hot weather
1.3.1 Potential problems for concrete in the freshly mixed state are likely to include:
- Increased water demand;
- Increased rate of slump loss and corresponding tendency to add water at the job site;
- Increased rate of setting, resulting in greater difficulty with handling, compacting, and finishing, and a greater risk of cold joints;
- Increased tendency for plastic-shrinkage cracking; and
- Increased difficulty in controlling entrained air content.

1.3.2 Potential deficiencies to concrete in the hardened state may include:
- Decreased 28-day and later strengths resulting from

either higher water demand, higher concrete temperature, or both at time of placement or during the first several days;
- Increased tendency for drying shrinkage and differential thermal cracking from either cooling of the overall structure, or from temperature differentials within the cross section of the member;
- Decreased durability resulting from cracking;
- Greater variability of surface appearance, such as cold joints or color difference, due to different rates of hydration or different water-cementitious material ratios (*w/cm*);
- Increased potential for reinforcing steel corrosion—making possible the ingress of corrosive solutions; and
- Increased permeability as a result of high water content, inadequate curing, carbonation, lightweight aggregates, or improper matrix-aggregate proportions.

1.4—Potential problems related to other factors

Other factors that should be considered along with climatic factors may include:
- Use of cements with increased rate of hydration;
- Use of high-compressive-strength concrete, which requires higher cement contents;
- Design of thin concrete sections with correspondingly greater percentages of steel, which complicate placing and consolidation of concrete;
- Economic necessity to continue work in extremely hot weather; and
- Use of shrinkage-compensating cement.

1.5—Practices for hot weather concreting

Any damage to concrete caused by hot weather can never be fully alleviated. Good judgment is necessary to select the most appropriate compromise of quality, economy, and practicability. The procedures selected will depend on: type of construction; characteristics of the materials being used; and experience of the local industry in dealing with high ambient temperature, high concrete temperatures, low relative humidity, wind speed, and solar radiation.

The most serious difficulties occur when personnel placing the concrete lack experience in constructing under hot weather conditions or in doing the particular type of construction. Last-minute improvisations are rarely successful. Early preventive measures should be applied with the emphasis on materials evaluation, advanced planning and purchasing, and coordination of all phases of work. Planning in advance for hot weather involves detailed procedures for mixing, placing, protection, curing, temperature monitoring, and testing of concrete. Precautions to avoid plastic-shrinkage cracking are important. The potential for thermal cracking, either from overall volume changes or from internal restraint, should be anticipated. Methods to control cracking include: proper use of joints, increased amounts of reinforcing steel or fibers, limits on concrete temperature, reduced cement content, low-heat-of-hydration cement, increased form-stripping time, and selection and dosage of appropriate chemical and mineral admixtures.

The following list of practices and measures to reduce or avoid the potential problems of hot weather concreting are discussed in detail in Chapters 2, 3, and 4:
- Select concrete materials and proportions with satisfactory records in hot weather conditions;
- Cool the concrete;
- Use a concrete consistency that permits rapid placement and effective consolidation;
- Minimize the time to transport, place, consolidate, and finish the concrete;
- Plan the job to avoid adverse exposure of the concrete to the environment; schedule placing operations during times of the day or night when weather conditions are favorable;
- Protect the concrete from moisture loss during placing and curing periods; and
- Schedule a preplacement conference to discuss the requirements of hot weather concreting.

CHAPTER 2—EFFECTS OF HOT WEATHER ON CONCRETE PROPERTIES

2.1—General

2.1.1 Properties of concrete that make it an excellent construction material can be affected adversely by hot weather, as defined in Chapter 1. Harmful effects are minimized by control procedures outlined in this report. Strength, impermeability, dimensional stability, and resistance of the concrete to weathering, wear, and chemical attack all depend on the following factors: selection and proper control of materials and mixture proportioning; initial concrete temperature; wind speed; solar radiation; ambient temperature; and humidity condition during the placing and curing period.

2.1.2 Concrete mixed, placed, and cured at elevated temperatures normally develops higher early strengths than concrete produced and cured at lower temperatures, but strengths are generally lower at 28 days and later ages. The data in Fig. 2.1.2 shows that with increasing curing temperatures, 1-day strength will increase, and 28-day strength decreases (Klieger 1958; Verbeck and Helmuth 1968). Some researchers conclude that a relatively more uniform microstructure of the hydrated cement paste can account for higher strength of concrete mixtures cast and cured at lower temperatures (Mehta 1986).

2.1.3 Laboratory tests have demonstrated the adverse effects of high temperatures with a lack of proper curing on concrete strength (Bloem 1954). Specimens molded and cured in air at 73 F (23 C), 60% relative humidity and at 100 F (38 C), 25% relative humidity produced strengths of only 73 and 62%, respectively, of that obtained for standard specimens moist-cured at 73 F (23 C) for 28 days. The longer the delay between casting the cylinders and placing into standard moist storage, the greater the strength reduction. The data illustrate that inadequate curing in combination with high placement temperatures impairs the hydration process and reduces strength. The tests were made on plain concrete without admixtures or pozzolans that might have improved its performance at elevated temperatures. Other researchers determined that insufficient curing is more detri-

Table 2.1.5—Typical concrete temperatures for various relative humidities potentially critical to plastic-shrinkage cracking

Concrete temperature, F (C)	Air temperature, F (C)	Critical evaporation rate			
		0.2 lb/ft²/h (1.0 kg/m²/h)	0.15 lb/ft²/h (0.75 kg/m²/h)	0.10 lb/ft²/h (0.50 kg/m²/h)	0.05 lb/ft²/h (0.25 kg/m²/h)
		Relative humidity, %*			
105 (41)	95 (35)	85	100	100	100
100 (38)	90 (32)	80	95	100	100
95 (35)	85 (29)	75	90	100	100
90 (32)	80 (27)	60	85	100	100
85 (29)	75 (24)	55	80	95	100
80 (27)	70 (21)	35	60	85	100
75 (24)	65 (19)	20	55	80	100

*Relative humidity, % which evaporation rate will exceed the critical values shown, assuming air temperature is 10 F (6 C) cooler than concrete temperature and a constant wind speed of 10 mph (16 km/h), measured at 20 in. (0.5 m) above the evaporating surface.
Note: Based on NRMCA-PCA nomograph (Fig. 2.1.5), results rounded to nearest 5%.

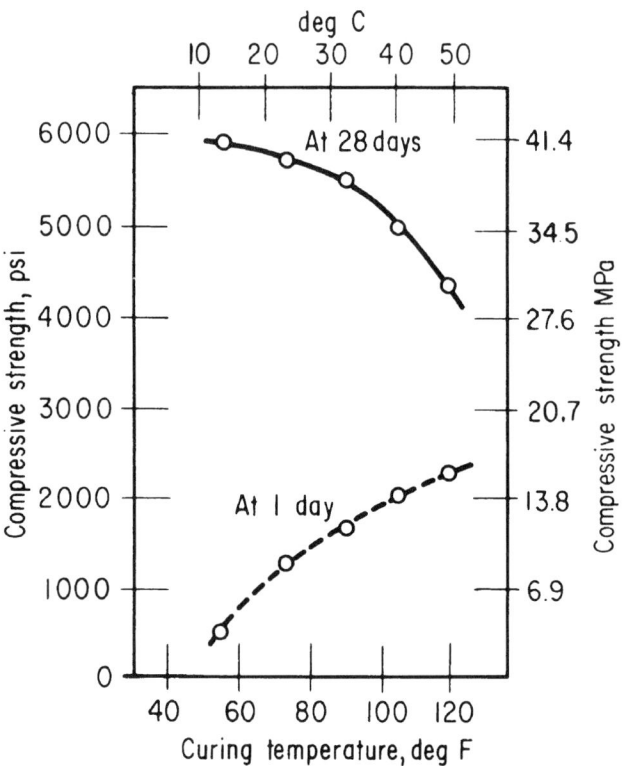

Fig 2.1.2—Effects of curing temperature on compressive strength of concrete (Verbeck and Helmuth 1968).

evaporation rate is given in Section 5.1.3. High concrete temperatures, high wind speed, and low humidity, alone or in combination, cause rapid evaporation of surface water. The rate of bleeding, on the other hand, depends on concrete mixture ingredients and proportions, on the depth of the member being cast, and on the type of consolidation and finishing. Because surface drying is initiated when evaporation rate exceeds bleeding rate, the probability of plastic-shrinkage cracking therefore increases whenever the environmental conditions increase evaporation, or when the concrete has a reduced bleeding rate. For example, concrete mixtures incorporating fly ash, silica fume, or fine cements frequently have a low to negligible bleeding rate, making such mixtures highly sensitive to surface drying and plastic shrinkage, even under moderately evaporative conditions (ACI 234R).

2.1.5 Plastic-shrinkage cracking is seldom a problem in hot-humid climates where relative humidity is rarely less than 80%. Table 2.1.5 shows, for various relative humidities, the concrete temperatures that may result in critical evaporation rate levels, and therefore increase the probability of plastic-shrinkage cracking. The table is based on the assumption of a 10 mph (16 km/h) wind speed and an air temperature of 10 F (6 C) cooler than the concrete temperature.

The nomograph in Fig. 2.1.5 is based on common hydrological methods for estimating the rate of evaporation of water from lakes and reservoirs, and is therefore the most accurate when estimating the rate of evaporation from the surface of concrete while that surface is covered with bleed water. When the concrete surface is not covered with bleed water, the nomograph and its underlying mathematical expression tends to overestimate the actual rate of water loss from the concrete surface by as much as a factor of 2 or more (Al-Fadhala 1997). The method is therefore the most useful in estimating the evaporation potential of the ambient conditions, and not as an estimator of the actual rate of water loss from the concrete. Early in the bleeding process, however, and at rates of evaporation less than or equal to 0.2 lb/ft²/h (1.0 kg/m²/h), the method has been shown to be in good agreement with water loss measurements, as long as the temperature, humidity, and wind speed have been measured

mental than high temperatures (Cebeci 1986), and also that required strength levels can be maintained by the proper use of either chemical or mineral admixtures are used in the concrete (Gaynor et al 1985; Mittelacher 1985 & 1992).

2.1.4 Plastic-shrinkage cracking is frequently associated with hot weather concreting in arid climates. It occurs in exposed concrete, primarily in flatwork, but also in beams and footings, and may develop in other climates when the surface of freshly cast concrete dries and subsequently shrinks. Surface drying is initiated whenever the evaporation rate is greater than the rate at which water rises to the surface of recently placed concrete by bleeding. A method to estimate

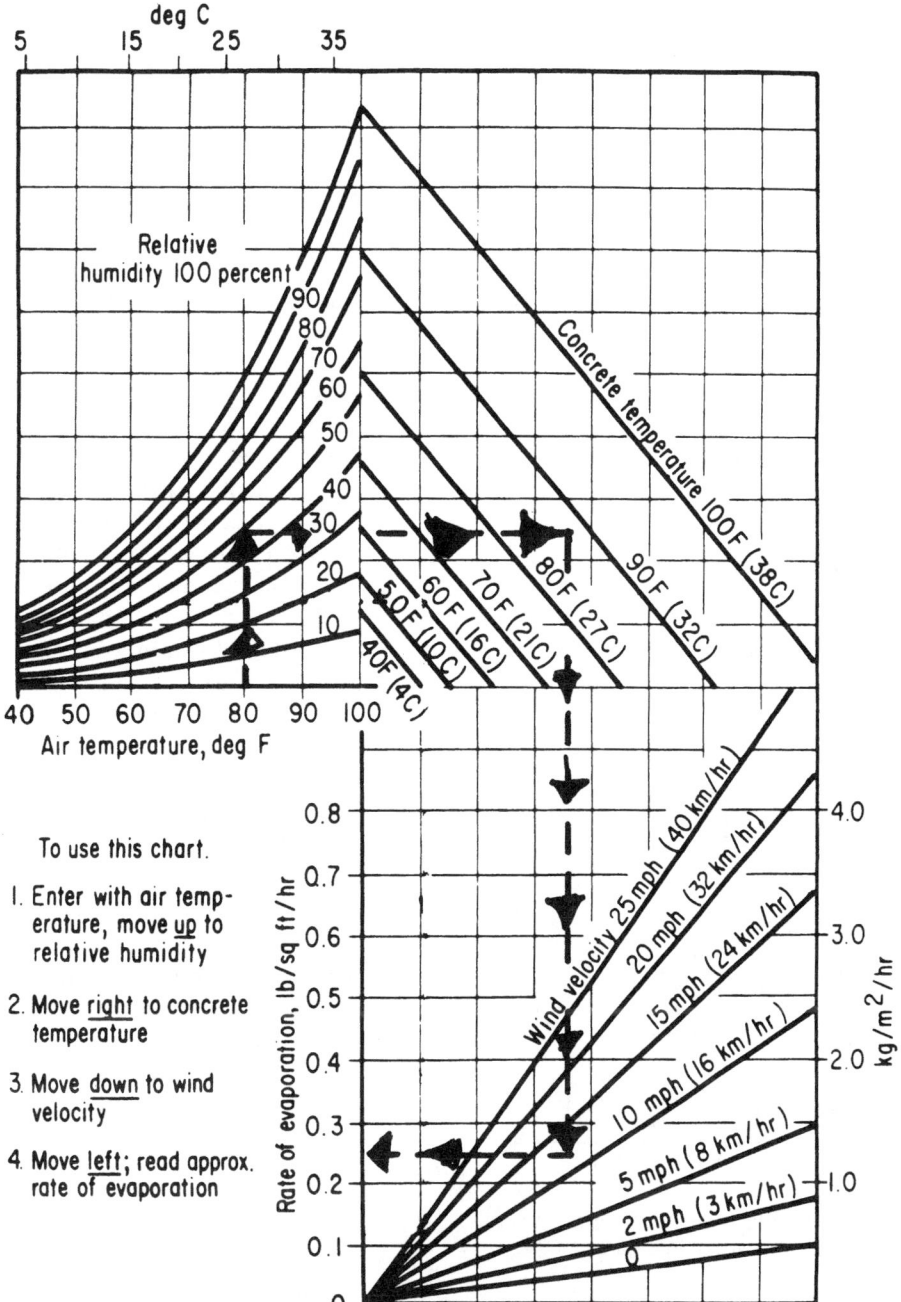

Fig. 2.1.5—Effect of concrete and air temperatures, relative humidity, and wind speed on the rate of evaporation of surface moisture from concrete. This chart provides a graphic method of estimating the loss of surface moisture for various weather conditions. To use this chart, follow the four steps outlined above. If the rate of evaporation approaches 0.2 lb/ft²/h (1 kg/m²/h), precautions against plastic-shrinkage cracking are necessary (Lerch 1957). Wind speed is the average horizontal air or wind speed in mph (km/h) and should be measured at a level approximately 20 in. (510 mm) higher than the evaporating surface. Air temperature and relative humidity should be measured at a level approximately 4 to 6 ft (1.2 to 1.8 m) higher than the evaporating surface on its windward side shielded from the sun's rays (PCA Journal 1957).

as described in the text below Fig. 2.1.5. It is especially critical that wind speed be monitored at 20 in. (0.5 m) above the evaporating surface. This is because wind speed increases rapidly with height above the surface, and wind measurements taken from higher than the prescribed height used in developing the nomograph will overestimate evaporation rate. Note also that wind speed varies tremendously over time, and estimates should not be based on transient gusts of wind. Use of Fig. 2.1.5 provides evaporation rate estimates based on environmental factors of temperature, humidity, and wind speed that contribute to plastic-shrinkage cracking. The graphic method of the chart also yields ready information on the effect of changes in one or more of these factors. For example, it shows that concrete at a temperature of 70 F

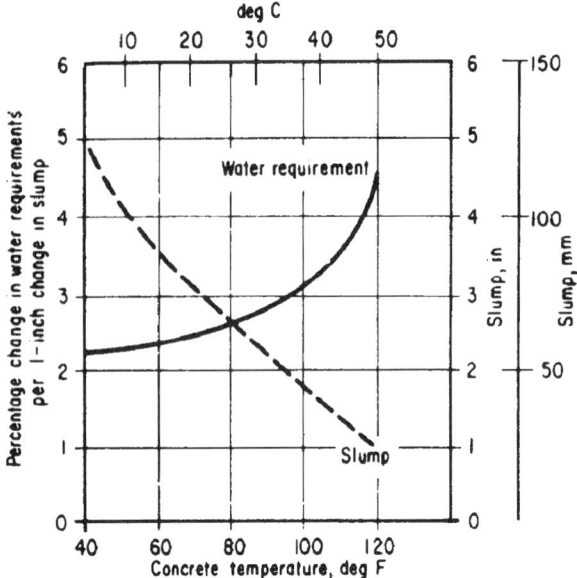

Fig. 2.2.1(a)—Effect of concrete temperature on slump and on water required to change slump (average data for Type I and II cements) (Klieger 1958).

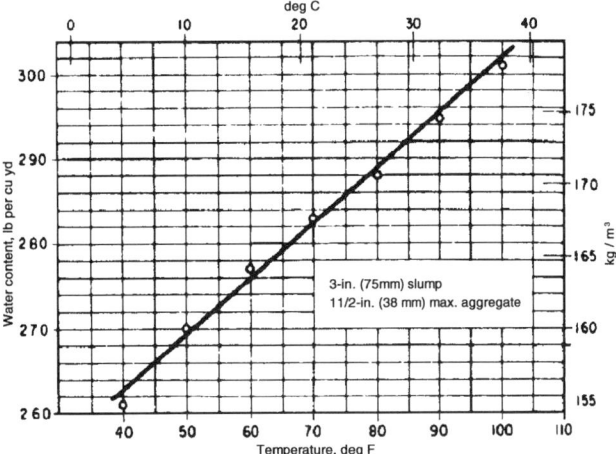

Fig 2.2.1(b)—Effect of temperature increase on the water requirement of concrete (U.S. Bureau of Reclamation 1975).

(21 C) placed at an air temperature of 70 F (21 C), with a relative humidity of 50% and a moderate wind speed of 10 mph (16 km/h), will have six times the evaporation rate of the same concrete placed when there is no wind.

2.1.6 When evaporation rate is expected to approach the bleeding rate of the concrete, precautions should be taken, as explained in detail in Chapter 4. Because bleeding rates vary from zero to over 0.2 lb/ft^2/h (1.0 kg/m^2/h), over time, and are not normally measured, it is common to assume a value for the critical rate of evaporation. The most commonly quoted value is 0.2 lb/ft^2/h (1.0 kg/m^2/h). More recent experience with bridge deck overlays containing silica fume has led to specified allowable evaporation rates of only 0.05 lb/ft^2/h (0.025 kg/m^2/h) (Virginia Department Of Transportation). Construction specifications for the State of New York and the City of Cincinnati are intermediate evaporation rates of 0.15 and 0.10 lb/ft^2/h (0.75 and 0.50 kg/m^2/h), respectively.

The probability for plastic-shrinkage cracks to occur may be increased if the setting time of the concrete is delayed due to the use of slow-setting cement, an excessive dosage of retarding admixture, fly ash as a cement replacement, or cooled concrete. Fly ash is also likely to reduce bleeding and may thereby contribute to a cracking tendency (ACI 226.3R). Plastic-shrinkage cracks are difficult to close once they have occurred (see Section 4.3.5).

2.2—Temperature of concrete

2.2.1 Unless measures are taken to control concrete performance at elevated temperatures, by the selection of suitable materials and proportions as outlined in Sections 2.3 through 2.9, increases in concrete temperature will have the following adverse effects. Other adverse effects are listed in Section 1.3.

- The amount of the water required to produce a given slump increases with the time. For constant mixing time, the amount of water required to produce a given slump also increases with the temperature, as shown in Fig. 2.2.1(a) and 2.2.1(b);
- Increased water content will create a decrease in strength and durability, if the quantity of cementitious material is not increased proportionately;
- Slump loss will be evident earlier after initial mixing and at a more rapid rate, and may cause difficulties with handling and placing operations;
- In an arid climate, plastic-shrinkage cracks are more probable;
- In sections of large dimensions, there will be an increased rate of hydration and heat evolution that will increase differences in temperature between the interior and the exterior concrete. This may cause thermal cracking (ACI 207.1R);
- Early curing is critical and lack of it increasingly detrimental as temperatures rise.

2.3—Ambient conditions

2.3.1 In the more general types of hot weather construction (as defined in Section 1.2), it is impractical to recommend a maximum ambient or concrete temperature because the humidity and wind speed may be low, permitting higher ambient and concrete temperatures. A maximum ambient or concrete temperature that will serve a specific case may be unrealistic in others. Accordingly, the committee can only provide information about the effects of higher temperatures in concrete as mentioned in Sections 1.3 and 2.2.1, and advise that at some temperature between approximately 75 and 100 F (24 and 38 C) there is a limit that will be found to be most favorable for best results in each hot weather operation, and such a limit should be determined for the work. Practices for hot weather concreting should be discussed during the preplacement conference.

Trial batches of concrete for the job should be made at the limiting temperature selected, or at the expected job site high temperature, rather than the 68 to 86 F (20 to 30 C) range given in ASTM C 192. Procedures for testing of concrete batches at temperatures higher than approximately 70 F (21 C) are given in Section 2.9.

2.4—Water requirements

2.4.1 Water, as an ingredient of concrete, greatly influences many of its significant properties, both in the freshly mixed and hardened state. High water temperatures cause higher concrete temperatures, and as the concrete temperature increases, more water is needed to obtain the same slump. Fig. 2.2.1(b) illustrates the possible effect of concrete temperature on water requirements. Unless the amount of cementitious material is increased proportionately, the extra water increases the water-cementitious material ratio and will decrease the strength, durability, watertightness, and other related properties of the concrete. This extra water must be accounted for during mix proportioning. Although pertinent to concrete placed under all conditions, this points to the special need to control the use of additional water in concrete placed under hot weather conditions; see Section 2.3.1.

2.4.2 Fig. 2.2.1(a) illustrates the general effects of increasing concrete temperature on slump of concrete when the amount of mixing water is held constant. It indicates that an increase of 20 F (11 C) in temperature may be expected to decrease the slump by about 1 in. (25 mm). Fig. 2.2.1(a) also illustrates changes in water requirement that may be necessary to produce a 1 in. (25 mm) increase in slump at various temperature levels. For 70 F (21 C) concrete, about 2-1/2% more water is required to increase slump 1 in. (25 mm); for 120 F (50 C) concrete, 4-1/2% more water is needed for the 1 in. slump increase. The original mixing water required to change slump may be less if a water-reducing, midrange water-reducing, or high-range water-reducing admixture is used.

2.4.3 Drying shrinkage generally increases with total water content (Portland Cement Association Design and Control of Control Mixtures 1992). Rapid slump loss in hot weather often increases the demand for water, increasing total water content, and therefore, increasing the potential for subsequent drying shrinkage. Concrete cast in hot weather is also susceptible to thermal-shrinkage as it subsequently cools. The combined thermal and drying shrinkage can lead to more cracking than observed for the same concrete placed under milder conditions.

2.4.4 Because water has a specific heat of about four to five times that of cement or aggregates, the temperature of the mixing water has the greatest effect per unit weight on the temperature of concrete. The temperature of water is easier to control than that of the other components. Even though water is used in smaller quantities than the other ingredients, cooled water will reduce the concrete placing temperature, but usually by not more than approximately 8 F (4.5 C) (Fig. 2.4.4). The quantity of cooled water should not exceed the batch water requirement, which will depend on the mixture proportions and the moisture content of aggregates. In general, lowering the temperature of the batch water by 3.5 to 4 F (2.0 to 2.2 C) will reduce the concrete temperature approximately 1 F (0.5 C). Efforts should therefore be made to obtain cold water. To keep it cold, tanks, pipes, or trucks used for storing or transporting water should be either insulated, painted white, or both. Water can be cooled to as low as 33 F (1 C) using water chillers, ice, heat pump technology, or liquid nitrogen. These methods and their effectiveness are discussed further.

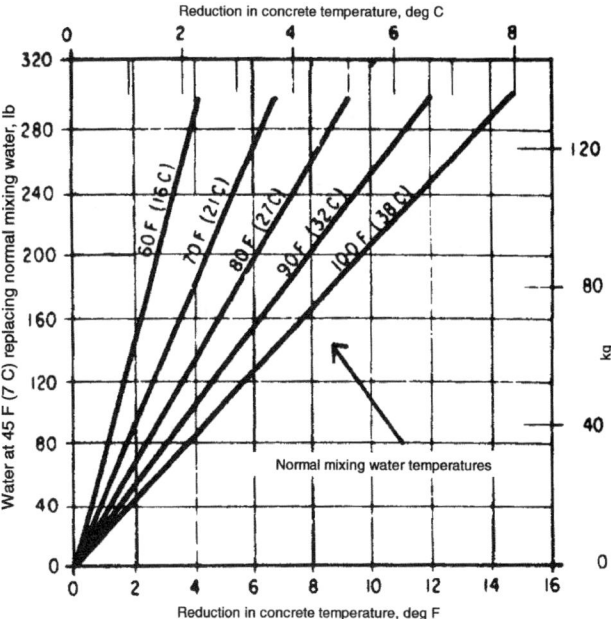

Fig 2.4.4—General effects of cooled mixing water on concrete temperature (National Ready Mixed Concrete Association 1962).

2.4.5 Using ice as part of the mixing water has remained a major means of reducing concrete temperature. On melting, ice absorbs heat at the rate of 144 Btu/lb (335 J/g). To be most effective, the ice should be crushed, shaved, or chipped when placed directly into the mixer as part of the mixing water. For maximum effectiveness, the ice should not be allowed to melt before it is placed in the mixer in contact with other ingredients, however, but it must melt completely prior to the completion of mixing of the concrete. For a more rapid blending of materials at the beginning of mixing, not all of the available batch water should be added in the form of ice. Its quantity may have to be limited to approximately 75% of the batch water requirement. To maximize amounts of ice or cold mixing water, aggregates should be well-drained of free moisture, permitting a greater quantity of ice or cold mixing water to be used. Fig. 2.4.5 illustrates potential reductions in concrete temperature by substituting varying amounts of ice at 32 F (0 C) for mixing water at the temperatures shown. Mixing should be continued until the ice is melted completely. Crushed ice should be stored at a temperature that will prevent lumps from forming by refreezing of particles.

2.4.6 The temperature reduction can also be estimated by using Eq. (A-4) or (A-5) in Appendix A. For most concrete, the maximum temperature reduction with ice is approximately 20 F (11 C). When greater temperature reductions are required, cooling by injection of liquid nitrogen into the mixer holding mixed concrete may be the most expedient means. See Appendix B for additional information. Liquid injected nitrogen does not affect the mixing water requirement except by reducing concrete temperature.

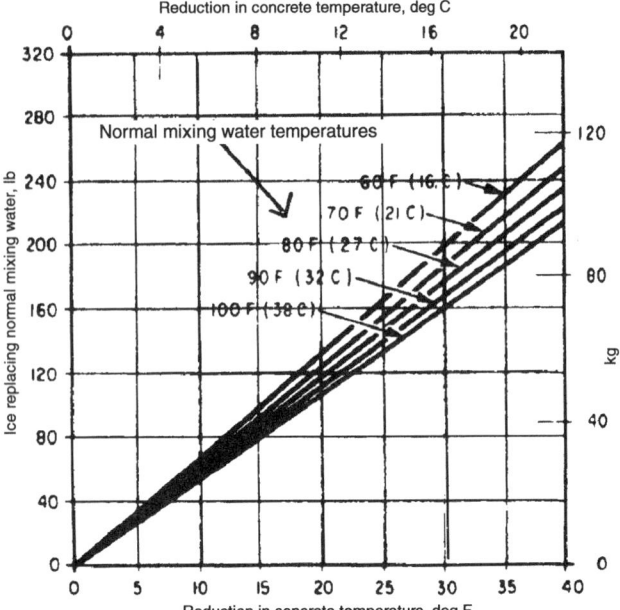

Fig. 2.4.5—General effects of ice in mixing water on concrete temperature. Temperatures are normal mixing water temperatures (National Ready Mixed Concrete Association 1962).

2.5—Effect of cement

2.5.1 High concrete temperature increases the rate of hydration (Fig. 2.5.2). As a result, concrete stiffens more rapidly and requires more water to produce or maintain the desired slump. The higher water content will cause strength loss and increase the cracking tendency of the concrete unless offset by measures described in Sections 2.6.1 and 2.7.

2.5.2 Selection of a particular cement may have a decided effect on the hot weather performance of concrete, as illustrated in Fig. 2.5.2. Although the curves are based on limited data from mixtures using different cements in combination with a set-retarding admixture, they show, for example, that when tested at 100 F (38 C), the concrete with the slowest setting cement reaches time of final setting 2-1/2 h later than the concrete with the fastest setting cement. The concrete that sets slowest at 100 F (38 C) was the fastest-setting cement when tested at 50 F (10 C). Fig. 2.5.2 is a good example of the difficulty of predicting performance of concrete at different temperatures. In general, use of a normally slower-setting Type II portland cement (ASTM C 150) or Type IP or IS blended cement (ASTM C 595) may improve the handling characteristics of concrete in hot weather (ACI 225R). Concrete containing the slower setting cements will be more likely to exhibit plastic-shrinkage cracking.

2.5.3 When using slower hydrating cements, the slower rate of heat development and the simultaneous dissipation of heat from the concrete result in lower peak temperatures. There will be less thermal expansion, and the risk of thermal cracking upon cooling of the concrete will be reduced. This is an important consideration for slabs, walls, and mass concretes, as discussed in ACI 207.1R and ACI 207.2R. The temperature increase from hydration of cement in a given concrete mixture is proportional to its cement content. Therefore, the cement content should be limited to that required to provide strength and durability. Concrete mixtures that obtain high strength at an early age will develop high concrete temperature during initial curing. These concrete mixtures should be provided thermal protection to ensure gradual cooling at a rate that will not cause them to crack; see Section 4.4.1.

2.5.4 Cement may be delivered at relatively high temperatures. This is not unusual for newly manufactured cement that has not had an opportunity to cool after grinding of the component materials. Concrete mixtures will consist of approximately 10 to 15% cement. This will increase concrete temperature approximately 1 F (0.5 C) for each 8 F (4 C) increase in cement temperature.

2.6—Supplementary cementitious materials

2.6.1 Materials in this category include fly ash and other pozzolans (ASTM C 618) and ground granulated blast-furnace slag (ASTM C 989). Each are widely used as partial replacements for portland cement; they may impart a slower rate of setting and of early strength gain to the concrete, which is desirable in hot weather concreting, as explained in Section 2.5.2. Faster setting cements or cements causing a rapid slump loss in hot weather may perform satisfactorily in combination with these materials (Gaynor et al 1985). The use of fly ash may reduce the rate of slump loss of concrete under hot weather conditions (Ravina 1984; Gaynor et al 1985).

2.7—Chemical admixtures

2.7.1 Various types of chemical admixtures (ASTM C 494) have been found beneficial in offsetting some of the undesirable characteristics of concrete placed during periods of high ambient temperatures (see also ACI 212.3R). The benefits may include lower mixing water demand, extended periods of use, and strengths comparable with, or higher than, concrete without admixtures placed at lower temperatures. Their effectiveness depends on the chemical reactions of the cement with which they are used in the concrete. Admixtures without a history of satisfactory performance at the expected hot weather conditions should be evaluated before their use, as explained in Section 2.7.5. Chemical admixtures affect the properties of concrete as described in the following.

2.7.2 Retarding admixtures meeting ASTM C 494, Type D requirements have both water-reducing and set-retarding properties, and are used widely under hot weather conditions. They can be included in concrete in varying proportions and in combination with other admixtures so that, as temperature increases, higher dosages of the admixture may be used to obtain a uniform time of setting. Their water-reducing properties largely offset the higher water demand resulting from increases in concrete temperature. Because water-reducing retarders generally increase concrete strength, they can be used, with proper mixture adjustments, to avoid strength losses that would otherwise result from high concrete temperatures (Gaynor et al 1985; Mittelacher 1985 and 1992). Compared with concrete without admixture, a concrete mixture that uses a water-reducing and retarding admixture may have a higher rate of slump loss. The net water reduction and other benefits remain substan-

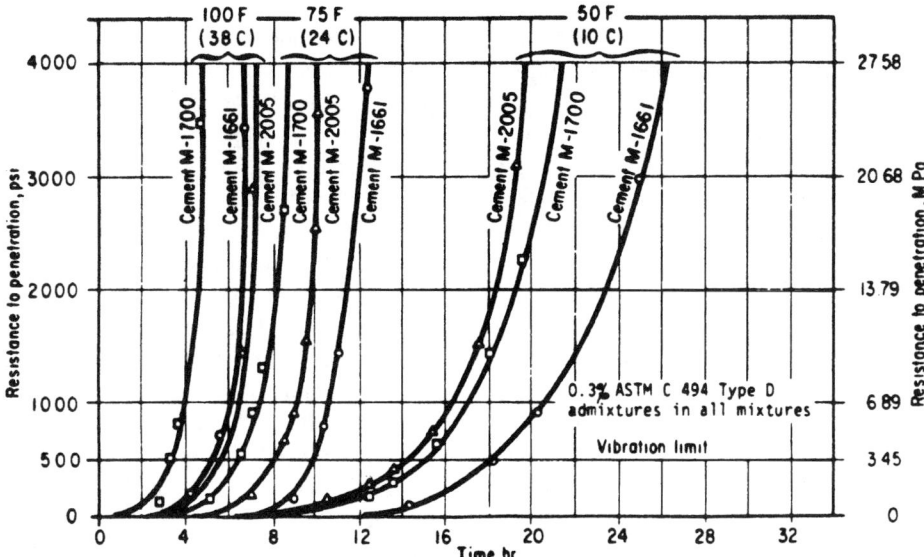

Fig 2.5.2—Effect of temperature and brand of cement on setting time characteristics of concrete mortars (Tuthill and Cordon 1955).

tial even after the initial slump is increased to compensate for slump loss.

2.7.3 Admixtures of the hydroxylated carboxylic acid type (ACI 212.3R, Class 3) and some types meeting ASTM C 494, Type D requirements may increase the early bleeding and rate of bleeding of concrete. This admixture-induced early bleeding may be helpful in preventing drying of the surface of concrete placed at high ambient temperature and low humidity. Concrete that is prone to bleeding generally should be reconsolidated after most of the bleeding has taken place. Otherwise, differential settling may occur that can lead to cracks over reinforcing steel and other inserts in near-surface locations. This cracking is more likely in cool weather with slower setting concretes than hot weather. If the admixture reduces the tensile strength and tensile strain capacity, however, plastic-shrinkage tendencies may be increased (Ravina and Shalon 1968). Other admixtures (ACI 212.3R, Classes 1 and 2) may reduce bleeding rate. If drying conditions are such that crusting of the surface blocks bleed water from reaching the surface, continued bleeding may cause scaling. Under such conditions, fog sprays, evaporation retardants (materials that retard the evaporation of bleeding water of concrete), or both, should be used to prevent crusting.

2.7.4 Some high-range, water-reducing and retarding admixtures (ASTM C 494, Type G), and plasticizing and retarding admixtures (ASTM C 1017, Type II), often referred to as superplasticizers, can provide significant benefits under hot weather conditions when used to produce flowing concrete. At higher slumps, heat gain from internal friction during mixing of the concrete will be less (see ASTM STP 169C and ACI 207.4R). The improved handling characteristics of flowing concrete permit more rapid placement and consolidation, and the period between mixing and initial finishing can therefore be reduced. The rate of slump loss of flowing concrete may also be less at higher temperatures than in concrete using conventional retarders (Yamamoto and Kobayashi 1986). Concrete strengths are generally found to be substantially higher than those of comparable concrete without admixture and with the same cement content. Certain products may cause significant bleeding, which may be beneficial in many instances, but may require some precautions in others (see Section 2.7.3). Air-content tests will be needed before placement to assure maintenance of proper air content. Assurance also may be needed that the air-void system is not impaired if it is required for the freezing and thawing resistance of the concrete. This can be determined by requiring hardened air analysis or ASTM C 666 freezing and thawing testing. Some high-range water-reducing retarders can maintain the necessary slump for extended periods at elevated concrete temperatures (Collepardi et al 1979; Hampton 1981; Guennewig 1988). These will be of particular benefit in the event of delayed placements or deliveries over greater distances. Other high-range water-reducing admixtures may greatly accelerate slump loss, particularly when initial slumps are less than 3 to 4 in. (75 to 100 mm). Some water-reducing admixtures can cause the concrete to extend its working time by a couple of hours, followed by acceleration of strength gain.

2.7.5 Since the early 1990s, the use of midrange water-reducing admixtures in hot weather has increased. Midrange water-reducing admixtures provide up to 15% water reduction, which is higher than conventional water-reducing admixtures, but lower water reduction than high-range water-reducing admixtures. Although at present there is no ASTM classification, midrange water-reducing admixtures comply with the requirements of ASTM C 494, Type A admixtures, and in some cases, Type F admixtures. These admixtures will not delay the setting time of the concrete significantly. At higher dosages, conventional water-reducing admixtures can achieve this water reduction, but with significant increase in the setting time of the concrete. The pumping and finishing characteristics of concrete containing midrange water-reducing admixtures are improved when compared with concrete containing conventional Type A water reducers. The use of midrange water reducers is particularly beneficial in cases where aggregate properties contribute to poor workability or finishing difficulties. The surface appearance of concrete containing a midrange water reducer could be changed, thereby requiring a change of the timing of finishing operations. Also available are midrange water-reducing and retarding admixtures that comply with ASTM C 494 requirements for Type D admixtures.

2.7.6 The use of extended set-control admixtures to stop the hydration process of freshly mixed concrete (freshly batched or returned plastic concrete that normally would be disposed), and concrete residue (washwater) in ready-mix truck drums has gained increased acceptance in hot weather environments since their introduction in 1986. Some extended set-control admixtures comply with ASTM C 494 requirements for Type B, retarding admixtures, and Type D, water-reducing and retarding mixtures. Extended set-control admixtures differ from conventional retarding admixtures in that they stop the hydration process of both the silicate and aluminate phases in portland cement. Regular retarding admixtures only act on the silicate phases, which extend (not stop) the hydration process. The technology of extended set-control admixtures may also be used to stop the hydration process of freshly batched concrete for hauls requiring extended time periods or slow placement methods during transit. For this application, the extended set-control admixture is added during or immediately after the batching process. Proper dosage rates of extended set-control admixtures should be determined by trial mixtures incorporating project time requirements in this way ensuring that the concrete will achieve the required setting time. Additional admixtures are not required to restart hydration.

2.7.7 The qualifying requirements of ASTM C 494 afford a valuable screening procedure for the selection of admixture products. Admixtures without a performance history pertaining to the concrete material selected for the work should be first evaluated in laboratory trial batches at the expected high job temperature, using one of the procedures described in Section 2.9. Some high-range, water-reducing retarders may not demonstrate their potential benefits when used in small laboratory batches. Further testing may then be required in production-size concrete batches. During preliminary field use, concrete containing admixture should be evaluated for consistency of performance in regard to the desired characteristics in hot weather construction. When evaluating admixtures, properties such as workability, pumpability, early strength development, placing and finishing characteristics, appearance, and effect on reuse of molds and forms should be considered in addition to the basic properties of slump retention, setting time, and strength. These characteristics may influence selection of an admixture and its dosage more than properties usually covered by most specifications.

2.8—Aggregates

2.8.1 Aggregates are the major constituent of concrete, as they account for 60 to 80% of the volume of normalweight concrete used in most structures. Therefore, the properties of the aggregate affect the quality of concrete significantly. The size, shape, and grading of the aggregate are three of the principal factors that affect the amount of water required to produce concrete at a given slump. Aggregate properties desirable in hot weather concreting include the following:

- Gradation, particle shape, and the absence of undersized material are very important in minimizing water demand (ACI 221R). Crushed coarse aggregate also contributes to higher water demand, but is reported to provide better resistance to cracking than rounded gravels (ACI 224R). The blending of three or more aggregate sizes may reduce the mixing water requirements and improve workability at a given slump (Shilstone, Sr. and Shilstone, Jr. 1993).

2.8.2 With coarse aggregate being the ingredient of greatest mass in concrete, changes in its temperature have a considerable effect on concrete temperatures. For example, a moderate 1.5 to 2 F (0.8 to 1.1 C) temperature reduction will lower the concrete temperature 1 F (0.5 C). Cooling the coarse aggregate may be an effective supplementary means to achieve desired lower concrete temperature (see Appendix B).

2.9—Proportioning

2.9.1 Mixture proportions may be established or adjusted on the basis of field-performance records in accordance with ACI 318/318R (ACI 318/318RM), provided the records indicate the effect of expected seasonal temperatures and delivery times.

2.9.2 Selection of ingredients and their proportions should be guided by their contribution to satisfactory performance of the concrete under hot weather conditions (ACI 211.1 and 211.2). Cement content should be kept as low as possible but sufficient to meet strength and durability requirements. Inclusion of supplementary cementitious materials, such as fly ash or ground granulated blast-furnace slag, should be considered to delay setting and to mitigate the temperature rise from heat of hydration. The use of various types of water-reducing admixtures can offset increased water demand and strength loss that could otherwise be caused by higher concrete temperatures. High-range, water-reducing retarders formulated for extended slump retention should be considered if longer delivery periods are anticipated. Unless required otherwise, concrete should be proportioned for a slump of not less than 3 in. (75 mm) to permit prompt placement and effective consolidation in the form.

2.9.3 The performance of the concrete mixtures proposed for the work should be verified under conditions approximating the delivery time and hot weather environment expected at the project. Trial batches used to select proportions are normally prepared in accordance with ASTM C 192. The method requires concrete materials to be at room temperature [in the range of 68 to 86 F (20 to 30 C)]. Trial batches, however, should also be performed at the expected maximum placing temperature with consideration of using a mixing and agitating period longer than that required in ASTM C 192 to help define the performance to be expected.

2.9.4 In determining mixture proportions using laboratory trial batches, a procedure for estimating the slump loss during the period between first mixing of the concrete and its placement in the form is suggested in Procedures A and B, below, adopted from ACI 223, Section 4.5.2 on shrinkage-compensating concrete. These procedures from ACI 223 were found to produce a rate of slump loss similar to that expected for a 30 to 40 min delivery time.

Procedure A—

1. Prepare the batch using ASTM C 192 procedures, but add 10% additional water over that normally required;

2. Mix initially in accordance with ASTM C 192 (3 min mixing followed by a 3 min rest and 2 min remixing);

3. Determine the slump and record as initial slump;

4. Continue mixing for 15 min;

5. Determine the slump and record as estimated placement slump. Experience has shown this slump correlates with that expected for 30 to 40 min delivery time. If this slump does not meet the specification limits, either discard and repeat the procedure with an appropriate water adjustment or add water to give the required slump and then test the concrete; and

6. Determine other properties of fresh concrete (temperature, air content, unit weight), and mold strength test specimens.

Procedure B—

1. Prepare the batch using ASTM C 192 procedures for the specified slump;

2. Mix in accordance with ASTM C 192 (3 min mixing, 3 min rest, and 2 min remixing) and confirm the slump;

3. Stop the mixer and cover the batch with wet burlap;

4. After 20 min, remix 2 min, adding water to produce the specified slump. The total water (initial water plus the remixing water) can be expected to equal that required at the batch plant to give the required job site slump; and

5. Determine other properties of fresh concrete (temperature, air content, unit weight), and mold-strength test specimens.

2.9.5 As an alternative method, use of full-size production batches may be considered for verification of mixture proportions, provided the expected high temperature levels of the concrete can be attained. This may be the preferred method when using admixtures selected for extended slump retention. It requires careful recording of batch quantities at the plant and of water added for slump adjustment before sampling. Sampling procedures of ASTM C 172 should be strictly observed.

CHAPTER 3—PRODUCTION AND DELIVERY
3.1—General

Production facilities and procedures should be capable of providing the required quality of concrete under hot weather conditions at production rates required by the project. Satisfactory control of production and delivery operations should be assured. Concrete plant and delivery units should be in good operating condition. Intermittent stoppage of deliveries due to equipment breakdown can be much more serious under hot weather conditions than in moderate weather. In hot weather concreting operations, concrete placements may be scheduled at times other than during daylight hours, such as during the coolest part of the morning. Night-time production requires good planning and good lighting.

3.2—Temperature control of concrete

3.2.1 Concrete can be produced in hot weather without maximum limits on placing temperature and will perform satisfactorily if proper precautions are observed in proportioning, production, delivery, placing, and curing. As part of these precautions, an effort should be made to keep the concrete temperature as low as practical. Using the relationships given in Appendix A, it can be shown, for example, that the temperature of concrete of usual proportions can be reduced by 1 F (0.5 C) if any of the following reductions are made in material temperatures:

- 8 F (4 C) reduction in cement temperature;
- 4 F (2 C) reduction in water temperature; or
- 2 F (1 C) reduction in the temperature of the aggregates.

3.2.2 Fig. 3.2.2 shows the influence of the temperature of concrete ingredients on concrete temperature. As the greatest portion of concrete is aggregate, reduction of aggregate temperature brings about the greatest reduction in concrete temperature. Therefore, all practical means should be employed to keep the aggregates as cool as possible. Shaded storage of fine and coarse aggregates, and sprinkling and fog spraying of coarse aggregates stock-piles under arid conditions will help. Sprinkling of coarse aggregates with cool water can reduce aggregate temperature by evaporation and direct cooling (Lee 1987). Passing water through a properly sized evaporative cooling tower will chill the water to the wet bulb temperature. This procedure will have greater effects in areas that have low relative humidity. Wetting of aggregates, however, tends to cause variations in surface moisture and thereby complicates slump control. Above-ground storage tanks for mixing water should be provided with shade and thermal insulation. Silos and bins will absorb less heat if coated with heat-reflective paints. Painting mixer surfaces white to minimize solar heat gain will be of some help. Based on 1 h delivery time on a hot, sunny day, concrete in a clean white mixer drum should be 2 to 3 F (1 to 1.5 C) cooler than in a black or red mixer drum, and 0.5 F (0.3 C) cooler than in a cream-colored drum. If an empty mixer drum stands in the sun for an extended period before concrete is batched, the heat stored in the metal drum would produce concrete temperatures 0.5 to 1 F (0.3 to 0.5 C) lower for a white mixer drum than a yellow or red mixer drum. Spraying the exterior of the mixer drum with water before batching or during delivery has been suggested as a means of minimizing concrete temperature, but it can be expected to be of only marginal benefit.

3.2.3 Setting up the means for cooling sizeable amounts of concrete production requires planning well in advance of placement and installation of specialized equipment. This can include chilling of batch water by water chillers or heat pump technology as well as other methods, such as substituting crushed or flaked ice for part of the mixing water, or cooling by liquid nitrogen. Delivery of the required quantity of cooling materials should be assured for each placement. Details for estimating concrete temperatures are provided in Appendix A. Various cooling methods are described in Appendix B. The general influence of the temperature of

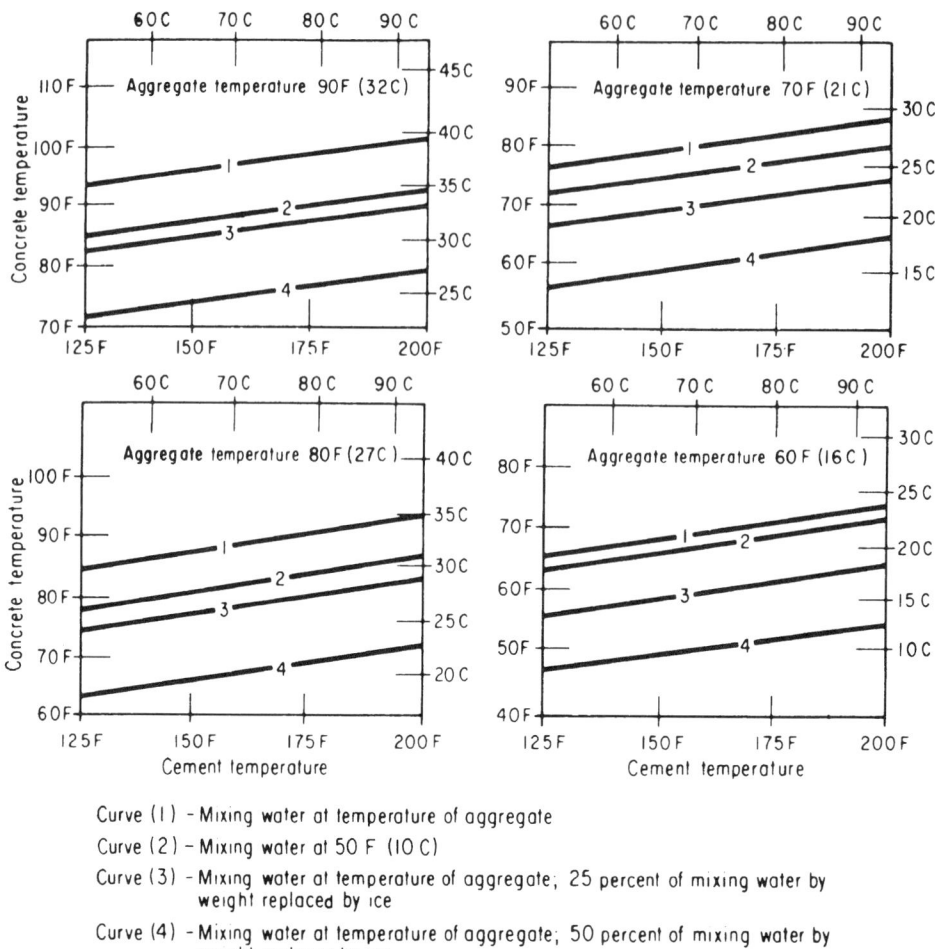

Fig. 3.2.2—*Influence of temperature of concrete ingredients on concrete temperature. Calculated from equations in Appendix A.*

concrete ingredients on concrete temperature is calculated from the equations in Appendix A, and shown in Fig. 3.2.2.

3.3—Batching and mixing

3.3.1 Batching and mixing is described in ACI 304R. Procedures under hot weather conditions are no different from good practices under normal weather conditions. Producing concrete of the correct slump and other specified properties to confirm with applicable specifications is essential. An interruption in the concrete placement due to rejection may cause the formation of a cold joint or serious problems in finishing. Testing of concrete must be diligent and accurate so that results represent the true condition of the concrete.

3.3.2 For truck-mixed concrete, initial mixing of approximately 70 revolutions at the batch plant prior to transporting will allow an accurate verification of the condition of the concrete, primarily its slump and air content. Generally, centrally mixed concrete can be inspected visually as it is being discharged into the transportation unit. Slump can easily change due to minor changes in materials and concrete characteristics. For example, an undetected change of only 1.0% moisture content of the fine and coarse aggregates could change slump by 1 to 2 in. (25 to 50 mm) (ACI 211.1). An error range of approximately 0.5% in the determination of aggregate moisture complicates moisture control, even with advanced systems. Operators often batch concrete in a drier condition than desired to avoid producing a slump higher than specified; a small water addition may be needed at the job site.

3.3.3 Hot weather conditions and extended hauling time may indicate a need to split the batching process by batching the cement at the job site, or layering the materials in the mixer drum at the plant to keep some of the cement dry and then mixing the concrete after arrival at the job site. This may not, however, contribute to concrete uniformity between loads. These methods may, on occasion, offer the best solution under existing conditions. A better controlled concrete can usually be provided when all materials are batched at the concrete production facility. By using some effective retarding admixtures at appropriate dosages, preferably in combination with cementitious material of slow-setting characteristics, concrete can be maintained in a placeable condition for extended periods even in hot weather (see Section 2.7). Field experience indicates that concrete set retardation can be extended further by separately batching the retarding admixture with a small portion of mixing water, 1 to 2 gal/yd^3 (5 to 10 L/m^3), after the concrete has been mixed for several minutes. These admixtures, together with the cementitious materials and other ingredients proposed for the project, should be evaluated in the field for desired properties. Should the slump be lower than required, the use of midrange water-reducing or high-range water-reducing admixtures is recommended to increase the concrete slump.

3.3.4 Under hot weather conditions, the amount of mixing at mixing speed of the mixer should be held to a minimum to avoid any unnecessary heat gain of the concrete (ACI 207.4R). For efficient mixing, mixers should be free of buildup of hardened concrete and excessive wear of mixer blades. As soon as the concrete has been mixed to a homogeneous condition, all further drum rotation should be at the lowest agitating speed of the unit (generally one revolution per min). The drum should not be stopped for extended periods of time. There is potential for false setting problems causing the concrete to stiffen rapidly or set in the drum, or for flattening of the mixer rollers.

3.3.5 Specifications governing the total number of revolutions of the drum usually set a limit of 300 revolutions for truck mixers. This limit should be waived for conditions that require further thorough mixing of the concrete:
- Separate addition of high-range, water-reducing admixtures;
- Direct addition of liquid injected nitrogen into the mixer as a means of lowering the concrete temperature; and
- If the concrete retains its workability without the addition of water.

3.4—Delivery
Cement hydration, temperature rise, slump loss, aggregate grinding, and either loss or, occasionally, gain of air content all occur with the passage of time while the concrete is in the mixer; thus, the period between start of mixing to start of placement of the concrete should be minimized. Coordinating the dispatching of mixer trucks with the rate of concrete placement avoids delays in arrival or waiting periods until discharge. On major concrete placements, provisions should be made to have good communications between the job site and concrete-production facility. Major placements should be scheduled during periods of lower urban traffic loads. When placement is slow, consideration should be given to reducing load size, using set-retarding admixture, or using cooled concrete.

3.5—Slump adjustment
Fresh concrete is subject to slump loss with time, whether it is used in moderate or hot weather. With given materials and mixture proportions, the slump change characteristics between plant and job site should be established. With the limitations on accurately predicting slump, as explained in Section 3.3.2, uncertainty in traffic, and the timing of placing operations, operators need to batch concrete in a drier condition to avoid a slump higher than specified. If, on arrival at the job site, the slump is less than the specified maximum, additional water may be added if the maximum allowable water content is not exceeded. When water is added to bring the slump within required limits, the drum or blades must be turned an additional 30 revolutions or more, if necessary, at mixing speed. For expeditious placement and effective consolidation, structural concrete should have a minimum slump of 3 or 4 in. (75 or 100 mm). Slump increases should be allowed when chemical admixtures are used, providing the admixture-treated concrete has the same or lower water-cementitious materials ratio *(w/cm)* and does not exhibit segregation potential.

3.6—Properties of concrete mixtures
The proposed mixtures should be suitable for expected job conditions. This is particularly important when there are no limits on placing temperatures, as is the case in most general construction in the warmer regions. Use of cements or cementitious materials that perform well under hot weather conditions, in combination with water-reducing and retarding admixtures, can provide concrete of required properties (Mittelacher 1985). When using high-range, water-reducing and retarding admixtures, products should be selected that provided extended slump retention in hot weather (Collepardi et al 1979; Guennewig 1988). In dry and windy conditions, the setting rate of concrete used in flatwork should be adjusted to minimize plastic-shrinkage cracking or crusting of the surface, with the lower layer still in a plastic condition. The type of adjustment depends on local climatic conditions, timing of placements, and concrete temperatures. A change in admixture dosage or formulation can often provide the desired setting time.

3.7—Retempering
Retempering is defined as "additions of water and remixing of concrete, or mortar which has lost enough workability to become unplaceable or unsaleable" (ACI 116R). Laboratory research, as well as field experience, shows that strength reduction and other detrimental effects are proportional to the amount of retempering water added. Therefore, water additions in excess of the proportioned maximum water content or *w/cm* to compensate for loss of workability should be prohibited. Adding chemical admixtures, particularly high-range water-reducing admixtures, may be very effective to maintain workability.

CHAPTER 4—PLACING AND CURING
4.1—General
4.1.1 The requirements for good results in hot weather concrete placing and curing are no different than in other seasons. The same necessities exist:
- Concrete be handled and transported with a minimum of segregation and slump loss;
- Concrete be placed where it is to remain;
- Concrete be placed in layers shallow enough to assure vibration well into the layer below and that the elapsed time between layers be minimized to avoid cold joints;
- Construction joints outlined in ACI 224.3R be made on sound, clean concrete;
- Finishing operations and their timing be guided only by the readiness of the concrete for them, and nothing else; and
- Curing be conducted so that at no time during the pre-

scribed period will the concrete lack ample moisture and temperature control to permit full development of its potential strength and durability.

4.1.2 Details of placing, consolidation, and curing procedures are described in ACI 304R, 308R, and 309R. It is the purpose of this chapter to point out the factors peculiar to hot weather that can affect these operations and the resulting concrete and to recommend what should be done to prevent or offset their influence.

4.2—Preparations for placing and curing

4.2.1 *Planning hot weather placements*—Prior to the start of the project, plans should be made for minimizing the exposure of the concrete to adverse conditions. Whenever possible, placing of slabs should be scheduled after roof structure and walls are in place to minimize problems associated with drying winds and direct sunlight. This will also reduce thermal shock from rapid temperature drops caused by wide day and night temperature differences or cool rain on concrete heated by the sun earlier in the day.

Under hot weather conditions, scheduling concrete placements at other-than-normal hours may be advisable. Pertinent considerations include ease of handling and placing, and avoiding the risk of plastic-shrinkage and thermal cracking.

4.2.2 *Preparing for ambient conditions*—Personnel in charge of concrete construction should be aware in advance of the damaging combinations of high air temperature, direct sunlight, drying winds, and high concrete temperature. Monitoring of local weather reports and routine recording of conditions at the site, including air temperature, sun exposure, relative humidity, and prevailing winds, can be conducted locally. These data, together with projected or actual concrete temperatures, enable supervisory personnel through reference to Fig. 2.1.5 to determine and prepare the required protective measures. Equipment should also be available at the site for measuring the evaporation rate in accordance with Section 5.1.3.

4.2.3 *Expediting placements*—Preparations must be made to transport, place, consolidate, and finish the concrete at the fastest possible rate. Delivery of concrete to the job should be scheduled so it will be placed promptly on arrival, particularly the first batch. Many concrete placements get off to a bad start because the concrete was ordered before the job was ready and slump control was lost at this most critical time. Traffic arrangements at the site should ensure easy access of delivery units to the unloading points over stable roadways. Site traffic should be coordinated for a quick turn-around of concrete mixer trucks. If possible, large or critical placements should be scheduled during periods of low urban traffic loads.

4.2.4 *Placing equipment*—Equipment for placing the concrete shall be of suitable design and have ample capacity to perform its functions efficiently. All equipment should have adequate power for the work and be in first-class operating condition. Breakdowns or delays that stop or slow the placement can seriously affect the quality and appearance of the work. Arrangements should be made for readily available backup equipment. Concrete pumps, if used, must be capable of pumping the specified class of concrete through the length of line and elevation at required rates per h. If placement is by crane and buckets, wide-mouth buckets with steep-angled walls should be used to permit rapid and complete discharge of bucket contents. Adequate means of communication between bucket handlers and placing crew should be provided to assure that concrete is charged into buckets only if the placing crew is ready to use the concrete without delay. Concrete should not be allowed to rest exposed to the sun and high temperature before placing it into the form. To minimize the heat gain of the concrete during placement, delivery units, conveyors, pumps, and pump lines should be kept in the shade if possible. In addition, pump lines should be painted white. Lines can also be cooled by covering with damp burlap, kept wet with a soaker hose, or similar means.

4.2.5 *Consolidation equipment*—There should be ample vibration equipment and manpower to consolidate the concrete immediately as it is received in the form. Procedures and equipment are described in ACI 309R. Provision should be made for an ample number of standby vibrators—at least one standby for each three vibrators in use. On sites subject to occasional power outages, portable generators should be available for uninterrupted vibrator operation. Apart from the unsightliness of poorly consolidated concrete, insufficient compaction in the form may seriously impair the durability and structural performance of reinforced concrete.

4.2.6 *Preparations for protecting and curing the concrete*—Ample water should be available at the project site for moistening the subgrade, as well as for fogging forms and reinforcement prior to concrete placement, and for moist curing if applicable. Fog nozzles should produce a fog blanket. They should not be confused with common garden-hose nozzles, which generate an excessive washing spray. Pressure washers with a suitable nozzle attachment may be a practical means for fogging on smaller jobs. Materials and means should be on hand for erecting temporary windbreaks and shades as needed to protect against drying winds and direct sunlight. Plastic sheeting or sprayable compounds for applying temporary moisture-retaining films should be available to reduce evaporation from flatwork between finishing passes. If concrete placed under hot weather conditions is exposed to rapid temperature drops, thermal protection should be provided to protect the concrete against thermal-shrinkage cracking. Finally, curing materials should be readily available at the project site to permit prompt protection of all exposed surfaces from premature drying upon completion of the placement.

4.2.7 *Preparing incidental work*—Due to faster setting and hardening of the concrete in hot weather, the timing of various final operations as saw-cutting joints and applying surface retarders becomes more critical; therefore, these operations must be planned in advance. Plans should be made for the timely sawing of contraction joints in flatwork to minimize cracking due to excessive tensile stresses. Typically, joints that are cut using the conventional wet or dry process are made within 4 to 12 h after the slab has been finished; 4 h in hot weather to 12 h in cold weather. For early entry dry-cut saws, the waiting period will typically vary from 1 h in hot weather to 4 h in cold weather (ACI 302.1R).

4.3—Placement and finishing

4.3.1 *General*—Speed-up of placement and finishing materially reduces hot weather difficulties. Delays increase slump loss and invite the addition of water to offset it. Each operation in finishing should be carried out promptly when the concrete is ready for it. The concrete should not be placed faster than it can be properly consolidated and finished. If the placing rate is not coordinated with the available work force and equipment, the quality of the work will be marred by cold joints, poor consolidation, and uneven surface finishes.

4.3.2 *Placing formed concrete*—In hot weather, it is usually necessary to place concrete in shallower layers than those used in moderate weather to assure coverage of the lower layer while it will still respond readily to vibration. The interval between monolithic wall and deck placements becomes very short in hot weather. This interval may be extended by the judicious use of set-retarding admixtures.

4.3.3 *Placement of flatwork*—During the depositing of concrete for flatwork on the ground, the subgrade should be moist, yet free of standing water and soft spots. In placing concreting slabs of any kind, it may be necessary in hot weather to keep the operation confined to a small area and to proceed on a front having a minimum amount of exposed surface to which concrete is to be added. A fog nozzle should be used to cool the air, to cool any forms and steel immediately ahead, and to lessen rapid evaporation from the concrete surface before and after each finishing operation. Excessive fog application (which would wash the fresh concrete surface or cause surplus water to cling to reinforcement or stand on the concrete surface during floating and troweling) must be avoided. Other means of reducing moisture loss include spreading and removing impervious sheeting or application of sprayable moisture-retaining (monomolecular) films one or more times as needed, between the various finishing operations. Finishing of flatwork should commence after the surface sheen of the (monomolecular) film has disappeared. These products should not be used as finishing aids or worked into the surface, as concrete durability may be reduced. Contact the product manufacturer for information on proper application and dosage. These procedures may cause a slight increase of the concrete temperature in place due to reduced evaporative cooling. Generally, the benefit from reduced moisture evaporation is more important than the increase of in-place concrete temperature (Berhane 1984).

4.3.4 *Plastic-shrinkage cracks*—Without protection against moisture loss, plastic-shrinkage cracks may occur, as described in Section 2.1.5. In relatively massive placements, revibration before floating can sometimes close this type of cracking. Before the concrete reaches final set, the cracks can frequently be closed by striking the surface on each side of the crack with a float. The affected area is then retroweled to level finish.

It serves no lasting purpose to merely trowel a slurry over the cracks, because these are likely to reappear if not firmly closed and immediately covered to avoid evaporation.

4.4—Curing and protection

4.4.1 *General*—After completing placing and finishing operations, efforts must continue to protect the concrete from high temperature, direct sunlight, low humidity, and drying winds. If possible, the work should be kept in a uniformly moderate temperature condition to allow the concrete to develop its full strength potential. High initial curing temperatures are detrimental to the ultimate strength to a greater degree than high placing temperatures (Bloem 1954; Barnes et al 1977; Gaynor et al 1985). Procedures for keeping exposed surfaces from drying must be promptly commenced, with ample coverage and continued without interruption. Failure to do so may result in excessive shrinkage and cracking, and will impair the surface durability and strength of the concrete. Curing should be continued for at least the first 7 days. If a change in curing method is made during this period, it should be done only after the concrete is 3 days old. The concrete surface should not be permitted to become dry during the transition. The various methods of curing are described in ACI 308R. The concrete should also be protected against thermal-shrinkage cracking from rapid temperature drops, particularly during the first 24 h. This type of cracking is usually associated with a cooling rate of more than 5 F (3 C) per h, or more than 50 F (28 C) in a 24 h period for concrete with a least dimension less than about 12 in. (300 mm). Concrete exposed to rapid cooling has a lower tensile strain capacity and is more susceptible to cracking than concrete that is allowed to cool at a slower rate (ACI 207.4R). Hot weather patterns likely to cause thermal cracking include wide day and night temperature differences and cold rain. Under these conditions, the concrete should be protected by placing several layers of waterproof paper over the concrete, or by using other insulating methods and materials described in ACI 306R.

4.4.2 *Moist-curing of flatwork*—Of the different curing procedures, moist-curing is the best method for developing the strength of concrete and minimizing early drying shrinkage. It can be provided by ponding, covering with clean sand kept continuously wet, or continuous sprinkling. This will require an ample water supply and disposal of the runoff. When sprinkling is used, care must be taken that erosion of the surface does not occur. A more practical method of moist-curing is that of covering the prewetted concrete with impervious sheeting or application of absorptive mats or fabric kept continuously wet with a soaker hose or similar means. Suitable coverings are described in ACI 308R. These materials should be kept in contact with the concrete surface at all times. Alternate cycles of wetting and drying must be avoided because this may result in pattern cracking. The temperature of water used for curing must be as close as possible to that of the concrete to avoid thermal shock.

4.4.3 *Membrane curing of flatwork*—Use of liquid membrane-forming compounds is the most practical method of curing where job conditions are not favorable for moist-curing. The membranes restrict the loss of moisture from the concrete, thereby allowing the development of strength, durability, and abrasion resistance of flatwork. On concrete surfaces exposed to the sun, heat reflecting white pigmented compounds should be used. The capability for moisture retention varies considerably between products. For use under hot weather conditions, a material should be selected that provides better moisture retention than required by ASTM C 309. It limits the moisture loss in a 72 h period to not more than 9 lb/yd^3 (0.55 kg/m^3) when tested in accordance with ASTM C 156. Some agencies have set a more restrictive limit of 0.39 kg/m^2 of moisture loss in a 72 h period. On flatwork, application should be started immediately after disappearance of the surface water sheen after the final finishing pass. When applied by spraying, the spray nozzles should be held or positioned sufficiently close to the surface to assure the correct application rate and prevent wind-blown dispersion. Manual application should be in two passes, with the second pass proceeding at right angles to the first application. Most curing compounds should not be used on any surface against which additional concrete or other materials are to be bonded, unless the curing material will not reduce bond strength or unless removal of the curing material is assured before subsequent bonded construction.

4.4.4 *Curing of concrete in forms*—Forms should be covered and kept continuously moist during the early curing period. Formed concrete requires early access to ample external curing water for strength development. This is particularly important when using high-strength concrete having a *w/cm* less than approximately 0.40 (ACI 363R). The forms should be loosened as soon as this can be done without damage to the concrete, and provisions made for the curing water to run down inside them. Cracking may occur when the concrete cools rapidly from a high peak temperature and is restrained from contracting. In more massive members, and if the internal temperature rise cannot be controlled by available means, the concrete should be given thermal protection so that it will cool gradually at a rate that will not cause the concrete to crack. After form removal, form tie holes can be filled and any necessary repairs made by uncovering a small portion of the concrete at a time to carry on this work. These repairs should be completed in the first few days after stripping so the repairs and tie-hole fillings can cure with the surrounding concrete. At the end of the curing period (7 days should be minimum; 10 days is better), the covering should be left in place without wetting for several days (4 days is suggested) so that the concrete surface will dry slowly and be less subject to surface shrinkage cracking. The effects of drying can also be minimized by applying a sprayable curing compound at the end of the moist-curing period.

CHAPTER 5—TESTING AND INSPECTION
5.1—Testing

5.1.1 Tests on the fresh concrete sample should be conducted and specimens prepared in accordance with applicable ASTM Standards. Tests should be performed by an ACI certified concrete technician. The sample should be as representative as possible of the potential strength and other properties of the concrete as delivered. High temperature, low relative humidity, and drying winds are particularly detrimental to the sample of fresh concrete used for making tests and molding specimens. Leaving the sample of fresh concrete exposed to sun, wind, or dry air will invalidate test results.

5.1.2 It is sometimes desirable in hot weather to conduct tests such as slump, air content, ambient and concrete temperature, relative humidity, and unit weight more frequently than for normal conditions.

5.1.3 The most important factor affecting plastic shrinkage is the evaporation rate, which can be estimated from Fig. 2.1.5 with the prevailing temperature, relative humidity, and wind speed. The evaporation rate can be determined more accurately by evaporating water from a cake pan having an area of approximately 1 ft^2 (0.093 m^2). The pan is filled with water and the mass determined every 15 to 20 min to determine the evaporation rate, which is equal to the loss of water mass from the pan. A balance of at least 5.5 lb (2500 g) capacity is satisfactory.

5.1.4 Particular attention should be given to the protection and curing of strength test specimens used as a basis for acceptance of concrete. Due to their small size in relation to most parts of the structure, test specimens are influenced more readily by changes in ambient temperatures. Extra effort is needed in hot weather to maintain strength test specimens at a temperature of 60 to 80 F (16 to 27 C) and to prevent moisture loss during the initial curing period, in accordance with ASTM C 31/C 31M. If possible, the specimens should be provided with an impervious cover and placed in a temperature-controlled job facility immediately after molding. If stored outside, exposure to the sun should be avoided and the cooling effect of evaporating water should be used to help provide the required curing condition. The following methods for nonpotentially absorptive test molds have been found practical:

- Embedding in damp sand. Care should be taken to maintain sand in continuously moist conditions (not to be used for cardboard molds);
- Covering with wet burlap. Care should be taken to maintain burlap in a continuously moist condition and out of contact with the concrete;
- Continuous fog sprays. Care should be taken to prevent interruptions of the fog spray; and
- Total immersion in water (not to be used for cardboard molds). Specimens may be immersed immediately in saturated limewater after molding. Because specimens are made with hydraulic cement, which hardens under water, specimen cylinders need not be covered with a cap, but generally they are, as a precautionary measure to prevent external damage.

5.1.5 Molds must not be of a type that is potentially absorptive and expands when in contact with moisture or when immersed in water. Molds should meet the requirements of ASTM C 470. Merely covering the top of the molded test cylinder with a lid or plate is usually not sufficient in hot weather to prevent loss of moisture and to maintain the required initial curing temperature. During the transfer to the testing facility, the specimens should be kept moist and also be protected and handled carefully. They should then be stored in a moist condition at 73 ± 2 F (23 ± 1.7 C) until the moment of test.

5.1.6 Specimens in addition to those required for acceptance may be made and cured at the job site to assist in determining when forms can be removed, when shoring can be removed, and when the placement can be placed in service. Unless specimens used for these purposes are cured at the same place and as nearly as possible under the same conditions as the placement, results of the tests can be misleading. Alternative test methods for determining in place concrete strength are described in ASTM C 900 and ASTM C 918.

5.2—Inspection

5.2.1 The numerous details to be looked after in concrete construction are covered in ACI 311.1R and 311.4R. The particular effects of hot weather on concrete performance and the precautions to be taken to minimize adverse effects have been discussed previously. Project inspection of concrete is necessary to ensure compliance with these additional precautions and procedures. Adequate inspection is also necessary to verify and document this compliance. The need for such measures as spraying of forms and subgrade, cooling concrete, providing sunshades, windscreens, or fogging and the like, and minimizing delays in placement and curing should be anticipated.

5.2.2 Air temperature, concrete temperature (ASTM C 1064), general weather conditions (clear, cloudy), wind speed, relative humidity, and evaporation rate should be recorded at frequent intervals. In addition, the following should be recorded and identified with the work in progress so that conditions relating to any part of the concrete construction can be identified at a later date:

- All water added to the mixture with corresponding mixing times;
- Time batched, time discharge started, and time discharge completed;
- Concrete temperature at time of delivery and after placing concrete;
- Observations on the appearance of concrete as delivered and after placing in forms;
- Slump of concrete as delivered;
- Slump of concrete as discharged; and
  - Protection and curing:
  - Method;
  - Time of application;
  - Rate of application;
  - Visual appearance of concrete; and
  - Duration of curing.

These observations should be included in the permanent project records.

CHAPTER 6—REFERENCES
6.1—Referenced standards and reports

The documents of the various standards-producing organizations referred to in this document are listed below with their serial designation.

American Concrete Institute

116R	Cement and Concrete Terminology
201.2R	Guide to Durable Concrete
207.1R	Mass Concrete
207.2R	Effect of Restraint, Volume Change, and Reinforcement on Cracking of Massive Concrete
207.4R	Cooling and Insulating Systems for Mass Concrete
211.1R	Standard Practice for Selecting Proportions for Normal, Heavyweight, and Mass Concrete
211.2	Standard Practice for Selecting Proportions for Structural Lightweight Concrete
211R	Guide for the Use of Normal Weight Aggregates in Concrete
212.3R	Chemical Admixtures for Concrete
221R	Guide for Use of Normal Weight Aggregates in Concrete
223R	Standard Practice for the Use of Shrinkage-Compensating Concrete
224R	Control of Cracking in Concrete Structures
224.3R	Joints in Concrete Construction
225R	Guide to the Selection and Use of Hydraulic Cements
226.3R	Use of Fly Ash in Concrete
234R	Guide for the Use of Silica Fume in Concrete
301	Specifications for Structural Concrete for Buildings
302.1R	Guide for Concrete Floor and Slab Construction
304R	Guide for Measuring, Mixing, Transporting, and Placing Concrete
306R	Cold Weather Concreting
308R	Standard Practice for Curing Concrete
309R	Guide for Consolidation of Concrete
311.1R/ (SP-2)	Manual of Concrete Inspection
311.4R	Guide for Concrete Inspection Programs
318/ 318R	Building Code Requirements for Structural Concrete and Commentary
318M/ 318RM	Building Code Requirements for Structural Concrete and Commentary
363R	State-of-the-Art Report on High Strength Concrete
E4-96	Chemical and Air-Entraining Admixtures for Concrete

ASTM International

C 31/C 31M	Standard Practice for Making and Curing Concrete Test Specimens in the Field
C 156	Standard Test Method for Water Retention by Concrete Curing Materials
C 172	Standard Practice for Sampling Freshly Mixed Concrete
C 192	Standard Practice for Making and Curing Concrete Test Specimens in the Laboratory
C 309	Standard Specification for Liquid Membrane-Forming Compounds for Curing Concrete
C 494	Standard Specification for Chemical Admixtures for Concrete
C 595	Standard Specification for Blended Hydraulic Cements
C 618	Standard Specification for Fly Ash and Raw or Calcined Natural Pozzolan for Use as a Mineral Admixture in Portland Cement Concrete

C 989	Standard Specification for Ground Granulated Blast-Furnace Slag for Use in Concrete and Mortars
C 1017	Standard Specification for Chemical Admixtures for Use in Producing Flowing Concrete
C 1064	Standard Test Method for Temperature of Freshly Mixed Portland-Cement Concrete
STP 169 C	Significance of Tests and Properties of Concrete and Concrete-Making Materials, 1994, 571 pp.

These publications may be obtained from the following organizations:

American Concrete Institute
P.O. Box 9094
Farmington Hills, Mich. 48333-9094

ASTM International
100 Barr Harbor Drive
West Conshohocken, Pa. 19428

6.2—Cited references

American Concrete Institute, 1996, "Practitioner's Guide to Hot Weather Concreting," ACI PP-1.

Barnes, B. D.; Orndorff, R. L.; and Roten, J. E., 1977, "Low Initial Curing Temperature Improves the Strength of Concrete Test Cylinders," ACI JOURNAL, *Proceedings* V. 74, No. 12, Dec., pp. 612-615.

Berhane, Z., 1984, "Evaporation of Water from Fresh Mortar and Concrete at Different Environmental Conditions," ACI JOURNAL, *Proceedings* V. 81, No. 6, Nov.-Dec., pp. 560-565.

Bloem, D., 1954, "Effect of Curing Conditions on Compressive Strengths of Concrete Cylinders," *Publication No. 53*, National Ready Mixed Concrete Association, Dec., 15 pp.

Cebeci, O. Z., 1986, "Hydration and Porosity of Cement Paste in Warm and Dry Environment," *8th International Congress on the Chemistry of Cement, Rio de Janeiro*, V. III, pp. 412-416; 423-424.

Cebeci, O. Z., 1987, "Strength of Concrete in Warm and Dry Environment," *Materials and Structures, Research and Testing* (RILEM, Paris), V. 20, No. 118, July pp. 270-272.

Collepardi, M.; Corradi, M.; and Valente, M., 1979, "Low-Slump-Loss Superplasticized Concrete," *Transportation Research Record 720*, Transportation Research Board, Washington, D.C., Jan., pp. 7-12,

Gaynor, R. D.; Meininger, R. C.; and Khan, T. S., 1985, "Effects of Temperature and Delivery Time on Concrete Proportions," *Temperature Effects on Concrete*, STP-858, ASTM, Philadelphia, pp. 68-87.

Guennewig, T., 1988, "Cost-Effective Use of Superplasticizers," *Concrete International: Design & Construction*, V. 10, No. 3, Mar., pp. 31-34.

Hampton, J. S., 1981, "Extended Workability of Concrete Containing High-Range Water-Reducing Admixtures in Hot Weather," *Developments in the Use of Superplasticizers*, SP-68, V. M. Malhotra, ed., American Concrete Institute, Farmington Hills, Mich., pp. 409-422.

Klieger, P., 1958, "Effect of Mixing and Curing Temperature on Concrete Strength," ACI JOURNAL, *Proceedings* V. 54, No. 12, June, pp. 1063-1081. Also, *Research Department Bulletin 103*, Portland Cement Association.

Lee, M., 1987, "New Technology in Concrete Cooling," *Concrete Products*, V. 89, No. 7, July, pp. 24-26, 36.

Lerch, William, 1957, "Plastic Shrinkage," ACI JOURNAL, *Proceedings*, V. 53, No. 8, Feb., 797-802.

Mentha, P. K., 1986, "Concrete Structure, Properties and Materials", pp. 56-57

Mittelacher, M., 1985, "Effect of Hot Weather Conditions on the Strength Performance of Set-Retarded Field Concrete," *Temperature Effects on Concrete,* STP 858, ASTM, Philadelphia, pp. 88-106.

Mittelacher, M., 1992, "Compressive Strength and The Rising Temperature of Field Concrete," *Concrete International*, V.14, No.12, Dec., pp 29–33.

National Ready Mixed Concrete Association, 1962, "Cooling Ready Mixed Concrete," *Publication* No. 106, Silver Spring, June, 7 pp.

Olivieri, E., and Martin, I., 1963, "Curing of Concrete in Puerto Rico," *Revista*, Colegio de Agricultura y Artes Mecanicas, Universidad de Puerto Rico, Mayaguez.

Portland Cement Association, 1992, "Design And Control of Concrete Mixtures," Thirteenth Edition, p. 80.

Ravina, D., 1984, "Slump Loss of Fly Ash Concrete," *Concrete International: Design & Construction*, *Proceedings*, V. 6, No. 4, Apr., pp. 35-39.

Ravina, D., and Shalon, R., 1968a, "Shrinkage of Fresh Mortars Cast under and Exposed to Hot Dry Climatic Conditions," *Proceedings*, Colloquium on Shrinkage of Hydraulic Concrete, RILEM/Cembureau, Paris, V. 2, (published by Instituto Eduardo Torroja, Madrid).

Ravina, D., and Shalon, R., 1968b, "Plastic Shrinkage and Cracking," ACI JOURNAL, *Proceedings* V. 65, No. 4, Apr., pp. 282-291.

Shilstone J., Sr., and Shilstone J., Jr., 1993, "High-Performance Concrete Mixtures For Durability," *High-Performance Concrete in Severe Environments*, SP–140, P. Zia, ed. American Concrete Institute, Farmington Hills, Mich., pp. 281-305.

Tuthill, L. H., and Cordon, W. A., 1955, "Properties and Uses of Initially Retarded Concrete," ACI JOURNAL, *Proceedings* V. 52, No. 3, Nov., pp. 273-286.

U.S. Bureau of Reclamation, 1952, "Effect of Initial Curing Temperatures on the Compressive Strength and Durability of Concrete," *Report* No. C-625, Denver, 7 pp.

U.S. Bureau of Reclamation, 1975, *Concrete Manual*, 8th Edition., Denver, 627 pp.

Verbeck, G. J., and Helmuth, R. H., 1968, "Structure and Physical Properties of Cement Pastes," *Proceedings*, Fifth International Symposium on the Chemistry of Cement, Tokyo, V. III, pp. 1-32.

Yamamoto, Y. and Kobayashi, S., 1986, "Effect of Temperature on the Properties of Superplasticized Concrete," ACI JOURNAL. *Proceedings* V. 83, No. 1, Jan-Feb., pp. 80-87.

APPENDIX A—ESTIMATING CONCRETE TEMPERATURE

A1—Equations for estimating temperature T of freshly mixed concrete are shown in the following.

Without ice (in. – lb and SI units)

$$T = \frac{0.22(T_a W_a + T_c W_c) + T_w W_w + T_a W_{wa}}{0.22(W_a + W_c) + W_w + W_{wa}} \quad \text{(A-1)}$$

With ice (in. – lb units)

$$T = \frac{0.22(T_a W_a + T_c W_c) + T_w W_w + T_a W_{wa} - 112 W_i}{0.22(W_a + W_c) + W_w + W_i + W_{wa}} \quad \text{(A-2)}$$

With ice (SI units)

$$T = \frac{0.22(T_a W_a + T_c W_c) + T_w W_w + T_a W_{wa} 79.6 W_w}{0.22(W_a + W_c) + W_w + W_w + W_i + W_{wa}} \quad \text{(A-3)}$$

where
T_a = temperature of aggregate
T_c = temperature of cement
T_w = temperature of batched mixing water from normal supply excluding ice
T_i = temperature of ice. (Note: The temperature of free and absorbed water on the aggregate is assumed to be the same temperature as the aggregate. All temperatures are in F or C.)
W_a = dry mass of aggregate
W_c = mass of cement
W_i = mass of ice
W_w = mass of batched mixing water
W_{wa} = mass of free and absorbed moisture in aggregate at T_a. (Note: All masses are in lb or kg.)

A2—Eq. (A-2) and (A-3), for estimating the temperature of concrete with ice in U. S. customary or SI units, assume that the ice is at its melting point. A more exact approach would be to use Eq. (A-4) or (A-5), which includes the temperature of the ice.

With ice (in.-lb units)

$$T = \frac{0.22(T_a W_a + T_c W_c) + T_w W_w}{0.22(W_a + W_c) + W_w + W_i + W_{wa} 79.6 W_w} \quad \text{(A-4)}$$

$$+ \frac{T_a W_{wa} + - W_i(128 - 0.5 T_i)}{0.22(W_a + W_c) + W_w + W_i + W_{wa}}$$

With ice (SI units)

$$T = \frac{0.22(T_a W_a - T_c W_c) T_w W_w}{0.22(W_a + W_c) + W_w + W_i + W_{wa}} \quad \text{(A-5)}$$

$$+ \frac{T_a W_{wa} + - W_i(79.6 - 0.5 T_i)}{0.22(W_a + W_c) + W_w + W_i + W_{wa}}$$

APPENDIX B—METHODS FOR COOLING FRESH CONCRETE

The summary is limited to a description of methods suitable for most structural uses of concrete. Methods for the cooling of mass concrete are explained in ACI 207.4R.

B1—Cooling with chilled mixing water

Concrete can be cooled to a moderate extent by using chilled mixing water; the maximum reduction in concrete temperature that can be obtained is approximately 10 F (6 C). The quantity of cooled water cannot exceed the mixing water requirement, which will depend upon the moisture content of aggregates and mixture proportions. The method involves a significant investment in mechanical refrigeration equipment and insulated water storage large enough for the anticipated hourly and daily production rates of cooled concrete. Available systems include one that is based on heat-pump technology, which is usable for both cooling and heating of concrete. Apart from its initial installation cost, this system appears to offer cooling at the lowest cost of available systems for cooling mixing water.

B2—Liquid nitrogen cooling of mixing water

Mixing water can be chilled rapidly through injection of liquid nitrogen into an insulated holding tank. This chilled water is then dispensed into the batch. Alternatively, the mixing water may be turned into ice slush by liquid nitrogen injection into the mixing water stream as it is discharged into the mixer. The system enables cooling by as much as 20 F (11 C). The ratio of ice-to-water in the slush must be adjusted to produce the temperature of concrete desired. Installation of this system requires insulated mixing water storage, a nitrogen supply vessel, batch controls, and auxiliary equipment. Apart from installation costs, there are operating expenses from liquid nitrogen usage and rental fees for the nitrogen supply vessel. The method differs from that by direct liquid nitrogen injection into mixed concrete described in B4.

B3—Cooling concrete with ice

Concrete can be cooled by using ice for part of the mixing water. The amount of cooling is limited by the amount of mixing water available for ice substitution. For most concrete, the maximum temperature reduction is approximately 20 F (11 C). For correct proportioning, the ice must be weighed. Cooling with block ice involves the use of a crusher/slinger unit, which can finely crush a block of ice and blow it into the mixer. A major obstacle to the use of block ice in many areas is insufficient supply. Costs of using block ice are: the cost of ice including transportation, refrigerated storage, handling and crushing equipment, additional labor, and if required, provisions for weighing the ice. An alternative to using block ice is to set up an ice plant near the concrete plant. As the ice is produced, it is weighed, crushed, and conveyed into the mixer. It may also be produced and used as flake ice. This system requires a large capital investment.

B4—Cooling mixed concrete with liquid nitrogen

B4.1 Injection of liquid nitrogen into freshly mixed concrete is an effective method for reduction of concrete temperature. The practical lower limit of concrete temperature is reached when concrete nearest the injection nozzle forms into a frozen lump; this is likely to occur when the desired concrete temperature is less than 50 F. The method has been successfully used in a number of major concrete placements. The performance of concrete was not affected adversely by its exposure to large amounts of liquid nitrogen. Cost of this method is relatively high, but it may be justified on the basis of practical considerations and overall effectiveness.

B4.2 Installations of the system consist of a nitrogen supply vessel and injection facility for central mixers, or one or more injection stations for truck mixers. The system can be set up at the construction site for last-minute cooling of the concrete before placement. This reduces temperature gains of cooled concrete in transit between the concrete plant and job site. Coordination is required in the dispatching of liquid nitrogen tanker trucks to injection stations for the timely replenishing of gas consumed in the cooling operations. The quantity of liquid nitrogen required will vary according to mixture proportions and constituents, and the amount of temperature reduction. The use of 135 ft^3 (48 m^3) of liquid nitrogen will usually reduce concrete temperature 1 F (0.5 C).

B5—Cooling of coarse aggregates

B5.1 An effective method of lowering the temperature of the coarse aggregate is by cool water spraying or inundation. Coarse aggregate has the greatest mass in a typical concrete mixture. Reducing the temperature of the aggregate approximately 2 ± 0.5 F (1 ± 0.5 C) will lower the final concrete temperature approximately 1 F (0.5 C). To use this method, the producer must have available large amounts of chilled water and the necessary water-cooling equipment for production requirements. This method is most effective when adequate amounts of coarse material are contained in a silo or bin so that cooling can be accomplished in a short period of time. Care must be taken to evenly inundate the material so that slump variation from load to load is minimized.

B5.2 Cooling of coarse aggregate can also be accomplished by blowing air through the moist aggregate. The air flow will enhance evaporative cooling and can bring the coarse aggregate temperature within 2 F (1 C) of wet bulb temperature. Effectiveness of the method depends on ambient temperature, relative humidity, and velocity of air flow. The added refinement of using chilled air instead of air at ambient temperature can reduce the coarse aggregate temperature to as low as 45 F (7 C). This method, however, involves a relatively high installation cost.

ACI 306R-88
(Reapproved 2002)

This document has been approved for use by agencies of the Department of Defense and for listing in the DoD Index of Specifications and Standards.

Cold Weather Concreting

Reported by ACI Committee 306

Nicholas J. Carino, Chairman*

Fred A. Anderson*	Gilbert J. Haddad	Albert W. Knott	Derle Thorpe*
Peter Antonich	Don B. Hill	William F. Perenchio	Valery Tokar*
George R. U. Burg*	Jules Houde	John M. Scanlon*	Harry H. Tormey
Oleh B. Ciuk	David A. Hunt	Michael L. Shydlowski*	Lewis H. Tuthill*
Douglas J. Haavik*	Robert A. Kelsey	Bruce A. Suprenant	Harold B. Wenzel

The general requirements for producing satisfactory concrete during cold weather are discussed, and methods for satisfying these requirements are described. One of the objectives of cold weather concreting practice is to provide protection of the concrete at early ages to prevent damage from freezing. For many structural concretes, protection considerably in excess of that required to prevent damage by early freezing is needed to assure development of adequate strength.

The following items are discussed in the report: recommended temperature of concrete, temperature records, temperature of materials, preparations prior to placement, duration of protection period, methods for determining in-place strength, form removal, protective insulating covers, heated enclosures, curing methods, and accelerating admixtures. References are included that provide supplementary data on the effects of curing temperature on concrete strength.

Keywords: accelerating admixtures; age; aggregates; calcium chloride; **cold weather**; compressive strength; **concrete construction**; concretes; curing; durability; form removal; formwork (construction); freeze-thaw durability; heating; **in-place testing; insulation**; materials handling; **protection**; subgrade preparation; **temperature**.

CONTENTS

Chapter 1 — Introduction, p. 306R-1
 1.1 — Definition of cold weather
 1.2 — Standard specification
 1.3 — Objectives
 1.4 — Principles
 1.5 — Economy

Chapter 2 — General requirements, p. 306R-3
 2.1 — Planning
 2.2 — Protection during fall and spring
 2.3 — Concrete temperature
 2.4 — Temperature records
 2.5 — Heated enclosures
 2.6 — Exposure to freezing and thawing
 2.7 — Concrete slump

Chapter 3 — Temperature of concrete as mixed and placed and heating of materials, p. 306R-5
 3.1 — Placement temperature
 3.2 — Mixing temperature
 3.3 — Heating mixing water
 3.4 — Heating aggregates
 3.5 — Steam heating of aggregates
 3.6 — Overheating of aggregates
 3.7 — Calculation of mixture temperature
 3.8 — Temperature loss during delivery

Chapter 4 — Preparation before concreting, p. 306R-7
 4.1 — Temperature of surfaces in contact with fresh concrete
 4.2 — Metallic embedments
 4.3 — Removal of snow and ice
 4.4 — Condition of subgrade

Chapter 5 — Protection against freezing and protection for concrete not requiring construction supports, p. 306R-7
 5.1 — Protection to prevent early-age freezing
 5.2 — Need for additional protection
 5.3 — Length of protection period
 5.4 — Stripping of forms
 5.5 — Temperature drop after removal of protection
 5.6 — Allowable temperature differential

Chapter 6 — Protection for structural concrete requiring construction supports, p. 306R-9
 6.1 — Introduction
 6.2 — Tests of field-cured specimens
 6.3 — In-place testing
 6.4 — Maturity method
 6.5 — Attainment of design strength
 6.6 — Increasing early strength
 6.7 — Cooling of concrete
 6.8 — Estimating strength development
 6.9 — Removal of forms and supports
 6.10 — Requirements

ACI Committee Reports, Guides, Standard Practices, and Commentaries are intended for guidance in planning, designing, executing, and inspecting construction. This document is intended for the use of individuals who are competent to evaluate the significance and limitations of its content and recommendations and who will accept responsibility for the application of the material it contains. The American Concrete Institute disclaims any and all responsibility for the stated principles. The Institute shall not be liable for any loss or damage arising therefrom.

Reference to this document shall not be made in contract documents. If items found in this document are desired by the Architect/Engineer to be a part of the contract documents, they shall be restated in mandatory language for incorporation by the Architect/Engineer.

---

*Task force member.
ACI 306R-88 supersedes ACI 306R-78 (Revised 1983).
Copyright © 1988, American Concrete Institute.
All rights reserved including rights of reproduction and use in any form or by any means, including the making of copies by any photo process, or by electronic or mechanical device, printed, written, or oral, or recording for sound or visual reproduction or for use in any knowledge or retrieval system or device, unless permission in writing is obtained from the copyright proprietors.

Chapter 7 — Materials and methods of protection, p. 306R-13
7.1 — Introduction
7.2 — Insulating materials
7.3 — Selection of insulation
7.4 — Enclosures
7.5 — Internal electric heating
7.6 — Covering after placement
7.7 — Temporary removal of protection
7.8 — Insulated forms

Chapter 8 — Curing requirements and methods, p. 306R-20
8.1 — Introduction
8.2 — Curing during the protection period
8.3 — Curing following the protection period

Chapter 9 — Acceleration of setting and strength development, p. 306R-21
9.1 — Introduction
9.2 — Calcium chloride as an accelerating admixture
9.3 — Other accelerating admixtures

Chapter 10 — References, p. 306R-22
10.1 — Recommended references
10.2 — Cited references
10.3 — Selected references

CHAPTER 1 — INTRODUCTION
1.1 — Definition of cold weather

This report describes construction procedures which, if properly followed, can result in concrete placed in cold weather of sufficient strength and durability to satisfy intended service requirements. Concrete placed during cold weather will develop these qualities only if it is properly produced, placed, and protected. The necessary degree of protection increases as the ambient temperature decreases.

Cold weather is defined as a period when, for more than 3 consecutive days, the following conditions exist: 1) the average daily air temperature is less than 40 F (5 C) and 2) the air temperature is not greater than 50 F (10 C) for more than one-half of any 24-hr period.* The average daily air temperature is the average of the highest and the lowest temperatures occurring during the period from midnight to midnight. Cold weather, as defined in this report, usually starts during fall and usually continues until spring.

1.2 — Standard specification

If requirements for cold weather concreting are needed in specification form, ACI 306.1 should be referenced; if necessary, appropriate modifications should be added to the contract documents after consulting the specification checklist.

1.3 — Objectives

The objectives of cold weather concreting practices are to:

1.3.1 — prevent damage to concrete due to freezing at early ages. When no external water is available, the degree of saturation of newly placed concrete decreases as the concrete gains maturity and the mixing water combines with cement during hydration. Under such conditions, the degree of saturation falls below the critical level (the degree of water saturation where a single cycle of freezing would cause damage) at approximately the time that the concrete attains a compressive strength of 500 psi (3.5 MPa) (Powers 1962). At 50 F (10 C), most well-proportioned concrete mixtures reach this strength during the second day.

1.3.2 — assure that the concrete develops the required strength for safe removal of forms, for safe removal of shores and reshores, and for safe loading of the structure during and after construction.

1.3.3 — maintain curing conditions that foster normal strength development without using excessive heat and without causing critical saturation of the concrete at the end of the protection period.

1.3.4 — limit rapid temperature changes, particularly before the concrete has developed sufficient strength to withstand induced thermal stresses. Rapid cooling of concrete surfaces or large temperature differences between exterior and interior members of the structure can cause cracking, which can be detrimental to strength and durability. At the end of the required period, insulation or other means of protection should be removed gradually so that the surface temperature decreases gradually during the subsequent 24-hr period (see Section 5.5).

1.3.5 — provide protection consistent with the intended serviceability of the structure. Concrete structures are intended for a useful life of many years. The attainment of satisfactory strength for 28-day, standard-cured cylinders is irrelevant if the structure has corners damaged by freezing; dehydrated areas; and cracking from overheating because of inadequate protection, improper curing, or careless workmanship. Similarly, early concrete strength achieved by indiscriminate use of excessive calcium chloride is of no avail if the concrete bcomes excessively cracked in later years because of the likelihood of disruptive internal expansion due to alkali-aggregate reaction or of possible corrosion of reinforcement (see Section 9.2). Short-term construction economy should not be obtained at the expense of long-term durability.

1.4 — Principles

This report presents recommendations to achieve the objectives listed in Section 1.3 (Schnarr and Young 1934a and 1934b). The practices and procedures described in this report stem from the following principles concerning cold weather concreting:

1.4.1 — Concrete that is protected from freezing until it has attained a compressive strength of at least 500 psi (3.5 MPa) will not be damaged by exposure to a single freezing cycle (Powers 1962).

1.4.2 — Concrete that is protected as in Section 1.4.1 will mature to its potential strength despite subsequent exposure to cold weather (Malhotra and Berwanger 1973). No further protection is necessary unless a certain strength must be attained in less time.

---
*The values in SI units are direct conversions of the in.-lb. values. They do not necessarily represent common metric ranges or sizes. For practical application, the user should adjust them to conform with local practice.

Table 3.1—Recommended concrete temperatures

Line	Air temperature	Section size, minimum dimension, in. (mm)			
		< 12 in. (300 mm)	12-36 in. (300-900 mm)	36-72 in. (900-1800 mm)	>72 in. (1800 mm)
		Minimum concrete temperature as placed and maintained			
1	—	55 F (13 C)	50 F (10 C)	45 F (7 C)	40 F (5 C)
		Minimum concrete temperature as mixed for indicated air temperature*			
2	Above 30 F (−1 C)	60 F (16 C)	55 F (13 C)	50 F (10 C)	45 F (7 C)
3	0 to 30 F (−18 to −1 C)	65 F (18 C)	60 F (16 C)	55 F (13 C)	50 F (10 C)
4	Below 0 F (−18 C)	70 F (21 C)	65 F (18 C)	60 F (16 C)	55 F (13 C)
		Maximum allowable gradual temperature drop in first 24 hr after end of protection			
5	—	50 F (28 C)	40 F (22 C)	30 F (17 C)	20 F (11 C)

*For colder weather a greater margin in temperature is provided between concrete as mixed and required minimum temperature of fresh concrete in place.

1.4.3 — Where a specified concrete strength must be attained in a few days or weeks, protection at temperatures above 50 F (10 C) is required. See Chapters 5 and 6.

1.4.4 — Except within heated protective enclosures, little or no external supply of moisture is required for curing during cold weather. See Chapter 8.

1.4.5 — Under certain conditions, calcium chloride should not be used to accelerate setting and hardening because of the increased chances of corrosion of metals embedded in concrete or other adverse effects. See Chapter 9.

Times and temperatures given in this report are not exact values for all situations and they should not be used as such. The user should keep in mind the primary intent of these recommendations and should use discretion in deciding what is adequate for each particular circumstance.

1.5 — Economy

Experience has shown that the overall costs of adequate protection for cold weather concreting are not excessive, considering what is required and the resulting benefits. The owner must decide whether the extra costs involved in cold weather concreting operations are a profitable investment or if it is more cost effective to wait for mild weather. Neglect of protection against early freezing can cause immediate destruction or permanently weakened concrete. Therefore, if cold weather concreting is performed, adequate protection from low temperatures and proper curing are essential.

CHAPTER 2 — GENERAL REQUIREMENTS
2.1 — Planning

It is recommended that the concrete contractor, concrete supplier, and owner (or architect/engineer) meet in a preconstruction conference to define in clear terms how cold weather concreting methods will be used. This report provides a basis for the contractor to select specific methods to satisfy the minimum requirements during cold weather concreting.

Plans to protect fresh concrete from freezing and to maintain temperatures above the recommended minimum values should be made well before freezing temperatures are expected to occur. Necessary equipment and materials should be at the work site before cold weather is likely to occur, not after concrete has been placed and its temperature begins to approach the freezing point.

2.2 — Protection during fall and spring

During periods not defined as cold weather, such as in fall or spring, but when heavy frost or freezing is forecast at the job site,* all concrete surfaces should be protected from freezing for at least the first 24 hr after placement. Concrete protected in this manner will be safe from damage by freezing at an early age. If the concrete is air entrained and properly cured, the ultimate strength and durability of the concrete will be unimpaired. Protection from freezing during the first 24 hr does not assure a satisfactory rate of strength development, particularly when followed by considerably colder weather. Protection and curing should continue long enough — and at a temperature sufficiently above freezing — to produce the strength required for form removal or structural safety (see Chapters 5 and 6).

2.3 — Concrete temperature

During cold weather, the concrete temperature at the time of placement should not be lower than the values given in Chapter 3. In addition, to prevent freezing at early ages, the concrete temperature should be maintained at not less than the recommended placement temperature for the length of time given in Chapter 5. This length of time depends on the type and amount of cement, whether an accelerating admixture is used, and the service category.

The recommended minimum placement temperatures given in Table 3.1 in Chapter 3 apply to normal weight

---
*Charts showing mean dates of freezing weather in the United States may be obtained from the National Climatic Center, Federal Building, Ashville, NC 28801.

concrete. Experience indicates that freshly mixed lightweight concrete loses heat more slowly than freshly mixed normal-weight concrete. Lighter weight insulating concretes lose heat even more slowly. However, when exposed to freezing temperatures, such concretes are more susceptible to damage from surface freezing.

The temperature of concrete at the time of placement should always be near the minimum temperatures given in Chapter 3, Table 3.1. Placement temperatures should not be higher than these minimum values by more than 20 F (11 C). One should take advantage of the opportunity provided by cold weather to place low-temperature concrete. Concrete that is placed at low temperatures [40 to 55 F (5 to 13 C)] is protected against freezing and receives long-time curing, thus developing a higher ultimate strength (Klieger 1958) and greater durability. It is, therefore, less subject to thermal cracking than similar concrete placed at higher temperatures. Placement at higher temperatures may expedite finishing in cold weather, but it will impair long-term concrete properties.

2.4 — Temperature records

The actual temperature at the concrete surface determines the effectiveness of protection, regardless of air temperature. Therefore, it is desirable to monitor and record the concrete temperature. Temperature recording and monitoring must consider the following:

2.4.1 — The corners and edges of concrete are more vulnerable to freezing and usually are more difficult to maintain at the required temperature, therefore, their temperature should be monitored to evaluate and verify the effectiveness of the protection provided.

2.4.2 — Inspection personnel should keep a record of the date, time, outside air temperature, temperature of concrete as placed, and weather conditions (calm, windy, clear, cloudy, etc.). Temperatures of concrete and the outdoor air should be recorded at regular time intervals but not less than twice per 24-hr period. The record should include temperatures at several points within the enclosure and on the concrete surface, corners, and edges. There should be a sufficient number of temperature measurement locations to show the range of concrete temperatures. Temperature measuring devices embedded in the concrete surface are ideal, but satisfactory accuracy and greater flexibility of observation can be obtained by placing thermometers against the concrete under temporary covers of heavy insulating material until constant temperatures are indicated.

2.4.3 — Maximum and minimum temperature readings in each 24-hr period should be recorded. Data recorded should clearly show the temperature history of each section of concrete cast. A copy of the temperature readings should be included in the permanent job records. It is preferable to measure the temperature of concrete at more than one location in the section cast and use the lowest reading to represent the temperature of that section. Internal temperature of concrete should be monitored to insure that excessive heating does not occur (see Section 7.4). For this, expendable thermistors or thermocouples cast in the concrete may be used.

2.5 — Heated enclosures

Heated enclosures must be strong enough to be windproof and weatherproof. Otherwise, proper temperatures at corners, edges, and in thin sections may not be maintained despite high energy consumption. Combustion heaters should be vented and they should not be permitted to heat or to dry the concrete locally. Fresh concrete surfaces exposed to carbon dioxide, resulting from the use of salamanders or other combustion heaters that exhaust flue gases into an enclosed area, may be damaged by carbonation of the concrete. Carbonation may result in soft surfaces or surface crazing depending on the concentration of carbon dioxide, the concrete temperature, and the relative humidity (see Section 7.4). Carbon monoxide, which can result from partial combustion, and high levels of carbon dioxide are potential hazards to workers.

In addition, strict fire prevention measures should be enforced. Fire can destroy the protective enclosures as well as damage the concrete. Concrete can be damaged by fire at any age. However, at a very early age additional damage can occur by subsequent freezing of the concrete before new protective enclosures are provided.

2.6 — Exposure to freezing and thawing

If, during construction, it is likely that the concrete will be exposed to cycles of freezing and thawing while it is in a saturated condition, it should be properly air entrained even though it will not be exposed to freezing and thawing in service. The water-cement ratio should not exceed the limits recommended in ACI 201.2R, and the concrete should not be allowed to freeze and thaw in a saturated condition before developing a compressive strength of 3500 psi (24 MPa). Therefore, new sidewalks and other flatwork exposed to melting snow during daytime and freezing during nighttime should be air entrained and protected from freezing until a strength of at least 3500 psi (24 MPa) has been attained.

2.7 — Concrete slump

Concrete with a slump lower than normal [less than 4 in. (100 mm)] is particularly desirable in cold weather for flatwork; bleeding of water is minimized and setting occurs earlier. During cold weather, bleed water may remain on the surface for such a long period that it interferes with proper finishing. If the bleed water is mixed into the concrete during trowelling, the resulting surface will have a lower strength and may be prone to dusting and subsequent freeze-thaw damage if exposed. Thus, during cold weather, the concrete mixture should be proportioned so that bleeding is minimized as much as practicable. If bleedwater is present on flatwork, it should be skimmed off prior to trowelling by using a rope or hose.

CHAPTER 3 — TEMPERATURE OF CONCRETE AS MIXED AND PLACED AND HEATING OF MATERIALS

3.1 — Placement temperature

During cold weather, the concrete mixing temperature should be controlled as described in Section 3.2 so that when the concrete is placed its temperature is not below the values shown in Line 1 of Table 3.1. The placement temperature of concrete should be determined according to ASTM C 1064. The more massive the concrete section, the less rapidly it loses heat; therefore, lower minimum placement temperatures are recommended as concrete sections become larger. For massive structures, it is especially beneficial to have low placement temperatures (see ACI 207.1R). Concrete temperatures that are much higher than the values in Line 1 do not result in a proportionally longer protection against freezing because the rate of heat loss is greater for larger temperature differentials.

In addition, higher temperatures require more mixing water, increase the rate of slump loss, may cause quick setting, and increase thermal contraction. Rapid moisture loss from exposed surfaces of flatwork may cause plastic shrinkage cracks. Rapid moisture loss can occur from surfaces exposed to cold weather because the warm concrete heats the surrounding cold air and reduces its relative humidity (see ACI 302.1R). Therefore, the temperature of concrete as placed should be kept as close to the recommended minimum value as is practicable. Placement temperatures should not be higher than these minimum values by more than 20 F (11 C).

3.2 — Mixing temperature

The recommended minimum temperature of concrete at the time of mixing is shown in Lines 2, 3, and 4 of Table 3.1. As the ambient air temperature decreases, the concrete temperature during mixing should be increased to offset the heat lost in the interval between mixing and placing. The mixing temperature should not be more than 15 F (8 C) above the recommended values in Lines 2, 3, and 4. While it is difficult to heat aggregates uniformly to a predetermined temperature, the mixing water temperature can be adjusted easily by blending hot and cold water to obtain a concrete temperature within 10 F (5 C) of the required temperature.

3.3 — Heating mixing water

Mixing water should be available at a consistent, regulated temperature, and in sufficient quantity to avoid appreciable fluctuations in temperature of the concrete from batch to batch. Since the temperature of concrete affects the rate of slump loss and may affect the performance of admixtures, temperature fluctuations can result in variable behavior of individual batches.

Premature contact of very hot water and concentrated quantities of cement has been reported to cause flash set and cement balls in truck mixers. When water above 140 F (80 C) is used, it may be necessary to adjust the order in which ingredients are blended. It may be helpful to add the hot water and coarse aggregate ahead of the cement and to stop or slow down the addition of water while the cement and aggregate are loaded.

If the cement is batched separately from the aggregate, mixing may be more difficult. To facilitate mixing, about three-fourths of the added hot water should be placed in the drum either ahead of the aggregates or with them. To prevent packing at the end of the mixer, coarse aggregate should be added first. The cement should be added after the aggregates. As the final ingredient, the remaining one-fourth of the mixing water should be placed into the drum at a moderate rate.

Water with a temperature as high as the boiling point may be used provided that resulting concrete temperatures are within the limits discussed in Section 3.2 and no flash setting occurs. If loss of effectiveness of the air-entraining admixture is noted due to an initial contact with hot water, the admixture must be added to the batch after the water temperature has been reduced by contact with the cooler solid materials.

3.4 — Heating aggregates

When aggregates are free of ice and frozen lumps, the desired temperature of the concrete during mixing can usually be obtained by heating only the mixing water, but when air temperatures are consistently below 25 F (−4 C), it is usually necessary to also heat the aggregates. Heating aggregates to temperatures higher than 60 F (15 C) is rarely necessary if the mixing water is heated to 140 F (60 C). If the coarse aggregate is dry and free of frost, ice, and frozen lumps, adequate temperatures of freshly mixed concrete can be obtained by increasing the temperature of only the sand, which seldom has to be above about 105 F (40 C), if mixing water is heated to 140 F (60 C). Seasonal variations must be considered, as average aggregate temperatures can be substantially higher than air temperature during autumn, while the reverse may occur during spring.

3.5 — Steam heating of aggregates

Circulating steam in pipes is recommended for heating aggregates. For small jobs, aggregates may be thawed by heating them carefully over culvert pipes in which fires are maintained. When aggregates are thawed or heated by circulating steam in pipes, exposed surfaces of aggregate should be covered with tarpaulins as much as is practicable to maintain a uniform distribution of heat and to prevent formation of ice crusts. Steam jets liberated in aggregate may cause troublesome moisture variation, but this method is the most thermally efficient procedure to heat aggregate. If steam is confined in a pipe-heating system, difficulties from variable moisture in aggregates are avoided, but the likelihood of localized hot, dry spots is increased. Wear and corrosion of steam pipes in aggregates will eventually cause leaks, which may lead to the same moisture variation problem caused by steam jets. Periodic inspection of the pipes and replacement as necessary are recommended.

When conditions require thawing of substantial quantities of extremely low temperature aggregates, steam jets may be the only practicable means of providing the necessary heat. In such a case, thawing must be done as far in advance of batching as is possible to achieve substantial equilibrium in both moisture content and temperature. After thawing is completed, the steam supply can be reduced to the minimum that will prevent further freezing, thereby reducing to some extent the problems arising from variable moisture content. Nevertheless, under such conditions, mixing water control must be largely on an individual batch adjustment basis. Dry hot air instead of steam has been used to keep aggregates ice free.

3.6 — Overheating of aggregates

Aggregates should be heated sufficiently to eliminate ice, snow, and frozen lumps of aggregate. Often 3-in. (76-mm) frozen lumps will survive mixing and remain in the concrete after placing. Overheating should be avoided so that spot temperatures do not exceed 212 F (100 C) and the average temperature does not exceed 150 F (65 C) when the aggregates are added to the batch. Either of these temperatures is considerably higher than is necessary for obtaining desirable temperatures of freshly mixed concrete. Materials should be heated uniformly since considerable variation in their temperature will significantly vary the water requirement, air entrainment, rate of setting, and slump of the concrete.

Extra care is required when batching the first few loads of concrete following a prolonged period of steaming the aggregates in storage bins. Many concrete producers recycle the first few tons of very hot aggregates. This material is normally discharged and recycled by placing it on top of the aggregates in the storage bins.

3.7 — Calculation of mixture temperature

If the weights and temperatures of all constituents and the moisture content of the aggregates are known, the final temperature of the concrete mixture may be estimated from the formula

$$T = \frac{[0.22(T_s W_s + T_a W_a + T_c W_c) + T_w W_w + T_s W_{ws} + T_a W_{wa}]}{[0.22(W_s + W_a + W_c) + W_w + W_{wa} + W_{ws}]} \quad (3\text{-}1)$$

where

T = final temperature of concrete mixture (deg F or C)
T_c = temperature of cement (deg F or C)
T_s = temperature of fine aggregate (deg F or C)
T_a = temperature of coarse aggregate (deg F or C)
T_w = temperature of added mixing water (deg F or C)
W_c = weight of cement (lb or kg)
W_s = saturated surface-dry weight of fine aggregate (lb or kg)
W_a = saturated surface-dry weight of coarse aggregate (lb or kg)
W_w = weight of mixing water (lb or kg)
W_{ws} = weight of free water on fine aggregate (lb or kg)
W_{wa} = weight of free water on coarse aggregate (lb or kg)

Eq. (3-1) is derived by considering the equilibrium heat balance of the materials before and after mixing and by assuming that the specific heats of the cement and aggregates are equal to 0.22 Btu/(lb F) [0.22 kcal/(kg C)].

If the temperature of one or both of the aggregates is below 32 F (0 C), the free water will be frozen, and Eq. (3-1) must be modified to take into account the heat required to raise the temperature of the ice to 32 F (0 C), to change the ice into water, and to raise the temperature of the free water to the final mixture temperature. The specific heat of ice is 0.5 Btu/(lb F) [0.5 kcal/(kg C)] and the heat of fusion of ice is 144 Btu/lb (80 kcal/kg). Thus Eq. (3-1) is modified by substituting the following expressions for $T_s W_{ws}$ or $T_a W_{wa}$, or both, depending on whether the fine aggregate or coarse aggregate, or both, are below 32 F (0 C).

For in.-lb units

$$\text{for } T_s W_{ws} \text{ substitute } W_{ws}(0.50T_s - 128) \quad (3\text{-}2)$$

$$\text{for } T_a W_{wa} \text{ substitute } W_{wa}(0.50T_a - 128) \quad (3\text{-}3)$$

For SI units

$$\text{for } T_s W_{ws} \text{ substitute } W_{ws}(0.50T_s - 80) \quad (3\text{-}4)$$

$$\text{for } T_a W_{wa} \text{ substitute } W_{wa}(0.50T_a - 80) \quad (3\text{-}5)$$

In these equations, the numbers 128 and 80 are obtained from the heat of fusion needed to melt the ice, the specific heat of the ice, and the melting temperature of ice.

3.8 — Temperature loss during delivery

The Swedish Cement and Concrete Research Institute (Petersons 1966) performed tests to determine the expected decrease in concrete temperature during delivery in cold weather. Their studies included revolving drum mixers, covered-dump bodies, and open-dump bodies. Approximate temperature drop for a delivery time of 1 hr can be computed using Eq. (3-6)-(3-8). For revolving drum mixers

$$T = 0.25(t_r - t_a) \quad (3\text{-}6)$$

For covered-dump body

$$T = 0.10(t_r - t_a) \quad (3\text{-}7)$$

For open-dump body

$$T = 0.20(t_r - t_a) \quad (3\text{-}8)$$

where

T = temperature drop to be expected during a 1-hr delivery time, deg F or C. (This value must be added to t_r to determine the required temperature of concrete at the plant.)
t_r = concrete temperature required at the job, deg F or C.
t_a = ambient air temperature, deg F or C.

The values from these equations are proportionally adjusted for delivery times greater than or less than one hour.

3.9.1 — The following examples illustrate the application of these approximate equations:

1. Concrete is to be continuously agitated in a revolving drum mixer during a 1-hr delivery period. The air temperature is 20 F and the concrete at delivery must be at least 50 F. From Eq. (3-6)

$$T = 0.25 (50 - 20) = 7.5 \text{ F}$$

Therefore, allowance must be made for a 7.5-deg temperature drop, and the concrete at the plant must have a temperature of at least (50 + 7.5 F), or about 58 F.

2. For the same temperature conditions given in Example 1, the concrete will be delivered within 1 hr and the drum will not be revolved except for initial mixing and again briefly at the time of discharge. Assuming that Eq. (3-7) represents this situation best, the temperature drop is

$$T = 0.10 (50 - 20) = 3 \text{ F}$$

Thus provisions must be made for a concrete temperature of (50 + 3 F), or 53 F, at the plant.

The advantage of covered dump bodies over revolving drums suggests that temperature losses can be minimized by not revolving the drum more than is absolutely necessary during delivery.

CHAPTER 4 — PREPARATION BEFORE CONCRETING

4.1 — Temperature of surfaces in contact with fresh concrete

Preparation for concreting, other than mentioned in Section 2.1, consists primarily of insuring that all surfaces that will be in contact with newly placed concrete are at temperatures that cannot cause early freezing or seriously prolong setting of the concrete. Ordinarily, the temperatures of these contact surfaces, including subgrade materials, need not be higher than a few degrees above freezing, say 35 F (2 C), and preferably not more than 10 F (5 C) higher than the minimum placement temperatures given in Line 1 of Table 3.1.

4.2 — Metallic embedments

The placement of concrete around massive metallic embedments that are at temperatures below the freezing point of the water in concrete may result in local freezing of the concrete at the interface. If the interface remains frozen beyond the time of final vibration, there will be a permanent decrease in the interfacial bond strength. Whether freezing will occur, the volume of frozen water, and the duration of the frozen period depend primarily upon the placement temperature of concrete, the relative volumes of the concrete and the embedment, and the temperature of the embedment. An analytical study, using the finite element method to solve the heat flow problem, has been reported (Suprenant and Basham 1985). Two cases were investigated: a No. 9 bar in a slab and a square steel tube filled with concrete. Based on that limited study, it was suggested that steel embedments having a cross-sectional area greater than 1 in.^{2} (650 mm^2) should have a temperature of at least 10 F (−12 C) immediately before being surrounded by fresh concrete at a temperature of at least 55 F (13 C). Additional study is required before definitive recommendations can be formulated. The engineer/architect should determine whether the structure contains large embedments that pose potential problems. If heating is required, the heating process should not alter the mechanical or metallurgical properties of the metal. The contractor should submit the plan for heating to the engineer for approval.

4.3 — Removal of snow and ice

All snow, ice, and frost must be removed so that it does not occupy space intended to be filled with concrete. Hot-air jets can be used to remove frost, snow, and ice from forms, reinforcement, and other embedments. Unless the work area is housed, this work should be done immediately prior to concrete placement to prevent refreezing.

4.4 — Condition of subgrade

Concrete should not be placed on frozen subgrade material. The subgrade sometimes can be thawed acceptably by covering it with insulating material for a few days before the concrete placement, but in most cases external heat must be applied. Experimenting at the site will show what combinations of insulation and time causes subsurface heat to thaw the subgrade material. If necessary, the thawed material should be recompacted.

CHAPTER 5 — PROTECTION AGAINST FREEZING AND PROTECTION FOR CONCRETE NOT REQUIRING CONSTRUCTION SUPPORTS

5.1 — Protection to prevent early-age freezing

To prevent early-age freezing, protection must be provided immediately after concrete placement. Arrangements for covering, insulating, housing, or heating newly placed concrete should be made before placement. The protection that is provided should be adequate to achieve, in all sections of the concrete cast, the temperature and moisture conditions recommended in this report. In cold weather, the temperature of the newly placed concrete should be kept close to the val-

Table 5.1—Length of protection period required to prevent damage from early-age freezing of air-entrained concrete

Line	Exposure	Protection period at temperature indicated in Line 1 of Table 3.1, days*	
		Type I or II cement	Type III cement, or accelerating admixture, or 100 lb/yd³ (60 kg/m³) of additional cement
1	Not exposed	2	1
2	Exposed	3	2

*A day is a 24-hr period.

Table 5.3—Length of protection period for concrete placed during cold weather

Line	Service category	Protection period at temperature indicated in Line 1 of Table 3.1, days*	
		Type I or II cement	Type III cement, or accelerating admixture, or 100 lb/yd³ (60 kg/m³) of additional cement
1	1—no load, not exposed	2	1
2	2—no load, exposed	3	2
3	3—partial load, exposed	6	4
4	4—full load	See Chapter 6	

*A day is a 24-hr period.

ues shown in Line 1 of Table 3.1 for the lengths of time indicated in Table 5.1 for protection against early-age freezing. The length of the protection period may be reduced by: (1) using Type III cement; (2) using an accelerating admixture; or (3) using 100 lb/yd³ (60 kg/m³) of cement in excess of the design cement content. Line 1 of Table 5.1 refers to concrete that will be exposed to little or no freezing and thawing in service or during construction, such as in foundations and substructures. Line 2 refers to concrete that will be exposed to the weather in service or during construction.

It has been shown that when there is no external source of curing water, concrete that has attained a strength of 500 psi will not be damaged by one cycle of freezing and thawing (Powers 1962; Hoff and Buck 1983). The protection periods given in Table 5.1 may be reduced if it is verified that the concrete, including corners and edges, has attained an in-place compressive strength of at least 500 psi (3.5 MPa), and will not be expected to be exposed to more than one cycle of freezing and thawing before being buried or backfilled. Techniques for estimating the in-place strength are discussed in Chapter 6. To protect massive concrete against thermal cracking, a longer protection period than given in Table 5.1 is required. For concrete with a low cement content, a longer protection period may be needed to reach a strength of 500 psi (3.5 MPa).

5.2 — Need for additional protection

The comparatively short periods of protection shown in Table 5.1 are for air-entrained concrete having the air content recommended in ACI 211.1. These are the minimum protection requirements to prevent damage from one early cycle of freezing and thawing* and thereby assure that there is no impairment to the ultimate durability of the concrete. These short periods are permissible only when: (1) there is sufficient subsequent curing (see Chapter 8) and protection to develop the required safe strength for the specific service category (see Section 5.3); and (2) the concrete is not subject to freezing in a critically saturated condition. When there are early-age strength requirements, it is necessary to extend the protection period beyond the minimum duration given in Table 5.1.

5.3 — Length of protection period

The length of the required protection period depends on the type and amount of cement, whether an accelerating admixture is used, and the service category (Sturrup and Clendening 1962). Table 5.3 gives the minimum length of the protection period at the temperatures given in Line 1 of Table 3.1. These minimum protection periods are recommended unless the in-place strength of the concrete has attained a previously established value. The service categories are as follows:

5.3.1 *Category 1: No load, not exposed* — This category includes foundations and substructures that are not subject to early load, and, because they are buried deep within the ground or are backfilled, will undergo little or no freezing and thawing in service. For concrete in this service category, conditions are favorable for continued natural curing. This concrete requires only the protection time recommended for Category 1 (Line 1) in Table 5.3. It is seen that for Category 1, the length of the protection period in Table 5.3 is the same as the requirement for protection against early-age freezing given in Line 1 of Table 5.1. Thus, for this service category, only protection against early-age freezing is necessary.

5.3.2 *Category 2: No load exposed* — This category includes massive piers and dams that have surfaces exposed to freezing and weathering in service but have no early strength requirements. Interior portions of these structures are self-curing. Exterior surfaces will continue to cure when natural conditions are favorable. To provide initial curing and insure durability of surfaces and edges, the concrete should receive at least the length of protection recommended for Category 2 (Line 2) in Table 5.3. It is seen that for Category 2, the length of the protection period in Table 5.3 is the same as the requirement for protection against early-age freezing given in Line 2 of Table 5.1. Thus, for this service category, only protection against early-age freezing is necessary.

5.3.3 *Category 3: Partial load, exposed* — The third category includes structures exposed to the weather that may be subjected to small, early-age loads compared

---

*Since non-air-entrained concrete should not be used where freezing and thawing occur, this concrete is not covered in the recommendations. However, the limited durability potential of non-air-entrained concrete is best achieved by using a protection period that is at least twice that indicated in Table 5.1.

Table 5.5—Maximum allowable temperature drop during first 24 hr after end of protection period

Section size, minimum dimensions, in. (mm)			
< 12 in. (< 300 mm)	12 to 36 in. (300 to 900 mm)	36 to 72 in. (900 to 1800 mm)	> 72 in. (> 1800 mm)
50 F (28 C)	40 F (22 C)	30 F (17 C)	20 F (11 C)

with their design strengths and will have an opportunity for additional strength development prior to the application of design loads. In such cases, the concrete should have at least the length of protection recommended for Category 3 in Table 5.3.

5.3.4 *Category 4: Full load* — This category includes structural concrete requiring temporary construction supports to safely resist construction loads. Protection requirements for this category are discussed in Chapter 6.

5.4 — Stripping of forms

During cold weather, protection afforded by forms, except those made of steel, is often of great significance. In heated enclosures, forms serve to evenly distribute the heat. In many cases, if suitable insulation or insulated forms are used, the forms, including those made of steel, would provide adequate protection without supplemental heating. Thus it is often advantageous to keep forms in place for at least the required minimum period of protection. However, an economical construction schedule often dictates their removal at the earliest practicable time. In such cases, forms can be removed at the earliest age that will not cause damage or danger to the concrete. Refer to Chapter 6 and ACI 347 for additional information on form removal.

If wedges are used to separate forms from young concrete, they should be made of wood. Usually, if the concrete is sufficiently strong, corners and edges will not be damaged during stripping. The minimum time before stripping can best be determined by experience, since it is influenced by several job factors, including type and amount of cement and other aspects of the concrete mixture, curing temperature, type of structure, design of forms, and skill of workers. After removal of forms, concrete should be covered with insulating blankets or protected by heated enclosures for the time recommended in Table 5.3. If internal heating by embedded electrical coils is used, concrete should be covered with an impervious sheet and heating continued for the recommended time.

In the case of retaining walls, basement walls, or other structures where one side could be subjected to hydrostatic pressure, hasty removal of forms while the concrete is still relatively young may dislodge the form ties and create channels through which water can flow.

5.5 — Temperature drop after removal of protection

At the end of the protection period, concrete should be cooled gradually to reduce crack-inducing differential strains between the interior and exterior of the structure. The temperature drop of concrete surfaces should not exceed the rates indicated in Table 5.5. This can be accomplished by slowly reducing sources of heat, or by allowing insulation to remain until the concrete has essentially reached equilibrium with the mean ambient temperatures. Insulated forms, however, can present some difficulties in lowering the surface temperatures. Initial loosening of forms away from the concrete and covering with polyethylene sheets to allow some air circulation can alleviate the problem. As shown in Table 5.5, the maximum allowable cooling rates for surfaces of mass concrete are lower than for thinner members.

5.6 — Allowable temperature differential

Although concrete should be cooled to ambient temperatures to avoid thermal cracking, a temperature differential may be permitted when protection is discontinued. For example, Fig. 5.6 can be used to determine the maximum allowable difference between the concrete temperature in a wall and the ambient air temperature (winds not exceeding 15 mph [24 km/h]). These curves compensate for the thickness of the wall and its shape restraint factor, which is governed by the ratio of wall length to wall height.

CHAPTER 6 — PROTECTION FOR STRUCTURAL CONCRETE REQUIRING CONSTRUCTION SUPPORTS

6.1 — Introduction

For structural concrete, where a considerable level of design strength must be attained before safe removal of forms and shores is permitted, additional protection time must be provided beyond the minimums given in Table 5.1, since these minimum times are not sufficient to allow adequate strength gain. The criteria for removal of forms and shores from structural concrete should be based on the in-place strength of the concrete rather than on an arbitrary time duration. The recommendations in this chapter are based on job conditions meeting the requirements listed in Section 6.10.

6.2 — Tests of field-cured specimens

One method used to verify attainment of sufficient in-place strength before support is reduced, changed, or removed, and before curing and protection are discontinued, is to cast at least six field-cured test specimens from the last 100 yd^3 (75 m^3) of concrete. However, at least three specimens should be cast for each 2 hr of the entire placing time, or for each 100 yd^3 (75 m^3) of concrete, whichever provides the greater number of specimens. The specimens should be made in accordance with ASTM C 31, following the procedures given for "Curing Cylinders for Determining Form Removal Time or When a Structure May be Put into Service." The specimens should be protected immediately from the cold weather until they can be placed under the same protection provided for the parts of the structure they represent. After demolding, the cylinders should

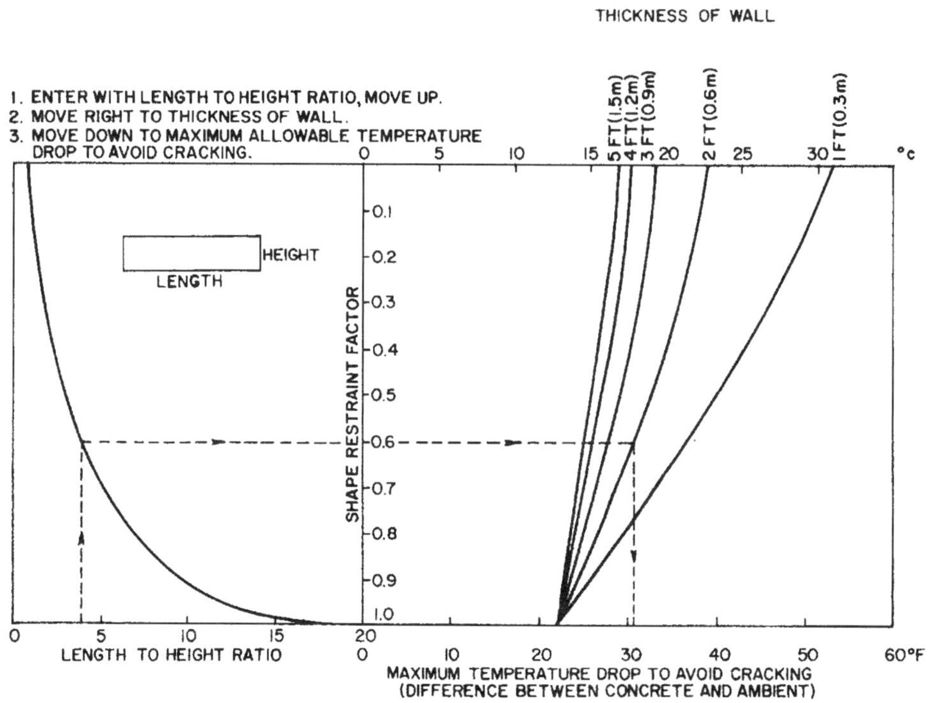

Fig. 5.6—Graphical determination of safe differential temperature for walls (Mustard and Ghosh 1979)

be capped and tested in accordance with the applicable sections of ASTM C 31 and ASTM C 39.

For flatwork, field-cured test specimens can be obtained by using special cylindrical molds that are positioned in the formwork and filled during the placement of concrete in the structure (ASTM C 873). Since the test specimens are cured in the structure, they experience the same temperature history as the structure. When a strength determination is required, the molds are extracted from the structure and the cylinder is prepared for testing according to ASTM C 39. The holes remaining in the structure would be filled with concrete.

6.3 — In-place testing

In-place and nondestructive concrete strength testing (Malhotra 1976), when correlated with field-cured and standard-cured (ASTM C 192) cylinder test results, is another method that can be used to verify attainment of strength. These tests are performed on the concrete in the structure using portable, hand-held instruments, and they offer advantages compared with testing field-cured specimens. For example, in-place testing eliminates the difficulty of trying to prepare test specimens that truly experience the same temperature history as the concrete in the structure. Hence, they are usually preferable to testing field-cured specimens prepared according to ASTM C 31. Applicable in-place test methods include the probe penetration method (ASTM C 803) and the pullout test method (ASTM C 900). The architect/engineer should review and accept the proposed method, including appropriate correlation data, for estimating in-place strength.

6.4 — Maturity method

Since strength gain of concrete is a function of time and temperature, estimation of strength development of concrete in a structure also can be made by relating the time-temperature history of field concrete to the strength of cylinders of the same concrete mixture cured under standard conditions in a laboratory. This relationship has been established (Bergstrom 1953) by use of a maturity factor M expressed as

$$M = \sum (T - T_o) \Delta t \qquad (6\text{-}1)$$

where
M = maturity factor, deg-hr
T = temperature of concrete, deg F (C)
T_o = datum temperature, deg F (C)
Δt = duration of curing period at temperature T, hr

When concrete temperature is constant, as in laboratory curing methods, the summation sign in Eq. (6-1) is not necessary. The appropriate value for the datum temperature T_o depends on the type of cement, the type and quantity of admixture, and the range of the curing temperature. A value of 23 F (-5 C) is suggested (Carino 1984) for concrete made with Type I cement and cured within the range of 32 to 70 F (0 to 20 C). This value may not be applicable to other types of cements or to Type I cement in combination with liquid or mineral admixtures. A procedure for experimental determination of the datum temperature is given in ASTM C 1074.

6.4.1 — The principle of the maturity method is that the strength of a given concrete mixture can be related

Table 6.4.4 — Calculation of maturity factor and estimated in-place strength

1	2	3	4	5	6	7	8	9	10
			Average temperature in structure					Maturity factor	Corresponding compressive
Date	Elapsed time h, hr	Temperature in structure, F	F	C	Col. 5 + 5 C, C	Time interval h, hr	Col. 6 x Col. 7, C hr	Σ Col. 8, C hr	strength, psi
Sept. 1	0	50							
			50	10	15	12	180		
	12	50						180	—
			50	10	15	12	180		
2	24	50						360	—
			48	8.9	13.9	6	83		
	30	46						443	400
			47	8.3	13.3	18	240		
3	48	48						683	1080
			47	8.3	13.3	12	160		
	60	46						843	1400
			45	7.2	12.2	12	146		
4	72	44						989	1600
			43	6.1	11.1	96	1065		
8	168	42						2054	2600
			42	5.6	10.6	72	763		
11	240	42						2817	3100
			42	5.6	10.6	72	763		
14	312	42						3580	3400

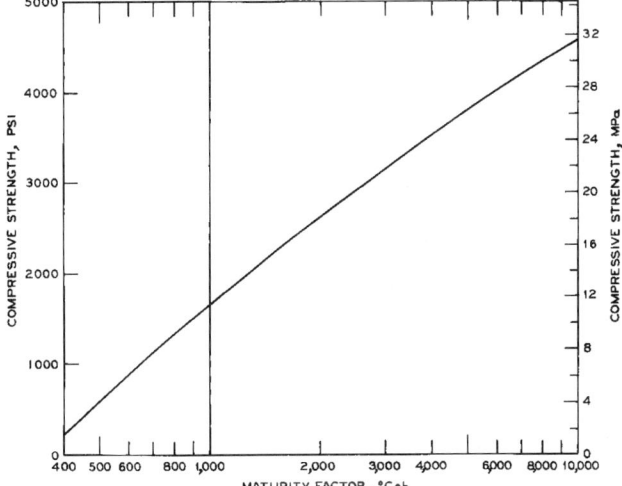

Fig. 6.4.4—Example of a strength-maturity factor relationship for laboratory-cured cylinders (22.8 C)

to the value of the maturity factor. To use this technique, a strength versus maturity factor curve is established by performing compressive strength tests at various ages on a series of cylinders made with concrete similar to that which will be used in construction. The specimens are usually cured at room temperature and the temperature history of the concrete is recorded to compute the maturity factor at the time of testing. The average cylinder strengths and corresponding maturity factors at each test age are plotted, and a smooth curve is fitted to the data.

6.4.2 — The in-place strength of properly cured concrete at a particular location and at a particular time is predicted by determining the maturity factor at that time and reading the corresponding strength on the strength-maturity factor curve. The in-place maturity factor at a particular location is determined by measuring the temperature of the concrete at closely spaced time intervals and using Eq. (6-1) to sum the successive products of the time intervals and the corresponding average concrete temperature above the datum temperature.

Temperatures can be measured with expendable thermistors or thermocouples cast in the concrete. The temperature sensors should be embedded in the structure at critical locations in terms of severity of exposure and loading conditions. Electronic instruments known as maturity meters are available that permit direct and continuous determination of the maturity factor at a particular location in the structure. These instruments use a probe inserted into a tube embedded in the concrete to measure the temperature, and they automatically compute and display the maturity factor in degree-hours.

6.4.3 — Strength prediction based on the maturity factor assumes that the in-place concrete has the same strength potential as the concrete used to develop the strength-maturity factor curve. Prior to removing forms or shores, it is necessary to determine whether the in-place concrete has the assumed strength potential. This may be done by additional testing of the concrete in question, such as by testing standard-cured cylinders at early ages, by using accelerated strength tests as described in ASTM C 684, by testing field-cured cylinders for which the maturity factor has been monitored, or by using one of the in-place tests listed in 6.3.

6.4.4 Example — The following example illustrates the use of the maturity factor method:

In anticipation of cold weather, a contractor installed thermocouples at critical locations in a concrete wall placed at 9:00 A.M. on Sept. 1. A history of the strength gain for the particular concrete mixture to be used in the wall had been developed under laboratory conditions, and the strength-maturity factor curve, which is shown in Fig. 6.4.4, had been established. A record of the in-place concrete temperature was maintained as indicated in Columns 2 and 3 of Table 6.4.4.

After 3 days (72 hr), the contractor needed to know the in-place strength of the concrete in the wall. Using the temperature record, the contractor calculated the average temperature (Column 5) during the various time intervals and the cumulative maturity factors at different ages (Column 9). Based on the strength-maturity factor curve (Fig. 6.4.4), the predicted in-place

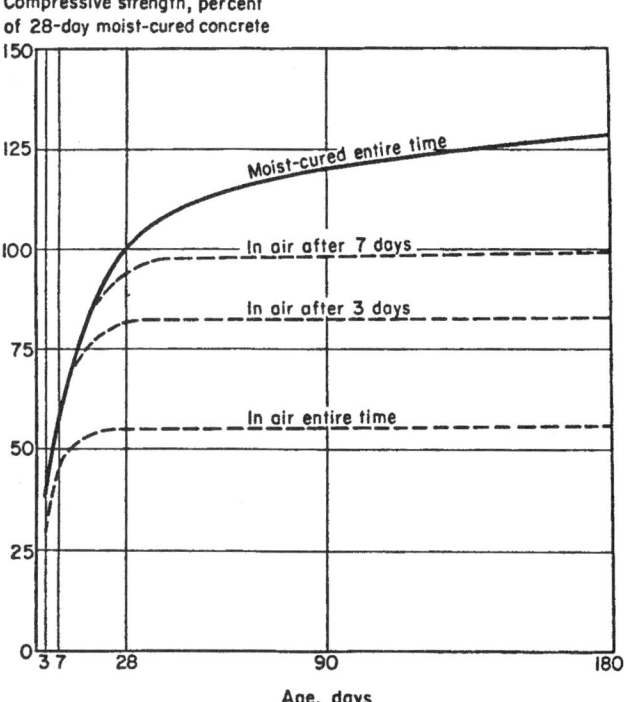

Fig. 6.5—Compressive strength of concrete dried in laboratory air after preliminary moist curing (Price 1951)

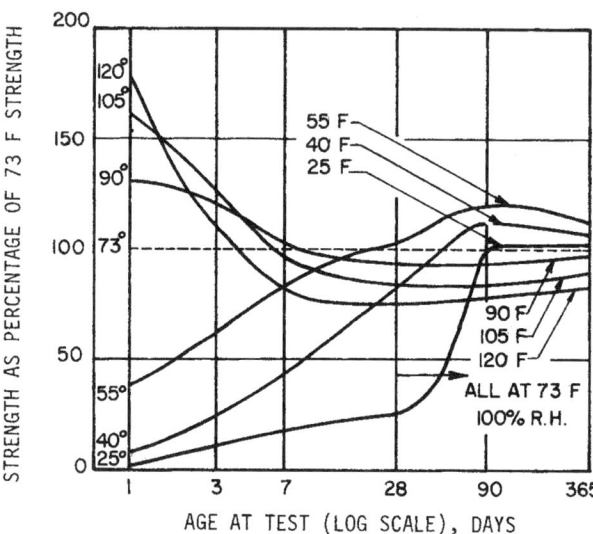

Fig. 6.6.1—Effect of temperature conditions on the strength development of concrete (Type I cement) (Kleiger 1958)

strength (Column 10) at 72 hr is 1600 psi (11.0 MPa). By continuing the procedure, strength at later ages can be predicted.

6.5 — Attainment of design strength

Generally there is little opportunity for further curing of structural concrete beyond that provided initially. Fig. 6.5 illustrates the strength development of concrete specimens removed from moist curing at various ages and subsequently exposed to laboratory air. It is seen that as the specimens dry out, strength gain ceases. For this reason, early strengths high enough to assure later attainment of design strength must be attained before temporarily supported structural concretes can be safely released from cold weather protection and exposed to freezing weather.

6.6 — Increasing early strength

The time needed for concrete to attain the strength required for safe removal of shores is influenced by many factors. Most important among them are those that affect the rate and level of strength development, such as the initial temperature of the concrete when placed, the temperature at which the concrete is maintained after placing, the type of cement, the type and amount of accelerating admixture or other admixtures used, and the conditions of protection and curing. Economic considerations may dictate an accelerated construction schedule even though the resulting concrete may be of lesser quality in terms of reduced long-term ultimate strength or increased thermal cracking. In such cases, the duration of protection may be substantially reduced by:

6.6.1 — holding the temperature during protection and curing at a higher level than that indicated in Line 1 of Table 3.1. Fig. 6.6.1 illustrates the effects of curing temperature on strength development, where strength is expressed as a percentage of the strength at the same age for curing at 73 F (23 C). Type I and III cements give somewhat higher strengths than Type II at early ages. Because of variations in the performance of any one given cement, the data in Fig. 6.6.1 should be used only as a guide.

6.6.2 — using types and compositions of cement that exhibit higher earlier strength development and by using a higher cement content with a lower water-cement ratio (see Section 9.1).

6.6.3 — using an accelerating admixture conforming to ASTM C 494, Type C (accelerating), or Type E (water-reducing and accelerating). The several items for concern mentioned in Chapter 9 should be reviewed before using calcium chloride or Type C or Type E admixtures containing calcium chloride. Due to variation in performance with different brands and types of cement, tests should be performed in advance at the anticipated curing temperature using the cement, aggregates, and admixtures proposed for use.

6.7 — Cooling of concrete

To lower the likelihood of cracking due to thermal stresses, precautions should be taken to assure gradual cooling of concrete surfaces at the termination of the protection period (see Section 5.5).

6.8 — Estimating strength development

There may be times when conservative estimates of concrete strength must be made when adequate curing and protection was provided but no actions were taken to determine the level of strength development. In such

Table 6.8 — Duration of recommended protection for percentage of standard-cured 28-day strength*

Percentage of standard-cured 28-day strength	At 50 F (10 C), days			At 70 F (21 C), days		
	Type of cement			Type of cement		
	I	II	III	I	II	III
50	6	9	3	4	6	3
65	11	14	5	8	10	14
85	21	28	16	16	18	12
95	29	35	26	23	24	20

*The data in this table were derived from concretes with strengths from 3000 to 5000 psi (20.7 to 34.4 MPa) after 28 days of curing at 70 ± 3 F (21 ± 1.7 C). The 28-day strength for each type of cement was considered as 100 percent in determining the times to reach various percentages of this strength for curing at 50 F (10 C) and 70 F (21 C). These times are only approximate, and specific values should be obtained for the concrete used on the job.

cases, Table 6.8 may be used as a guide to determine the recommended duration of curing and protection at 50 or 70 F (10 or 21 C) to achieve different percentages of the standard-cured 28-day strength.

6.9 — Removal of forms and supports

The removal of forms and supports and the placing and removal of reshores should be in accordance with the recommendations of ACI 347:

6.9.1 — The in-place strength of concrete that is required to permit removal of forms and shores should be specified by the engineer/architect.

6.9.2 — Suitable tests of field-cured concrete specimens or of concrete in place should be made (see Sections 6.2 and 6.3).

6.9.3 — Methods for evaluating the results of concrete strength tests and the minimum strength required for form and shore removal should be completely prescribed in the specifications.

6.9.4 — The results of all tests, as well as records of weather conditions and other pertinent information, should be recorded and used by the engineer/architect in deciding when to remove forms and shores.

6.9.5 — The reshoring procedure, which is one of the most critical operations in formwork, should be planned in advance and reviewed by the engineer. This operation should be performed so that large areas of new construction are not subjected to combined dead and construction loads in excess of their capacity, as determined by the in-place concrete strength at the time of form removal and reshoring.

6.10 — Requirements

The recommendations made in this chapter and in Table 6.8 are based on job conditions meeting the following requirements:

6.10.1 — The internal concrete temperature is at least 50 F (10 C) after the concrete is in place. To reduce subsequent thermal contractions, this temperature should be exceeded as little as is practicable.

6.10.2 — Facilities are available to maintain the concrete temperature throughout the structure at 50 F (10 C) or above until protection can be safely discontinued. Such facilities should incorporate, as required, the following:

   a. suitable protection from wind and loss of heat

   b. effective and sufficient heating equipment and personnel to maintain all parts of the concrete at the required temperature

   c. necessary fire protection equipment

   d. protection and heating to include the top surface of newly placed slabs or floors

   e. venting and circulation to maintain an even temperature at the top and the bottom of vertical units such as walls, piers, and columns

6.10.3 — Reshores are left in place as long as is necessary to safeguard all members of the structure. The number of tiers reshored below the tier being placed and how long such reshores remain in place should be based on reliable evidence that sufficient strength exists to safely carry the applied loads.

6.10.4 — The concrete is made with Type I, II, or III portland cement.

6.10.5 — Proper curing is used, particularly to avoid drying of the concrete in heated enclosures.

6.10.6 — Inspections are performed to check compliance with the construction requirements of ACI 301, ACI 318, and other ACI standards for construction practices.

CHAPTER 7 — MATERIALS AND METHODS OF PROTECTION

7.1 — Introduction

Concrete placed during cold weather should be maintained as close as possible to the recommended temperatures in Line 1 of Table 3.1 and for the length of times recommended in Table 5.3 unless the in-place strength has reached a previously established target value. The specific protection system required to maintain the recommended temperatures depends on such factors as the ambient weather conditions, the geometry of the structure, and the concrete mixture proportions. In some cases, it may only be necessary to cover the concrete with insulating materials and use the natural heat of hydration to maintain the temperature at the recommended levels. In more extreme cases, it may be necessary to build enclosures and use heating units to maintain the desired temperatures.

7.2 — Insulating materials

Since most of the heat of hydration of the cement is generated during the first 3 days, heating from external sources may not be required to prevent freezing of the concrete if the generated heat is retained. Heat of hydration may be retained by using insulating blankets on unformed surfaces and by using insulating forms (Tuthill, et al. 1951; Wallace 1954; Mustard and Ghosh 1979). Insulation must be kept in close contact with the concrete or the form surface to be effective. Some commonly used insulating materials include:

7.2.1 *Polystyrene foam sheets* — These may be cut to shape and wedged between the studs of the forms or glued into place.

7.2.2 *Urethane foam* — This foam may be sprayed onto the surface of forms making a continuous insulating layer. A good weather-resistant enamel should be sprayed over urethane foam to reduce water absorption and protect it from the deteriorating effect of ultraviolet rays. Urethane foam should be used with caution because it generates highly noxious fumes when exposed to fire.

7.2.3 *Foamed vinyl blankets* — These materials are pliable blankets of foamed vinyl with an extruded vinyl backing. They also may have electric wires embedded in the foam to provide additional heat. Blankets without embedded wires are available in rolls of standard widths. Heated blankets must be custom ordered.

7.2.4 *Mineral wool or cellulose fibers* — Generally, mineral wool or cellulose fibers are encased in heavy polyethylene liners and formed into large mats or rolls. The plastic liners are sometimes given a rough surface finish to reduce the risk of slipping. They may be laid flat to cover slabs or they may be draped over other structural elements.

7.2.5 *Straw* — Straw is still popular although it is not as effective as blankets or mats. Disadvantages of straw are its bulk, its flammability, and the need to protect it from moisture. Tarpaulins, polyethylene sheets, or waterproof paper must be used as a protective cover to inhibit wind convection and to keep the straw dry and in place.

7.2.6 *Blanket or batt insulation* — Commercial blanket or batt insulation must be adequately protected from wind, rain, snow, or other moisture by means of tough, moistureproof cover material because wetting will impair its insulating value. Closed-cell material is particularly advantageous because of its resistance to wetting.

7.3 — Selection of insulation

Concrete temperature records reveal the effectiveness of different amounts or kinds of insulation and of other methods of protection for various types of concrete work under different weather conditions. Using these temperature records, appropriate modifications can be made to the protection method or selected materials. Additional methods for estimating the temperatures that can be maintained by various insulation arrangements under given weather conditions have been published (Tuthill et al. 1951; Mustard and Ghosh 1979).

As was mentioned in Section 7.2, the heat of hydration is high during the first few days after placement and it then gradually decreases with age. Thus, to maintain a specified temperature throughout the protection period, the amount of required insulation is greater for a long protection period than for a shorter period. Conversely, for a given insulation system, concrete that is to be protected for a short period, such as 3 days, can be exposed to a lower ambient temperature than concrete that is to be protected for a longer period, such as 7 days.

Based on the requirements of this report, Tables 7.3.1, 7.3.2, 7.3.3, and 7.3.4 and Fig. 7.3.1, 7.3.2, 7.3.3, and 7.3.4 indicate the minimum ambient air temperatures to which concrete walls or slabs of different thicknesses may be exposed for different values of thermal resistance R, for different cement contents, and for protection periods of 3 or 7 days. For protection periods less than 3 days, the tables or figures for 3 days should be used. For protection periods between 3 and 7 days, interpolation between the values in the tables or figures should be used. For these figures and tables it is assumed that the concrete temperature as placed is 50 F (10 C).

These tables and figures can also be used to determine the required thermal resistance R under different conditions. The thickness of the chosen insulating material that is needed to obtain the required thermal resistance can be calculated using the insulation values in Table 7.3.5. The thermal resistance of the various insulating materials have been calculated under the assumption that the insulation is applied to the exterior surfaces of steel forms. When $\tfrac{3}{4}$-in. (20-mm) plywood forms are used, an R value of 0.94 (0.17) for the plywood form should be added to the R-value of the added insulation. The values shown in the tables and figures are based on the assumption that wind speeds are not greater than 15 mph (24 km/h). With higher wind speeds, the effectiveness of a given thickness of insulation diminishes. However, at a wind speed of 30 mph (48 km/h), the decrease in the effective thermal resistance amounts to less than an R-value of 0.1. Thus, for all practical purposes, the effects of wind speed may be neglected in determining the required thickness of added insulation.

Corners and edges are particularly vulnerable during cold weather. Therefore, the thickness of insulation for these parts should be about three times the thickness that is required for walls or slabs. In addition, the tables and figures are for cement having a heat of hydration similar to Type I portland cement. For Type II cement and blended hydraulic cements with moderate heat of hydration, the insulation requirements given in the tables and figures should be increased by about 30 percent. Where other types of cements or blends of cement and other cementitious materials are used,* similar proportional adjustments should be made to the amount of insulation (Tuthill et al. 1951; Mustard and Ghosh 1979). Insulation beyond the required amount should not be used because it may raise the internal temperature of the concrete above recommended levels, which will lengthen the gradual cooling period, increase thermal shrinkage, and increase the risk of cracking due to thermal shock.

7.3.1 *Example* — The following example illustrates how to determine the required thickness of insulation for a given condition.

---

*Typical heat of hydration curves for various cements can be found in "Concrete for Massive Structures," *Bulletin No. IS128T*, Portland Cement Association, Skokie, IL.

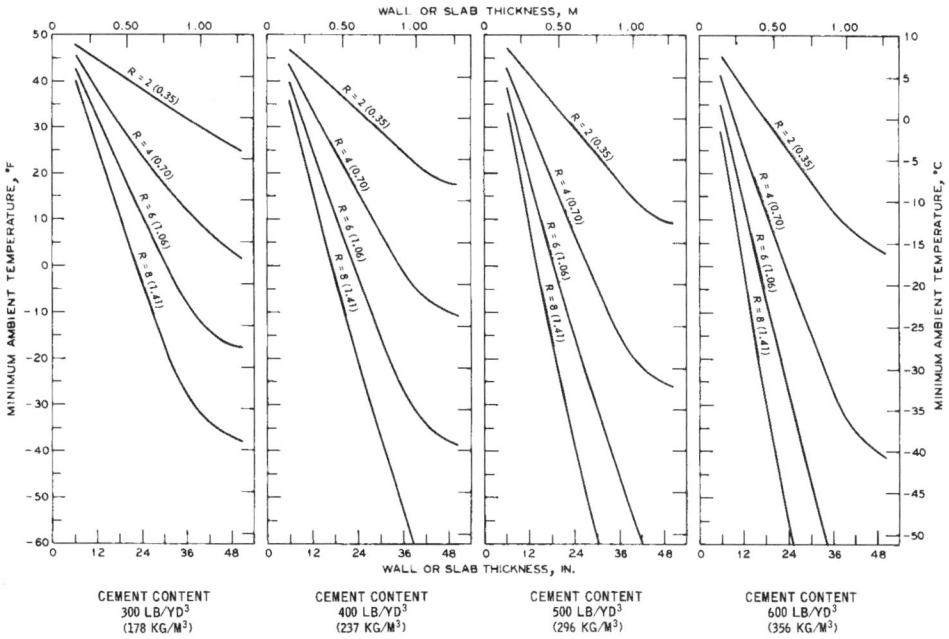

Fig. 7.3.1—Minimum exposure temperatures for concrete slabs above ground and walls as a function of member thickness, R-value, and cement content. Concrete placed and surface temperature maintained at 50 F (10 C) for 7 days

Table 7.3.1—Minimum exposure temperatures for concrete slabs above ground and walls for concrete placed and surface temperature maintained at 50 F (10 C) for 7 days

Wall or slab thickness, in. (m)	Minimum ambient air temperature, deg F (C) allowable when insulation having these values of thermal resistance R, hr·ft²·F/Btu (m²·K/W), is used			
	$R = 2$ (0.35)	$R = 4$ (0.70)	$R = 6$ (1.06)	$R = 8$ (1.41)
Cement content = 300 lb/yd³ (178 kg/m³)				
6 (0.15)	48 (9)	46 (8)	43 (6)	40 (4)
12 (0.30)	45 (7)	39 (4)	32 (0)	25 (−4)
18 (0.46)	41 (5)	31 (−1)	21 (−6)	11 (−12)
24 (0.61)	38 (3)	24 (−4)	10 (−12)	−2 (−19)
36 (0.91)	32 (0)	12 (−11)	−8 (−22)	−28 (−33)
48 (1.2)	26 (−3)	3 (−16)	−17 (−27)	−37 (−38)
60 (1.5)	26 (−3)	3 (−16)	−17 (−27)	−37 (−38)
Cement content = 400 lb/yd³ (237 kg/m³)				
6 (0.15)	47 (8)	44 (7)	40 (4)	36 (2)
12 (0.30)	43 (6)	35 (2)	26 (−3)	17 (−8)
18 (0.46)	39 (4)	25 (−4)	11 (−12)	−2 (−19)
24 (0.61)	34 (1)	16 (−8)	−2 (−19)	−20 (−29)
36 (0.91)	25 (−4)	−1 (−18)	−27 (−31)	−53 (−47)
48 (1.2)	18 (−8)	−10 (−23)	−38 (−39)	*
60 (1.5)	18 (−8)	−10 (−23)	−38 (−39)	*
Cement content = 500 lb/yd³ (296 kg/m³)				
6 (0.15)	47 (8)	43 (6)	38 (3)	33 (1)
12 (0.30)	42 (6)	31 (−1)	20 (−7)	9 (−13)
18 (0.46)	36 (2)	19 (−7)	2 (−17)	−15 (−26)
24 (0.61)	30 (−1)	7 (−14)	−16 (−27)	−39 (−39)
36 (0.91)	18 (−8)	−15 (−26)	−46 (−43)	−79 (−62)
48 (1.2)	10 (−12)	−25 (−32)	−60 (−51)	*
60 (1.5)	10 (−12)	−25 (−32)	*	*
Cement content = 600 lb/yd³ (356 kg/m³)				
6 (0.15)	46 (8)	41 (5)	35 (2)	29 (−2)
12 (0.30)	40 (4)	28 (−2)	14 (−10)	0 (−18)
18 (0.46)	33 (1)	13 (−11)	−7 (−22)	−29 (−34)
24 (0.61)	26 (−3)	−1 (−18)	−28 (−33)	−55 (−48)
36 (0.91)	12 (−11)	−27 (−31)	−66 (−54)	*
48 (1.2)	4 (−16)	−40 (−40)	*	*
60 (1.5)	4 (−16)	−40 (−40)	*	*

*<< −60 F (−51 C).

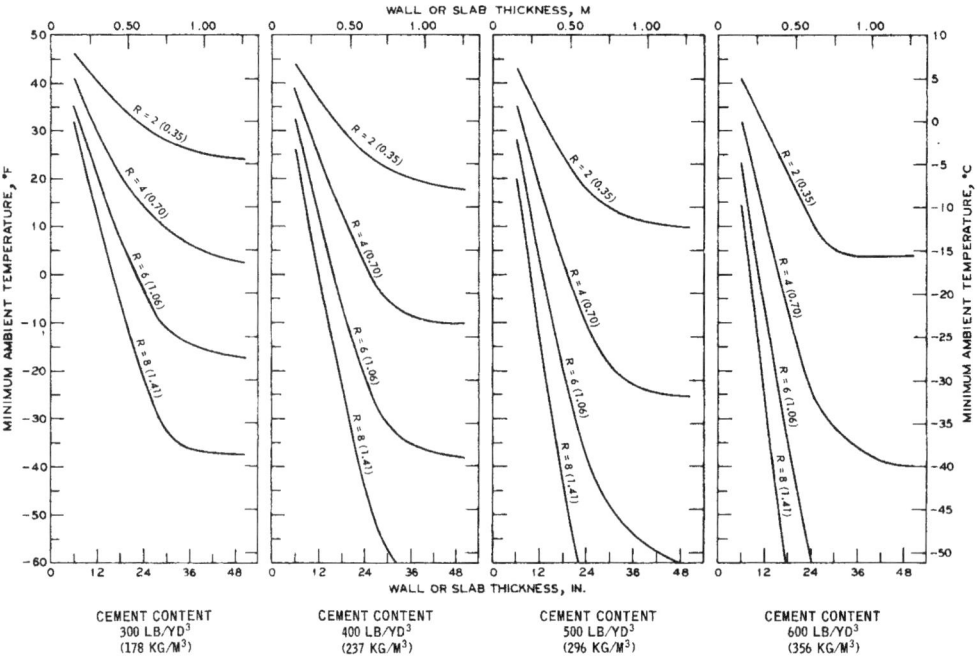

Fig. 7.3.2—Minimum exposure temperatures for concrete slabs above ground and walls as a function of member thickness, R-value, and cement content. Concrete placed and surface temperature maintained at 50 F (10 C) for 3 days

Table 7.3.2—Minimum exposure temperatures for concrete slabs above ground and walls for concrete placed and surface temperature maintained at 50 F (10 C) for 3 days

Wall or slab thickness, in. (m)	Minimum ambient air temperature, deg F (C) allowable when insulation having these values of thermal resistance R, hr·ft²·F/Btu (m²·K/W), is used			
	$R = 2$ (0.35)	$R = 4$ (0.70)	$R = 6$ (1.06)	$R = 8$ (1.41)
Cement content = 300 lb/yd³ (178 kg/m³)				
6 (0.15)	46 (8)	41 (5)	36 (2)	32 (0)
12 (0.30)	41 (5)	31 (−1)	21 (−6)	11 (−12)
18 (0.46)	36 (2)	21 (−6)	8 (−13)	−5 (−21)
24 (0.61)	31 (−1)	14 (−10)	−3 (−19)	−21 (−29)
36 (0.91)	26 (−3)	8 (−13)	−14 (−26)	−36 (−38)
48 (1.2)	26 (−3)	3 (−16)	−17 (−27)	−37 (−38)
60 (1.5)	26 (−3)	3 (−16)	−17 (−27)	−37 (−38)
Cement content = 400 lb/yd³ (237 kg/m³)				
6 (0.15)	44 (7)	38 (3)	32 (0)	26 (−3)
12 (0.30)	37 (3)	24 (−4)	12 (−11)	0 (−18)
18 (0.46)	30 (−1)	12 (−11)	−6 (−21)	−24 (−31)
24 (0.61)	25 (−4)	2 (−17)	−21 (−29)	−44 (−42)
36 (0.91)	20 (−7)	−9 (−23)	−36 (−38)	−63 (−53)
48 (1.2)	18 (−8)	−10 (−23)	−38 (−39)	*
60 (1.5)	18 (−8)	−10 (−23)	−38 (−39)	*
Cement content = 500 lb/yd³ (296 kg/m³)				
6 (0.15)	43 (6)	35 (2)	28 (−2)	20 (−7)
12 (0.30)	34 (1)	18 (−8)	3 (−16)	−12 (−24)
18 (0.46)	25 (−4)	2 (−16)	−21 (−29)	−44 (−42)
24 (0.61)	18 (−8)	−10 (−23)	−38 (−39)	−68 (−56)
36 (0.91)	12 (−11)	−23 (−31)	−60 (−51)	*
48 (1.2)	10 (−12)	−25 (−32)	*	*
60 (1.5)	10 (−12)	−25 (−32)	*	*
Cement content = 600 lb/yd³ (356 kg/m³)				
6 (0.15)	41 (5)	32 (0)	23 (−5)	14 (−10)
12 (0.30)	31 (−1)	12 (−11)	−7 (−22)	−26 (−32)
18 (0.46)	21 (−6)	−7 (−22)	−35 (−37)	−63 (−53)
24 (0.61)	11 (−12)	−24 (−31)	−59 (−51)	*
36 (0.91)	4 (−16)	−36 (−38)	*	*
48 (1.2)	4 (−16)	−40 (−40)	*	*
60 (1.5)	4 (−16)	−40 (−40)	*	*

*<< −60 F (−51 C).

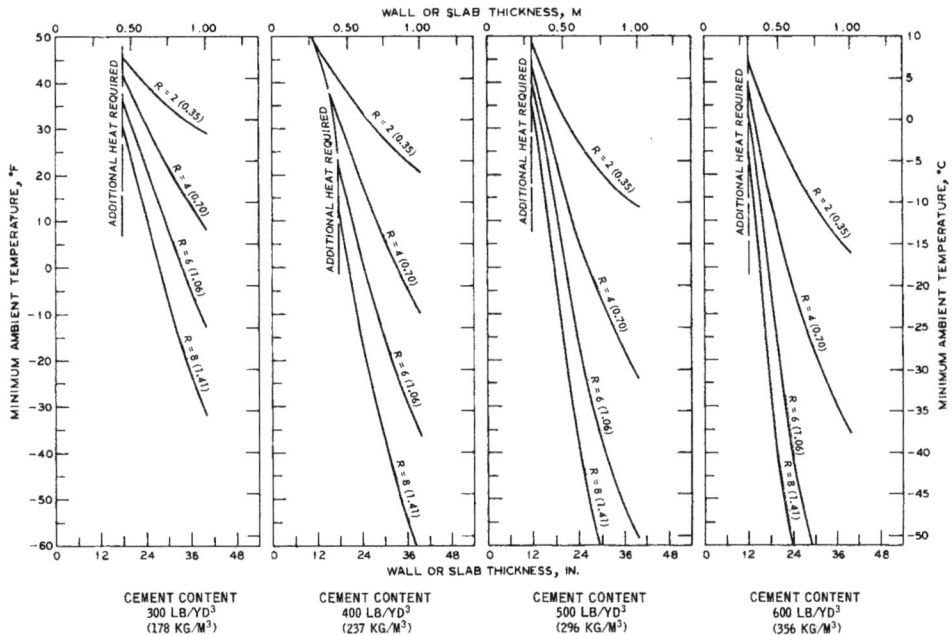

Fig. 7.3.3—Minimum exposure temperatures for concrete flatwork placed on the ground as a function of member thickness, R-value, and cement content. Concrete placed and surface temperature maintained at 50 F (10 C) for 7 days on ground at 35 F (2 C)

Table 7.3.3—Minimum exposure temperatures for concrete flatwork placed on the ground for concrete placed and surface temperature maintained at 50 F (10 C) for 7 days on ground at 35 F (2 C)

Slab thickness, in. (m)	Minimum ambient air temperature, deg F (C) allowable when insulation having these values of thermal resistance R, hr·ft²·F/Btu (m²·K/W), is used			
	$R = 2$ (0.35)	$R = 4$ (0.70)	$R = 6$ (1.06)	$R = 8$ (1.41)
Cement content = 300 lb/yd³ (178 kg/m³)				
4 (0.10)	*	*	*	*
8 (0.20)	*	*	*	*
12 (0.31)	*	*	*	*
18 (0.46)	46 (8)	42 (6)	36 (2)	30 (−1)
24 (0.61)	40 (4)	31 (−1)	22 (−6)	11 (−12)
30 (0.76)	35 (2)	22 (−6)	7 (−14)	−8 (−22)
36 (0.91)	31 (−1)	13 (−11)	−5 (−21)	−23 (−31)
Cement content = 400 lb/yd³ (237 kg/m³)				
4 (0.10)	*	*	*	*
8 (0.20)	*	*	*	*
12 (0.31)	*	*	*	50 (10)
18 (0.46)	41 (5)	32 (0)	22 (−6)	12 (−11)
24 (0.61)	35 (2)	19 (−7)	−1 (−17)	−15 (−26)
30 (0.76)	28 (−2)	8 (−13)	−14 (−26)	−36 (−38)
36 (0.91)	23 (−5)	−4 (−20)	−29 (−34)	−54 (−48)
Cement content = 500 lb/yd³ (296 kg/m³)				
4 (0.10)	*	*	*	*
8 (0.20)	*	*	*	*
12 (0.31)	48 (9)	44 (7)	40 (4)	36 (2)
18 (0.46)	36 (2)	22 (−6)	8 (−13)	−6 (−21)
24 (0.61)	28 (−2)	6 (−14)	−16 (−27)	−38 (−39)
30 (0.76)	22 (−6)	−7 (−22)	−36 (−38)	−64 (−53)
36 (0.91)	16 (−9)	−18 (−28)	−50 (−46)	†
Cement content = 600 lb/yd³ (356 kg/m³)				
4 (0.10)	*	*	*	*
8 (0.20)	*	*	*	*
12 (0.31)	44 (7)	38 (3)	32 (0)	26 (−3)
18 (0.46)	31 (−1)	14 (−10)	−5 (−21)	−24 (−31)
24 (0.61)	22 (−6)	−5 (−21)	−32 (−36)	−61 (−52)
30 (0.76)	14 (−10)	−19 (−28)	−67 (−55)	†
36 (0.91)	7 (−14)	−30 (−34)	†	†

* > 50 F (10 C): additional heat required.
† < < −60 F (−51 C).

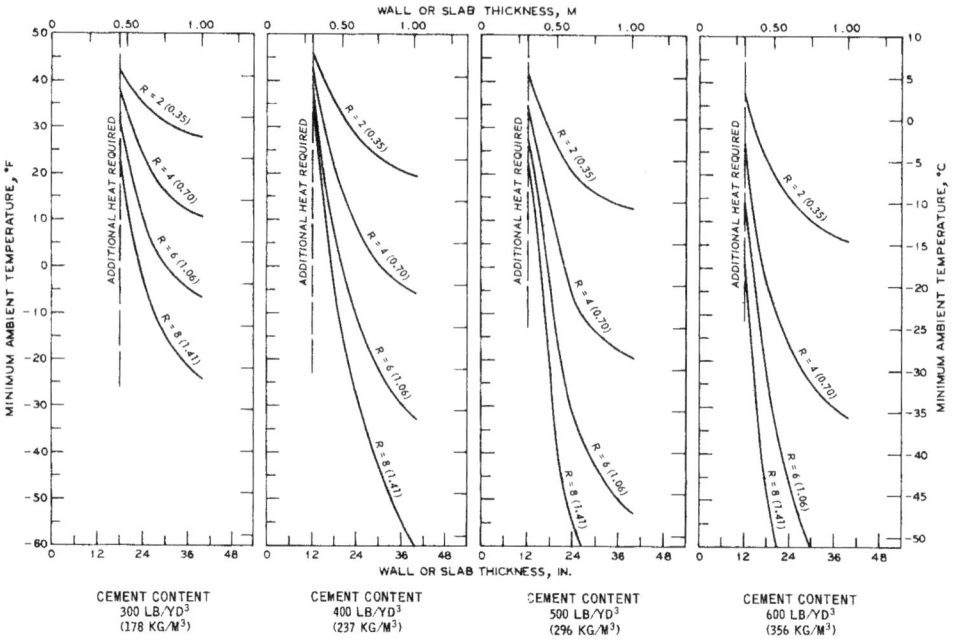

Fig. 7.3.4—Minimum exposure temperatures for concrete flatwork placed on the ground as a function of member thickness, R-value, and cement content. Concrete placed and surface temperature maintained at 50 F (10 C) for 3 days on ground at 35 F (2 C)

Table 7.3.4—Minimum exposure temperatures for concrete flatwork placed on the ground for concrete placed and surface temperature maintained at 50 F (10 C) for 3 days on ground at 35 F (2 C)

Slab thickness, in. (m)	Minimum ambient air temperature, deg F (C) allowable when insulation having these values of thermal resistance R, hr·ft²·F/Btu (m²·K/W), is used			
	$R = 2$ (0.35)	$R = 4$ (0.70)	$R = 6$ (1.06)	$R = 8$ (1.41)
Cement content = 300 lb/yd³ (178 kg/m³)				
4 (0.10)	*	*	*	*
8 (0.20)	*	*	*	*
12 (0.31)	*	*	*	*
18 (0.46)	42 (6)	38 (3)	32 (0)	26 (−3)
24 (0.61)	37 (3)	25 (−4)	11 (−12)	−3 (−19)
30 (0.76)	31 (−1)	15 (−9)	−1 (−18)	−17 (−27)
36 (0.91)	31 (−1)	12 (−11)	−5 (−21)	−22 (−30)
Cement content = 400 lb/yd³ (237 kg/m³)				
4 (0.10)	*	*	*	*
8 (0.20)	*	*	*	*
12 (0.31)	46 (8)	44 (7)	42 (6)	40 (4)
18 (0.46)	36 (2)	22 (−6)	8 (−13)	−6 (−21)
24 (0.61)	28 (−2)	9 (−13)	−10 (−23)	−29 (−34)
30 (0.76)	21 (−6)	0 (−18)	−21 (−29)	−42 (−41)
36 (0.91)	21 (−6)	−4 (−20)	−29 (−34)	−50 (−46)
Cement content = 500 lb/yd³ (296 kg/m³)				
4 (0.10)	*	*	*	*
8 (0.20)	*	*	*	*
12 (0.31)	42 (6)	36 (2)	30 (−1)	24 (−4)
18 (0.46)	30 (−1)	12 (−11)	−6 (−21)	−22 (−30)
24 (0.61)	21 (−6)	−5 (−21)	−31 (−35)	−50 (−46)
30 (0.76)	16 (−9)	−10 (−23)	−42 (−41)	−74 (−59)
36 (0.91)	16 (−9)	−18 (−28)	−50 (−46)	†
Cement content = 600 lb/yd³ (356 kg/m³)				
4 (0.10)	*	*	*	*
8 (0.20)	*	*	*	*
12 (0.31)	38 (3)	26 (−3)	14 (−10)	2 (−17)
18 (0.46)	24 (−4)	0 (−18)	−24 (−31)	−48 (−44)
24 (0.61)	14 (−10)	−16 (−27)	−46 (−43)	−82 (−63)
30 (0.76)	10 (−12)	−20 (−29)	−62 (−52)	†
36 (0.91)	7 (−14)	−30 (−34)	†	†

* > 50 F (10 C): additional heat required.
† << −60 F (−51 C).

Table 7.3.5—Thermal resistance of various insulating materials

Insulating material	Thermal resistance R for these thicknesses of material*	
	1 in., hr·ft²·F/Btu	10 mm, m²·K/W
Boards and slabs		
Expanded polyurethane (R-11 exp.)	6.25	0.438
Expanded polystyrene extruded (R-12 exp.)	5.00	0.347
Expanded polystyrene extruded, plain	4.00	0.277
Glass fiber, organic bonded	4.00	0.277
Expanded polystyrene, molded beads	3.57	0.247
Mineral fiber with resin binder	3.45	0.239
Mineral fiber board, wet felted	2.94	0.204
Sheathing, regular density	2.63	0.182
Cellular glass	2.63	0.182
Laminated paperboard	2.00	0.139
Particle board (low density)	1.85	0.128
Plywood	1.25	0.087
Blanket		
Mineral fiber, fibrous form processed from rock, slag, or glass	3.23	0.224
Loose fill		
Wood fiber, soft woods	3.33	0.231
Mineral fiber (rock, slag, or glass)	2.50	0.173
Perlite (expanded)	2.70	0.187
Vermiculite (exfoliated)	2.20	0.152
Sawdust or shavings	2.22	0.154

*Values from *ASHRAE Handbook of Fundamentals*, 1977, American Society of Heating, Refrigerating, and Air-Conditioning Engineers, New York.

Problem — A contractor anticipates having to place an 18 in. (0.46 m) thick concrete wall when the ambient temperature will be about 0 F (−18 C). The concrete will have a cement content of 500 lb/yd³ (296 kg/m³). There will be no early-age strength requirements for the concrete; therefore, a 3-day protection period will be used. The forms are made of ¾-in. plywood. The contractor would like to use plain expanded polystyrene boards for insulation. What should be the thickness of the polystyrene boards?

Solution — Table 7.3.2 or Fig. 7.3.2 may be used to solve the problem. According to Table 7.3.2, an R-value of 4 (0.7) is sufficient for an ambient temperature of 2 F (−16 C), which will be assumed to be close enough to the expected ambient temperature. Since the plywood forms will provide some of this insulation, the required added insulation must have an R-value of $4 - 0.94 = 3.06$ $(0.70 - 0.17 = 0.53)$. Referring to Table 7.3.5, it is seen that a 1 in. (25 mm) plain polystyrene board has an R-value of 4 (0.7). Therefore, ¾-in. (19 mm) thick polystyrene boards will provide the needed additional insulation.

7.4 — Enclosures

Enclosures are probably the most effective means of protection, but they are also the most expensive. The need for enclosures is dependent on the nature of the structure and the weather conditions (wind and snow). Experience has shown that they are generally required for placing operations when the air temperature is less than −5 F (−20 C). Builders attempting to place concrete at such low temperatures without enclosures will encounter manpower and equipment difficulties that will result in inferior constructions.

Enclosures block the wind, keep out the cold air, and conserve heat. They can be made with any suitable material such as wood, canvas, building board, or plastic sheet. Enclosures made with flexible materials are less expensive and easier to build and remove. Enclosures built with rigid materials are more effective in blocking wind and maintaining perimeter temperatures. Enclosures should be capable of withstanding wind and snow loads and be reasonably airtight. Sufficient space should be provided between the concrete and the enclosure to permit free circulation of the warmed air. Sufficient headroom should be provided, so that workers can work efficiently.

Heat can be supplied to the enclosures by live steam, forced hot air, or combustion heaters of various types. Although live steam heating provides an ideal curing environment, it offers less than ideal working conditions and can cause icing problems around the perimeter of the enclosure. Heaters and ducts should be positioned so as not to cause areas of overheating or drying of the concrete surface. During the protection period, concrete surfaces should not be exposed to air at more than 20 F (11 C) above the minimum placement temperatures given in Line 1 of Table 3.1 unless higher values are requird by an accepted curing method.

Combustion heaters should be vented to prevent reaction of the carbon dioxide in the exhaust gases with the exposed surfaces of newly placed concrete. Carbon dioxide reacts with calcium hydroxide in the concrete and forms calcium carbonate (Kauer and Freeman 1955) resulting in a weak dusty surface.

Combustion heaters called vented heaters have a heat exchanger in the firebox so that none of the products of combustion are blown into the heated area. The exhaust gases are vented from the firebox to the outside atmosphere. Even though they use energy inefficiently compared to direct firebox heaters, vented heaters must be used to maintain concrete floor finish quality and for safety. The fact that a conventional combustion heater is located outside the enclosed concreting area and hot air is blown into the enclosure through ducts does not mean that the heater is vented, since the hot air may contain carbon dioxide.

Operation of combustion heaters should be supervised continuously, and adequate fire fighting equipment should be available at the job site at all times.

7.5 — Internal electric heating

Concrete can be heated internally by using embedded coiled and insulated electrical resistors. Low voltage current is passed through the coils, which are embedded near the surfaces of the section in a predetermined pattern. Internal concrete temperature may be raised to any required level by selecting the appropriate spacing or pitch of the coils. Gradual cooling may be controlled by intermittently interrupting the current passing through the coils. Heating usually begins after a presetting period of 4 to 5 hr, depending on the setting characteristics of the concrete. Therefore, insulated forms are required only to prevent freezing of concrete during the presetting period and to minimize heat dissipation from surfaces where coils are not used. Moisture loss due to evaporation from unformed surfaces should be prevented by covering the surfaces with plastic sheets. Concrete temperatures must be monitored so that the recommended values are not exceeded.

7.6 — Covering after placement

Tarpaulins or other easily movable coverings, supported on sawhorses or on other framework, should follow closely the finishing of the concrete. The tarpaulins should be arranged so that heated air can circulate freely on both the top and, where exposed, the bottom of the slab. Layers of insulating material placed directly on the concrete are also effective in preventing freezing. This protection is particularly important in the case of lightweight structural concrete. The lower rate of heat diffusion of lightweight concrete allows more rapid freezing of surfaces to occur compared to normal weight concrete.

7.7 — Temporary removal of protection

Housing and enclosures should be left in place for the entire recommended protection period. Sections may be temporarily removed to permit placing additional forms or concrete, but scheduling of this work must insure that the previously placed concrete does not freeze. Sections that are removed should be replaced as soon as forms or concrete are in their final position. The time during which protection is temporarily removed should not be considered as part of the protection period. Time lost from the required period of protection should be made up with twice the number of lost degree-hours before protection is discontinued. For example, if protection was temporarily removed for 6 hr and the surface temperature dropped 15 F (8 C) below the minimum value given in Table 3.1, the deficiency in protection would be 90 F-hr (48 C-hr). Therefore, the protection period should be extended for 180 F-hr (96 C-hr).

7.8 — Insulated forms

When insulated forms are used in addition to heated enclosures, the interior temperature as well as the surface temperature of the concrete should be monitored to insure that the concrete is not heated more than is necessary. This recommendation applies particularly to mass concrete.

CHAPTER 8 — CURING REQUIREMENTS AND METHODS

8.1 — Introduction

Newly placed concrete must be protected from drying so that adequate hydration can occur. Normally, measures must be taken to prevent evaporation of moisture from concrete. During cold weather, when the air temperature is below 50 F (10 C), atmospheric conditions in most areas will not cause excessive drying. However, new concrete is vulnerable to freezing when it is in a critically saturated condition. Therefore, if concrete has been saturated during the protection period, it should be allowed to undergo some drying before being exposed to freezing temperatures.

8.2 — Curing during the protection period

Concrete exposed to cold weather is not likely to dry at an undesirable rate; however, this may not be true for concrete that is being protected from cold weather. As long as forms remain in place, concrete surfaces adjacent to the forms will retain adequate moisture. On the other hand, exposed horizontal surfaces, particularly finished floors, are prone to rapid drying in a heated enclosure.

When concrete that is warmer than 60 F (16 C) is exposed to air at 50 F (10 C) or higher, it is essential that measures be taken to prevent drying. The preferred technique is to use steam for both heating and preventing excessive evaporation. When dry heating is used, the concrete should be covered with an approved impervious material or a curing compound meeting the requirements of ASTM C 309, or it may be water cured. However, water curing is not recommended and is the least desirable method, since during periods of subfreezing temperatures it produces icing problems where water runs out of the enclosure or where there is a poor seal. It also increases the likelihood of the concrete freezing in a nearly saturated condition when protection is removed. If water or steam curing are used, they should be terminated 12 hr before the end of the temperature protection period, and the concrete should be permitted to dry prior to and during the period of gradual adjustment to ambient cold weather conditions, as discussed in Section 5.5.

When the air temperature within the enclosure has fallen to 50 F (10 C), the concrete can be exposed to the air provided the relative humidity is not less than 40 percent. During very cold weather, it is always necessary to add moisture to heated air to maintain this humidity. For example, if the outside temperature is 10 F (−12 C), the relative humidity of the air within the heated enclosure will be less than 20 percent if no moisture is added.

8.3 — Curing following the protection period

Following the removal of the temperature protection, it is usually not necessary to provide measures to prevent excessive drying as long as the air temperature remains below 50 F (10 C), except when concrete is placed in extremely arid regions. If a curing compound is applied during the first period of above-freezing temperature after protection is removed, the need to conduct further curing operations if the temperature should rise above 50 F (10 C) is eliminated.

The severity of drying is dependent on four factors: (1) the temperature of the concrete, (2) the temperature of the air, (3) the wind speed, and (4) the relative humidity of the air. For example, drying will be excessive if concrete at 50 F (20 C) is exposed to air having a temperature of 50 F (10 C) and a relative humidity less than 40 percent. As the air temperature decreases, the air requires a higher relative humidity to prevent excessive drying. For example, if concrete at 50 F (10 C) is exposed to air having a temperature of 40 F (5 C), the air must have a relative humidity greater than 60 percent to prevent excessive drying. However, when the concrete temperature has dropped to 40 F (5 C), an ambient air temperature of 40 F (5 C) with a relative humidity of 11 percent can be tolerated. For air temperatures above 50 F (10 C), the rate of drying increases rapidly.

If excessive drying is anticipated, concrete may be water cured when no freezing is expected. Otherwise, the use of curing compounds or an impervious cover is the preferred procedure. During cold weather periods when freezing occurs, occasional peak temperatures above 50 F (10 C) should not be a cause of concern. However, when temperatures above 50 F (10 C) occur during more than half of any 24-hr period for three consecutive days, the concrete should no longer be regarded as cold weather concrete and normal curing practice should apply (see ACI 308).

CHAPTER 9 — ACCELERATION OF SETTING AND STRENGTH DEVELOPMENT

9.1 — Introduction

If proper precautions are taken, accelerating admixtures, Type III cement (high early strength), or additional cement can be used to shorten the times needed to achieve setting and required strength. The reduction in setting time and the acceleration of strength gain often result in savings due to a shorter protection period, faster reuse of forms, earlier removal of shores, or less labor in finishing flatwork (see ACI 302). The strength development of concrete in massive structures, in which internal temperature rise is critical, should not be accelerated.

Methods used to obtain high early strength concrete increase the heat of hydration, which can be favorable in some instances. When resistance to sulfate attack is not a concern, cement having higher percentages of tricalcium silicate and tricalcium aluminate may be advantageous in cold weather because these compounds contribute to earlier strength development and higher heat of hydration. Cements of the same type and even of the same brand name, however, can have wide variations in their setting times and rates of strength development. Tests on concrete made with the cements to be used for a particular job are recommended to show which of the alternatives will produce the desired properties. Additional information on the subject of accelerating setting and strength development may be found in ACI 212.

9.2 — Calcium chloride as an accelerating admixture

9.2.1 — Calcium chloride is a popular and widely used accelerating admixture that reduces the setting time and increases the rate of early-age strength development. The use and effects of calcium chloride are discussed in ACI 212, ACI 201, and Shideler (1952). The maximum chloride ion content of concrete is prescribed in ACI 318, and these limits govern where the code is in use.

Calcium chloride should not be used under certain conditions; these restrictions are as follows:

  a. Calcium chloride should not be used in prestressed concrete because of its potential for accelerating the rate of stress corrosion (Monfore and Verbeck 1960).

  b. The presence of chlorides has been associated with galvanic corrosion between galvanized and plain steel when galvanizing is used on permanent forms for roof decks or embedded parts. Calcium chloride should not be used in such construction (Wright and Jencks 1963).

  c. Galvanic corrosion of metals embedded in concrete is intensified by addition of calcium chloride to the concrete. Examples of this occurrence are as follows (Wright and Jencks 1963):

   1. When aluminum and steel sheets embedded in concrete were connected externally, the galvanic corrosion of the aluminum was proportional to the concentration of calcium chloride.

   2. Where large slabs of reinforced concrete were cast with aluminum conduit electrically connected to embedded steel, the results showed that calcium chloride promoted galvanic corrosion.

  d. Calcium chloride should not be used where sulfate resisting concrete is required.

  e. Calcium chloride has been found to increase the expansion caused by cement alkali-silica reaction (Shideler 1952).

  f. Varying the calcium chloride content of concrete in a single structural unit because of changes in weather conditions may create concentration cells and, in the presence of moisture, may promote corrosion of reinforcing steel in the areas of higher chloride ion concentration. This practice is dangerous and should not be permitted in structures where continuity of reinforcement exists.

  g. Even where a uniform dosage of calcium chloride has been used throughout a structural element, galvanic cells can develop within the reinforcing network due solely to differences in moisture contents. The relatively dry portions can act as cathodes, due to greater amounts of available oxygen, and the relatively damp portions can act as anodes where corrosion occurs. The degree of dampness required to support anodic activity has not been established definitively; it is believed that relative humidities above 70 percent may be sufficient (Tuutti 1982).

9.2.2 — Calcium chloride, and most other chemical admixtures, added to the concrete mixture in permissible quantities will not lower the freezing point of water in concrete to any significant degree. To avoid misplaced confidence in such a practice, and to avoid use of harmful materials, any attempt to lower the freezing point of the water in concrete by the use of calcium chloride should not be permitted.

9.3 — Other accelerating admixtures

9.3.1 — Some water-reducing accelerating admixtures, conforming to Type E in ASTM C 494, have been found to accelerate setting and strength gain at ambient temperatures of 50 F (10 C) and below, and also to reduce the required water content of the mixture. Some Type E admixtures contain small percentages of calcium chloride, usually amounting to less than 0.2 percent by weight of the cement when used at recommended dosage rates (see Section 9.3.2). This amount by itself has a limited effect on early strength development of the concrete. Tests indicate, however, that other Type E water-reducing accelerating admixtures, meeting the requirements of ASTM C 494, will substantially improve the 24-hr strength of concrete maintained at 50 F (10 C). The strength of concrete containing these admixtures approaches the strength obtained by using 2 percent calcium chloride and will be appreciably greater than for some, but not all, concretes made with Type III cements. The data also indicate that water-reducing accelerating admixtures may reduce setting time as well as produce substantial strength increases at all later ages.

9.3.2 — If adequate information or past performance records are not available, tests should be made to evaluate the effect of a particular admixture or admixtures on the concrete for the job. These tests should be carried out at the expected job temperatures using the materials approved for the job. The manufacturers of many proprietary admixtures claim accelerated setting and strength development properties. It should be determined whether the admixture under consideration contains chloride. If chloride is present, the percentage of chloride, by weight of cement, that would be introduced into the concrete if the admixture were to be used should be determined and compared with the permissible limits given in ACI 201, or ACI 318. The admixture should not be used if the limits are exceeded.

Not all accelerating admixtures contain added chlorides. If job requirements prohibit the use of admixtures containing added chloride, noncorrosive accelerating admixtures are available.

CHAPTER 10 — REFERENCES

10.1 — Recommended references

The documents of the various standards-producing organizations referenced in this report are listed with their serial designation, including year of adoption or revision. The documents listed were the latest effort at the time this report was revised. Since some of these documents are revised frequently, usually in minor detail only, the user of this report should check directly with the sponsoring group if it is desired to refer to the latest revision.

American Concrete Institute

201.2R-77	Guide to Durable Concrete
207.1R-70 (Reapproved 1980)	Mass Concrete for Dams and Other Massive Structures
211.1-81	Standard Practice for Selecting Proportions for Normal, Heavyweight, and Mass Concrete
212.2R-81	Guide for Use of Admixtures in Concrete
301-84	Specifications for Structural Concrete for Buildings
302.1R-80	Guide for Concrete Floor and Slab Construction
306.1-87	Standard Specification for Cold Weather Concreting
308-81	Standard Practice for Curing Concrete
318-83	Building Code Requirements for Reinforced Concrete
347-78 (Reapproved 1984)	Recommended Practice for Concrete Formwork

ASTM

C 31-87a	Standard Practice for Making and Curing Concrete Test Specimens in the Field
C 39-86	Standard Test Method for Compressive Strength of Cylindrical Concrete Specimens
C 192-81	Standard Method of Making Curing Concrete Test Specimens in the Laboratory
C 309-81	Standard Specification for Liquid Membrane-Forming Compounds for Curing Concrete
C 494-82	Standard Specification for Chemical Admixtures for Concrete
C 684-81	Standard Method of Making, Accelerated Curing, and Testing of Concrete Compression Test Specimens
C 803-82	Standard Test Method for Penetration Resistance of Hardened Concrete
C 873-85	Standard Test Method for Compressive Strength of Concrete Cylinders Cast in Place in Cylindrical Molds
C 900-87	Standard Test Method for Pullout Strength of Hardened Concrete
C 1064-86	Standard Test Method for Temperature of Freshly Mixed Portland-Cement Concrete
C 1074-87	Standard Practice for Estimating Concrete Strength by the Maturity Method

These publications are available from the following organizations:

American Concrete Institute
P.O. Box 9094
Farmington Hills, Mich. 48333-9094

ASTM International
100 Barr Harbor Dr.
West Conshohocken, Pa. 19428

10.2 — Cited references

Bergström, Sven C., Dec. 1953, "Curing Temperature, Age, and Strength of Concrete," *Magazine of Concrete Research* (London), V. 5, No. 14, pp. 61-66.

Carino, Nicholas, J., Winter 1984, "The Maturity Method—Theory and Application," *Cement, Concrete, and Aggregates,* V. 6, No. 2, pp. 61-73.

Hoff, George C., and Buck, Alan D., Sept.-Oct. 1983, "Considerations in the Prevention of Damage to Concrete Frozen at Early Ages," ACI JOURNAL, *Proceedings* V. 80, No. 5, pp. 371-376.

Kauer, J. A., and Freeman, R. L., Dec. 1955, "Effect of Carbon Dioxide on Fresh Concrete," ACI JOURNAL, *Proceedings* V. 52, No. 4, pp. 447-454.

Klieger, Paul, June 1958, "Effect of Mixing and Curing Temperature on Concrete Strength," ACI JOURNAL, *Proceedings* V. 54, No. 12, pp. 1063-1082.

Malhotra, V. M., 1976, *Testing Hardened Concrete: Nondestructive Methods,* ACI Monograph No. 9, American Concrete Institute/Iowa State University Press, Detroit, 204 pp.

Malhotra, V. M., and Berwanger, Carl, 1973, "Effect of Below Freezing Temperatures on Strength Development of Concrete," *Behavior of Concrete Under Temperature Extremes,* SP-39, American Concrete Institute, Detroit, pp. 37-58.

Monfore, G. E., and Verbeck, G. J., Nov. 1960, "Corrosion of Prestressed Wire in Concrete," ACI JOURNAL, *Proceedings* V. 57, No. 5, pp. 491-516.

Mustard, J. Neil, and Ghosh, Ram S., Jan. 1979, "Minimum Protection and Thermal Stresses in Winter Concreting," *Concrete international: Design & Construction,* V. 1, No. 1, pp. 96-101.

Petersons, N., 1966, "Concrete Quality Control and Authorization of Ready Mixed Concrete Factories in Sweden," *Reprint* No. 43, Swedish Cement and Concrete Research Institute, Stockholm.

Powers, T. C., Mar. 1962, "Prevention of Frost Damage to Green Concrete," *RILEM Bulletin* (Paris), No. 14, pp. 120-124. Also, *Research Department Bulletin* No. 148, Portland Cement Association.

Price, Walter H., Feb. 1951, "Factors Influencing Concrete Strength," ACI JOURNAL, *Proceedings* V. 47, No. 7, pp. 417-432.

Schnarr, Wilfrid, and Young, R. B., Mar.-Apr. 1934a, "Cold Weather Protection of Concrete," ACI JOURNAL, *Proceedings* V. 30, No. 4, pp. 292-304. Also, Discussion, *Proceedings* V. 31, No. 1, Sept.-Oct. 1934, pp. 47-51.

Schnarr, Wilfrid, and Young, R. B., Mar.-Apr. 1934b, "Manufacturing Concrete During Cold Weather," ACI JOURNAL, *Proceedings* V. 30, No. 4, pp. 279-291. Also, Discussion, *Proceedings* V. 31, No. 1, Sept.-Oct. 1934, pp. 47-51.

Shideler, J. J., Mar. 1952, "Calcium Chloride in Concrete," ACI JOURNAL, *Proceedings* V. 48, No. 7, pp. 537-560.

Sturrup, V. R., and Clendenning, T. G., 1962, "Technical Factors Involved in Winter Concreting," *Ontario Hydro Research News* (Toronto), V. 14, First Quarter, pp. 1-9.

Suprenant, B. A., and Basham, K. D., 1985, "Effects of Cold Embedments on the Temperature of Fresh Concrete," *Proceedings,* 3rd RILEM International Symposium on Winter Concreting, Technical Research Center of Finland, (VTT), Espoo.

Tuthill, L. H.; Glover, R. E.,; Spencer, C. H.; and Bierce, W. B., Nov. 1951, "Insulation for Protection of New Concrete in Winter," ACI JOURNAL, *Proceedings* V. 48, No. 3, pp. 253-272.

Tuutti, K., 1982, "Corrosion of Steel in Concrete," *CBI Research* No. 4.82, Swedish Cement and Concrete Research Institute, Stockholm.

Wallace, George B., Nov. 1954, "New Winter Concreting Techniques Successful on Bureau of Reclamation Jobs," *Western Construction,* V. 29, No. 11.

Wright, T. E., and Jenks, I. H., Oct. 1963, "Galvanic Corrosion in Concrete Containing Calcium Chloride," *Proceedings,* ASCE, V. 89, ST5, pp. 117-132.

10.3 — Selected references

ACI Committee 604, Sept. 1948, selected references from "Recommended Practice for Winter Concreting Methods (ACI 604-48)," ACI JOURNAL, *Proceedings* V. 45, No. 1, Sept. 1948, pp. 13-18.

Clear, K. C., Feb. 1974, "Evaluation of Portland Cement Concrete for Permanent Bridge Deck Repair," *Report* No. FHWA-RD-74-5, Federal Highway Administration, Washington, D.C.

Clemmer, H. F., Oct. 1939, "When and Where to Use Calcium Chloride in Concrete Construction," *Concrete,* V. 47, p. 14.

National Ready Mixed Concrete Association, 1968, "Cold Weather Ready-Mixed Concrete," *Publication* No. 130, Silver Spring, 19 pp.

Neville, Adam A., 1971, *Hardened Concrete: Physical and Mechanical Aspects,* ACI Monograph No. 6, American Concrete Institute/Iowa State University Press, Detroit, 260 pp.

Nisbet, E. G., and Maitland, S. T., Mar. 1976, "Mass Concrete Sections and the Maturity Concept," *Canadian Journal of Civil Engineering* (Ottawa), V. 3, No. 1, pp. 47-57.

Racey, H. J., Nov. 1957, "Cold Weather Concrete Failures," *Civil Engineering—ASCE,* V. 27, No. 11, p. 57.

"RILEM Recommendations for Winter Concreting," Dec. 1963, *RILEM Bulletin* (Paris), No. 21, pp. 3-31.

Shideler, Joseph J.; Brewer, Harold W.; and Chamberlin, Wilbur H., Feb. 1951, "Entrained Air Simplifies Winter Concreting," ACI JOURNAL, *Proceedings* V. 47, No. 6, pp. 449-460.

U.S. Bureau of Reclamation, 1975, *Concrete Manual,* 8th Edition, Denver, 627 pp.

This report was submitted to letter ballot of the committee and was approved in accordance with ACI balloting procedures.

ACI 308R-01

Guide to Curing Concrete

Reported by ACI Committee 308

	Steven H. Gebler Chairman		Cecil L. Jones Secretary	
Don Brogna		Gene D. Hill, Jr.		Aimee Pergalsky
Joseph Cabrera[†]		Edward P. Holub		William S. Phelan
James N. Cornell II		R. Doug Hooton		Robert E. Price[†]
Ronald L. Dilly		Kenneth C. Hover[*]		Larry R. Roberts
Jonathan E. Dongell		John C. Hukey		Phillip Smith
Ben E. Edwards		Frank A. Kozeliski		Luke M. Snell
Derek Firth		James A. Lee		Joel Tucker
Jerome H. Ford		Daryl Manuel		Patrick M. Watson
Sid Freedman		Bryant Mather		John B. Wojakowski
Gilbert J. Haddad		Calvin W. McCall		
Samuel B. Helms		H. Celik Ozyildirim		

[*]Chair of document subcommittee
[†]Deceased

The term "curing" is frequently used to describe the process by which hydraulic-cement concrete matures and develops hardened properties over time as a result of the continued hydration of the cement in the presence of sufficient water and heat. While all concrete cures to varying levels of maturity with time, the rate at which this development takes place depends on the natural environment surrounding the concrete, and the measures taken to modify this environment by limiting the loss of water, heat, or both, from the concrete, or by externally providing moisture and heat. The word "curing" is also used to describe the action taken to maintain moisture and temperature conditions in a freshly placed cementitious mixture to allow hydraulic-cement hydration and, if applicable, pozzolanic reactions to occur so that the potential properties of the mixture may develop. Current curing techniques are presented; commonly accepted methods, procedures, and materials are described. Methods are given for curing pavements and other slabs on ground, for structures and buildings, and for mass concrete. Curing methods for several specific categories of cement-based products are discussed in this document. Curing measures, in general, are specified in ACI 308.1. Curing measures directed toward the maintenance of satisfactory concrete temperature under specific environmental conditions are addressed in greater detail by Committees 305 and 306 on Hot and Cold Weather Concreting, respectively, and by ACI Committees 301 and 318.

Keywords: cold weather; concrete; curing; curing compound; hot weather construction; mass concrete; reinforced concrete; sealer; shotcrete; slab-on-ground.

CONTENTS
Chapter 1—Introduction, p. 308R-2
1.1—Introduction
1.2—Definition of curing
1.3—Curing and the hydration of portland cement
 1.3.1—Hydration of portland cement
 1.3.2—The need for curing
 1.3.3—Moisture control and temperature control
1.4—When deliberate curing procedures are required
 1.4.1—Natural conditions
 1.4.2—Sequence and timing of curing steps for unformed surfaces
 1.4.3—When curing is required for formed surfaces
 1.4.4—When curing is required: cold and hot weather
 1.4.5—Duration of curing
1.5—The curing-affected zone
1.6—Concrete properties influenced by curing

ACI Committee Reports, Guides, Standard Practices, and Commentaries are intended for guidance in planning, designing, executing, and inspecting construction. This document is intended for the use of individuals who are competent to evaluate the significance and limitations of its content and recommendations and who will accept responsibility for the application of the material it contains. The American Concrete Institute disclaims any and all responsibility for the stated principles. The Institute shall not be liable for any loss or damage arising therefrom.

Reference to this document shall not be made in contract documents. If items found in this document are desired by the Architect/Engineer to be a part of the contract documents, they shall be restated in mandatory language for incorporation by the Architect/Engineer.

ACI 308R-01 became effective August 14, 2001.
Copyright © 2001, American Concrete Institute.
All rights reserved including rights of reproduction and use in any form or by any means, including the making of copies by any photo process, or by electronic or mechanical device, printed, written, or oral, or recording for sound or visual reproduction or for use in any knowledge or retrieval system or device, unless permission in writing is obtained from the copyright proprietors.

CHAPTER 1—INTRODUCTION
1.1—Introduction
This guide reviews and describes the state of the art for curing concrete and provides guidance for specifying curing procedures. Curing practices, procedures, materials, and monitoring methods are described. Although the principles and practices of curing discussed in this guide are applicable to all types of concrete construction, this document does not specifically address high-temperature or high-pressure accelerated curing.

1.2—Definition of curing
The term "curing" is frequently used to describe the process by which hydraulic-cement concrete matures and develops hardened properties over time as a result of the continued hydration of the cement in the presence of sufficient water and heat. While all concrete cures to varying levels of maturity with time, the rate at which this development takes place depends on the natural environment surrounding the concrete and on the measures taken to modify this environment by limiting the loss of water, heat, or both, from the concrete, or by externally providing moisture and heat. The term "curing" is also used to describe the action taken to maintain moisture and temperature conditions in a freshly placed cementitious mixture to allow hydraulic-cement hydration and, if applicable, pozzolanic reactions to occur so that the potential properties of the mixture may develop (ACI 116R and ASTM C 125). (A mixture is properly proportioned and adequately cured when the potential properties of the mixture are achieved and equal or exceed the desired properties of the concrete.) The curing period is defined as the time period beginning at placing, through consolidation and finishing, and extending until the desired concrete properties have developed. The objectives of curing are to prevent the loss of moisture from concrete and, when needed, supply additional moisture and maintain a favorable concrete temperature for a sufficient period of time. Proper curing allows the cementitious material within the concrete to properly hydrate. Hydration refers to the chemical and physical changes that take place when portland cement reacts with water or participates in a pozzolanic reaction. Both at depth and near the surface, curing has a significant influence on the properties of hardened concrete, such as strength, permeability, abrasion resistance, volume stability, and resistance to freezing and thawing, and deicing chemicals.

1.3—Curing and the hydration of portland cement
1.3.1 *Hydration of portland cement*—Portland cement concrete is a composite material in which aggregates are bound in a porous matrix of hardened cement paste. At the microscale, the hardened paste is held together by bonds that develop between the products of the reaction of cement with water. Similar products are formed from the reactions between cement, water, and other cementitious materials.

The cement-water reaction includes both chemical and physical processes that are collectively known as the hydration of the cement (Taylor 1997).[1] As the hydration process continues, the strength of the interparticle bonding increases,

Chapter 2—Curing methods and materials, p. 308R-12
2.1—Scope
2.2—Use of water for curing concrete
2.3—Initial curing methods
 2.3.1—Fogging
 2.3.2—Liquid-applied evaporation reducers
2.4—Final curing measures
 2.4.1—Final curing measures based on the application of water
 2.4.2—Final curing methods based on moisture retention
2.5—Termination of curing measures
2.6—Cold-weather protection and curing
 2.6.1—Protection against rapid drying in cold weather
 2.6.2—Protection against frost damage
 2.6.3—Rate of concrete strength development in cold weather
 2.6.4—Removal of cold-weather protection
2.7—Hot-weather protection and curing
2.8—Accelerated curing
2.9—Minimum curing requirements
 2.9.1—General
 2.9.2—Factors influencing required duration of curing

Chapter 3—Curing for different types of construction, p. 308R-19
3.1—Pavements and other slabs on ground
 3.1.1—General
 3.1.2—Curing procedures
 3.1.3—Duration of curing
3.2—Buildings, bridges, and other structures
 3.2.1—General
 3.2.2—Curing procedures
 3.2.3—Duration of curing
3.3—Mass concrete
 3.3.1—General
 3.3.2—Methods and duration of curing
 3.3.3—Form removal and curing formed surfaces
3.4—Curing colored concrete floors and slabs
3.5—Other constructions

Chapter 4—Monitoring curing and curing effectiveness, p. 308R-22
4.1—General
4.2—Evaluating the environmental conditions in which the concrete is placed
 4.2.1—Estimating evaporation rate
4.3—Means to verify the application of curing
4.4—Quantitative measures of the impact of curing procedures on the immediate environment
4.5—Quantitative measures of the impact of curing procedures on moisture and temperature
4.6—Maturity method
4.7—Measuring physical properties of concrete affected by temperature and moisture control to assess curing effectiveness

Chapter 5—References, p. 308R-26
5.1—Referenced standards and reports
5.2—Cited references

and the interparticle porosity decreases. Figure 1.1 shows particles of unhydrated portland cement observed through a scanning electron microscope. In contrast to Fig. 1.1, Fig. 1.2 shows the development of hydration products and interparticle bonding in partially hydrated cement. Figure 1.3 shows a single particle of partially hydrated portland cement. The surface of the particle is covered with the products of hydration in a densely packed, randomly oriented mass known as the cement gel. In hydration, water is required for the chemical formation of the gel products and for filling the micropores that develop between the gel products as they are being formed (Powers and Brownyard 1947; Powers 1948). The rate and extent of hydration depend on the availability of water. Parrott and Killoh (1984) found that as cement paste comes to equilibrium with air at successively lower relative humidity (RH), the rate of cement hydration dropped significantly. Cement in equilibrium with air at 80% RH hydrated at only 10% the rate as companion specimens in a 100% RH curing environment. Therefore, curing procedures ensure that sufficient water is available to the cement to sustain the rate and degree of hydration necessary to achieve the desired concrete properties at the required time.

The water consumed in the formation of the gel products is known as the chemically bound water, or hydrate water, and its amount varies with cement composition and the conditions of hydration. A mass fraction of between 0.21 to 0.28 of chemically bound water is required to hydrate a unit mass of cement (Powers and Brownyard 1947; Copeland, Kantro, and Verbeck 1960; Mills 1966). An average value is approximately 0.25 (Kosmatka and Panarese 1988; Powers 1948).

As seen in Fig. 1.2 and 1.3, the gel that surrounds the hydrated cement particles is a porous, randomly oriented mass. Besides the hydrate water, additional water is adsorbed onto the surfaces and in the interlayer spaces of the layered gel structure during the hydration process. This is known as physically bound water, or gel water. Gel water is typically present in all concrete in service, even under dry ambient conditions, as its removal at atmospheric pressure requires heating the hardened cement paste to 105 C (221 F) (Neville 1996). The amount of gel water adsorbed onto the expanding surface of the hydration products and into the gel pores is "about equal to the amount that is (chemically) combined with the cement" (Powers 1948). The amount of gel water has been calculated more precisely to be a mass fraction of about 0.20 of the mass of hydrated cement (Powers 1948; Powers and Brownyard 1947; Cook 1992; Taylor 1997).

Both the hydrate water and physically adsorbed gel water are distinct in the microstructure of the hardened cement paste, yet both are required concurrently as portland cement cures. Neville (1996) writes that continued hydration of the cement is possible "only when sufficient water is available both for the chemical reactions and for the filling of the gel pores being formed." The amount of water consumed in the hydration of portland cement is the sum of the water incorporated physically onto the gel surfaces plus the water incorporated

Fig 1.1—*Unhydrated particles of portland cement—magnification 2000× (photo credit Fig. 1.1-1.3, Eric Soroos).*

Fig 1.2—*Multiple particles of partially hydrated portland cement—magnification 4000×.*

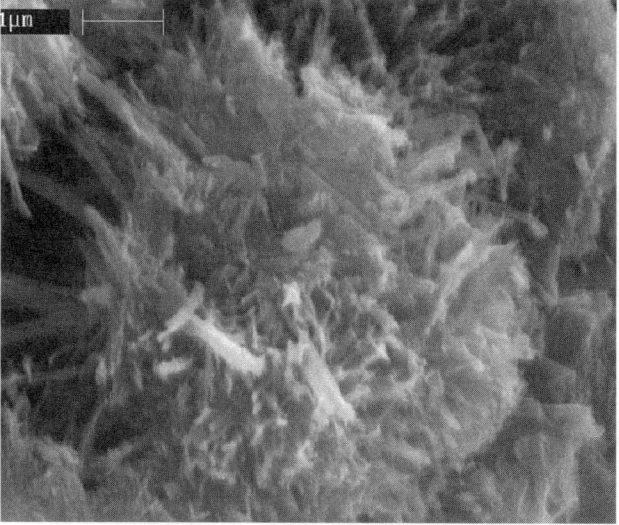

Fig 1.3—*Close-up of a single particle of hydrated cement—magnification 11,000×.*

---

[1]"In cement chemistry the term 'hydration' denotes the totality of the changes that occur when an anhydrous cement, or one of its constituent phases, is mixed with water" (Taylor 1997).

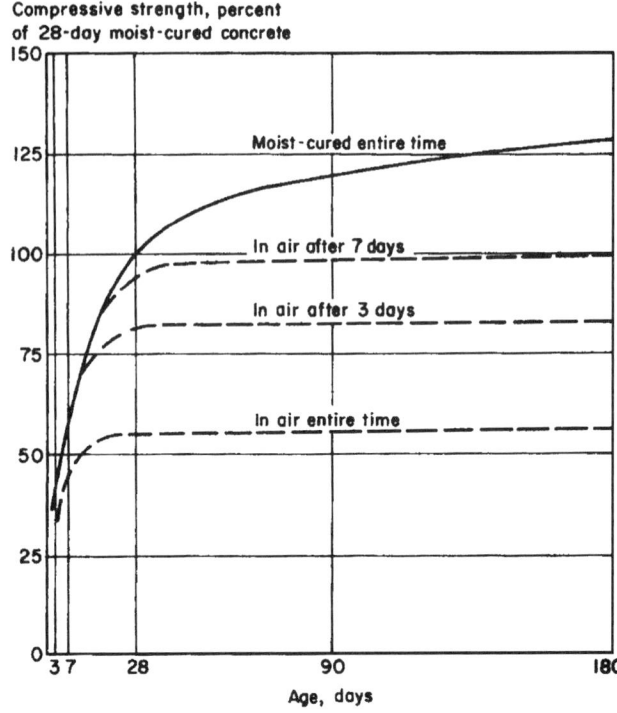

Fig 1.4—Compressive strength of 150 x 300 mm (6 x 12 in.) cylinders as a function of age for a variety of curing conditions (Kosmatka and Panarese 1988).

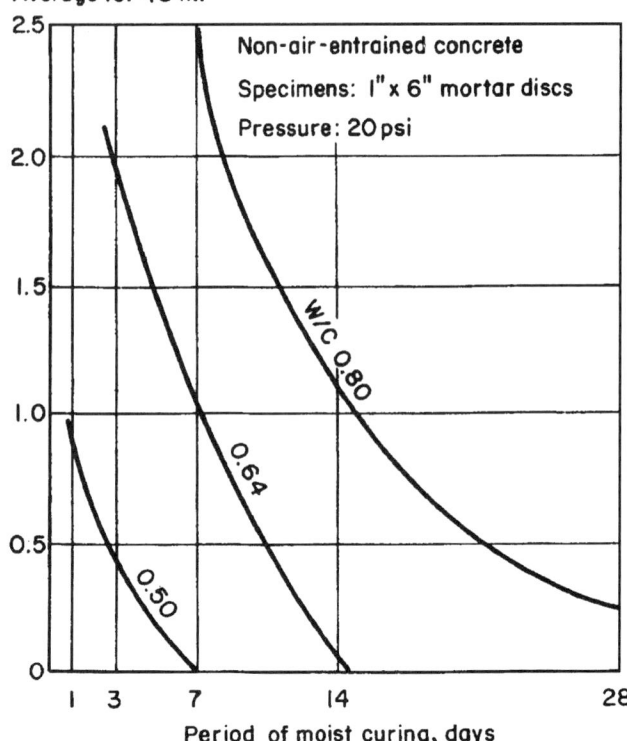

Fig 1.5—Influence of curing on the water permeability of mortar specimens (Kosmatka and Panarese 1988).

chemically into the hydrate products themselves. (Neville 1996; Powers and Brownyard 1947; Mindess and Young 1981; Taylor 1997.) Because hydration can proceed only in saturated space, the total water requirement for cement hydration is "about 0.44 g of water per gram of cement,[2] plus the curing water that must be added to keep (the capillary pores of) the paste saturated" (Powers 1948). As long as sufficient water is available to form the hydration products, fill the interlayer gel spaces and ensure that the reaction sites remain water-filled, the cement will continue to hydrate until all of the available pore space is filled with hydration products or until all of the cement has hydrated.

The key to the development of both strength and durability in concrete, however, is not so much the degree to which the cement has hydrated but the degree to which the pores between the cement particles have been filled with hydration products (Powers and Brownyard 1947, Powers 1948). This is evident from the microperspective seen in Fig. 1.2 and from the macrobehavior illustrated in Fig. 1.4 and 1.5, in which it can be seen that the continued pore-filling accompanying sustained moist-curing leads to a denser, stronger, less-permeable concrete. The degree to which the pores are filled, however, depends not only on the degree to which the cement has hydrated, but also on the initial volume of pores in the paste, thus the combined importance of the availability of curing water and the initial water-cement ratio (w/c).

The pore volume between cement particles seen in Fig. 1.2 (darker areas of the photograph) was originally occupied in the fresh paste by the mixing water. As the volume of mixing water decreases relative to the volume of the cement, the initial porosity of the paste decreases as well. For this reason, pastes with lower w/c have a lower initial porosity, requiring a reduced degree of hydration to achieve a given degree of pore-filling. This is clearly demonstrated in Fig. 1.5, which shows the combined effects of curing and w/c. For the particular mortar specimens tested, a leakage rate of 2.4 kg/m^2/h (0.5 lb/ft^2/h) was achieved after 21 days of moist curing for a w/c of 0.80. The same level of permeability, and same degree of pore-filling, was reached after 10 days for w/c = 0.64, and 2.5 days for w/c = 0.50.

This interaction of curing and w/c in developing the microstructure of hardened cement paste is potentially confusing. On one hand, it is important to minimize the volume of mixing water to minimize the pore space between cement particles. This is done by designing concrete mixtures with a low w/c. On the other hand, it is necessary to provide the cement with sufficient water to sustain the filling of those pores with hydration products. While a high w/c may provide sufficient water to promote a high degree of hydration, the net result would be a low degree of pore-filling due to the high initial paste porosity. The more effective way to achieve a high degree of pore-filling is to minimize initial paste porosity with a low w/c and then to foster hydration by preventing loss of the internal mixing water, or externally applying curing water to promote the maximum possible degree of hydration.

---

[2]Other sources place this approximate value at 0.42 to 0.44 g of water for each gram of dry cement (Powers 1947; Taylor 1997; Neville 1996).

The maximum degree of hydration achievable is a function of both *w/c* and the availability of water (Mills 1966).

1.3.2 *The need for curing*—If the amount of water initially incorporated into the concrete as mixing water will sustain sufficient hydration to develop the desired properties for a given concrete mixture, curing measures are required to ensure that this water remains in the concrete until the desired properties are achieved. At lower initial water contents, where advantage is being taken of lower *w/c* and lower initial porosity, it may be necessary to use curing measures that provide additional water to sustain hydration to the degree of pore-filling required to achieve desired concrete properties. In 1948, Powers demonstrated that concrete mixtures with a *w/c* less than approximately 0.50 and sealed against loss of moisture cannot develop their full potential hydration due to lack of water. Such mixtures would therefore benefit from externally applied curing water (Powers 1948). Powers also pointed out, however, that not all mixtures need to reach their full hydration potential to perform satisfactorily, and externally applied curing water is not always required for mixtures with *w/c* less than 0.50.

A related issue in concrete with a low w/c^3 is that of self-desiccation, which is the internal drying of the concrete due to consumption of water by hydration (Neville 1996; Parrott 1986; Patel et al. 1988; Spears 1983). As the cement hydrates, insufficient mixing water remains to sustain further hydration. Low *w/c* mixtures, sealed against water loss or water entry, can dry themselves from the inside. This problem is most commonly associated with mixtures with a *w/c* around 0.40 or less (Powers 1948; Mills 1966; Cather 1994; Meeks and Carino 1999) and is responsible for an almost negligible long-term strength gain in many low *w/c* mixtures. Given that water also interacts with cementitious materials such as fly ash, slag, and silica fume, self-desiccation can also arise with mixtures having low water-cementitious materials ratios (*w/cm*).

Self-desiccation can be remedied near the concrete surface by externally providing curing water to sustain hydration. At such low values of *w/c*, however, the permeability of the paste is normally so low that externally applied curing water will not penetrate far beyond the surface layer (Cather 1994; Meeks and Carino 1999). Conversely, the low permeability of low *w/c* mixtures prevents restoration of moisture lost in drying at the surface by migration of moisture from the interior. The surface of low *w/c* concrete can therefore dry quickly, calling attention to the critical need to rapidly provide curing water to the surface of low *w/c* concrete (Aïtcin 1999). This also means that surface properties, such as abrasion resistance and scaling resistance, can be markedly improved by wet-curing low *w/c* concrete, while bulk properties, such as compressive strength, can be considerably less sensitive to surface moisture conditions. (See Sections 1.5 and 1.6.)

1.3.3 *Moisture control and temperature control*—Curing procedures that address moisture control ensure that sufficient water is available to the cement to sustain the degree of hydration necessary to achieve the desired concrete properties. The hydration process—a series of chemical reactions—is thermally dependent—the rate of reaction approximately doubles for each 10 C (18 F) rise in concrete temperature. Curing procedures should also ensure that the concrete temperature will sufficiently sustain hydration. As early-age concrete temperatures increase, however, the rate of hydration can become so rapid as to produce concrete with diminished strength and increased porosity, thus requiring temperature control measures (see ACI 305R). Curing measures directed primarily toward the maintenance of satisfactory concrete temperature under specific environmental conditions are addressed in greater detail by ACI Committees 305 and 306 on Hot and Cold Weather Concreting, respectively, and by ACI Committees 301 and 318.

1.4—When deliberate curing procedures are required

Deliberate curing measures are required to add or retain moisture whenever the development of desired concrete properties will be unacceptably delayed or arrested by insufficient water being available to the cement or cementitious materials. Curing measures are required as soon as the concrete is at risk of drying and when such drying will damage the concrete or inhibit the development of required properties. Curing measures should be maintained until the drying of the surface will not damage the concrete, and until hydration has progressed so that the desired properties have been obtained, or until it is clear that the desired properties will develop in the absence of deliberate curing measures.

1.4.1 *Natural conditions*—Whether action is required to maintain an adequate moisture content and temperature in the concrete depends on the ambient weather conditions, the concrete mixture, and on desired properties of the hardened concrete. Under conditions that prevent excessive moisture loss from the concrete, or when the required performance criteria for the concrete are not compromised by early moisture loss, it is entirely possible that no deliberate action needs to be taken to protect the concrete. Guidance for predicting the impact of ambient conditions on the behavior of fresh concrete is found in Section 1.4 and in Chapters 2 through 4. The best source for guidance on the impact of ambient conditions on hardened concrete properties would be field experience with environmental conditions and the concrete mixture in question. Note that in most environments it is unlikely that favorable, natural conditions will exist for the duration of the curing period. The contractor should therefore be prepared to initiate curing measures as soon as ambient conditions change.

1.4.2 *Sequence and timing of curing steps for unformed surfaces*—Curing has traditionally been considered to be a single-step process, conducted some time after the concrete has been placed and finished. Adequate control of moisture, however, can require that several different procedures be initiated in sequence, culminating in a last step that is defined herein as final curing. This section will describe three stages

---

[3]Because the discussion focuses on the hydration of portland cement and not on the related reactions involving materials such as fly ash, ground-generated slag, or silica fume, the appropriate terminology is water-cement ratio (*w/c*) rather than the more generic water-cementitious materials ratio (*w/cm*).

of curing procedures, defined by the techniques used and the time at which they are initiated.

Initial curing refers to procedures implemented anytime between placement and final finishing of the concrete to reduce moisture loss from the surface. Examples of initial curing measures include fogging and the use of evaporation reducers.

Intermediate curing is sometimes necessary and refers to procedures implemented when finishing is completed, but before the concrete has reached final set. During this period, evaporation may need to be reduced, but the concrete may not yet be able to tolerate the direct application of water or the mechanical damage resulting from the application of fabric or plastic coverings. Spray-applied, liquid membrane-forming curing compounds can be used effectively to reduce evaporation until a more substantial curing method can be implemented, if required.

Final curing refers to procedures implemented after final finishing and after the concrete has reached final set. Examples of final curing measures include application of wet coverings such as saturated burlap, ponding, or the use of spray-applied, liquid membrane-forming curing compounds.

Curing procedures and their time of application vary depending on when the surface of the concrete begins to dry and how far the concrete has advanced in the setting process. Curing measures should be coordinated with the sequence and timing of placing and finishing operations.

1.4.2.1 *Timing of placing and finishing operations*—Transport, placing, consolidation, strike off, and bull-floating of unformed concrete surfaces, such as slabs, all take place before the concrete reaches initial setting. Time of initial set is also known as the vibration limit, indicating that the concrete cannot be properly consolidated after reaching initial set (Tuthill and Cordon 1955; Dodson 1994). Surface texturing can begin at initial set but should be completed by the time the concrete has reached final set. Both initial and final set are defined on the basis of the penetration-resistance test (ASTM C 403/C 403 M) for mortar sieved from concrete (Kosmatka 1994; Dodson 1994). This concept is defined similarly for concrete (ACI 302.1R; Suprenant and Malisch 1998a,b,e; Abel and Hover 2000), as indicated in Fig. 1.6(a).

Surface finishing (beyond bull-floating) should not be initiated before initial set nor before bleed water has disappeared from the concrete surface. Before initial set, the concrete is not stiff enough to hold a texture nor stiff enough to support the weight of a finisher or finishing machine.

Furthermore, bleeding of the concrete also controls the timing of finishing operations. Bleed water rises to the surface of freshly cast concrete because of the settling of the denser solid particles in response to gravity and accumulates on the surface until it evaporates or is removed by the contractor (Section 1.4.2.2.2). Bleed water is evident by the sheen on the surface of freshly cast concrete, and its amount can be measured by ASTM Test Method C 232 (Suprenant and Malisch 1998a,e). Finishing the concrete surface before settlement and bleeding has ended can trap the residual bleed water below a densified surface layer, resulting in a weakened zone just below the surface. Finishing before the bleed water fully disappears remixes accumulated bleed water back into the concrete surface, thus increasing the *w/cm* and decreasing strength and durability in this critical near-surface region. Remixing bleed water can also decrease air content at the surface, further reducing durability. Proper finishing should not start until bleeding has ceased and the bleed water has disappeared or has been removed. In most cases, the concrete surface is drying while it is being finished.

The presence of bleed water is detected visually. The appearance of the concrete surface can be misleading, however, when the rate of evaporation equals or exceeds the rate of bleeding. In this case, the apparently dry surface would suggest that bleeding has stopped and that finishing can begin. In reality, however, finishing may yet be premature as bleed water is still rising to the surface. When it is necessary to evaluate this situation more carefully, a clear plastic sheet can be placed over a section of the concrete to block evaporation and to allow observation of bleeding.

Surface finishing should be completed before the concrete attains the level of stiffness (or penetration resistance measured by ASTM C 403) characterized by having reached final set (Abel and Hover 2000). Attempts to texture the concrete beyond final set usually require the addition of water to the surface. This practice should not be allowed because of the loss of surface strength and durability that results from the addition of water to the concrete surface. ACI 302.1R has coined the phrase "window of finishability" to denote the time period between initial and final set (Fig. 1.6(a)) (Suprenant and Malisch 1998e; Abel and Hover 2000).

1.4.2.2 *Timing of curing procedures*—Curing measures should be initiated when the concrete surface begins to dry, which starts as soon as the accumulated bleed water evaporates faster than it can rise to the concrete surface (Lerch 1957; Kosmatka 1994; Al-Fadhala and Hover 2001). The time at which drying and the need for curing begins depends not only on the environment and the resulting rate of evaporation, but also on the bleeding characteristics of the concrete, as shown schematically in Fig. 1.7. The figure illustrates the cumulative bleeding of three different mixtures, measured as a function of time since concrete placement. Superimposed on this diagram is the cumulative loss of surface water due to evaporation arising from three different environments, characterized by high, medium, and low evaporation rates (Rates 1, 2, and 3). Given that surface drying begins as soon as cumulative evaporation catches up with cumulative bleeding, it can be seen that there is a wide divergence in the time at which curing measures are required to control such drying.

1.4.2.2.1 *Evaporation*—The rate of evaporation is influenced by air and concrete temperatures, relative humidity, wind, and radiant energy from direct sunshine. The driving force for evaporation of water from the surface of concrete is the pressure difference between the water vapor at the surface and the water vapor in the air above that surface; the greater the pressure difference, the faster the evaporation. Vapor pressure at the concrete surface is related to the temperature of the water, which is generally assumed to be the same as the concrete surface temperature. The higher the concrete surface temperature, the higher the surface water-vapor pressure.

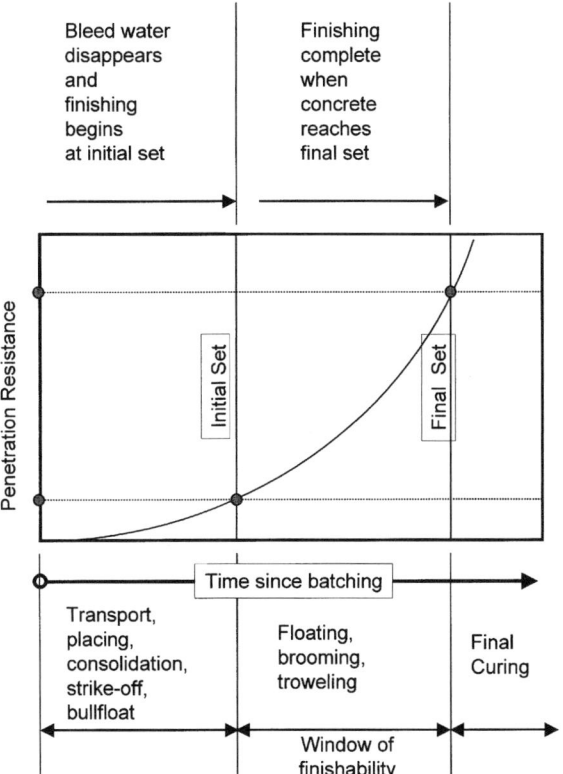

Fig. 1.6(a)—Conventional construction operations under ideal conditions: (1) Initial set coincides with the cessation of bleeding and all bleed water has just evaporated at the beginning of finishing operations; and (2) final set coincides with the completion of finishing. Final curing can begin immediately after finishing with final set.

Because evaporation is driven by the difference between vapor pressure at the surface and in the air, factors that lower water-vapor pressure in the air will increase evaporation. While low humidity in the air increases evaporation rate, it is not as well known that low air temperature, especially in combination with low humidity, increases evaporation. Evaporation rate is high in hot, dry weather because the concrete temperature rises, not because the air is warm. Wind speed becomes a factor as well, because wind moves water vapor away from the surface as it evaporates. In still air, evaporation slows with time due to the accumulation of water vapor (increased humidity) in the air immediately over the evaporating surface. Direct sunlight also accelerates evaporation by heating the water on the surface.

There have been multiple attempts to mathematically estimate evaporation rate based on these factors, dating back to 1802 (Dalton). The most commonly used evaporation rate predictor in the concrete industry is that introduced by Menzel (1954) but developed from 1950 to 1952 by Kohler (1955) for hydrological purposes, as reported by Veihemeyer (1964) and Uno (1998). Most well-known is the evaporation rate nomograph that was reformatted from Menzel's earlier version in 1960 by the National Ready Mixed Concrete Association (NRMCA) (1960). The use, limitations, and accuracy of this tool for estimating rate of evaporation are discussed in detail in Chapter 4, Sections 4.2.1.1 to 4.2.1.3.

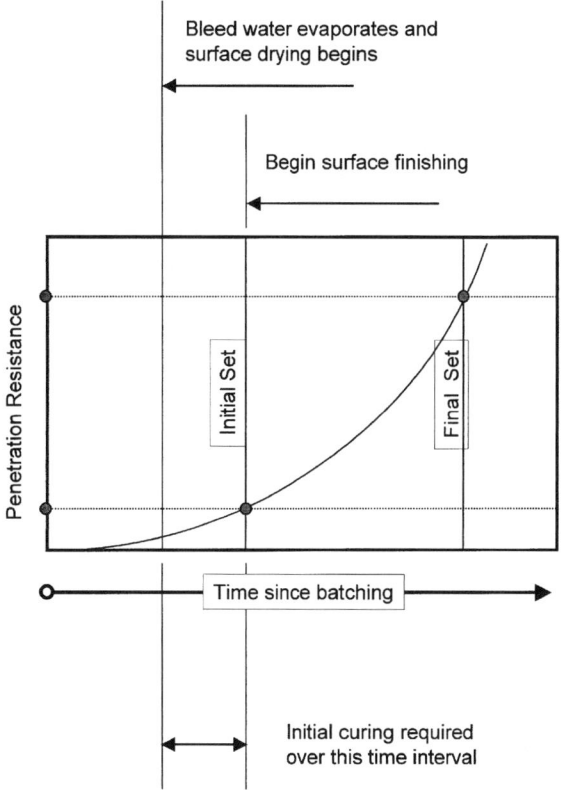

Fig. 1.6(b)—Bleed water disappears and surface drying commences at some time before beginning finishing. Initial curing is required to minimize moisture loss before and during finishing operations.

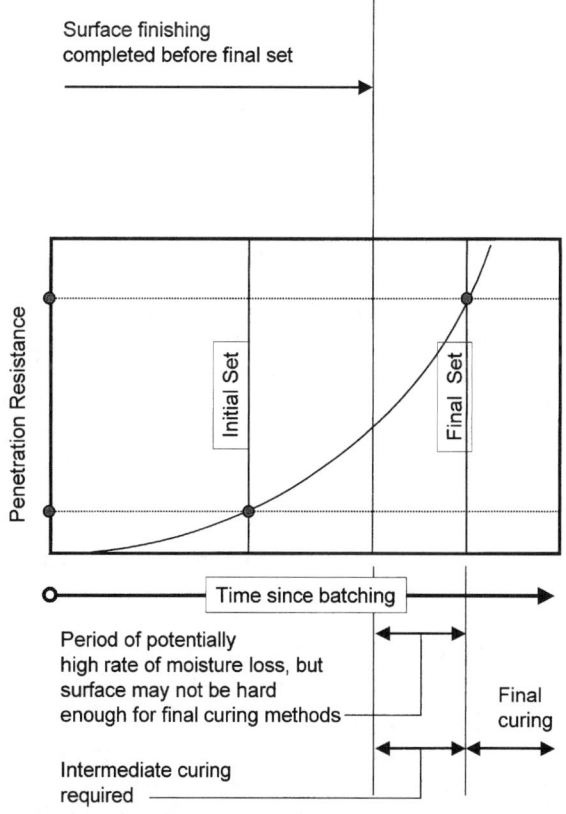

Fig. 1.6(c)—Surface finishing has been completed before the concrete surface has reached final set.

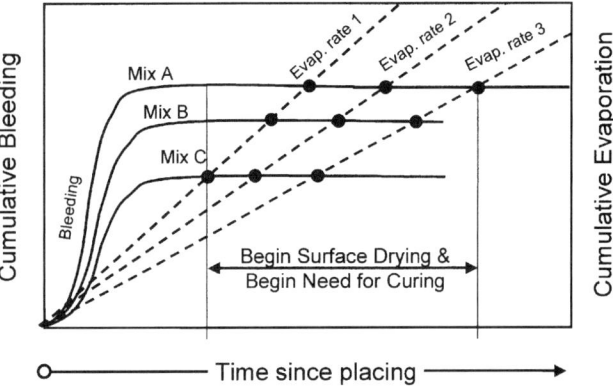

Fig. 1.7—Schematic illustration showing the combined influence of bleeding characteristics and evaporation in determining the time at which the surface of concrete begins to dry.

1.4.2.2.2 *Bleeding*—Both the rate and duration of bleeding depend on the concrete mixture, the depth or thickness of the concrete, and the method of consolidation (Kosmatka 1994; Suprenant and Malisch 1998a). Although water content and *w/cm* are the primary compositional factors, cement, cementitious materials, aggregates, admixtures, and air content all influence bleeding. Thorough vibration brings bleed water to the surface earlier, and deep members tend to show increased bleeding compared with shallow members (Kosmatka 1994). The rate of bleeding diminishes as setting takes place, even in the absence of surface drying, so that surface drying will ultimately occur even under benign evaporation conditions (Al-Fadhala and Hover 2001). Mixtures with a low to negligible bleeding rate are particularly susceptible to surface drying early in the placing and finishing process. Such concrete mixtures often incorporate silica fume, fine cements, or other fine cementitious materials, low *w/cm*, high air contents, or water-reducing admixtures.

1.4.2.2.3 *Initial curing*—For mixtures with a low to zero bleeding rate, or in the case of aggressively evaporative environments, or both, surface drying can begin well before initial set and well before initiation of finishing operations, as indicated in Fig. 1.6(b). Under such conditions, it is necessary to reduce moisture loss by one or more initial curing techniques, such as fogging, the use of evaporation reducers, or by modifying the environment with sunshades, windscreens, or enclosures (Section 2.3). Because finishing can involve several separate and time-consuming operations, initial curing measures may need to be continued or reapplied until finishing is complete.

Initial curing measures should be applied immediately after the bleed water sheen has disappeared, because the concrete surface is protected against drying as long as it is covered with bleed water. When finishing begins immediately after the disappearance of the bleed water, it is unnecessary to apply initial curing measures. When the concrete exhibits a reduced tendency to bleed, when evaporative conditions are severe, or both, the concrete can begin to dry immediately after placing. Under such conditions, initial curing measures, such as fog-spraying to increase the humidity of the air or the application of a liquid-applied evaporation reducer, should be initiated immediately after strike-off, and in some cases, before bull floating. Such initial curing measures should be continuously maintained until more substantial curing measures can be initiated. Excess water from a fog spray or an evaporation reducer should be removed or allowed to evaporate before finishing the surface. (Refer to ACI 302.1R.)

Application of initial curing measures is also frequently required for concretes that exhibit low or negligible bleeding. Such concrete mixtures often incorporate silica fume, fine cements, or other fine cementitious materials, low *w/cm*, high air contents, or water-reducing admixtures. Initial curing measures are frequently required immediately upon placing such concrete to minimize plastic-shrinkage cracking. Plastic shrinkage is initiated by surface drying, which begins when the rate of evaporative water loss from the surface exceeds the rate at which the surface is moistened by bleed water. Refer to ACI 305R for further discussion on plastic shrinkage, and to ACI 234R for further discussion on curing concrete incorporating silica fume.

1.4.2.2.4 *Final curing*—The concrete surface should be protected against moisture loss immediately following the finisher or finishing machine. Significant surface-drying can occur when curing measures are delayed until the entire slab is finished because the peak rate of evaporation from a concrete surface often occurs immediately after the last pass of the finishing tool, as tool pressure brings water to the surface (Al-Fadhala and Hover 2001; Shaeles and Hover 1988). This is especially true when the finished texture has a high surface area such as a broomed or tined surface (Shariat and Pant 1984). Therefore, it is necessary to control moisture loss immediately after finishing (Transportation Research Board 1979).

When the conclusion of finishing operations coincides with the time of final set, as indicated in Fig. 1.6(a), final curing is applied at exactly the right time to reduce the peak rate of moisture loss. A delay in final curing can result in considerable water loss (Al-Fadhala and Hover 2001). Under some conditions, however, applying final curing measures immediately after completion of finishing can be deleterious. These conditions are described in the next section.

1.4.2.2.5 *Conditions under which intermediate curing is recommended*—Intermediate curing measures are required whenever the concrete surface has been finished before the concrete has reached final set. This can happen when the desired surface texture is rapidly achieved, when setting is delayed, or both.

A freshly finished concrete surface is not only vulnerable to the deleterious loss of moisture, but can be vulnerable to damage from the early application of curing materials. The need to protect against moisture loss can conflict with the need to prevent damage to the surface immediately following finishing. Of particular concern is concrete that has been surface-finished before the concrete has reached final set, as shown in Fig. 1.6(c).

Before reaching final set, the concrete surface is susceptible to marring by applying wet burlap, plastic sheets, or other curing materials. Furthermore, the bonds between the cement particles can be easily broken and the particles displaced by water added to the concrete surface and forced between the cement particles, resulting in weakening normally associated with the premature addition of water. For the reason that the earlier water is applied as a final curing measure, the more gently it should be applied to avoid displacement of cement particles. (Fogging is an example of a gentle application, as long as accumulated water is not finished into the surface.) As setting progresses with an increased strength of cement particle bonding, water can be applied to the surface more aggressively. In laboratory and field tests of this principle (Falconi 1996), application of wet burlap to concrete surfaces immediately after finishing reduced resistance to deicer salt scaling. When concrete slabs of the same mixture were lightly covered with plastic sheets immediately after finishing, and the plastic replaced with wet burlap when the concrete had reached final set (measured by ASTM C 403), wet-curing was consistently beneficial in increasing scaling resistance.

Intermediate curing methods can be a continuation of initial curing measures, such as evaporation reducers, or fogging, maintained until the final curing is applied. Membrane-forming curing compounds meeting the requirements of ASTM C 309 or C 1315 can be applied from a power sprayer, making it unnecessary to walk on the concrete surface, and can be applied immediately behind the final pass of the finishing tool or machine. Curing compounds have the advantage of being applicable before final set, as well as being a frequently acceptable final curing method. Curing compounds, therefore, can be an effective intermediate curing method or precursor to other final curing methods, such as water curing or protective coverings, minimizing water loss during the last stages of the setting process.

The combination of a curing compound as an intermediate curing method followed by water-saturated coverings as a final curing method is more common in bridge construction than in building construction (Krauss and Rogalla 1996). The curing compound can be spray-applied to the concrete surface from the perimeter of the bridge deck immediately behind the finishing machine or from the finishers' work bridge. After the curing compound has dried, the wet burlap or similar material is applied and soaker hoses or plastic sheets are installed. This is not a dual or redundant application of two equivalent curing methods. Curing compounds and so-called "breathable sealers" meeting the requirements of ASTM C 309 and C 1315, permit moisture transmission and have a variable capacity to retard moisture loss, depending on the quality of the product used, field application, and field conditions. Wet curing by ponding, sprinkling, or the application of saturated burlap not only prevents water loss but also supplies additional curing water to sustain cement hydration, which is important for low w/cm mixtures that can self-desiccate (Powers 1948; Mills 1966; Mindess and Young 1981; Neville 1996; Persson 1997; Carino and Meeks 1999).

1.4.2.3 *Preparation for casting and curing*—Curing procedures have to be initiated as soon as possible when the concrete surface begins to dry or whenever evaporative conditions become more severe. The curing measures to be used should be anticipated so that the required materials are available on site and ready to use if needed. Water or curing chemicals, coverings, and application equipment and accessories need to be ready, particularly when harsh environmental conditions may require rapid action. To be effective, sunshades or windbreaks (Section 2.7) should be erected in advance of concrete placing operations. Actions such as dampening the subgrade, forms, or adjacent construction, or cooling reinforcing steel or formwork are likewise required in advance of concrete placement. See ACI 301, 302.1R, 305R, and 306R for other commentary on preparedness.

1.4.3 *When curing is required for formed surfaces*—Moisture loss is a concern for both formed and unformed surfaces. Forms left in place reduce moisture loss if the forms are not water-absorbent. Dry, absorbent forms will extract water from the concrete surface. In addition, concrete usually shrinks from the form near the top of the section and it is not unusual to find dry concrete surfaces immediately after removing forms. After form removal, formed surfaces can benefit from curing (Section 3.3.3).

1.4.4 *When curing is required: cold and hot weather*—The environment dictates the need for curing and influences the effectiveness and logistical difficulty in applying the curing methods. For example, use of a fog spray as an initial curing method in freezing weather is impractical and may be of little value despite the critical need to limit surface evaporation under such conditions. Similarly, in hot, arid environments there is a critical need to prevent loss of water from the concrete surface. Such factors often influence the choice of curing methods in hot or cold weather. This choice should be made with consideration of not only the logistical and economic issues, but also of the relative effectiveness of the curing methods proposed in terms of surface strength, resistance to abrasion or deicer scaling, surface permeability, or other factors. The influence of the curing method on the desired properties of the concrete should be given first consideration in such decisions. See Sections 2.6 and 2.7 for details.

1.4.5 *Duration of curing*—The required duration of curing depends on the composition and proportions of the concrete mixture, the values to be achieved for desired concrete properties, the rate at which desired properties are developing while curing measures are in place, and the rates at which those properties will develop after curing measures are terminated. Tests have shown that the duration of wet curing required to bring pastes of different w/c to an equivalent permeability varied, from 3 days for low w/c, to 1 year for high w/c (Powers, Copeland, and Mann 1959). The duration of curing is sensitive to the w/c of the pastes because a lower w/c results in closer initial spacing of the cement particles, requiring less hydration to fill interparticle spaces with hydration products.

Curing should be continued until the required concrete properties have developed or until there is a reasonable assurance that the desired concrete properties will be achieved after the curing measures have been terminated and the concrete is exposed to the natural environment. Most likely, the continued rate of development of the concrete properties will be slower after curing measures have been terminated. Figure 1.4 shows the compressive strength of 150 x 300 mm (6 x 12 in.) cylinders for a particular con-

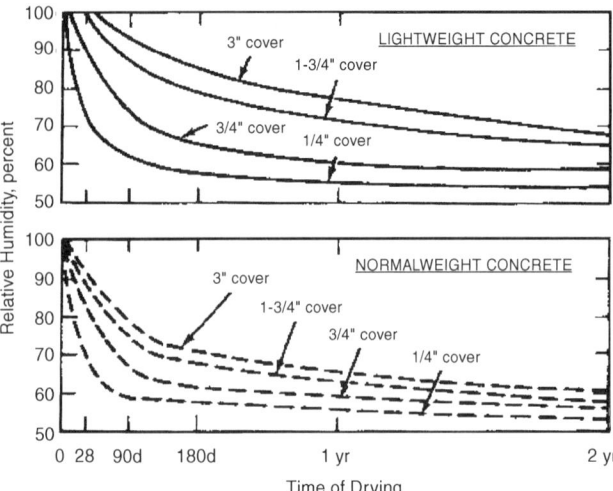

Fig. 1.8—Example of variation of internal relative humidity with depth from surface of concrete cylinder (Hanson 1968) [1 in. = 25.4 mm].

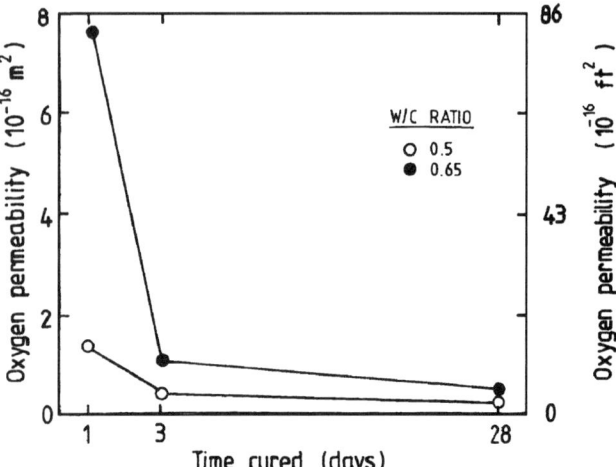

Fig. 1.10—The effect of curing on reducing the oxygen permeability of a concrete surface (Grube and Lawrence 1984; Gowriplan et al. 1990).

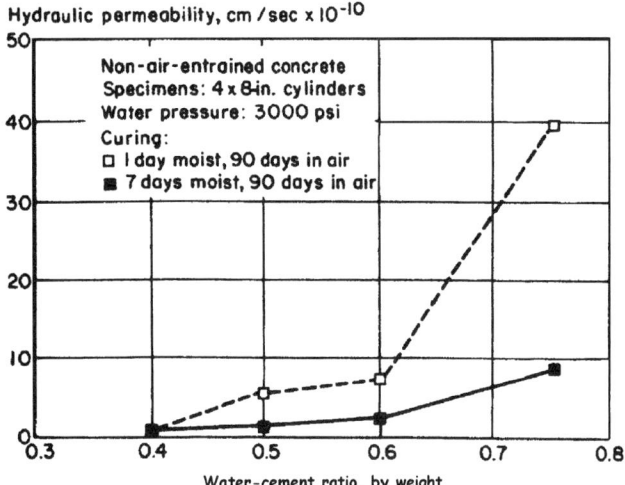

Fig. 1.9—Influence of curing on the water permeability of concrete (Kosmatka and Panarese 1988) (1 cm/s = 0.39 in./s).

crete mixture as a function of curing time for a variety of curing conditions. The figure demonstrates that the rate of continued strength development decreases sharply after curing procedures are terminated. This postcuring rate of continued development should be considered in approving the termination of curing anytime before full attainment of specified concrete properties. For example, it is common to permit termination of curing measures when the compressive strength of the concrete has reached 70% of the specified strength. This is a reasonable practice if the anticipated postcuring conditions allow the concrete to continue to develop to 100% of the specified strength within the required time period. When postcuring conditions are not likely to allow the required further development of concrete properties, it may be more reasonable to require curing until the concrete has developed the full required properties.

In determining the appropriate duration of curing, concrete properties that are desired in addition to compressive strength should be considered. For example, if both high compressive strength and low permeability are required concrete performance characteristics, then the curing needs to be long enough to develop both properties to the specified values. The appropriate duration of curing will depend on the property that is the slowest to develop. Other considerations in determining the specified duration of curing include the cost of applying and subsequently maintaining various curing measures, and the risk and costs associated with not achieving the necessary concrete properties if curing is insufficient. See Section 2.8 for details on required duration of curing.

1.5—The curing-affected zone

Concrete is most sensitive to moisture loss, and therefore, most sensitive and responsive to curing at its surface, where it is in contact with dry, moving air or absorptive media such as a dry subgrade or porous formwork. Figure 1.8 shows an example how internal relative humidity varies with depth from the surface for a 150 x 300 mm (6 x 12 in.) concrete cylinder (Hanson 1968). (Concrete with an internal RH of 70%, for example, would gain or lose no moisture when placed in air at an RH of 70%.) The cylinder specimens had been moist cured for 7 days and then dried at 23 C (73 F) and 50% RH. For the specimen cast with normalweight aggregate, the humidity at 6.4 mm (1/4 in.) depth was approximately 70% at an age of 28 days, while the humidity was about 95% at the center of the cylinder. At 28 days, cement in the outer 6.4 mm (1/4 in.) would have ceased to hydrate, while that in the center of the cylinder would have continued to hydrate (Section 1.3).

Cather (1992) defined the curing-affected zone as that portion of the concrete most influenced by curing measures. This zone extends from the surface to a depth varying from approximately 5 to 20 mm (1/4 to 3/4 in.), depending on the characteristics of the concrete mixture, such as *w/cm* and permeability and the ambient conditions (Carrier 1983; Spears 1983). Concrete properties in the curing-affected zone will be strongly influenced by curing effectiveness, while properties further from the surface will be less susceptible to moisture loss.

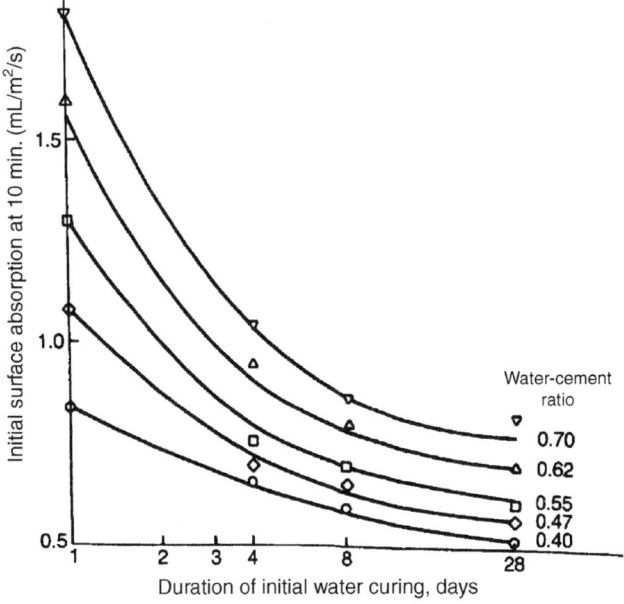

Fig 1.11—Surface absorption is reduced by 50% by extending water curing from 1 to 4 days (Dhir, Hewlett, and Chan 1987) ($1\ mL/m^2/s = 0.74\ lb/ft^2/h$).

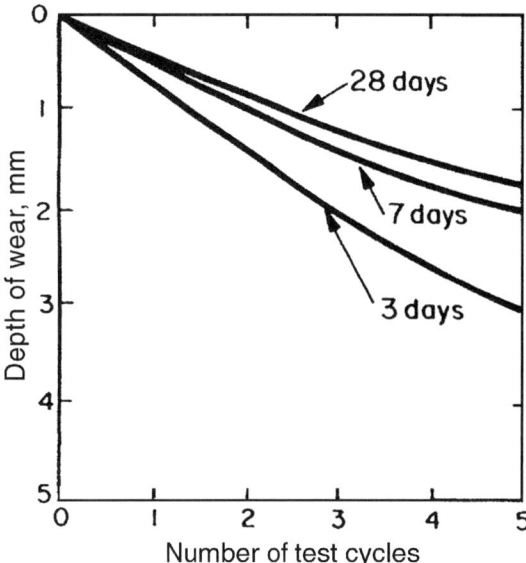

Fig. 1.12—Sawyer (1957) demonstrated the effects of curing on abrasion resistance (1 mm = 0.04 in).

The lower the permeability of the concrete, the more slowly moisture moves between the surface and the interior. Similarly, the lower the permeability, the less-readily water from the interior can replenish water removed from the surface by evaporation (Pihlajavaara 1964, 1965). In such low-permeability concrete, surface-drying can inhibit the development of surface properties, while interior or bulk properties may develop more fully.

Surface hardness, abrasion resistance, scaling resistance, surface permeability and absorption, flexural tension strength (modulus of rupture), surface cracking, surface strain capacity, and similar surface-type properties are strongly influenced by curing. Further, the results of tests for such properties can be useful indicators of curing effectiveness (Chapter 4). Conventional compression tests of cores, cylinders, or cubes are useful as indicators of concrete strength within the bulk of the specimen, but are not necessarily representative of the surface properties. While tests of compressive strength have traditionally been used to demonstrate the effects of curing (Fig. 1.4), such tests are actually not as representative of curing effectiveness as tests of the surface properties listed above. This is because the curing-affected zone is not critical with regard to the compressive strength of cylinders or core, which fail away from their ends. For example, drilled cores can be misleading indicators of curing effectiveness when the curing-affected zone includes only the top 12 mm (1/2 in.) or so of the core sample (Montgomery, Basheer and Long 1992). In a typical core test, the concrete in the curing-affected zone is covered or reinforced with neoprene caps or capping compound, ground smooth, or cutoff altogether. Core tests, therefore, are not consistently reliable indicators of concrete performance in the curing-affected zone, nor are core tests necessarily reliable indicators of curing effectiveness as related to surface properties and performance.

1.6—Concrete properties influenced by curing

Because curing directly affects the degree of hydration of the cement, curing has an impact on the development of all concrete properties. The impact of curing on a broad range of concrete properties is illustrated by the following collection of data from various sources.

As seen previously, Fig. 1.4 indicates the influence of curing on compressive strength, and Fig. 1.5 and 1.9 indicate the influence of curing on the water permeability of hardened concrete. Figure 1.10 shows a 50% reduction in permeability achieved by extending the duration of moist curing from 1 to 3 days and a similar improvement achieved by further increasing the curing period to seven days. Figure 1.9 shows a similar trend.

Figure 1.10 shows the effect of curing on the reduction of the oxygen permeability on a concrete surface (Grube 1984; Gowriplan 1990). The significant reduction in permeability that accompanies a curing extension from one to three days is apparent.

The data in Fig. 1.11 indicate that surface absorption is reduced by about 50% by extending water curing from 1 to 4 days (Dhir, Hewlett, and Chan 1987).

Sawyer (1957) demonstrated the effects of curing on abrasion resistance (Fig. 1.12) and the effects of a 24 h delay in curing (Fig. 1.13). The number of test cycles is the number of successive applications of the abrasive wear test device. Dhir (1991) demonstrated a similar relationship between abrasion resistance and curing, as shown in Fig. 1.14.

Murdock, Brook, and Dewar (1991) showed a relationship between the duration of wet curing and the resistance to freezing and thawing of air-entrained concrete, as shown in Fig. 1.15. For a w/c of 0.45 (ACI 318 maximum value for concrete exposed to freezing while moist), resistance to

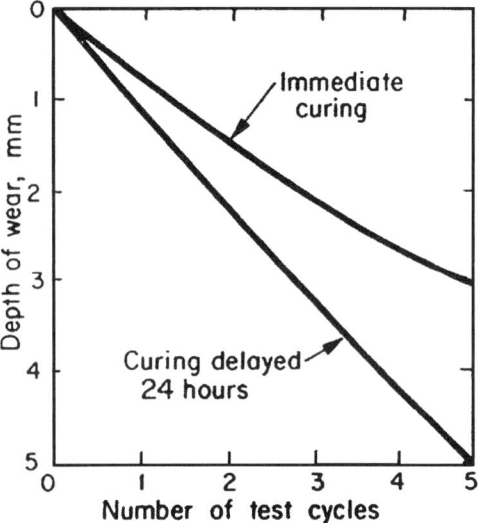

Fig. 1.13—Sawyer (1957) demonstrated the effects of delaying curing on abrasion resistance (1 mm = 0.04 in).

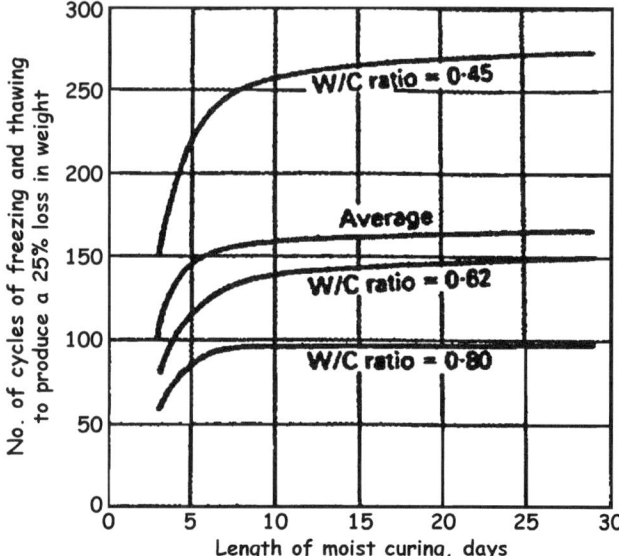

Fig. 1.15—Influence of duration of moist-curing time on freezing and thawing durability of concrete, also as a function of w/c (Murdock, Brook, and Dewar 1991).

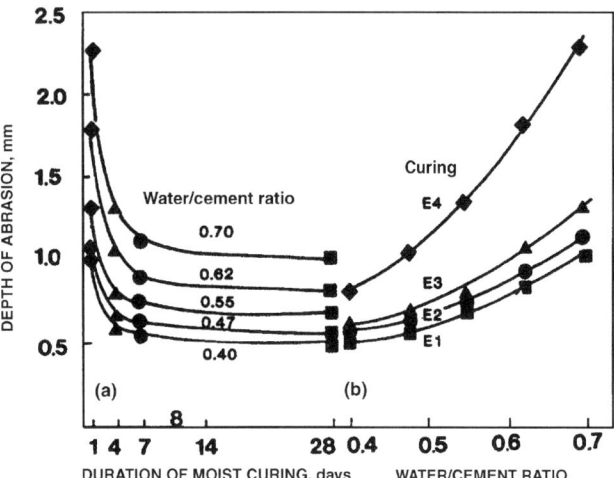

Fig. 1.14—Dhir, Hewlett, and Chan (1991) demonstrated the relationship between abrasion resistance and curing (1 mm = 0.04 in.). Note: E1= 24 h wet burlap followed by 27 days immersion curing in water at 20 C (68 F); E2= 24 h wet burlap followed by 6 days immersion curing in water at 20 C (68 F) and 21 days in air at 20 C (68 F) and 55% RH; E3= 24 h wet burlap followed by 3 days immersion curing in water at 20 C (68 F) and 24 days in air at 20 C (68 F) and 55% RH; and E4= 24 h wet burlap followed by 27 days in air at 20 C (68 F) and 55% RH.

freezing and thawing continues to develop over the entire 28-day curing period.

Gowriplan et al. (1990) demonstrated a relationship between curing and oxygen permeability at various depths from the surface. This work included comparisons of various methods for curing (Fig. 1.16).

CHAPTER 2—CURING METHODS AND MATERIALS
2.1—Scope

Regardless of the materials or methods used for curing concrete, the concrete should maintain a satisfactory moisture content and temperature so that its properties develop.

While there are many methods for controlling the temperature and moisture content of freshly placed concrete, not all such methods are equal in price, appropriateness, or effectiveness. The means and methods to be used will depend on the demands of each set of circumstances. The economics of the particular method of curing selected should be evaluated for each job, because the availability of water or other curing materials, labor, control of runoff (if water is continuously applied), and subsequent construction, such as the application of floor coverings or other treatments, will influence price and feasibility.

The two general systems for maintaining adequate moisture content vary in effectiveness depending on the concrete mixture and curing methods and materials used, the details of construction operations, and the ambient weather conditions. These two systems are the continuous or frequent application of water through ponding, fogging, steam, or saturated cover materials such as burlap or cotton mats, rugs, sand, and straw or hay, and the minimization of water loss from the concrete by use of plastic sheets or other moisture-retaining materials placed over the exposed surfaces, or by the application of a membrane-forming compound (commonly referred to as a curing compound) meeting the requirements of ASTM C 309 or C 1315.

This guide does not address all of the safety concerns, if any, associated with the use of curing materials. It is the responsibility of the user of this guide to establish appropriate safety and health practices and to determine the applicability of regulatory limitations before its use.

2.2—Use of water for curing concrete

The method of water curing selected should provide a complete and continuous cover of water that is free of harmful amounts of deleterious materials. Several methods of water curing are described as follows.

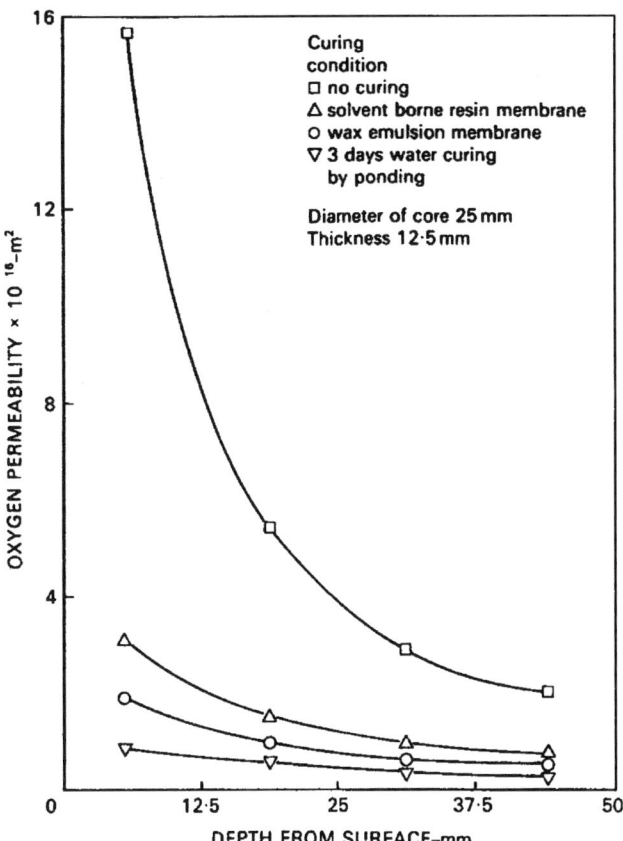

Fig. 1.16—The influence of various curing types on oxygen permeability at various depths (Gowriplan et al., 1990) (1 mm = 0.04 in; 1 m² = 10.76 ft²).

The curing water should be free of "aggressive impurities that would be capable of attacking or causing deterioration of the concrete" (Pierce 1994). In general, water that is potable and satisfactory as mixing water, meeting the requirements of ASTM C 94, is acceptable as curing water. Where appearance is a factor, the water should be free of harmful amounts of substances that will stain or discolor the concrete. Dissolved iron or organic impurities may cause staining, and the potential staining ability of curing water can be evaluated by means of CRD-C401 (U.S. Army Corps of Engineers 1975). The use of seawater as curing water is controversial, as is the use of seawater as mixing water. The potential effects are discussed by Eglington (1998).

Care needs to be taken to avoid thermal shock or excessively steep thermal gradients due to use of cold curing water. Curing water should not be more than 11 C (20 F) cooler than the internal concrete temperature to minimize stresses due to temperature gradients that could cause cracking (Kosmatka and Panarese 1988). A sudden drop in concrete temperature of about 11 C (20 F) can produce a strain of about 100 millionths, which approximates the typical strain capacity of concrete. (See also discussion in Mather [1987].)

2.3—Initial curing methods

As discussed in Section 1.4, initial curing refers to procedures implemented anytime between placement and finishing of concrete to reduce the loss of moisture from the concrete surface.

2.3.1 *Fogging*—Fogging provides excellent protection against surface drying when applied properly and frequently and when the air temperature is well above freezing. Fogging requires the use of an inexpensive but specially designed nozzle that atomizes the water into a fog-like mist. The fog spray should be directed above, not at, the concrete surface, as its primary purpose is to increase the humidity of the air and reduce the rate of evaporation. This effect lasts only as long as the mist is suspended in the air over the slab. This means frequent or continuous fogging is necessary, and the frequency of application should be increased as wind velocity increases over the concrete surface. (The droplets are fine enough and the application is continuous enough when a visible fog is suspended over the concrete surface.) Fogging is also useful for reducing the tendency for a crust to form on the surface of the freshly cast concrete. Fogging can precipitate water on the concrete surface and is not deleterious as long as the water from the sprayer does not mar or penetrate the surface. Water from fogging should not be worked into the surface in subsequent finishing operations. Water from fogging should be removed or allowed to evaporate before finishing.

2.3.2 *Liquid-applied evaporation reducers*—Evaporation reducers (Cordon and Thorpe 1965) are solutions of organic chemicals in water that are capable of producing a monomolecular film over the bleed water layer that rises to the top surface of concrete. If present in sufficient concentration, these chemicals form an effective film that reduces the rate of evaporation of the bleed water from the concrete surface.

Evaporation reducers can be sprayed onto freshly placed concrete to reduce the risk of shrinkage when the evaporation rate equals or exceeds the bleeding rate. Evaporation reducers are not to be used for the purpose of making it easier to finish concrete surfaces (materials designed for such purpose are often referred to as finishing aids), and should be used only in accordance with manufacturer's instructions.

2.4—Final curing measures

As discussed in Section 1.4, final curing refers to procedures implemented after final finishing and when the concrete has reached final set. As discussed in detail in that section, it may be necessary to use an intermediate curing technique when the concrete surface has been finished before the concrete reaches final set, as premature application of final curing may damage the freshly cast concrete.

2.4.1 *Final curing measures based on the application of water*

2.4.1.1 *Sprinkling the surface of the concrete*—Fogging or sprinkling with nozzles or sprays provides excellent curing when the air temperature is above freezing. Lawn sprinklers are effective after the concrete has reached final set and where water runoff is not a concern. A disadvantage of sprinkling is the cost of the water in regions where an ample supply is not readily available. Intermittent sprinkling should not be used if the concrete surface is allowed to dry between periods of wetting. Soaker hoses are useful, especially on surfaces that are vertical or nearly so. Care should be taken to avoid erosion of the surface.

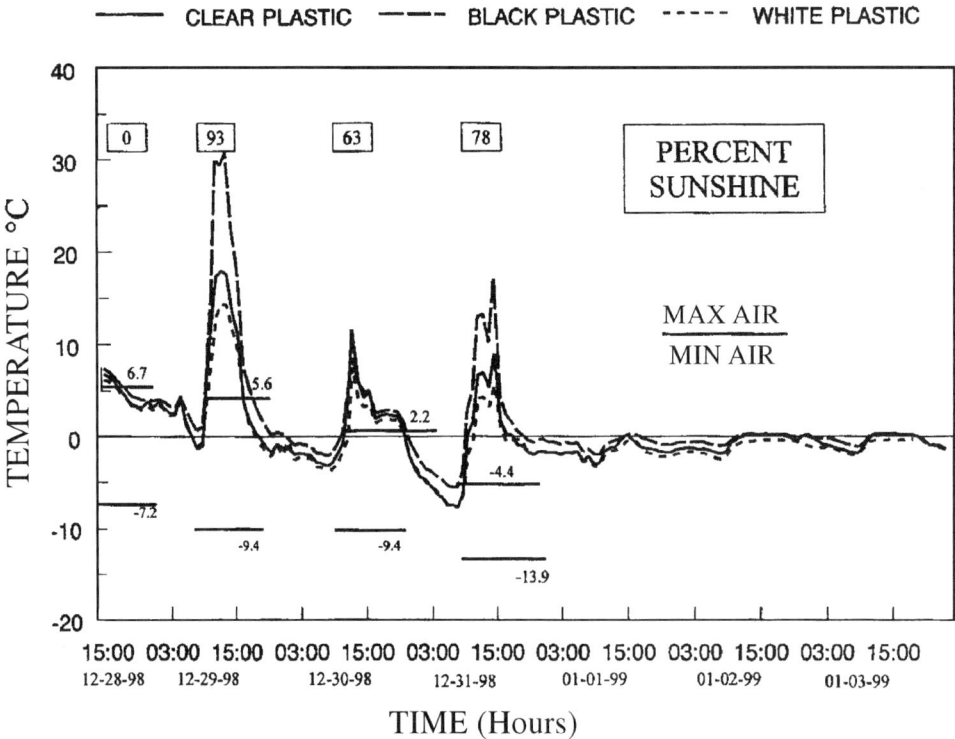

Fig. 2.1—Temperature variations under clear, black, and white plastic (Wojokowski 1999).

2.4.1.2 *Ponding or immersion*—Though seldom used, the most thorough method of water curing consists of immersion of the finished concrete in water. Ponding is sometimes used for slabs, such as culvert floors or bridge decks, pavements, flat roofs, or wherever a pond of water can be created by a ridge, dike, or other dam at the edge of the slab. (Van Aardt documented dilution of the paste and weakening of the surface resulting from premature application of ponding [Van Aardt 1953]).

2.4.1.3 *Burlap, cotton mats, and other absorbent materials*—Burlap, cotton mats, and other coverings of absorbent materials can hold water on horizontal or vertical surfaces. These materials should be free of harmful substances, such as sugar or fertilizer, or substances that may discolor the concrete. To remove soluble substances, burlap should be thoroughly rinsed in water before placing it on the concrete. Burlap that has been treated to resist rot and fire is preferred for use in curing concrete. Burlap should also be dried to prevent mildew when it is to be stored between jobs. The thicker the burlap, the more water it will hold and the less frequently it will need to be wetted. Double thickness may be used advantageously. Lapping the strips by half widths when placing will give greater moisture retention and aid in preventing displacement during high wind or heavy rain. A continuous supply of moisture is required when high temperature, low humidity, or windy conditions prevail. The concrete surface should remain moist throughout the curing period. When burlap is permitted to dry, it can draw moisture from the surface of the concrete.

Absorbent mats made of cotton or similar fibers can be applied much the same as burlap, except that due to their greater mass, application to a freshly finished surface should be delayed until the concrete has hardened to a greater degree than for burlap.

Whenever concrete slabs are so large that the workers have to walk on the freshly placed concrete to install the curing materials, it will be necessary to wait until the concrete has sufficiently hardened to permit such operations without marring the surface.

2.4.1.4 *Sand curing*—Wet, clean sand can be used for curing provided it is kept saturated throughout the curing period. The sand layer should be thick enough to hold water uniformly over the entire surface to be cured. Sand should meet ASTM C 33 or similar requirements for deleterious materials in fine aggregate to minimize the risk of damage to the concrete surface from deleterious materials.

2.4.1.5 *Straw or hay*—Wet straw or hay can be used for wet-curing small areas, but there is the danger that wind might displace it unless it is held down with screen wire, burlap, or other means. There is also the danger of fire if the straw or hay is allowed to become dry. Such materials may discolor the surface for several months after removal. If these materials are used, the wetted layer should be at least 150 mm (6 in.) deep and kept wet throughout the curing period.

2.4.2 *Final curing methods based on moisture retention*—Curing materials are sheets or liquid membrane-forming compounds placed on concrete to reduce evaporative water loss from the concrete surface.

These materials have several advantages:

1) They do not need to be kept wet to ensure that they do not absorb moisture from the concrete;

2) They are easier to handle than burlap, sand, straw or hay; and

3) They can often be applied earlier than water-curing methods.

As discussed in Section 1.4.2.2.5, curing materials can be applied immediately after finishing without the need to wait for final setting of the concrete.

2.4.2.1 *Plastic film*—Plastic film has a low mass per unit area and is available in clear, white, or black sheets. Plastic film should meet the requirements of ASTM C 171, which specifies a minimum thickness of 0.10 mm (0.004 in.). White film minimizes heat gain by absorption of solar radiation. Clear and black sheeting have advantages in cold weather by absorption of solar radiation but should be avoided during warm weather except for shaded areas. The general effect of the color of plastic sheeting on concrete surface temperature is shown in Fig. 2.1 (Wojokowski 1999). Wojokowski placed clear, black, and white plastic over a hardened concrete surface and measured the temperature over a period of one week during winter. Within the week four separate 12-h periods were evaluated for minimum and maximum air temperature, and for percentage of the period in which the sun was shining. Sunshine varied from 0 to 93%, and the outside air temperatures varied over the time periods as shown by the pairs of horizontal lines indicating maximum and minimum values. The concrete under plastic was as much as 25 C (45 F) degrees warmer than the air. The temperature under black plastic was approximately 15 C (27 F) warmer than under clear plastic and almost 20 C (36 F) warmer than under white plastic. The temperature difference was negligible in the absence of sunshine. (When plastic covers freshly cast concrete, heat is also provided from the hydration of the cement.)

Care should be taken to avoid tearing or otherwise interrupting the continuity of the film. Plastic film reinforced with glass or other fibers is more durable and less likely to be torn.

Where the as-cast appearance is important, concrete should be cured by means other than moisture-retaining methods because the use of smooth plastic film often results in a mottled appearance due to variations in temperature, moisture content, or both (Greening and Landgren 1966). Curing methods requiring the application of water may be necessary because mottling can be minimized or prevented by occasional flooding under the film. Combinations of plastic film bonded to absorbent fabric help to retain and more evenly distribute moisture between the plastic film and the concrete surface have been effective in reducing mottling.

Plastic film should be placed over the surface of the fresh concrete as soon as possible after final finishing without marring the surface and should cover all exposed surfaces of the concrete. It should be placed and weighted so that it remains in contact with the concrete during the specified duration of curing. On flat surfaces such as pavements, the film should extend beyond the edges of the slab at least twice the thickness of the slab. The film should be placed flat on the concrete surface, avoiding wrinkles, to minimize mottling. Windrows of sand or earth, or pieces of lumber should be placed along all edges and joints in the film to retain moisture and prevent wind from getting under the film and displacing it. Alternatively, it is acceptable and generally more economical to use a narrow strip of plastic film along the vertical edges, placing it over the sheet on the horizontal surface and securing all edges with windrows or strips of wood. To remove the covering after curing, the strip can be pulled away easily, leaving the horizontal sheet to be rolled up without damage from tears or creases.

2.4.2.2 *Reinforced paper*—Composed of two layers of kraft paper cemented together with a bituminous adhesive and reinforced with fiber, reinforced paper should comply with ASTM C 171. Most papers used for curing have been treated to increase tear resistance when wetted and dried. The sheets of reinforced paper can be cemented together with bituminous cement to meet width requirements.

Paper sheets with one white surface to give reflectance and reduce absorption of solar radiation are available. A reflectance requirement is included in ASTM C 171. Reinforced paper is applied in the same manner as plastic film (Section 2.4.2.1). Reinforced paper can be reused as long as it is effective in retaining moisture on the concrete surface. Holes and tears should be repaired with a patch of paper cemented with a suitable glue or bituminous cement. Pin holes, resulting from walking on the paper or from deterioration of the paper through repeated use, are evident if the paper is held up to the light. When the paper no longer retains moisture, it should be discarded or used in double thickness.

2.4.2.3 *Liquid membrane-forming compounds*—Liquid membrane-forming compounds for curing concrete should comply with the requirements of ASTM C 309 or C 1315 when tested at the rate of coverage to be used on the job. Compounds formulated to meet the requirements of ASTM C 309 include: Type 1, clear; Type 1D, clear with fugitive dye; Type 2, white pigmented; Class A, unrestricted composition (usually used to designate wax-based products); and Class B, resin-based compositions. A note in ASTM C 309 states: "Silicate solutions are chemically reactive rather than membrane-forming, therefore, they do not meet the intent of this specification." ASTM C 156 is the test method used to evaluate water-retention capability of liquid membrane-forming compounds. ASTM C 1151 provides an alternative laboratory test for determining the efficiency of liquid membrane-forming compounds.

Membrane-forming curing compounds that meet the requirements of ASTM C 309 permit the loss of some moisture and have a variable capacity to reduce moisture loss from the surface, depending on field application and ambient conditions (Mather 1987, 1990; Shariat and Pant 1984; Senbetta 1988).

Compounds formulated to meet the requirements of ASTM C 1315 have special properties, such as alkali resistance, acid resistance, adhesion-promoting qualities, and resistance to degradation by ultraviolet light, in addition to their moisture-retention capability as measured by ASTM C 156. Curing compounds are classed according to their tendency to yellow or change color with age and exposure and by whether they are clear or pigmented. Products meeting the requirements of ASTM C 1315 are often referred to as "breathable membrane sealers" after they have performed the function of a curing membrane used during the final curing. When these products are tested in accordance with ASTM C 156, the allowable moisture loss is 0.40 kg/m^2 (0.08 lb/ft^2) in 72 h when applied at a curing compound coverage rate of 5.0 or 7.4 m^2/L (200 or 300 ft^2/gal.) for Type I or Type II compounds, respectively. When performing the ASTM C 156 test in a laboratory to verify compliance with either ASTM C 309 or C 1315, the surface of the test specimen is relatively smooth, and the laboratory-applied coverage rates specified in the test method can be readily obtained. Achieving the same coverage, and therefore achieving the same moisture-retention effectiveness in the field, is more difficult, given the variable surface texture and the need to evenly cover the concrete. ASTM C 1315 notes that agencies can require a substantially greater application rate on deeply textured surfaces. Further, the specified evaporative environment for the ASTM C 156 test corresponds to an evaporation rate from a free water surface varying between about 1/2 to 1 kg/m^2/h (0.1 to 0.2 lb/ft^2/h), which may be less severe than encountered in a given construction environment. For these reasons, the water loss experienced by concrete in the field may vary from the values specified by ASTM for compliance with the requirements of C 309 or C 1315.

Curing materials can be composed of wax or other organic material thinned with a solvent. The solvent can make the use of the curing compound subject to various restrictions or regulations governing the transport, storage, or use of hazardous materials. When appropriate, adequate ventilation should be provided and other safety precautions should be taken when using solvent-based compounds. Other similar curing compounds based on water-soluble solids or a water emulsion are available. When the concrete surface is to receive paint, finishes, or toppings that require positive bond to the concrete, it is critical that the curing procedures and subsequent coatings, finishes, or toppings be compatible to achieve the necessary bond. Curing compounds meeting the requirements of ASTM C 1315 have been formulated to promote such adhesion, and ASTM C 1315 includes references to test methods for evaluating the bonding of tile and other floor coverings. Testing to establish compatibility among the curing compound, subsequent surface treatments, concrete moisture content, and the actual finished surface texture of the concrete is recommended when performance is critical. Such testing is beyond the scope of this document, but useful references include Suprenant and Malisch (1998d,1999a,b).

Alternatives to ensuring compatibility of the curing compound with subsequent surface treatments include deliberate removal of the curing compound in accordance with manufacturer's recommendations, the use of an alternative water-retention curing method, or water curing.

Some organic resin-based curing compounds will degrade over time with exposure to ultraviolet light (direct sunlight). Dissipation is accelerated in the presence of water. A self-dissipating resin compound is expected to break down some time after application and can be used on troweled or floated surfaces receiving subsequent finishes. Some of these products may not dissipate fast enough in interior or shaded locations and are not recommended if dissipation is the expected method of removal. Mechanical or chemical removal should be specified when self-dissipating compounds are used in interior applications.

For floors and slabs designed for high wear resistance, optimum top surface strength development, durability, and minimal cracking, the curing compound should meet or exceed the moisture retention requirements of ASTM C 1315.

In accordance with the manufacturer recommendations, liquid membrane-forming curing compounds should be stirred or agitated before use and applied uniformly at the manufacturer's recommended rate. The curing compound should be applied in two applications at right angles to each other to ensure uniform and more complete coverage. On very deeply textured surfaces, the surface area to be treated can be at least twice the surface area of a trowelled or floated surface (Shariat and Pant 1984). In such cases, two separate applications may be needed, each at 5 m^2/L (200 ft^2/gal.), with the first being allowed to become tacky before the second is applied. A curing compound can be applied by hand or power sprayer, using appropriate wands and nozzles with pressure usually in the range of 0.2 to 0.7 MPa (25 to 100 psi). If the job size is large, application by power sprayer is preferred because of speed and uniformity of distribution. For very small areas such as repairs, the compound can be applied with a wide, soft-bristled brush or paint roller. Application rates are most readily verified by recording the number of containers of compound used, or the number of sprayer tanks or buckets of compound applied to the surface.

For maximum beneficial effect, liquid membrane-forming compounds should be applied immediately after the disappearance of the surface water sheen following final finishing. Delayed application of these materials not only allows drying of the surface during the period of peak water loss, but also increases the likelihood that the liquid curing compound will be absorbed into the concrete and hence not form a membrane.

When the evaporation rate exceeds the rate of bleeding of the concrete, the surface will appear dry even though bleeding is still occurring. Under such conditions, finishing the concrete or the application of curing compound, or both, can be detrimental because bleed water can be consequently trapped just below the concrete surface. Clearly identifying this condition in the field is difficult, and it is always risky to delay finishing and curing. In such cases where it is important to diagnose this problem, a transparent plastic sheet can be placed over the unfinished, uncured concrete to shield the test area from evaporation, and any bleed water can be seen accumulating under the plastic.

Another consequence of applying curing compound to a freshly cast concrete surface that appears dry is that evaporation will be temporarily stopped, but bleeding might continue, resulting in map cracking of the membrane film with the subsequent reduction in moisture-retention capability. This situation would require reapplication of the curing compound.

When using curing compounds to reduce moisture loss from formed surfaces, the exposed surface should be wetted immediately after form removal and kept moist until the curing compound is applied. The concrete should be allowed to reach a uniformly damp appearance with no free water on the surface, and then application of the compound should begin at once. As with flatwork, dampening the concrete prevents absorption of the curing compound, which would prevent the formation of a membrane.

2.4.2.4 *Linseed oil-based curing compounds*—Various linseed oil emulsions have been successfully used as curing compounds by some highway agencies (Sedgwick County 1999; Minnesota Department of Transportation 1998). Although not strictly a membrane-forming compound, these products should conform with Formulation 6325-39-2 of the Oilseed Crops Laboratory, U.S.D.A., and meet the requirements of ASTM C 309 (AASHTO M 148) with the exception of the four hour maximum drying time requirement.

These products are used as an ASTM C 309 Type A curing compound for retaining moisture in freshly cast concrete and in some cases are used as a surface treatment to increase the durability of hardened concrete. This latter use is beyond the scope of this document. When used as a curing compound, however, linseed oil-based compounds are intended for application directly after final texturing. Some retardation of setting time may result from the use of these products.

2.5—Termination of curing measures

During the curing period, saturated cover materials should not be allowed to dry and absorb water from the concrete. At the end of the required wet-curing period, cover materials should be allowed to dry thoroughly before removal to provide uniform, slow drying of the concrete surface. Controlled and gradual termination of wet or moist curing is particularly critical in cold weather when there is a risk of freezing the freshly exposed and water-saturated surface. ACI 306R and ACI 301 contain recommendations for gradual termination of curing and protection.

2.6—Cold-weather protection and curing

Cold weather puts immature concrete at risk in at least four ways. First, the rate of evaporation of water from the surface of concrete can be higher in cold weather than in warm weather, particularly when the concrete is warm and the humidity is low. Second, if the concrete temperature greatly decreases, pore water can freeze in the pores of the concrete, leading to frost damage. Third, cold concrete temperature slows the rate of hydration of the cement, slowing the rate at which all concrete properties develop. Fourth, when protection is removed at the end of the curing period, there is a risk of rapid drying, and a rapid drop in temperature can crack the concrete. In cold weather, control of moisture should be accompanied by control of temperature. Wet-curing in freezing weather can be as beneficial to the long-term serviceability of the concrete as it is in warm weather, but only if the moist concrete is kept from freezing. Air-entrained concrete should be used whenever the concrete is expected to be subjected to freezing temperature while in a moist condition.

Water curing in cold weather can require the construction of a heated, temporary enclosure. For these reasons cold-weather curing procedures frequently include evaporation reducers as an initial curing, followed by membrane-forming curing compounds, or by covering with dry plastic sheets (or similar coverings), or by both a membrane-forming curing compound and a covering in place of curing by the direct application of water. These techniques may or may not produce the equivalent concrete surface properties as providing added water, but the risk of freezing damage to the concrete surface is reduced.

Cold-weather curing further requires that water supplies, the water distribution system, and the runoff water also should be kept from freezing, and workers should be protected against ice-related hazards and cold injury such as frostbite.

ACI 306.1 provides specification requirements for cold-weather concreting, and additional information is found in ACI 306R.

2.6.1 *Protection against rapid drying in cold weather*—The rate of evaporation from a freshly placed concrete surface can be greater in cold weather than in hot weather, especially when the concrete has been heated by the addition of hot water that evaporates into cold, dry air. Further, cold weather is often accompanied by faster average wind speeds. Senbetta and Brury (1991) have discussed the likelihood of developing plastic shrinkage cracking in cold weather. To minimize plastic shrinkage cracking and to sustain hydration of the cement, concrete placed in cold weather should be maintained at a high moisture content at the surface.

2.6.2 *Protection against frost damage*—Maintenance of a high moisture content at the surface when the air temperature may drop below freezing increases the risk of frost damage, scaling, and aggregate pop-outs. If freezing and thawing is anticipated at any time during construction or in service, when either the cement paste or the aggregates are critically saturated near the surface of the concrete, air-entrained concrete should be used. For such conditions, properly air-entrained concrete containing the air contents required by ACI 318 or ACI 301, or recommended by ACI 201.2R, should be used. Damage from freezing at early ages should be prevented by protecting the concrete from freezing, and by curing without the addition of external water until a compressive strength of at least 3.5 MPa (500 psi) is developed. If it is likely that the concrete will be critically saturated when subsequently exposed to freezing and thawing temperatures, protection and curing should be continued until a compressive strength of about 28 MPa (4000 psi) is reached. For resistance to deicer scaling, a compressive strength of 31 MPa (4500 psi) should be attained before such exposure is permitted (Klieger 1956; Powers 1962; Mather 1990). These general rules may not apply for concretes incorporating special cements or other special ingredients or for concrete that is cured at high temperature (Pfeifer and Marusin 1982; Heinz and Ludwig 1987; Kelham 1996).

2.6.3 *Rate of concrete strength development in cold weather*—When added-water curing is required in freezing weather, ACI 306 and ACI 318 require that the temperature of the moist concrete be kept above 10 C (50 F). Although concrete continuously maintained at a curing temperature of 10 C (50 F) in the field will be protected against freezing, such concrete will develop compressive strength at about half the rate of a companion cylinder cured in the lab at 23 C (73 F). This reduced rate of strength gain can have significant impact on construction operations such as form and shore removal, or the introduction of construction or service loads. When rate of strength development is critical in cold weather, it may be necessary to increase the curing temperature on the basis of tests with the specific concrete mixture. In-place tests may be necessary.

2.6.4 *Removal of cold-weather protection*—When water curing is used, either by retention or application of water, and the ambient air temperature is or will be below freezing, the concrete should be allowed to dry for at least 12 h before discontinuing or removing the temperature protection to minimize the likelihood of a nearly saturated surface condition when the concrete is exposed to freezing temperatures. Otherwise, a moist and perhaps warm concrete surface can be rapidly cooled and dried, resulting in freezing and, in some cases, cracking. Therefore, cold-weather curing coverings should be removed in stages to slow the rates of cooling and drying.

2.7—Hot-weather protection and curing

Hot weather can include warm, humid environments like summer along the Gulf of Mexico or within large river valleys that can be relatively benign in regard to concrete curing, or the more hostile warm and dry environments like arid regions of the west or southwest U.S. In these dry environments, it is critical to maintain adequate moisture content in the concrete, and under such conditions, the added curing water itself can evaporate so quickly that it requires constant replacement. This is complicated by the limited availability of curing water in such environments.

Hot weather is defined as any combination of high air temperature, low relative humidity, and wind velocity that impairs the quality of fresh or hardened concrete, or otherwise results in undesirable concrete properties. Because hot weather can lead to rapid drying of concrete, protection and curing are critical. Additional information about curing concrete in hot weather is contained in ACI 305R.

Hot-weather curing starts before the concrete is placed, with steps taken to ensure that the subgrade, adjacent concrete, or formwork do not absorb water from the freshly placed concrete. This problem can be minimized by spraying the formwork, existing concrete, reinforcement, and subgrade with water before placement, which can also lower the temperature of those surfaces. Quality-control measures should be used to avoid standing or ponded water on these surfaces while placing the concrete.

During hot weather, initial curing methods should be used immediately after placing, and before and during the finishing process. Steps for initial curing should be taken to slow the evaporation of the bleed water or to replenish the bleed water. Evaporation rate is reduced by windscreens or sunscreens that block wind and radiant energy, and by fogging that temporarily increases the humidity of the air above the concrete. Some of the fog droplets fall to the concrete surface and augment the bleed water. Training, judgment, and quality-control measures are required to replace the evaporated bleed water. At no time is it proper to mix surface water with the top layer of cement paste in subsequent finishing operations. Mixing of water into the paste increases the *w/cm* at the surface, reducing strength and durability in this critical portion of the concrete. When high temperatures with wind, low humidity, or both, prevail, an evaporation-reducing film may need to be applied one or more times during the finishing operation to reduce the risk of plastic shrinkage cracking and crusting (Section 2.3.2).

Final curing methods can be used once the concrete surface will not be damaged by the application of curing materials or water. The need for continuous curing is greatest during the first few days after placement of the concrete in hot weather. During hot weather, provided that favorable moisture conditions are continuously maintained, concrete can attain a high degree of maturity in a short time. Water-curing, if used, should be continuous to avoid volume changes due to alternate wetting and drying. Liquid membrane-forming compounds with white (Type II) pigments should be used to reflect solar radiation.

2.8—Accelerated curing

A variety of proprietary products and specialized curing procedures have been developed to rapidly cure concrete products. These include insulating the concrete to accelerate curing by retaining heat of hydration or the addition of heat via steam or other methods. Such procedures, alone or in combination, are used to reduce the total time required for the concrete to achieve sufficient strength to permit handling. High-temperature, high-pressure, or steam curing are beyond the scope of this document. See Pfeifer and Marusin (1982), Heinz and Ludwig (1987), and Kelham (1996).

2.9—Minimum curing requirements

2.9.1 *General*—Curing should be continued long enough to ensure that 100% of the specified value for concrete properties will be developed in a reasonable time period after deliberate curing measures have been terminated, especially for mechanical properties such as strength or modulus of elasticity, and for durability-related properties, such as low permeability, abrasion or scaling resistance, initial surface absorption, or resistance to freezing and thawing. After curing measures are terminated and the concrete is fully exposed to the natural environment, the rate at which mechanical- or durability-related properties continue to develop could be reduced significantly. In the case of concrete properties in the curing-affected zone, further development may cease altogether upon drying of the near surface. For these reasons, it is always best to maintain deliberate curing until the desired in-place properties have been achieved. Termination of deliberate curing at some time short of full development of desired properties may be reasonable when based on experience with a given concrete mixture in a given environment. Thus, when strength is the essential performance criterion, it is common to maintain curing measures until a minimum of 70% of the specified 28-day strength f_c', has been achieved (ACI 301). When the structure's performance requires that the in-place strength or other concrete property reaches 100% of the specified value, curing should be extended until tests prove that the specified property has been reached. The temperature and moisture content of small, field-cured cylinders can differ significantly from that of the larger concrete placement that they are meant to represent, however. The in-place strength can be verified by tests such as penetration resistance (ASTM C 803), pullout tests (ASTM C 900), maturity measurement (ASTM C 1074), or tests of cast-in-place cylinder specimens (ASTM C 873). Also refer to ACI 228.1R for procedures to implement these in-place tests.

When performance criteria, such as surface hardness, abrasion resistance, resistance to freezing and thawing, surface absorption, or permeability are required, curing may need to be extended until the required values for such properties are achieved. It may be necessary to perform laboratory tests to evaluate the effect of curing on various concrete properties. Useful standard tests for this purpose may include C 666, C 642, C 1151, C 1202, C 944, C 418, C 779, and C 1138. (See also Liu [1994].) In-place tests for surface penetrability are discussed in 228.2R and in Chapter 4 of this document.

2.9.2 *Factors influencing required duration of curing*—The duration of curing required to achieve the desired levels of strength, durability, or both, depends on the chemical composition and fineness of the cementitious materials, *w/cm*, mixture proportions, aggregate characteristics, chemical and mineral admixtures, the temperature of the concrete, and the effectiveness of the curing method in retaining moisture in the concrete. This complex set of factors makes it difficult to confidently state the minimum curing time required to achieve the desired level of performance with the particular mixture in question. For concrete with and without pozzolans and chemical admixtures, a 7-day minimum duration of curing will often be sufficient to attain approximately 70% of the specified compressive strength. It is not necessarily true, however, that durability characteristics, such as abrasion resistance or surface absorption, will reach satisfactory levels in the same minimum time. Certain cement and admixture combinations, and high temperature are likely to reduce the time required to less than 7 days, while other combinations of materials, cooler concrete temperatures, or both, will extend the time required.

In general, when the development of a given strength or durability is critical to the performance of the concrete during construction or in service, the minimum duration of curing should be established on the basis of tests of the required properties performed with the concrete mixture in question. It is the responsibility of the designer and specifier to determine which properties are critical to the performance of the concrete under the intended service conditions and to develop a testing program to verify that the curing has been maintained long enough so that such properties have been achieved.

When natural weather conditions of temperature, humidity, and precipitation combine to cause zero net evaporation from the surface of the concrete, no curing measures are required to maintain adequate moisture content for as long as those natural conditions remain. In most climates, however, such conditions can change hourly, or daily, and rarely persist for the time required to foster development of the required concrete properties. It is therefore necessary to take steps to protect the concrete against loss of moisture. When no data are available from experience, values are not specified for concrete strength or durability, and when testing is not performed to verify in-place strength, concrete should be maintained above 10 C (50 F) and kept moist for the minimum curing periods shown in Table 2.1. Table 2.1 is not intended to apply to accelerated curing under high temperature, high pressure, or both.

Table 2.1—Recommended minimum duration of curing for concrete mixtures[*]

	Minimum curing period
ASTM C 150 Type I	7 days
ASTM C 150 Type II	10 days
ASTM C 150 Type III or when accelerators are used to achieve results demonstrated by test to be comparable to those achieved using ASTM C 150 Type III cement	3 days
ASTM C 150 Type IV or Type V cement	14 days
Blended cement, combinations of cement and other cementitious materials of various types in various proportions in accordance with ASTM C 595, C 845, and C 1157	Variable. See section 2.9.

[*]with various cement types when no testing is performed and no concrete properties are specified

CHAPTER 3—CURING FOR DIFFERENT TYPES OF CONSTRUCTION
3.1—Pavements and other slabs on ground

3.1.1 *General*—Slabs on ground include highway pavements, airfield pavements, canal linings, parking lots, driveways, walkways, and floors. Slabs have a high ratio of exposed surface area to volume of concrete. Without preventive measures, the early loss of moisture due to evaporation from the concrete surface could be so large and rapid as to result in plastic shrinkage cracking. Continued loss of moisture, and the accompanying decrease in the degree of hydration, would have a deleterious effect on strength, abrasion resistance, and durability. When moisture loss is predominantly at the top surface of the concrete, the gradient in moisture content leads to greater shrinkage at the top than at the bottom, which in turn leads to an upwards curling of the slab (Ytterberg 1987a,b,c). Alternatively, moisture can be lost from the bottom surface due to absorption into a dry subgrade, causing the opposite moisture gradient if the top surface is kept moist. This also leads to distortion of the slab. To minimize the development of such gradients in moisture content, both the top and bottom of slabs on ground should be uniformly moist or uniformly dry. Uniformly moist conditions are usually required if the properties of the concrete surface are important for the performance or appearance of the slab. This is achieved by prewetting the subgrade, and minimizing moisture loss at the top surface through initial, intermediate, and final curing as described in Chapter 1. Similarly, when an impervious membrane or vapor barrier is installed below the slab, maintaining the top surface in a moist condition is imperative to minimize curling. Placement of a 100 mm (4 in.) compacted, drainable fill on top of membranes and vapor retarders helps to dry the bottom of the slab so that curling is reduced while both the top and bottom surfaces dry (ACI 302.1R). The final tendency for distortion of the slab due to differential volume change, however, will depend on the moisture gradient after curing measures have been terminated.

3.1.2 *Curing procedures*—To maintain a satisfactory moisture content and temperature, the entire surface of the newly placed concrete should be treated in accordance with one of the water-curing methods (Section 2.2), one of the curing material methods (Section 2.4.2), or a combination thereof, beginning as soon as possible after finishing operations, without marring the surface.

To avoid plastic-shrinkage cracks, protective measures such as sun shields, wind breaks, evaporation reducers, or fog spraying should be initiated immediately to reduce evaporation. Exposed surfaces of the slab should be entirely covered and kept wet until the required concrete properties have developed to the desired level.

Mats used for curing can either be left in place and kept saturated for completion of the curing, or can be subsequently replaced by a liquid membrane-forming curing compound, plastic sheeting, reinforced paper, straw, or water. If the concrete has been kept continuously moist since casting and finishing, drying of the concrete with its accompanying shrinkage can begin only when the curing procedures are discontinued. Therefore, the surface should be protected against rapid loss of moisture upon the termination of curing by replacing wet burlap with plastic sheets until the surface has dried under the sheets.

3.1.3 *Duration of curing*—When the average ambient daily temperature (computed as the average of the highest and lowest temperature from midnight to midnight) is above 5 C (40 F), the recommended minimum period of maintenance of moisture and temperature for all procedures is as shown in Table 2.1 (Section 2.9.2) or it is the time necessary to attain an in-place compressive strength of the concrete of at least 70% of the specified compressive or flexural strength, whichever period is longer. If testing is not performed to verify in-place strength, concrete should be maintained above 10 C (50 F) and kept moist for the time periods shown in Table 2.1, unless otherwise directed in the specifications. Strength-based criteria should be replaced or augmented with durability-related criteria when appropriate. When concrete is placed at an average daily temperature of 5 C (40 F) or lower, precautions should be taken to prevent damage by freezing as recommended in ACI 306R. These general-purpose recommendations can be insufficient if durability-related surface properties are required.

3.2—Buildings, bridges, and other structures

3.2.1 *General*—Concrete in structures and buildings includes cast-in-place walls, columns, slabs, beams, and all other portions of buildings except slabs-on-grade, that are covered in Section 3.1. It also includes small footings, piers, retaining walls, tunnel linings, and conduits. Not included are mass concrete (see Section 3.3), precast concrete, and other constructions as discussed in Section 3.4.

3.2.2 *Curing procedures*—Under usual placing conditions, curing should be accomplished by one or a combination of methods discussed in Chapters 1 and 2.

Additional curing should be provided after the removal of forms when the surface strength or durability of underside surfaces is deemed important, or when it is necessary to minimize dusting. Additional curing is done by either applying a liquid membrane-forming curing compound or by promptly applying sufficient water to keep the surface continuously moist. Water curing of vertical surfaces can be done by using wet burlap covered with polyethylene. Water curing of the bottom of slabs and beams is not recommended and is rarely effective. Form removal should be done when curing has been sufficient.

After the concrete has hardened and while the forms are still in place on vertical and other formed surfaces, form ties may be loosened when damage to the concrete will not occur and water applied to run down on the inside of the form to keep the concrete wet. Care should be taken to prevent thermal shock and cracks when using water that is significantly cooler than the concrete surface. Curing water should not be more than about 11 C (20 F) cooler than the concrete (Section 2.2.1). Immediately following form removal, the surfaces should be kept continuously wet by a water spray or water-saturated fabric or until the membrane-forming curing compound is applied. Curing measures should include treatment of top surfaces.

3.2.3 *Duration of curing*—When the daily mean ambient temperature is above 5 C (40 F), curing should be continuous for the time periods shown in Table 2.1, or for the time necessary to attain a minimum of 70% of specified compressive (or flexural strength if appropriate), whichever period is longer. If concrete is placed with daily mean ambient temperatures at 5 C (40 F) or lower, precautions should be taken as recommended in ACI 306R. Strength-based criteria should be replaced or augmented with durability-related criteria when appropriate (See Section 2.9.1 and Chapter 4).

3.3—Mass concrete

3.3.1 *General*—Mass concrete is any volume of cast-in-place concrete with dimensions large enough to require measures be taken to cope with the generation of heat and attendant volume change and to minimize cracking. It is most frequently encountered in piers, abutments, dams, heavy footings, and similar massive constructions; although, the impact of temperature rise and thermal gradients should be considered in all concrete, whether the concrete is reinforced or not. Such problems are exacerbated where high strength and high cementitious materials contents are required. Recommendations for the control of temperature and thermal gradients in mass concrete are found in ACI 207.1R and ACI 207.2R.

3.3.2 *Methods and duration of curing*—Mass concrete is often cured with water for the additional cooling benefit in warm weather; however, this can be counterproductive when the temperature gradient between the warmer interior and the cooler surface generates stress in the concrete. Horizontal or sloping unformed surfaces of mass concrete can be maintained continuously wet by water spraying, wet sand, or water-saturated fabrics. For vertical and other formed surfaces, after the concrete has hardened and the forms are still in place, the form ties may be loosened and water supplied to run down the inside of the form to keep the concrete wet (Section 3.2.2). Immediately following form removal, the surfaces can be kept continuously wet by a water spray or water-saturated fabric. Curing water should not be more than approximately 11 C (20 F) cooler than the concrete, because induced surface strains may cause cracking.

Liquid membrane-forming curing compounds may be the best alternative in some instances. During cold weather, for example, after the initial protection period from freezing, application of a liquid membrane-forming curing compound, in lieu of spraying surfaces with water, will adequately reduce drying and provide satisfactory curing conditions without icing problems. The use of curing compounds may be permitted if the surface is not a construction joint or if the membrane is removed before placing adjacent concrete. A self-dissipating membrane-forming curing compound can be used on concrete surfaces that are to receive an additional layer of concrete or other bonded surface treatment. The surface should be cleaned before the new concrete is placed. Use of the membrane-forming curing compound, however, may alter the appearance of the concrete surface.

Curing should start as soon as the concrete has hardened sufficiently to prevent surface damage. For unreinforced massive sections not containing ground granulated blast-furnace slag or pozzolan, curing should be continued for not less than 2 weeks. Where ground granulated blast-furnace slag or pozzolan is included in the concrete, the minimum time for curing should be not less than 3 weeks. For reinforced mass concrete, curing should be continuous for a minimum of 7 days or until 70% of the specified compressive strength is obtained, if strength is the key concrete performance criterion. For construction joints, curing should be continued until resumption of concrete placement or until the required curing period is completed.

3.3.3 *Form removal and curing formed surfaces*—Forms for mass concrete can be removed as soon as removal operations can be safely performed without damage to the concrete or impairment to the serviceability of the structure. During cold weather, the protection afforded by forms can make it advantageous to leave the forms in place until the end of the minimum protection period or even longer. When forms are removed and protection is discontinued, the concrete should be cooled gradually to ambient temperature at rates not exceeding 14 C (25 F) in 24 h. The concrete can be cooled gradually by replacing the forms with coverings that retain less heat when the forms are removed. When the temperature differential between the concrete surface and the ambient air is less than 14 C (25 F), forms can be removed and protection discontinued without the need for gradual cooling.

3.4—Curing colored concrete floors and slabs

Concrete can be colored by applying a dry-shake hardener or using integral coloring pigments. The goal is normally to obtain a colored surface with minimal variations. It is imperative that the curing process be prompt, continuous, and uniform.

Table 3.1—Curing for specialty concrete

Specialty concrete	ACI committee report
Refractory concrete	547.1R
Insulating concrete	523.1R
Expansive cement concrete	223
Roller-compacted concrete	207.5R
Architectural concrete	303R
Shotcrete	506.2
Fiber-reinforced concrete	544.3R
Vertical slipform construction	313

The following methods have been used successfully to provide satisfactory moisture retention, adequate strength development for the wearing surface, and to minimize cracking:

- Application of a clear membrane-forming curing and sealing compound meeting ASTM C 1315, Type I, Class A. (Note that even nonyellowing compounds will discolor over time.) For colored industrial floors subjected to moderate or heavy traffic, the curing compound should limit moisture loss to 0.040 kg/m^2 (0.008 lb/ft^2) at a coverage rate of 7.4 m^2/L (300 ft^2/gal.);

- Application of a matching pigmented membrane-forming curing compound. (Note that a significant color difference can be expected when the curing compound wears off);

- Application of a removable curing compound. The removal process should be thoroughly discussed before applying these materials. The timing of the removal process is affected by many factors. One such factor is the rate of top surface strength development;

- Application of an approved nonstaining sheet membrane. This membrane has plastic on the outer surface and felt (or similar absorptive, nonstaining material) on the inner surface. This membrane should be placed flat on the concrete surface to minimize mottling due to differential moisture loss. The use of polyethylene alone is not recommended because the contact between the polyethylene and the surface of the concrete is variable, resulting in a mottled appearance; and

- Application of ponding or other equivalent moist-curing or water-retention methods. The surface should be kept continuously moist for 7 days or longer without periodic drying. Ponding can affect the appearance of the colored concrete. Check the water source for minerals or compounds that can stain or modify the color of the concrete. Also check any water-retention coverings that can discolor the concrete.

Placement of a test slab is recommended to visually assess the appearance achieved by the combination of concrete coloring and curing methods. The test slab should be larger than 10 m^2 (100 ft^2) and should be prepared using the concrete mixture and finishing and curing techniques planned for the project. The environmental conditions should be the same or similar to those expected for the project. Several different curing methods can be used on the test slab(s).

3.5—Other constructions

Previous chapters and sections have addressed curing for normal cast-in-place concrete. Curing for specialty concrete and special construction techniques is referenced in Table 3.1.

CHAPTER 4—MONITORING CURING AND CURING EFFECTIVENESS
4.1—General

Most specifications for curing freshly placed concrete prescribe a curing method or acceptable alternatives combined with a specified duration over which the methods must be used. Monitoring the effectiveness of the curing methods used or evaluating the environment in which the concrete has been placed can be of value but is rarely done. Several of the techniques currently available for such monitoring are listed as follows and are discussed in detail later in this chapter. Some of these techniques will likely be developed further to evaluate the need for curing, the effectiveness of the curing methods used, and compliance with applicable specifications. The following actions can be taken to evaluate curing and curing effectiveness:

- Monitor the environmental conditions in which the concrete is placed to evaluate the need for temperature and moisture control;
- Verify that the specified curing procedures have been used;
- Monitor the quantitative changes in the immediate environment as a result of curing procedures;
- Monitor the moisture content and temperature in the concrete; and
- Monitor the physical properties of the concrete, as influenced by the application of curing procedures. Properties of the concrete near to the surface are the most sensitive to curing and are often the most useful or reliable indicators of curing effectiveness.

4.2—Evaluating the environmental conditions in which the concrete is placed

The need for moisture control in freshly placed concrete depends on the rate of moisture loss from that concrete. Moisture loss depends on the water content, the ease with which water can move through the fresh concrete, the rate of bleeding, the rate of absorption into forms or subgrade, and the rate of evaporation of water from the exposed surfaces of the fresh concrete. The rate of evaporation from the surface of concrete further depends on the temperature and other properties of the mixture, construction operations, surface texture, and ambient environmental conditions (Section 1.4.2.2.1). Finally, evaporation from the surface depends on whether the surface is directly exposed to the air or is covered with bleed water, curing water or chemical treatments, or with surface coverings.

Evaporation depends on environmental factors that include the temperature of the concrete, the temperature of the water on the surface of the concrete, the temperature and RH of the air above the surface of the concrete, and the wind speed close to the concrete surface (Section 1.4.2.2.1). These factors combine with the characteristics of the concrete mixture and surface texture to promote or hinder evaporation of water from the concrete surface.

4.2.1 *Estimating evaporation rate*—Approximate methods for estimating the rate of evaporation of water from a water-covered surface have been proposed since the early 1800s (Brutsaert 1982; Veihmeyer 1964; Uno 1998). Each of these methods has in common the estimation of rate of evaporation on the basis of measurements of air temperature and relative humidity, water temperature, and wind speed. The most common of these approximate methods used by the concrete industry is the relationship adopted from hydrological applications (Menzel 1954; Veihmeyer 1964; Uno 1998). This was subsequently reformatted by the National Ready Mixed Concrete Association (NMRCA 1960) to produce the nomograph shown in Fig. 4.1.

The nomograph is most commonly used to estimate evaporation rate for the purpose of evaluating the risk of plastic shrinkage cracking (Lerch 1957; NRMCA 1960). This is based on Lerch's supposition that the surface begins to dry when the evaporation rate exceeds the bleeding rate. Because bleeding rates vary for different mixtures from 0 to over 1.0 kg/m^2/h (0.2 lb/ft^2/h), vary over time after casting, and are not normally measured in the field, a value is most often assumed for the bleeding rate that then becomes an implicitly assumed value for critical rate of evaporation. The most commonly quoted value is 1.0 kg/m^2/h (0.2 lb/ft^2/h) based on work originally reported in 1954 and 1955 (Menzel 1954; Lerch 1957). Recent experience with high-performance bridge deck overlays containing silica fume that exhibited a sharply reduced bleeding rate has led to specified maximum allowable evaporation rates of only 0.25 kg/m^2/h (0.05 lb/ft^2/h) for overlays (Virginia DOT 1997). Other specifications reflecting a reduced bleeding rate with modern concretes vary from 0.50 to 0.75 kg/m^2/h (0.10 to 0.15 lb/ft^2/h) (Krauss and Rogalla 1996). When the concrete mixture has a zero bleeding rate, the critical evaporation rate above which the surface will dry is zero. Zero bleeding rates are characteristic of dense mixtures incorporating fly ash, silica fume, ground-granulated blast-furnace slag or other pozzolans, high cement contents, low w/cm, fine cements, or high air contents. Further, all concrete mixtures exhibit a diminishing bleeding rate during setting with an eventual bleeding rate of zero.

4.2.1.1 *The Menzel/NRMCA nomograph*—The nomograph[4] (Fig. 4.1) and its underlying equations characterize the evaporative environment and are not intended to estimate the actual rate at which water is being lost from the concrete surface. The nomograph provides an estimate of "evaporativity," which is the maximum rate "at which the atmosphere can vaporize water from a free water surface" (Jalota and Prihar 1998; Uno 1998). The nomograph is therefore most useful as a means of approximately characterizing the environment into which the concrete is being placed. This can be helpful for forecasting the need for curing and protection measures and for estimating the likely effects of changes in air or concrete temperature, humidity, or wind speed on evaporation. The limitations inherent in the method are discussed in Section 4.2.1.2.

4.2.1.2 *Limitations and accuracy*—Because the nomograph was derived from an experimental fit of actual evaporation rates

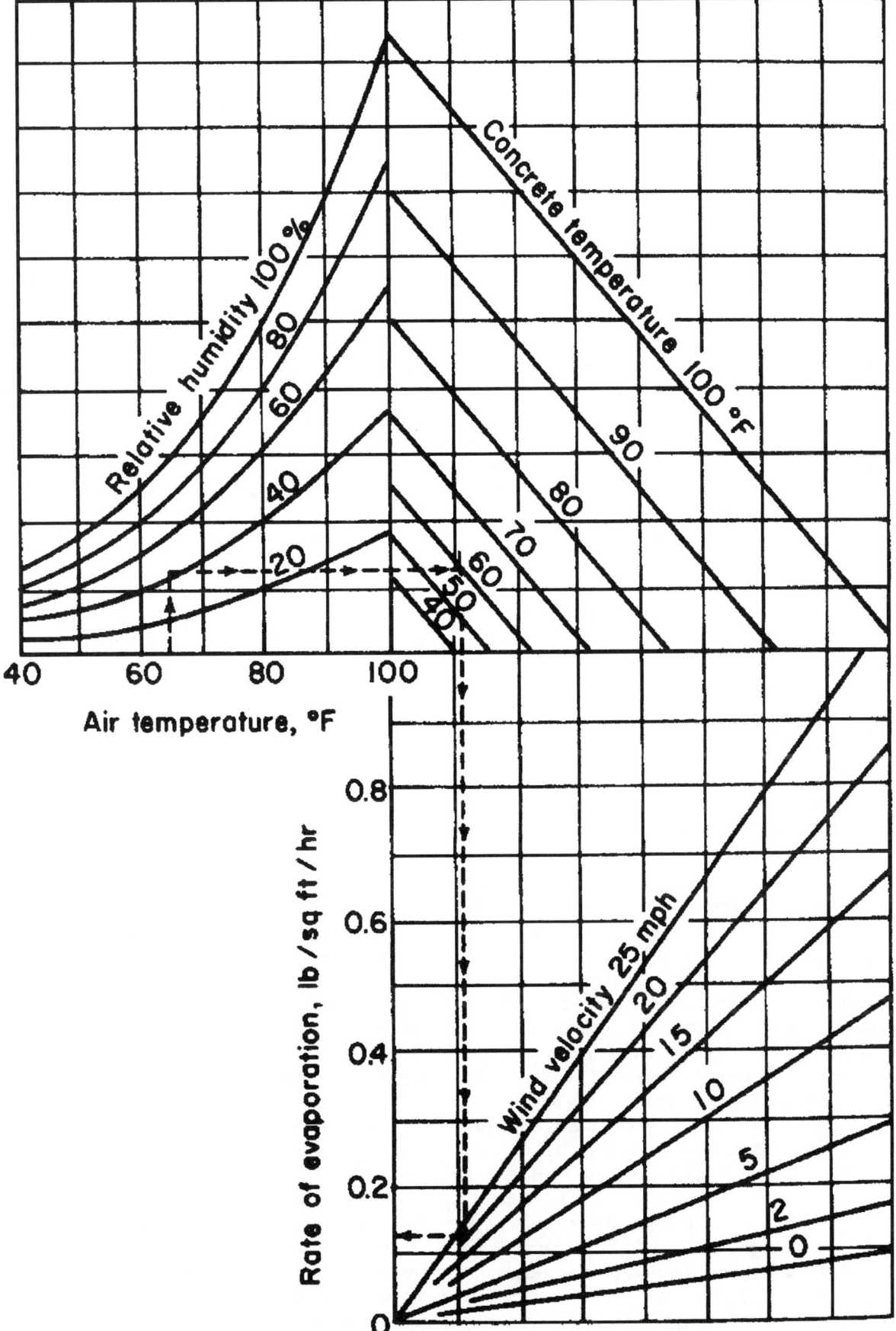

Fig. 4.1—Nomograph for estimating the maximum potential rate of evaporation of the environment, assuming a water-covered surface in which the water temperature is equal to the concrete temperature (Menzel 1954; NRMCA 1960).

with measurements of wind speed, temperature, and humidity, environmental measurements in the field should be taken as in the original experiments, as follows:
- The air temperature is to be taken 1.2 to 1.8 m (4 to 6 ft) above the evaporating surface, in the shade;
- The temperature of the water being evaporated at the surface is equal to the temperature of the concrete, and direct sunlight is not contributing to evaporation;
- The relative humidity should be measured in the shade, on the windward side (upwind) of, and 1.2 to 1.8 m (4 to 6 ft) above the evaporating surface;
- The wind speed should be measured at a height of 0.5 m (20 in) above the surface of the concrete. It is further assumed that the wind velocity profile (variation of wind speed with height above the evaporating surface) is identical to that which prevailed in the experiments that led to development of the equation.

These measurement conditions were defined in the original work by Menzel but were inadvertently deleted from the nomograph in 1960 and in all subsequent versions.

Uno (1998), Hover (1992), Shaeles and Hover (1988), and Al-Fadhala and Hover (2001) have discussed the accuracy of the nomograph and its sensitivity to the measuring protocols. In wind-tunnel testing under controlled conditions, Al-Fadhala and Hover (2001) demonstrated that when measurements were taken as originally defined in Menzel's paper (1954), estimates based on the nomograph were within ± 25% of actual evaporation rates up to 1.0 kg/m^2/h (0.2 lb/ft^2/h). At higher evaporation rates the nomograph consistently overestimated the actual rate by up to 50% at 1.8 kg/m^2/h (0.36 lb/ft^2/h).

The most common error in using the nomograph is to measure wind speed at other than 0.5 m (20 in.) above the evaporating surface in question. Because the surface air speed defines the evaporative environment and is highly variable based on ground clutter and the prevailing wind velocity profile, measurements taken from nearby weather stations at heights varying from 2 to 12 m (6 to 36 ft) will almost always lead to overestimates of evaporation rate. Furthermore, because wind speed fluctuates widely over even a short period of time, the average evaporation rate over the time required to cast a concrete slab, for example, is related to an average wind speed over that period. Therefore, a spot measurement of wind speed can be misleading in estimating evaporation rate.

Because the output of the nomograph is the rate of water loss from a water-covered surface under the same environmental conditions (such as a lake, reservoir, or a water-filled evaporation pan), the computed result approximates the water loss from concrete only when the concrete surface is covered with bleed water. Al-Fadhala and Hover showed that for the short time that concrete specimens were covered with bleed water, the rate of evaporation from the concrete was fairly well-approximated by the value obtained from the nomograph. The actual rate of water loss from the concrete surface decreased to approximately 50% of the free-water evaporation rate at 3 h after batching, however, diminishing to 10% at approximately 8 h. This is similar to the results obtained by Berhane (1984) and is due to the reduction of paste porosity that accompanies hydration of the cement, which in turn hinders the movement of water out of the concrete. No standardized technique is available to estimate the rate of evaporation from a concrete surface not covered with water (Berhane 1984, 1985; Mather 1984; Shaeles and Hover 1988; Al-Fadhala and Hover 2001).

Given the inherent variability in fluctuating environmental conditions and the empirical nature of the nomograph itself, this method is primarily useful for characterizing the environment in an approximate manner. The method correctly identifies the factors controlling evaporation and allows the user to predict how the various factors interact to either increase or decrease the severity of the evaporative environment. When field measurements are taken, the nomograph produces an estimate that is likely to be within 25% of the actual evaporativity of the construction environment. Alternative evaporation predictors are available in the hydrologic literature. Several methods for estimating evaporation rate are reviewed by Uno (1998), Brutsaert (1987), and Shaeles and Hover (1988).

4.2.1.3 *Alternative technologies for estimating evaporation rate*—Remote sensing devices have been developed for agricultural applications to estimate the rate of evaporation from the surface of field crops. In limited tests, one such instrument based on the horticultural work of Idso (1968) and Idso et al. (1969) was useful for estimating the rate of evaporation from the surface of concrete.

More accurate estimates of the actual rate of evaporation at the surface of the concrete can be made by typical agricultural or meteorological methods of evaporation pans and recording devices. A modification of these methods was used in which containers of the concrete mixture were exposed to evaporative conditions and periodically weighed to estimate the rate of evaporation from the concrete surface (Shaeles and Hover 1988; Al-Fadhala and Hover 2001; Krauss and Rogalla 1996).

Regardless of the method used to estimate evaporative conditions, the key requirement is to recognize that the rate of evaporation can vary considerably over a given localized area, and that the most significant evaporation rate is that measured at the concrete surface at the time of concrete placing, finishing, and curing. Data obtained from nearby weather stations, airports, or furnished by a local news service are questionable when estimating the conditions at a particular job site, or over a particular concrete surface on that site, and can lead to significant errors in calculating the rate of evaporation (Falconi 1996).

4.3—Means to verify the application of curing

When a particular curing process is specified, one can assess the application of curing in accordance with those specifications. For example, if the specifications require a 3-day water cure, one can monitor the duration over which the concrete surfaces were continuously exposed to water. Similarly, if covering with plastic or other materials is required, one simply observes whether this was done or not, and whether the concrete surface remained continuously covered.

On the other hand, when a liquid-applied curing membrane is used and a particular rate of application is required (either directly by the specifications or by reference to manufacturer's instructions), an average rate of application can be determined by monitoring the volume of liquid applied over a measured area.

While this can be done by counting empty drums of curing material at the conclusion of several concrete placements, more reliable estimates are made by first-hand observation of the application of the material. When power spray equipment is used, it is sometimes possible to monitor the flow rate from the hose or nozzle. Verification of curing-compound coverage rate is more difficult for deeply textured surfaces, however, where the contact surface area can be more than twice the projected horizontal area of a slab (Shariat and Pant 1984). This is true not only for slabs, but also for architecturally formed surfaces with bold relief.

To develop a sense for the degree of saturation or coverage corresponding to the required rate of application, apply the material in question to a small measured trial area. For example, at heavy rates of application, the liquid-applied curing material often forms a visible sheen over the surface and will sometimes stand in puddles at low spots. Once it is determined that this is the appropriate appearance for a sufficiently heavy application, it would become obvious that a light coating that is barely visible would not comply with the applicable specification.

4.4—Quantitative measures of the impact of curing procedures on the immediate environment

Some curing procedures, such as using a fog spray, erecting wind screens, or heating the air around the concrete being cast, modify the immediate environment. The effectiveness of such procedures can be directly monitored by temperature, humidity, or wind speed measurements as discussed previously.

The effect of curing procedures on the rate of evaporation can be assessed either by estimating evaporative conditions before and after the application of curing, or by other indirect methods. Indirect assessment of evaporation rate comes from recognition of the impact of evaporative cooling on the surface temperature of concrete. Thus, a drop in surface temperature can be the result of an increase in the rate of cooling. In the absence of other factors causing a change in concrete surface temperature, Folliard[4] observed that the application of a high-solids curing compound immediately reduced the evaporation rate and resulted in a rapid increase in surface temperature. This experiment was performed indoors without incident sunlight.

4.5—Quantitative measures of the impact of curing procedures on moisture and temperature

Curing reduces the loss of internal and surface moisture, and controls the temperature of concrete to permit sufficient hydration of the cementitious materials. Therefore, monitoring the effectiveness of the curing procedures can be performed by monitoring either the degree to which the temperature and internal moisture content have been controlled, or the degree to which hydration has progressed and developed improved concrete properties.

Direct measurements of concrete temperature are readily performed by a variety of techniques. Typical insertion thermometers used to measure the temperature of fresh concrete are of little value after setting, while embedded thermocouples are an inexpensive means of monitoring multiple locations and various depths in both fresh and hardened concrete (ACI 306R). Specially designed surface thermometers are useful. Pyrometers and infrared devices can be used to measure surface temperature remotely.

Internal moisture content or internal relative humidity can be measured with instruments of a variety of levels of sophistication. Relatively inexpensive, embeddable moisture gages are available, as are electronic moisture and humidity gages. (Refer to Section 4.7 for references on use of gages and interpretation of results.)

Through the use of such devices, one can comply with a performance specification for curing that requires that the concrete remain within a particular range of temperature and internal moisture content for a specified period of time.

Such instrumentation may also be valuable in assessing the tendency for undesirable phenomena such as shrinkage, cracking, and curling. Such phenomena are related to mixture composition and proportions, internal temperature and moisture content, and thermal and moisture gradients within the concrete.

4.6—Maturity method

The maturity method has been developed as a means of estimating the cumulative effect of time and temperature during curing on the development of concrete properties. The method is based on the influence of temperature on the rate of the reaction between cement and water, and assumes that the higher the concrete temperature, the more rapidly the cement hydrates, and therefore, the more rapid the development of strength and related properties. Concrete matures as the degree of hydration increases.

Application of the maturity method requires monitoring in-place concrete temperature in a structure over time and computation of the effect of that time-temperature history on the maturity of the in-place concrete. Accompanying laboratory work on the same concrete mixture correlates maturity to strength (or other property of interest) by testing standard specimens after exposure to various curing temperatures for varying lengths of time. In-place strength in the structure is inferred from the level of maturity determined in the field and from the strength-maturity relationship developed in the lab. Throughout this process, it is necessary to make sure that the concrete mixture used in the structure is the same as that used in the specimens for developing the correlation between maturity and strength, and that there is sufficient water for hydration. The maturity method is described in detail in ASTM C 1074 and in multiple references (Carino 1991; ACI 306R; ACI 228.1R).

The temperature record required for maturity calculations directly indicates the effectiveness of curing measures for controlling concrete temperature. Moisture control is not recorded in most current applications of maturity because it is assumed that the moisture condition of the structure and that of the test specimens is the same such that temperature differences alone control the relative rates of hydration. An error can be introduced into the maturity approach when an inadequate supply of moisture inhibits hydration of the cement. For example, a strength-maturity

---

[4] Folliard, K., and Hover, K. C., 1989, "National Science Foundation Research for Undergraduates Program, Cornell University (NSFREU)," *Report of Activities*, Aug., 1989.

Table 4.1—Concrete characteristics in the curing-affected zone

Near surface property affected by curing	Reference
Degree of hydration	Wainwright et al. 1990
Pore size distribution mercury-intrusion porosimetry	Wainwright et al. 1990
Oxygen or air permeability	Wainwright et al. 1990
	Ballim 1993
	Kollek 1989
	Nolan et al. 1997
	Basheer et al. 1990
	Dhir et al. 1995
	Ben-Othman and Buenfeld 1990
	Parrott 1995
	Dinku and Reinhardt 1997
	Figg 1973
	Tang and Nilson 1986
	Montgomery et al. 1992
Initial surface absorption	BS 1881: Part 5, 1970; Part 122, 1983; Part 208, 1996
	ASTM C 1151; Senbetta and Scholer 1984
	Hooton et al. 1993
	DeSouza et al. 1997
	DeSouza et al. 1998
Surface permeability/absorption test devices	Nolan et al. 1997
	Basheer et al. 1990
	McCarter et al. 1996
	Balayssac et al. 1993
	Ballim 1993
	Figg 1973
	Concrete Society 1988
	Dhir et al. 1987
	McCarter 1995
	Price and Bamforth 1991
	Sabir et al. 1998
	Balayssac et al. 1998
	Montgomery et al. 1992
Internal moisture content	Parrott 1995
	McCarter et al. 1996
	Nolan et al. 1997
	Persson 1997
Tension strength of surface concrete pull-off testing	Nolan et al. 1997
	Long and Murray 1984
	Montgomery et al. 1992
Depth of carbonation	Dinku and Reinhardt 1997
	RILEM Recommendations CPC-18 1988
	Nolan et al. 1997
	Balayssac et al. 1993
	Bier 1987
Abrasion resistance	Montgomery et al. 1992
	Sawyer 1957

relationship developed on the basis of wet-cured test specimens would lead to an over-estimate of strength for in-place concrete that was allowed to dry even though it was kept warm. Future developments in the maturity theory would allow the quantitative evaluation of the effects of both temperature and moisture on the development of concrete properties, and would direct a new interest in comprehensively monitoring curing effectiveness.

4.7—Measuring physical properties of concrete affected by temperature and moisture control to assess curing effectiveness

Because the ultimate goal of proper curing is the development of appropriate concrete properties, the final test of the effectiveness of curing is whether those properties are developed (Wainwright, Cabrera, and Gowriplan 1990). While virtually all concrete properties are sensitive to curing, the adequacy of curing is most readily observed in the properties of concrete at the curing-affected zone.

The curing-affected zone (Cather 1992) will vary in thickness depending on the properties of the concrete, the severity of the ambient conditions, and the curing time involved. For example, in low w/c concrete with a specified 28-day cylinder compressive strength of 55.2 MPa (8000 psi), the resulting curing-affected zone was observed to extend from 6 to 10 mm (1/4 to 3/8 in.) from the surface after a period of 1 year of continuous wet curing (Hover 1984). In more conventional concretes, this zone may be 10 to 13 mm (3/8 to 1/2 in.) deep at an age of 28 days (Hover 1984; Cather 1992; Dhir et al. 1986-92). Regardless of the shallowness of the zone, however, the concrete properties in this zone are most frequently those that determine the durability and serviceability of the concrete. In the case of an industrial floor, for example, the top 6 mm (1/4 in.) is the most critical part of the entire concrete installation and is vital to satisfactory performance.

Given the shallowness of the curing-affected zone, test techniques that evaluate the properties of the concrete at depth, such as measuring the compressive strength of drilled cores, have a limited sensitivity to the effectiveness of curing. Concrete characteristics in the curing-affected zone that are likely to be more sensitive to curing effectiveness, along with a listing of references for test methods for these properties, are shown in Table 4.1.

CHAPTER 5—REFERENCES
5.1—Referenced standards and reports

The standards and reports listed below were the latest editions at the time this document was prepared. Because these documents are revised frequently, the reader is advised to contact the proper sponsoring group if it is desired to refer to the latest version.

AASHTO Material Standards
M148   Liquid Membrane Forming Curing Compounds
M182   Burlap Cloth Made From Jute or Kenaf
T26   Quality of water to be used in concrete

ACI Standards and Reports
116R   Cement and Concrete Terminology
201.2R   Guide to Durable Concrete
207.1R   Mass Concrete

207.2R	Effect of Restraint, Volume Change, and Reinforcement on Cracking of Massive Concrete
207.5R	Roller-Compacted Mass Concrete
223	Shrinkage-Compensating Concrete
228.1R	In-Place Methods to Estimate Concrete Strength
232.2R	Use of Fly Ash in Concrete
233R	Ground Granulated Blast-Furnace Slag as a Cementitious Constituent in Concrete
234R	Guide for Use of Silica Fume in Concrete
301	Standard Specification for Structural Concrete
302.1R	Guide for Concrete Floor and Slab Construction
303R	Cast-in-Place Architectural Concrete
305R	Hot Weather Concreting
306R	Cold Weather Concreting
306.1	Standard Specification for Cold Weather Concreting
308	Standard Practice for Curing Concrete
308.1	Standard Specification for Curing Concrete
313	Design and Construction of Concrete Silos and Stacking Tubes for Storing Granular Materials
318	Building Code Requirements for Structural Concrete (318-99) and Commentary (318R-99)
506.2	Specification for Shotcrete
523.1R	Guide for Cast-in-Place Low-Density Concrete
544.3R	Guide for Specifying, Proportioning, Mixing, Placing and Finishing Steel Fiber Reinforced Concrete
547.1R	Refractory Plastics and Ramming Mixes
548.1R	Polymers in Concrete

ASTM Standards

C 33	Specifications for Concrete Aggregates
C 94	Specification for Ready Mixed Concrete
C 125	Terminology Relating to Concrete and Concrete Aggregate
C 156	Test for Water Retention by Concrete Curing Materials
C 171	Specification for Sheet Materials for Curing Concrete
C 232	Test Method for Bleeding of Concrete
C 309	Specification for Liquid Membrane-Forming Compounds for Curing Concrete
C 403/C 403M	Test Method for Time of Setting of Concrete Mixtures by Penetration Resistance
C 418	Test method for Abrasion Resistance of Concrete by Sandblasting
C 666	Test Method for Resistance of Concrete to Rapid Freezing and Thawing
C 672/C 672M	Test Method for Scaling Resistance of Concrete Surfaces Exposed to Deicing Chemicals
C 779	Test Method for Abrasion Resistance of Horizontal Concrete Surfaces
C 803/C 803M	Test Method for Penetration Resistance of Hardened Concrete
C 805	Test for Rebound Number of Hardened Concrete
C 873	Test for Compressive Strength of Concrete Cylinders Cast-in-Place in Cylindrical Molds
C 900	Test for Pullout Strength of Hardened Concrete
C 944	Test Method for Abrasion Resistance of Concrete or Mortar Surfaces by Rotating-Cutter Method
C 1074	Practice for Estimating Concrete Strength by the Maturity Method
C 1138	Test Method for Abrasion Resistance of Concrete (Underwater Method)
C 1151	Test Methods for Evaluating the Effectiveness of Materials for Curing Concrete
C 1202	Test Method for Electrical Indication of Concrete's Ability to Resist Chloride Ion Penetration
C 1315	Specification for Liquid Membrane-Forming Compounds Having Special Properties for Curing and Sealing Concrete

5.2—Cited references

Abel, J. D., and Hover, K. C., 2000, "Field Study of the Setting Behavior of Fresh Concrete," *Cement, Concrete, and Aggregates*, V. 22, No. 2, Dec., pp. 95-102.

Aïtcin, P.-C., 1999, "Does Concrete Shrink or Does it Swell?" *Concrete International*, V. 21, No. 12, Dec. pp. 77-80.

Al-Fadhala, M., and Hover, K. C., 2001, "Rapid Evaporation from Freshly Cast Concrete and the Gulf Environment," *Construction and Building Materials*, V. 15, No. 1, Jan., pp. 1-7.

Balayssac, J. P.; Detriche, C. H.; and Diafat, N., 1998, "Effect of Wet Curing Duration upon Cover Concrete Characteristics," *Materials and Structures*, V. 31, June, pp. 325-328.

Balayssac, J. P.; Detriche, C. H.; and Grandet, J., 1993, "Validity of the Water Absorption Test for Characterizing Cover Concrete," *Materials and Structures*, V. 26, May, pp. 226-230.

Ballim, Y., 1993, "Curing and the Durability of OPC, Fly Ash, and Blast-Furnace Slag Concretes," *Materials and Structures*, V. 26, May, pp. 238-244.

Basheer, P. A. M.; Long, A. E.; and Montgomery, F. R., 1994, "The Autoclam—a New Test for Permeability," *Concrete*, July-Aug., pp. 27-29.

Basheer, P. A. M.; Montgomery, F. R.; Long, A. E.; and Batayneh, M., 1990, "Durability of Surface Treated Concrete," *Protection of Concrete*, Proceedings of the International Conference, Dundee, Scotland, R. K. Dhir and J. W. Green, eds., E&FN Spon, London, pp. 211-221.

Ben-Othman, B., and Buenfeld, N.R., 1990, "Oxygen Permeability of Structural Lightweight Aggregate Concrete," *Protection of Concrete*, Proceedings of the International Conference, Dundee, Scotland, R. K. Dhir and J. W. Green, eds., E&FN Spon, London, pp. 725-736.

Berhane, Z., 1984, "Evaporation of Water from Fresh Mortar and Concrete at Different Environmental Conditions," ACI JOURNAL, *Proceedings* V. 81, No. 6, Nov.-Dec. pp. 560-564.

Berhane, Z., 1985, Closure to discussion on paper, "Evaporation of Water from Fresh Mortar and Concrete at Different Environmental Conditions," ACI JOURNAL, *Proceedings* V. 82, No. 6, Nov.-Dec., pp. 930-933.

Bier, T. A., 1987, "Influence of Type of Cement and Curing on Carbonation Progress and Pore Structure of Hydrated Cement Paste," *Materials Research Society Symposium*, V. 85, pp. 123-134.

British Standards Institute, "Testing Concrete—Methods of Testing Hardened Concrete For Other Than Strength," BS 1881-5:1970, London, 36 pp.

British Standards Institute, "Testing Concrete—Method for Determination of Water Absorption," BS 1881-122:1983, London, 4 pp.

British Standards Institute, "Testing Concrete—Recommendations for the Determination of the Initial Surface Absorption of Concrete," BS 1881-208:1996, London, 14 pp.

Brutsaert, W., 1982, *Evaporation Into the Atmosphere: Theory, History, and Applications*, D. Reidel Publishing Co., Dordrecht, Holland, 299 pp.

Brutsaert, W., 1987, *Evaporation into the Atmosphere: Theory, History and Application*, D. Reidel Publishing Co., Boston, Mass., 299 pp.

Bungey, J. H., 1990, "Measurement of Quality," *Protection of Concrete*, Proceedings of the International Conference, Dundee, Scotland, R. K. Dhir and J. W. Green, eds., E&FN Spon, London, pp. 833-845.

Bungey, J. H., 1992, "Near to Surface Strength Testing," *Concrete*, Sept.-Oct., pp. 34-36.

Cabrera, J. G.; Gowriplan, N.; and Wainwright, P. J., 1989, "An Assessment of Concrete Curing Efficiency using Gas Permeability, *Magazine of Concrete Research*, V. 41, No. 149, Dec., pp 193-198.

Cabrera, J. G., and Lynsdaale, C. J., 1988, "A New Gas Permeameter for Measuring the Permeability of Mortar and Concrete," *Magazine of Concrete Research*, V. 40, No. 144, Sept., pp. 177-182.

Carino, N. J., 1991, "The Maturity Method," *CRC Handbook on Nondestructive Testing of Concrete*, V. M. Malhotra and N. J. Carino, ed., CRC Press, Boca Raton, Fla., pp. 101-146.

Carrier, R. E., 1983, "Concrete Curing Tests," *Concrete International*, V. 5, No. 4, Apr., pp. 23-26.

Cather, R., 1992, "How to Get Better Curing," *Concrete, The Journal of the Concrete Society*, London, V. 26., No. 5, Sept.-Oct., pp. 22-25.

Cather, R., 1994, "Curing: the True Story?" *Magazine of Concrete Research*, V. 46. No. 168, Sept., pp. 157-161.

Concrete Society, 1988, "Permeability of Concrete—A Review of Testing and Experience," *Technical Report 31*, London, 95 pp.

Cook, R. A., 1992, "A Generalized Form of the Powers Model of Cement Microstructure and its Evaluation via Mercury Porosimetry," PhD dissertation, Cornell University, Ithaca, N.Y.

Copeland, L. E., and, Bragg, R. H., 1955, "Self-Desiccation in Portland Cement Pastes," *ASTM Bulletin* No. 204, Feb., 34 pp.

Copeland, L. E.; Kantro, D. L.; and Verbeck, G., 1960, "Chemistry of Hydration of Portland Cement," *Proceedings of the Fourth International Symposium on the Chemistry of Cement*, Washington, D.C.

Cordon, W. A., and Thorpe, J. D., 1965, "Control of Rapid Drying of Fresh Concrete by Evaporation Control," ACI JOURNAL, *Proceedings* V. 62, No. 8, Aug., pp. 977-985.

Dalton, J., 1802, "Experimental Essays on Evaporation," *Proceedings*, V. 5, Manchester Literary and Philosophical Society, pp. 536-602.

DeSouza, S. J.; Hooton, R. D.; and Bickley, J. A., 1997, "Evaluation of Laboratory Drying Procedures Relevant to Field Conditions for Concrete Sorptivity Measurements," *Cement Concrete and Aggregates*, V. 19, No. 2, Dec., pp. 59-63.

DeSouza, S. J.; Hooton, R. D.; and Bickley, J. A, 1998, "A Field Test for Evaluating High Performance Concrete Covercrete Quality," *Canadian Journal of Civil Engineering*, V. 25, No. 3, June, pp. 551-556.

Dhir, R. K.; Hewlett, P. C.; Byars, E. A.; and Shaaban, I. G., 1995, "A New Technique for Measuring the Air Permeability of Near-Surface Concrete," *Magazine of Concrete Research*, V. 47, No. 171, June, pp. 167-176.

Dhir, R. K.; Hewlett, P. C.; and Chan, Y. N., 1986, "Near Surface Characteristics and Durability of Concrete: An Initial Appraisal," *Magazine of Concrete Research*, V. 38, pp. 54-56.

Dhir, R. K.; Hewlett, P. C.; and Chan, Y. N., 1987, "Near Surface Characteristics and Durability of Concrete: Assessment and Development of In Situ Test Methods," *Magazine of Concrete Research*, V. 39, No. 141, Dec., pp. 183-195.

Dhir, R.K.; Hewlett, P. C.; and Chan, Y. N., 1989, "Near Surface Characteristics and Durability of Concrete: Intrinsic Permeability," *Magazine of Concrete Research*, V. 41, No. 148, Sept., pp. 87-97.

Dhir, R. K.; Hewlett, P. C.; and Chan, Y. N., 1989, "Near Surface Characteristics and Durability of Concrete: Prediction of Carbonation," *Magazine of Concrete Research*, V. 41, No. 148, Sept., pp. 137-144.

Dhir, R. K.; Hewlett, P. C.; and Chan, Y. N., 1991, "Near Surface Characteristics of Concrete: Abrasion Resistance," *Materials and Structures/Matériaux et Constructions*, V. 24, Mar., pp. 122-128.

Dhir, R. K.; Levitt, M.; and Wang, J., 1989, "Membrane Curing of Concrete: Water Vapour Permeability of Curing Membranes," *Magazine of Concrete Research*, V. 41, Dec., pp. 221-229.

Dinku, A., and Reinhardt, H. W., 1997, "Gas Permeability Coefficient of Cover Concrete as a Performance Control," *Materials and Structures*, V. 30, Sept., pp. 387-393.

Dodson, V. H., 1994, "Time of Setting, Significance of Tests and Properties of Concrete and Concrete-Making Materials," ASTM STP 169C, P. Klieger and J. Lamond, eds., ASTM, West Conshohocken, Pa., pp. 77-87.

Eglington, M., 1998, "Resistance of Concrete to Destructive Agencies," *Lea's Chemistry of Cement and Concrete*, 4th Edition, P. C. Hewlett, ed., Arnold Press, London, pp. 299-340.

Falconi, M. I., 1996, "An Investigation of the Scaling Resistance and Other Properties of Slag-Cement Concrete," Master's thesis, Cornell University, Ithaca, N.Y.

Figg, J. W., 1973, "Methods of Measuring the Air and Water Permeability of Concrete," *Magazine of Concrete Research*, V. 25, No. 85, Dec., pp. 213-219.

Gowriplan, N.; Cabrera, J. G.; Cusens, A. R.; and Wainwright, P. J., 1990, "Effect of Curing on Durability," *Concrete International*, V. 12, No. 2, Feb., pp. 47-54.

Greening, N. R., and Landgren, R., 1966, "Surface Discoloration of Concrete Flatwork," *Journal of the PCA Research and Development Laboratory*, V. 8, No. 3, Sept., pp. 34-50.

Grube, H., and Lawrence, C. D., 1984, "Permeability of Concrete to Oxygen," *Proceedings,* RILEM Seminar on the Durability of Concrete under Normal Outdoor Exposure, Hanover University, Mar., pp. 68-79.

Hanson, J. A., 1968, "Effects of Curing and Drying Environments on the Splitting Tensile Strength of Concrete," *PCA Development Bulletin D141.*

Heinz, D., and Ludwig, U., 1987, "Mechanism of Secondary Ettringite Formation in Mortars and Concretes Subjected to Heat Treatment," *Concrete Durability: Katharine and Bryant Mather International Conference*, SP-100, J. Scanlon, ed., American Concrete Institute, Farmington Hills, Mich., pp. 2059-2071.

Hong, C. Z., and Parrott, L., 1989, *Air Permeability of Cover Concrete and the Effect of Curing*, British Cement Association, Wexham Springs, Slough, England, Oct.

Hooton, R. D.; Mesic, T.; and Beal, D. L., 1993, "Sorptivity Testing of Concrete as an Indicator of Concrete Durability and Curing Efficiency," *Proceedings*, Third Canadian Symposium on Cement and Concrete, Ottawa, Aug.

Hover, K. C., 1992, "Evaporation of Surface Moisture: A Problem in Concrete Technology and Human Physiology," *Concrete in Hot Climates: Proceedings of the Third International RILEM Conference on Concrete in Hot Climates*, M. J. Walker, ed., E&FN Spon, London, pp. 13-24.

Hover, K. C., 1984, "The Influence of Moisture on the Physical Properties of Hardened Concrete and Mortar," PhD dissertation, Cornell University, Ithaca, N.Y., Aug.

Idso, S., 1968, *Analysis of Environment-Plant Relationships*, St. Paul Agricultural Experiment Station, Minn.

Idso, S. B.; Jackson, R. D.; Ehrler, W. L.; and Mitchell, S. T., 1969, "A Method for Determination of Infrared Emittance of Leaves," *Ecology*, V. 50, May, pp. 899-902.

Jalota, S. K., and Prihar, S. S., 1998, *Reducing Soil Water Evaporation with Tillage and Straw Mulching*, Iowa State University Press, Ames, Iowa, 142 pp.

Kelham, S., 1996, "The Effect of Cement Composition and Fineness on Expansion Associated with Delayed Ettringite Formation," *Cement and Concrete Composites*, V. 18, June, pp. 171-179.

Klieger, P., 1956, "Curing Requirements for Scale Resistance of Concrete," *Bulletin* No. 150, Highway Research Board, Jan.

Kohler, M. A.; Nordenson, T. J.; and Fox, W. E., 1955, "Evaporation from Pans and Lakes," *Research Paper* No. 38, U.S. Department of Commerce, Washington, D.C., May.

Kollek, J. J., 1989, "The Determination of the Permeability of Concrete to Oxygen by the Cembureau Method—a Recommendation," *Materials and Structures*, V. 22, No. 129, pp. 225-230.

Kosmatka, S. H., 1994, "Bleeding," *Significance of Tests and Properties of Concrete and Concrete-Making Materials,* ASTM STP 169C, P. Klieger and J. Lamond, eds., ASTM, West Conshohocken, Pa., pp. 88-111.

Kosmatka, S. H., and Panarese, W. C., 1988, *Design & Control of Concrete Mixtures*, Portland Cement Association, Skokie Ill., 13th Edition.

Krauss, P. D., and Rogalla, E. A., 1996, "Transverse Cracking in Newly Constructed Bridge Decks," *NCHRP Report 380*, National Cooperative Highway Research Program, Transportation Research Board, National Academy Press, Washington, D.C., 126 pp.

Lerch, W., 1957, "Plastic Shrinkage," ACI JOURNAL, *Proceedings* V. 53, Feb., pp. 797-802.

Liu, T. C., 1994, "Abrasion Resistance," *Significance of Tests and Properties of Concrete and Concrete-Making Materials,* ASTM STP 169C, P. Klieger and J. Lamond, eds., ASTM, West Conshohocken, Pa., pp. 182-191.

Long, A. E., and Murray, A. M., 1984, "The Pull-Off Partially Destructive Test for Concrete," *In-Situ/Nondestructive Testing of Concrete*, SP-82, V. M. Malhotra, ed., American Concrete Institute, Farmington Hills, Mich., pp. 327-350.

Mather, B., 1984, Discussion on a paper by Z. Berhane, "Evaporation of Water from Fresh Mortar and Concrete at Different Environmental Conditions," ACI JOURNAL, *Proceedings* V. 81, No. 6, Nov.-Dec., pp. 930-933.

Mather, B., 1987, "Curing of Concrete," *Lewis H. Tuthill International Symposium on Concrete and Concrete Construction*, SP-104, G. T. Halvorsen, ed., American Concrete Institute, Farmington Hills, Mich., pp. 145-159.

Mather, B., 1990, "Curing Compounds," *Concrete International*, V. 12, No. 2, Feb., pp. 40-41.

McCarter, W. J., 1995, "Properties of Concrete in the Cover Zone: Developments in Monitoring Techniques," *Magazine of Concrete Research*, V. 47, No. 172, pp. 243-251.

McCarter, W. J.; Ezirim, H.; and Emerson, M., 1996, "Properties of Concrete in the Cover Zone: Water Penetration, Sorptivity and Ionic Ingress," *Magazine of Concrete Research*, V. 48, No. 176, pp. 149-156.

Meeks, K. W., and Carino, N. J., 1999, *Curing of High Performance Concrete: Report of the State-of-the-Art*, NISTR 6295, National Institute of Standards and Technology, Building and Fire Research Laboratory, Gaithersburg, Md., Mar., 191 pp.

Menzel, C. A., 1954, "Causes and Prevention of Crack Development in Plastic Concrete," *Proceedings*, Portland Cement Association Annual Meeting, pp. 130-136.

Mills, R. H., 1966, "Factors Influencing the Cessation of Hydration in Water Cured Pastes," *Symposium on Structure of Portland Cement Paste and Concrete*, HRB SR 90, Highway Research Board, Washington, pp. 406-424.

Mindess, S., and Young, J. F., 1981, *Concrete*, Prentice Hall, Englewood Cliffs, N.J., 671 pp.

Minnesota Department of Transportation, 1998, "Standard Specifications for Road and Bridge Construction," Extreme Service Membrane Curing Compound, St. Paul, Minn.

Montgomery, F. R.; Basheer, P. A. M.; and Long, A. E., 1992, "Influence of Curing Conditions on the Durability Related Properties of Near Surface Concrete and Cement Mortars," *Durability of Concrete: G. M. Idorn International*

Symposium, SP-131, J. Holm and M. Geiker, eds., American Concrete Institute, Farmington Hills, Mich., pp. 127-138.

Murdock, L. J.; Brook, K. M.; and Dewar, J. D., 1991, *Concrete Materials and Practice*, 6th Edition, Edward Arnold Press, London, 470 pp.

National Ready Mixed Concrete Association, 1960, "Plastic Cracking of Concrete," *Engineering Information*, NRMCA, Silver Spring, Md., July, 2 pp.

Neville, A. M., 1996, *Properties of Concrete*, 4th Edition, Wiley, 844 pp.

Neville, A. M., and Brooks, J. J., 1987, *Concrete Technology*, Longman Scientific & Technical Press, Harlow, England, 438 pp.

Nolan, E.; Ali, M. A.; Basheer, P. A. M.; and Marsh, B. K., 1997, "Testing the Effectiveness of Commonly-Used Site Curing Regimes," *Materials and Structures*, V. 30, Jan.-Feb., pp. 53-60.

Parrott, L. J., 1995, "Influence of Cement Type and Curing on the Drying and Air Permeability of Cover Concrete," *Magazine of Concrete Research*, V. 47, No. 171, pp. 103-111.

Parrott, L. J., and Killoh, D. C., 1984, *Proceedings of the British Ceramic Society*, No. 35, pp. 41-53.

Parrott, L. J.; Killoh, D. C.; and Patel, R. G., 1986, "Cement Hydration under Partially Saturated Conditions," *Proceedings*, 8th Congress on Chemistry of Cement, Rio de Janeiro, V. 3, pp. 46-50.

Patel, R. G.; Killoh, D. C.; Parrott, L. J.; and Gutteridge, W. A., 1988, "Influence of Curing at Different Relative Humidities upon Compound Reactions and Porosity in Portland Cement Paste," *Materials and Structures*, V. 21, May, pp. 192-197.

Patel, R. G.; Parrott, L. J.; Martin, J. A.; and Killoh, D. C., 1985, "Gradients in Microstructure and Diffusion Properties in Cement Paste Caused by Drying," *Cement and Concrete Research*, V. 15, No. 2, pp. 343-356.

Persson, B., 1997, "Self-Desiccation and its Importance in Concrete Technology," *Materials and Structures*, V. 30, June, pp. 293-305.

Pfeifer D., and Marusin, S., 1982, "Energy Efficient Accelerated Curing of Concrete: A State of the Art Review," *Prestressed Concrete Institute Technical Report* No. 1, Chicago, Ill.

Pierce, J. S., 1994, "Mixing and Curing Water for Concrete," *Significance of Tests and Properties of Concrete and Concrete-Making Materials*, STP 169C, P. Klieger and J. Lamond, eds., ASTM, West Conshohocken, Pa., pp. 473-477.

Pihlajavaara, S. E, 1964, "On the Interrelation of the Moisture Content and the Strength of Mature Concrete and its Reversibility," *Report Series III*, State Institute for Technical Research, Finland, Building 76.

Pihlajavaara, S. E., 1965, "Estimation of the Drying of Concrete," *Proceedings,* RILEM Symposium on the Problem of Moisture in Buildings, Helsinki, Finland, Aug.

Portland Cement Association, 1955, "Prevention of Plastic Cracking in Concrete," *Concrete Information*, ST 80, Structural Bureau, Skokie, Ill.

Powers, T. C.; Copeland, L. E.; and Mann, H. M., 1959, "Capillary Continuity or Discontinuity in Cement Pastes," *Journal of the Portland Cement Association Research and Development Laboratories*, V. 1, No. 2, May, pp. 38-48.

Powers, T. C., 1948, "A Discussion of Cement Hydration in Relation to the Curing of Concrete," *Proceedings,* Highway Research Board, V. 27, pp. 178-188.

Powers, T. C., and Brownyard, T. L., 1947, "Studies of the Physical Properties of Hardened Portland Cement Paste," *Research Laboratories of the Portland Cement Association, Bulletin 22*, Mar., 992 pp.

Powers, T. C., 1962, "Prevention of Frost Damage to Green Concrete," *Bulletin* No. 14, RILEM, Mar., pp. 120-124.

Price, W. F., and Bamforth, P. B., 1991, "Initial Surface Absorption of Concrete: Examination of Modified Test Apparatus for Obtaining Uniaxial Flow," *Magazine of Concrete Research*, V. 43, No. 155, June, pp. 93-104.

RILEM Recommendations CPC-18, 1998, "Measurement of Hardened Concrete Carbonation Depth," *Materials and Structures*, V. 21, No. 126, pp. 453-455.

Sabir, B. B.; Wild, S.; and O'Farrell, M., 1998, "A Water Sorptivity Test for Mortar and Concrete," *Materials and Structures*, V. 31, Oct., pp. 568-574.

Sawyer, J. L., 1957, "Wear Test on Concrete Using the German Standard Method of Test Machine," *Proceedings*, ASTM, V. 57, July, pp. 1145-1153.

Sedgwick County, Kansas, 1998, "Special Provision, Linseed Oil Emulsion Treatment," *Standard Specifications for Road and Bridge Construction*.

Senbetta, E., 1988, "Concrete Curing Practices in the United States," *Concrete International*, V. 10, No. 11, Nov., pp. 64-67.

Senbetta, E., and Brury, M. A., 1991, "Control of Plastic Shrinkage Cracking in Cold Weather," *Concrete International*, V. 13, No. 3, Mar., pp. 49-53.

Senbetta, E., and Scholer, C. F., 1984, "A New Approach for Testing Concrete Curing Efficiency," ACI JOURNAL, *Proceedings* V. 81, No. 1, Jan.-Feb., pp. 82-86.

Shaeles, C. A., 1986, *Plastic Shrinkage Cracking in Mortar and Concrete*, Master's thesis, Cornell University, Ithaca, N.Y., Aug.

Shaeles, C. A., and Hover, K. C., 1988, "The Influence of Construction Operations on the Plastic Shrinkage Cracking of Thin Slabs," *ACI Materials Journal*, V. 85, No. 6, Nov.-Dec., pp. 495-504.

Shariat, S. M. S., and Pant, P. D., 1984, "Curing and Moisture Loss of Grooved Concrete Surfaces," *Transportation Research Record* No. 986, pp. 4-8.

Soroos, E., 1994, "Scanning Electron Microscope Images," *JSM 135 SEM*, School of Materials Science and Engineering, Cornell University, Ithaca, N.Y., May.

Spears, R.E., 1983, "The 80% Solution to Inadequate Curing Problems," *Concrete International*, V. 5, No. 4, Apr., pp. 15-18.

Suprenant, B., and Malisch, W., 1998, "Diagnosing Slab Delaminations (Part II)," *Concrete Construction*, Feb., pp. 169-175.

Suprenant, B., and Malisch, W., 1998, "Diagnosing Slab Delaminations (Part III)," *Concrete Construction*, Mar., pp. 169-175.

Suprenant, B., and Malisch, W., 1998, "Where Should You Place the Vapor Retarder?" *Concrete Construction*, May, pp. 427-433.

Suprenant, B., and Malisch, W., 1998, "Are Your Slabs Dry Enough for Floor Coverings?" *Concrete Construction*, Aug., pp. 671-677.

Suprenant, B., and Malisch, W., 1998, "The True Window of Finishability," *Concrete Construction*, Oct., pp. 859-863.

Suprenant, B., and Malisch, W., 1999, "Effect of Water Vapor Emissions on Floor Covering Adhesives," *Concrete Construction*, Jan., pp. 27-32.

Suprenant, B., and Malisch, W., 1999, "Why Won't the Concrete Dry?" *Concrete Construction*, July, pp. 29-33.

Tang, L., and Nilson, L. O., 1986, "Effect of Drying at an Early Age on Moisture Distributions in Concrete Specimens Used for Air Permeability Tests," *Nordic Concrete Research Establishment Report*, Watford, UK, 19 pp.

Taylor, H. F. W., 1997, *Cement Chemistry*, 2nd Edition, Thomas Telford, London, 459 pp.

Transportation Research Board, 1979, "Curing of Concrete Pavements," *Transportation Research Circular*, No. 208, June, ISSN 0097-8515, Transportation Research Board, Washington, D.C., 13 pp.

Tuthill, L. H., and Cordon, W. A., 1955, "Properties and Uses of Initially Retarded Concrete," ACI JOURNAL, *Proceedings* V. 52, pp. 273-286.

U.S. Army Corps of Engineers, 1975, "Method of Test for the Staining Properties of Water," CRD C 401, *Handbook for Concrete and Cement*, Vicksburg, Miss.

Uno, P. J., 1998, "Plastic Shrinkage Cracking and Evaporation Formulas," *ACI Materials Journal*, V. 95, No. 4, July-Aug., pp. 365-375.

Van Aardt, J. H. P., 1953, "Discussion to F. E. Jones on The Physical Structure of Cement Products and its Effect on Durability," *Proceedings of the Third International Symposium on the Chemistry of Cement*, London, pp. 433.

Veihmeyer, F. J., 1964, "Evapotranspiration," *Handbook of Applied Hydrology*, V. T. Chow, ed., McGraw Hill, pp. 11-1 to 11-38.

Virginia Department of Transportation, 1997, "Specifications for Highway and Bridge Construction."

Wainwright, P. J.; Cabrera, J. G.; and Gowriplan, N., 1990, "Assessment of the Efficiency of Chemical Membranes to Cure Concrete, Protection of Concrete," *Proceedings of the International Conference, Dundee, Scotland*, R. K. Dhir and J. W. Green, eds., E&FN Spon, London, pp. 907-920.

Wojokowski, J., 1999, *Report to ACI Committee 308 on Effect of Plastic Covering on Variations in Concrete Surface Temperature*, Kansas Department of Transportation, Topeka, Nov. 1.

Ytterberg, R., 1987, "Shrinkage and Curling of Slabs on Grade (Part I)," *Concrete International*, V. 9, No 4., Apr., pp. 22-31.

Ytterberg, R., 1987, "Shrinkage and Curling of Slabs on Grade (Part II)," *Concrete International*, V. 9, No 5., May, pp. 54-61.

Ytterberg, R., 1987, "Shrinkage and Curling of Slabs on Grade (Part III)," *Concrete International*, V. 9, No 6., June, pp. 72-81.

ACI 309R-96

Guide for Consolidation of Concrete

Reported by ACI Committee 309

H. Celik Ozyildirim Chairman	Richard E. Miller, Jr. Subcommittee Chairman

Dan A. Bonikowsky	Roger A. Minnich
Neil A. Cumming	Mikael P. J. Olsen
Timothy P. Dolen	Larry D. Olson
Jerome H. Ford	Sandor Popovics
Steven H. Gebler	Steven A. Ragan
Kenneth C. Hover	Donald L. Schlegel
Gary R. Mass	Bradley K. Violetta
Bryant Mather	

Consolidation is the process of removing entrapped air from freshly placed concrete. Several methods and techniques are available, the choice depending mainly on the workability of the mixture, placing conditions, and degree of air removal desired. Some form of vibration is usually employed.

This guide includes information on the mechanism of consolidation, and gives recommendations on equipment, characteristics, and procedures for various classes of construction.

Keywords: admixtures; air; air entrainment; amplitude; centrifugal force; concrete blocks; concrete construction; concrete pavements; concrete pipes; concrete products; concrete slabs; concretes; consistency; consolidation; floors; formwork (construction); heavyweight concretes; inspection; lightweight aggregate concretes; maintenance; mass concrete; mixture proportioning; placing; plasticizers; precast concrete; quality control; reinforced concrete; reinforcing steels; segregation; surface defects; tamping; vacuum-dewatered concrete; vibration; vibrators (machinery); water-reducing admixtures; workability.

CONTENTS

Chapter 1—General, p. 309R-2

Chapter 2—Effect of mixture properties on consolidation, p. 309R-3
2.1—Mixture proportions
2.2—Workability and consistency
2.3—Workability requirements

Chapter 3—Methods of consolidation, p. 309R-4
3.1—Manual methods
3.2—Mechanical methods
3.3—Methods used in combinations

Chapter 4—Consolidation of concrete by vibration, p. 309R-5
4.1—Vibratory motion
4.2—Process of consolidation

Chapter 5—Equipment for vibration, p. 309R-7
5.1—Internal vibrators
5.2—Form vibrators
5.3—Vibrating tables
5.4—Surface vibrators
5.5—Vibrator maintenance

Chapter 6—Forms, p. 309R-14
6.1—General
6.2—Sloping surfaces
6.3—Surface defects
6.4—Form tightness
6.5—Forms for external vibration

ACI Committee Reports, Guides, Standard Practices, and Commentaries are intended for guidance in planning, designing, executing, and inspecting construction. This document is intended for the use of individuals who are competent to evaluate the significance and limitations of its content and recommendations and who will accept responsibility for the application of the material it contains. The American Concrete Institute disclaims any and all responsibility for the stated principles. The Institute shall not be liable for any loss or damage arising therefrom.

Reference to this document shall not be made in contract documents. If items found in this document are desired by the Architect/Engineer to be a part of the contract documents, they shall be restated in mandatory language for incorporation by the Architect/Engineer.

ACI 309R-96 became effective May 24, 1996. This report supersedes ACI 309R-87.
Copyright © 1997, American Concrete Institute.
All rights reserved including rights of reproduction and use in any form or by any means, including the making of copies by any photo process, or by electronic or mechanical device, printed, written, or oral, or recording for sound or visual reproduction or for use in any knowledge or retrieval system or device, unless permission in writing is obtained from the copyright proprietors.

In addition to the members of ACI Committee 309, the following individuals contributed significantly to the development of this report: George R. U. Burg, Lars Forssblad, John C. King, Kenneth L. Saucier, and C. H. Spitler. Their contribution is sincerely appreciated.

Chapter 7—Recommended vibration practices for general construction, p. 309R-16
    7.1—Procedure for internal vibration
    7.2—Judging the adequacy of internal vibration
    7.3—Vibration of reinforcement
    7.4—Revibration
    7.5—Form vibration
    7.6—Consequences of improper vibration

Chapter 8—Structural concrete, p. 309R-21
    8.1—Design and detailing prerequisites
    8.2—Mixture requirements
    8.3—Internal vibration
    8.4—Form vibration
    8.5—Tunnel

Chapter 9—Mass concrete, p. 309R-22
    9.1—Mixture requirements
    9.2—Vibration equipment
    9.3—Forms
    9.4—Vibration practices
    9.5—Roller-compacted concrete

Chapter 10—Normal weight concrete floor slabs, p. 309R-25
    10.1—Mixture requirements
    10.2—Equipment
    10.3—Structural slabs
    10.4—Slabs on grade
    10.5—Heavy-duty industrial floors
    10.6—Vacuum dewatering

Chapter 11—Pavements, p. 309R-27
    11.1—Mixture requirements
    11.2—Equipment
    11.3—Vibration procedures
    11.4—Special precautions

Chapter 12—Precast products, p. 309R-30
    12.1—Mixture requirements
    12.2—Forming material
    12.3—Production technique
    12.4—Other factors affecting choice of consolidation method
    12.5—Placing methods

Chapter 13—Lightweight concrete, p. 309R-31
    13.1—Mixture requirements
    13.2—Behavior of lightweight concrete during vibration
    13.3—Consolidation equipment and procedures
    13.4—Floors

Chapter 14—High density concrete, p. 309R-32
    14.1—Mixture requirements
    14.2—Placing techniques

Chapter 15—Quality control and inspection, p. 309R-33
    15.1—General
    15.2—Adequacy of equipment and procedures
    15.3—Checking equipment performance

Chapter 16—Consolidation of test specimens, p. 309R-35
    16.1—Strength tests
    16.2—Unit weight tests
    16.3—Air content tests
    16.4—Consolidating very stiff concrete in laboratory specimens

Chapter 17—Consolidation in congested areas, p. 309R-36
    17.1—Common placing problems
    17.2—Consolidation techniques

Chapter 18—Information sources, p. 309R-37
    18.1—Specified and/or recommended references
    18.2—Cited references

Appendix A—Fundamentals of vibration, p. 309R-38
    A.1—Principles of simple harmonic motion
    A.2—Action of a rotary vibrator
    A.3—Vibratory motion in the concrete

CHAPTER 1—GENERAL

A mass of freshly placed concrete is usually honeycombed with entrapped air. If allowed to harden in this condition, the concrete will be nonuniform, weak, porous, and poorly bonded to the reinforcement. It will also have a poor appearance. The mixture must be consolidated if it is to have the properties normally desired and expected of concrete.

Consolidation is the process of inducing a closer arrangement of the solid particles in freshly mixed concrete or mortar during placement by the reduction of voids, usually by vibration, centrifugation, rodding, tamping, or some combination of these actions; it is also applicable to similar manipulation of other cementitious mixtures, soils, aggregates, or the like.

Drier and stiffer mixtures require greater effort to achieve proper consolidation. By using certain chemical admixtures, consistencies requiring reduced consolidation effort can be achieved at a lower water content. As the water content of the concrete is reduced, concrete quality (strength, durability, and other properties) improves, provided it is properly consolidated. Alternatively, the cement content can be lowered, reducing the cost while maintaining the same quality. If adequate consolidation is not provided for these drier or stiffer mixtures, the quality of the inplace concrete drops off rapidly.

Equipment and methods are now available for fast and efficient consolidation of concrete over a wide range of placing conditions. Concrete with a relatively low water content can be readily molded into an unlimited variety of shapes, making it a highly versatile and economical construction material. When good consolidation practices are combined with good formwork, concrete surfaces have a highly pleasing appearance [see Fig. 1(a) through 1(c)].

Fig. 1(a)—Pleasing appearance of concrete in church construction.

Fig. 1(b)—Pleasing appearance of concrete in utility building construction.

Fig. 1(c)—Close-ups of surfaces resulting from good consolidation.

CHAPTER 2—EFFECT OF MIXTURE PROPERTIES ON CONSOLIDATION

2.1—Mixture proportions

Concrete mixtures are proportioned to provide the workability needed during construction and the required properties in the hardened concrete. Mixture proportioning is described in detail in documents prepared by ACI Committee 211, as listed in Chapter 18.1.

2.2—Workability and consistency

Workability of freshly mixed concrete is that property that determines the ease and homogeneity with which it can be mixed, placed, consolidated, and finished. Workability is a function of the rheological properties of the concrete.

As shown in Fig. 2.2, workability may be divided into three main aspects:

1. Stability (resistance to bleeding and segregation).
2. Ease of consolidation.
3. Consistency, affected by the viscosity and cohesion of the concrete and angle of internal friction.

Workability is affected by grading, particle shape, proportions of aggregate and cement, use of chemical and mineral admixtures, air content, and water content of the mixture.

Consistency is the relative mobility or ability of freshly mixed concrete to flow. It also largely determines the ease with which concrete can be consolidated. Once the materials and proportions are selected, the primary control over work-

Table 2.1—Consistencies used in construction**

Consistency description	Slump, in. (mm)	Vebe time, sec	Compacting factor average	Thaulow drop table revolutions
Extremely dry	—	32 to 18	—	112-56
Very stiff	—	18 to 10	0.70	56-28
Stiff	0 to 1* (0 to 25)	10 to 5	0.75	28 to 14
Stiff plastic	1 to 3 (25 to 75)	5 to 3	0.85	14-7
Plastic	3 to 5 (75 to 125)	3 to 0*	0.90	<7
Highly plastic	5 to 7$^1/_2$ (125 to 190)	—	—	—
Flowing	7$^1/_2$ plus (190 plus)	—	0.95	—

*Test method is of limited value in this range.
**ACI 211.3 Table 2.3.1 (a)

ability is through changes in the consistency brought about by minor variations in the water content.

The slump test (ASTM C 143) is widely used to indicate consistency of mixtures used in normal construction. The Vebe test is generally recommended for stiffer mixtures.

Values of slump, compacting factor, drop table, and Vebe time for the entire range of consistencies used in construction are given in Table 2.1.

Other measures of consistency such as the Powers remolding test and Kelly ball are available. These are not used as frequently as slump. Information on various consistency tests has been discussed by Neville (1981), Vollick (1966), and Popovics (1982).

2.3—Workability requirements

The concrete should be sufficiently workable so that consolidation equipment, properly used, will give adequate consolidation. A high degree of flowability may be undesirable because it may increase the cost of the mixture and may reduce the quality of the hardened concrete. Where such a high degree of flowability is the result of too much water in the mixture, the mixture will generally be unstable and will probably segregate during the consolidation process.

Mixtures having moderately high slump, small maximum-size aggregate, and excessive fine aggregate are frequently used because the high degree of flowability means less work in placing.

At the other extreme, it is inadvisable to use mixtures that are too stiff for conditions of consolidation. They will require great consolidation effort and even then may not be adequately consolidated. Direction and guidance are often required to achieve the use of mixtures of lower slump or fine aggregate content, or a larger maximum size aggregate, so as to give a more efficient use of the cement.

Concrete containing certain chemical admixtures may be placed in forms with less consolidation effort. Refer to reports of ACI Committee for additional information on chemical admixtures. The use of fly ash, slag, or silica fume may also affect the consolidation of concrete by permitting placement with less consolidation effort. Refer to reports of ACI Committee 226 for more information regarding these materials. The amount of consolidation effort required with or without the use of admixtures can best be determined by trial mixtures under field conditions.

It is the workability of the mixture in the form that determines the consolidation requirements. Workability may be considerably less than at the mixer because of slump loss due to high temperature, false set, delays, or other cause.

CHAPTER 3—METHODS OF CONSOLIDATION

The consolidation method should be compatible with the concrete mixture, placing conditions, form intricacy, amount of reinforcement, etc. Many manual and mechanical methods are available.

3.1—Manual methods

Some consolidation is caused by gravity as the concrete is deposited in the form. This is particularly true for well proportioned flowing mixtures where less additional consolidation effort is required.

Plastic or more flowable mixtures may be consolidated by rodding. Spading is sometimes used at formed surfaces—a flat tool is repeatedly inserted and withdrawn adjacent to the form. Coarse particles are shoved away from the form and movement of air voids and water pockets toward the top surface is facilitated, thereby reducing the number and size of bugholes in the formed concrete surface.

Hand tamping may be used to consolidate stiff mixtures. The concrete is placed in thin layers, and each layer is carefully rammed or tamped. This is an effective consolidation method, but laborious and costly.

The manual consolidation methods are generally only used on smaller nonstructural concrete placement.

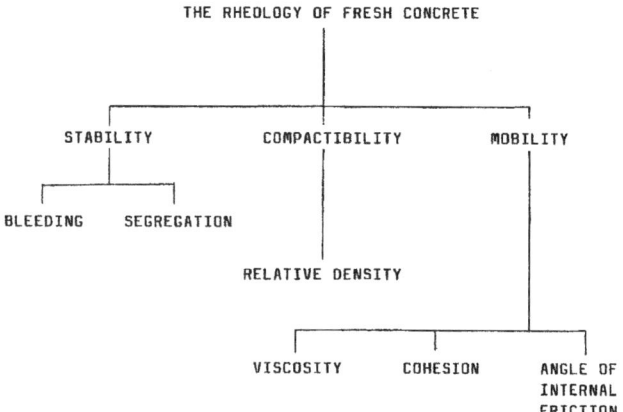

Fig. 2.2—Parameters of the rheology of fresh concrete.

Fig. 4—Internal vibrator "liquifying" low-slump concrete

3.2—Mechanical methods

The most widely used consolidation method is vibration. It will receive the most attention in this guide. Vibration may be either internal, external, or both.

Power tampers may be used to compact stiff concrete in precast units. In addition to the ramming or tamping effect, there is a low-frequency vibration that aids in the consolidation.

Mechanically operated tamping bars are suitable for consolidating stiff mixtures for some precast products, including concrete blocks.

Equipment that applies static pressures to the top surface may be used to consolidate thin concrete slabs of plastic or flowing consistency. Concrete is literally squeezed into the mold, and entrapped air and part of the mixing water is forced out.

Centrifugation (spinning) is used to consolidate concrete in concrete pipe, piles, poles, and other hollow sections.

Many types of surface vibrators are available for slab construction, including vibrating screeds, vibratory roller screeds, plate and grid vibratory tampers, and vibratory finishing tools.

Shock tables, sometimes called drop tables, are suitable for consolidating low-slump concrete. The concrete is deposited in thin lifts in sturdy molds. As the mold is filled, it is alternately raised a short distance and dropped on to a solid base. The impact causes the concrete to be rammed into a dense mass. Frequencies are 150 to 250 drops per min., and the free fall is $1/8$ to $1/2$ in. (3 to 13 mm).

3.3—Methods used in combination

Under some conditions, a combination of two or more consolidation methods gives the best results.

Internal and external vibration can often be combined to advantage in precast work and occasionally in cast-in-place concrete. One scheme uses form vibrators for routine consolidation and internal vibrators for spot use at critical, heavily reinforced sections prone to voids or poor bond with the reinforcement. Conversely, in sections where the primary consolidation is by internal vibrators, form vibration may also be applied to achieve the desired surface appearance.

Vibration may be simultaneously applied to the form and top surface. This procedure is frequently used in making precast units on vibrating tables. The mold is vibrated while a vibratory plate or screed working on the top surface exerts additional vibratory impulses and pressure.

Vibration of the form is sometimes combined with static pressure applied to the top surface. Vibration under pressure is particularly useful in concrete block production where the very stiff mixtures do not react favorably to vibration alone.

Centrifugation, vibration, and rolling may be combined in the production of concrete pipe and other hollow sections.

CHAPTER 4—CONSOLIDATION OF CONCRETE BY VIBRATION

Vibration consists of subjecting freshly placed concrete to rapid vibratory impulses which liquefy the mortar (see Fig. 4) and drastically reduce the internal friction between aggregate particles. While in this condition, concrete settles under the action of gravity (sometimes aided by other forces). When vibration is discontinued, friction is reestablished.

4.1—Vibratory motion

A concrete vibrator has a rapid oscillatory motion that is transmitted to the freshly placed concrete. Oscillating motion is basically described in terms of frequency (number of oscillations or cycles per unit of time) and amplitude (deviation from point of rest).

Rotary vibrators follow an orbital path caused by rotation of an unbalanced weight or eccentric inside a vibrator casing. The oscillation is essentially simple harmonic motion, as explained in the Appendix. Acceleration, a measure of intensity of vibration, can be computed from the frequency and amplitude when they are known. It is usually expressed by g,

Table 5.1.5—Range of characteristics, performance, and applications of internal* vibrators

Column 1	2	3	4	5	6	7	8	9
			\multicolumn{3}{c}{Suggested values of}	\multicolumn{2}{c}{Approximate values of}	
Group	Diameter of head, in. (mm)	Recommended frequency, vibrations per min (Hz)	Eccentric moment, in. lb mm-kg(10^{-3})	Average amplitude, in. (mm)	Centrifugal force, lb (kg)	Radius of action, in. (mm)	Rate of concrete placement, yd	Application
1	3/4-1 1/2 (2-4) (20-40)	9000-15,000 (150-200)	0.03-0.10 (0.035-0.12) (3.5-12)	0.015-0.03 (0.04-0.08) (0.4-0.8)	100-400 (45-180)	3-6 (8-15) (80-150)	1-5 (0.8-4)	Plastic and flowing concrete in very thin members and confined places. May be used to supplement larger vibrators, especially in prestressed work where cables and ducts cause congestion in forms. Also used for fabricating laboratory test specimens.
2	1 1/4-2 1/2 (3-6) (30-60)	8500-12,500 (140-210)	0.08-0.25 (0.09-0.29) (9-29)	0.02-0.04 (0.05-0.10) (0.5-1.0)	300-900 (140-400)	5-10 (13-25) (130-250)	3-10 (2.3-8)	Plastic concrete in thin walls, columns, beams, precast piles, thin slabs, and along construction joints. May be used to supplement larger vibrators in confined areas.
3	2-3 1/2 (5-9) (50-90)	8000-12,000 (130-200)	0.20-0.70 (0.23-0.81) (23-81)	0.025-0.05 (0.06-0.13) (0.6-1.3)	700-2000 (320-900)	7-14 (18-36) (180-360)	6-20 (4.6-15)	Stiff plastic concrete (less than 3-in. [80-mm] slump) in general construction such as walls, columns, beams, prestressed piles, and heavy slabs. Auxiliary vibration adjacent to forms of mass concrete and pavements. May be gang mounted to provide full-width internal vibration of pavement slabs.
4	3-6 (8-15) (80-150)	7000-10,500 (120-180)	0.70-2.5 (0.81-2.9) (81-290)	0.03-0.06 (0.08-0.15) (0.8-1.5)	1500-4000 (680-1800)	12-20 (30-51) (300-510)	(15-40) (11-31)	Mass and structural concrete of 0 to 2-in. (50 mm) slump deposited in quantities up to 4 yd^3 (3m^3) in relatively open forms of heavy construction (powerhouses, heavy bridge piers, and foundations). Also auxiliary vibration in dam construction near forms and around embedded items and reinforcing steel.
5	5-7 (13-18) (130-150)	5500-8500 (90-140)	2.25-3.50 (2.6-4.0) (260-400)	0.04-0.08 (0.10-0.20) (1.0-2.0)	2500-6000 (1100-2700)	16-24 (40-61) (400-610)	25-50 (19-38)	Mass concrete in gravity dams, large piers, massive walls, etc. Two or more vibrators will be required to operate simultaneously to mix and consolidate quantities of concrete of 4 yd^3 (3 m^3) or more deposited at one time in the form.

Column 3—While vibrator is operating in concrete.
Column 4—Computed by formula in Fig. A.2 in Appendix A.
Column 5—Computed or measured as described in Section 15.3.2. This is peak amplitude (half the peak-to-peak value), operating in air.
Column 6—Computed by formula in Fig. A.2 in Appendix, using frequency of vibrator while operating in concrete.
Column 7—Distance over which concrete is fully consolidated.
Column 8—Assumes the insertion spacing is 1 1/2 times the radius of action, and that vibrator operates two-thirds of time concrete is being placed.
Columns 7 and 8—These ranges reflect not only the capability of the vibrator but also differences in workability of the mix, degree of deaeration desired, and other conditions experienced in construction.
*Generally, extremely dry or very stiff concrete (Table 2.1) does not respond well to internal vibrators.

which is the ratio of the vibration acceleration to the acceleration of gravity. Acceleration is a useful parameter for external vibration, but not for internal vibration where the amplitude in concrete cannot be measured readily.

For vibrators other than the rotary type, reciprocating vibrators for example, the principles of harmonic motion do not apply. However, the basic concepts described here are still useful.

4.2—Process of consolidation

When low-slump concrete is deposited in the form, it is in a honeycombed condition, consisting of mortar-coated coarse-aggregate particles and irregularly distributed pockets of entrapped air. Reading (1967) stated that the volume of entrapped air depends on the workability of the mixture, size and shape of the form, amount of reinforcing steel and other items of congestion, and method of depositing the concrete. It is generally in the range of 5 to 20 percent. The purpose of consolidation is to remove practically all of the entrapped air because of its adverse effect on concrete properties and surface appearance.

Consolidation by vibration is best described as consisting of two stages—the first comprising subsidence or slumping of the concrete, and the second a deaeration (removal of entrapped air bubbles). The two stages may occur simultaneously, with the second stage under way near the vibrator before the first stage has been completed at greater distances (Kolek 1963).

When vibration is started, impulses cause rapid disorganized movement of mixture particles within the vibrator's radius of influence. The mortar is temporarily liquefied. Internal friction, which enabled the concrete to support itself in its original honeycombed condition, is reduced drastically. The mixture becomes unstable, and seeks a lower level and denser condition. It flows laterally to the form and around the reinforcing steel and embedments.

At the completion of this first stage, honeycomb has been eliminated; the large voids between the coarse aggregate are now filled with mortar. The concrete behaves somewhat like a liquid containing suspended coarse-aggregate particles. However, the mortar still contains many entrapped air bubbles, ranging up to perhaps 1 in. (25 mm) across and amounting to several percent of the concrete volume.

After consolidation has proceeded to a point where the coarse aggregate is suspended in the mortar, further agitation of the mixture by vibration causes entrapped air bubbles to rise to the surface. Large air bubbles are more easily removed than small ones because of their greater buoyancy.

Also those near the vibrator are released before those near the outer fringes of the radius of action.

The vibration process should continue until the entrapped air is reduced sufficiently to attain a concrete density consistent with the intended strength and other requirements of the mixture. To remove all of the entrapped air with standard vibrating equipment is usually not practical.

The mechanism and principles involved in vibration of fresh concrete are described in detail in ACI 309.1R.

CHAPTER 5—EQUIPMENT FOR VIBRATION

Concrete vibrators can be divided into two main classes—internal and external. External vibrators may be further divided into form vibrators, surface vibrators, and vibrating tables.

5.1—Internal vibrators

Internal vibrators, often called spud or poker vibrators, have a vibrating casing or head. The head is immersed in and acts directly against the concrete. In most cases, internal vibrators depend on the cooling effect of the surrounding concrete to prevent overheating.

All internal vibrators presently in use are the rotary type (see Section 4.1). The vibratory impulses emanate at right angles to the head.

5.1.1 *Flexible shaft type*—This type of vibrator is probably the most widely used. The eccentric is usually driven by an electric or pneumatic motor, or by a portable internal combustion engine [see Fig. 5.1.1(a)].

For the electric motor-driven type, a flexible drive shaft leads from the electric motor into the vibrator head where it turns the eccentric weight. The motor generally has universal, 120 (occasionally 240) volt, single-phase, 60 Hz alternating-current characteristics. Fifty Hz AC current is used in some countries. The frequency of this type of vibrator is quite high when operating in air—generally in the range of 12,000 to 17,000 vibrations per min (200 to 283 Hz) (the higher values being for the smaller head sizes). However, when operating in concrete, the frequency is usually reduced by about one-fifth. In this report, frequency is expressed in vibrations per min to conform to current industry practice in the United States; however, frequency is given in hertz in the Appendix to agree with textbook formulas.

For the engine-driven types, both gasoline and diesel, the engine speed is usually about 3600 revolutions per min (60 Hz). A V-belt drive or gear transmission is used to step up this speed to an acceptable frequency level. Another type of unit uses a 2-cycle gasoline engine operating at a no-load speed of 12,000 RPM [Fig. 5.1.1.(b)], so the need for a step-up transmission is eliminated. This unit is portable and is usually carried on a back pack. Again a flexible shaft leads into the vibrator head. While larger and more cumbersome than electric motor-driven vibrators, engine-driven vibrators are attractive where commercial power is not readily available.

For most flexible-shaft vibrators, the frequency is the same as the speed of the shaft. However, the roll-gear (conical-pendulum) type is able to achieve high vibrator frequency with modest electric motor and flexible shaft speeds. The end of the

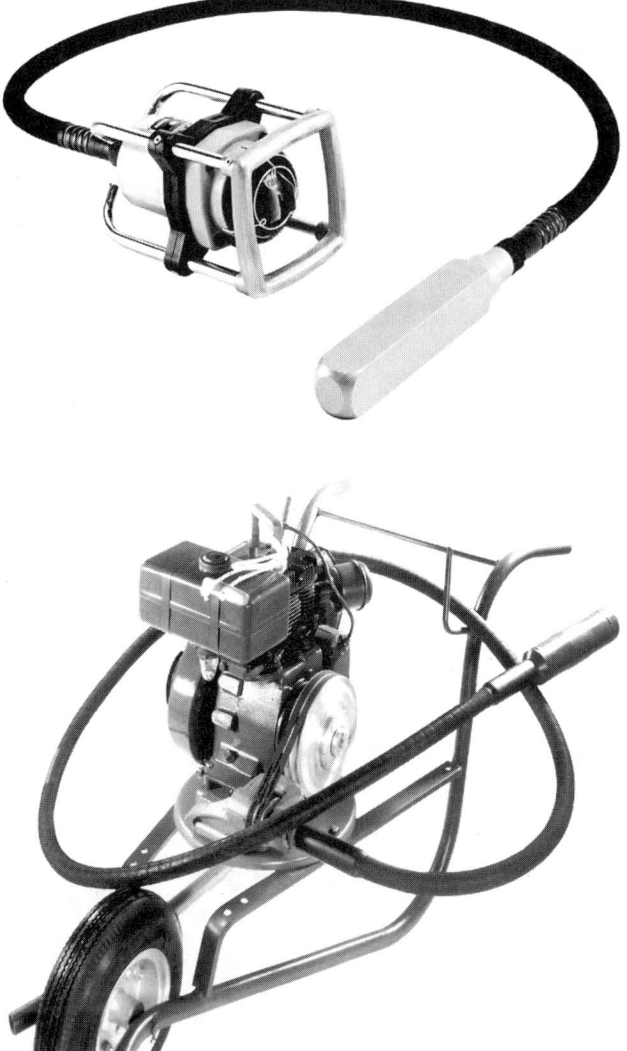

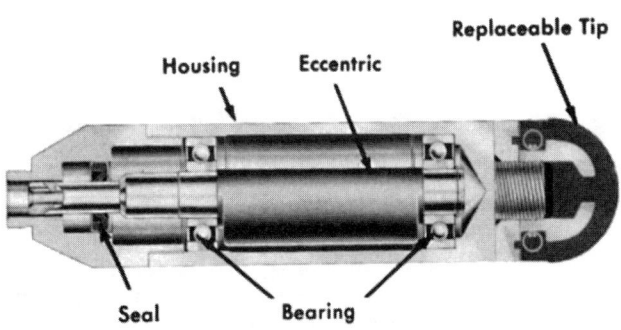

Fig. 5.1.1(a)—Flexible shaft vibrators; electric motor-driven type (top); gasoline engine-driven type (middle); and cross section through head (bottom).

pendulum strikes the inner housing in a star-shaped pattern, giving the vibrator head a frequency higher than the shaft driving it. Motor speeds are usually about 3600 revolutions per min with 60 Hz current (about 3000 revolutions per min with 50 Hz current). A single induction or three-phase squirrel-cage motor

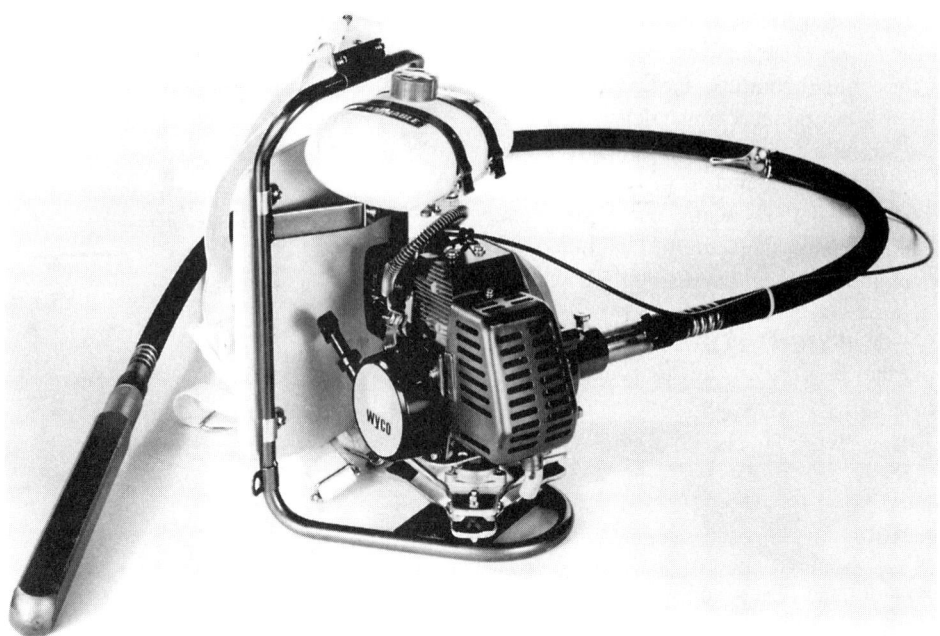

Fig. 5.1.1(b)—Back pack two-cycle gasoline engine-driven vibrator

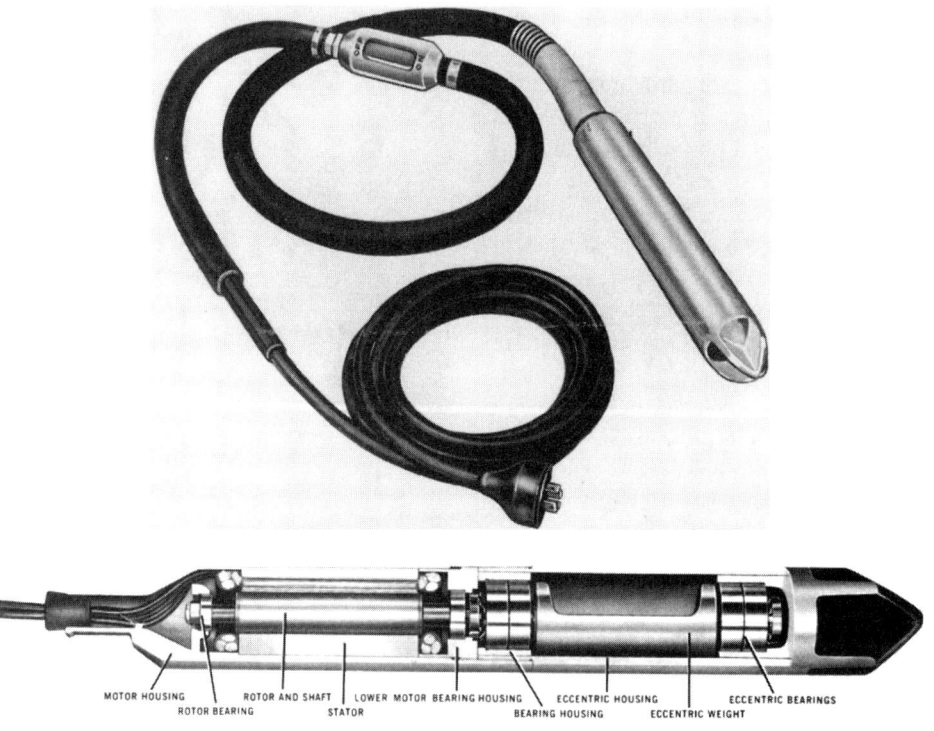

Fig. 5.1.2—Electric motor-in-head vibrator; external appearance (top) and internal construction of head (bottom).

is generally used. The low speed of the flexible shaft is favorable from the standpoint of maintenance.

5.1.2 *Electric motor-in-head type*—Electric motor-in-head vibrators have increased in popularity in recent years (see Fig. 5.1.2). Since the motor is in the vibrator head, there is no separate motor and flexible drive to handle. A substantial electrical cable, which also acts as a handle, leads into the head. Electric motor-in-head vibrators are generally at least 2 in. (50 mm) in diameter.

This type of vibrator is available in two designs. One uses a universal motor and the other a 180 Hz (high-cycle) three-phase motor. In the latter, the energy is usually supplied by a portable gasoline engine-driven generator; however, commercial power passed through a frequency converter may be used. The design uses an induction-type motor that has little dropoff in speed when immersed in concrete. It can rotate a heavier eccentric weight and develops a greater centrifugal force than current universal motor-in-head models of the

Table 5.5.1—Sample service log for flexible shaft vibrator

Model _____ Serial No. _____
Date purchased _____
Date checked out from equipment pool _____
Estimated use, hr per day _____

Item	Frequency of preventive maintenance		
	Clean and inspect	Lubricate	Replace
	Electric motor		
Filter	___	___	___
Brushes	___	___	___
Switch	___	___	___
Armature and field	___	___	___
Bearings	___	___	___
	Flexible shaft		
Shaft	___	___	___
	Vibrator head		
Seals	___	___	___
Bearings	___	___	___
Oil change	___	___	___

same diameter. Vibrator motors operating on 150 or 200 Hz current are used in some countries.

5.1.3 *Pneumatic vibrators*—Pneumatic vibrators (see Fig. 5.1.3) are operated by compressed air, the pneumatic motor generally being inside the vibrator head. The vane type has been the most common, with both the motor and the eccentric elements supported on bearings. Bearingless models, which generally require less maintenance, are also available. A few flexible-shaft pneumatic models, with the air motor outside the head, are also available.

Pneumatic vibrators are attractive where compressed air is the most readily available source of power. The frequency is highly dependent on the air pressure, so the air pressure should always be maintained at the proper level, usually that recommended by the manufacturer. In some cases, it is desirable to vary the air pressure to obtain a different frequency.

5.1.4 *Hydraulic vibrators*—Hydraulic vibrators, using a hydraulic gear motor, are popular on paving machines. Here the vibrator is connected to the paver's hydraulic system by means of high-pressure hoses. The frequency of vibration can be regulated by varying the rate of flow of hydraulic fluid through the vibrator. The efficiency of the vibrator is dependent on the pressure and flow rate of the hydraulic fluid. It is, therefore, important that the hydraulic system be checked frequently.

5.1.5 *Selecting an internal vibrator for the job*—The principal requirement for an internal vibrator is effectiveness in consolidating concrete. It should have an adequate radius of action, and it should be capable of flattening and de-aerating the concrete quickly. Insofar as possible, the vibrator should also be reliable in operation, easy to handle and manipulate, resistant to wear, and be such that it does not damage embedded items. Some of these requirements are mutually opposed, so compromises are necessary. However, some of the problems can be minimized or eliminated by careful vibrator design. For example, it is known that very high frequencies and high centrifugal force tend to increase maintenance requirements and reduce the life of vibrators.

Evidence strongly indicates that the effectiveness of an internal vibrator depends mainly on the head diameter, frequency, and amplitude. The amplitude is largely a function of the eccentric moment and head mass, as explained in the Appendix.

Fig. 5.1.3—Air vibrators for ordinary construction (top) and for mass concrete (bottom).

Frequency may be readily determined (see Section 15.3.1), but there is no simple method for determining amplitude of a vibrator operating in concrete. It is therefore necessary to use the amplitude as determined while the vibrator is operating in air, which is somewhat greater than the amplitude in concrete. This amplitude may be either measured or computed, as described in Section 15.3.2.

While not strictly correct for internal vibrators, the centrifugal force may be used as a rough overall measure of the output of a vibrator. Fig. A.2 in the Appendix explains how it is computed.

The radius of action, and hence the insertion spacing, depends not only on the characteristics of the vibrator, but also on the workability of the mixture and degree of congestion.

Table 5.1.5 gives the ordinary range of characteristics, performance, and applications of internal vibrators. (Some special-purpose vibrators fall outside these ranges.) Recommended frequencies are given, along with suggested values of eccentric moment, average amplitude, and centrifugal force.

Approximate ranges are also given for the radius of action and rate of concrete placement. These are empirical values based mainly on previous experience.

Equally good results can usually be obtained by selecting a vibrator from the next larger group, provided suitable adjustments are made in the spacing and time of the insertions. In selecting the vibrator and vibration procedures, consideration should be given to the vibrator size relative to the form size. Crazing of formed concrete surfaces is due to drying shrinkage that occurs in the high concentration of cement paste brought to the surface by a vibrator too large for the application.

The values in Table 5.1.5 are not to be considered as a guarantee of performance under all conditions. The best measure of vibrator performance is its effectiveness in consolidating job concrete.

5.1.6 *Special shapes of vibrator heads*—The recommendations in Table 5.1.5 assume round vibrators. Other shapes of vibrator head (square or other polygonal shapes, fluted, finned, etc.), have a different surface area and have a different distribution of force between the vibrator and the concrete (see Fig. 5.1.6).

The effect of shape on vibrator performance has not been thoroughly evaluated. For the purpose of this guide, it is recommended that the equivalent diameter of a specially shaped vibrator be considered as that of a round vibrator having the same perimeter.

5.1.7 *Data to be supplied by manufacturer*—The vibrator manufacturer's catalog should include the physical dimensions (length and diameter) and total mass of the vibrator head, eccentric moment, frequency in air and approximate frequency in concrete, and centrifugal force at these two frequencies.

The catalog should also include certain other data needed for proper hookup and operation of the vibrators. Voltage and current requirements and wire sizes (depending on the length of run) for electric vibrators should be given. For pneumatic vibrators, compressed air pressure and flow capacity should be stated, as well as size of piping or hose (also depending on the length of run). Speed should be given for gasoline-engine driven units.

Information for hydraulic vibrators should include recommended operating pressures and a chart showing frequency, at various flow rates.

5.2—Form vibrators

5.2.1 *General description*—Form vibrators are external vibrators attached to the outside of the form or mold. They vibrate the form, which in turn transmits the vibration to the concrete. Form vibrators are self-cooling and may be of either the rotary or reciprocating type.

Concrete sections as thick as 24 in. (600 mm) and up to 30 in. (750 mm) deep have been effectively vibrated by form vibrators in the precast concrete industry. For walls and deeper placements, it may be necessary to supplement a form vibrator with internal vibration for sections thicker than 12 in. (300 mm).

5.2.2 *Types of form vibrators*

   5.2.2.1 *Rotary*—Rotary form vibrators produce essentially simple harmonic motion. The impulses have components both perpendicular to and in the plane of the form. This type may be pneumatically, hydraulically, or electrically driven (see Fig. 5.2.2.1).

In the pneumatically and hydraulically driven models, centrifugal force is developed by a rotating cylinder or revolving eccentric mass (similar to internal vibrators). These vibrators generally work at frequencies of 6000 to 12,000 vibrations per min (100 to 200 Hz). The frequency may be varied by adjusting the air pressure on the pneumatic models or the fluid pressure on the hydraulic models.

The electrically driven models have an eccentric mass attached to each end of the motor shaft. Generally, these masses are adjustable. In most cases, induction motors are used and the frequency is 3600 vibrations per min (60 Hz AC, or 3000 vibrations per min for 50 Hz AC). Higher frequency vibrators operating at 7200 or 10,800 vibrations per min (120 or 180 Hz) are also available (6000, 9000, or

Fig. 5.1.6—Several of the different sizes and shapes of vibrator heads available. From left to right: short head, round head, square head, hexagonal head, and rubber-tipped head.

12,000 vibrations per min [100, 150, or 200 Hz] in Europe). These higher frequency vibrators require a frequency converter. There are also electric form vibrators with frequencies of 6000 to 9000 vibrations per min (100 to 150 Hz) that are powered by single-phase universal motors.

The manufacturer's catalog should include physical dimensions, mass, and eccentric moment. For pneumatically driven models, frequency in air and approximate frequency under load should be given. For electric models, the frequency at the rated electric load should be stated. The centrifugal force at the given frequency values should be provided. In addition, the catalog should provide data needed for proper hookup of the vibrators (as in Section 5.1.7).

5.2.2.2 *Reciprocating*—In reciprocating vibrators, a piston is accelerated in one direction, stopped (by impacting against a steel plate), and then accelerated in the opposite direction (see Fig. 5.2.2.2). This type is pneumatically driven, and frequencies are usually in the range of 1000 to 5000 vibrations per min (20 to 80 Hz).

These vibrators produce impulses acting perpendicular to the form. The principles of simple harmonic motion do not apply.

5.2.2.3 *Other types*—Other types of form vibrators, less commonly used, include:

a. Electromagnetic, which usually develops a combination sinusoidal-saw-tooth wave form.

b. Pneumatic or electric hand-held hammers, which are sometimes used to assist in consolidating small concrete units.

5.2.3 *Selecting external vibrators for vertical forms*—Low-frequency high-amplitude vibration is normally preferred for stiffer mixtures. High frequency, low amplitude vibration generally results in better consolidation and better surfaces (fewer bugholes) for more plastic consistencies. In this guide, the dividing line between high and low frequency for external vibration is arbitrarily taken as 6000 vibrations per min (100 Hz), and between high and low amplitude 0.005 in. (0.13 mm).

The effectiveness of form vibrators is largely a function of the acceleration imparted to the concrete by the form. Accelerations in the range of 1 to 2 g are generally recommended for plastic mixtures and 3 to 5 g for stiff mixtures. In addition, the amplitude should not be less than 0.001 in. (0.025 mm) for plastic mixtures or 0.002 in. (0.050 mm) for stiff mixtures.

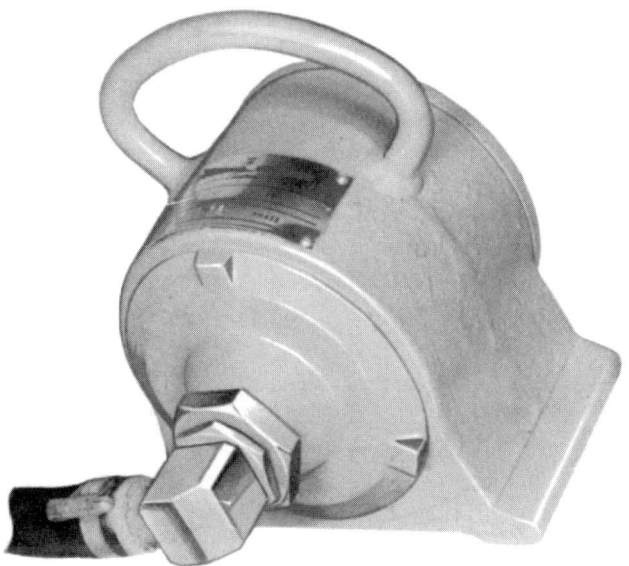

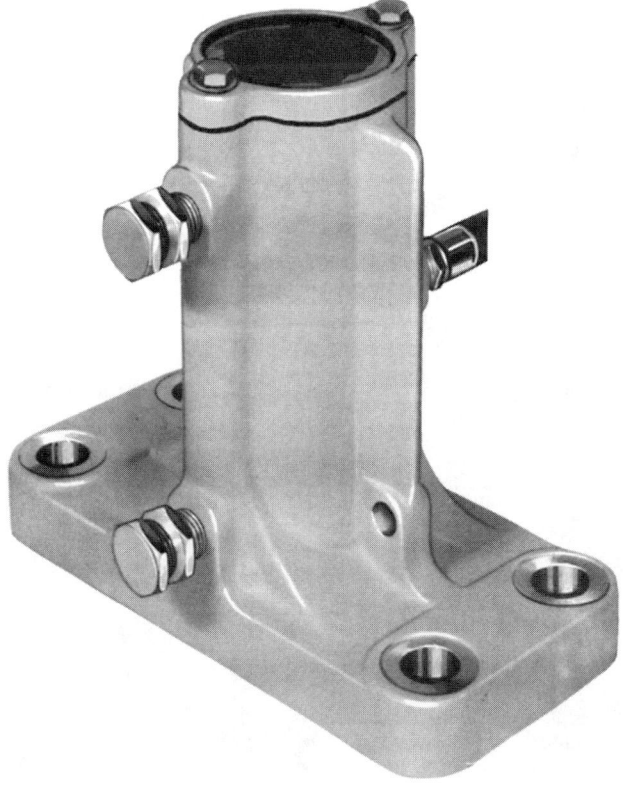

Fig. 5.2.2.1—Rotary form vibrators; pneumatically driven (top) and electrically driven (bottom).

Fig. 5.2.2.2—Reciprocating form vibrator.

The acceleration of a form is a function of the centrifugal force of the vibrators as related to the mass of form and concrete activated. The following empirical formulas recommended by Forssblad (1971) have been found useful in estimating the centrifugal force of form vibrators needed to provide adequate consolidation:

1. For plastic mixtures in beam and wall forms: Centrifugal force = 0.5 [(mass of form) + 0.2 (concrete mass)].

2. For stiff mixtures in pipe and other rigid forms: Centrifugal force = 1.5 [(weight of form) + 0.2 (concrete weight)].

Formulas should be checked against field experience. The prospective user should submit drawings of the structure to be vibrated to the vibrator manufacturer and should solicit recommendation as to size, quantity, and location of vibrator units. The proper distance between form vibrators is normally within the range of 5 to 8 ft. (1.5 to 2.5 m) and supplemental internal vibration may be required for sections thicker than 12 in. (300 mm).

Frequency and amplitude should be checked at several points on the form with a vibrograph or other suitable device (see Sections 7.5 and 15.3.3). From these values, the actual acceleration may be computed using the formula in Fig. A.1 in Appendix A.

When external vibration employs electrically operated vibrators on thin form membranes, caution should be used to prevent burning out these vibrators.

5.3—Vibrating tables

A vibrating table normally consists of a steel or reinforced concrete table with external vibrators rigidly mounted to the supporting frame (see Fig. 5.3). The table and frame are isolated from the base by steel springs, neoprene isolation pads, or other means.

Fig. 5.3—Vibrating table.

The table itself can be part of the mold. However, a separate mold usually rests on top of the table. Vibration is transmitted from the table to the mold and thence to the concrete. There is a difference of opinion as to the advisability of fastening the mold to the table.

Low frequency (below 6000 vibrations per min [100 Hz]), high amplitude (over 0.005 in. [0.13 mm]) vibration is normally preferred, at least for stiffer mixtures.

The effectiveness of table vibration is largely a function of the acceleration imparted to the concrete by the table. Accelerations in the range of 3 to 10 g (30 to 100 m/sec^2) are generally recommended, the higher values being needed for the stiffer mixtures. In addition, the amplitude should not be less than 0.001 in. (0.025 mm) for plastic mixtures, or 0.002 in. (0.050 mm) for stiff mixtures.

Acceleration of the table is a function of the vibrational force as related to the mass of form and concrete activated. The following empirical formulas have been useful in estimating the required centrifugal force of the vibrators (Forssblad 1971):

1. Rigid vibrating table or vibrating beams, with form placed loosely on the table: Centrifugal force = (2 to 4) [(mass of table) + 0.2 (mass of form) + 0.2 (mass of concrete)].

2. Rigid vibrating table, with form attached to the table: Centrifugal force = (2 to 4) [(mass of table) + (mass of form) + 0.2 (mass of concrete)].

3. Flexible vibrating table, continuous over several supports: Centrifugal force = (0.5 to 1) [(mass of table + 0.2 (mass of concrete)].

The choice of vibrators and spacing should be based on the preceding formulas and previous experience. Frequency and amplitude should be checked at several points on the table, with a vibrograph or other suitable device. The actual acceleration may then be computed. The vibrators should be moved around until dead spots are eliminated and the most uniform vibration is attained.

When concrete sections of different sizes are to be vibrated, the table should have a variable amplitude. Variable frequency is an added advantage.

If the vibrating table has a vibrating element containing only one eccentric, a circular vibrational motion may be obtained which imparts an undesirable rotational movement to the concrete. This may be prevented by mounting two vibrators side by side in such a manner that their shafts rotate in opposite directions. This neutralizes the horizontal component of vibration, so the table is subjected to a simple harmonic motion in the vertical direction only. Very high amplitudes may be obtained in this manner.

To achieve good consolidation of very stiff mixtures, it is frequently necessary to apply pressure to the top surface during vibration.

5.4—Surface vibrators

Surface vibrators are applied to the top surface and consolidate the concrete from the top down by maintaining a head of concrete in front of them. Their leveling effect assists the finishing operation. They are used mainly in slab construction.

There are three principal types of surface vibrators:

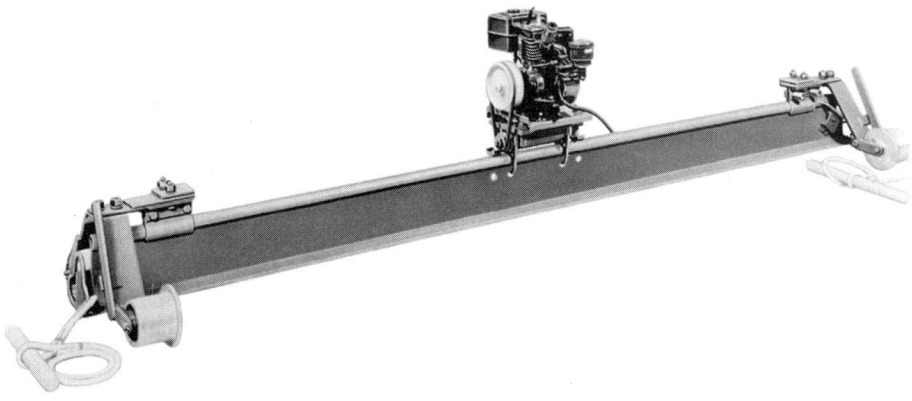

Fig. 5.4(a)—Vibrating screed for small jobs. Single beam type.

a. *Vibrating screed*—This consists of a single or double beam spanning the slab width [see Fig. 5.4(a) and (b)]. Vibrating screeds are most suited for horizontal or nearly horizontal surfaces. Caution should be exercised in using vibrating screeds on sloping surfaces. One or more eccentrics, depending on the screed length, are attached to the top. The eccentrics are driven by an internal combustion engine, or by electric or pneumatic power. The beam is supported on the forms or suitable rails; this controls the screed elevation so that it acts not only as a compactor but also provides the final finish. Vibratory screeds are usually hand drawn on small jobs and power towed on larger ones.

Vibration produced by oscillation of the beam is transmitted to the concrete near the vibrating member. A large amplitude is needed, especially for stiffer consistencies, to attain a considerable depth of consolidation. Frequencies of 3000 to 6000 vibrations per min (50 to 100 Hz) have been found to be satisfactory. Vibrating screeds usually work best with accelerations of about 5 g. Research by Kirkham (1963) has shown that consolidation is proportional to the mass times the amplitude times the frequency divided by the machine's forward speed.

$$Consolidation \propto \frac{Mass \cdot Amplitude \cdot Frequency}{Speed}$$

b. *Plate or grid vibratory tampers*—This consists of a small vibrating plate or grid, usually a few square feet (about 0.2 m^2) in area, that is moved over the slab surface. These vibrators work best on relatively stiff concrete.

c. *Vibratory roller screed*—This unit strikes off as well as consolidates. One model consists of three rollers in which the front acts as an eccentric and is the vibrating roller, rotating at 100 to 400 revolutions per min (1.7 to 6.7 Hz) (regulated according to the consistency of the mixture) in a direction opposite to the direction of movement. It knocks down, screeds, and provides mild vibration. This equipment is suitable for plastic mixtures.

Vibratory hand floats or trowels are also available. Small vibratory devices, electrically or pneumatically powered, attached to standard finishing tools provide for easier finishing.

Fig. 5.4(b)—Vibrating screed for small jobs. Double beam type.

5.5—Vibrator maintenance

Vibration equipment uses an eccentric or out-of-balance mass; therefore, higher-than-normal loads are imposed on parts such as bearings.

Regardless of vibrator type, care should be given to its maintenance. The manufacturers usually issue manuals giving instructions for servicing their machines. Nevertheless, stand-by vibrators should always be on hand.

For electrical vibrators, precautions should be taken to prevent accidental electrical shock.

Periodic measurements of energy input to the vibrator system (motor, flex shaft [if used], and vibrator head) should be taken under no load to determine free-load losses. This can be useful to indicate pending failure.

Preventive maintenance is a system of planned inspections, adjustments, repairs, and overhauls. Preventive maintenance of vibratory equipment is necessary for it to operate at full effectiveness and to avoid production shutdowns. Certain items need daily attention, while others require less frequent care, as recommended by the vibrator manufacturer.

Usually, the contractor is responsible for vibrator maintenance. Sometimes, however (especially in the case of certain mass-concrete vibrators), the contractor performs only the daily maintenance, with other servicing left to the manufacturer.

5.5.1 *Preventive maintenance program*—A file should be established with data on use and servicing requirements for each vibrator. Servicing requirements are obtained mainly from the manufacturer's service manual and spare parts list. The file might contain some or all of the following:

a. Make, serial number, and date of purchase.

b. Line voltage and amperage requirements for electrical vibrators, air volume consumed by air units, minimum cable or pipe sizes, and other pertinent information.

c. Spare parts that are apt to wear out quickly. If these are difficult to procure, they should be carried in stock.

d. Log giving a breakdown of service requirements, from the power source to the vibrator tip. Items of wear, items to lubricate and inspect in each stage, and the recommended lubricants and frequency of lubrication are listed.

Table 5.5.1 is a service log that might be used for a flexible-shaft vibrator. Starting with the date that the vibrator is checked out from the equipment pool, an actual calendar schedule can be set up for the items listed. For best results this program should be handled by a separate maintenance division rather than the operating line.

CHAPTER 6—FORMS

Formwork, form release agents, mixture design, and consolidation are some key factors in establishing the appearance of concrete work. The concrete surface appearance is a reflection of the form surface, provided that consolidation is properly accomplished. Since repairs to a defective surface are costly and seldom fully satisfactory, they should be avoided by establishing and maintaining quality forming and consolidation procedures.

6.1—General

Form strength, design, and other requirements are covered in ACI 347R and ACI SP-4, *Formwork for Concrete* (Hurd 1989). These publications deal mainly with forms for concrete that is internally vibrated. Very little guidance is given on the design of forms for external vibration.

6.2—Sloping surfaces

It is difficult to consolidate concrete that has a sloping top surface. When the slope is approximately 1:4 (vertical to horizontal) or steeper, consolidation is best assured by providing a temporary holding form or slipform screed to prevent sag or flow of concrete during vibration. An advantage of the temporary holding form or slipform screed is elimination of the need to strike off the top surface (Tuthill 1967). The holding form can be removed before the concrete has reached its final set so that surface blemishes can be removed by hand. When the sloping form cannot be removed before the concrete has set, the form should be removed as soon as possible to permit filling of the blemishes.

6.3—Surface defects

Some surface defects are related to a combination of the consolidation process and formwork details. Formwork considerations are addressed by ACI 347R, while ACI 303R provides information on the use of form release agents.

The formed concrete finish should be observed when the form is stripped so that appropriate corrective measures can be expeditiously implemented. Additional information concerning surface defects may be found in ACI 309.2R.

6.4—Form tightness

Form joints should be mortar-tight for all concrete construction and should be taped to prevent leakage where appearance is important. If holes, open joints, or cracks occur in the form sheathing, hydrostatic pressure will cause mortar to flow out when vibration momentarily converts it to a fluid consistency. Such loss of mortar will cause rock pockets or sand streaks at these locations (see Fig. 6.4). Also, air may sometimes be sucked into the form at points of leakage, causing additional voids in the concrete surface. These imperfections seriously impair surface appearance and in some cases may weaken the structure. Moreover, it is practically impossible to make repairs that are inconspicuous.

Forms may also lose mortar at the bottom during vibration if the bottom plate does not fit the base tightly. The forms may cause this leakage by floating upward during vibration, especially if one or both sides are battered. Forms must be securely tied down and tightly caulked if this leakage is to be prevented.

Fig. 6.4—Sand streaks caused by mortar leak

Fig. 6.5.3—Mounting of vibrators; wood wall form and pipe form (inset).

A 1 by 4 in. (25 by 100 mm) closed-cell rubber or polyvinylchloride foam strip tacked to the underside of the plate is quite effective in stopping this leakage. It is very helpful to secure flat, straight surfaces on which to set the plate.

Mortar leakage at form joints between form panels and at the bottom of wall forms can be minimized by extending the form sheathing about $1/8$ in. (3 mm), or more in some cases, beyond the form-framing members. This arrangement allows the relatively thin edges of the sheathing to conform more easily and tightly to adjacent surfaces than wide and unyielding faces of form-framing members. When it is desired to disguise the joints, rustication strips should be used.

ACI 347R and SP-4 (Hurd 1989) suggest a 1 in. (25 mm) or less overlap for form sheathing. Otherwise forms spread and promote loss of mortar. The wales should overlap the casting below and should be held tightly to the previous casting by form ties. Anchors or bolts in the previous placement are recommended.

6.5—Forms for external vibration

6.5.1 *General*—Forms must withstand the lateral pressure of the vibrating liquefied concrete. Forms for external vibration must also be able to stand up under the repeated, reversing stresses induced by vibrators attached to the forms. Furthermore, they must be capable of transmitting the vibration over a considerable area in a uniform manner. Form design and vibration requirements should be coordinated before purchasing the forms.

The low-frequency, high-amplitude type of vibration has a greater impact and is harder on forms than the high-frequency, low-amplitude type. Extremely rugged forms are required where high-frequency, high-amplitude vibration is used.

6.5.2 *Forming material*—Steel is the preferred forming material because it has good structural strength and fatigue properties, is well suited for attachment of vibrators, and when properly reinforced provides good, uniform transmission of vibration. Wood, plastic, or reinforced concrete forms are generally less suitable, but will give satisfactory results if their limitations are understood and proper allowances are made.

6.5.3 *Design and construction*—Forms should be designed to resist the pressure of concrete without excessive deflection and to transmit the vibratory impulses to the concrete. A steel plate, 3/16 to 3/8 in. (5 to 10 mm) or thicker, stiffened with vertical and/or horizontal ribs, will perform these functions. Oscillation (flexing) of the steel plate between the stiffeners is normally somewhat greater than for the stiffeners themselves, but it should not be excessive if the stiffeners are closely spaced. Special attention should be directed to attachments when external vibration is anticipated to insure that excessive form deflections do not occur.

Special members, such as steel I-beams or channels, should be placed next to the plate, passing through the stiffeners in a continuous run. It is generally desirable to weld the stiffeners to these members.

The vibrators should be rigidly attached to the special members (see Fig. 6.5.3). Damage to the form and vibrator will occur if the vibrator shakes loose.

When rotary electric units are used, the rigidity of mounting required can readily be determined by measuring the amperage draw. If it exceeds the nameplate rating, the support is not strong enough. Air units cannot be evaluated as easily, but observing the movement of the form gives an indication of the rigidity. It is essential that the form hardware be securely fastened. Since wedges have a tendency to work loose under vibration, bolting is more dependable. Special attention should be paid to the strength of welds.

Vertical forms should be placed on rubber pads or other resilient base material to prevent transmission and loss of vibration to the supporting foundation as well as leakage of mortar.

It is difficult to attain and maintain form tightness when vibration is of the external type; even minute openings in the form will permit loss of mortar. Rubber or other suitable seals may be used to prevent grout loss through steel forms.

Attaching external vibrators directly to the form is generally unsatisfactory because the skin may flutter or develop a diaphragm action. This movement causes the vibrational force to be highly localized, and sometimes results in early form failure. However, portable vibrators mounted to brackets on metal forms have been successfully used in precast work and occasionally in general construction. One or more vibrators are moved from bracket to bracket over the form as placing progresses. This method should be used with extreme caution, and only with units having low amplitude and high frequency.

CHAPTER 7—RECOMMENDED VIBRATION PRACTICES FOR GENERAL CONSTRUCTION

After proper vibration equipment has been selected (see Chapter 5), it should be operated by conscientious, well-trained operators. The vibrator operator should have developed, through experience, the ability to determine the time necessary for the vibrator to remain in the concrete to insure proper consolidation. By a systematic review of the operator's previous work, the operator and supervisor should be able to determine the vibrator spacing and the vibration time needed to produce dense concrete without segregation.

Internal vibration is generally best suited for ordinary construction, provided the section is large enough for the vibrator to be effectively used. However, external vibration or consolidation aids may be needed to supplement internal vibration in areas congested with reinforcement or otherwise inaccessible (See Chapter 17). In many thin sections, especially in precast work and slabs, external vibration should be the primary method of consolidation.

7.1—Procedure for internal vibration

Concrete should be deposited in layers compatible with the work being done. In large mats and heavy pedestals, the maximum layer depth should be limited to 20 in. (500 mm). The depth should be nearly equal to the vibrator head length. In walls and columns, the layer depths should generally not exceed 20 in. (500 mm). The layers should be as level as possible so that the vibrator is not used to move the concrete laterally, since this could cause segregation. Fairly level surfaces can be obtained by depositing the concrete in the form at close intervals; the use of elephant trunks is frequently helpful.

Even though the concrete has been carefully deposited in the form, there are likely to be some small mounds or high spots. Some minor leveling can be accomplished by inserting the vibrator into the center of these spots to flatten them. Excessive movement should be avoided, particularly through reinforced structural elements.

After the surface is leveled, the vibrator should be inserted vertically at a uniform spacing over the entire placement area. The distance between insertions should be about $1\frac{1}{2}$ times the radius of action, and should be such that the area visibly affected by the vibrator overlaps the adjacent just-vibrated area. In slabs, a standard length vibrator should be sloped towards the vertical, or a short stubby 5-inch-long vibrator should be held vertically. Both should be kept 2 in. (50 mm) away from the bottom if the slab is a tilt-up panel and when a tilt-up panel slab has an architectural bottom face. The vibration should be sufficient to close the bottom edges of the placed concrete layers.

An alternate method that has been successfully used is as follows. The vibrator should penetrate rapidly to the bottom of the layer and at least 6 in. (150 mm) into the preceding layer. The vibrator should be manipulated in an up and down motion, generally for 5 to 15 sec, to knit the two layers together. The vibrator should then be withdrawn gradually with a series of up and down motions. The down motion should be a rapid drop to apply a force to the concrete which, in turn, increases internal pressure in the freshly placed mixture.

Rapidly extract the vibrator from the concrete when the head becomes only partially immersed in the concrete. The concrete should move back into the space vacated by the vibrator. For dry mixtures where the hole does not close during the withdrawal, sometimes reinserting the vibrator within 1/2 influence radius will solve the problem; if this is not effective, the mixture or vibrator should be changed.

Thin slabs supported on beams should be vibrated in two stages: first, after beam concrete has been placed, and again when the concrete is brought to finished grade.

The vibrator exerts forces outward from the shaft. Air pockets at the same level as, or located below, the head tend to be trapped. Therefore, air pockets should be worked upward in front of the vibrator.

When the placement consists of several layers, concrete delivery should be scheduled so that each layer is placed while the preceding one is still plastic to avoid cold joints. If the underlying layer has stiffened just beyond the point where it can be penetrated by the vibrator, bond can still be obtained by thoroughly and systematically vibrating the new concrete into contact with the previously placed concrete; however, an unavoidable joint line will show on the surface when the form is removed.

7.2—Judging the adequacy of internal vibration

Presently, there is no quick and fully reliable indicator for determining the adequacy of consolidation of the freshly placed concrete. Adequacy of internal vibration is judged mainly by the surface appearance of each layer. The principal indicators of well consolidated concrete are:

1. Embedment of large aggregate. Except in architectural concrete with exposed aggregate surfaces, general batch leveling, blending of the batch perimeter with concrete previously placed, a thin film of mortar on the top surface, and cement paste showing at the junction of the concrete and form.

2. General cessation in escape of large entrapped air bubbles at the top surface. Thicker layers require more vibration time than thin layers, because it takes longer for deep-seated bubbles to make their way to the surface.

Sometimes the pitch or tone of the vibrator is a helpful guide. When an immersion vibrator is inserted in concrete, the frequency usually drops off, then increases, and finally becomes constant when the concrete is free of entrapped air. An experienced operator also learns the proper feel of a vibrator when consolidation is complete.

There is a tendency for inexperienced vibrator operators to merely flatten the batch. Complete consolidation is assured only when the other items evidencing adequate vibration are sought and attained.

7.3—Vibration of reinforcement

When the concrete cannot be reached by the vibrator, such as congested reinforcement areas, it may be helpful to vibrate exposed portions of reinforcing bars. Some engineers have suggested possible degradation in concrete-to-steel bond from vibration carried down through reinforcement to partially set concrete in the lower layers of a placement. Careful examination of hardened concrete consolidated in this manner has uncovered no grounds for such fears. When the concrete is still mobile, this vibration actually increases the concrete-to-steel bond through the removal of entrapped air and water from underneath the reinforcing bars.

A form vibrator, attached to the reinforcing steel with a suitable fitting, should be used for this purpose. Binding an immersion vibrator to a reinforcing bar may damage the vibrator.

7.4—Revibration

Revibration is the process of vibrating concrete that was vibrated some time earlier. Actually most concrete is revibrated unintentionally when, in placing successive layers of concrete, the vibrator extends down into the underlying layer (which was previously vibrated). However, the term revibration as used here refers to an intentional, systematic revibration some time after placing is completed (Vollick 1958).

Revibration can be accomplished any time the running vibrator will sink under its own weight into the concrete and liquefy it momentarily. This revibration has generally been considered to be most effective when performed just prior to the time of initial setting of the concrete for mixtures with slumps of 3 in. (75 mm) or more.

Revibration generally results in improved compressive strength of standard cylinders. The effect of revibration on concrete-to-steel bond strength is not as clear. Revibration appears to improve bond strength for top reinforcing steels placed in high-slump concrete. Revibration may, however, severely damage bond strength for reinforcing steel in well-consolidated, low-slump concrete. Revibration is almost universally detrimental to the bond strength of bottom reinforcing steel. Overall, revibration tends to reduce the differences in bond strength caused by differences in slump and position (Altowaiji, Darwin, and Donahey 1984).

Revibration is most beneficial in the top few feet (0.5 to 1 m) of a placement, where air and water voids are most prevalent. Revibration of the tops of walls normally results in a more uniform appearance of vertical surfaces.

Revibration can be very effective in minimizing cracks at the top of doorways, arches, major boxouts, etc. The procedure is to delay additional concrete placement for 1 to 2 hr, depending upon temperature, after reaching the springline of arches or headline of doors, boxouts, or joints between column and floor, etc., to permit settlement shrinkage to occur before revibration of the materials in place and the resumption of placement.

7.5—Form vibration

The size and spacing of form vibrators should be such that the proper intensity of vibration is distributed over the desired area of form. The spacing is a function of the type and shape of the form, depth, and thickness of the concrete, force output per vibrator, workability of the mixture, and vibrating time.

The recommended approach is to start with a spacing, generally in the range of 4 to 8 ft (1.2 to 2.4 m), based on the guidelines in Section 5.2.3 and previous experience. If this pattern does not produce adequate and uniform vibration, the vibrators should be relocated as necessary until proper results are obtained. Achieving optimum spacing requires knowledge of the distribution of frequency and amplitude over the form, and an understanding of the workability and compactibility of the mixture.

The frequency can readily be determined by a vibrating reed tachometer (see Section 15.3.1). However, the small amplitudes associated with form vibration have been difficult to measure in the past. Inadequate amplitudes cause poor consolidation, while excessive local amplitudes are not only wasteful of vibrator power but can also cause the concrete to roll and tumble so that it does not consolidate properly.

Moving one's hand over the form will locate areas of strong or weak vibration (high or low amplitude) or dead spots. The vibrating reed tachometer can provide slightly more reliable information; the difference in oscillation of the reed at various points gives a rough indication of the difference in amplitude.

The vibrograph makes it possible to get reliable values of the amplitude at various locations on forms vibrated externally. The frequency and wave form are also generally provided.

Concrete compacted by form vibration should be deposited in layers 10 to 15 in. (250 to 400 mm) thick. Each layer should be vibrated separately. Vibration times are considerably longer than for internal vibration, frequently as much as 2 min and as much as 30 min or more in some deep sections.

Another procedure which has given good results in precast work involves continuously placing ribbons of concrete 2 to 4 in. (50 to 100 mm) thick, accompanied by continuous vibration. It can produce surfaces nearly free of bugholes.

It is desirable to be able to vary the frequency and amplitude of the vibrators. On electrically driven external vibrators, amplitudes can be adjusted to different fixed values quite readily. The frequency of air-driven external vibrators can be adjusted by varying the air pressure, while the amplitude can be altered by changing the eccentric mass.

Since most of the movement imparted by form vibrators is perpendicular to the plane of the form, the form tends to act as a vibrating membrane, with an oil-can effect. This is particularly true if the vibration is of the high-amplitude type, and the plate is too thin or lacks adequate stiffeners. This in-and-out movement can cause the forms to pump air into the

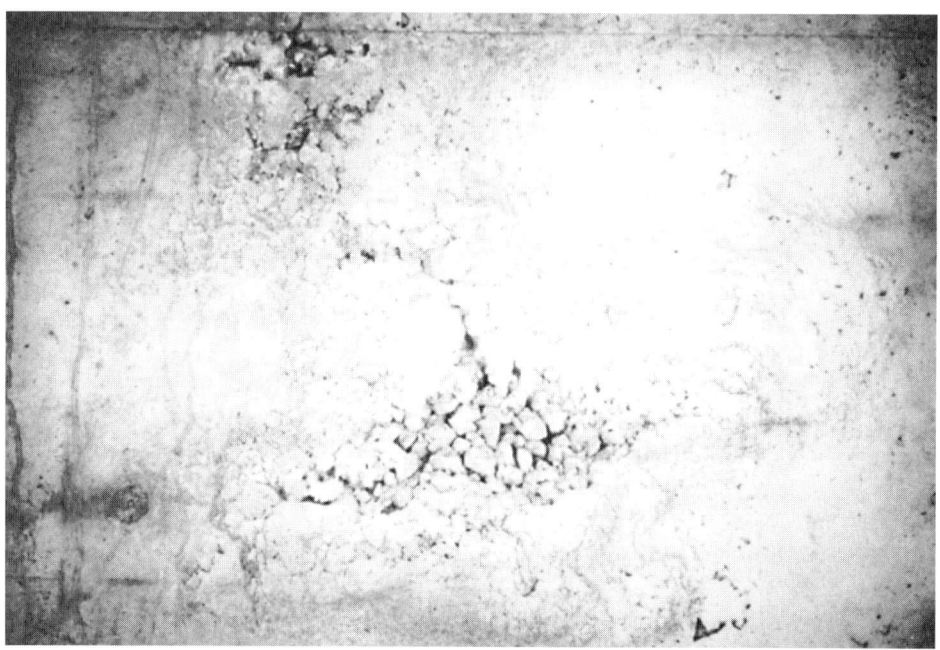

Fig. 7.6.1(a)—Honeycomb.

Fig. 7.6.1(b)—Haphazard procedure may result in mortar accumulation at the surface and leave rock pockets below, particularly at batch perimeters.

concrete, especially in the top few feet (0.5 to 1 m) of a wall or column lift, creating a gap between the concrete and the form. Here there are no subsequent layers of concrete to assist in closing the gap. It is therefore often advisable to use an internal vibrator in this region.

Form vibration during stripping is sometimes beneficial. The minute movement of the entire form surface helps to loosen it from the concrete and permit easy removal without damage to the concrete surface.

7.6—Consequences of improper vibration

The most serious defects resulting from undervibration are honeycomb, excessive entrapped air voids (bugholes), sand streaks, subsidence cracking, and placement lines.

7.6.1 *Honeycomb*—Honeycomb occurs [see Fig. 7.6.1(a)] when the mortar does not fill the space between the coarse aggregate particles. The presence of honeycomb indicates that the first stage of consolidation (see Section 4.2) has not been completed at these locations. When it shows on the surface, it is necessary to chip out the area and make a repair. Such repairs should be kept to a minimum, mainly because they mar the appearance and reduce the concrete strength. Honeycomb is generally caused by using improper or faulty vibrators, improper placement procedures, poor vibration procedures, inappropriate concrete mixtures, or congested reinforcement. Unsystematic insertions of internal vibrators at haphazard angles are likely to cause an accumulation of mortar at the top surface, while the lower portion of the layer may be undervibrated [Fig. 7.6.1(b)].

Guidance on proper placing techniques to minimize separation of coarse aggregate from mortar can be obtained from Chapter 9 of ACI *Manual of Concrete Inspection*, SP-2.

Concrete properties contributing to honeycomb are insufficient paste to fill the voids between the aggregate, improper ratio of fine to total aggregate, poor aggregate grading, or

Fig. 7.6.1(c)—Poorly designed, congested reinforcement which will make good consolidation extremely difficult.

improper slump for the placing conditions. Insufficient clearance between the reinforcing steel is an important cause of honeycomb [see Fig. 7.6.1(c)]. In establishing steel spacing, both the designer and builder must keep in mind that the concrete must be consolidated.

7.6.2 *Excessive entrapped-air voids*—Concrete that is free of honeycomb still contains entrapped air voids because complete removal of entrapped air is rarely feasible (See Section 4.2). The amount of entrapped air remaining in the concrete after vibration is largely a function of the vibratory equipment and procedure, but it is also affected by concrete mixture constituents, the properties of the concrete mixture, location in the placement, and other factors (Samuelsson 1970). When proper equipment or procedures are not used, or other unfavorable conditions occur, the entrapped-air content will be high and surface voids (commonly called bugholes) are likely to be excessive (see Fig. 7.6.2).

Fig. 7.6.2—Excessive air voids on formed surface.

To reduce air voids in concrete surfaces, the distance between internal vibrator insertions should be reduced, and the time at each insertion increased. Use of a more powerful vibrator may help for some situations. Also there should be a row of insertions close to the form, but without touching it. When form contact is almost unavoidable, the vibrator should be rubber tipped; even then, any such contact should be avoided if possible because this may mar the form and disfigure the concrete surface. It is critical that the locations of vibrator insertions be such that zones of influence overlap.

Form coatings of high viscosity or those that are applied in overly thick applications tend to hold air bubbles and should be avoided.

Form vibrators tend to draw mortar to the form, and when used in combination with internal vibrators have proved effective in reducing the size and number of air voids on the surface.

For difficult conditions and when the concrete appearance is quite important, spading next to the form has been helpful in reducing air voids.

It is nearly impossible to eliminate air voids from inwardly sloping formed surfaces, and designers should recognize this fact. However, these voids can be minimized if sticky, oversanded mixtures are avoided, the concrete is deposited in shallow layers of 1 ft. (0.3 m) or less, and the vibrator is inserted as closely as possible to the form. By attaching an external vibrator to the sloping form and reducing the layer thickness to 6 in. (150 mm), voids can be considerably reduced.

7.6.3 *Sand streaking*—Sand streaking is caused by heavy bleeding and mortar loss along the form, resulting from the character and proportions of the materials and method of depositing the concrete (see Fig. 7.6.3). Harsh, wet mixtures that are deficient in cement and contain poorly graded aggregates—particularly those deficient in the No. 50 to 100 (300 to 150 μm) and minus No. 100 (150 μm) fractions—may cause sand streaking, as well as other problems. Dropping concrete through reinforcing steel and depositing it in thick lifts without adequate vibration may also cause streaking, as well as honeycomb. Another cause of sand streaking is form

Fig. 7.6.3—Sand streaking caused by heavy bleeding along form.

Fig. 7.6.4—"Pour" lines.

Table 12.1—Consolidation methods for precast concrete products

Products	Mix Classification (Section 12.1)	Forming material	Conveying and placing method	Consolidation method
Concrete pipe	a to d	Steel	Pumping, conveyors, or bucket (thin layers)	Tamping; internal or external vibration; centrifugation; vacuum; pressure
Concrete piles and poles	c, d	Steel	Pumped, or conveyed by mixer trucks	Centrifugation; internal or external, high frequency, low amplitude vibration; roller packed
Concrete block	b	Steel	Machine hopper	Low frequency, high amplitude vibration plus pressure
Slab and beam sections	b, c	Steel	Traveling hopper, mixer trucks, belt conveyors	External vibration with or without roller compactions; internal vibration with surface vibrating screed
Wall panels	a to c	Reinforced concrete, steel, or wood	Buckets and belt conveyors (continuous ribbon feed)	Tampers; internal and external vibration

vibrators that are attached to leaky forms that have a pumping action with a resulting loss of fines or an indrawing of air at the joints.

7.6.4 *Placement lines*—Placement lines are dark lines (see Fig. 7.6.4) on the formed surface at the boundary between adjacent batches of concrete. Generally, they indicate that the vibrator was not lowered far enough to penetrate the layer below the one being vibrated.

7.6.5 *Cold joints*—Delays in concreting can result in cold joints. To avoid cold joints, placing should be resumed substantially before the surface hardens. For unusually long delays during concreting, the concrete should be kept live by periodically re-vibrating it. Concrete should be vibrated at approximately 15-min intervals or less depending upon job conditions. However, concrete should not be overvibrated to the point of causing segregation. Furthermore, should the concrete approach time of initial setting, vibration should be discontinued and the concrete should be allowed to harden. A cold joint will result and suitable surface preparation measures should be applied.

7.6.6 *Subsidence cracking*—Subsidence cracking results from the development of tension when the concrete mechanically settles at or near initial setting time. To eliminate this type of cracking, the concrete should be revibrated at the latest time at which the vibrator will sink into the concrete under its own mass.

7.6.7—Undervibration is far more common than overvibration. Normal weight concretes that are well proportioned and have adequate consistency are not readily susceptible to overvibration. Consequently, if there is any doubt as to the adequacy of consolidation, it should be resolved by additional vibration.

7.6.8—Overvibration can occur if, due to careless operation or use of grossly oversized equipment, vibration is many times the recommended amount. This overvibration may result in:

a. *Segregation*—The mechanics of segregation come into play when the forces of gravity and vibration are given sufficient time to interact. With excessive vibration time, the cohesive forces within the concrete are overcome by gravity and vibration causes the heavier aggregates in the mixture to settle and the lighter aggregates to work upward borne by the paste matrix. Examination during or after this type of placement will show a layer of laitance, a layer of mortar containing a minor proportion of large aggregate, and an accumulation of large aggregate in the bottom of the placement layer. This condition is more likely with wet mixtures with large differences in the densities of the aggregates and the mortar and when mixtures having too high a proportion of mortar to coarse aggregate. Lightweight aggregate is a problem all its own unrelated to mortar proportion. Proper control of consistency will minimize the problem.

b. *Sand streaks*—They are most likely with harsh, lean mixtures and with concrete moved horizontally with the vibrator.

c. *Loss of entrained air in air-entrained concrete*—This can reduce the concrete's resistance to cycles of freezing and thawing. The problem generally occurs in mixtures with excessive water contents. If the concrete originally contained the amount of entrained air recommended by ACI Committee 211 (see Chapter 18.1) and the slump is in the proper range, serious loss of entrained air is highly unlikely. However, too many insertions of the vibrator too close together in concrete can cause a coalescing of the entrained-air system, which may cause a reduction in freeze-thaw durability.

d. *Excessive form deflections or form damage*—These are most likely with external vibration.

e. *Form failure*—Excessive internal pressures that may cause form failure can occur by allowing the vibrator to be immersed too long in the concrete at the same location. Pressure caused by excessive depth (deeper than the designed rate of rise per hour) of fresh concrete, augmented by the dynamic forces of prolonged vibration, may cause the form to fail instantaneously.

CHAPTER 8—STRUCTURAL CONCRETE

8.1—Design and detailing prerequisites

In designing structural members and detailing formwork and reinforcement, consideration should be given to depositing the freshly mixed concrete as closely as possible to its final position in such a way that segregation, honeycombing, and other surface and internal imperfections are minimized. Also, the method of consolidation should be carefully considered when detailing reinforcement and formwork. For example, for internal vibration, openings in the reinforcement must be provided to allow insertion of vibrators. Typically, 4 by 6-in. (100 by 150-mm) openings at 24-in. (600-mm) centers are required.

These items require that special attention be directed to member size, reinforcing steel size, location, spacing, and other factors that influence the placing and consolidation of concrete. This is particularly true in structures designed for seismic loads, where the reinforcement often becomes extremely congested and effective concrete consolidation using conventional mixtures and procedures becomes impossible.

The designer should communicate with the constructor during the early structural design. Problem areas should be recognized in time to take appropriate remedial measures such as staggering splices, bundling reinforcing steel, modifying stirrup spacing, and increasing section size. When conditions contributing to substandard consolidation exist, one or more of the following actions should be taken: redesign the member, redesign the reinforcing steel, modify the mixture, utilize mock-up tests to develop a procedure, and alert the constructor to critical conditions.

The placing of concrete in congested areas is discussed in more detail in Chapter 17.

8.2—Mixture requirements

Structural concrete mixtures should be proportioned to give the placeability, durability, strength, and other properties required with proper regard to placement conditions. The concrete should work readily into the form corners and around reinforcement by the consolidation methods employed, without segregation or excessive free water collecting on the surface. Some guidance on proportioning may be found in Chapter 2, and ACI 301 covers this subject in detail. In areas of congested reinforcement, the procedures in Chapter 17 should be considered. Also, consideration should be given to using mechanical connections for the reinforcement to minimize congestion.

A 3-in. (75-mm) slump is normally ample for properly vibrated structural concrete in forms. What may be regarded as a need for higher slump concrete in many quarters is better satisfied by more thorough vibration. Actually, concrete for heavy structural members can often be satisfactorily placed at a 2 in. (50 mm) maximum slump when effectively vibrated.

In those areas where thorough vibration cannot be achieved due to congested reinforcement or other obstructions, it may be desirable to increase the slump by using admixtures to produce a flowing concrete that can be more effectively consolidated (ACI 309.3R). However, it is important to note that the use of flowing concrete does not preclude the need for vibration.

8.3—Internal vibration

For most structural concrete, vibration is most effectively performed by means of standard immersion vibrators meeting the guidelines in Table 5.1.5. It is important that the vibrator selected be suitable for the mixture and placing conditions.

The recommended procedure for internal vibration is described in Section 7.1. In walls and beams, two vibrators should generally be used, one for leveling the mixture immediately after placement and the other for further consolidation. On larger and more critical jobs, a third unit, which may be less powerful than the other two, may be useful. It should be used in a row of closely spaced insertions within a few inches (several centimeters) of the form, and also in the top layer of the placement, to assist air bubbles to rise and escape.

Slabs placed monolithically with joists or beams should be constructed in the following manner: all joists and beams should be placed and vibrated before the slab itself. A time interval of about an hour will permit settlement and consequent bleeding to take place in these elements prior to placing the concrete in the slab section. The slab concrete should be placed and vibrated prior to the beam concrete taking its initial set. Vibrators should penetrate through the slab into the previously placed beam concrete to consolidate and bond the structural elements.

8.4—Form vibration

Form vibration is suitable for many thin sections and is a useful supplement to internal vibration at locations where steel is unusually congested, where concrete cannot be directly placed but must flow into position, or where an internal vibrator cannot be inserted. However, form vibration can result in form pressures substantially higher than normal, and particular consideration should be given to formwork design.

Procedures for form vibration are described in Section 7.5. In any use of form vibration, it is important to avoid excessive vibration at any given location. The vibrators should be moved, as necessary, to keep them operating just below the top surface of the concrete, not on unfilled areas of forms.

8.5—Tunnel

Form vibrators are used for concrete consolidation in tunnel linings. Frequently, form vibration is supplemented by immersion vibrators that are used behind the form or through access windows in the form. Tunnel-lining concrete is most commonly placed by pumping, with pump lines positioned in the sidewalls and crown. It is important to have a workable yet cohesive mixture that will respond well to vibration. The slump should be about 5 in. (130 mm) at the discharge end of the pumpline.

When the level of concrete behind the form reaches the crown, an advancing slope of fresh concrete is produced. This advancing slope will generally vary from $2^1/_2$ to 1 to as much as 5 to 1, horizontal to vertical. Form vibrators should be operated within a few feet (about one meter) of the advancing slope and should be frequently moved forward horizontally. Special attention should be given to form vibration in the crown so that concrete that has been pumped into the highest points within the form is not drawn down by vibration. As the placement proceeds, the withdrawal of the pumpline and position and timing of vibration must insure maximum filling of the form.

CHAPTER 9—MASS CONCRETE

Mass concrete is defined as any volume of concrete with dimensions large enough to require that measures be taken to cope with generation of heat from hydration of the cement and attendant volume change to minimize cracking. To reduce the heat rise and to achieve economy, low cement contents and large aggregates are used and low slumps are maintained. These measures generally require special attention in consolidation.

9.1—Mixture requirements

Proper proportioning and optimum use of chemical admixtures, fly ash, and slag in mass concrete facilitate proper consolidation. Refer to ACI 211.1 for information on mixture proportioning. Additional information on mass concrete is found in ACI 207.1R.

9.2—Vibration equipment

Mass concrete containing aggregate larger than $1^1/_2$ in. (38 mm) and low cement contents presents a unique vibration problem when low slump consistencies are used. This

Fig. 9.4(a)—Stepped construction used for mass concrete construction (Photo courtesy U.S. Bureau of Reclamation).

Fig. 9.4(b)—Flattening a pile of mass concrete just deposited in form.

condition requires that powerful equipment meeting the requirements of Group 5 in Table 5.1.5 be available for proper consolidation. Pneumatically driven vibrators are generally used in the United States. The air supply must be ample and the force at the vibrator must be sufficient for adequate consolidation. In heavily reinforced areas, vibrators with small diameters may be needed to penetrate between the bars and achieve proper consolidation.

9.3—Forms

For economy of forms and better control of temperature, mass concrete is placed in fairly shallow lifts—usually 5 to 10 ft. (1.5 to 3.0 m) thick. In addition to normal form requirements (see Chapter 6), forms for mass concrete are often dependent on anchors embedded in concrete for their strength and security of position. Embedment depth for these anchors should provide anchorage sufficient to withstand the impact of fast dumping from high-line or gantry buckets as well as the ordinary concrete pressures during vibration.

9.4—Vibration practices

The lifts should be built up with multiple layers 12 to 20 in. (300 to 500 mm) thick, depending on the aggregate size. Such lifts can be reliably consolidated with some penetration of the vibrator into lower layers. Heavily reinforced sections may need thinner layers and proper attention to insure the encasement of reinforcement by concrete.

Each layer is constructed in strips 6 to 12 ft (1.8 to 3.6 m) wide. The forward edge of each upper layer should be held back 4 to 5 ft (1.2 to 1.5 m) from the one below so that it will not move when vibrating the adjacent strip of lower-layer batches placed along the edge. This procedure produces a stair-step effect of the layers [see Fig. 9.4(a)]. The placement is thus completed to full thickness and area with minimum surface exposure. This practice minimizes warming of pre-cooled concrete and cold joint problems between layers in warm weather. It also makes the placement easier in wet weather. Details for manufacture and placement of mass concrete may be found elsewhere (*U.S. Bureau of Reclamation Concrete Manual*, 1981; ACI 207.1R).

For effective consolidation of mass concrete, the vibrator crew should follow a systematic procedure. The crew should work closely together and move as a unit, rather than each operator working separately with widely spaced, random insertions. The vibrators should be inserted nearly vertically into the tops of the deposited piles at fairly uniform spacings and then reinserted as necessary to flatten the pile to the proper depth and spread it to the area it should occupy [see Fig. 9.4(b)]. Then the subsequent placements should be systematically vibrated with the vibrator penetrating the full depth of the layer and into the preceding layer, but staying away from the forward edges [see Fig. 9.4(c)]. The edges in contact with the previous strip and previous batch should be very thoroughly knitted together. Each vibrator operator should have his particular area of attention.

Vibration at each point should continue until entrapped air ceases to escape. Depending on mixture and slump, this time will usually range from 10 to 15 sec. The insertions must be spaced and timed to achieve thorough consolidation, not only near the surface but for the full depth of the layer and below it.

The completed top surface of the block should be left fairly even and free of footprints and vibrator holes, to facilitate the subsequent joint cleanup. The final vibration should be done by a vibrator operator on plywood snowshoes using a small

Fig. 9.4(c)—Systematic vibration of concrete layer.

vibrator if necessary. When consolidation is completed, the top of the coarse aggregate should be approximately at the level of the concrete surface.

The amount of concrete that can be handled by one vibrator will depend on the capability of the vibrator, the experience and diligence of the operator, and the response to vibration of the particular concrete mixture being used. Under optimum conditions, an efficient crew may handle as much as 50 yd^3 (40 m^3) per hr per vibrator. Around embedded items and in complicated formwork, the amount handled might be less than half this amount.

In Europe, Japan, and Canada, successful use has been made of gang vibrators using bulldozers, cranes, and hydraulic hoists. One bulldozer spreads and levels the concrete ready for consolidation. This is followed by systematic consolidation across the freshly spread concrete by three or more vibrators mounted on a frame. Successful use of this procedure requires an open form with a minimum of form ties. When a bulldozer is used to manipulate the frame, care is required in turning so that the tracks of the dozer do not dig into the concrete.

9.5—Roller-compacted concrete

Mass concrete can be compacted with vibratory rollers. Roller-compacted concrete (RCC) is a concrete of zero slump consistency that is transported, placed, and compacted in horizontal layers using the same equipment that is used for highway construction and earth and rockfill construction. Since the consolidation phase of RCC construction is performed by equipment of the sort used in earthwork, the soils term *compaction* has been used in place of the concrete term *consolidation*. Detailed information on RCC can be found in ACI 207.5R.

Roller-compacted concrete placed in the United States is generally placed and spread in 8 to 12-in. (200 to 300-mm) layers, although layers up to 3 ft. (1 m) thick have been used in some applications. For layers thicker than 12 in. (300 mm), the concrete should be deposited and spread in several thin layers prior to compaction. In open areas, layers are compacted by smooth-drum vibratory rollers with a static linear mass of 1200 to 3000 lb/ft. (1800 to 4500 kg/m) of drum width. In some applications, finish rolling has been accomplished with pneumatic-tired rollers with a static mass of up to 26 tons (24,000 kg). In tight areas and areas adjacent to walls and other obstructions, smaller walk-behind rollers and mechanical tampers can be used to compact the RCC. When using this equipment, care should be taken to place the RCC in thinner layers to assure compaction. Placement and rolling is generally done on horizontal layers. However, RCC has been placed and compacted on moderate slopes where a winch line has been used to assist the travel of the roller on the slope.

Generally, for richer and more plastic mixtures, the first pass by the roller is in the static mode (no vibration), followed by repeated passes in the vibratory mode. A delayed finish rolling approximately 1 hr after initial compaction has been effective in reducing surface cracking. Operators should insure a minimum of 6 in. (150 mm) overlap between adjacent rolling lanes and at the end of each run. Careful attention should be given to compaction of the joint along placing lanes, particularly if the concrete in the previous lane has reached its time of initial setting. This has been achieved by rolling the edges of lanes on a 2-to-1 slope or cutting back a vertical edge into well-compacted concrete with a grader.

Selection of vibratory rollers is not yet fully understood and equipment selection should be established through field-test procedures. Vibratory rollers generally fall under two categories:

1. *High-frequency, low-amplitude rollers*—1800 to 3200 vibrations per minute (30 to 50 Hz), 0.015 to 0.03 in. (0.38 to 0.75 mm)—are used for asphalt compaction.

2. *Lower-frequency, higher-amplitude rollers*—1200 to 1800 vibrations per minute (20 to 30 Hz), 0.03 to 0.06 in. (0.75 to 1.5 mm)—are used in earth and rockfill compaction.

Construction parameters, such as lift thickness, and characteristics of the concrete mixture, nominal maximum size of aggregate and water content, may influence selection of rollers.

Special care should be taken in proportioning the RCC mixture and in placing techniques to avoid segregation or contamination over the previously placed lift to assure a well-bonded, low permeability lift joint. When freshly mixed RCC concrete is placed on a hardened lift surface, the surface should be clean, and a thin layer of mortar or several inches (±100 mm) of a more plastic bedding mixture should be placed on the surface before covering with the regular RCC mixture. Generally, 4 to 6 passes with a properly sized vibratory roller are sufficient to produce a dense, well-compacted concrete. However, increased lift thickness and stiffer-consistency RCC mixtures may require more passes. Field trials should be conducted to determine the number of roller passes required to achieve full compaction.

CHAPTER 10—NORMAL WEIGHT CONCRETE FLOOR SLABS

10.1—Mixture requirements

Concrete for slab construction should be proportioned to give the required placeability, finishability, abrasion resistance, strength, and durability. ACI 302.1R covers recommended procedures for floor and slab construction.

Stiffer mixtures are commonly used for durable, abrasion-resistant surfaces. These require consolidation by vibration or other effective means. Recommendations in this guide are primarily for this class of construction.

10.2—Equipment

Surface vibration is recommended for consolidating slabs up to 6 in. (150 mm) thick, provided they are unreinforced or contain only light mesh. Vibrating screeds, supported on the forms, screed boards, or rails, are the most common means. They should be low-frequency (3000 to 6000 vibrations per min [50 to 100 Hz]) and high-amplitude to minimize machine wear and provide adequate depth of consolidation without creating an objectionable layer of fines at the surface. Use of the high-frequency, low-amplitude type is acceptable when applied solely to accommodate the finishing operation. Unreinforced slabs 6 to 8 in. (150 to 200 mm) thick may be consolidated by either internal or surface vibration.

Fig. 10.6—Vacuum dewatering of concrete slab is shown just behind the floor finishing operation

Internal vibration, using equipment described in Table 5.1.5, is recommended for all slabs more than 8 in. (200 mm) thick. It is also recommended for slabs of lesser thickness that contain reinforcement or other embedments, such as conduit. Internal vibration should also be provided adjacent to load transfer devices and forms.

10.3—Structural slabs

Structural slabs that contain reinforcement and conduit and should be internally vibrated. Vibrating screeds are also used frequently to facilitate finishing; a high-frequency, low-amplitude type may be used in this case.

Often, the slab will contain projecting columns, conduit, or reinforcing bars that prevent setting forms or screed boards needed for a vibrating screed. Such floors must be screeded by hand and slumps in excess of 2 in. (50 mm) are required. At these slumps, adequate consolidation will be obtained by internal vibration and the hand-screeding and finishing operations.

10.4—Slabs on grade

The procedures described in Chapter 11 should be followed on large jobs when practical. However, many floor slabs are small, odd-shaped, or on nonuniform sections so that highly mechanized procedures cannot be used. Such construction is covered by the procedures given in this chapter.

10.4.1 *Internal vibration*—The vibrator head should be completely immersed during vibration. For thick slabs, it will be possible to insert the vibrator vertically, while for thinner slabs it should be inserted at an angle, or even horizontally. Contact of the vibrator with the subgrade should be kept to a minimum since this might contaminate the concrete with foreign material.

The use of vibrating screeds, when edge forms or screed rails can be used, will facilitate strikeoff operations after the slab has been consolidated by internal vibration. By using a vibrating screed, one can use concrete of lower slump.

10.4.2 *Surface vibration*—Slumps in the range of 1 to 2 in. (25 to 50 mm) are generally recommended for concrete consolidated by vibrating screeds. For slumps in excess of 3 in. (75 mm) vibrating screeds should be used with care, since such concrete will have an accumulation of mortar on the finished surface after vibration.

Vibrating screeds strike off and straightedge the concrete in addition to providing consolidation. To perform significant consolidation, the leading edge of the shoe must be at an angle to the surface and the proper surcharge (height of uncompacted concrete required to produce a finished surface at the proper elevation) must be carried in front of the leading straightedge.

When it is impractical to set screed boards or forms for vibrating screeds or other surface vibrators, the slump will have to be increased to between 3 and 4 in. (75 and 100 mm) and the primary consolidation obtained through the straightedging and finishing operations. Spading or internal vibration will be required to consolidate concrete adequately around reinforcing steel, load-transfer devices, keyways, and the edges of forms.

10.5—Heavy-duty industrial floors

The wearing surface of heavy-duty industrial floors should be of a high-quality concrete. For information regarding the various floor classifications and requirements, refer to Table 1.1 in ACI 301.1R. Many industrial floors are placed as two courses, with conventional concrete in the bottom course and a higher quality concrete in the top course. The top course should preferably be placed before the bottom course has attained final set. The use of two course floor systems provides economy and a more efficient use of materials.

The top surface should be struck off slightly above the finish grade. The wearing course should then be compacted by rolling, tamping, or other surface vibration. The use of a power-disc float with hammers will provide additional consolidation of the near-surface region. In these concretes, the disc float must be used soon after the screeding operation if sufficient mortar cannot be brought to the surface to adequately fill the surface voids.

Chemical admixtures may be used to increase the workability of mixtures to make consolidation easier.

10.6—Vacuum dewatering

The vacuum process is a method of improving the concrete quality near the surface by removing part of the mixing water after the concrete has been placed; however, some reconsolidation is involved (see Fig. 10.6). Mats are applied to the surface after the normal consolidation has been completed and they are connected to vacuum pumps. The suction applied by the pumps and the atmospheric pressure (a consolidating force), acting simultaneously on the mats, remove water and entrapped air from the region near the surface and close up the spaces formerly occupied by the water.

CHAPTER 11—PAVEMENTS

Highway and airfield pavement jobs include applications such as continuously reinforced pavements and bridge decks and may use concrete at rates in excess of 500 yd^3 (400 m^3) per hr. Automated equipment capable of handling 1 to 2-in. (25 to 50-mm) slump concrete is generally used for placing and finishing. At the other extreme, residential developments may require less than 100 yd^3 (80 m^3) of concrete per day. Considerable hand-work is frequently used, necessitating slumps in the range of 2 to 4 in. 50 to 100 mm).

This guide is aimed at highway and airfield construction. The procedures described generally apply either to fixed-form or slipformed pavements. Zero-slump concrete pavements are placed by the roller compaction process as described in Section 9.5.

11.1—Mixture requirements

The concrete mixture should have adequate placeability and finishability to achieve the desired consolidation and finish. The slump should be 2 in. (50 mm) or less to keep segregation and loss of entrained air to a minimum and to maintain the quality of the concrete.

The concrete received at the placing point should be uniform. Variations in the mixture may result in segregation or inadequate consolidation, causing the pavement to have poor riding qualities and poor durability. For fiber-reinforced concrete, internal vibrators must be used at a closer spacing and for a longer period of time to obtain satisfactory results (see ACI 544.1R).

11.2—Equipment

11.2.1 *Selection of equipment*—All pavements should be consolidated by full-width vibration. The type of vibration—internal or surface—is determined by the slab thickness, the rate of production, consistency, and other characteristics of the concrete mixture.

Internal vibrators, usually gang-mounted spud vibrators, meeting the guideline in Table 5.1.5 should be used when pavement thicknesses are 8 in. (200 mm) or more. When equipment moves rapidly over slabs to attain high production rates, internal vibration may be needed in pavements as thin as 4 in. (100 mm). Hydraulic vibrators have increased rapidly in popularity in recent years, mainly because the frequency is adjustable and maintenance requirements are low.

Surface vibrators may be used for pavements less than 8 in. (200 mm) thick and have been successfully used for pavements up to 10 in. (250 mm) thick using greater vibrational effort. However, the production rate will be lower than that obtained with internal vibrators. Also, surface vibration in combination with striking off, screeding, and floating can bring excess fine material to the surface. This can happen as a result of improper mixture proportions or over-working the surface, or both.

The speed of the paving train controls the time of vibration, and the equipment and mixture proportions must be selected accordingly.

11.2.2 *General requirements*—Both surface and internal vibrators should be controlled by an automatic on-off switch that operates the vibrators simultaneously, and only when the machine is in forward motion.

The ability to vary frequency is desirable to permit adjustment for the job conditions and materials being used.

Standby vibrator units should be available for replacement or if needed for additional vibration.

11.2.3 *Internal vibrators*—In addition to the usual internal vibrators described in Chapter 5, L-shaped spuds are also available for pavement construction. The latter are especially adapted for consolidating the thinner slabs and for operating above the mesh in reinforced pavements.

The vibrators are usually gang-mounted on a horizontal frame (see Fig. 11.2.3) that should be located immediately in front of the first screed or extrusion plate. The frame should be adjustable forward and backward to compensate for differences in concrete consistency from job to job.

The frame should be capable of spacing 10 to 14 vibrators over a 24 ft (7.3 m) paving width. It should also be capable of vertical movement so that the spuds can be completely withdrawn from the concrete or lowered to the exact position in the concrete required for optimum vibration.

Fig. 11.2.3—Gang-mounted spud vibrators for consolidating pavement concrete.

Fig. 11.2.4(a)—Pan-type surface vibrator for pavement construction.

Fig. 11.2.4(b)—Older screeds with trucks permitting cam-like action to raise screed to clear concrete surface when moving for second pass.

The vibrators should be capable of angular adjustment that can be maintained during vibration.

The vibrator frequency should be adjustable between 8000 and 12,000 vibrations per min (130 and 200 Hz). The frequency from vibrator to vibrator should be uniform.

Hand-held immersion vibrators of the type used in consolidating structural concrete may be useful along forms or in irregular areas.

11.2.4 *Surface vibrators*—Three types of surface vibrators are used in concrete pavement construction—vibratory screed, vibratory pan, and vibratory-roller screed (see Section 5.3).

The vibratory screed is a dual-purpose unit that consolidates the concrete and strikes off the surface. The ends of the screed generally are not equipped with trucks (wheel assemblies) [Fig. 11.2.4(a)]. Older equipment may have trucks [Fig. 11.2.4(b)] with a cam-like action, so that the screed may be raised clear of the concrete surface when moving it back for a second pass. Small screeds may be lifted by hand. A unit is normally required for each lane of width. Vibratory screeds should be capable of varying the frequency from 3000 to 8000 vibrations per min (50 to 130 Hz).

The vibratory pan is the only surface vibrator used strictly for consolidation. The pan should be mounted on a horizontal frame capable of raising it clear of the concrete or holding it at the surface as desired. The pan vibrator should be adjustable in frequency from 3000 to 6000 vibrations per min (50 and 100 Hz).

The final type of surface vibrator is the vibratory-roller screed, which strikes off as well as consolidates. The frequency should be adjustable; the range for the most widely used current model is 100 to 400 vibrations per min (2 to 7 Hz). This equipment requires a concrete slump of more than 2 in. (50 mm) and its use should be limited to irregular areas and hand placements.

11.3—Vibration procedures

11.3.1 *Internal vibrations using gang-mounted vibrators*—The centrifugal force and vibrator spacing should be based upon the aggregate to be used, mixture characteristics, rate of concrete delivery, method of reinforcement placement, and paver speed. Vibrators with a centrifugal force near the low end of the range shown in Group 3 in Table 5.1.5 should be used for mixtures with small coarse aggregates and high fine aggregate contents. Normally, the trial spacing should be 20 to 30 in. (500 to 750 mm). The lower the centrifugal force and the shallower the slab, the closer the spacing. The location of the outside vibrators is critical, especially in slipform paving.

When nonuniformity or mortar streaking occurs in vibrator paths while operating at normal paving speeds, the vibrators should be lowered in the concrete, their angularity changed, the frequency increased or decreased, the amplitude changed (usually by changing the eccentric mass), or additional vibrators added until the streaking is eliminated. Proper consolidation is generally achieved when the concrete surface has a uniform texture and sheen, with coarse-aggregate particles barely visible on or immediately below the surface.

For pavements less than 10 in. (250 mm) thick, the vibrators should be operated parallel with, or at a slight angle to, the subbase. For thicker nonreinforced pavements, the vibrators should be close to the vertical, with the vibrator tip preferably about 2 in. (50 mm) from the subbase, and the top of the vibrator a few inches below the pavement surface.

A 4 to 6-in. (100 to 150-mm) surcharge of concrete should be carried over the vibrators during the placing operation. Greater surcharge loads are likely to cause surging behind the screed or extrusion plate and prevent full release of entrapped air.

For reinforced pavement with thicknesses less than 10 in. (250 mm), the vibrators should be parallel with the subbase above and as near as practical to the reinforcement but at least two vibrator diameters below the surface. When the reinforcement is close to the surface, the concrete should be placed in multiple passes to permit consolidation. If inadequate consolidation is discovered at the bottom of the slab under the steel, space the vibrators closer together, increase the vibratory effort, or decrease the paver speed. Since it is common practice to attach the vibratory unit to the equipment carrying the first transverse screed, the proper adjustment of the vibrators will depend on the forward speed of this equipment.

Reinforced slabs in which the reinforcement is placed by vibration after full-depth concrete placement require initial consolidation prior to steel placement. In continuously reinforced pavements where the steel is placed on chairs prior to concrete placement, care should be taken to insure that the concrete below the reinforcing steel is receiving adequate consolidation. For reinforcement placed with a mesh depressor, less vibration will normally be required than for mesh placed on chairs or for concrete placed in two courses. For reinforced slabs placed in two courses, the vibrators should be used in both courses.

Olsen, Winn, and Ledbetter et al. (1984) provide additional information on consolidation of concrete pavements.

11.3.2 *Surface vibration*

11.3.2.1 The vibratory-pan unit should be positioned behind the surface strikeoff equipment. The vibration frequency should be set in accordance with the forward speed of the equipment on which it is mounted. A surcharge should not be allowed to build up in front of the pan because it will dampen the vibrations. An internal spud vibrator may assist in consolidating concrete along each form.

11.3.2.2 It is usually advisable to make two passes of the screed or roller. The first strikes off and consolidates the concrete, and the second provides the surface finish. Maximum frequency should be used on the first pass and a reduced frequency on the second. In this case, surface appearance is not a satisfactory criterion of the adequacy of consolidation. An understanding of the effectiveness of consolidation below the surface is required.

11.3.3 *Manual vibration*—Hand-held immersion vibrators should be used adjacent to all headers (bulkheads) and joint assemblies, unless a vibratory dowel installer or full-width internal vibration is used. They should also be used in other areas where gang-mounted vibrators are not practicable. The vibrator head should be completely immersed in as near a vertical position as practicable to avoid segregation and mortar streaking. The concrete should be vibrated to the required depth by systematic vibration of overlapping areas. The insertion spacing should generally be 20 to 30 in. (500 to 750 mm) or about $1\frac{1}{2}$ times the effective radius of action. It is better to space the insertions too closely than too far apart.

The vibrator should operate in one location until the concrete is consolidated thoroughly, then it should be withdrawn slowly to insure closing the hole resulting from the vibrator insertion. The length of time to effect thorough consolidation will vary with the concrete workability and the centrifugal force of the vibrator. Vibration time may be as short as 5 sec or as long as 20 sec per point of application.

11.4—Special precautions

When placing air entrained concrete, air content of the consolidated concrete in place should be checked. Certain methods of consolidating and finishing pavements will effect the characteristics of the air void system. When air entrainment is required for frost resistance, the air void parameters in the hardened concrete should be verified. If the air content falls below the specified level, changes should be made in the vibrating procedures or in the amount or type of air-entraining admixture being used. The depth and location of reinforcing steel should be checked behind the vibrators to assure that the reinforcement has not been dislocated.

When fixed forms are used, the pavement edge should be examined after form removal to determine the effectiveness of the vibrators. If honeycomb is observed, one or more of the following changes should be made to prevent its recurrence: (1) position vibrators closer to the forms, (2) increase frequency or amplitude of the vibrators, or (3) reduce the forward speed of the paving equipment.

In slipform paving, the equipment should move forward as continuously as possible, especially in warm weather. Delays, and starting and stopping the paver, may produce tearing of the surface and edges of already consolidated concrete. Tearing can extend to a depth of 6 to 8 inches (150 to 200 mm) and result in a loss of consolidation. The condition is caused by the development of excessive friction between the top or side form of the paver and concrete. Factors that can contribute to tearing include thickness of the slab, use of concrete with too low a slump, concrete temperature, wind and humidity, mixture proportions, particle shape of the aggregates, rate of slump loss, and adjustment and operation of the slipform paver. Once tearing has occurred, the only means of restoring integrity to the concrete is to use immersion vibrators and revibrate the affected area. If tearing is near or on the edge, installation of side forms may be required to retain the concrete during vibration.

Cores should be taken periodically to check the adequacy of consolidation. Those taken to check pavement thickness may be suitable for this purpose. The top surface of cores should be examined to determine the thickness of the mortar layer above the coarse aggregate. Mortar thicknesses over coarse aggregate in excess of 1/8 in. (3 mm) indicate overvibration or overfinishing, which can result in reduced abrasion resistance. This also indicates an over-mortared mixture. The inspector should record locations of breakdowns, delays, or other unusual events and should request cores from these areas.

The density of fresh concrete immediately after vibration can be determined by the use of nuclear gages. These gages measure relative density, which is the plastic mass per unit volume measured in the normal manner (ASTM C 138). This can provide a useful means for indicating when the desired degree of consolidation has been achieved. Useful results can be obtained on large jobs where the cost can be justified, where testing personnel are available, the instrument is properly calibrated, and the concrete mixture is reasonably uniform.

Excessive entrapped-air voids in the cores indicate a need for additional vibration, or a change in the location or spacing of vibrators. Intrusion of subbase material into the concrete may result from internal vibrators set too low or at an incorrect angle.

Changing job conditions such as weather, rate of progress, changes in equipment, and slump may necessitate a change in the characteristics or position of the vibrators. The inspector should watch for nonuniformity behind the vibrators. Nonuniformity caused by improper use of gang vibrators has been known to produce lines of weakness that can develop into longitudinal cracks.

CHAPTER 12—PRECAST PRODUCTS

The consolidation method for precast products should be selected on the basis of the end use of the product, concrete mixture, forming material, and production technique so that the entire operation can be efficiently planned and coordinated. Table 12.1 summarizes pertinent data for some precast concrete products.

12.1—Mixture requirements

The workability of the mixture is an important consideration in selecting the consolidation method for precast work. In precast work, generally the following consistencies are used:

a. Stiff mixtures. These are harsh, zero-slump mixtures that exhibit little cohesiveness when squeezed in the hand. Because of their low water content, moist curing is generally used to achieve adequate cement hydration.

b. Stiff plastic mixtures. These mixtures have some cohesiveness and are slightly plastic, usually with less than 1 in. (25 mm) slump.

c. Uniformly or gap-graded mixtures having slump in the 1- to 4-in. (25- to 100-mm) range. These mixtures are cohesive and plastic.

d. Mixtures having over 4-in. (100-mm) slump that flow readily, and have a high potential for segregation if mechanical vibration is applied.

In precast work, it may be necessary to adjust the mixture proportions, within reasonable limits, to provide compatibility with the available precasting equipment.

12.2—Forming material

The consolidation method should be compatible with the form or mold material. Steel, wood, and reinforced concrete are

Fig. 12.4—Form vibration used in precast beam construction

generally preferred. Forms may be lined with fiberglass or other plastics to produce special surfaces. Rubber has also been used.

Care should be taken to prevent form damage during consolidation. For example, internal vibrators should have a rubber tip and contact between the vibrator and form should be avoided.

12.3—Production technique

For products that have been standardized, well developed methods are generally available. Machinery is available for manufacturing the following standardized products:

   a. Concrete pipe;
   b. Concrete block and lintels;
   c. Floor slab units;
   d. Small paving slabs (patio block, etc.);
   e. Building units such as load bearing wall panels.

Custom-built products present more difficult problems. Experience in mixture proportioning, mold design, and other factors lead to the best casting and consolidation method. The number of units to be cast should also be considered.

The information in Chapter 5 should be helpful. Previous experience and experimentation are frequently employed to arrive at the final solution.

12.4—Other factors affecting choice of consolidation method

External form vibration (see Fig. 12.4) or vibrating tables are generally preferred over internal vibration in the precast industry. They give more uniform control and allow more economical techniques to be adopted in day-to-day production of similar units. When the section involves large concrete masses remote from external vibrators, supplemental internal vibration should be provided.

Tamping is an effective method of consolidating stiff concrete placed in thin layers.

Pressure vibration is suitable for stiff mixtures. Here a given concrete volume is placed in a mold and a force is applied to the top concurrent with the vibration.

The curing method may affect the choice and operation of consolidation equipment. External form vibrators that are not removable and are exposed to steam and moisture are likely to have high maintenance costs, especially if they are electrically powered.

12.5—Placing methods

The method of depositing concrete in the forms is important to consolidation. To expel the maximum amount of entrapped air and to keep the voids on formed surfaces at a minimum, vibration should be continuous during concrete placement.

Dumping concrete in intermittent heaps should be avoided. Portable mixers or mixer trucks should discharge in a continuous moving ribbon directly into the form, rather than discharging into a bucket and intermittently dumping the concrete in heaps.

When using vibrating or drop tables, a uniform concrete layer should be placed in the mold before the table is placed in operation. When shallow slabs are manufactured, the form should be completely filled before vibration starts. If the depth exceeds 12 in. (300 mm), it is best to use two or more layers. The concrete consistency and desired surface appearance will also affect the method employed; the lower the water-cement ratio, the shallower the lift that should be used.

CHAPTER 13—LIGHTWEIGHT CONCRETE

Concrete made with lightweight aggregate is used to reduce dead loads resulting in smaller structural members and foundation sizes. Lightweight concrete is also used to provide better fire resistance and to serve as insulation against sound and heat transmission.

13.1—Mixture requirements

Most commercially available lightweight coarse aggregates have a nominal maximum size of 1/2 or 3/4 in. (13 or 19 mm). The fine aggregate may be either normal weight or lightweight, or a combination of both, providing the concrete meets density and strength requirements.

A slump of 2 to 3 in. (50 to 75 mm) is adequate for normal construction. With higher slumps the larger pieces of lightweight aggregate may float to the top surface during vibration. Stiffer mixtures are frequently used in precast work.

Air entrainment is highly desirable in lightweight concrete. It imparts cohesiveness to the mortar so that the coarser particles have less tendency to float during vibration.

13.2—Behavior of lightweight concrete during vibration

During vibration, the entrapped air bubbles are brought to the surface through buoyancy and dissipated as for normal weight concrete. However, the lower density of the mixture results in somewhat less buoyancy for the air bubbles. It is important to allow enough vibrating time to remove the air bubbles, while noting that with lengthy vibration times much of the entrained air may be lost and some of the lightweight aggregate particles may float.

Segregation of concrete mixture components during vibration is caused by differences in their specific gravities. In normal weight concrete, the coarse aggregate is heavier than the mortar and therefore tends to sink during vibration. In lightweight concrete, the reverse is true, although the tendency for the coarse aggregate to float is less when the mortar contains lightweight fine aggregate. Dry mixtures will not segregate as rapidly under vibratory action as wet ones.

13.3—Consolidation equipment and procedures

Equipment recommended for consolidating normal-weight concrete is also suitable for lightweight concrete.

As for normal-weight concrete, lightweight concrete should be placed as closely to its final position as practicable to avoid segregation. Vibrators should not be used to move the concrete laterally. Shovels are frequently helpful in depositing or moving the concrete.

Most practices used for vibrating normal-weight concrete can be followed with lightweight concrete. However, due to the reduced buoyancy of entrapped air bubbles in lightweight concrete, the layer depths should be reduced to approximately 80 percent of those given in Section 7.1. The vibrators should be inserted at close intervals and should penetrate the previously placed layer. Sufficient time, usually about 10 sec, should be given at each insertion to get adequate consolidation. Stiffer mixtures may require a few additional seconds.

On walls where surface air voids are objectionable, the following procedure is suggested. Each layer should be vibrated in the normal manner, and then revibrated immediately prior to placing the succeeding lift. If a period of about 30 min (or as long as practical) is allowed between vibration operations, this procedure can be quite effective. As an alternative to the second vibration, which may require additional vibrators, hand spading or spudding against the form surface is moderately effective.

13.4—Floors

Consolidation and finishing operations should receive particular attention when lightweight concrete is used in floor construction. While most of the recommendations in Chapter 10 are applicable, some additional precautions are helpful.

Air entrainment and minimal slump are both very desirable. These will assist in preventing the lightweight coarse aggregate particles from coming to the top surface.

Best consolidation is obtained by dragging the vibrator through the concrete in a nearly horizontal position at about the same spacing as used for vertical insertions. Dragging at a constant velocity will give more uniform vibration than jerking motions. In lieu of internal vibrators, vibrating screeds may be used for thin floors where there are no obstructions to impede their use.

Where segregation has occurred, a hand-operated grid tamper or mesh roller may be used to depress the floating lightweight coarse aggregates slightly below the top surface.

CHAPTER 14—HIGH DENSITY CONCRETE

Concrete made with high density aggregates is primarily used for radiation shielding and counterweights. For radiation shielding, it is absolutely essential that the concrete be dense, practically free of voids and cracks, and homogeneous.

14.1—Mixture requirements

Aggregates for high density concrete comprise iron products (specific gravity 7.5 to 8.0), heavy slags (specific gravity over 5.0), and hydrous or mineral ores (specific gravity 3.5 to 4.8). These materials may be used individually or in combination to obtain concrete densities from about 160 to over 380 lb/ft.^{3} (2600 to over 6100 kg/m^3). (See ASTM C637 or C638.)

Normal mixture proportions range between 1:6 and 1:10 by mass of cement to combined fine and coarse aggregate. The water-cement ratio is usually between 0.45 and 0.65.

Settlement can generally be minimized by proper proportioning and incorporation of suitable chemical admixtures.

14.2—Placing techniques

Heavyweight concrete is fabricated by conventional mixing and placing methods, by aggregate immersion (puddling), or by preplaced aggregate construction (ACI 304.3R). Formwork should receive careful attention, because heavyweight concrete exerts considerably higher pressures on forms than normal weight concrete. Form pressure can be reduced by placing concrete in slowly rising lifts. Care must be taken to avoid excessive loads on the concrete handling equipment due to the higher density of heavyweight concrete. It is common practice to reduce concrete truck and bucket loads by half.

14.2.1 *Conventional placing techniques*—Conventional placement methods may be used for concrete containing high density aggregates, provided the mixture is workable and the forms are relatively free of embedded items. However, such concrete presents special problems due to the tendency of the high density aggregate particles to segregate. Segregation is greatest where the aggregates are not uniform in grading or density, the mixture contains excessive moisture, or the slump is excessive. Concrete slump should generally be between 1.5 and 3 in. (40 and 75 mm) for high density mineral aggregate mixtures. Placement and consolidation must be closely controlled to insure uniform density and freedom from segregation.

Internal vibration is often supplemented with external vibration, but extra care must be taken when the heavy aggregates are friable and easily broken down. Vibrator frequencies used for normal-weight concrete are usually satisfactory for heavyweight concrete. However, somewhat higher frequencies—about 11,000 vibrations per min (180 Hz)—together with shorter vibration periods have sometimes been found to reduce the tendency for segregation, especially when steel punchings or other very high density aggregates are used. The potential for overvibration is increased with the use of high density aggregates, which can result in the settlement of the heavy particles. The radius of action of a vibrator in heavyweight concrete is less than in conventional concrete, so a closer spacing of insertions is required.

14.2.2 *Special placing techniques*—When segregation cannot be avoided or when embedded items or restrictions prohibit conventional placement, the preplaced or postplaced aggregate methods may be employed.

In the preplaced aggregate method (ACI 207.1R, ACI 304.1R, and 304.3R), embedments such as heavy reinforcement, pipes, and conduits may be vibrated during aggregate placement to minimize unfilled pockets. When vibration of embedded items cannot be tolerated, the aggregate may be hand placed or rodded into position. Vibration during grout pumping should be avoided except where a superior surface finish is desired. Hurd (1989) indicates that forms may be lightly vibrated near the grout surface.

Postplaced aggregate is a rarely used technique in which up to one foot (300 mm) of high density grout is placed in the form and heavy aggregate is embedded into it. The coarse aggregate is worked into place by rodding. Internal vibration should be avoided, especially where the grout contains high-density fine aggregates.

CHAPTER 15—QUALITY CONTROL AND INSPECTION

15.1—General

Good consolidation is the result of:

1. Good specifications and enforcement;

2. Good design relative to geometry and reinforcing steel;

3. Good mixture proportions;

4. Use of proper equipment, and maintenance practices to keep it in good working order;

5. Proper field procedures. Workers should understand why they are consolidating the concrete and the consequences if it is improperly done;

6. Quality control procedures implemented by the contractor;

7. Quality assurance and testing to see that proper quality control procedures are followed.

15.2—Adequacy of equipment and procedures

Concrete workability is not constant, even with the best of control. Variations in aggregate grading and in consistency due to slump loss between the mixer and form should be compensated for by slight changes in the consolidation procedure. There should be sufficient flexibility—in vibration time, vibrator spacing, and sometimes vibrator properties—to adjust to these changed conditions.

Slumps should be as low as practical for the working conditions. Properly sized vibrators in good operating condition are essential. Use of the recommended layer depth, vibrator spacings, timing, and penetration depth are also important to the quality of the final product.

Spare vibrators should be available at the point of placement to maintain production in the event of a breakdown, or when vibrators are taken out of service for routine maintenance and repair.

Mechanical consolidation equipment cannot operate properly unless adequate power is available. With electric vibrators, voltage can be expected to vary appreciably and should be regularly checked. With pneumatic vibrators, the air pressure at the vibrator should be regularly checked, either by installing an ordinary dial gage in the line, or by inserting a needle gage in the air hose.

Since internal vibrators are used in wet (conductive) locations, all electric units should be grounded to the power source. Power generators should also be grounded to maintain continuity of the grounding system. Units operating at less than 50 volts, or that are protected by an approved double insulation system, are excepted. In the United States, electric vibrators are subject to Article 250-45 of the National Electric Code (1990).

15.3—Checking equipment performance

All vibratory units should be checked prior to starting the work, and periodically during construction, to verify that they are working properly.

15.3.1 *Frequency of internal vibrators*—The vibrating reed tachometer (see Fig. 15.3.1) is a simple device for checking the frequency of an internal vibrator. The frequency should be occasionally determined while the vibrator is operating in air, but it is the frequency while operating in concrete that is most important and requires regular checking. The latter can be determined by holding the device against the back end of the vibrator while it is almost submerged; for a pneumatic vibrator, holding the device against the hose is equally satisfactory. This measurement should be taken just before the vibrator is withdrawn, and is always the fastest speed while it is operating in concrete. The resonant reed tachometer is a more expensive instrument that gives more accurate values of frequency.

15.3.2 *Amplitude of internal vibrators*—The amplitude of an internal vibrator varies linearly along the head with the maximum value occurring at the tip. The average amplitude of most internal vibrators while operating in air may be approximately computed by the formula given in Fig. A.2 in the Appendix.

With care, this device is capable of an accuracy of about 0.005 in. (0.13 mm).

The actual amplitude should also be determined by measurement. This will serve as a check on the manufacturer's data and will indicate whether the vibrator is working properly. It will also provide other useful data, for example, the maximum amplitude and the distribution of amplitude along the head. A visual-effect scale (optical wedge) may be used for this purpose. Several vibrator firms have prepared scales on stickers which may readily be attached to the vibrator head. See Figure 15.3.2.

For flexible-shaft electric and most pneumatic vibrators, a measurement should be taken near the tip and another near the back end of the head, and these results averaged.

For the motor-in-head and pendulum vibrators, where the eccentric is near the tip, the amplitude will generally be relatively large at the tip. It will decrease rapidly until a node (point of zero amplitude) is reached near the back end, and the amplitude will increase to a relatively small value at the extreme back end. The node can be verified and located by moving one's hand over the vibrator surface. If the node is less than one-fifth of the head length away from the back end, the average amplitude may be taken as one-half the measured tip amplitude. If the node point is at a greater distance from the back end, a second measurement (probably near the back end) should be taken. The average amplitude can then be determined as the mean of the two measurements.

15.3.3 *Frequency and amplitude for external vibration*—The frequency and amplitude of vibrating forms and vibrating tables should be determined at sufficient points to establish their distribution over the surface.

The frequency may be determined by a vibrating reed or resonant reed tachometer.

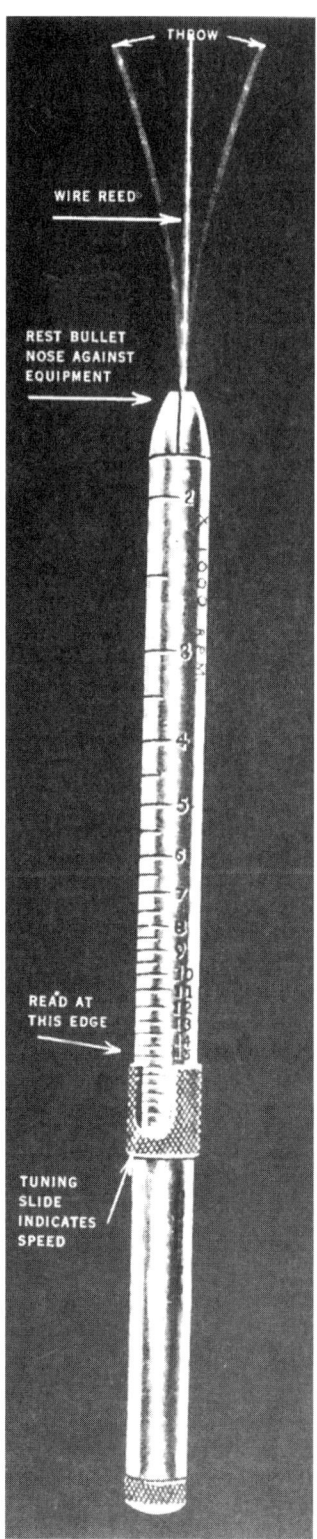

Fig. 15.3.1—Vibrating reed tachometer.

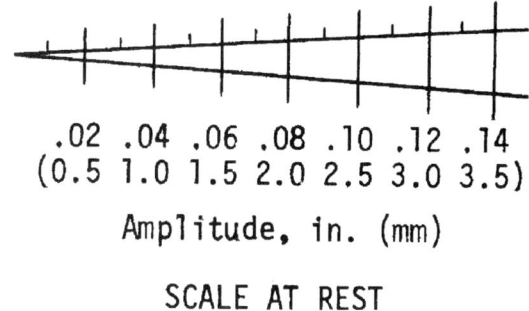

SCALE AT REST

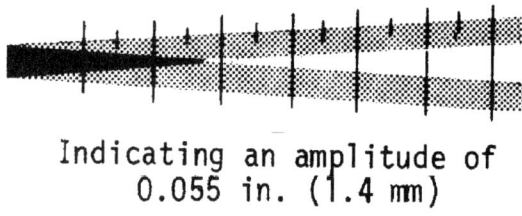

Indicating an amplitude of 0.055 in. (1.4 mm)

SCALE DURING VIBRATION

INSTRUCTIONS FOR USE

Attach scale to vibrator head at point where amplitude is desired, with center line of "V" parallel to axis of vibrator.

With the head vibrating, a black triangle forms at the apex of the "V." The scale reading at the tip of the triangle is the peak amplitude (peak-to-peak total displacement). A hand reading glass (2 to 3x) improves the accuracy of the reading.

With care, this device is capable of an accuracy of about 0.005 in. (0.13 mm).

Fig. 15.3.2—Visual effect scale for measuring amplitude of vibrator operating in air.

Fig. 15.3.3—Using vibrograph to determine amplitude and frequency of vibrating form.

The amplitude may be determined by using a vibrograph. The model shown in Fig. 15.3.3 measures amplitude within an accuracy of about 0.0005 in. (0.013 mm). It also records the wave form, which is frequently of interest, and provides the frequency. It is quite portable.

CHAPTER 16—CONSOLIDATION OF TEST SPECIMENS

16.1—Strength tests

In current ASTM standards (C 31, C 192, and C 1018) for making control specimens for strength tests:

a. Rodding is required for concrete with slumps of more than 3 in. (75 mm). Vibration is prohibited because of the danger of removing excessive entrained air and causing segregation.

b. Either rodding or vibration is permitted for slumps in the 1 to 3 in. (25 to 75 mm) range.

c. For slumps less than 1 in. (25 mm), vibration is required.

d. For concrete of very low water content, external table or plank vibration combined with superimposed load, or tamping is required.

e. For concrete containing fiber reinforcement, external vibration is required per ASTM C 1018. It is understood that extremely low-slump fiber concrete cannot be well consolidated.

For internal vibrators, ASTM requires a minimum frequency of 7000 vibrations per min (120 Hz) and head diameter between 0.75 and 1.5 in. (20 and 40 mm). Table 5.1.5 recommends a minimum of 9000 vibrations per min (150 Hz) for internal vibrators in thin members. For vibrating tables, a minimum frequency of 3600 vibrations per min (60 Hz) is required, with higher frequencies suggested.

The intensity and time of vibration for laboratory specimens is not closely regulated. The standards merely suggest that consolidation has been achieved as soon as the specimen's surface is smooth. Entrained air may be unintentionally removed from small specimens. The concrete strength is increased about 5 percent for each percent of air removed.

Normally the consolidation of test specimens is not required to match that in construction. If it is desired to match field concrete in the laboratory, suitable consolidation procedures must be followed. Some prefer core strengths or the strength of cubes cut from the concrete obtained from the structure as a means for estimating the strength of concrete in the structure.

16.2—Density tests

Tests for density of freshly mixed concrete (ASTM C 138) are widely used to determine the mass of the concrete per unit volume, which is used to compute the cement and air content or as a method of controlling the density of hardened lightweight concrete. The density of fresh concrete is closely related to the total air content and hence to the degree of consolidation.

ASTM C 138 requires consolidation in accordance with Section 16.1. For measures less than 0.4 ft.^{3} (.01 m^3), rodding is required. For slumps in excess of 3 in. (75 mm), the rodding procedure should produce essentially complete consolidation, but for lower slumps the degree of consolidation may be less than in a structure where the concrete is compacted by vibration.

16.3—Air content tests

ASTM C 231 provides for consolidation by rodding for slumps greater than 3 in. (75 mm) and by rodding or vibration when slumps are 3 in. (75 mm) or less. ASTM C 173 provides for consolidation only by hand rodding.

It would appear more reasonable to follow the consolidation procedures recommended in Section 16.1. Internal vibrators should be satisfactory when the slump is greater than about 1/2 in. (13 mm). Although no specific test data are available, it would appear that the pressure method, ASTM C 231, will not work properly on very harsh or low-slump mixtures. With such mixtures, the application of pressure to the surface of the concrete may not result in the expected compression of the air in the void system. The volumetric method, ASTM C 173, is not subject to this limitation and should produce accurate results on even extremely dry concrete.

ASTM C 1170 gives a method of determining the density of stiff to extremely dry concrete mixtures using a vibrating table with or without a 50 lb (22.7 kg) surcharge to consolidate the sample. The CRD C 160 method uses a 27.5 lb (12.5 kg) surcharge. These methods can be adapted to use a standard pressure air meter to determine the air content of the concrete.

16.4—Consolidating very stiff concrete in laboratory specimens

Cylinders consolidated under surcharge using ASTM C 1176 have also been used to determine the density of stiff to extremely dry mixtures. This method uses a 20 lb (9.1 kg) surcharge. Other non-standard methods have been used to consolidate cylinders by tamping equipment or vibrating compaction hammers.

It is important that the density of the laboratory concrete be close to the density of the concrete in the structure being represented. This may require a modification of the consolidation effort. During the early stages of a project it may be

Fig. 17.1.1—Congestion due to reinforcing details.

desirable to compare cylinder densities to core densities to determine the correct amount of consolidation to use.

CHAPTER 17—CONSOLIDATION IN CONGESTED AREAS*

Congested areas are areas where the lateral movement of freshly placed concrete is unduly restricted or hindered. To achieve structurally sound and esthetically pleasing concrete, special consideration must be given to select techniques that will allow proper consolidation in congested areas. Some common problems and remedial measures are described here.

17.1—Common placing problems

17.1.1 *Congestion of reinforcement*—Reinforcing steel congestion occurs in a variety of ways; for example, structural and seismic design requires multiple ties at the top and bottom of columns. Where design requirements override consolidation considerations, the horizontal tie spacing is often reduced so that the largest aggregate in the mixture is restricted from moving horizontally to the form face. Reinforcing steel congestion also occurs in areas where there is additional reinforcement around formed openings, particularly in thin wall sections, or columns intersecting with other elements (see Fig. 17.1.1).

17.1.2 *Electrical conduit, pipe sleeves and other embedded items*—Electrical designers often specify a multiple of 1 to 6-in. (25 to 150-mm) diameter conduits in localized areas for powerfeeds and cable trays. Pipe sleeves and complex structural embedments also can create barriers that affect concrete placement and consolidation (see Fig. 17.1.2).

17.1.3 *Boxouts*—Formed boxouts within walls and slabs can create congested zones because the concrete flow is restricted under the boxouts and between adjacent formed openings. This situation can be alleviated by adding construction joints or by adding access openings within the boxouts (see Fig. 17.1.3).

Fig. 17.1.2—Congestion due to a pipe passing through a concrete floor.

Fig. 17.1.3—Large blockout within a wall with pipes through the formed blockout to permit access for concrete placement and vibration.

17.2—Consolidation techniques

Consolidation in congested areas can be enhanced by special attention to construction practices in three specific areas:
1. Placing and consolidation techniques;
2. Use of admixtures;
3. Use of modified mixtures.

---

*See "Guide to Consolidation of Concrete in Congested Areas," ACI 309.3R.

17.2.1 *Placing and consolidation techniques*—The first principle of good consolidation in congested areas is to place the concrete as close to its final location as possible before consolidation. In crane and bucket applications, the use of hoppers and trunks should be considered. When using concrete pumps, wire-reinforced rubber hose attached to the boom pipe is an excellent method of getting concrete close to its final location. In extreme cases, the use of lie-flat hose is recommended. The hose will conform to the varying clearances through the reinforcement. The hose can be cut off to facilitate removal as the placement rises in the form.

In congested wall sections, the provision of placing ports in one side of the wall form insures good consolidation. The ports are located on grids patterned to address the congested areas and need to be about 2 ft. (0.6 m) square. As the concrete reaches the first set of ports, the ports are closed off and vibrators raised to the next row of ports. Additional visual access may be provided by using a transparent plastic plate as a form face in congested areas. This allows the placement crew to take additional steps to remedy problems if necessary in areas of congestion.

To achieve proper concrete consolidation in congested areas by internal vibration, obstruction-free vertical runs of 4 by 6-in. (100 by 150-mm) minimum cross section are needed to permit vibrator insertion. The horizontal spacing of these vertical runs should not exceed 24 in. (610 mm) or $1^1/_2$ times the radius of action indicated in Table 5.1.5. Also, these openings should not be more than 12 in. (300 mm) or $^3/_4$ times the radius of action from the form. If such runs cannot be provided without compromising structural integrity, the engineer should specify construction details and procedures to achieve proper consolidation.

17.2.2 *Use of chemical admixtures*—Proper consolidation in congested areas can generally be improved by increasing the flowability of the mixture by the judicious use of concrete admixtures. They provide high-slump concrete without altering the proportioned water-cementitious material ratio. Additional information on the use of admixtures to achieve flowing concrete can be found in the report of ACI Committee 212.3R.

It must be understood that the use of chemical admixtures does not replace the requirement for good consolidation by vibration as outlined in Chapter 7.

17.2.3 *Use of modified mixtures*—In situations where it cannot be guaranteed that the proportioned mixture will be able to flow to the form face due to congestion, the use of modified mixtures is recommended. The modified mixture containing aggregate of a reduced nominal maximum size can be used to obtain highly plastic or flowing concrete that falls into Groups 1 and 2 of Table 5.1.5 for vibrator selection. The modified mixture should generally be proportioned to have a strength equal to or greater than the original mixture.

17.2.4 *Conclusion*—The previously discussed techniques provide the designer, contractor, and supplier with methods to improve consolidation while maintaining quality. The need for quality flowable concrete is especially required in situations where extreme congestion exists and is unavoidable.

CHAPTER 18—INFORMATION SOURCES

18.1—Specified and/or recommended references

The documents of the various standards-producing organizations referred to in this document are listed with their serial designations.

American Concrete Institute

116R	Cement and Concrete Terminology, SP-19
207.1R	Mass Concrete for Dams and Other Massive Structure
207.5R	Roller Compacted Concrete
211.1	Standard Practice for Selecting Proportions for Normal, Heavyweight, and Mass Concrete
211.2	Standard Practice for Selecting Proportions for Structural Lightweight Concrete
211.3	Standard Practice for Selecting Proportions for No-Slump Concrete
213.1R	Chemical Admixtures for Concrete
226.1R	Ground Granulated Blast-Furnace Slag as a Cementitious Constituent in Concrete
226.3R	Use of Fly Ash in Concrete
301	Specifications for Structural Concrete for Buildings
302.1R	Guide for Concrete Floor and Slab Construction
303R	Guide to Cast-in-Place Architectural Concrete Practice
304R	Guide for Measuring, Mixing, Transporting, and Placing Concrete
304.3R	Heavyweight Concrete: Measuring, Mixing, Transporting, and Placing
309R	Standard Practice for Consolidation of Concrete
309.1R	Behavior of Fresh Concrete During Vibration
309.2R	Identification and Control of Consolidation-Related Surface Defects in Formed Concrete
309.3R	Guide for Consolidation of Concrete in Congested Areas
318	Building Code Requirements for Reinforced Concrete
347	Recommended Practice for Concrete Formwork
544.1R	State-of-the-Art Report on Fiber Reinforced Concrete
SP-2	ACI Manual of Concrete Inspection

ASTM International

C 31	Standard Method of Making and Curing Concrete Test Specimens in the Field
C 138	Standard Test Method for Unit Weight, Yield, and Air Content (Gravimetric) of Concrete
C 143	Standard Test Method for Slump of Portland Cement Concrete
C 173	Standard Test Method for Air Content of Freshly Mixed Concrete by Volumetric Method
C 192	Standard Method of Making and Curing Concrete Test Specimens in the Laboratory
C 231	Standard Test Method for Air Content of Freshly Mixed Concrete by the Pressure Method
C 637	Standard Specification for Aggregates for Radiation-Shielding Concrete
C 638	Descriptive Nomenclature of Constituents of Aggregates for Radiation-Shielding Concrete

C 1018   Standard Test Method for Flexural Toughness and First-Crack Strength of Fiber-Reinforced Concrete (Using Beam with Third-Point Loading)

C 1170   Standard Test Methods for Determining Consistency and Density of Roller-Compacted Concrete Using a Vibrating Table

C 1176   Standard Practice for Making Roller-Compacted Concrete in Cylinder Molds Using a Vibrating Table

U.S. Army Corps of Engineers

CRD C 160   Standard Practice for Making Roller-Compacted Concrete in Cylinder Molds Using a Vibrating Table

U.S. Bureau of Reclamation Concrete Manual

These publications may be obtained from the following organizations:

American Concrete Institute
P.O. Box 9094
Farmington Hills, MI 48333-9094

ASTM International
100 Barr Harbor Drive
West Conshohocken, PA 19428

18.2—Cited references

1. Altowaiji, Wisam A. K.; Darwin, David; and Donahey, Rex C., "Preliminary Study of the Effect of Revibration on Concrete-Steel Bond Strength," SL *Report* No. 84-2, University of Kansas Center for Research, Lawrence, Nov. 1984, 29 pp.

2. Ersoy, Sedad, "Investigations on the Consolidation Effect of Immersion Vibrators (Untersuchungen uber die Verdichtungswirkung von Tauchruttlern)," Technische Hochschule, Aachen, 1962.

3. Eyman, Krystian, "Pulses in Concrete Technology," ACI JOURNAL, *Proceedings* V. 77, No. 2, Mar.-Apr. 1980, pp. 78-81.

4. Forssblad, Lars, "Investigations of Internal Vibration of Concrete," Acta Polytechnica Scandinavica, *Civil Engineering and Building Construction Series* No. 29, 1969, Stockholm.

5. Forssblad, Lars, "Concrete Compaction in the Manufacture of Concrete Products and Prefabricated Building Units," The Swedish Association, Malmo, 1971.

6. Hurd, M. K., *Formwork for Concrete*, SP-4, 4th Edition, American Concrete Institute, Farmington Hills, Mich., 1989, 464 pp.

7. Kirkham, R. H. H., "The Compaction of Concrete by Surface Vibration," *Reports*, Conference on Vibration-Compaction Techniques, Budapest, 1963, pp. 251-268.

8. Kolek, J., "Research on the Vibration of Fresh Concrete," *Reports*, Conference on Vibration-Compaction Techniques, Budapest, 1963, pp. 61-76.

9. National Fire Protection Association, "National Electrical Code," (70 P-84), Quincy, 1984, 751 pp.

10. Neville, A. M., *Properties of Concrete*, 3rd Edition, Pitman Publishing, Inc., Marshfield, Chapter 4, 1981.

11. Olsen, M. P. J.; Winn, D. P.; and Ledbetter, W. B., "Consolidation of Concrete Pavement," *Research Report* No. 341-1, Texas Transportation Institute, Texas A & M University, College Station, Aug. 1984.

12. Popovics, Sandor, *Fundamentals of Portland Cement Concrete: A Quantitative Approach*, V. 1, Fresh Concrete, John Wiley & Sons, New York, 1982, 477 pp.

13. Reading, Thomas J., "What You Should Know about Vibration," *Concrete Construction*, V. 12, No. 6, June 1967, pp. 213-217.

14. Rebut, P., "Practical Guide to Vibration of Concrete (Guide Pratique de la Vibration des Betons)," Eyrolles, Paris, 1962, 418 pp.

15. Samuelsson, Paul, "Voids in Concrete Surfaces," ACI JOURNAL, *Proceedings* V. 67, No. 11, Nov. 1970, pp. 868-874.

16. Stamenkovic, Hrista, "Prevention and Repair of Voids Around Congested Reinforcement," ACI JOURNAL, *Proceedings* V. 81, No. 1, Jan.-Feb. 1984, pp. 40-46.

17. Tuthill, L. H., "How the California Water Project Endeavors to Get Uniformly Excellent Concrete," *Civil Engineering—ASCE*, V. 37, No. 7, July 1967, pp. 43-44.

18. U.S. Bureau of Reclamation, *Concrete Manual*, 8th Edition, Denver, 1981, 627 pp.

19. Vollick, C. A., "Effects of Revibrating Concrete," ACI JOURNAL, *Proceedings* V. 54, No. 9, Mar. 1958, pp. 721-732.

20. Vollick, C. A., "Uniformity and Workability," *Significance of Tests and Properties of Concrete and Concrete-Making Materials*, STP-169A, ASTM, Philadelphia, Apr. 1966, pp. 73-89.

21. Walz, Kurt, *Vibrated Concrete (Ruttelbeton)*, 3rd Edition, Wilhelm Ernst und Sohn, Berlin, 1960.

22. Wilde, Robert L., *Be Your Own Vibration Expert*, Koehring, Dart Division, Denver, 1970.

APPENDIX—FUNDAMENTALS OF VIBRATION

A.1—Principles of simple harmonic motion

The movement of an internal rotary concrete vibrator is essentially harmonic motion, characterized by a sinusoidal wave form, as shown in Fig. A.1. (Actually, harmonics are often superimposed, but it has been found that the assumption of simple harmonic motion is reasonably consistent with experimental data.) This figure shows the path of any point on the head of an operating vibrator and the relationship between frequency, amplitude, and acceleration.

A.2—Action of a rotary vibrator

Rotating the eccentric inside the vibrator head or casing causes the head to revolve in an orbit; that is, any point on the casing follows a circular path whose radius is the amplitude of the vibrator. Fig. A.2 shows the action of a rotary vibrator and gives the significant parameters, for example, mass, eccentric moment, frequency, centrifugal force, and computed average amplitude.

The centrifugal force computed in this manner is not strictly correct, since it is for the hypothetical case where the vibrator shell has zero amplitude while the rotor (eccentric) turns in its bearings. In spite of these limitations, however, the values thus obtained are useful as a rough indicator of the relative effectiveness of different vibrators.

A.3—Vibratory motion in the concrete

When immersed in concrete, the orbiting head (now under load) has a somewhat lesser amplitude than when operating in air. The concrete is subjected to vibratory impulses which produce wave motion emanating at right angles to the head. These pressure waves are mainly responsible for the consolidation.

The waves decay rapidly with distance from the source because of the expanding area of the wave front and the absorption of energy (damping) by the concrete. This decay (reduction in amplitude) causes a reduction in the acceleration (intensity of vibration). Where the acceleration in the concrete is less than about 1 g for plastic mixes, or about 3 gs for stiff mixes, the vibration is no longer effective. A considerable amplitude at the vibrator is required to attain a satisfactory radius of action.

The response of fresh concrete to vibration is largely a function of its rheological (flow) properties. Much more research is needed on this subject.

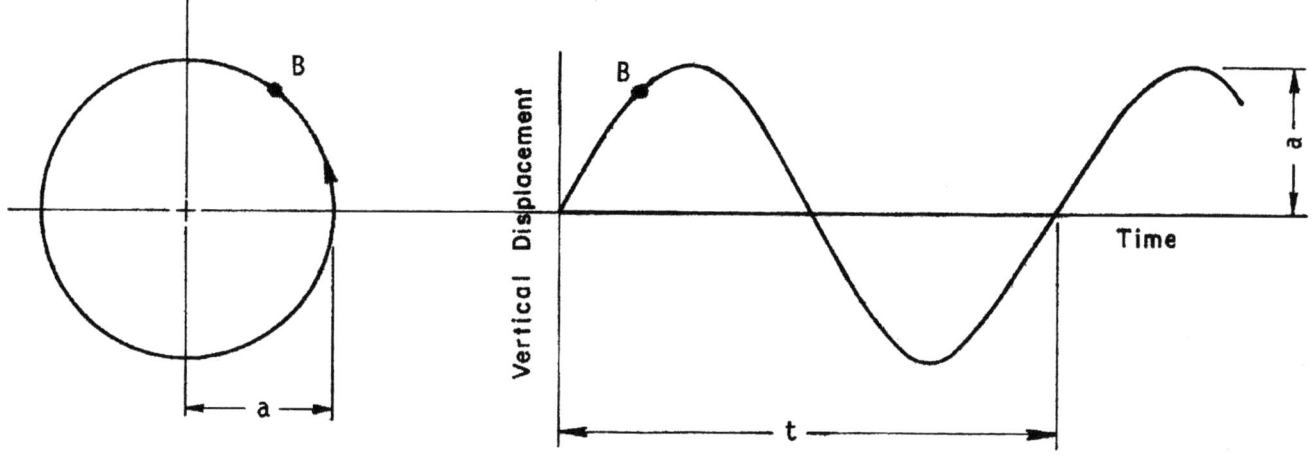

B	= random point on vibrator spud
t	= time for one complete revolution or vibration cycle, sec
n	= $1/t$ = frequency, vibration cycles or vibrations per sec (Hz)
a	= amplitude (deviation from point of rest),* in. (mm)
A	= $4\pi^2 n^2 a$ = acceleration, in. per sec^2 (mm/sec^2)

Acceleration, $gs, = \dfrac{4\pi^2 n^2 a}{g}$, where g is 386 in.

* It should be noted that amplitude as used here (and elsewhere in this report) is peak amplitude, which is half the peak-to-peak amplitude or displacement used by some in describing vibrations.

Fig. A.1—Principles of simple harmonic motion applied to rotary vibrator.

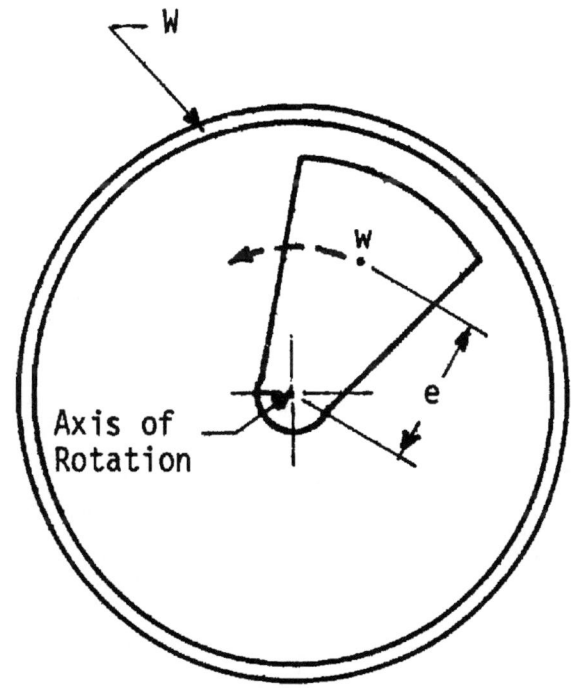

W	= weight of shell and other nonmoving parts, lb (kg)
w	= weight of eccentric, lb (kg)
$W + w$	= total weight of vibrator
e	= eccentricity, i.e., distance from center of gravity of eccentric to its center of rotation, in. (mm)
we	= eccentric moment, in.-lb (mm-kg)
n	= frequency, cycles per sec (Hz)
F	= $\dfrac{w}{g} 4\pi^2 n^2 e$ = centrifugal force, lb (kN)
a'	= $w \dfrac{e}{W + w}$ = computed average amplitude, in. (mm)

Fig. A.2—Action of a rotary vibrator.

Chapter 3 through Chapter 7 of:

BUILDING CODE REQUIREMENTS FOR STRUCTURAL CONCRETE AND COMMENTARY (ACI 318-05)

REPORTED BY ACI COMMITTEE 318

ACI Committee 318
Structural Building Code

James K. Wight
Chair

Basile G. Rabbat
Secretary

Sergio M. Alcocer	Luis E. Garcia	Dominic J. Kelly	Myles A. Murray
Florian G. Barth	S. K. Ghosh	Gary J. Klein	Julio A. Ramirez
Roger J. Becker	Lawrence G. Griffis	Ronald Klemencic	Thomas C. Schaeffer
Kenneth B. Bondy	David P. Gustafson	Cary S. Kopczynski	Stephen J. Seguirant
John E. Breen	D. Kirk Harman	H. S. Lew	Roberto Stark
James R. Cagley	James R. Harris	Colin L. Lobo	Eric M. Tolles
Michael P. Collins	Neil M. Hawkins	Leslie D. Martin	Thomas D. Verti
W. Gene Corley	Terence C. Holland	Robert F. Mast	Sharon L. Wood
Charles W. Dolan	Kenneth C. Hover	Steven L. McCabe	Loring A. Wyllie
Anthony E. Fiorato	Phillip J. Iverson	W. Calvin McCall	Fernando V. Yanez
Catherine E. French	James O. Jirsa	Jack P. Moehle	

Subcommittee Members

Neal S. Anderson	Juan P. Covarrubias	Michael E. Kreger	Vilas S. Mujumdar	Guillermo Santana
Mark A. Aschheim	Robert J. Frosch	Daniel A. Kuchma	Suzanne D. Nakaki	Andrew Scanlon
John F. Bonacci	Harry A. Gleich	LeRoy A. Lutz	Theodore L. Neff	John F. Stanton
JoAnn P. Browning	R. Doug Hooton	James G. MacGregor	Andrzej S. Nowak	Fernando R. Stucchi
Nicholas J. Carino	Javier F. Horvilleur[†]	Joe Maffei	Randall W. Poston	Raj Valluvan
Ned M. Cleland	L. S. Paul Johal	Denis Mitchell	Bruce W. Russell	John W. Wallace
Ronald A. Cook				

Consulting Members

C. Raymond Hays   Richard C. Meininger   Charles G. Salmon

[†]Deceased

ACI 318-05 is deemed to satisfy ISO 19338, "Performance and Assessment Requirements for Design Standards on Structural Concrete," Reference Number ISO 19338.2003(E). Also Technical Corrigendum 1: 2004.

CHAPTER 3 — MATERIALS

CODE

3.1 — Tests of materials

3.1.1 — The building official shall have the right to order testing of any materials used in concrete construction to determine if materials are of quality specified.

3.1.2 — Tests of materials and of concrete shall be made in accordance with standards listed in 3.8.

3.1.3 — A complete record of tests of materials and of concrete shall be retained by the inspector for 2 years after completion of the project, and made available for inspection during the progress of the work.

3.2 — Cements

3.2.1 — Cement shall conform to one of the following specifications:

(a) "Standard Specification for Portland Cement" (ASTM C 150);

(b) "Standard Specification for Blended Hydraulic Cements" (ASTM C 595), excluding Types S and SA which are not intended as principal cementing constituents of structural concrete;

(c) "Standard Specification for Expansive Hydraulic Cement" (ASTM C 845);

(d) "Standard Performance Specification for Hydraulic Cement" (ASTM C 1157).

3.2.2 — Cement used in the work shall correspond to that on which selection of concrete proportions was based. See 5.2.

COMMENTARY

R3.1 — Tests of materials

R3.1.3 — The record of tests of materials and of concrete should be retained for at least 2 years after completion of the project. Completion of the project is the date at which the owner accepts the project or when the certificate of occupancy is issued, whichever date is later. Local legal requirements may require longer retention of such records.

R3.2 — Cements

R3.2.2 — Depending on the circumstances, the provision of 3.2.2 may require only the same type of cement or may require cement from the identical source. The latter would be the case if the sample standard deviation[3.1] of strength tests used in establishing the required strength margin was based on a cement from a particular source. If the sample standard deviation was based on tests involving a given type of cement obtained from several sources, the former interpretation would apply.

CODE

3.3 — Aggregates

3.3.1 — Concrete aggregates shall conform to one of the following specifications:

(a) "Standard Specification for Concrete Aggregates" (ASTM C 33);

(b) "Standard Specification for Lightweight Aggregates for Structural Concrete" (ASTM C 330).

Exception: Aggregates that have been shown by special test or actual service to produce concrete of adequate strength and durability and approved by the building official.

3.3.2 — Nominal maximum size of coarse aggregate shall be not larger than:

(a) 1/5 the narrowest dimension between sides of forms, nor

(b) 1/3 the depth of slabs, nor

(c) 3/4 the minimum clear spacing between individual reinforcing bars or wires, bundles of bars, individual tendons, bundled tendons, or ducts.

These limitations shall not apply if, in the judgment of the engineer, workability and methods of consolidation are such that concrete can be placed without honeycombs or voids.

3.4 — Water

3.4.1 — Water used in mixing concrete shall be clean and free from injurious amounts of oils, acids, alkalis, salts, organic materials, or other substances deleterious to concrete or reinforcement.

3.4.2 — Mixing water for prestressed concrete or for concrete that will contain aluminum embedments, including that portion of mixing water contributed in the form of free moisture on aggregates, shall not contain deleterious amounts of chloride ion. See 4.4.1.

3.4.3 — Nonpotable water shall not be used in concrete unless the following are satisfied:

3.4.3.1 — Selection of concrete proportions shall be based on concrete mixes using water from the same source.

COMMENTARY

R3.3 — Aggregates

R3.3.1 — Aggregates conforming to the ASTM specifications are not always economically available and, in some instances, noncomplying materials have a long history of satisfactory performance. Such nonconforming materials are permitted with special approval when acceptable evidence of satisfactory performance is provided. Satisfactory performance in the past, however, does not guarantee good performance under other conditions and in other localities. Whenever possible, aggregates conforming to the designated specifications should be used.

R3.3.2 — The size limitations on aggregates are provided to ensure proper encasement of reinforcement and to minimize honeycombing. Note that the limitations on maximum size of the aggregate may be waived if, in the judgment of the engineer, the workability and methods of consolidation of the concrete are such that the concrete can be placed without honeycombs or voids.

R3.4 — Water

R3.4.1 — Almost any natural water that is drinkable (potable) and has no pronounced taste or odor is satisfactory as mixing water for making concrete. Impurities in mixing water, when excessive, may affect not only setting time, concrete strength, and volume stability (length change), but may also cause efflorescence or corrosion of reinforcement. Where possible, water with high concentrations of dissolved solids should be avoided.

Salts or other deleterious substances contributed from the aggregate or admixtures are additive to the amount which might be contained in the mixing water. These additional amounts are to be considered in evaluating the acceptability of the total impurities that may be deleterious to concrete or steel.

CODE

3.4.3.2 — Mortar test cubes made with nonpotable mixing water shall have 7-day and 28-day strengths equal to at least 90 percent of strengths of similar specimens made with potable water. Strength test comparison shall be made on mortars, identical except for the mixing water, prepared and tested in accordance with "Standard Test Method for Compressive Strength of Hydraulic Cement Mortars (Using 2-in. or [50-mm] Cube Specimens)" (ASTM C 109).

3.5 — Steel reinforcement

3.5.1 — Reinforcement shall be deformed reinforcement, except that plain reinforcement shall be permitted for spirals or prestressing steel; and reinforcement consisting of structural steel, steel pipe, or steel tubing shall be permitted as specified in this code.

3.5.2 — Welding of reinforcing bars shall conform to "Structural Welding Code — Reinforcing Steel," ANSI/AWS D1.4 of the American Welding Society. Type and location of welded splices and other required welding of reinforcing bars shall be indicated on the design drawings or in the project specifications. ASTM reinforcing bar specifications, except for ASTM A 706, shall be supplemented to require a report of material properties necessary to conform to the requirements in ANSI/AWS D1.4.

COMMENTARY

R3.5 — Steel reinforcement

R3.5.1 — Fiber reinforced polymer (FRP) reinforcement is not addressed in this code. ACI Committee 440 has developed guidelines for the use of FRP reinforcement.[3.2, 3.3]

Materials permitted for use as reinforcement are specified. Other metal elements, such as inserts, anchor bolts, or plain bars for dowels at isolation or contraction joints, are not normally considered to be reinforcement under the provisions of this code.

R3.5.2 — When welding of reinforcing bars is required, the weldability of the steel and compatible welding procedures need to be considered. The provisions in ANSI/AWS D1.4 Welding Code cover aspects of welding reinforcing bars, including criteria to qualify welding procedures.

Weldability of the steel is based on its chemical composition or carbon equivalent (CE). The Welding Code establishes preheat and interpass temperatures for a range of carbon equivalents and reinforcing bar sizes. Carbon equivalent is calculated from the chemical composition of the reinforcing bars. The Welding Code has two expressions for calculating carbon equivalent. A relatively short expression, considering only the elements carbon and manganese, is to be used for bars other than ASTM A 706 material. A more comprehensive expression is given for ASTM A 706 bars. The CE formula in the Welding Code for A 706 bars is identical to the CE formula in the ASTM A 706 specification.

The engineer should realize that the chemical analysis, for bars other than A 706, required to calculate the carbon equivalent is not routinely provided by the producer of the reinforcing bars. For welding reinforcing bars other than A 706 bars, the design drawings or project specifications should specifically require results of the chemical analysis to be furnished.

The ASTM A 706 specification covers low-alloy steel reinforcing bars intended for applications requiring controlled tensile properties or welding. Weldability is accomplished in the A 706 specification by limits or controls on chemical composition and on carbon equivalent.[3.4] The producer is required by the A 706 specification to report the chemical composition and carbon equivalent.

CODE

3.5.3 — Deformed reinforcement

3.5.3.1 — Deformed reinforcing bars shall conform to the requirements for deformed bars in one of the following specifications:

(a) "Standard Specification for Deformed and Plain Carbon-Steel Bars for Concrete Reinforcement" (ASTM A 615);

(b) "Standard Specification for Low-Alloy Steel Deformed and Plain Bars for Concrete Reinforcement" (ASTM A 706);

(c) "Standard Specification for Rail-Steel and Axle-Steel Deformed Bars for Concrete Reinforcement" (ASTM A 996). Bars from rail-steel shall be Type R.

COMMENTARY

The ANSI/AWS D1.4 Welding Code requires the contractor to prepare written welding procedure specifications conforming to the requirements of the Welding Code. Appendix A of the Welding Code contains a suggested form that shows the information required for such a specification for each joint welding procedure.

Often it is necessary to weld to existing reinforcing bars in a structure when no mill test report of the existing reinforcement is available. This condition is particularly common in alterations or building expansions. ANSI/AWS D1.4 states for such bars that a chemical analysis may be performed on representative bars. If the chemical composition is not known or obtained, the Welding Code requires a minimum preheat. For bars other than A 706 material, the minimum preheat required is 300 F for bars No. 6 or smaller, and 400 F for No. 7 bars or larger. The required preheat for all sizes of A 706 is to be the temperature given in the Welding Code's table for minimum preheat corresponding to the range of CE "over 45 percent to 55 percent." Welding of the particular bars should be performed in accordance with ANSI/AWS D 1.4. It should also be determined if additional precautions are in order, based on other considerations such as stress level in the bars, consequences of failure, and heat damage to existing concrete due to welding operations.

Welding of wire to wire, and of wire or welded wire reinforcement to reinforcing bars or structural steel elements is not covered by ANSI/AWS D1.4. If welding of this type is required on a project, the engineer should specify requirements or performance criteria for this welding. If cold drawn wires are to be welded, the welding procedures should address the potential loss of yield strength and ductility achieved by the cold working process (during manufacture) when such wires are heated by welding. Machine and resistance welding as used in the manufacture of welded plain and deformed wire reinforcement is covered by ASTM A 185 and A 497, respectively, and is not part of this concern.

R3.5.3 — Deformed reinforcement

R3.5.3.1 — ASTM A 615 covers deformed carbon-steel reinforcing bars that are currently the most widely used type of steel bar in reinforced concrete construction in the United States. The specification requires that the bars be marked with the letter *S* for type of steel.

ASTM A 706 covers low-alloy steel deformed bars intended for applications where controlled tensile properties, restrictions on chemical composition to enhance weldability, or both, are required. The specification requires that the bars be marked with the letter *W* for type of steel.

Deformed bars produced to meet both ASTM A 615 and A 706 are required to be marked with the letters *S* and *W* for type of steel.

CODE

3.5.3.2 — Deformed reinforcing bars shall conform to one of the ASTM specifications listed in 3.5.3.1, except that for bars with f_y exceeding 60,000 psi, the yield strength shall be taken as the stress corresponding to a strain of 0.35 percent. See 9.4.

3.5.3.3 — Bar mats for concrete reinforcement shall conform to "Standard Specification for Welded Deformed Steel Bar Mats for Concrete Reinforcement" (ASTM A 184). Reinforcing bars used in bar mats shall conform to ASTM A 615 or A 706.

3.5.3.4 — Deformed wire for concrete reinforcement shall conform to "Standard Specification for Steel Wire, Deformed, for Concrete Reinforcement" (ASTM A 496), except that wire shall not be smaller than size D4 and for wire with f_y exceeding 60,000 psi, the yield strength shall be taken as the stress corresponding to a strain of 0.35 percent.

3.5.3.5 — Welded plain wire reinforcement shall conform to "Standard Specification for Steel Welded Wire Reinforcement, Plain, for Concrete" (ASTM A 185), except that for wire with f_y exceeding 60,000 psi, the yield strength shall be taken as the stress corresponding to a strain of 0.35 percent. Welded intersections shall not be spaced farther apart than 12 in. in direction of calculated stress, except for welded wire reinforcement used as stirrups in accordance with 12.13.2.

COMMENTARY

Rail-steel reinforcing bars used with this code are required to conform to ASTM A 996 including the provisions for Type R bars, and marked with the letter R for type of steel. Type R bars are required to meet more restrictive provisions for bend tests.

R3.5.3.2 — ASTM A 615 includes provisions for Grade 75 bars in sizes No. 6 through 18.

The 0.35 percent strain limit is necessary to ensure that the assumption of an elasto-plastic stress-strain curve in 10.2.4 will not lead to unconservative values of the member strength.

The 0.35 strain requirement is not applied to reinforcing bars having specified yield strengths of 60,000 psi or less. For steels having specified yield strengths of 40,000 psi, as were once used extensively, the assumption of an elasto-plastic stress-strain curve is well justified by extensive test data. For steels with specified yield strengths, up to 60,000 psi, the stress-strain curve may or may not be elasto-plastic as assumed in 10.2.4, depending on the properties of the steel and the manufacturing process. However, when the stress-strain curve is not elasto-plastic, there is limited experimental evidence to suggest that the actual steel stress at ultimate strength may not be enough less than the specified yield strength to warrant the additional effort of testing to the more restrictive criterion applicable to steels having specified yield strengths greater than 60,000 psi. In such cases, the ϕ-factor can be expected to account for the strength deficiency.

R3.5.3.5 — Welded plain wire reinforcement should be made of wire conforming to "Standard Specification for Steel Wire, Plain, for Concrete Reinforcement" (ASTM A 82). ASTM A 82 has a minimum yield strength of 70,000 psi. The code has assigned a yield strength value of 60,000 psi, but makes provision for the use of higher yield strengths provided the stress corresponds to a strain of 0.35 percent.

CODE	COMMENTARY
3.5.3.6 — Welded deformed wire reinforcement shall conform to "Standard Specification for Steel Welded Wire Reinforcement, Deformed, for Concrete" (ASTM A 497), except that for wire with f_y exceeding 60,000 psi, the yield strength shall be taken as the stress corresponding to a strain of 0.35 percent. Welded intersections shall not be spaced farther apart than 16 in. in direction of calculated stress, except for welded deformed wire reinforcement used as stirrups in accordance with 12.13.2.	**R3.5.3.6** — Welded deformed wire reinforcement should be made of wire conforming to " Standard Specification for Steel Welded Wire Reinforcement, Deformed, for Concrete (ASTM A 497)." ASTM A 497 has a minimum yield strength of 70,000 psi. The code has assigned a yield strength value of 60,000 psi, but makes provision for the use of higher yield strengths provided the stress corresponds to a strain of 0.35 percent.
3.5.3.7 — Galvanized reinforcing bars shall comply with "Standard Specification for Zinc-Coated (Galvanized) Steel Bars for Concrete Reinforcement" (ASTM A 767). Epoxy-coated reinforcing bars shall comply with "Standard Specification for Epoxy-Coated Steel Reinforcing Bars" (ASTM A 775) or with "Standard Specification for Epoxy-Coated Prefabricated Steel Reinforcing Bars" (ASTM A 934). Bars to be galvanized or epoxy-coated shall conform to one of the specifications listed in 3.5.3.1.	**R3.5.3.7** — Galvanized reinforcing bars (A 767) and epoxy-coated reinforcing bars (A 775) were added to the 1983 code, and epoxy-coated prefabricated reinforcing bars (A 934) were added to the 1995 code recognizing their usage, especially for conditions where corrosion resistance of reinforcement is of particular concern. They have typically been used in parking decks, bridge decks, and other highly corrosive environments.
3.5.3.8 — Epoxy-coated wires and welded wire reinforcement shall comply with "Standard Specification for Epoxy-Coated Steel Wire and Welded Wire Reinforcement" (ASTM A 884). Wires to be epoxy-coated shall conform to 3.5.3.4 and welded wire reinforcement to be epoxy-coated shall conform to 3.5.3.5 or 3.5.3.6.	
3.5.4 — Plain reinforcement	**R3.5.4 — Plain reinforcement**
3.5.4.1 — Plain bars for spiral reinforcement shall conform to the specification listed in 3.5.3.1(a) or (b).	Plain bars and plain wire are permitted only for spiral reinforcement (either as lateral reinforcement for compression members, for torsion members, or for confining reinforcement for splices).
3.5.4.2 — Plain wire for spiral reinforcement shall conform to "Standard Specification for Steel Wire, Plain, for Concrete Reinforcement" (ASTM A 82), except that for wire with f_y exceeding 60,000 psi, the yield strength shall be taken as the stress corresponding to a strain of 0.35 percent.	
3.5.5 — Prestressing steel	**R3.5.5 — Prestressing steel**
3.5.5.1 — Steel for prestressing shall conform to one of the following specifications:	**R3.5.5.1** — Because low-relaxation prestressing steel is addressed in a supplement to ASTM A 421, which applies only when low-relaxation material is specified, the appropriate ASTM reference is listed as a separate entity.
(a) Wire conforming to "Standard Specification for Uncoated Stress-Relieved Steel Wire for Prestressed Concrete" (ASTM A 421);	
(b) Low-relaxation wire conforming to "Standard Specification for Uncoated Stress-Relieved Steel Wire for Prestressed Concrete" including Supplement "Low-Relaxation Wire" (ASTM A 421);	

CODE

(c) Strand conforming to "Standard Specification for Steel Strand, Uncoated Seven-Wire for Prestressed Concrete" (ASTM A 416);

(d) Bar conforming to "Standard Specification for Uncoated High-Strength Steel Bars for Prestressing Concrete" (ASTM A 722).

3.5.5.2 — Wire, strands, and bars not specifically listed in ASTM A 421, A 416, or A 722 are allowed provided they conform to minimum requirements of these specifications and do not have properties that make them less satisfactory than those listed in ASTM A 421, A 416, or A 722.

3.5.6 — Structural steel, steel pipe, or tubing

3.5.6.1 — Structural steel used with reinforcing bars in composite compression members meeting requirements of 10.16.7 or 10.16.8 shall conform to one of the following specifications:

(a) "Standard Specification for Carbon Structural Steel" (ASTM A 36);

(b) "Standard Specification for High-Strength Low-Alloy Structural Steel" (ASTM A 242);

(c) "Standard Specification for High-Strength Low-Alloy Columbium-Vanadium Structural Steel" (ASTM A 572);

(d) "Standard Specification for High-Strength Low-Alloy Structural Steel with 50 ksi (345 MPa) Minimum Yield Point to 4 in. (100 mm) Thick" (ASTM A 588);

(e) "Standard Specification for Structural Steel Shapes" (ASTM A 992).

3.5.6.2 — Steel pipe or tubing for composite compression members composed of a steel encased concrete core meeting requirements of 10.16.6 shall conform to one of the following specifications:

(a) Grade B of "Standard Specification for Pipe, Steel, Black and Hot-Dipped, Zinc-Coated Welded and Seamless" (ASTM A 53);

(b) "Standard Specification for Cold-Formed Welded and Seamless Carbon Steel Structural Tubing in Rounds and Shapes" (ASTM A 500);

(c) "Standard Specification for Hot-Formed Welded and Seamless Carbon Steel Structural Tubing" (ASTM A 501).

COMMENTARY

CODE

3.6 — Admixtures

3.6.1 — Admixtures to be used in concrete shall be subject to prior approval by the engineer.

3.6.2 — An admixture shall be shown capable of maintaining essentially the same composition and performance throughout the work as the product used in establishing concrete proportions in accordance with 5.2.

3.6.3 — Calcium chloride or admixtures containing chloride from other than impurities from admixture ingredients shall not be used in prestressed concrete, in concrete containing embedded aluminum, or in concrete cast against stay-in-place galvanized steel forms. See 4.3.2 and 4.4.1.

3.6.4 — Air-entraining admixtures shall conform to "Standard Specification for Air-Entraining Admixtures for Concrete" (ASTM C 260).

3.6.5 — Water-reducing admixtures, retarding admixtures, accelerating admixtures, water-reducing and retarding admixtures, and water-reducing and accelerating admixtures shall conform to "Standard Specification for Chemical Admixtures for Concrete" (ASTM C 494) or "Standard Specification for Chemical Admixtures for Use in Producing Flowing Concrete" (ASTM C 1017).

3.6.6 — Fly ash or other pozzolans used as admixtures shall conform to "Standard Specification for Coal Fly Ash and Raw or Calcined Natural Pozzolan for Use in Concrete" (ASTM C 618).

3.6.7 — Ground granulated blast-furnace slag used as an admixture shall conform to "Standard Specification for Ground Granulated Blast-Furnace Slag for Use in Concrete and Mortars" (ASTM C 989).

COMMENTARY

R3.6 — Admixtures

R3.6.3 — Admixtures containing any chloride, other than impurities from admixture ingredients, should not be used in prestressed concrete or in concrete with aluminum embedments. Concentrations of chloride ion may produce corrosion of embedded aluminum (e.g., conduit), especially if the aluminum is in contact with embedded steel and the concrete is in a humid environment. Serious corrosion of galvanized steel sheet and galvanized steel stay-in-place forms occurs, especially in humid environments or where drying is inhibited by the thickness of the concrete or coatings or impermeable coverings. See 4.4.1 for specific limits on chloride ion concentration in concrete.

R3.6.7 — Ground granulated blast-furnace slag conforming to ASTM C 989 is used as an admixture in concrete in much the same way as fly ash. Generally, it should be used with portland cements conforming to ASTM C 150, and only rarely would it be appropriate to use ASTM C 989 slag with an ASTM C 595 blended cement that already contains a pozzolan or slag. Such use with ASTM C 595 cements might be considered for massive concrete placements where slow strength gain can be tolerated and where low heat of hydration is of particular importance. ASTM C 989 includes appendices which discuss effects of ground granulated blast-furnace slag on concrete strength, sulfate resistance, and alkali-aggregate reaction.

CODE

3.6.8 — Admixtures used in concrete containing ASTM C 845 expansive cements shall be compatible with the cement and produce no deleterious effects.

3.6.9 — Silica fume used as an admixture shall conform to ASTM C 1240.

3.7 — Storage of materials

3.7.1 — Cementitious materials and aggregates shall be stored in such manner as to prevent deterioration or intrusion of foreign matter.

3.7.2 — Any material that has deteriorated or has been contaminated shall not be used for concrete.

3.8 — Referenced standards

3.8.1 — Standards of ASTM International referred to in this code are listed below with their serial designations, including year of adoption or revision, and are declared to be part of this code as if fully set forth herein:

A 36/ A 36M-04	Standard Specification for Carbon Structural Steel
A 53/ A 53M-02	Standard Specification for Pipe, Steel, Black and Hot-Dipped, Zinc-Coated, Welded and Seamless
A 82-02	Standard Specification for Steel Wire, Plain, for Concrete Reinforcement
A 184/ A 184M-01	Standard Specification for Welded Deformed Steel Bar Mats for Concrete Reinforcement
A 185-02	Standard Specification for Steel Welded Wire Reinforcement, Plain, for Concrete
A 242/ A 242M-04	Standard Specification for High-Strength Low-Alloy Structural Steel
A 307-04	Standard Specification for Carbon Steel Bolts and Studs, 60,000 psi Tensile Strength
A 416/ A 416M-02	Standard Specification for Steel Strand, Uncoated Seven-Wire for Prestressed Concrete
A 421/ A 421M-02	Standard Specification for Uncoated Stress-Relieved Steel Wire for Prestressed Concrete

COMMENTARY

R3.6.8 — The use of admixtures in concrete containing ASTM C 845 expansive cements has reduced levels of expansion or increased shrinkage values. See ACI 223.[3.5]

R3.8 — Referenced standards

The ASTM standard specifications listed are the latest editions at the time these code provisions were adopted. Since these specifications are revised frequently, generally in minor details only, the user of the code should check directly with the sponsoring organization if it is desired to reference the latest edition. However, such a procedure obligates the user of the specification to evaluate if any changes in the later edition are significant in the use of the specification.

Standard specifications or other material to be legally adopted by reference into a building code should refer to a specific document. This can be done by simply using the complete serial designation since the first part indicates the subject and the second part the year of adoption. All standard documents referenced in this code are listed in 3.8, with the title and complete serial designation. In other sections of the code, the designations do not include the date so that all may be kept up-to-date by simply revising 3.8.

ASTM standards are available from ASTM, 100 Barr Harbor Drive, West Conshohocken, Pa., 19428.

CODE

A 496-02   Standard Specification for Steel Wire, Deformed, for Concrete Reinforcement

A 497/ A 497M-02   Standard Specification for Steel Welded Wire Reinforcement, Deformed, for Concrete

A 500-03a   Standard Specification for Cold-Formed Welded and Seamless Carbon Steel Structural Tubing in Rounds and Shapes

A 501-01   Standard Specification for Hot-Formed Welded and Seamless Carbon Steel Structural Tubing

A 572/ A 572M-04   Standard Specification for High-Strength Low-Alloy Columbium-Vanadium Structural Steel

A 588/ A 588M-04   Standard Specification for High-Strength Low-Alloy Structural Steel with 50 ksi [345 MPa] Minimum Yield Point to 4-in. [100-mm] Thick

A 615/ A 615M-04b   Standard Specification for Deformed and Plain Carbon Steel Bars for Concrete Reinforcement

A 706/ A 706M-04b   Standard Specification for Low-Alloy Steel Deformed and Plain Bars for Concrete Reinforcement

A 722/ A 722M-98(2003)   Standard Specification for Uncoated High-Strength Steel Bars for Prestressing Concrete

A 767/ A 767M-00b   Standard Specification for Zinc-Coated (Galvanized) Steel Bars for Concrete Reinforcement

A 775/ A 775M-04a   Standard Specification for Epoxy-Coated Steel Reinforcing Bars

A 884/ A 884M-04   Standard Specification for Epoxy-Coated Steel Wire and Welded Wire Reinforcement

A 934/ A 934M-04   Standard Specification for Epoxy-Coated Prefabricated Steel Reinforcing Bars

A 992/ A 992M-04   Standard Specification for Structural Steel Shapes

A 996/ A 996M-04   Standard Specification for Rail-Steel and Axle-Steel Deformed Bars for Concrete Reinforcement

COMMENTARY

Type R rail-steel bars are considered a mandatory requirement whenever ASTM A 996 is referenced in the code.

CODE		**COMMENTARY**

C 31/ C 31M-03a	Standard Practice for Making and Curing Concrete Test Specimens in the Field
C 33-03	Standard Specification for Concrete Aggregates
C 39/ C 39M-03	Standard Test Method for Compressive Strength of Cylindrical Concrete Specimens
C 42/ C 42M-04	Standard Test Method for Obtaining and Testing Drilled Cores and Sawed Beams of Concrete
C 94/ C 94M-04	Standard Specification for Ready-Mixed Concrete
C 109/ C 109M-02	Standard Test Method for Compressive Strength of Hydraulic Cement Mortars (Using 2-in. or [50-mm] Cube Specimens)
C 144-03	Standard Specification for Aggregate for Masonry Mortar
C 150-04a	Standard Specification for Portland Cement
C 172-04	Standard Practice for Sampling Freshly Mixed Concrete
C 192/ C 192M-02	Standard Practice for Making and Curing Concrete Test Specimens in the Laboratory
C 260-01	Standard Specification for Air-Entraining Admixtures for Concrete
C 330-04	Standard Specification for Lightweight Aggregates for Structural Concrete
C 494/ C 494M-04	Standard Specification for Chemical Admixtures for Concrete
C 496/ C 496M-04	Standard Test Method for Splitting Tensile Strength of Cylindrical Concrete Specimens
C 567-04	Standard Test Method for Determining Density of Structural Lightweight Concrete
C 595-03	Standard Specification for Blended Hydraulic Cements

CODE

C 618-03	Standard Specification for Coal Fly Ash and Raw or Calcined Natural Pozzolan for Use in Concrete
C 685/ C 685M-01	Standard Specification for Concrete Made by Volumetric Batching and Continuous Mixing
C 845-04	Standard Specification for Expansive Hydraulic Cement
C 989-04	Standard Specification for Ground Granulated Blast-Furnace Slag for Use in Concrete and Mortars
C 1017/ C 1017M-03	Standard Specification for Chemical Admixtures for Use in Producing Flowing Concrete
C 1157-03	Standard Performance Specification for Hydraulic Cement
C 1218/ C 1218M-99	Standard Test Method for Water-Soluble Chloride in Mortar and Concrete
C 1240-04	Standard Specification for Silica Fume Used in Cementitious Mixtures

3.8.2 — "Structural Welding Code—Reinforcing Steel" (ANSI/AWS D1.4-98) of the American Welding Society is declared to be part of this code as if fully set forth herein.

3.8.3 — Section 2.3.3 Load Combinations Including Flood Loads and 2.3.4 Load Combinations Including Atmospheric Ice Loads of "Minimum Design Loads for Buildings and Other Structures" (SEI/ASCE 7-02) is declared to be part of this code as if fully set forth herein, for the purpose cited in 9.2.4.

3.8.4 — "Specification for Unbonded Single Strand Tendons (ACI 423.6-01) and Commentary (423.6R-01)" is declared to be part of this code as if fully set forth herein.

3.8.5 — Articles 9.21.7.2 and 9.21.7.3 of Division I and Article 10.3.2.3 of Division II of AASHTO "Standard Specification for Highway Bridges" (AASHTO 17th Edition, 2002) are declared to be a part of this code as if fully set forth herein, for the purpose cited in 18.15.1.

3.8.6 — "Qualification of Post-Installed Mechanical Anchors in Concrete (ACI 355.2-04)" is declared to be part of this code as if fully set forth herein, for the purpose cited in Appendix D.

COMMENTARY

R3.8.3 — SEI/ASCE 7 is available from ASCE Book Orders, Box 79404, Baltimore, MD, 21279-0404.

R3.8.5 — The 2002 17th Edition of the AASHTO "Standard Specification for Highway Bridges" is available from AASHTO, 444 North Capitol Street, N.W., Suite 249, Washington, DC, 20001.

R3.8.6 — Parallel to development of the ACI 318-05 provisions for anchoring to concrete, ACI 355 developed a test method to define the level of performance required for post-installed anchors. This test method, ACI 355.2, con-

CODE

3.8.7 — "Structural Welding Code—Steel (AWS D 1.1/D1.1M-2004)" of the American Welding Society is declared to be part of this code as if fully set forth herein.

3.8.8 — "Acceptance Criteria for Moment Frames Based on Structural Testing (ACI T1.1-01)," is declared to be part of this code as if fully set forth herein.

COMMENTARY

tains requirements for the testing and evaluation of post-installed anchors for both cracked and uncracked concrete applications.

Notes

CHAPTER 4 — DURABILITY REQUIREMENTS

CODE

4.1 — Water-cementitious material ratio

COMMENTARY

R4.1 — Water-cementitious material ratio

Chapters 4 and 5 of earlier editions of the code were reformatted in 1989 to emphasize the importance of considering durability requirements before the designer selects f_c' and cover over the reinforcing steel.

Maximum water-cementitious material ratios of 0.40 to 0.50 that may be required for concretes exposed to freezing and thawing, sulfate soils or waters, or for preventing corrosion of reinforcement will typically be equivalent to requiring an f_c' of 5000 to 4000 psi, respectively. Generally, the required average compressive strengths, f_{cr}', will be 500 to 700 psi higher than the specified compressive strength, f_c'. Since it is difficult to accurately determine the water-cementitious material ratio of concrete during production, the f_c' specified should be reasonably consistent with the water-cementitious material ratio required for durability. Selection of an f_c' that is consistent with the water-cementitious material ratio selected for durability will help ensure that the required water-cementitious material ratio is actually obtained in the field. Because the usual emphasis on inspection is for strength, test results substantially higher than the specified strength may lead to a lack of concern for quality and production of concrete that exceeds the maximum water-cementitious material ratio. Thus an f_c' of 3000 psi and a maximum water-cementitious material ratio of 0.45 should not be specified for a parking structure, if the structure will be exposed to deicing salts.

The code does not include provisions for especially severe exposures, such as acids or high temperatures, and is not concerned with aesthetic considerations such as surface finishes. These items are beyond the scope of the code and should be covered specifically in the project specifications. Concrete ingredients and proportions are to be selected to meet the minimum requirements stated in the code and the additional requirements of the contract documents.

4.1.1 — The water-cementitious material ratios specified in Tables 4.2.2 and 4.3.1 shall be calculated using the weight of cement meeting ASTM C 150, C 595, C 845, or C 1157 plus the weight of fly ash and other pozzolans meeting ASTM C 618, slag meeting ASTM C 989, and silica fume meeting ASTM C 1240, if any, except that when concrete is exposed to deicing chemicals, 4.2.3 further limits the amount of fly ash, pozzolans, silica fume, slag, or the combination of these materials.

R4.1.1 — For concrete exposed to deicing chemicals the quantity of fly ash, other pozzolans, silica fume, slag, or blended cements used in the concrete is subject to the percentage limits in 4.2.3. Further, in 4.3 for sulfate exposures,[4.1] the pozzolan should be Class F by ASTM C 618, or have been tested by ASTM C 1012[4.2] or determined by service record to improve sulfate resistance.

CODE

4.2 — Freezing and thawing exposures

4.2.1 — Normalweight and lightweight concrete exposed to freezing and thawing or deicing chemicals shall be air-entrained with air content indicated in Table 4.2.1. Tolerance on air content as delivered shall be ± 1.5 percent. For f_c' greater than 5000 psi, reduction of air content indicated in Table 4.2.1 by 1.0 percent shall be permitted.

TABLE 4.2.1—TOTAL AIR CONTENT FOR FROST-RESISTANT CONCRETE

Nominal maximum aggregate size, in.*	Air content, percent	
	Severe exposure	Moderate exposure
3/8	7.5	6
1/2	7	5.5
3/4	6	5
1	6	4.5
1-1/2	5.5	4.5
2†	5	4
3†	4.5	3.5

* See ASTM C 33 for tolerance on oversize for various nominal maximum size designations.

† These air contents apply to total mix, as for the preceding aggregate sizes. When testing these concretes, however, aggregate larger than 1-1/2 in. is removed by handpicking or sieving and air content is determined on the minus 1-1/2 in. fraction of mix (tolerance on air content as delivered applies to this value.). Air content of total mix is computed from value determined on the minus 1-1/2 in. fraction.

4.2.2 — Concrete that will be subject to the exposures given in Table 4.2.2 shall conform to the corresponding maximum water-cementitious material ratios and minimum f_c' requirements of that table. In addition, concrete that will be exposed to deicing chemicals shall conform to the limitations of 4.2.3.

TABLE 4.2.2—REQUIREMENTS FOR SPECIAL EXPOSURE CONDITIONS

Exposure condition	Maximum water-cementitious material ratio*, by weight, normalweight concrete	Minimum f_c', normal-weight and light-weight concrete, psi*
Concrete intended to have low permeability when exposed to water	0.50	4000
Concrete exposed to freezing and thawing in a moist condition or to deicing chemicals	0.45	4500
For corrosion protection of reinforcement in concrete exposed to chlorides from de-icing chemicals, salt, salt water, brackish water, seawater, or spray from these sources.	0.40	5000

* When both Table 4.3.1 and Table 4.2.2 are considered, the lowest applicable maximum water-cementitious material ratio and highest applicable minimum f_c' shall be used.

COMMENTARY

R4.2 — Freezing and thawing exposures

R4.2.1 — A table of required air contents for frost-resistant concrete is included in the code, based on **"Standard Practice for Selecting Proportions for Normal, Heavyweight, and Mass Concrete"** (ACI 211.1).[4.3] Values are provided for both severe and moderate exposures depending on the exposure to moisture or deicing salts. Entrained air will not protect concrete containing coarse aggregates that undergo disruptive volume changes when frozen in a saturated condition. In Table 4.2.1, a severe exposure is where the concrete in a cold climate may be in almost continuous contact with moisture prior to freezing, or where deicing salts are used. Examples are pavements, bridge decks, sidewalks, parking garages, and water tanks. A moderate exposure is where the concrete in a cold climate will be only occasionally exposed to moisture prior to freezing, and where no deicing salts are used. Examples are certain exterior walls, beams, girders, and slabs not in direct contact with soil. Section 4.2.1 permits 1 percent lower air content for concrete with f_c' greater than 5000 psi. Such high-strength concretes will have lower water-cementitious material ratios and porosity and, therefore, improved frost resistance.

R4.2.2 — Maximum water-cementitious material ratios are not specified for lightweight concrete because determination of the absorption of these aggregates is uncertain, making calculation of the water-cementitious material ratio uncertain. The use of a minimum specified compressive strength, f_c', will ensure the use of a high-quality cement paste. For normalweight concrete, use of both minimum strength and maximum water-cementitious material ratio provide additional assurance that this objective is met.

CODE

4.2.3 — For concrete exposed to deicing chemicals, the maximum weight of fly ash, other pozzolans, silica fume, or slag that is included in the concrete shall not exceed the percentages of the total weight of cementitious materials given in Table 4.2.3.

TABLE 4.2.3—REQUIREMENTS FOR CONCRETE EXPOSED TO DEICING CHEMICALS

Cementitious materials	Maximum percent of total cementitious materials by weight*
Fly ash or other pozzolans conforming to ASTM C 618	25
Slag conforming to ASTM C 989	50
Silica fume conforming to ASTM C 1240	10
Total of fly ash or other pozzolans, slag, and silica fume	50†
Total of fly ash or other pozzolans and silica fume	35†

* The total cementitious material also includes ASTM C 150, C 595, C 845, and C 1157 cement.
The maximum percentages above shall include:
(a) Fly ash or other pozzolans present in Type IP or I(PM) blended cement, ASTM C 595, or ASTM C 1157;
(b) Slag used in the manufacture of a IS or I(SM) blended cement, ASTM C 595, or ASTM C 1157;
(c) Silica fume, ASTM C 1240, present in a blended cement.
† Fly ash or other pozzolans and silica fume shall constitute no more than 25 and 10 percent, respectively, of the total weight of the cementitious materials.

4.3 — Sulfate exposures

4.3.1 — Concrete to be exposed to sulfate-containing solutions or soils shall conform to requirements of Table 4.3.1 or shall be concrete made with a cement that provides sulfate resistance and that has a maximum water-cementitious material ratio and minimum f_c' from Table 4.3.1.

COMMENTARY

R4.2.3 — Section 4.2.3 and Table 4.2.3 establish limitations on the amount of fly ash, other pozzolans, silica fume, and slag that can be included in concrete exposed to deicing chemicals.[4.4-4.6] Research has demonstrated that the use of fly ash, slag, and silica fume produce concrete with a finer pore structure and, therefore, lower permeability.[4.7-4.9]

R4.3 — Sulfate exposures

R4.3.1 — Concrete exposed to injurious concentrations of sulfates from soil and water should be made with a sulfate-resisting cement. Table 4.3.1 lists the appropriate types of cement and the maximum water-cementitious material ratios and minimum specified compressive strengths for various exposure conditions. In selecting a cement for sulfate resistance, the principal consideration is its tricalcium aluminate (C_3A) content. For moderate exposures, Type II cement is limited to a maximum C_3A content of 8.0 percent under ASTM C 150. The blended cements under ASTM C 595 with the MS designation are appropriate for use in moderate sulfate exposures. The appropriate types under ASTM C 595 are IP(MS), IS(MS), I(PM)(MS), and I(SM)(MS). For severe exposures, Type V cement with a maximum C_3A content of 5 percent is specified. In certain areas, the C_3A

TABLE 4.3.1—REQUIREMENTS FOR CONCRETE EXPOSED TO SULFATE-CONTAINING SOLUTIONS

Sulfate exposure	Water soluble sulfate (SO_4) in soil, percent by weight	Sulfate (SO_4) in water, ppm	Cement type	Maximum water-cementitious material ratio, by weight, normalweight concrete*	Minimum f_c', normalweight and lightweight concrete, psi*
Negligible	$0.00 \leq SO_4 < 0.10$	$0 \leq SO_4 < 150$	—	—	—
Moderate†	$0.10 \leq SO_4 < 0.20$	$150 \leq SO_4 < 1500$	II, IP(MS), IS(MS), P(MS), I(PM)(MS), I(SM)(MS)	0.50	4000
Severe	$0.20 \leq SO_4 \leq 2.00$	$1500 \leq SO_4 \leq 10,000$	V	0.45	4500
Very severe	$SO_4 > 2.00$	$SO_4 > 10,000$	V plus pozzolan‡	0.45	4500

* When both Table 4.3.1 and Table 4.2.2 are considered, the lowest applicable maximum water-cementitious material ratio and highest applicable minimum f_c' shall be used.
† Seawater.
‡ Pozzolan that has been determined by test or service record to improve sulfate resistance when used in concrete containing Type V cement.

CODE

4.3.2 — Calcium chloride as an admixture shall not be used in concrete to be exposed to severe or very severe sulfate-containing solutions, as defined in Table 4.3.1.

4.4 — Corrosion protection of reinforcement

4.4.1 — For corrosion protection of reinforcement in concrete, maximum water soluble chloride ion concentrations in hardened concrete at ages from 28 to 42 days contributed from the ingredients including water, aggregates, cementitious materials, and admixtures shall not exceed the limits of Table 4.4.1. When testing is performed to determine water soluble chloride ion content, test procedures shall conform to ASTM C 1218.

COMMENTARY

content of other available types such as Type III or Type I may be less than 8 or 5 percent and are usable in moderate or severe sulfate exposures. Note that sulfate-resisting cement will not increase resistance to some chemically aggressive solutions, for example ammonium nitrate. The project specifications should cover all special cases.

Using fly ash (ASTM C 618, Class F) also has been shown to improve the sulfate resistance of concrete.[4.9] Certain Type IP cements made by blending Class F pozzolan with portland cement having a C_3A content greater than 8 percent can provide sulfate resistance for moderate exposures.

A note to Table 4.3.1 lists seawater as moderate exposure, even though it generally contains more than 1500 ppm SO_4. In seawater exposures, other types of cement with C_3A up to 10 percent may be used if the maximum water-cementitious material ratio is reduced to 0.40.

ASTM test method C 1012[4.2] can be used to evaluate the sulfate resistance of mixtures using combinations of cementitious materials.

In addition to the proper selection of cement, other requirements for durable concrete exposed to concentrations of sulfate are essential, such as, low water-cementitious material ratio, strength, adequate air entrainment, low slump, adequate consolidation, uniformity, adequate cover of reinforcement, and sufficient moist curing to develop the potential properties of the concrete.

R4.4 — Corrosion protection of reinforcement

R4.4.1 — Additional information on the effects of chlorides on the corrosion of reinforcing steel is given in **"Guide to Durable Concrete"** reported by ACI Committee 201[4.10] and **"Corrosion of Metals in Concrete"** reported by ACI Committee 222.[4.11] Test procedures should conform to those given in ASTM C 1218. An initial evaluation may be obtained by testing individual concrete ingredients for total chloride ion content. If total chloride ion content, calculated on the basis of concrete proportions, exceeds those permitted in Table 4.4.1, it may be necessary to test samples of the hardened concrete for water-soluble chloride ion content described in the ACI 201 guide. Some of the total chloride ions present in the ingredients will either be insoluble or will react with the cement during hydration and become insoluble under the test procedures described in ASTM C 1218.

CODE

TABLE 4.4.1—MAXIMUM CHLORIDE ION CONTENT FOR CORROSION PROTECTION OF REINFORCEMENT

Type of member	Maximum water soluble chloride ion (Cl⁻) in concrete, percent by weight of cement
Prestressed concrete	0.06
Reinforced concrete exposed to chloride in service	0.15
Reinforced concrete that will be dry or protected from moisture in service	1.00
Other reinforced concrete construction	0.30

4.4.2 — If concrete with reinforcement will be exposed to chlorides from deicing chemicals, salt, salt water, brackish water, seawater, or spray from these sources, requirements of Table 4.2.2 for maximum water-cementitious material ratio and minimum f_c', and the minimum concrete cover requirements of 7.7 shall be satisfied. See 18.16 for unbonded tendons.

COMMENTARY

When concretes are tested for soluble chloride ion content the tests should be made at an age of 28 to 42 days. The limits in Table 4.4.1 are to be applied to chlorides contributed from the concrete ingredients, not those from the environment surrounding the concrete.

The chloride ion limits in Table 4.4.1 differ from those recommended in ACI 201.2R[4.10] and ACI 222R.[4.11] For reinforced concrete that will be dry in service, a limit of 1 percent has been included to control total soluble chlorides. Table 4.4.1 includes limits of 0.15 and 0.30 percent for reinforced concrete that will be exposed to chlorides or will be damp in service, respectively. These limits compare to 0.10 and 0.15 recommended in ACI 201.2R.[4.10] ACI 222R[4.11] recommends limits of 0.08 and 0.20 percent by weight of cement for chlorides in prestressed and reinforced concrete, respectively, based on tests for acid soluble chlorides, not the test for water soluble chlorides required here.

When epoxy or zinc-coated bars are used, the limits in Table 4.4.1 may be more restrictive than necessary.

R4.4.2 — When concretes are exposed to external sources of chlorides, the water-cementitious material ratio and specified compressive strength f_c' of 4.2.2 are the minimum requirements that are to be considered. The designer should evaluate conditions in structures where chlorides may be applied, in parking structures where chlorides may be tracked in by vehicles, or in structures near seawater. Epoxy- or zinc-coated bars or cover greater than the minimum required in 7.7 may be desirable. Use of slag meeting ASTM C 989 or fly ash meeting ASTM C 618 and increased levels of specified strength provide increased protection. Use of silica fume meeting ASTM C 1240 with an appropriate high-range water reducer, ASTM C 494, Types F and G, or ASTM C 1017 can also provide additional protection.[4.12] The use of ASTM C 1202[4.13] to test concrete mixtures proposed for use will provide additional information on the performance of the mixtures.

Notes

CHAPTER 5 — CONCRETE QUALITY, MIXING, AND PLACING

CODE

5.1 — General

5.1.1 — Concrete shall be proportioned to provide an average compressive strength, f'_{cr}, as prescribed in 5.3.2 and shall satisfy the durability criteria of Chapter 4. Concrete shall be produced to minimize the frequency of strength tests below f'_c, as prescribed in 5.6.3.3. For concrete designed and constructed in accordance with the code, f'_c shall not be less than 2500 psi.

5.1.2 — Requirements for f'_c shall be based on tests of cylinders made and tested as prescribed in 5.6.3.

5.1.3 — Unless otherwise specified, f'_c shall be based on 28-day tests. If other than 28 days, test age for f'_c shall be as indicated in design drawings or specifications.

5.1.4 — Where design criteria in 9.5.2.3, 11.2, and 12.2.4 provide for use of a splitting tensile strength value of concrete, laboratory tests shall be made in accordance with "Standard Specification for Lightweight Aggregates for Structural Concrete" (ASTM C 330) to establish a value of f_{ct} corresponding to f'_c.

5.1.5 — Splitting tensile strength tests shall not be used as a basis for field acceptance of concrete.

COMMENTARY

R5.1 — General

The requirements for proportioning concrete mixtures are based on the philosophy that concrete should provide both adequate durability (Chapter 4) and strength. The criteria for acceptance of concrete are based on the philosophy that the code is intended primarily to protect the safety of the public. Chapter 5 describes procedures by which concrete of adequate strength can be obtained, and provides procedures for checking the quality of the concrete during and after its placement in the work.

Chapter 5 also prescribes minimum criteria for mixing and placing concrete.

The provisions of 5.2, 5.3, and 5.4, together with Chapter 4, establish required mixture proportions. The basis for determining the adequacy of concrete strength is in 5.6.

R5.1.1 — The basic premises governing the designation and evaluation of concrete strength are presented. It is emphasized that the average compressive strength of concrete produced should always exceed the specified value of f'_c used in the structural design calculations. This is based on probabilistic concepts, and is intended to ensure that adequate concrete strength will be developed in the structure. The durability requirements prescribed in Chapter 4 are to be satisfied in addition to attaining the average concrete strength in accordance with 5.3.2.

R5.1.4 — Sections 9.5.2.3 (modulus of rupture), 11.2 (concrete shear strength) and 12.2.4 (development of reinforcement) require modification in the design criteria for the use of lightweight concrete. Two alternative modification procedures are provided. One alternative is based on laboratory tests to determine the relationship between average splitting tensile strength f_{ct} and specified compressive strength f'_c for the lightweight concrete. For a lightweight aggregate from a given source, it is intended that appropriate values of f_{ct} be obtained in advance of design.

R5.1.5 — Tests for splitting tensile strength of concrete (as required by 5.1.4) are not intended for control of, or acceptance of, the strength of concrete in the field. Indirect control will

CODE

5.2 — Selection of concrete proportions

5.2.1 — Proportions of materials for concrete shall be established to provide:

(a) Workability and consistency to permit concrete to be worked readily into forms and around reinforcement under conditions of placement to be employed, without segregation or excessive bleeding;

(b) Resistance to special exposures as required by Chapter 4;

(c) Conformance with strength test requirements of 5.6.

5.2.2 — Where different materials are to be used for different portions of proposed work, each combination shall be evaluated.

5.2.3 — Concrete proportions shall be established in accordance with 5.3 or, alternatively, 5.4, and shall meet applicable requirements of Chapter 4.

5.3 — Proportioning on the basis of field experience or trial mixtures, or both

COMMENTARY

be maintained through the normal compressive strength test requirements provided by 5.6.

R5.2 — Selection of concrete proportions

Recommendations for selecting proportions for concrete are given in detail in **"Standard Practice for Selecting Proportions for Normal, Heavyweight, and Mass Concrete"** (ACI 211.1).[5.1] (Provides two methods for selecting and adjusting proportions for normalweight concrete: the estimated weight and absolute volume methods. Example calculations are shown for both methods. Proportioning of heavyweight concrete by the absolute volume method is presented in an appendix.)

Recommendations for lightweight concrete are given in **"Standard Practice for Selecting Proportions for Structural Lightweight Concrete"** (ACI 211.2).[5.2] (Provides a method of proportioning and adjusting structural grade concrete containing lightweight aggregates.)

R5.2.1 — The selected water-cementitious material ratio should be low enough, or in the case of lightweight concrete the compressive strength high enough to satisfy both the strength criteria (see 5.3 or 5.4) and the special exposure requirements (Chapter 4). The code does not include provisions for especially severe exposures, such as acids or high temperatures, and is not concerned with aesthetic considerations such as surface finishes. These items are beyond the scope of the code and should be covered specifically in the project specifications. Concrete ingredients and proportions are to be selected to meet the minimum requirements stated in the code and the additional requirements of the contract documents.

R5.2.3 — The code emphasizes the use of field experience or laboratory trial mixtures (see 5.3) as the preferred method for selecting concrete mixture proportions.

R5.3 — Proportioning on the basis of field experience or trial mixtures, or both

In selecting a suitable concrete mixture there are three basic steps. The first is the determination of the sample standard deviation. The second is the determination of the required average compressive strength. The third is the selection of mixture proportions required to produce that average strength, either by conventional trial mixture procedures or by a suitable experience record. Fig. R5.3 is a flow chart outlining the mixture selection and documentation procedure.

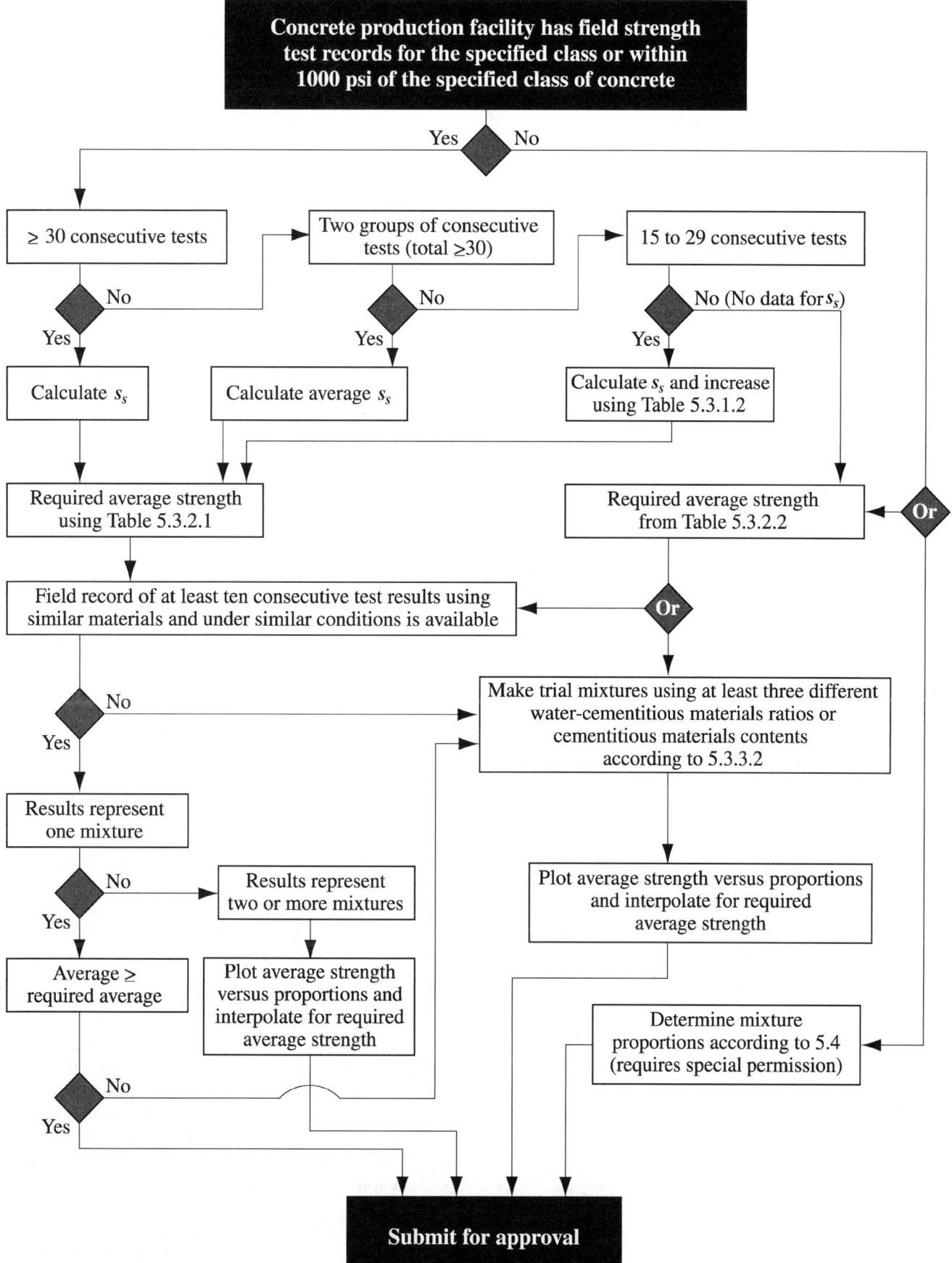

Fig. R5.3—Flow chart for selection and documentation of concrete proportions

CODE

5.3.1 — Sample standard deviation

5.3.1.1 — Where a concrete production facility has test records, a sample standard deviation, s_s, shall be established. Test records from which s_s is calculated:

(a) Shall represent materials, quality control procedures, and conditions similar to those expected and changes in materials and proportions within the test records shall not have been more restricted than those for proposed work;

(b) Shall represent concrete produced to meet a specified compressive strength or strengths within 1000 psi of f_c';

(c) Shall consist of at least 30 consecutive tests or two groups of consecutive tests totaling at least 30 tests as defined in 5.6.2.4, except as provided in 5.3.1.2.

5.3.1.2 — Where a concrete production facility does not have test records meeting requirements of 5.3.1.1, but does have a record based on 15 to 29 consecutive tests, a sample standard deviation s_s shall be established as the product of the calculated sample standard deviation and modification factor of Table 5.3.1.2. To be acceptable, test records shall meet requirements (a) and (b) of 5.3.1.1, and represent only a single record of consecutive tests that span a period of not less than 45 calendar days.

TABLE 5.3.1.2—MODIFICATION FACTOR FOR SAMPLE STANDARD DEVIATION WHEN LESS THAN 30 TESTS ARE AVAILABLE

No. of tests[*]	Modification factor for sample standard deviation[†]
Less than 15	Use table 5.3.2.2
15	1.16
20	1.08
25	1.03
30 or more	1.00

[*] Interpolate for intermediate numbers of tests.
[†] Modified sample standard deviation, s_s, to be used to determine required average strength, f_{cr}', from 5.3.2.1.

COMMENTARY

The mixture selected should yield an average strength appreciably higher than the specified strength f_c'. The degree of mixture over design depends on the variability of the test results.

R5.3.1 — Sample standard deviation

When a concrete production facility has a suitable record of 30 consecutive tests of similar materials and conditions expected, the sample standard deviation, s_s, is calculated from those results in accordance with the following formula:

$$s_s = \left[\frac{\Sigma(x_i - \bar{x})^2}{(n-1)}\right]^{1/2}$$

where:

s_s = sample standard deviation, psi
x_i = individual strength tests as defined in 5.6.2.4
$\bar{x}$ = average of n strength test results
n = number of consecutive strength tests

The sample standard deviation is used to determine the average strength required in 5.3.2.1.

If two test records are used to obtain at least 30 tests, the sample standard deviation used shall be the statistical average of the values calculated from each test record in accordance with the following formula:

$$\bar{s}_s = \left[\frac{(n_1-1)(s_{s1})^2 + (n_2-1)(s_{s2})^2}{(n_1+n_2-2)}\right]^{1/2}$$

where:

$\bar{s}_s$ = statistical average standard deviation where two test records are used to estimate the sample standard deviation
s_{s1}, s_{s2} = sample standard deviations calculated from two test records, 1 and 2, respectively
n_1, n_2 = number of tests in each test record, respectively

If less than 30, but at least 15 tests are available, the calculated sample standard deviation is increased by the factor given in Table 5.3.1.2. This procedure results in a more conservative (increased) required average strength. The factors in Table 5.3.1.2 are based on the sampling distribution of the sample standard deviation and provide protection (equivalent to that from a record of 30 tests) against the possibility that the smaller sample underestimates the true or universe population standard deviation.

The sample standard deviation used in the calculation of required average strength should be developed under conditions "similar to those expected" [see 5.3.1.1(a)]. This requirement is important to ensure acceptable concrete.

CODE

5.3.2 — Required average strength

5.3.2.1 — Required average compressive strength f_{cr}' used as the basis for selection of concrete proportions shall be determined from Table 5.3.2.1 using the sample standard deviation, s_s, calculated in accordance with 5.3.1.1 or 5.3.1.2.

TABLE 5.3.2.1—REQUIRED AVERAGE COMPRESSIVE STRENGTH WHEN DATA ARE AVAILABLE TO ESTABLISH A SAMPLE STANDARD DEVIATION

Specified compressive strength, psi	Required average compressive strength, psi
$f_c' \leq 5000$	Use the larger value computed from Eq. (5-1) and (5-2) $f_{cr}' = f_c' + 1.34 s_s$ (5-1) $f_{cr}' = f_c' + 2.33 s_s - 500$ (5-2)
$f_c' > 5000$	Use the larger value computed from Eq. (5-1) and (5-3) $f_{cr}' = f_c' + 1.34 s_s$ (5-1) $f_{cr}' = 0.90 f_c' + 2.33 s_s$ (5-3)

COMMENTARY

Concrete for background tests to determine sample standard deviation is considered to be "similar" to that required if made with the same general types of ingredients under no more restrictive conditions of control over material quality and production methods than on the proposed work, and if its specified strength does not deviate more than 1000 psi from the f_c' required [see 5.3.1.1(b)]. A change in the type of concrete or a major increase in the strength level may increase the sample standard deviation. Such a situation might occur with a change in type of aggregate (i.e., from natural aggregate to lightweight aggregate or vice versa) or a change from non-air-entrained concrete to air-entrained concrete. Also, there may be an increase in sample standard deviation when the average strength level is raised by a significant amount, although the increment of increase in sample standard deviation should be somewhat less than directly proportional to the strength increase. When there is reasonable doubt, any estimated sample standard deviation used to calculate the required average strength should always be on the conservative (high) side.

Note that the code uses the sample standard deviation in pounds per square inch instead of the coefficient of variation in percent. The latter is equal to the former expressed as a percent of the average strength.

Even when the average strength and sample standard deviation are of the levels assumed, there will be occasional tests that fail to meet the acceptance criteria prescribed in 5.6.3.3 (perhaps 1 test in 100).

R5.3.2 — Required average strength

R5.3.2.1 — Once the sample standard deviation has been determined, the required average compressive strength, f_{cr}', is obtained from the larger value computed from Eq. (5-1) and (5-2) for f_c' of 5000 psi or less, or the larger value computed from Eq. (5-1) and (5-3) for f_c' over 5000 psi. Equation (5-1) is based on a probability of 1-in-100 that the average of three consecutive tests may be below the specified compressive strength f_c'. Equation (5-2) is based on a similar probability that an individual test may be more than 500 psi below the specified compressive strength f_c'. Equation (5-3) is based on the same 1-in-100 probability that an individual test may be less than $0.90 f_c'$. These equations assume that the sample standard deviation used is equal to the population value appropriate for an infinite or very large number of tests. For this reason, use of sample standard deviations estimated from records of 100 or more tests is desirable. When 30 tests are available, the probability of failure will likely be somewhat greater than 1-in-100. The additional refinements required to achieve the 1-in-100 probability are not considered necessary, because of the uncertainty inherent in assuming that conditions operating when the test record was accumulated will be similar to conditions when the concrete will be produced.

CODE

5.3.2.2 — When a concrete production facility does not have field strength test records for calculation of s_s meeting requirements of 5.3.1.1 or 5.3.1.2, f'_{cr} shall be determined from Table 5.3.2.2 and documentation of average strength shall be in accordance with requirements of 5.3.3.

TABLE 5.3.2.2—REQUIRED AVERAGE COMPRESSIVE STRENGTH WHEN DATA ARE NOT AVAILABLE TO ESTABLISH A SAMPLE STANDARD DEVIATION

Specified compressive strength, psi	Required average compressive strength, psi
$f'_c < 3000$	$f'_{cr} = f'_c + 1000$
$3000 \leq f'_c \leq 5000$	$f'_{cr} = f'_c + 1200$
$f'_c > 5000$	$f'_{cr} = 1.10 f'_c + 700$

5.3.3 — Documentation of average compressive strength

Documentation that proposed concrete proportions will produce an average compressive strength equal to or greater than required average compressive strength f'_{cr} (see 5.3.2) shall consist of a field strength test record, several strength test records, or trial mixtures.

5.3.3.1 — When test records are used to demonstrate that proposed concrete proportions will produce f'_{cr} (see 5.3.2), such records shall represent materials and conditions similar to those expected. Changes in materials, conditions, and proportions within the test records shall not have been more restricted than those for proposed work. For the purpose of documenting average strength potential, test records consisting of

COMMENTARY

R.5.3.3 — Documentation of average compressive strength

Once the required average compressive strength f'_{cr} is known, the next step is to select mixture proportions that will produce an average strength at least as great as the required average strength, and also meet special exposure requirements of Chapter 4. The documentation may consist of a strength test record, several strength test records, or suitable laboratory or field trial mixtures. Generally, if a test record is used, it will be the same one that was used for computation of the standard deviation. However, if this test record shows either lower or higher average compressive strength than the required average compressive strength, different proportions may be necessary or desirable. In such instances, the average from a record of as few as 10 tests may be used, or the proportions may be established by interpolation between the strengths and proportions of two such records of consecutive tests. All test records for establishing proportions necessary to produce the average compressive strength are to meet the requirements of 5.3.3.1 for "similar materials and conditions."

For strengths over 5000 psi where the average compressive strength documentation is based on laboratory trial mixtures, it may be appropriate to increase f'_{cr} calculated in Table 5.3.2.2 to allow for a reduction in strength from laboratory trials to actual concrete production.

CODE

less than 30 but not less than 10 consecutive tests are acceptable provided test records encompass a period of time not less than 45 days. Required concrete proportions shall be permitted to be established by interpolation between the strengths and proportions of two or more test records, each of which meets other requirements of this section.

5.3.3.2 — When an acceptable record of field test results is not available, concrete proportions established from trial mixtures meeting the following restrictions shall be permitted:

(a) Materials shall be those for proposed work;

(b) Trial mixtures having proportions and consistencies required for proposed work shall be made using at least three different water-cementitious materials ratios or cementitious materials contents that will produce a range of strengths encompassing f'_{cr};

(c) Trial mixtures shall be designed to produce a slump within ± 0.75 in. of maximum permitted, and for air-entrained concrete, within ± 0.5 percent of maximum allowable air content;

(d) For each water-cementitious materials ratio or cementitious materials content, at least three test cylinders for each test age shall be made and cured in accordance with "Standard Practice for Making and Curing Concrete Test Specimens in the Laboratory" (ASTM C 192). Cylinders shall be tested at 28 days or at test age designated for determination of f'_c;

(e) From results of cylinder tests a curve shall be plotted showing the relationship between water-cementitious materials ratio or cementitious materials content and compressive strength at designated test age;

(f) Maximum water-cementitious material ratio or minimum cementitious materials content for concrete to be used in proposed work shall be that shown by the curve to produce f'_{cr} required by 5.3.2, unless a lower water-cementitious materials ratio or higher strength is required by Chapter 4.

5.4 — Proportioning without field experience or trial mixtures

5.4.1 — If data required by 5.3 are not available, concrete proportions shall be based upon other experience or information, if approved by the registered design professional. The required average compressive strength f'_{cr} of concrete produced with materials

COMMENTARY

R5.4 — Proportioning without field experience or trial mixtures

R5.4.1 — When no prior experience (5.3.3.1) or trial mixture data (5.3.3.2) meeting the requirements of these sections is available, other experience may be used only when special permission is given. Because combinations of different ingredients may vary considerably in strength level, this

CODE

similar to those proposed for use shall be at least 1200 psi greater than f_c'. This alternative shall not be used if f_c' is greater than 5000 psi.

5.4.2 — Concrete proportioned by this section shall conform to the durability requirements of Chapter 4 and to compressive strength test criteria of 5.6.

5.5 — Average compressive strength reduction

As data become available during construction, it shall be permitted to reduce the amount by which the required average concrete strength, f_{cr}', must exceed f_c', provided:

(a) Thirty or more test results are available and average of test results exceeds that required by 5.3.2.1, using a sample standard deviation calculated in accordance with 5.3.1.1; or

(b) Fifteen to 29 test results are available and average of test results exceeds that required by 5.3.2.1 using a sample standard deviation calculated in accordance with 5.3.1.2; and

(c) Special exposure requirements of Chapter 4 are met.

5.6 — Evaluation and acceptance of concrete

5.6.1 — Concrete shall be tested in accordance with the requirements of 5.6.2 through 5.6.5. Qualified field testing technicians shall perform tests on fresh concrete at the job site, prepare specimens required for curing under field conditions, prepare specimens required for testing in the laboratory, and record the temperature of the fresh concrete when preparing specimens for strength tests. Qualified laboratory technicians shall perform all required laboratory tests.

COMMENTARY

procedure is not permitted for f_c' greater than 5000 psi and the required average compressive strength should exceed f_c' by 1200 psi. The purpose of this provision is to allow work to continue when there is an unexpected interruption in concrete supply and there is not sufficient time for tests and evaluation or in small structures where the cost of trial mixture data is not justified.

R5.6 — Evaluation and acceptance of concrete

Once the mixture proportions have been selected and the job started, the criteria for evaluation and acceptance of the concrete can be obtained from 5.6.

An effort has been made in the code to provide a clear-cut basis for judging the acceptability of the concrete, as well as to indicate a course of action to be followed when the results of strength tests are not satisfactory.

R5.6.1 — Laboratory and field technicians can establish qualifications by becoming certified through certification programs. Field technicians in charge of sampling concrete; testing for slump, unit weight, yield, air content, and temperature; and making and curing test specimens should be certified in accordance with the requirements of ACI Concrete Field Testing Technician—Grade 1 Certification Program, or the requirements of ASTM C 1077,[5.3] or an equivalent program. Concrete testing laboratory personnel should be certified in accordance with the requirements of ACI Concrete Laboratory Testing Technician, Concrete Strength Testing Technician, or the requirements of ASTM C 1077.

Testing reports should be promptly distributed to the owner, registered design professional responsible for the design, contractor, appropriate subcontractors, appropriate suppliers, and building official to allow timely identification of either compliance or the need for corrective action.

CODE

5.6.2 — Frequency of testing

5.6.2.1 — Samples for strength tests of each class of concrete placed each day shall be taken not less than once a day, nor less than once for each 150 yd^3 of concrete, nor less than once for each 5000 ft^2 of surface area for slabs or walls.

5.6.2.2 — On a given project, if total volume of concrete is such that frequency of testing required by 5.6.2.1 would provide less than five strength tests for a given class of concrete, tests shall be made from at least five randomly selected batches or from each batch if fewer than five batches are used.

5.6.2.3 — When total quantity of a given class of concrete is less than 50 yd^3, strength tests are not required when evidence of satisfactory strength is submitted to and approved by the building official.

5.6.2.4 — A strength test shall be the average of the strengths of two cylinders made from the same sample of concrete and tested at 28 days or at test age designated for determination of f_c'.

5.6.3 — Laboratory-cured specimens

5.6.3.1 — Samples for strength tests shall be taken in accordance with "Standard Practice for Sampling Freshly Mixed Concrete" (ASTM C 172).

5.6.3.2 — Cylinders for strength tests shall be molded and laboratory-cured in accordance with "Standard Practice for Making and Curing Concrete Test Specimens in the Field" (ASTM C 31) and tested in accordance with "Standard Test Method for Compressive Strength of Cylindrical Concrete Specimens" (ASTM C 39).

COMMENTARY

R5.6.2 — Frequency of testing

R5.6.2.1 — The following three criteria establish the required minimum sampling frequency for each class of concrete:

(a) Once each day a given class is placed, nor less than

(b) Once for each 150 yd^3 of each class placed each day, nor less than

(c) Once for each 5000 ft^2 of slab or wall surface area placed each day.

In calculating surface area, only one side of the slab or wall should be considered. Criteria (c) will require more frequent sampling than once for each 150 yd^3 placed if the average wall or slab thickness is less than 9-3/4 in.

R5.6.2.2 — Samples for strength tests are to be taken on a strictly random basis if they are to measure properly the acceptability of the concrete. To be representative, the choice of times of sampling, or the batches of concrete to be sampled, are to be made on the basis of chance alone, within the period of placement. Batches should not be sampled on the basis of appearance, convenience, or other possibly biased criteria, because the statistical analyses will lose their validity. Not more than one test (average of two cylinders made from a sample, 5.6.2.4) should be taken from a single batch, and water may not be added to the concrete after the sample is taken.

ASTM D 3665[5.4] describes procedures for random selection of the batches to be tested.

R5.6.3 — Laboratory-cured specimens

CODE

5.6.3.3 — Strength level of an individual class of concrete shall be considered satisfactory if both of the following requirements are met:

(a) Every arithmetic average of any three consecutive strength tests equals or exceeds f_c';

(b) No individual strength test (average of two cylinders) falls below f_c' by more than 500 psi when f_c' is 5000 psi or less; or by more than **0.10f_c'** when f_c' is more than 5000 psi.

5.6.3.4 — If either of the requirements of 5.6.3.3 is not met, steps shall be taken to increase the average of subsequent strength test results. Requirements of 5.6.5 shall be observed if requirement of 5.6.3.3(b) is not met.

COMMENTARY

R5.6.3.3 — A single set of criteria is given for acceptability of strength and is applicable to all concrete used in structures designed in accordance with the code, regardless of design method used. The concrete strength is considered to be satisfactory as long as averages of any three consecutive strength tests remain above the specified f_c' and no individual strength test falls below the specified f_c' by more than 500 psi if f_c' is 5000 psi or less, or falls below f_c' by more than 10 percent if f_c' is over 5000 psi. Evaluation and acceptance of the concrete can be judged immediately as test results are received during the course of the work. Strength tests failing to meet these criteria will occur occasionally (probably about once in 100 tests) even though concrete strength and uniformity are satisfactory. Allowance should be made for such statistically expected variations in deciding whether the strength level being produced is adequate.

R5.6.3.4 — When concrete fails to meet either of the strength requirements of 5.6.3.3, steps should be taken to increase the average of the concrete test results. If sufficient concrete has been produced to accumulate at least 15 tests, these should be used to establish a new target average strength as described in 5.3.

If fewer than 15 tests have been made on the class of concrete in question, the new target strength level should be at least as great as the average level used in the initial selection of proportions. If the average of the available tests made on the project equals or exceeds the level used in the initial selection of proportions, a further increase in average level is required.

The steps taken to increase the average level of test results will depend on the particular circumstances, but could include one or more of the following:

(a) An increase in cementitious materials content;

(b) Changes in mixture proportions;

(c) Reductions in or better control of levels of slump supplied;

(d) A reduction in delivery time;

(e) Closer control of air content;

(f) An improvement in the quality of the testing, including strict compliance with standard test procedures.

Such changes in operating and testing procedures, or changes in cementitious materials content, or slump should not require a formal resubmission under the procedures of 5.3; however, important changes in sources of cement, aggregates, or admixtures should be accompanied by evidence that the average strength level will be improved.

Laboratories testing cylinders or cores to determine compliance with these requirements should be accredited or inspected for conformance to the requirement of ASTM C 1077[5.3] by a recognized agency such as the American Associ-

CODE

5.6.4 — Field-cured specimens

5.6.4.1 — If required by the building official, results of strength tests of cylinders cured under field conditions shall be provided.

5.6.4.2 — Field-cured cylinders shall be cured under field conditions in accordance with "Practice for Making and Curing Concrete Test Specimens in the Field" (ASTM C 31).

5.6.4.3 — Field-cured test cylinders shall be molded at the same time and from the same samples as laboratory-cured test cylinders.

5.6.4.4 — Procedures for protecting and curing concrete shall be improved when strength of field-cured cylinders at test age designated for determination of f_c' is less than 85 percent of that of companion laboratory-cured cylinders. The 85 percent limitation shall not apply if field-cured strength exceeds f_c' by more than 500 psi.

5.6.5 — Investigation of low-strength test results

5.6.5.1 — If any strength test (see 5.6.2.4) of laboratory-cured cylinders falls below f_c' by more than the values given in 5.6.3.3(b) or if tests of field-cured cylinders indicate deficiencies in protection and curing (see 5.6.4.4), steps shall be taken to assure that load-carrying capacity of the structure is not jeopardized.

5.6.5.2 — If the likelihood of low-strength concrete is confirmed and calculations indicate that load-carrying capacity is significantly reduced, tests of cores drilled from the area in question in accordance with "Standard Test Method for Obtaining and Testing Drilled Cores and Sawed Beams of Concrete" (ASTM C 42) shall be permitted. In such cases, three cores shall be taken for each strength test that falls below the values given in 5.6.3.3(b).

COMMENTARY

ation for Laboratory Accreditation (A2LA), AASHTO Materials Reference Laboratory (AMRL), National Voluntary Laboratory Accreditation Program (NVLAP), Cement and Concrete Reference Laboratory (CCRL), or their equivalent.

R5.6.4 — Field-cured specimens

R5.6.4.1 — Strength tests of cylinders cured under field conditions may be required to check the adequacy of curing and protection of concrete in the structure.

R5.6.4.4 — Positive guidance is provided in the code concerning the interpretation of tests of field-cured cylinders. Research has shown that cylinders protected and cured to simulate good field practice should test not less than about 85 percent of standard laboratory moist-cured cylinders. This percentage has been set as a rational basis for judging the adequacy of field curing. The comparison is made between the actual measured strengths of companion job-cured and laboratory-cured cylinders, not between job-cured cylinders and the specified value of f_c'. However, results for the job-cured cylinders are considered satisfactory if the job-cured cylinders exceed the specified f_c' by more than 500 psi, even though they fail to reach 85 percent of the strength of companion laboratory-cured cylinders.

R5.6.5 — Investigation of low-strength test results

Instructions are provided concerning the procedure to be followed when strength tests have failed to meet the specified acceptance criteria. For obvious reasons, these instructions cannot be dogmatic. The building official should apply judgment as to the significance of low test results and whether they indicate need for concern. If further investigation is deemed necessary, such investigation may include nondestructive tests, or in extreme cases, strength tests of cores taken from the structure.

Nondestructive tests of the concrete in place, such as by probe penetration, impact hammer, ultrasonic pulse velocity or pull out may be useful in determining whether or not a portion of the structure actually contains low-strength concrete. Such tests are of value primarily for comparisons within the same job rather than as quantitative measures of strength. For cores, if required, conservatively safe acceptance criteria are provided that should ensure structural ade-

CODE

5.6.5.3 — Cores shall be prepared for transport and storage by wiping drilling water from their surfaces and placing the cores in watertight bags or containers immediately after drilling. Cores shall be tested no earlier than 48 hours and not later than 7 days after coring unless approved by the registered design professional.

5.6.5.4 — Concrete in an area represented by core tests shall be considered structurally adequate if the average of three cores is equal to at least 85 percent of f_c' and if no single core is less than 75 percent of f_c'. Additional testing of cores extracted from locations represented by erratic core strength results shall be permitted.

5.6.5.5 — If criteria of 5.6.5.4 are not met and if the structural adequacy remains in doubt, the responsible authority shall be permitted to order a strength evaluation in accordance with Chapter 20 for the questionable portion of the structure, or take other appropriate action.

5.7 — Preparation of equipment and place of deposit

5.7.1 — Preparation before concrete placement shall include the following:

(a) All equipment for mixing and transporting concrete shall be clean;

(b) All debris and ice shall be removed from spaces to be occupied by concrete;

(c) Forms shall be properly coated;

(d) Masonry filler units that will be in contact with concrete shall be well drenched;

(e) Reinforcement shall be thoroughly clean of ice or other deleterious coatings;

COMMENTARY

quacy for virtually any type of construction.[5.5-5.8] Lower strength may, of course, be tolerated under many circumstances, but this again becomes a matter of judgment on the part of the building official and design engineer. When the core tests fail to provide assurance of structural adequacy, it may be practical, particularly in the case of floor or roof systems, for the building official to require a load test (Chapter 20). Short of load tests, if time and conditions permit, an effort may be made to improve the strength of the concrete in place by supplemental wet curing. Effectiveness of such a treatment should be verified by further strength evaluation using procedures previously discussed.

A core obtained through the use of a water-cooled bit results in a moisture gradient between the exterior and interior of the core being created during drilling. This adversely affects the core's compressive strength.[5.9] The restriction on the commencement of core testing provides a minimum time for the moisture gradient to dissipate.

Core tests having an average of 85 percent of the specified strength are realistic. To expect core tests to be equal to f_c' is not realistic, since differences in the size of specimens, conditions of obtaining samples, and procedures for curing, do not permit equal values to be obtained.

The code, as stated, concerns itself with assuring structural safety, and the instructions in 5.6 are aimed at that objective. It is not the function of the code to assign responsibility for strength deficiencies, whether or not they are such as to require corrective measures.

Under the requirements of this section, cores taken to confirm structural adequacy will usually be taken at ages later than those specified for determination of f_c'.

R5.7 — Preparation of equipment and place of deposit

Recommendations for mixing, handling and transporting, and placing concrete are given in detail in **"Guide for Measuring, Mixing, Transporting, and Placing Concrete"** reported by ACI Committee 304.[5.10] (Presents methods and procedures for control, handling and storage of materials, measurement, batching tolerances, mixing, methods of placing, transporting, and forms.)

Attention is directed to the need for using clean equipment and for cleaning forms and reinforcement thoroughly before beginning to deposit concrete. In particular, sawdust, nails, wood pieces, and other debris that may collect inside the forms should be removed. Reinforcement should be thoroughly cleaned of ice, dirt, loose rust, mill scale, or other coatings. Water should be removed from the forms.

CODE

(f) Water shall be removed from place of deposit before concrete is placed unless a tremie is to be used or unless otherwise permitted by the building official;

(g) All laitance and other unsound material shall be removed before additional concrete is placed against hardened concrete.

5.8 — Mixing

5.8.1 — All concrete shall be mixed until there is a uniform distribution of materials and shall be discharged completely before mixer is recharged.

5.8.2 — Ready-mixed concrete shall be mixed and delivered in accordance with requirements of "Standard Specification for Ready-Mixed Concrete" (ASTM C 94) or "Standard Specification for Concrete Made by Volumetric Batching and Continuous Mixing" (ASTM C 685).

5.8.3 — Job-mixed concrete shall be mixed in accordance with the following:

(a) Mixing shall be done in a batch mixer of approved type;

(b) Mixer shall be rotated at a speed recommended by the manufacturer;

(c) Mixing shall be continued for at least 1-1/2 minutes after all materials are in the drum, unless a shorter time is shown to be satisfactory by the mixing uniformity tests of "Standard Specification for Ready-Mixed Concrete" (ASTM C 94);

(d) Materials handling, batching, and mixing shall conform to applicable provisions of "Standard Specification for Ready-Mixed Concrete" (ASTM C 94);

(e) A detailed record shall be kept to identify:

(1) number of batches produced;

(2) proportions of materials used;

(3) approximate location of final deposit in structure;

(4) time and date of mixing and placing.

5.9 — Conveying

5.9.1 — Concrete shall be conveyed from mixer to place of final deposit by methods that will prevent separation or loss of materials.

COMMENTARY

R5.8 — Mixing

Concrete of uniform and satisfactory quality requires the materials to be thoroughly mixed until uniform in appearance and all ingredients are distributed. Samples taken from different portions of a batch should have essentially the same unit weight, air content, slump, and coarse aggregate content. Test methods for uniformity of mixing are given in ASTM C 94. The necessary time of mixing will depend on many factors including batch size, stiffness of the batch, size and grading of the aggregate, and the efficiency of the mixer. Excessively long mixing times should be avoided to guard against grinding of the aggregates.

R5.9 — Conveying

Each step in the handling and transporting of concrete needs to be controlled to maintain uniformity within a batch and from batch to batch. It is essential to avoid segregation of the coarse aggregate from the mortar or of water from the other ingredients.

CODE	COMMENTARY
5.9.2 — Conveying equipment shall be capable of providing a supply of concrete at site of placement without separation of ingredients and without interruptions sufficient to permit loss of plasticity between successive increments.	The code requires the equipment for handling and transporting concrete to be capable of supplying concrete to the place of deposit continuously and reliably under all conditions and for all methods of placement. The provisions of 5.9 apply to all placement methods, including pumps, belt conveyors, pneumatic systems, wheelbarrows, buggies, crane buckets, and tremies.
	Serious loss in strength can result when concrete is pumped through pipe made of aluminum or aluminum alloy.[5.11] Hydrogen gas generated by the reaction between the cement alkalies and the aluminum eroded from the interior of the pipe surface has been shown to cause strength reduction as much as 50 percent. Hence, equipment made of aluminum or aluminum alloys should not be used for pump lines, tremies, or chutes other than short chutes such as those used to convey concrete from a truck mixer.

5.10 — Depositing

R5.10 — Depositing

5.10.1 — Concrete shall be deposited as nearly as practical in its final position to avoid segregation due to rehandling or flowing.

Rehandling concrete can cause segregation of the materials. Hence the code cautions against this practice. Retempering of partially set concrete with the addition of water should not be permitted, unless authorized. This does not preclude the practice (recognized in ASTM C 94) of adding water to mixed concrete to bring it up to the specified slump range so long as prescribed limits on the maximum mixing time and water-cementitious materials ratio are not violated.

5.10.2 — Concreting shall be carried on at such a rate that concrete is at all times plastic and flows readily into spaces between reinforcement.

5.10.3 — Concrete that has partially hardened or been contaminated by foreign materials shall not be deposited in the structure.

Section 5.10.4 of the 1971 code contained a requirement that "where conditions make consolidation difficult or where reinforcement is congested, batches of mortar containing the same proportions of cement, sand, and water as used in the concrete, shall first be deposited in the forms to a depth of at least 1 in." That requirement was deleted from the 1977 code since the conditions for which it was applicable could not be defined precisely enough to justify its inclusion as a code requirement. The practice, however, has merit and should be incorporated in job specifications where appropriate, with the specific enforcement the responsibility of the job inspector. The use of mortar batches aids in preventing honeycomb and poor bonding of the concrete with the reinforcement. The mortar should be placed immediately before depositing the concrete and should be plastic (neither stiff nor fluid) when the concrete is placed.

5.10.4 — Retempered concrete or concrete that has been remixed after initial set shall not be used unless approved by the engineer.

5.10.5 — After concreting is started, it shall be carried on as a continuous operation until placing of a panel or section, as defined by its boundaries or predetermined joints, is completed except as permitted or prohibited by 6.4.

5.10.6 — Top surfaces of vertically formed lifts shall be generally level.

5.10.7 — When construction joints are required, joints shall be made in accordance with 6.4.

5.10.8 — All concrete shall be thoroughly consolidated by suitable means during placement and shall be thoroughly worked around reinforcement and embedded fixtures and into corners of forms.

Recommendations for consolidation of concrete are given in detail in **"Guide for Consolidation of Concrete"** reported by ACI Committee 309.[5.12] (Presents current information on the mechanism of consolidation and gives recommendations on equipment characteristics and procedures for various classes of concrete.)

CODE

5.11 — Curing

5.11.1 — Concrete (other than high-early-strength) shall be maintained above 50 F and in a moist condition for at least the first 7 days after placement, except when cured in accordance with 5.11.3.

5.11.2 — High-early-strength concrete shall be maintained above 50 F and in a moist condition for at least the first 3 days, except when cured in accordance with 5.11.3.

5.11.3 — Accelerated curing

5.11.3.1 — Curing by high-pressure steam, steam at atmospheric pressure, heat and moisture, or other accepted processes, shall be permitted to accelerate strength gain and reduce time of curing.

5.11.3.2 — Accelerated curing shall provide a compressive strength of the concrete at the load stage considered at least equal to required design strength at that load stage.

5.11.3.3 — Curing process shall be such as to produce concrete with a durability at least equivalent to the curing method of 5.11.1 or 5.11.2.

5.11.4 — When required by the engineer or architect, supplementary strength tests in accordance with 5.6.4 shall be performed to assure that curing is satisfactory.

COMMENTARY

R5.11 — Curing

Recommendations for curing concrete are given in detail in **"Guide to Curing Concrete"** reported by ACI Committee 308.[5.13] (Presents basic principles of proper curing and describes the various methods, procedures, and materials for curing of concrete.)

R5.11.3 — Accelerated curing

The provisions of this section apply whenever an accelerated curing method is used, whether for precast or cast-in-place elements. The compressive strength of steam-cured concrete is not as high as that of similar concrete continuously cured under moist conditions at moderate temperatures. Also the modulus of elasticity E_c of steam-cured specimens may vary from that of specimens moist-cured at normal temperatures. When steam curing is used, it is advisable to base the concrete mixture proportions on steam-cured test cylinders.

Accelerated curing procedures require careful attention to obtain uniform and satisfactory results. Preventing moisture loss during the curing is essential.

R5.11.4 — In addition to requiring a minimum curing temperature and time for normal- and high-early-strength concrete, the code provides a specific criterion in 5.6.4 for judging the adequacy of field curing. At the test age for which the compressive strength is specified (usually 28 days), field-cured cylinders should produce strength not less than 85 percent of that of the standard, laboratory-cured cylinders. For a reasonably valid comparison to be made, field-cured cylinders and companion laboratory-cured cylinders should come from the same sample. Field-cured cylinders should be cured under conditions identical to those of the structure. If the structure is protected from the elements, the cylinder should be protected.

Cylinders related to members not directly exposed to weather should be cured adjacent to those members and provided with the same degree of protection and method of curing. The field cylinders should not be treated more favorably than the elements they represent. (See 5.6.4 for additional information.) If the field-cured cylinders do not provide satisfactory strength by this comparison, measures should be taken to improve the curing. If the tests indicate a possible serious deficiency in strength of concrete in the structure, core tests may be required, with or without supplemental wet curing, to check the structural adequacy, as provided in 5.6.5.

CODE	COMMENTARY
5.12 — Cold weather requirements	**R5.12 — Cold weather requirements**
5.12.1 — Adequate equipment shall be provided for heating concrete materials and protecting concrete during freezing or near-freezing weather.	Recommendations for cold weather concreting are given in detail in **"Cold Weather Concreting"** reported by ACI Committee 306.[5.14] (Presents requirements and methods for producing satisfactory concrete during cold weather.)
5.12.2 — All concrete materials and all reinforcement, forms, fillers, and ground with which concrete is to come in contact shall be free from frost.	
5.12.3 — Frozen materials or materials containing ice shall not be used.	
5.13 — Hot weather requirements	**R5.13 — Hot weather requirements**
During hot weather, proper attention shall be given to ingredients, production methods, handling, placing, protection, and curing to prevent excessive concrete temperatures or water evaporation that could impair required strength or serviceability of the member or structure.	Recommendations for hot weather concreting are given in detail in **"Hot Weather Concreting"** reported by ACI Committee 305.[5.15] (Defines the hot weather factors that effect concrete properties and construction practices and recommends measures to eliminate or minimize the undesirable effects.)

… STRUCTURAL CONCRETE BUILDING CODE AND COMMENTARY (ACI 318-05) … 597

CHAPTER 6 — FORMWORK, EMBEDDED PIPES, AND CONSTRUCTION JOINTS

CODE

6.1 — Design of formwork

6.1.1 — Forms shall result in a final structure that conforms to shapes, lines, and dimensions of the members as required by the design drawings and specifications.

6.1.2 — Forms shall be substantial and sufficiently tight to prevent leakage of mortar.

6.1.3 — Forms shall be properly braced or tied together to maintain position and shape.

6.1.4 — Forms and their supports shall be designed so as not to damage previously placed structure.

6.1.5 — Design of formwork shall include consideration of the following factors:

(a) Rate and method of placing concrete;

(b) Construction loads, including vertical, horizontal, and impact loads;

(c) Special form requirements for construction of shells, folded plates, domes, architectural concrete, or similar types of elements.

6.1.6 — Forms for prestressed concrete members shall be designed and constructed to permit movement of the member without damage during application of prestressing force.

6.2 — Removal of forms, shores, and reshoring

6.2.1 — Removal of forms

Forms shall be removed in such a manner as not to impair safety and serviceability of the structure. Concrete exposed by form removal shall have sufficient strength not to be damaged by removal operation.

6.2.2 — Removal of shores and reshoring

The provisions of 6.2.2.1 through 6.2.2.3 shall apply to slabs and beams except where cast on the ground.

6.2.2.1 — Before starting construction, the contractor shall develop a procedure and schedule for removal of shores and installation of reshores and for calculating the loads transferred to the structure during the process.

COMMENTARY

R6.1 — Design of formwork

Only minimum performance requirements for formwork, necessary to provide for public health and safety, are prescribed in Chapter 6. Formwork for concrete, including proper design, construction, and removal, demands sound judgment and planning to achieve adequate forms that are both economical and safe. Detailed information on formwork for concrete is given in: **"Guide to Formwork for Concrete,"** reported by Committee 347.[6.1] (Provides recommendations for design, construction, and materials for formwork, forms for special structures, and formwork for special methods of construction. Directed primarily to contractors, the suggested criteria will aid engineers and architects in preparing job specifications for the contractors.)

Formwork for Concrete[6.2] prepared under the direction of ACI Committee 347. (A how-to-do-it handbook for contractors, engineers, and architects following the guidelines established in ACI 347R. Planning, building, and using formwork are discussed, including tables, diagrams, and formulas for form design loads.)

R6.2 — Removal of forms, shores, and reshoring

In determining the time for removal of forms, consideration should be given to the construction loads and to the possibilities of deflections.[6.3] The construction loads are frequently at least as great as the specified live loads. At early ages, a structure may be adequate to support the applied loads but may deflect sufficiently to cause permanent damage.

Evaluation of concrete strength during construction may be demonstrated by field-cured test cylinders or other procedures approved by the building official such as:

(a) Tests of cast-in-place cylinders in accordance with "Standard Test Method for Compressive Strength of Concrete Cylinders Cast-in-Place in Cylindrical Molds" (ASTM C 873[6.4]). (This method is limited to use in slabs where the depth of concrete is from 5 to 12 in.);

CODE

(a) The structural analysis and concrete strength data used in planning and implementing form removal and shoring shall be furnished by the contractor to the building official when so requested;

(b) No construction loads shall be supported on, nor any shoring removed from, any part of the structure under construction except when that portion of the structure in combination with remaining forming and shoring system has sufficient strength to support safely its weight and loads placed thereon;

(c) Sufficient strength shall be demonstrated by structural analysis considering proposed loads, strength of forming and shoring system, and concrete strength data. Concrete strength data shall be based on tests of field-cured cylinders or, when approved by the building official, on other procedures to evaluate concrete strength.

6.2.2.2 — No construction loads exceeding the combination of superimposed dead load plus specified live load shall be supported on any unshored portion of the structure under construction, unless analysis indicates adequate strength to support such additional loads.

6.2.2.3 — Form supports for prestressed concrete members shall not be removed until sufficient prestressing has been applied to enable prestressed members to carry their dead load and anticipated construction loads.

COMMENTARY

(b) Penetration resistance in accordance with "Standard Test Method for Penetration Resistance of Hardened Concrete" (ASTM C 803[6.5]);

(c) Pullout strength in accordance with "Standard Test Method for Pullout Strength of Hardened Concrete" (ASTM C 900[6.6]);

(d) Maturity factor measurements and correlation in accordance with ASTM C 1074.[6.7]

Procedures (b), (c), and (d) require sufficient data, using job materials, to demonstrate correlation of measurements on the structure with compressive strength of molded cylinders or drilled cores.

Where the structure is adequately supported on shores, the side forms of beams, girders, columns, walls, and similar vertical forms may generally be removed after 12 h of cumulative curing time, provided the side forms support no loads other than the lateral pressure of the plastic concrete. Cumulative curing time represents the sum of time intervals, not necessarily consecutive, during which the temperature of the air surrounding the concrete is above 50 F. The 12-h cumulative curing time is based on regular cements and ordinary conditions; the use of special cements or unusual conditions may require adjustment of the given limits. For example, concrete made with Type II or V (ASTM C 150) or ASTM C 595 cements, concrete containing retarding admixtures, and concrete to which ice was added during mixing (to lower the temperature of fresh concrete) may not have sufficient strength in 12 h and should be investigated before removal of formwork.

The removal of formwork for multistory construction should be a part of a planned procedure considering the temporary support of the whole structure as well as that of each individual member. Such a procedure should be worked out prior to construction and should be based on a structural analysis taking into account the following items, as a minimum:

(a) The structural system that exists at the various stages of construction and the construction loads corresponding to those stages;

(b) The strength of the concrete at the various ages during construction;

(c) The influence of deformations of the structure and shoring system on the distribution of dead loads and construction loads during the various stages of construction;

(d) The strength and spacing of shores or shoring systems used, as well as the method of shoring, bracing, shore removal, and reshoring including the minimum time

CODE

COMMENTARY

intervals between the various operations;

(e) Any other loading or condition that affects the safety or serviceability of the structure during construction.

For multistory construction, the strength of the concrete during the various stages of construction should be substantiated by field-cured test specimens or other approved methods.

6.3 — Conduits and pipes embedded in concrete

R6.3 — Conduits and pipes embedded in concrete

6.3.1 — Conduits, pipes, and sleeves of any material not harmful to concrete and within limitations of 6.3 shall be permitted to be embedded in concrete with approval of the engineer, provided they are not considered to replace structurally the displaced concrete, except as provided in 6.3.6.

R6.3.1 — Conduits, pipes, and sleeves not harmful to concrete can be embedded within the concrete, but the work should be done in such a manner that the structure will not be endangered. Empirical rules are given in 6.3 for safe installations under common conditions; for other than common conditions, special designs should be made. Many general building codes have adopted ANSI/ASME piping codes B 31.1 for power piping[6.8] and B 31.3 for chemical and petroleum piping.[6.9] The specifier should be sure that the appropriate piping codes are used in the design and testing of the system. The contractor should not be permitted to install conduits, pipes, ducts, or sleeves that are not shown on the plans or not approved by the engineer or architect.

For the integrity of the structure, it is important that all conduit and pipe fittings within the concrete be carefully assembled as shown on the plans or called for in the job specifications.

6.3.2 — Conduits and pipes of aluminum shall not be embedded in structural concrete unless effectively coated or covered to prevent aluminum-concrete reaction or electrolytic action between aluminum and steel.

R6.3.2 — The code prohibits the use of aluminum in structural concrete unless it is effectively coated or covered. Aluminum reacts with concrete and, in the presence of chloride ions, may also react electrolytically with steel, causing cracking and/or spalling of the concrete. Aluminum electrical conduits present a special problem since stray electric current accelerates the adverse reaction.

6.3.3 — Conduits, pipes, and sleeves passing through a slab, wall, or beam shall not impair significantly the strength of the construction.

6.3.4 — Conduits and pipes, with their fittings, embedded within a column shall not displace more than 4 percent of the area of cross section on which strength is calculated or which is required for fire protection.

6.3.5 — Except when drawings for conduits and pipes are approved by the structural engineer, conduits and pipes embedded within a slab, wall, or beam (other than those merely passing through) shall satisfy 6.3.5.1 through 6.3.5.3.

6.3.5.1 — They shall not be larger in outside dimension than 1/3 the overall thickness of slab, wall, or beam in which they are embedded.

CODE

6.3.5.2 — They shall not be spaced closer than 3 diameters or widths on center.

6.3.5.3 — They shall not impair significantly the strength of the construction.

6.3.6 — Conduits, pipes, and sleeves shall be permitted to be considered as replacing structurally in compression the displaced concrete provided in 6.3.6.1 through 6.3.6.3.

6.3.6.1 — They are not exposed to rusting or other deterioration.

6.3.6.2 — They are of uncoated or galvanized iron or steel not thinner than standard Schedule 40 steel pipe.

6.3.6.3 — They have a nominal inside diameter not over 2 in. and are spaced not less than 3 diameters on centers.

6.3.7 — Pipes and fittings shall be designed to resist effects of the material, pressure, and temperature to which they will be subjected.

6.3.8 — No liquid, gas, or vapor, except water not exceeding 90 F nor 50 psi pressure, shall be placed in the pipes until the concrete has attained its design strength.

6.3.9 — In solid slabs, piping, unless it is for radiant heating or snow melting, shall be placed between top and bottom reinforcement.

6.3.10 — Concrete cover for pipes, conduits, and fittings shall not be less than 1-1/2 in. for concrete exposed to earth or weather, nor less than 3/4 in. for concrete not exposed to weather or in contact with ground.

6.3.11 — Reinforcement with an area not less than 0.002 times area of concrete section shall be provided normal to piping.

6.3.12 — Piping and conduit shall be so fabricated and installed that cutting, bending, or displacement of reinforcement from its proper location will not be required.

6.4 — Construction joints

6.4.1 — Surface of concrete construction joints shall be cleaned and laitance removed.

COMMENTARY

R6.3.7 — The 1983 code limited the maximum pressure in embedded pipe to 200 psi, which was considered too restrictive. Nevertheless, the effects of such pressures and the expansion of embedded pipe should be considered in the design of the concrete member.

R6.4 — Construction joints

For the integrity of the structure, it is important that all construction joints be defined in construction documents and constructed as required. Any deviations should be approved by the engineer or architect.

CODE

6.4.2 — Immediately before new concrete is placed, all construction joints shall be wetted and standing water removed.

6.4.3 — Construction joints shall be so made and located as not to impair the strength of the structure. Provision shall be made for transfer of shear and other forces through construction joints. See 11.7.9.

6.4.4 — Construction joints in floors shall be located within the middle third of spans of slabs, beams, and girders.

6.4.5 — Construction joints in girders shall be offset a minimum distance of two times the width of intersecting beams.

6.4.6 — Beams, girders, or slabs supported by columns or walls shall not be cast or erected until concrete in the vertical support members is no longer plastic.

6.4.7 — Beams, girders, haunches, drop panels, and capitals shall be placed monolithically as part of a slab system, unless otherwise shown in design drawings or specifications.

COMMENTARY

R6.4.2 — The requirements of the 1977 code for the use of neat cement on vertical joints have been removed, since it is rarely practical and can be detrimental where deep forms and steel congestion prevent proper access. Often wet blasting and other procedures are more appropriate. Because the code sets only minimum standards, the engineer may have to specify special procedures if conditions warrant. The degree to which mortar batches are needed at the start of concrete placement depend on concrete proportions, congestion of steel, vibrator access, and other factors.

R6.4.3 — Construction joints should be located where they will cause the least weakness in the structure. When shear due to gravity load is not significant, as is usually the case in the middle of the span of flexural members, a simple vertical joint may be adequate. Lateral force design may require special design treatment of construction joints. Shear keys, intermittent shear keys, diagonal dowels, or the shear transfer method of 11.7 may be used whenever a force transfer is required.

R6.4.6 — Delay in placing concrete in members supported by columns and walls is necessary to prevent cracking at the interface of the slab and supporting member caused by bleeding and settlement of plastic concrete in the supporting member.

R6.4.7 — Separate placement of slabs and beams, haunches, and similar elements is permitted when shown on the drawings and where provision has been made to transfer forces as required in 6.4.3.

Notes

CHAPTER 7 — DETAILS OF REINFORCEMENT

CODE

7.1 — Standard hooks

The term standard hook as used in this code shall mean one of the following:

7.1.1 — 180-deg bend plus $4d_b$ extension, but not less than 2-1/2 in. at free end of bar.

7.1.2 — 90-deg bend plus $12d_b$ extension at free end of bar.

7.1.3 — For stirrup and tie hooks

(a) No. 5 bar and smaller, 90-deg bend plus $6d_b$ extension at free end of bar; or

(b) No. 6, No. 7, and No. 8 bar, 90-deg bend plus $12d_b$ extension at free end of bar; or

(c) No. 8 bar and smaller, 135-deg bend plus $6d_b$ extension at free end of bar.

7.1.4 — Seismic hooks as defined in 21.1.

7.2 — Minimum bend diameters

7.2.1 — Diameter of bend measured on the inside of the bar, other than for stirrups and ties in sizes No. 3 through No. 5, shall not be less than the values in Table 7.2.

7.2.2 — Inside diameter of bend for stirrups and ties shall not be less than $4d_b$ for No. 5 bar and smaller. For bars larger than No. 5, diameter of bend shall be in accordance with Table 7.2.

7.2.3 — Inside diameter of bend in welded wire reinforcement for stirrups and ties shall not be less than $4d_b$ for deformed wire larger than D6 and $2d_b$ for all other wires. Bends with inside diameter of less than $8d_b$ shall not be less than $4d_b$ from nearest welded intersection.

COMMENTARY

R7.1 — Standard hooks

Recommended methods and standards for preparing design drawings, typical details, and drawings for the fabrication and placing of reinforcing steel in reinforced concrete structures are given in the *ACI Detailing Manual*, reported by ACI Committee 315.[7.1]

All provisions in the code relating to bar, wire, or strand diameter (and area) are based on the nominal dimensions of the reinforcement as given in the appropriate ASTM specification. Nominal dimensions are equivalent to those of a circular area having the same weight per foot as the ASTM designated bar, wire, or strand sizes. Cross-sectional area of reinforcement is based on nominal dimensions.

R7.1.3 — Standard stirrup and tie hooks are limited to No. 8 bars and smaller, and the 90-deg hook with $6d_b$ extension is further limited to No. 5 bars and smaller, in both cases as the result of research showing that larger bar sizes with 90-deg hooks and $6d_b$ extensions tend to pop out under high load.

R7.2 — Minimum bend diameters

Standard bends in reinforcing bars are described in terms of the inside diameter of bend since this is easier to measure than the radius of bend. The primary factors affecting the minimum bend diameter are feasibility of bending without breakage and avoidance of crushing the concrete inside the bend.

R7.2.2 — The minimum $4d_b$ bend for the bar sizes commonly used for stirrups and ties is based on accepted industry practice in the United States. Use of a stirrup bar size not greater than No. 5 for either the 90-deg or 135-deg standard stirrup hook will permit multiple bending on standard stirrup bending equipment.

R7.2.3 — Welded wire reinforcement can be used for stirrups and ties. The wire at welded intersections does not have the same uniform ductility and bendability as in areas which were not heated. These effects of the welding temperature are usually dissipated in a distance of approximately four wire diameters. Minimum bend diameters permitted

CODE

TABLE 7.2—MINIMUM DIAMETERS OF BEND

Bar size	Minimum diameter
No. 3 through No. 8	$6d_b$
No. 9, No. 10, and No. 11	$8d_b$
No. 14 and No. 18	$10d_b$

7.3 — Bending

7.3.1 — All reinforcement shall be bent cold, unless otherwise permitted by the engineer.

7.3.2 — Reinforcement partially embedded in concrete shall not be field bent, except as shown on the design drawings or permitted by the engineer.

7.4 — Surface conditions of reinforcement

7.4.1—At the time concrete is placed, reinforcement shall be free from mud, oil, or other nonmetallic coatings that decrease bond. Epoxy-coating of steel reinforcement in accordance with standards referenced in 3.5.3.7 and 3.5.3.8 shall be permitted.

7.4.2 — Except for prestressing steel, steel reinforcement with rust, mill scale, or a combination of both shall be considered satisfactory, provided the minimum dimensions (including height of deformations) and weight of a hand-wire-brushed test specimen comply with applicable ASTM specifications referenced in 3.5.

COMMENTARY

are in most cases the same as those required in the ASTM bend tests for wire material (ASTM A 82 and A 496).

R7.3 — Bending

R7.3.1 — The engineer may be the design engineer or architect or the engineer or architect employed by the owner to perform inspection. For unusual bends with inside diameters less than ASTM bend test requirements, special fabrication may be required.

R7.3.2 — Construction conditions may make it necessary to bend bars that have been embedded in concrete. Such field bending should not be done without authorization of the engineer. The engineer should determine whether the bars should be bent cold or if heating should be used. Bends should be gradual and should be straightened as required.

Tests[7.2,7.3] have shown that A 615 Grade 40 and Grade 60 reinforcing bars can be cold bent and straightened up to 90 deg at or near the minimum diameter specified in 7.2. If cracking or breakage is encountered, heating to a maximum temperature of 1500 F may avoid this condition for the remainder of the bars. Bars that fracture during bending or straightening can be spliced outside the bend region.

Heating should be performed in a manner that will avoid damage to the concrete. If the bend area is within approximately 6 in. of the concrete, some protective insulation may need to be applied. Heating of the bar should be controlled by temperature-indicating crayons or other suitable means. The heated bars should not be artificially cooled (with water or forced air) until after cooling to at least 600 F.

R7.4 — Surface conditions of reinforcement

Specific limits on rust are based on tests,[7.4] plus a review of earlier tests and recommendations. Reference 7.4 provides guidance with regard to the effects of rust and mill scale on bond characteristics of deformed reinforcing bars. Research has shown that a normal amount of rust increases bond. Normal rough handling generally removes rust that is loose enough to injure the bond between the concrete and reinforcement.

CODE

7.4.3 — Prestressing steel shall be clean and free of oil, dirt, scale, pitting and excessive rust. A light coating of rust shall be permitted.

7.5 — Placing reinforcement

7.5.1 — Reinforcement, including tendons, and post-tensioning ducts shall be accurately placed and adequately supported before concrete is placed, and shall be secured against displacement within tolerances permitted in 7.5.2.

7.5.2 — Unless otherwise specified by the registered design professional, reinforcement, including tendons, and post-tensioning ducts shall be placed within the tolerances in 7.5.2.1 and 7.5.2.2.

7.5.2.1 — Tolerance for d and minimum concrete cover in flexural members, walls, and compression members shall be as follows:

	Tolerance on d	Tolerance on minimum concrete cover
$d \leq 8$ in.	±3/8 in.	−3/8 in.
$d > 8$ in.	±1/2 in.	−1/2 in.

except that tolerance for the clear distance to formed soffits shall be minus 1/4 in. and tolerance for cover shall not exceed minus 1/3 the minimum concrete cover required in the design drawings and specifications.

7.5.2.2 — Tolerance for longitudinal location of bends and ends of reinforcement shall be ± 2 in., except the tolerance shall be ± 1/2 in. at the discontinuous ends of brackets and corbels, and ± 1 in. at the

COMMENTARY

R7.4.3 — Guidance for evaluating the degree of rusting on strand is given in Reference 7.5.

R7.5 — Placing reinforcement

R7.5.1 — Reinforcement, including tendons, and post-tensioning ducts should be adequately supported in the forms to prevent displacement by concrete placement or workers. Beam stirrups should be supported on the bottom form of the beam by positive supports such as continuous longitudinal beam bolsters. If only the longitudinal beam bottom reinforcement is supported, construction traffic can dislodge the stirrups as well as any prestressing tendons tied to the stirrups.

R7.5.2 — Generally accepted practice, as reflected in **"Standard Specifications for Tolerances for Concrete Construction and Materials,"** reported by ACI Committee 117.[7.6] has established tolerances on total depth (formwork or finish) and fabrication of truss bent reinforcing bars and closed ties, stirrups, and spirals. The engineer should specify more restrictive tolerances than those permitted by the code when necessary to minimize the accumulation of tolerances resulting in excessive reduction in effective depth or cover.

More restrictive tolerances have been placed on minimum clear distance to formed soffits because of its importance for durability and fire protection, and because bars are usually supported in such a manner that the specified tolerance is practical.

More restrictive tolerances than those required by the code may be desirable for prestressed concrete to achieve camber control within limits acceptable to the designer or owner. In such cases, the engineer should specify the necessary tolerances. Recommendations are given in Reference 7.7.

R7.5.2.1 — The code specifies a tolerance on depth d, an essential component of strength of the member. Because reinforcing steel is placed with respect to edges of members and formwork surfaces, the depth d is not always conveniently measured in the field. Engineers should specify tolerances for bar placement, cover, and member size. See ACI 117.[7.6]

CODE

discontinuous ends of other members. The tolerance for minimum concrete cover of 7.5.2.1 shall also apply at discontinuous ends of members.

7.5.3 — Welded wire reinforcement (with wire size not greater than W5 or D5) used in slabs not exceeding 10 ft in span shall be permitted to be curved from a point near the top of slab over the support to a point near the bottom of slab at midspan, provided such reinforcement is either continuous over, or securely anchored at support.

7.5.4 — Welding of crossing bars shall not be permitted for assembly of reinforcement unless authorized by the engineer.

7.6 — Spacing limits for reinforcement

7.6.1 — The minimum clear spacing between parallel bars in a layer shall be d_b, but not less than 1 in. See also 3.3.2.

7.6.2 — Where parallel reinforcement is placed in two or more layers, bars in the upper layers shall be placed directly above bars in the bottom layer with clear distance between layers not less than 1 in.

7.6.3 — In spirally reinforced or tied reinforced compression members, clear distance between longitudinal bars shall be not less than $1.5d_b$ nor less than 1-1/2 in. See also 3.3.2.

7.6.4 — Clear distance limitation between bars shall apply also to the clear distance between a contact lap splice and adjacent splices or bars.

7.6.5 — In walls and slabs other than concrete joist construction, primary flexural reinforcement shall not be spaced farther apart than three times the wall or slab thickness, nor farther apart than 18 in.

7.6.6 — Bundled bars

7.6.6.1 — Groups of parallel reinforcing bars bundled in contact to act as a unit shall be limited to four in any one bundle.

7.6.6.2 — Bundled bars shall be enclosed within stirrups or ties.

7.6.6.3 — Bars larger than No. 11 shall not be bundled in beams.

7.6.6.4 — Individual bars within a bundle terminated within the span of flexural members shall terminate at different points with at least $40d_b$ stagger.

COMMENTARY

R7.5.4 — "Tack" welding (welding crossing bars) can seriously weaken a bar at the point welded by creating a metallurgical notch effect. This operation can be performed safely only when the material welded and welding operations are under continuous competent control, as in the manufacture of welded wire reinforcement.

R7.6 — Spacing limits for reinforcement

Although the minimum bar spacings are unchanged in this code, the development lengths given in Chapter 12 became a function of the bar spacings since the 1989 code. As a result, it may be desirable to use larger than minimum bar spacings in some cases. The minimum limits were originally established to permit concrete to flow readily into spaces between bars and between bars and forms without honeycomb, and to ensure against concentration of bars on a line that may cause shear or shrinkage cracking. Use of nominal bar diameter to define minimum spacing permits a uniform criterion for all bar sizes.

R7.6.6 — Bundled bars

Bond research[7.8] showed that bar cutoffs within bundles should be staggered. Bundled bars should be tied, wired, or otherwise fastened together to ensure remaining in position whether vertical or horizontal.

A limitation that bars larger than No. 11 not be bundled in beams or girders is a practical limit for application to building size members. (The **Standard Specifications for Highway Bridges**[7.9] permits two-bar bundles for No. 14 and No. 18 bars in bridge girders.) Conformance to the crack control requirements of 10.6 will effectively preclude bundling of bars larger than No. 11 as tensile reinforcement. The code phrasing "bundled in contact to act as a unit," is intended to

CODE

7.6.6.5 — Where spacing limitations and minimum concrete cover are based on bar diameter, d_b, a unit of bundled bars shall be treated as a single bar of a diameter derived from the equivalent total area.

7.6.7 — Tendons and ducts

7.6.7.1 — Center-to-center spacing of pretensioning tendons at each end of a member shall be not less than $4d_b$ for strands, or $5d_b$ for wire, except that if specified compressive strength of concrete at time of initial prestress, f'_{ci}, is 4000 psi or more, minimum center-to-center spacing of strands shall be 1-3/4 in. for strands of 1/2 in. nominal diameter or smaller and 2 in. for strands of 0.6 in. nominal diameter. See also 3.3.2. Closer vertical spacing and bundling of tendons shall be permitted in the middle portion of a span.

7.6.7.2 — Bundling of post-tensioning ducts shall be permitted if shown that concrete can be satisfactorily placed and if provision is made to prevent the prestressing steel, when tensioned, from breaking through the duct.

7.7 — Concrete protection for reinforcement

7.7.1 — Cast-in-place concrete (nonprestressed)

The following minimum concrete cover shall be provided for reinforcement, but shall not be less than required by 7.7.5 and 7.7.7:

	Minimum cover, in.
(a) Concrete cast against and permanently exposed to earth	3
(b) Concrete exposed to earth or weather:	
No. 6 through No. 18 bars	2
No. 5 bar, W31 or D31 wire, and smaller	1-1/2
(c) Concrete not exposed to weather or in contact with ground:	
Slabs, walls, joists:	
No. 14 and No. 18 bars	1-1/2
No. 11 bar and smaller	3/4

COMMENTARY

preclude bundling more than two bars in the same plane. Typical bundle shapes are triangular, square, or L-shaped patterns for three- or four-bar bundles. As a practical caution, bundles more than one bar deep in the plane of bending should not be hooked or bent as a unit. Where end hooks are required, it is preferable to stagger the individual bar hooks within a bundle.

R7.6.7 — Tendons and ducts

R7.6.7.1 — The allowed decreased spacing in this section for transfer strengths of 4000 psi or greater is based on Reference 7.10, 7.11.

R7.6.7.2 — When ducts for prestressing steel in a beam are arranged closely together vertically, provision should be made to prevent the prestressing steel from breaking through the duct when tensioned. Horizontal disposition of ducts should allow proper placement of concrete. A clear spacing of one and one-third times the size of the coarse aggregate, but not less than 1 in., has proven satisfactory. Where concentration of tendons or ducts tends to create a weakened plane in the concrete cover, reinforcement should be provided to control cracking.

R7.7 — Concrete protection for reinforcement

Concrete cover as protection of reinforcement against weather and other effects is measured from the concrete surface to the outermost surface of the steel to which the cover requirement applies. Where minimum cover is prescribed for a class of structural member, it is measured to the outer edge of stirrups, ties, or spirals if transverse reinforcement encloses main bars; to the outermost layer of bars if more than one layer is used without stirrups or ties; or to the metal end fitting or duct on post-tensioned prestressing steel.

The condition "concrete surfaces exposed to earth or weather" refers to direct exposure to moisture changes and not just to temperature changes. Slab or thin shell soffits are not usually considered directly exposed unless subject to alternate wetting and drying, including that due to condensation conditions or direct leakage from exposed top surface, run off, or similar effects.

Alternative methods of protecting the reinforcement from weather may be provided if they are equivalent to the additional concrete cover required by the code. When approved by the building official under the provisions of 1.4, reinforcement with alternative protection from the weather may

CODE

Beams, columns:
　Primary reinforcement, ties,
　stirrups, spirals 1-1/2

Shells, folded plate members:
　No. 6 bar and larger.............................. 3/4
　No. 5 bar, W31 or D31 wire,
　and smaller 1/2

7.7.2 — Cast-in-place concrete (prestressed)

The following minimum concrete cover shall be provided for prestressed and nonprestressed reinforcement, ducts, and end fittings, but shall not be less than required by 7.7.5, 7.7.5.1, and 7.7.7:

Minimum cover, in.

(a) Concrete cast against and
permanently exposed to earth 3

(b) Concrete exposed to earth or weather:

　Wall panels, slabs, joists................. 1
　Other members......................... 1-1/2

(c) Concrete not exposed to
weather or in contact with ground:

　Slabs, walls, joists 3/4
　Beams, columns:
　　Primary reinforcement........................... 1-1/2
　　Ties, stirrups, spirals 1

　Shells, folded plate members:
　　No. 5 bar, W31 or D31 wire,
　　and smaller 3/8
　　Other reinforcement .... d_b but not
　　less than 3/4

7.7.3 — Precast concrete (manufactured under plant control conditions)

The following minimum concrete cover shall be provided for prestressed and nonprestressed reinforcement, ducts, and end fittings, but shall not be less than required by 7.7.5, 7.7.5.1, and 7.7.7:

Minimum cover, in.

(a) Concrete exposed to earth or weather:

Wall panels:
　No. 14 and No. 18 bars, prestressing
　tendons larger than 1-1/2 in.
　diameter............................... 1-1/2
　No. 11 bar and smaller, prestressing
　tendons 1-1/2 in. diameter and smaller,
　W31 and D31 wire and smaller............. 3/4

COMMENTARY

have concrete cover not less than the cover required for reinforcement not exposed to weather.

The development length given in Chapter 12 is now a function of the bar cover. As a result, it may be desirable to use larger than minimum cover in some cases.

R7.7.3 — Precast concrete (manufactured under plant control conditions)

The lesser cover thicknesses for precast construction reflect the greater convenience of control for proportioning, placing, and curing inherent in precasting. The term "manufactured under plant control conditions" does not specifically imply that precast members should be manufactured in a plant. Structural elements precast at the job site will also qualify under this section if the control of form dimensions, placing of reinforcement, quality control of concrete, and curing procedure are equal to that normally expected in a plant.

Concrete cover to pretensioned strand as described in this section is intended to provide minimum protection against weather and other effects. Such cover may not be sufficient to transfer or develop the stress in the strand, and it may be necessary to increase the cover accordingly.

CODE

Other members:
    No. 14 and No. 18 bars, prestressing
    tendons larger than 1-1/2 in. diameter 2
    No. 6 through No. 11 bars, prestressing
    tendons larger than 5/8 in. diameter
    through 1-1/2 in. diameter 1-1/2
    No. 5 bar and smaller, prestressing
    tendons 5/8 in. diameter and smaller,
    W31 and D31 wire, and smaller 1-1/4

(b) Concrete not exposed to weather or in contact with ground:

Slabs, walls, joists:
    No. 14 and No. 18 bars, prestressing
    tendons larger than 1-1/2 in.
    diameter 1-1/4
    Prestressing tendons 1-1/2 in.
    diameter and smaller 3/4
    No. 11 bar and smaller,
    W31 or D31 wire, and smaller 5/8

Beams, columns:
    Primary reinforcement .... d_b but not less
    than 5/8 and need not
    exceed 1-1/2
    Ties, stirrups, spirals 3/8

Shells, folded plate members:
    Prestressing tendons 3/4
    No. 6 bar and larger 5/8
    No. 5 bar and smaller,
    W31 or D31 wire, and smaller 3/8

7.7.4 — Bundled bars

For bundled bars, minimum concrete cover shall be equal to the equivalent diameter of the bundle, but need not be greater than 2 in.; except for concrete cast against and permanently exposed to earth, where minimum cover shall be 3 in.

7.7.5 — Corrosive environments

In corrosive environments or other severe exposure conditions, amount of concrete protection shall be suitably increased, and denseness and nonporosity of protecting concrete shall be considered, or other protection shall be provided.

COMMENTARY

R7.7.5 — Corrosive environments

Where concrete will be exposed to external sources of chlorides in service, such as deicing salts, brackish water, seawater, or spray from these sources, concrete should be proportioned to satisfy the special exposure requirements of Chapter 4. These include minimum air content, maximum water-cementitious materials ratio, minimum strength for normal weight and lightweight concrete, maximum chloride ion in concrete, and cement type. Additionally, for corrosion protection, a minimum concrete cover for reinforcement of 2 in. for walls and slabs and 2-1/2 in. for other members is recommended. For precast concrete manufactured under plant control conditions, a minimum cover of 1-1/2 and 2 in., respectively, is recommended.

CODE

7.7.5.1 — For prestressed concrete members exposed to corrosive environments or other severe exposure conditions, and which are classified as Class T or C in 18.3.3, minimum cover to the prestressed reinforcement shall be increased 50 percent. This requirement shall be permitted to be waived if the precompressed tensile zone is not in tension under sustained loads.

7.7.6 — Future extensions

Exposed reinforcement, inserts, and plates intended for bonding with future extensions shall be protected from corrosion.

7.7.7 — Fire protection

When the general building code (of which this code forms a part) requires a thickness of cover for fire protection greater than the minimum concrete cover specified in 7.7, such greater thicknesses shall be used.

7.8 — Special reinforcement details for columns

7.8.1 — Offset bars

Offset bent longitudinal bars shall conform to the following:

7.8.1.1 — Slope of inclined portion of an offset bar with axis of column shall not exceed 1 in 6.

7.8.1.2 — Portions of bar above and below an offset shall be parallel to axis of column.

7.8.1.3 — Horizontal support at offset bends shall be provided by lateral ties, spirals, or parts of the floor construction. Horizontal support provided shall be designed to resist 1-1/2 times the horizontal component of the computed force in the inclined portion of an offset bar. Lateral ties or spirals, if used, shall be placed not more than 6 in. from points of bend.

7.8.1.4 — Offset bars shall be bent before placement in the forms. See 7.3.

7.8.1.5 — Where a column face is offset 3 in. or greater, longitudinal bars shall not be offset bent. Separate dowels, lap spliced with the longitudinal bars adjacent to the offset column faces, shall be provided. Lap splices shall conform to 12.17.

COMMENTARY

R7.7.5.1 — Corrosive environments are defined in sections 4.4.2 and R4.4.2. Additional information on corrosion in parking structures is given in ACI 362.1R-97,[7.12] "Design of Parking Structures," pp. 21-26.

R7.8 — Special reinforcement details for columns

CODE

7.8.2 — Steel cores

Load transfer in structural steel cores of composite compression members shall be provided by the following:

7.8.2.1 — Ends of structural steel cores shall be accurately finished to bear at end bearing splices, with positive provision for alignment of one core above the other in concentric contact.

7.8.2.2 — At end bearing splices, bearing shall be considered effective to transfer not more than 50 percent of the total compressive stress in the steel core.

7.8.2.3 — Transfer of stress between column base and footing shall be designed in accordance with 15.8.

7.8.2.4 — Base of structural steel section shall be designed to transfer the total load from the entire composite member to the footing; or, the base shall be designed to transfer the load from the steel core only, provided ample concrete section is available for transfer of the portion of the total load carried by the reinforced concrete section to the footing by compression in the concrete and by reinforcement.

7.9 — Connections

7.9.1 — At connections of principal framing elements (such as beams and columns), enclosure shall be provided for splices of continuing reinforcement and for anchorage of reinforcement terminating in such connections.

7.9.2 — Enclosure at connections shall consist of external concrete or internal closed ties, spirals, or stirrups.

7.10 — Lateral reinforcement for compression members

7.10.1 — Lateral reinforcement for compression members shall conform to the provisions of 7.10.4 and 7.10.5 and, where shear or torsion reinforcement is required, shall also conform to provisions of Chapter 11.

7.10.2 — Lateral reinforcement requirements for composite compression members shall conform to 10.16. Lateral reinforcement requirements for tendons shall conform to 18.11.

7.10.3 — It shall be permitted to waive the lateral reinforcement requirements of 7.10, 10.16, and 18.11 where tests and structural analysis show adequate strength and feasibility of construction.

COMMENTARY

R7.8.2 — Steel cores

The 50 percent limit on transfer of compressive load by end bearing on ends of structural steel cores is intended to provide some tensile capacity at such splices (up to 50 percent), since the remainder of the total compressive stress in the steel core are to be transmitted by dowels, splice plates, welds, etc. This provision should ensure that splices in composite compression members meet essentially the same tensile capacity as required for conventionally reinforced concrete compression members.

R7.9 — Connections

Confinement is essential at connections to ensure that the flexural capacity of the members can be developed without deterioration of the joint under repeated loadings.[7.13,7.14]

R7.10 — Lateral reinforcement for compression members

R7.10.3 — Precast columns with cover less than 1-1/2 in., prestressed columns without longitudinal bars, columns smaller than minimum dimensions prescribed in earlier code editions, columns of concrete with small size coarse aggregate, wall-like columns, and other special cases may require special designs for lateral reinforcement. Wire, W4, D4, or larger, may be used for ties or spirals. If such special

CODE

7.10.4 — Spirals

Spiral reinforcement for compression members shall conform to 10.9.3 and to the following:

7.10.4.1 — Spirals shall consist of evenly spaced continuous bar or wire of such size and so assembled to permit handling and placing without distortion from designed dimensions.

7.10.4.2 — For cast-in-place construction, size of spirals shall not be less than 3/8 in. diameter.

7.10.4.3 — Clear spacing between spirals shall not exceed 3 in., nor be less than 1 in. See also 3.3.2.

7.10.4.4 — Anchorage of spiral reinforcement shall be provided by 1-1/2 extra turns of spiral bar or wire at each end of a spiral unit.

7.10.4.5 — Spiral reinforcement shall be spliced, if needed, by any one of the following methods:

(a) Lap splices not less than the larger of 12 in. and the length indicated in one of (1) through (5) below:
    (1) deformed uncoated bar or wire .... $48d_b$
    (2) plain uncoated bar or wire .... $72d_b$
    (3) epoxy-coated deformed bar or wire ... $72d_b$
    (4) plain uncoated bar or wire with a standard stirrup or tie hook in accordance with 7.1.3 at ends of lapped spiral reinforcement. The hooks shall be embedded within the core confined by the spiral reinforcement .... $48d_b$
    (5) epoxy-coated deformed bar or wire with a standard stirrup or tie hook in accordance with 7.1.3 at ends of lapped spiral reinforcement. The hooks shall be embedded within the core confined by the spiral reinforcement .... $48d_b$

(b) Full mechanical or welded splices in accordance with 12.14.3.

7.10.4.6 — Spirals shall extend from top of footing or slab in any story to level of lowest horizontal reinforcement in members supported above.

COMMENTARY

columns are considered as spiral columns for load strength in design, the volumetric reinforcement ratio for the spiral, ρ_s, is to conform to 10.9.3.

R7.10.4 — Spirals

For practical considerations in cast-in-place construction, the minimum diameter of spiral reinforcement is 3/8 in. (3/8 in. round, No. 3 bar, or equivalent deformed or plain wire). This is the smallest size that can be used in a column with 1-1/2 in. or more cover and having concrete compressive strengths of 3000 psi or more if the minimum clear spacing for placing concrete is to be maintained.

Standard spiral sizes are 3/8, 1/2, and 5/8 in. diameter for hot rolled or cold drawn material, plain or deformed.

The code allows spirals to be terminated at the level of lowest horizontal reinforcement framing into the column. However, if one or more sides of the column are not enclosed by beams or brackets, ties are required from the termination of the spiral to the bottom of the slab or drop panel. If beams or brackets enclose all sides of the column but are of different depths, the ties should extend from the spiral to the level of the horizontal reinforcement of the shallowest beam or bracket framing into the column. These additional ties are to enclose the longitudinal column reinforcement and the portion of bars from beams bent into the column for anchorage. See also 7.9.

Spirals should be held firmly in place, at proper pitch and alignment, to prevent displacement during concrete placement. The code has traditionally required spacers to hold the fabricated spiral cage in place but was changed in 1989 to allow alternate methods of installation. When spacers are used, the following may be used for guidance: For spiral bar or wire smaller than 5/8 in. diameter, a minimum of two spacers should be used for spirals less than 20 in. in diameter, three spacers for spirals 20 to 30 in. in diameter, and four spacers for spirals greater than 30 in. in diameter. For spiral bar or wire 5/8 in. diameter or larger, a minimum of three spacers should be used for spirals 24 in. or less in diameter, and four spacers for spirals greater than 24 in. in diameter. The project specifications or subcontract agreements should be clearly written to cover the supply of spacers or field tying of the spiral reinforcement. In the 1999 code, splice requirements were modified for epoxy-coated and plain spirals and to allow mechanical splices.

CODE

7.10.4.7 — Where beams or brackets do not frame into all sides of a column, ties shall extend above termination of spiral to bottom of slab or drop panel.

7.10.4.8 — In columns with capitals, spirals shall extend to a level at which the diameter or width of capital is two times that of the column.

7.10.4.9 — Spirals shall be held firmly in place and true to line.

7.10.5 — Ties

Tie reinforcement for compression members shall conform to the following:

7.10.5.1 — All nonprestressed bars shall be enclosed by lateral ties, at least No. 3 in size for longitudinal bars No. 10 or smaller, and at least No. 4 in size for No. 11, No. 14, No. 18, and bundled longitudinal bars. Deformed wire or welded wire reinforcement of equivalent area shall be permitted.

7.10.5.2 — Vertical spacing of ties shall not exceed 16 longitudinal bar diameters, 48 tie bar or wire diameters, or least dimension of the compression member.

7.10.5.3 — Ties shall be arranged such that every corner and alternate longitudinal bar shall have lateral support provided by the corner of a tie with an included angle of not more than 135 deg and no bar shall be farther than 6 in. clear on each side along the tie from such a laterally supported bar. Where longitudinal bars are located around the perimeter of a circle, a complete circular tie shall be permitted.

7.10.5.4 — Ties shall be located vertically not more than one-half a tie spacing above the top of footing or slab in any story, and shall be spaced as provided herein to not more than one-half a tie spacing below the lowest horizontal reinforcement in slab or drop panel above.

COMMENTARY

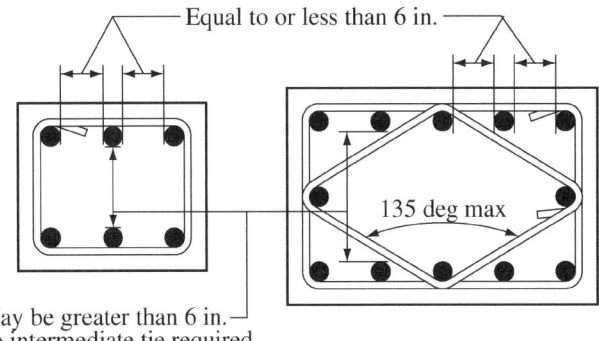

Fig. R7.10.5—Sketch to clarify measurements between laterally supported column bars

R7.10.5 — Ties

All longitudinal bars in compression should be enclosed within lateral ties. Where longitudinal bars are arranged in a circular pattern, only one circular tie per specified spacing is required. This requirement can be satisfied by a continuous circular tie (helix) at larger pitch than required for spirals under 10.9.3, the maximum pitch being equal to the required tie spacing (see also 7.10.4.3).

The 1956 code required "lateral support equivalent to that provided by a 90-deg corner of a tie," for every vertical bar. Tie requirements were liberalized in 1963 by increasing the permissible included angle from 90 to 135 deg and exempting bars that are located within 6 in. clear on each side along the tie from adequately tied bars (see Fig. R7.10.5). Limited tests[7.15] on full-size, axially-loaded, tied columns containing full-length bars (without splices) showed no appreciable difference between ultimate strengths of columns with full tie requirements and no ties at all.

Since spliced bars and bundled bars were not included in the tests of Reference 7.15, is prudent to provide a set of ties at each end of lap spliced bars, above and below end-bearing splices, and at minimum spacings immediately below sloping regions of offset bent bars.

Standard tie hooks are intended for use with deformed bars only, and should be staggered where possible. See also 7.9.

Continuously wound bars or wires can be used as ties provided their pitch and area are at least equivalent to the area and spacing of separate ties. Anchorage at the end of a continuously wound bar or wire should be by a standard hook as for separate bars or by one additional turn of the tie pattern. A circular continuously wound bar or wire is considered a spiral if it conforms to 7.10.4, otherwise it is considered a tie.

CODE

7.10.5.5 — Where beams or brackets frame from four directions into a column, termination of ties not more than 3 in. below lowest reinforcement in shallowest of such beams or brackets shall be permitted.

7.10.5.6 — Where anchor bolts are placed in the top of columns or pedestals, the bolts shall be enclosed by lateral reinforcement that also surrounds at least four vertical bars of the column or pedestal. The lateral reinforcement shall be distributed within 5 in. of the top of the column or pedestal, and shall consist of at least two No. 4 or three No. 3 bars.

7.11 — Lateral reinforcement for flexural members

7.11.1 — Compression reinforcement in beams shall be enclosed by ties or stirrups satisfying the size and spacing limitations in 7.10.5 or by welded wire reinforcement of equivalent area. Such ties or stirrups shall be provided throughout the distance where compression reinforcement is required.

7.11.2 — Lateral reinforcement for flexural framing members subject to stress reversals or to torsion at supports shall consist of closed ties, closed stirrups, or spirals extending around the flexural reinforcement.

7.11.3 — Closed ties or stirrups shall be formed in one piece by overlapping standard stirrup or tie end hooks around a longitudinal bar, or formed in one or two pieces lap spliced with a Class B splice (lap of $1.3\ell_d$) or anchored in accordance with 12.13.

7.12 — Shrinkage and temperature reinforcement

7.12.1 — Reinforcement for shrinkage and temperature stresses normal to flexural reinforcement shall be provided in structural slabs where the flexural reinforcement extends in one direction only.

7.12.1.1 — Shrinkage and temperature reinforcement shall be provided in accordance with either 7.12.2 or 7.12.3.

7.12.1.2 — Where shrinkage and temperature movements are significantly restrained, the requirements of 8.2.4 and 9.2.3 shall be considered.

COMMENTARY

R7.10.5.5 — With the 1983 code, the wording of this section was modified to clarify that ties may be terminated only when elements frame into all four sides of square and rectangular columns; for round or polygonal columns, such elements frame into the column from four directions.

R7.10.5.6 — Provisions for confinement of anchor bolts that are placed in the top of columns or pedestals were added in the 2002 code. Confinement improves load transfer from the anchor bolts to the column or pier for situations where the concrete cracks in the vicinity of the bolts. Such cracking can occur due to unanticipated forces caused by temperature, restrained shrinkage, and similar effects.

R7.11 — Lateral reinforcement for flexural members

R7.11.1 — Compression reinforcement in beams and girders should be enclosed to prevent buckling; similar requirements for such enclosure have remained essentially unchanged through several editions of the code, except for minor clarification.

R7.12 — Shrinkage and temperature reinforcement

R7.12.1 — Shrinkage and temperature reinforcement is required at right angles to the principal reinforcement to minimize cracking and to tie the structure together to ensure its acting as assumed in the design. The provisions of this section are intended for structural slabs only; they are not intended for soil-supported slabs on grade.

R7.12.1.2 — The area of shrinkage and temperature reinforcement required by 7.12 has been satisfactory where shrinkage and temperature movements are permitted to occur. For cases where structural walls or large columns provide significant restraint to shrinkage and temperature movements, it may be necessary to increase the amount of reinforcement normal to the flexural reinforcement in 7.12.1.2 (see Reference 7.16). Top and bottom reinforcement are both effective in controlling cracks. Control strips during

CODE

7.12.2 — Deformed reinforcement conforming to 3.5.3 used for shrinkage and temperature reinforcement shall be provided in accordance with the following:

7.12.2.1 — Area of shrinkage and temperature reinforcement shall provide at least the following ratios of reinforcement area to gross concrete area, but not less than 0.0014:

(a) Slabs where Grade 40 or 50
deformed bars are used 0.0020

(b) Slabs where Grade 60
deformed bars or welded wire
reinforcement are used.................................. 0.0018

(c) Slabs where reinforcement
with yield stress exceeding 60,000
psi measured at a yield strain
of 0.35 percent is used .... $\dfrac{0.0018 \times 60{,}000}{f_y}$

7.12.2.2 — Shrinkage and temperature reinforcement shall be spaced not farther apart than five times the slab thickness, nor farther apart than 18 in.

7.12.2.3 — At all sections where required, reinforcement to resist shrinkage and temperature stresses shall develop f_y in tension in accordance with Chapter 12.

7.12.3 — Prestressing steel conforming to 3.5.5 used for shrinkage and temperature reinforcement shall be provided in accordance with the following:

7.12.3.1 — Tendons shall be proportioned to provide a minimum average compressive stress of 100 psi on gross concrete area using effective prestress, after losses, in accordance with 18.6.

7.12.3.2 — Spacing of tendons shall not exceed 6 ft.

7.12.3.3 — When spacing of tendons exceeds 54 in., additional bonded shrinkage and temperature reinforcement conforming to 7.12.2 shall be provided between the tendons at slab edges extending from the slab edge for a distance equal to the tendon spacing.

COMMENTARY

the construction period, which permit initial shrinkage to occur without causing an increase in stresses, are also effective in reducing cracks caused by restraint.

R7.12.2 — The amounts specified for deformed bars and welded wire reinforcement are empirical but have been used satisfactorily for many years. Splices and end anchorages of shrinkage and temperature reinforcement are to be designed for the full specified yield strength in accordance with 12.1, 12.15, 12.18, and 12.19.

R7.12.3 — Prestressed reinforcement requirements have been selected to provide an effective force on the slab approximately equal to the yield strength force for nonprestressed shrinkage and temperature reinforcement. This amount of prestressing, 100 psi on the gross concrete area, has been successfully used on a large number of projects. When the spacing of tendons used for shrinkage and temperature reinforcement exceeds 54 in., additional bonded reinforcement is required at slab edges where the prestressing forces are applied in order to adequately reinforce the area between the slab edge and the point where compressive stresses behind individual anchorages have spread sufficiently such that the slab is uniformly in compression. Application of the provisions of 7.12.3 to monolithic cast-in-place post-tensioned beam and slab construction is illustrated in Fig. R7.12.3.

Tendons used for shrinkage and temperature reinforcement should be positioned vertically in the slab as close as practicable to the center of the slab. In cases where the shrinkage and temperature tendons are used for supporting the principal tendons, variations from the slab centroid are permissible; however, the resultant of the shrinkage and temperature tendons should not fall outside the kern area of the slab.

CODE

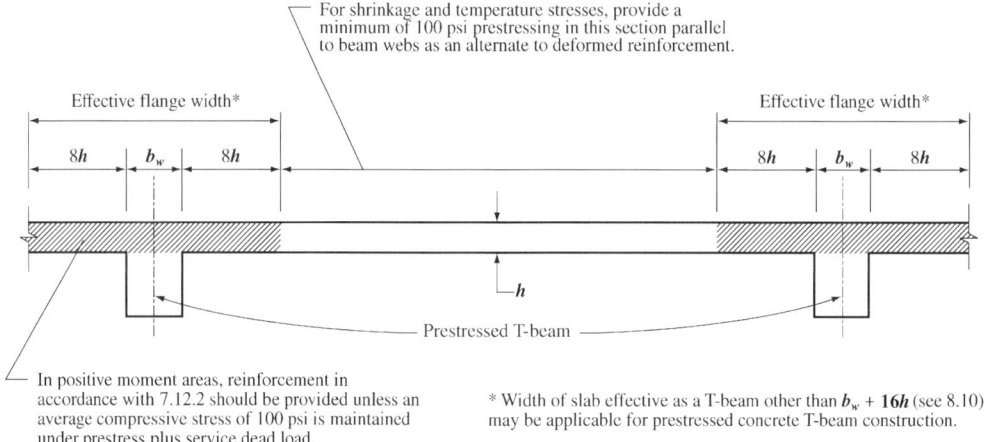

Fig. R7.12.3—Prestressing used for shrinkage and temperature

7.13 — Requirements for structural integrity

7.13.1 — In the detailing of reinforcement and connections, members of a structure shall be effectively tied together to improve integrity of the overall structure.

7.13.2 — For cast-in-place construction, the following shall constitute minimum requirements:

7.13.2.1 — In joist construction, at least one bottom bar shall be continuous or shall be spliced with a Class A tension splice or a mechanical or welded splice satisfying 12.14.3 and at noncontinuous supports shall be terminated with a standard hook.

7.13.2.2 — Beams along the perimeter of the structure shall have continuous reinforcement consisting of:

(a) at least one-sixth of the tension reinforcement required for negative moment at the support, but not less than two bars; and

(b) at least one-quarter of the tension reinforcement required for positive moment at midspan, but not less than two bars.

7.13.2.3 — Where splices are needed to provide the required continuity, the top reinforcement shall be spliced at or near midspan and bottom reinforcement

COMMENTARY

The designer should evaluate the effects of slab shortening to ensure proper action. In most cases, the low level of prestressing recommended should not cause difficulties in a properly detailed structure. Special attention may be required where thermal effects become significant.

R7.13 — Requirements for structural integrity

Experience has shown that the overall integrity of a structure can be substantially enhanced by minor changes in detailing of reinforcement. It is the intent of this section of the code to improve the redundancy and ductility in structures so that in the event of damage to a major supporting element or an abnormal loading event, the resulting damage may be confined to a relatively small area and the structure will have a better chance to maintain overall stability.

R7.13.2 — With damage to a support, top reinforcement that is continuous over the support, but not confined by stirrups, will tend to tear out of the concrete and will not provide the catenary action needed to bridge the damaged support. By making a portion of the bottom reinforcement continuous, catenary action can be provided.

Requiring continuous top and bottom reinforcement in perimeter or spandrel beams provides a continuous tie around the structure. It is not the intent to require a tensile tie of continuous reinforcement of constant size around the entire perimeter of a structure, but simply to require that one half of the top flexural reinforcement required to extend past the point of inflection by 12.12.3 be further extended and spliced at or near midspan. Similarly, the bottom reinforcement required to extend into the support by 12.11.1 should be made continuous or spliced with bottom reinforcement from the adjacent span. If the depth of a continuous beam changes at a support, the bottom reinforcement in the deeper member should be terminated with a standard hook and bottom reinforcement in the shallower member should be extended into and fully developed in the deeper member.

CODE

shall be spliced at or near the support. Splices shall be Class A tension splices or mechanical or welded splices satisfying 12.14.3. The continuous reinforcement required in 7.13.2.2(a) and 7.13.2.2(b) shall be enclosed by the corners of U-stirrups having not less than 135-deg hooks around the continuous top bars, or by one-piece closed stirrups with not less than 135-deg hooks around one of the continuous top bars. Stirrups need not be extended through any joints.

7.13.2.4 — In other than perimeter beams, when stirrups as defined in 7.13.2.3 are not provided, at least one-quarter of the positive moment reinforcement required at midspan, but not less than two bars, shall be continuous or shall be spliced over or near the support with a Class A tension splice or a mechanical or welded splice satisfying 12.14.3, and at noncontinuous supports shall be terminated with a standard hook.

7.13.2.5 — For two-way slab construction, see 13.3.8.5.

7.13.3 — For precast concrete construction, tension ties shall be provided in the transverse, longitudinal, and vertical directions and around the perimeter of the structure to effectively tie elements together. The provisions of 16.5 shall apply.

7.13.4 — For lift-slab construction, see 13.3.8.6 and 18.12.6.

COMMENTARY

In the 2002 code, provisions were added to permit the use of mechanical or welded splices for splicing reinforcement, and the detailing requirements for the longitudinal reinforcement and stirrups in beams were revised. Section 7.13.2 was revised in 2002 to require U-stirrups with not less than 135-deg hooks around the continuous bars, or one-piece closed stirrups, because a crosstie forming the top of a two-piece closed stirrup is ineffective in preventing the top continuous bars from tearing out of the top of the beam.

R7.13.3 — The code requires tension ties for precast concrete buildings of all heights. Details should provide connections to resist applied loads. Connection details that rely solely on friction caused by gravity forces are not permitted.

Connection details should be arranged so as to minimize the potential for cracking due to restrained creep, shrinkage and temperature movements. For information on connections and detailing requirements, see Reference 7.17.

Reference 7.18 recommends minimum tie requirements for precast concrete bearing wall buildings.

Notes

ACI 347-04

Guide to Formwork for Concrete
An ACI Standard

Reported by ACI Committee 347

Pericles C. Stivaros	Kevin L. Wheeler
Chair	Secretary

Rodney D. Adams	Samuel A. Greenberg	H. S. Lew
Kenneth L. Berndt	R. Kirk Gregory	Donald M. Marks
Randolph H. Bordner	Awad S. Hanna	Robert G. McCracken
Ramon J. Cook	G. P. Jum Horst	William R. Phillips
James N. Cornell	Mary K. Hurd	Douglas J. Schoonover
Jack L. David	David W. Johnston	W. Thomas Scott
William A. Dortch	Roger S. Johnston	Aviad Shapira
Jeffrey C. Erson	Dov Kaminetzky	Rolf A. Spahr
N. John Gardner	Harry B. Lancelot	

Objectives of safety, quality, and economy are given priority in these guidelines for formwork. A section on contract documents explains the kind and amount of specification guidance the engineer/architect should provide for the contractor. The remainder of the report advises the formwork engineer/contractor on the best ways to meet the specification requirements safely and economically. Separate chapters deal with design, construction, and materials for formwork. Considerations peculiar to architectural concrete are also outlined in a separate chapter. Other sections are devoted to formwork for bridges, shells, mass concrete, and underground work. The concluding chapter on formwork for special methods of construction includes slipforming, preplaced-aggregate concrete, tremie concrete, precast, and prestressed concrete.

Keywords: anchors; architectural concrete; coatings; concrete; construction; falsework; form ties; forms; formwork; foundations; quality control; reshoring; shoring; slipform construction; specifications; tolerances.

ACI Committee Reports, Guides, Standard Practices, and Commentaries are intended for guidance in planning, designing, executing, and inspecting construction. This document is intended for the use of individuals who are competent to evaluate the significance and limitations of its content and recommendations and who will accept responsibility for the application of the material it contains. The American Concrete Institute disclaims any and all responsibility for the stated principles. The Institute shall not be liable for any loss or damage arising therefrom.

Reference to this document shall not be made in contract documents. If items found in this document are desired by the Architect/Engineer to be a part of the contract documents, they shall be restated in mandatory language for incorporation by the Architect/Engineer.

It is the responsibility of the user of this document to establish health and safety practices appropriate to the specific circumstances involved with its use. ACI does not make any representations with regard to health and safety issues and the use of this document. The user must determine the applicability of all regulatory limitations before applying the document and must comply with all applicable laws and regulations, including but not limited to, United States Occupational Safety and Health Administration (OSHA) health and safety standards.

CONTENTS
Preface, p. 347-2

Chapter 1—Introduction, p. 347-2
1.1—Scope
1.2—Definitions
1.3—Achieving economy in formwork
1.4—Contract documents

Chapter 2—Design, p. 347-5
2.1—General
2.2—Loads
2.3—Unit stresses
2.4—Safety factors for accessories
2.5—Shores
2.6—Bracing and lacing
2.7—Foundations for formwork
2.8—Settlement

Chapter 3—Construction, p. 347-9
3.1—Safety precautions
3.2—Construction practices and workmanship
3.3—Tolerances
3.4—Irregularities in formed surfaces
3.5—Shoring and centering
3.6—Inspection and adjustment of formwork
3.7—Removal of forms and supports
3.8—Shoring and reshoring of multistory structures

Chapter 4—Materials, p. 347-16
4.1—General

ACI 347-04 supersedes ACI 347R-03 and became effective October 15, 2004.
Copyright © 2004, American Concrete Institute.
All rights reserved including rights of reproduction and use in any form or by any means, including the making of copies by any photo process, or by electronic or mechanical device, printed, written, or oral, or recording for sound or visual reproduction or for use in any knowledge or retrieval system or device, unless permission in writing is obtained from the copyright proprietors.

4.2—Properties of materials
4.3—Accessories
4.4—Form coatings and release agents

Chapter 5—Architectural concrete, p. 347-18
5.1—Introduction
5.2—Role of the architect
5.3—Materials and accessories
5.4—Design
5.5—Construction
5.6—Form removal

Chapter 6—Special structures, p. 347-22
6.1—Discussion
6.2—Bridges and viaducts, including high piers
6.3—Structures designed for composite action
6.4—Folded plates, thin shells, and long-span roof structures
6.5—Mass concrete structures
6.6—Underground structures

Chapter 7—Special methods of construction, p. 347-26
7.1—Recommendations
7.2—Preplaced-aggregate concrete
7.3—Slipforms
7.4—Permanent forms
7.5—Forms for prestressed concrete construction
7.6—Forms for site precasting
7.7—Use of precast concrete for forms
7.8—Forms for concrete placed under water

Chapter 8—References, p. 347-30
8.1—Referenced standards and reports
8.2—Cited references

PREFACE
Before the formation of ACI Committee 347 (formerly ACI Committee 622) in 1955, there had been an increase in the use of reinforced concrete for longer span structures, multistoried structures, and increased story heights.

The need for a formwork standard and increased knowledge concerning the behavior of formwork was evident from the rising number of failures, sometimes resulting in the loss of life. The first report by the committee, based on a survey of current practices in the United States and Canada, was published in the ACI JOURNAL in June 1957.[1.1] The second committee report was published in the ACI JOURNAL in August 1958.[1.2] This second report was an in-depth review of test reports and design formulas for determining lateral pressure on vertical formwork. The major result of this study and report was the development of a basic formula establishing form pressures to be used in the design of vertical formwork.

The first standard was ACI 347-63. Subsequent revisions were ACI 347-68 and ACI 347-78. Two subsequent revisions, ACI 347R-88 and ACI 347R-94, were committee reports because of changes in the ACI policy on the style and format of standards. ACI 347-01 returned the guide to the standardization process.

A major contribution of the committee has been the sponsorship and review of *Formwork for Concrete*[1.3] by M. K. Hurd, first published in 1963 and currently in its sixth edition. Now comprising more than 490 pages, this is the most comprehensive and widely used document on this subject. (The Japan National Council on Concrete has published a Japanese translation.)

The paired values stated in inch-pound and SI units are usually not exact equivalents. Therefore, each system is to be used independently of the other. Combining values from the two systems may result in nonconformance with this document.

CHAPTER 1—INTRODUCTION
1.1—Scope
This guide covers:
- A listing of information to be included in the contract documents;
- Design criteria for horizontal and vertical forces on formwork;
- Design considerations, including safety factors, to be used in determining the capacities of formwork accessories;
- Preparation of formwork drawings;
- Construction and use of formwork, including safety considerations;
- Materials for formwork;
- Formwork for special structures;
- Formwork for special methods of construction; and
- Qualification of personnel for inspection and testing.

This guide is based on the premise that layout, design, and construction of formwork should be the responsibility of the formwork engineer/contractor. This is believed to be fundamental to the achievement of safety and economy of formwork for concrete.

1.2—Definitions
The following definitions will be used in this guide. Many of the terms can also be found in ACI 116R:

backshores—shores placed snugly under a concrete slab or structural member after the original formwork and shores have been removed from a small area at a time, without allowing the slab or member to deflect; thus, the slab or other member does not yet support its own weight or existing construction loads from above.

bugholes—surface air voids: small regular or irregular cavities, usually less than 0.6 in. (15 mm) in diameter, resulting from entrapment of air bubbles in the surface of formed concrete during placement and consolidation. Also called blowholes.

centering—specialized temporary support used in the construction of arches, shells, and space structures where the entire temporary support is lowered (struck or decentered) as a unit to avoid introduction of injurious stresses in any part of the structure.

climbing form—a form that is raised vertically for succeeding lifts of concrete in a given structure.

diagonal bracing—supplementary formwork members designed to resist lateral loads.

engineer/architect—the engineer, architect, engineering firm, architectural firm, or other agency issuing project plans and specifications for the permanent structure, administering the work under contract documents, or both.

flying forms—large prefabricated, mechanically handled sections of formwork designed for multiple reuse; frequently including supporting truss, beam, or shoring assemblies completely unitized. Note: Historically, the term has been applied to floor forming systems.

form—a temporary structure or mold for the support of concrete while it is setting and gaining sufficient strength to be self-supporting.

formwork—total system of support for freshly placed concrete, including the mold or sheathing that contacts the concrete and all supporting members, hardware, and necessary bracing.

formwork engineer/contractor—engineer of the formwork system, contractor, or competent person in charge of designated aspects of formwork design and formwork operations.

ganged forms—large assemblies used for forming vertical surfaces; also called gang forms.

horizontal lacing—horizontal bracing members attached to shores to reduce their unsupported length, thereby increasing load capacity and stability.

preshores—added shores placed snugly under selected panels of a deck-forming system before any primary (original) shores are removed. Preshores and the panels they support remain in place until the remainder of the complete bay has been stripped and backshored, a small area at a time.

reshores—shores placed snugly under a stripped concrete slab or other structural member after the original forms and shores have been removed from a large area, requiring the new slab or structural member to deflect and support its own weight and existing construction loads to be applied before installation of the reshores.

scaffold—a temporary elevated platform (supported or suspended) and its supporting structure used for supporting workers, tools, and materials; adjustable metal scaffolding can be used for shoring in concrete work, provided its structure has the necessary load-carrying capacity and structural integrity.

shores—vertical or inclined support members designed to carry the weight of the formwork, concrete, and construction loads above.

slipform—a form that is pulled or raised as concrete is placed; may move in a horizontal direction to lay concrete for concrete paving or on slopes and inverts of canals, tunnels, and siphons; or may move vertically to form walls, bins, or silos.

1.3—Achieving economy in formwork

The engineer/architect can help overall economy in the structure by planning so that formwork costs are minimized. The cost of formwork in the United States can be as much as 60% of the total cost of the completed concrete structure in place and sometimes greater. This investment requires careful thought and planning by the engineer/architect when designing and specifying the structure and by the formwork engineer/contractor when designing and constructing the formwork.

Formwork drawings, prepared by the formwork engineer/contractor, can identify potential problems and should give project site employees a clear picture of what is required and how to achieve it. The following guidelines show how the engineer/architect can plan the structure so that formwork economy may best be achieved:

- To simplify and permit maximum reuse of formwork, the dimensions of footings, columns, and beams should be of standard material multiples, and the number of sizes should be minimized;
- When interior columns are the same width as or smaller than the girders they support, the column form becomes a simple rectangular or square box without boxouts, and the slab form does not have to be cut out at each corner of the column;
- When all beams are made one depth (beams framing into beams as well as beams framing into columns), the supporting structures for the beam forms can be carried on a level platform supported on shores;
- Considering available sizes of dressed lumber, plywood, and other ready-made formwork components and keeping beam and joist sizes constant will reduce labor time;
- The design of the structure should be based on the use of one standard depth wherever possible when commercially available forming systems, such as one- or two-way joist systems, are used;
- The structural design should be prepared simultaneously with the architectural design so that dimensions can be better coordinated. Room sizes can vary a few inches to accommodate the structural design;
- The engineer/architect should consider architectural features, depressions, and openings for mechanical or electrical work when detailing the structural system, with the aim of achieving economy. Variations in the structural system caused by such items should be shown on the structural plans. Wherever possible, depressions in the tops of slabs should be made without a corresponding break in elevations of the soffits of slabs, beams, or joists;
- Embedments for attachment to or penetration through the concrete structure should be designed to minimize random penetration of the formed surface; and
- Avoid locating columns or walls, even for a few floors, where they would interfere with the use of large formwork shoring units in otherwise clear bays.

1.4—Contract documents

The contract documents should set forth the tolerances required in the finished structure but should not attempt to specify the manner in which the formwork engineer/contractor designs and builds the formwork to achieve the required tolerances.

The layout and design of the formwork and its construction should be the responsibility of the formwork engineer/contractor. This approach gives the necessary freedom to use skill, knowledge, and innovation to safely construct an economical structure. By reviewing the formwork drawings, the engineer/architect can understand how the formwork engineer/contractor has interpreted the contract documents. Some local areas have legal requirements defining the specific responsibilities of the engineer/architect in formwork design, review, or approval.

1.4.1 *Individual specifications*—The specification writer is encouraged to refer to this guide as a source of recommendations that can be written into the proper language for contract documents.

The specification for formwork will affect the overall economy and quality of the finished work; therefore, it should be tailored for each particular job, clearly indicate what is expected of the contractor, and ensure economy and safety.

A well-written formwork specification tends to equalize bids for the work. Unnecessarily exacting requirements can make bidders question the specification as a whole and make it difficult for them to understand exactly what is expected. They can be overly cautious and overbid or misinterpret requirements and underbid.

A well-written formwork specification is of value not only to the owner and the contractor, but also to the field representative of the engineer/architect, approving agency, and the subcontractors of other trades. Some requirements can be written to allow discretion of the contractor where quality of finished concrete work would not be impaired by the use of alternative materials and methods.

Consideration of the applicable general requirements suggested herein will not be sufficient to make a complete specification. Requirements should be added for actual materials, finishes, and other items peculiar to and necessary for the individual structure. The engineer/architect can exclude, call special attention to, strengthen, or make more lenient any general requirement to best fit the needs of the particular project. Helpful and detailed information is given in *Formwork for Concrete*.[1.3]

1.4.2 *Formwork materials and accessories*—If the particular design or desired finish requires special attention, the engineer/architect can specify in the contract documents the formwork materials and such other features necessary to attain the objectives. If the engineer/architect does not call for specific materials or accessories, the formwork engineer/contractor can choose any materials that meet the contract requirements.

When structural design is based on the use of commercially available form units in standard sizes, such as one-way or two-way joist systems, plans should be drawn to make use of available shapes and sizes. Some latitude should be permitted for connections of form units to other framing or centering to reflect the tolerances and normal installation practices of the form type anticipated.

1.4.3 *Finish of exposed concrete*—Finish requirements for concrete surfaces should be described in measurable terms as precisely as practicable. Refer to Section 3.4 and Chapter 5.

1.4.4 *Design, inspection, review, and approval of formwork*—Although the safety of formwork is the responsibility of the contractor, the engineer/architect or approving agency may, under certain circumstances, decide to review and approve the formwork, including drawings and calculations. If so, the engineer/architect should call for such review or approval in the contract documents.

Approval might be required for unusually complicated structures, structures whose designs were based on a particular method of construction, structures in which the forms impart a desired architectural finish, certain post-tensioned structures, folded plates, thin shells, or long-span roof structures.

The following items should be clarified in the contract documents:
- Who will design the formwork;
- Who will inspect the specific feature of formwork and when will the inspection be performed; and
- What reviews, approvals, or both will be required—
  a. For formwork drawings;
  b. For the formwork before concreting and during concreting; and
  c. Who will give such reviews, approvals, or both.

1.4.5 *Contract documents*—The contract documents should include all information about the structure necessary for the formwork engineer/contractor to design the formwork and prepare formwork drawings, such as:
- Number, location, and details of all construction joints, contraction joints, and expansion joints that will be required for the particular job or parts of it;
- Sequence of concrete placement, if critical;
- Tolerances for concrete construction;
- The live load and superimposed dead load for which the structure is designed and any live-load reduction used. This is a requirement of ACI 318;
- Intermediate supports under stay-in-place forms, such as metal deck used for forms and permanent forms of other materials; supports, bracing, or both, required by the structural engineer's design for composite action; and any other special supports;
- The location and order of erection and removal of shores for composite construction;
- Special provisions essential for formwork for special construction methods and for special structures such as shells and folded plates. The basic geometry of such structures, as well as their required camber, should be given in sufficient detail to permit the formwork engineer/contractor to build the forms;
- Special requirements for post-tensioned concrete members. The effect of load transfer and associated movements during tensioning of post-tensioned members can be critical, and the contractor should be advised of any special provisions that should be made in the formwork for this condition;
- Amount of required camber for slabs or other structural members to compensate for deflection of the structure. Measurements of camber attained should be made at the soffit level after initial set and before removal of formwork supports;
- Where chamfers are required or prohibited on beam soffits or column corners;
- Requirements for inserts, waterstops, built-in frames for openings and holes through concrete; similar requirements where the work of other trades will be

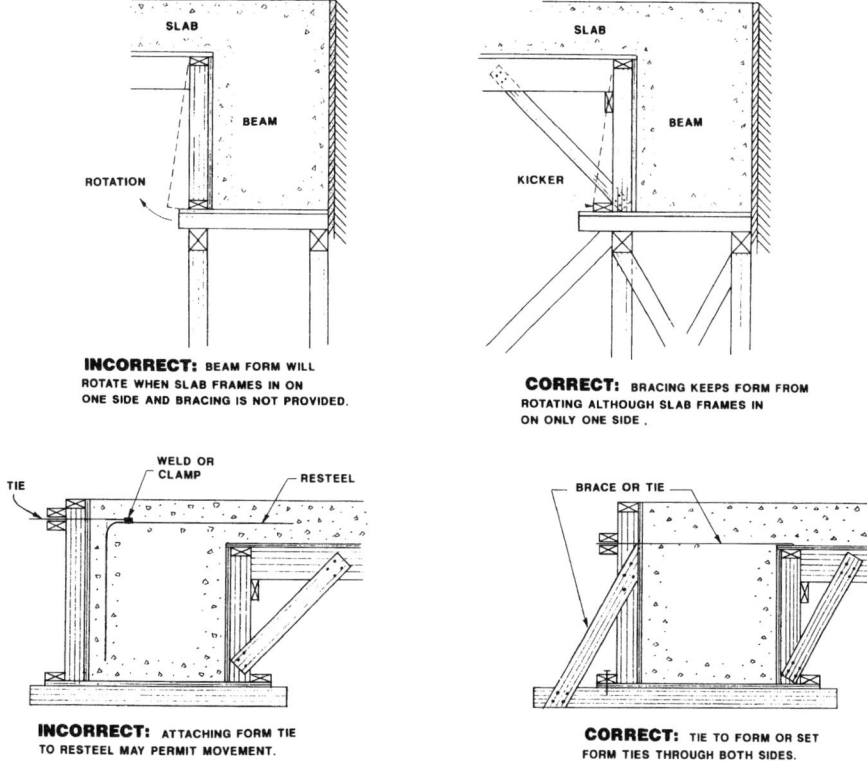

Fig. 2.1—Prevention of rotation is important where the slab frames into the beam form on only one side.

attached to, supported by, or passed through formwork;
- Where architectural features, embedded items, or the work of other trades could change the location of structural members, such as joists in one- or two-way joist systems, such changes or conditions should be coordinated by the engineer/architect; and
- Locations of and details for architectural concrete. When architectural details are to be cast into structural concrete, they should be so indicated or referenced on the structural plans because they can play a key role in the structural design of the form.

CHAPTER 2—DESIGN
2.1—General

2.1.1 *Planning*—All formwork should be well planned before construction begins. The amount of planning required will depend on the size, complexity, and importance (considering reuses) of the form. Formwork should be designed for strength and serviceability. System stability and member buckling should be investigated in all cases.

2.1.2 *Design methods*—Formwork is made of many different materials, and the commonly used design practices for each material are to be followed (refer to Chapter 4). For example, wood forms are designed by working-stress methods recommended by the American Forest and Paper Association. When the concrete structure becomes a part of the formwork support system, as in many multistory buildings, it is important for the formwork engineer/contractor to recognize that the concrete structure has been designed by the strength method. Accordingly, in communication of the loads, it should be clear whether they are service loads or factored loads.

Throughout this guide, the terms design, design load, and design capacity are used to refer to design of the formwork. Where reference is made to design load for the permanent structure, structural design load, structural dead load, or some similar term is used to refer to unfactored service loads on the structure.[*]

2.1.3 *Basic objectives*—Formwork should be designed so that concrete slabs, walls, and other members will have the correct dimensions, shape, alignment, elevation, and position within established tolerances. Formwork should also be designed so that it will safely support all vertical and lateral loads that might be applied until such loads can be supported by the concrete structure. Vertical and lateral loads should be carried to the ground by the formwork system or by the in-place construction that has adequate strength for that purpose. Responsibility for the design of the formwork rests with the contractor or the formwork engineer hired by the contractor to design and be responsible for the formwork.

2.1.4 *Design deficiencies*—Some common design deficiencies that can lead to failure are:
- Lack of allowance in design for loadings such as wind, power buggies, placing equipment, and temporary material storage;
- Inadequate reshoring;
- Overstressed reshoring;
- Inadequate provisions to prevent rotation of beam forms where the slabs frame into them on only one side (Fig. 2.1);
- Insufficient anchorage against uplift due to battered form faces;

---
[*]As defined by ACI 318, both dead load and live load are unfactored loads.

- Insufficient allowance for eccentric loading due to placement sequences;
- Failure to investigate bearing stresses in members in contact with shores or struts;
- Failure to provide proper lateral bracing or lacing of shoring;
- Failure to investigate the slenderness ratio of compression members;
- Inadequate provisions to tie corners of intersecting cantilevered forms together;
- Failure to account for loads imposed on form hardware anchorages during closure of form panel gaps when aligning formwork; and
- Failure to account for elastic shortening during post-tensioning.

2.1.5 *Formwork drawings and calculations*—Before constructing forms, the formwork engineer/contractor may be required to submit detailed drawings, design calculations, or both of proposed formwork for review and approval by the engineer/architect or approving agency. If such drawings are not approved by the engineer/architect or approving agency, the formwork engineer/contractor should make such changes as may be required before the start of construction of the formwork.

The review, approval, or both of the formwork drawings does not relieve the contractor of the responsibility for adequately constructing and maintaining the forms so that they will function properly. If reviewed by persons other than those employed by the contractor, the review or approval indicates that no exception is taken by the reviewer to the assumed design loadings in combination with design stresses shown; the proposed construction methods; the placement rates, equipment, and sequences; the proposed form materials; and the overall scheme of formwork. All major design values and loading conditions should be shown on formwork drawings. These include assumed values of live load; the compressive strength of concrete for formwork removal and for application of construction loads; rate of placement, minimum temperature, height, and drop of concrete; weight of moving equipment that can be operated on formwork; foundation pressure; design stresses; camber diagrams; and other pertinent information, if applicable.

In addition to specifying types of materials, sizes, lengths, and connection details, formwork drawings should provide for applicable details, such as:
- Procedures, sequence, and criteria for removal of forms, shores, and reshores;
- Design allowance for construction loads on new slabs when such allowance will affect the development of shoring, reshoring schemes, or both (refer to Sections 2.5 and 3.8 for shoring and reshoring of multistory structures);
- Anchors, form ties, shores, lateral bracing, and horizontal lacing;
- Field adjustment of forms;
- Waterstops, keyways, and inserts;
- Working scaffolds and runways;
- Weepholes or vibrator holes, where required;
- Screeds and grade strips;
- Location of external vibrator mountings;
- Crush plates or wrecking plates where stripping can damage concrete;
- Removal of spreaders or temporary blocking;
- Cleanout holes and inspection openings;
- Construction joints, contraction joints, and expansion joints in accordance with contract documents (also refer to ACI 301);
- Sequence of concrete placement and minimum elapsed time between adjacent placements;
- Chamfer strips or grade strips for exposed corners and construction joints;
- Camber;
- Mudsills or other foundation provisions for formwork;
- Special provisions, such as safety, fire, drainage, and protection from ice and debris at water crossings;
- Formwork coatings;
- Notes to formwork erector showing size and location of conduits and pipes projecting through formwork; and
- Temporary openings or attachments for climbing crane or other material handling equipment.

2.2—Loads

2.2.1 *Vertical loads*—Vertical loads consist of dead and live loads. The weight of formwork plus the weight of the reinforcement and freshly placed concrete is dead load. The live load includes the weight of the workers, equipment, material storage, runways, and impact.

Vertical loads assumed for shoring and reshoring design for multistory construction should include all loads transmitted from the floors above as dictated by the proposed construction schedule. Refer to Section 2.5.

The formwork should be designed for a live load of not less than 50 lb/ft^2 (2.4 kPa) of horizontal projection. When motorized carts are used, the live load should not be less than 75 lb/ft^2 (3.6 kPa).

The design load for combined dead and live loads should not be less than 100 lb/ft^2 (4.8 kPa) or 125 lb/ft^2 (6.0 kPa) if motorized carts are used.

2.2.2 *Lateral pressure of concrete*—Unless the conditions of Section 2.2.2.1 or 2.2.2.2 are met, formwork should be designed for the lateral pressure of the newly placed concrete given in Eq. (2.1a) or (2.1b). Minimum values given for other pressure formulas do not apply to Eq. (2.1a) and (2.1b).

$$p = wh \text{ (lb/ft}^2\text{)} \qquad (2.1a)$$

$$p = \rho g h \text{ (kPa)} \qquad (2.1b)$$

where
- p = lateral pressure, lb/ft^2 (kPa);
- w = unit weight of concrete, lb/ft^3;
- ρ = density of concrete, kg/m^3;
- g = gravitational constant, 9.81 N/kg; and
- h = depth of fluid or plastic concrete from top of placement to point of consideration in form, ft (m).

The set characteristics of a mixture should be understood, and using the rate of placement, the level of fluid concrete

Table 2.1—Unit weight coefficient C_w

Inch-pound version		SI version	
Unit weight of concrete	C_w	Density of concrete	C_w
Less than 140 lb/ft^3	$C_w = 0.5[1 + (w/145 \text{ lb/ft}^3)]$ but not less than 0.80	Less than 2240 kg/m^3	$C_w = 0.5[1 + (w/2320 \text{ kg/m}^3)]$ but not less than 0.80
140 to 150 lb/ft^3	1.0	2240 to 2400 kN/m^3	1.0
More than 150 lb/ft^3	$C_w = w/145 \text{ lb/ft}^3$	More than 2400 kg/m^3	$C_w = w/2320 \text{ kg/m}^3$

can be determined. For columns or other forms that can be filled rapidly before stiffening of the concrete takes place, h should be taken as the full height of the form or the distance between horizontal construction joints when more than one placement of concrete is to be made. When working with mixtures using newly introduced admixtures that increase set time or increase slump characteristics, such as self-consolidating concrete, Eq. (2.1a) [(2.1b)] should be used until the effect on formwork pressure is understood by measurement.

2.2.2.1 *Inch-pound version*—For concrete having a slump of 7 in. or less and placed with normal internal vibration to a depth of 4 ft or less, formwork can be designed for a lateral pressure as follows, where p_{max} = maximum lateral pressure, lb/ft^2; R = rate of placement, ft/h; T = temperature of concrete during placing, °F; C_w = unit weight coefficient per Table 2.1; and C_c = chemistry coefficient per Table 2.2.[2.1]

For columns:

$$p_{max} = C_w C_c [150 + 9000R/T] \qquad (2.2)$$

with a minimum of $600C_w$ lb/ft^2, but in no case greater than wh.

For walls with a rate of placement of less than 7 ft/h and a placement height not exceeding 14 ft

$$p_{max} = C_w C_c [150 + 9000R/T] \qquad (2.3)$$

with a minimum of $600C_w$ lb/ft^2, but in no case greater than wh.

For walls with a placement rate less than 7 ft/h where placement height exceeds 14 ft, and for all walls with a placement rate of 7 to 15 ft/h

$$p_{max} = C_w C_c [150 + 43{,}400/T + 2800R/T] \qquad (2.4)$$

with a minimum of $600C_w$ lb/ft^2, but in no case greater than wh.

2.2.2.1 *SI version*—For concrete having a slump of 175 mm or less and placed with normal internal vibration to a depth of 1.2 m or less, formwork can be designed for a lateral pressure as follows, where p_{max} = maximum lateral pressure, kPa; R = rate of placement, m/h; T = temperature of concrete during placing, °C; C_w = unit weight coefficient per Table 2.1; and C_c = chemistry coefficient per Table 2.2.[2.1]

For columns

Table 2.2—Chemistry coefficient C_c

Cement type or blend	C_c
Types I, II, and III without retarders*	1.0
Types I, II, and III with a retarder	1.2
Other types or blends containing less than 70% slag or 40% fly ash without retarders*	1.2
Other types or blends containing less than 70% slag or 40% fly ash with a retarder*	1.4
Blends containing more than 70% slag or 40% fly ash	1.4

*Retarders include any admixture, such as a retarder, retarding water reducer, retarding midrange water-reducing admixture, or high-range water-reducing admixture (super-plasticizer), that delays setting of concrete.

$$p_{max} = C_w C_c \left[7.2 + \frac{785R}{T + 17.8} \right] \qquad (2.2)$$

with a minimum of $30C_w$ kPa, but in no case greater than $\rho g h$.

For walls with a rate of placement of less than 2.1 m/h and a placement height not exceeding 4.2 m

$$p_{max} = C_w C_c \left[7.2 + \frac{785R}{T + 17.8} \right] \qquad (2.3)$$

with a minimum of $30C_w$ kPa, but in no case greater than $\rho g h$.

For walls with a placement rate less than 2.1 m/h where placement height exceeds 4.2 m, and for all walls with a placement rate of 2.1 to 4.5 m/h

$$p_{max} = C_w C_c \left[7.2 + \frac{1156}{T + 17.8} + \frac{244R}{T + 17.8} \right] \qquad (2.4)$$

with a minimum of $30C_w$ kPa, but in no case greater than $\rho g h$.

2.2.2.1.1—The unit weight coefficient C_w is determined from Table 2.1.

2.2.2.1.2—The chemistry coefficient C_c is determined from Table 2.2.

2.2.2.1.3—For the purpose of applying the pressure formulas, columns are defined as vertical elements with no plan dimension exceeding 6.5 ft (2 m). Walls are defined as vertical elements with at least one plan dimension greater than 6.5 ft (2 m).

2.2.2.2—Alternatively, a method based on appropriate experimental data can be used to determine the lateral pressure used for form design (References 2.2 to 2.7).

Table 2.3—Minimum safety factors of formwork accessories*

Accessory	Safety factor	Type of construction
Form tie	2.0	All applications
Form anchor	2.0	Formwork supporting form weight and concrete pressures only
	3.0	Formwork supporting weight of forms, concrete, construction live loads, and impact
Form hangers	2.0	All applications
Anchoring inserts used as form ties	2.0	Precast-concrete panels when used as formwork

*Safety factors are based on the ultimate strength of the accessory when new.

2.2.2.3—If concrete is pumped from the base of the form, the form should be designed for full hydrostatic head of concrete wh plus a minimum allowance of 25% for pump surge pressure. In certain instances, pressures can be as high as the face pressure of the pump piston.

2.2.2.4—Caution is necessary and additional allowance for pressure should be considered when using external vibration or concrete made with shrinkage compensating or expansive cements. Pressures in excess of the equivalent hydrostatic head can occur.

2.2.2.5—For slipform lateral pressures, refer to Section 7.3.2.4.

2.2.3 *Horizontal loads*—Braces and shores should be designed to resist all horizontal loads such as wind, cable tensions, inclined supports, dumping of concrete, and starting and stopping of equipment. Wind loads on enclosures or other wind breaks attached to the formwork should be considered in addition to these loads.

2.2.3.1—For building construction, the assumed value of horizontal load due to wind, dumping of concrete, inclined placement of concrete, and equipment acting in any direction at each floor line should be not less than 100 lb/linear ft (1.5 kN/m) of floor edge or 2% of total dead load on the form distributed as a uniform load per linear foot (meter) of slab edge, whichever is greater.

2.2.3.2—Wall form bracing should be designed to meet the minimum wind load requirements of the local building code or ANSI/SEI/ASCE-7 with adjustment for shorter recurrence interval as provided in SEI/ASCE 37. For wall forms exposed to the elements, the minimum wind design load should be not less than 15 lb/ft^2 (0.72 kPa). Bracing for wall forms should be designed for a horizontal load of at least 100 lb/linear ft (1.5 kN/m) of wall length, applied at the top.

2.2.3.3—Wall forms of unusual height or exposure should be given special consideration.

2.2.4 *Special loads*—The formwork should be designed for any special conditions of construction likely to occur, such as unsymmetrical placement of concrete, impact of machine-delivered concrete, uplift, concentrated loads of reinforcement, form handling loads, and storage of construction materials. Form designers should provide for special loading conditions, such as walls constructed over spans of slabs or beams that exert a different loading pattern before hardening of concrete than that for which the supporting structure is designed.

Imposition of any construction loads on the partially completed structure should not be allowed, except as specified in formwork drawings or with the approval of the engineer/architect. Refer to Section 3.8 for special conditions pertaining to multistory work.

2.2.5 *Post-tensioning loads*—Shores, reshores, and backshores need to be analyzed for both concrete placement loads and for all load transfer that takes place during post-tensioning.

2.3—Unit stresses

Unit stresses for use in the design of formwork, exclusive of accessories, are given in the applicable codes or specifications listed in Chapter 4. When fabricated formwork, shoring, or scaffolding units are used, manufacturer's recommendations for allowable loads can be followed if supported by engineering calculations, test reports of a qualified and recognized testing agency, or successful experience records. For formwork materials that will experience substantial reuse, reduced values should be used. For formwork materials with limited reuse, allowable stresses specified in the appropriate design codes or specifications for temporary structures or for temporary loads on permanent structures can be used. Where there will be a considerable number of formwork reuses or where formwork is fabricated from materials such as steel, aluminum, or magnesium, the formwork should be designed as a permanent structure carrying permanent loads.

2.4—Safety factors for accessories

Table 2.3 shows recommended minimum factors of safety for formwork accessories, such as form ties, form anchors, and form hangers. In selecting these accessories, the formwork designer should be certain that materials furnished for the job meet these minimum ultimate-strength safety requirements.

2.5—Shores

Shores and reshores or backshores (as defined in Section 1.2) should be designed to carry all loads transmitted to them. A rational analysis should be used to determine the number of floors to be shored, reshored, or backshored and to determine the loads transmitted to the floors, shores, and reshores or backshores as a result of the construction sequence.

The analysis should consider, but should not necessarily be limited to:
- Structural design load of the slab or member including live load, partition loads, and other loads for which the engineer of the permanent structure designed the slab. Where the engineer included a reduced live load for the design of certain members and allowances for construction loads, such values should be shown on the structural plans and be taken into consideration when performing this analysis;
- Dead load weight of the concrete and formwork;
- Construction live loads, such as placing crews and equipment or stored materials;

- Design strength of specified concrete;
- Cycle time between the placement of successive floors;
- Strength of concrete at the time it is required to support shoring loads from above;
- The distribution of loads between floors, shores, and reshores or backshores at the time of placing concrete, stripping formwork, and removal of reshoring or backshoring;[1.3, 2.8, 2.9, 2.10]
- Span of slab or structural member between permanent supports;
- Type of formwork systems, that is, span of horizontal formwork components, and individual shore loads; and
- Minimum age of concrete where appropriate.

Commercially available load cells can be placed under selected shores to monitor actual shore loads to guide the shoring and reshoring during construction.[2.11]

Field-constructed butt or lap splices of timber shoring are not recommended unless they are made with fabricated hardware devices of demonstrated strength and stability. If plywood or lumber splices are made for timber shoring, they should be designed to prevent buckling and bending of the shoring.

Before construction, an overall plan for scheduling of shoring and reshoring or backshoring, and calculation of loads transferred to the structure, should be prepared by a qualified and experienced formwork designer. The structure's capacity to carry these loads should be reviewed or approved by the engineer/architect. The plan and responsibility for its execution remain with the contractor.

2.6—Bracing and lacing

The formwork system should be designed to transfer all horizontal loads to the ground or to completed construction in such a manner as to ensure safety at all times. Diagonal bracing should be provided in vertical and horizontal planes where required to resist lateral loads and to prevent instability of individual members. Horizontal lacing can be considered in design to hold in place and increase the buckling strength of individual shores and reshores or backshores. Lacing should be provided in whatever directions are necessary to produce the correct slenderness ratio l/r for the load supported, where l = unsupported length and r = least radius of gyration. The braced system should be anchored to ensure stability of the total system.

2.7—Foundations for formwork

Proper foundations on ground, such as mudsills, spread footings, or pile footings, should be provided. If soil under mudsills is or may become incapable of supporting superimposed loads without appreciable settlement, it should be stabilized or other means of support should be provided. No concrete should be placed on formwork supported on frozen ground.

2.8—Settlement

Formwork should be designed and constructed so that vertical adjustments can be made to compensate for take-up and settlements.

CHAPTER 3—CONSTRUCTION
3.1—Safety precautions

Contractors should follow all state, local, and federal codes, ordinances, and regulations pertaining to forming and shoring. In addition to the very real moral and legal responsibility to maintain safe conditions for workmen and the public, safe construction is, in the final analysis, more economical than any short-term cost savings from cutting corners on safety provisions.

Attention to safety is particularly significant in formwork construction that supports the concrete during its plastic state and until the concrete becomes structurally self-sufficient. Following the design criteria contained in this guide is essential for ensuring safe performance of the forms. All structural members and connections should be carefully planned so that a sound determination of loads may be accurately made and stresses calculated.

In addition to the adequacy of the formwork, special structures, such as multistory buildings, require consideration of the behavior of newly completed beams and slabs that are used to support formwork and other construction loads. It should be kept in mind that the strength of freshly cast slabs or beams is less than that of a mature slab.

Formwork failures can be attributed to substandard materials and equipment, human error, and inadequacy in design. Careful supervision and continuous inspection of formwork during erection, concrete placement, and removal can prevent many accidents.

Construction procedures should be planned in advance to ensure the safety of personnel and the integrity of the finished structure. Some of the safety provisions that should be considered are:
- Erection of safety signs and barricades to keep unauthorized personnel clear of areas in which erection, concrete placing, or stripping is under way;
- Providing experienced form watchers during concrete placement to ensure early recognition of possible form displacement or failure. A supply of extra shores or other material and equipment that might be needed in an emergency should be readily available;
- Provision for adequate illumination of the formwork and work area;
- Inclusion of lifting points in the design and detailing of all forms that will be crane-handled. This is especially important in flying forms or climbing forms. In the case of wall formwork, consideration should be given to an independent work platform bolted to the previous lift;
- Incorporation of scaffolds, working platforms, and guardrails into formwork design and all formwork drawings;
- Incorporation of provisions for anchorage of alternative fall protection devices, such as personal fall arrest systems, safety net systems, and positioning device systems; and
- A program of field safety inspections of formwork.

3.1.1 *Formwork construction deficiencies*—Some common construction deficiencies that can lead to formwork failures are:
- Failure to inspect formwork during and after concrete

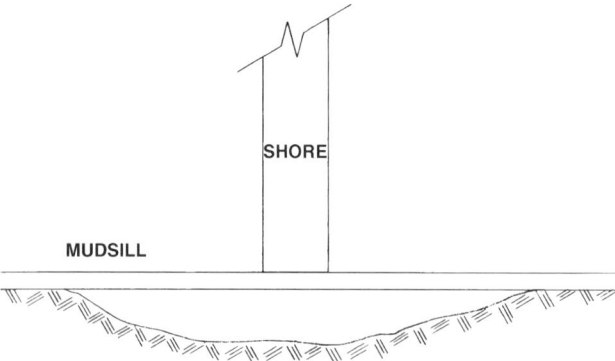

Fig. 3.1—Inadequate bearing under mudsill.

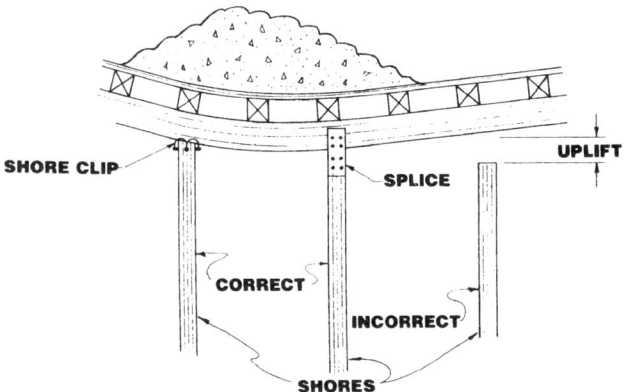

Fig. 3.2—Uplift of formwork. Connection of shores to joists and stringers should hold shores in place when uplift or torsion occurs. Lacing to reduce the shore slenderness ratio can be required in both directions.

placement to detect abnormal deflections or other signs of imminent failure that could be corrected;
- Insufficient nailing, bolting, welding, or fastening;
- Insufficient or improper lateral bracing;
- Failure to comply with manufacturer's recommendations;
- Failure to construct formwork in accordance with the form drawings;
- Lack of proper field inspection by qualified persons to ensure that form design has been properly interpreted by form builders; and
- Use of damaged or inferior lumber having lower strength than needed.

3.1.1.1 *Examples of deficiencies in vertical formwork*—Construction deficiencies sometimes found in vertical formwork include:
- Failure to control rate of placing concrete vertically without regard to design parameters;
- Inadequately tightened or secured form ties or hardware;
- Form damage in excavations resulting from embankment failure;
- Use of external vibrators on forms not designed for their use;
- Deep vibrator penetration of earlier semihardened lifts;
- Improper framing of blockouts;
- Improperly located or constructed pouring pockets;
- Inadequate bulkheads;
- Improperly anchored top forms on a sloping face;
- Failure to provide adequate support for lateral pressures on formwork; and
- Failure to provide adequate bracing resulting in attempts to plumb forms against concrete pressure force.

3.1.1.2 *Examples of deficiencies in horizontal formwork*—Construction deficiencies sometimes found in horizontal forms for elevated structures include:
- Failure to properly regulate the rate and sequence of placing concrete horizontally to avoid unanticipated loadings on the formwork;
- Shoring not plumb, thus inducing lateral loading and reducing vertical load capacity;
- Locking devices on metal shoring not locked, inoperative, or missing. Safety nails missing on adjustable two-piece wood shores;
- Failure to account for vibration from adjacent moving loads or load carriers;
- Inadequately tightened or secured shore hardware or wedges;
- Loosening or premature removal of reshores or back-shores under floors below;
- Premature removal of supports, especially under cantilevered sections;
- Inadequate bearing area or unsuitable soil under mudsills (Fig. 3.1);
- Mudsills placed on frozen ground subject to thawing;
- Connection of shores to joists, stringers, or wales that are inadequate to resist uplift or torsion at joints (refer to Fig. 3.2);
- Failure to consider effects of load transfer that can occur during post-tensioning (refer to Section 3.8.7); and
- Inadequate shoring and bracing of composite construction.

3.2—Construction practices and workmanship

3.2.1—*Fabrication and assembly details*

3.2.1.1—Studs, wales, or shores should be properly spliced.

3.2.1.2—Joints or splices in sheathing, plywood panels, and bracing should be staggered.

3.2.1.3—Shores should be installed plumb and with adequate bearing and bracing.

3.2.1.4—Specified size and capacity of form ties or clamps should be used.

3.2.1.5—All form ties or clamps should be installed and properly tightened as specified. All threads should fully engage the nut or coupling. A double nut may be required to develop the full capacity of the tie.

3.2.1.6—Forms should be sufficiently tight to prevent loss of mortar from the concrete.

3.2.1.7—Access holes may be necessary in wall forms or other high, narrow forms to facilitate concrete placement.

3.2.2—*Joints in the concrete*

3.2.2.1—Contraction joints, expansion joints, control joints, construction joints, and isolation joints should be installed as specified in the contract documents (refer to Fig. 3.3) or as requested by the contractor and approved by the engineer/architect.

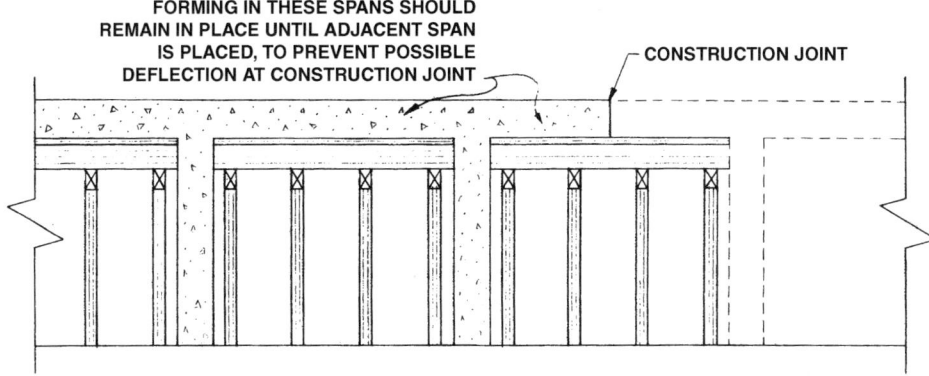

Fig. 3.3—Forming and shoring restraints at construction joints in supported slabs.

3.2.2.2—Bulkheads for joints should preferably be made by splitting the bulkhead along the lines of reinforcement passing through the bulkhead. By doing this, each portion can be positioned and removed separately. When required on the engineer/architect's plans, beveled inserts at control joints should be left undisturbed when forms are stripped and removed only after the concrete has been sufficiently cured. Wood strips inserted for architectural treatment should be kerfed to permit swelling without causing pressure on the concrete.

3.2.3 *Sloping surfaces*—Sloped surfaces steeper than 1.5 horizontal to 1 vertical should be provided with a top form to hold the shape of the concrete during placement, unless it can be demonstrated that the top forms can be omitted.

3.2.4 *Inspection*—The inspection should be performed by a person certified as an ACI Concrete Construction Inspector or a person having equivalent formwork training and knowledge.

3.2.4.1—Forms should be inspected and checked before the reinforcing steel is placed to confirm that the dimensions and the location of the concrete members will conform to the structural plans.

3.2.4.2—Blockouts, inserts, sleeves, anchors, and other embedded items should be properly identified, positioned, and secured.

3.2.4.3—Formwork should be checked for camber when specified in the contract documents or shown on the formwork drawings.

3.2.5—*Cleanup and coatings*

3.2.5.1—Forms should be thoroughly cleaned of all dirt, mortar, and foreign matter and coated with a release agent before each use. Where the bottom of the form is inaccessible from within, access panels should be provided to permit thorough removal of extraneous material before placing concrete. If surface appearance is important, forms should not be reused if damage from previous use would cause impairment to concrete surfaces.

3.2.5.2—Form coatings should be applied before placing of reinforcing steel and should not be used in such quantities as to run onto bars or concrete construction joints.

3.2.6—*Construction operations on the formwork*

3.2.6.1—Building materials, including concrete, should not be dropped or piled on the formwork in such a manner as to damage or overload it.

3.2.6.2—Runways for moving equipment should be provided with struts or legs as required and should be supported directly on the formwork or structural member. They should not bear on or be supported by the reinforcing steel unless special bar supports are provided. The formwork should be suitable for the support of such runways without significant deflections, vibrations, or lateral movements.

3.2.7 *Loading new slabs*—Overloading of new slabs by temporary material stockpiling or by early application of permanent loads should be avoided. Loads, such as aggregate, lumber, reinforcing steel, masonry, or machinery should not be placed on new construction in such a manner as to damage or overload it.

3.3—Tolerances

Tolerance is a permissible variation from lines, grades, or dimensions given in contract documents. Suggested tolerances for concrete structures can be found in ACI 117.

The contractor should set and maintain concrete forms, including any specified camber, to ensure completed work is within the tolerance limits.

3.3.1 *Recommendations for engineer/architect and contractor*—Tolerances should be specified by the engineer/architect so that the contractor will know precisely what is required and can design and maintain the formwork accordingly. Specifying tolerances more exacting than needed can increase construction costs.

Contractors should be required to establish and maintain control points and benchmarks in an undisturbed condition until final completion and acceptance of a project. Both should be adequate for the contractor's use and for reference to establish tolerances. This requirement can become even more important for the contractor's protection when tolerances are not specified or shown. The engineer/architect should specify tolerances or require performance appropriate to the type of construction. Specifying tolerances more stringent than commonly obtained for a specific type of construction should be avoided, as this usually results in disputes among the parties involved. For example, specifying permitted irregularities more stringent than those allowed for a Class C surface (Table 3.1) is incompatible with most concrete one-way joist construction techniques. Where a project involves features sensitive to the cumulative effect of tolerances on

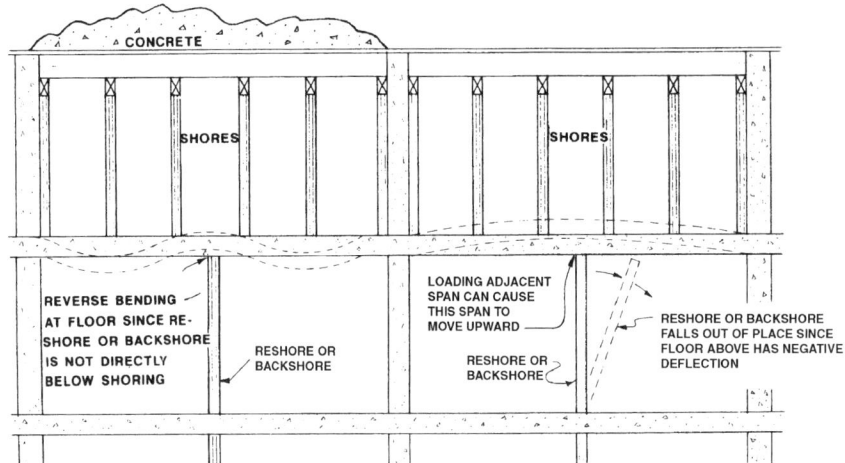

Fig. 3.4—Reshore installation. Improper positioning of shore from floor to floor can create bending stresses for which the slab was not designed.

Table 3.1—Permitted abrupt or gradual irregularities in formed surfaces as measured within a 5 ft (1.5 m) length with a straightedge

Class of surface			
A	B	C	D
1/8 in. (3 mm)	1/4 in. (6 mm)	1/2 in. (13 mm)	1 in. (25 mm)

individual portions, the engineer/architect should anticipate and provide for this effect by setting a cumulative tolerance. Where a particular situation involves several types of generally accepted tolerances on items such as concrete, location of reinforcement, and fabrication of reinforcement, which become mutually incompatible, the engineer/architect should anticipate the difficulty and specify special tolerances or indicate that governs. The project specifications should clearly state that a permitted variation in one part of the construction or in one section of the specifications should not be construed as permitting violation of the more stringent requirements for any other part of the construction or in any other such specification section.

The engineer/architect should be responsible for coordinating the tolerances for concrete work with the tolerance requirements of other trades whose work adjoins the concrete construction. For example, the connection detail for a building's façade should accommodate the tolerance range for the lateral alignment and elevation of the perimeter concrete member.

3.4—Irregularities in formed surfaces

This section provides a way of evaluating surface variations due to forming quality but is not intended for evaluation of surface defects, such as bugholes (blowholes) and honeycomb, attributable to placing and consolidation deficiencies. The latter are more fully explained by ACI 309.2R. Allowable irregularities are designated either abrupt or gradual. Offsets and fins resulting from displaced, mismatched, or misplaced forms, sheathing, or liners, or from defects in forming materials are considered abrupt irregularities. Irregularities resulting from warping and similar uniform variations from planeness or true curvature are considered gradual irregularities.

Gradual irregularities should be checked with a straightedge for plane surfaces or a shaped template for curved or warped surfaces. In measuring irregularities, the straightedge or template can be placed anywhere on the surface in any direction.

Four classes of formed surface are defined in Table 3.1. The engineer/architect should indicate which class is required for the work being specified or indicate other irregularity limits where needed, or the concrete surface tolerances as specified in ACI 301 should be followed.

Class A is suggested for surfaces prominently exposed to public view where appearance is of special importance. Class B is intended for coarse-textured, concrete-formed surfaces intended to receive plaster, stucco, or wainscoting. Class C is a general standard for permanently exposed surfaces where other finishes are not specified. Class D is a minimum-quality requirement for surfaces where roughness is not objectionable, usually applied where surfaces will be permanently concealed. Special limits on irregularities can be needed for surfaces continuously exposed to flowing water, drainage, or exposure. If permitted irregularities are different from those given in Table 3.1, they should be specified by the engineer/architect.

3.5—Shoring and centering

3.5.1 *Shoring*—Shoring should be supported on satisfactory foundations, such as spread footings, mudsills, or piling, as discussed in Section 2.7.

Shoring resting on intermediate slabs or other construction already in place need not be located directly above shores or reshores below, unless the slab thickness and the location of its reinforcement are inadequate to take the reversal of stresses and punching shear. The reversal of stresses results from the reversal of bending moments in the slab over the shore or reshore below as shown in Fig. 3.4. Where the conditions are questionable, the shoring location should be approved by the engineer/architect. If reshores do not align with the shores above, then calculate for reversal stresses. Generally, the dead load stresses are sufficient to compensate for reversal stresses caused by reshores. Reshores should be prevented from falling.

All members should be straight and true without twists or bends. Special attention should be given to beam and slab or one- and two-way joist construction to prevent local overloading when a heavily loaded shore rests on the thin slab.

Multitier shoring, single-post shoring in two or more tiers, is a dangerous practice and is not recommended.

Where a slab load is supported on one side of the beam only (refer to Fig. 2.1), edge beam forms should be carefully planned to prevent tipping of the beam due to unequal loading.

Vertical shores should be erected so that they cannot tilt and should have a firm bearing. Inclined shores should be braced securely against slipping or sliding. The bearing ends of shores should be square. Connections of shore heads to other framing should be adequate to prevent the shores from falling out when reversed bending causes upward deflection of the forms (refer to Fig. 3.2).

3.5.2 *Centering*—When centering is used, lowering is generally accomplished by the use of sand boxes, jacks, or wedges beneath the supporting members. For the special problems associated with the construction of centering for folded plates, thin shells, and long-span roof structures, refer to Section 6.4.

3.5.3 *Shoring for composite action between previously erected steel or concrete framing and cast-in-place concrete*—Refer to Section 6.3.

3.6—Inspection and adjustment of formwork

Helpful information about forms before, during, and after concreting can be found in Reference 1.3 and ACI 311.1R.

3.6.1—*Before concreting*

3.6.1.1—Telltale devices should be installed on shores or forms to detect formwork movements during concreting.

3.6.1.2—Wedges used for final alignment before concrete placement should be secured in position before the final check.

3.6.1.3—Formwork should be anchored to the shores below so that movement of any part of the formwork system will be prevented during concreting.

3.6.1.4—Additional elevation of formwork should be provided to allow for closure of form joints, settlements of mudsills, shrinkage of lumber, and elastic shortening and dead load deflections of form members.

3.6.1.5—Positive means of adjustment (wedges or jacks) should be provided to permit realignment or readjustment of shores if settlement occurs.

3.6.2 *During and after concreting*—During and after concreting, but before initial set of the concrete, the elevations, camber, and plumbness of formwork systems should be checked using telltale devices.

Formwork should be continuously watched so that any corrective measures found necessary can be promptly made. Form watchers should always work under safe conditions and establish in advance a method of communication with placing crews in case of emergency.

3.7—Removal of forms and supports

3.7.1 *Discussion*—Although the contractor is generally responsible for design, construction, and safety of formwork, criteria for removal of forms or shores should be specified by the engineer/architect.

3.7.2—*Recommendations*

3.7.2.1—The engineer/architect should specify the minimum strength of the concrete to be attained before removal of forms or shores. The strength can be determined by tests on job-cured specimens or on in-place concrete. Other concrete tests or procedures (refer to ACI 228.1R) can be used such as the maturity method, rebound numbers, penetration resistance, or pullout tests, but these methods should be correlated to the actual concrete mixture used in the project, periodically verified by job-cured specimens, and approved by the engineer/architect. The engineer/architect should specify who will make the specimens and who will make the tests. Results of such tests, as well as records of weather conditions and other pertinent information, should be recorded by the contractor. Depending on the circumstances, a minimum elapsed time after concrete placement can be established for removal of the formwork.

Determination of the time of form removal should be based on the resulting effect on the concrete.[*] When forms are stripped there should be no excessive deflection or distortion and no evidence of damage to the concrete due to either removal of support or to the stripping operation (Fig. 3.5). When forms are removed before the specified curing is completed, measures should be taken to continue the curing and provide adequate thermal protection for the concrete. Supporting forms and shores should not be removed from beams, floors, and walls until these structural units are strong enough to carry their own weight and any approved superimposed load. In no case should supporting forms and shores be removed from horizontal members before the concrete has achieved the strength specified by the engineer/architect.

As a general rule, the forms for columns and piers can be removed before forms for beams and slabs. Formwork and shoring should be constructed so each can be easily and safely removed without impact or shock and permit the concrete to carry its share of the load gradually and uniformly.

3.7.2.2—The removal of forms, supports, and protective enclosures, and the discontinuance of heating and curing should follow the requirements of the contract documents. When standard beam or cylinder tests are used to determine stripping times, test specimens should be cured under conditions that are not more favorable than the most unfavorable conditions for the concrete the test specimens represent. The curing records can serve as the basis on which the engineer/architect will determine the review or approval of form stripping.

3.7.2.3—Because the minimum stripping time is a function of concrete strength, the preferred method of determining stripping time is using tests of job-cured cylinders or concrete in place. When the contract documents do not

---

[*]Helpful information on strength development of concrete under varying conditions of temperature and with various admixtures can be found in ACI 305R and ACI 306R.

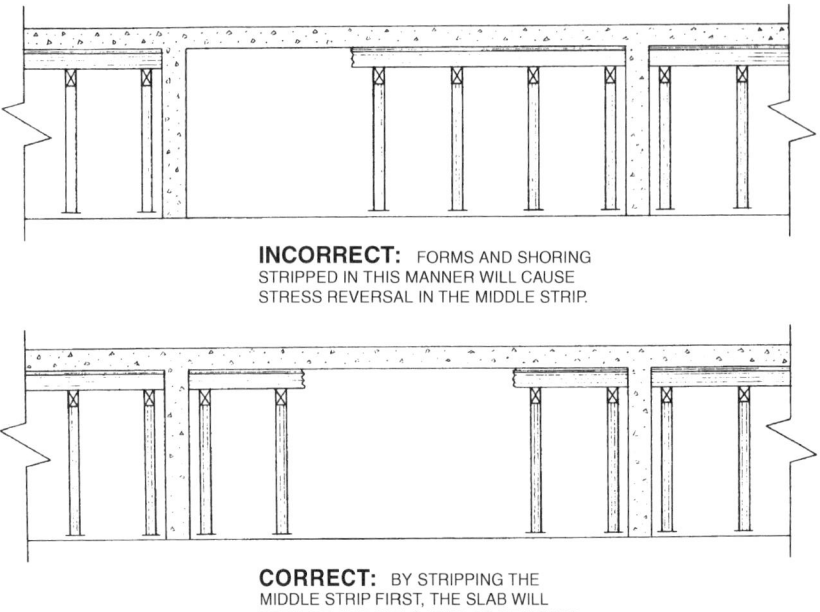

Fig. 3.5—*Stripping sequence for two-way slabs.*

specify the minimum strength required of concrete at the time of stripping, however, the following elapsed times can be used. The times shown represent a cumulative number of days, or hours, not necessarily consecutive, during which the temperature of the air surrounding the concrete is above 50 °F (10 °C). If high early-strength concrete is used, these periods can be reduced as approved by the engineer/architect. Conversely, if ambient temperatures remain below 50 °F (10 °C), or if retarding agents are used, then these periods should be increased at the discretion of the engineer/architect. Shorter stripping times listed for live load to dead load ratios greater than 1.0 are the result of more reserve strength being available for dead load in absence of live load at time of stripping.

Walls[*] .... 12 h
Columns[*] .... 12 h
Sides of beams and girders[*] .... 12 h
Pan joist forms[†]
   30 in. (760 mm) wide or less .... 3 days
   Over 30 in. (760 mm) wide .... 4 days

	Structural live load less than structural dead load	Structural live load more than structural dead load
Arch centers	14 days	7 days
Joist, beam or girder soffits		
Under 10 ft (3 m) clear span between structural supports	7 days[‡]	4 days
10 to 20 ft (3 to 6 m) clear span between structural supports	14 days[‡]	7 days
Over 20 ft (6 m) clear span between structural supports	21 days[‡]	14 days
One-way floor slabs		
Under 10 ft (3 m) clear span between structural supports	4 days[‡]	3 days
10 to 20 ft (3 to 6 m) clear span between structural supports	7 days[‡]	4 days
Over 20 ft (6 m) clear span between structural supports	10 days[‡]	7 days

Two-way slab systems[§]........Removal times are contingent on reshores where required, being placed as soon as practicable after stripping operations are complete but not later than the end of the working day in which stripping occurs. Where reshores are required to implement early stripping while minimizing sag or creep (rather than for distribution of superimposed construction loads as covered in Section 3.8), capacity and spacing of such reshores should be designed by the formwork engineer/contractor and reviewed by the engineer/architect.

Post-tensioned slab system[§]........As soon as full post-tensioning has been applied.

3.8—Shoring and reshoring of multistory structures

3.8.1 *Discussion*—The following definitions apply for purposes of this discussion:

shores—vertical or inclined support members designed to carry the weight of formwork, concrete, and construction loads.

reshores—shores placed snugly under a stripped concrete slab or structural member after the original forms and shores have been removed from a large area. This requires the new

---

[*]Where such forms also support formwork for slab or beam soffits, the removal times of the latter should govern.

[†]Of the type that can be removed without disturbing forming or shoring.

[‡]Where forms can be removed without disturbing shores, use half of values shown but not less than 3 days.

[§]Refer to Section 3.8 for special conditions affecting number of floors to remain shored or reshored.

slab or structural member to deflect and support its own weight and existing construction loads applied before the installation of the reshores. It is assumed that the reshores carry no load at the time of installation. Afterward, additional construction loads will be distributed among all members connected by reshores.

Multistory work represents special conditions, particularly in relation to the removal of forms and shores. Reuse of form material and shores is an obvious economy. Furthermore, the speed of construction in this type of work permits other trades to follow concreting operations from floor to floor as closely as possible. The shoring that supports freshly placed and low-strength early-age concrete, however, is supported by lower floors that were not originally designed specifically for these loads. The loads imposed must not exceed the safe capacity of each floor providing support. For this reason, shoring or reshoring should be provided for a sufficient number of floors to distribute the imposed construction loads to several slab levels without causing excessive stresses, excessive slab deflections, or both.[1.3,2.8,2.9,2.10] Reshoring is used to distribute construction loads to the lower floors.

In a common method of analysis, while reshoring remains in place at grade level, each level of reshores carries the weight of only the new slab plus other construction live loads. The weight of intermediate slabs is not included because each slab carries its own weight before reshores are put in place.

Once the tier of reshores in contact with grade has been removed, the assumption is made that the system of slabs behaves elastically. The slabs interconnected by reshores will deflect, equally during addition or removal of loads. Loads will be distributed among the slabs in proportion to their developed stiffness. The deflection of concrete slabs can be considered elastic, that is, neglecting shrinkage and creep. Caution should also be taken when a wood compressible system is used. Such systems tend to shift most of the imposed construction loads to the upper floors, which have less strength. Addition or removal of loads may be due to construction activity or to removing shores or reshores in the system. Shore loads are determined by equilibrium of forces at each floor level.

3.8.2 *Advantages of reshoring*—Stripping formwork is more economical if all the material can be removed at the same time and moved from the area before placing reshores. Slabs are allowed to support their own weight, reducing the load in the reshores. Combination of shores and reshores usually requires fewer levels of interconnected slabs, thus freeing more areas for other trades.

3.8.3 *Other methods*—Other methods of supporting new construction are less widely used and involve leaving the original shores in place or replacing them individually (backshoring and preshoring), which prevents the slab from deflecting and carrying its own weight. These methods are not recommended unless performed under careful supervision by the formwork engineer/contractor and with review by the engineer/architect because excessively high slab and shore stresses can develop.

3.8.4 *Design*—Refer to Chapter 2.

3.8.5 *Placing reshores*—When used in this section, the word shore refers to either reshores or the original shores.

Reshoring is one of the most critical operations in formwork; consequently, the procedure should be planned in advance by the formwork engineer/contractor and should be reviewed or approved by the engineer/architect. Operations should be performed so that areas of new construction will not be required to support combined dead and construction loads in excess of their capability, as determined by design load and developed concrete strength at the time of stripping and reshoring.

Shores should not be located so as to alter the pattern of stress determined in the structural analysis or induce tensile stresses where reinforcing bars are not provided. Size and number of shores, and bracing, if required, should provide a supporting system capable of carrying any loads that could be imposed on it.

Where possible, shores should be located in the same position on each floor so that they will be continuous in their support from floor to floor. When shores above are not directly over shores below, an analysis should be made to determine whether or not detrimental stresses are produced in the slab. This condition seldom occurs in reshoring because the bending stresses normally caused by the offset reshores are not large enough to overcome the stress resulting from the slab carrying its own dead load. Where slabs are designed for light live loads or on long spans where the loads on the shores are heavy, care should be used in placing the shores so that the loads on the shores do not cause excessive punching shear or bending stress in the slab.

While reshoring is under way, no construction loads should be permitted on the new construction unless the new construction can safely support the construction loads.

When placing reshores, care should be taken not to preload the lower floor and not to remove the normal deflection of the slab above. The reshore is simply a strut and should be tightened only to the extent necessary to achieve good bearing contact without transferring load between upper and lower floors.

3.8.6 *Removal of reshoring*—Shores should not be removed until the supported slab or member has attained sufficient strength to support itself and all applied loads. Removal operations should be carried out in accordance with a planned sequence so that the structure supported is not subject to impact or loading eccentricities.

3.8.7 *Post-tensioning effects on shoring and reshoring*—The design and placement of shores and reshores for post-tensioned construction requires more consideration than for normal reinforced concrete. The stressing of post-tensioning tendons can cause overloads to occur in shores, reshores, or other temporary supports. The stressing sequence has the greatest effect. When a slab is post-tensioned, the force in the tendon produces a downward load at the beam. If the beam is shored, the shoring should carry this added load. The magnitude of the load can approach the dead load of 1/2 the slab span on both sides of the beam. If the floor slab is tensioned before the supporting beams and girders, a careful analysis of the load transfer to the beam or girder shores or reshores will be required.

Similar load transfer situations occur in post-tensioned bridge construction.

CHAPTER 4—MATERIALS
4.1—General
The selection of materials suitable for formwork should be based on the price, safety during construction, and the quality required in the finished product. Approval of formwork materials by the engineer/architect, if required by the contract documents, should be based on how the quality of materials affects the quality of finished work. Where the concrete surface appearance is critical, the engineer/architect should give special notice and make provision for preconstruction mock-ups. Refer to Chapter 5 for architectural concrete provisions.

4.2—Properties of materials
4.2.1 *General*—*Formwork for Concrete*[1.3] describes the formwork materials commonly used in the United States and provides extensive related data for form design. Useful specification and design information is also available from manufacturers and suppliers. Table 4.1 indicates specific sources of design and specification data for formwork materials.

This tabulated information should not be interpreted to exclude the use of any other materials that can meet quality and safety requirements established for the finished work.

4.2.2 *Sheathing*—Sheathing is the supporting layer of formwork closest to the concrete. It can be in direct contact with the concrete or separated from it by a form liner. Sheathing consists of wood, plywood, metal, or other materials capable of transferring the load of the concrete to supporting members, such as joists or studs. Liners are made of wood, plastic, metal, cloth, or other materials selected to alter or enhance the surface of the finished concrete.

In selecting and using sheathing and lining materials, important considerations are:
- Strength;
- Stiffness;
- Release;
- Reuse and cost per use;
- Surface characteristics imparted to the concrete, such as wood grain transfer, decorative patterns, gloss, or paintability;
- Absorptiveness or ability to drain excess water from the concrete surface;
- Resistance to mechanical damage, such as from vibrators and abrasion from slipforming;
- Workability for cutting, drilling, and attaching fasteners;
- Adaptability to weather and extreme field conditions, temperature, and moisture; and
- Weight and ease of handling.

4.2.3 *Structural supports*—Structural support systems carry the dead and live loads that have been transferred through the sheathing. Important considerations are:
- Strength;
- Stiffness;
- Dimensional accuracy and stability;
- Workability for cutting, drilling, and attaching fasteners;
- Weight;
- Cost and durability; and
- Flexibility to accommodate varied contours and shapes.

4.3—Accessories
4.3.1 *Form ties*—A form tie is a tensile unit used to hold concrete forms against the active pressure of freshly placed plastic concrete. In general, it consists of an inside tensile member and an external holding device, both made to specifications of various manufacturers. These manufacturers also publish recommended working loads on the ties for use in form design. There are two basic types of tie rods: the one-piece prefabricated rod or band type and the threaded internal disconnecting type. Their suggested working loads range from 1000 to more than 50,000 lb (4.4 to more than 220 kN).

4.3.2 *Form anchors*—Form anchors are devices used to secure formwork to previously placed concrete of adequate strength. The devices normally are embedded in the concrete during placement. The actual load-carrying capacity of the anchors depends on their shape and material, the strength and type of concrete in which they are embedded, the area of contact between concrete and anchor, and the depth of embedment and location in the member. Manufacturers publish design data and test information to assist in the selection of proper form anchor devices.

4.3.3 *Form hangers*—Form hangers are devices used to suspend formwork loads from structural steel, precast concrete, or other members.

4.3.4 *Side form spacers*—A side form spacer is a device that maintains the desired distance between a vertical form and reinforcing bars. Both factory-made and job-site fabricated devices have been successfully used. Advantages and disadvantages of the several types are explained in References 1.3, 4.1, and 4.2.

4.3.5—*Recommendations*

4.3.5.1—The recommended factors of safety for ties, anchors, and hangers are given in Section 2.4.

4.3.5.2—The rod- or band-type form tie, with a supplemental provision for spreading the forms and a holding device engaging the exterior of the form, is the common type used for light construction.

The threaded internal disconnecting type of tie (also called through tie) is more often used for formwork on heavy construction, such as heavy foundations, bridges, power houses, locks, dams, and architectural concrete.

Removable portions of all ties should be of a type that can be readily removed without damage to the concrete and that leaves the smallest practicable holes to be filled. Removable portions of the tie should be removed unless the contract documents permit their remaining in place.

A minimum specification for form ties should require that the bearing area of external holding devices be adequate to prevent excessive bearing stress in form lumber.

4.3.5.3—Form hangers should support the dead load of forms, weight of concrete, and construction and impact loads. Form hangers should be symmetrically arranged on the supporting member and loaded, through proper sequencing of

Table 4.1—Form materials with data sources for design and specification

Materials	Principal uses	Data sources
Sawn lumber	Form framing, sheathing, and shoring	"American Softwood Lumber Standard," PS 20-94 *Wood Handbook*, Reference 4.3 *Manual for Wood Frame Construction*, Reference 4.4 *National Design Specification for Wood Construction*, ANSI/AF&PA NDS-1997, Reference 4.7 *Timber Construction Manual*, Reference 4.6 *Structural Design in Wood*, Reference 4.5
Engineered wood*	Form framing and shoring	*Engineered Wood Products*, Reference 4.21 "Code for Engineering Design in Wood," (Canada) CAN3-086 "Engineering Design in Wood (Limit States Design)," CAN/CSA-096.1-94
Plywood	Form sheathing and panels	"Construction and Industrial Plywood," PSI-95 APA *Plywood Design Specification*, Reference 4.8 APA *Concrete Forming*, Reference 4.20
Steel	Panel framing and bracing Heavy forms and falsework	*Specification for Structural Steel Buildings—Allowable Stress Design and Plastic Design*, Reference 4.9 *Specification for Design of Cold Formed Steel Structural Members*, Reference 4.10
Steel	Column and joist forms	*Forms for One-Way Joist Construction*, ANSI A48.1 *Forms for Two-Way Concrete Joist Construction*, ANSI A48.2 *Recommended Industry Practice for Concrete Joist Construction*, part of Reference 4.1
Steel	Stay-in-place deck forms	ASTM A 446 (galvanized steel)
Steel	Shoring	*Recommended Safety Requirements for Shoring Concrete Formwork*, Reference 4.19
Steel	Steel joists used as horizontal shoring	*Recommended Horizontal Shoring Beam Erection Procedure*, Reference 4.18
Steel	Expanded metal bulkheads, single-sided forms	*Standard Specification and Load Tables for Open Web Steel Joists*, Reference 4.17 *Expand Your Forming Options*, Reference 4.16
Aluminum†	Form panels and form framing members Horizontal and vertical shoring and bracing	*Aluminum Construction Manual*, Reference 4.11
Reconstituted wood panel products‡	Form liners and sheathing	*Mat Formed Wood Particle Board*, ANSI A208.1 *Hardboard Concrete Form Liners*, LLB-810a *Performance Standard for Wood-Based Structural Use Panels*, PS2-92
Insulation materials • Wood fiber or glass fiber • Other commercial products	Stay-in-place form liners or sheathing Cold-weather protection for fresh concrete	ASTM C 532 (insulating form board)
Fiber or laminated paper pressed tubes or forms	Column and beam forms Void forms for slabs, beams, girders and precast piles	
Corrugated cardboard	Internal and under-slab void forms Void forms in beams and girders (normally used with internal "egg-crate" stiffeners)	*A Study of Cardboard Voids for Prestressed Concrete Box Slabs*, Reference 4.12

Note: Manufacturers' recommendations, when supported by test data and field experience, are a primary source for many form materials. In addition, the handbooks, standards, specifications, and other data sources cited herein are listed in more detail in *Formwork for Concrete* and in the references for Chapter 4 and Chapter 8 of this document. Be sure to check cautionary footnotes for engineered wood, aluminum, and panel products made of reconstituted wood.

*Structural composite lumber products are proprietary and unique to a particular manufacturer. They cannot be interchanged because industry-wide common grades have not been established to serve as a basis for equivalence.

†Should be readily weldable and protected against galvanic action at the point of contact with steel. If used as a facing material in contact with fresh concrete, should be nonreactive to concrete or concrete containing calcium chloride.

‡Check surface reaction with wet concrete.

Table 4.1 (cont.)—Form materials with data sources for design and specification

Materials	Principal uses	Data sources
Concrete	Stay-in-place forms	*Building Code Requirements for Structural Concrete and Commentary,* ACI 318
	Molds for precast units	*Precast Concrete Units Used as Form for Cast-in-Place Concrete,* ACI 347.1R
Glass fiber-reinforced plastic	Ready-made column forms	*Using Glass-Fiber Reinforced Forms,* Reference 4.13
	Domes and pans for concrete joist construction	
	Custom-made forms for special architectural effects	*Nonmetallic Form Ties,* Reference 4.14
	Form ties	
Cellular plastics	Form lining and insulation	*Cellular Plastics in Construction,* Reference 4.15
	Stay-in-place wall forms	Insulating Concrete Forms Association
Other plastics, including ABS, polypropylene, polyethylene, polyvinyl chloride, polyurethane	Form liners, both rigid and flexible, for decorative concrete	*Plastic Form Liners,* Reference 4.22
	Chamfer and rustication formers	
Rubber and rubberized or architectural fabrics	Form lining and void forms	Monolithic Dome Institute
	Inflatable forms for dome and culvert construction	
Form ties, anchors, and hangers	Hold formwork secure against loads and pressures from concrete and construction activities	Safety factors recommended in Section 2.4
		Also refer to Reference 4.14
Side form spacers	Maintain correct distance between reinforcement and form to provide specified concrete cover for steel	*Side Form Spacers,* Reference 4.2
Plaster	Waste molds for architectural concrete	
Release agents and protective form coatings	Help preserve form facing and facilitate release	*Choosing and Using a Form Release Agent,* Reference 4.23

Note: Manufacturers' recommendations, when supported by test data and field experience, are a primary source for many form materials. In addition, the handbooks, standards, specifications, and other data sources cited herein are listed in more detail in *Formwork for Concrete* and in the references for Chapter 4 and Chapter 8 of this document. Be sure to check cautionary footnotes for engineered wood, aluminum, and panel products made of reconstituted wood.

the concrete placement, to minimize twisting or rotation of the hanger or supporting members. Form hangers should closely fit the flange or bearing surface of the supporting member so that applied loads are transmitted properly.

4.3.5.4—Where the concrete surface is exposed and appearance is important, the proper type of form tie or hanger will not leave exposed metal at the surface. Otherwise, noncorrosive materials should be used when tie holes are left unpatched, exposing the tie to the elements.

4.4—Form coatings and release agents

4.4.1 *Coatings*—Form coatings or sealers are usually applied in liquid form to contact surfaces either during manufacture or in the field to serve one or more of the following purposes:
- Alter the texture of the contact surface;
- Improve the durability of the contact surface;
- Facilitate release from concrete during stripping; or
- Seal the contact surface from intrusion of moisture.

4.4.2 *Release agents*—Form release agents are applied to the form contact surfaces to prevent bond and thus facilitate stripping. They can be applied permanently to form materials during manufacture or applied to the form before each use. When applying in the field, be careful to avoid coating adjacent construction joint surfaces or reinforcing steel.

4.4.3 *Manufacturers' recommendations*—Manufacturers' recommendations should be followed in the use of coatings, sealers, and release agents, but independent investigation of their performance is recommended before use. When concrete surface color is critical, effects of the coating, sealing, and release agents should be evaluated. Where surface treatments such as paint, tile adhesive, sealers, or other coatings are to be applied to formed concrete surfaces, be sure that adhesion of such surface treatments will not be impaired or prevented by use of the coating, sealers, or release agent. Also, consider bonding requirements of subsequent concrete placements.

CHAPTER 5—ARCHITECTURAL CONCRETE
5.1—Introduction
5.1.1 *Objective*—The general requirements for formwork presented in preceding chapters for the most part also apply to architectural concrete. Additional information is available in ACI 301 and ACI 303R.

This chapter identifies and emphasizes additional factors that can have a critical influence on formwork for cast-in-place architectural concrete. Tilt-up and precast architectural concrete are not considered here. Concrete receiving coatings or plasters that hide the surface color and texture is not considered architectural.

5.1.2 Definition—ACI Committee 303 defines architectural concrete as concrete that is exposed as an interior or exterior surface in the completed structure, contributes to its visual character, and is specifically designated as such in the contract documents. Particular care should be taken in the selection of materials, design, and construction of the formwork, and placing and consolidation of the concrete to eliminate bulges, offsets, or other unsightly features in the finished surface and to maintain the integrity of the surface texture or configuration. The character of the concrete surface to be produced should also be considered when the form materials are selected. Special attention should be given to closure techniques, concealment of joints in formwork materials, and to the sealing of forms to make them watertight.

5.1.3 Factors in addition to formwork—Many factors other than formwork affect the architectural effects achieved in concrete surfaces. They start at the design stage and carry through to the completed project. Factors affecting the concrete can also include the mixture proportions or aggregate, the method of placing the concrete, the consolidation technique, and the curing procedure. Chemicals can have an effect on the final product, whether used as additives in the mixture; applied directly to the concrete, such as curing compounds; or applied indirectly, such as form release agents. Even after the structure is completed, weather and air pollution will affect the appearance of the concrete. These and other influencing factors should be identified and their effects evaluated during the initial design stages. The single most important factor for the success of an architectural concrete job is good workmanship.

5.1.4 Uniform construction procedures—Architectural concrete should have a uniform color and surface finish. The best way for the contractor to achieve this uniformity is to be consistent in all construction practices. Forming materials should be kept the same, and release agents should be applied uniformly and consistently. Placement and consolidation of the concrete should be standardized so that uniform density is achieved. Stripping and curing sequences should be kept constant throughout the work to control color variations.

5.2—Role of the architect

5.2.1 Preplanning—Much architectural concrete is also structural, but surface quality generally desired for architectural concrete is higher than what is typically satisfactory for structural concrete, and is more costly. The architect can use the latest information available in the art of forming and concrete technology during the design process to keep plans in line with the budget for the structure. However, intricacies and irregularities can raise the budget to a point that outweighs the architectural concrete's aesthetic contributions. The architect can make form reuse possible by standardizing building elements, such as columns, beams, and windows, and by making uninterrupted form areas the same size wherever possible to facilitate the use of standard form gangs or modules. The increased size of these uninterrupted areas will contribute to forming economy. A prebid conference with qualified contractors will bring out many practical considerations before the design is finalized.

5.2.2 Contract documents and advance approvals—The architect should prepare contract documents that fully instruct the bidder as to the location and desired appearance of architectural surfaces, as well as other specific requirements listed in Sections 5.2.3 to 5.2.7. On major work, this is frequently achieved by specifying a preconstruction mockup prepared and finished by the contractor for approval by the architect, using proposed form materials, jointing techniques, and form surface treatments, such as wetting, oiling, or lacquering. Once such a mockup has been completed to the satisfaction of the architect, it remains at the site for the duration of the work as a standard with which the rest of the work should comply.

Design reference samples, which are smaller specimens of concrete with the proposed surface appearance, may also be created for approval by the architect. Small samples like these, kept at the job site for reference, are not as good as a full-scale mockup but can be helpful. The samples should be large enough to adequately represent the surface of the concrete. If the samples are to be used as a basis for acceptance, several should be made to represent the variation that can occur in the finish.

In the absence of physical mockups or reference samples, it can be helpful to specify viewing conditions under which the concrete surfaces will be evaluated for compliance with the specifications.

5.2.3 Tolerances—The architect should specify dimensional tolerances considered essential to successful execution of the design. ACI 117 can be consulted, but the architect should realize that the tolerances therein are for concrete construction in general, and more restrictive tolerances can be required for architectural work. No numerical limits are suggested herein because the texture, lighting, and configuration of surfaces will all have an influence. ACI Committee 347R notes, however, that concrete construction tolerances of 1/2 those called for in ACI 117 are considered the achievable limit.

5.2.4 Camber—The contractor should camber formwork to compensate for deflection of the formwork during concrete placement. The architect should, however, specify any additional camber required to compensate for structural deflection or optical sag (the illusion that a perfectly horizontal long-span member is sagging). The architect should be aware that horizontal members are checked for compliance with tolerances and camber before the removal of the forms and shores.

5.2.5 Joints and details—Location, number, and details of items such as openings, contraction joints, construction joints, and expansion joints should be shown on the design plans or the architect should specify a review of the proposed location of all of these details as shown on the formwork drawings. (Some guidance on joint locations can be found in ACI 224R, 303R, and 332R.) Because it is impossible to disguise the presence of joints in the form face, it is important for their positions to be predetermined and, if possible, planned as part of the architectural effect.

The architect can plan joint locations between surface areas on a scale and module suitable to the size of available materials and prevailing construction practices. If this is not aesthetically satisfactory, dummy joints can be introduced to give a smaller pattern. Actual joints between sheathing materials can be masked by means of rustication strips (splayed fillets) attached to the form face. Rustication strips at horizontal and vertical construction joints can also create crisp edges accented by shadow lines instead of the potential ragged edge of a construction joint left exposed to full view. Special care should be taken during placement and vibration to minimize bugholes and honeycombing that form when air is trapped beneath horizontal rustications.

Sometimes construction joints in beams can be concealed above the support columns and joints in floors above their supporting beams instead of in the more customary regions of low shear.

5.2.6 *Ties and inserts*—Form ties and accompanying tie holes are an almost inescapable part of wall surfaces. Architects frequently integrate tie holes into the visual design quality of the surface. If this is planned and any effects or materials other than those provided in Section 5.3.4 are desired, they should be clearly specified as to both location and type.

Where tie holes are to be patched or filled, the architect should specify the treatment desired, unless it has been shown on the preconstruction mockup.

5.2.7 *Cover over reinforcing steel*—Adequate cover over reinforcement, as required by codes, is needed for protection of steel and long-term durability of the concrete. Reinforcement that is properly located is important in the control of surface cracking. For positive assurance of maintaining required cover, the architect can specify appropriate side form spacers as defined in Section 4.3.4.

There is no advantage in specifying more cover than required by code because excessive cover can contribute to increased cracking. The architect should specify sufficient cover to allow for any reduction that will result from the incorporation of grooves or indented details and from surface treatments, such as aggregate exposure and tooling. The maximum thickness of any material to be removed should be added to basic required cover.

5.3—Materials and accessories

5.3.1 *Sheathing or form facing*—Architectural concrete form sheathing should be of appropriate quality to maintain the uniformity of concrete surfaces through multiple uses and to control deflection within appropriate limits. Plywood, steel, glass fiber-reinforced plastic, and aluminum can all be suitable as sheathing or facing materials. Select the grade or class of material needed for pressure, framing, and deflection requirements. Be sure that the chosen material meets the specification requirements for the concrete surface texture. Procedures for controlling the rusting of steel should be carefully followed.

5.3.2 *Structural framing*—Form facing can be supported with lumber, steel, or aluminum members straight and rigid enough to meet the architectural specifications.

5.3.3 *Form liners*—A form liner is a material attached to the inside face of the form to alter or improve surface texture or quality of the concrete. It is not required structurally.

Wood, rigid plastic, elastomeric materials, and glass fiber-reinforced plastics are all suitable liner materials when carefully detailed and fabricated. Plastics should be handled and assembled with care to avoid distortion caused by daily temperature cycles at the job site.

5.3.4 *Form ties*—Form-tie assemblies for architectural concrete should permit tightening of forms and leave no metal closer to the surface than 1-1/2 in. (38 mm) for steel ties and 1 in. (25 mm) for stainless steel ties. The ties should not be fitted with lugs, cones, washers, or other devices that will leave depressions in the concrete less than the diameter of the device, unless specified. Ties should be tight fitting or tie holes in the form should be sealed to prevent leakage at the holes in the form. If textured surfaces are to be formed, ties should be carefully evaluated as to fit, pattern, grout leakage, and aesthetics.

5.3.5 *Side form spacers*—Side form spacers, as defined in Section 4.3.4, are particularly important in architectural concrete to maintain adequate cover over reinforcing steel and to prevent development of rust streaking on concrete surfaces. Plastic, plastic-protected, rubber-tipped, or other noncorroding spacers should be attached to the reinforcing bar so that they do not become dislodged during concrete placement and vibration. The number and location of the side form spacers should be adequate for job conditions.

5.4—Design

5.4.1 *Special considerations*—The general procedure will follow the principles outlined in Chapter 2. The formwork engineer/contractor, however, will frequently have limitations imposed by the architectural design. Some of these considerations are: tie spacing and size, form facing preferences, location and special treatment of form joints, special tolerances, and use of admixtures. Because these factors can influence form design, they should be fully reviewed at the beginning of the form design process.

5.4.2 *Lateral pressure of concrete*—Architectural concrete can be subjected to external vibration, revibration, set retardants, high-range water-reducing admixtures, and slumps greater than those assumed for determining the lateral pressure as noted in Section 2.2.2. Particular care should be exercised in these cases to design the forms for the increased lateral pressures resulting from sources noted in Section 2.2.2.

5.4.3 *Structural considerations*—Because deflections in the contact surface of the formwork reflect directly in finished surfaces under varying light conditions, forms for architectural concrete should be designed carefully to minimize deflections. In most cases, deflections govern design rather than bending (flexural stress) or horizontal shear. Deflections of sheathing, studs, and wales should be designed so that the finished surface meets the architectural specifications. Limiting these deflections to $l/400$, where l is the clear span between supports, is satisfactory for most architectural formwork. Forms bow with reuse; therefore, more bulging will be reflected in the surface formed after several uses. This effect should be considered when designing forms.

When tie size and spacing are limited by the architect, the formwork engineer/contractor may have to reverse the usual procedure to arrive at a balanced form design. Given the capacity of the available tie and the area it supports, the formwork engineer/contractor can find the allowable pressure, design supporting members, and establish a rate of concrete placing.

Where wood forms are used, stress-graded lumber (or equivalent) free of twists and warps should be used for structural members. Form material should be sized and positioned to prevent deflections detrimental to the surfaces formed. Joints of sheathing materials should be backed with structural members to prevent offsets.

5.4.4 *Tie and reanchor design*—Tie layout should be planned. If the holes are to be exposed as part of the architectural concrete, tie placement should be symmetrical with the member formed. If tie holes are not to be exposed, ties should be located at rustication marks, control joints, or other points where the visual effect will be minimized.

Externally braced forms can be used instead of any of the aforementioned methods to avoid objectionable blemishes in the finished surface. Externally braced forms, however, can be more difficult and more costly to build.

Consideration should be given to reanchoring forms in preceding or adjacent placements to achieve a tight fit and prevent grout leakage at these points. Ties should be located as close as possible to the construction joint to facilitate reanchoring the form to adjacent placements.

5.4.5 *Joints and details*—In architectural concrete, joints should, where feasible, be located at the junction of the formwork panels. At contraction or construction joints, rustication strips should be provided and fastened to the face of forms.

Corners should be carefully detailed to prevent grout leakage. Sharp corners should, wherever possible, be eliminated by the use of chamfer strips except when prohibited by project specifications.

5.4.6 *Tolerances*—The formwork engineer/contractor should check for dimensional tolerances specified by the architect that can have a bearing on the design of the forms. If no special tolerances are given, the formwork engineer/ contractor can use ACI 117 tolerances for structural concrete.

5.5—Construction

5.5.1 *General*—Forms should be carefully built to resist the pressures to which they will be subjected and to limit deflections to a practicable minimum within the tolerances specified. Joints in structural members should be kept to a minimum and, where necessary, should be suitably spliced or otherwise constructed to maintain continuity.

Pour pockets for vibrating or placing concrete should be planned to facilitate careful placement and consolidation of the concrete to prevent segregation, honeycomb, sanding, or cold joints in the concrete. The location of pour pockets should be coordinated with the architect.

Attachment of inserts, rustication strips, and ornamental reliefs should be planned so that forms can be removed without exerting pressure on these attachments.

Where special forming systems are specified by the engineer of the project for structural purposes (such as one- and two-way joist systems) in areas that are considered architectural, the architect and engineer should coordinate their requirements to be sure the architectural effect is consistent with the forming method and material specified.

Forms that will be reused should be carefully inspected after each use to ensure that they have not become damaged, distorted, disassembled, or otherwise unable to perform as designed.

5.5.2 *Sheathing and jointing*—Contact surfaces of the formwork should be carefully installed to produce neat and symmetrical joint patterns, unless otherwise specified. Joints should be either vertical or horizontal and, where possible, should be staggered to maintain structural continuity.

Nailing should be done with care using hammers with smooth and well-dressed heads to prevent marring of the form surfaces. When required, box nails should be used on the contact surface and should be placed in a neat pattern.

Wherever possible, sheathing or panel joints should be positioned at rustication strips or other embedded features that can conceal or minimize the joint.

Construction joints should be formed with a grade strip attached to the form to define a clean straight line on the joint of the formed surface. Formwork should be tightened at a construction joint before the next placement to prevent seepage of water between the form and previously placed concrete surfaces.

Architectural concrete forms should be designed to resist water leakage and avoid discoloration. One method to prevent water loss from the concrete at the joints between sections of the formwork and at construction joints is to attach a gasket of flexible material to the edge of each panel. The gasket is compressed when the formwork is assembled or placed against the existing concrete. Caulk, tape, joint compound, or combinations of these can be used to seal joints. In all cases, unsupported joints between sheathing sheets should be backed by framing. Water-tight forms require more care during vibration to remove entrapped air that can cause bug holes.

Textured surfaces on multilift construction should be separated with rustication strips or broad reveals because accumulation of construction tolerances, random textures, or both, prevent texture matching. Furthermore, the grout seal between the bottom of a textured liner and the top of the previous placement is impractical without the rustication strip.

5.5.3 *Cleaning, coating, and release agents*—Form coatings or releasing agents should be applied before reinforcing steel is placed and should be applied carefully to avoid contacting adjacent construction joints or reinforcing. No form coating should be used unless it can be demonstrated not to stain the concrete or impair the adhesion of paints or other surface treatments.

Form sealers should be tested to ensure that they will not adversely affect the texture of the form lining material.

Ties that are to be pulled from the wall should be coated with nonstaining bond breaker or encased in sleeves to facilitate removal.

Forms should be carefully cleaned and repaired between uses to prevent deterioration of the quality of surface formed. Film or splatter of hardened concrete should be thoroughly removed.

5.5.4 *Ornamental liners and detail*—Ornamental concrete is usually formed by elastomeric molds or wood, plastic, or plaster waste molds. Members making up wood molds should be kerfed on the back wherever such members can become wedged between projections in the ornament. Molds should be constructed so that joints will not be opened by slight movement or swelling of the wood. Joints in the molds should be made inconspicuous by pointing.

The molds should be carefully set in the forms and securely held in position to reproduce the design shown on the plans. Where wood forms adjoin molds, the wood should be neatly fitted to the profile of the mold and all joints should be carefully pointed. The molds and the adjacent wood forms should be detailed so that the wood forms can be stripped without disturbing the molds. The edge of the mold or pattern strip should be tapered to a slight draft to permit removing the detail material without damaging the concrete. Special provisions should be made for early form removal, retardation, or both when sandblasting, wire brushing, or other treatments are required.

Form liners should be attached securely with fasteners or glue recommended by the manufacturer. The form behind the liner should hold the fasteners. The surfaces should be cleaned and dried thoroughly so that the glue will bond. Do not use glue at temperatures lower than those recommended by the manufacturer.

5.6—Form removal
5.6.1 *Avoiding damage*—When concrete surfaces are to be left as cast, it is important not to damage or scar the concrete face during stripping. Forms should be supported so that they do not fall back or against the architectural surface. The use of pry bars and other stripping tools should be strictly supervised. In no case should pry bars be placed directly against the concrete. Even the use of wood or plastic wedges does not ensure that damage will not occur. Once formwork is removed, the architectural surfaces should be protected from continuing construction operations.

5.6.2 *Concrete strength*—It is desirable for architectural concrete to have a higher compressive strength than normal for stripping. This can be accomplished by adjusting the mixture proportions or leaving forms in place longer. If concrete is not strong enough to overcome the adhesion between the form surface and the concrete, concrete can scale or spall. Therefore, a good quality surface might require the forms to stay in place longer. The longer the forms stay in place, however, the darker the concrete will become. The engineer/architect should specify what concrete strength is required before stripping can take place.

5.6.3 *Uniformity*—To ensure surface quality, uniformity in stripping time and curing practices are essential. Where the objective is to produce as consistent an appearance as possible, it is beneficial to protect the concrete by leaving the formwork in place somewhat longer than normal. Early exposure of concrete to the air affects the manner in which the surface dries. The ambient conditions can influence the eventual color of the concrete.

5.6.4 *Avoiding thermal shock*—Cold-weather concreting requires that special attention be paid to the sudden temperature change of concrete. To avoid thermal shock and consequent crazing of the concrete surface, the change in temperature of the concrete should be controlled within the limits outlined in ACI 303R. This can be accomplished by heating the work area, leaving the forms in place to contain the heat of hydration, or by insulating the concrete after the forms have been removed (refer to ACI 306R). Positive steps should be taken to inspect, record, and document the procedures used to cure the concrete.

CHAPTER 6—SPECIAL STRUCTURES
6.1—Discussion
Formwork for all structures should be designed, constructed, and maintained in accordance with recommendations in Chapters 1 to 4. This section deals with the additional requirements for formwork for several special classes of work. ACI 344R contains information on design and construction of circular prestressed-concrete structures.

6.2—Bridges and viaducts, including high piers
6.2.1 *Discussion*—The construction and removal of formwork should be planned in advance. Forms and supports should be sufficiently rigid to ensure that the finished structure will fulfill its intended structural function and that exposed concrete finishes will present a pleasing appearance to the public.

6.2.2 *Shoring and centering*—Recommended practice in Sections 3.5 and 3.7 for erection and removal should be followed. In continuous structures, support should not be released in any span until the first and second adjoining spans on each side have reached the specified strength. For post-tensioned bridges, the shore design should consider the resulting redistribution of loads on the shores similar to the effects discussed in Section 3.8.7.

6.2.3 *Forms*—Forms can be of any of a large number of materials but most commonly are wood or metal. They should be built mortar-tight of sound material strong enough to prevent distortion during placing and curing of the concrete.

6.3—Structures designed for composite action
6.3.1 *Recommendations*—Structures or members that are designed so that the concrete acts compositely with other materials or with other parts of the structure present special forming problems that should be anticipated in the design of the structure. Requirements for shoring or other deflection control of the formwork should be clearly presented by the engineer/architect in the specifications. Where successive placements are to act compositely in the completed structure, deflection control becomes extremely critical.

Shoring, with or without cambering portions of the structure during placement and curing of the concrete, should be analyzed separately for the effects of dead load of newly placed concrete and for the effect of other construction loads that can be imposed before the concrete attains its design strength.

6.3.2 *Design*—Formwork members and shores should be designed to limit deflections to a practical minimum consistent with the structural member being constructed. Where camber is specified for previously installed components of the structure, allowance should be made for the resultant preloading of the shores before application of the dead load of concrete.

In members constructed in several successive placements, such as box-girder structures, formwork components should be sized, positioned, supported, or all three to minimize progressive increases in deflection of the structure that would excessively preload the reinforcing steel or other portions of the composite member.

In multistory work where shoring of composite members is required, consideration should be given to the number of stories of shores necessary, in conjunction with the speed of construction and concrete strengths, to minimize deflections due to successive loadings. Distinction should be made in such analyses for shores posted to relatively unyielding support, such as foundations, instead of to structures or members already in elastic support (refer to Section 3.8).

Composite construction can have beams of relatively light cross sections that are fully adequate when construction is complete. During construction, these beams may not be laterally supported by the formwork, thus leaving them with a high slenderness ratio and reduced beam strength. The engineer/architect should alert the contractor to this problem in general notes on the structural plans or in notes on applicable plans when this condition exists. The formwork engineer/contractor should be alert to this possibility and provide shoring or lateral support where needed.

6.3.3 *Erection*—Construction, erection of formwork, or both for composite construction follows basic recommendations contained in Chapter 3. Shoring of members that will act compositely with the concrete to be placed should be done with great care to ensure sufficient bearing, rigidity, and tightness to prevent settlement or deflections beyond allowable limits. Wedges, shims, and jacks should be provided to permit adjustment if required before or during concreting, as well as to permit removal without jarring or impacting the completed construction. Provision should be made for readily checking the accuracy of position and grade during placement. Even though adjustment of forms can be possible during or after placing, it is not recommended. Any required adjustment should be made before the initial set of the concrete.

Where camber is required, a distinction should be made between that part which is an allowance for settlement or deflection of formwork or shoring and that which is provided for design loadings. The former should generally be the responsibility of the formwork engineer/contractor who designs the forms and supports unless such camber is stipulated by the engineer/architect. Measurement of camber provided for structural design loadings should be made after hardening of the concrete but before removal of the supports (also refer to Section 1.4.5). This is because the structural deflection occurring upon removal of the supports is a function of the structural design and cannot be controlled by the contractor.

6.3.4 *Removal*—In addition to meeting the provisions of Section 3.7, forms, supports, or both should be removed only after tests and specified curing operations indicate to the satisfaction of the engineer/architect that the most recently placed concrete has attained the strength required to develop composite action, and then only after approval of the engineer/architect. The sequence of such removal should be approved by the engineer/architect.

6.4—Folded plates, thin shells, and long-span roof structures

6.4.1 *Discussion*—For long-span and space structures requiring a complex, three-dimensional design analysis and presenting three-dimensional problems in formwork design, erection, and removal, formwork planning should be done by formwork engineers having the necessary special qualifications and experience. These formwork engineers should consult and cooperate with the engineer/architect to make sure that the resulting surfaces will conform to his or her design.

6.4.2—*Design*
- The engineer/architect should specify limiting values and directions of the reactive forces when the falsework is supported by the permanent structure.
- When applicable, the engineer/architect should include a decentering sequence plan with the bidding documents as a basis for the design of the forming and support system to be used by the contractor.
- *Lateral loads*—In determining the lateral forces acting on the formwork, the wind load should be calculated on the basis of a minimum of 15 lb/ft^2 (0.72 kPa) of projected vertical area as specified for wall forms in Section 2.2.3. For structures such as domes, negative forces due to suction created by the wind on the leeward side of the structure should be considered.
- *Analysis*—The provisions of Sections 2.1.1 and 2.3 should be adhered to in formwork planning.

Assumed design loads should be shown on the formwork drawings. Complete stress analyses should be prepared by competent structural engineers, and the maximum and minimum values of stress, including reversal of stress, should be shown for each member for the most severe loading conditions. Consideration should be given to unsymmetrical or eccentric loadings that might occur during concrete placement and during erection, decentering, or moving of travelers. The vertical or lateral deflection of the moving forms or travelers, as well as the stability under various loads, should be investigated to confirm that the formwork will function satisfactorily and that the concrete tolerances will be met.

Particular care should be taken in the design and detailing of individual members and connections. Where truss systems are used, connections should be designed to keep eccentricities as small as possible to minimize deflections or distortions.

Because the weight of the formwork can be equal to or greater than the design live load of the structure, form details should be designed to avoid the formwork hanging up and overloading the structure during decentering.

Due to the special shapes involved, tolerances based on functions of these shapes should be specified by the engineer/architect in the bidding documents.

6.4.3 *Drawings*—When required, the formwork engineer/contractor should submit detailed drawings of the formwork for approval of the engineer/architect.

These drawings should show the proposed concrete placing sequence and the resulting loads. To ensure that the structure can assume its deflected shape without damage, the decentering or handling sequence of the formwork should be shown on the drawings. The formwork design, drawings, and procedures should comply with federal and local safety laws, as well as the contract documents.

Deflection of these structures can cause binding between the form and the concrete during decentering. Formwork drawings and form details should be planned to prevent binding and facilitate stripping of forms. Drawings should show such details as type of inserts and joints in sheathing where spreading of the form can result in the form becoming keyed into the concrete.

6.4.4 *Approval*—The engineer/architect should review the design and drawings for the formwork and the procedures for construction to ensure the structural integrity of the permanent structure. The engineer/architect should approve in writing the loads imposed by the formwork, the sequence of the concrete placing operations, and the timing and procedures of decentering and stripping.

6.4.5 *Construction*—In planning and erecting formwork, provisions should be made for adequate means of adjustment during placing where necessary. Telltales should be installed to check alignment and grade during placement. Where the forming system is based on a certain placing sequence, that sequence should be clearly defined and adhered to in the field.

6.4.6 *Removal of formwork*—Formwork should be removed and decentered in accordance with the procedure and sequence specified on the form drawings or on the contract documents. Decentering methods used should be planned to prevent any concentrated reaction on any part of the permanent structure. Due to the large deflections and the high dead load-to-live load ratio common to this type of structure, decentering and form removal should not be permitted until specified tests demonstrate that the concrete strength and the modulus of elasticity specified in contract documents have been reached. Moduli of elasticity can determine time of decentering, although required compressive strengths may already have been attained. Decentering should begin at points of maximum deflection and progress toward points of minimum deflection, with the decentering of edge members proceeding simultaneously with the adjoining shell.

6.5—Mass concrete structures

6.5.1 *Discussion*—ACI 116R defines mass concrete as "any volume of concrete with dimensions large enough to require that measures be taken to cope with generation of heat from hydration of the cement and attendant volume change to minimize cracking." Mass concrete occurs in heavy civil engineering construction, such as in gravity dams, arch dams, gravity-retaining walls, lock walls, power-plant structures, and large building foundations (ACI 207.1R). Special provisions are usually made to control the temperature rise in the mass by the use of cement or cementitious material combinations possessing low or moderate heat-generating characteristics, by postcooling (cooling the fresh concrete) or by placing sequence. Heat rise in mass concrete is most often controlled by replacement of cement with pozzolans, particularly fly ash.

Formwork for mass concrete falls into two distinct categories, namely, low and high lift. Low-lift formwork, for heights of 5 to 10 ft (1.5 to 3 m), usually consists of multiuse steel cantilever form units that incorporate their own work platforms and, on occasion, lifting devices. High-lift formwork is comparable with the single-use wood forms used extensively for structural concrete.

6.5.2 *Lateral pressure of concrete*—The lateral pressure formulas for concrete placed in walls can be used for mass concrete (refer to Section 2.2.2). The formwork engineer needs to carefully review the concrete mixture proportion to determine the appropriate formula from Section 2.2.2. Concrete additives or cement substitutes can improve heat generation characteristics, but the same materials can increase concrete set time and increase lateral pressures.

Consideration should be given to placing sequence in the determination of pressure. Frequently, concrete is layered in such a way that the fresh concrete rate of placement locally is substantially greater than the average rate of placement. Local lateral pressures can be greater than would be estimated on the basis of the average rate of placement. In addition, the use of large concrete buckets with rapid discharge of concrete can cause high impact loads near the forms.

6.5.3 *Design consideration*—Mass concrete forming can require special form tie and anchor design.

6.5.3.1—Forming sloping surfaces requires ties or anchors to resist pressure forces that are perpendicular to the face of the form. Using horizontal ties will leave the vertical component of pressure untied. Vertical (hold down) anchors are required.

6.5.3.2—Forms tied or anchored to a rock face require particular care. Often, rock anchors are placed before the forms are erected. This requires the form designer to accommodate tie and anchor misalignment. Rock anchors should be checked to ensure that the anchor can resist the tie forces.

6.5.3.3—Bending and welding of high tensile steel tie rods should not be permitted without the approval of the tie manufacturer. Any approved welding should be by a certified welder using approved written welding procedures.

6.5.3.4—The capacity of anchors and form ties embedded in previously placed concrete is dependent on the strength of the concrete, which is very low at early ages. The embedded strength should be sufficient to sustain design loadings from the new placement and initial bolting stresses.

6.5.4 *Tolerances*—Refer to Section 3.3 and ACI 117.

6.6—Underground structures

6.6.1 *Discussion*—Underground structures differ from corresponding surface installations in that the construction takes place inside an excavation instead of in the open, providing unique problems in handling and supporting formwork and in the associated concrete placing. As a result, four factors usually make the design of formwork for underground structures entirely different than for their above-ground counterparts. First, concrete to fill otherwise inaccessible areas can be placed pneumatically or by positive displacement pump and pipeline. Second, rock is sometimes used as a form backing, permitting the use of rock anchors and tie rods instead of external bracing and shores. Third, the limits of the excavation demand special handling equipment that adds particular emphasis to the removal and reuse of forms. Fourth, rock surfaces can sometimes be used for attaching hoisting devices.

When placement is done by pneumatic or positive displacement pump and pipeline methods, the plastic concrete is forced under pressure into a void, such as the crown of a tunnel lining. For more information on the pumping process, refer to ACI 304.2R.

6.6.2—*Design loads*

6.6.2.1 *Vertical loads*—Vertical and construction loads assumed in the design of formwork for underground structures are similar to those for surface structures, with the exception of unusual vertical loads occurring near the crown of arch or tunnel forms and flotation or buoyancy effect beneath tunnel forms.

In placing concrete in the crowns of tunnel forms, pressures up to 3000 lb/ft^2 (144 kPa) have been induced in areas of overbreak and near vertical bulkheads from concrete placed pneumatically or by positive displacement pump. Overbreak is the excess removal of rock or other escavated material above the forms beyond the required tunnel lining thickness. Until more definite recommendations can be made, the magnitude and distribution of pressure should be determined by the formwork engineer. The assumed pressure should not be less than 1500 lb/ft^2 (72 kPa) acting normally to the form plus the dead weight of the concrete placed pneumatically or by pump.

6.6.2.2 *Lateral loads*—For shafts and exterior walls against rock, the values listed in Section 2.2.2 should apply.

When the shaft form relies on the single shear value of embedded anchors in the previous placement as a means of support, the minimum time lapse between successive placements (or minimum concrete strength) and maximum allowable loading additional to the dead weight of the form should be specified.

For arch forms and portions of tunnel forms above the maximum horizontal dimension or spring line of the form, the pressure should be compatible with the pressures discussed under vertical loads in Section 6.6.2.1.

6.6.3 *Drawings*—In addition to the provisions of Chapters 1, 2, and 3, the following data should be included on the drawings for specialized formwork and formwork for tunnels:

- All pressure diagrams used in the design of the form, including diagrams for uplift, for unbalanced lateral or vertical loads, for pressurized concrete, or for any other load applicable to the particular installation;
- Recommended method of supplemental strutting or bracing to be employed in areas where form pressures can exceed those listed due to abnormal conditions;
- Handling diagrams and procedures showing the proposed method of handling the form during erection or installation for concrete placement plus the method of bracing and anchorage during normal operation;
- Concrete placement method and, for tunnel arch forms, whether the design is based on the unit or bulkhead system of concrete placement or the continuously advancing slope method; and
- The capacity and working pressure of the pump and the size, length, and maximum embedment of the discharge line when placement by pumping is anticipated.

6.6.4 *Construction*—The two basic methods of placing a tunnel arch entail problems in the construction of the formwork that require special provisions to permit proper reuse. These two basic methods are commonly known as the bulkhead method and the continuously advancing slope method.

The former is used exclusively where poor ground conditions exist, requiring the lining to be placed concurrently with tunnel driving operations. It is also used when some factor, such as the size of the tunnel, the introduction of reinforcing steel, or the location of construction joints, precludes the advancing slope method. The advancing slope method, a continuous method of placement, is usually preferred for tunnels driven through competent rock, ranging between 10 and 25 ft (3 and 8 m) in diameter and at least 1 mi (1.6 km) in length.

The arch form for the bulkhead method is usually fabricated into a single unit between 50 and 150 ft (15 and 45 m) long, which is stripped, moved ahead, and re-erected using screw jacks or hydraulic rams. These are permanently attached to the form and supporting traveling gantry. The arch form for the continuously advancing slope method usually consists of eight or more sections that range between 15 and 30 ft (5 and 9 m) in length. These are successively stripped or collapsed, telescoped through the other sections, and re-erected using a form traveler.

Although the minimum stripping time for tunnel arch forms is usually established on the basis of experience, it can be safely predetermined by tests. At the start of a tunnel arch concreting operation, the recommended minimum stripping time is 12 h for exposed surfaces and 8 h for construction joints. If the specifications provide for a reduced minimum stripping time based on site experience, such reductions should be in time increments of 30 min or less and should be established by laboratory tests and visual inspection and surface scratching of sample areas exposed by opening the form access covers. Arch forms should not be stripped prematurely when unvented groundwater seepage could become trapped between the rock surface and the concrete lining.

6.6.5 *Materials*—The choice of materials for underground formwork is typically predicated on the shape, degree of reuse and mobility of the form, and the magnitude of pump or pneumatic pressures to which it is subjected. Usually, tunnel and shaft forms are made of steel or a composite of wood and steel. Experience is important in the design and fabrication of a satisfactory tunnel form due to the nature of the pressures developed by the concrete, placing techniques, and the high degree of mobility required.

When reuse is not a factor, plywood and tongue-and-groove lumber are sometimes used for exposed surface finishes. High humidity in underground construction alleviates normal shrinkage and warping.

CHAPTER 7—SPECIAL METHODS OF CONSTRUCTION

7.1—Recommendations

The applicable provisions of Chapters 2, 3, and 4 also apply to the work covered in this chapter.

7.2—Preplaced-aggregate concrete

7.2.1 *Discussion*—Preplaced-aggregate concrete is made by injecting (intruding) mortar into the voids of a preplaced mass of clean, graded aggregate. For normal construction, the preplaced aggregates are vibrated thoroughly into forms and around reinforcing and then wetted and kept wet until the injection of mortar into the voids is completed. In underwater construction, the mortar displaces the water and fills the voids. In both types of construction, this process can create concrete with a high content of coarse aggregate.

The injected mortar contains water, fine sand, portland cement, pozzolan, and a chemical admixture designed to increase the penetration and pumpability of the mortar. The structural coarse aggregate is similar to coarse aggregate for conventional concrete. It is well washed and graded from 1/2 in. (13 mm) to the largest size practicable. After compaction in the forms, it usually has a void content ranging from 35 to 45%. Refer to ACI 304.1R.

7.2.2 *Design considerations*—Due to the method of placement, the lateral pressures on formwork are considerably different from those developed for conventional concrete as given in Section 2.2.2. The formwork engineer/contractor should be alerted to the unique problems created by preplaced-aggregate, by mass placings where heat of hydration and drying shrinkage are critical, and by differential pressures in the form structure when mortar injection varies greatly from one form face to another. For additional information, refer to ACI 359, ACI 207.1R, and ACI SP-34. Because of the pressure created during aggregate packing and mortar pumping, forms that mortar is injected through should be anchored and braced far more securely than for ordinary concrete. Particular attention should be paid to uplift pressures created in battered forms. Provision should be made to prohibit even the slightest uplift of the form. Injection pipes spaced 5 to 6 ft (1.5 to 1.8 m) apart, penetrating the face of the form, require that the form be checked for structural integrity as well as a means of plugging or shutting off the openings when the injection pipes are removed. Some of these problems are reduced where mortar can be injected vertically in open top forms.

Forms, ties, and bracing should be designed for the sum of:

a) The lateral pressure of the coarse aggregate as determined from the equivalent fluid lateral pressure of the dry aggregate using the Rankine or Coulomb theories for granular materials; or a reliable bin action theory (refer to theories and references presented in ACI 313 and ACI 313R); and

b) The lateral pressure of the injected mortar; as an equivalent fluid the mortar normally weighs 130 lb/ft^3 (21 kN/m^3), but can weigh as much as 200 lb/ft^3 (32 kN/m^3) for high-density mortars.

The time required for the initial set of the fluidized mortar (from 1 to 2 h) and the rate of rise should be ascertained. The maximum height of fluid to be assumed in determining the lateral pressure of the mortar is the product of the rate of rise (ft/h) and the time of initial set in hours. The lateral pressure for the design of formwork at any point is the sum of the pressures determined from Steps (a) and (b) for the given height.

7.2.3 *Construction*—In addition to the provisions of Chapter 3, the forms should be mortar-tight and effectively vented because preplaced-aggregate concrete entails forcing mortar into the voids around the coarse aggregate.

Where increased lateral pressures are expected, the workmanship and details of formwork should be of better quality than formwork for conventional concrete.

7.2.4 *Materials for formwork*—For unexposed surfaces, mortar-tight forms of steel or plywood are acceptable. Absorptive form linings are not recommended because they permit the coarse aggregate to indent the lining and form an irregular surface. Form linings, such as hardboard on common sheathing, are not successful because they do not transmit the external form vibration normally used for ensuring a void-free finished surface. Where external vibration is used, added strength is needed in the form.

7.3—Slipforms

7.3.1 *Discussion*—Refer to ACI 313 for silo construction. Slipforming is a quasicontinuous forming process in which a special form assembly slips or moves in the appropriate direction leaving the formed concrete in place. The process is, in some ways, similar to an extrusion process. Plastic concrete is placed in the forms, and the forms can be thought of as moving dies to shape the concrete. The rate of movement of the forms is regulated so the forms leave the concrete only after it is stiff enough to retain its shape while supporting its own weight and the lateral forces caused by wind and equipment. Formwork of this type can be used for vertical structures, such as silos, storage bins, building cores, bearing wall buildings, piers, chimneys, shaft linings, communication and observation towers, nuclear shield walls, and similar structures.

Horizontal slipforming lends itself to concrete structures, such as tunnel linings, water conduits, drainage channels, precast elements, canal linings, highway median barriers, pavements, curbs, shoulder barriers, and retaining walls.

Vertical slipforms, concreted while rising, are usually moved in small increments by jacks that propel themselves on smooth steel rods or tubing embedded in or attached to the hardened concrete. Horizontal slipforms generally move on a rail system, tractor treads, wheels, and other similar means resting on a shaped berm. Working and storage decks and finisher's scaffolding are attached to and carried by the moving formwork.

The vertical or horizontal movement of forms can be a continuous process or a planned sequence of finite placements.

Slipforms used on structures such as tunnels and shafts should comply with the applicable provisions of Section 6.6. Slipforms used on mass concrete structures, such as dams, should comply with the applicable provisions of Section 6.5.

7.3.2—Vertical slipforms

7.3.2.1—A vertical slipform system has five main components: sheathing, wales, yokes, jacks and jackrods, and working or storage decks and scaffolding.

The sheathing or vertical forms can be wood staves, plywood, metal, glass fiber-reinforced plastic, wood, or a combination of these materials. The function of the sheathing is to contain and shape the concrete.

Wales have three main functions:
- Support and hold the sheathing in place;
- Transmit the lifting force from the yokes to the sheathing and to the other elements of the form; and
- Provide support for various platforms and scaffolding.

Yokes support the wales at regular intervals with their legs, transmit the lifting forces from the jacks to the wales, and resist the lateral force of plastic concrete within the form.

The jacks, installed on the yoke's beams, climb up the jackrods and provide the force needed to raise the entire slipform system.

Various platforms, decks, and scaffolding complete the slipform system. They provide space for storing concrete, reinforcing steel, and embedments, and serve as a working area for placing and finishing.

7.3.2.2 *Design and construction considerations*—Slipforms should be designed by engineers familiar with slipform construction. Construction of the slipform and slipping should be carried out under the immediate supervision of a person experienced in slipform work. Drawings should be prepared by a slipform engineer employed by the contractor. The drawings must show the jack layout, formwork, working decks, and scaffolds. A developed elevation of the structure should be prepared, showing the location of all openings and embedments. The slipform engineer should be experienced in the use of the exact brand of equipment to be used by the contractor because there is significant variation in equipment between manufacturers.

7.3.2.3 *Vertical loads*—In addition to dead loads, live loads assumed for the design of decks should not be less than the following:

Sheathing and joists .... 75 lb/ft² (3.6 kPa)
or concentrated buggy wheel loads, whichever is greater

Beams, trusses, and wales .... 50 lb/ft² (2.4 kPa)
Light-duty finishers' scaffolding .... 25 lb/ft² (1.2 kPa)

7.3.2.4 *Lateral pressure of concrete*—The lateral pressure of fresh concrete to be used in designing forms, bracing, and wales can be calculated as follows.

Inch-pound version:

$$p = c_1 + \frac{6000R}{T}$$

where
c_1 = 100;
p = lateral pressure, lb/ft²;
R = rate of concrete placement, ft/h; and
T = temperature of concrete in the forms, °F.

SI version:

$$p = c_1 + \frac{524R}{T + 17.8}$$

where
c_1 = 4.8;
p = lateral pressure, kPa;
R = rate of concrete placement, m/h; and
T = temperature of concrete in the forms, °C.

c_1 = 100 lb/ft² (4.8 kPa) is justified because vibration is slight in slipform work because the concrete is placed in shallow layers of 6 to 10 in. (150 to 250 mm) with no revibration. For some applications, such as gas-tight or containment structures, additional vibration can be required to achieve maximum density of the concrete. In such cases, the value of c_1 should be increased to 150 lb/ft² (7.2 kPa).

7.3.2.5 *Tolerances*—Prescribed tolerances for slipform construction of building elements are listed in ACI 117.

7.3.2.6 *Sliding operation*—The maximum rate of slide should be limited by the rate for which the forms are designed. In addition, both maximum and minimum rates of slide should be determined by an experienced slipform supervisor to accommodate changes in weather, concrete slump, initial set of concrete, workability, and the many exigencies that arise during a slide and cannot be accurately predicted beforehand. A person experienced in slipform construction should be present on the deck at all times during the slide operation.

During the initial placing of the concrete in the slipform, the placing rate should not exceed that for which the form was designed. Ideally, concrete should be placed in approximately 6 to 8 in. (150 to 200 mm) lifts throughout the slipform operation.

The level of hardened concrete in the form should be checked frequently by the use of a probing rod to establish safe lifting rates. Forms should be leveled before they are filled and should be maintained level unless otherwise required for out-of-tolerance corrections. Care should be taken to prevent drifting of the forms from alignment or designed dimensions and to prevent torsional movement.

Experience has shown that a plumb line, optical plummet, laser, or combination of these used in conjunction with a water level system is effective in maintaining the form on line and grade and for positioning openings and embedded items.

The alignment and plumbness of a structure should be checked at least once during every 4 h that the slide is in operation and preferably every 2 h. In work that is done in separate intermittent slipping operations, a check of alignment and plumbness should be made at the beginning of each slipping operation.

More frequent readings should be taken on single tall structures with relatively small plan sections, as the form system in these structures tends to twist and go out of plumb more readily.

Sufficient checks of plumbness should be provided to readily detect and evaluate movements of the form for all slipformed structures so that appropriate adjustment can be made in sufficient time by experienced personnel.

Detailed records of both vertical and lateral form movements should be maintained throughout the slipform operation.

7.3.3 *Horizontal slipforms*—The general provisions of Section 2.1.4 should be met and the formwork engineer/contractor should submit drawings of the slipform for review and approval by the engineer/architect. These drawings should show the handling diagrams, the placing procedure, and the provisions for ensuring attainment of the required concrete surfaces.

7.4—Permanent forms

7.4.1 *Discussion*—Permanent forms, or stay-in-place forms, are forms left in place that may or may not become an integral part of the structural frame. These forms can be rigid—such as metal deck, precast concrete, wood, plastics, and various types of fiberboard—or the flexible type—such as reinforced, water-repellent, corrugated paper, or wire mesh with waterproof paper backing.

When the permanent form is used as a deck form, it is supported from the main structural frame with or without an intermediate system of temporary supports. If temporary supports are required under, or to provide structural stability for, the structural frame members to support the weight of the fresh concrete without causing excessive deflection or member instability, such information should be specified by the engineer/architect.

7.4.2 *Design considerations*—If the stay-in-place form is not covered in the contract specifications because it has no function in the finished structure, the form manufacturer's specifications should be used; the manufacturer's recommended practice should be followed for size, span, fastenings, and other special features pertinent to this type of form, such as being water repellent and protected against chemical attack from wet concrete; and the minimum requirements of Chapters 2 and 3 should be followed. Particular care should be taken in the design of such forms by the formwork engineer/contractor to minimize distortion or deformation of the form or supporting members under the construction loads.

The engineer/architect who specifies the use of permanent rigid forms should consider in the structural analysis both the construction dead and live loads on the form as well as the structure's stability during construction, in addition to consideration of the form's performance in the finished structure.

When metal deck to become an integral part of the structure is used as a permanent form, its shape, depth, gage, coating, physical dimensions, properties, and intermediate temporary support should be as called for in contract documents. If structural continuity is assumed in the design of the form, the engineer of the permanent structure should specify the required number of permanent supports over which the form material should be continuous.

7.4.3—*Installation*

7.4.3.1 *Shop drawings*—The formwork engineer/contractor should submit fully detailed shop drawings for all permanent deck forms to the engineer/architect for review, approval, or both. Shop drawings should show all form thicknesses, metal gauges, physical dimensions and properties, accessories, finishes, methods of attachment to the various classes of the work, and temporary shoring requirements.

7.4.3.2 *Fastenings*—The permanent deck form should be properly fastened to supporting members and to adjacent sections of deck form and properly lapped, in accordance with manufacturers' recommendations, to provide a tight joint that will prevent loss of mortar during the placement of concrete. Where required, end closures for corrugated or fluted forms should be provided, together with fill pieces where a tight fit is required. To prevent buckling, allow for expansion of metal deck forms after fastening and before concrete placement.

Flexible types of forms (those that depend for lateral stiffness on supporting members) should be drawn tight for proper installation. Adequate temporary bracing or anchors should be provided in the plane of the top chord of the supporting members to prevent lateral buckling and rotation of these supports and to maintain the required tension in the flexible form.

Paper or metal forms used to form voids in concrete construction should be properly placed and anchored to the reinforcement and to side or deck forms with wire ties or other approved methods to prevent displacement or flotation during placing of concrete. Water should be prevented from entering voids. Where water intrusion is possible, weep holes should be provided to reduce its entrapment.

7.4.4 *Deflections*—The vertical and lateral deflections of the permanent form between supports under the load of fresh concrete should be investigated by the engineer/architect. Temporary supports, such as shoring and stringers, should be specified, if necessary, to keep deflection within desired tolerances.

7.5—Forms for prestressed concrete construction

7.5.1 *Discussion*—The engineer/architect should indicate in the contract documents any special requirements for prestressed concrete construction.

It may be necessary to provide appropriate means of lowering or removing the formwork before full prestress is applied to prevent damage due to upward deflection of resilient formwork.

Pretensioning or post-tensioning of strands, cables, or rods can be done with or without side forms of the member in place, in accordance with Section 7.5.2. Bottom forms and supporting shores or falsework should remain in place until the member is capable of supporting its dead load and anticipated construction loads, as well as any formwork carried by the member.

The concreting sequence for certain structures should also be planned so that concrete is not subjected to bending stress caused by deflection of the formwork.

7.5.2—*Design*

7.5.2.1—Where the side forms cannot be conveniently removed from the bottom or soffit form after concrete has set, such forms should be designed with slip joints or with added panel and connection strength for additional axial or bending loads that can be superimposed on them during the prestressing operation.

7.5.2.2—Side forms that remain in place during the transfer of prestressing force should be designed to allow for vertical and horizontal movements of the cast member during the prestressing operation. The form should be designed to minimize the restraint to elastic shortening in the prestressing operation. For example, small components or wrecking strips should be planned that can be removed or destroyed to relieve load on side forms as well as to eliminate their restraint during prestressing. In all cases, the restraint to shrinkage of concrete should be kept to a minimum, and the deflections of members due to prestressing force and the elastic deformation of forms or falsework should be considered in the design and removal of the forms.

7.5.2.3—For reasons of safety, when using post-tensioned, cast-in-place elevated slabs, the contractor should be careful to ensure that supporting shores do not fall out due to lifting of the slab during tensioning. For large structures where the dead load of the member remains on the formwork during prestressing, displacement of the dead load toward end supports should be considered in design of the forms and shoring, including sills or other foundation support.

7.5.3 *Construction accessories*—Hold-down or push-down devices for deflected cables or strands should be provided in the casting bed or forms. All openings, offsets, brackets, and all other items required in the concrete work should be provided for in the formwork. Bearing plates, anchorage assemblies, prestressing steel, conduits, tube enclosures, and lifting devices shown or specified to be set in concrete should be accurately located with formwork templates and anchored to remain within the tolerances given on contract documents. Quality and strength of these accessories should be as specified.

7.5.4 *Tolerances*—Prescribed ranges of tolerances for job site precast and plant manufactured precast-prestressed concrete members are given in ACI 117 and the PCI report[7.1] on tolerances.

7.5.5 *Special provisions for curing and for safety of workers*—Where necessary to allow early reuse of forms, provisions should be made to use accelerated curing processes such as steam curing, vacuum processing, or other approved methods.

Safety shields should be provided at end anchorages of prestressing beds or where necessary for the protection of workmen or equipment against possible breakage of prestressing strands, cables, or other assemblies during prestressing or casting operation.

7.6—Forms for site precasting

7.6.1 *Discussion*—Forms for site precasting are used for precast concrete items that can be either load- or nonload-bearing members for structural or architectural uses.

7.6.2 *Construction*—Exterior braces only should be used when exposed metal or filled-in pockets resulting from the use of metal ties would present an objectionable appearance.

To ensure uniformity of appearance in the cast members or units, particularly in adjacent units where differences in texture, color, or both would be visible, care should be taken that the contact surfaces of forms or form liners are of uniform quality and texture.

Form oil or retardant coatings (nonstaining, if required) should be applied uniformly and in accordance with manufacturers' recommendations for this particular class of work.

7.6.3 *Accessories*—It is particularly important in this class of work that positive and rigid devices be used to ensure proper location of reinforcement. All openings, cutouts, offsets, inserts, lift rings, and connection devices required to be set in concrete should be accurately located and securely anchored in the formwork.

The finished surfaces of members should be free of lift rings and other erection items where it will be exposed, interfere with the proper placing of precast members or other materials, or be subject to corrosion. Such items should be removed so that no remaining metal will be subject to corrosion.

The quality and strength of these accessories should be as required by the contract documents, but the lifting devices or other accessories not called for in the contract documents are the responsibility of the contractor.

7.6.4 *Tolerances*—Prescribed tolerances for precast-concrete construction are listed in ACI 117.

7.6.5 *Removal of forms*—Precast members or units should be removed from forms only after the concrete has reached a specified strength, as determined by the field-cured test cylinders or beams and job history of concrete curing.

Where required to allow early reuse of forms, provisions can be made to use accelerated curing processes, such as steam curing, or other approved methods. Methods of lifting precast units from forms should be approved by the engineer/architect.

7.7—Use of precast concrete for forms

7.7.1 *Discussion*—Precast concrete panels or molds have been used as forms for cast-in-place and precast concrete, either as permanent forms, integrated forms, or as removable, reusable forms. They have been used for both structural and architectural concrete, designed as structurally composite with the cast-in-place material or to provide a desired quality of outer surface and, in some cases, to serve both of these purposes. Concrete form units can be plain, reinforced, or prestressed, and either cast in the factory or at the job site. The most common use of precast concrete form units has been for elevated slabs acting compositely with topping concrete, as in bridge and commercial or institutional construction. Precast units are also common as ground holding systems in tunneling and as stay-in-place forms for rehabilitation of navigation lock walls.

7.7.2—*Design*

7.7.2.1 *Responsibility for design*—Where the integrated form is to act compositely with the structure concrete, the form panel should be designed by the engineer/architect who should also indicate what additional external support is required for the permanent forms. For permanent forms intended to achieve a desired architectural effect, the engineer/architect can specify surface finish and desired minimum thickness of architectural material. Design and layout of temporary forms and supporting systems should normally be the responsibility of the formwork engineer/contractor.

7.7.2.2 *Connections*—Connection details should be planned to overcome problems of mating precast members to each other and to the existing or cast-in-place structure.

7.7.2.3 *Bonding concrete form to concrete structure*—Effective bond between precast form unit and the concrete structure is essential and can be achieved by: 1) special treatment, such as grooving or roughening the form face in contact with the structure concrete; 2) use of anchoring devices extending across the interface between form panel and structure concrete; 3) a combination of 1) and 2); and 4) use of paint-on or spray-on bonding chemicals. Lifting hooks in a form unit can be designed to serve also as anchors or shear connectors.

7.7.2.4 *Code requirements*—Precast concrete forms used in composite design with cast-in-place concrete in buildings should be designed in accordance with ACI 318.

7.7.3—*During and after concreting*

7.7.3.1 *Vibration*—Thorough consolidation of site-cast concrete is required to prevent voids that would interrupt the bond of the form to structure concrete, but sufficient care should be used to prevent damage of concrete panels by contact with vibrators.

7.7.3.2 *Protection of architectural finish*—Care should be taken to avoid spilling fresh concrete on exposed surfaces, and any spilled or leaked concrete should be thoroughly removed before it has hardened. After concreting, protection of precast architectural concrete form facings may need to be considered.

7.8—Forms for concrete placed under water

7.8.1 *Discussion*—There are two basic approaches to the problem of placing concrete under water: the concrete can be mixed in the conventional manner and then placed by special methods, or the preplaced aggregate method can be used.

In the first approach, placement can be made by either pump, underwater bucket, or tremie. The tremie is a steel pipe, suspended vertically in the water, with a hopper attached to the upper end above the water surface. The lower end of the pipe, with an ejectable plug, extends to the bottom of the area to be concreted. This pipe is charged with concrete from the surface. Once the pipe is filled with concrete, it is kept full and its bottom should be kept immersed in the fresh concrete.

In the second approach, the forms are filled with coarse aggregate, which is then grouted so that the voids around the aggregate are filled as discussed in Section 7.2. The grout is introduced at the bottom and the water is displaced upward as the grout rises.

7.8.2—*Underwater bucket and tremie*

7.8.2.1 *Design*—Forms for underwater concreting are designed with the same considerations as other forms covered in Section 2.2, except that the density of the submerged concrete can be reduced by the weight of the water displaced. Because of large local pressures that can develop due to the head of concrete in the tremie, the location of the tremie and possible resulting loads on the form should be evaluated by experienced personnel. Some designers have ignored the effects of submergence because this results in a practical design that is sturdy enough to withstand the extra rigors of underwater conditions.

In tidal zones, forms should be designed for the lowest possible water level. Changes in construction schedules can transform a planned submerged placement to one made above water, thus losing the offsetting water pressure.

7.8.2.2 *Construction*—Underwater forms should be built on the surface in large units because final positioning and fitting when done under water by divers is slow and costly. For this reason, foundations should be kept simple in shape, and forms should be free of complex bracing and connection details. Through-ties, which could interfere with the concrete placing, should be avoided. Forces imposed on preassembled forms during lifting should be considered in the form design.

Forms should be carefully fitted and secured to adjacent materials or construction to avoid loss of mortar under pressure developed. If there is any water current flow past the form, small openings in the form should be avoided as they will permit washing or scouring of the fresh concrete.

When it is intended to permit concrete to overflow the form and screed it off to grade, it is essential that the form is positioned to the proper grade and is detailed so that the overflow will not interfere with the proposed method and devices for stripping.

Forms should be well detailed, and such details should be scrupulously followed so that divers employed to remove the form can visualize and plan their work before descending.

Multiuse forms can have special devices for positioning forms from above water and special stripping devices, such as hydraulic jacks, that permit releasing the form from the surface.

7.8.3—*Preplaced aggregate*

7.8.3.1 *Design*—The formwork should be designed with the same considerations as mentioned previously in Section 7.2.2.

7.8.3.2 *Construction*—It is important to ensure that silt is excluded from the forms because silt chokes the voids in the aggregate and interferes with the flow of grout. Silt, if left adhering to the aggregate, can reduce the bond between the aggregate and the grout.

The inspection of the forms before concrete placement should verify that the perimeters of the forms are effectively sealed against the leakage of grout or the intrusion of silt or other fines.

CHAPTER 8—REFERENCES
8.1—Referenced standards and reports

The standards and reports listed as follows were the latest editions at the time this document was prepared. Because these documents are revised frequently, the reader is advised to contact the proper sponsoring group if it is desired to refer to the latest version.

American Concrete Institute

116R	Cement and Concrete Terminology
117	Standard Specifications for Tolerances for Concrete Construction and Materials
207.1R	Mass Concrete
224R	Control of Cracking in Concrete Structures

228.1R	In-Place Methods to Estimate Concrete Strength
301	Specifications for Structural Concrete
303R	Guide to Cast-in-Place Architectural Concrete Practice
304.1R	Guide for the Use of Preplaced-Aggregate Concrete for Structural and Mass Concrete Applications
304.2R	Placing Concrete by Pumping Methods
305R	Hot Weather Concreting
306R	Cold Weather Concreting
309.2R	Identification and Control of Consolidation-Related Surface Defects in Formed Concrete
311.1R	Manual of Concrete Inspection
313	Standard Practice for Design and Construction of Concrete Silos and Stacking Tubes for Storing Granular Materials
313R	Commentary on Standard Practice for Design and Construction of Concrete Silos and Stacking Tubes for Storing Granular Materials
318	Building Code Requirements for Reinforced Concrete
332R	Guide to Residential Cast-in-Place Concrete Construction
344R	Design and Construction of Circular Prestressed Concrete Structures
347.1R	Precast Concrete Units Used as Forms for Cast-in-Place Concrete
359	Code for Concrete Reactor Vessels and Containments
SP-34	Concrete for Nuclear Reactors

American Forest & Paper Association
National Design Specification for Wood Construction
Load and Resistance Factor Manual for Engineered Wood Construction

American National Standards Institute
ANSI/SEI/ ASCE 7	Minimum Design Loads for Buildings and Other Structures
A48.1	Forms for One-Way Concrete Joist Construction
A48.2	Forms for Two-Way Concrete Joist Construction
A208.1	Mat-Formed Wood Particle Board

American Society of Civil Engineers
SEI/ASCE 37 Design Loads on Structures During Construction

APA—The Engineered Wood Association
Plywood Design Specification and supplements, 1997

ASTM International
A 446	Standard Specification for Steel Sheet, Zinc-Coated (Galvanized) by the Hot-Dip Process, Structural (Physical) Quality
C 532	Standard Specification for Structural Insulating Formboard (Cellulosic Fiber)
E 329	Specification for Agencies Engaged in the Testing and/or Inspection of Materials Used in Construction

Canadian Standards Association
CAN3-O86-M80	Code for Engineering Design in Wood
CAN/CSA-O96.1.94	Engineered Design in Wood (Limit States Design)

U.S. Department of Commerce
LLB-810a	Hardboard Concrete Form Liners (Simplified Practice Recommendation)
PS 1-95	Construction and Industrial Plywood
PS20-94	American Softwood Lumber

These publications may be obtained from the following organizations:

American Concrete Institute
P.O. Box 9094
Farmington Hills, MI 48333-9094

American Forest & Paper Association
American Wood Council
1111 19th St., NW
Washington, DC 20036

American National Standards Institute
11 W. 42nd St.
New York, NY 10036

APA—The Engineered Wood Association
P.O. Box 11700
Tacoma, WA 98411

ASTM International
100 Barr Harbor Dr.
West Conshohocken, PA 19428

CSA International
178 Rexdale Blvd.
Etobicoke (Toronto) ON
M9W 1R3 Canada

U.S. Department of Commerce publications available from:

U.S. Government Printing Office
Washington, DC 20402

8.2—Cited references
CHAPTER 1 REFERENCES
1.1. ACI Committee 622, "Form Construction Practices," ACI JOURNAL, *Proceedings* V. 53, No. 12, May 1957, pp. 1105-1118.

1.2. ACI Committee 622, "Pressures on Formwork," ACI JOURNAL, *Proceedings* V. 55, No. 2, Aug. 1958, pp. 173-190.

1.3. Hurd, M. K., *Formwork for Concrete*, SP-4, 6th Edition, American Concrete Institute, Farmington Hills, Mich., 1995, 492 pp.

CHAPTER 2 REFERENCES
2.1. Barnes, J. M., and Johnston, D. W., "Modification Factors for Improved Prediction of Fresh Concrete Lateral

Pressures on Formwork," Institute of Construction, Department of Civil Engineering, North Carolina State University, Raleigh, N.C., Oct., 1999, 90 pp.

2.2. Gardner, N. J., "Pressure of Concrete Against Formwork," ACI JOURNAL, *Proceedings* V. 77, No. 4, July-Aug. 1980, pp. 279-286; and discussion, *Proceedings* V. 78, No. 3, May-June 1981, pp. 243-246.

2.3. Gardner, N. J., and Ho, P. T.-J., "Lateral Pressure of Fresh Concrete," ACI JOURNAL, *Proceedings* V. 76, No. 7, July 1979, pp. 809-820.

2.4. Clear, C. A., and Harrison, T. A., 1985, "Concrete Pressure on Formwork," *CIRIA Report* No. 108, Construction Industry Research and Information Association, London, 32 pp.

2.5. "Pressure of Concrete on Vertical Formwork (Frischbeton auf Lotrechte Schalungen)," DIN 18218, Deutsches Institut für Normung e.V., Berlin, 1980, 4 pp.

2.6. Gardner, N. J., "Pressure of Concrete on Formwork—A Review," ACI JOURNAL, *Proceedings* V. 82, No. 5, July-Aug. 1985, pp. 744-753.

2.7. British Cement Association, "Hi-Rib Permanent Formwork Trials," Report and Appendix, RE1.031.01.1 BCA, Slough, UK, Feb. and July 1992, 22 and 9 pp.

2.8. Grundy, P., and Kabaila, A., "Construction Loads on Slabs with Shored Formwork in Multistory Buildings," ACI JOURNAL, *Proceedings* V. 60, No. 12, Dec. 1963, pp. 1729-1738.

2.9. Agarwal, R. K., and Gardner, N. J., "Form and Shore Requirements for Multistory Flat Slab Type Buildings," ACI JOURNAL, *Proceedings* V. 71, No. 11, Nov. 1974, pp. 559-569.

2.10. Stivaros, P. C., and Halvorsen, G. T., "Shoring/Reshoring Operations for Multistory Buildings," *ACI Structural Journal*, V. 87, No. 5, Sept.-Oct. 1990, pp. 589-596.

2.11. Noble, J., "Stop Guessing at Reshore Loads—Measure Them," *Concrete Construction*, V. 20, No. 7, 1975, pp. 277-280.

CHAPTER 4 REFERENCES

4.1. *Manual of Standard Practice*, 27th Edition, Concrete Reinforcing Steel Institute, Schaumburg, Ill., 2001, 116 pp.

4.2. Randall, F. A., Jr., and Courtois, P. D., "Side Form Spacers," ACI JOURNAL, *Proceedings* V. 73, No. 2, Feb. 1976, pp. 116-120.

4.3. "Wood Handbook: Wood as an Engineering Material," *Agriculture Handbook 72*, Forest Products Society, U. S. Department of Agriculture, Madison, Wisc, 1998, 464 pp.

4.4. *Manual for Wood Frame Construction*, National Forest Products Association (now American Forest & Paper Association), Washington, D.C., 1988.

4.5. Stalnaker, J. J., and Harris, E. C., *Structural Design in Wood*, 2nd Edition, Chapman & Hall, 1997, 448 pp.

4.6. American Institute of Timber Construction, *Timber Construction Manual*, 4th Edition, John Wiley & Sons, New York, 1994, 928 pp.

4.7. "National Design Specification for Wood Construction (ANSI/AF&PA NDS-1997)," American Forest & Paper Association, Washington, D.C., 1997, 174 pp.

4.8. "Plywood Design Specification," APA—The Engineered Wood Association, Tacoma, Wash., 1997, 32 pp.

4.9. "Specification for Structural Steel Buildings—Allowable Stress Design and Plastic Design," American Institute of Steel Construction, Chicago, Ill., 1989, 220 pp.

4.10. "Specification for the Design of Cold-Formed Steel Structural Members," American Iron and Steel Institute, Washington, D.C., 1986, 82 pp.

4.11. *Aluminum Design Manual: Specifications & Guidelines for Aluminum Structures*, The Aluminum Association, Washington, D.C., 1994.

4.12. Ziverts, G. J., "A Study of Cardboard Voids for Prestressed Concrete Box Slabs," *PCI Journal*, V. 9, No. 3, 1964, pp. 66-93, and V. 9, No. 4, pp. 33-68.

4.13. Hurd, M. K., "Using Glass-Fiber-Reinforced-Plastic Forms," *Concrete Construction*, V. 42, No. 9, 1997, 689 pp.

4.14. Hurd, M. K., "Nonmetallic Form Ties," *Concrete Construction*, V. 38, No. 10, 1993, pp. 695-699.

4.15. Building Materials Committee, *Cellular Plastics in Construction*, Cellular Plastics Division, Society of the Plastics Industry, Washington, D.C.

4.16. Hurd, M. K., "Expand Your Forming Options," *Concrete Construction*, V. 42, No. 9, Sept. 1997, pp. 725-728.

4.17. "Standard Specifications and Load Tables for Open Web Steel Joists," Steel Joist Institute, Myrtle Beach, S.C, 1994, 96 pp.

4.18. "Recommended Horizontal Shoring Beam Erection Procedure," Scaffolding, Shoring, and Forming Institute, Cleveland, Ohio, 1983.

4.19. "Recommended Safety Requirements for Shoring Concrete Formwork," Scaffolding, Shoring, and Forming Institute, Cleveland, Ohio, 1990.

4.20. "Concrete Forming," V345, APA—The Engineered Wood Association, Tacoma, Wash., 1998, 28 pp.

4.21. Smulski, S., ed., *Engineered Wood Products: A Guide for Specifiers, Designers, and Users*, PFS Research Foundation, Madison, Wisc., 1997, 330 pp.

4.22. Hurd, M. K., "Plastic Form Liners," *Concrete Construction*, V. 39, No. 11, Nov. 1994, pp. 847-853.

4.23. Hurd, M. K., "Choosing and Using a Form Release Agent," *Concrete Construction*, V. 41, No. 10, 1996, pp. 732-736.

CHAPTER 7 REFERENCES

7.1. PCI Committee on Tolerances, "Tolerances for Precast and Prestressed Concrete," *PCI Journal*, V. 30, No. 1, Jan.-Feb. 1985, pp. 26-112.

ACI 533.1R-02

Design Responsibility for Architectural Precast-Concrete Projects

Reported by ACI Committee 533

Benjamin Lavon
Chair

Donald F. Meinheit
Secretary

Robert B. Austin
George F. Baty
Harry A. Chambers
Sidney Freedman[*]

Edward M. Frisbee
Harry A. Gleich
Thomas J. Grisinger
Theodore W. Hunt[†]
Allan R. Kenney

Navin N. Pandya
James B. Quinn
Ralph C. Robinson[†]
Joseph R. Tucker

[*]Primary author.
[†]Deceased.

Architectural precast concrete is a unique subcategory of concrete construction that involves the same basic entities as other construction. This document outlines the responsibilities for various parties of the design/construction team for architectural precast-concrete projects.

Keywords: Architect; architectural concrete; construction; design; design responsibility; general contractor; precast concrete.

CONTENTS
Chapter 1—Introduction, p. 533.1R-1
 1.1—Background

Chapter 2—General responsibilities, p. 533.1R-2
 2.1—Architect
 2.2—Engineer of record
 2.3—General contractor
 2.4—Precaster (manufacturer)
 2.5—Erector
 2.6—Inspection

Chapter 3—Forms of contracts, p. 533.1R-4
 3.1—Negotiated versus competitive bid
 3.2—Single-source responsibility
 3.3—Mockups

Chapter 4—Conclusion, p. 533.1R-5

Chapter 5—References, p. 533.1R-5
 5.1—Referenced standards and reports

Appendix—Prebid process, p. 533.1R-6

CHAPTER 1—INTRODUCTION

Design and construction of structures is a complex process. Defining the scope of work and the responsibilities of the involved parties by means of the contract documents is necessary to achieve a high-quality structure. This is a guide document for all parties involved in a precast-concrete project and defines the responsibilities of each party. This document does not specifically address the inspection functions but provides direction on who should be conducting inspections. These responsibilities are subject to relationships between the parties defined in the contract documents.

1.1—Background
Practices regarding the assignment and acceptance of responsibility in design and construction vary throughout North America. In many cases, there has been confusion regarding the responsibility of the various parties. This situation has sometimes led to protracted legal proceedings. The first consensus document to attempt to define essential

---

ACI Committee Reports, Guides, Standard Practices, and Commentaries are intended for guidance in planning, designing, executing, and inspecting construction. This document is intended for the use of individuals who are competent to evaluate the significance and limitations of its content and recommendations and who will accept responsibility for the application of the material it contains. The American Concrete Institute disclaims any and all responsibility for the stated principles. The Institute shall not be liable for any loss or damage arising therefrom.
Reference to this document shall not be made in contract documents. If items found in this document are desired by the Architect/Engineer to be a part of the contract documents, they shall be restated in mandatory language for incorporation by the Architect/Engineer.

ACI 533.1R-02 became effective June 26, 2002.
Copyright © 2002, American Concrete Institute.
All rights reserved including rights of reproduction and use in any form or by any means, including the making of copies by any photo process, or by electronic or mechanical device, printed, written, or oral, or recording for sound or visual reproduction or for use in any knowledge or retrieval system or device, unless permission in writing is obtained from the copyright proprietors.

roles was *Quality in the Constructed Project,* published by the American Society of Civil Engineers (ASCE) in 1990 (ASCE 1990).

In 1987, ACI organized a standing board committee on Responsibility in Concrete Construction. The document, *Guidelines for Authorities and Responsibilities in Concrete Design and Construction,* was published by the committee (ACI Committee on Responsibility in Concrete Construction 1995).

One of the basic principles of the construction industry is that responsibility and authority should go hand in hand. Another principle is that every entity should be responsible for its own work. These principles are frequently violated. For example, an architect/engineer (A/E) can require that certain tasks not be undertaken by the contractor without the A/E's approval, but the A/E may not wish to accept responsibility for problems that develop resulting from requiring those tasks. This is a case of requiring compliance without accepting responsibility. There have also been cases where owners have sued architects and engineers for approving poor-quality construction but gave them no contract to monitor the work as it progressed. Safety enforcement agencies and plaintiffs' lawyers have often charged engineers or architects with the responsibility for construction accidents. These last two situations typically are cases of responsibility without authority, although there could be instances where a design professional's work can affect jobsite safety. If the designers are involved with construction-management functions, they could be making decisions affecting worker safety as well as quality of construction.

Construction has reached a level of complexity today where it is essential to have design input from the subcontractors. This input, whether submitted as value engineering proposals, in response to performance requirements, or simply offered as design alternatives, has a legitimate place in construction.

Panels are designed for stripping, handling, and installation loads, usually by the panel manufacturer. Service loads are set by the governing building code and are multiplied by the appropriate load factor. Minimum load factors are set by the governing building code.

Panel design should consider dead, live, and environmental loads including wind, earthquake (if applicable), temperature, and moisture effects. Service loads set by the governing building code should be considered only as minimum requirements.

Frequently, a precaster wants to change some items in the design to make a fabrication or erection operation easier or more economical. In approving the changes, the A/E still has responsibility for proper interfacing with other materials in contact with precast concrete. This notion of responsibility is presented in the ASCE document (ASCE 1990) and in the Precast/Prestressed Concrete Institute's guidelines (PCI Ad-Hoc Committee for Responsibility for Design of Precast-Concrete Structures 1998).

The engineer of record (EOR) always has to take overall responsibility for the structural design of the complete structure. Often, certain aspects of the design are delegated to specialty engineers working for the material suppliers or subcontractors. When any of this structural design work for that portion of the structure involves engineering (as opposed to simply detailing), then the design work should be under the control of a professional engineer licensed in the same state as the project who takes responsibility for the work done. One state, Florida, has formal legal procedures for this process. Local regulatory authorities should be consulted for their specific requirements. Contract documents often require that structural design be the responsibility of a professional engineer, regardless of government mandate.

CHAPTER 2—GENERAL RESPONSIBILITIES
2.1—Architect

The architect develops the design concept, overall structure geometry, selects the cladding material for appearance, provides details for weatherproofing, selects tolerances for proper interfacing with other materials, and specifies performance and quality characteristics and inspection and testing requirements in the project specifications.

The architect and EOR should have responsibility for all aspects of the precast-concrete design. The architect can specify in the contract documents that design services for portions of the work are to be provided by the precaster. Such design services should be performed for the precaster by a licensed precast engineer who can be an employee of the precaster or an independent structural engineer. The architect and EOR should review these designs, including structural calculations. This review does not relieve the precaster and the specialty engineer of their design responsibility. The contract and the design documents should state clearly the scope of both the precast design and review responsibilities, and the responsibilities of others providing design services.

The contract drawings prepared by the architect/engineer should provide the overall geometry of the structure and typical connection concepts to permit design, estimating, and bidding. Frequently, the architect's drawings will only show joints, reveals, or panel articulation. This lets the precaster determine panel sizes. In the prebid process, the precaster and erector should discuss their approach to panelize and subsequently connect the units to the building frame within the architectural and structural concepts of the project. In addition, the contract documents (design drawings and specifications) also should provide the general performance criteria, including concrete strength requirements for loading, deflection requirements, temperature considerations, and any tolerance or clearance requirements for proper interfacing with other parts of the structure.

The contract documents should clearly define:
- Precast-concrete components that are the design responsibility of the precaster (who takes responsibility for elements at interfaces with other parts of the structure, such as the secondary steel bracing to prevent rotation of beams or panels);
- Details or concepts of supports, connections, and clearances that are part of the structure designed by the architect and that will interface with the precast-concrete components; and
- Permissible load transfer points and indicate connection types to avoid having the precaster make assumptions on connection types and piece counts during bidding.

The architect and EOR should review designs, calculations, and shop drawings submitted for conformance with design criteria, loading requirements, and design concepts as specified in the design documents. This review, however, does not relieve the precaster and the precast engineer of their design responsibilities.

Key design issues for the architect—Buildings using architectural precast panels are becoming increasingly complex. The architect should understand the issues that affect a precaster's bid and make sure the contract documents address these items clearly. For preparation of shop drawings, all items interfacing with other materials should be defined. Contract documents that lack detail generally require numerous requests for information. While such documents are easier and less expensive to produce, they may ultimately result in disputes, delays, and additional costs.

The contract drawings developed by the architect should provide a clear interpretation of the configurations and dimensions of individual units and their relationship to the structure as a whole. To do this, the contract documents should supply the following data:

- Elevations, wall sections, and dimensions necessary to define the sizes and shapes of each different type of wall panel;
- Locations and sizes of all joints, both real (functional) and false (aesthetic). Joints between units should be completely detailed;
- Required materials and finishes for all surfaces, with a clear indication of which surfaces are to be exposed to view when in place;
- Corner details;
- Details for jointing and interfacing with other materials (coordinated with the general contractor);
- Details for special or unusual conditions, including fire requirements;
- Governing building code and design loads;
- Deflection limitations; and
- Specified dimensional tolerances for the precast concrete and the supporting structure, location tolerances for the contractor's hardware, clearance requirements, and erection tolerances for the precast concrete. Any exceptions to industry tolerances specified in the contract documents are not recommended.

Lack of detail will extend shop drawing time and potentially lead to disputes over work scope, schedule delays, or both. Also, lack of detail can lead to unanticipated changes that will impact cost. Poor detailing of panel cross sections within the contract documents will often lead to disputes.

Ideally, the assembly (or erection) drawing process should be as simple as submitting elevations showing panel sizes, surface features, and panel relationships; detail sheets showing panel cross sections and special edge conditions, and feature details; and connections showing mechanisms and locations of force transfer to the supporting structure. The review of shop drawings by the architect should be performed within the time specified in the contract documents. These drawings should be reviewed and any minor revisions made so that production can start. Verification of minor dimensional and detailing revisions is anticipated. When major shop drawing revisions are required, however, it can indicate a lack of planning or detailing within the contract documents.

It is the architect's responsibility to establish the standards of acceptability for surface finish, color range, and remedial procedures for defects and damage. This can be best accomplished by the precaster producing at least three sample panels a minimum of 4 x 4 ft before the initial production to establish the range of acceptability with respect to color and texture variations, surface blemishes, and overall appearance. When the units have returns, the same size return should appear in the sample panels.

2.2—Engineer of record

The EOR has responsibility for describing loading on precast element inserts and loading criteria (combinations, wind, seismic) for the structural design of the complete structure and the effects of the precast erection sequences on individual structural members, for example, steel spandrel beams when numerous concentrated panel loads are placed on them. The EOR should anticipate these loadings and provide means to support them. Responsibility for the precast-concrete design can be delegated to someone else, such as the precaster or precast engineer. The EOR should consider the consequences of the weight and eccentricity of the panels when designing the supporting structure. The EOR should also determine where, when, and what type of loading is to be assigned to the panels and the structure. The EOR has the responsibility of reviewing the delegated design work for compatibility with the overall structural design. This does not, however, relieve the preparer of the design work of the responsibility for doing it correctly.

Panels typically span column to column and are supported on or near the column. The EOR should determine and show on the contract documents the locations for supporting the gravity and lateral loads, including midpoint lateral (tie-back) connections, if necessary. The panel loads are routinely provided by the precaster to the EOR. The EOR determines during assembly (erection) drawing review whether or not the structure is adequate, within defined deflection limitations, to resist the loads and forces. The EOR should also provide sufficient information on seismic detailing. It is important that preliminary meeting(s) with the architect, EOR, and precasters be held before structural members are ordered, fabricated, or both, so that panel sizes, shapes, and basic connections, as well as their locations can be established (refer to the Appendix).

The EOR will know whether or not a spandrel panel is designed to transfer load to the columns, but the exact location of the load transfer can vary from precaster to precaster. For example, spandrel panel loads can be transferred from near the panel bottom or from near the top. The gravity supports of precast-concrete panels are almost always eccentric to the centerline of the supporting steel or concrete member. A concrete member is generally stiff and strong enough in torsion so that this is not a problem. Because the precaster does not design the columns or beams, the EOR should design to prevent excessive deflection and rotation of the supporting structure during and after erection of the precast concrete, as well as determine the need for diagonal bracing or web stiffening (CASE National Guidelines Committee 1994; CASE Task Group on Specialty Engineering 1996). The contract documents should address the issue of reimbursements to the EOR for engineering the bracing or reinforcement of the structure if the precaster's panelization changes the EOR's designed connection locations. In some areas, the precaster is responsible for designing the bracing and may also supply the secondary steel. These responsibilities should be clearly addressed in the contract documents and discussed in a prebid meeting.

2.3—General contractor

The general contractor (GC) typically has responsibility and authority for implementing the design intent of the contract documents, which includes furnishing materials, equipment, labor, maintaining specified quality requirements, and coordinating all trades. The GC is responsible for construction means, methods, techniques, sequences, and construction procedures. Also, the GC should initiate, maintain, and supervise all safety procedures and programs on the construction site. Site access to the structure for erection of the precast elements can become a problem. The responsibility to provide and maintain access roads should be clearly stated in the contract documents.

The GC has no design responsibility. The GC, however, does have considerable impact on the design process through its coordination role. The GC receives the different shop drawing submittals from the various trades and puts them together to form the completed design. The GC is normally responsible for project schedule, grid dimensions (which include control points, benchmarks, and lines on the building), quantities, and dimensional interfacing of the precast concrete with other trades, and maintenance of specified tolerances of the structure to ensure proper fit. During shop drawing review, the GC should notify the precaster when as-built conditions (dimensions) vary beyond tolerances on the contract drawings. In particular, dimensional tolerances between interfacing materials, such as precast concrete and glazing, should be considered.

The GC should be a party to direct communication between the precaster, EOR, and the architect. Communication channels should be established among the parties. The GC needs to be informed in writing, particularly if decisions affect the GC's activities.

Typically, the GC is responsible for placing embedments in cast-in-place concrete and coordinating steel attachments with the steel fabricator. In most instances, the most economical approach is to have gravity haunch connection hardware attached to steel columns by the steel fabricator. This necessitates awarding the precast-concrete contract in a timely manner.

2.4—Precaster (manufacturer)

To achieve practical and economical construction, the precaster first designs panelization and then connections. Ideally, a precaster performs value engineering as early as during preliminary design (in a partnering relationship) to improve economics, structural soundness, and performance. The precaster should request clarification of ambiguities, in writing from the architect, through contractual channels on special conditions not clearly defined in the contract documents. When the construction schedule demands a rapid turnaround time for review of drawing submittals, the precaster should notify all concerned of their obligations to review and return submitted drawings within the requested time period. At that time, the architect's and EOR's cooperation is needed to expeditiously review submitted documents.

The precaster or precaster's specialty engineer prepares detailed assembly, or erection drawings and design calculations that are usually signed and sealed. These drawings, calculations, or both, should show all design criteria, identify each material, show how precast panels interface with each other and the structure, and indicate the magnitude and location of all design loads imparted to the structure by the connections.

The precaster or specialty engineer designs the precast panels for the specified loads and is responsible for selecting, designing, and locating hardware, and panel reinforcement or items associated with the precasters methods of handling, storing, shipping, and erecting precast-concrete units. If necessary, this also includes an erection and bracing sequence developed in conjunction with the erector, EOR, and GC to maintain stability of the structure during erection.

Any additional design responsibility vested with the precast-concrete manufacturer should be defined clearly in the contract documents prepared by the architect. Most precast-concrete work is covered in Option II, Table 1. Option I in Table 1 has considerable liability for the architect and EOR.

2.5—Erector

The responsibility for erection of architectural precast concrete is usually determined by the GC. The contract documents rarely require that the erection be part of the precast-concrete manufacturer's work, be performed by the precaster's workers, or be subcontracted to specialized erection firms. Fabrication and erection included into one contract is preferred by some precasters because this improves coordination and reduces vulnerability to backcharges. The GC, however, may choose to issue separate contracts for fabrication and erection.

Erectors and precasters coordinate development of efficient connections for each project based on their equipment and expertise.

2.6—Inspection

Quality control for product manufacturing will be provided by the precaster according to provisions contained in a comprehensive quality system manual developed by the precaster. The quality system manual will be available to the owner and EOR for review.

Quality assurance will be provided through the precaster's participation in a recognized industry quality certification program. One such program is the PCI Plant Certification Program. Additional inspections may be required, by specification, through the owner's quality-assurance agency.

Installation quality assurance will be provided by adherence to industry standards such as the *PCI Erectors' Manual*. Additional quality assurance can be provided by requiring installation by an industry-qualified erector.

CHAPTER 3—FORMS OF CONTRACTS
3.1—Negotiated versus competitive bid

The price of architectural precast concrete is a relatively small percentage of the total building cost (usually less than 10%). Therefore, the possible difference in the price between the lowest precast bidder and the precaster who is ultimately awarded the contract will have a minor impact on the overall cost.

In a negotiated project, the precaster can become part of the building team at the very early design stage and be more effective in providing valuable expertise on panel design, performance, and economics (refer to Appendix).

3.2—Single-source responsibility

The architect or owner sometimes prefers single-source responsibility for wall cladding (which includes windows, precast, and all related sealants) responsibility for the following reasons: to enhance technical and aesthetic coordination between building systems; to establish a single-source warranty; to centralize control for erection and problem-solving issues; and generally, to provide for a single source of knowledge and total understanding of the entire system. In

Table 1—Design responsibilities

Contract information supplied by designer	Responsibility of manufacturer of precast concrete
Option I	
Provide complete drawings and specifications detailing all aesthetic, functional and structural requirements plus dimensions.	The manufacturer should make shop drawings (erection and production drawings) as required, with details as shown by the designer. Modifications may be suggested that, in manufacturer's estimation, would improve the economics, structural soundness, or performance of the precast-concrete installation. The manufacturer should obtain specific approval for such modifications. Full responsibility for the precast-concrete design, including such modifications, remains with the designer. Alternative proposals from a manufacturer should match the required quality and remain within the parameters established for the project. It is particularly advisable to give favorable consideration to such proposals if the modifications are suggested so as to conform to the manufacturer's normal and proven procedures.
Option II	
Detail all aesthetic and functional requirements but specify only the required structural performance of the precast-concrete units. Specified performance should include all limiting combinations of loads together with their points of application. This information should be supplied in such a way that all details of the unit can be designed without reference to the behavior of other parts of the structure. The division of responsibility for the design should be clearly stated in the contract.	The manufacturer has two alternatives: (a) Submit erection and shape drawings with all necessary details and design information for the approval and ultimate responsibility of the designer. (b) Submit erection and shape drawings for general approval and assume responsibility for part of the structural design; that is, the individual units, but not their effect on the building. Firms accepting this practice may either stamp (seal) drawings themselves, or commission engineering firms to perform the design and stamp the drawings. The choice between the alternatives (a) and (b) should be decided between the designer and the manufacturer prior to bidding, with either approach clearly stated in the specifications for proper allocation of design responsibility. Experience has shown that divided design responsibility can create contractual problems. It is essential that the allocation of design responsibility is understood and clearly expressed in the contract documents.

single-source responsibility, detailing issues are delegated to the contractor and material suppliers of the wall.

Single-source responsibility usually puts the precaster in a position of a broker/contractor without having the management and engineering skills to work out interfacing details for the window system, insulation, interior finishes, and sealants. A more logical entity for single-source responsibility is the GC. In some cases, the GC may prefer that single-source responsibilities be separated for greater profit potential.

Generally, the precast-concrete industry tends to avoid single-source responsibility, not only because of technical concerns regarding materials whose quality they do not control directly but also due to economics. It can be uneconomical to package everything under the precaster's construction umbrella because this additional responsibility requires additional compensation. The notion of a single-source responsibility for exterior enclosure can break down if that single source is not clearly defined.

3.3—Mockups

Panel-to-panel joint design is necessary to prevent air and water infiltration, and to properly seal windows and other openings. The architect is responsible for these designs. Because precast concrete is inherently watertight and impermeable, the panel joints and interfacing performance become the primary concerns. If testing is desired, it should be specified in the contract documents. Shipping a full-scale mockup to a testing lab for a wind-driven rain test, although costly and time consuming, is one way to satisfy these concerns. Also, aesthetics can be refined during this process. The cost of these tests needs to be identified in the bid documents and included in the precast budget.

A mockup will help determine how the total facade is assembled. Also, it will help in establishing the actual field-construction techniques. If a leak develops, which usually occurs at the window to precast-concrete interface, the details need to be examined and modified. The contract documents should require that the same sealant contractor seal both the precast-to-precast panel joints and the window interface to avoid sealant incompatibility.

CHAPTER 4—CONCLUSION

A successful precast-concrete project requires teamwork. This means close cooperation between and coordination of all participants, including the owner, architect, engineer, precast-concrete manufacturer, erector, general contractor or construction manager, and all trades. The precast work scope and the responsibilities of each party (usually defined by the contract documents) should be established at an early stage in the development of a project to achieve the desired results and schedule (refer to Table 1). Each party has the responsibility for communicating with all other parties through the GC/CM to achieve optimum efficiency during construction, and quality in the completed structure. When authority and responsibility roles are correctly and properly defined by the contract documents and communicated, responsibility issues are easily resolved.

CHAPTER 5—REFERENCES
5.1—Referenced standards and reports

ACI Committee on Responsibility in Concrete Construction, 1995, "Guidelines for Authorities and Responsibilities in Concrete Design and Construction," *Concrete International*, V. 17, No. 9, Sept., pp. 66-69.

American Society of Civil Engineers (ASCE), 1990, "Quality in the Constructed Project," *Manuals and Reports on Engineering Practice* No. 73, American Society of Civil Engineers, New York, N.Y.

CASE National Guidelines Committee, 1994 "National Practice Guidelines for the Structural Engineer of Record," 2nd Edition, Coalition of American Structural Engineers, Washington, D.C., 15 pp.

CASE Task Group on Specialty Engineering, 1996, "National Practice Guidelines for Specialty Structural

Engineers," Coalition of American Structural Engineers, Washington, D.C., 12 pp.

PCI Ad-Hoc Committee for Responsibility for Design of Precast-Concrete Structures, 1998, "Recommendations on Responsibility for Design and Construction of Precast-Concrete Solutions," *PCI Journal*, July-Aug., Chicago, Ill.

APPENDIX
Prebid process

Where the selection of a precaster cannot be negotiated or controlled by owner or architect but is governed by an open-bid situation, the following prebid process is desirable.

Step 1: Verification of architect's concepts and systems—A preview of the proposed precast-concrete assumptions during the design development stage (50% complete) of the architectural contract documents should be arranged with at least one local precaster. This review confirms or modifies the concept so that a realistic approach is presented on the architect's bid drawings. Those attending the meeting should:

- Discuss panelization and piece and joint sizes, determine what can be made, and what can be shipped and erected efficiently;
- Discuss architect's concept for structural support of precast concrete so that the architect can communicate to the EOR what support requirements are needed;
- Review desired finish(es) and continue or finalize the sample process; and
- Review the architect's intent for any interfaces with adjacent systems, such as roofing, windows, or building entrances.

Step 2: The prebid conference—This should be a mandatory meeting for all precasters wanting to bid the project and is usually held at least 3 weeks before the bid date. The architect presents the project's precast-concrete concepts with the intent of communicating straightforward information so bids will be prepared on a comparable basis. Questions can be asked or clarifications made at this time. Items to be discussed include:

- How and where the project's precast concrete will be structurally attached to the building frame;
- Specifications and any special provisions;
- Design responsibilities and lines of communication;
- The architect's approved finish samples with information on the type and size of aggregates and cement used, where applicable;
- The finish acceptance criteria and inspection (who, what, when, and where);
- Prebid submittal requirements such as proposal drawings and finish samples;
- Mockups, if applicable;
- Potential problems, discrepancies, or both found in the contract documents; and
- Special erection needs and logistics.

Step 3: Information submitted with bid or after-bid award—This submittal allows a review of each precaster's intent and confirms the precaster's ability to conform to concepts and finish requirements. (Realistically, only the three low bidders are required to provide this submission.) This material should include:

- Proposal drawings expressing the panelization and structural concepts;
- Small-size finish samples;
- The history of the precaster's organization as well as confirmation of the plant's quality assurance (plant certification) program;
- A list of comparable projects, references, and financial capability;
- Key schedule items such as mockup panels, shop drawings and design submittals, mold production, scheduled start of manufacture, and production schedules; and
- Qualifications to the bid that can be listed and reviewed.

If bidders are limited to a small group (two to four) by prequalification or other means, then all precasters are contacted during the development process; then limited prebid meetings, bid submittals, or both are needed.

If the project will allow for a negotiated precast contract and the precaster is brought onto the project team in the initial stages of development, then prebid and bid submittal information can be eliminated.

Step 4: The preconstruction conference—A preconstruction conference should be held at the job site as soon as possible after award of the precast concrete and erection contracts. The GC/CM should conduct job site meetings frequently to coordinate the precast-concrete erection with the work of other trades and to facilitate the erection process. These meetings should include those subcontractors whose work affects or is affected by the precast-concrete erection.

The coordinating meetings should consider all details of loading, delivery sequences and schedules, types of transportation, routes of ingress and egress for delivery trucks and erection cranes, handling techniques and devices, connections, erection methods and sequences, effects of temporary bracing on other trades, and on-site storage and protection. Questions regarding access, street use, sidewalk permits, oversized loads, lighting, or working hours should be addressed at this time.

VERMONT STATE COLLEGES

0 0003 0753775 3

DISCARD

Hartness Library
Vermont Technical College
One Main St.
Randolph Center, VT 05061